Australian Vegetation

Third Edition

Australian Vegetation has been an essential reference for students and researchers in botany, ecology and natural resource management for over 35 years. Now fully updated and with a new team of authors, the third edition presents the latest insights on the patterns and processes that shaped the vegetation of Australia. The first part of the book provides a synthesis of ecological processes that influence vegetation traits throughout the continent, using a new classification of vegetation. New chapters examine the influences of climate, soils, fire regimes, herbivores and aboriginal people on vegetation, in addition to completely revised chapters on evolutionary biogeography, quaternary vegetation history and alien plants. The book's second half presents detailed ecological portraits for each major vegetation type and offers data-rich perspectives and comparative analysis presented in tables, graphs, maps and colour illustrations. This authoritative book will inspire readers to learn and explore first-hand the vegetation of Australia.

David A. Keith is Professor of Botany in the Centre for Ecosystem Science at the University of New South Wales Australia and the NSW Office of Environment and Heritage. He won the Australian Ecology Research Award in 2013 and led the team awarded the 2015 Eureka Prize for Environmental Research. He is the author of the award-winning book *Ocean shores to desert dunes: the native vegetation of New South Wales and the ACT.*

Australian Vegetation

DAVID A. KEITH
University of New South Wales, Sydney
New South Wales Office of Environment and Hertiage, Sydney
Australian Long Term Ecological Research Network, Canberra

Third Edition

CAMBRIDGE
UNIVERSITY PRESS

CAMBRIDGE
UNIVERSITY PRESS

Shaftesbury Road, Cambridge CB2 8EA, United Kingdom

One Liberty Plaza, 20th Floor, New York, NY 10006, USA

477 Williamstown Road, Port Melbourne, VIC 3207, Australia

314–321, 3rd Floor, Plot 3, Splendor Forum, Jasola District Centre, New Delhi – 110025, India

103 Penang Road, #05–06/07, Visioncrest Commercial, Singapore 238467

Cambridge University Press is part of Cambridge University Press & Assessment, a department of the University of Cambridge.

We share the University's mission to contribute to society through the pursuit of education, learning and research at the highest international levels of excellence.

www.cambridge.org
Information on this title: www.cambridge.org/9781107118430

DOI: 10.1017/9781316339701

First published 1981
Second edition 1994
Reprinted 2001
Third edition 2017

A catalogue record for this publication is available from the British Library

Library of Congress Cataloging-in-Publication data
Names: Keith, David (David Andrew), author.
Title: Australian vegetation / David A. Keith.
Description: New York, NY : Cambridge University Press, 2017. |
Includes bibliographical references and index.
Identifiers: LCCN 2016049394 | ISBN 9781107118430 (hardback : alk. paper)
Subjects: LCSH: Plants–Australia.
Classification: LCC QK431 .K45 2017 | DDC 581.994–dc23 LC record available at https://lccn.loc.gov/2016049394

ISBN 978-1-107-11843-0 Hardback

To Roger Carolin, Peter Myerscough
and Doug Benson,
who inspired appreciation and learning in botany,
ecology and conservation.

CONTENTS

CONTRIBUTORS

TONY D. AULD
Centre for Ecosystem Science, School of Biological, Earth and Environmental Sciences, University of New South Wales, Sydney, NSW, Australia; Centre for Sustainable Ecosystem Solutions, School of Biological Sciences, University of Wollongong, Wollongong, NSW, Australia; and NSW Office of Environment and Heritage, Hurstville, NSW, Australia

JAYNE BALMER
Department of Primary Industries, Water, Parks and Environment, Hobart, TAS, Australia; and School of Land and Food, University of Tasmania, Hobart, TAS, Australia

PHILIP BARTON
Fenner School of Environment and Society, The Australian National University, Canberra, ACT, Australia

LINDA J. BEAUMONT
Department of Biological Sciences, Macquarie University, Sydney, NSW, Australia

LAUREN T. BENNETT
School of Ecosystem and Forest Sciences, The University of Melbourne, Melbourne, VIC, Australia

ANDREW J.W. BIGGS
Department of Natural Resources and Mines, Toowoomba, QLD, Australia

WILLIAM J. BOND
South African Environmental Observation Network, National Research Foundation, Claremont, South Africa; and Department of Biological Sciences, University of Cape Town, Rondebosch, South Africa

PAUL BOON
Institute for Sustainability and Innovation, Victoria University, Melbourne, VIC, Australia

ROSS A. BRADSTOCK
Centre for Sustainable Ecosystem Solutions, School of Biological Sciences, University of Wollongong, Wollongong, NSW, Australia

MARK C. BRUNDRETT
Department of Parks and Wildlife, Swan Region, WA, Australia and School of Plant Biology, University of Western Australia, Perth, WA, Australia

DON W. BUTLER
Queensland Herbarium, Toowong, QLD, Australia

SAMANTHA J. CAPON
Australian Rivers Institute, Griffith University, Nathan, QLD, Australia

JANE A. CATFORD
Centre for Biological Sciences, University of Southampton, Southampton, UK; and Department of Ecology, Evolution and Behavior, University of Minnesota, St Paul, MN, USA

PHILIP CLARKE
Griffith and Federation Universities, Brisbane, QLD, Australia and South Australian Museum, Adelaide, SA, Australia

GARRY D. COOK
CSIRO Ecosystem Sciences, Winnellie, NT, Australia

MICHAEL M. DOUGLAS
School of Earth and Environment, University of Western Australia, Perth, WA, Australia

DAVID J. ELDRIDGE
Centre for Ecosystem Science, School of Biological, Earth and Environmental Sciences, University of New South Wales, Sydney, NSW, Australia

EMILIE ENS
Department of Environmental Sciences, Macquarie University, Sydney, NSW, Australia

AGUSTÍN F. FACELLI
Department of Environment, Water and Natural Resources of South Australia, Adelaide, SA, Australia

JOSÉ M. FACELLI
School of Biological Sciences, Department of Ecology and Environmental Sciences, The University of Adelaide, Adelaide, SA, Australia

ROD J. FENSHAM
School of Biological Sciences, University of Queensland, St Lucia, QLD, Australia; and Queensland Herbarium, Toowong, QLD, Australia

PAUL W. FOREMAN
Department of Ecology, Environment and Evolution, La Trobe University, Bundoora, VIC, Australia

RAY H. FROEND
School of Natural Sciences, Edith Cowan University, Joondalup, WA, Australia

RACHAEL V. GALLAGHER
Department of Biological Sciences, Macquarie University, Sydney, NSW, Australia

LOUISE GILFEDDER
Discipline of Geography and Spatial Sciences, School of Land and Food, University of Tasmania, Hobart, TAS, Australia

ROBERT GODFREE
CSIRO, National Collections and Marine Infrastructure, Canberra, ACT, Australia

MEGAN GOOD
Federation University, Ballarat, VIC, Australia

CARL R. GOSPER
CSIRO Land and Water, Floreat, WA, Australia; and Science and Conservation Division, Department of Parks and Wildlife, Kensington, WA, Australia

PETER T. GREEN
Department of Ecology, Environment and Evolution, La Trobe University, Bundoora, VIC, Australia; and Long Term Ecological Research Network, Terrestrial Ecosystem Research Network, Canberra, ACT, Australia

TONY GRICE
School of Marine and Tropical Biology, James Cook University, Townsville, QLD, Australia

PAULINE F. GRIERSON
School of Biological Sciences, University of Western Australia, Perth, WA, Australia

SIMON HABERLE
Department of Archaeology and Natural History, Australian National University, Canberra, ACT, Australia

TOM D. HARWOOD
CSIRO Land and Water, Canberra, ACT, Australia

STEPHEN D. HOPPER
Centre of Excellence in Natural Resource Management and School of Biological Sciences, The University of Western Australia, Albany, WA, Australia

ISAAC B. JANSENS
NSW Office of Environment and Heritage, Hurstville, NSW, Australia

PETER JOBSON
Flora and Fauna Division, Department of Land Resource Management, Northern Territory Government, Alice Springs, NT, Australia

GREGORY J. JORDAN
School of Plant Science, University of Tasmania, Hobart, TAS, Australia

DAVID A. KEITH
Centre for Ecosystem Science, School of Biological, Earth and Environmental Sciences, University of New South Wales, Kensington, NSW, Australia; NSW Office of Environment and Heritage, Hurstville, NSW, Australia; and Long Term Ecological Research Network, Terrestrial Ecosystem Research Network, Canberra, ACT, Australia

A. PETER KERSHAW
School of Earth, Atmosphere and Environment, Monash University, Clayton, VIC, Australia

JAMIE KIRKPATRICK
Discipline of Geography and Spatial Sciences, School of Land and Food, University of Tasmania, Hobart, TAS, Australia

BYRON B. LAMONT
Department of Environment and Agriculture, Curtin University, Perth, WA, Australia

MICHELLE R. LEISHMAN
Department of Biological Sciences, Macquarie University, Sydney, NSW, Australia

ADAM C. LIEDLOFF
CSIRO Land and Water, Winnellie, NT, Australia

CATHERINE LOVELOCK
School of Biological Sciences, University of Queensland, St Lucia, QLD, Australia

HARRY J. MACDERMOTT
School of Biological Sciences, University of Queensland, St Lucia, QLD, Australia

ADRIAN D. MANNING
Fenner School of Environment and Society, The Australian National University, Canberra, ACT, Australia

KEITH MCDOUGALL
Department of Ecology, Environment and Evolution, La Trobe University, Bundoora, VIC, Australia

DANIEL J. METCALFE
CSIRO Landscape Management, Brisbane, QLD Australia; and Long Term Ecological Research Network, Terrestrial Ecosystem Research Network, Canberra, ACT, Australia

BEN P. MILLER
Biodiversity Conservation Centre, Botanic Gardens and Parks Authority, Kings Park, WA, Australia

SCOTT D. MOONEY
School of Biological, Earth and Environmental Sciences, University of New South Wales Australia, NSW, Australia

JOHN W. MORGAN
Department of Ecology, Environment and Evolution, La Trobe University, Bundoora, VIC, Australia; and Long Term Ecological Research Network, Terrestrial Ecosystem Research Network, Canberra, ACT, Australia

BRETT P. MURPHY
Research Institute for Livelihoods and the Environment, Charles Darwin University, Darwin, NT, Australia

CATHERINE NANO
Flora and Fauna Division, Department of Land Resource Management, Northern Territory Government, Alice Springs, NT, Australia

JOHN NELDNER
Department of Science, Information Technology and Innovation, Queensland Herbarium, Toowong, QLD, Australia

MARK K.J. OOI
Centre for Ecosystem Science, School of Biological, Earth and Environmental Sciences, University of New South Wales, Sydney, NSW, Australia; and Centre for Sustainable Ecosystem Solutions, School of Biological Sciences, University of Wollongong, Wollongong, NSW, Australia

NEIL PETTIT
University of Western Australia, Perth, WA, Australia; and Edith Cowan University, Joondalup, WA, Australia

SUZANNE M. PROBER
CSIRO Land and Water, Floreat, WA, Australia

JANE ROBERTS
Institute of Applied Ecology, University of Canberra, Canberra, ACT, Australia

JESSICA ROE
School of Biological, Earth and Environmental Sciences, University of New South Wales, Sydney, NSW, Australia; and UNSW Learning, Design and Development Service, University of New South Wales, Sydney, NSW, Australia

KERRYLEE ROGERS
School of Earth and Environmental Sciences, University of Wollongong, Wollongong, NSW, Australia

NEIL SAINTILAN
Department of Environmental Sciences, Macquarie University, Sydney, NSW, Australia

SAMANTHA SETTERFIELD
Faculty of Science, University of Western Australia, Perth, Australia

CHRISTOPHER C. SIMPSON
NSW Office of Environment and Heritage, Hurstville, NSW, Australia; and Centre for Ecosystem Science, School of Biological, Earth and Environmental Sciences, University of New South Wales, Kensington, NSW, Australia

RHIANNON SMITH
University of New England, Armidale, NSW, Australia

KALE SNIDERMAN
School of Earth Sciences, University of Melbourne, Melbourne, VIC, Australia

KEVIN R. THIELE
School of Plant Biology, The University of Western Australia, Crawley, WA, Australia

MARK G. TOZER
NSW Office of Environment and Heritage, Hurstville, NSW, Australia; and Centre for Ecosystem Science, School of Biological, Earth and Environmental Sciences, University of New South Wales, Kensington, NSW, Australia

SAMANTHA K. TRAVERS
Centre for Ecosystem Science, School of Biological, Earth and Environmental Sciences, University of New South Wales, Sydney, NSW, Australia

SUSANNA VENN
Evolution, Ecology and Genetics, Research School of Biology, The Australian National University, Canberra, ACT, Australia

FIONA WALSH
CSIRO, Land and Water, Alice Springs, NT, Australia

NEVILLE WALSH
Royal Botanic Gardens Victoria, South Yarra, VIC, Australia

GRANT WARDELL-JOHNSON
School of Science, Curtin University, Perth, WA, Australia

GLENDA M. WARDLE
Desert Ecology Research Group, School of Life and Environmental Sciences, University of Sydney, NSW, Australia; and Long Term Ecological Research Network, Terrestrial Ecosystem Research Network, Canberra, ACT, Australia

CHRISTOPHER J. WESTON
School of Ecosystem and Forest Sciences, The University of Melbourne, Melbourne, VIC, Australia

PETER H. WESTON
National Herbarium of New South Wales, Royal Botanic Gardens and Domain Trust, Sydney, NSW, Australia

JENNIE WHINAM
Discipline of Geography and Spatial Sciences, School of Land and Food, University of Tasmania, Hobart, TAS, Australia

KRISTEN J. WILLIAMS
CSIRO Land and Water, Canberra, ACT, Australia

RICHARD J. WILLIAMS
CSIRO Land and Water, Winnellie, NT, Australia; and Long Term Ecological Research Network, Terrestrial Ecosystem Research Network, Canberra, ACT, Australia

SARAS M. WINDECKER
School of BioSciences, The University of Melbourne, Melbourne, VIC, Australia

COLIN J. YATES
Science and Conservation Division, Department of Parks and Wildlife, Kensington, WA, Australia

FOREWORD TO THE THIRD EDITION

Richard H. Groves

Attempts to classify and name the various major types of Australian vegetation have a long history. The Aboriginal peoples who first settled the continent recognised some different plant assemblages and assigned them descriptive words in their own languages, such as brigalow, kwongan, mallee, mulga and wallum, all of which remain in general use by Australians today. On 6 August 1699, William Dampier wrote of the vegetation at Shark's Bay in Western Australia:

> The Grass grows in great Tufts, as big as a Bushel, here and there a Tuft: Being intermix'd with much Heath. [. . .] Of Trees or Shrubs here are divers Sorts; but none above 10 Foot high: There Bodies about 3 Foot about, and 5 or 6 Foot high. [. . .] There were also beside some Plants, Herbs, and tall Flowers, some very small Flowers, growing on the Ground, that were sweet and beautiful, and for the most part unlike any I had seen elsewhere.

Spencer (1981, p. 108)

This is most likely the first description of an Australian vegetation type. Subsequent early European collectors, such as Robert Brown, or visitors such as Charles Darwin, were also intrigued by the different types of Australian vegetation they encountered, and especially by the eucalypt forests. The predominant emphasis over this early period was, however, on collecting, naming and describing the indigenous flora. It was really only in the late nineteenth century with the development of the new discipline of oecology in the northern hemisphere that European botanists and plant geographers began to describe vegetation types and attempt to classify them systematically (see e.g. Schimper 1903; Diels 1906; Warming 1909). Resident Australian botanists soon followed this development (Osborn 1914; Andrews 1916; Adamson and Osborn 1924) and some notable regional accounts of natural and modified vegetation were published subsequently (Wood 1937; Beadle 1948; Moore 1953a, b; Costin 1954).

To map the vegetation at a national scale, however, a consensual and rigorous classification of vegetation types was necessary; a terminological need met initially by Wood (1939), Beadle and Costin (1952) and Wood and Williams (1960), and subsequently by Specht (1970, 1981). Specht's classification used a matrix of two structural aspects – namely, life-form and foliage projective cover of the tallest stratum – to characterise 28 major vegetation types of Australia (Table F.1). It was this 1970 classification that Carnahan (1976) used to map the natural vegetation of Australia and subsequently to map the present vegetation types (Carnahan 1990). Specht's 1970 classification also largely formed the basis of accounts of the major vegetation types in the two earlier editions of this book (Groves 1981, 1994), although less so in other accounts, either national (Beadle 1981) or regional (Beard 1976; Kirkpatrick and Dickinson 1984; Keith 2004).

Table F.1 **Structural formations of Australian vegetation omitting the littoral and alpine complexes, and salt lakes (abridged from Specht 1970).**

	Foliage projective cover of tallest stratum			
Life form and height of tallest stratum	100–70%	70–30%	30–10%	<10%
Trees[a] > 30 m	Tall closed-forest	Tall open-forest	(Tall woodland)	(Tall open-woodland)
Trees 10–30 m	Closed-forest	Open forest	Woodland	Open-woodland
Trees < 10 m	Low closed-forest	Low open forest	Low woodland	Low open-woodland
Shrubs[a] > 2 m	Closed-scrub	Open-scrub	Tall shrubland	Tall open-shrubland
Shrubs 0.25–2 m[b]	Closed-heath	Open-heath	Low shrubland	Low open-shrubland
Shrubs < 0.25 m[b]	–	–	Fellfield	Open-fellfield
Hummock grasses < 2 m	–	–	Hummock grassland	Open-hummock grassland
Graminoids, herbs, ferns, etc.	Closed-grassland	Grassland	Open-grassland	Ephemeral herbland

[a] A *tree* is defined as a woody plant usually with a single stem; a *shrub* is a woody plant with many stems arising at or near the base.
[b] These categories may be subdivided into (1) sclerophyllous and (2) semi-succulent/succulent.

Current descriptions of major Australian vegetation types reflect this century-long history of attempts to classify vegetation types and usually comprise a mixture of structural and floristic nomenclature, as implicit in terms such as chenopod shrublands (Chapter 24). In this current edition, some environmental descriptors are added, e.g. 'dry' sclerophyll forest, even when this assemblage of sclerophyllous trees and shrubs may grow in moderately moist conditions, though without the prominence of ferns which seems to characterise so-called 'wet' sclerophyll forest (see Chapters 12 and 13). Both 'dry' and 'wet' sclerophyll forests are open forests in Specht's 1970 terminology (Table F.1). So-called 'desert' vegetation is common over extensive areas of central Australia, but is really a mixture of hummock grassland (Chapter 26), *Acacia* open-woodland and shrubland (Chapter 25), and low eucalypt open-woodland (Chapter 23) in varying proportions and is not an ecologically valid term for this admixture of several vegetation types that characterise Australian arid environments. At the other extreme from Specht's structural classification, Bridgewater (1994) presented a system for classifying Australian vegetation on a solely floristic basis. Each of his 26 divisions was defined by a suite of 'character genera' (see Groves (1999) for a further discussion of this floristic classification). Whatever the nomenclatural usages and bases for classification by individual botanists, Australian vegetation has increasingly captured the attention of plant ecologists over the last century or more, especially since the first edition of this book was published in 1981.

Despite the contemporary strength of Australian vegetation studies (as evidenced in the chapters that follow), I wish to introduce a few advances in vegetation science over the 35-year period since the earlier editions of this book were published. I shall also point to some deficiencies in coverage of earlier strengths as I perceive them.

The first advance seems to me to be the linking of vegetation type and functionality, as implicit in the title to Part I of this book and Chapter 1. An initially descriptive science of studying vegetation patterns at a continental scale now includes a consideration of functional aspects, such as seasonal water usage and nutrient cycling, as drivers of the dynamics of a vegetation type and, ultimately, its distribution nationally.

A second and allied advance is the enhanced study of fire effects on the distribution of a range of vegetation types. Gill (1975) introduced the concept of a 'fire regime' to begin to refine studies of fire ecology in Australia and elsewhere, but especially in relation to adaptive traits shown by plant species and, indeed, by sclerophyll vegetation generally, in responding to fires in Mediterranean-climate ecosystems (Gill 1977). The concept of the fire regime (a function of the frequency, intensity and season of fire) unified understanding of previous local descriptions of wild fires and enabled a recent attempt to predict the effects of changing fire regimes on the responses of woody plants at a continental scale (Clarke *et al.* 2015; Chapter 5).

Mapping and classification of Australian vegetation types has advanced greatly with the development of computers and geographic information systems (GIS). The initial intuitive classification of Australian rainforests of Webb (1959) was developed further only 10 or more years later into a computerised numerical analysis of field data (Webb *et al.* 1970, 1976) that enabled classification of a range of structural subtypes of rainforest and their different environmental relationships. So too the use of GIS methods to generate computer-enabled maps of Australian vegetation, such as that of the National Vegetation Information System (NVIS) of the Department of the Environment and Energy (2007), the products of which now replace the earlier hard-copy maps of Carnahan (1976, 1990). The latest NVIS version (4.2) maps 79 subgroups of native vegetation types in terms of both structure and floristics, e.g. *Banksia* woodlands, *Callitris* open woodlands, and 6 subgroups of non-native vegetation.

Some earlier strengths of Australian vegetation science have not advanced commensurately as the three topics I have pointed to above, however. In the first edition of this book, I noted that pollination biology was an aspect 'greatly in need of more research' and this need continues 35 years later. A further deficiency is the apparent lack of newly initiated long-term field studies of vegetation change of the type initiated by the Botany Department of the University of Adelaide at Koonamore (Hall *et al.* 1964; Sinclair 2005) in 1926 or on the Bogong High Plains initiated by the Botany Department of the University of Melbourne in 1947 (Carr and Turner 1959; Wahren *et al.* 1994). Although these are ongoing (Foulkes *et al.* 2014; Williams *et al.* 2014), some important long-term studies have fallen dormant, such as vegetation monitoring over 20 years the since cessation of grazing of alpine and sub-alpine plots in Kosciusko National Park (Wimbush and Costin 1979a, b). A notable exception to this apparent decline in the effects of time on grazed vegetation is the recent account of floristic changes over a 46-year period in mulga (*Acacia aneura*) vegetation in arid Western Australia by Davies *et al.* (2015).

Each edition of *Australian Vegetation* contains a chapter on the vegetation history of Australia (see Chapters 2 and 3). And yet it seems to me that currently there is a decline in studies of the important and longer-term effects of time on the dynamics of Australian vegetation types. The current absence of studies of long-term changes in vegetation type of the sort evidenced by the earlier studies of Singh at Lake George, New South Wales, and Kershaw on the Atherton Tableland, Queensland is especially significant (Singh *et al.* 1981). The results of both studies show an alternation between sclerophyll and rainforest vegetation with changing climates and the incidences of fires over a period as long as the last 100 000 years. Such studies offer valuable perspectives on recent vegetation history over much of eastern Australia and the effects of changing climates on vegetation distribution; they also provide clues to the resilience of such vegetation to human-induced climate change in future time.

I conclude that Australian studies in vegetation science have advanced considerably, both in number and in quality, since 1980. There have been some obvious strengths and weaknesses in research coverage and I have pointed to a few examples in each category. Previous editions of this book provide markers of where the science was at in both 1981 and in 1994, as seen through the eyes of the many contributors to both editions and this former editor. So too will the chapters in this third edition and the personal predilections of the present editor and the authors of the 27 chapters that follow this Foreword. I trust this third edition will continue to provide stimulus to graduate students in botany and visitors to this country who wish to learn more of the structural, floristic and functional aspects of Australian vegetation. I can attest personally to a continuing fascination with the diversity and beauty of the vegetation of Australia, especially as evoked by its eucalypt forests, its sclerophyll shrublands (heaths) and its natural grasslands.

R. H. Groves
CSIRO Honorary Research Fellow, Canberra

References

Adamson, R.S. and Osborn, T.G.B. (1924) The ecology of the *Eucalyptus* forests of the Mount Lofty Ranges (Adelaide district), South Australia. *Transactions of the Royal Society of South Australia* 48, 87–144.
Andrews, E.C. (1916) The geological history of the Australian flowering plants. *American Journal of Science* 42, 171–232.
Beadle, N.C.W. (1948) *The Vegetation and Pastures of Western New South Wales with Special Reference to Soil Erosion.* Government Printer: Sydney.
Beadle, N.C.W. (1981) *The Vegetation of Australia.* Cambridge University Press: Cambridge.
Beadle, N.C.W. and Costin, A.B. (1952) Ecological classification and nomenclature. *Proceedings of the Linnean Society of New South Wales* 77, 61–82.
Beard, J.S. (1976) *Vegetation Survey of Western Australia, Murchison.* University of Western Australia Press: Nedlands.

Bridgewater, P.B. (1994) Conspectus of major vegetation units for Australia. *Phytocoenologia* 24, 283–297.

Carnahan, J.A. (1976) Natural vegetation. In '*Atlas of Australian Resources.' Second Series*. Department of Natural Resources: Canberra.

Carnahan, J.A. (1990) *Atlas of Australian Resources. Third Series. Volume 6. Vegetation*. Australian Surveying and Land Information Group: Canberra.

Carr, S.G.M. and Turner, J.S. (1959) The ecology of the Bogong High Plains. I. The environmental factors and the grassland communities. *Australian Journal of Botany* 7, 12–33.

Clarke, P.J., Lawes, M.J., Murphy, B.P. *et al.* (2015) A synthesis of postfire recovery traits of woody plants in Australian ecosystems. *Science of the Total Environment* 534, 31–42.

Costin, A.B. (1954) *A Study of the Ecosystems of the Monaro Region of New South Wales*. Government Printer: Sydney.

Davies, S.J.J.F., Kenny, S.A. and Walsh, T.F.M. (2015) The population dynamics of some arid zone plants during 46 years of grazing on Mileura Station, Western Australia. *The Rangeland Journal* 37, 367–377.

Department of the Environment and Energy (2007) *NVIS data products 4.2*. Australian Government: Canberra. Available at: http://www.environment.gov.au/land/native-vegetation/national-vegetation-information-system/data-products (accessed 10 May 2016).

Diels, L. (1906) *Die Vegetation der Erde. VII. Die Planzenwelt von west-Australien sudlich des Wandekraises*. W. Engelmann: Leipzig.

Foulkes, J., de Preu, N., Sinclair, R. *et al.* (2014) Chenopod and acacia shrublands. In *Biodiversity and Environmental Change: Monitoring, Challenges and Direction*, ed. D. Lindenmayer, E. Burns, N. Thurgate, A. Lowe. CSIRO Publishing: Melbourne, pp. 379–438.

Gill, A.M. (1975) Fire and the Australian flora: a review. *Australian Forestry* 38, 1–25.

Gill, A.M. (1977) Plant traits adaptive to fires in Mediterranean land systems. In *Proceedings of the Symposium on the Environmental Consequences of Fire and Fuel Management in Mediterranean Ecosystems*. USDA Forest Service General Technical Report WO-3: Washington.

Groves, R.H. (1981) *Australian Vegetation*. Cambridge University Press: Cambridge.

Groves, R.H. (1994) *Australian Vegetation*. 2nd edition. Cambridge University Press: Cambridge.

Groves, R.H. (1999) Present vegetation types. In *Flora of Australia*, Vol. 1, 2nd edition. ABRS/CSIRO Australia: Melbourne.

Hall, E.A.A., Specht, R.L. and Eardley, C.M. (1964) Regeneration of the vegetation on Koonamore vegetation reserve, 1926–1962. *Australian Journal of Botany* 12, 205–264.

Keith, D. (2004) *Ocean Shores to Desert Dunes: The Native Vegetation of New South Wales and the ACT*. Department of Environment and Conservation (NSW): Hurstville.

Kirkpatrick, J.B. and Dickinson, K.J.M. (1984) *Vegetation of Tasmania*. Forestry Commission of Tasmania: Hobart.

Moore, C.W.E. (1953a) The vegetation of the south-eastern Riverina, New South Wales. I. The climax communities. *Australian Journal of Botany* 1, 485–547.

Moore, C.W.E. (1953b) The vegetation of the south-eastern Riverina, New South Wales. II. The disclimax communities. *Australian Journal of Botany* 1, 548–567.

Osborn, T.G.B. (1914) Sketches of vegetation at home and abroad. VIII. Notes on the flora around Adelaide, South Australia. *New Phytologist* 13, 109–121.

Schimper, A.F.W. (1903) *Plant Geography upon a Physiological Basis*. Oxford University Press: Oxford.

Sinclair, R. (2005) Long-term changes in vegetation, gradual and episodic, on the TGB Osborn Vegetation Reserve, Koonamore, South Australia (1926–2002). *Australian Journal of Botany* 53, 283–296.

Singh, G., Kershaw, A.P. and Clark, R. (1981) Quaternary vegetation and fire history of Australia. In *Fire and the Australian Biota. I*. Australian Academy of Science: Canberra.

Specht, R.L. (1970) Vegetation. In *'The Australian Environment'*, 4th edition. CSIRO Australia and Melbourne University Press: Melbourne.

Specht, R.L. (1981) Major vegetation formations in Australia. In *Ecological Biogeography of Australia*. Dr W. Junk: The Hague.

Spencer, J. (1981). *William Dampier 1652–1715. A Voyage to New Holland: the English voyage of discovery to the South Seas in 1699*. Alan Sutton: Gloucester, UK.

Wahren, C.-H.A., Papst, W.A. and Williams, R.J. (1994) Long-term vegetation change in relation to cattle grazing in sub-alpine grassland and heathland on the Bogong High Plains. An analysis of vegetation records from 1945 to 1994. *Australian Journal of Botany* 42, 607–639.

Warming, E. (1909) *Oecology of Plants*. Oxford University Press: Oxford.

Webb, L.J. (1959) A physiognomic classification of Australian rainforests. *Journal of Ecology* 40, 551–570.

Webb, L.J., Tracey, J.G., Williams, W.T. and Lance, G.N. (1970) Studies in the numerical analysis of complex rainforest communities. V. A comparison of the properties of floristic and physiognomic-structural data. *Journal of Ecology* 58, 203–232.

Webb, L.J., Tracey, J.G. and Williams, W.T. (1976) The value of structural features in tropical rainforest typology. *Australian Journal of Ecology* 1, 3–28.

Wimbush, D.J. and Costin, A.B. (1979a) Trends in vegetation at Kosciusko. II. Sub-alpine range transects 1959–1978. *Australian Journal of Botany* 27, 789–831.

Wimbush, D.J. and Costin, A.B. (1979b) Trends in vegetation at Kosciusko. III. Alpine range transects 1959–1978. *Australian Journal of Botany* 27, 833–871.

Wood, J.G. (1937) *The Vegetation of South Australia*. Government Printer: Adelaide.

Wood, J.G. (1939) Ecological concepts and nomenclature. *Transactions of the Royal Society of South Australia* 63, 215–223.

Wood, J.G. and Williams, R.J. (1960) Vegetation. In *The Australian Environment*. 3rd edition. CSIRO Australia: Melbourne.

ACKNOWLEDGEMENTS

I am indebted to Richard Groves, who first suggested that I might continue the valuable legacy that he established with two seminal editions of *Australian Vegetation* in 1981 and 1994. Some considerable time later, Alan Crowden pursued me through conference crowds to secure an agreement. Both Richard and Alan provided invaluable and realistic advice on the challenge ahead. I am grateful to Lindsey Tate, Judith Shaw and Geetha Williams who managed production, copy editing and typesetting, respectively. Max McMaster produced a meticulous and comprehensive index.

It's been a great pleasure and privilege to work with 80 of Australia's leading botanists who responded to the call and produced something special in these pages. A much larger network of our colleagues were influential on the contents of the chapters through their published work and discussion of ideas. I especially thank the lead authors who responded to my suggestions, criticisms and threats in good spirit. Equally important, are those who contributed peer reviews of submitted manuscripts. Thanks to all those listed below for making this a better work.

The compilation of enormous volumes of map data has been one of the pivotal, but largely unseen tasks that differentiates this edition from previous ones. Chris Simpson provided indispensable GIS expertise in making order from many thousands of pieces that were sent down the wire. In the end, he fled to the ski fields.

Chantel Benbow proofread the entire manuscript and maintained sanity, composure and a truly meticulous eye for consistency and detail to the last word.

I am deeply grateful to my family for their support and understanding when I disappeared into the place in my head where this book took shape, and for their faith that 'he'll be back some day'.

Finally, to the subject itself. How lucky we are to live amongst it.

Scientific reviewers

Associate Professor Paul Adam, University of New South Wales
Dr Alan Andersen, CSIRO Land and Water
Dr Tony Auld, New South Wales Office of Environment and Heritage
Dr Teresa Ayre, Queensland Herbarium
Dr Jayne Balmer, Department of Primary Industries, Water and Environment, Tasmania
Dr Margaret Brock, University of Tasmania
Dr Elisabeth Bui, CSIRO Land and Water
Associate Professor Tina Bell, University of Sydney
Dr Margaret Byrne, Department of Parks and Wildlife, Western Australia
Dr James Camac, Macquarie University

Dr Samantha Capon, Griffith University
Dr Jane Catford, University of Minnesota
Professor Carla Catterall, Griffith University
Dr Will Cornwell, University of New South Wales
Professor John Dodson, Australian Nuclear Science and Technology Organisation
Mr Michael Doherty, CSIRO Land and Water
Dr Josh Dorrough, Natural Regeneration Australia
Professor David Eldridge, University of New South Wales
Dr Rod Fensham, Queensland Herbarium
Dr Rachael Gallagher, Macquarie University
Dr Neil Gibson, Department of Parks and Wildlife, Western Australia
Dr Carl Gosper, CSIRO Land and Water and Department of Parks and Wildlife, Western Australia
Professor Bob Hill, University of Adelaide
Professor Rod Keenan, University of Melbourne
Professor Peter Kershaw, Monash University
Professor Richard Kingsford, University of New South Wales
Associate Professor Andrew Le Brocque, University of Southern Queensland
Dr Steven Leonard, La Trobe University
Mr Berin MacKenzie, New South Wales Office of Environment and Heritage
Dr Sue McIntyre, Australian National University
Dr John Morgan, La Trobe University
Dr Steve Morton, CSIRO Land and Water
Professor Laco Mucina, University of Western Australia
Professor Adrienne Nicotra, Australian National University
Dr Mark Ooi, University of New South Wales
Dr Neil Pettit, University of Western Australia
Dr Owen Price, University of Wollongong
Dr Lynda Prior, University of Tasmania
Dr Suzanne Prober, CSIRO Land and Water
Dr Maurizio Rossetto, National Herbarium of New South Wales
Professor Peter Saenger, Southern Cross University
Dr David Tng, James Cook University
Mr Mark Tozer, New South Wales Office of Environment and Heritage
Associate Professor Peter Vesk, University of Melbourne
Dr John Virtue, Primary Industries and Regions, South Australia
Dr Paul Williams, James Cook University
Dr Boyd Wright, Alice Springs Herbarium

PART I Ecological Processes

1 Girt: A Continental Synthesis of Australian Vegetation

DAVID A. KEITH AND MARK G. TOZER

Australian vegetation, much like that on other continents, is characterised by evergreen forests in humid climates, savannas in the seasonal tropics, desert shrublands and grasslands in arid climates, and wetland vegetation where water accumulates. While continental plant diversity is intermediate on a global scale, Australian vegetation is characterised by high endemism at species and genus levels, a number of biodiversity hotspots and some structurally and physiognomically unique forms. We propose a continental model of vegetation assembly in which autochthonous and intrusive species pools are exposed to resource filters dominated by water and nutrients, other abiotic drivers such as temperature and disturbance regimes (particularly fire and floods), as well as local biotic filters, most notably competition and predation. We present a new high-level ecological typology of Australian vegetation and evaluate how ecological processes and legacies shape and differentiate major types of vegetation across the continent. Some key plant traits apparently enable persistence and turnover of species groups that characterise vegetation types inhabiting different environments. We conclude that Australian vegetation reflects a broad ensemble of evolutionary filters, varying combinations of which may come to the fore under particular conditions. Advances in Australian plant ecology over more than a century increasingly reflect the complex relationships between ecological legacy, directional selection by contrasting agents under varying environments, and stochasticity.

1.1 Introduction

The Australian continent and associated islands support a rich tapestry of vegetation composed of forests, woodlands, shrublands, grasslands and wetlands. In a broad sense, these patterns and the major processes that shape them are common features shared by other continents (Walter 1979). Like other continents, evergreen forests dominate Australia's high-rainfall regions, savannas dominate the seasonal tropics, desert shrublands and grasslands dominate the arid subtropics, and wetland vegetation is found where water accumulates (Beadle 1981). Even cursory examination, however, reveals unique compositional features and ecological behaviours.

Published estimates suggest more than 20 000 described vascular plant species are native to Australia, of which more than 90% are endemic (Chapman 2009). Although these are likely substantial underestimates due to ongoing discovery, high levels of endemism are unequivocal and hotspots of diversity occur in contrasting environments that span wet tropical and seasonal temperate climates, but largely exclude arid climates (Laffan and Crisp 2003).

Citation: Keith, D.A. and Tozer, M.G. (2017). Girt: A Continental Synthesis of Australian Vegetation. In D. A. Keith, ed., *Australian Vegetation*, 3rd edn. Cambridge: Cambridge University Press, pp. 3–39.

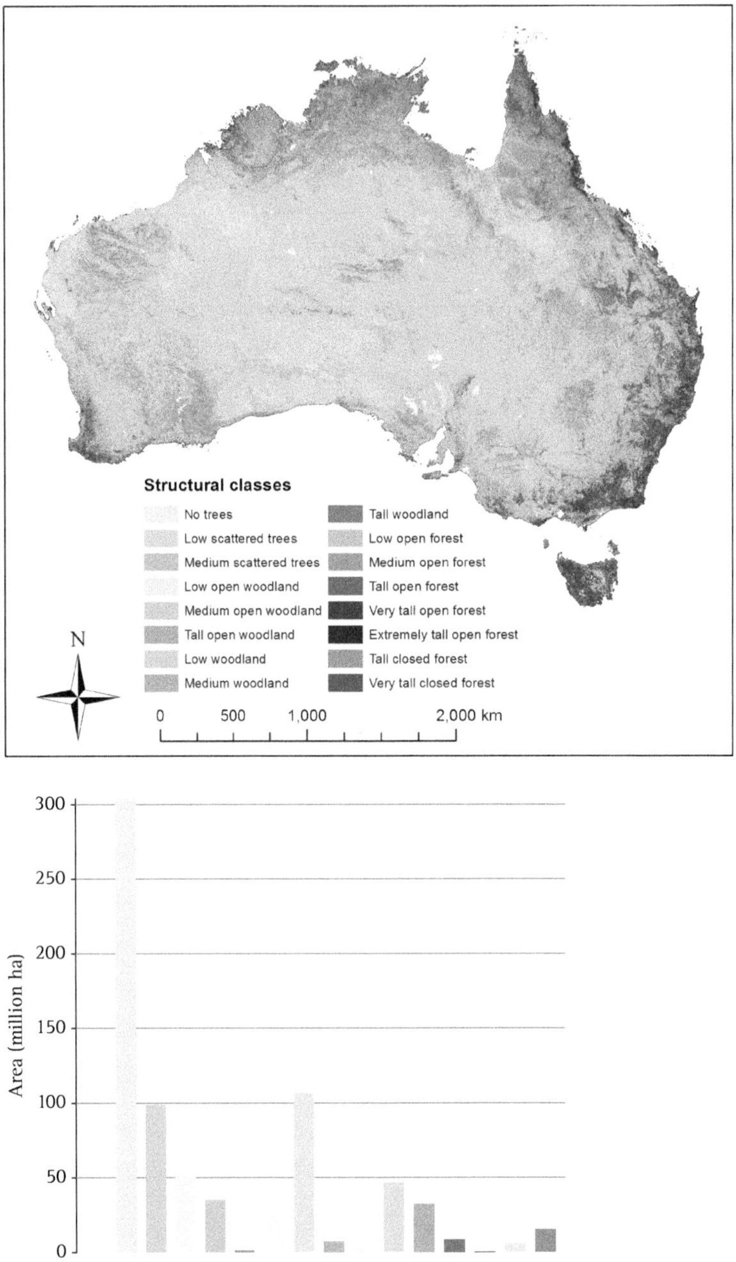

Figure 1.1 Map of vegetation structure of Australia derived from ALOS-1 PALSAR, ICESat/GLAS and Landsat imagery time series for 2003–2009. Histogram shows areas of structural types with a vast core of low open woodland and scattered trees, girt by a semi-continuous band of medium woodland and narrow coastal fringe of forest.
Data from Auscover (2016). (A black and white version of this figure will appear in some formats. For the colour version, please refer to the plate section.)

Australia also includes structurally unique forms of vegetation including open-canopy (eucalypt) forests and woodlands, and hummock grasslands. The forests and woodlands of more humid climes fringe an extensive arid biome of low open vegetation that extends to the west coast (Figure 1.1). Relative to other land masses, Australia has an overrepresentation of sclerophylly and underrepresentation of succulence, while cold deciduous forests and exclusively coniferous forests are virtually absent (Box 1996; Orians and Milewski 2007).

In explaining the uniqueness and analogues of Australian vegetation, researchers have pointed to the biogeographical legacies of long isolation and intermittent long-distance dispersal (Barlow 1994; Crisp and Cook 2013; Weston and Jordan, Chapter 2, this volume); the widespread pressures of seasonal and extended droughts and temperature extremes (Nix 1982; Byrne *et al.* 2008); landscape age and climatic stability (Hopper 2009; Mucina and Wardell-Johnson 2011); nutritional poverty of soils (Lambers *et al.* 2010); and the pervasive influence of wildland fires (Bond *et al.* 2005; Bradstock 2010). The chapters in Part I of this book examine some of the most important ecological processes that shape Australian vegetation in both historical and contemporary time frames. A continental synthesis is needed to explain and predict where particular processes have strong influences on vegetation, where they do not, and why. Such a synthesis requires: (i) a conceptual model that provides a theoretical basis for explanation and prediction; and (ii) a descriptive framework for comparative analysis and expression of ecological relationships.

In this chapter, we first build on community assembly theory (Keddy 1992) to develop a generic conceptual model for understanding how ecological processes shape vegetation at a continental scale. The model emphasises, on one hand, species pools and the selection pressures to which they are exposed and, on the other hand, the resulting expressions of ecological traits, processes and ecosystem behaviour. Second, we propose a general ecological typology of Australian vegetation, defining the units in a way that is spatially explicit at continental scale and reflects the ecological traits of their component species. We briefly review and build on previous descriptive work in this area. We then use the conceptual model and typology to discuss how legacies and ecological processes differentiate major types of vegetation across the continent, focussing on some key plant traits that enable persistence and turnover of component species groups. In doing so, we orientate our interpretations to existing theories on the distribution and character of vegetation at global and continental scales. We conclude by reflecting on the progressive development of continental-scale ecology of Australian vegetation.

1.2 A model of vegetation assembly

Ecological communities are shaped by their natural capital of biological variation and the selective forces that act upon it. The outcomes of this filtering process are reflected in the spectra of ecological traits exhibited by the suite of component species that characterise a given assemblage (Keddy 1992; Grime 2001; Shipley *et al.* 2006; Craine *et al.* 2012). Hence the observed properties of any two assemblages may differ because: (i) selective forces on a common species pool differ in type, direction or relative strength; (ii) a similar suite of selective forces act upon different species pools; or (iii) both species pools and selective forces differ. Neutral processes, such as stochastic arrivals and extinctions (Crisp and Cook 2013), stochastic variation in the filtering process and equivalence of competitive abilities may also play a role in differentiating ecological assemblages, even where species pools, traits and selection pressures are similar (Hubbell 2005; Chase 2010).

The relative influence of niche differentiation and neutral processes in structuring local communities depends on the balance between environmental variation (both spatial and temporal) and fitness differences (Adler *et al.* 2007), as well as the degree of resource partitioning and effects of natural enemies (Adler *et al.* 2013). All of these elements vary with spatial scale, with different suites of selective forces and neutral processes shaping biomes at continental and global scales (Box 1995) and ecological communities at local scales (Keddy 1992). In this chapter, we focus on continental-scale vegetation types, and propose a conceptual model to represent key elements and processes involved in their assembly (Figure 1.2).

The species pool is central to biome assembly (Figure 1.2), as it defines the evolutionary capital on which filters, stabilising mechanisms and neutral processes operate. It reflects legacies of historical biogeography (Crisp and Cook 2013; Weston and Jordan, Chapter 2, this volume) as well as antecedent filtering and stochastic events. Four main groups of drivers operate on traits represented within the species pool to shape the assemblage in a particular area at a particular time. We separate the supply of primary resources (Craine *et al.* 2012) from other abiotic drivers that influence their availability or the ability to acquire them (Figure 1.2). The latter include ambient factors such as temperature and salinity, as well as naturally occurring toxins such as aluminium (an inhibitor of root growth) and heavy metals. Another group of environmental filters includes disturbance regimes, notably sequences of fire, flood and storm events, which interrupt resource acquisition and indeed liberate resources from existing biota or

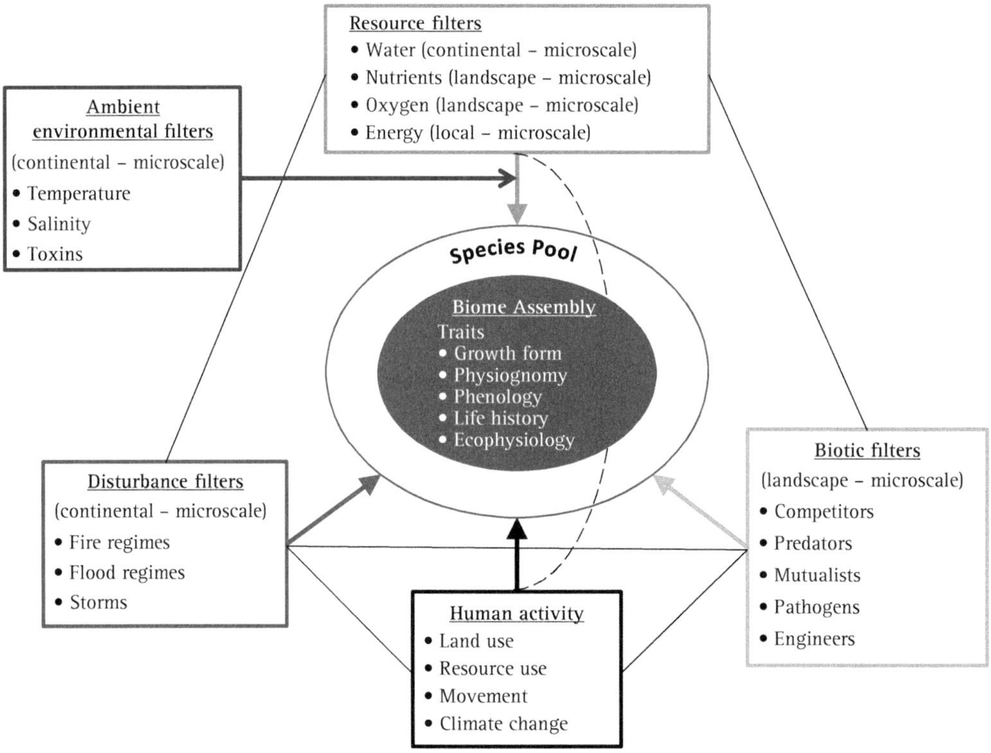

Figure 1.2 Conceptual model of vegetation assembly showing influence of, and interactions between, the main groups of ecosystem drivers and the spatial scales over which they vary (see text for details).
Note: Interactions between human activity and resource drivers are present but not shown in graphic display. Ambient environmental filters also influence biotic processes, disturbance and human activity (arrows not shown).

other sources. The biotic drivers include filters, such as interspecific competition and pathogens, as well as a range of stabilising or compensatory mechanisms that may promote coexistence (Figure 1.2). We distinguish human activity from other biotic drivers to emphasise its influence on biome assembly through exploitation of resources (including biota), movement of biota and modification of global circulation systems.

Spatial and temporal heterogeneity regulate the influence of the four main groups of drivers on biome assembly. Spatial heterogeneity is critical to niche partitioning, with different elements of each driver varying at a range of spatial scales. For example, the availability of water varies with climate at global scales and also with variations in microtopography. Other drivers vary over a narrower range of scales, such as nutrient availability, which is strongly influenced by substrate distributions and topography at landscape scales. Temporal variability in biome drivers is also critical to coexistence by disrupting competitive elimination (Tilman 1988; Keddy 1989) or through mechanisms such as storage effects (Chesson 2000). Some elements, such as water availability, vary over a broad range of temporal scales from days to decades, while others, such as energy influx, vary on a daily and seasonal basis. Disturbance, by definition, involves temporal variation expressed through the duration of events and intervals between them (e.g. Bunn *et al.* 2006).

Finally, interactions among drivers may inflate or moderate their influences on biome assembly (Figure 1.2). For example, an overabundance of water may generate anoxia in plant substrates (Vartapetian and Jackson 1997). Herbivore activity may influence fire regimes through their effects on available fine fuels (Keeley *et al.* 2012). Human activities may be major mediators of other drivers. Water extraction and storage on industrial scales, for example, limit availability of a key resource and transform hydrological flows and flood regimes critical to the persistence of some biota (Bunn and Arthington 2002).

1.3 An ecological typology of Australian vegetation

The direct outcomes of assembly processes are expressed in the composition and trait spectra of ecological communities. Ecological classifications are abstract descriptive tools for summarising patterns in these essentially continuous outcomes and for projecting them collectively across real landscapes (Whittaker 1975). We propose a new typology of Australian vegetation that enables an ecological interpretation of vegetation assembly across the continent and associated islands. Our aim is to build on existing, primarily descriptive typologies to produce a framework that not only represents spatial patterns in species pools, but also reflects similarities and contrasts among plant traits represented in different assemblages and characterises ecological drivers that shape and sustain the biota (Figure 1.2). The framework is hierarchical to enable representation of ecological and evolutionary relationships at different scales. While finer levels of the typology are under development (Keith and Pellow 2015), here, we present the upper level of the hierarchy comprising 16 'vegetation formations' (Table 1.1) based largely on growth form and physiognomic traits that are interpreted as outcomes of ecological drivers and selection processes. There are also strong patterns in floristic and structural features that help identify the units (Table 1.1), although these properties are not the primary basis for classification at the formation level.

We developed a continental view of spatial vegetation patterns at the formation level (Figure 1.3) by re-interpreting national and regional mapping (Department of the Environment and Water Resources 2007). Such mapping is enabled in part by distinctive remote sensing signatures that represent combinations of physiognomic, energetic and structural features of vegetation (Keith 2004). As well, relatively strong environmental relationships among the formations enable their environmental niches to be characterised (Table 1.1; Figure 1.3)

Table 1.1 **Characteristics of Australian vegetation formations and their relationships to other national and international vegetation typologies.**

Australian vegetation formations	Description	Environmental niches	Distribution
Rainforests (Chapter 11)	Closed-canopy non-eucalypt forests, vine scrubs and monsoon forests	Humid–semi-humid, tropical–cool temperate macroclimates and mesoclimates, eutrophic–mesotrophic substrates, rarely fire-prone	Tropics, east coast lowlands and ranges
Wet sclerophyll forests (Chapter 12)	Tall eucalypt forests over mesomorphic shrubs, ferns or herbs	Humid, subtropical–cool temperate macro and mesoclimates, mesotrophic substrates, prone to warm-season century-scale crown fires	East coast lowlands, ranges, far south-west outlier
Dry sclerophyll forests and woodlands (Chapter 13)	Eucalypt forests and woodlands over scleromorphic shrub understories	Humid–semi-humid, aseasonal subtropical–cool temperate macroclimates, mesotrophic–oligotrophic substrates, prone to warm-season decadal fires	East coast lowlands, ranges and slopes, south-west outlier
Heathlands (Chapter 14)	Closed–open scleromorphic shrublands, some with emergent eucalypts	Humid–semi-humid, tropical–cool temperate macroclimates, oligotrophic substrates, prone to warm-season decadal fires	South-west and east coast lowlands and ranges, southern sandplains, tropical plateaus and sandplains
Savanna (Chapters 15, 16)	Eucalypt and other tree genera over C_4 grasses and scattered shrubs	Winter – dry, humid tropical–subtropical macroclimates, eutrophic–mesotrophic substrates, prone to cool-season subdecadal fires	Tropics, eastern subtropics
Temperate subhumid woodlands (Chapter 17)	Eucalypt woodland over mixed C_3/C_4 grasses and forbs with variable shrub cover	Winter – wet or aseasonal, humid–subhumid temperate macroclimates, eutrophic–mesotrophic substrates, prone to warm-season decadal surface fires	South-west hinterlands, south-east coastal valleys, tablelands and western slopes
Tussock grasslands (Chapter 18)	Perennial tussock grasses with interstitial herbs and few or no woody plants	Humid–semi-arid, tropical–temperate macroclimates, eutrophic–mesotrophic periodically damp clay substrates, prone to decadal–subdecadal surface fires	Tropical and arid plains, south-east hinterlands and plateaus, coast and offshore islands

Beadle and Costin (1952); Specht and Specht (2001)	Whittaker (1975)	Faber-Langendoen *et al.* (2014)
Rainforest (temperate, subtropical, tropical and monsoon)	1 Tropical rainforest, 3 Temperate rainforest, 5a Temperate evergreen forest–broadleaf	1.A.2 Tropical lowland humid forest, 1.A.3 Tropical montane humid forest, 1.B.1 Warm temperate forest and Woodland, 1.B.2 Cool Temperate Forest and Woodland
Sclerophyll forest (wet)	5c Temperate evergreen forest–sclerophyll [Mediterranean]	1.B.1 Warm temperate forest and woodland, 1.B.2 Cool temperate forest and woodland
Sclerophyll forest (dry)	5c Temperate evergreen forest–sclerophyll [Mediterranean], 10b Temperate woodland–sclerophyll	1.B.1 Warm temperate forest and woodland, 1.B.2 Cool temperate forest and woodland
Heath, mallee (wet), scrub (wet)	11b Temperate shrubland–heath, 11c Temperate shrubland–sclerophyll [Mediterranean], 29 Marine rocky shores	2.A.3 Tropical scrub and herb coastal vegetation, 2.B.1 Mediterranean scrub and grassland, 2.B.2 Temperate grassland and shrubland, 6.A.1 Tropical cliff, scree and other rock vegetation, 6.B.1 Temperate and boreal cliff, Scree and other rock vegetation
Savannah (shrub, tree, mallee)	2 Tropical seasonal forest, 8 Tropical broadleaf woodland	1.A.1 Tropical dry forest and woodland, 2.A.1 Tropical lowland grassland, savanna and shrubland
Woodland (savannah and sub-alpine)	5c Temperate evergreen forest–sclerophyll [Mediterranean], 10b Temperate woodland–sclerophyll	1.B.1 Warm temperate forest and woodland, 1.B.2 Cool temperate forest and woodland
Grassland (dry and wet tussock)	12 Savanna [tropical grassland], 13 Temperate grassland, 30 Marine sandy beaches	2.A.1 Tropical lowland grassland, savanna and shrubland, 2.B.2 Temperate grassland and shrubland, 3.A.2 Warm desert and semi-desert scrub and grassland

Table 1.1 (cont.)

Australian vegetation formations	Description	Environmental niches	Distribution
Alpine (cryophilic) herbfields and shrublands (Chapter 19)	Alpine forbs and grasses with or without low shrubs tolerant of snow burial	Humid, alpine–subpolar macroclimates with seasonal snow cover, eutrophic–mesotrophic periodically damp substrates, prone to century-scale surface fires	South-east mountains, sub-Antarctic islands
Freshwater wetlands (Chapter 20)	Perennial and ephemeral wetlands with helophytic shrubs, forbs and/or graminoids	Diverse macroclimates on depositional fluvial or groundwater-influenced landforms with mesotrophic–eutrophic substrates, periodically inundated by streams, overland flows or rising groundwater	Throughout tropical and temperate regions
Floodplain (hydrogenic) forests, woodlands and shrublands (Chapter 21)	Riparian and flood-prone forests, woodlands and thickets of *Eucalyptus*, *Melaleuca* or *Casuarina* with helophytic understorey	Diverse macroclimates on depositional fluvial landforms with mesotrophic–eutrophic substrates, periodically inundated by perennial or ephemeral streams or overland flows	Tropical, temperate and semi-arid floodplains and rivers
Saline wetlands (Chapter 22)	Halophytic forests, scrubs, succulent forblands or grasslands	Diverse macroclimates on depositional flats in coastal estuaries or drainage pans with saline substrates, periodically inundated by tides, overland flows or rising groundwater	Tropical and temperate tidal coasts, arid inland depressions
Semi-arid acacia/casuarina woodlands (Chapter 16, 25)	Woodlands of *Acacia*, *Casuarina* or *Callitris* with reduced leaves (leaf blades absent) and mixed perennial/ephemeral ground layer	Semi-arid–arid aseasonal tropical–temperate macroclimates, mesotrophic–eutrophic fine–medium-textured substrates, mostly depositional landforms, rarely fire-prone	Subtropical and temperate semi-arid inland
Semi-arid eucalypt woodlands (Chapter 23)	Eucalypt woodlands including mallee, with xerophytic shrubs and mixed perennial/ephemeral ground layer	Semi-arid–arid aseasonal subtropical–temperate macroclimates on mostly oligotrophic coarse-textured substrates on erosional landforms, occasionally fire-prone	Subtropical and temperate semi-arid inland plains and ranges

Beadle and Costin (1952); Specht and Specht (2001)	Whittaker (1975)	Faber-Langendoen *et al.* (2014)
Alpine herbfield (tall and short), grassland (sod), Fjeldmark	11d Temperate shrubland–sub-alpine, needleleaf, 11e Temperate shrubland–sub-alpine, broadleaf, 14 Alpine shrubland, 15 Alpine grassland	4.B.1 Temperate and boreal alpine dwarf-shrub and grassland
Fen, bog (valley and raised bog)	22 Cool temperate bog	2.C.2 Temperate to polar bog and fen, 2.C.3 Tropical freshwater marsh, wet meadow and shrubland, 2.C.4 Temperate to polar freshwater marsh, wet meadow and shrubland, 5.B.1 Tropical freshwater aquatic vegetation, 5.B.2 Temperate to polar freshwater aquatic vegetation
Sclerophyll forest (swamp), woodland (tall and swamp)	24 Temperate freshwater swamp forest, 23 Tropical freshwater swamp forest	1.A.4 Tropical flooded and swamp forest, 1.B.3 Temperate flooded and swamp forest
Scrub (mangrove)	25 Mangrove swamp, 25 Saltmarsh, 26, 27, 29, 31 Marine mudflats	1.A.5 Mangrove, 2.C.5 Salt marsh, 5.A.1 Floating and suspended macroalgae saltwater vegetation, 5.A.2 Benthic macroalgae saltwater vegetation, 5.A.3 Benthic vascular saltwater vegetation
Woodland (shrub)		
Mallee (dry), woodland (shrub)		

Table 1.1 (cont.)

Australian vegetation formations	Description	Environmental niches	Distribution
Arid subsucculent chenopod shrublands (Chapter 24)	Chenopod shrublands of *Atriplex*, *Maireana* or *Sclerolaena* with tussock grasses	Arid – semi-arid aseasonal temperate macroclimates with eutrophic–mesotrophic fine-textured substrates, rarely fire-prone	Temperate arid inland
Arid sclerophyll shrublands (Chapter 25)	*Acacia* and other xeromorphic shrublands with ephemeral forbs and grasses	Arid–semi-arid aseasonal subtropical–temperate macroclimates with mesotrophic–oligotrophic coarse-textured substrates, prone to occasional fire	Subtropical and temperate arid inland
Hummock grasslands (Chapter 26)	Deserts dominated by hummock *Triodia* with scattered shrubs and ephemeral forbs	Arid aseasonal subtropical–temperate macroclimates with oligotrophic substrates, prone to occasional fire	Subtropical and warm temperate arid inland

Global formations not represented within Australia. Whittaker (1975): 4 Temperate deciduous forest; 5b Temperate evergreen forest–needleleaf; 6 Taiga and subarctic–sub-alpine needle-leaved forest; 7 Elfin woodland; 9a Thornwood–woodland; 9b Thornwood–scrub; 10a Temperate woodland–needleleaf; 10c Temperate woodland–deciduous broadleaf; 11a Temperate shrubland–deciduous; 16 Tundra; 18a Cool semi-desert–open scrub; 18b Cool semi-desert–dry grassland; 19 Arctic–alpine semi-desert; 20 True desert; 21 Arctic–alpine desert (Note: 16, 19 and 21 on sub-Antarctic islands).

and their distributions to be spatially modelled and temporally projected (Prober *et al.* 2012; Keith *et al.* 2014). Part II of this book (Chapters 11–26 of this book; cross-referenced in Table 1.1) elaborate on floristic, structural and ecological features of the formations and their variation across the continent.

1.3.1 Historical context and relationships to other classifications

The first significant typological synthesis of Australia's vegetation (Diels 1906) recognised elements with major physiognomic contrasts such as rainforest, savanna, sclerophyll forest, brigalow, mallee and desert, but there are a number of omissions and anomalies in the broad typology and accompanying map (Beard 2001). Beadle and Costin (1952) advanced a more comprehensive typology of physiognomic formations and subformations. Many of these correspond well with units of our proposed typology (Table 1.1). Beadle (1981) later elaborated on continental-scale floristic variation at the alliance level and provided a generalised continental map. This filled many of the gaps in his formation-level typology, but did not explicitly relate alliances to formations.

More recent typologies of Australian vegetation focus on structural and taxonomic attributes of vegetation. Specht (1970) proposed a factorial classification based on growth form, height and cover of the tallest vegetation stratum, and later revised it and developed a detailed classification of floristic associations (Specht *et al.* 1995). A scheme that defined dominant plant taxa based on cover, irrespective of their vertical stratum, was first applied in Western Australia (Beard and Webb 1974) and adapted in a spatially explicit continental synthesis

Beadle and Costin (1952); Specht and Specht (2001)	Whittaker (1975)	Faber-Langendoen *et al.* (2014)
Saltbush	17 Warm semi-desert scrub	3.A.2 Warm desert and semi-desert scrub and grassland
Scrub (dry)	17 Warm semi-desert scrub	3.A.2 Warm desert and semi-desert scrub and grassland
Grassland (hummock)		3.A.2 Warm desert and semi-desert scrub and grassland

Faber-Langendoen *et al.* (2014): 1.B.4 Boreal forest and woodland; 1.B.5 Boreal flooded and swamp forest; 2.A.2 Tropical montane grassland and shrubland; 2.B.3 Boreal grassland and shrubland; 2.B.4 Temperate to polar scrub and herb coastal vegetation; 2.C.1 Tropical bog and fen; 3.A.1 Tropical thorn woodland; 3.B.1 Cool semi-desert scrub and grassland; 4.A.1 Tropical high montane scrub and grassland; 4.B.2 Polar tundra and barrens; 5.A.4 Benthic lichen saltwater vegetation.

of vegetation (Carnahan 1976) considerably more detailed than Diels' (1906) synthesis.

Proliferation of vegetation surveys and mapping by state agencies after 1970 eventually lead to the establishment of a spatially explicit national repository of vegetation information (National Land and Water Resources Audit 2001), with an accompanying typology of Major Vegetation Groups and Subgroups. These units are founded on structural and taxonomic features of the tallest vegetation stratum and have undergone a series of refinements (e.g. Department of the Environment and Water Resources 2007). The recent proliferation and development of remote sensing technology and computing capacity now enables continental-scale mapping of vegetation structure at a spatial resolution in the order of metres (Figure 1.1). While structural characteristics of vegetation reflect its biomass and primary productivity (Specht 1970), the emphasis in our typology on physiognomic attributes and ecological processes (Table 1.1) reflects their fundamental importance for understanding vegetation dynamics and managing and conserving biodiversity. A revision of Major Vegetation Subgroups (Keith and Pellow 2015) will help to resolve the relationships between the two schemes.

Global typologies of vegetation help place Australia's vegetation in context with that on other continents. Many of these give primary emphasis to vegetation–climate relationships, perhaps influenced by Clements' (1916) concept of the climatic climax which posits that all vegetation develops to climatically defined limits unless constrained by local factors. Walter (1979) later developed the zonobiome concept as a major classification unit of global ecosystems based on climate as the primary independent environmental factor structuring vegetation. Box (1981, 1996) advanced

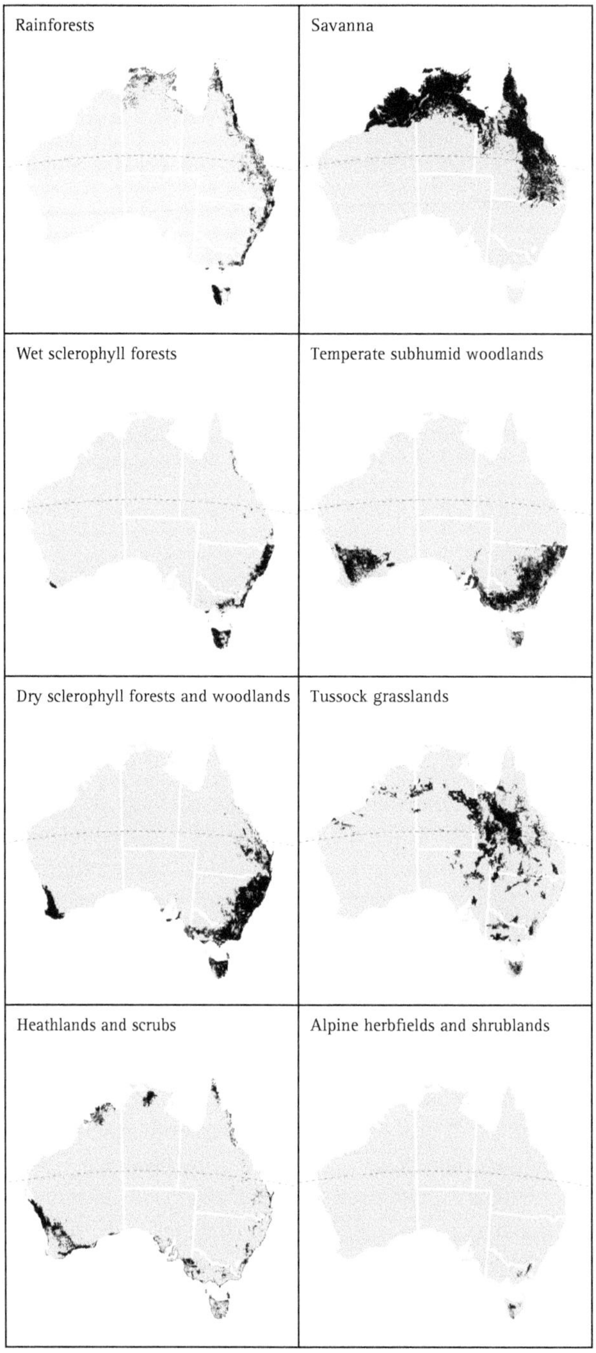

Figure 1.3 Indicative distributions of 16 Australian vegetation formations based on interpretation of state agency vegetation maps held in NVIS 4.2 (Department of the Environment and Energy 2007).

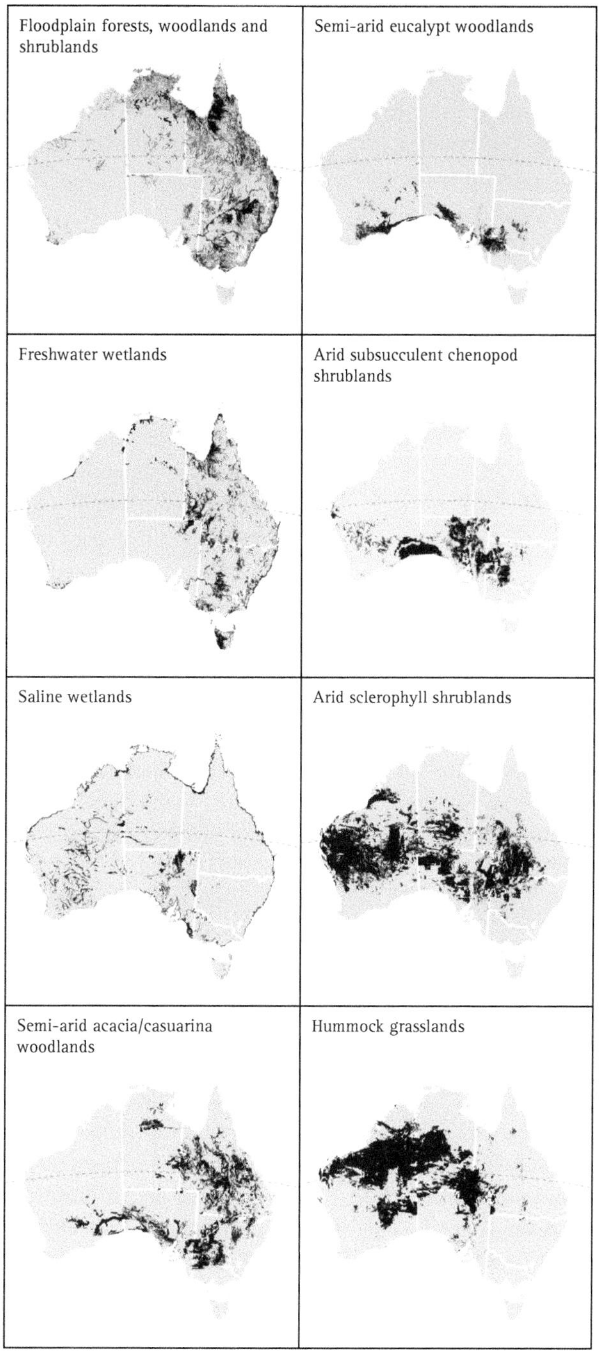

Figure 1.3 (cont.)

understanding of vegetation–climate relationships by developing a mechanistic approach to classification, using ecophysiological and physiognomic traits to identify tree types capable of net growth under different climatic conditions. This enabled prediction of vegetation types on a global scale and established capacity to examine global vegetation distributions under alternative climate scenarios (Box 1995).

The emphasis on climate, if not Clements' concept of the climatic climax, continues in a recent global classification scheme that grew from North American typologies (Faber-Langendoen *et al.* 2014). The limited correspondence between the formations of that scheme with Australian vegetation formations (Table 1.1) recalls the difficulties that early descriptive ecologists encountered in applying Clementsian concepts to Australian vegetation. Beadle and Costin (1952) noted that 'the greatest confusion over classification has arisen because of the erroneous impression that climate is [always] the controlling factor in determining vegetation'. This conflict reflects particular evolutionary influences on vegetation that have limited expression over much of the globe but come to the fore on the Australian continent (Orians and Milewski 2007; Hopper 2009; Mucina and Wardell-Johnson 2011). We return to these influences later in the chapter.

Whittaker's (1975) global vegetation formations or biomes are based on physiognomic and phenological traits as well as vegetation–climate relationships, and thus provide a useful framework for interpreting Australian vegetation in a global context (Table 1.1). Strong Australian representation of evergreen sclerophyll forests and shrublands is notable, although their distribution extends well beyond Mediterranean climates in Australia (Tozer *et al.*, Chapter 13, this volume, cf. Whittaker 1975). A striking feature emerging from cross-referencing of Australian vegetation formations with Whittaker's (1975) global formations is the lack of any substantial representation in Australia of coniferous (boreal) forests, tropical montane grasslands, tropical bogs and fens, thorn scrubs, tundra or cold deserts, while Australian semi-arid woodlands and hummock grasslands seem to lack direct analogues on other continents (Table 1.1).

1.4 Assembly of Australian biomes

1.4.1 Species pools

The phytogeographic elements thought to contribute to species pools of Australian ecosystems (see Weston and Jordan, Chapter 2, this volume) developed from those proposed by Hooker (1860) with advances in tectonic theory (Hess 1962) and molecular technology (Crisp and Cook 2013). Here, we adopt Barlow's (1994) interpretation of Gondwanan and intrusive elements, respectively, comprising species derived from ancestors present in Australia prior to the Gondwanan break-up, and those arriving after Australia's separation from the rest of Gondwana. Barlow (1994) further divided the Gondwanan element into relict and autochthonous subelements, respectively, based on whether they were largely unaltered survivors of the humid Gondwanan flora or radiating in isolation through multiple climatic cycles. Analyses of molecular phylogenies suggest that diversification of clades traditionally regarded as autochthonous began some millions of years before Australia's connection to South America through Antarctica was finally severed, but after its separation from other parts of Gondwana (Crisp and Cook 2013). Barlow (1994) subdivided the intrusive element into three groups: taxa dispersed from tropical Malesia; widely dispersed cosmopolitan taxa; and neoaustral taxa of northern hemisphere origin. For our purposes, we recognise a fourth 'modern' subelement including taxa that have arrived recently through human movement (a subgroup of the cosmopolitan element).

There are strong patterns in the contributions of different phytogeographic elements to the species pools of different vegetation formations (Table 1.2). Gondwanan relicts are largely confined to rainforests, though some are found in the understories of wet sclerophyll forests. Gondwanan autochthonous elements are pervasive through most Australian ecosystems and particularly dominate dry sclerophyll forests and heathlands, almost to the exclusion of other elements, as well as the sclerophyllous formations of the semi-arid and arid zones (Table 1.2). Most of the autochthonous taxa are sclerophyllous and apparently underwent multiple episodes of diversification within Australia, initially during a fire-prone period in the late Cretaceous (Crisp *et al.* 2011; Lamont and He 2012), again during the Miocene aridification (Barlow 1994; Crisp *et al.* 2011; Toon *et al.* 2015) and most recently during the Pleistocene climatic oscillations (Hopper 1979; Byrne 2007). Eucalypts and some other autochthonous taxa exhibit strong phylogenetic structure across the formations (Wardell-Johnson *et al.* 1997; Keith 2004).

Intrusive elements are particularly dominant in savannas, tussock grasslands, freshwater and saline wetlands and arid subsucculent shrublands (Table 1.2). Strong representation of tropical intrusives in the savannas reflects the relatively recent emergence of this biome under seasonally dry tropical conditions (Crisp and Cook 2013). Tropical intrusives are also well represented in rainforests, but their incursion diminishes into the temperate latitudes reflecting the limited time for dispersal since the Australian tectonic plate came within range of Malesian sources (Sniderman and Jordan 2011).

Cosmopolitan intrusives dominate grassy, wetland and some arid ecosystems (Table 1.2), reflecting the dispersal abilities of taxa such as Amaranthaceae, Asteraceae, Brassicaceae and Poaceae. Several progenitors of the woody chenopods (Amaranthaceae), for example, seem to have arrived in coastal saline habitats and radiated within the arid zone during the Miocene aridification (Kadereit *et al.* 2010; Kadereit and Freitag 2011). Similarly, there have been multiple arrivals of C_4 grasses, with major expansions across savannas, grasslands, temperate and semi-arid woodlands and arid shrublands since the mid-Miocene, reflecting a global episode of expansion (Keeley and Rundel 2005). In contrast, their incursion into rainforests, sclerophyll forests and alpine systems has been limited. Cosmopolitan elements dominate most freshwater and saline wetlands (Table 1.2) reflecting strong hydrochorous and zoochorous dispersal capability, the latter primarily through migratory birds. Some autochthonous Gondwanan elements occur in freshwater wetlands, but are mostly confined to mire ecosystems that exhibit relationships with heathland floras (Lamont and Keith, Chapter 14, this volume).

The neoaustral intrusive subelement is comparatively small and focused on alpine ecosystems. These have composite species pools that include cosmopolitan and autochthonous elements, and even some Gondwanan relics, representing multiple sources of colonisation into recently uplifted, and even more recently glaciated mountains (Barlow 1989).

The modern introduced flora of Australia is substantial, but incursions have been selective (Leishman *et al.*, Chapter 9 this volume). Sclerophyll vegetation formations remain relatively resilient to plant invasions; although mesophyllous woody species (especially vines) have established in disturbed rainforests and wet sclerophyll forests, perennial and ephemeral grasses and forbs are conspicuous components of temperate and arid grassy systems, and a range of aquatic and amphibious forbs have invaded wetlands and floodplain woodlands (Table 1.2).

1.4.2 Resource filters

Of the four major resource groups (Figure 1.2), the most significant filters shaping Australian vegetation appear to be water and nutrients. In Figure 1.4, we propose a simplified conceptual model showing the influence of these resource gradients on the distribution of vegetation formations. This necessitates simplification of complex multidimensional

Table 1.2 **Species pools of Australian vegetation formations. Species pools follow Barlow's (1994) phytogeographic elements and subelements, with the addition of an intrusive modern subelement, whose dispersal to the continent has been facilitated by humans. Dominant elements are those that make up a large portion of taxa or biomass within the vegetation.**

Australian vegetation formation	Gondwanan relict	Gondwanan autochthonous	Intrusive tropical	Intrusive cosmopolitan	Intrusive neoaustral	Intrusive modern
Rainforests	dominant		dominant			
Wet sclerophyll forests	subdominant	dominant				subdominant
Dry sclerophyll forests and woodlands		dominant				
Heathlands		dominant				
Savannas		codominant	codominant			
Temperate subhumid woodlands		codominant		codominant		subdominant
Tussock grasslands		subdominant		dominant		subdominant
Alpine herbfields and shrublands		dominant		subdominant	subdominant	
Floodplain forests, woodlands and shrublands		subdominant		dominant		subdominant
Freshwater wetlands		subdominant		dominant		
Saline wetlands				dominant		
Semi-arid acacia/casuarina woodlands		dominant		subdominant		subdominant
Semi-arid eucalypt woodlands		dominant		subdominant		
Arid subsucculent chenopod shrublands				dominant		subdominant
Arid sclerophyll shrublands		dominant		subdominant		
Hummock grasslands		dominant		subdominant		

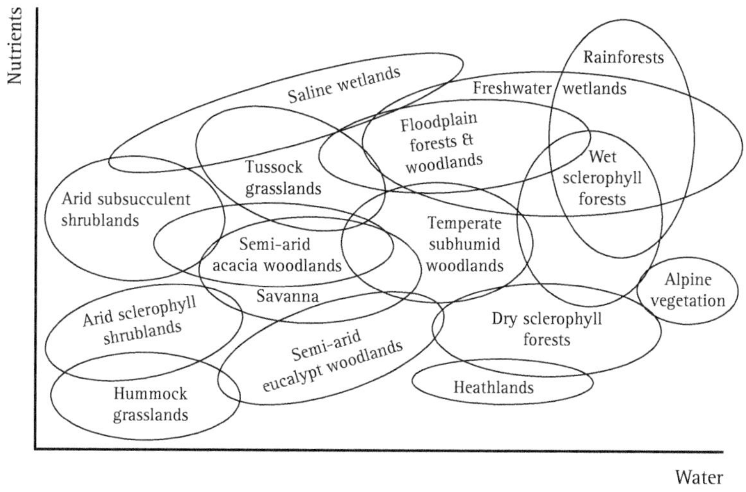

Figure 1.4 Idealised model of niche segregation of Australian vegetation formations (Table 1.1) with respect to simplified resource gradients for water and nutrients.

gradients, as any of several essential nutrients may be limiting and water may be available from contrasting sources (rainfall, surface flows, groundwater) that vary on different time scales. The model aims to simply portray our hypotheses about the relationships between the most limiting aspects (source, type, timing) of these resources and the occurrence of respective formations. Water is postulated to be most abundant in wetlands, some rainforests and alpine vegetation, and most limiting in arid shrublands and hummock grasslands. Nutrients are postulated to be most abundant in some rainforests, wetlands and tussock grasslands, and most limiting in arid sclerophyll shrublands, hummock grasslands and heathlands (Figure 1.4).

Although there are markedly contrasting expressions of vegetation at the formation level along major water and nutrient gradients, contrasting selection pressures may generate remarkably convergent expressions of particular plant traits (Table 1.3). Sclerophylly (Table 1.3), for example, is one of several functional traits expressed in many taxa exposed to severe drought, yet it is also common among taxa exposed to nutrient deprivation in well-watered environments (Lamont *et al.* 2002; Tozer *et al.*, Chapter 13, this volume), and thus is well represented in most formations except those in the top right and top left of Figure 1.4.

Grime's (2001) C–S–R model makes instructive predictions about plant traits across the resource space in Figure 1.4. The relative constancy of nutrient deprivation (along the bottom of Figure 1.4) should promote proliferation of traits associated with stress tolerance (Table 1.3). In contrast, drought-prone environments (on the left of Figure 1.4) undergo periods of water deprivation invariably punctuated by episodes of abundant water supply and may be populated by organisms with ruderal traits (R) that allow rapid resource exploitation and stress avoidance, as well as those with stress tolerance traits (S) (Table 1.3). Conversely, environments with more abundant supply of both water and nutrients (top right of Figure 1.4) are dominated by traits that promote competitive ability (C).

Such relationships suggest strong interplay between contrasting drivers and perhaps a finite range of evolutionary solutions to natural selection. While many factors may have a role in shaping

Table 1.3 **Differentiation of plant traits among Australian vegetation formations. Growth forms adapted from Hnatiuk *et al.* (2009); leaf types based on Box (1996) and Berry and Roderick (2002), who describe characteristic leaf trait values in their Table 1; leaf sizes follow Raunkaier (1934); seedbank types based on Keith *et al.* (2002); leaf turnover estimates are in years; dispersal modes follow Keith (2012); CSR strategy after Grime (2001) (C competitor, S stress tolerator, R ruderal).**

Formation	Growth forms	Leaf types	Leaf sizes	Leaf turnover	Seedbank types	Dispersal modes	CSR strategy
Rainforests	Closed-canopy trees, vines, ferns	Mesomorphic, turgor, laurophyll	Micro–megaphyll	Most <2	Transient, persistent	Endozoochory, anemochory	C,(R)
Wet sclerophyll forests	Open-canopy trees, shrubs, ferns, forbs	Scleromorphic, mesomorphic, turgor	Noto–mesophyll	Most <2	Persistent, transient	Anemochory, endozoochory	C,R
Dry sclerophyll forests and woodlands	Open-canopy trees, shrubs, graminoids	Scleromorphic	Noto–mesophyll	>2	Persistent, serotinous	Myrmecochory, anemochory	S
Heathlands	Shrubs, graminoids	Scleromorphic	Lepto–microphyll	>2	Persistent, serotinous	Myrmecochory, anemochory	S,(R)
Savannas	Open-canopy trees, summergreen (C_4) grasses	Mesomorphic, turgor, scleromorphic	Noto–mesophyll	Most <2	Transient, persistent (?)	Anemochory, epizoochory, endozoochory	C,R
Temperate subhumid woodlands	Open-canopy trees, wintergreen (C_3) and summergreen (C_4) grasses	Scleromorphic, coriaceous, turgor	Noto–mesophyll	Woody spp. >2, some non-woody spp. <2	Transient	Anemochory, epizoochory	S,R
Tussock grasslands	Wintergreen (C_3) and summergreen (C_4) grasses	Turgor, mesomorphic	Nano–microphyll	<2	Transient	Anemochory, epizoochory	C,R
Alpine herbfields and shrublands	Grasses, forbs, low shrubs	Turgor, coriaceous	Pico–nanophyll	Most >2, some non-woody <2	Transient, persistent	Anemochory	S,R
Floodplain forests woodlands and shrublands	Open-canopy trees, graminoids, forbs	Coriaceous, turgor, mesomorphic	Nano–mesophyll	Most <2	Transient, persistent	Hydrochory	C,S

Formation	Growth forms	Leaf types	Leaf sizes	Leaf turnover	Seedbank types	Dispersal modes	CSR strategy
Freshwater wetlands	Graminoids, forbs, shrubs	Turgor, coriaceous	Nano–microphyll	<2	Transient, persistent	Hydrochory	C,S,R
Saline wetlands	Forbs, grasses, closed-canopy trees	Succulent, turgor	Lepto–notophyll	<2	Transient	Hydrochory	S
Semi-arid acacia/ casuarina woodlands	Open-canopy trees, shrubs, tussock grasses, forbs	Reduced to phyllodes or scales, coriaceous, scleromorphic	Pico–microphyll	Woody spp. >2, some non-woody spp. <2	Persistent, transient	Anemochory, endozoochory	S,R
Semi-arid eucalypt woodlands	Open-canopy trees, shrubs, hummock grasses, forbs	Coriaceous	Micro–mesophyll	Woody spp. >2, some non-woody spp. <2	Persistent	Anemochory, epizoochory, myrmecochory	S,R
Arid subsucculent chenopod shrublands	Shrubs, ephemeral forbs, tussock grasses	Subsucculent, turgor, coriaceous	Lepto–nanophyll	Woody spp. > 2, some non-woody spp. <2	Transient	Anemochory, endozoochory	S,R
Arid sclerophyll shrublands	Shrubs, ephemeral forbs	Scleromorphic, coriaceous	Nano–microphyll	Woody spp. >2, some non-woody spp. <2	Persistent	Anemochory, myrmecochory	S,R
Hummock grasslands	Hummock grasses, ephemeral forbs and grasses	Scleromorphic, coriaceous	Nanophyll	Woody spp. >2, some non-woody spp. <2	Persistent	Anemochory, myrmecochory	S,R

vegetation in particular environments, particular assembly filters come to the fore as major influences that sometimes override most of the others (Figure 1.2).

1.4.2.1 Water

While Australia is the world's second driest continent after Antarctica in terms of precipitation, water availability gradients are complex and expressed in multiple dimensions and scales. At continental scale, physiognomically defined Australian vegetation formations (Table 1.1) are strongly differentiated in their ability to tolerate drought and their ability to exploit available moisture to produce biomass (Berry and Roderick 2002). Tolerance and exploitative ability may be explored at continental scale with the aid of spatially interpolated vegetation and climate data (Figure 1.5). The overall inverse relationship between these two hydrological factors (Figure 1.5) illustrates the trade-offs involved in the selection of plant functional traits in response to opposing selection pressures associated with this key plant resource. It also reveals greater complexity in hydrological relationships among

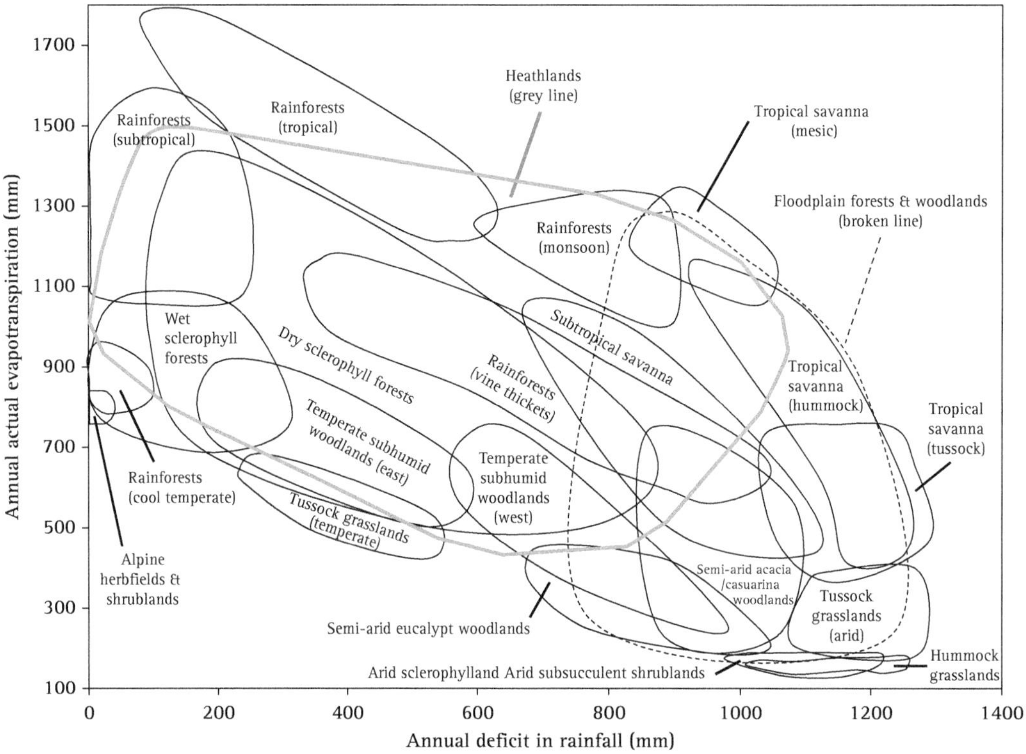

Figure 1.5 Continental-scale hydrological niches of Australian vegetation formations (and parts thereof) in relation to productivity (as estimated by actual evapotranspiration) and drought tolerance (annual rainfall deficit). Ellipses represent hydrological niches drafted from the values derived from intersections of vegetation maps in Figure 1.3 with spatial outputs of a water-balance model (Dyer 2009) compiled at 9' resolution (i.e. 9 minutes of latitude and longitude) from temperature and rainfall spatial data (6' resolution) obtained from the Bureau of Meteorology (n.d.) and spatial data for soil available water capacity (3' resolution) obtained from the Commonwealth Scientific and Industrial Research Organisation (n.d.).

vegetation types than can be represented in a simple one-dimensional hydrological model implicit in Figure 1.4.

Rainforests inhabit environments with a low water deficit, exploiting available moisture with rapid plant growth (high evapotranspiration, ET), although vine thickets extend to environments with surprisingly high rainfall deficit (Figure 1.5). Sclerophyll forests are characterised by intermediate ET and mostly low drought tolerance, particularly during summer. Savannas exhibit the best trade-off with intermediate ET in the face of high water deficit, particularly during the winter dry season. Temperate woodlands exhibit relatively low ET but span a very wide range of water deficit, with a marked segregation of humid woodlands in south-eastern Australia from semi-arid woodlands in the south-west. Shrubland and hummock grassland formations of the arid zone are poorly differentiated within a narrow hydrological niche at the extremes of low ET and high water deficit (Figure 1.5).

Water resource trade-offs are best known in leaf type and size traits (Wright *et al.* 2004). Specific leaf area, leaf thickness, carbon:nitrogen (C:N) ratio and leaf longevity all generally increase from left to right and top to bottom in Figure 1.5 and may be generalised via representation of leaf types (Table 1.3), while leaf size shows the reverse trend (Table 1.3). Other key plant traits for drought tolerance include seasonal leaf phenology (Eamus 1999), leaf:sapwood area ratio (Gleason *et al.* 2013), CAM and C_4 photosynthetic pathways (Murphy and Bowman 2007), stomatal regulation and exposure (Toon *et al.* 2015), canopy height (Falster and Westoby 2005), robust xylem structure and extensive root architecture (Mitchell *et al.* 2008).

Mean annual rainfall varies continentally from 150 mm in the Lake Eyre region of central Australia, where subsucculent and sclerophyllous arid shrublands dominate, to more than 4000 mm in the rainforest-dominated region of tropical north-eastern Australia, with hydraulic limitation exerting a major influence on vegetation structure along regional moisture gradients (Givnish *et al.* 2014). Rainfall seasonality may also exert a major selection pressure (e.g. McPherson *et al.* 2004; Prior *et al.* 2004), with savannas dominating winter-drought climates of the tropical north, and heathlands and temperate subhumid forests and woodlands dominating the summer-drought Mediterranean climates of the south.

Landscape-scale and local water gradients have a profound influence on the distribution of wetland vegetation, which is concentrated along corridors of surface water movement and at sites of accumulation. Local variations in the water regime (depth, duration, season and extent of inundation or waterlogging) are needed to resolve distribution of wetland types (Catford *et al.*, Chapter 20, this volume), but currently are not feasible to model at continental scales (cf. Figure 1.5). Groundwater has also emerged as a major driver of regional and local ecosystems (Eamus *et al.* 2015). Interannual variation in rainfall appears to be at least as influential in shaping Australian ecosystems as long-term averages; the area of ephemeral wetlands greatly exceeds that of perennial wetlands and Australian rivers include those with the most variable flows on Earth (Puckridge *et al.* 1998). Rainfall variability also drives major trophic boom–bust cycles in both wetland and dryland ecosystems (Bunn *et al.* 2006; Letnic and Dickman 2006).

1.4.2.2 Nutrients

Nutritional filters appear to operate on Australian vegetation more widely and to greater extremes than on other continents (Orians and Milewski 2007). These function at landscape and local scales determined by substrate and topographic variation. Nutrient availability is determined by substrate characteristics, lateral movement and sorting processes, and in some cases (mainly coastal) aerosol inputs. Nonetheless, continental-scale modelling of major nutrients is feasible and provides insights into relationships between vegetation formations. Higher-nutrient environments are characterised by certain types of rainforests, grass-dominated vegetation or wetlands, while lower-nutrient environments are dominated by arid sclerophyll shrublands and hummock grasslands (Figure 1.6). Nutrient-rich ecosystems are relatively restricted to alluvial plains and relictual mafic volcanics and their derivatives. The extent of scleromorphic vegetation in humid and arid climates suggests widespread nutrient deficiency across the continent (Beadle 1981). Comparisons among vegetation formations suggest contrasting responses to phosphorus and nitrogen. Arid subsucculent shrublands and tussock grasslands inhabit soils that are richer in phosphorus than those supporting arid sclerophyll shrublands and hummock grasslands, but all four formations occupy soils that are very low in nitrogen (Figure 1.6). Forests in humid climates do not occupy soils with low nitrogen levels, due to the interplay between soil nitrogen, organic matter, decomposition rates and climatic humidity (Brundrett *et al.*, Chapter 7, this volume), but may occupy soils with a wide range of phosphorus levels (Figure 1.6). These results, like those for water availability, suggest that more complex resource gradients exist than can be represented in one-dimensional models (Figure 1.4).

Australian plants have evolved a wide range of leaf, root and life-history traits in response to nutritional selection filters (Brundrett *et al.*, Chapter 7, this volume). These include: (i) high production rates of low-mineral carbohydrates such as cellulose, lignin and sugars; (ii) the ability to regulate metabolic activity; and (iii) root structures and biotic associations to enhance nutrient capture. Nutrient-related patterns in plant traits are most conspicuous

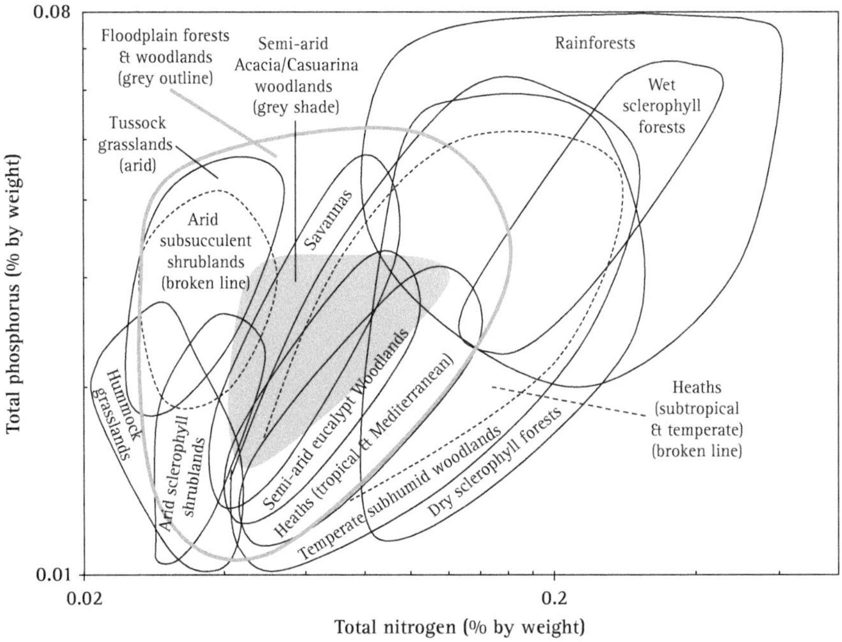

Figure 1.6 Continental-scale soil nutrient niches of Australian vegetation formations (and parts thereof) in relation to total phosphorus and nitrogen (log axes). Ellipses represent hydrological niches drafted from the values derived from intersections of vegetation maps in Figure 1.3 with spatial models of soil total phosphorus and total nitrogen (3' resolution) obtained from the Commonwealth Scientific and Industrial Research Organisation (http://www.clw.csiro.au/aclep/soilandlandscapegrid/).

in environments where water is less limiting. For example, heathlands and sclerophyll woodlands replace rainforests and wet sclerophyll forests with declining nutritional status along local edaphic gradients in eastern Australia (Beadle 1966), although more subtle trends are also evident in the arid zone where sclerophyllous shrublands and hummock grasslands replace subsucculent shrublands and tussock grasslands with increasing soil phosphorus (Figures. 1.3, 1.6). Specialised root structures are particularly well represented in heathlands on the least fertile soils, reflecting selection for enhanced nutrient extraction (Lambers *et al.* 2010). Specific leaf area, C:N ratio and leaf longevity all increase with nutrient deprivation (top to bottom on Figure 1.3; Wright *et al.* 2004), reflecting stronger expression of sclerophylly and nutrient conservation (Table 1.3). Endozoochory (dispersal of fleshy fruits by ingestion) also tends to diminish in favour of myrmecochory (ant dispersal) with nutrient deprivation, similarly reflecting selection for nutrient conservation (Hughes *et al.* 1993). Lamont and Keith (Chapter 14, this volume) review several other traits associated with nutrient poverty.

Like water, nutrient gradients can be viewed as multidimensional, with different elements or their ratios potentially limiting plant survival, growth and/or reproduction. Calcium and its ratio to magnesium, for example, appears to be important in the distribution of different arid zone plant communities, as are sodium and potassium (Leigh 1994). Overall, however, phosphorus is believed to be the most widely limiting nutrient and a principle driver of scleromorphy because of its limited concentration in many substrates, its limited solubility and its influence on nitrogen acquisition (Beadle 1966; Lambers *et al.* 2010).

1.4.2.3 Light and oxygen

Light and oxygen appear to be important resource filters in Australian biomes at some local scales, but not at continental scales. Continental patterns in light regime are influenced more by cloud cover than latitude, with energy uptake unlikely to be limited even on the sub-Antarctic islands (cf. polar ecosystems; Box 1996). Cloud cover is more influential on water availability (through both precipitation and evapotranspiration) than photosynthesis (Nix 1982). Latitudinal light gradients could be important at local scales where photosynthetically active radiation (PAR) may become limiting due to interception by tall competitors in vertically structured vegetation (Lusk *et al.* 2011). Notably, plant height increases with both water and nutrient availability (Figure 1.4; Table 1.3), reflecting competitive ability (Falster and Westoby 2005). In these situations a range of leaf traits have evolved to maximise light capture, including photosynthetic capacity, respiration rate, leaf longevity and mechanical strength and plasticity in those traits (Lusk *et al.* 2010; Pérez-Harguindeguy *et al.* 2013).

While plants are net oxygen producers, they are also consumers through respiration. Oxygen deprivation, particularly to roots, becomes influential on vegetation only in local sites subject to prolonged inundation of the surface and substrate (Vartapetian and Jackson 1997). These sites are occupied by aquatic or peaty vegetation that contain plants with aerenchymatous tissues, surface roots and other traits that enhance the diffusion and internal transport of oxygen (Finlayson 2005; Catford *et al.*, Chapter 20, this volume).

1.4.3 Other abiotic filters

1.4.3.1 Thermal niche

Temperature is a major correlate of vegetation patterns in Australia (Nix 1982), although its influence on vegetation assembly operates indirectly through its effect on resource acquisition and utilisation in metabolic functions. For example, temperature effects on water availability through evapotranspiration may account for some of this influence (Tozer *et al.*, Chapter 13, this volume). Physiological temperature limits and optima for resource acquisition and growth may vary markedly among plant taxa

(Wardlaw 1979). Patterns in growth of eucalypt seedlings at different experimental temperatures suggest their distribution is related to tolerance of extreme conditions rather than maximum growth rate under less severe conditions (Paton 1980). Seasonally cold temperatures and snow cover are defining physical features of cryophilic ecosystems in Australia's restricted alpine landscapes and sub-Antarctic islands (Venn *et al.*, Chapter 19, this volume), although only two of Box's (1995) four cold-tolerant tree types are well represented in Australia. The selection pressures include limitations on the length of the growing season, reduced growth rates, and chill and frost injuries (Briceño *et al.* 2014).

Plants with large soft leaf laminas or C_4 photosynthetic pathways are more prone to chill injury, have growth optima at higher temperatures and reduced propensity for cold acclimation, and thus tend to be distributed in warmer climates than species with converse traits. As on other continents, there are strong latitudinal and altitudinal gradients in grass distribution, with tropical savannas and arid shrublands dominated by C_4 species and temperate humid woodlands and grasslands dominated by C_3 species (Table 1.3; Murphy and Bowman 2007). In cool temperate and sub-alpine environments, low temperatures interact with overabundance of PAR to cause photoinhibition, which apparently limits the distribution of trees during their recruitment phase (Ball *et al.* 1991; Holly *et al.* 1994).

Heat shock is likely to be a significant selection agent discriminating different ecosystems (Larcher 2003), with plant responses depending on intensity, duration and rates of change in temperature as well as developmental stage and growing conditions prior to exposure (Larkindale *et al.* 2005). While many species of the temperate northern hemisphere suffer fatal heat stress with prolonged exposure to temperatures over 45°C, near-surface temperatures routinely exceed 50°C in the Australian arid zone (Ooi *et al.* 2009). Consequently, selection in hot climates is likely to favour traits that reduce exposure or increase tolerance to high temperatures, such as increased transpiration (in humid climates), vertical orientation of leaves, reduced leaf lamina size, and vestiture or waxed leaf surfaces (Angus *et al.* 2010). High temperatures also inhibit seed germination, a response likely to mediate seasonal patterns of recruitment, as well as the range of climates in which different species are able to sustain populations (Ooi *et al.* 2004).

1.4.3.2 Salinity

Substrate salinity is an important abiotic driver of coastal ecosystems and in parts of the southern and central inland. In these environments, high salt concentrations are maintained either by tidal inundation or accumulation in landscape sinks from groundwater sources or sea salt aerosols. On the east coast, periodically high rates of salt deposition associated with strong winds and high seas sometimes trigger episodes of dieback in sclerophyll vegetation (e.g. Keith 2002). Substrate salinity exceeding 0.1% by weight disrupts osmotic function and enzyme structure in cells of most glycophytes. However, the halophytes that occupy saline wetlands and subsucculent semi-arid shrublands (Table 1.1) possess a range of ecophysiological traits that enable them to tolerate substrate salinity that is more than an order of magnitude greater than that level (Wainwright 1981). Indeed some halophytic species (e.g. *Atriplex*, *Tecticornia*) reach optimal growth rates at salinities that exceed the tolerance thresholds of many glycophytic species (Atwell *et al.* 1999).

The saline wetlands and subsucculent semi-arid shrublands include virtually all of Australia's modest succulent terrestrial flora (Adam 2009), including numerous genera from Amaranthaceae (chenopods) and Aizoaceae. However, halophytic species that inhabit these ecosystems, including non-succulents, belong to a wide range of taxa with diverse representation in other biomes, suggesting multiple independent colonisations of, and adaptation to, saline environments from contrasting ancestral stock.

1.4.3.3 Metal toxins

A third group of abiotic drivers include essential micronutrients or non-essential metals that may reach toxic concentrations in plant substrates due to soil pH. The solubility of many toxic metals is so closely mediated by pH that it is difficult to disentangle the limiting factors. Acid soils are widespread in Australia and typified by high levels of aluminium and iron below pH 5. Aluminium suppresses root growth, but tolerance varies between plant species and toxicity may be moderated by organic chelates derived from humus or root exudates (Delhaize and Ryan 1995). Some Gondwanan relicts (e.g. Cunoniaceae, Proteaceae) and likely many Gondwanan autochthonous taxa characteristic of sclerophyllous vegetation formations (Table 1.3) have apparently evolved mechanisms to tolerate toxic conditions associated with acid soils. In some cases these involve mycorrhizal associations that selectively increase root uptake of deficient essential nutrients and reduce uptake of aluminium (Egerton-Warburton *et al.* 1993).

A contrasting condition, toxic levels of heavy metals associated with high pH, occurs in soils derived from ultramafic (or 'serpentine') substrates, which support a wide range of plant communities spanning rainforest, sclerophyll woodland and xeric shrubland formations (Table 1.1). These are rich in chromium, cobalt, nickel, iron and magnesium, but severely deficient in nitrogen, phosphorus, potassium and calcium. While far more restricted in Australia than acid soils, plant communities on ultramafic soils may be distinctive and include a number of endemic (or near-endemic) species (van der Ent *et al.* 2015). Some of these are hyper-accumulators of nickel, while others may have physiological traits that enable them to tolerate or exclude high concentrations of heavy metals (van der Ent *et al.* 2015). In general, however, there seems little differentiation of vegetation on ultramafic substrates from that surrounding it (Gibson *et al.* 2012). Further research is needed to resolve the mechanisms by which ubiquitous species tolerate potentially toxic metal concentrations.

1.4.4 Disturbance regimes

Disturbance regimes are a special class of abiotic filter due to their ability to kill living biomass, liberate resources and trigger life-history processes in particular species. Fire and surface water flows are the most important disturbance regimes across the Australian continent (Table 1.4), although high winds associated with storms may exert an important influence on tropical and east coast forests, woodlands and estuarine vegetation (e.g. Metcalfe *et al.* 2008).

1.4.4.1 Fires

Fires are pervasive through all Australian vegetation formations except rainforests, saline wetlands and subsucculent arid shrublands (Table 1.4), however, the frequency, intensity, season and type vary greatly across the continent (Miller and Murphy, Chapter 5 this volume). Generally there is a shift from frequent low-intensity cool-season fires in the tropical north to less frequent high-intensity warm-season fires in the temperate south (Bradstock 2010; Murphy *et al.* 2013), but regional and local nuances exist (Table 1.4).

Plants within fire-prone vegetation exhibit a suite of life-history syndromes including: (i) the ability to resprout from fire-resistant organs; (ii) seedling regeneration from serotinous or soil seedbanks; (iii) delayed regeneration from seed produced by resprouting shoots; (iv) delayed regeneration from seed produced by adults that escape surface fires; and (v) regeneration from seed dispersed from sources outside burnt areas (Keith 1996; Keeley *et al.* 2012). Occurrence of these syndromes varies greatly between different vegetation formations that likely reflect different species pools and selection pressures exerted by respective fire regimes. For example, resprouting is more frequent in savanna

Table 1.4 **Disturbance drivers of Australia vegetation formations (see also Murphy *et al.* 2013).**

Formation	Disturbance drivers
Rainforests	Very rarely fire-prone. Decadal-scale cyclonic storms.
Wet sclerophyll forests	Century-scale warm-season canopy fires, occasional understorey fires.
Dry sclerophyll forests and woodlands	Decadal-scale warm-season canopy fires, rarely understorey fires.
Heathlands and scrubs	Decadal-scale canopy fires, mostly warm-season (cool-season in tropics). Coastal storms.
Savannas	Subdecadal–decadal cool-season understorey fires, occasional canopy fires.
Temperate subhumid woodlands	Decadal-scale warm-season understorey fires, rarely canopy fires.
Tussock grasslands	Subdecadal warm- and cool-season canopy fires (varying with latitude).
Alpine herbfields and shrublands	Century–millennial-scale warm-season canopy and substrate fires. Seasonal snow cover.
Floodplain forests, woodlands and shrublands	Rarely prone to understorey or substrate fires. Annual–decadal inundation by fluvial waters.
Freshwater wetlands	Most types rarely fire-prone, sclerophyll mires prone to decadal–century-scale warm-season canopy and substrate fires. Permanent–decadal inundation or waterlogging by fluvial and ombrogenous waters.
Saline wetlands	Rarely fire-prone. Daily–decadal inundation by saline tidal waters, spring tides or local run-on.
Semi-arid acacia/casuarina woodlands	Very rarely fire-prone.
Semi-arid eucalypt woodlands	Decadal–century-scale warm-season canopy fires, rarely understorey fires.
Arid subsucculent chenopod shrublands	Very rarely fire-prone. Some types occasionally inundated by brief surface flows.
Arid sclerophyll shrublands	Irregular century-scale mostly warm-season canopy fires.
Hummock grasslands	Irregular decadal–century-scale mostly warm-season canopy fires.

floras than in heathland floras (Clarke *et al.* 2015), but the converse is true of serotiny (Table 1.4).

Fossil evidence and recent molecular dating of fire-related traits suggest that some of these syndromes have a deep evolutionary history in flowering plants dating back to the late Cretaceous (60–80 million years ago) before Australia's separation from other southern land masses (Crisp *et al.* 2011; Lamont and He 2012), and probably longer in some gymnosperms. At the community level, there is evidence that fires promote diversity by interrupting competitive elimination, invoking storage effects and lottery recruitment (Keith 2012).

1.4.4.2 Floods

Inundation and waterlogging regimes are critical to sustainability and turnover of freshwater and saline wetlands and to floodplain forests, woodlands and shrublands (Catford *et al.*, Chapter 20, this volume; Good *et al.*, Chapter 21, this volume; Rogers *et al.*, Chapter 22, this volume). The frequency of inundation varies from daily in tidal wetlands to multi-decadal in some arid ephemeral wetlands (Table 1.4). The flows of Australian arid zone rivers are renowned for having the most variable flows in the world (Puckridge *et al.* 1998) and some rivers of the tropics and temperate south-east also have high-volume, high-velocity flow events that generate major disturbances (Warfe *et al.* 2011).

Recurring inundations by surface flows are primary dispersal vectors of seeds and vegetative propagules (Nilsson *et al.* 2010). Wetting associated with inundation also triggers germination in many wetland plant species (Brock *et al.* 2003). This, together with the role of floods as agents of resource enrichment, initiates boom–bust cycles for which Australian ephemeral wetland systems are renowned (Bunn *et al.* 2006). The flow of resources initiates successive episodes of growth and reproduction through trophic networks of producers, primary consumers and predators. After recession of water from the landscape, populations of many plants gradually return to low levels or dormant phases (Grime's ruderal strategy, Table 1.3), while a few long-lived perennials persist with reduced rates of metabolic activity (Grime's stress tolerator strategy, Table 1.3) until the next inundation event (Brock *et al.* 2003). In contrast, permanent wetlands of temperate humid climates tend to be dominated by strong competitors (Table 1.3) and tolerators of oxygen deprivation. During intervals between major flow events, the residual wetlands and perennial streams become important refuges for a range of biota. The extent of inundation events, microtopographic variation and the movement patterns of water therefore determine the distribution of plants and the spatial structure of microhabitats within these ecosystems (Capon 2005).

In contrast to fire-prone ecosystems, which are characterised by Gondwanan autochthonous elements, especially in the temperate and arid zones, flood-prone ecosystems are dominated by intrusive cosmopolitan and modern elements (Table 1.2). In part, this reflects the colonisation abilities of intrusive species and the predisposition of flood-prone environments to invasions, but may also be a consequence of varying resource availability that selects for fast life histories when these systems are in resource-rich states.

1.4.5 Biotic drivers

A wide range of biotic filters operate on Australian vegetation, mostly at local spatial scales. Here, we focus on competition and predation, but note that poorly understood facilitation processes are potentially important in structuring communities, especially in low-resource or high-stress environments (Jarrad *et al.* 2012; James *et al.* 2015). Pathogens are also emerging as agents of vegetation change, for example, in temperate sclerophyll systems where plant root pathogens are transforming the structure and composition of diverse plant communities (Cahill *et al.* 2008).

1.4.5.1 Competition

Competitive interactions are important over distances at which neighbouring plants can interact. The most striking effects are evident above ground in vertically structured communities, where established tall plants deprive smaller plants of light, including recruits of taller species, limiting photosynthesis and growth, while promoting selection of shade-tolerance leaf traits (Lusk *et al.* 2010). Light intensity may vary by several orders of magnitude from above the canopy to the soil surface, not only in forests, but also heathlands and grasslands (Keith and Myerscough 1993).

In rainforests and wet sclerophyll forests, understorey plants and recruits are equipped with plastic responses that enable them to adjust leaf form and growth rates to tolerate low light conditions and also take advantage of opportunistic canopy gaps (Lusk *et al.* 2010). Equivalent data are scarce from sclerophyll communities, but less plastic responses may explain strong inverse relationships between canopy cover and understorey diversity (Specht and Specht 1989; Keith and Bradstock 1994).

Below-ground competitive interactions are poorly studied but may also be important in structuring local communities (Coomes and Grubb 2000). The degree of functional divergence in root architecture and uptake mechanisms suggests the existence of significant biotic filters in Australian ecosystems where availability of water and nutrients is limited (Brundrett *et al.*, Chapter 5, this volume; Lamont and Keith, Chapter 14, this volume).

1.4.5.2 Predation

Plant predation interactions (including herbivory, granivory and frugivory) may operate over landscape or local scales, depending on movement abilities of the predators. Different vegetation formations display contrasting signatures of predator selection pressures. Invertebrate herbivory has an important role in structuring Australian rainforests and some eucalypt communities (Lowman 1992; Rinker and Lowman 2004) and is likely to maintain diversity of these systems by countering density-dependent competition (Bagchi *et al.* 2014). Regional outbreaks of herbivorous insects have had profound impacts of Australian temperate eucalypt forests and woodlands (Stone and Coops 2004).

In contrast, vertebrate herbivores strongly influence ground-layer composition and diversity in grasslands and some of the woodland formations, and may mediate the dominance of woody plants (Eldridge *et al.* 2016; Eldridge *et al.*, Chapter 6, this volume). Relative to grassy systems, vertebrate herbivory seems to be less important in structuring Australian sclerophyll forests, woodlands and heathlands, as foraging activity may be limited by poor nutritional status and high cellulose content of foliage (Orians and Milewski 2007). Chemical and physical traits such as leaf C:N ratio, secondary metabolites, spinescence, resprouting and growth form are expressed in a wide range of Australian plants, potentially in response to herbivory (Díaz *et al.* 2006; Eldridge *et al.*, Chapter 6, this volume). Invertebrate granivory appears to influence assembly through recruitment, as seeds are relatively nutrient- and energy-rich resources in these systems (Campbell and Clarke 2006), and some species have evolved mass seed release or masting traits reducing the impact of granivores on seed availability for recruitment (O'Dowd and Gill 1984; Wright and Zuur 2014).

1.5 Ecological synthesis in historical context

A long and interesting chronology of syntheses of the ecology and evolution of Australian vegetation shows progressive development from simple biogeographic narratives to complex evolutionary models. It began with Hooker's (1860) essay in *Florae Tasmaniae*. Hooker's emphasis was on biogeography and although he recognised the endemism within the Australian flora, the relationships with other southern floras led him to deduce historical connections with the species pools of other continents a century before geological evidence finally emerged (Weston and Jordan, Chapter 2, this volume). This was a visionary synthesis of isolation and connections for the biota of an island continent, at a time when his colleague Darwin had just published evolutionary theory. Hooker's ideas have since been updated in the light of developments in plate tectonics; cladistics and molecular phylogeny (Barlow 1994; Crisp and Cook 2013; Weston and Jordan, Chapter 2, this volume).

The next significant development was Diels' (1906) account of the plant world of Western Australia, in which he published the first vegetation map of the Australian continent, recognising an arid core of desert and mulga, surrounded by a ring of 'scrub' and 'savanna' grassland, with forests and woodlands occupying humid and subhumid coastal regions. Diels distinguished tropical and temperate rainforest, both confined to the east coast, but more importantly following Schimper (1903), he drew a distinction between two major types of eucalypt forest and woodland, depending on whether they were characterised by 'sklerophyllen' or 'savannen' understories (Beard 2001).

While sclerophylly was, and still is, widely interpreted as a form of xeromorphy related to aridity (Tozer *et al.*, Chapter 13, this volume), Andrews (1916) was the first to propose a nutritional explanation. His ideas were later developed and tested experimentally by Beadle (1966), who posited that soil phosphorus was a key limiting factor in the distribution and form of Australian plant communities. While recent work has reinforced the importance of nutrition (Lambers *et al.* 2010), theories continued to develop about hydrological influences and energetics as key drivers of form and function alongside nutritional constraints in Australian vegetation (Specht and Specht 2001). In his descriptive treatise on vegetation of Australia, Beadle (1981) identified historic, climatic and edaphic factors as the three key elements affecting the patterning of plant communities on the Australian continent. Within that descriptive structure, he identified many of the selection filters incorporated in our model of biome assembly (Figure 1.2).

Another important advance came when Orians and Milewski (2007) linked nutritional poverty to fire regimes and other ecosystem processes. Until that time, fire regimes had been largely treated as marginal to continental-scale ecological patterns and evolutionary processes (Specht and Specht 2001). Orians and Milewski (2007) were no doubt influenced by the thinking of Bond *et al.* (2005), who proposed a major role for fire in global biogeography through positive feedbacks, as opposed to bottom-up resource and resource-mediating factors. In essence, they proposed that environments with abundant light and periodically adequate moisture, but limited supplies of nutrients, promote plants that synthesise excess carbohydrate that is channelled into production of cellulose, lignin, starch, cuticular waxes and sugar-rich nectars, rather than nutrient-demanding photosynthetic or reproductive tissues. The high concentration of cellulose and lignin and low concentration of salts in foliage enhances flammability and reduces herbivory in the system, while 'intense' fires reinforce nutritional poverty through volatilisation. The interplay between productivity and fire regimes as well as their ecological effects was subsequently emphasised by Bradstock (2010). Although presented as a continental ecology, Orians and Milewski's (2007) implicit focus was on humid and subhumid sclerophyll systems, and consequently their theory ignored the continental diversity of fire regimes (Murphy *et al.* 2013), other fire-prone systems such as savanna, and low-flammability resource-rich systems. Nonetheless, highlighting the importance of feedbacks between disturbance regimes and biotic interactions between plants and animals was a major outcome of their work.

A broader synthesis came with Hopper's (2009) theory that all terrestrial ecosystems could be placed on a continuum from old climatically buffered infertile landscapes (OCBILs) to young often disturbed fertile landscapes (YODFELs). He posited that due to their landscape age and stability, OCBILs should be characterised by long-lived organisms with poor dispersal ability, high endemism, persistence of old lineages, genetic mechanisms to cope with insularity, nutritional and other biological traits to cope with infertility and susceptibility to soil disturbance. Conversely, YODFELs should be populated with organisms that are widely dispersed and capable of exploiting resource-rich environments after landscape rejuvenation by major geological disturbances such as glaciation, orogeny, volcanic eruptions, marine transgressions and episodes of fluvial expansion and deposition. Hopper (2009) and Lambers *et al.* (2010) showed that sclerophyll ecosystems of south-western Australia and South Africa matched the predicted biotic, geophysical and historical profile of extreme OCBILs, while YODFELs were best exemplified by post-glacial ecosystems of northern Europe and America.

Mucina and Wardell-Johnson (2011) extended the OCBIL–YODFEL continuum into three dimensions representing landscape age, climatic stability and fire. They identified a need to consider the evolutionary influence of landscape aging and rejuvenation over multiple spatial and temporal scales. They pointed to the evolutionary significance of different manifestations of climatic stability including annual-scale buffering effects of maritime influence and the predictability of seasonal variability, each of which may be expressed over different evolutionary time scales. Finally, they suggested that fire-driven evolution could be viewed in terms of fire regime stability over different evolutionary time scales.

By bringing Australian vegetation under the lens of community assembly theory, we aim to promote thinking about a broad ensemble of evolutionary filters, varying combinations of which may come to the fore under particular conditions (Figure 1.2). Rather than focussing on any single gradient, or suite of correlated gradients, community assembly theory helps to illuminate different key processes that influence selection of plant traits and shape ecosystems. A new ecological classification of vegetation formations (Table 1.1) is central to understanding the outcomes of contrasting species pools and environmental conditions, as well as the ongoing maintenance and development of ecological processes across Australia. In this vein, an improved understanding of ecological processes contributing to assembly of Australian vegetation is an essential foundation for the conservation of its spectacular, but increasingly fragile diversity (Keith and Auld, Chapter 26, this volume). Girt (no word could better capture the essence of Australia) by a throng of past, present and future human activities (Hunt 2013), Australian vegetation is undergoing its greatest transformation since Miocene aridification. The subsequent chapters in this book examine the major drivers and their influence on the full range of vegetation formations across the continent.

Acknowledgements

We thank Chris Simpson, Chantel Benbow and Belinda Pellow for assistance in preparing the vegetation classification and maps presented in this chapter.

References

Adam, P. (2009) Australian saltmarshes in a global context. In *Australian Saltmarsh Ecology*, ed. N.J. Saintilan. CSIRO Publishing: Melbourne, pp. 1–22.

Adler, P.B., HilleRisLambers, J. and Levine, J.M. (2007) A niche for neutrality. *Ecology Letters* 10, 95–104.

Adler, P.B., Fajardo, A., Kleinhesselink, A.R. and Kraft, N.J.B. (2013) Trait-based tests of coexistence mechanisms. *Ecology Letters* 16, 1294–1306.

Andrews, E.C. (1916) The geological history of the Australian flowering plants. *American Journal of Science* 249, 171–232.

Angus, J., Atkin, O., Brummell, D. *et al.* (2010) Temperature and acclimation. In *Plants in Action*, 2nd edition, ed. R. Munns. Australian Society of Plant Scientists: Brisbane. Available at: http://plantsinaction.science.uq.edu.au (accessed 23 June 2016).

Atwell, B.J., Kriedemann, P.E., Turnbull, C.G.N. (1999) *Plants in Action: Adaptation in Nature, Performance in Cultivation*, 1st edition Macmillan: Melbourne.

Auscover (2016) Auscover Remote Sensing Data Facility, Terrestrial Ecosystem Research Network. Available at: http://www.auscover.org.au/?s=vegetation+height (accessed 8 August 2016).

Bagchi, R., Gallery, R.E., Gripenberg, S. *et al.* (2014) Pathogens and insect herbivores drive rainforest plant diversity and composition. *Nature* 506, 85–88.
Ball, M.C., Hodges, V.S. and Laughlin, G.P. (1991) Cold-induced photoinhibition limits regeneration of snow gums at tree-line. *Functional Ecology* 5, 663–668.
Barlow, B.A. (1989) The alpine flora: autochthones and peregrines. In *The Scientific Significance of the Australian Alps*, ed. R. Good. The Australian Alps Liaison Committee: Canberra, pp. 69–78.
Barlow, B.A. (1994) Phytogeography of the Australian region. In *Australian vegetation*, 2nd edition, ed. R.H. Groves. Cambridge University Press: Cambridge, pp. 3–36.
Beadle, N.C.W. (1966) Soil phosphate and its role in molding segments of the Australian flora and vegetation, with special reference to xeromorphy and sclerophylly. *Ecology* 47, 992–1007.
Beadle, N.C.W. (1981) *The Vegetation of Australia*. Cambridge University Press: Cambridge.
Beadle, N.C.W. and Costin, A. (1952) Ecological classification and nomenclature. *Proceedings of the Linnean Society of New South Wales* 77, 61–82.
Beard, J.S. (2001) A historic vegetation map of Australia. *Austral Ecology* 26, 441–443.
Beard, J.S. and Webb, M.J. (1974) The vegetation survey of Western Australia: its aims, objects and methods. In *Great Sandy Desert, 1:1,000,000 Vegetation Series: Explanatory Notes to Sheet 2*. University of Western Australia Press: Nedlands.
Berry, S.L. and Roderick, M.L. (2002) Estimating mixtures of leaf functional types using continental-scale satellite and climatic data. *Global Ecology and Biogeography* 11, 23–39.
Bond, W.J., Woodward, F.I. and Midgley, G.F. (2005) The global distribution of ecosystems in a world without fire. *New Phytologist* 165, 525–538.
Box, E.O. (1981) *Macroclimate and Plant Forms: An Introduction to Predictive Modelling in Phytogeography. Tasks for Vegetation Science*, Vol. 1. Junk: The Hague.
Box, E.O. (1995) Factors determining distributions of tree species and plant functional types. *Vegetatio* 121, 101–116.
Box, E.O. (1996) Plant functional types and climate at the global scale. *Journal of Vegetation Science* 7, 309–320.
Bradstock, R.A. (2010) A biogeographic model of fire regimes in Australia: current and future implications. *Global Ecology and Biogeography* 19, 145–158.
Briceño, V.F., Harris-Pascal, D., Nicotra, A.B. *et al.* (2014) Variation in snow cover drives differences in frost resistance in seedlings of the alpine herb *Aciphylla glacialis*. *Environmental and Experimental Botany* 106, 174–181.
Brock, M.A., Nielsen, D.L., Shiel, R.J. *et al.* (2003) Drought and aquatic community resilience: the role of eggs and seeds in sediments of temporary wetlands. *Freshwater Biology* 48, 1207–1218.
Bunn, S.E. and Arthington, A.H. (2002) Basic principles and ecological consequences of altered flow regimes for aquatic biodiversity. *Environmental Management* 30, 492–507.
Bunn, S.E., Thoms, M.C., Hamilton, S.K. and Capon, S.J. (2006) Flow variability in dryland rivers: boom, bust and the bits in between. *River Research and Applications* 22, 179–186.
Bureau of Meteorology (n.d.) Gridded climate data. Canberra: Australian Government. Available at: http://www.bom.gov.au/climate/averages/climatology/gridded-data-info/gridded-climate-data.shtml
Byrne, M. (2007) Phylogeography provides an evolutionary context for the conservation of a diverse and ancient flora. *Australian Journal of Botany* 55, 316–325.

Byrne, M., Yeates, D.K., Joseph, L. *et al.* (2008) Birth of a biome: insights into the assembly and maintenance of the Australian arid zone biota. *Molecular Ecology* 17, 4398–4417.

Cahill, D.M., Rookes, J.E., Wilson, B.A. *et al.* (2008) Turner Review No. 17: *Phytophthora cinnamomi* and Australia's biodiversity: impacts, predictions and progress towards control. *Australian Journal of Botany* 56, 279–310.

Campbell, M.L. and Clarke, P.J. (2006) Seed dynamics of resprouting shrubs in grassy woodlands: seed rain, predators and seed loss constrain recruitment potential. *Austral Ecology* 31, 1016–1026.

Capon, S.J. (2005) Flood variability and spatial variation in plant community composition and structure on a large arid floodplain. *Journal of Arid Environments* 60, 283–302.

Carnahan, J.A. (1976) Natural vegetation. In *Atlas of Australian Resources*, 2nd series. Department of Natural Resources: Canberra.

Chapman, A.D. (2009) *Numbers of Living Species in Australia and the World*, 2nd edition. Australian Biological Resources Study: Canberra. Available at: https://www.environment.gov.au/system/files/pages/2ee3f4a1-f130-465b-9c7a-79373680a067/files/nlsaw-2nd-complete.pdf (accessed 23 June 2016).

Chase, J.M. (2010) Stochastic community assembly causes higher biodiversity in more productive environments. *Science* 328, 1388–1391.

Chesson, P. (2000) Mechanisms of maintenance of species diversity. *Annual Review of Ecology and Systematics* 31, 343–366.

Clarke, P.J., Lawes, M.J., Murphy, B.P. *et al.* (2015) A synthesis of postfire recovery traits of woody plants in Australian ecosystems. *Science of the Total Environment* 534, 31–42.

Clements, F.E. (1916) *Plant succession: an analysis of the development of vegetation*. Carnegie Institution of Washington: Washington DC.

Coomes, D.A. and Grubb, P.J. (2000) Impacts of root competition in forests and woodlands: a theoretical framework and review of experiments. *Ecological Monographs* 70, 171–207.

Craine, J.M., Engelbrecht, B.M.J., Lusk, C.H. *et al.* (2012) Resource limitation, tolerance, and the future of ecological plant classification. *Frontiers in Plant Science* 3, 246.

Crisp, M.D. and Cook, L.G. (2013) How was the Australian flora assembled over the last 65 million years? A molecular phylogenetic perspective. *Annual Review of Ecology, Evolution and Systematics* 44, 303–324.

Crisp, M.D., Burrows, G.E., Cook, L.G. *et al.* (2011) Flammable biomes dominated by eucalypts originated at the Cretaceous–Palaeogene boundary. *Nature Communications* 2, 193.

Commonwealth Scientific and Industrial Research Organisation (CSIRO) (n.d.) *Soil and Landscape Grid of Australia*. TERN: Canberra. Available at: http://www.clw.csiro.au/aclep/soilandlandscapegrid/

Delhaize, E. and Ryan, P.R. (1995) Aluminum toxicity and tolerance in plants. *Plant Physiology* 107, 315–321.

Department of the Environment and Energy (2007) NVIS data products 4.2. Australian Government: Canberra. Available at: http://www.environment.gov.au/land/native-vegetation/national-vegetation-information-system/data-products (accessed 10 May 2016).

Department of the Environment and Water Resources (2007) *Australia's Native Vegetation: A Summary of Australia's Major Vegetation Groups, 2007*. Australian Government: Canberra. Available at: https://www.environment.gov.au/system/files/resources/a9897cf2-9d38-4201-bea2-13dadf3af9a8/files/major-veg-summary.pdf (accessed 24 June 2016).

Díaz, S., Lavorel, S., McIntyre, S. *et al.* (2006) Plant trait responses to grazing: a global synthesis. *Global Change Biology* 13, 313–341.

Diels, L. (1906) Die pflanzenwelt von West-Australien südlich des wendekreises: mit einer einleitung über die pflanzenwelt Gesamt-Australiens in grundzügen. In *Die Vegetation der Erde*. Vol. 7. Verlag Von Wilhelm Engelmann: Leipzig.

Dyer, J.M. (2009) Assessing topographic patterns in moisture use and stress using a water-balance approach. *Landscape Ecology* 24, 391–403.

Eamus, D. (1999) Ecophysiological traits of deciduous and evergreen woody species in the seasonally dry tropics. *Trends in Ecology and Evolution* 14, 11–16.

Eamus, D., Zolfaghar, S., Villalobos-Vega, R. *et al.* (2015) Groundwater-dependent ecosystems: recent insights from satellite and field based studies. *Hydrology and Earth System Sciences* 19, 4229–4256.

Egerton-Warburton, L.M., Kuo, J., Griffin, B.J. and Lamont, B.B. (1993) The effect of aluminium on the distribution of calcium, magnesium and phosphorus in mycorrhizal and non-mycorrhizal seedlings of Eucalyptus rudis: a cryo-microanalytical study. *Plant Soil* 155, 481–484.

Eldridge, D.J., Poore, A.G.B., Ruiz-Colmenero, M. *et al.* (2016) Ecosystem structure, function and composition in rangelands are negatively affected by livestock grazing. *Ecological Applications* 26, 1273–1283.

Faber-Langendoen, D., Keeler-Wolf, T., Meidinger, D. *et al.* (2014) EcoVeg: a new approach to vegetation description and classification. *Ecological Monographs* 84, 533–561.

Falster, D.S. and Westoby, M. (2005) Alternative height strategies among 45 dicot rain forest species from tropical Queensland, Australia. *Journal of Ecology* 93, 521–535.

Finlayson, C.M. (2005) Plant ecology of Australia's tropical floodplain wetlands: a review. *Annals of Botany* 96, 541–555.

Gibson, N., Meissner, R., Markey, A.S. and Thompson, W.A. (2012) Patterns of plant diversity in ironstone ranges in arid south western Australia. *Journal of Arid Environments* 77, 25–31.

Givnish, T.J., Wong, C.S., Stuart-Williams, H. *et al.* (2014) Determinants of maximum tree height in *Eucalyptus* species along a rainfall gradient in Victoria, Australia. *Ecology* 95, 2991–3007.

Gleason, S.M., Butler, D.W. and Waryszak, P. (2013) Shifts in leaf and stem hydraulic traits across aridity gradients in eastern Australia. *International Journal of Plant Sciences* 174, 1292–1301.

Grime, J.P. (2001) *Plant Strategies, Vegetation Processes, and Ecosystem Properties*, 2nd edition. John Wiley & Sons: Chichester.

Hess, H.H. (1962) History of ocean basins. In *Petrologic Studies*, ed. A.E.J. Engel, H.L. James and B.F. Leonard. Geological Society of America: Boulder, CO, pp. 599–620.

Hnatiuk, R.J., Thackway, R. and Walker, J. (2009) Vegetation. In *Australian Soil and Land Survey Field Handbook*, 3rd edition, ed. National Committee on Soil and Terrain, Australia. CSIRO Publishing: Melbourne, pp. 73–126.

Holly, C., Laughlin, G.P. and Ball, M.C. (1994) Cold-induced photoinhibition and design of shelters for establishment of eucalypts in pasture. *Australian Journal of Botany* 42, 139–147.

Hooker, J.D. (1860) Introductory essay. In *Flora Tasmaniae. Volume 1. Part III. Dicotyledones*, ed. J.D. Hooker. Lovell Reeve: London, pp. i–cxxviii.

Hopper, S.D. (1979) Biogeographical aspects of speciation in the southwest Australian flora. *Annual Review of Ecology and Systematics* 10, 399–422.
Hopper, S.D. (2009) OCBIL theory: towards an integrated understanding of the evolution, ecology and conservation of biodiversity on old, climatically buffered, infertile landscapes. *Plant Soil* 322, 49–86.
Hubbell, S.P. (2005) Neutral theory in community ecology and the hypothesis of functional equivalence. *Functional Ecology* 19, 166–172.
Hughes, L., Westoby, M. and Johnson, A.D. (1993) Nutrient costs of vertebrate- and ant-dispersed fruits. *Functional Ecology* 7, 54–62.
Hunt, D. (2013) *Girt: The Unauthorised History of Australia*. Black Inc.: Melbourne.
Jarrad, F.C., Walshe, T., Chee, Y.E. and Burgman, M.A. (2012) A test of biotic interactions among two alpine plant species in Australia. *Austral Ecology* 37, 90–96.
James, C.S., Capon, S.J. and Quinn, G.P. (2015) Nurse plant effects of a dominant shrub (*Duma florulenta*) on understorey vegetation in a large, semi-arid wetland in relation to flood frequency and drying. *Journal of Vegetation Science* 26, 985–994.
Kadereit, G. and Freitag, H. (2011) Molecular phylogeny of Camphorosmeae (Camphorosmoideae, Chenopodiaceae): implications for biogeography, evolution of C_4-photosynthesis and taxonomy. *Taxon* 60, 51–78.
Kadereit, G., Mavrodiev, E., Zacharias, E.H. and Sukhorukov, A.P. (2010) Molecular phylogeny of Atripliceae (Chenopodioideae, Chenopodiaceae): implications for systematics, biogeography, flower and fruit evolution, and the origin of C_4 photosynthesis. *American Journal of Botany* 97, 1664–1687.
Keddy, P.A. (1989) *Competition*. Chapman and Hall: London.
Keddy, P.A. (1992) Assembly and response rules: two goals for predictive community ecology. *Journal of Vegetation Science* 3, 157–164.
Keeley, J.E. and Rundel, P.W. (2005) Fire and the Miocene expansion of C_4 grasslands. *Ecology Letters* 8, 683–690.
Keeley, J.E., Bond, W.J., Bradstock, R.A. *et al.* (2012) *Fire in Mediterranean Ecosystems: Ecology, Evolution and Management*. Cambridge University Press: Cambridge.
Keith, D.A. (1996) Fire-driven mechanisms of extinction in vascular plants: a review of empirical and theoretical evidence in Australian vegetation. *Proceedings of the Linnean Society of New South Wales* 116, 37–78.
Keith, D.A. (2002) Population dynamics of an endangered heathland shrub, *Epacris stuartii* (Epacridaceae): recruitment, establishment and survival. *Austral Ecology* 27, 67–76.
Keith, D.A. (2004) *Ocean Shores to Desert Dunes: The Native Vegetation of New South Wales and the ACT*. NSW Department of Environment and Conservation: Sydney
Keith, D.A. (2012) Functional traits: their roles in understanding and predicting biotic responses to fire regimes from individuals to landscapes. In *Flammable Australia: Fire Regimes, Biodiversity and Ecosystems in a Changing World*, 2nd edition, ed. R.A. Bradstock, R.J. Williams and A.M. Gill. CSIRO Publishing: Melbourne, pp. 97–125.
Keith, D.A. and Bradstock, R.A. (1994) Fire and competition in Australian heath: a conceptual model and field investigations. *Journal of Vegetation Science* 5, 347–354.
Keith, D.A. and Myerscough, P.J. (1993) Floristics and soil relations of upland swamp vegetation near Sydney. *Australian Journal of Ecology* 18, 325–344.
Keith, D.A. and Pellow, B.J. (2015) *Review of Australia's Major Vegetation Classification and Description*. Centre for Ecosystem Science: Sydney. Available at: https://www.environment

.gov.au/system/files/pages/c838080e-2efb-4718-9342-472bf9da70e6/files/review-australias-major-vegetation-classification-descriptions.pdf (accessed 23 June 2016).
Keith, D.A., McCaw, W.L. and Whelan, R.J. (2002) Fire regimes in Australian heathlands and their effects on plants and animals. In *Flammable Australia: The Fire Regimes and Biodiversity of a Continent*, ed. R.A. Bradstock, J.E. Williams and A.M. Gill. Cambridge University Press: Cambridge, pp. 199–237.
Keith, D.A., Elith, J. and Simpson, C.C. (2014) Predicting distribution changes of a mire ecosystem under future climates. *Diversity and Distributions* 20, 440–454.
Laffan, S.W. and Crisp, M.D. (2003) Assessing endemism at multiple spatial scales, with an example from the Australian vascular flora. *Journal of Biogeography* 30, 511–520.
Lambers, H., Brundrett, M.C., Raven, J.A. and Hopper, S.D. (2010) Plant mineral nutrition in ancient landscapes: high plant species diversity on infertile soils is linked to functional diversity for nutritional strategies. *Plant and Soil* 334, 11–31.
Lamont, B.B. and He, T. (2012) Fire-adapted Gondwanan angiosperm floras evolved in the Cretaceous. *BMC Evolutionary Biology* 12, 223.
Lamont, B.B., Groom, P.K. and Cowling, R.M. (2002) High leaf mass per area of related species assemblages may reflect low rainfall and carbon isotope discrimination rather than low phosphorus and nitrogen concentrations. *Functional Ecology* 16, 403–412.
Larcher, W. (2003) '*Physiological plant ecology: ecophysiology and stress physiology of functional groups*'. 4th edition. (Springer: Berlin).
Larkindale, J., Mishkind, M. and Vierling, E. (2005) Plant responses to high temperature. In *Plant Abiotic Stress*, ed. M.A. Jenks and P.M. Hasegawa. Blackwell Publishing: Oxford, pp. 100–144.
Leigh, J.H. (1994) Chenopod shrublands. In *Australian Vegetation*, ed. R.H. Groves. Cambridge University Press: Cambridge, pp. 345–368.
Letnic, M. and Dickman, C.R. (2006) Boom means bust: interactions between the El Niño/Southern Oscillation (ENSO), rainfall and the processes threatening mammal species in arid Australia. *Biodiversity and Conservation* 15, 3847–3880.
Lowman, M.D. (1992) Leaf growth dynamics and herbivory in five species of Australian rainforest canopy trees. *Journal of Ecology* 80, 433–447.
Lusk, C.H., Onoda, Y., Kooyman, R. and Gutiérrez-Girón, A. (2010) Reconciling species-level vs. plastic responses of evergreen leaf structure to light gradients: shade leaves punch above their weight. *New Phytologist* 186, 429–438.
Lusk, C.H., Sendall, K. and Kooyman, R. (2011) Latitude, solar elevation angles and gap-regenerating rain forest pioneers. *Journal of Ecology* 99, 491–502.
McPherson, S., Eamus, D. and Murray, B.R. (2004) Seasonal impacts on leaf attributes of several tree species growing in three diverse ecosystems of south-eastern Australia. *Australian Journal of Botany* 52, 293–301.
Metcalfe, D.J., Bradford, M.G. and Ford, A.J. (2008) Cyclone damage to tropical rain forests: species- and community-level impacts. *Austral Ecology* 33, 432–441.
Mitchell, P.J., Veneklaas, E.J., Lambers, H. and Burgess, S.S.O. (2008) Using multiple trait associations to define hydraulic functional types in plant communities of south-western Australia. *Oecologia* 158, 385–397.
Mucina, L. and Wardell-Johnson, G.W. (2011) Landscape age and soil fertility, climatic stability, and fire regime predictability: beyond the OCBIL framework. *Plant and Soil* 341, 1–23.
Murphy, B.P. and Bowman, D.M.J.S. (2007) Seasonal water availability predicts the relative abundance of C_3 and C_4 grasses in Australia. *Global Ecology and Biogeography* 16, 160–169.

Murphy, B.P., Bradstock, R.A., Boer, M.M. *et al.* (2013) Fire regimes of Australia: a pyrogeographic model system. *Journal of Biogeography* 40, 1048–1058.

National Land and Water Resources Audit (2001) *Australian Native Vegetation Assessment 2001*. National Land and Water Resources Audit: Canberra.

Nilsson, C., Brown, R.L., Jansson, R. and Merritt, D.M. (2010) The role of hydrochory in structuring riparian and wetland vegetation. *Biological Reviews* 85, 837–858.

Nix, H.A. (1982) Environmental determinants of biogeography and evolution in Terra Australis. In *Evolution of the Flora and Fauna of Arid Australia*, ed. W.R. Barker and P.J.M. Greenslade. Peacock: Frewville, pp. 47–66.

O'Dowd, D.J. and Gill, A.M. (1984) Predator satiation and site alteration following fire: mass reproduction of alpine ash (*Eucalyptus delegatensis*) in southeastern Australia. *Ecology* 65, 1052–1066.

Ooi, M.K.J., Auld, T.D. and Whelan, R.J. (2004) Delayed post-fire seedling emergence linked to season: a case study with *Leucopogon* species (Epacridaceae). *Plant Ecology* 174, 183–196.

Ooi, M.K.J., Auld, T.D. and Denham, A.J. (2009) Climate change and bet-hedging: interactions between increased soil temperatures and seed bank persistence. *Global Change Biology* 15, 2375–2386.

Orians, G.H. and Milewski, A.V. (2007) Ecology of Australia: the effects of nutrient-poor soils and intense fires. *Biological Reviews* 82, 393–423.

Paton, D.M. (1980) Eucalyptus physiology. II. Temperature responses. *Australian Journal of Botany* 28, 555–566.

Pérez-Harguindeguy, N., Díaz, S., Garnier, E. *et al.* (2013) New handbook for standardised measurement of plant functional traits worldwide. *Australian Journal of Botany* 61, 167–234.

Prior, L.D., Bowman, D.M.J.S. and Eamus, D. (2004) Seasonal differences in leaf attributes in Australian tropical tree species: family and habitat comparisons. *Functional Ecology* 18, 707–718.

Prober, S.M., Hilbert, D.W., Ferrier, S. *et al.* (2012) Combining community-level spatial modelling and expert knowledge to inform climate adaptation in temperate grassy eucalypt woodlands and related grasslands. *Biodiversity Conservation* 21, 1627–1650.

Puckridge, J.T., Sheldon, F., Walker, K.F. and Boulton, A.J. (1998) Flow variability and the ecology of large rivers. *Marine and Freshwater Research* 49, 55–72.

Raunkaier, C. (1934) *The Life Forms of plants*. Translated from the original published in Danish. Oxford University Press: Oxford.

Rinker, H.B. and Lowman, M.D. (2004) Insect herbivory in tropical forests. In *Forest Canopies*, 2nd edition, ed. M. Lowman and H.B. Rinker. Elsevier: Burlington, pp. 359–386.

Schimper, A.F.W. (1903) *Plant Geography Upon a Physiological Basis*. English translation by W.R. Fisher, ed. P. Groom, I.B. Balfour. Clarendon Press: Oxford.

Shipley, B., Vile, D. and Garnier, E. (2006) From plant traits to plant communities: a statistical mechanistic approach to biodiversity. *Science* 314, 812–824.

Sniderman, J.M.K. and Jordan, G.J. (2011) Extent and timing of floristic exchange between Australian and Asian rainforests. *Journal of Biogeography* 38, 1445–1455.

Specht, R.L. (1970) Vegetation. In *The Australian Environment*, 4th edition, ed. G.W. Leeper. CSIRO Australia & Melbourne University Press: Melbourne.

Specht, R.L. and Specht, A. (1989) Species richness of sclerophyll (heathy) plant communities in Australia: the influence of overstorey cover. *Australian Journal of Botany* 37, 337–350.

Specht, R.L. and Specht, A. (2001) *Australian Plant Communities: Dynamics of Structure, Growth and Biodiversity*. Oxford University Press: Melbourne.

Specht, R.L., Specht, A., Whelan, M.B. and Hegarty, E.E. (1995) *Conservation Atlas of Plant Communities in Australia*. Southern Cross University Press: Lismore.

Stone, C. and Coops, N.C. (2004) Assessment and monitoring of damage from insects in Australian eucalypt forests and commercial plantations. *Australian Journal of Entomology* 43, 283–292.

Tilman, D. (1988) *Plant Strategies and the Dynamics and Structure of Plant Communities*. Princeton University Press: Princeton.

Toon, A., Crisp, M.D., Gamage, H. *et al.* (2015) Key innovation or adaptive change? A test of leaf traits using *Triodiinae* in Australia. *Scientific Reports* 5, 1–12.

van der Ent, A., Jaffré, T., L'Huillier, L. *et al.* (2015) The flora of ultramafic soils in the Australia–Pacific region: state of knowledge and research priorities. *Australian Journal of Botany* 63, 173–190.

Vartapetian, B.B. and Jackson, M.B. (1997) Plant adaptations to anaerobic stress. *Annals of Botany* 79, 3–20.

Wainwright, S.J. (1981) Plants in relation to salinity. *Advances in Botanical Research* 8, 221–261.

Walter, H. (1979) *Vegetation of the Earth and Ecological Systems of the Geo-biosphere*. Springer-Verlag: New York.

Walter, H. and Box, E.O. (1976) Global classification of natural terrestrial ecosystems. *Vegetatio* 32, 75–81.

Wardell-Johnson, G., Williams, J.E., Hill, K.D. and Cumming, R. (1997) Evolutionary biogeography and contemporary distribution of eucalypts. In *Eucalypt Ecology: Individuals to Ecosystems*, ed. J.E. Williams and J.C.Z. Woinarski. Cambridge University Press: Cambridge, pp. 92–128.

Wardlaw, I.F. (1979) The physiological effects of temperature on plant growth. *Proceedings of the Agronomy Society of New Zealand* 9, 39–48.

Warfe, D.M., Pettit, N.E., Davies, P.M. *et al.* (2011) The 'wet–dry' in the wet-dry tropics drives river ecosystem structure and processes in northern Australia. *Freshwater Biology* 56, 2169–2195.

Whittaker, R.H. (1975) *Communities and Ecosystems*, 2nd edition. MacMillan: New York.

Wright, B.R. and Zuur, A.F. (2014) Seedbank dynamics after masting in mulga (*Acacia aptaneura*): implications for post-fire regeneration. *Journal of Arid Environments* 107, 10–17.

Wright, I.J., Reich, P.B., Westoby, M. *et al.* (2004) The worldwide leaf economics spectrum. *Nature* 428, 821–827.

2 Evolutionary Biogeography of the Australian Flora in the Cenozoic Era

PETER H. WESTON AND GREGORY J. JORDAN

The dominant historical pattern of the Australian flora over the last 65 million years is one of contraction of a widespread mesic flora that contained both rainforest and sclerophyll components, and expansion and diversification of dry climate and fire tolerant groups, especially sclerophyllous groups. Many species (perhaps the majority) are from lineages that have existed continuously in Australia since the final break-up of Gondwana. However, this autochthonous flora has been overlain by a complex pattern of migration into and out of Australia, including trans-oceanic dispersal among temperate southern hemisphere landmasses and immigration across land (or island hopping) from Asia. Some major historical biogeographic patterns within Australia seem fairly clear (such as the separation of mesic zones of southwestern and southeastern Australia). However, many other biogeographic patterns and processes within Australia are still very poorly understood.

2.1 Introduction

Evolutionary history is important for understanding how biomes were assembled, why floras from physically similar environments in different parts of the world are composed of strikingly different taxa and why some regions have more species, endemic species or endemic clades than others.

In this chapter, we aim to provide a synthetic, testable narrative of the evolution of the flora of Australia (including southern New Guinea) in space and time, concentrating on the most influential formative years prior to and after the isolation of the continent in the late Eocene. We will first review previous ideas on the evolution of the Australian flora. We will then consider the evidence that is now available on the evolution and historical biogeography of major Australian plant taxa, the major biomes to which they belonged, and the evolution of the landscapes in which they have lived.

2.2 A history of histories

2.2.1 Joseph Hooker's nineteenth century insights

Hooker (1853) recognised that many genera and other 'well marked' taxa were shared by New Zealand, temperate South America and south-eastern Australia (including Tasmania), but were found nowhere else. He believed that this 'Antarctic flora' demanded a general explanation and argued that a previously continuous, ancestral Antarctic flora had been

Australian Vegetation, Third Edition, ed. David A. Keith. Published by Cambridge University Press. © Cambridge University Press 2017.

Citation: Weston, P.H. and Jordan, G.J. (2017). Evolutionary Biogeography of the Australian Flora in the Cenozoic Era. In D. A. Keith, ed., *Australian Vegetation*, 3rd edn. Cambridge: Cambridge University Press, pp. 40–62.

fragmented by 'geological and climatic causes'. This established the idea of vicariance (disjunctions caused by breaking up of past distributions, rather than dispersal across oceans or other barriers) and anticipated the idea of Gondwanic vicariance by many years. With no concept of continental drift, Hooker postulated that previous intercontinental land connections may have sunk beneath the waves due to geological uplift (supported, for example, by fossils of extant marine molluscs found hundreds of metres above sea level in South America), subsidence and erosion.

Charles Darwin might have been Hooker's closest scientific confidant but he disagreed strongly with Hooker's biogeographic explanation, asserting that dispersal of individual taxa across vast oceanic gaps was more likely than the movement of whole floras along hypothetical land bridges or stepping stones (Darwin 1859). Since then, Darwinian long-distance dispersal and Hookerian movement of whole floras followed by vicariance have competed for the favour of biogeographers as explanations of discontinuous distributions.

Hooker (1860) then analysed the biogeography of the Australian flora. By counting the genera unique to Australia and those shared with surrounding continents he identified other major components of the flora: 'peculiarly Indian' genera in the monsoon tropics, 'Malayan' and 'Polynesian' genera in the wet tropics; a South African connection in the south-west; a European contingent in montane south-eastern Australia; and a large group of characteristically Australian, predominantly sclerophyllous genera (later dubbed the 'autochthonous element'). Hooker pointedly noted that the non-Australian elements were concentrated in parts of the continent that were geographically closest to their exotic relatives and that many more 'peculiarly Indian' genera occurred in Australia than 'Australian genera' in Asia. Hooker also noted striking differences between the floras of south-eastern and south-western Australia, especially the very high species richness and endemism in the south-west. He speculated that the autochthonous element originated in South Africa, dispersed to south-western Australia and later to the rest of the continent.

2.2.2 Elaborations of Hooker's biogeography, 1860–1960

In the hundred years following Hooker's analyses, several botanists used similar methods and improved knowledge of the extant flora, palaeobotany and landscapes of remote parts of the continent to flesh out his ideas, often accompanied by hypotheses of centres of origin of floristic elements. Some of these ideas have endured, such as the hypothesis that a barrier to dispersal of sclerophyllous plants across what is now the Nullarbor Plain formed in the Miocene and effectively isolated the mesic floras of south-eastern and south-western Australia. However, most of the hypotheses form an outdated 'invasion theory' that assumes that all major plant groups originated elsewhere and dispersed into Australia over a stable configuration of continents and oceans (Barlow 1981, 1994; Crisp *et al.* 1999).

Early palaeobotanists working on Australian specimens used 'picture matching' similarity of fossil leaves, flowers, fruit and wood to living taxa to infer that Australia contained an implausible mixture of northern hemisphere and modern Australian groups of plants. Some of these fossil identifications have stood the test of time, but none should be accepted without critical appraisal (Hill 1988). In the 1950s, Isabel Cookson transformed Australian palaeobotany by systematically studying fossil pollen (e.g. Cookson and Pike 1954) and leaf cuticles (e.g. Cookson and Duigan 1950), backed up by detailed comparisons with living species. However, this work was constrained by the poor capacity to determine the ages of the fossils, and limits to the understanding of the living flora. Overall, by the 1960s, it was clear that some important modern Australian groups, such as *Banksia* (Figure 2.1a, b), *Eucalyptus*, *Nothofagus*, conifers, Santalaceae and Casuarinaceae had long histories on the continent.

A century after Hooker's seminal biogeographic analysis, Burbidge (1960) produced a monumental update, using similar methods, but armed with greatly improved knowledge of the Australian flora, especially in tropical and arid regions. She divided the continent into four floristic zones, three

A. Fossil *Banksia cooksoniae* from the Latrobe Valley Coal. Isabel Cookson described this fossil as *Banksieaephyllum acuminatum*. It demonstrated the antiquity of *Banksia*.

B. Fossil of the unmistakeable pollen of *Banksia*. Cookson showed that this type of pollen, and many others were present in Australia deep in the Cenozoic.

C. Paleogene fossil of an extinct podocarp species (possibly *Acmopyle*)

D. Early Eocene fossil *Araucaria* from Tasmania

E. Scanning electron micrograph of cuticle extracted from the fossil *Araucaria* in D

F. Fossil leaf of *Nothofagus gunnii* alongside a leaf of the extant species.

G. *Campynema* is a relictual relative of lilies found in alpine Tasmania

H. *Isophysis tasmanica* from alpine Tasmania is sister to all other irises

Figure 2.1 Macrofossils and pollen in dated sediments provide independent calibration data for phylogenies and evolutionary narratives. Phylogenetically isolated lineages also provide historical signals.
Photos: Greg Jordan. (A black and white version of this figure will appear in some formats. For the colour version, please refer to the plate section.)

interzones and three focal areas of particular biogeographic interest. By counting the numbers of families and genera endemic to these regions, and those shared between the regions and with other landmasses, she identified floristic relationships or distributional tracks. Burbidge's tracks were consistent with Hooker's elements but she speculated in more detail about the historical processes that might have produced the tracks. For example, she argued that Eremaean taxa that have their closest relatives in the northern hemisphere invaded central Australia as strand or dune plants on the coastlines of Cretaceous epicontinental seas, and adapted to arid environments much later, during the Pleistocene.

2.2.3 Reinterpretation in the light of plate tectonic theory, 1960–1980

The scientific revolution that replaced the model of fixed continents with one of mobile tectonic plates (Hess 1962), thus corroborating Wegener's (1929) idea of continental drift, forced paradigm shifts on historical biogeography. Vicariance replaced long-distance dispersal as the default hypothesis for explaining disjunct intercontinental distributions and Australia quickly came to be seen as a fragment of Gondwana, not an eternally isolated island continent. It was finally conceded that major plant taxa might have originated in Australia rather than having invaded it. A turning point in Australian biogeography was the work of Raven and Axelrod (1974), who fitted biogeographic narratives for many angiosperm families to a model of drifting continents. Using an essentially Darwinian approach, they drew dispersal routes from centres of origin over new, dynamic paleomaps of the globe, and concluded that angiosperms originated in West Gondwana during the early Cretaceous and invaded the rest of the world from there.

Barlow (1981, 1994) then created a new synthesis of Australian phytogeography that incorporated new information about paleoclimatology and geomorphology. He concluded that the Antarctic and autochthonous elements of the Australian flora were both parts of a Gondwanan flora inherited from the disintegrating supercontinent, and only differing in the degree to which they had adapted to Australia's changing environment over the past 65 million years. Thus, he considered that the autochthonous element had diversified and adapted to live in the expanding non-rainforest biomes, whereas the Antarctic element remained in the shrinking rainforests. He explained the Indo-Malayan element as coming from the collision of the Australian Plate with the Sunda Shelf in the early Miocene, with ensuing floristic exchange strongly biased in favour of Asian invaders over Australian emigrants.

Barlow was supported by other biogeographers, notably Johnson and Briggs (1975, 1981), who reinterpreted the history of Proteaceae, Myrtaceae and Restionaceae in the light of Gondwanic fragmentation and inferred their phylogenetic relationships. They concluded that most of the numerous taxa with overlapping distributions across Gondwanic fragments had vicariant histories, having dispersed with the landmasses themselves as the supercontinent broke up. They postulated that sclerophyllous members of the Proteaceae and Myrtaceae evolved from rainforest ancestors, starting in the early Cenozoic.

2.2.4 The impact of phylogenetic methodology

The phylogenetic revolution in biological systematics initiated by Hennig (1966) had profound impacts on historical biogeographic research. Until the 1980s, phylogenetic background knowledge was provided almost entirely by taxonomy and the distributions and characteristics of genera or species. Groups that taxonomists chose not to name were usually ignored by biogeographers. This imposed substantial sources of error – the named taxa might not be monophyletic; only a small proportion of potential information about phylogenetic relationships was being considered by biogeographers; and genera were treated as units in quantitative analyses, even though genera are rarely equivalent in age, number of species or morphological diversity.

Phylogenies can estimate every speciation event in a clade's evolutionary history, especially when fuelled by spectacular technological advances in molecular biology. These phylogenies have enabled systematists at last to test phylogenetic hypotheses rigorously, and more recently, to estimate the ages of nodes in phylogenetic trees (Ho 2014). This in turn was the basis for the development of a number of potentially powerful ways to use this goldmine of historical information to test biogeographic hypotheses (Crisp *et al.* 2011b).

2.2.5 Palaeobotanical progress since 1960

Recent palaeobotanical studies have presented a moderately clear impression of the vegetation and flora of south-eastern Australia, as well as useful snapshots of other parts of the continent, for the Late Paleocene to the mid-Miocene (~59 to 12 Ma [million years ago]). The subsequent period until the late Pliocene (~3.6 Ma) represents a conspicuous gap with very little plant fossil evidence, which poses problems for interpreting the fossil record because dry climates became prevalent in Australia during this period (Byrne *et al.* 2011).

Stover and Partridge's (1973) stratigraphic framework for fossil pollen and spores finally provided a means for determining the age of most plant fossil assemblages in south-eastern Australia. These ages are sometimes validated by the marine fossil record, palaeomagnetism, radioactive isotopes or other methods. With refinements, this system has been tentatively extended to other parts of Australia (see Macphail 2007).

Interpreting and identifying fossils has also improved through elaborations of Cookson's use of microfossils and the cuticles of fossil leaves, the incorporation of phylogenetic evidence and logic (e.g. Jordan and Hill 1999; Sauquet *et al.* 2009), and the use of non-taxonomic (physiognomic) fossil signals of climate (e.g. Greenwood *et al.* 2003) or vegetation structure (e.g. Jordan *et al.* 2014). Using phylogenies in identifying Australian fossils has helped overcome the problem that fossil identifications may be erroneous when based on ancestral characteristics shared among species that are not closely related (symplesiomorphies). Phylogenetic approaches are becoming more common but many fossils are still identified on the basis of overall similarity. The potential cost of demanding phylogenetically informed identifications is that many fossils would be ignored (e.g. Sauquet *et al.* 2009) except when only used as physiognomic proxies.

2.2.6 Recent meta-analyses and syntheses

Our knowledge of the evolutionary phytogeography of Australia has consolidated in the new millennium, with the publication of reviews and meta-analyses that have used greatly enhanced palaeobotanical and phylogenetic knowledge to test biogeographic hypotheses (Sanmartín and Ronquist 2004; Hopper 2009; Bowman *et al.* 2010; Byrne *et al.* 2008, 2011; Sniderman and Jordan 2011; Crisp and Cook 2007, 2013; Kooyman *et al.* 2014; Crayn *et al.* 2015). These studies found support for some of the general ideas postulated by earlier biogeographers but, not surprisingly, found many of them to be simplistic. The much greater accuracy and precision that detailed phylogenetic analyses provide over intuitively generated taxonomies and fossil identifications have allowed complex biogeographic patterns and processes to be disentangled to an extent that was unimaginable as recently as the 1980s.

2.3 Earth history

This section focuses on geological and climatic changes for the Paleogene and Neogene (~65 to

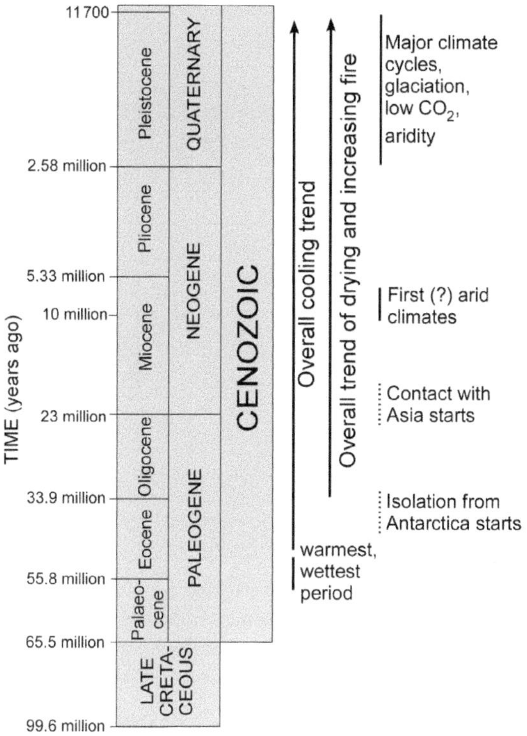

Figure 2.2 Geological time scale indicating the approximate ages of relevant geological periods and epochs, as well as major environmental events, as described in the text.

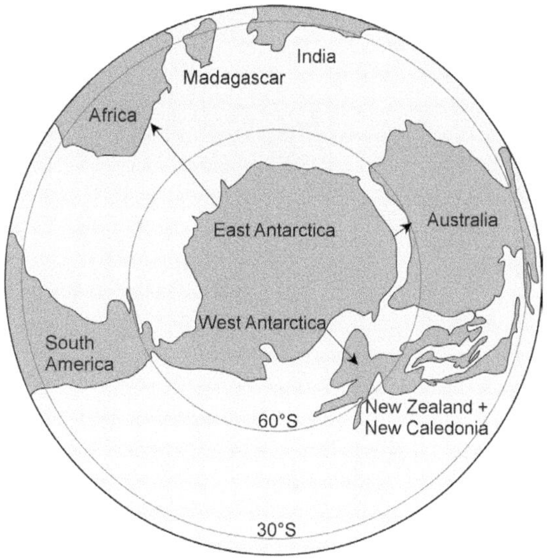

Figure 2.3 Gondwanan landmasses during the latest Cretaceous (~ 70 Ma). Arrows indicate prior rifting from the rest of Gondwana. Figure based on Cantrill and Poole (2012).

2.58 Ma) (Figure 2.2). The subsequent period of ice ages (the Quaternary) is described by Mooney *et al.* (Chapter 3, this volume).

2.3.1 Historical geology

Since the beginning of the twenty-first century, there has been general agreement about the very broad picture of the break-up of Gondwana (see McLoughlin 2001). Approximately 182 Ma, the major southern hemisphere landmasses, plus India, Sri Lanka, northern Borneo and some other fragments were connected to form Gondwana. Rifting then led to the early departure of India and Madagascar, followed by Africa in the mid-Cretaceous. Australia remained connected through Tasmania with a widening rift valley to the west. This rift valley was progressively flooded starting from the western end, creating a seaway south of Western Australia perhaps as early as the mid-Cretaceous, and increasingly isolating Australia from Antarctica except through a diminishing connection via Tasmania. That connection was severed at least 37 Ma (Exon *et al.* 2004), although a shallow water gap may have formed as early as 48 Ma (Figure 2.3). On the opposite side of Gondwana, South America remained connected or close to Antarctica until approximately the same time (Exon *et al.* 2004). The plates that now make up New Zealand, New Caledonia and a large area of land now submerged were connected to each other and attached to West Gondwana and the eastern margin of Australia. 'Zealandia' probably

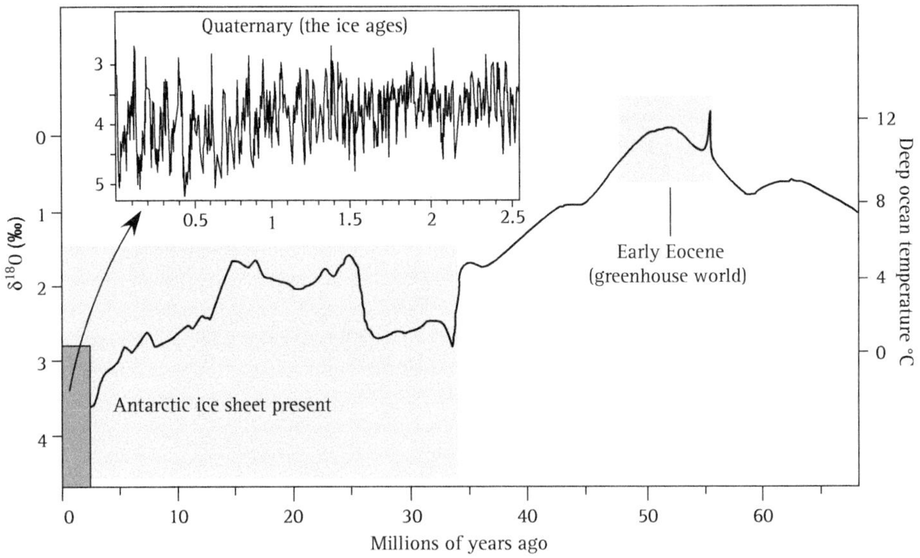

Figure 2.4 The oxygen isotope from fossil shells in drill cores from the deep ocean floor shows major trends in climate through time. Low values (i.e. high on the vertical axis) of $\delta^{18}O$ indicate a combination of globally warm temperatures and small amounts of water trapped in ice, high values indicate low temperatures and/or large ice caps. The main figure shows $\delta^{18}O$ for the last 65 million years, based on Zachos *et al.* (2001), and the inset shows $\delta^{18}O$ for the Quaternary (the last 2.58 million years, which includes the Pleistocene and Holocene), based on Tidemaan *et al.* (1994).

split from West Antarctica and southern Australia c. 84 Ma, but may have remained connected with north-eastern Australia until c. 50 Ma (Ladiges and Cantrill 2007). Australia's move northward led to a collision with Southeast Asia c. 25 Ma. Since then, Australia and Asia may well have been intermittently connected by land or separated by narrow seaways.

Uplift of the highlands of eastern and south-eastern Australia commenced in the early Cretaceous c. 120 Ma and continued at varying rates at different latitudes until the present (Müller *et al.* 2016). In the north, the rate of uplift was almost constant, in contrast to that in the central eastern highlands and south-east, where it slowed dramatically in the mid-Cretaceous before accelerating in the south-east during the Eocene to reach alpine altitudes in the Neogene. Uplift was accompanied by sporadic volcanism through most of the Cenozoic along an arc more or less following the Great Dividing Range from north-east Queensland to Victoria and extending to Mt Gambier in South Australia. Different regions experienced volcanic activity at different times, with eruptions until a few thousand years ago. The volcanic activities may have created disturbance regimes that facilitated the survival of disturbance-dependant taxa in periods with little or no fire. However, the major impact was the provision of some areas with fertile soils that differ dramatically from the extremely depauperate soils found across most of the continent.

2.3.2 Paleoclimatology

Oxygen isotopes in deep-sea sediments (Figure 2.4) provide a framework for understanding changes in Australia's climate over the Cenozoic. They indicate an increase in global temperatures during the Paleocene (65–56 Ma), peaking briefly at the Paleocene/Eocene boundary and then again for an extended period through the early Eocene (c. 56–48 Ma). Temperatures have trended downwards ever since, with an abrupt drop at the Eocene/Oligocene boundary (c. 34 Ma) and partial recoveries in the late Oligocene to mid-Miocene (c. 26–15 Ma) and mid-Pliocene (c. 3 Ma).

Most evidence for local terrestrial palaeoclimates comes from fossil pollen and spores (summarised by Macphail 2007). Based on this evidence, southern Australia broadly followed the global temperature trends described above. Thus, there is evidence that, even at high palaeolatitudes, climates were nearly as warm as current lowland tropical climates during the early Eocene (Carpenter *et al.* 2012; Pross *et al.* 2012). In addition, the major cooling event at the end of the Eocene was sufficient to create glaciers to relatively low altitudes in Tasmania (Macphail *et al.* 1993). Northern Australia may have shown less overall cooling since climate change was strongest in polar regions (Macphail 2007).

The long-held view that precipitation changed in parallel with temperature still holds well for south-eastern Australia and moderately well for the rest of the continent. Thus, wet climates were present widely across Australia from the Paleocene to the Oligocene, followed by a general trend of drying (Byrne *et al.* 2008). Palaeoclimatic evidence for Western, Central and Northern Australia is very patchy, but the presence of mesic taxa (such as the conifers *Dacrycarpus* and *Dacrydium*) indicate that Paleocene to Oligocene climates were more-or-less mesic (although possibly less so than south-eastern Australia). Fossil *Banksia* leaves, combined with silcrete formation, suggest that south-western Australia had become at least seasonally dry by the late Eocene (Carpenter *et al.* 2014). Some patchy palaeobotanical evidence from north-west and central Australia (Macphail 2007) suggests that such seasonality may have been quite widespread. Geomorphological evidence shows considerable aridification of central Australia (and most of the rest of Australia) during the late Miocene. A brief reversion to wet climates sufficient to support distinctly mesic vegetation on the Nullarbor Plain occurred in the mid-Pliocene (c. 3.5 Ma) (Sniderman *et al.* 2016).

2.4 The evolution of dominant and diverse taxa

Different taxa have responded to opportunities and selective pressures in various ways, depending on luck and the features that they inherited, so it pays to consider the history of particular clades as well as that of whole biotas. Here we concentrate on dominant or particularly diverse taxa.

2.4.1 Conifers

The Cenozoic history of Australia's conifers can be summarised as early dominance and diversity, followed by massive depletion induced by fire and dry climates over the last 30 million years. Most groups were driven either into mesic refugia or, more often, to regional or global extinction. Indeed, conifers reached their highest species and phylogenetic diversity in the Australian fossil record in the early Oligocene, peaking at levels possibly greater than any region of the modern world.

Fossil pollen indicates that multiple lineages of Podocarpaceae (*Dacrydium*, *Microcachrys*, *Lagarostrobos* and others that probably included *Podocarpus*) occurred in all major regions of Australia in the Eocene and probably Oligocene (Macphail 2007). Macrofossils (mostly foliage; e.g. Figure 2.1c) show that many more genera were present in the Eocene and Oligocene than at present, especially in Tasmania (Hill and Brodribb 1999). The extent of the subsequent restriction of the family is indicated by the fact that most of the phylogenetic diversity in Podocarpaceae now resides in the wettest regions of the continent – western Tasmania and the rainforests of north-eastern Queensland.

Molecular dating (e.g. Biffin *et al.* 2011) generally estimates that the most recent common ancestors of extant members of podocarp genera are considerably younger than the oldest fossils of these genera. For instance, molecular dates suggest that *Phyllocladus aspleniifolius* from Tasmania diverged from its congeners in New Zealand in the late Miocene, long after the separation of Zealandia from Australia. However, unmistakeable fossil foliage of *Phyllocladus* occurs in much older sediments in both Australia (Hill and Brodribb 1999) and New Zealand (Jordan *et al.* 2011), implying either that the genus became extinct in one of these regions and then recolonised, or that the molecular dates substantially underestimate the age of the group.

Like the podocarps, Araucariaceae show Gondwanan vicariance, long-distance dispersal and fossil evidence (Figure 2.1d, e) of a late Cenozoic retreat to scattered mesic refuges in eastern Australia (Hill *et al.* 1999; Biffin *et al.* 2010). Cupressaceae, on the other hand, include clades with contrasting stories. The inferred timing and cladistic patterns of the family's diversification are consistent with Pangaean fragmentation (Mao *et al.* 2012). The Australian taxa that are restricted to fire-free environments (*Papuacedrus, Athrotaxis, Diselma*) are relictual, showing extremely low net diversification rates. The only clade that occupies fire-prone and arid environments is *Callitris–Actinostrobus*. This clade has diversified into 20 species since 50 Ma and extends over vast areas of the Australian arid zone.

2.4.2 Nothofagaceae

Numerous authors have used *Nothofagus* as a test group for understanding the biogeography of the southern hemisphere. In recent decades, well-supported phylogenetic trees, cladistic biogeographic analysis and molecular dating have been used to test ideas about the biogeography of the genus. Cladistic biogeographic analysis shows Australian species clustering closer to New Zealand taxa than to South American ones in two clades (Ladiges 1998). This pattern is incongruent with the understood sequence of Gondwanic fragmentation, which predicts that Australian and South American taxa should be closest relatives. Most recently, Sauquet *et al.* (2012) used relaxed clock molecular dating to estimate ages of clades of *Nothofagus* under a wide range of calibration strategies for molecular dating. All their strategies except those assuming vicariance yielded chronograms in which disjunctions between South America and Australasia were temporally consistent with vicariance between those continents but disjunctions between Australia and New Zealand were much younger than the Tasman Sea, strongly suggesting dispersal across the Tasman Sea. Other taxa that have similar histories to *Nothofagus* include the Atherospermataceae, Monimiaceae (Renner 2005), Winteraceae (Thomas *et al.* 2014) and Elaeocarpaceae (Crayn *et al.* 2006).

Nothofagus has a rich fossil record going back to the late Cretaceous, based on morphologically distinctive fossils of its abundant, wind-dispersed and distinctive pollen, leaves (Figure 2.1f) and fruit. The genus was widespread across Australia during the Paleogene. Pollen grains of subgenus *Brassospora* (which is now extinct on mainland Australia) and subgenus *Lophozonia* formed a conspicuous component of fossil pollen floras of the Murray–Darling Basin as recently as the mid-Miocene, when both disappeared from the record at a time of sharply drying climate (Macphail and Truswell 1989), although *Lophozonia* returned briefly in the early Pliocene. The south-eastern Australian rainforest species *Nothofagus cunninghamii* retreated to numerous moist refugia during glacial maxima of the Pleistocene, expanding during interglacials (Worth *et al.* 2009).

2.4.3 Proteaceae

The Proteaceae exemplify Barlow's (1981) idea that the autochthonous and Antarctic elements in the Australian flora are one and the same. Thus, the Proteaceae contain both elements in the form of a suite of small rainforest genera, many of which have close relatives on other continents, and a species-rich set of largely endemic sclerophyll genera. However, both groups are old. The tribes of Proteaceae, except perhaps for the Proteeae, were probably all present on what is now Australia by the Paleocene (Sauquet *et al.* 2009, fig. S1). *Telopea*, *Alloxylon*, and Australian *Lomatia* and *Orites* are 'Antarctic elements' with surviving sister groups in South America. All are Paleocene or Eocene age, but some 'autochthonous' groups, such as *Banksia*, are at least as old. Fossil *Banksia* from New Zealand (Carpenter *et al.* 2010) emphasise the tenuous nature of the distinction between autochthonous and Antarctic elements.

Like *Nothofagus*, Proteaceae occurs on Gondwanan fragments and was long seen as showing Gondwanan vicariance (Johnson and Briggs 1975). However, again like *Nothofagus*, this view has been challenged by molecular dating analyses (Barker *et al.* 2007; Mast *et al.* 2008, 2015; Sauquet *et al.* 2009) that suggest that long-distance dispersal over water may have established many disjunct distributions between Australia and other continental fragments: South Africa (Leucadendreae, Petrophileeae, Macadamiinae), South America (Macadamiinae, Gevuininae), New Caledonia (Persoonieae, Virotiinae, Gevuininae, Hakeinae), New Zealand (Persoonieae) and Madagascar (Malagasiinae). In contrast, *Lomatia* and subtribes Embothriinae and Roupalinae are estimated to be the right age to have diversified by vicariance with the rifting of Australia from Antarctica–South America (Sauquet *et al.* 2009; Milner *et al.* 2015). Similarly, molecular dating is consistent with *Helicia* invading South East Asia over land and narrow sea gaps.

The cohort of angiosperm groups that came to dominate Australia's vegetation in the Paleocene included at least 27 extant clades of Proteaceae (Sauquet *et al.* 2009) as well as many extinct groups (Hill *et al.* 1995). Some groups that are now restricted to rainforests of the Queensland wet tropics and/or Burbidge's (1960) MacPherson–Macleay Overlap were much more widespread during the Paleogene and Miocene than they are now. These groups include subtribe Musgraveinae (sister of *Banksia*), *Athertonia* and *Eidothea*. A fossil inflorescence of Musgraveinae was found in an Eocene deposit in southern Victoria (Christophel 1984), *Athertonia* occurred in several places in central Queensland and south-eastern Victoria in the Oligo-Miocene (Rozefelds 1995) and *Eidothea* is known from several fossil sites in Victoria and on the Bass Straight Islands (Rozefelds *et al.* 2005). *Embothrium*, which is now endemic to South America, occurred in many parts of Australia (Macphail 2007).

The fates of proteaceous clades in Australia since the Cenozoic climatic optimum in the early Eocene are mostly stories of either adaptation to progressively more xeric conditions, often associated with rapid diversification, or ecological conservatism as rainforest trees with very low net diversification rates and distributional contraction (Figure 2.5). In *Banksia*, the only large clade of Proteaceae for which an almost comprehensive species-level phylogeny is available, the rate of diversification accelerated in the Neogene and decelerated from the early Miocene to the Holocene (Cardillo and Pratt 2013). Two counteracting processes may have shaped this diversification. Aridification must have driven extinction in most lineages of Proteaceae, including many sclerophyllous and xeromorphic taxa. However, this aridification also left survivors in disconnected mesic refugia, which would have facilitated allopatric speciation between the inhabitants of geographically isolated habitats. The only clades that penetrated deep into the arid zone are several groups of *Hakea* and *Grevillea*. These groups diversified to a moderate extent, with some species establishing very widespread distributions (Mast *et al.* 2015).

2.4.4 Myrtaceae

According to the most recently published molecular chronogram of the Myrtaceae (Thornhill *et al.* 2015), the family diverged from its African sister group, the Vochysiaceae, in the mid-Cretaceous,

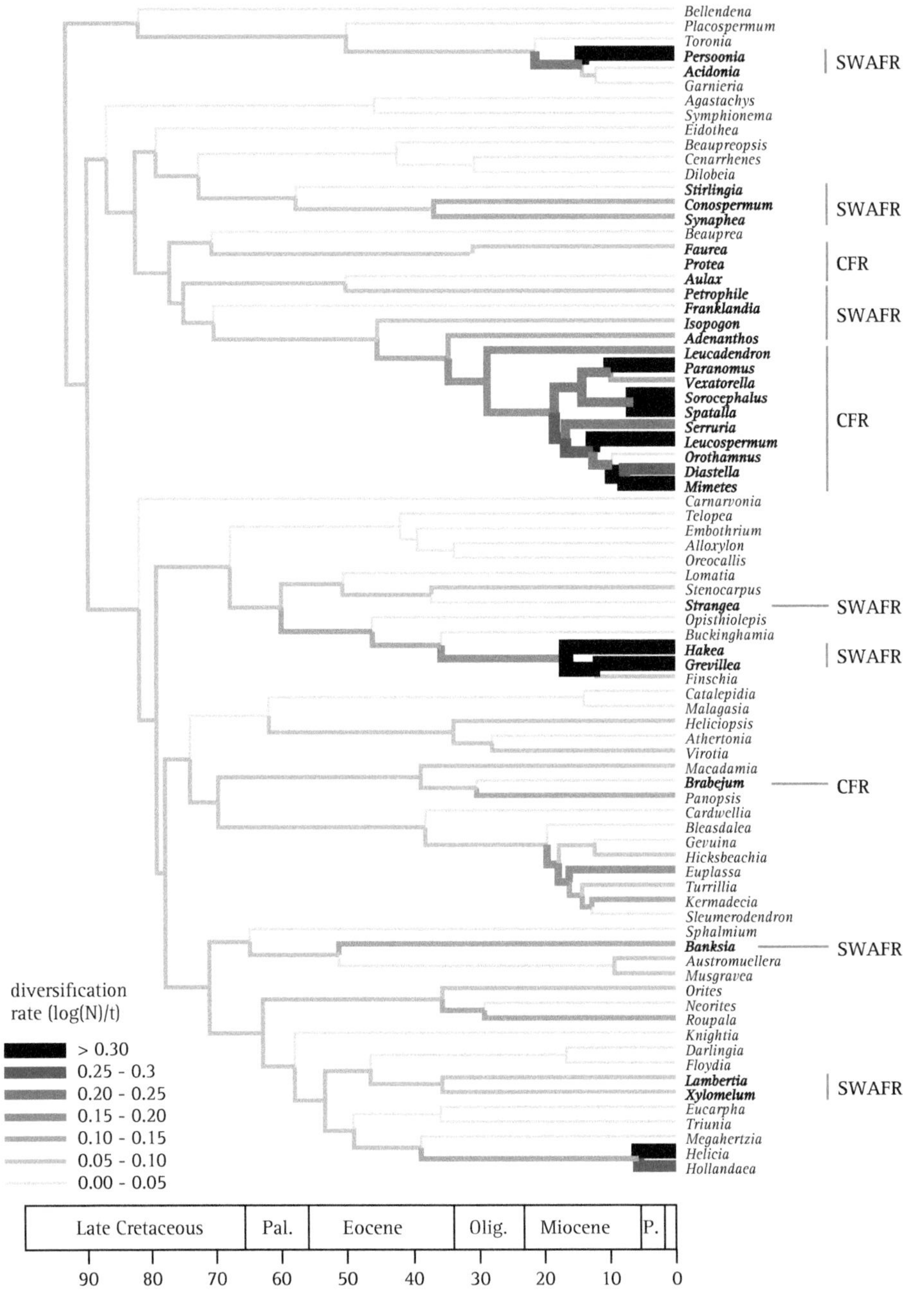

Figure 2.5 Molecular chronogram for the genera of Proteaceae, produced using the Bayesian uncorrelated lognormal method (implemented in BEAST) from an alignment of nucleotide sequence data for eight loci (Sauquet *et al.* 2009, Fig. 1, redrawn with permission of the senior author). Branches are toned according to absolute net diversification rate by stem age of their subtending clade. Taxa present in Mediterranean biodiversity hotspots are identified with either SWA (south-western Australia) or CFR (Cape Floristic Region). Absolute ages are in million years.

as Africa rifted from the remainder of Gondwana. Myrtaceae largely differentiated on the fragmenting continental block of Zealandia–South America–Antarctica–Australia during the Cenozoic. A beautiful demonstration of this is provided by exquisitely preserved and precisely dated macrofossils of *Eucalyptus* of Early Eocene age from Patagonia (Hermsen *et al.* 2012).

Five extant Australian myrtaceous lineages survived the Cretaceous–Paleogene boundary (Thornhill *et al.* 2015). The family diversified quickly during the Eocene, by the end of which 34 extant Australian clades had differentiated, including 5 major clades of eucalypts (Thornhill *et al.* 2015). Diversification of *Eucalyptus* accelerated through the Oligocene (Crisp and Cook 2013), as Australia's climate cooled and dried, and then – in a similar pattern to that in *Banksia* – decelerated from the mid-Miocene to the present when much of the continent became dry, presumably eliminating many species from central Australia and forcing a few lineages to adapt. Comparing the number of extant species of the two main clades of the eucalypts illustrates the strikingly higher net diversification of the sclerophyllous clade (about 930 species) over its rainforest sister group (4 species) by the end of the Cenozoic, in a drying and fire-prone environment. The fossil record is consistent with these patterns, but at present the capacity to differentiate fossils of different groups is limited (Thornhill and Crisp 2012).

2.4.5 Fabaceae

Two large, endemic or nearly endemic clades dominate the legume flora of Australia: the egg and bacon peas (tribes Mirbelieae and Bossiaeeae) and *Acacia*. However, the flora contains a number of other smaller clades of Fabaceae with diverse histories (in genera such as *Archidendron*, *Senna*, *Indigofera*, *Glycine* and *Desmodium*).

The age of the diversification of the Mirbelieae–Bossiaeeae clade was estimated by Lavin *et al.* (2005) to be 48.4 ± 1.3 Ma, old enough to have inhabited the Australia–Antarctica–South America supercontinent, although the geographic origin of the clade is difficult to determine. Lineage through time plots for the clade are consistent with diversification through the Eocene, followed by a mass extinction at about the Eocene–Oligocene boundary (Crisp and Cook 2009), when the climate cooled dramatically (Exon *et al.* 2004). The clade is now widespread across Australia, but most diverse in the temperate to subtropical south.

Acacia is much younger than Mirbelieae–Bossiaeeae, having started diversifying in the early Miocene, 23 Ma (Miller *et al.* 2013). Its closest relative is *Paraserianthes lophantha*, which has one subspecies restricted to south-western Australia and one in Indonesia (Brown *et al.* 2011). Since diverging from *Paraserianthes*, *Acacia* has proliferated into almost 1000 species. This striking disparity between genera can be partly explained by numerous lineages of *Acacia* adapting to desert environments at various times during the Neogene and the failure of *Paraserianthes* to adapt to aridity. However, *Acacia* has also diversified spectacularly in the sclerophyllous communities of south-western Australia, where *Paraserianthes* is native. It is not clear whether the characteristic phyllodes of *Acacia* acted as a 'key innovation' in this explosive radiation.

2.4.6 Asterales

Phylogenetic evidence indicates that Goodeniaceae diverged from Asteraceae plus Calyceraceae clade in the Cretaceous or Paleocene, when Australia and South America were still connected to Antarctica by dry land, and started to diversify in the Eocene (Beaulieu *et al.* 2013; Jabaily *et al.* 2014). The Goodeniaceae seems to have been restricted to Australia until the late Miocene, after which two lineages of *Goodenia* and four of *Scaevola* dispersed out of Australia. Diversification accelerated in the family in the Miocene with many events of colonisation of the arid zone, which may even have served as a 'stepping stone' allowing expansion of clades between the mesic regions of south-eastern and south-western Australia via ancestors in the arid zone.

The Asteraceae first diversified in South America, then a lineage dispersed to Africa, probably in the mid-Eocene, and diversified explosively there (Funk *et al.* 2009; Barres *et al.* 2013). Almost half of Australia's 1000 or so species of Asteraceae belong to a single clade within the tribe Gnaphalieae that dispersed from Africa less than 16 Ma (Nie *et al.* 2016). The other major clade of Asteraceae in the Australian flora is the tribe Astereae, which includes about 300 native species, including *Brachyscome* and the polyphyletic *Olearia*. These taxa diversified from at least two dispersal events into Australia from unknown locations (Cross *et al.* 2002; Jafari *et al.* 2015). Both Gnaphalieae and Astereae have adapted to all the terrestrial biomes of Australia and New Zealand. Some smaller Australian clades also appear to have originated in Africa. The widely overlapping distributions and relatively young age of many closely related genera within Gnaphalieae, Astereae and other clades suggest very complex histories within this group, involving successive rounds of range expansion and allopatric diversification and dispersal to (and possibly back from) New Zealand.

2.4.7 Poaceae

Molecular studies agree that the most recent common ancestor of the Poaceae lived in a moist, closed habitat during the Paleocene or late Cretaceous (Bouchenak-Khelladi *et al.* 2010; Christin *et al.* 2014). Multiple clades successfully adapted to live in dry, open habitats during the late Eocene to Miocene (Bouchenak-Khelladi *et al.* 2010). Diversification accelerated and C_4 photosynthesis evolved at least 15 times between 32 and 3 Ma, as climates cooled and dried and open habitats expanded (Christin *et al.* 2008).

Ancestral area reconstructions suggest that Australian grasses arrived by multiple dispersal events during the Miocene, either from Asia, as Australia approached and collided with it, or across wide ocean gaps (Bouchenak-Khelladi *et al.* 2010). This result is consistent with the Australian fossil record, in which grass pollen (some of which may be misidentified Restionaceae) is extremely rare until the late Oligocene (Macphail and Hill 2002). After this, it becomes steadily more abundant, and by the Pliocene and Pleistocene is often the dominant pollen type, especially in dry regions (Macphail and Hill 2002). Phylogenies of two clades that are important in Australia indicate immigration into Australia via long-distance dispersal. Thus, the small, warm temperate subfamily Danthonioideae apparently followed an intricate scenario of repeated instances of long-distance dispersal across the Pacific and Southern Oceans during the Miocene, with *Rytidosperma* being derived from long-distance dispersal from southern Africa (Linder *et al.* 2013). The Australian species of *Poa* fall within a clade that otherwise occurs from Alaska to Chile (Gillespie *et al.* 2007). Their divergence has not been dated but inspection of branch lengths suggests that the Australian and American taxa are too recently diverged to be vicariant.

Triodia, the dominant understorey taxon in over 30% of arid and monsoon tropical Australia, diverged from its closest relatives in Asia, Europe and Africa in the early Miocene and started to diversify during the mid-Miocene aridification of Australia (Toon *et al.* 2015). *Triodia* may well represent a case of an arid-adapted grass that immigrated initially into southern Australia then spread northwards as the climate dried.

2.4.8 Amaranthaceae

The family Amaranthaceae (including Chenopodiaceae) has a similar history to the grasses in that it originated around the Cretaceous–Paleogene boundary (Magallón *et al.* 2015), had multiple origins of C_4 photosynthesis, and only really diversified in Australia after the onset of severe aridity in the mid-Miocene. Cabrera *et al.* (2011) and Kadereit *et al.* (2005, 2006, 2010) suggest that at least 10 lineages dispersed independently into

Australia, into arid and/or saline habitats similar to those occupied by their colonising ancestors. Only the monotypic genus *Scleroblitum* is old enough, according to molecular dating analyses, to have originated from its closest relatives by vicariance with South America. However, its biogeographic history remains obscure because its closest relatives are widely distributed. Other Australian clades range in age from 23 to 1.7 Ma. Pollen grains consistent with the Amaranthaceae did not become common in Australian palynofloras until the late Miocene, when the Australian clades of *Atriplex* and tribe Camphorosmeae were rapidly diversifying.

The largest Australian group (12 genera and 147 species) is a clade that includes *Maireana* and *Sclerolaena* (Cabrera *et al.* 2011). This clade arrived in Australia in the mid-Miocene, probably from semi-arid to arid parts of Eurasia but did not start to diversify until the late Miocene, about 7 Ma. Ancestral area analysis suggests the clade first took root in south-western Australia and then spread to central, then eastern, then northern Australia (Cabrera *et al.* 2011). Most of Australia's 57 native species of *Atriplex*, all of which have C_4 photosynthesis, diversified rapidly from a species that arrived from central Asia in the Late Miocene, although a few species evolved from a slightly earlier immigrant (Kadereit *et al.* 2010).

2.5 Synthesis: a historical narrative

Although the biogeographic histories of individual clades often include some striking idiosyncrasies, several general historical patterns shine through. These generalisations fit in well with our knowledge of Earth history and are the subject of our concluding section.

2.5.1 The antiquity of rainforests, sclerophylls and fire adaptation

A range of earlier authors asserted that when Australia was part of Gondwana, the landscape was dominated by rainforest. However, fossils from the Paleocene (65–55 Ma) show that Australia carried predecessors of some conspicuously different components of the modern flora, including groups characteristic of rainforest (e.g. leafy Lauraceae), groups characteristic of sclerophyll communities (e.g. *Banksia* and probably *Eucalyptus*) and groups that span these habitats (e.g. Casuarinaceae and Loranthaceae). In fact, the oldest compelling fossil evidence for 'modern' vegetation types is somewhat older (latest Cretaceous) and shows distinctly sclerophyll vegetation (Carpenter *et al.* 2015).

Eucalyptus and Proteaceae shed specific light on the deeper history of rainforest and sclerophyll components of the flora. Meristematic strands deeply embedded in xylem enable most eucalypts to resprout after fire, and are likely to contribute to the eucalypts' current success in fire-prone Australia. However, the very early evolution of this innovation challenges the view that fire only became significant in the last 25 million years (e.g. Kershaw *et al.* 1994; Martin 1996). Dated ancestral state reconstruction implies that the *Eucalyptus–Corymbia–Angophora–Arillastrum* clade had these buds at least 60 Ma, during the Paleocene (Crisp *et al.* 2011a). This corresponds well with evidence of abundant charcoal in Cretaceous deposits, mainly of the northern hemisphere (Bond and Scott 2010). *Eucalyptus* may well have been in Australia at that time – some ambiguous pollen fossils (Thornhill and Crisp 2012) suggest that this is so, but impressive fossils show that it was in South America in the early Eocene (Hermsen *et al.* 2012).

Ancestral character state reconstructions have shown that, within the Proteaceae, xeromorphic adaptations, adaptations to fire, and anatomical structures that make leaves sclerophyllous all evolved repeatedly (Jordan *et al.* 2005, 2008; Lamont and He 2012). Lamont and He (2012) further argued that the pattern of evolution of fire adaptations suggested that at least some Proteaceae became adapted to fire in the Cretaceous, long before the generally accepted increase in fire in the Neogene. These inferences have been confirmed by fossilised sclerophyllous, proteaceous leaves from a late Cretaceous deposit in central Australia that contains charcoal. The fossils closely match leaves of *Banksia, Stirlingia, Isopogon, Conospermum* and the tribe Persoonieae (Carpenter *et al.* 2015). However, ancestral state reconstructions of the preferred habitat of the most recent common ancestor of the Proteaceae, which lived in the mid-Cretaceous, are equivocal.

In contrast, the very warm and wet early Eocene does show widespread fossil evidence for rainforest and little for sclerophyll floras. Widespread rainforest continued to exist even as temperatures declined during the middle and late Eocene, though extinctions did occur – such as numerous Proteaceae and the global last gasp of Corystosperms (extinct gymnosperm 'seed ferns') (McLoughlin *et al.* 2008). In addition, some of the dominant groups through this period are no longer in Australia, but are now found in places such as New Zealand, New Guinea and/or New Caledonia (e.g. a suite of conifers, *Nothofagus* subgenus *Brassospora, Beauprea* and Strasburgeriaceae). Lactoridaceae represents a remarkable case of being stranded on a desert island (Robinson Crusoe Island).

Figure 2.6 Vicariance and long-distance dispersal among the austral floras, derived from the data set of Crisp *et al.* (2009). The first value gives the number of disjunctions in sampled taxa assigned to vicariance, the second the number of disjunctions attributed to dispersal. The line thickness indicates the proportion of disjunctions attributed to vicariance. Figure redrawn with permission from H. Peter Linder.

2.5.2 Immigration and dispersal

Although much of the mesic flora of Australia is descended from lineages inherited from Gondwana (Sanmartín and Ronquist 2004; Kooyman *et al.* 2014), many disjunct clades with apparently Gondwanan distributions seem to have dispersed around the southern hemisphere (including dispersal in and out of Australia) across large ocean gaps after Australia's connection to Antarctica was severed (Crisp *et al.* 2009; Figures 2.6 and 2.7). In addition to the examples described above, there is evidence that, although the disjunctions between Australasian and South American epacrids resulted from vicariance, many clades of Ericaceae have dispersed between Australia, New Zealand and New Caledonia quite recently (Wagstaff *et al.* 2010; Puente-Lelièvre *et al.* 2013; Schwery *et al.* 2015). This included regional extinctions and recolonisations (Jordan *et al.* 2010; Puente-Lelièvre *et al.* 2013). Dispersal of tropical taxa from Asia into Australia's wet tropical and monsoonal biomes dates back at least to 33 Ma but the rate of immigration accelerated about 12 Ma as the Sunda and Sahul shelves came in to close proximity, with a strong bias towards immigration over emigration (Sniderman and Jordan 2011; Crayn *et al.* 2015).

2.5.3 Aridification and the increase in fire

Although there has been a long-held view that aridification of Australia commenced in the Miocene, parts of Australia appear to have had relatively dry

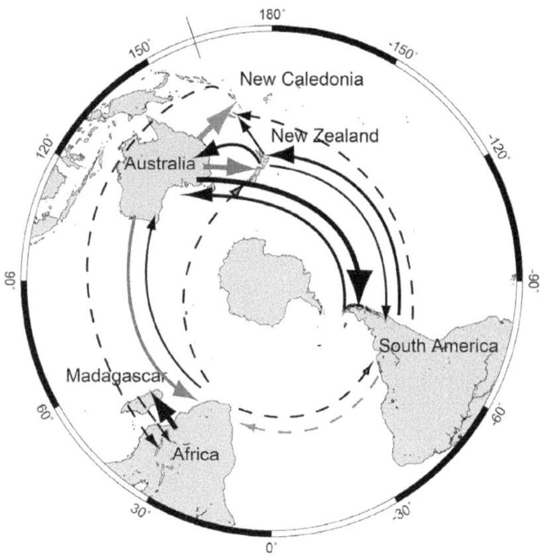

Figure 2.7 The number of colonisation events among the austral floras, derived from the data set of Crisp *et al.* (2009). Arrows indicate the number of colonisation events: broken lines, fewer than 2 events, thin solid lines 2–5 events, next thickness 5–10, then 10–20, and thickest lines more than 20 colonisation events. Figure redrawn with permission from H. Peter Linder.

climates earlier than this, with late Eocene rocks in south-western Australia containing fossil leaves of *Banksia* that are clearly xeromorphic and not just scleromorphic (Carpenter *et al.* 2014). The scarcity of plant fossils outside south-eastern and eastern Australia may also be a sign that other parts of Australia had dry climates unsuited to preserving plant fossils. Thus, some habitats in central, northern and western Australia may well have been dry enough for xeric lineages to persist through even the very wet early Eocene.

Notwithstanding the previous comments, the Australian fossil record indicates that climates dried, fire-dependent groups (such as eucalypts, Asteraceae and grasses) and charcoal increased, and rainforest contracted (Kershaw *et al.* 1994) during the Neogene. A wide range of molecular evidence supports the idea that the arid zone expanded and Eremaean taxa diversified in that period (Byrne *et al.* 2008). This evolution of the arid zone flora included *in situ* adaptation, dispersal from more mesic environments and long-distance dispersal of already xeric taxa from other continents (Byrne *et al.* 2008).

Rainforest and other mesic groups became increasingly restricted to a sparse archipelago of well-watered areas separated by barriers to dispersal and gene flow (Byrne *et al.* 2011). Although this fragmentation probably induced some evolution and speciation in rainforest clades, rates of diversification (net speciation) are considerably greater in relatively dry habitats in the South West Floristic Region, eastern Australia and the arid zone (Figure 2.5; Crisp *et al.* 2009; Byrne *et al.* 2011). Thus, relative to their ages, dry climate and sclerophyll lineages are generally more species rich than rainforest lineages (Crayn *et al.* 2006; Sauquet *et al.* 2009).

2.5.4 Geographic splits within the sclerophyll flora

The most spectacular biogeographic barrier in Australia is the Nullarbor Plain. This barrier appears to explain many of the numerous east–west disjunctions, especially in sclerophyllous plant groups, with a distinct peak in age estimates of disjunctions between 10 and 15 million years ago, the inferred time of formation of the barrier (Figure 2.8; also Crisp and Cook 2007). A few disjunctions, such as that between the major eastern Australian clade of *Eucalyptus* subgenus *Eucalyptus* and its closest western relatives (Crisp *et al.* 2004), and in *Banksia* (Crisp and Cook 2007), are older (Oligocene). Ladiges *et al.* (2011) argue that such distributions (including disjunctions in *Corymbia* and *Eucalyptus* subgenus *Eudesmia*) may have been bisected by drying at the Oligocene–Miocene boundary. The barrier is not complete, with numerous (and presumably relatively recent) disjunctions within species across the barrier, and fossil evidence for relatively mesic habitat in the heart of the Nullarbor since the main formation of the barrier (Sniderman *et al.* 2016).

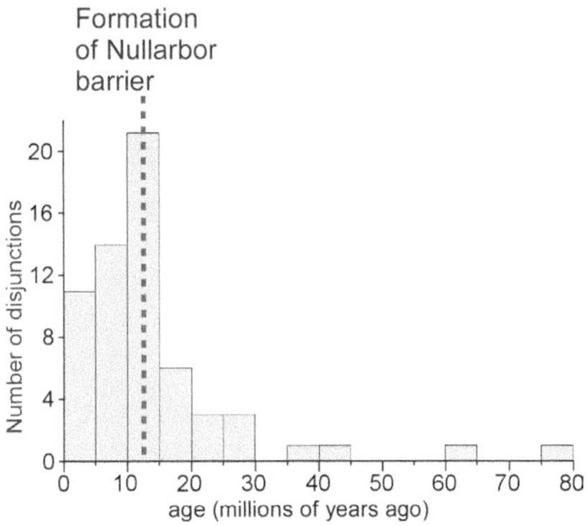

Figure 2.8 Dated disjunctions in plant taxa between mesic areas of south-western Australia and eastern Australia. Note the spike in number of disjunctions at the time of formation of the Nullarbor Plain barrier, and also that many disjunctions are younger than this. Data come from published chronograms of *Grevillea/Hakea*, *Banksia*, Fabaceae, Ericaceae, Restionaceae, Goodeniaceae, Rhamnaceae, Colchicaceae, Casuarinaceae, conifers, Asparagaceae, Anarthriaceae and Haemodoraceae.

Similarly, some disjunctions between northern and southern/eastern Australia seem to relate to aridification. In both *Corymbia* and *Eucalyptus* subgenus *Eudesmia*, species endemic to the Top End and Kimberley regions are more closely related to eastern species than to those in the south-west (Ladiges *et al.* 2011). Also, both *Corymbia* and *Eudesmia* show intricate patterns of diversification within the arid zone, implying that the arid zone is not homogenous in biogeographic terms, but acts as a highly dissected landscape.

2.5.5 Other biomes

The history of the alpine flora can be broadly split into very distinct groups. One small component of the alpine flora focused on the western Tasmanian mountains has very deep history, containing a globally significant suite of palaeoendemic clades, including early diverging endemic, or nearly endemic genera of conifers, Proteaceae and monocots (Figure 2.1g, h; Jordan *et al.* 2016). However, most alpine species belong to young and very widespread clades, often being distributed widely across the southern hemisphere, or even globally. Examples include several clades of Asteraceae (e.g. Wagstaff *et al.* 2006), genera of Cyperaceae and other herbs, such as *Poa* and Danthonioid grasses (e.g. Gillespie *et al.* 2007; Linder *et al.* 2013). Some of these alpine plants appear to represent *in situ* evolution from lowland clades (e.g. *Grevillea* species; Mast *et al.* 2015). This recent origin probably reflects the relatively recent creation of alpine habitat in Australia by cooling climates combined with geological uplift (Müller *et al.* 2016).

The history of the monsoon biome remains relatively unclear, with most evidence summarised by Bowman *et al.* (2010). It appears to have included multiple immigrations of monsoon clades, especially from Asia (Crayn *et al.* 2015), although some clades (e.g. within Proteaceae, Mast *et al.* 2015) are autochthonous. There is some evidence for a mid-Miocene origin of several Australian savanna clades (Toon *et al.* 2015).

2.6 Conclusions

Many of the dominant patterns and processes postulated in our historical narrative include ideas that have evolved directly from hypotheses first developed in the nineteenth century. However others have emerged strongly in the last few decades in response to discoveries in plant systematics, paleobotany and Earth history. These seem likely to withstand further testing. However, many details of the story, such as historical relationships between disjunct mesic refugia within Australia, remain poorly understood. Progress in closing these gaps should be rapid over the next few years, given the availability of rapidly improving molecular and analytical techniques. The evolutionary biogeography of the Australian flora remains an exciting area of scientific research.

References

Barker, N.P., Weston, P.H., Rutschmann, F. *et al.* (2007) Molecular dating of the 'Gondwanan' plant family Proteaceae is only partially congruent with the timing of Gondwanan break-up. *Journal of Biogeography* 34, 2012–2027.

Barlow, B.A. (1981) The Australian flora: its origin and evolution. *Flora of Australia* 1, 25–75.

Barlow, B.A. (1994) Phytogeography of the Australian region. In *Australian Vegetation*, 2nd edition, ed. R.H. Groves. Cambridge University Press: Cambridge, pp. 3–35.

Barres, L., Sanmartín, I., Anderson, C.L. *et al.* (2013) Reconstructing the evolution and biogeographic history of tribe Cardueae (Compositae). *American Journal of Botany* 100, 867–882.

Beaulieu, J.M., Tank, D.C. and Donoghue, M.J. (2013) A southern hemisphere origin for campanulid angiosperms, with traces of the break-up of Gondwana. *BMC Evolutionary Biology* 13, 80.

Biffin, E., Hill, R.S. and Lowe, A.J. (2010) Did Kauri (*Agathis*: Araucariaceae) really survive the Oligocene drowning of New Zealand? *Systematic Biology* 59, 594–602.

Biffin, E., Conran, J.G. and Lowe, A.J. (2011) Podocarp evolution: a molecular phylogenetic perspective. In *Ecology of Podocarpaceae in Tropical Forests. Smithsonian Contributions to Botany No. 95*, ed. B. Turner and L. Cernusak. Smithsonian Scholarly Press: Washington, pp. 1–20.

Bond, W.J. and Scott, A.C. (2010) Fire and the spread of flowering plants in the Cretaceous. *New Phytologist* 188, 1137–1150.

Bouchenak-Khelladi, Y., Verboom, G.A., Savolainen, V. *et al.* (2010) Biogeography of the grasses (Poaceae): a phylogenetic approach to reveal evolutionary history in geographical space and geological time. *Botanical Journal of the Linnean Society* 162, 543–557.

Bowman, D.M.J.S., Brown, G.K., Braby, M.F. *et al.* (2010) Biogeography of the Australian monsoon tropics. *Journal of Biogeography* 37, 201–216.

Brown, G.K., Murphy, D.J. and Ladiges, P.Y. (2011) Relationships of the Australo-Malesian genus *Paraserianthes* (Mimosoideae: Leguminosae) identifies the sister group of *Acacia* sensu stricto and two biogeographical tracks. *Cladistics* 27, 380–390.

Burbidge, N.T. (1960) The phytogeography of the Australian region. *Australian Journal of Botany* 8, 75–211.

Byrne, M., Yeates, D.K., Joseph, L. *et al.* (2008) Birth of a biome: insights into the assembly and maintenance of the Australian arid zone biota. *Molecular Ecology* 17, 4398–4417.

Byrne, M., Steane, D., Joseph, L. *et al.* (2011) Decline of a biome: evolution, contraction, fragmentation, extinction and invasion of the Australian mesic zone biota. *Journal of Biogeography* 38, 1635–1656.

Cabrera, J., Jacobs, S.W.L. and Kadereit, G. (2011) Biogeography of *Camphorosmeae* (Chenopodiaceae): tracking the Tertiary history of Australian aridification. *Telopea* 13, 313–326.

Cantrill, D.J. and Poole, I. (2012) *The Vegetation of Antarctica through Geological Time.* Cambridge University Press: Cambridge.

Cardillo, M. and Pratt, R. (2013) Evolution of a hotspot genus: geographic variation in speciation and extinction rates in *Banksia* (Proteaceae). *BMC Evolutionary Biology* 13, 155.

Carpenter, R.J., Jordan, G.J., Lee, D.E. *et al.* (2010) Leaf fossils of *Banksia* (Proteaceae) from New Zealand: an Australian abroad. *American Journal of Botany* 97, 288–297.

Carpenter, R.J., Jordan, G.J., Macphail, M.K. *et al.* (2012) Near-tropical early Eocene terrestrial temperatures at the Australo-Antarctic margin, western Tasmania. *Geology* 40, 267–270.

Carpenter, R.J., McLoughlin, S., Hill, R.S. *et al.* (2014) Early evidence of xeromorphy in angiosperms: stomatal encryption in a new Eocene species of *Banksia* (Proteaceae) from Western Australia. *American Journal of Botany* 101, 1486–1497.

Carpenter, R.J., Macphail, M.K., Jordan, G.J. *et al.* (2015) Fossil evidence for open, Proteaceae-dominated heathlands and fire in the Late Cretaceous of Australia. *American Journal of Botany* 102, 2092–2107.

Christin, P.A., Besnard, G., Samaritani, E. *et al.* (2008) Oligocene CO_2 decline promoted C_4 photosynthesis in grasses. *Current Biology* 18, 37–43.

Christin, P.A., Spriggs, E., Osborne, C.P. *et al.* (2014) Molecular dating, evolutionary rates, and the age of the grasses. *Systematic Biology* 63, 153–165.

Christophel, D.C. (1984) Early Tertiary Proteaceae: the first floral evidence for the Musgraveinae. *Australian Journal of Botany* 32, 177–186.

Cookson, I.C. and Duigan, S.L. (1950) Fossil Banksieae from Yallourn, Victoria, with notes on the morphology and anatomy of living species. *Australian Journal of Scientific Research, Series B* 3, 133–165.

Cookson, I.C. and Pike, K.M. (1954) Some dicotyledonous pollen types from Cainozoic deposits in the Australian region. *Australian Journal of Botany* 2, 197–219.

Crayn, D.M., Rossetto, M. and Maynard, D.J. (2006) Molecular phylogeny and dating reveals an Oligomiocene radiation of dry-adapted shrubs (former Tremandraceae) from rainforest tree progenitors (Elaeocarpaceae) in Australia. *American Journal of Botany* 93, 1328–1342.

Crayn, D.M., Costion, C. and Harrington, M.G. (2015) The Sahul–Sunda floristic exchange: dated molecular phylogenies document Cenozoic intercontinental dispersal dynamics. *Journal of Biogeography* 42, 11–24.

Crisp, M.D. and Cook, L.G. (2007) A congruent molecular signature of vicariance across multiple plant lineages. *Molecular Phylogenetics and Evolution* 43, 1106–1117.

Crisp, M.D. and Cook, L.G. (2009) Explosive radiation or cryptic mass extinction? Interpreting signatures in molecular phylogenies. *Evolution* 63, 2257–2265.

Crisp, M.D. and Cook, L.G. (2013) How was the Australian flora assembled over the last 65 million years? A molecular phylogenetic perspective. *Annual Review of Ecology, Evolution and Systematics* 44, 303–324.

Crisp, M.D., West, J.G. and Linder, H.P. (1999) Biogeography of the terrestrial flora. In *Flora of Australia Volume 1 – Introduction*, 2nd edition. CSIRO Publishing: Melbourne, pp. 321–367.

Crisp, M.D., Cook, L.G. and Steane, D. (2004) Radiation of the Australian flora: what can comparisons of molecular phylogenies across multiple taxa tell us about the evolution of diversity in present-day communities? *Philosophical Transactions of the Royal Society London B* 359, 1551–1571.

Crisp, M.D., Arroyo, M.T.K., Cook, L.G. *et al.* (2009) Phylogenetic biome conservatism on a global scale. *Nature* 458, 754–758.

Crisp, M.D., Burrows, G.E., Cook, L.G. *et al.* (2011a) Flammable biomes dominated by eucalypts originated at the Cretaceous–Palaeogene boundary. *Nature Communications* 2, 193.

Crisp, M.D., Trewick, S.A. and Cook, L.G. (2011b) Hypothesis testing in biogeography. *Trends in Ecology and Evolution* 26, 66–72.

Cross, E.W., Quinn, C.J. and Wagstaff, S.J. (2002) Molecular evidence for the polyphyly of *Olearia* (Astereae: Asteraceae). *Plant Systematics and Evolution* 235, 99–120.

Darwin, C. (1859) *The Origin of Species*. John Murray: London.

Exon, N.F., Kennett, J.P. and Malone, M.J. (2004) Leg 189 synthesis: Cretaceous–Holocene history of the Tasmanian gateway. *Proceedings of the Ocean Drilling Program, Scientific Results* 189, 1–37.

Funk, V.A., Susanna, A., Stuessy, T.F. and Robinson, H. (2009) Classification of Compositae. In *Systematics, Evolution, and Biogeography of Compositae*, ed. V.A. Funk, A. Susanna, T.F. Stuessy and R.F. Bayer. International Association for Plant Taxonomy: Bratislava, pp. 171–189.

Gillespie, L.J., Archambault, A. and Soreng, R.J. (2007) Phylogeny of *Poa* (Poaceae) based on trnT–trnF sequence data: major clades and basal relationships. *Aliso* 23, 420–434.

Greenwood, D.R., Moss, P.T., Rowett, A.I. *et al.* (2003) Plant communities and climate change in southeastern Australia during the early Paleogene. *Geological Society of America* 369, 365–380.

Hennig, W. (1966) *Phylogenetic Systematics.* University of Illinois Press: Urbana, IL.

Hermsen, E.J., Gandolfo, M.A. and del Carmen Zamaloa, M. (2012) The fossil record of *Eucalyptus* in Patagonia. *American Journal of Botany* 99, 1356–1374.

Hess, H.H. (1962) History of ocean basins. In *Petrologic Studies: A Volume to Honor A. F. Buddington*, ed. A.E.J. Engel, H.L. James, B.F. Leonard. Geological Society of America: New York, pp. 599–620.

Hill, R.S. (1988) Australian Tertiary angiosperm and gymnosperm leaf remains: an updated catalogue. *Alcheringa* 12, 207–219.

Hill, R.S. and Brodribb, T.J. (1999) Southern conifers in time and space. *Australian Journal of Botany* 47, 639–696.

Hill, R.S., Scriven, L.J. and Jordan, G.J. (1995) The fossil record of Australian Proteaceae. *Flora of Australia* 16, 21–30.

Hill, R.S., Truswell, E.M., McLoughlin, S. and Dettman, M.E. (1999) Evolution of the Australian flora: fossil evidence. In *Flora of Australia Volume 1: Introduction*, 2nd edition. CSIRO Publishing: Melbourne, pp. 251–320.

Ho, S.Y.W. (2014) The changing face of the molecular evolutionary clock. *Trends in Ecology and Evolution* 29, 496–503.

Hooker, J.D. (1853) Introductory essay. In *The Botany of the Antarctic Voyage of the H.M. Discovery Ships* Erebus *and* Terror *in the years 1839–1843, Vol. 2, Flora Novæ-Zealandiæ*. Lovell Reeve: London.

Hooker, J.D. (1860) Introductory essay. In *The Botany of the Antarctic Voyage of the H.M. Discovery Ships* Erebus *and* Terror *in the years 1839–1843, Vol. 3, Flora Tasmaniae*. Lovell Reeve: London.

Hopper, S.D. (2009) OCBIL theory: towards an integrated understanding of the evolution, ecology and conservation of biodiversity on old, climatically buffered, infertile landscapes. *Plant Soil* 322, 49–86.

Jabaily, R.S., Shepherd, K.A., Gardner, A.G. *et al.* (2014) Historical biogeography of the predominantly Australian plant family Goodeniaceae. *Journal of Biogeography* 41, 2057–2067.

Jafari, F., Osaloo, S.K. and Mozffarian, V. (2015) Molecular phylogeny of the tribe Astereae (Asteraceae) in SW Asia based on nrDNA ITS and cpDNA psbA-trnH sequences. *Willdenowia* 45, 77–92.

Johnson, L.A.S. and Briggs, B.G. (1975) On the Proteaceae: the evolution and classification of a southern family. *Botanical Journal of the Linnean Society* 70, 83–182.

Johnson, L.A.S. and Briggs, B.G. (1981) Three old southern families: Myrtaceae, Proteaceae and Restionaceae. In *Ecological Biogeography of Australia*, ed. A Keast. Junk: The Hague, pp. 427–469.

Jordan, G.J. and Hill, R.S. (1999) The phylogenetic affinities of *Nothofagus* (Nothofagaceae) leaf fossils based on combined molecular and morphological data. *International Journal of Plant Sciences* 160, 1177–1188.

Jordan, G.J., Dillon, R.A. and Weston, P.H. (2005) Solar radiation as a factor in the evolution of scleromorphic leaf anatomy in Proteaceae. *American Journal of Botany* 92, 789–796.

Jordan, G.J., Weston, P.H., Carpenter, R.J. *et al.* (2008) The evolutionary relations of sunken, covered, and encrypted stomata to dry habitats in Proteaceae. *American Journal of Botany* 95, 521–530.

Jordan, G.J., Bannister, J.M., Mildenhall, D.C. *et al.* (2010) Fossil Ericaceae from New Zealand: deconstructing the use of fossil evidence in historical biogeography. *American Journal of Botany* 97, 59–70.

Jordan, G.J., Carpenter, R.J., Bannister, J.M. *et al.* (2011) High conifer diversity in Oligo-Miocene New Zealand. *Australian Systematic Botany* 24, 121–136.

Jordan, G.J., Carpenter, R.J. and Brodribb, T.J. (2014) Using fossil leaves as evidence for open vegetation. *Palaeogeography, Palaeoclimatology, Palaeoecology* 395, 168–175.

Jordan, G.J., Harrison, P.A., Worth, J.R.P. *et al.* (2016) Palaeoendemic plants provide evidence for persistence of open, well-watered vegetation since the Cretaceous. *Global Ecology and Biogeography* 25, 127–140.

Kadereit, G., Gotzek, D., Jacobs, S. and Freitag, H. (2005) Origin and age of Australian Chenopodiaceae. *Organisms, Diversity & Evolution* 5, 59–80.

Kadereit, G., Mavrodiev, E.V., Zacharias, E.H. and Sukhorukov, A.P. (2010) Molecular phylogeny of *Atripliceae* (Chenopodioideae, Chenopodiaceae): implications for systematics, biogeography, flower and fruit evolution, and the origin of C_4 photosynthesis. *American Journal of Botany* 97, 1664–1687.

Kadereit, G., Mucina, L. and Freitag, H. (2006) Phylogeny of *Salicornioideae* (Chenopodiaceae): diversification, biogeography, and evolutionary trends in leaf and flower morphology. *Taxon* 55, 617–642.

Kershaw, A.P., Martin, H.A. and McEwen Mason, J.R.C. (1994) The Neogene: a period of transition. In *History of the Australian Vegetation: Cretaceous to Recent*, ed. R.S. Hill. Cambridge University Press: Cambridge, pp. 299–327.

Kooyman, R.M., Wilf, P., Barreda, V.D. *et al.* (2014) Paleo-Antarctic rainforest into the modern old world tropics: the rich past and threatened future of the 'southern wet forest survivors'. *American Journal of Botany* 101, 2121–2135.

Ladiges, P.Y. (1998) Biogeography after Burbidge. *Australian Systematic Botany* 11, 231–242.

Ladiges, P.Y. and Cantrill, D. (2007) New Caledonia–Australian connections: biogeographic patterns and geology. *Australian Systematic Botany* 20, 383–389.

Ladiges, P.Y., Parra-O, C., Gibbs, A. *et al.* (2011) Historical biogeographical patterns in continental Australia: congruence among areas of endemism of two major clades of eucalypts. *Cladistics* 27, 29–41.

Lamont, B.B. and He, T. (2012) Fire-adapted Gondwanan angiosperm floras evolved in the Cretaceous. *BMC Evolutionary Biology* 12, 223.

Lavin, M., Herendeen, P.S. and Wojciechowski, M.F. (2005) Evolutionary rates analysis of Leguminosae implicates a rapid diversification of lineages during the Tertiary. *Systematic Biology* 54, 575–594.

Linder, H.P., Antonelli, A., Humphreys, A.M. *et al.* (2013) What determines biogeographical ranges? Historical wanderings and ecological constraints in the danthonioid grasses. *Journal of Biogeography* 40, 821–834.

McLoughlin, S. (2001) The breakup history of Gondwana and its impact on pre-Cenozoic floristic provincialism. *Australian Journal of Botany* 49, 271–300.

McLoughlin, S., Carpenter, R.J., Jordan, G.J. *et al.* (2008) Seed ferns survived the end-Cretaceous mass extinction in Tasmania. *American Journal of Botany* 95, 465–471.

Macphail, M.K. (2007) *Australian Palaeoclimates: Cretaceous to Tertiary. A Review of Palaeobotanical and Related Evidence to the Year 2000.* CRC LEME: Bentley.

Macphail, M.K., Colhoun, E.A., Kiernan, K. *et al.* (1993) Glacial climates in the Antarctic region during the late Paleogene: evidence from northwest Tasmania, Australia. *Geology* 21, 145–148.

Macphail, M.K. and Hill, R.S. (2002) Paleobotany of the Poaceae. *Flora of Australia* 43, 37–70.

Macphail, M.K. and Truswell, E.M. (1989) Palynostratigraphy of the central west Murray Basin. *BMR Journal of Australian Geology and Geophysics* 11, 301–331.

Magallón, S., Gómez-Acevedo, S., Sánchez-Reyes, L.L. and Hernández-Hernández, T. (2015) A metacalibrated time-tree documents the early rise of flowering plant phylogenetic diversity. *New Phytologist* 207, 437–453.

Mao, K., Milne, R.I., Zhang, L. *et al.* (2012) Distribution of living Cupressaceae reflects the breakup of Pangea. *Proceedings of the National Academy of Sciences of the USA.* 109, 7793–7798.

Martin, H.A. (1996) Wildfires in past ages. *Proceedings of the Linnean Society of New South Wales* 116, 3–18.

Mast, A.R., Olde, P., Makinson, R.O. *et al.* (2015) Paraphyly changes understanding of timing and tempo of diversification in subtribe Hakeinae (Proteaceae), a giant Australian plant radiation. *American Journal of Botany* 102, 1634–1646.

Mast, A.R., Willis, C.L., Jones, E.H. *et al.* (2008) A smaller *Macadamia* from a more vagile tribe: inference of phylogenetic relationships and divergence times in *Macadamia* and relatives (tribe Macadamieae; Proteaceae). *American Journal of Botany* 95, 843–870.

Miller, J.T., Murphy, D.J., Ho, S.Y.W. *et al.* (2013) Comparative dating of *Acacia*: combining fossils and multiple phylogenies to infer ages of clades with poor fossil records. *Australian Journal of Botany* 61, 436–445.

Milner, M.L., Weston, P.H., Rossetto, M. *et al.* (2015) Biogeography of the Gondwanan genus *Lomatia* (Proteaceae): vicariance at continental and intercontinental scales. *Journal of Biogeography* 42, 2440–2451.

Müller, R.D., Flament, N., Matthews, K.J. *et al.* (2016) Formation of Australian continental margin highlands driven by plate–mantle interaction. *Earth and Planetary Science Letters* 441, 60–70.

Nie, Z.L., Funk, V.A., Meng, Y. *et al.* (2016) Recent assembly of the global herbaceous flora: evidence from the paper daisies (Asteraceae: Gnaphalieae). *New Phytologist* 209, 1795–1806.

Pross, J., Contreras, L., Bijl, P.K. *et al.* (2012) Persistent near-tropical warmth on the Antarctic continent during the early Eocene epoch. *Nature* 487, 73–77.

Puente-Lelièvre, C., Harrington, M.G., Brown, E.A. *et al.* (2013) Cenozoic extinction and recolonization in the New Zealand flora: the case of the fleshy-fruited epacrids (Styphelieae, Styphelioideae, Ericaceae). *Molecular Phylogenetics and Evolution* 66, 203–214.

Raven, P.H. and Axelrod, D.I. (1974) Angiosperm biogeography and past continental movements. *Annals of the Missouri Botanical Garden* 61, 539–673.

Renner, S.S. (2005) Variation in diversity among Laurales, Early Cretaceous to present. *Biologiske Skrifter* 55, 441–458.

Rozefelds, A.C. (1995) Miocene *Wilkinsonia* fruits (Hicksbeachiinae, Proteaceae) from the base of the Yallourn formation, Latrobe valley, Victoria. *Papers and Proceedings of the Royal Society of Tasmania* 129, 59–62.

Rozefelds, A.C., Dettmann, M.E. and Clifford, H.T. (2005) *Xylocaryon lockii* F.Muell. (Proteaceae) fruits from the Cenozoic of south-eastern Australia. *Kanunnah* 1, 91–102.

Sanmartín, I. and Ronquist, F. (2004) Southern hemisphere biogeography inferred by event-based models: plant versus animal patterns. *Systematic Biology* 53, 216–243.

Sauquet, H., Weston, P.H., Anderson, C.L. *et al.* (2009) Contrasted patterns of hyperdiversification in Mediterranean hotspots. *Proceedings of the National Academy of Sciences of the USA* 106, 221–225.

Sauquet, H., Ho, S.Y.W., Gandolfo, M.A. *et al.* (2012) Testing the impact of calibration on molecular divergence times using a fossil-rich group: the case of *Nothofagus* (Fagales). *Systematic Biology* 61, 289–313.

Schwery, O., Onstein, R.E., Bouchenak-Khelladi, Y. *et al.* (2015) As old as the mountains: the radiations of the Ericaceae. *New Phytologist* 207, 355–367.

Sniderman, J.M.K. and Jordan, G.J. (2011) Extent and timing of floristic exchange between Australian and Asian rain forests. *Journal of Biogeography* 38, 1445–1455.

Sniderman, J.M.K., Woodhead, J., Hellstrom, J. *et al.* (2016) Pliocene reversal of late Neogene aridification. *Proceedings of the National Academy of Sciences of the USA* 113, 1994–2004.

Stover, L.E. and Partridge, A.D. (1973) Tertiary and Late Cretaceous spores and pollen from the Gippsland Basin, southeastern Australia. *Proceedings of the Royal Society of Victoria* 85, 237–286.

Thomas, N., Bruhl, J.J., Ford, A. and Weston, P.H. (2014) Molecular dating of Winteraceae reveals a complex biogeographic history involving both ancient Gondwanan vicariance and long-distance dispersal. *Journal of Biogeography* 41, 894–904.

Thornhill, A.H. and Crisp, M.D. (2012) Phylogenetic assessment of pollen characters in Myrtaceae. *Australian Systematic Botany* 25, 171–187.

Thornhill, A.H., Ho, S.Y.W., Külheim, C. *et al.* (2015) Interpreting the modern distribution of Myrtaceae using a dated molecular phylogeny. *Molecular Phylogenetics and Evolution* 93, 29–43.

Tiedemann, R., Sarnthein, M. and Shackleton, N.J. (1994) Astronomic timescale for the Pliocene Atlantic δ18O and dust flux records of Ocean Drilling Program Site 659. *Paleoceanography* 9, 619–638.

Toon, A., Crisp, M.D., Gamage, H. *et al.* (2015) Key innovation or adaptive change? A test of leaf traits using *Triodiinae* in Australia. *Scientific Reports* 5, 12398.

Wagstaff, S., Breitwieser, I. and Swenson, U. (2006) Origin and relationships of the austral genus *Abrotanella* (Asteraceae) inferred from DNA sequences. *Taxon* 55, 95–106.

Wagstaff, S., Dawson, M.I., Venter, S. *et al.* (2010) Origin, diversification, and classification of the Australasian genus *Dracophyllum* (Richeeae, Ericaceae). *Annals of the Missouri Botanical Garden* 97, 235–258.

Wegener, A. (1929) *The Origin of Continents and Oceans*, 4th edition. Methuen: London.

Worth, J.R.P., Jordan, G.J., McKinnon, G.E. *et al.* (2009) The major Australian cool temperate rainforest tree *Nothofagus cunninghamii* withstood Pleistocene glacial aridity within multiple regions: evidence from the chloroplast. *New Phytologist* 182, 519–532.

Zachos, J., Pagani, H., Sloan, L. *et al.* (2001) Trends, rhythms, and aberrations in global climate 65 Ma to present. *Science* 292, 686–693.

3 Quaternary Vegetation in Australia

SCOTT D. MOONEY, KALE SNIDERMAN, A. PETER KERSHAW, SIMON HABERLE AND JESSICA ROE

The Quaternary, the last 2.6 million years of Geologic Time, is a period of considerable environmental change. Widespread evidence within ocean sediments indicates more than 50 cold (glacial) and warm (interglacial) climatic episodes during this time and these were responsible for, and coupled with, other considerable environmental changes. Sediments accumulating in lakes, swamps and marine environments may preserve biological material, including pollen, spores and macrofossils, and these offer a window through which we can investigate past vegetation and climate and the dynamic nature of our vegetation communities. The early Quaternary in southern Australia supported a rainforest/sclerophyll mosaic and the closed communities included taxa that are now restricted to more northerly locations. We describe a case study for the early Quaternary vegetation using pollen and macrofossils from Stony Creek Basin, located in western Victoria. This reveals a rhythmic change every ~20,000 years, representing the expansion and contraction of closed/open forest plant communities within an upland vegetation mosaic. Importantly, this record suggests that both the closed rainforest and open sclerophyll-dominated vegetation communities have undergone significant change in the last ~1.5 million years, perhaps resulting from the novel climate extremes that influenced south-eastern Australia. The mid-to-late Quaternary is dominated by glacial cycles that occur at eccentricity (~100,000 year) periodicities: we consider the vegetation of the last glacial maximum and use a case study from north-eastern Queensland to demonstrate considerable vegetation change in the postglacial period. This case study also demonstrates that the vegetation of the modern environment was only really established in the last 8,000 – 9,000 years.

3.1 Introduction

In the last few million years of the Earth's long history, major climatic oscillations have been a significant driver of environmental change. These, coupled with the emergence of anatomically modern *Homo sapiens*, the human colonisation of most terrestrial environments and, more recently, rapid increases in global population and consumption means that the Quaternary (the last ~2.6 million years) is a period of significant change. Such environmental change has driven and has indeed been coupled with, dramatic vegetation change across the globe.

Previous publications that describe the Quaternary vegetation of Australia at a continental scale based on palynological records include Kershaw (1980, 1981), Singh (1982), Singh *et al.* (1981), Kershaw

Australian Vegetation, Third Edition, ed. David A. Keith. Published by Cambridge University Press.

Citation: Mooney, S.D. *et al.* (2017). Quaternary Vegetation in Australia. In D. A. Keith, ed., *Australian Vegetation*, 3rd edn. Cambridge: Cambridge University Press, pp. 63–88.

et al. (1991b), Hope (1994) and, in the second edition of this monograph (Groves 1994), the chapter 'Quaternary vegetation history' (Dodson 1994). Bowler (1982), Markgraf *et al.* (1992), Harrison and Dodson (1993), Hubbard (1995a, 1995b), Kershaw (1997) and Dodson (1998) also consider aspects of Quaternary vegetation in Australia, albeit at smaller spatial scales or during critical time intervals of the late Quaternary. Hope *et al.* (2004) reviewed terrestrial and marine pollen records from across the Austral-Asian region, placing vegetation changes recorded in the Quaternary in Australia into a wider context. Pickett *et al.* (2004) represents the only quantitative ('biomisation') approach to reconstructing the Australian vegetation at a continental scale, and focuses on two time periods, at 6000 and 18 000 years ago.

Given that more than a decade has elapsed since the last of these publications and more than 15 years since the last of the continent-wide summaries was published, it is timely to reconsider the vegetation of Australia during the Quaternary. We now have a better understanding of the nature of the Quaternary and the drivers of vegetation change, including temperature, moisture availability and atmospheric CO_2 concentration, during this period and our database has also expanded in the intervening years.

This chapter describes the nature of the Quaternary and the tools at our disposal to consider the vegetation of Australia over the last few million years. We have emphasised the transition from the pre-Quaternary, and hence we naturally follow on from Weston and Jordan (Chapter 2, this volume), and consider recent evidence for the vegetation of the early part of the Quaternary. We describe the late Pleistocene vegetation of northern Australia using long pollen records from marine sediments and then focus on the vegetation of the period during and after the last glacial maximum. We conclude by considering how our examples of Quaternary vegetation history offer insight into the contemporary Australian vegetation.

3.2 The Quaternary

The early Cenozoic, before about 37 million years ago (mega annum or Ma) has traditionally been associated with a 'pan-Austral closed-canopy forest', or rainforest (Lange 1982), largely based on evidence from south-eastern Australia. The nature and timing of the decline of this mesic vegetation and its replacement by more open (and sclerophyll) vegetation is described in detail by Weston and Jordan (Chapter 2, this volume). The history of many Australian open vegetation communities is complex but their increasing dominance is perhaps most simply related to the increasing aridification of Australia from about the mid-to-late Cenozoic. This drying of the continent is often attributed to a series of interrelated processes starting with the separation of Australia from Antarctica at about 40 Ma, the development of the circum-Antarctic current and concomitant changes to atmospheric circulation systems. Together these resulted in a steepening of the meridional temperature gradient and the growth of significant ice sheets in Antarctica, which then further consolidated these oceanic and atmospheric changes.

For Australia, the influence of the strengthening subtropical high-pressure belt, increasing seasonality and the physical position of the continent had important influences on its climate and vegetation. As an example, Bowler (1982) suggested that by about 6 Ma southern Australia would have been under the influence of the subtropical high-pressure belt, resulting in the development of a strongly seasonal climate. At the other end of the continent, northern Australia resisted the global trend towards cooling in the late Cenozoic due to its northward movement into the tropics (Truswell 1993).

Although slow global cooling and the growth of the Antarctic ice sheet characterises the late Cenozoic (Zachos *et al.* 2001), the lower boundary of the Quaternary, set at 2.588 Ma (Gibbard *et al.* 2009), relates to the onset of extensive northern hemisphere glaciation (Haug *et al.* 2005). Why Earth's climate descended into the glaciated state remains a question of lively debate, with some highlighting tectonic processes (e.g. uplift of the Himalayan–Tibetan Plateau, the formation of the Panama Isthmus, and other ocean basin changes that saw the Arctic become increasingly land-locked and the restriction of the Indonesian throughflow) that 'sensitised' the system (Ruddiman and Raymo 1988; Raymo 1994; Ravelo *et al.* 2004).

To equate the Quaternary only with glaciation and cold conditions ignores the significant climatic and other environmental fluctuations of the period (Figure 3.1). The tempo of the characteristic (and at least during the last ~1 Ma, dramatic) glacial–interglacial cycles is driven by subtle changes in the orbital parameters of the Earth, which alter the seasonal and spatial distribution of incoming solar radiation (the Milankovitch–Croll hypothesis). This in turn triggers a number of within-Earth processes and feedbacks within the climate system (e.g. accumulation and ablation of continental ice sheets, changes in greenhouse gas concentrations, ocean circulation, land cover and albedo), which together determine the magnitude of the climatic cycles.

The climatic oscillations of the Quaternary are recorded in many natural archives, but the best in terms of temporal depth and spatial coverage are the changing balance of the two stable isotopes

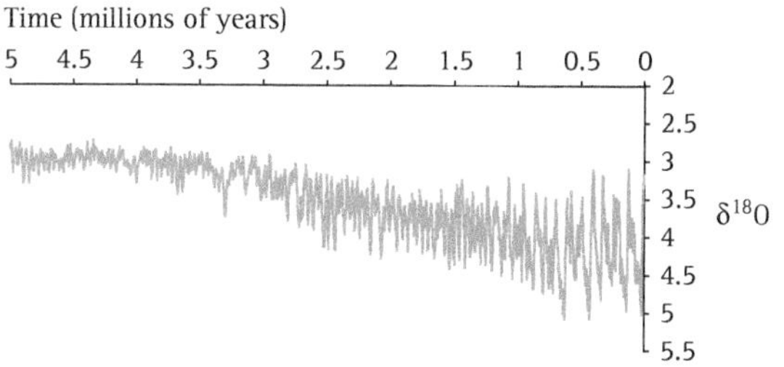

Figure 3.1 The climate of Earth over the last five million years, as recorded by changes in the oxygen isotope content of carbonate tests within marine sediment cores. In this diagram up is warmer (and/or less ice on Earth) and down is colder/more ice on Earth. The magnitude of the glacial–interglacial cycles increases over the last million years or so.
Source: Lisiecki and Raymo 2005.

of oxygen, ^{18}O and ^{16}O, preserved in the carbonate tests of Foraminifera deposited in ocean sediments. These marine oxygen isotope records, expressed by convention against a standard using the delta notation $\delta^{18}O$, are a 'proxy' of past changes in global ice volume and temperatures, and hence reveal much about both the lead up to the Quaternary and the Quaternary itself (Figure 3.1). The importance of oxygen isotope data is revealed by the widespread use of marine isotope stages (MIS) to describe the alternating warm and cold periods on Earth. Ice cores, from polar ice sheets (e.g. EPICA 2004) and high altitude glaciers also provide important sources of palaeoclimatic and palaeoenvironmental information, recorded via the accumulation of past snowfall and in the gas bubbles and other inclusions (Figure 3.2).

Before about 3.2 Ma, average $\delta^{18}O$ is low and oscillations are relatively small in amplitude (Figure 3.1). This reflects relatively warm temperatures across the globe and the lack of major ice sheets in the northern hemisphere. The situation changes dramatically in the period between about 3.0 and 2.5 Ma, reflected in a major transition in the $\delta^{18}O$ record at 2.74 Ma, when the mean changes and the oscillations are larger with a ~41 ka periodicity. The causes of the step-like nature and hence the rapidity of this and subsequent changes are a little uncertain: it is possible that these relatively rapid changes result from non-linear responses to gradual external forcing, however a gradual build-up of continental ice is perhaps more plausible.

In the period from 1.9 to 1.2 Ma, the glacial–interglacial cycles are fairly uniform in duration (of about 41 000 years) and amplitude (Figure 3.1). After about 1.2 Ma, the tempo of the climatic oscillations slows and their amplitude increases. This change is somewhat complex, but by the later Quaternary a periodicity approximating 100 000 years is apparent and the amplitude of the oxygen isotope record is greater than any other time of the Quaternary. The rhythm of glacial and

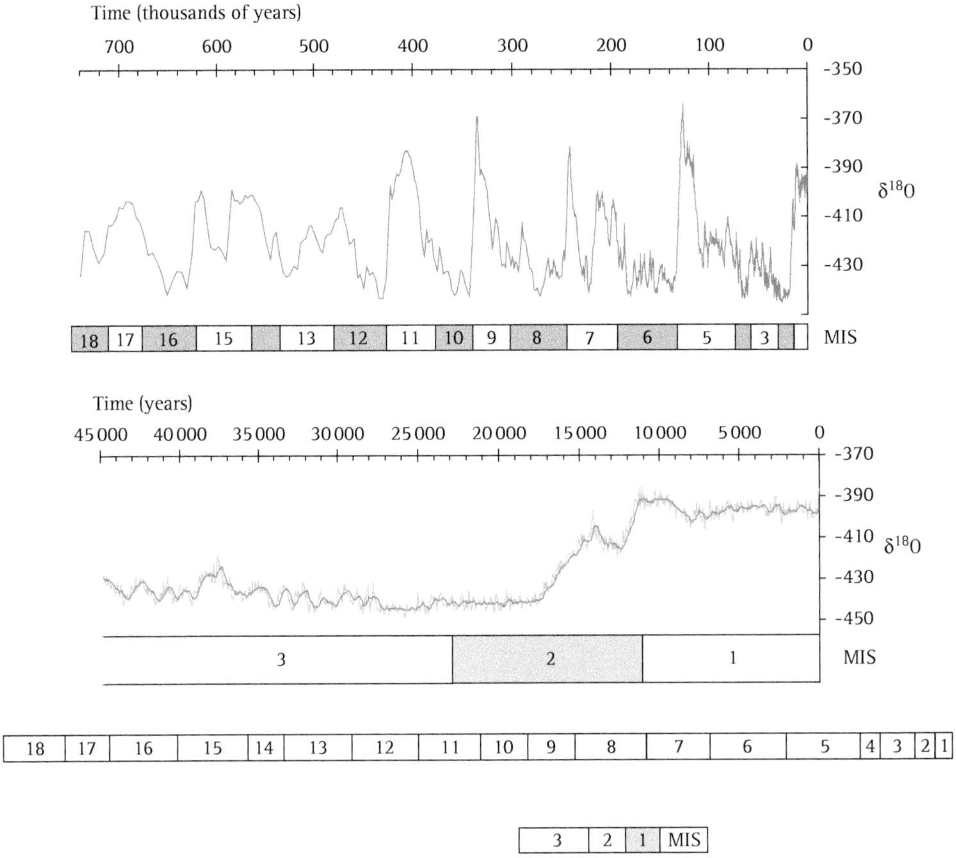

Figure 3.2 Antarctic ice core records of the late Quaternary. In these diagrams up is warmer and down represents colder temperatures. The upper panel shows the EPICA Dome C ice core deuterium data against mean age, revealing the glacial cycles of the last 740 000 years (source: Jouzel 2004). The lower panel shows the high-resolution EPICA Dome C ice core deuterium data of the last ~45 000 years.
Source: Röthlisberger et al. 2005.

interglacial cycles in the late Quaternary seemingly reflects eccentricity, one of the Earth's orbital parameters, but this is somewhat implausible as variations in eccentricity have a relatively small impact on insolation. The mid-Pleistocene transition to these longer and more severe cycles, at about 800 000 years ago, has also been linked to an abrupt increase in Antarctic ice (Elderfield *et al.* 2012) and coincides with a distinct trend towards drier conditions across Australia.

It should be emphasised that, although the glacial cycles of the last 1 or 1.5 million years have some characteristics in common (see Figure 3.1), each cycle has distinct features (e.g. Figure 3.2 upper panel). In the late Quaternary a 'full glacial cycle' begins with rapid deglaciation (a 'termination'), followed by an interglacial of about 10 ka, then slow cooling lasting for 70–90 ka, culminating with full glacial conditions (a glacial maximum). The glacial conditions are separated by (briefer) interglacials (Figures 3.1 and 3.2) with warmer, generally wetter climates, enhanced greenhouse gases and higher sea levels. The number of secondary fluctuations, representing relatively warmer (interstadials) and cooler (stadials) periods varies within each glacial cycle, but is commonly between 4 and 6.

The Quaternary period includes two epochs: the Pleistocene (2.588 Ma to 11.7 ka) and the Holocene, representing the current interglacial period, which started 11 700 years before present (BP). In comparison to the Holocene, the Pleistocene is an epoch where change is the norm and stability is relatively rare and brief. At high latitudes and altitudes, the cycles were responsible for the accumulation and wasting of ice, as either sea ice or as glaciers or ice sheets on land. During sustained cold periods sea levels were significantly lower as large volumes of water were tied up in continental ice sheets, and so land areas were larger and connections existed between otherwise isolated land areas if the intervening sea was relatively shallow. Ups and downs in sea level during the Quaternary climatic oscillations meant that Australia, New Guinea and Tasmania were repeatedly joined (as 'Sahul') and separated during interglacials. Glacial intervals are also associated with reduced moisture availability and aridity, lowered atmospheric CO_2 and other greenhouse gases, stronger winds and increased fluxes of dust. In contrast, during interglacials and interstadials, warmer and wetter climatic conditions allowed biogeographic patterns approximating those of the Holocene (before significant human impacts).

The variability of climate and environmental conditions during the Quaternary drove significant biotic change, through changes in species abundance, local extinctions, and migration with responses dependent on the tolerances and vagility of individual species. When viewed across the Quaternary, these changes and the expansion/contraction of vegetation advantaged/disadvantaged by the glacial/interglacial conditions resulted in considerable 'shuffling' that is often detectable in patterns of community composition (Box 3.1) and genetic diversity within and between species (Hewitt 2000).

Box 3.1 Tools for reconstructing Quaternary vegetation: pollen and plant macrofossils

Palynology, the study of pollen, spores and related 'palynomorphs', emerged as the principle technique for the reconstruction of vegetation history in the early-to-mid-twentieth century. Pollen grains are small (*c.* 10–80 μm), produced in abundance, easily incorporated into sediments, resistant to decay in anaerobic conditions, and are generally well mixed, such that the 'pollen rain' in an area tends to be uniform and is a function of the composition of the vegetation. Any change in vegetation through time is registered via the temporal sequence of pollen spectra (the number or proportion of each type) and a spatial dimension can be gained by examining spectra of the same age from sites across a landscape.

Although palynology is primarily concerned with the history of vegetation of an area, inferences about events that have affected this history can be drawn: the composition of vegetation provides information about climate, but it can also be used to examine human

Box 3.1 (cont.)

impacts. In the late-twentieth century, fine-resolution palynology, in which sediments are sampled contiguously at small increments in sample depth and chronological control is greater, was increasingly applied to questions of ecological significance (Green 1983).

Our knowledge of the nature and dynamics of Quaternary vegetation is overwhelmingly biased towards the late Quaternary, primarily because older sediments are relatively rare. Our understanding of late Quaternary vegetation depends on the analysis of the pollen contained in organic-rich sediments preserved mostly in shallow lakes and mires (colloquially swamps). Many of these depositional environments have organic-rich, basal dates of early Holocene age, beneath which clastic, inorganic Pleistocene/glacial sediments typically preserve little pollen. While similar depositional features must have existed throughout the Quaternary, especially during earlier interglacial phases, the sediments are poorly preserved over multiple glacial–interglacial cycles. This introduces another bias in our pollen database: lakes and mires (and hence pollen records) are more often found in the well-watered continental margin of Australia, which are also more accessible to major research centres.

Our understanding of Quaternary vegetation, especially beyond the late Quaternary, has also been hindered by the limited taxonomic resolution of fossil pollen (Jackson and Overpeck 2000), reflecting the evolutionary conservativeness of pollen as a plant organ. Australian rainforest, because it includes many species-poor genera and families that produce morphologically distinctive pollen, can be richly documented with palynology (Kershaw *et al.* 2007); but Australia's much more widespread, open-canopied sclerophyll vegetation tends to generate pollen assemblages dominated by a small number of coarsely resolved types (principally *Eucalyptus*/*Corymbia*/*Angophora*, Casuarinaceae, Asteraceae, Poaceae and Amaranthaceae including chenopods) (Kershaw *et al.* 2000), each of which may represent up to hundreds of ecologically and climatically divergent species.

Longer pollen records, extending back beyond the last interglacial, are available from marine sediment cores (e.g. Moss and Kershaw 2000; Kershaw *et al.* 2003), lakes generated from meteorite impact (Colhoun and van der Geer 1988; Dodson and Ramrath 2001; Dodson and Macphail 2004) or through tectonic (Singh and Geissler 1985; McEwan Mason 1991) and volcanic (Kershaw *et al.* 1991a; Wagstaff *et al.* 2001; Kershaw *et al.* 2004, 2007; Sniderman *et al.* 2007, 2011) processes. Australia has been relatively inactive geologically during the Quaternary and hence tectonically derived sites are relatively few in number, and dry climates associated with repeated glacial intervals have inhibited the preservation of organic materials within many sites. Small lakes, even if very deep, fill with sediment within a few hundred thousand years, after which they become difficult to identify in the landscape (Wagstaff *et al.* 2001; Kershaw *et al.* 2004). Deriving a chronology for older sediments is also difficult: radiocarbon (^{14}C) dating is a common, relatively simple and commercially available technique for dating organic sediments, but it has a ~50 ka limit and this may discourage examination of older sediments.

Marine pollen records, in which palynology is applied to sediment cores sampled from offshore marine environments, are playing an increasingly important part in describing the Quaternary vegetation of Australia. They offer a number of advantages, as they are generally continuous, long and well-dated (usually by comparison to $\delta^{18}O$ or magnetic profiles) records that are useful to assess regional vegetation histories due to their large pollen catchment areas. Marine pollen records have proven to be particularly useful for northern Australia from the Queensland Wet Tropics to northwestern Australia (e.g. Kershaw and van der Kaars 2012) and they are often the only means of obtaining data for a large part of semi-arid Australia. Existing records have demonstrated, for example, that the regional responses of vegetation to orbital forcing are variable (Kershaw and van der Kaars 2012).

Importantly, the temporal depth of marine pollen records from northern Australia means that they often allow the consideration of vegetation across most or all of the last glacial cycle or even multiple cycles. This means that they allow inferences about vegetation dynamics across eccentricity periodicities, which are the defining feature of the late Quaternary. The continuity of marine sediment cores also means that they provide reliable data about the nature of the last glacial maximum (LGM) when terrestrial records might be discontinuous. On the negative side, marine pollen records are unable to provide data on short-term dynamics of individual plant communities. As well, changing land–ocean geographies associated with glacial-induced sea-level variations may distort climate and vegetation patterns preserved in marine sediments.

As noted, pollen grains are small in size and so they require substantial magnification for identification. In contrast, macroscopic organic remains, including plant organs such as flowers, fruits, seeds, leaves and wood, might be preserved in sediments under the right circumstances. When available, plant macrofossils can sometimes be identified to species level, particularly in the case of leaves and woody fruits (Jordan 1995a), which is rarely achievable with fossil pollen. Because intact plant macrofossils are generally too large and fragile to have been transported far, their presence typically testifies to the local presence of source plants, in contrast to the often regional, and uncertain, area of vegetation integrated by fossil pollen assemblages.

Box 3.1 (cont.)

This means that plant macrofossils offer the potential to document past vegetation with more taxonomic detail than is possible with pollen. In exceptional circumstances very good preservation might also allow aspects of the vegetation community to be considered, allowing further insight into the environmental conditions of the time. As an example, Sniderman *et al.* (2013) used the diversity of fossil leaves in the early Quaternary sediments of the small palaeolake Stony Creek Basin (described in detail in the case study below) to infer patterns in species richness at that time.

3.3 The vegetation of the early Quaternary in south-eastern Australia

Due to the limitations discussed in Box 3.1, and hence a paucity of data, we can only consider the nature of the vegetation during the earliest Quaternary at a regional scale in south-eastern Australia. Aspects of this regional vegetation can be inferred from sediment cores drilled in the Murray-Darling Basin (Martin 1978; Macphail 1997, 1999), in the offshore Gippsland Basin (Macphail 1997), as well as from Lake George, a tectonic lake on the southern Tablelands of New South Wales (Kershaw *et al.* 1991a). Because the catchments of these records can be very large, their fossil assemblages integrate vegetation changes over large, and potentially subcontinental regions. Some age control is provided by the presence of globally correlated marine faunas in the offshore sediments, but the ages of terrestrial records are often only estimated or inferred by magnetic stratigraphy.

Dodson and Ramrath (2001) and Dodson and Macphail (2004) described a lacustrine record from an impact crater at Yallalie in Western Australia, spanning ~3.6 to 2.5 Ma, just prior to the Quaternary. The pollen record at this site indicated a high diversity of Myrtaceae and Proteaceae, families that dominate highly biodiverse sclerophyll shrublands in the region today. In addition, pollen with affinities to rainforest taxa was present, albeit in low quantities, suggesting that the vegetation was a woodland/heath and rainforest mosaic, perhaps with rainforest confined to small, presumably largely fire-protected enclaves within a mostly sclerophyll landscape. Pollen or spores with affinities to rainforest include the fern *Lygodium* (Lygodiaceae), the conifers *Araucaria* and *Agathis* (Araucariaceae), *Dacrydium*, *Dacrycarpus* and *Phyllocladus* (Podocarpaceae), and the angiosperm taxa *Nothofagus* subgenus *Fuscospora* and subgenus *Brassospora* (Nothofagaceae), Musgraveinae (Proteaceae), Sapotaceae, Ebenaceae and Vitaceae (Dodson and Macphail 2004).

At Lake George in New South Wales, an association of *Nothofagus* subgenus *Brassospora*, Podocarpaceae and Casuarinaceae was apparent prior to 2.8 Ma, when taxa of open vegetation are poorly represented (Macphail *et al.* 2015). This, arguably represented the last remnants of closed-canopy forests in the Lake George region, but unfortunately the sediments are not continuous and the transition to 'Quaternary' vegetation is missing. Nonetheless, a major change at Lake George at 2.8 Ma sees the start of alternating phases of pollen-rich and pollen-poor sediments probably representing increasingly drier conditions especially during the glacial intervals.

These, and other early Quaternary pollen records (Martin 1987; Kershaw *et al.* 1991b; Macphail 1997, 1999), have been interpreted as revealing a transition from forests in which rainforest taxa such as Araucariaceae and Podocarpaceae were present, to open vegetation communities in which Casuarinaceae and Asteraceae are much more common. This transition is generally assumed to represent the vegetation response to the onset of the Quaternary, although their coarse spatial, temporal and taxonomic resolution means they provide only a broad outline of vegetation change. It is also apparent that the climates of the Quaternary occur within longer trends, without a distinct onset (Figure 3.1); hence these trends in Australian vegetation are also likely to have predated the Quaternary (Weston and Jordan, Chapter 2, this volume).

Greater spatial resolution is provided by early to mid-Pleistocene sediments from sites in western Tasmania, which have been dated using palaeomagnetism and by comparisons of the pollen floras with those in the offshore Gippsland Basin. At Regatta Point, in Macquarie Harbour, western Tasmania, macrofossil floras (leaves, fruits, etc.) have been recovered from fine-grained clasts preserved within fluvial gravels. These floras, interpreted as early Pleistocene in age, include combinations of extant western Tasmanian species and extinct taxa (Jordan 1997a). The extinct taxa include conifers such as *Dacrycarpus carpenterii* (Podocarpaceae) and *Callitris strahanensis* (Cupressaceae) (Jordan 1995b), and angiosperms from rainforest, such as *Nothofagus pachyphylla* (Jordan 1999), *Quintinia tasmanensis* (Paracryphiaceae) and *Laurophyllum australum* (Lauraceae) (Jordan 1997b), as well as from open communities, for example the extinct species *Oxylobium pungens*, *Acacia bulbosa* (Fabaceae), *Rubus nebuloides* (Rosaceae), *Banksia kingii* and *B. strahanensis*, *Orites truncata* and *Telopea strahanensis* (Proteaceae) (Jordan and Hill 1991; Jordan 1995a, 1997b). Associated pollen records (Macphail *et al.* 1993) demonstrate the presence of additional rainforest taxa that no longer occur in Tasmania, such as *Symplocos* (Symplocaceae).

The mixture of extant and extinct taxa at Regatta Point presents a confusing picture of the climate at the time of deposition. Today, the nearest living relatives of several of the extinct taxa at Regatta Point are found in warmer climates than those occupied by the extant species. In explaining this contradiction, Jordan (1997a) suggested that many of the extinct early Pleistocene Tasmanian taxa were cool temperate microtherms, using the bioclimatic classification of Nix (1991), even though most or all of their surviving relatives are now confined to tropical and subtropical climates (mesotherms). These fossil floras strongly suggest that western Tasmania has undergone a complex history of floral impoverishment during the Quaternary, affecting both its rainforest and sclerophyll floras.

3.3.1 The pollen and macrofossil record from Stony Creek Basin, Victoria and its implications for the early Quaternary vegetation of Australia

Evidence from Stony Creek Basin (SCB), a small (*c.* 300 m diameter) in-filled lake of probable volcanic origin, at *c.* 500 m a.s.l. in Victoria's western uplands, reinforces the view that early Quaternary vegetation differed substantially from that of today. SCB preserves a 40 m thick sediment record. This appears to have been deposited between about 1.55 and 1.83 Ma based on magnetic polarity, fission track dating of zircons and annually deposited sediment laminations (Sniderman *et al.* 2007).

The SCB pollen record (Figure 3.3) alternates, rhythmically, between dominance by rainforest taxa (chiefly several Podocarpaceae, and several angiosperms, some now extinct in southern Australia, e.g. *Ilex*) and by the sclerophyll trees (*Eucalyptus*, Casuarinaceae or *Callitris*), though neither rainforest nor sclerophylls ever entirely exclude the other. In a small basin such as SCB, pollen is predominantly locally derived (Prentice 1985). Hence the rhythmic

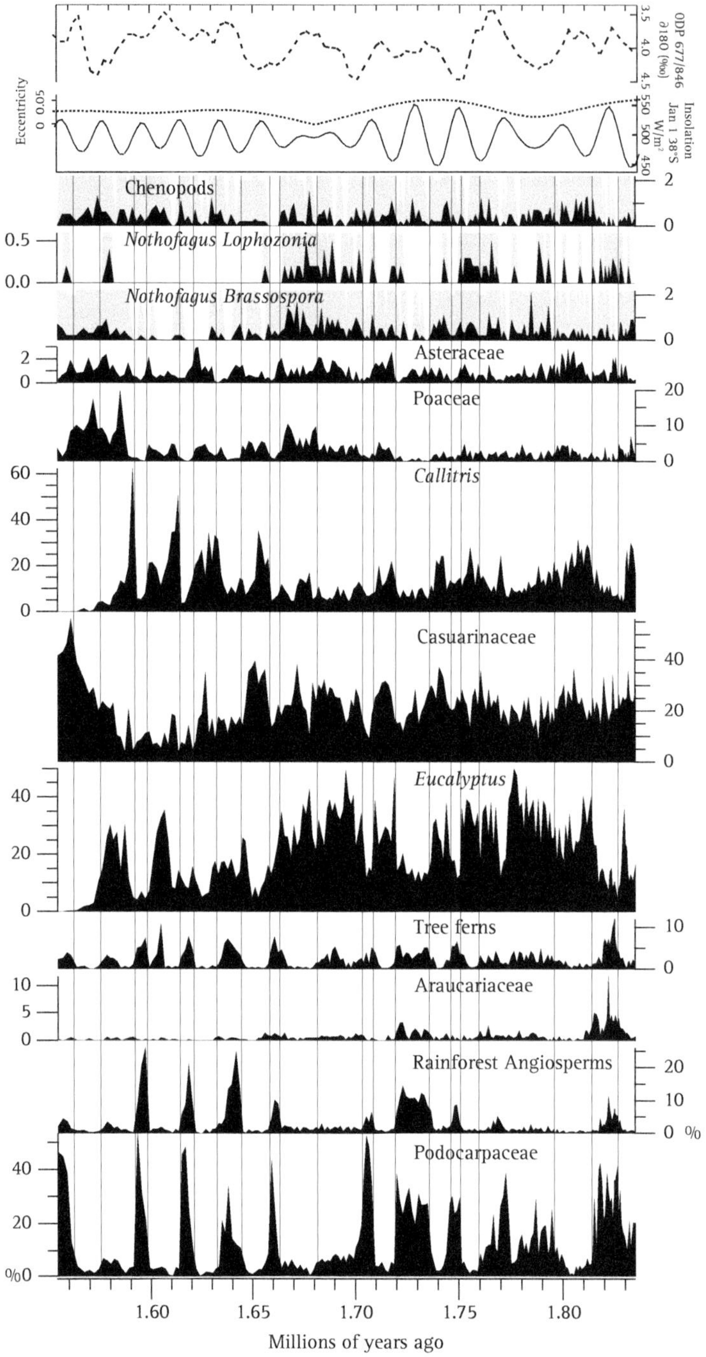

Figure 3.3 The early Quaternary pollen record from Stony Creek Basin. The top two panels show the oxygen isotope record and the January insolation curve at the latitude of the site resulting from the Earth's orbital parameters, respectively, while the lower panels show the palynology data.

changes in pollen abundance suggest expansion and contraction of closed and open forest within an upland vegetation mosaic. The marine isotope record of the time suggests that the SCB vegetation record is responding to orbital precessional (~20 ka) cycles, rather than obliquity cycles (~40 ka). Dominance by this precessional cycle is not easily explained because this is a feature of contemporary tropical regions largely unaffected by the seasonal tilt of the Earth.

In the SCB record, there is little sign of open-canopied vegetation communities dominated by Casuarinaceae and Asteraceae and it includes little contribution from herbs or grasses, unlike the lowland and offshore records for the early Pleistocene (Kershaw *et al.* 1991b). Although Casuarinaceae is consistently abundant at SCB, it is unlikely to signal open vegetation: macrofossils demonstrate the presence of two species, one with close affinities to *Allocasuarina torulosa* (Sniderman *et al.* 2013), which primarily inhabits moist open forests of the eastern coast (Ashton and Attiwill 1994).

Two taxa of contrasting ecological and climatic affinity are consistently recorded at SCB as small percentages of the pollen sum (Sniderman 2011). The Chenopodiaceae (Amaranthaceae *sensu lato*) is adapted to the driest climates, and *Nothofagus* subgenus *Brassospora* is now extinct in Australia but occurs in New Caledonia and New Guinea where annual precipitation exceeds 1500 mm (Read *et al.* 2005). Both are wind-pollinated taxa (Macphail 2007) and are strongly overrepresented, relative to the size, or even local presence, of source populations. Their limited representation in the SCB pollen record suggests that neither taxon was likely to be present in the vegetation immediately surrounding the site but their pollen may have been dispersed from vegetation growing in drier and wetter climates, respectively, along a regional precipitation gradient in south-eastern Australia. Pollen records from the Murray–Darling (Macphail and Truswell 1993) and offshore Gippsland (Macphail 1997) Basins corroborate the existence of extensive open vegetation with a substantial chenopod component further inland from SCB by ~1.8 Ma.

The composition of rainforest represented at SCB is quite different from extant southern Australian cool- (microtherm) or warm temperate/subtropical- (mesotherm) rainforests. The geographically closest modern rainforest analogues, in southern and central Victoria, are microtherm forests dominated by Myrtle Beech, *Nothofagus cunninghamii* (Busby and Brown 1994). *Nothofagus* subgenus *Lophozonia*, to which *N. cunninghamii* belongs, produces a very distinctive pollen type (Macphail *et al.* 1994), which was only present at SCB in small quantities (Sniderman 2011). Although this implies that source plants were present in the western uplands, they were unlikely to have been important in the local vegetation community. Instead, the SCB rainforest of the early to mid-Quaternary was probably more analogous, structurally and floristically, to podocarp-rich forests extant in New Zealand, New Caledonia and New Guinea (Sniderman and Haberle 2012). Relics of this Australian 'conifer/broadleaf rainforest' (Ogden and Stewart 1995) persist in western Tasmania, as small isolated patches in eastern Australia and in north-eastern Queensland, where podocarps remained an ecologically dominant part of the flora until the late Pleistocene (Kershaw *et al.* 2007).

The diversity of extinct rainforest taxa in the SCB record implies that modern rainforest patches in southern Australia may have developed their current floristics and distribution only after the pattern of modern cool-season rainfall was initiated, some time after deposition ceased at SCB, perhaps between 1.5 and 1.0 Ma (Sniderman *et al.* 2009; McLaren and Wallace 2010). Prior to this time throughout much of the Cenozoic, south-eastern Australian rainforests apparently supported a wide range of Podocarpaceae and *Nothofagus* subgenus *Brassospora*, even if by the early Pleistocene these were confined to the wettest sites. It should be noted, however, that the evidence from SCB that *Brassospora* survived in south-eastern Australia until c. 1.55 Ma contrasts with its disappearance in the offshore Gippsland Basin sediments at approximately 3.6 Ma (Macphail 1997), although this date is not well constrained.

The apparent spatial juxtaposition of sclerophyll and rainforest vegetation at SCB during the early Quaternary, and at Yallalie during the Pliocene, is consistent with contemporary mosaics of rainforest and sclerophyll communities in many parts of eastern Australia. Charcoal records from SCB suggest that fire was a regular feature of both the sclerophyll community and the angiosperm-dominated phases of rainforest, occurring at least once every few centuries, but were very rare during gymnosperm-dominated phases (Sniderman and Haberle 2012). Podocarp forests are sensitive to fire on landmasses where they are still ecologically dominant and widespread, such as New Zealand (Ogden and Stewart 1995) and vegetation boundaries are often controlled by fire.

The sclerophyll flora at SCB in the early Pleistocene is documented in some detail by macrofossils including at least 69 dicot morphotaxa. These include some fruits and flowers but mostly mummified and carbonised leaves, mainly of sclerophyllous, microphyllous angiosperms interpreted as components of open forest communities, though microphyllous leaves of *Callitris*, several Podocarpaceae and of Araucariaceae were also found (R. Paull, G.J. Jordan, R.S. Hill, unpublished data). Sniderman *et al.* (2013) argued that the macrofossil flora represented an ecological 'snapshot' of a single plant community because they were recovered from sediments representing no more than 2500 years of deposition and they probably were transported from a relatively small area of vegetation (< c. 50 ha).

Sniderman *et al.* (2013) concluded that the richness of sclerophyll species at SCB during the early Pleistocene was comparable to that of contemporary sclerophyll plant communities in south-western Australia and in the Cape Floristic Region of South Africa, regions known for their extraordinary plant diversity. Twenty distinct morphospecies of Styphelioideae (Ericaceae), for example, have been identified from the site (Jordan *et al.* 2007; Sniderman *et al.* 2013).

The apparent decline in diversity of sclerophyll flora at SCB some time after 1.5 Ma might be indicative of more widespread declines in south-eastern Australia. In contrast, the extraordinary plant diversity of south-western Australia has been maintained (but rainforest communities have disappeared). Under the relatively subdued climatic variability of the early Pleistocene, diverse sclerophyll floras may have been widespread in southern Australia and after c. 1 Ma, perhaps with the development of an east–west gradient in the amplitude of glacial–interglacial climate variability, many eastern species within both rainforest and sclerophyll vegetation communities were driven extinct by novel climate extremes. This is consistent with a suggestion that regional rates of extinction are correlated with the amplitude of glacial–interglacial climate variability (Jansson and Dynesius 2002).

The coarse taxonomic resolution of palynology, and the Neogene history of rainforest contraction and sclerophyll expansion evident in pollen records (Kershaw 1988; Kershaw *et al.* 1994; Martin 1978, 1994), together may have encouraged a view that the diversity of the sclerophyll flora has never been higher than it is today. The macrofossil record from SCB indicates that this is not correct, and that, at least for eastern Australia, diversity within the sclerophyll biome probably peaked during, or no later than, the early Pleistocene.

3.4 The climate and vegetation of late Quaternary Australia

As we come forward in time, information about past environmental conditions and vegetation tends to have better resolution and chronological control. In this section, we describe the late Pleistocene and then focus on the vegetation immediately preceding and during the last glacial maximum in Australia. The section concludes with a case study describing a high-resolution record of vegetation change since the last glacial maximum from Lake Euramoo in north-eastern Australia.

In the mid to late Quaternary glacial and interglacial cycles at eccentricity periodicities (~100 ka) are the defining feature and determine much of the dynamics of Australian vegetation. The summary pollen diagram from Core MD98-2167, in the Timor Sea off the Kimberley coast (Figure 3.4), is used here as an example of long marine pollen records. This record, covering approximately 250 000 years, is especially useful as it is located in a region where the very seasonal climate is not conducive to the accumulation of terrestrial polleniferous sediments. Figure 3.4 is a summary diagram, which masks some of the detail, but the overall importance of the 100 000-year eccentricity cycle is clearly evident. When individual pollen types are considered other cycles also become apparent: *Eucalyptus* pollen,

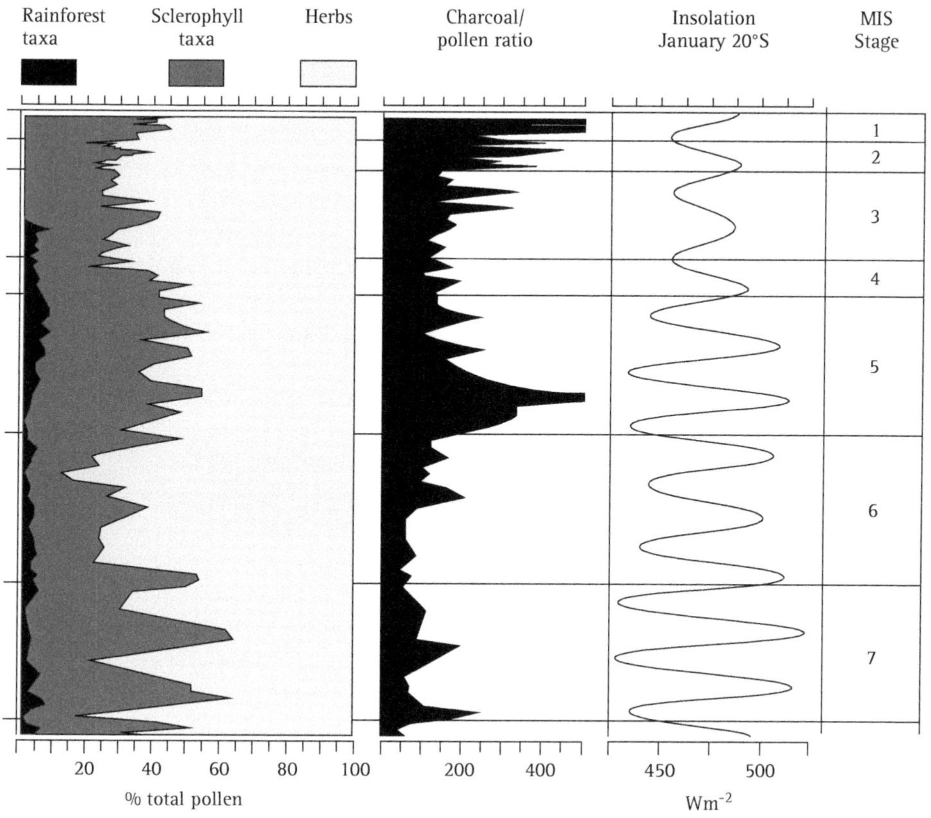

Figure 3.4 A summary diagram of the long pollen record (left panel) from marine sediment core MD982167, located off the Kimberley coast in the Timor Sea. This allows the reconstruction of vegetation and environments over the most recent glacial–interglacial cycle, from about 250 000 years ago (marine isotope stages (MIS) 7 to 1). The charcoal/pollen ratio (middle panel) indicates past fire activity. An insolation curve for January (right panel) shows the influence of variation in Earth's orbital parameters as potential long-term drivers of vegetation change.

reflecting the regional importance of savanna woodland, shows a ~20 000 year precessional cycle following summer insolation. This can be best explained by changes to the southern hemisphere monsoon. A general trend indicating increasingly dry conditions or increasing disturbance, characteristic of a number of records, is suggested by the charcoal curve and the possible extinction of a rainforest taxon, cf. *Olea*, some 45 000 years ago.

As is common with all late Quaternary climatic cycles, the last glacial cycle shows a progressive decline in temperatures and the build-up of ice over the last ~100 000 years, albeit in stages (Figure 3.2). This progressive cooling resulted in two distinct glaciations in the Snowy Mountains Australia (Barrows *et al.* 2001): the 'early' Kosciuszko glaciation, at about 59 ka was the most extensive, and this seems part of a larger spatial pattern as the high-resolution ice core data from Antarctica (Figure 3.2 upper panel) also reveals that MIS 4 (~70 to 57 ka BP) almost matched the last glacial maximum in intensity. It appears that inland Australia had significantly enhanced effective precipitation during some intervals of MIS 5 and in MIS 4 (Cohen *et al.* 2012). Cool and humid conditions also dominated MIS Stage 3 (57 to 29 ka BP) with enhanced fluvial discharge, generally fuller, larger and/or more permanent lakes (e.g. Bowler *et al.* 2012; Cohen *et al.* 2012; Reeves *et al.* 2013a).

The first glacier advance of the 'late Kosciuszko' series, at about 32 ka BP (Barrows *et al.* 2001), was a harbinger of the approaching cold and dry glacial maximum. Although humid conditions continued in southern temperate parts of Australia until ~25 ka BP, there is widespread evidence in other parts of Australia of near full glacial conditions by ~28 ka BP. Environmental conditions then deteriorated towards the LGM, at 21 ± 3 ka BP, but this trend was not always unidirectional (Reeves *et al.* 2013a).

Australia appears to have been about 4–8°C cooler than present on average during the LGM. Fletcher and Thomas (2010) calculated that lowland Tasmania was 4.2°C colder at the LGM than at present, which is less than has been assumed (e.g. 9°C in south-eastern Australia by Galloway 1965). It is also possible that any LGM reduction in temperature was negligible in parts of the tropics: this latitudinal gradient is evident in sea surface temperatures (SSTs), which were <3°C colder in the tropical western Pacific at the LGM compared to the present (Visser *et al.* 2003) and 3–6°C colder in the Southern Ocean (Barrows *et al.* 2007).

At the LGM atmospheric circulation was different partially due to changes in SST gradients and altered ocean currents, for example, related to expanded sea ice around Antarctica (e.g. Bostok *et al.* 2006). Northern Australia experienced drier conditions at the LGM, due to relatively weak south-east trade winds and also a weaker monsoon system (Kuhnt *et al.* 2015). The reorganisation of pressure systems and wind at the LGM meant that the westerlies, which are important sources of winter moisture in southern Australia, were in a more northerly position (by up to 5°C) during summer compared to their contemporary position (e.g. Hesse 1994).

Other environmental changes were also important. Based on the Vostok Antarctic ice core record, CO_2 was 25–30% lower at the LGM (180 ppm) compared to the average during the Holocene (270 ppm). Sea level during the LGM was ~125 m lower than current (pre-industrial) levels (Lewis *et al.* 2013), and this increased the size of Australia, enhanced continentality and so contributed to the aridity of some parts of the continent.

Primarily based on evidence from eastern and southern Australia, the vegetation at the LGM can be described as a semi-arid grassland steppe dominated by Asteraceae and Poaceae (Kershaw 1980, 1981; Singh *et al.* 1981; Singh 1982; Kershaw *et al.* 1991b; Dodson 1994; Hope 1994). In contrast, Pickett *et al.* (2004) suggested that xerophytic shrub, heath and woodland communities dominated by myrtaceous shrubs, Asteraceae and Chenopodiaceae characterised southern Australia during the LGM.

Although patches of more mesic vegetation communities are alluded to in pollen spectra, they are absent from the LGM. As an example, there is no record of rainforest taxa during any of the glacial maximums (Kershaw 1994) and sclerophyll forest is only apparent during the glacial maximums in Queensland. It has thus generally been concluded that the expansion of the arid core of central Australia saw trees increasingly confined to sheltered refugia, including on the continental shelves exposed by the lower sea levels at the LGM (e.g. Harle 1997) and treed communities disassociated.

These suggestions and their inferred climatic conditions are strangely juxtaposed by enhanced fluvial activity in south-eastern Australia during the LGM. This includes evidence on the riverine plain of the Murray–Darling Basin (e.g. Kemp and Rhodes 2010) and hints elsewhere of at least periodically wet conditions (see Reeves *et al.* 2013a). As an example the lakes of the Willandra system were full at the LGM, and after this time began a slow decline culminating in presently dry or ephemeral systems (Bowler *et al.* 2012). These anomalous records of significant moisture have been associated with the more northerly position of the westerlies or seasonal snowmelt from the eastern highlands but they are an enigma and difficult to reconcile with the traditional view of a largely treeless, arid LGM in eastern Australia.

Other studies suggest evidence of tree-dominated vegetation during the LGM at Barrington Tops (Sweller and Martin 2001), on the southern fringes of the Sydney Basin (Black *et al.* 2006) and at Mountain Lagoon in the Blue Mountains west of Sydney (Robbie and Martin 2007). The Little Llangothlin Lagoon on the New England tableland also apparently has a significant presence of rainforest taxa at the LGM (Daniel Ellerton, Jamie Shulmeister, Craig Woodward *pers. comm.*). This could reflect refuges on the New England escarpment, where extant wet forests are extensive. However, uncertainty about LGM vegetation stems from the taxonomic imprecision of palynology: these works include pollen types that relate to taxa that include both tree and non-tree species.

The LGM vegetation of Australia is often described as a 'no-analogue' mixture of Casuarinaceae and Asteraceae. These could represent riparian and ephemeral species, respectively, given the enhanced fluvial activity described above in an otherwise dry landscape, or an expansion of current 'dryland' *Allocasuarina* (e.g. *A. verticillata*) in response to the environmental conditions of the LGM. Alternatively, the aridity of the LGM may have been overemphasised and we need a better explanation for the supposedly 'arid' vegetation of the time. Some of this uncertainty stems from the LGM sites sampled thus far, most of which have condensed or only partial chronological sequences and this is accentuated by difficulties relating to radiocarbon dating of sediments with little organic content. Despite this uncertainty, it is clear that at a regional scale sclerophyll forests contracted during the LGM and humid and tropical vegetation were reduced to isolated and potentially small refugia.

Post-glacial increases in temperature were evident across Australia by 17 ka BP (Reeves *et al.* 2013a). Despite this relatively fast recovery in temperature after the LGM the post-glacial recovery of tree-dominated vegetation communities in temperate Australia was not initiated until after 15 ka and as late as 13 ka in some places (Dodson 1998). See the Lake Euramoo case study (Section 3.4.1 below) for more details. This means that arid and semi-arid vegetation communities were widespread in south-eastern Australia in the period of post-glacial warming. Where the spatial density of pollen records is high (e.g. the Atherton Tableland Queensland or western Victoria), the data suggest local variability the rates of woodland/forest recovery after the LGM.

Generalisations regarding palaeoclimate become more difficult across the continent in the late Pleistocene (Reeves *et al.* 2013a). The Antarctic Cold Reversal (ACR, from ~14.5 to 12.5 ka BP) influenced the Australian sector of the Southern Ocean, with declining SST and probably wetter conditions in southern Australia (Fletcher and

Moreno 2011; Petherick *et al.* 2013). These southern conditions are in contrast to northern Australia where consistently warmer SSTs are evident from 15 ka BP (Reeves *et al.* 2013b). The northern monsoon crosses a significant threshold at 13.4 ka BP (Kuhnt *et al.* 2015) reflecting southern hemisphere warming and the flooding of the shallow Sahul shelf. Prior to 13.4 ka BP 'glacial' conditions meant that winter rainfall was more important in northern Australia than in the modern summer-dominated regime.

After the cold and dry conditions of the late Pleistocene abated, vegetation patterns approaching those of modern day began to develop. As an example, Dodson (1998) suggested that the vegetation of temperate Australia underwent considerable change in the period from 15 to 14 ka BP, and forest cover increased rapidly between 12 and 9.5 ka BP. From this it is obvious that some species assemblages that are observable today may not necessarily have had a very long history.

The lower panel of Figure 3.2 shows that maximum post-glacial warmth in Antarctica occurred between ~11.5 and 9 ka BP. This warmth was also apparent in our region, with high lake levels and SST approaching modern values (Reeves *et al.* 2013a). In north-west Australia maximum fluvial runoff occurred between about 11.5 and 11 ka BP (Kuhnt *et al.* 2015) but this precipitation maximum extends until ~8 ka BP across much of northern Australia (Reeves *et al.* 2013b), leading to suggestions that the monsoon extended further south (e.g. to 22°S) in the early Holocene. A return to drier conditions then is apparent in northern Australia at about 8 ka BP. In southern Australia SST peaks at about 6 ka BP and warmer and wetter conditions are apparent from 6 to 5.5 ka BP (Petherick *et al.* 2013). This mid-Holocene peak in effective moisture contrasts with the late Holocene, during which climatic variability increased, as the El Niño phase of ENSO increased in frequency and/or strength (e.g. Moy *et al.* 2002). This ENSO variability is also an important control on the vegetation of north-eastern Australia, as described in the Lake Euramoo case study below.

3.4.1 A high-resolution pollen record from Lake Euramoo demonstrating vegetation dynamics after the Last Glacial Maximum

A high-resolution record of pollen and charcoal from Lake Euramoo covering the last 23 000 years, exemplifies how vegetation changed from the arid LGM through to the more humid Holocene. Lake Euramoo is located on the Atherton Tableland, a region that includes long, glacial and interglacial pollen records of vegetation (e.g. Kershaw 1994; Kershaw *et al.* 2007). The high-resolution record from Lake Euramoo provides detail that aids in the interpretation of these more coarsely resolved but long records. The dynamics of the vegetation relates to both climate and human activity primarily through the application of fire (Haberle 2005).

The pollen record from Lake Euramoo (Figure 3.5) has been divided into five significant zones (Haberle 2005):

1. 23 000–16 800 calibrated years before present (cal BP) dry sclerophyll woodland;
2. 16 800–8600 cal BP sclerophyll woodland with marginal rainforest in protected pockets with advancing rainforest from 16 800-10 900 cal BP and a brief reversal to greater sclerophyll woodland between 10 900-8600 cal BP;
3. 8600–4800 cal BP cool lower montane rainforest dominated by Cunoniaceae and Urticaceae/Moraceae. This rainforest formation is more common at higher altitudes in the region today;
4. 4800–70 cal BP, drier sub-montane rainforest dominated by Elaeocarpaceae; and
5. 70 cal BP–AD 1999, degraded sub-montane rainforest with increasing influence of invasive species and fire.

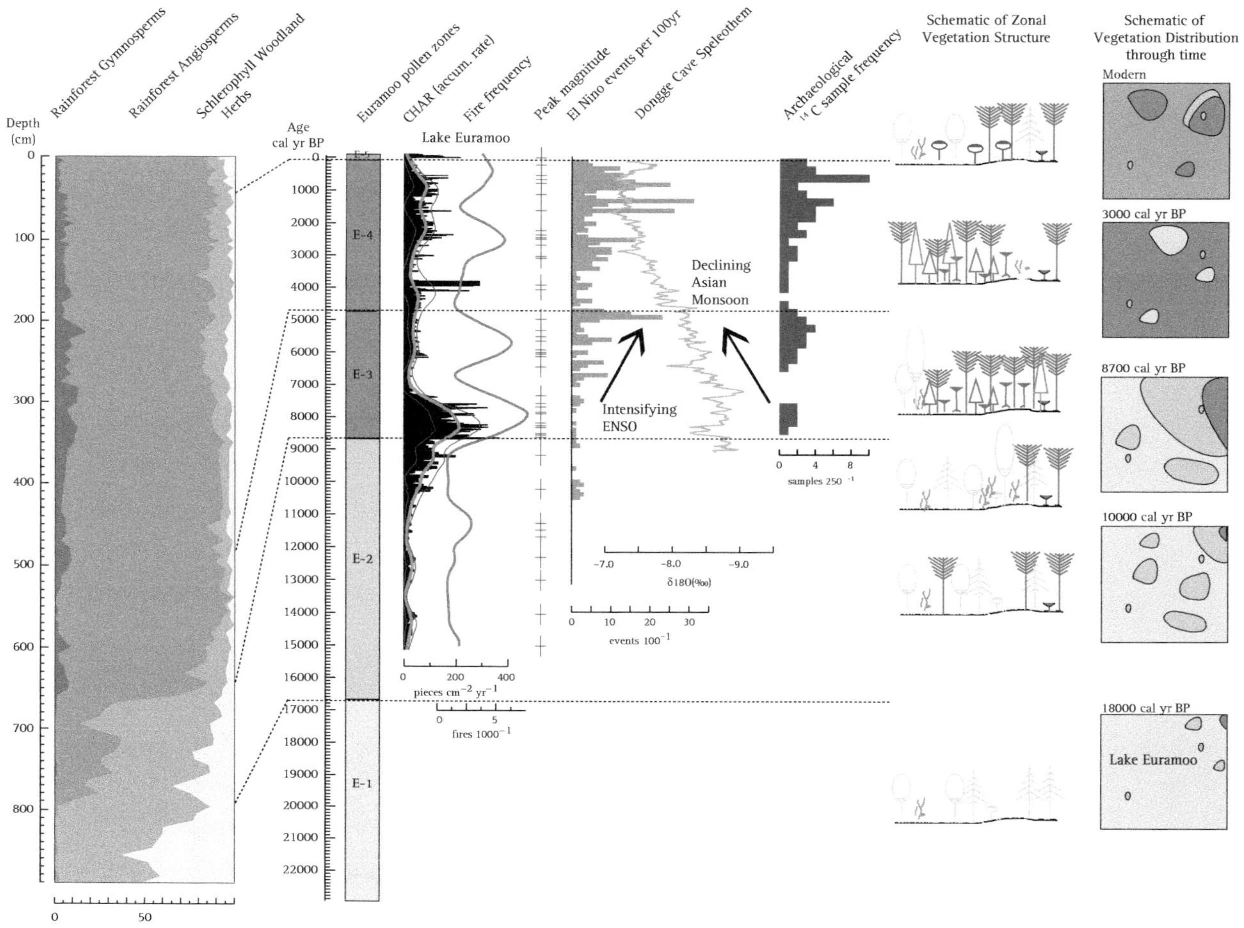

Figure 3.5 A summary diagram of the high-resolution pollen record from Lake Euramoo on the Atherton Tableland in north-eastern Queensland. The reconstruction of fire frequency, based on charcoal accumulation (CHAR), extends back to 15 000 years ago, as prior to that time charcoal abundance is very low. Some of the possible drivers of vegetation and fire change in this landscape are shown for comparison including the total number of strong El Niño events per century interpreted from flood deposits in Laguna Pallcacocha, Ecuador (orange bars; Moy *et al.* 2002) and past precipitation in China from the Dongge Cave speleothem δ18O record (blue line; Wang *et al.* 2005).
Diagrams: Simon Haberle. (A black and white version of this figure will appear in some formats. For the colour version, please refer to the plate section.)

In contrast to earlier suggestions, the sclerophyll woodland close to and immediately after the LGM is associated with low levels of charcoal and hence probably best reflects the (dry, cool) climatic conditions rather than high frequency fire. The woodland at the time was dominated by *Casuarina*/*Allocasuarina* with a grassy understorey. The first indications of rainforest taxa in the post-glacial period are at 16 800 cal BP, and the peak development of lower montane rainforest was at 7300 cal BP, indicating that the transition from the LGM sclerophyll woodland to rainforest was extremely protracted.

This development of rainforest at Lake Euramoo occurred through several phases, starting as a limited suite of angiosperm and gymnosperm taxa from rainforest and wet sclerophyll that expanded from glacial refugia at 16 800 cal BP. The development also included some reversals of trends, for example *Casuarina* returned in the period 12 600 to 9600 cal BP and *Agathis* and *Podocarpus*, both rainforest gymnosperms, disappeared temporarily from the record. The diversity of species within the rainforest increased dramatically during the late glacial transition as new species arrived and sclerophyll woodland taxa were gradually excluded. Closed-canopy rainforest was then established between 9600 and 8700 cal BP, perhaps in response to the early Holocene increase in precipitation that was described in the previous section.

Haberle (2005) argued that the process of rainforest development at Lake Euramoo appears to be at least partly controlled by the precession orbital parameter, though more local environmental variables and human activity were also significant factors. As an example, the protracted advance of rainforest across the Atherton Tableland during the early Holocene could have been partly due to the efforts of humans using fire to maintain a mosaic of open and forested vegetation communities. At Lake Euramoo the peak development of lower montane rainforest at 7300 cal BP was achieved despite the persistent recurrence of fires.

Figure 3.5 also includes a reconstruction of the frequency of past fire events, derived from the statistical de-trending of charcoal accumulation into peak events (e.g. see Higuera *et al.* 2010). This record of fire extends to about 15 ka, as prior to that time the influx of charcoal is very low. The Lake Euramoo record shows that past fire frequencies in rainforest environments fluctuated between 2 and 5 events per 1000 years. Peak fire frequency occurred at *c.* 8000 cal BP and was associated with the final rapid transition to rainforest dominance.

During the mid-Holocene (6 to 4 ka), warm temperate rainforest, typical of rainforest at lower altitudes under wetter and warmer climates, covered the region and fire frequencies were low. A combination of low interannual climate variability (ENSO-related), declining strength of the Asian monsoon and low density of Aboriginal occupation in the rainforests are likely causes of low charcoal quantities and fire frequencies. It was not until around 4000 years ago that charcoal quantities and fire frequencies began to rise across the region, peaking in the last 2000 years. El Niño activity has been highest in the last 2000 years (Moy *et al.* 2002) and may have been a significant cause of drought and greater potential fire ignition over this time. Importantly, however, this time period also coincides with evidence for increased Aboriginal site occupation, the adoption of more complex food harvesting strategies and intensive use of rainforest resources (Turney and Hobbs 2006; Cosgrove *et al.* 2007).

The Lake Euramoo pollen record is significant in illustrating the dynamics of the World Heritage rainforests of the Wet Tropics of north-east Queensland. Although often described as of great antiquity (e.g. ‘ancient rainforests’; WTMA 2016), this and other palynological records on the Atherton Tableland have demonstrated that rainforests had a dramatically smaller range during the colder episodes of the Quaternary, expanding out of refugia and re-assembling during warmer periods.

3.5 Lessons from the Quaternary

Despite the limitations to the tools at our disposal for the reconstruction of Quaternary vegetation in Australia (Box 3.1), several important conclusions can be drawn from the case studies in this chapter. First, there has been a pervasive view that there were relatively few plant extinctions during the Quaternary, and that these were mainly of local scale. The evidence instead indicates that, at least in southern Australia, rainforests at the dawn of the Quaternary included a range of tree taxa that are now extinct or restricted to more northerly rainforests in Australia. Some of these taxa, such as *Nothofagus* subgenus *Brassospora*, may have been eliminated from Australia well before ~1 Ma. For other taxa, such as the Podocarpaceae, there is evidence of progressive decline extending to the late Quaternary, in both the south and the north of the continent. This pattern emphasises the importance of protecting remaining Australian rainforests that include Podocarpaceae.

The modern rainforest patches in southern Australia communities seem to have developed their current floristics, diversity and distribution only after the modern cool-season rainfall patterns were initiated, perhaps between 1.5 and 1.0 Ma. Quaternary extinctions appear to have significantly reduced the southern Australian microtherm flora and mesotherm taxa became increasingly confined to their current northern (Australian) location. Similarly, macrofossil floras and palynology indicate that sclerophyll communities in some parts of south-eastern Australia have also undergone a complex history of floral impoverishment during the Quaternary.

Significant plant extinctions in the Quaternary are perhaps not surprising given the large glacial–interglacial changes in moisture availability, especially apparent since the mid to late Quaternary. Accommodating these changes suggests resilience, whereas extinctions necessitate an alternative perspective, in which at least some elements were fragile. This has implications for Quaternary Australian megafaunal extinctions, which Dodson (1989) hypothesised were more likely to be human mediated, citing the lack of evidence (at that time) of coincident plant extinctions which would be expected if climate was a primary cause.

Early to mid-Quaternary vegetation change at SCB has emphasised the interpretation of vegetation as a *mosaic*, for example, with the inclusion of both rainforest and sclerophyll in the local environment, in the context of indicators of a drier regional climate. Detailed pollen analysis provides insight into previous conundrums regarding the late survival of Cenozoic elements in the vegetation (e.g. *Nothofagus*, *Araucaria*, *Phyllocladus*), which were previously believed to be abruptly eliminated. The evidence suggests a rainforest/sclerophyll mosaic during the early to mid-Quaternary, not unlike modern mosaics, for example, along the temperate eastern Australian escarpment.

In the last few million years Australian vegetation has experienced at least two major agents of environmental change: the climatic inconstancy of glacial–interglacial cycles; and the entry of humans into the region. The former forced repeated changes in vegetation communities in response to changes in temperature, atmospheric CO_2, sea level and other climate-mediated disturbances. Local changes in species abundance must have also been a significant feature of this period.

During Quaternary glaciations, the denudation or severe reduction of vegetation beneath extensive continental ice sheets and around their margins is well known in Europe and North America. In Australia, however, there is little evidence that landscapes were entirely devegated. Instead there is palynological evidence of a glacial biota, resolved at the generic level but not the species level. Many taxa that dominate Australian humid/mesic vegetation must have been confined to refugia during the unfavourable environmental conditions of glacials/stadials, but some other taxa underwent significant range expansion during these times. A simple model of 'retreat' and 'recovery' of tree-dominated vegetation, fails to explain responses to the environmental conditions of the majority of the Quaternary, which were far from benign.

3.6 Human impact on Australian vegetation

This chapter has largely ignored human impacts on Australian vegetation, especially prehistoric human impacts associated with Australia's First People. Current evidence indicates that people first entered the continent at about 50 ka (Bird *et al.* 2004), but the extent of their influence on the vegetation is uncertain. Earlier, suggestions that humans significantly altered the vegetation of Australia, for example through burning activities that result in sclerophyll dominance (e.g. Singh *et al.* 1981) or landcover changes (e.g. Miller *et al.* 2005a, 2005b), are yet to be rigorously tested. Impacts associated with the use of fire by prehistoric people in Australia at regional or continental spatial scales are contentious and unresolved (e.g. Mooney *et al.* 2011; Williams *et al.* 2015).

The question of human impact in Australia is also difficult to separate from polarised views on Pleistocene megafauna extinctions. One view argues that megafauna were driven to extinction exclusively, or at least primarily, through human agency, relatively soon after the first human arrival. Another view argues that humans either were not involved, having arrived too late; were incapable of entirely eliminating large vertebrates at a continental scale; or were only responsible for a small subset of the extinctions. Proponents of this latter view suggest that the palaeontological record is incomplete or conflates the timing of extinction of many of these species (Field *et al.* 2008), and that extinctions were more likely to be staggered throughout much of the last few hundred thousand years, perhaps in response to gradually deteriorating climatic conditions (Wroe *et al.* 2013; cf. Saltré *et al.* 2016).

Regardless of which scenario is more accurate, it seems clear that Australia's ecosystems lost their largest vertebrate consumers and predators relatively recently in evolutionary terms. Although cascading influences of megafaunal extinctions on vegetation, soil and fire dynamics seem likely, the coarse taxonomic resolution of fossil pollen, in part, make it difficult to distinguish these extinctions from other possible explanations for vegetation and floristic changes (Rule *et al.* 2012).

References

Ashton, D.H. and Attiwill, P.M. (1994) Tall open-forests. In *Australian Vegetation*, ed. R.H. Groves. Cambridge University Press: Cambridge, pp. 157–196.

Barrows, T.T., Stone, J.O., Fifield, L.K. and Cresswell, R.G. (2001) Late Pleistocene glaciation of the Kosciuszko Massif, Snowy Mountains, Australia. *Quaternary Research* 55, 179–189.

Barrows, T.T., Juggins, S., De Deckker, P. *et al.* (2007) Long-term sea surface temperature and climate change in the Australian-New Zealand region. *Paleoceanography* 22, PA 2215.

Bird, M.I., Hope, G. and Taylor, D. (2004) Populating PEP II: the dispersal of humans and agriculture through Austral-Asia and Oceania. *Quaternary International* 118–119, 145–163.

Black, M.P., Mooney, S.D. and Martin, H.A. (2006) A > 43 000-year vegetation and fire history from Lake Baraba, New South Wales, Australia. *Quaternary Science Reviews*, 25, 3003–3016.

Bowler, J.M. (1982) Aridity in the late Tertiary and Quaternary of Australia. In *Evolution of the Flora and Fauna of arid Australia*, ed. W.R. Barker and P.J.M. Greenslade. Peacock Publications: Australia, pp. 35–44.

Bostock, H.C., Opdyke, B.N., Gagan, M.K., Kiss, A.E. and Fifield, L.K. (2006) Glacial/inter-glacial changes in the East Australian current. *Climate Dynamics* 26, 645–659.

Bowler, J.M., Gillespie, R., Bolkjovac, K. and Johnson, H. (2012) Wind v water: glacial maximum records from the Willandra lakes. In *Peopled Landscapes: Archaeological and Biogeographic Approaches to Landscapes*, ed. S.G. Haberle and B. David. ANU E-Press: Canberra, pp. 271–296.

Busby, J.R. and Brown, M.J. (1994) Southern rainforests. In *Australian Vegetation*, ed. R.H. Groves. Melbourne: Cambridge University Press, pp. 131–155.

Cohen, T.J., Nanson, G.C., Jansen, J.D. *et al.* (2012) Late Quaternary mega-lakes fed by the northern and southern river systems of central Australia: varying moisture sources and increased continental aridity. *Palaeogeography, Palaeoclimatology, Palaeoecology* 356–357, 89–108.

Colhoun, E. and van der Geer, G. (1988) Darwin Crater, the King and Linda valleys. In *Cainzoic Vegetation of Tasmania*, ed. E.A. Colhoun. University of Newcastle: Newcastle, pp. 30–71.

Cosgrove, R., Field, J. and Ferrier, Å. (2007) The archaeology of Australia's tropical rainforests. *Palaeogeography, Palaeoclimatology, Palaeoecology* 251, 150–73.

Dodson, J.R. (1989) Late Pleistocene vegetation and environmental shifts in Australia and their bearing on faunal extinctions. *Journal of Archaeological Science* 16, 207–17.

Dodson, J.R. (1994) Quaternary vegetation history. In *Australian Vegetation*, ed. R.H. Groves. Melbourne: Cambridge University Press, pp. 37–56.

Dodson, J.R. (1998) Timing and response of vegetation change to Milankovitch forcing in temperate Australia and New Zealand. *Global and Planetary Change* 19, 161–174.

Dodson, J.R. and Macphail, M.K. (2004) Palynological evidence for aridity events and vegetation change during the Middle Pliocene, a warm period in southwestern Australia. *Global and Planetary Change* 41, 285–307.

Dodson, J. and Ramrath, A. (2001) An Upper Pliocene lacustrine environmental record from south-Western Australia: preliminary results. *Palaeogeography, Palaeoclimatology, Palaeoecology* 167, 309–320.

Elderfield, H., Ferretti, P., Greaves, M. *et al.* (2012) Evolution of ocean temperature and ice volume through the mid-Pleistocene climate transition. *Science* 337, 704–709.

European Project for Ice Coring in Antarctica (EPICA) (2004) Eight glacial cycles from an Antarctic ice core. *Nature* 429, 623–628.

Field, J., Fillios, M. and Wroe, S. (2008) Chronological overlap between humans and megafauna in Sahul (Pleistocene Australia–New Guinea): a review of the evidence. *Earth Science Reviews* 89, 97–115.

Fletcher, M.-S. and Moreno, P.I. (2011) Zonally symmetric changes in the strength and position of the Southern Westerlies drove atmospheric CO_2 variations over the past 14 k.y. *Geology* 39, 419–422.

Fletcher, M.-S. and Thomas, I. (2010) A quantitative Late Quaternary temperature reconstruction from western Tasmania, Australia. *Quaternary Science Reviews* 29, 2351–2361.

Galloway, R.W. (1965) Late Quaternary climates in Australia. *Journal of Geology* 73, 603–618.

Gibbard, P.L., Head, M.J. and Walker, M.J,C., The Subcommission on Quaternary Stratigraphy (2009) Formal ratification of the Quaternary System/Period and the Pleistocene Series/Epoch with a base at 2.58 Ma. *Journal of Quaternary Science* 25, 96–102.
Green, D.G. (1983) The ecological interpretation of fine resolution pollen records. *New Phytologist* 94, 459–77.
Groves, R.H. (1994) *Australian Vegetation.* Cambridge University Press: Cambridge.
Haberle, S.G. (2005) A 23 000-yr pollen record from Lake Euramoo, Wet Tropics of NE Queensland, Australia. *Quaternary Research* 64, 343–356.
Harle, K.J. (1997) Late Quaternary vegetation and climate change in southeastern Australia: palynological evidence from marine core E55-6. *Palaeogeography, Palaeoclimatology, Palaeoecology* 131, 465–483.
Harrison, S.P. and Dodson, J. (1993) Climates of Australia and New Guinea since 18 000 yr B.P. In *Global Climates Since the Last Glacial Maximum*, ed. H.E. Wright *et al.* University Minnesota Press: Minneapolis, pp. 265–293.
Haug, G.H., Ganopolski, A., Sigman, D.M. *et al.* (2005) North Pacific seasonality and the glaciation of North America 2.7 million years ago. *Nature* 422, 821–825.
Hesse, P.P. (1994) The record of continental dust from Australia in Tasman Sea sediments. *Quaternary Science Reviews* 13, 257–272.
Hewitt, G. (2000) The genetic legacy of the Quaternary ice ages. *Nature* 405, 907–913.
Higuera, P.E., Gavin, D.G., Bartlein, P.J. and Hallett, D.J. (2010) Peak detection in sediment-charcoal records: impacts of alternative data analysis methods on fire history interpretations. *International Journal of Wildland Fire* 19, 996–1014.
Hope, G.S. (1994) Quaternary vegetation. In *History of the Australian Vegetation: Cretaceous to Recent*, ed. R.S. Hill. Cambridge University Press: Cambridge, pp. 368–389.
Hope, G.S., Kershaw, A.P., van der Kaars, S. *et al.* (2004) History of vegetation and habitat change in the Austral-Asian region. *Quaternary International* 118–119, 103–126.
Hubbard, N.N. (1995a) In search of regional paleoclimates: Australia, 18 000 yr BP. *Palaeogeography, Palaeoclimatology, Palaeoecology* 116, 167–188.
Hubbard, N.N. (1995b) An integrated method for reconstructing regional paleoclimates: Australia (18,000 yr bp). *Palaeogeography, Palaeoclimatology, Palaeoecology* 116, 141–166.
Jackson, S.T. and Overpeck, J.T. (2000) Responses of plant populations and communities to environmental changes of the Late Quaternary. *Paleobiology* 26, 194–220.
Jansson, R. and Dynesius, M. (2002) The fate of clades in a world of recurrent climatic change: Milankovitch oscillations and evolution. *Annual Review of Ecology and Systematics* 33, 741–777.
Jordan, G.J. (1995a) Early-Middle Pleistocene leaves of extinct and extant Proteaceae from western Tasmania, Australia. *Botanical Journal of the Linnean Society* 118, 19–35.
Jordan, G.J. (1995b) Extinct conifers and conifer diversity in the Early Pleistocene of western Tasmania. *Review of Palaeobotany and Palynology* 84, 375–387.
Jordan, G.J. (1997a) Contrasts between the climatic ranges of fossil and extant taxa: causes and consequences for palaeoclimatic estimates. *Australian Journal of Botany* 45, 465–474.
Jordan, G.J. (1997b) Evidence of Pleistocene plant extinction and diversity from Regatta Point, western Tasmania, Australia. *Botanical Journal of the Linnean Society* 123, 45–71.
Jordan, G.J. (1999) A new Early Pleistocene species of Nothofagus and the climatic implications of co-occurring Nothofagus fossils. *Australian Systematic Botany* 12, 757–765.

Jordan, G.J. and Hill, R.S. (1991) Two new Banksia species from Pleistocene sediments in western Tasmania. *Australian Systematic Botany* 4, 499–511.
Jordan, G.J., Bromfield, K.E., Sniderman, J.M.K. and Crayn, D. (2007) Diverse fossil epacrids (Styphelioideae; Ericaceae) from Early Pleistocene sediments at Stony Creek Basin, Victoria, Australia. *International Journal of Plant Sciences* 168, 1359–1376.
Jouzel, J. (2004) *EPICA Dome C Ice Cores Deuterium Data. IGBP PAGES/World Data Center for Paleoclimatology Data Contribution Series # 2004-038*. NOAA/NGDC Paleoclimatology Program, Boulder CO, USA.
Kemp, J. and Rhodes, E. (2010) Episodic fluvial activity of inland rivers in southeastern Australia: palaeochannel systems and terraces of the Lachlan River. *Quaternary Science Reviews* 29, 732–752.
Kershaw, A.P. (1980) Evidence for vegetation and climatic change during the Quaternary. In *The Geology and Geophysics of Northeastern Australia*, ed. R.A. Henderson and P.J. Stephenson. Geological Society of Australia: Brisbane, pp. 398–402.
Kershaw, A.P. (1981) Quaternary vegetation and environments. In *Ecological Biogeography in Australia*, ed. A. Keast. Dr Junk: The Hague, pp. 83–101.
Kershaw, A.P. (1988) Australasia. In *Vegetation History*, ed. B. Huntley and T.I. Webb. Kluwer Academic: New York, pp. 237–306.
Kershaw, A.P. (1994) Pleistocene vegetation of the humid tropics of northeastern Queensland, Australia. *Palaeogeography, Palaeoclimatology, Palaeoecology* 109, 399–412.
Kershaw, A.P. (1997) Environments of mainland southeastern Australia at the climatic extremes of the last glacial cycle: evidence from pollen. In *Climates of the Past*, ed. J. Mecco and N. Petit-Maire. Servicio de Publicaciones: Canary Islands, pp. 115–122.
Kershaw, A.P. and van der Kaars, S. (2012) Australia and the southwest Pacific. In *Quaternary Environmental Change in the Tropics*, ed. S.E. Metcalfe and D.J. Nash. Wiley-Blackwell: Chichester, pp. 236–262.
Kershaw, A.P., Baird, J.G., D'Costa, D.M. *et al.* (1991a) A comparison of long Quaternary pollen records from the Atherton and Western Plains volcanic provinces, Australia. In *The Cainozoic in Australia: A Re-appraisal of the Evidence*, ed. M.A.J. Williams, P. De Deckker and A.P. Kershaw. Geological Society of Australia: Sydney, pp. 288–301.
Kershaw, A.P., D'Costa, D.M., McEwan Mason, J.R.C. and Wagstaff, B.E. (1991b) Palynological evidence for Quaternary vegetation and environments of mainland southeastern Australia. *Quaternary Science Reviews* 10, 391–404.
Kershaw, A.P., Martin, H.A. and McEwan Mason, J.R.C. (1994) The Neogene: a period of transition. In *History of the Australian Vegetation: Cretaceous to Recent*, ed. R.S. Hill. Cambridge University Press: Cambridge, pp. 299–327.
Kershaw, A.P., Quilty, P.G., David, B. *et al.* (2000) Palaeobiogeography of the Quaternary of Australasia. *Memoir of the Association of Australasian Palaeontologists* 23, 471–516.
Kershaw, A.P., van der Kaars, S. and Moss, P.T. (2003) Late Quaternary Milankovitch-scale climatic change and variability and its impact on monsoonal Australasia. *Marine Geology* 201, 81–95.
Kershaw, A.P., D'Costa, D.M., Tibby, J. *et al.* (2004) The last million years around Lake Keilambete, western Victoria. *Proceedings of the Royal Society of Victoria* 116, 93–104.
Kershaw, A.P., Bretherton, S.C. and van der Kaars, S. (2007) A complete pollen record of the last 230 ka from Lynch's Crater, north-eastern Australia. *Palaeogeography, Palaeoclimatology, Palaeoecology* 251, 23–45.

Kuhnt, W., Holbourn, A., Xu, J. *et al.* (2015) Southern hemisphere control on Australian monsoon variability during the late deglaciation and Holocene. *Nature Communications* 6, 5916.

Lange, R.T. (1982) Australian Tertiary vegetation. In *A History of Australasian Vegetation*, ed. J.M.B. Smith. McGraw-Hill: Sydney, pp. 44–89.

Lewis, S.E., Sloss, C.R., Murray-Wallace, C.V. *et al.* (2013) Post-glacial sea-level changes around the Australian margin: a review. *Quaternary Science Reviews* 74, 115–138.

Lisiecki, L.E. and Raymo, M.E. (2005) A Pliocene-Pleistocene stack of 57 globally distributed benthic 18O records. *Paleoceanography* 20, PA1003.

Macphail, M.K. (1997) Late Neogene climates in Australia: fossil pollen- and spore-based estimates in retrospect and prospect. *Australian Journal of Botany* 45, 425–464.

Macphail, M.K. (1999) Palynostratigraphy of the Murray Basin, inland southeastern Australia. *Palynology* 23, 197–240.

Macphail, M.K. (2007) Australian Palaeoclimates: Cretaceous to Tertiary – a review of palaeobotanical and related evidence to the year 2000. *CRC LEME Special Volume Open File Report* 151, 266.

Macphail, M.K. and Truswell, E.M. (1993) Palynostratigraphy of the Bookpurnong Beds and related late Miocene-early Pliocene facies in the central West Murray Basin; Part 2, spores and pollens. *AGSO Journal of Australian Geology and Geophysics* 14, 383–409.

Macphail, M.K., Jordan, G.J. and Hill, R.S. (1993) Key periods in evolution of the flora and vegetation in western Tasmania I. the Early–Middle Pleistocene. *Australian Journal of Botany* 41, 673–707.

Macphail, M.K., Alley, N.F., Truswell, E.M. and Sluiter, I.R.K. (1994) Early Tertiary vegetation: evidence from spores and pollen. In *History of the Australian Vegetation: Cretaceous to Recent*, ed. R.S. Hill. Cambridge University Press: Cambridge, pp. 189–261.

Macphail, M.K., Fifield, L.K., Pillans, B. *et al.* (2015) Lake George revisited: new evidence for the origin and evolution of a large closed lake, Southern Tablelands, NSW, Australia. *Australian Journal of Earth Sciences* 6, 853–871.

Markgraf, V., Dodson, J.R., Kershaw, A.P. *et al.* (1992) Evolution of late Pleistocene and Holocene climates in the circum-South Pacific land areas. *Climate Dynamics* 6, 193–211.

Martin, H.A. (1978) Evolution of the Australian flora and vegetation through the Tertiary: evidence from pollen. *Alcheringa* 2, 181–202.

Martin, H.A. (1987) Cainozoic history of the vegetation and climate of the Lachlan River region, New South Wales. *Proceedings of the Linnean Society of New South Wales* 109, 213–257.

Martin, H.A. (1994) Australian Tertiary phyogeography: evidence from palynology. In *History of the Australian Vegetation: Cretaceous to Recent*, ed. R.S. Hill. Cambridge University Press: Cambridge, pp. 104–142.

McEwan Mason, J.R.C. (1991) The late Cainozoic magnetostratigraphyh and preliminary palynology of Lake George, New South Wales. In *The Cainozoic of Australia: a re-appraisal of the evidence*, ed. M.A.J. Williams, P. De Deckker and A.P. Kershaw. Geological Society of Australia: Sydney, pp. 195–209.

McLaren, S. and Wallace, M.W. (2010) Plio-Pleistocene climate change and the onset of aridity in southeastern Australia. *Global and Planetary Change* 71, 55–72.

Miller, G.H., Fogel, M.L., Magee, J.W. *et al.* (2005a) Ecosystem collapse in Pleistocene Australia and a human role in megafaunal extinction. *Science* 309, 287–290.

Miller, G.H., Mangan, J., Pollard, D. *et al.* (2005b) Sensitivity of the Australian monsoon to insolation and vegetation: implications for human impact on continental moisture balance. *Geology* 33, 65–68.

Mooney, S.D., Harrison, S.P., Bartlein, P.J. *et al.* (2011) Late Quaternary fire regimes in Australia. *Quaternary Science Reviews* 30, 28–46.

Moss, P.T. and Kershaw, A.P. (2000) The last glacial cycle from the humid tropics of northeastern Australia: comparison of a terrestrial and a marine record. *Palaeogeography, Palaeoclimatology, Palaeoecology* 155, 155–176.

Moy, C.M., Seltzer, G.O., Rodbell, D.T. and Anderson, D.M. (2002) Variability of El Niño/Southern Oscillation activity at millennial timescales during the Holocene epoch. *Nature* 420, 162–165.

Nix, H.A. (1991) An environmental analysis of Australian rainforests. In *The Rainforest Legacy: Australian National Rainforests Study. Volume 2. Flora and Fauna of the Rainforests*, ed. G. Werren and A.P. Kershaw. Australian Government Publishing Service: Canberra, pp. 1–11.

Ogden, J. and Stewart, G.H. (1995) Community dynamics of the New Zealand conifers. In *Ecology of the Southern Conifers*, ed. N.J. Enright and R.S. Hill. Melbourne University Press: Melbourne, pp. 81–119.

Petherick, L., Bostock, H., Cohen, T. *et al.* (2013) Climatic records over the past 30 ka from temperate Australia: a synthesis from the Oz-INTIMATE workgroup. *Quaternary Science Reviews* 74, 57–77.

Pickett, E.J., Harrison, S.P., Hope, G. *et al.* (2004) Pollen-based reconstructions of biome distributions for Australia, South East Asia and the Pacific (SEAPAC region) at 0, 6000 and 18 000 14C yr B.P. *Journal of Biogeography* 30, 1381–1444.

Prentice, I.C. (1985) Pollen representation, source area, and basin size: toward a unified theory of pollen analysis. *Quaternary Research* 23, 76–86.

Ravelo, A.C., Andreasen, D.H., Lyle, M. *et al.* (2004) Regional climate shifts caused by gradual global cooling in the Pliocene epoch. *Nature* 429, 263–267.

Raymo, M.E. (1994) The initiation of northern-hemisphere glaciation. *Annual Review of Earth and Planetary Sciences* 22, 353–383.

Read, J., Hope, G.S. and Hill, R.S. (2005) Phytogeography and climate analysis of *Nothofagus* subgenus *Brassospora* in New Guinea and New Caledonia. *Australian Journal of Botany* 53, 297–312.

Reeves, J.M., Barrows, T.T., Cohen, T.J. *et al.* (2013a) Climate variability over the last 35 000 years recorded in marine and terrestrial archives in the Australian region: an OZ-INTIMATE compilation. *Quaternary Science Reviews* 74, 21–34.

Reeves, J.M., Bostock, H.C., Ayliffe, L.K. *et al.* (2013b) Palaeoenvironmental change in tropical Australasia over the last 30 000 years: a synthesis by the OZ-INTIMATE group. *Quaternary Science Reviews* 74, 97–114.

Robbie, A. and Martin, H.A. (2007) The history of the vegetation from the last glacial maximum at Mountain Lagoon, Blue Mountains, New South Wales. *Proceedings of the Linnean Society of New South Wales* 128, 57–80.

Röthlisberger, R., Bigler, M., Wolff, E.W. *et al.* (2005) *EPICA Dome C Ice Core nss-Ca and Na Data. IGBP PAGES/World Data Center for Paleoclimatology Data Contribution Series #2005-046*. NOAA/NGDC Paleoclimatology Program, Boulder CO, USA.

Ruddiman, W.F. and Raymo, M.E. (1988) Northern hemisphere climate regimes during the past 3-Ma: possible tectonic connections. *Philosophical Transactions of the Royal Society of London Series B, Biological Sciences* 318, 411–430.

Rule, S., Brook, B.W., Haberle, S.G. *et al.* (2012) The aftermath of megafaunal extinction: ecosystem transformation in Pleistocene Australia. *Science* 335, 1483–1486.

Saltré, F., Rodríguez-Rey, M., Brook, B.W. *et al.* (2016) Climate change not to blame for late Quaternary megafauna extinctions in Australia. *Nature Communications* 7, 10511.

Singh, G. (1982) Environmental upheaval: the vegetation of Australasia during the Quaternary. In *A History of Australasian Vegetation*, ed. J.M.B. Smith. McGraw-Hill: Sydney, pp. 90–108.

Singh, G. and Geissler, E.A. (1985) Late Cainzoic history of vegetation, fire, lake levels and climate, at Lake George, New South Wales, Australia. *Philosophical Transactions of the Royal Society of London B* 311, 379–447.

Singh, G., Kershaw, A.P. and Clark, R. (1981) Quaternary vegetation and fire history in Australia. In *Fire and the Australian Biota*, ed. A.M. Gill, R.H. Groves and I.R. Noble. Australian Academy of Science: Canberra, pp. 23–54.

Sniderman, J.M.K. (2011) Early Pleistocene vegetation change in upland south-eastern Australia. *Journal of Biogeography* 38, 1456–1470.

Sniderman, J.M.K. and Haberle, S.G. (2012) Fire and vegetation change during the Early Pleistocene in southeastern Australia. *Journal of Quaternary Science* 27, 307–317.

Sniderman, J.M.K., Pillans, B., O'Sullivan, P.B. and Kershaw, A.P. (2007) Climate and vegetation in southeastern Australia respond to southern hemisphere insolation forcing in the late Pliocene-early Pleistocene. *Geology* 35, 41–44.

Sniderman, J.M.K., Porch, N. and Kershaw, A.P. (2009) Quantitative reconstruction of Early Pleistocene climate in southeastern Australia and implications for atmospheric circulation. *Quaternary Science Reviews* 28, 3185–3196.

Sniderman, J.M.K., Jordan, G.J. and Cowling, R.M. (2013) Fossil evidence for a hyperdiverse sclerophyll flora under a non–Mediterranean-type climate. *Proceedings of the National Academy of Science* 110, 3423–3428.

Sweller, S. and Martin, H.A. (2001) A 40 000 year vegetation history and climatic interpretations of Burraga Swamp, Barrington Tops, New South Wales. *Quaternary International* 83–85, 233–244.

Truswell, E.M. (1993) Vegetation changes in the Australian Tertiary in response to climatic and phytogeographyic forcing factors. *Australian Systematic Botany* 6, 533–558.

Turney, C.S.M. and Hobbs, D. (2006) ENSO influence on Holocene Aboriginal populations in Queensland, Australia. *Journal of Archaeological Science* 33, 1744–1748.

Visser, K., Thunell, R. and Stott, L. (2003) Magnitude and timing of temperature change in the Indo-Pacific warm pool during deglaciation. *Nature* 421, 152–155.

Wagstaff, B.E., Kershaw, A.P., O'Sullivan, P.B. *et al.* (2001) An Early to Middle Pleistocene palynological record from the volcanic crater of Pejark Marsh, Western Plains of Victoria, southeastern Australia. *Quaternary International* 83–5, 211–232.

Wang, Y.J., Cheng, H., Edwards, R.L. *et al.* (2005) The Holocene Asian monsoon: links to solar changes and north Atlantic climate. *Science* 308, 854–857.

Williams, A.N., Mooney, S.D., Scisson, S.A. and Marlon, J. (2015) Exploring the relationship between Aboriginal population indices and fire in Australia over the last 20 000 years. *Palaeogeography, Palaeoclimatology, Palaeoecology* 432, 49–57.

Wroe, S., Field, J.H., Archer, M. *et al.* (2013) Climate change frames debate over the extinction of megafauna in Sahul (Pleistocene Australia–New Guinea). *Proceedings of the National Academy of Sciences of the United States of America* 110, 8777–8781.

WTMA (2016) *Wet Tropics Management Authority homepage*. Wet Tropics Management Authority: Cairns. Available at: http://www.wettropics.gov.au (accessed February 2016).

Zachos, J., Pagani, M., Sloan, L.C. *et al.* (2001) Trends, rhythms, and aberrations in global climate 65 Ma to present. *Science* 292, 686–693.

4 Aboriginal People and Australia's Vegetation: Past and Current Interactions

EMILIE ENS, FIONA WALSH AND PHILIP CLARKE

Plants have had significant and multiple roles in past and present Aboriginal people's lives. Aboriginal people extracted the food, medicine and material resources they needed from their immediate environment. Seasonal changes in plants and animals were markers in regional calendars and today they continue to influence some Aboriginal people's resource use, environmental management and movements. The first section of this chapter presents a selection of the many Aboriginal uses of plants from across Australia. Aboriginal people also secured resources from distant environments through travel and trade networks. Like people the world over and today, they manipulated their resources and their environments at local to landscape scales, for example through fire and harvesting practices, to supply material needs and suit their tastes and preferences. The second section of this chapter presents multidisciplinary evidence that illustrates similarities and differences in modes and intent of plant resource procurement, "management" and production across Australia. Today, Aboriginal people maintain a variety of links to their ancestral estates. These links are influenced by policy, legislation and locally-driven aspirations. Contemporary influences on Australia's vegetation occur through: continued customary practices for resource, ceremonial and social purposes; involvement in government or industry supported environmental management programs; and small-scale commercial bush foods, art and craft enterprises. An overview of some contemporary practices is presented in the third section of the chapter to present a summary of Aboriginal people's past to present uses and influences on Australia's vegetation.

4.1 Aboriginal demography and vegetation change

Before European settlement, Aboriginal people lived in loose tribal groups that clustered in land-holding clans (ancestral estates). Groups had responsibilities to maintain the resources essential for survival. Over 250 distinct languages were spoken, but now, most traditional languages are considered endangered or threatened (McConvell and Thieberger 2001). Knowledge related to plant species is encoded in spoken and now written languages.

The past, present and future influences of Aboriginal people on Australia's landscapes are integral to any account of Australia's vegetation (Altman *et al.* 2007; Ens *et al.* 2014). The distribution and density of pre-colonial Aboriginal occupation would have had local to landscape-scale impacts on Australia's vegetation. 'Traditional' sociopolitical structures have radically changed through post-colonial times but

Australian Vegetation, Third Edition, ed. David A. Keith. Published by Cambridge University Press. © Cambridge University Press 2017.

Citation: Ens, E., Walsh, F. and Clarke, P. (2017). Aboriginal People and Australia's Vegetation: Past and Current Interactions. In D. A. Keith, ed., *Australian Vegetation*, 3rd edn. Cambridge: Cambridge University Press, pp. 89–112.

many elements persist and continue to influence Australia's extant vegetation composition and change. The role of Aboriginal rangers and Indigenous Protected Areas in contemporary conservation has become a significant and increasing influence in contemporary conservation.

Traditional Australian Aboriginal belief systems recognise that human origins lie in 'Dreamings' that tell of when people and environments were created by the ancestral beings (Rose 1996; Thorne 2005). Archaeological evidence shows that the human ancestors of modern Aboriginal people arrived in Australia more than 50 000 years ago, during an era of low sea level, and successive waves of immigration followed (Mulvaney and Kamminga 1999; Hiscock 2007). From this time on, analysis of Aboriginal impacts on Australia's vegetation is complicated by climatic, environmental and human variations.

Reconstructions of population fluctuations based on radiocarbon analyses suggest founder populations of 1000–2000 people. Densities may have remained low until the relative climatic stability of the early Holocene (about 12 000 years ago). Archaeological evidence shows intensification of Aboriginal occupation during the late Holocene in regions that are now desert (Veth *et al.* 2005) and rainforest (Ferrier and Cosgrove 2012). Pre-colonial population estimates of approximately 500 000 people (Taylor 2005) may underestimate numbers due to the disease-related mortalities that preceded European records. Some evidence suggests that Australian Aboriginal populations may have peaked at about 1.2 million people in AD 1500 (Williams 2013).

In 2011, Australia's Indigenous population was estimated at 670 000; 3% of Australia's total population (ABS 2011). This figure is projected to rise to 900 000 by 2026. Aboriginal people continue to occupy remote, regional and urban places with a trend of increasing urbanisation. Today, over 33% of Australia's land surface is owned by Aboriginal people (mostly in remote central and northern Australia) and up to 75% of land has declared Indigenous interests (Altman and Markham 2014).

4.2 Aboriginal uses of plants

Across Australia, about 4000 plant species were, and some still are, used by Aboriginal people as food and medicine, equating to about 20% of named Australian vascular plants (Isaacs 1987). Many more species are known to have uses, including as utensils, shelter and other materials, food for game and totems (Clarke 2007). For example, on Arrente lands, over 30% vascular plants have uses (Walsh and Douglas 2011). Aboriginal people, such as Senior Marra expert Betty Roberts (*pers. comm.*) have said, 'Every plant has a use!' Such statements reflect both Aboriginal people's resourcefulness and understanding of the ecological connections between species.

European explorers in Australia were interested in Australian plants as food to survive and as potential future crops. Plant collectors often relied physically and intellectually on Aboriginal guides (Clarke 2008b, 2010). The involvement of Aboriginal people in the botanical 'discovery' of Australia's plants is illustrated by the common use of many plant words derived from Aboriginal language (Dixon *et al.* 2006). For instance, the word waratah (*Telopea speciosissima*) is derived from *Warada* in the Dharug language of the Sydney region, and mallee (*Eucalyptus* species) is probably from *Mali* in the Wemba Wemba language of western Victoria (Dixon *et al.* 2006; see also Yates *et al.*, Chapter 23, this volume). Often Aboriginal people and their knowledge were critical to the European botanical 'discovery' of Australia's plants and plant populations.

4.2.1 Biogeographic regionalisation of bush food plants

Dominant vegetation types and, hence, Aboriginal plant food types, differ across arid, temperate and tropical Australia. Spinifex, tussock grass and *Acacia* shrublands dominate arid and semi-arid ecosystems (Nano *et al.*, Chapter 25, this volume; Wardle and Nano, Chapter 26, this volume). Unpredictable rainfall shapes the arid zone flora with a predominance of species that rapidly set seed after rain (Morton *et al.* 2011). Correspondingly, seed harvesting dominated the food production systems of Aboriginal people in inland Australia (Smith 2013). In arid regions, seed species commonly made into edible dampers included wild millet (*Panicum decompositum*) and native woolly butt (*Eragrostis eriopoda*) (Latz 1995). The seeds of many wattle (*Acacia*) species were roasted to eat as a ground paste and some were eaten as green beans (Clarke 2007). The sporocarps of the nardoo or water fern (*Marsilea drummondii*) were heavily utilised as food during droughts in eastern central Australia but they are highly toxic and so require extensive preparation (Clarke 2013). The use of nardoo as food source, in the absence of Aboriginal knowledge and experience, led to the deaths of the trans-Australian explorers Burke and Willis in 1861.

By contrast, southern Australia, where humid climates produce more standing and flowing water, has a higher diversity of species with tuberous and underground organs for starch storage and both vegetative and sexual reproduction. According to Gott (2008), in south-eastern Australia, the diet of Aboriginal people was root-dominated and seeds comprised a minor dietary component. Root species commonly eaten in south-eastern Australia include the yam daisy (*Microseris lanceolata*) and cumbungi (*Typha domingensis*) (Gott 1982). In south-eastern Queensland and north-eastern New South Wales, the bunya pine (*Araucaria bidwillii*) seasonally produced enormous crops of edible nuts that enabled Aboriginal groups from a wide area to gather for feasts and ceremonies (Clarke 2007).

In tropical northern Australia, storage roots, fruits and edible nuts of many species were and in some cases still are a major part of Aboriginal diets (Crawford 1982; Russell-Smith *et al.* 1997). Many different types of yams and underground storage organs are still eaten today, including cheeky yam (*Dioscorea bulbifera*), long yam (*Dioscorea transversa*) and water chestnut (*Eleocharis dulcis*) (Yunupingu *et al.* 1994; Raymond *et al.* 1999). Fruits include the green plum (*Buchanania obovata*) and billy goat plum (*Terminalia ferdinandiana*) (Yunupingu *et al.* 1994; Raymond *et al.* 1999). Across the monsoonal tropics, edible nuts of the cycads (*Cycas* and *Bowenia* species) were widely used and stored, although they required processing and leaching of poisonous toxins before consumption. According to Pedley (1993), at the time of colonisation, 10–14% of Aboriginal wet–dry rainforest people's diet comprised edible toxic nuts including black walnut (*Endiandra palmerstonii*) and black bean (*Castanospermum australe*). Similar toxic nuts requiring processing were also used in southern Australia including the cycads from the *Macrozamia* and *Lepidozamia* genera (Asmussen 2011).

While biogeographic regions typically had dominant food groups, people sought a diversity of food types from local sources or through trade. For example, despite desert regions being seed-food dominated, tubers, such as those of the bush onion (*Cyperus bulbosus*) and bush potato (*Ipomoea costata*), and fruits (e.g. *Solanum* spp.) were also favoured. Despite tuber-dominated diets in temperate regions, in south-eastern Australia, fruits such as muntry (*Kunzea pomifera*) were prized (Clarke 2013). Additionally, a variety of nectar and honey, green food, edible gum exudates and other minor but nutritionally important plant food groups occurred in each bioregion. Consumption of Australian native truffles and other edible fungi (Kalotas 1996) may have been significant and requires further study.

4.2.2 Bush medicine plants

Aboriginal groups had a developed and refined pharmacopoeia of medicinal plants. There is much documentation of Aboriginal bush medicines from the language group (e.g. Nabarula *et al.* 1978; Rose 1987; Everard *et al.* 2002) to national (e.g. Webb 1969; Isaacs 1987; Low 1990; Clarke 2008a) scale. Plant-based medicines treated fever, congestion, headache, skin sores, swollen or aching limbs, digestive problems and more (Box 4.1). Medicinal

Box 4.1 Blending traditional and scientific knowledge to create local Aboriginal plant-based enterprises

Yirralka Miyalk (women) rangers

The Yolngu people of north-east Arnhem Land maintain cultural knowledge, including medicinal knowledge of the native plants. Our ancestors have passed this knowledge down to us through the generations. Our native plants are also entwined in our kinship system and have a moiety, either Dhuwa or Yirritja. This demonstrates the interconnectedness between people and Country. The plants are in our song lines, and certain species are the responsibility of particular clan groups. We know of the plants that can for example heal skin wounds, make a cold or flu pass, and heal stings and wounds from insects and animals.

Being Yirralka rangers, we miyalk (women) rangers have responsibility for certain projects. We do jobs like care for plant nurseries, develop our herbarium and record Indigenous Ecological Knowledge. One of our plant projects is the transfer of our medicinal knowledge into products that we can sell.

We have chosen three plants on which to base our 'Yirralka Rangers Bush Products'. The first is Nambarra (*Melaleuca dealbata*) – a paperbark that is common to the floodplains of this region. The Dhudi djapu clan is responsible for this plant, which belongs to the Dhuwa moiety. This is what we use for a cold or flu infection, as its strong scent helps clear the nose and mouth. The second plant we use in our products is gadayka (*Eucalyptus tetrodonta*), which is widespread across tropical northern Australia (Figure 4.1). The clan responsible for gadayka is the Marrakulu clan, also from the Dhuwa moiety. This is one of the plants we use to rid our skin of bad irritations, such as scabies. The third plant we call Butjiringanging, a Yirritja moiety monsoon vine forest plant with the Latin name *Litsea glutinosa*. This plant we can apply to help skin conditions and other complaints such as sore muscles.

To make the products, we boil the plant matter and infuse it into products that balanda (non-Indigenous people) recognise and often use like beeswax and oils. We sell lip balms, bath scrubs, soap and body rubs.

Figure 4.1 Senior Yirralka Miyalk ranger Gurrundul Marika cutting bark of gadayka (*Eucalyptus tetrodonta*).
Photo: Emilie Ens.

We hold with our minds and our tongues the traditional way of preparing these plants. We learn, know, use and pass on this knowledge. The enterprise allows us to share what we are willing to show people who are not Yolngu. We are building an enterprise from this knowledge and experience. It is also a way to maintain this knowledge for future Yolngu. This enterprise also provides a way for us to learn about balanda knowledge systems and ways of business too.

concoctions included washes, drinks, rubs, salves and aromatherapy. For example, drinks were heated water with plant additives, now referred to in Aboriginal English as 'tea'. Some plant materials were heated, ground then rubbed or massaged into swollen parts of the patient's body. Some plant extracts were suspended or activated in animal fats. Alternately, patients inhaled aromatic and volatile oils in the steam and smoke of purposeful hearth fires. For example, the apple-bush (*Pterocaulon serrulatum*) of northern Australia is a widespread aromatic inhalant. Rock fuchsia-bush (*Eremophila freelingii*) produced medicinal salve and herbal teas in central Australia (Barr *et al.* 1988) and old man beard (*Clematis microphylla*) was made into a poultice in southern Australia (Clarke 2015). Narcotics and stimulants were highly prized, widely sought and traded. They included Solanaceae species such as pituri (*Duboisia hopwoodii*) and the wild tobaccos (*Nicotiana* spp.). Pharmacological research has revealed the bioactive properties and active constituents of some species (Barr *et al.* 1988; Ghisalberti 1994). While some species are under, or warrant, further investigation, there are serious considerations of informed consent, intellectual property and benefit sharing to be negotiated (Davis *et al.* 2009).

4.2.3 Other plant-based economic materials

Aboriginal tool-makers chose their timber from a wealth of different tree species. They selected timbers with specific properties to enhance the functioning of the type of tool being made (Clarke 2007, 2012). For instance, in the tropics the lightweight timber of bean tree (*Erythrina vespertilio*) was preferred for dugout canoe construction. The heavy dense wood of the northern ironwood (*Erythrophleum chlorostachys*) made strong digging tools and harpoon points (Yunupingu *et al.* 1994; Raymond *et al.* 1999). Across Australia, bark and broadleaf sheeting cut from *Eucalyptus* and paperbark (*Melaleuca* and *Leptospermum*) species were used to construct shelters or make useful containers (Clarke 2012).

Plant fibre was and remains a fundamental artefact-making material across Aboriginal Australia. Fibres are woven into bags, basket, mats, fishing and game nets, harpoon lines, ornaments, bindings and watercraft lashings. In the tropics, a strong weather-resistant string was made from the inner bark of fig (*Ficus* species) aerial roots. Leaf fibre for baskets and mats is still commonly stripped from the leaf shoots of the screw palm (*Pandanus spiralis*) and sand palm (*Livistona humilis*) (Yunupingu *et al.* 1994; Raymond *et al.* 1999). In the arid zone, strong fibre from New Holland rattle-pod (*Crotalaria novae-hollandiae*) was used to make string. Fibre sandals were made from the bark of bird-flower bush (*Crotalaria cunninghamii*) (Latz 1995). Aboriginal basket-making traditions in south-eastern and eastern Australia utilised Cyperaceae stems, such as the spiny-headed sedge (*Cyperus gymnocaulos*) and the many-flowered mat-rush (*Lomandra multiflora*) (Clarke 2015). Due to its importance to recent Indigenous fibre craft traditions, the spiny-headed sedge is now cultivated on the New South Wales south coast at sites where it did not occur in early pre-colonial times (Nash 2012).

Artists derived many of their art materials from plant sources. For instance, on the Adelaide Plains in South Australia the red paint for bark shields was made from the tuber skin of scented sundew (*Drosera whittakeri*) (Clarke 2013). Across Arnhem Land, the crushed roots of scarlet bloodroot (*Haemodorum coccineum*) yielded a dye to colour palm fibre red and purple (Raymond *et al.* 1999; Hamby 2005). Ornaments were decorated with a variety of colourful seeds including the red-brown seeds of the bean tree (*Erythrina vespertilio*) and the red and black seeds of gidgee-gidgee (*Abrus precatorius*) in the tropics (Yunupingu *et al.* 1994; Clarke 2012).

Adhesive cements prepared from plant gums and resins were also crucial to production of Aboriginal tools and artefacts such as spears and spear throwers. Adhesive cements were essential to join objects together, to caulk water-holding vessels and watercraft and to sculpt. Across the arid zone, gum gathered from spinifex (*Triodia* species) was intensively used for these purposes (Pitman and Wallis 2012). In temperate Australia, grass tree resin was preferred and in the tropics, northern ironwood resin was utilised (Clarke 2012). The ease to which organic materials can be shaped and absorb shock set them apart from mineral substances.

Plants also had many ephemeral uses. For example, the juice from the bracket tree orchid (*Dendrobium affine*) was utilised as paint fixative, and the crushed bark of the freshwater mangrove (*Barringtonia acutangula*) used to make fish poison (Raymond *et al.* 1999). The compendium of useful resource plants is diverse and varies across bioregions, but has some consistent features across the continent.

Aboriginal people sought to maintain both the diversity of useful resource species available to them and the volume of certain species needed in large quantities. Species use is the basis for understanding the impacts of Aboriginal activities upon Australia's vegetation. Aboriginal people consciously or sometimes inadvertently influenced the proliferation of certain plants species. It is also likely they may have contributed to the decline of other species through displacement or range reduction, although more research is required to substantiate the widespread anecdotal reports.

4.3 Aboriginal manipulation and 'management' of plants and vegetation in pre-European times

The intensity of Aboriginal people's pre-colonial influence on populations of plants, animals and ecosystems is still debated. Some scholars argue that European settlers ignored the influence of Aboriginal people on Australian ecosystems or assumed it was minimal. Perpetuation of the myth of *terra nullius* allowed for unconstrained pastoral expansions (Langton 1998; Pascoe 2014). However, a growing body of evidence has revealed that Aboriginal people manipulated ecosystems to enhance food and resource production (Kimber 1976; Bowman 1998; Gerritson 2008; Gammage 2011; Pascoe 2014). It is likely that species relocations, burning, water management and other Aboriginal activities shaped Australian pre-colonial environments to varying degrees in time and space, although much more remains to be investigated.

Aboriginal deference to practices encoded in Dreaming beliefs reflects another form of intent. Cultural strategies related to ceremonies, site visits and protocols around plants as sentient beings are believed by Aboriginal people to shape the production and abundance of some species (Keen 2004). We respect these cultural practices but they are not discussed further here. Like many human endeavours, it is likely there was a dynamic variety of intent in space and time.

The intensity of people's manipulation of resource species and the effects of Aboriginal practices on species, vegetation structure and composition can be identified from the late Holocene to pre-European times using archaeological, biological, historical and ethnographic evidence. Aboriginal occupation patterns, technologies and economies were a function of the carrying capacity of the ecosystems inhabited. Human processes interactively affected and were affected by weather, water availability and plant and animal densities (Ulm 2013).

4.3.1 Species-scale manipulation

4.3.1.1 Species harvest

Plant harvest had different impacts depending on the species' ecology and the particular human actions. In pre-European times, Aboriginal people gathered food species in bulk quantities by the coolamon full (e.g. native woolly butt, *Eragrostis eriopoda*). For species that required intensive processing (e.g. yandying, threshing, washing, leaching, grinding, pounding), people preferred to gather a species when abundant rather than from sparsely productive patches. Sometimes patches were avoided to allow for replenishment.

There are few reconstructive assessments of plant harvest in relation to population size. Smith (2013) used early contact records to quantify harvest of the prized narcotic, pituri (*Duboisia hopwoodii*). In the central and western deserts, individual men each carried about 32 kg of dried leaves and stems on round trips of 800–1200 km. This equated to 900–1600 kg or about 2000 plants transported annually. Comparing this to a modern population of 24 000 plants, Smith (2013) concluded that this harvest was likely to be within the pituri population's capacity.

Other highly desirable species may have been both intensively harvested and managed. For example, root food species across Australia were manipulated to maintain or even promote populations. There is evidence from Arnhem Land and Cape York that Aboriginal people harvested yams (e.g. *Dioscorea* spp.) by removing the basal part of the tuber and leaving the top portion with its emergent vine *in situ* (Hynes and Chase 1982; Russell-Smith *et al.* 1997). Such practices are referred to as 'domiculture' (Kimber 1976) and areas such as the littoral forest habitat of *Dioscorea sativa* var. *elongata* were described in modern creole as a 'garden for yams' (Hynes and Chase 1982).

In south-eastern Australia, about 10% of the flora in grasslands and woodlands consisted of root species. Thirteen genera and numerous species including yam-daisies (*Microseris* species) were dug for edible tubers, corms and bulbs (Gott 2008; Gott *et al.* 2015). Digging thinned patch densities, aerated and loosened soil, and also turned in litter, which increased soil nutrients. Digging and burning maintained the genetic diversity of yam daisy species (Gott 2008). In arid and semi-arid Australia, tuberous species were scarce and their harvest was also managed. The bush onion (*Cyperus bulbosus*) was an important plant food of the floodplains (Latz 1995). The combination of soil pounding, tilling then tuber thinning increased the vegetative reproduction of tubers and also created softer aerated soil beds suited to water filtration and seed-derived germination. These practices were widespread. For example, Arrernte and Manjilyjarra Aboriginal groups more than 1200 km apart each had practices specifying bush onion beds were hand pounded with a round flat stone rather than dug with a stick (Dobson *et al.* 2008; Walsh 2008). Pounding reduced fragmentation of the fine rhizomes with terminal tubers.

There is ethnographic evidence that various fruit species were also carefully harvested with protocols including taking portions not whole crops, not damaging unripe fruit and not breaking plant limbs (e.g. Russell-Smith *et al.* 1997). Certain biological characteristics reduced the risk of overharvest to some species, e.g. asynchronous ripening or both vegetative and sexual reproduction.

4.3.1.2 Species dispersal

There is evidence for pre-colonial translocations of plants by Australian Aboriginal people. Entire plants, vegetative parts or reproductive parts were moved. Plant species were gifted, traded or collected across hundreds of kilometres. These relocations ranged from campsites, across regions and the continent.

The local movement of useful plant parts contributed to increased species patch densities, diversity and proximity to some Aboriginal occupation sites. In tropical rainforests of eastern Australia, groves of high-value trees with edible fruit and nuts were found near Aboriginal camps and ceremonial grounds (Hynes and Chase 1982). Within coastal vegetation communities, midden sites near Stony Point in north-eastern Queensland exhibited exceptional diversity of high-value plant food species. Thirty-seven species including the favoured lady apple (*Syzygium suborbiculare*) occurred near middens due to domiculture practices (Hynes and Chase 1982). These researchers also proposed that humans influenced plant population phenotypes by preferencing certain tastes or sizes. From Aboriginal perspectives, the co-occurrence of ancestral camping sites and stands of important food species is evidence of the co-development between people, plants and sentient spiritual beings. Yanyuwa descriptions of habitats integrate landform, camping sites and plant resource associations (Bradley *et al.* 2006).

In plant biogeography studies, inter- and intra-continental human dispersal of plants may explain some disjunct Australian occurrences of useful species. Australian Aboriginal people's plant trade with maritime traders and explorers is at least 3000 years old (Bean 2007). Macassan traders planted tamarind (*Tamarindus indica*) in the eighteenth and nineteenth centuries which have been archaeologically linked to trepang (sea cucumber) processing sites in northern Australia (Russell 2004).

Linguistic and genetic research of boab (*Adansonia gregorii*) populations across the Kimberley showed high similarity and gene flow between geographically separated populations. Humans are proposed as gene vectors of boabs within the Kimberley and from Africa (Bell *et al.* 2014; Rangan *et al.* 2015). Similarly, microsatellite analyses of *Livistona* showed relationships between *L. mariae* in a central Australian river bed and its nearest phylogenetic relative *L. rigida* in the Top End (Kondo *et al.* 2012). With the small genetic divergence but long distance between these two *Livistona* species, early humans were suggested to have translocated this food resource species in the late Pleistocene (about 15 000 BP). Smith (2013) questioned whether *Livistona* was of sufficient value to warrant translocation, however, other *Livistona* species are well-known as principal food resources in the northern Kimberley (*L. eastonii*; Crawford 1982) and Top End (*L. benthamii* and *L. humilis*; Yunupingu *et al.* 1994). Similar phytogeographic studies are ongoing for several other large-fruited species in the Kimberley and eastern Australian rainforests. The results of these studies could blur distinctions between 'alien' and 'native' species and natural plant distributions, with implications for contemporary legislation and land management (Trudgen *et al.* 2012).

4.3.2 Ecosystem-scale manipulation

It is well established that Aboriginal people used fire across Australia for many purposes. Less well-known are the techniques of manipulating water sources and consequences for vegetation. Again, the spread, intensity and reasons for these manipulations by Aboriginal people changed through time and space. Generalisations are difficult and contested, requiring further enquiry.

4.3.2.1 Aboriginal burning regimes

There is an extensive patchwork of scholarship on the pattern and impacts of Aboriginal burning on Australia's ecosystems. These include national syntheses (Flannery 1994; Bowman 1998; Pyne 1998; Gammage 2011) and reviews for arid

(Latz 1995; Burrows *et al.* 2006; Hallam and Mulvaney 2014), tropical (Vigilante 2001; Hill *et al.* 2004; Russell-Smith *et al.* 2009) and temperate (Kohen 1996; Gott 2005) regions. Despite this, the precise nature of historical Aboriginal fire regimes, and the degree to which they varied from climatically determined regimes, remains uncertain for much of the continent (Mooney *et al.*, Chapter 3, this volume; Miller and Murphy, Chapter 5, this volume). Aboriginal fire–vegetation interactions need to be analysed at landscape to patch scales with consideration of inter- and intra-annual variability. Practices and consequences recorded in time and place cannot be assumed as transferable. The concepts of patchwork or mosaic burning emerged from arid regions (Latz and Griffin 1978) but they remain problematic concepts. The notion of mosaic burning across Australia's deserts is not universally supported as shown by explorer and early accounts of Aboriginal burning from the Simpson Desert and Lake Eyre Basin (Kimber and Friedel 2016). The occurrence of fire depended on vegetation and terrain type, weather, Aboriginal access to water and their reasons to use fire (e.g. communication) or not.

In arid regions, burning practices varied across desert systems. Western Desert peoples burnt for utilitarian purposes to create and maintain spinifex grasslands of different seral stages. They burnt to hunt game, promote preferred plant and animal food resources, open up walking access to Country, track animals in clearer sand and many other reasons (Kimber 1983; Burrows and Christensen 1991; Walsh 1990; Latz 2007). Fire was used to hunt sand goanna (*Varanus gouldii*) and small mammals and it provided habitat patchiness utilised by those species (Bird *et al.* 2013). Burning changed arid vegetation structure, at least in the short term (Burrows *et al.* 2006). For example, the alpha diversity of food resource plants was higher in recently burnt vegetation than older burns (Walsh 1990) with each seral or regeneration stage having a characteristic suite of resource species (Bird *et al.* 2005).

In the wet tropical rainforests of Queensland, Aboriginal burning modified forest structure or at least its boundaries (Metcalfe *et al.* 2014). In this vegetation system, research suggests that Aboriginal burns imposed a fine local-scale patterning on vegetation but regional-scale vegetation patterns were determined by environmental factors (Hill and Baird 2003). Within the rainforest, Kuku Yalanji people burnt and maintained small areas of open forest where *Cycas media*, *Dioscorea bulbifera* and other carbohydrate-rich resource species grew. These species and associated practices were likely to have persisted from Holocene times but affected only small areas. Hill and Baird (2003) challenged views that Aboriginal fires caused widespread contraction of rainforest areas in the Holocene.

In the mid-east coast subtropics, a vegetation mosaic of rainforest–eucalypt savanna–grassland is attributed to thousands of generations of skilful and persistent burning (Butler *et al.* 2014). The rainforest included bunya pine (*Araucaria bidwillii*) whose large seeds attracted huge gatherings of Aboriginal people. Grasslands in this area were apparently cultural creations maintained over millennia to provide game habitats and travel routes.

Across tropical Australian savanna, people operated a multi-faceted burning management system (Preece 2002; Russell-Smith *et al.* 2009; Vigilante *et al.* 2009). Analysis of explorer's records from the Kimberley, Top End and Cape York suggest that, generally speaking, burning in savanna ecosystems started not long after the monsoonal rains ceased throughout the dry season until the wet season began, although with more burning activity early rather than late in the dry season (Fensham 1997; Vigilante 2001; Preece 2002). Aboriginal fire regimes altered the structure and abundance of savanna woody vegetation to maintain productive plant food patches, accessible and habitable vegetation and pasture and habitat for kangaroos (Russell-Smith *et al.* 1997). Kangaroos utilise structural mosaics by feeding in open grassy plains or patches and resting in the shelter and shade of woodlands. Burning promoted grass health and reduced woody stems in hunting areas.

Similarly, some ethnohistorians interpret the extensive grassy woodland ecosystems in southern Australia at European settlement as deliberate creations of kangaroo habitat and hunting grounds by Aboriginal people (Kohen 1996; Gammage 2011). In south-east Australia, grasslands, where yam beds and root species were most abundant, could have been opened and maintained by Aboriginal burning strategies (Gott *et al.* 2015). However, these broad interpretations of human influence on landscape fire regimes and vegetation patterns overlook the strong association of grassy temperate grassy ecosystems with particular climatic and edaphic environments (Prober *et al.*, Chapter 17, this volume; Morgan *et al.*, Chapter 18, this volume), as well as the limited paleo-evidence of broad-scale changes in both fire and vegetation, independent of climate, after arrival of Aboriginal people (Mooney *et al.* 2011). In Tasmania, the pre-European distribution of sedgelands may have been the product of frequent Aboriginal burning (Bowman 1998). Colloff (2014) found limited evidence that Aboriginal burning structured the widespread and variable river red gum woodlands of south-eastern Australia. Yorta Yorta and other people combined fire, flooding and sustainable harvest to manage extensive reed beds along rivers within river red gum woodlands. The *Phragmites australis*, *Typha domingensis* and *T. orientalis* beds were highly valued resources for fibre, food carbohydrates, waterbird game, protection and other uses. Cumbungi (bullrush) beds were burnt to maintain young pliable reeds and shoots.

The biology and ecology of plant resource species and their responses to burn regimes shaped management practices by Aboriginal people. Specific burn strategies were applied for certain species depending on promotion, reduction or protection intent. In the sandy desert bioregions, at least eight *Solanum* species are early colonisers after fire and rainfall events. Clonal resprouting from rootstock was manipulated by Aboriginal burning (Walsh 1990; Latz 1998; Lee 2013). On Martu desert lands, burns were avoided near Quondong (*Santalum acuminatum*) trees and stands of *Grevillea eriostachya* when bearing nectar (Walsh 2008). Sometimes protective burning was dependent on individual plants or phenology, especially for high-value plant species (Russell-Smith *et al.* 1997). Across tropical savanna woodlands, groves of various fruit-bearing plant species were protected, for example, people back burnt near *Buchanania obovata* stands to prevent damage to tree structure and fruit production (Vigilante *et al.* 2004).

Aboriginal burn regimes were widespread, variable in relation to vegetation types and seasonally diverse. Mosaic patterns were a cumulative consequence of the strategies applied for many species if not a deliberately created patchwork, in combination with unplanned fires.

4.3.3 Wetland management

Aboriginal people 'managed' some wetlands to regulate water flows and influence aquatic animal and plant resources. They constructed weirs, diversions, ponds, traps and flood-on areas. Even in the arid zone, aquatic resources in riparian systems and small wetlands such as soaks, rock holes and springs were manipulated (Bayly 1999). Compared to burn regimes, there is less documented evidence for Aboriginal water management efforts because much of the infrastructure was destroyed, damaged or modified during early European exploration and development. As suggested by Colloff (2014), it is likely there were dynamic interactions between the manipulations of both water flows and fire regimes.

Better-documented wetland management systems include those by Gunditjmara people at Lake Condah in Victoria, along the Murray–Darling system (Humphries 2007; Weir 2009) and the Roper River in Northern Territory (Barber and Jackson 2014). Over at least 6500 years, Gunditjmara and their ancestors constructed fish traps to harvest eels, fish and aquatic plants including the water ribbon

(*Cygnogeton procera*, syn.*Triglochin procera*) with edible tubers (Gunditjmara people and Wettenhall 2010). On the Roper River, temporary weirs were constructed that created lagoons to retain fish and promote resource plants such as red lily (*Nelumbo nucifera*) with edible stems and seeds (Barber and Jackson 2014).

Aboriginal water management practices were likely to have had long-term landscape-scale effects (Barber and Jackson 2014). Flow regimes are a major driver of river and floodplain wetlands (Bunn and Arthington 2002) with a higher diversity of species after temporary flooding and structural change with sustained flooding. Aquatic ecologist Humphries (2007) argues for the need to better understand environmental effects of past Aboriginal water manipulation to inform contemporary restoration and management.

4.4 Contemporary Aboriginal uses and manipulation of vegetation

Following European colonisation, Aboriginal uses of plants and influences on vegetation were dramatically disrupted. With colonisation, many traditional practices were forcibly prohibited. Yet Aboriginal people often tried to maintain their customary economy and connections to ancestral estates, knowledge, practices and languages. The dominant settlers waged a consistent campaign to remove Aboriginal people from their lands to make way for broad-scale European style agriculture and settlements (Elder 2003). As a result, millennia of Aboriginal resource management practices were radically altered. In recent decades with new legislation, policy and funds, many Aboriginal groups and partner organisations are now working to revitalise and reapply blends of traditional and modern ways of managing Australia's vegetation (see examples in Boxes 4.1–4.3).

Europeans colonised the deserts and northern tropical parts of Australia later than southern Australia. These harsher environments were less conducive to early agriculture and settlement. Groups such as the Pintupi and Martu of the Western Desert continued hunter-gatherer-manager practices until the 1960s (Long 1971) and into the era of returning to homelands. The western and central deserts, Kimberley, Arnhem Land and Cape York remain strongholds of traditionally derived Aboriginal knowledge, practices and language (Ens *et al.* 2015).

After 150 years of colonisation, Aboriginal rights are becoming more widely recognised. These include the rights to be recognised as citizens, rights to vote and rights to claim land and/or gain access. Despite ongoing contests across most legislation and jurisdictions, there has been increasing support for Indigenous land management initiatives from government and Aboriginal agencies including programmes such as Landcare, the Natural Heritage Trust, 'Caring for Country' and 'Working on Country' (Kerins 2012). Aboriginal ranger groups and Indigenous Protected Areas are initiatives of the past 15–20 years that have gained momentum. Industry programmes have also contributed to investment in Aboriginal land management, such as through the carbon market (Russell-Smith *et al.* 2009; Robinson *et al.* 2014) and biodiversity and social offsets (Fitzsimons *et al.* 2012).

In 1992, the Australian Government initiated the Indigenous Protected Area (IPA) programme. By 2015, there were 60 IPAs, covering 48 million hectares of mainly remote Australia. IPA constitutes 36% of Australia's National Reserve System (Australian Government 2015a). Aboriginal people encourage the development of community ranger programmes as they appear to support custodial responsibilities for Country (Laudine 2009). In parallel, the Australian Government established the Working on Country programme in 2007 to support indigenous ranger jobs with 680 rangers in 2015 (Australian Government 2015b).

Recognition of the values of Indigenous biocultural knowledge and conservation partnerships between Indigenous and non-Indigenous

Australians has increased in recent decades (Ens *et al.* 2015). National legislation and international conventions including the *Environmental Protection and Biodiversity Conservation Act 1999* and the Convention on Biological Diversity Aichi Targets mandate the inclusion of Aboriginal knowledge and partnerships in environmental programmes (Ens *et al.* 2015).

Contemporary Aboriginal land management initiatives have been built on and informed by research and recognition of Aboriginal people's use and management of land and sea resources across Australia (e.g. Jones 1969; Meehan 1975; Kimber 1976; Horton 1982; Kohen 1986; Altman 1987). The continuity, changes and adaptation of Aboriginal practices were major themes through much of this research. Through the 1990s and early 2000s there was some convergence between Aboriginal people and conservation interests albeit with tensions or cautions that pivoted on the 'intentionality' of practices to maintain ecosystems or species production.

In relation to vegetation management, the Australian Government's 'Caring for Country' programmes typically direct Aboriginal rangers to undertake Western or science-derived management priorities. These focus largely on ecological threat abatement (through weed and feral animal control), fire suppression and threatened fauna species (May 2010; Kerins 2012). More recently, the IPA and Working on Country programmes have promoted traditionally derived or cultural forms of 'management' in a 'two-way' blend of Western scientific and traditional methods. This reflects that fact that many Aboriginal land and sea management groups practice two-way management (Ens *et al.* 2012a; Preuss and Dixon 2012; Hill *et al.* 2013; Holmes and Jampijinpa 2013; Walsh *et al.* 2013). For example, over the last decade in western Arnhem Land, Aboriginal Rangers have worked with scientists to mimic traditional fire management practice of early dry season low-intensity patch burns. These improve vegetation structural diversity and protect fire-sensitive vegetation such as monsoon rainforest, *Allosyncarpia ternata* and *Callitris intratropica* stands (Russell-Smith *et al.* 1997, 2009; Yibarbuk *et al.* 2001). Re-establishment of deliberate Indigenous burning practices in western Arnhem Land replaced decades of absence of people in the landscape (Cooke 2009). The absence of their practices caused intense late dry season wildfires that threatened fire-sensitive ecosystems (Preece 2002; Russell-Smith *et al.* 2009).

In Kakadu National Park, traditional floodplain burning has continued for hunting and gathering reasons but now also to minimise the invasion of new weed species (McGregor *et al.* 2010). Indigenous people and scientists also worked together to control grassy Weeds of National Significance on Nywaigi lands in north Queensland (Grice *et al.* 2012). In southern Arnhem Land, the Yugul Mangi rangers are creating multifunctional landscapes to serve a variety of purposes: they fenced culturally significant wetlands to exclude feral ungulates (horse, *Equus callabus*; buffalo, *Bubalis bubalis*; and pigs, *Sus scrofa*) and promote the regeneration of food plant species such as yarlbun (*Nymphaea* species), while allowing some feral ungulates to remain for food resources and potential income (Ens *et al.* 2016). Yarlbun has various edible parts, was a staple food and features in continued ceremonial practices (Ens *et al.* 2012b, 2016). Similarly in desert regions, contemporary Indigenous land management practices continue to influence certain plant species and vegetation types (Bird *et al.* 2005) for example through fire management (Figure 4.2).

Aboriginal people's use of plants for food, medicine, crafts and materials has also continued to contemporary times, albeit with modified purposes, intensities and extent across Australia (Clarke 2007, 2010). Some people in remote and regional locations continue to hunt for animals (sometimes using fire), and collect plant foods and medicines that are easily procured, particularly berries and yams (Clarke 2007).

Figure 4.2 Contemporary Aboriginal burning in the western Great Sandy Desert. In spinifex hummock grasslands with isolated *Melaleuca glomerata* stands, at least four post-fire age classes result from fires ignited by both Martu rangers and Martu families who burn to hunt near Punmu. *Photo: Gareth Catt of Kanyirninpa Jukurrpa. (A black and white version of this figure will appear in some formats. For the colour version, please refer to the plate section.)*

There is a paucity of research on the condition and change in resources important to Aboriginal people, including post-colonial impacts and response to climate change. In some cases, the distribution and condition of favoured fruit trees has been assessed at different phenological stages with information contributing to local land management decisions. For example, djutpi (*Antidesma ghaesembilla*) research in Arnhem Land urged harvesters to return to traditional collection methods of fruits only, not ripping entire branches or plants out (Ens *et al.* 2010). Similarly, known groves of glai (green plum, *Buchanania obovata*) and billy goat plum (*Terminalia ferdinandiana*) in the Kimberley that were managed in pre-European times to maintain fruit production are again purposefully managed through low-intensity early dry season fires to protect flowers and hence fruit production (Vigilante and Bowman 2004).

Across Australia, Aboriginal people and/or enterprise partners have trialled and developed a limited suite of plant species for small-scale commercial harvest. This harvest has included: timbers for art and craft enterprises (Griffiths *et al.* 2003; Koenig *et al.* 2007, 2011); plants for horticultural sale (e.g. Griffiths *et al.* 2006); seeds and fruits for food and mining rehabilitation sales (Walsh and Douglas 2011); and bush medicine-based products (Box 4.1). Attempts to sustain such enterprises in northern Australia have had relatively low success largely due to economic and social factors, rather than ecological limits (Cunningham *et al.* 2008).

Evidence for commercial harvests of plant species that are ecologically unsustainable by contemporary Aboriginal people is limited. In Arnhem Land there was little evidence that harvest of two high-value artefact timber species, *Brachychiton diversifolius* and *Bombax ceiba* threatened the viability of their populations (Koenig *et al.* 2007, 2011), although *B. ceiba* has a restricted habitat and is vulnerable to fire and feral animals. Thus, continued monitoring of harvest rates was recommended (Koenig and Griffiths 2012). Nevertheless, risks and concerns about localised declines due to commercial and customary harvest persist. Assessments of the sustainability of plant harvest and establishment of plant-based enterprises are one focus of the Cairns TIEC (Box 4.2), although more research is required.

In temperate Australia with dense European populations and freehold lands, contemporary Indigenous management of vegetation is restricted to small land tracts returned to Aboriginal management through collaborations with national parks, private landholders, Native Title or Indigenous Land Use Agreements (Hunt 2012). For example Booderee National Park on the New South Wales south coast is collaboratively managed by Parks Australia and local Aboriginal people (Farrier and Adams 2011). These areas are important focal places for Indigenous people to maintain and enhance their connection to Country and culture (for example Wattleridge IPA, Box 4.3).

In more populated regions of Australia, Indigenous enterprises focused on plant use, management and cultural maintenance are increasing. Bush tucker and medicine gardens feature in many government run botanic gardens and community gardens. Bush tucker and medicine enterprises are the focus of new crop research and industry development for restaurant, horticultural and pharmaceutical industries. Potentially, these will build opportunities for Indigenous people and help educate other Australians about the Aboriginal uses and significance of plants.

4.4.1 Synthesis

Most of the documented evidence for pre-colonial and contemporary influences of Aboriginal people on Australia's vegetation across arid, temperate and tropical zones is from the late Holocene. However, Aboriginal people harvested and manipulated plant species for over 50 000 years. These time depths are almost inconceivable. Extreme climatic, ecological and human changes complicate simple brief explanations of prehistoric human impacts upon Australian vegetation (Mooney *et al.*, Chapter 3, this volume).

Box 4.2 Tropical Indigenous Ethnobotany Centre, Cairns

Gerry Turpin

The Tropical Indigenous Ethnobotany Centre (TIEC) is a research and resource centre to support traditional owners in conservation, management, use and communication of their traditional plant knowledge. TIEC was the first Indigenous-driven research centre in Australia. Since being established in November 2010, TIEC has, in association with other research and non-government organisations, assisted about 40 Indigenous ranger and traditional owner groups.

IEC does work to: record and document traditional plant use and other knowledge; survey flora and weeds; train Indigenous people in plant identification and collection; collate and provide historical information on traditional plant use; and use modern technologies to bridge Traditional Ecological Knowledge and Western science to support management of Country. For example, the Mbabaram Medicinal Project is a collaboration between the Mbabaram Elders and traditional owners, the Australian Tropical Herbarium through TIEC, and the National Institute of Complimentary Medicine (University of Western Sydney). The project's purpose is to test the bioactive potential of medicinal plants to develop natural medicines and enterprise opportunities for Mbabaram people. Mbabaram botanical knowledge was recorded, documented and stored in a database. Another aim is to reinvigorate traditional plant knowledge in the community and train young land managers in plant identification and collection.

TIEC also acts as a conduit to bring research and other organisations to partner with traditional owner groups in collaborative research. It aims to develop and enhance community capability and strengthen community awareness of cultural biodiversity. TIEC guides the conduct of research in respectful and culturally appropriate ways. It provides a Traditional Knowledge toolkit that recommends prior informed consent, equitable benefit sharing and protection of intellectual property.

Figure 4.3 Gerry Turpin, ethnobotanist at Tropical Indigenous Ethnobotany Centre, recording Indigenous knowledge from Peter Wallace. *Photo: Darren Crayn.*

Box 4.3 Vegetation management at Wattleridge Indigenous Protected Area, New England Tableland, NSW

Tanya Elone, Lesley Patterson and Michelle McKemey

Wattleridge Indigenous Protected Area (IPA) is 35 km north-east of Guyra, NSW. In 1998 the Indigenous Land Corporation purchased the 650 ha property Wattleridge on behalf of Banbai Land Enterprises Limited and Banbai Traditional Owners of this region. After 2 years of consultation, Wattleridge was declared an IPA in 2001, the first IPA in New South Wales.

Wattleridge contains some of the most intact and least burnt patches of vegetation on the New England Granite outcrop. The IPA has a diversity of plant species no longer found elsewhere in the region. Over 25 orchid species and 2 vulnerable tree species – the dwarf cypress pine (*Callitris oblonga*) and the backwater grevillea (*Grevillea scortechinii* subsp. *sarmentosa*) – occur there. There are many culturally significant species and cultural sites.

The Banbai rangers have established a monitoring and management plan for these species. Senior woman ranger, Lesley Patterson, propagates the backwater grevillea and the rangers monitor regeneration of this species after fire. This monitoring programme is part of a broader burn plan implemented with assistance from the Firesticks Project (NCCNSW) and the Rural Fire Service. Research is conducted on biodiversity responses to compare small-scale fires in the long-unburnt Wattleridge IPA with the more frequently burnt adjacent Warra National Park.

Wattleridge IPA performs an important role in local plant conservation. It aids in identification of fire management strategies that will enhance natural and cultural values in the region.

Figure 4.4 Photo: Banbai Rangers Lesley Patterson and Merv Torrens supervising a burn in the Wattleridge IPA.
Photo: Michelle McKemey.

In pre-colonial times, Aboriginal people directly utilised up to a third of Australian vascular plant species, although all plants were used for something (e.g. fire). Aboriginal people collated intimate knowledge of Australia's ecosystems. They applied skills to procure and produce foods and many other resources. They were not passive hunter-gatherers but managed ecosystems to increase resource production and to facilitate wider travel and survival. Aboriginal practices span from local spatial scales (such as digging and planting of different species) to larger areas (such as burning and water flow manipulation). Cumulatively, in some (but not all) bioregions these practices aggregated to create vegetation mosaics of different scales and composition suited to key game species and/or favoured plant species. These mosaics were dynamic in space and time.

European settlement radically and irrevocably altered Aboriginal societies and Australian environments. Aboriginal beliefs, knowledge and practices related to flora and fauna use and management changed profoundly. Yet there are many elements of continuity. Plants for bush food and medicine continue to be important to many Aboriginal groups even in settled and urbanised contexts.

Across remote, rural and urban regions, Aboriginal groups maintain interests in land and sea resources. Legal and policy changes since the 1960s have granted Aboriginal people certain rights and opportunities, with some form of title to over 33% of Australia's land surface. Notionally, these areas can be managed by Aboriginal people. However, new cultures and technologies have influenced how Aboriginal people interact with Australia's vegetation. Some traditional practices have been maintained and some now have new intent (such as to generate income). Some new practices have been adopted such as the use of helicopters and fuel incendiaries to burn IPAs by government funded Aboriginal rangers. Nevertheless, Aboriginal people did and continue to have significant stakes and influences on Australia's vegetation.

References

ABS (Australian Bureau of Statistics) (2011) *Estimates of Aboriginal and Torres Strait Islander Australians, June 2011*. Available at: http://www.abs.gov.au/ausstats/abs@.nsf/mf/3238.0.55.001 (accessed 4 December 2016).

Altman, J.C. (1987) *Hunter-gatherers Today: An Aboriginal Economy in North Australia*. Australian Institute of Aboriginal Studies: Canberra.

Altman, J. and Markham, F. (2014) Submission 0136 to the Joint Select Committee on Northern Australia, Inquiry into the Development of Northern Australia. Parliament of Australia: Canberra.

Altman, J.C., Buchanan, G.J. and Larsen, L. (2007) *The Environmental Significance of the Indigenous Estate: Natural Resource Management as Economic Development in Remote Australia*. The Australian National University: Canberra.

Asmussen, B. (2011) 'There is likewise a nut. . .' a comparative ethnobotany of Aboriginal processing methods and consumption of Australian *Bowenia*, *Cycas*, *Lepidozamia* and *Macrozamia* species. *Technical Reports of the Australian Museum* 23, 147–163.

Australian Government (2015a) *Indigenous Protected Areas*. Commonwealth of Australia: Canberra. Available at: www.environment.gov.au/land/indigenous-protected-areas (accessed 7 July 2016).

Australian Government (2015b) *Working on Country*. Commonwealth of Australia: Canberra.

Barber, M. and Jackson, S. (2014) Autonomy and the intercultural: interpreting the history of Australian Aboriginal water management in the Roper River catchment, Northern Territory. *Journal of the Royal Anthropological Institute* 20, 670–693.

Barr, A., Knight, T., Andrews, M. and Alexander, V. (1988) *Traditional Bush Medicines: An Aboriginal Pharmacopoeia*. Greenhouse Publications: Richmond.

Bayly, I.A.E. (1999) Review of how Indigenous people managed for water in desert regions of Australia. *Journal of the Royal Society of Western Australia* 82, 17–25.

Bean, A.R. (2007) A new system for determining which plant species are indigenous in Australia. *Australian Systematic Botany* 20, 1–43.

Bell, K.L., Rangan, H., Fowler, R. *et al.* (2014) Genetic diversity and biogeography of the boab *Adansonia gregorii* (Malvaceae: Bombacoideae). *Australian Journal of Botany* 62, 164–174.

Bird, D.W., Bliege Bird, R. and Parker, C.H. (2005) Aboriginal burning regimes and hunting strategies in Australia's Western Desert. *Human Ecology* 33, 443–464.

Bird, R.B., Tayor, N., Codding, B.R. and Bird, D.W. (2013) Niche construction and Dreaming logic: Aboriginal patch mosaic burning and varanid lizards (*Varanus gouldii*) in Australia. *Proceedings of the Royal Society B: Biological Sciences* 280, 2013 2297.

Bowman, D.M.J.S. (1998) The impact of Aboriginal landscape burning on the Australian biota. *New Phytologist* 140, 385–410.

Bradley, J., Holmes, M., Marrngawi, D.N. *et al.* (2006) *Yumbulyumbulmantha ki-Awarawu = all kinds of things from Country: Yanyuwa ethnobiological classification*. Aboriginal and Torres Strait Islander Studies Unit: Brisbane.

Bunn, S.E. and Arthington, A.H. (2002) Basic principles and ecological consequences of altered flow regimes for aquatic biodiversity. *Environmental Management* 30, 492–507.

Burrows, N.D. and Christensen, P.E.S. (1991) A survey of Aboriginal fire patterns in the Western Desert of Australia. In *Fire and the Environment: Ecological and Cultural Perspectives*, ed. S.C. Nodvin and S.C. Waldrop. US Department of Agriculture, Forest Service, South Eastern Forest Experiment Station: Asheville, pp. 20–24.

Burrows, N.D., Burbidge, A.A., Fuller, P.J. and Behn, G. (2006) Evidence of altered fire regimes in the Western Desert region of Australia. *Conservation Science Western Australia* 5, 272–284.

Butler, D.W., Fensham, R.J., Murphy, B.P., Haberle, S.G., Bury, S.J. and Bowman, D.M.J.S. (2014) Aborigine-managed forest, savanna and grassland: biome switching in montane eastern Australia. *Journal of Biogeography* 41, 1492–1505.

Clarke, P.A. (2007) *Aboriginal People and Their Plants*. Rosenberg Publishing: Dural.

Clarke, P.A. (2008a) Aboriginal healing practices and Australian bush medicine. *Journal of the Anthropological Society of South Australia* 33, 3–38.

Clarke, P.A. (2008b) *Aboriginal Plant Collectors: Botanists and Australian Aboriginal People in the Nineteenth Century*. Rosenberg Publishing: Kenthurst.

Clarke, P.A. (2010) Economic botanists, ethnobotany and Aboriginal plant use. In *The Museum of Economic Botany at the Adelaide Botanic Garden*, ed. K.T. Emmett. Board of the Botanic Gardens and State Herbarium: Adelaide, pp. 62–65.

Clarke, P.A. (2012) *Australian Plants and Aboriginal Tools*. Rosenberg Publishing: Dural.

Clarke, P.A. (2013) The Aboriginal ethnobotany of the Adelaide region, South Australia. *Transactions of the Royal Society of South Australia* 137, 97–126.

Clarke, P.A. (2015) The Aboriginal ethnobotany of the south east of South Australia region. Part 2: foods, medicines and narcotics. *Transactions of the Royal Society of South Australia* 139, 247–272.

Colloff, M. (2014) Flooded forest and desert creek: ecology and history of the River red gum. CSIRO Publishing: Collingwood.

Cooke, P. (2009) Buffalo and tin, baki and Jesus: the creation of a modern wilderness. In *Culture, Ecology and Economy of Fire Management in North Australian Savannas: in the Tradition of Wurrk*, ed. J. Russell-Smith, P.J. Whitehead and P. Cooke. CSIRO Publishing: Collingwood, pp. 69–84.

Crawford, I.M. (1982) *Traditional Aboriginal Plant Resources in the Kalumburu Area: Aspects in Ethno-economics.* Western Australian Museum: Perth.

Cunningham, A.B., Garnett, S., Gorman, J. *et al.* (2008) Eco-enterprises and Terminalia ferdinandiana: 'best laid plans' and Australian policy lessons. *Economic Botany* 63, 16–28.

Davis, M., Holcombe, S. and Janke, T. (2009) *Maintain and Strengthen Your Culture: Handbook for Working with Indigenous Ecological Knowledge and Intellectual Property.* Natural Resources Management Board: Darwin.

Dixon, R.M., Moore, B., Ramson, W.S. and Thomas, M. (2006) *Australian Aboriginal Words in English: Their Origin and Meaning.* Oxford University Press: Melbourne.

Dobson, V.P., Walsh, F. and Sati, W. (2008) Anpernirrentye: *Relationships between Bush Foods, Creation Laws, People, Country and All Things, Illustrated by Three Plants (DVD).* Desert Knowledge CRC and CSIRO Publishing: Alice Springs.

Elder, B. (2003) *Blood on the Wattle: Massacres and Maltreatment of Aboriginal Australians since 1788*, 3rd edition. New Holland Publishers: Sydney.

Ens, E.J., Finlayson, M., Preuss, K. *et al.* (2012a) Australian approaches for managing 'country' using Indigenous and non-Indigenous knowledge. *Ecological Management and Restoration* 13, 100–107.

Ens, E.J., Towler, G.M., Daniels, C. *et al.* (2012b) Looking back to move forward: collaborative ecological monitoring in remote Arnhem Land. *Ecological Management and Restoration* 13, 26–35.

Ens, E.J., Burns, E., Russell-Smith, J. *et al.* (2014) The cultural imperative: broadening the vision of long-term ecological monitoring to enhance environmental policy and management outcomes. In *Biodiversity and Environmental Change: Monitoring, Challenges and Direction*, ed. D.B. Lindenmayer, E. Burns, N. Thurgate and A. Lowe. CSIRO Publishing: Collingwood, pp. 85–110.

Ens, E.J., Pert, P., Clarke, P.A. *et al.* (2015) Indigenous biocultural knowledge in ecosystem science and management: review and insight from Australia. *Biological Conservation* 181, 133–149.

Ens, E.J., Daniels, C., Nelson, E. *et al.* (2016) Creating multi-functional landscapes: using exclusion fences to frame feral ungulate management preferences in remote Aboriginal-owned northern Australia. *Biological Conservation* 197, 235–246.

Everard, P., Kalotas, A.C. and Goddard, C. (2002) *Punu: Yankunytjatjara Plant Use: Traditional Methods of Preparing Foods, Medicines, Utensils and Weapons from Native Plants.* Jukurrpa Books: Alice Springs.

Farrier, D. and Adams, M. (2011) Indigenous-government co-management of protected areas: Booderee National Park and the national framework in Australia. In *Guidelines for Protected Area Legislation*, ed. B. Lausche. IUCN: Gland, pp. 1–40.

Fensham, R. (1997) Aboriginal fire regimes in Queensland, Australia: analysis of the explorers' record. *Journal of Biogeography* 24, 11–22.

Ferrier, Å. and Cosgrove, R. (2012) Aboriginal exploitation of toxic nuts as a late-Holocene subsistence strategy in Australia's tropical rainforests. In *Peopled Landscapes: Archaeological and Biogeographic Approaches to Landscapes*, ed. S. Haberle and B. David. ANU E-Press: Canberra, pp. 103–120.

Fitzsimons, J., Russell-Smith, J., James, G. *et al.* (2012) Insights into the biodiversity and social benchmarking components of the Northern Australian fire management and carbon abatement programmes. *Ecological Management and Restoration* 13, 51–57.
Flannery, T.F. (1994) *The Future Eaters: An Ecological History of the Australasian Islands and People.* Reed Books: Chatswood.
Gammage, B. (2011) *The Biggest Estate on Earth: How Aborigines Made Australia.* Allen and Unwin: Melbourne.
Gerritson, R. (2008) *Australia and the Origins of Agriculture.* Archaeopress: Oxford.
Ghisalberti, E. (1994) The ethnopharmacology and phytochemistry of *Eremophila* species (Myoporaceae). *Journal of Ethnopharmacology* 44, 1–9.
Gott, B. (1982) Ecology of root use by the Aborigines of southern Australia. *Archaeology in Oceania* 17, 59–67.
Gott, B. (2005) Aboriginal fire management in south-eastern Australia: aims and frequency. *Journal of Biogeography* 32, 1203–1208.
Gott, B. (2008) Indigenous use of plants in south-eastern Australia. *Telopea* 12, 215–229.
Gott, B., Williams, N.S. and Antos, M. (2015) Humans and grasslands: a social history. In *Land of Sweeping Plains – Managing and Restoring the Native Grasslands of South-eastern Australia*, ed. N.S. Williams, A. Marshall and J. Morgan CSIRO Publishing: Clayton South, pp. 5–26.
Grice, A.C., Cassady, J. and Nicholas, D.M. (2012) Indigenous and non-Indigenous knowledge and values combine to support management of Nywaigi lands in the Queensland coastal tropics. *Ecological Management and Restoration* 13, 93–97.
Griffiths, A.D., Philips, A. and Godjuwa, C. (2003) Harvest of *Bombax ceiba* for the Aboriginal arts industry, central Arnhem Land, Australia. *Biological Conservation* 113, 295–305.
Griffiths, A.D., Schult, H.J. and Gorman, J. (2006) Wild harvest of *Cycas arnhemica* (Cycadaceae): impact on survival, recruitment and growth in Arnhem Land, northern Australia. *Australian Journal of Botany* 53, 771–779.
Gunditjmara People and Wettenhall, G. (2010) *The People of Budj Bim.* emPress: Victoria.
Hallam, S.J. and Mulvaney, J. (2014) *Fire and Hearth: A Study of Aboriginal Usage and European Usurpation in South-western Australia.* Western Australian UWA Publishing: Crawley.
Hamby, L. (2005) *Twined Together: kunmadj njalehnjaleken.* Injalak Arts and Crafts: Gunbalanya.
Hill, R. and Baird, A. (2003) Kuku-Yalanji Rainforest Aboriginal people and carbohydrate resource management in the Wet Tropics of Queensland, Australia. *Human Ecology* 30, 27–52.
Hill, R., Baird, A., Buchanan, D. *et al.* (2004) *Yalanji-Warranga Kaban: Yalanji People of the Rainforest Fire Management Book.* Little Ramsay Press: Cairns.
Hill, R., Pert, P., Davies, J. *et al.* (2013) *Indigenous Land Management in Australia: Extent, Scope, Diversity, Barriers and Success Factors.* CSIRO Ecosystem Sciences: Cairns.
Hiscock, P. (2007) *Archaeology of Ancient Australia.* Routledge: London.
Holmes, M.C.C. and Jampijinpa, W. (2013) Law for country: the structure of Warlpiri ecological knowledge and its application to natural resource management and ecosystem stewardship. *Ecology and Society* 18, 19.
Horton, D.R. (1982) The burning question: Aborigines, fire and Australian ecosystems. *Mankind* 13, 237–252.

Humphries, P. (2007) Historical Indigenous use of aquatic resources in Australia's Murray–Darling Basin, and its implications for river management. *Ecological Management and Restoration* 8, 106–113.
Hunt, J. (2012) North to south? In *People on Country, Vital Landscapes, Indigenous Futures*, ed. J. Altman and S. Kerins. Federation Press: Sydney, pp. 94–118.
Hynes, R. and Chase, A. (1982) Plants, sites and domiculture: Aboriginal influence upon plant communities in Cape York Peninsula. *Archaeology in Oceania* 17, 38–50.
Isaacs, J. (1987) *Bush Food: Aboriginal Food and Herbal Medicine.* The Rocks: Lansdowne.
Jones, R. (1969) Fire-stick farming. *Australian Natural History* 16, 224–228.
Kalotas, A. (1996) Aboriginal knowledge and use of fungi. *Fungi of Australia* 1, 269–298.
Keen, I. (2004) *Aboriginal Economy and Society: At the Threshold of Colonisation.* Oxford University Press: Melbourne.
Kerins, S. (2012) Caring for Country to working on Country. In *People on Country, Vital Landscapes, Indigenous Futures*, ed. J.C. Altman and S. Kerins. Federation Press: Sydney, pp. 26–44.
Kimber, R. (1976) Beginnings of farming? Some man–plant–animal relationships in Central Australia. *Mankind* 10, 142–150.
Kimber, R. (1983) Black lightning: Aborigines and fire in central Australia and the Western Desert. *Archaeology in Oceania* 18, 38–45.
Kimber, R. and Friedel, M. (2016) Challenging the concept of Aboriginal mosaic fire practices in the Lake Eyre Basin. *The Rangeland Journal* 37, 623–630.
Koenig, J., Altman, J.C., Griffiths, A.D. and Kohen, A. (2007) 20 years of Aboriginal woodcarving in Arnhem Land, Australia: using art sales records to examine the dynamics of sculpture production. *Forests, Trees and Livelihoods* 17, 43–60.
Koenig, J., Altman, J.C. and Griffiths, A.D. (2011) Artists as harvesters: natural resource use by Indigenous woodcarvers in Central Arnhem Land, Australia. *Human Ecology* 39, 407–419.
Koenig, J. and Griffiths, A. (2012) The population ecology of two tropical trees, *Brachychiton diversifolius* (Malvaceae) and *Bombax ceiba* (Bombaceae), harvested by Indigenous woodcarvers in Arnhem Land, Australia. *Environmental Management* 50, 555–565.
Kohen, J.L. (1986) *Prehistoric Settlement in the Western Cumberland Plain, Resources, Environment and Technology*. Macquarie University: Sydney.
Kohen, J.L. (1996) Aboriginal use of fire in southeastern Australia. *Proceedings of the Linnean Society of New South Wales* 116, 19–26
Kondo, T., Crisp, M.D., Linde, C. *et al.* (2012) Not an ancient relic: the endemic Livistona palms of arid central Australia could have been introduced by humans. *Proceedings of the Royal Society B: Biological Sciences* 279, 2652–2661.
Langton, M. (1998) *Burning questions: emerging environmental issues for indigenous peoples in northern Australia.* Centre for Indigenous Natural and Cultural Resource Management, Northern Territory University.
Latz, P.K. (1995) *Bushfires and Bushtucker: Aboriginal Plant Use in Central Australia.* I.A.D. Press: Alice Springs.
Latz, P.K. (1998) The desert raisin and other *Solanum. Australian Bushfoods Magazine* 5, 8–9.
Latz, P.K. (2007) *The Flaming Desert: Arid Australia – A Fire Shaped Landscape.* I.A.D. Press: Alice Springs.
Latz, P.K. and Griffin, G.F. (1978) Changes in Aboriginal land management in relation to fire and food plants in central Australia. In *Symposium on the Nutrition of Aborigines, Canberra (Australia), 23 Oct 1976*. Commonwealth Scientific and Industrial Research Organization: Canberra.

Laudine, C. (2009) *Aboriginal environmental knowledge: rational reverence*. Ashgate Publishing: Farnham.

Lee, L.S. (2013) Horticultural development of bush food plants and rights of Indigenous people as traditional custodians. The Australian bush tomato (*Solanum centrale*) example: a review. *The Rangeland Journal* 34, 359–373.

Long, J.P. (1971) Arid region Aborigines: the Pintubi. In *Aboriginal Man and Environment in Australia*, ed. D.J. Mulvaney and J. Golson. Australian National University Press: Canberra, pp. 262–270.

Low, T. (1990) *Bush Medicine: A Pharmacopoeia of Natural Remedies.* Angus and Robertson: North Ryde.

May, K. (2010) Government support for Indigenous cultural and natural resource management in Australia: the role of the Working on Country program. *Australian Journal of Social Issues* 45, 395–416.

McConvell, P. and Thieberger, N. (2001) *State of Indigenous Languages in Australia - 2001.* Department of the Environment and Heritage: Canberra. Available at: http://hdl.handle.net/11343/34091 (accessed 7 July 2016).

McGregor, S., Lawson, V., Christophersen, P. *et al.* (2010) Indigenous wetland burning: conserving natural and cultural resources in Australia's World Heritage-listed Kakadu National Park. *Human Ecology* 38, 721–729.

Meehan, B. (1975) Shell bed to shell midden. PhD Thesis, Australian National University, Canberra.

Metcalfe, D.J., Liddell, M.J., Bradford, M.G. and Green, P.T. (2014) Tropical rainforests of eastern Australia. In *Biodiversity and Environmental Change: Monitoring, Challenges and Direction*, ed. D. Lindenmayer, E. Burns, N. Thurgate and A. Lowe. CSIRO Publishing: Collingwood, pp. 111–165.

Mooney, S.D., Harrison, S.P., Bartlein, P.J. *et al.* (2011) Late Quaternary fire regimes in Australia. *Quaternary Science Reviews* 30, 28–46.

Morton, S., Smith, D.S., Dickman, C. *et al.* (2011) A fresh framework for the ecology of arid Australia. *Journal of Arid Environments* 75, 313–329.

Mulvaney, D.J. and Kamminga, J. (1999) *Prehistory of Australia.* Allen and Unwin: Sydney.

Nabarula, A.D., Nabarnardi, C.F., Nambijimba, E.R. and Nungari, S.P. (1978) Bush medicines used at Warrabri. *Aboriginal and Islander Health Worker Journal* 2, 12–14.

Nash, D. (2012) 'Heritage knowledge': Indigenous people and fibre plants on the NSW South Coast. *Artefact: The Journal of the Archaeological and Anthropological Society of Victoria* 35, 20–58.

Pascoe, B. (2014) *Dark Emu. Black Seeds: Agriculture or Accident?* Magabala Books: Broome.

Pedley, H. (1993) Plant detoxification in the rainforest: the processing of poisonous plant foods by the Jirrbal-Girramay people. Masters Thesis, James Cook University, Townsville.

Pitman, H.P. and Wallis, L.A. (2012) The point of spinifex: Aboriginal uses of spinifex grasses in Australia. *Ethnobotany Research and Applications* 10, 109–132.

Preece, N. (2002) Aboriginal fires in monsoonal Australia from historical accounts. *Journal of Biogeography* 29, 321–336.

Preuss, K. and Dixon, M. (2012) 'Looking after country two-ways': insights into Indigenous community-based conservation from the Southern Tanami. *Ecological Management and Restoration* 13, 2–15.

Pyne, S.J. (1998) Forged in fire: history, land, and anthropogenic fire. In *Advances in historical ecology*, ed. W. Balee. Columbia University Press: New York, pp. 64–103.

Rangan, H., Bell, K.L., Baum, D.A. *et al.* (2015) New genetic and linguistic analyses show ancient human influence on Baobab evolution and distribution in Australia. *PloS One* 10, e0119758.

Raymond, E., Blutja, J., Gingina, L. *et al.* (1999) *Wardaman Ethnobiology. Aboriginal Plant and Animal Knowledge from the Flora River and South-west Katherine Region, North Australia.* Parks and Wildlife Commission of the Northern Territory and Northern Territory University: Darwin.

Robinson, C.J., Gerrard, E., May, T. and Maclean, K. (2014) Australia's Indigenous carbon economy: a national snapshot. *Geographical Research* 52, 123–132.

Rose, D.B. (1987) *Bush Medicines: A Ngarinman and Bilinara Pharmacopoeia.* Australian Institute of Aboriginal Studies: Canberra.

Rose, D.B. (1996) *Nourishing Terrains: Australian Aboriginal Views of Landscape and Wilderness.* Australian Heritage Commission: Canberra.

Russell, D. (2004) Aboriginal-Makassan interactions in the eighteenth and nineteenth centuries in northern Australia and contemporary sea rights claims. *Australian Aboriginal Studies* 1, 3–17.

Russell-Smith, J., Lucas, D., Gapindi, M. *et al.* (1997) Aboriginal resource utilisation and fire management practice in western Arnhem Land, monsoonal northern Australia: notes for prehistory, lessons for the future. *Human Ecology* 25, 159–195.

Russell-Smith, J., Whitehead, P. and Cooke, P. (2009) *Culture, Ecology and Economy of Fire Management in North Australian Savannas: Rekindling the Wurrk Tradition.* CSIRO Publishing: Collingwood.

Smith, M. (2013) *The Archaeology of Australia's Deserts.* Cambridge University Press: Cambridge.

Taylor, J. (2005) *Population and Patterns of Residence.* Macquarie/Macmillan: Sydney.

Thorne, A. (2005) *Migration and Prehistory.* Macquarie/Macmillan: Sydney.

Trudgen, M.S., Webber, B.L. and Scott, J.K. (2012) Human-mediated introduction of Livistona palms into central Australia: conservation and management implications. *Proceedings of the Royal Society of London B: Biological Sciences* 279, 4115–4117.

Ulm, S. (2013) 'Complexity' and the Australian continental narrative: themes in the archaeology of Holocene Australia. *Quaternary International* 285, 182–192.

Veth, P., Smith, M. and Hiscock, P. (2005) *Desert Peoples: Archaeological Perspectives.* Blackwell Publishing: Oxford.

Vigilante, T. (2001) Analysis of explorers' records of Aboriginal landscape burning in the Kimberley region of Western Australia. *Australian Geographical Studies* 39, 135–155.

Vigilante, T. and Bowman, D.M.J.S. (2004) Effects of individual fire events on the flower production of fruit-bearing tree species, with reference to Aboriginal people's management and use, at Kalumburu, North Kimberley, Australia. *Australian Journal of Botany* 52, 405–415.

Vigilante, T., Murphy, B.P. and Bowman, D.M.J.S. (2009) Aboriginal fire use in Australian tropical savannas: ecological effects and management lessons. In *Tropical Fire Ecology*, ed. MA Cochrane. Springer: Chichester, pp. 143–167.

Walsh, F.J. (1990) An ecological study of traditional Aboriginal use of 'Country': Martu in the Great and Little Sandy Deserts. In *Australian Ecosystems: 200 Years of Utilization, Degradation and Reconstruction*, ed. D.A. Saunders, A.J. Hopkins and R.A. How. Surrey Beatty and Sons: Sydney, pp. 23–37.

Walsh, F.J. (2008) To hunt and to hold: Martu Aboriginal people's uses and knowledge of their country, with implications for co-management in Karlamilyi (Rudall River) National Park and the Great Sandy Desert, Western Australia. PhD Thesis, University of Western Australia, Perth.

Walsh, F. and Douglas, J. (2011) No bush foods without people: the essential human dimension to the sustainability of trade in native plant products from desert Australia. *Rangeland Journal* 33, 395–416.

Walsh, F.J., Dobson, P.V. and Douglas, J.C. (2013) Anpernirrentye: a framework for enhanced application of Indigenous ecological knowledge in natural resource management. *Ecology and Society* 18, 18.

Webb, L.J. (1969) The use of plant medicines and poisons by Australian Aborigines. *Mankind* 7, 137–146.

Weir, J.K. (2009) *Murray River Country An Ecological Dialogue with Traditional Owners.* Aboriginal Studies Press: Canberra.

Williams, A.N. (2013) A new population curve for prehistoric Australia, *Proceedings of the Royal Society B* 280, 20130486.

Yibarbuk, D., Whitehead, P.F., Russell-Smith, J. *et al.* (2001) Fire ecology and Aboriginal land management in central Arnhem Land, northern Australia: a tradition of ecosystem management. *Journal of Biogeography* 28, 325–343.

Yunupingu, B., Yunupingu-Marika, L., Marika, D. *et al.* (1994) *Rirratjingu ethnobotany: Aboriginal plant use from Yirrkala, Arnhem Land, Australia.* Conservation Commission of the Northern Territory: Darwin.

5 Fire and Australian Vegetation

BEN P. MILLER AND BRETT P. MURPHY

Fire has been a major force shaping the distribution and form of Australian vegetation through deep evolutionary time, and remains so today. While the role of fire in the evolution of the Australian flora is extremely difficult to demonstrate, most of Australia's most diverse and dominant plant genera have traits that appear to be adapted to living with fire, and these traits are likely to have contributed to their success and radiation. Fire regime components include fire season, intensity, frequency and type. Fire regime, fire pattern and size, are influenced by climate, net primary productivity, landscape features, and human activity. Attributes of vegetation such as vertical structure, horizontal pattern and connectivity, volatile oil production, moisture content and seasonal variation influence the behaviour of fire. At the same time, fire regime components interact with plant traits to filter the species capable of persisting at a site, and hence influence the structure, dynamics and composition of vegetation. Vegetation and fire represent a single dynamic system driven by climate and site attributes that vary in response to local and global drivers. These inherent dynamics now interact with a powerful set of anthropogenic impacts – fragmentation, invasive species, grazing, timber extraction, climate change and fire management – changing the attributes of Australia's vegetation. Here we review the evolutionary and ecological history of fire in Australia, patterns and drivers of modern Australian fire regimes, interactions between plant traits and fire regimes and the influence of anthropogenic change on fire–vegetation interactions.

5.1 Introduction

Australia is a megadiverse continent both in terms of vegetation and fire regimes. In some of the highest rainfall areas of southern Australia, fire in mountain ash forests may occur at intervals measured in centuries, yet the regeneration of mountain ash may be dependent on stand-replacing fires. These are triggered by extreme fire weather conditions following long periods of intense drought and typically burn the moist surface layers as well as the open eucalypt canopy over 90 m above the ground. These fires are perhaps the most intense experienced anywhere on Earth. At the other extreme, fire can occur annually in grasslands and savannas in the monsoonal tropics, but its intensity and impact vary with season of burn. While fire management of these tropical systems is increasingly returning to the traditional practices maintained by Aboriginal people over millennia, the pre-European fire regime of most temperate Australian grasslands is almost unknown. Climate, weather conditions, soils and

Australian Vegetation, Third Edition, ed. David A. Keith. Published by Cambridge University Press. © Cambridge University Press 2017.

Citation: Miller, B.P. and Murphy, B.P. (2017). Fire and Australian Vegetation. In D. A. Keith, ed., *Australian Vegetation*, 3rd edn. Cambridge: Cambridge University Press, pp. 113–34.

topography, human management, together with the response of plants to fire, their flammability and diverse functional traits all interact to create a complex pattern of fire and vegetation dynamics that in many places construct and define vegetation patterns within landscapes and across the continent.

5.2 Evolutionary and ecological history of fire in Australia

Evidence for the ecological significance of fire in the Australian geological record comes in three forms: (i) plant fossils demonstrating traits that are today associated with fire-prone ecosystems (e.g. serotiny in *Banksia*, epicormic resprouting in eucalypts); (ii) phylogenetic reconstructions of the presence of these traits through time; and (iii) charcoal, inertinite and related depositional records suggesting past fire. The first two of these lines of evidence assume that the plant traits that confer fitness in modern fire-prone environments evolved, or were retained through time, because fire existed in the ecological system. While this is plausible, the role of fire in the evolution of the Australian flora is extremely difficult to demonstrate (Bradshaw *et al.* 2011; Keeley *et al.* 2011). This line of reasoning provides dates for fire-dependent communities in Australia from ~60 million years ago (Ma) for both *Banksia* and *Eucalyptus* (Crisp *et al.* 2011; He *et al.* 2011). Globally, similar analyses indicate that ecologically relevant fire has a much longer record (to the Gondwanan Permian), associated with varying atmospheric O_2 or CO_2 levels (increasing flammability and severity of fire weather, respectively) (Belcher *et al.* 2010; Brown *et al.* 2012; Jasper *et al.* 2013; Glasspool *et al.* 2015).

Bond and colleagues (Bond and Scott 2010; Bond and Midgley 2012; Bond 2015) argue that the Cretaceous diversification of angiosperms, and the evolution of grasses, and particularly C_4 grasses, about 8 Ma, was facilitated by, and contributed to, changes in global fire regimes. They associate these changes with the greater propensity of grasses to burn, and the evolution of closed broadleaf forests which exclude fire. This dichotomy persists today, in the dynamism of savanna–rainforest boundaries, mediated by the flammability of grasses, on the one hand, and moist microclimates and low flammability of forest litter, on the other (Bowman and Wilson 1988; Little *et al.* 2012). Sedgeland–rainforest boundaries persisting in Tasmania may represent a similar, more ancient, dynamic (Jackson 1968).

Humans arrived in Australia around 45–50 thousand years ago (Bowler *et al.* 2003; Ens *et al.*, Chapter 4, this volume), but because anthropogenic landscape fire is near-invisible in the archaeological record, it is unclear when they first started deliberately burning the landscape. However, at the time of European arrival, Aboriginal people were making extensive use of fire and presumably had been burning some Australian landscapes for many millennia. The impacts of this burning on Australian fire regimes and biota is controversial (Bowman 1998; Gammage 2011; Mooney *et al.* 2011), but it must have had significant effects in at least some areas. A particularly controversial hypothesis of indirect impacts of early humans on Australian fire regimes claims that humans hunted the megafauna to extinction (by about 45 thousand years ago: Roberts *et al.* 2001) and the associated reduction of herbivory led to a dramatic increase in fire activity (Rule *et al.* 2012). Others have claimed that the signature of fire in charcoal deposits and $\delta^{13}C$ in emu eggshells indicates that shifts in patterns of fire occurrence and vegetation change can be explained by climate drivers alone (Daniau *et al.* 2012; Murphy *et al.* 2012).

There is no doubt that Aboriginal people used fire as a tool to manage landscapes, burning in different locations and seasons, at different times of the day, and under different weather conditions to achieve specific and various economic and cultural objectives (Bowman 1998; Yibarbuk *et al.* 2001; Hassell and Dodson 2002; Whitehead *et al.* 2003; Bowman *et al.* 2004; Burrows *et al.* 2006; Gammage 2011; Codding *et al.* 2014). Evidence from central and northern Australia, where ongoing practice and memory of traditional burning customs is strongest, suggests widespread and frequent burning in recent pre-European history, but also that this was not ubiquitous. Within these landscapes, certain localities and vegetation types (e.g. rainforests) were actively protected from fire, and others were deliberately burnt (e.g. grassy savannas). Elsewhere in Australia, the evidence is sparser. Except for the temperate grasslands (now almost extinct), even the most flammable vegetation types in the south and east are not able to support fire frequencies as high as can occur in the north. As such, it is not clear which vegetation types were more or less managed than others. Tall eucalypt forests (such as south-western karri and south-eastern mountain ash), temperate rainforests, as well as alpine regions, mulga and chenopod shrublands are examples of vegetation that perhaps were not deliberately burnt. Other vegetation types were more likely to have been managed with fire, but evidence for the extent, season, intensity or frequency of this imposed burning is sparse or confounded (Ward *et al.* 2001; Miller *et al.* 2007; O'Donnell *et al.* 2011).

Inference of the significance of Aboriginal burning in creating vegetation patterns is derived from studies across Australia that demonstrate vegetation change in the absence of traditional fire management (e.g. Lunt 1998b; Bowman *et al.* 2001, 2012b; Bowman and Prior 2004). Palynological studies also suggest a role for humans in modifying vegetation patterns in some systems such as Tasmanian moorlands (Fletcher and Thomas 2007) and subtropical grassland–rainforest landscapes (Butler *et al.* 2014), but not others (e.g. Blue Mountains forests and shrublands; Black and Mooney 2006). Other lines of evidence find no support for the contention that Aboriginal burning led to changes in fire regimes beyond those driven by varying climate (Daniau *et al.* 2012; Williams *et al.* 2015). Analyses of multiple systems, over varying time frames, with varying methods and resolution, leads to confusion: a more realistic interpretation is that Aboriginal people imposed different fire management regimes (including no deliberate burning) in different places, and the impacts of these activities varied among vegetation types and biomes. As occurs today, chance extreme weather and accidents would have added an unplanned element to any managed regime.

5.3 Patterns and drivers of Australian fire regimes

The 'fire regime' is one of the most fundamental concepts in fire ecology. Gill (1975) is widely credited with the classical definition, encompassing the core ecologically relevant components of fire activity for a point in a landscape – intensity, frequency, season and type of fire. Subsequently, the concept has been broadened, most notably to include a number of spatial components such as fire size and patchiness, as well as spatiotemporal variability (pyrodiversity) (Scott *et al.* 2013). Understanding the spatial distribution of fire regimes (pyrogeography), its underlying drivers and interactions with vegetation, have emerged as key challenges in fire and vegetation science in the last two decades.

5.3.1 Pyrogeographic models

For the purposes of fire fighting and fire prevention, the ingredients required for the occurrence of fire have long been described using the classic 'fire triangle' – the ingredients being fuel, oxygen and heat. This concept has been borrowed by pyrogeographers in recent years, with a number of conceptual models identifying key environmental 'switches' which must be activated for landscape fire to occur (Meyn *et al.* 2007; Bradstock 2010; Krawchuk and Moritz 2011). Descriptions of these switches vary, but Bradstock (2010), in his model of the drivers of fire activity in Australia, identified four: (i) sufficient biomass; (ii) its availability to burn (e.g. low fuel moisture, low bulk density); (iii) weather conditions conducive to fire spread (e.g. high wind speeds, high temperature); and (iv) ignitions.

The importance of each of Bradstock's (2010) four switches is difficult to evaluate empirically, because the switches themselves tend to be highly

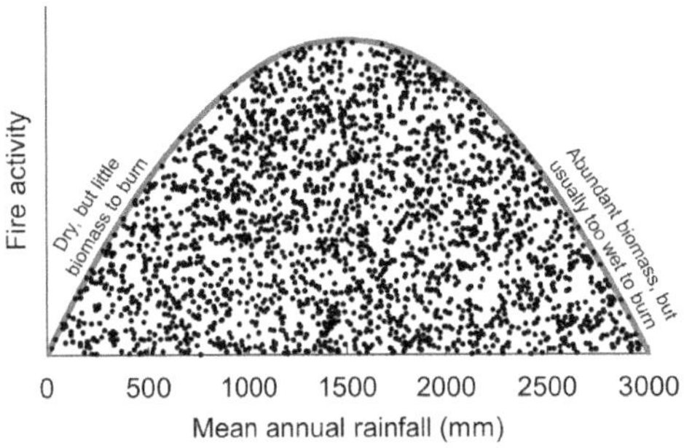

Figure 5.1 The fire–rainfall relationship, which has been postulated conceptually and demonstrated empirically at a range of spatial scales (Pausas and Bradstock 2007; van der Werf *et al.* 2008; Pausas and Ribeiro 2013; Bowman *et al.* 2014c; Figure 5.3). Water availability (e.g. mean annual rainfall) is best thought of as a correlate of the *upper bound* of fire activity (Murphy *et al.* 2013). Fire is strongly constrained in very dry and unproductive environments, due to lack of fuel, and in very wet environments, due to high fuel moisture. However, even at intermediate rainfall, fire activity can be very low.

correlated, or difficult to measure directly at large spatial scales. Regional- to global-scale analyses of fire activity have tended to focus on one or two of the key switches – such as biomass and its availability to burn, and even these are often expressed by a single proxy. For example, there is a well-known 'humped' relationship between the upper bound to fire activity and mean annual rainfall (Figure 5.1), which is assumed to primarily reflect productivity (i.e. the 'biomass' switch) and fuel moisture (i.e. the 'fuel availability to burn' switch). In arid environments, fire activity is very low because low productivity severely limits fuel accumulation. In wet environments, fire activity is also very low, because while biomass is abundant, it is typically too wet to burn. High fire activity is only possible at intermediate productivity and rainfall, where there is the most regular occurrence of abundant, dry biomass.

5.3.2 Climatic drivers

Australia is frequently described as the most fire-prone continent on Earth – an assertion which is difficult to verify. But there is no doubt that much of the continent, especially the far north, burns with remarkably high frequency (Figure 5.2), as does much of the seasonally wet–dry tropics globally. Other parts burn infrequently, but when they do burn, exceptionally high fire intensities are possible. Murphy *et al.* (2013) have argued that the geography of Australian fire regimes is a product of the strong latitudinal gradient in summer monsoon activity (Figure 5.3). Frequent low-intensity fires occur in the monsoonal north and infrequent, high-intensity fires in the temperate south, demonstrating a trade-off between frequency and intensity. That is, very high-intensity fires are only associated with very low-frequency fire regimes in the tall eucalypt forests of southern Australia. While these forests occasionally experience extremely intense fires (fire-line intensities >>50 000 kW m^{-1}, among the highest experienced anywhere on Earth), such regimes are exceptional, with most of the continent dominated by grass fuels, typically burning with lower-intensity (fire-line intensities <5000 kW m^{-1}).

Demonstrating the relative importance of different drivers of Australian fire regimes is challenging. Taking the example of the latitudinal gradient in fire frequency and intensity, it is difficult to know whether this is a product of latitudinal variation in rainfall seasonality, dry season temperatures or lightning activity. All of these climatic variables – closely related to the fire 'switches' of Bradstock (2010) – are close to their peak in monsoonal northern Australia (Figure 5.3). Monsoonal northern Australia is characterised by high rainfall but extreme rainfall seasonality, year-round high temperatures, high lightning activity (especially at the end of the dry season) – all of which make these landscapes 'primed' to burn each year, and annual burning does occur in some grass-dominated ecosystems (Figure 5.2a). Moving from north to south, fire frequency steadily declines, the predominant season of fire shifts from winter-spring to summer–autumn, and fire intensities (both typical and extreme) increase.

Part of this pattern in fire regime is due to vegetation patterns and part clearly due to climate. For instance, native temperate grasslands are now rare in southern Australia but in aseasonal and summer-rainfall areas of south-eastern Australia,

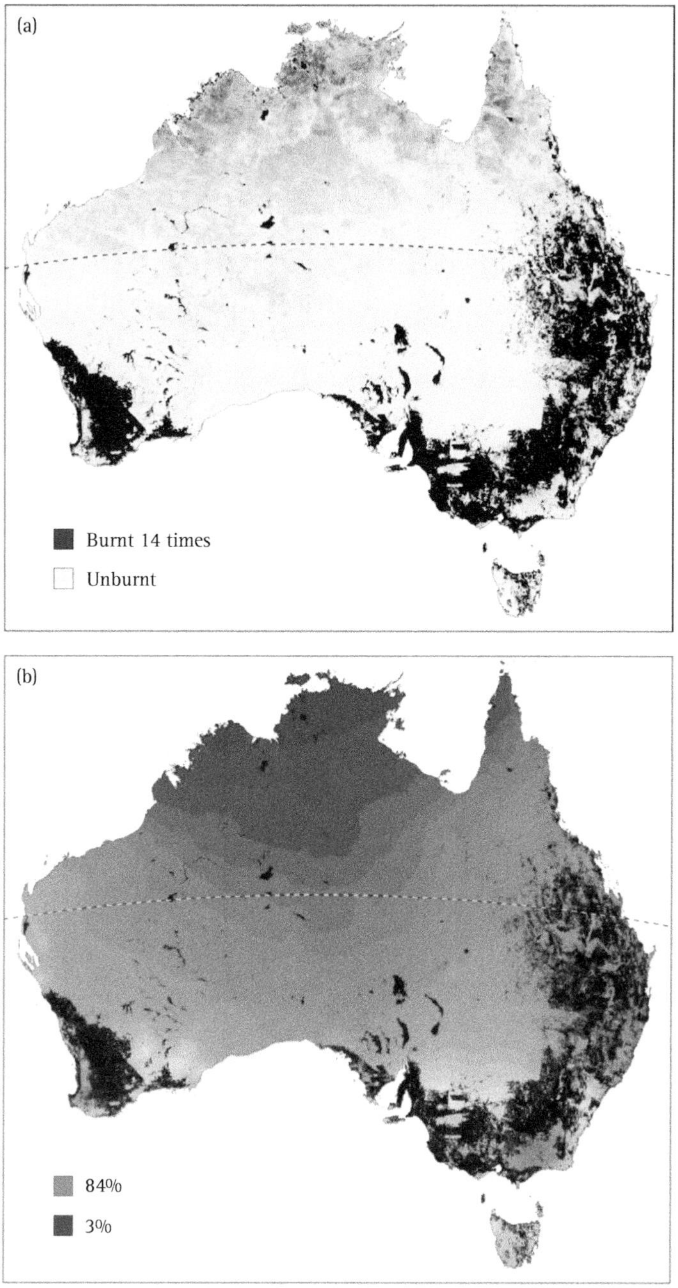

Figure 5.2 Australia-wide variation in: **(a)** fire frequency (number of times burnt in a 14-year period, 1997–2010), with fire scars derived from Advanced Very High Resolution Radiometer satellite imagery (Russell-Smith *et al.* 2007); and **(b)** the proportion of burning in the summer and autumn months of December–March, inclusive, inferred from the density of active fire detections by the Moderate-Resolution Imaging Spectroradiometer (MODIS) satellite sensors (Giglio *et al.* 2003). The surface shown in **(b)** is derived using interpolation with inverse distance weighting. Cleared areas, bare areas and waterbodies are shown in black (National Vegetation Information System 2012).
Maps: Brett Murphy. (A black and white version of this figure will appear in some formats. For the colour version, please refer to the plate section.)

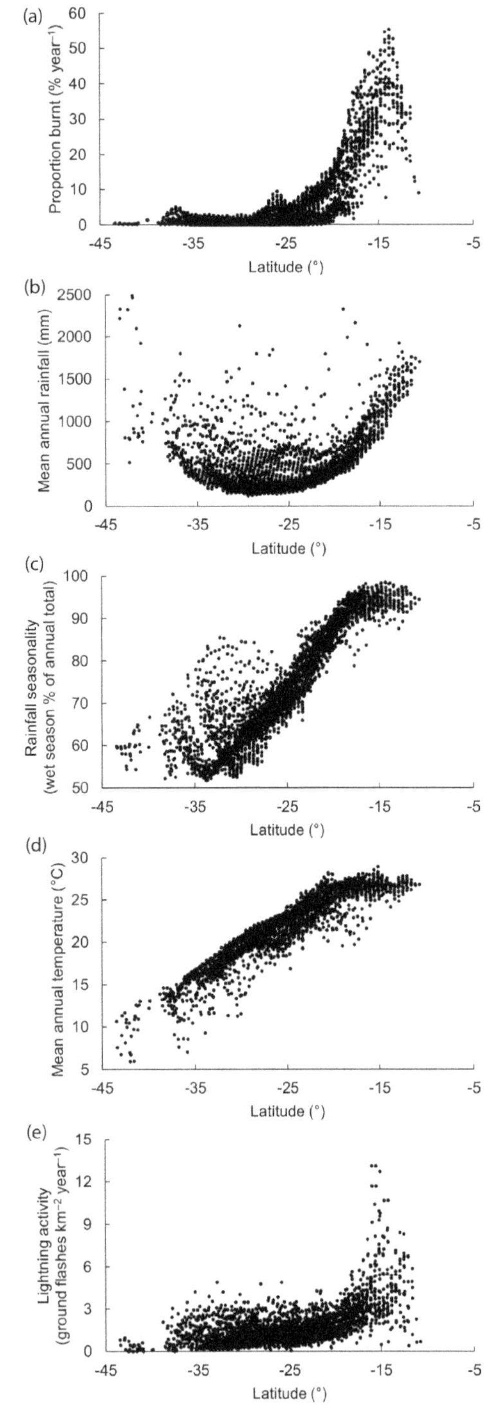

Figure 5.3 Australia's latitudinal gradient in (a) annual proportion burnt and (b–e) key climate variables likely to affect fire activity. Annual proportion burnt refers to the average proportion of 70 × 70 km cells (covering the Australian mainland and Tasmania) which is burnt each year in the 14-year period 1997–2010, with fire scars derived from Advanced Very High Resolution Radiometer satellite imagery (Russell-Smith *et al.* 2007). Climate data are long-term averages from the Australian Bureau of Meteorology (2015), and are averaged across the same 70 × 70 km cells. In (c), rainfall seasonality is defined as the proportion of annual rainfall occurring in the wettest six consecutive months.

remnant native *Themeda* grasslands can support annual burning (Morgan 1999). Fire frequency and seasonality is also similar in comparable temperate vegetation communities along the rainfall seasonality gradient that extends across southern Australia. For instance, mean fire return intervals in temperate sclerophyllous shrublands are similar in the Mediterranean-climate areas of south-western Australia (ranging from 17 to 38 (some >40) years) and summer rainfall areas of New South Wales (14 to 21 (and some >40) years (Enright *et al.* 2012)). Seasonally consistent or summer-dominated rainfall areas of south-eastern Australia appear to have a higher concentration of fire in summer than the predictable Mediterranean-climate of the southwest (Figure 5.2b). These patterns suggest a strong role for fire weather and summer drought, whether predictable or not, in driving temperate fire regimes (Nolan *et al.* 2016). Extensive fires in temperate wet or moist sclerophyllous ecosystems are less regular, and more closely tied to periodic drought and associated with very high summer temperatures (>40°C in forested areas). For example, Victoria's catastrophic bushfires of 1983 and 2009 ('Ash Wednesday' and 'Black Saturday', respectively) were each triggered by unusually high summer temperatures, extreme winds and timely ignitions, after a decade of drought.

5.3.3 Local-scale drivers

The relationship between regional climate and fire regimes can be severely blurred by local environmental factors – especially topography. Topographically enclosed areas such as river valleys tend to have higher soil moisture and experience more moderate fire weather conditions (e.g. lower temperatures, lower wind speeds, higher relative humidity) – and hence are much less likely to burn, and typically burn with lower intensity – than topographically exposed areas such as tops of ridges and slopes more exposed to the sun and prevailing winds in the fire season. Topographic features such as cliffs and areas of high surface roughness (i.e. topographic complexity) reduce the likelihood of fire spread. Hence, fire frequency tends to be reduced in rugged mountainous areas.

Vegetation itself plays a critical role in controlling fire regimes at local scales. In areas with skeletal soils, vegetation can be so sparse that fires can rarely be carried. Rocky areas, which tend to burn with much lower frequency than surrounding areas are a good example of this, such as the sparse Arnhem Plateau heathlands in northern Australia (Russell-Smith *et al.* 1998) or granite outcrops within temperate eucalypt forests and sandstone pavements in the Sydney Basin (Clarke 2002) and jarrah forests of the south-west (Yates *et al.* 2003); these areas tend to support a relatively high diversity of species sensitive to short fire intervals.

Vegetation can also act to reinforce topographic effects. For example, in low-lying, topographically protected areas such as valley floors and gullies, soil moisture is likely to be high and microclimate moderate, promoting the persistence of mesic vegetation – such as a perennial, mesic shrubby or ferny understorey, or rainforest canopy. Such vegetation may have higher moisture content (e.g. than an annual grassy or sclerophyll shrub understorey) and may moderate microclimatic conditions (Little *et al.* 2012), thereby reducing the fire-proneness of the site.

5.3.4 Feedbacks with vegetation

There are critical feedbacks between fire and vegetation which tend to be overlooked by pyrogeographers, paralleling the way that fire itself was overlooked as a determinant of vegetation distribution for many decades (Holdridge 1947; Whittaker 1975). Vegetation clearly affects fire regimes (by controlling fuel type, ground-level microclimate, biomass and its availability to burn); similarly, fire regimes affect vegetation by promoting or inhibiting the persistence of different species. Despite the fact that both vegetation and fire regimes tend to be well predicted by climate at continental and regional scales (e.g. Russell-Smith *et al.* 2007; Murphy and Bowman 2012), it is difficult to disentangle the direct effects of climate from stabilising feedbacks between fire and vegetation, which may themselves be in equilibrium with climate. Nonetheless, fire-mediated feedbacks and alternative stable states in vegetation patterns have been suggested for vegetation systems across Australia (Table 5.1). These include models for the persistence of less flammable vegetation (e.g. rainforest, from Tasmania to the tropics; arid *Acacia* woodlands; groves of fire-sensitive conifers) within a flammable (often grassy) matrix. In a number of cases, anthropogenic interference is thought to have played a key role in maintaining vegetation boundary feedback systems (Table 5.1).

The system for which the most solid evidence of strong stabilising feedbacks maintaining vegetation and fire regimes is the obligate-seeding tall eucalypt forests of south-eastern Australia – mountain ash (*Eucalyptus regnans*) and alpine ash (*Eucalyptus delegatensis*). Jackson (1968) proposed a groundbreaking alternative stable state model, whereby random variation in fire intervals determined whether the vegetation at a site was heath, sclerophyllous 'scrub', *Eucalyptus* forest or *Nothofagus* rainforest. Strong stabilising feedbacks between vegetation state and fire hazard meant that once a site was in a particular vegetation state, it was likely to remain in that state. There are two key feedbacks in Jackson's model. First, the development of a dense forest canopy moderates fire weather at ground level and maintains higher fuel moisture, thereby reducing the frequency of fire. Second, a forest canopy also enhances the nutrient status of the soil, which

Table 5.1 **Systems where fire–vegetation feedbacks have been documented or suggested, the role of fire regime in moving between states (switch) and mechanisms that reinforce their boundaries.**

Communities	Region	Switch	Mechanism	Reference
Athrotaxis cupressoides trees within a sub-alpine forest	Alpine Tasmania	fire severity	seedling herbivory and changed fuel profile	Holz *et al.* (2015)
Moorland–scrub–wet sclerophyll–rainforest	SW Tasmania	fire frequency	soil nutrition, hydrology and fuel availability mediated by aspect and geology	Jackson (1968); Wood *et al.* (2011)
Forested Cyperaceae-*Sphagnum* wetland–non-forested Restionaceae wetland	SW Tasmania	fire severity	changed hydrology: higher interception and transpiration versus water retaining rhizomatous mats	Fletcher *et al.* (2014)
Old-growth *Eucalyptus regnans*–regrowth dominated by *Acacia*	Central Victoria	fire frequency, drying climate, logging	tree density and fuel structure	Lindenmayer *et al.* (2011)
Treeless grasslands, sedgelands – eucalypt and rainforest	Northern Tasmania plateaus	fire frequency	fuel accumulation and availability: likely anthropogenically maintained	Bowman *et al.* (2012b)
Low diversity regeneration–old-growth *Eucalyptus salmonophloia* woodlands	SW Western Australia	fire interval	decreasing fuel availability in mature stands	Gosper *et al.* (2013)
Callitris glaucophylla patches within *Eucalyptus* forest	NE New South Wales	fire severity	difference in flammability of *Callitris* and altered fuel structure	Cohn *et al.* (2011)
Mulga shrubland–*Triodia* grassland	Central Northern Territory	fire frequency	non-fire soil–vegetation pattern feedbacks, fuel arrangement and flammability: partly anthropogenically maintained	Bowman *et al.* (2008)
Bunya mountain grassy balds–rainforest/sclerophyll forest	Southern Queensland	fire frequency	fuel accumulation and availability: likely anthropogenically maintained	Fensham and Fairfax (2006); Butler *et al.* (2014)
Callitris intratropica groves within savanna	Northern Territory tropics	fire severity	closed canopies inhibit fires by physically altering understorey fuel	Trauernicht *et al.* (2012)
Forest–savanna	Northern Territory tropics	fire occurrence	water, nutrients affecting tree growth rate, crown openness, humidity	Murphy and Bowman (2012); Clarke *et al.* (2014)
Rainforest–tall eucalypt forest–savanna	Northern Queensland	fire occurrence	microclimate, soil type, fertility, seasonal water stress	Warman and Moles (2009); Little *et al.* (2012)

means that the forest can recover from disturbance more easily than a forest establishing on a non-forest site. The key elements of Jackson's model, especially the vegetation–fire feedback, have stood the test of time (McCarthy *et al.* 1999; Wood and Bowman 2012). The extent to which vegetation–fire feedbacks can alter the distribution of fire regimes and vegetation at regional to continental scales is a major unresolved question in biogeography (Bond *et al.* 2005). For example, the existence of large tracts of fire-sensitive and fire-inhibiting *Acacia* communities in central Australia may well be relictual, reflecting a wider distribution of this community in the past when fire activity was less, and temperatures were cooler (Bowman *et al.* 2008). Given sustained burning, they could potentially be converted to fire-prone (and fire-tolerant) spinifex- (*Triodia* spp.) dominated communities (Bowman *et al.* 2008).

5.4 Interactions between plant traits and fire regime

5.4.1 Fire effects on plants and environment

Fire affects plants directly through heating, oxidation and desiccation, and is often lethal to plant tissues and whole plants. As well as the direct damaging effects of fire on plant tissues, short-term heat and chemical by-products of fire, such as in smoke, sometimes trigger morphological and physical changes in protected tissues to enable seed dispersal, overcome seed dormancy, cue seed germination and stimulate flowering in many species (Bell *et al.* 1993; Auld and Bradstock 1996; Lamont and Downes 2011; Flematti *et al.* 2013). Fire also alters soil, microclimate and biotic attributes (competitors, facilitators, herbivores) over a range of longer time frames, eliciting plant responses (York *et al.* 2012). How these influence individual plants and populations, the occurrence of plant species, the structure and composition of plant communities and the distribution of vegetation types depends on the nature of fire events and their environmental consequences, and more generally, by interactions between plant species' traits, or functional attributes, and the components of the fire regime (Keith 2012).

Fire remobilises nutrients stored in biomass and litter, increasing their availability in soils in the post-fire environment, but it can also volatilise macronutrients leading to their loss (Brundrett *et al.*, Chapter 7, this volume). While volatilised nitrogen can be replenished by nitrogen-fixing soil biota, soil phosphorous derives solely from weathering, and burning contributes to its incremental loss from soil and vegetation. Many Australian soils are both extraordinarily old and phosphorous depleted by global standards. A compelling argument may be made that, to the extent that phosphorous depletion has been exacerbated by the burning of vegetation over many millions of years, these plant traits, and additional vegetation attributes associated with nutrient impoverishment such as low productivity, low stature and sclerophylly, result from vegetation–fire feedbacks at geological timescales. For example, low-phosphorus content in foliage increases flammability (Orians and Milewski 2007; Scarff *et al.* 2012).

Fire changes the light, wind and temperature environment, the activity and abundance of competitors, herbivores and predators and the thermal, physical, hydrological, chemical, and biological properties of soils. Plants are sensitive to many of these changes, and also to by-products such as smoke, charcoal and ash. Facultative post-fire flowering may be stimulated or enhanced by the post-fire pulse in availability of nutrients, light or water (Lamont and Downes 2011). Obligate post-fire flowering may occur at a similar time to facultative flowering, but is ultimately triggered by fire-derived signals, such as ethylene (Lamont and Downes 2011). Similarly, chemicals associated with smoke, including karrikinolide and cyanide, stimulate germination in the post-fire environment (Flematti *et al.* 2013). Heat associated with fire drives structural changes that respectively overcome physical dormancy in soil-stored seeds with impermeable coats, and trigger post-fire release and dispersal of non-dormant seeds held in the woody structures (e.g. cones, nuts, etc.) of serotinous species. All of these processes result in enhanced seed germination in the seasons and years immediately following fire. In some systems (e.g. Mediterranean heathlands), seedling establishment is thought to occur almost exclusively following fire (Enright *et al.* 2007).

5.4.2 Plant responses to fire: traits and functional types

Plant species persist in fire-prone landscapes through three key groups of functional traits that: (i) enable individuals to survive fire; (ii) enable populations to regenerate *in situ*; or (iii) enable species to recolonise after fire (Noble and Gitay 1996; Keith 2012). Auld and Ooi (Chapter 10, this volume) discuss plant traits, including in relation to fire, in detail. The most important of these relate to fire survival and resprouting capacity, plant maturation time and longevity, the timing of recruitment in relation to fire (immediately following, delayed, independent) and seed persistence and dispersal traits (Noble and Slatyer 1980; Noble and Gitay 1996; Keith 2012; Miller and Dixon 2014). It is the interaction of these traits and the components of fire regimes that define the ability of species to persist under a given regime, and therefore the composition and structure of vegetation present.

While fires regime influence the distribution of species with particular plant traits, some of these traits might trade-off or be enabled – e.g. fire-stimulated flowering and resprouting (Lamont and Downes 2011) – by combination with other traits which are, in turn, influenced by other environmental variables. Resprouting is a dominant trait across Australia. It occurs in 61% of 2880 woody taxa recorded in samples of eight fire-prone ecosystems across Australia by Clarke *et al.* (2015). In those species recorded, resprouting varied strongly among growth form (>90% of climbers, arborescent monocots, cycads and ferns, 80% of trees, 50% of shrubs, 54% of subshrubs, 10% of parasites). Proportions also varied among vegetation types from 70 to 90% of savanna, spinifex grassland and rainforest species resprouting to ~50% of tussock grassland and heath species and 38% of mallee shrubland species (Clarke *et al.* 2015).

Resprouting also varies within regions: the proportion of resprouting species declines with temperature and rainfall in Sydney Basin dry sclerophyll forest (Hammill *et al.* 2016), but increases with rainfall and fire frequency in central Australian desert grassland and shrubland systems (Nano and Clarke 2011). In shrublands in south-western Australia resprouting varies from 62–74% of species in northern sandplains to 50% in the less-seasonal southern sandplains (Bell 2001).

Clarke *et al.* (2015) also examined the proportion of species that recruited from seed after fire: two-thirds of 2880 woody species stored their seeds (in canopy or soil seedbanks) between fire events and had fire-cued germination. Proportions were highest in ecosystems with crown fires across southern Australia (mallee 92%; heathlands 88%; eucalypt forests 77%) and lowest in rainforests (20%) and *Acacia* shrublands (27%). For species with canopy seed storage (serotiny), seed retention appears to be greater in less productive, more fire-prone environments. In New England, serotiny is highest in low-nutrient shrublands with high-severity, intermediate interval fire regimes and lowest in high nutrient, low-flammability forests (Clarke *et al.* 2013a). In south-western Australia, serotiny varies within *Banksia* species, being strong in drier, low shrubland communities that burn only in canopy fires to almost no serotiny in moister woodlands that may experience understorey or canopy fires (Cowling and Lamont 1985).

Taking a deductive approach, Noble and Gitay (1996) identified 13 categories of population dynamic responses to disturbance based on 27 feasible species types that combine, and are emergent from, plant traits. Other Australian fire-response functional type analyses disregard or combine uncommon combinations to identify fewer plant functional types (PFTs; Table 5.2). The most basic consider only the presence and absence of resprouting and seed bank persistence traits to derive four PFTs (e.g. Clarke *et al.* 2015; Hammill *et al.* 2016), others include longevity relative to fire interval and seed storage mode (soil or canopy stored) (Gill 1981; Keith 1996, 2012; Bradstock and Kenny 2003; Miller and Dixon 2014; Groom and Lamont 2015).

Table 5.2 **Varying plant functional type (PFT) sensitivity to fire regime elements for south-west Australian Mediterranean-climate shrubland plants. Compares two common PFT typologies and shows the number of fire-regime-driven demographic processes that threaten populations across PFTs, classified by longevity (L) in relation to mean fire interval (MFI), presence of resprouting (R+, R-), seed persistence (P+, P-) and storage mode (based on Miller and Dixon 2014). Threatening processes may or may not be shared across PFTs.**

Traits			Fire regime attribute→	Interval		Season	Intensity		Total	
L	R	P	Plant Functional Type	Short	Long	Aseasonal	High	Low		
<<MFI	R–	P+	Winter annuals		1	1	1		3	R- P+Obligate seeders
	R–	P+	Annual fire ephemerals		1	1	1	1	4	
	R–	P+	Longer-lived fire ephemerals	1	1	1	1	1	5	
=MFI	R–	P–	Seed non-persistent non-sprouters	1			1		2	R- P-Fire avoiders/ Obligate recolonisers
	R–	P+	Serotinous non-sprouters	1	1	1	1	1	5	
	R–	P+	Soil seedbank non-sprouters	1	1	1		1	4	R+ P- Obligate resprouters
>MFI	R+	P–	Seed non-persistent resprouters	3	2	2	1		8	
	R+	P+	Serotinous resprouters	4	2	1	2	1	10	R+ P+ Facultative resprouters
	R+	P+	Soil seedbank resprouters	4	2	2	2	1	11	
			Total	4	5	5	3	2	19	

Of course, reality is not simple: seedbank persistence is a variable, continuous trait and resprouting is not strictly binary. The capacity to resprout can vary among populations within species, and individuals within populations, and may be influenced by plant size, fire intensity or interval (Enright *et al.* 2011; Clarke *et al.* 2013b; Lawes *et al.* 2013).

PFTs represent different ways of surviving in a fire-prone landscape, but they interact with fire regimes in different ways. Noble and Gitay (1996) assessed 4 of their 13 population dynamics archetypes as sensitive to relatively long fire intervals, two to short intervals, and two to both. Miller and Dixon (2014) list 19 processes that theoretically threaten the persistence of sclerophyllous shrubland PFTs experiencing short or long fire intervals, aseasonal or high or low-intensity fire (Table 5.2). Only three of their nine PFTs are sensitive to all of these fire regime changes and none of the demographic or environmental processes affect all PFTs. The most widespread threat, affecting six PFTs, is that of populations failing to reach maturity or replacement levels of seed bank accumulation when successive fires occur at short intervals. PFTs not found in shrublands, and fire behaviour in other systems may be associated with other threatening processes: Keith (1996) analysed threats from fire-related processes to the persistence of plant populations generally and identified 20 processes. The power of these PFT analyses in relation to vegetation distribution is the potential to predict the response of species to varying fire regimes, and to guide fire management planning. Time to maturity and senescence of serotinous non-sprouting species is used as a guide for identification of tolerable fire intervals in several states (Bradstock and Kenny 2003; Burrows *et al.* 2008; Gosper *et al.* 2012).

Empirical tests of PFT-derived hypotheses for vegetation community responses to varying fire regime are rare. Keith *et al.* (2007) examined change in population size for 80 species in six PFTs of sclerophyllous wet heathland in New South Wales in relation to varying fire interval. Serotinous and non-serotinous obligate-seeder shrubs and fire ephemeral herbs all increased in abundance while resprouter shrubs, non-rhizomatous resprouting herbs and graminoids, and rhizomatous resprouting graminoids, herbs and ferns decreased; the changes were attributed to moderate fire intervals that favoured non-sprouters and the effects of competition from these PFTs on resprouter PFTs. A similar approach in south-western Australian shrublands (Enright *et al.* 2014) showed that non-sprouters were sensitive and resprouters resilient to shorter fire intervals, but that this relationship interacts with varying post-fire rainfall conditions. In the same system, however, Enright *et al.* (2011) also found that resprouter survival was sensitive to extremely short and extremely long fire intervals.

5.4.3 Plant traits and fire feedbacks

Plant traits that increase flammability (such as the presence of oils, fine leaves and twigs, and their retention on plants) or those that decrease it (such as large minimum branch diameters, open canopies and foliage with high water content) can influence fire regimes and enhance vegetation boundaries through feedback processes. For example, the succulent vegetation of arid salt flats dominated by *Tecticornia* and *Sarcostemma* species (Eldridge *et al.*, Chapter 24, this volume) have low productivity and low flammability, they burn very infrequently and resist encroachment of fire. Similarly, relatively low-flammability *Callitris* (cypress pine) trees change the local environment by modifying understorey vegetation and fuel structure to reduce fire intensity, which favours the persistence of *Callitris* (Cohn *et al.* 2011; Trauernicht *et al.* 2012).

Other dominant species and associated vegetation types can change environments in a manner that influences fire regimes by altering fuel types. For example, rainforest trees invading savanna exclude light and therefore reduce grassy fuel loads, while increasing biomass of litter fuels, which tend to have high bulk density and high water content, and are significantly less likely to burn. By altering the local environment to cooler and more humid conditions,

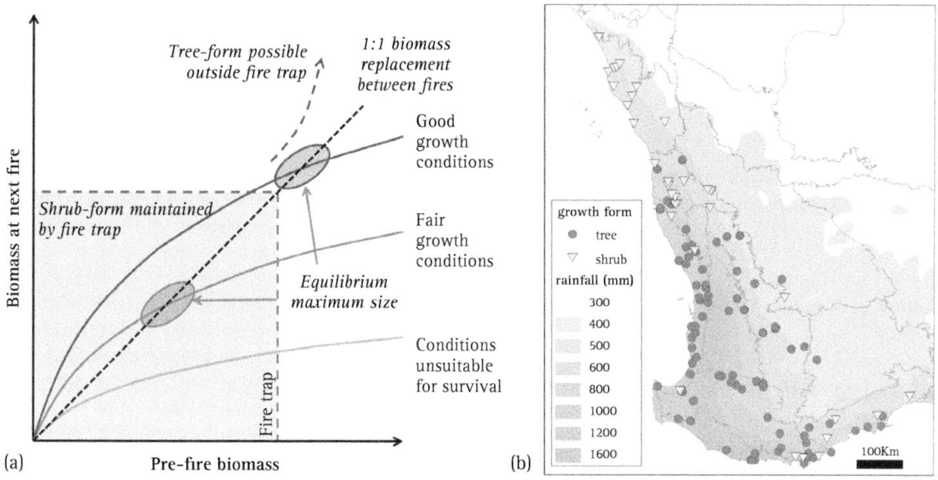

Figure 5.4 **(a)** Grady and Hoffmann (2012) hypothesise that varying growth rate with time since fire and mean fire return interval define the mean maximal size of woody resprouters at a site. The point at which the size-growth curve intercepts the self-replacement curve (where growth over the mean fire return interval ≡ pre-fire size) identifies a size that is difficult for plants to exceed under a given fire frequency. This point determines the structural vegetation height for resprouter populations and resprouter-dominated communities. The height of the growth curve varies with conditions for growth, leading to variation in equilibrial maximum height. In tree species subject to a fire trap (sensu Bond and Midgley 2001), where the equilibrial maximum height sits relative to the trap threshold determines whether individuals or populations can develop tree form. **(b)** Distribution of shrub and tree form *Banksia attenuata* in relation to rainfall (shading). *Data from Taylor and Hopper (1988).*

these vegetation communities also resist fire incursion under all but the most extreme conditions.

Resprouting can be epicormic (from buds or meristematic tissues buried in bark of above-ground stems), lignotuberous (arising from a more-or-less subsurface woody base), or from rhizomes or root suckers (Clarke *et al.* 2015). Epicormic resprouting enables the growth and persistence of trees in systems where fire intervals are shorter than the time required for tree seedlings or basally resprouting shoots to attain canopy height. The dominance of epicormic versus lignotuberous resprouting forms defines structural boundaries between basally resprouting mallee or heathland vegetation and adjacent sclerophyll forests and woodlands. *Banksia attenuata* and *B. menziesii* have two geographically distinct and genetically fixed forms: epicormically resprouting trees; and basally resprouting shrubs (Figure 5.4; Cowling and Lamont 1985; He 2014). The geographic distribution of this bud location trait within these dominant species defines the boundary between *Banksia* woodland and shrublands. This structural boundary is associated with, and leads to, differences in fire regime and understorey microclimate and community composition.

Few shrubs or emergent trees obtain sufficient height to escape canopy scorch when shrublands burn, but low-intensity surface fires that do not burn tree canopies can occur in *Banksia* woodlands, in addition to canopy fire (Bell *et al.* 1992). Tree-form juveniles resprout basally when burnt and, if repeatedly burnt will never attain tree form, either dying or perpetually persisting as resprouting juveniles – a situation described as a ‘fire trap’ (Bond and Midgley 2001). Grady and Hoffmann (2012) describe a process for woody resprouters where varying environmental conditions for growth, varying growth rate with time after fire and fire return intervals interact to define shrub and shrubland height (Figure 5.4a). Considering this process in relation to tree height thresholds required to escape fire traps may provide an explanation for *Banksia* woodland and other structural vegetation boundaries that reinforce edaphic or climatic thresholds and fire regime differences. If conditions for growth or fire frequency change, for instance under future climates, then the location of these vegetation boundaries may also be expected to change.

5.5 Consequences of contemporary anthropogenic change for fire–vegetation interactions

5.5.1 The transition from prehistoric to modern fire regimes

As noted above, interactions between Aboriginal people and fire regimes are uncertain but likely to have varied within climatic constraints at different spatial scales around the continent from active burning to negligible intervention to active fire exclusion. The transition from pre-human fire regimes to Aboriginal fire regimes is even less certain, and largely undetectable in many paleo-records (Mooney *et al.* 2011; Chapter 3, this volume). Nonetheless, widespread replacement of this traditional Aboriginal fire management following European invasion appears to have triggered various changes to fire regimes in different parts of the continent. In some tropical regions, perhaps relatively recently, there has been a replacement of frequent ignitions, throughout different times of year when vegetation will carry fire (Ens *et al.*, Chapter 4, this volume), to less frequent ignitions strongly concentrated in the time of year with peak lightning activity. In tropical areas with intensive pastoralism, overall fire frequencies are likely to have been reduced by the offtake of grass biomass by cattle. In other regions, however, there have been dramatic increases in the frequency of ignitions (Mooney *et al.* 2011; Bowman *et al.* 2012a). The outcomes of these changes vary between and within ecosystems.

5.5.2 Landscape fragmentation

Profound changes to fire regimes have been caused by land-use change and alteration of ignition patterns since Europeans settlement. Because fire is a spatial process, propagating across landscapes, the landscape context exerts strong control over a fire regime. This is particularly important for small patches of remnant vegetation within an agricultural matrix, where ignitions are unlikely to occur with sufficient frequency to maintain pre-European fire frequencies. Many remnant woodlands in southern Australia now experience fire relatively rarely, whereas in the past they would have been intermittently burnt by low-intensity fires (Yates and Ladd 2010). The consequences of such reductions in fire frequency can vary, ranging from regeneration failure (Gent and Morgan 2007) to woody thickening and turnover of dominant woody species (Lunt 1998a).

5.5.3 Invasive plant species

The introduction of exotic grasses is leading to dramatic changes to fire regimes in some Australian landscapes. Such grasses can initiate a 'grass–fire cycle' (D'Antonio and Vitousek 1992), by increasing fuel biomass and/or providing biomass which more readily burns and can burn with higher intensity (e.g. low bulk density, low moisture content). The consequent increases in fire frequency and intensity can lead to loss of woody cover, which further promotes increases in the biomass of the exotic grasses, leading to a positive feedback cycle. The most notable Australian examples of this process include gamba grass, veldt grass and buffel grass. Gamba (*Andropogon gayanus*) has become a serious problem in the high-rainfall savannas in the Darwin region. It dramatically increases fine fuel loads and fire intensities, profoundly altering ecosystem function, and often entirely removing the eucalypt canopy and *Callitris* stands within two decades (Ferdinands *et al.* 2006; Setterfield *et al.* 2010; Bowman *et al.* 2014a). It is rapidly spreading across the Top End. Veldt grass (*Ehrharta calycina*) is a serious weed in the *Banksia* woodlands of south-western Australia, and facilitates canopy loss due to high-severity fire (Fisher *et al.* 2009). Buffel grass (*Cenchrus ciliaris*) is increasing fire frequency and intensity over vast areas of central Australia (Miller *et al.* 2010), and facilitates the penetration of fire into fire-sensitive communities, which would otherwise burn rarely, if ever, including brigalow in central Queensland (Butler and Fairfax 2003). At the other extreme, fire-sensitive mesic species, such as *Pittosporum undulatum* (native to restricted areas in south-eastern Australia) can alter microclimates and fuel structures as they invade drier forest and other adjacent ecosystems to reduce fire spread, and change vegetation composition and structure (Rose and Fairweather 1997; Gleadow and Narayan 2007).

5.5.4 Climate change, interval squeeze and landscape traps

Climate change could profoundly alter some fire regimes, given the control exerted by longer-term climate and short-term weather. However, it is unlikely that there will be increases in either fire frequency or intensity 'across the board' due to a warming, drying climate. King *et al.* (2013) have suggested that fire activity is likely to decrease in central Australia, due to a decrease in fuel biomass, while increasing in high-rainfall parts of southern Australia, due to drier fuels and more extreme fire weather. Indeed, it seems likely that the regions most vulnerable to increasing fire frequency and/or intensity are the temperate forests and heathlands (Bradstock 2010; Enright *et al.* 2015). In these regions, infrequent high-intensity fires are associated with high summer temperatures and decadal drought – both of which are predicted to increase in frequency under future climates. Increased human population density and recreational activity could also lead to more frequent ignitions in these landscapes.

While it seems likely that climate change will directly impact many species and communities, those in landscapes where fire activity is likely to increase are particularly vulnerable – with fire potentially acting as a catalyst for the loss of species from a site. Enright *et al.* (2015) have recently proposed the 'interval squeeze' conceptual model to describe how climate change might cause rapid fire-driven decline of some woody species (especially obligate seeders). They identified two main processes associated with a changing climate, and climate–fire interactions, which could drive the loss of woody species from fire-prone sites: (i) *demographic shift*, where rates of plant growth, survival and propagule production are reduced; and (ii) *fire-interval shift*, where increased fire frequency reduces the time available to mature and reproduce between fires. In combination, these processes reduce the ability of woody species to quickly recover from fire, while at the same time reducing intervals between fires. Hence, sensitive species are 'squeezed' between declining demographic rates and shortening fire intervals. This conceptual model is most applicable in the parts of the continent where fire activity tends to be strongly limited by fuel moisture (i.e. much of the high-rainfall temperate zone).

The vulnerability of obligate-seeder forests to changing fire regimes has been highlighted recently with the development of the 'landscape trap' conceptual model by Lindenmayer *et al.* (2011). Focussing on the mountain ash (*Eucalyptus regnans*) forests of south-eastern Australia, they proposed that there is strong positive feedback between stand-replacing disturbance (e.g. high-intensity fire or clear-fell logging) and the risk of a subsequent stand-replacing wildfire. Once an old-growth forest is clear-felled or burnt by a stand-replacing fire, it has a significantly elevated risk of subsequent fires, which persists for several decades. Because obligate-seeder trees must grow from seed to reproductive maturity between high-severity fires, fires at intervals shorter than the maturation time of the individual trees (about 20 years) lead to loss of this obligate-seeder species and transformation of these forests to an alternative (possibly non-forest) state. The landscape trap model is likely to be equally applicable to alpine ash (*Eucalyptus delegatensis*) forests, which have a similar fire-driven regeneration cycle. Somewhat surprisingly, in their articulation of the landscape trap model, Lindenmayer *et al.* (2011) ignore the potential for extrinsic drivers of fire activity (e.g. climate and ignitions by humans) to increase wildfire incidence, thereby accelerating the fire-driven loss of obligate-seeder forests (Bowman *et al.* 2013, 2014b).

Acknowledgements

Murphy was supported by a fellowship from the Australian Research Council (DE130100434). Mark Ooi, Owen Price and David Keith provided constructive comments on the manuscript.

References

Auld, T.D. and Bradstock, R.A. (1996) Soil temperatures after the passage of a fire: do they influence the germination of buried seeds? *Austral Ecology* 21, 106–109.

Australian Bureau of Meteorology (2015) *Maps of Average Conditions*. Australian Bureau of Meteorology: Canberra. Available at: http://www.bom.gov.au/climate/averages/maps.shtml (accessed 8 July 2016).

Belcher, C.M., Mander, L., Rein, G. *et al.* (2010) Increased fire activity at the Triassic/Jurassic boundary in Greenland due to climate-driven floral change. *Nature Geoscience* 3, 426–429.

Bell, D.T. (2001) Ecological response syndromes in the flora of southwestern Western Australia: fire resprouters versus reseeders. *Botanical Review* 67, 417–440.

Bell, D.T., Loneragan, W.A., Ridley, W.J. *et al.* (1992) Response of tree canopy species of Kings Park, Perth, Western Australia to the severe summer wildfire of January 1989. *Journal of the Royal Society of Western Australia* 75, 35–39.

Bell, D.T., Plummer, J.A. and Taylor, S.K. (1993) Seed germination ecology in southwestern Western Australia. *The Botanical Review* 59, 24–73.

Black, M.P. and Mooney, S.D. (2006) Holocene fire history from the Greater Blue Mountains World Heritage Area, New South Wales, Australia: the climate, humans and fire nexus. *Regional Environmental Change* 6, 41–51.

Bond, W.J. (2015) Fires in the Cenozoic: a late flowering of flammable ecosystems. *Frontiers in Plant Science* 5.

Bond, W.J. and Midgley, J.J. (2001) Ecology of sprouting in woody plants: the persistence niche. *Trends in Ecology and Evolution* 16, 45–51.

Bond, W.J. and Midgley, J.J. (2012) Fire and the angiosperm revolutions. *International Journal of Plant Sciences* 173, 569–583.

Bond, W.J. and Scott, A.C. (2010) Fire and the spread of flowering plants in the Cretaceous. *New Phytologist* 188, 1137–1150.

Bond, W.J., Woodward, F.I. and Midgley, G.F. (2005) The global distribution of ecosystems in a world without fire. *New Phytologist* 165, 525–538.

Bowler, J.M., Johnston, H., Olley, J.M. *et al.* (2003) New ages for human occupation and climatic change at Lake Mungo, Australia. *Nature* 421, 837–840.

Bowman, D.M.J.S. (1998) The impact of Aboriginal landscape burning on the Australian biota. *New Phytologist* 140, 385–410.

Bowman, D.M.J.S. and Prior, L.D. (2004) Impact of Aboriginal landscape burning on woody vegetation in *Eucalyptus tetrodonta* savanna in Arnhem Land, northern Australia. *Journal of Biogeography* 31, 807–817.

Bowman, D.M.J.S. and Wilson, B.A. (1988) Fuel characteristics of coastal monsoon forests, Northern Territory, Australia. *Journal of Biogeography* 15, 807–817.

Bowman, D.M.J.S., Walsh, A. and Milne, D.J. (2001) Forest expansion and grassland contraction within a Eucalyptus savanna matrix between 1941 and 1994 at Litchfield National Park in the Australian monsoon tropics. *Global Ecology and Biogeography* 10, 535–548.

Bowman, D.M.J.S., Walsh, A. and Prior, L.D. (2004) Landscape analysis of Aboriginal fire management in Central Arnhem Land, north Australia. *Journal of Biogeography* 31, 207–223.

Bowman, D.M.J.S., Boggs, G.S. and Prior, L.D. (2008) Fire maintains an Acacia aneura shrubland–*Triodia* grassland mosaic in central Australia. *Journal of Arid Environments* 72, 34–47.

Bowman, D.M.J.S., Murphy, B.P., Burrows, G.E. and Crisp, M.D. (2012a) Fire regimes and the evolution of the Australian biota. In *Flammable Australia: Fire Regimes, Biodiversity and Ecosystems in a Changing World*, ed. R.A. Bradstock, A.M. Gill and R.J. Williams. CSIRO Publishing: Collingwood, pp. 27–47.

Bowman, D.M.J.S., Wood, S.W., Neyland, D. *et al.* (2012b) Contracting Tasmanian montane grasslands within a forest matrix is consistent with cessation of Aboriginal fire management. *Austral Ecology* 38, 627–638.

Bowman, D.M.J.S., Murphy, B.P., Boer, M.M. *et al.* (2013) Forest fire management, climate change, and the risk of catastrophic carbon losses. *Frontiers in Ecology and the Environment* 11, 66–68.

Bowman, D.M.J.S., MacDermott, H.J., Nichols, S.C. and Murphy, B.P. (2014a) A grass–fire cycle eliminates an obligate-seeding tree in a tropical savanna. *Ecology and Evolution* 4, 4185–4194.

Bowman, D.M.J.S., Murphy, B.P., Neyland, D.L.J. *et al.* (2014b) Abrupt fire regime change may cause landscape-wide loss of mature obligate-seeder forests. *Global Change Biology* 20, 1008–1015.

Bowman, D.M.J.S., Murphy, B.P., Williamson, G.J. and Cochrane, M.A. (2014c) Pyrogeographic models, feedbacks and the future of global fire regimes. *Global Ecology and Biogeography* 23, 821–824.

Bradshaw, S.D., Dixon, K.W., Hopper, S.D. *et al.* (2011) Little evidence for fire-adapted plant traits in Mediterranean-climate regions. *Trends in Plant Science* 16, 69–76.

Bradstock, R.A. (2010) A biogeographic model of fire regimes in Australia: current and future implications. *Global Ecology and Biogeography* 19, 145–158.

Bradstock, R.A. and Kenny, B.J. (2003) An application of plant functional types to fire management in a conservation reserve in southeastern Australia. *Journal of Vegetation Science* 14, 345–354.

Brown, S.A.E., Scott, A.C., Glasspool, I.J. and Collinson, M.E. (2012) Cretaceous wildfires and their impact on the Earth system. *Cretaceous Research* 36, 162–190.

Burrows, N.D., Burbidge, A.A., Fuller, P.J. and Behn, G. (2006) Evidence of altered fire regimes in the Western Desert region of Australia. *Conservation Science Western Australia* 5, 14–26.

Burrows, N.D., Wardell-Johnson, G. and Ward B. (2008) Post-fire juvenile period of plants in south-west Australia forests and implications for fire management. *Journal of the Royal Society of Western Australia* 91, 163–174.

Butler, D.W. and Fairfax, R.J. (2003) Buffel Grass and fire in a Gidgee and Brigalow woodland: a case study from central Queensland. *Ecological Management and Restoration* 4, 120–125.

Butler, D.W., Fensham, R.J., Murphy, B.P. *et al.* (2014) Aborigine-managed forest, savanna and grassland: biome switching in montane eastern Australia. *Journal of Biogeography* 41, 1492–1505.

Clarke, P.J. (2002) Habitat islands in fire-prone vegetation: do landscape features influence community composition? *Journal of Biogeography* 29, 677–684.

Clarke, P.J., Knox, K.E.J. and Butler, D. (2013a) Fire, soil fertility and delayed seed release: a community analysis of the degree of serotiny. *Evolutionary Ecology* 27, 429–443.

Clarke, P.J., Lawes, M.J., Midgley, J.J. *et al.* (2013b) Resprouting as a key functional trait: how buds, protection and resources drive persistence after fire. *New Phytologist* 197, 19–35.

Clarke, P.J., Prior, L.D., French, B.J. *et al.* (2014) Using a rainforest-flame forest mosaic to test the hypothesis that leaf and litter fuel flammability is under natural selection. *Oecologia* 176, 1123–1133.

Clarke, P.J., Lawes, M.J., Murphy, B.P. *et al.* (2015) A synthesis of postfire recovery traits of woody plants in Australian ecosystems. *Science of the Total Environment* 534, 31–42.

Codding, B.F., Bliege Bird, R., Kauhanen, P.G. and Bird, D.W. (2014) Conservation or co-evolution? Intermediate levels of Aboriginal burning and hunting have positive effects on kangaroo populations in Western Australia. *Human Ecology* 42, 1–11.

Cohn, J.S., Lunt, I.D., Ross, K.A. and Bradstock, R.A. (2011) How do slow-growing, fire-sensitive conifers survive in flammable eucalypt woodlands? *Journal of Vegetation Science* 22, 425–435.

Cowling, R.M. and Lamont, B.B. (1985) Variation in serotiny of three Banksia species along a climatic gradient. *Australian Journal of Ecology* 10, 345–350.

Crisp, M.D., Burrows, G.E., Cook, L.G. *et al.* (2011) Flammable biomes dominated by eucalypts originated at the Cretaceous–Palaeogene boundary. *Nature Communications* 2.

D'Antonio, C.M. and Vitousek, P.M. (1992) Biological invasions by exotic grasses, the grass/fire cycle, and global change. *Annual Review of Ecology and Systematics* 23, 63–87.

Daniau, A.L., Bartlein, P.J., Harrison, S.P. *et al.* (2012) Predictability of biomass burning in response to climate changes. *Global Biogeochemical Cycles* 26.

Department of the Environment and Energy (2007) *NVIS data products 4.2*. Australian Government: Canberra. Available at: http://www.environment.gov.au/land/native-vegetation/national-vegetation-information-system/data-products (accessed 10 May 2016).

Enright, N.J., Mosner, E., Miller, B.P. *et al.* (2007) Soil versus canopy seed storage and plant species coexistence in species-rich shrublands of southwestern Australia. *Ecology* 88, 2292–2304.

Enright, N.J., Fontaine, J.B., Westcott, V.C. *et al.* (2011) Fire interval effects on persistence of resprouter species in Mediterranean-type shrublands. *Plant Ecology* 212, 2071–2083.

Enright, N.J., Keith, D.A., Clarke, M.F. and Miller, B.P. (2012) Fire regimes in Australian sclerophyllous shrubby ecosystems: heathlands, heathy woodlands and mallee woodlands. In *Flammable Australia: Fire Regimes, Biodiversity and Ecosystems in a Changing World*, ed. R.A. Bradstock, A.M. Gill and R.J. Williams. CSIRO Publishing: Melbourne, pp. 215–234.

Enright, N.J., Fontaine, J.B., Lamont, B.B. *et al.* (2014) Resistance and resilience to changing climate and fire regime depend on plant functional traits. *Journal of Ecology* 102, 1572–1581.

Enright, N.J., Fontaine, J.B., Bowman, D.M.J.S. *et al.* (2015) Interval squeeze: altered fire regimes and demographic responses interact to threaten woody species persistence as climate changes. *Frontiers in Ecology and the Environment* 13, 265–272.

Fensham, R.J. and Fairfax, R.J. (2006) Can burning restrict eucalypt invasion on grassy balds? *Austral Ecology* 31, 317–325.

Ferdinands, K., Setterfield, S., Douglas, M. and Barratt, J. (2006) Africanising the tropical woodlands: canopy loss and tree death following gamba grass (Andropogon gayanus) invasion. In *Proceedings of the 15th Australian Weeds Conference*, ed. C. Preston, J.H. Watts and N.D. Crossman. Weed Management Society of South Australia: Adelaide, pp. 296.

Fisher, J.L., Loneragan, W.A., Dixon, K. *et al.* (2009) Altered vegetation structure and composition linked to fire frequency and plant invasion in a biodiverse woodland. *Biological Conservation* 142, 2270–2281.

Flematti, G.R., Waters, M.T., Scaffidi, A. *et al.* (2013) Karrikin and cyanohydrin smoke signals provide clues to new endogenous plant signaling compounds. *Molecular Plant* 6, 29–37.

Fletcher, M.S. and Thomas, I. (2007) Holocene vegetation and climate change from near Lake Pedder, south-west Tasmania, Australia. *Journal of Biogeography* 34, 665–677.

Fletcher, M.S., Wood, S.W. and Haberle, S.G. (2014) A fire-driven shift from forest to non-forest: evidence for alternative stable states? *Ecology* 95, 2504–2513.

Gammage, B. (2011) *The Biggest Estate on Earth: How Aborigines Made Australia*. Allen and Unwin: Crows Nest.

Gent, M.L. and Morgan, J.W. (2007) Changes in the stand structure (1975–2000) of coastal Banksia forest in the long absence of fire. *Austral Ecology* 32, 239–244.

Giglio, L., Descloitres, J., Justice, C.O. and Kaufman, Y.J. (2003) An enhanced contextual fire detection algorithm for MODIS. *Remote Sensing of Environment* 87, 273–282.

Gill, A.M. (1975) Fire and the Australian flora: a review. *Australian Forestry* 38, 4–25.

Gill, A.M. (1981) Adaptive responses of Australian vascular plant species to fires. In *Fire and the Australian Biota*, ed. A.M. Gill, R.H. Groves and I.R. Noble. Australian Academy of Science: Canberra, pp. 243–271.

Glasspool, I.J., Scott, A.C., Waltham, D. *et al.* (2015) The impact of fire on the late Paleozoic Earth System. *Frontiers in Plant Science* 6, 756 [doi:10.3389/fpls.2015.00756].

Gleadow, R.M. and Narayan, I. (2007) Temperature thresholds for germination and survival of *Pittosporum undulatum*: implications for management by fire. *Acta Oecologica* 31, 151–157.

Gosper, C.R., Yates, C.J. and Prober, S.M. (2012) Changes in plant species and functional composition with time since fire in two Mediterranean climate plant communities. *Journal of Vegetation Science* 23, 1071–1081.

Gosper, C.R., Prober, S.M. and Yates, C.J. (2013) Multi-century changes in vegetation structure and fuel availability in fire-sensitive eucalypt woodlands. *Forest Ecology and Management* 310, 102–109.

Grady, J.M. and Hoffmann, W.A. (2012) Caught in a fire trap: recurring fire creates stable size equilibria in woody resprouters. *Ecology* 93, 2052–2060.

Groom, P. and Lamont, B. (2015) *Plant Life of Southwestern Australia, Adaptations for Survival*. De Gruyter: Berlin.

Hammill, K., Penman, T. and Bradstock, R. (2016) Responses of resilience traits to gradients of temperature, rainfall and fire frequency in fire-prone, Australian forests: potential consequences of climate change. *Plant Ecology* 217, 1–17.

Hassell, C.W. and Dodson, J.R. (2002) Fire history of south-west Western Australia prior to European settlement in 1826–1829. In *Fire in Ecosystems of South-west Western Australia: Impacts and Management*, ed. I. Abbott and N. Burrows. Bachuys: Leiden, pp. 71–85.

He, T. (2014) Ecological divergence and evolutionary transition of resprouting types in *Banksia attenuata*. *Ecology and Evolution* 4, 3162–3174.

He, T., Lamont, B.B. and Downes, K.S. (2011) *Banksia* born to burn. *New Phytologist* 191, 184–196.

Holdridge, L.R. (1947) Determination of world plant formations from simple climatic data. *Science* 105, 367–368.

Holz, A., Wood, S.W., Veblen, T.T. and Bowman, D.M.J.S. (2015) Effects of high-severity fire drove the population collapse of the sub-alpine Tasmanian endemic conifer *Athrotaxis cupressoides*. *Global Change Biology* 21, 445–458.

Jackson, W.D. (1968) Fire, air, water and earth: an elemental ecology of Tasmania. *Proceedings of the Ecological Society of Australia* 3, 9–16.

Jasper, A., Guerra-Sommer, M., Abu Hamad, A.M.B. *et al.* (2013) The burning of Gondwana. Permian fires on the southern continent: a palaeobotanical approach. *Gondwana Research* 24, 148–160.

Keeley, J.E., Pausas, J.G., Rundel, P.W. *et al.* (2011) Fire as an evolutionary pressure shaping plant traits. *Trends in Plant Science* 16, 406–411.

Keith, D.A. (1996) Fire-driven extinction of plant populations: a synthesis of theory and review of evidence from Australian vegetation. *Proceedings of the Linnean Society of New South Wales* 116, 37–78.

Keith, D.A. (2012) Functional traits: their roles in understanding and predicting biotic responses to fire regimes from individuals to landscapes. In *Flammable Australia: Fire Regimes, Biodiversity and Ecosystems in a Changing World*, ed. R.A. Bradstock, A.M. Gill and R.J. Williams. CSIRO Publishing: Melbourne, pp. 97–125.

Keith, D.A., Holman, L., Rodoreda, S. *et al.* (2007) Plant functional types can predict decade-scale changes in fire-prone vegetation. *Journal of Ecology* 95, 1324–1337.

King, K.J., Cary, G.J., Bradstock, R.A. and Marsden-Smedley, J.B. (2013) Contrasting fire responses to climate and management: insights from two Australian ecosystems. *Global Change Biology* 19, 1223–1235.

Krawchuk, M.A. and Moritz, M.A. (2011) Constraints on global fire activity vary across a resource gradient. *Ecology* 92, 121–132.

Lamont, B.B. and Downes, K.S. (2011) Fire-stimulated flowering among resprouters and geophytes in Australia and South Africa. *Plant Ecology* 212, 2111–2125.

Lawes, M.J., Midgley, J.J. and Clarke, P.J. (2013) Costs and benefits of relative bark thickness in relation to fire damage: a savanna/forest contrast. *Journal of Ecology* 101, 517–524.

Lindenmayer, D.B., Hobbs, R.J., Likens, G.E. *et al.* (2011) Newly discovered landscape traps produce regime shifts in wet forests. *Proceedings of the National Academy of Sciences of the United States of America* 108, 15887–15891.

Little, J.K., Prior, L.D., Williamson, G.J. *et al.* (2012) Fire weather risk differs across rain forest–savanna boundaries in the humid tropics of north-eastern Australia. *Austral Ecology* 37, 915–925.

Lunt, I.D. (1998a) *Allocasuarina* (Casuarinaceae) invasion of an unburnt coastal woodland at Ocean Grove, Victoria: structural changes 1971–1996. *Australian Journal of Botany* 46, 649–656.

Lunt, I.D. (1998b) Two hundred years of land use and vegetation change in a remnant coastal woodland in southern Australia. *Australian Journal of Botany* 46, 629–647.

McCarthy, M.A., Malcolm Gill, A., Lindenmayer, D.B. (1999) Fire regimes in mountain ash forest: Evidence from forest age structure, extinction models and wildlife habitat. *Forest Ecology and Management* 124, 193–203.

Meyn, A., White, P.S., Buhk, C. and Jentsh, A. (2007) Environmental drivers of large, infrequent wildfires: the emerging conceptual model. *Progress in Physical Geography* 31, 287–312.

Miller, B.P. and Dixon, K.W. (2014) Plants and fire in Kwongan vegetation. In *Plant Life on the Sandplains in Southwest Australia, a Global Biodiversity Hotspot*, ed. H. Lambers. UWA Press: Perth, pp. 147–170.

Miller, B.P., Walshe, T., Enright, N.J. and Lamont, B.B. (2007) Error in the inference of fire history from grasstrees. *Austral Ecology* 32, 908–916.

Miller, G., Friedel, M., Adam, P. and Chewings, V. (2010) Ecological impacts of buffel grass (*Cenchrus ciliaris* L.) invasion in central Australia: does field evidence support a fire-invasion feedback? *The Rangeland Journal* 32, 353–365.

Mooney, S.D., Harrison, S.P., Bartlein, P.J. *et al.* (2011) Late Quaternary fire regimes of Australasia. *Quaternary Science Reviews* 30, 28–46.

Morgan, J.W. (1999) Defining grassland fire events and the response of perennial plants to annual fire in temperate grasslands of south-eastern Australia. *Plant Ecology* 144, 127–144.

Murphy, B.P. and Bowman, D.M.J.S. (2012) What controls the distribution of tropical forest and savanna? *Ecology Letters* 15, 748–758.

Murphy, B.P., Williamson, G.J. and Bowman, D.M.J.S. (2012) Did central Australian megafaunal extinction coincide with abrupt ecosystem collapse or gradual climate change? *Global Ecology and Biogeography* 21, 142–151.

Murphy, B.P., Bradstock, R.A., Boer, M.M. *et al.* (2013) Fire regimes of Australia: a pyrogeographic model system. *Journal of Biogeography* 6, 1048–1058.

Nano, C.E.M. and Clarke, P.J. (2011) How do drought and fire influence the patterns of resprouting in Australian deserts? *Plant Ecology* 212, 2095–2110.

Noble, I.R. and Gitay, H. (1996) A functional classification for predicting the dynamics of landscapes. *Journal of Vegetation Science* 7, 329–336.

Noble, I.R. and Slatyer, R.O. (1980) The use of vital attributes to predict successional changes in plant communities subject to recurrent disturbances. *Vegetatio* 43, 5–21.

Nolan, R.H., Boer, M.M., Resco de Dios, V. *et al.* (2016) Large-scale, dynamic transformations in fuel moisture drive wildfire activity across southeastern Australia. *Geophysical Research Letters* 43, 4229–4238.

O'Donnell, A.J., Boer, M.M., McCaw, W.L. and Grierson, P.F. (2011) Vegetation and landscape connectivity control wildfire intervals in unmanaged semi-arid shrublands and woodlands in Australia. *Journal of Biogeography* 38, 112–124.

Orians, G.H. and Milewski, A.V. (2007) Ecology of Australia: the effects of nutrient-poor soils and intense fires. *Biological Reviews* 82, 393–423.

Pausas, J. and Bradstock, R.A. (2007) Plant persistence fire traits along a productivity and disturbance gradient in Mediterranean shrublands of southeastern Australia. *Global Ecology and Biogeography* 16, 330–340.

Pausas, J.G. and Ribeiro, E. (2013) The global fire–productivity relationship. *Global Ecology and Biogeography* 22, 728–736.

Roberts, R.G., Flannery, T.F., Ayliffe, L.K. *et al.* (2001) New ages for the last Australian megafauna: continent-wide extinction about 46 000 ago. *Science* 292, 1888–1892.

Rose, S. and Fairweather, P.G. (1997) Changes in floristic composition of urban bushland invaded by *Pittosporum undulatum* in northern Sydney, Australia. *Australian Journal of Botany* 45, 123–149.

Rule, S., Brook, B.W., Haberle, S.G. *et al.* (2012) The aftermath of megafaunal extinction: ecosystem transformation in Pleistocene Australia. *Science* 335, 1483–1486.

Russell-Smith, J., Ryan, P.G., Klessa, D. *et al.* (1998) Fire regimes, fire-sensitive vegetation and fire management of the sandstone Arnhem Plateau, monsoonal northern Australia. *Journal of Applied Ecology* 35, 829–846.

Russell-Smith, J., Yates, C.P., Whitehead, P.J. *et al.* (2007) Bushfires 'down under': patterns and implications of contemporary Australian landscape burning. *International Journal of Wildland Fire* 16, 361–377.

Scarff, F.R., Gray, B.F. and Westoby, M. (2012) Exploring phosphate effects on leaf flammability using a physical chemistry model. *International Journal of Wildland Fire* 21, 1042–1051.

Scott, A.C., Bowman, D.M., Bond, W.J. *et al.* (2013) *Fire on Earth: An Introduction*. John Wiley and Sons: West Sussex.

Setterfield, S.A., Rossiter-Rachor, N.A., Hutley, L.B. *et al.* (2010) Turning up the heat: the impacts of *Andropogon gayanus* (gamba grass) invasion on fire behaviour in northern Australian savannas. *Diversity and Distributions* 16, 854–861.

Taylor, A. and Hopper, S.D. (1988) *The Banksia Atlas*. Australian Government Publishing Service: Canberra.

Trauernicht, C., Murphy, B.P., Portner, T.E. and Bowman, D.M.J.S. (2012) Tree cover-fire interactions promote the persistence of a fire-sensitive conifer in a highly flammable savanna. *Journal of Ecology* 100, 958–968.

van der Werf, G.R., Randerson, J.T., Giglio, L. *et al.* (2008) Climate controls on the variability of fires in the tropics and subtropics. *Global Biogeochemical Cycles* 22.

Ward, D.J., Lamont, B.B. and Burrows, C.L. (2001) Grasstrees reveal contrasting fire regimes in eucalypt forest before and after European settlement of southwestern Australia. *Forest Ecology and Management* 150, 323–329.

Warman, L. and Moles, A.T. (2009) Alternative stable states in Australia's Wet Tropics: a theoretical framework for the field data and a field-case for the theory. *Landscape Ecology* 24, 1–13.

Whitehead, P.J., Bowman, D.M.J.S., Preece, N. *et al.* (2003) Customary use of fire by indigenous peoples in northern Australia: its contemporary role in savanna management. *International Journal of Wildland Fire* 12, 415–425.

Whittaker, R. (1975) *Communities and Ecosystems*. MacMillan: New York.

Williams, A.N., Mooney, S.D., Sisson, S.A. and Marlon, J. (2015) Exploring the relationship between Aboriginal population indices and fire in Australia over the last 20 000 years. *Palaeogeography, Palaeoclimatology, Palaeoecology* 432, 49–57.

Wood, S.W. and Bowman, D.M.J.S. (2012) Alternative stable states and the role of fire-vegetation-soil feedbacks in the temperate wilderness of southwest Tasmania. *Landscape Ecology* 27, 13–28.

Wood, S.W., Murphy, B.P. and Bowman, D.M.J.S. (2011) Firescape ecology: how topography determines the contrasting distribution of fire and rain forest in the south-west of the Tasmanian Wilderness World Heritage Area. *Journal of Biogeography* 38, 1807–1820.

Yates, C.J. and Ladd, P.G. (2010) Using population viability analysis to predict the effect of fire on the extinction risk of an endangered shrub *Verticordia fimbrilepis* subsp. *fimbrilepis* in a fragmented landscape. *Plant Ecology* 211, 305–319.

Yates, C.J., Hopper, S.D., Brown, A. and Van Leeuwen, S. (2003) Impact of two wildfires on endemic granite outcrop vegetation in Western Australia. *Journal of Vegetation Science* 14, 185–194.

Yibarbuk, D., Whitehead, P.J., Russell-Smith, J. *et al.* (2001) Fire ecology and Aboriginal land management in central Arnhem Land, northern Australia: a tradition of ecosystem management. *Journal of Biogeography* 28, 325–343.

York, A., Bell, T.L. and Weston, C.J. (2012) Fire regimes and soil-based ecological processes: implications for biodiversity. In *Flammable Australia: Fire Regimes, Biodiversity and Ecosystems in a Changing World*, ed. R.A. Bradstock, A.M. Gill and R.J. Williams. CSIRO Publishing: Melbourne, pp. 127–148.

6 Direct and Indirect Effects of Herbivore Activity on Australian Vegetation

DAVID J. ELDRIDGE, SAMANTHA K. TRAVERS, ADRIAN D. MANNING AND PHILIP BARTON

Vertebrate and invertebrate herbivores influence vegetation worldwide. The Australian continent has a rich history of grazing by megafauna that went extinct during the Pleistocene, and by a suite of small to medium body-sized marsupial herbivores, many of which have become largely extinct on the mainland since European settlement. Herbivores introduced by Europeans over the past 200 years have caused substantial negative ecosystem impacts by altering vegetation structure and composition, reducing soil function, and impairing the capacity of ecosystems to recover from degradation. Australia has a relatively short history of grazing by domestic livestock, coincident with the decline in small-bodied herbivores, many of which had functional roles in ecosystems through their soil foraging activities. Australian vegetation also supports a rich suite of terrestrial, arboreal and folivorous invertebrates that have evolved specialised relationships with their plant hosts. In this chapter we outline the main effects of herbivore activity on Australian vegetation, adopting a broad definition that includes non-trophic engineering effects of animals on vegetation as well as the more direct effects through herbivory. We outline the distribution of vertebrate and invertebrate herbivores of forests, woodlands and grasslands, and describe their direct and indirect effects on plants, soils and other organisms. We use two case studies; termites as detritivores, and the link between psyllids and forest dieback, to reinforce the complex interactions that exist between herbivores, their hosts and ecosystem processes.

6.1 Introduction

The Australian continent has a long history of grazing (and browsing) by megafauna. Many of the megafauna were herbivores, and their sudden loss, potentially as a result of human activity, a changing climate or both (Rule *et al.* 2012), is likely to have caused substantial changes in vegetation. This vegetation shift may have been exacerbated by increased fire frequency. Today, Australia is grazed by free-ranging, low to moderate populations of smaller body-sized mammalian herbivores (e.g. wombats, macropods), omnivores (bilbies, bettongs) and arboreal folivores (e.g. possums, koalas), as well as a rich suite of invertebrate herbivores that have coevolved with major plant taxa such as eucalypts (Low 2014). However, the most pervasive and extensive herbivores are domestic livestock: cattle, sheep and goats.

The arrival of Europeans 200 years ago heralded a substantial change in Australian herbivore communities and their effects on plant communities. The introduction of domestic livestock coincided with invasions

Australian Vegetation, Third Edition, ed. David A. Keith. Published by Cambridge University Press.

Citation: Eldridge, D.J. *et al.* (2017). Direct and Indirect effects of Herbivore Activity on Australian Vegetation. In D. A. Keith, ed., *Australian Vegetation*, 3rd edn. Cambridge: Cambridge University Press, pp. 135–55.

by exotic pests and diseases, declines of indigenous mammalian fauna, alteration of fire regimes and initiation of European farming practices. Australian plants have not coevolved with domestic livestock, and grazing continues to be a major degradation process. Herbivore activity also has feedback effects on herbivory and detritivory by native arthropods, for example, dung beetles and the Australian plague locust.

In this chapter, we describe the effects of herbivores on Australian vegetation. We use a model of herbivore effects on vegetation and soil to discuss the trophic effects of vertebrate herbivory and the non-trophic effects, through habitat disturbance and engineering, on Australian plant communities. We discuss the distribution of the main vertebrate and invertebrate herbivores in Australia and use continent-wide data on livestock activity to describe the net effects of sheep and cattle on plant processes. To illustrate the complex feedback effects of herbivory on plant communities, we describe two case studies: one of detritivory by termites and a second the interactions between psyllids and bell miners. Our aim is to show that herbivore activity in a broad sense, by both vertebrates and invertebrates, influences key ecological processes that have structured the present vegetation communities of Australia.

6.2 A model of herbivore activity on vegetation and soils

In this chapter, we adopt a broad definition of 'herbivore activity' as a synthesis of two separate, but related groups of processes: trophic-level effects associated with herbivory (consumption of plant material) and non-trophic, engineering effects associated with changes in the physical environment of the soil or plant structure. Thus we take a broader view than normally considered under the term 'grazing'. We use a model (Figure 6.1) to outline the direct and indirect effects of both herbivory and engineering on plant communities. Direct effects are not mediated by any other variables. Conversely, indirect effects are those that are mediated by an intermediary variable. The classic, oft-quoted example of indirect effects is the trophic cascade among wolves and willows in North America. Wolves have no direct effect on willows, but they have a strong indirect effect by controlling large herbivores, such as elk, that remove willows.

6.2.1 Herbivory

Both vertebrates and invertebrates influence plant communities directly (direct effect) by consuming plant material or by moving or caching seeds (Murphy *et al.* 2005). Herbivores are typically selective, preferentially consuming phytomass from some species more than others, causing shifts in composition of the plant community. For example, macropods (kangaroos) tend to graze preferentially on short green grass (Bayliss 1987). Similarly, domestic livestock grazing favours annual plants at the expense of perennials (Dorrough *et al.* 2004; Lunt *et al.* 2007). This process is illustrated by **pathway 1** (Figure 6.1). Herbivores can also alter vegetation structure directly (**pathway 2**) by reducing plant size and therefore the connectivity of plant patches. This selects for short-stature plants (Díaz *et al.* 2006). Herbivore-induced structural changes in vegetation may subsequently induce changes in species composition (**pathway 6 via 2**). This is an example of

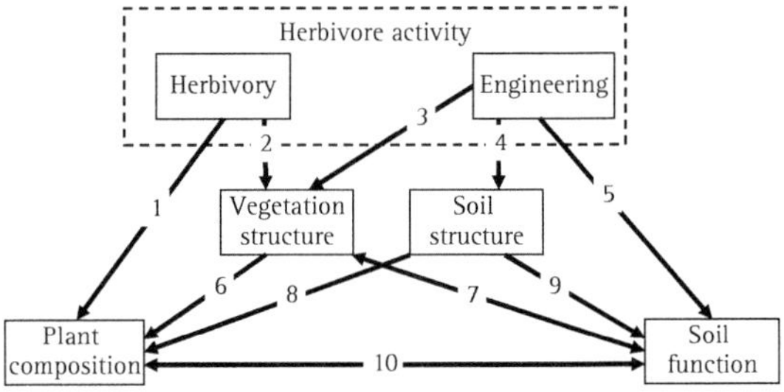

Figure 6.1 Model of the effects of herbivore activity on vegetation structure and composition, soil structure and function (the capacity of the soil to cycle nutrients, withstand deformation and infiltrate water). Herbivore activity is represented as a composite of two different processes; one associated with biomass consumption (i.e. herbivory, a trophic effect) and the other an engineering (non-trophic) effect. Arrows represent the direction of effects. Numbers refer to examples of effects. See text for explanation of the numbered effects.

an indirect effect, where vegetation structure mediates the effect of herbivores on plant composition. For example, changes in grazing regime may lead to the conversion of grassland to shrubland, which has been shown to alter the composition of understorey (protégé) plants through complex facilitative mechanisms (Soliveres *et al.* 2012). Similarly, insect galls can alter host plant architecture (Blanche 2012) and invertebrate leaf defoliators (e.g. sawfly larvae, Chrysomelid beetles) can potentially alter plant composition by killing their hosts and opening the canopy.

6.2.2 Engineering

Trampling of groundstorey vegetation or shrubs by large terrestrial vertebrates, or hollow formation by termite activity and fungal attack (Perry *et al.* 1985) will alter vegetation structure (**pathway 3**). Depending on the severity of these effects, plants may recover, but plant death will result in long-term structural change. Vertebrate engineers can alter soil physical and chemical properties by depositing dung and urine, and invertebrates, such as termites, can create organic-rich soils from the breakdown of plant tissue (**pathway 5**). Engineers can also alter soil function indirectly by altering soil structure (**pathways 4 and 9**), for example, by increasing soil macroporosity, thereby enhancing soil water flow and soil function (Eldridge 1994; Eldridge *et al.* 2015). Changes in soil structure can influence plant composition (**pathway 8**) by controlling soil moisture availability, for example preventing water extraction by woody plants (Harrington *et al.* 1984) and leading to dominance by forbs and grasses.

Figure 6.2 (a) Contrasting biomass, soil exposure and species composition of the vegetation ground-layer in semi-arid woodland either side of experimental herbivore exclosure fence, at Yathong Nature Reserve, central western New South Wales. Feral goats, rabbits and native kangaroos have access to the left hand side of the fence. Note dense growth of *Austrostipa scabra* and undershrubs on right-hand side of fence. (b) Piosphere effect with livestock tracks to water, severe denudation of vegetation, soil compaction and scalding near Bourke, north-west New South Wales. (c) Legacies of overgrazing, including soil compaction, inhibition of ground-layer vegetation and shrub encroachment persist 15 years after livestock removal near Enngonia Bourke, north-west New South Wales.
Photos: David Eldridge. (A black and white version of this figure will appear in some formats. For the colour version, please refer to the plate section.)

Shifts in vegetation structure caused by herbivore activity can influence soil function (**pathway 7**). For example, grazing by domestic livestock may: (i) alter shrub density, which in turn enhances the capture of windborne sediments and nutrients in shrublands (Eldridge and Soliveres 2015); or (ii) open up unvegetated patches that can be colonised by biological soil crusts. Plant composition has been shown to alter the composition of microbial communities and therefore soil function (**pathway 10**), however positive feedback effects of increasing soil function on plant community composition are likely (Maestre *et al.* 2012).

6.3 Vertebrate herbivores

Prior to European settlement, the mammalian fauna of Australia was dominated by a diverse range of small marsupials, which are now largely extinct or restricted to small areas on offshore islands (Johnson 2006). Many of these were herbivores or omnivores that foraged in the soil, with substantial non-trophic effects on ecosystems.

Currently, the most widespread native herbivores are the macropods (Table 6.1), which have increased in density and distribution, along with a rich suite of exotic herbivores, since European settlement. Currently four species of macropods (*Macropus rufus, M. robustus, M. giganteus, M. fuliginosus*) are distributed widely across the continent. Their increase has been attributed to a change in vegetation structure and composition brought about by the introduction of grazing by domestic livestock, land clearing for agriculture and timber removal associated with closer settlement. This has resulted from four largely related processes: (i) an increase in the area of grassland for grazers or regrowth forest for browsers (see Fensham *et al.*, Chapter 16, this volume; Prober *et al.*, Chapter 17, this volume); (ii) vegetation fragmentation, which has exacerbated the juxtaposition between open foraging area and wooded sheltered sites; (iii) increased water availability in landscapes where it was previously more limiting (e.g. Fensham and Fairfax 2008); and (iv) persecution of the main macropod predator, the dingo (*Canis dingo dingo*). Despite the widespread abundance of macropods, relatively few studies have assessed their roles as herbivores, and most of these have examined differences in dietary preference between macropods and livestock in semi-arid areas where they compete directly with livestock (Howland *et al.* 2014). Studies have indicated that kangaroo diets are not as broad as those of livestock (Squires 1981), and that dietary overlap tends to diminish as the availability of forage plants declines (Noble and Tongway 1986).

Other smaller macropods (e.g. wallabies) have substantial impacts on vegetation in more temperate climates, often by browsing young trees. Marsupial herbivores include small- to medium-sized animals, some of which forage in the soil (e.g. *Bettongia* spp., *Isoodon* spp., *Parameles* spp.) as well as three species of wombats (Table 6.1). Apart from domestic livestock (sheep and cattle), Australia has a number of exotic herbivores (Table 6.1). Herbivory by the European rabbit (*Oryctolagus cuniculus*) is a key process responsible for restricting establishment of shrub recruits in arid (Lange and Graham 1983; Tiver and Andrew 1997) and semi-arid (Auld 1990) woodlands and coastal heathlands (Cooke 1987). Larger-bodied feral herbivores include goats (*Caprus hircus*; Tiver and Andrew 1997), deer (*Cervus* spp., *Axis* spp.) and camels (*Camelus* spp.) (Table 6.1), which compete directly with macropods for habitat and biomass.

6.3.1 Direct and indirect effects of vertebrate herbivores

Vertebrate herbivores influence plant communities in a number of different ways, both trophic (**pathways 1, 2 and 6**) and non-trophic (ecosystem engineering; **pathways 3–5, 9**; Figure 6.1). These effects can also be direct or indirect. While often considered together, differentiating the type of pathway can help to understand how herbivores influence a given ecosystem. The direct effects of herbivore activity associated with the removal of plant material ('intrinsic effects', **pathways 1 and 2**) are strongly influenced by interactions between plant traits and herbivore characteristics (see next section). Indirect

Table 6.1 **Distribution, habitat preference, density and diet of key vertebrate herbivore groups in Australia**

Herbivore group	Distribution	Habitat	Diet	Key species	Density (animals/ha)	References
Large macropods	Widespread	Grasslands to open woodlands and forests	Herbivorous; grasses and some forbs	*Macropus giganteus, M. fuliginosus, M. rufus*	0.1–7.0	Fletcher (2007) Bayliss (1987)
Wombats	Widespread across southern Australia	Temperate and riparian forests in south-eastern Australia	Herbivorous; native and exotic grasses	*Vombatus ursinus*	0.3–2.6	Taylor (1993) Borchard and Eldridge (2012) McIlroy *et al.* (2012)
	Yorke and Eyre Peninsulas (SA)	Open semi-arid open woodlands	Herbivorous; nutrient-poor native grasses	*Lasiorhinus latifrons*	0.2	Taggart and Temple-Smith (2008)
	Epping State Forest (Qld)	Open semi-arid open woodlands	Herbivorous; nutrient-poor native grasses	*Lasiorhinus krefftii*	0.2	Taggart and Temple-Smith (2008)
Medium and small ground marsupials	Eastern and south-eastern Australia, coastal SA and WA	Coastal and temperate forest and woodlands	Omnivorous; insectivorous; bulbs	*Parameles nasuta* *Potorus* spp.	Variable, largely unknown	
Range-restricted fossorial marsupials	Predominantly offshore islands and conservation reserves	Semi-arid and arid woodlands, grasslands	Omnivorous; insectivorous; bulbs, chenopod shrub foliage, grasses	*Bettongia penicillata* *B. lesueur, Macrotis lagotis*	1.0–3.0 (in fenced reserves)	AWC (2016)
Arboreal folivorous mammals	Widespread across eastern Australia	Temperate forests and woodlands	Folivorous (*Eucalyptus*-specific); occasionally opportunistic generalist (possum, glider)	*Phascolarctos cinereus* *Petauroides volans* *Pseudocheirus peregrinus*	Spatially variable (koala); 2.9–13.5 (possum)	Finlayson *et al.* (2010)

Table 6.1 (cont.)

Herbivore group	Distribution	Habitat	Diet	Key species	Density (animals/ha)	References
Emus	Widespread	Semi-arid to arid grasslands and woodlands	Omnivorous; berries, fruit, insects, forbs	*Dromaius novaehollandiae*	0.1–1.6	Pople *et al.* (2000)
Camels	Semi-arid central and western Australia	Hummock grasslands, arid woodlands	Generalist herbivore; tree and shrub foliage, grasses	*Camelus dromadarius*	0.196	Ninti One (2013) Dorges and Heucke (2008)
Cattle	Widespread, but lower densities in arid areas	All communities excluding rainforest	Herbivorous; grasses, some shrub and tree foliage	*Bos taurus* *B. indicus*	Variable 0.1–4.0[a]	Bastin and ACRIS (2008)
Deer	Widespread and increasing in range	Temperate and semi-arid woodlands, forests	Herbivorous; fruit, tree and shrub foliage	*Cervus* spp. *Axis* spp.	<0.01 but increasing	Bayne *et al.* (2000)
Sheep	Widespread across eastern and southern Australia; inside the dingo fence	All communities excluding rainforest	Herbivorous; grasses, forbs, shrub foliage	*Ovis aries*	Variable 0.1–4.0[a]	Bastin and ACRIS (2008)
Goats	Widespread and increasing in density	Temperate and semi-arid woodlands, forests	Herbivorous; grasses, shrub and tree foliage	*Capra hircus*	0.29	Squires (1981) Tiver and Andrew (1997)
Pigs	Widespread	Riparian areas	Omnivorous; opportunistic; bulbs, roots, foliage, grasses	*Sus scrofa*	0.1–0.2 in wetlands	Saunders and Kay (1991) McIlroy and Saillard (1989)
Lagomorphs	Widespread; temperate and tropical areas	Grasslands, woodlands, open forest	Herbivorous; grasses, bulbs, roots	*Oyctolagus cuniculus* *Lepus europaeus*	Variable; 2.0–4.0	Cooke *et al.* (2010)

[a]Data include the sum of sheep and cattle.

(unassociated) effects on plant communities occur when herbivores alter the physical environment without consuming plant material (**pathway 3–5**; termed 'ecosystem engineering'; Jones *et al.* 1994).

Engineering effects may simply be a by-product of the presence of herbivores in the environment. These are often the effects associated with 'damage' to soils from 'overgrazing', such as soil compaction, removal of biological soil crusts, reductions in surface roughness, increases in soil particle mobilisation, loss of fine material through wind and water erosion, or alterations to the flow of surface water (**pathways 4 and 5**; Tongway *et al.* 2003; Eldridge and Whitford 2009). Direct effects of surface trampling, degradation of soil structure and reductions in nutrient concentrations are commonly reported effects of grazing worldwide (e.g. Fuhlendorf and Smeins 1999; Daryanto *et al.* 2013). In shrub-dominated rangelands, continued trampling in the shrub interspaces can reinforce the redistribution of resources from the interspaces to the shrub hummocks (Schlesinger *et al.* 1990) leading to reduced connectivity (Eldridge *et al.* 2015). Livestock activity can also enhance nutrient return to the soil through animal urine and faeces (Noble and Tongway 1986) or through the creation of carrion when animals die (Barton *et al.* 2013b).

Although the engineering effects described above are largely negative, there can be many positive effects of herbivore engineering, for example, by increasing soil surface heterogeneity and providing novel niches for plants and animals. In Australia, small fossorial (digging) mammals create pits in the soil while foraging for food such as invertebrates and plant tubers (**pathways 4 and 9**). Foraging pits of the short-beaked echidna (*Tachyglossus aculeatus*) have been shown to retain greater levels of soil moisture, support more soil carbon, nitrogen and plant litter, and have higher respiration rates than adjacent non-pit surfaces (Eldridge and Mensinga 2007). Even pits that were designed to mimic those created by greater bilbies (*Macrotis lagotis*) and burrowing bettongs (*Bettongia lesueur*) were cooler and held water for longer than adjacent non-pit surfaces (James *et al.* 2010). These pits even supported more plant germinants; an indirect engineering effect on plant composition (**pathways 4 and 8**).

There are numerous other examples in which localised ecosystem engineering by herbivores are important mechanisms for maintaining landscape heterogeneity. Depressions created by resting kangaroos at the base of tree trunks ('hip holes') have been shown to increase local sources of nitrogen and carbon by 47 and 68%, respectively (Eldridge and Rath 2002). Wombats, which construct large burrows in the sides of streambanks, have a strong engineering effect on soils (Borchard and Eldridge 2012). Associational engineering effects could also include 'transport engineering' i.e. the inadvertent or deliberate movement of seeds (e.g. scatter-hoarding) by animals. For example, the woylie (*Bettongia penicillata*) scatter-hoards the seeds of sandalwood *Santalum spicatum*, and research indicates that this has a significant role in regeneration of this species and by extension, the vegetation community (Murphy *et al.* 2005; **pathways 3 and 6**). Omnivorous emus also appear to have an important role in the movement of plant propagules over substantial distances (Calviño-Cancela *et al.* 2006).

6.4 Plant responses to herbivory: traits and defences

Plants have developed a range of different mechanisms to cope with herbivory, and their ability to grow and reproduce under grazing (grazing resistance; Briske 1996) depends on the extent to which they can avoid grazing (e.g. by being grazed less severely), tolerate grazing (by recovering more rapidly from defoliation) or both (Briske 1996).

A number of studies have examined how Australian plant communities respond to herbivory, largely by examining plant traits (e.g. McIntyre and Lavorel 1994, 2001; Pettit *et al.* 1995; Rogers 1995; Navie and Rogers 1997; Fensham 1998; Vesk and Westoby 2001; Vesk *et al.* 2004). Traits that have been shown to reflect plant responses to grazing include life history, plant height, habit and growth form, palatability, leaf characteristics, relative growth rate, seed production and seed bank persistence (Rogers 1993; McIntyre *et al.* 1999; McIntyre and Lavorel 2001; Dorrough *et al.* 2004; Vesk *et al.* 2004; Díaz *et al.* 2006).

Despite patterns derived from global synthesis (Díaz *et al.* 2006), studies from specific communities in Australia suggest that plant trait responses are highly species- and context-specific (e.g. environment) (Westoby 1999), and explain a relatively small amount of the variation in grazing responses (e.g. Vesk *et al.* 2004; Fensham *et al.*, Chapter 16, this volume; Eldridge *et al.*, Chapter 24, this volume). Below, we describe the main physical, morphological and chemical traits that plants employ to respond to herbivores in the Australian context.

6.4.1 Life history

Annual plants invest very few resources in the maintenance of a substantial plant structure. Instead, they invest their resources in rapidly producing a large number of seeds. Unlike perennial plants, plants with annual life histories are generally not disadvantaged by grazing, which is consistent with studies from environments with a short evolutionary history of grazing (Dorrough *et al.* 2004; Vesk *et al.* 2004; Díaz *et al.* 2006).

6.4.2 Growth form

The prediction that the abundance of certain growth forms (e.g. annual plants, shrubs) increases with grazing is not supported by global analyses. One explanation for this is that these methods of classifying plant growth form are extremely broad and are poorly related to potential grazing effects. For example, morphologically similar species within the same genus may show markedly different responses to grazing (e.g. *Austrodanthonia* or *Austrostipa*; Dorrough *et al.* 2011). These differences may relate to differences in palatability, crude protein or secondary metabolites, whether they are C_3 or C_4 plants, and whether they are tussock or stoloniferous grasses species (Díaz *et al.* 2006). Studies based on more narrowly defined growth form traits (e.g. erect cf. rosette forbs) illustrate considerable variability in grazing responses (Tremont and McIntyre 1994). In their global synthesis, Díaz *et al.* (2006) showed that the shrub growth form shows contrasting responses to grazing. This may at first seem inconsistent with the notion of woody encroachment (see following section), but in Australia, shrubs in the same genera (e.g. *Maireana* spp.) can show marked differences in their tolerance to grazing (Eldridge *et al.* 1990).

6.4.3 Plant height

Globally, grazing is thought to favour short plants irrespective of climate and grazing history. However, plant height may be less important in Australia because much of the Australian landscape has relatively open vegetation structure, largely due to low rainfall and low productivity (Vesk *et al.* 2004). Thus, there are likely to be fewer taller species providing protection for shorter understorey species. In temperate, high-productivity grasslands, grazing may release smaller annual plants and forbs from competitive exclusion by reducing biomass of large tussock grasses such as *Themeda*, allowing shorter grazing-avoiding or grazing-tolerant species to occupy inter-tussock spaces (McIntyre *et al.* 2003; Fynn *et al.* 2004; Frank 2005).

6.4.4 Plant habit and architecture

Grazing may increase the abundance of stoloniferous and rosette plant forms. Having a prostrate habit (stoloniferous plants) might increase the plant's ability to evade detection by herbivores (Dorrough *et al.* 2004). However, having a large surface area (e.g. rosette structure) confers advantages on plants such as the ability to be more competitive and to invest more resources in creating a suitable micro-environment. Thus, larger plants are likely to trap more resources such as nutrients and water (Eldridge and Koen 1993) and have higher levels of plant decomposition.

6.4.5 Ability to resprout

The ability to resprout and the presence of subterranean regenerative buds, root suckers or horizontal rhizomes are clear responses to herbivory. The presence of regenerating buds on native perennial grasses has been shown to increase their persistence under grazing compared with native forbs that lack these morphological features (Clarke 2003). The extent to which these features confer an advantage on some plant groups will be context-dependent. For example, in low productivity systems, grazed plants may not regrow due to extreme resource limitation (Proulx and Mazumder 1998; Frank 2005).

6.4.6 Palatability

Plant palatability represents a trade-off between plant allocation to the production of new photosynthetic tissue and anti-herbivore defences. The evidence for palatability as an anti-herbivore trait in Australia is relatively weak compared to systems with a long evolutionary history of grazing. Indeed the stronger the historical impact of grazing, the greater the likely investment in grazing avoidance mechanisms (Herms and Mattson 1992). Globally, evidence suggests that plant palatability should increase in response to grazing in humid systems (grazing tolerance, involving fast regrowth of high-quality tissue with very low structural defences), and should decline in dry systems (grazing avoidance, involving slow-growing tissue, rich in structural defence). However, the few studies that have examined palatability have shown no evidence that it influences herbivory, at least in dryland environments (Vesk *et al.* 2004).

An excellent example of plant secondary defences against herbivory is demonstrated by vertebrate folivores of *Eucalyptus* spp. Koalas (*Phascolarctos cinereus*) are highly specialised folivores that are specific to about 20 of the more than 500 species of eucalypts. Eucalypts are low in nitrogen but high in plant secondary metabolites such as tannins and phenolic compounds (Moore and Foley 2005). Along with the greater glider (*Petauroides volans*) and common ringtail possum (*Pseudocheirus peregrinus*), the koala has complex morphological and physiological adaptations to cope with a phenol- and fibre-rich diet (Foley and Hume 1987). Despite the presence of plant secondary metabolites in eucalypts, these folivores and browsers such as wallabies and kangaroos (O'Reilly-Wapstra *et al.* 2002) may reduce the fitness of eucalypts by extensive defoliation, which often kills the trees (Martin 1985).

In south-western Western Australia a significant number of plant species from the genus *Gastrolobium* contain the compound sodium monofluoroacetic acid, which is known to be lethal to non-native animals but is tolerated in low doses by native animals that are endemic to that area. This compound has been used commercially (1080 poison) to control pest animals (e.g. canids) in Australia and New Zealand.

6.5 Net effects of vertebrate herbivore activity on plant communities

The effects of herbivore activity on vegetation are typically more complex than those on individual species, often resulting from a combination of direct and indirect effects, with multiple feedback processes on plant composition and structure mediated by changes in soil structure and function. Here we illustrate the net effects of livestock activity with analyses of plant data from a continental study of grazing, and a discussion of woody encroachment.

6.5.1 Continental-scale effects of livestock activity on vegetation

The net effects of herbivore activity, through direct and indirect herbivory and engineering effects, are likely to vary among different communities and different herbivore species. Here we use a meta-analysis to report on the net effects of sheep and cattle on plant richness, abundance, cover and biomass (Figure 6.3). Using published and unpublished data from Australian studies, we calculated a log response ratio for different

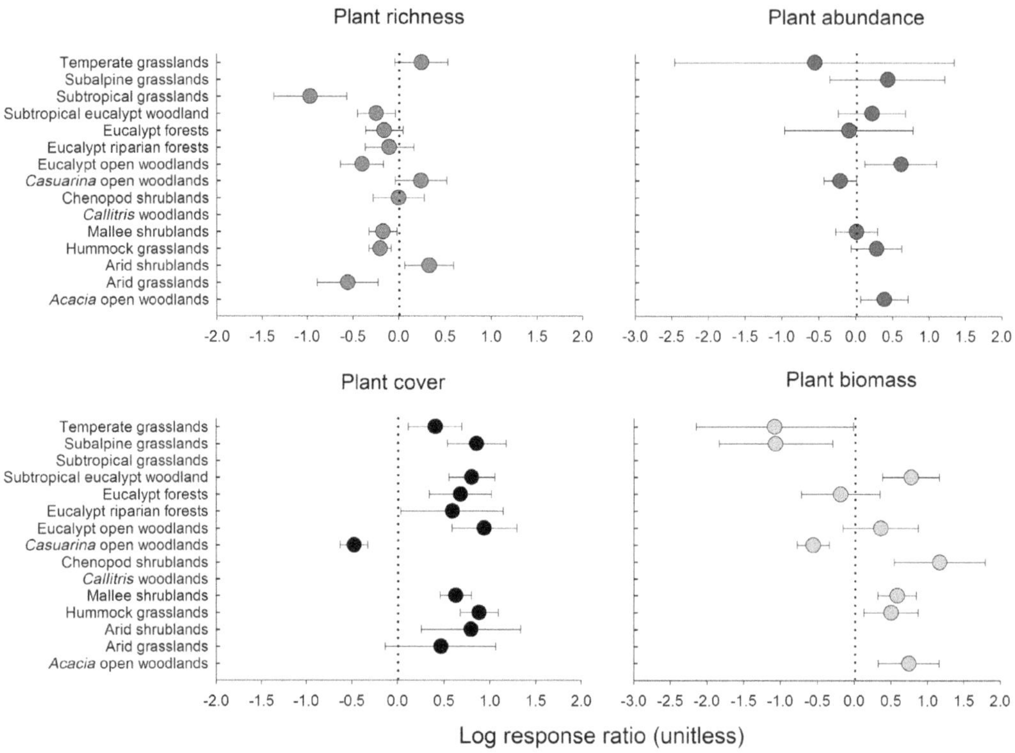

Figure 6.3 Mean (± 95% CI) log response ratio of the effects of grazing by sheep and cattle on plant richness, plant abundance, plant cover and plant biomass across 15 plant community types in Australia. Plant communities have been ranked from the least arid sites (i.e. temperate grasslands), to the most arid (*Acacia* open woodlands) based on the Food and Agriculture Organisation's (FAO) Aridity Index. The log response ratio is a measure of the average grazing response ($\log_{10}$ grazed minus $\log_{10}$ ungrazed). Thus positive values represent an increase in that attribute in response to increasing intensity of grazing (Eldridge *et al.* 2016).

combinations of grazing intensity such that a ratio greater than zero indicates an increase in that attribute with increasing grazing and a ratio less than zero indicates a decline. Grazing effects for different communities were sorted in relation to aridity (rainfall in relation to evaporation). The results indicate that grazing generally increases plant cover in most communities and biomass in half of the communities (i.e. log response ratio > 0). Positive effects of grazing on plant cover and biomass can be explained by a number of mechanisms: (i) stimulation of vigorous plant growth by removing dead or apical plant tissue (**pathway 2**; Figure 6.1); (ii) competitive release of large unpalatable forbs by selective consumption of their palatable competitors (**pathway 1**; Díaz *et al.* 2006); or (iii) grazing-induced surface disturbance which gives exotic species with a large soil seed bank a competitive advantage over less dominant natives (**pathways 8 and 4**).

Plant richness, however, generally declined with grazing but there were no clear trends in relation to aridity. Thus, grazing might lead to a change in composition with a loss of grazing-sensitive species or the replacement of one suite of species with another less diverse suite of species. The average effects of grazing on above-ground biomass, through herbivory, varied markedly among vegetation communities, with some strong reductions in biomass (e.g. sub-alpine grasslands) matched by relatively marked increases due to grazing (e.g. chenopod shrublands). Some of the increase in biomass in the latter community is the result of a grazing-induced change in composition from shrubs and forbs to derived grassland (Benson *et al.* 1997).

6.5.2 Reinforcing the encroachment of woody plants

A global example of the net effects of herbivore activity on vegetation is the transition from grasslands or open woodlands to shrublands (termed 'woody encroachment'; Eldridge *et al.* 2011). In Australia, encroachment has occurred in many vegetation communities ranging from semi-arid woodlands (Eldridge *et al.* 2013; Eldridge and Soliveres 2015) to coastal forests, temperate and riverine woodlands (Lunt *et al.* 2010), tropical savanna woodlands and grasslands (Murphy *et al.* 2013), eucalypt lowland forest (Hradsky *et al.* 2015) and alpine grasslands (Wahren *et al.* 2001; Fensham *et al.*, Chapter 16, this volume; Prober *et al.*, Chapter 17, this volume).

Heavy and sustained grazing by high densities of domestic and feral herbivores and macropods is a key driver of woody encroachment (Eldridge and Soliveres 2015; Hradsky *et al.* 2015). The mechanisms underpinning this phenomenon are complex, but involve a number of indirect effects. Sustained herbivory removes groundstorey plants (mainly grasses; **pathway 1**, Figure 6.1), altering vegetation structure by reducing plant biomass (**pathway 2**) and allowing woody plants to use soil moisture at the expense of grasses (**pathway 7**; Peltzer and Kochy 2001). Changes in soil function brought about by trampling can alter the distribution of soil nutrients around woody plant canopies (McCulley *et al.* 2004; **pathway 5**) and increase the cover of bare soil surfaces, which are likely to favour the dispersal and survival of the propagules of encroaching species (Archer *et al.* 2001; **pathway 10**). The net effect of these direct and indirect effects is a reduction in fire intensity and frequency, which would otherwise regulate woody encroachment (Archer *et al.* 2001). Woody plants, however, may signal ecosystem repair and protection from grazing by providing habitat that allows understorey (protégé) species to escape from herbivory (**pathway 6**) or to benefit from greater levels of soil nutrients beneath the canopy (Warnock *et al.* 2007; **pathway 7**). Under extremely high levels of livestock grazing, engineering effects will predominate (**pathway 5**) so that any positive effects of shrubs on soil function will diminish (Eldridge *et al.* 2015).

6.6 Invertebrate herbivores

Australian vegetation communities support a rich community of terrestrial, arboreal and folivorous invertebrates (Lowman 1985; Tassone and Majer 1997). The herbivorous insect fauna inhabiting Australian trees is generally well studied because many are important consumers of plant biomass (Ohmart and Edwards 1991). Many have evolved varying degrees of specialisation on their plant hosts ranging from monophagous (feeding on one species of plant), oligophagous (feeding on a small range of plants, often in one family), or polyphagous (feeding on a wide range of plants). Correspondingly, plants may host a distinct insect assemblage that has coevolved with the plant's defences, such as its suite of secondary metabolites and tannins or mechanical adaptations that reduce digestibility (Ohmart and Edwards 1991).

Grassland invertebrates are probably prolific, but apart from high profile pests or rare species, have been poorly studied. Threatened species include the morabine grasshopper (*Keyacris scurra*) and the wood cricket (*Cooraboorama canberrae*) in the ACT (Driscoll 1994), and the golden sun moth (*Synemon plana*), which is closely associated with *Austrodanthonia* grasslands in eastern Australia. Grassland species are particularly vulnerable to herbivore activity through trampling and habitat destruction by livestock. Grazing has been shown to reduce beetle scarabs and their larvae, ants, Lepidopteran adults and larvae, Hemiptera, Dipterans and Orthopterans (Driscoll 1994). Below we present two case studies of invertebrate herbivory; one of termites and the other the relationship between psyllids, bell miners and forest dieback.

6.6.1 Direct and indirect effects of invertebrate herbivores

Invertebrate herbivores have direct and indirect effects on vegetation. Direct effects result from herbivory, such as the extensive herbivory on the annual grass *Dactyloctenium radulans* by the Australian plague locust (*Chortoicetes terminifera*) following summer rainfall in arid and semi-arid Australia (Hunter *et al.* 2001). Woodland and forest eucalypts are also predated heavily by leaf-eating beetles (Coleoptera; pathway 1, Figure 6.1; Table 6.2). Canopy dieback in the New England

Table 6.2 **Distribution, habitat preference, diet, key taxa and ecosystem impacts of some key invertebrate herbivore groups in Australia**

Broad group	Distribution	Habitat	Diet	Key taxa	Ecosystem impacts	References
Larvae: Coleoptera, Lepidoptera, Hemiptera	Widespread; eastern Australia	Temperate forests	Herbivorous	Sawfly larvae (Symphyta); psyllids (Psylloidea)	Tree dieback; tree death	Ohmart and Edwards (1991)
Adults: Coleoptera, Lepidoptera, Hemiptera	Widespread; eastern Australia	Temperate forests	Herbivorous; folivorous	Scarab beetles (Chrysomelidae)	Soil aeration, tree dieback	
Orthoptera	Widespread	Arid grasslands to temperate forests	Herbivorous	Australian plague locust (*Chortoicetes terminifera*); mole crickets (family Gryllotalpidae)	Widespread tree defoliation; increased hydrological function	Rentz (1996)
Isoptera	Widespread	Dry sclerophyll forest, woodland and savanna	Detritivorous; polyphagous (wood, soil, plant, dung)	*Drepanotermes* spp. *Mastotermes* spp. *Nasutitermes* spp.	Organic matter decomposition; soil aeration; nutrient increase; hollow formation	Prior and Werner (2007) Watson *et al.* (1973) Whitford and Eldridge (2013)

Tablelands is an interesting example of direct and indirect interactions among vertebrates, invertebrate herbivores and trees. Overgrazing by livestock along with land clearing creates soil conditions favouring leaf beetles (Chrysomelidae), scarab beetles (Scarabaeidae) and sawflies (Symphyta). Repeated herbivory and psyllid (Psylloidea) attack lead to cycles of defoliation, reduction in canopy size, tree cover and often death (**pathway 2**). Death of paddock trees can lead to marked changes in soil nutrient concentrations (Eldridge and Wong 2004; **pathway 7**). Recent research has also revealed other notable associations between insects and woodland trees; the ecological role of termites as herbivores, and trophic links between psyllids, their bird predators, and forest dieback, which we detail below.

6.7 Case studies of herbivory in invertebrates

6.7.1 Termites as herbivores and detritivores

Termites are important detritivores and herbivores and have been shown to remove substantial quantities of above- and below-ground plant tissue across a range of communities (Whitford and Eldridge 2013). Termites convert recalcitrant cellulose to readily digestible sugars and starches due to a mutualistic association with bacteria and protozoa in their guts. Some species have specialised diets as either plant-, litter-, wood-, soil- or dung-feeders, while others consume a variety of substrates as polyphagous feeders. Consumption depends on availability and vegetation community. For example, *Drepanotermes tamminensis* in Western Australia has been shown to harvest bark and leaves of eucalypts in woodlands, but it harvests shoots of shrubs in *Allocasuarina campestris*-dominated shrublands (Park *et al.* 1993). In the northern Australian savannas, the biomass of termites is similar to that of vertebrate herbivores such as livestock and macropods (10–15 kg ha^{-1}; Watson *et al.* 1973). At these levels, termites can consume about 100 kg ha^{-1} annually, thus competing with domestic livestock; again reinforcing the relative contribution of invertebrates compared to vertebrates.

The greatest termite diversity in Australia occurs in dry sclerophyll forests, woodlands and savannas, with a diversity hotspot in south-west Western Australia (Watson and Abbey 1993; Abensperg-Traun and Steven 1997). Unlike trends globally, where termite diversity decreases with latitude (Collins 1989; Abensperg-Traun 1998) or reduced productivity (Eggleton *et al.* 1994), species diversity in Australia changes little from the tropics to the arid interior (Abensperg-Traun and Steven 1997), and only four of the 260 Australian species are restricted to tropical rainforests. In the Australian arid zone, termites are one of the most abundant invertebrates and, therefore, are a reliable prey item for reptiles and mammals (Noble and Tongway 1986; Abensperg-Traun and Steven 1997).

Termites affect vegetation communities directly by harvesting plant material, thereby reducing plant size (**pathway 2**, Figure 6.1), but their engineering effects on soil nutrients, soil porosity and flow of water (**pathways 10 via 5, or 8 via 4**) are likely to be more substantial than their herbivory effects (Whitford and Eldridge 2013). Termite tunnels act as sites of enhanced infiltration (Eldridge 1994; Tongway *et al.* 2003). Higher densities of macropores associated with termite mounds and surface pavements (**pathway 4**) increase infiltration rates around these structures (**pathway 9**). Further, runoff water shed from termite pavements has been shown to support a greater cover and biomass of plants in the annular zone surrounding the pavements (**pathway 7**; Spain and McIvor 1988). As central-place foragers, termites concentrate organic material in and around their nest and alter its distribution in the landscape (Whitford and Eldridge 2013). Termite mounds are also high in inorganic nitrogen, available phosphorus and exchangeable calcium, magnesium and potassium (Decaëns *et al.* 2001; Whitford and Eldridge 2013) and in semi-arid woodlands their activity may contribute up to 20% of the mineralised carbon and greater microbial biomass (Lee and Wood 1971; Holt 1998).

6.7.2 Psyllids, bell miners and forest dieback

Forest dieback occurs as a consequence of outbreaks of invertebrates and a series of complex ecological interactions involving avian communities and other ecosystem components. Related processes operate in woodlands, although our focus here is on eucalypt forests (see Wardell-Johnson *et al.*, Chapter 12, this volume, for a perspective on dieback as a threat to wet sclerophyll forest biodiversity and its management implications). Psyllids are an important group of insect herbivores in Australia's woodlands and forests (Ohmart and Edwards 1991). Psyllids are Hemiptera or 'true bugs' that feed on plant sap and liquids. Psyllids are involved with bell miners (*Manorina melanophrys*) in a form of insect-mediated dieback that has resulted in the loss of crown foliage over more than 2.5 million hectares of Australian eucalypt forest (Wardell-Johnson *et al.* 2006). The most susceptible forest communities are those dominated by either Dunn's white gum (*Eucalyptus dunnii*), Sydney blue gum (*E. saligna*), flooded gum (*E. grandis*), grey ironbarks (*E. siderophloia, E. paniculata*), narrow-leaved white mahogany (*E. acmenoides*) and/or grey gum (*E. punctata*) (Bell Miner Associated Dieback Working Group 2004). The most widely accepted mechanism for this dieback involves the removal of lerp by bell miners, leaving the psyllid to secrete more lerp to maintain its shelter. The bell miners aggressively exclude smaller bird species, which are natural predators of the psyllids. In response to psyllid attack, the trees stimulate nutrient-rich epicormic growth, further encouraging psyllid activity in a positive feedback. Psyllids can reach densities high enough to support bell miner colonies of several hundred birds, and in extreme cases, may kill entire stands of the host trees.

While bell miners have been strongly implicated in this dieback, it is the regrowth of foliage that is the most likely driver of increasing psyllid populations. Recent research suggests that increasing bell miner populations may only be a reinforcing feedback, not a cause of declining tree health (Steinbauer *et al.* 2015). Observations of bell miners switching prey types, from *Glycaspis* to *Cardiaspina* psyllid species, suggests that they might use different species opportunistically (te Marvelde *et al.* 2009), further reinforcing the notion that they are indicators, rather than drivers, of dieback.

Psyllid genera have different leaf preferences, and this affects the composition of their lerp and the response of trees (Steinbauer *et al.* 2015). *Cardiaspina* and *Creiis* species cause leaf damage, which increases canopy dieback. This can shift the age composition of eucalypt leaves in forests experiencing dieback and may cause further shifts in psyllid communities, from older to younger leaf specialists. The effect of psyllids on eucalypts and acacias may also be influenced by ants, which will tend psyllids in order to feed on their sugary excretions. This valuable resource for ants may encourage greater abundance of psyllids, potentially further damaging their tree hosts (Gibb and Cunningham 2009; Barton *et al.* 2013a).

6.8 Conclusions

We have identified here some of the direct and indirect effects of herbivore activity on Australian vegetation. Three generalisations arise from our work. First, the two constituent processes comprising herbivore activity, herbivory and ecosystem engineering, cannot be viewed in isolation. Australian vegetation is replete with many examples of the additive effects of these two processes on plant communities, and feedback effects on the herbivores themselves. Second, herbivore activity interacts with other drivers of ecosystem change such as climate, fire, flooding and anthropogenic change such as land clearing. While the net effect is often clear, it is not always possible to separate out the specific effects of herbivore activity. Third, the impacts of both herbivory and engineering are likely to change over the next century with projected changes in climate. The challenge in the next few decades will be to encourage land managers to adopt sustainable grazing practices for large areas of the continent where pastoralism is a major land use.

References

Abensperg-Traun, M. (1998) Termites (Isoptera) in Western Australia: present and future directions of ecological research. *Journal of the Royal Society of Western Australia* 81, 13–142.

Abensperg-Traun, M. and Steven, D. (1997) Latitudinal gradients in the species richness of Australian termites (Isoptera). *Austral Ecology* 22, 471–476.

Archer, S., Boutton, T.W. and Hibbard, K.A. (2001) Trees in grasslands: biogeochemical consequences of woody plant expansion. In *Global Biogeochemical Cycles in the Climate System*, ed. E.D. Schulze, S.P. Harrison, M. Heimann, E.A. Holland, J. Lloyd, I.C. Prentice and D. Schimel. Academic Press: San Diego, pp. 115–130.

Auld, T.D. (1990) Regeneration in populations of the arid zone plants *Acacia carnei* and *A. oswaldii*. *Proceeding of the Ecological Society of Australia* 16, 267–272.

AWC (2016) *Scotia*. Australian Wildlife Conservancy: Western Australia. Available at: http://www.australianwildlife.org/sanctuaries/scotia-sanctuary.aspx (accessed 8 March 2016).

Barton, P.S., Colloff, M.J., Pullen, K.R. and Cunningham, S.A. (2013a) Arthropod assemblages in a focal tree species (*Eucalyptus microcarpa*) depends on the species mix in restoration plantings. *Biodiversity and Conservation* 22, 2091–2110.

Barton, P.S., Cunningham, S.A., Lindenmayer, D.B. and Manning, A.D. (2013b) The role of carrion in maintaining biodiversity and ecological processes in terrestrial ecosystems. *Oecologia* 171, 761–772.

Bastin, G., ACRIS Management Committee (2008) *Taking the Pulse*. National Land and Water Resources Audit: Canberra. Available at: https://www.environment.gov.au/system/files/resources/a8015c25-4aa2-4833-ad9c-e98d09e2ab52/files/rangelands08-pulse.pdf (accessed 26 July 2016).

Bayliss, P. (1987) Kangaroo dynamics. In *Kangaroos their Ecology and Management in the Sheep Rangelands of Australia*, ed. G. Caughley, N. Shepherd and J. Short. Cambridge University Press: Cambridge, pp. 119–133.

Bayne, P., Harden, B., Pines, K. and Taylor, U. (2000) Controlling feral goats by shooting from a helicopter with and without the assistance of ground-based spotters. *Wildlife Research* 27, 517–523.

Bell Miner Associated Dieback Working Group (2004) *Bell Miner Associated Dieback Strategy*. Department of Environment and Conservation: Coffs Harbour. Available at: http://www.environment.nsw.gov.au/resources/nature/BMADStrategy.pdf (accessed 26 July 2016).

Benson, J.S., Ashby, E.M. and Porteners, M.F. (1997) The native grasslands of the riverine plain, New South Wales. *Cunninghamia* 5, 1–48.

Blanche, R. (2012) *Life in a Gall: The Biology and Ecology of Insects that Live in Plant Galls*. CSIRO Publishing: Collingwood.

Borchard, P. and Eldridge, D.J. (2012) Vegetation changes associated with cattle (*Bos taurus*) and wombat (*Vombatus ursinus*) activity in a riparian forest. *Applied Vegetation Science* 15, 62–70.

Briske, D. (1996) Strategies of plant survival in grazed systems: a functional interpretation. In *The Ecology and Management of Grazed Systems*, ed. J. Hodgson and A. Illius. CAB International: Wallingford, pp. 37–67.

Calviño-Cancela, M., Dunn, R.R., van Etten, E.J.B. and Lamont, B.B. (2006) Emus as non-standard seed dispersers and their potential for long-distance dispersal. *Ecography* 29, 632–640.

Clarke, P.J. (2003) Composition of grazed and cleared temperate grassy woodlands in eastern Australia: patterns in space and inferences in time. *Journal of Vegetation Science* 14, 5–14.

Collins, N.M. (1989) Termites. In *Ecosystems of the World. 14B. Tropical Rainforest Ecosystems*, ed. H. Lieth and M.J.A. Werger. Elsevier: Amsterdam, pp. 455–471.

Cooke, B.D. (1987) The effects of rabbit grazing on regeneration of sheoaks, *Allocasuarina verticilliata* and saltwater ti-trees, *Melaleuca halmaturorum*, in the Coorong National Park, South Australia. *Australian Journal of Ecology* 13, 11–20.

Cooke, B.D., Jones, R. and Gong, W. (2010) An economic decision model of wild rabbit, *Oryctolagus cuniculus*, control to conserve Australian native vegetation. *Wildlife Research* 37, 558–565.

Daryanto, S., Eldridge, D.J. and Throop, H.L. (2013) Managing semi-arid woodlands for carbon storage: grazing and shrub effects on above- and belowground carbon. *Agriculture Ecosystems and Environment* 169, 1–11.

Decaëns, T., Galvis, J.H. and Amézquita, E. (2001) Propriétés des structures produites par les ingénieures écologiques ála surface du sol d'une savane colombienne. *Comptes Rendus Academie Sciences Paris, Sciences de la vie* 324, 465– 478.

Díaz, S., Lavorel, S., McIntyre, S. *et al.* (2006) Plant trait responses to grazing: a global synthesis. *Global Change Biology* 13, 313–341.

Dorges, B. and Heucke, J. (2008) One-humped camel, *Camelus dromedarius*. In *The Mammals of Australia*, 3rd edition, ed. S. Van Dyck and R. Strahan. Reed New Holland: Sydney, pp. 758–760.

Dorrough, J., Ash, J. and McIntyre, S. (2004) Plant responses to livestock grazing frequency in an Australian temperate grassland. *Ecography* 27, 798–810.

Dorrough, J., McIntyre, S. and Scroggie, M.P. (2011) Individual plant species responses to phosphorus and livestock grazing. *Australian Journal of Botany* 59, 670–681.

Driscoll, D.A. (1994) *Invertebrates of lowland native grasslands in the Australian Capital Territory: conservation and research strategies for a recovery plan. Technical report 9.* ACT Parks and Conservation Service: Tuggeranong. Available at: http://www.environment.act.gov.au/__data/assets/pdf_file/0008/576809/Technical_Report_9.pdf (accessed 26 July 2016).

Eggleton, P., Williams, P.H. and Gaston, K.J. (1994) Explaining global termite diversity: productivity or history? *Biodiversity and Conservation* 3, 318–330.

Eldridge, D.J. (1994) Nests of ants and termites influence infiltration in a semi-arid woodland. *Pedobiologia* 38, 481–492.

Eldridge, D.J. and Koen, T.B. (1993) Runoff and sediment removal on a semi-arid soil in eastern Australia. II. Some variations in hydrological properties along a gradient in soil surface condition. *The Rangeland Journal* 15, 234–246.

Eldridge, D.J. and Mensinga, A. (2007) Foraging pits of the short-beaked echidna (*Tachyglossus aculeatus*) as small-scale patches in a semi-arid Australian box woodland. *Soil Biology and Biochemistry* 39, 1055–1065.

Eldridge, D.J. and Rath, D. (2002) Hip holes: kangaroo (*Macropus* spp.) resting sites modify the physical and chemical environment of woodland soils. *Austral Ecology* 27, 527–536.

Eldridge, D.J. and Soliveres, S. (2015) Are shrubs really a sign of declining ecosystem function? Disentangling the myths and truths of woody encroachment in Australia. *Australian Journal of Botany* 62, 594–608.
Eldridge, D.J. and Whitford, W.G. (2009) Soil disturbance by native animals along grazing gradients in an arid grassland. *Journal of Arid Environments* 73, 1144–1148.
Eldridge, D.J. and Wong, V.N.L. (2004) Clumped and isolated trees influence soil nutrient levels in an Australian temperate box woodland. *Plant and Soil* 270, 331–342.
Eldridge, D.J., Westoby, M. and Stanley, R.J. (1990) Population dynamics of the arid zone shrubs *Atriplex vesicaria*, *Maireana astrotricha* and *Maireana pyramidata* under grazing, 1980–1987. *Journal of Applied Ecology* 27, 502–512.
Eldridge, D.J., Bowker, M.A., Maestre, F.M. *et al.* (2011) Impacts of shrub encroachment on ecosystem structure and functioning: towards a global synthesis. *Ecology Letters* 14, 709–722.
Eldridge, D.J., Soliveres, S., Bowker, M.A. and Val, J. (2013) Grazing dampens the positive effects of shrub encroachment on ecosystem functions in a semi-arid woodland. *Journal of Applied Ecology* 50, 1028–1038.
Eldridge, D.J., Beecham, G. and Grace, J.B. (2015) Do shrubs reduce the adverse effects of grazing on soil properties? *Ecohydrology* 8, 1503–1513.
Eldridge, D.J.,Poore, A.G.B., Ruiz-Colmenero, M. *et al.* (2016) Ecosystem structure, function and composition in rangelands are negatively affected by livestock grazing. *Ecological Applications* 26, 1273–1283.
Fensham, R.J. (1998) The grassy vegetation of the Darling Downs, south-eastern Queensland, Australia. Floristics and grazing effects. *Biological Conservation* 84, 301–310.
Fensham, R.J. and Fairfax, R.J. (2008) Water-remoteness for grazing relief in Australian arid lands. *Biological Conservation* 141, 1447–1460.
Finlayson, G.R., Diment, A.N., Mitrovski, P. *et al.* (2010) Estimating western ringtail possum (*Pseudocheirus occidentalis*) density using distance sampling. *Australian Mammology* 32, 197–200.
Fletcher, D. (2007) Managing eastern grey kangaroos *Macropus giganteus* in the Australian Capital Territory: reducing overabundance – of opinion. In *Pest or Guest: The Zoology of Overabundance*, ed. D. Lunney, P. Eby, P. Hutchings and S. Burgin. Royal Society of New South Wales: Mosman, pp. 117–128.
Foley, W.J. and Hume, I.D. (1987) Digestion and metabolism of high-tannin *Eucalyptus* foliage by the brushtail possum (*Trichosurus vulpecula*) (Marsupialia: Phalangeridae). *Journal of Comparative Physiology* 157, 67–76.
Frank, D.A. (2005) The interactive effects of grazing ungulates and aboveground production on grassland diversity. *Oecologia* 143, 629–634.
Fuhlendorf, S.D. and Smeins, F.E. (1999) Scaling effects of grazing in a semi-arid grassland. *Journal of Vegetation Science* 10, 731–738.
Fynn, R.W.S., Morris, C.D. and Edwards, T.J. (2004) Effect of burning and mowing on grass and forb diversity in a longterm grassland experiment. *Journal of Vegetation Science* 7, 1–10.
Gibb, H. and Cunningham, S.A. (2009) Does the availability of arboreal honeydew determine the prevalence of ecologically dominant ants in restored habitats? *Insectes Sociaux* 56, 405–412.
Harrington, G.N., Wilson, A.D. and Young, M.D. (1984) *Management of Australia's Rangelands.* CSIRO Division of Wildlife and Rangelands Research: Canberra.

Herms, D.A. and Mattson, W.J. (1992) The dilemma of plants: to grow or defend. *The Quarterly Review of Biology* 67, 283–335.

Holt, J.A. (1998) Microbial activity in the mounds of some Australian termites. *Applied Soil Ecology* 9, 183–187.

Howland, B., Stojanovic, D., Gordon, I.J. *et al.* (2014) Eaten out of house and home: impacts of grazing on ground-dwelling reptiles in Australian grasslands and grassy woodlands. *PLoS ONE* 9, e105966.

Hradsky, B.A., Loschiavo, J., Hradsky, M. and Di Stefano, J. (2015) Shrub expansion alters forest structure but has little impact on native mammal occurrence. *Austral Ecology* 40, 611–625.

Hunter, D.M., Walker, P.W. and Elder, R.J. (2001) Adaptations of locusts and grasshoppers to the low and variable rainfall of Australia. *Journal of Orthoptera Research* 10, 347–351.

James, A.I., Eldridge, D.J. and Moseby, K.E. (2010) Foraging pits, litter, and plant germination in an arid shrubland. *Journal of Arid Environments* 74, 516–520.

Johnson, C.N. (2006) *Australia's Mammal Extinctions: A 50 000 Year History.* Cambridge University Press: Port Melbourne.

Jones, C.G., Lawton, J.H. and Shachak, M. (1994) Organisms as ecosystem engineers. *Oikos* 69, 373–386.

Lange, R.T. and Graham, C.R. (1983) Rabbits and the failure of regeneration in Australian arid zone *Acacia*. *Australian Journal of Ecology* 8, 377–381.

Lee, K.E. and Wood, T.G. (1971) Physical and chemical effects on soils of some Australian termites, and their pedological significance. *Pedobiologia* 11, 376–409.

Low, T. (2014) *Where Song Began. Australia's Birds and How they Changed the World.* Penguin Books: Australia.

Lowman, M.D. (1985) Temporal and spatial variability in insect grazing of the canopies of five Australian rain forest tree species. *Australian Journal of Ecology* 10, 7–24.

Lunt, I.D., Eldridge, D.J., Morgan, J.W. and Witt, G.B. (2007) A framework to predict the effects of livestock grazing and grazing exclusion on conservation values in natural ecosystems in Australia. *Australian Journal of Botany* 55, 401–415.

Lunt, I.D., Winsemius, L.M., McDonald, S.P. *et al.* (2010) How widespread is woody plant encroachment in temperate Australia? Changes in woody vegetation cover in lowland woodland and coastal ecosystems in Victoria from 1989 to 2005. *Journal of Biogeography* 37, 722–732.

Maestre, F.T., Quero, J.L., Gotelli, N.J. *et al.* (2012) Plant species richness and ecosystem multifunctionality in global drylands. *Science* 335, 214–218.

Martin, R.W. (1985) Overbrowsing, and decline of a population of the koala, *Phascolarctos cinereus*, in Victoria. I. Food preference and food tree defoliation. *Australian Wildlife Research* 12, 355–365.

McCulley, R.L., Archer, S.R., Boutton, T.W. *et al.* (2004) Soil respiration and nutrient cycling in wooded communities developing in grassland. *Ecology* 85, 2804–2817.

McIlroy, J., Collins, J. and Borchard, P. (2012) A revised method for estimating population densities of common wombats (*Vombatus ursinus*). *Australian Mammalogy* 34, 170–174.

McIlroy, J.C. and Saillard, R.J. (1989) The effect of hunting with dogs on the numbers and movements of feral pigs, *Sus scrofa*, and the subsequent success of poisoning exercises in Namadgi National Park, A.C.T. *Australian Wildlife Research* 16, 353–363.

McIntyre, S. and Lavorel, S. (1994) Predicting richness of native, rare, and exotic plants in response to habitat and disturbance variables across a variegated landscape. *Conservation Biology* 8, 521–531.

McIntyre, S. and Lavorel, S. (2001) Livestock grazing in subtropical pastures: steps in the analysis of attribute response and plant functional types. *Journal of Ecology* 89, 209–226.

McIntyre, S., Lavorel, S., Landsberg, J. and Forbes, T.D.A. (1999) Disturbance response in vegetation: towards a global perspective on functional traits. *Journal of Vegetation Science* 10, 621–630.

McIntyre, S., Heard, K.M. and Martin, T.G. (2003) The relative importance of cattle grazing in subtropical grasslands: does it reduce or enhance plant biodiversity. *Journal of Applied Ecology* 40, 445–457.

Moore, B.D. and Foley, W.J. (2005) Tree use by koalas in a chemically complex landscape. *Nature* 432, 488–490.

Murphy, H.T., Bradford, M.G., Dalongeville, A. *et al.* (2013) No evidence for long-term increases in biomass and stem density in the tropical rain forests of Australia. *Journal of Ecology* 101, 1589–1597.

Murphy, M.T., Garkaklis, M.J. and Hardy, G.E.S. (2005) Seed caching by woylies *Bettongia penicillata* can increase sandalwood *Santalum spicatum* regeneration in Western Australia. *Austral Ecology* 30, 747–755.

Navie, S.C. and Rogers, R.W. (1997) The relationship between attributes of plants represented in the germinable seed bank and stocking pressure in a semi-arid subtropical rangeland. *Australian Journal of Botany* 45, 1055–1071.

Ninti One (2013) *Managing the impacts of feral camels across remote Australia. Final report of the Australian feral camel management project*, ed. M. McGregor, Q. Hart and R. Davies. Australian Government: Canberra. Available at: http://www.nintione.com.au/resource/ManagingImpactsFeralCamels_FinalReportAFCMP.pdf (accessed 26 July 2016).

Noble, J.C. and Tongway, D.J. (1986) Herbivores in arid and semi-arid rangelands. In *Australian Soils: The Human Impact*, ed. J.S. Russell and R.F. Isbell. University of Queensland Press: St Lucia, pp. 243–270.

O'Reilly-Wapstra, J.M., McArthur, C. and Potts, B.M. (2002) Genetic variation in resistance of *Eucalyptus globulus* to marsupial browsers. *Oecologia* 130, 289–296.

Ohmart, C.P. and Edwards, P.B. (1991) Insect herbivory on *Eucalyptus*. *Annual Review of Entomology* 36, 637–657.

Park, H.C., Majer, J.D., Hobbs, T.R.J. and Bae, T.U. (1993) Harvesting rate of the termite, *Drepanotermes tamminensis* (Hill) within native woodland and shrubland of the Western Australian wheatbelt. *Ecological Research* 8, 269–275.

Peltzer, D.A. and Kochy, M. (2001) Competitive effects of grasses and woody plants in mixed-grass prairie. *Journal of Ecology* 89, 519–527.

Perry, D.H., Lenz, M. and Watson, J.A.L. (1985) Relationships between fire, fungal rots and termite damage in Australian forest trees. *Australian Forestry* 48, 46–53.

Pettit, N.E., Froend, R.H. and Ladd, P.G. (1995) Grazing in remnant woodland vegetation: changes in species composition and life form groups. *Journal of Vegetation Science* 6, 121–130.

Pople, A.R., Griff, G.C., Cairns, S.C. *et al.* (2000) Trends in the numbers of red kangaroos and emus on either side of the South Australian dingo fence: evidence for predator regulation? *Wildlife Research* 27, 269–276.

Prior, L. and Werner, P. (2007) Tree-piping termites negatively correlated with growth and survival of host trees in savanna woodland of north Australia. *Journal of Tropical Ecology* 23, 611–622.

Proulx, M. and Mazumder, A. (1998) Reversal of grazing impact on plant species richness in nutrient-poor vs. nutrient-rich ecosystems. *Ecology* 79, 2581–2592.

Rentz, D. (1996) *Grasshopper Country.* UNSW Press: Sydney.

Rogers, R.W. (1993) Selection for grasses with different carbon assimilation pathways under different sheep stocking pressures. *Australian Journal of Botany* 41, 327–331.

Rogers, R.W. (1995) Vegetative attributes, taxonomy, and character syndromes in grasses around a sheep camp on the northern Tablelands of New South Wales. *Australian Journal of Botany* 43, 317–325.

Rule, S., Brooke, B.W., Haberle, S.G. *et al.* (2012) The aftermath of megafaunal extinction: ecosystem transformation in Pleistocene Australia. *Science* 335, 1483–1486.

Saunders, G. and Kay, B. (1991) Movements of feral pigs (*Sus scrofa*) at Sunny Corner, New South Wales. *Wildlife Research* 18, 49–61.

Schlesinger, W.H., Reynolds, J.F., Cunningham, G.L. *et al.* (1990) Biological feedbacks in global desertification. *Science* 247, 1043–1048.

Soliveres, S., Eldridge, D.J., Hemmings, F. and Maestre, F.T. (2012) Nurse plant effects on plant species richness in drylands: the role of grazing, rainfall and species specificity. *Perspectives in Plant Ecology Evolution and Systematics* 14, 402–410.

Spain, A.V. and McIvor, J.G. (1988) The nature of herbaceous vegetation associated with 596 termitaria in north-eastern Australia. *Journal of Ecology* 76, 181–191.

Squires, V.R. (1981) *Livestock Management in the Arid Zone.* Inkata Press: Melbourne.

Steinbauer, M.J., Sinai, K.M.J., Anderson, A. *et al.* (2015) Trophic cascades in bell miner-associated dieback forests: quantifying relationships between leaf quality, psyllids and *Psyllaephagus* parasitoids. *Austral Ecology* 40, 77–89.

Taggart, D.A. and Temple-Smith, P.D. (2008) Southern hairy-nosed wombat: *Lasiorhinus latifrons*. In *The Mammals of Australia*, 3rd edition, ed. S. Van Dyck and R. Strahan. Reed New Holland: Sydney, pp. 204–206.

Tassone, R.A. and Majer, J.D. (1997) Abundance of arthropods in tree canopies of *Banksia* woodland on the Swan Coastal Plain. *Journal of the Royal Society of Western Australia* 80, 281–286.

Taylor, R.J. (1993) Observations on the behaviour and ecology of the common wombat *Vombatus ursinus* in northeast Tasmania. *Australian Mammalogy* 16, 1–17.

te Marvelde, L., McDonald, P.G., Kazem, A.J.N. and Wright, J. (2009) Do helpers really help? Provisioning biomass and prey type effects on nestling growth in the cooperative bell miner. *Animal Behaviour* 77, 727–735.

Tiver, F. and Andrew, M.H. (1997) Relative effects of herbivory by sheep, rabbits, goats and kangaroos on recruitment and regeneration of shrubs and trees in eastern South Australia. *Journal of Applied Ecology* 34, 903–914.

Tongway, D.J., Sparrow, A.D. and Friedel, M.H. (2003) Degradation and recovery processes in arid grazing lands of central Australia. Part 1: soil and land resources. *Journal of Arid Environments* 55, 301–326.

Tremont, R.M. and McIntyre, S. (1994) Natural grassy vegetation and native forbs in temperate Australia: structure, dynamics and life histories. *Australian Journal of Botany* 42, 641–58.

Vesk, P.A. and Westoby, M. (2001) Predicting plant species' responses to grazing. *Journal of Applied Ecology* 38, 897–909.
Vesk, P.A., Leishman, M.R. and Westoby, M. (2004) Simple traits do not predict grazing response in Australian dry shrublands and woodlands. *Journal of Applied Ecology* 41, 22–31.
Wahren, C.H.A., Williams, R.J. and Papst, W.A. (2001) Vegetation change and ecological processes in alpine and subalpine sphagnum bogs of the Bogong High Plains, Victoria, Australia. *Arctic, Antarctic, and Alpine Research* 33, 357–68.
Wardell-Johnson, G., Stone, C., Recher, H. and Lynch, A.J.J. (2006) Bell miner associated dieback (BMAD) independent scientific literature review: a review of eucalypt dieback associated with bell miner habitat in north-eastern New South Wales, Australia. Occasional Paper 2006/116 (DEC: Hurstville)
Warnock, A.D., Westbrooke, M.E., Florentine, S.K. and Hurst, C.P. (2007) Does *Geijera parviflora* Lindl. (Rutaceae) facilitate understorey species in semi-arid Australia? *The Rangeland Journal* 29, 207–216.
Watson, J.A.L. and Abbey, H.M. (1993) *Atlas of Australian Termites*. CSIRO Division of Entomology: Canberra.
Watson, J.A.L., Lendon, C. and Low, B.S. (1973) Termites in mulga lands. *Tropical Grasslands* 7, 121–126.
Westoby, M. (1999) The LHS strategy scheme in relation to grazing and fire. In *Proceedings of the VIth International Rangeland Congress,* Vol. 2, ed. D. Eldridge and D. Freudenberger. International Rangeland Congress: Townsville, pp. 893–896.
Whitford, W.G. and Eldridge, D.J. (2013) Effects of ants and termites on soil geomorphological processes. In *Treatise on Geomorphology*. Academic Press: Saint Louis, pp. 281–292.

7 Soils and the Below-ground Interactions that Shape Australian Vegetation

MARK C. BRUNDRETT, PAULINE F. GRIERSON, LAUREN T. BENNETT AND CHRISTOPHER J. WESTON

Vegetation and plant diversity patterns are strongly linked to soil properties, such as fertility, texture and depth in all parts of Australia. This results because many parts of Australia have relatively old and highly weathered soil profiles compared to most other regions. However, interactions between soils and vegetation are very complex because soils result from weathering processes that are determined by substrate type, topography, climate and vegetation, as well as the relocation of materials by wind and water. In a global context, Australian soils in many habitats are highly infertile, with low availability of phosphorus, nitrogen and trace elements. These infertile soils are often linked to shrub-dominated ecosystems, where it is assumed that soil fertility is insufficient to support tree growth, but plant productivity is also severely limited by water availability in many habitats. Low soil nutrient supply is counteracted by plant adaptations to growth in infertile soils, as well as efficient nutrient cycling, especially after fire. In Australia the majority of plants have mycorrhizal associations that assist in nutrient uptake and this includes an exceptionally high diversity of plants with ectomycorrhizal associations, as well as plants with nitrogen fixing symbioses. However, nonmycorrhizal plants are also exceptionally diverse in many Australian vegetation types and include many species with specialised means of nutrition, including carnivores, parasites and cluster-rooted species.

7.1 Introduction

Water and nutrients are key resources that shape Australia's vegetation (Keith and Tozer, Chapter 1, this volume). Most Australian plants obtain these resources from soils through their roots or rhizoids and/or fungal associates. The composition, structure and functioning of Australian vegetation are strongly regulated by the availability of soil nutrients and water, which in turn reflect the diverse geological and geomorphological features of the continent. The continental diversity of geological substrates encompasses exposed shields of ancient cratons, sedimentary sequences of varied age and texture, igneous intrusions and extrusions of varied age, texture and mineral composition, as well as expanses of relatively young unconsolidated aeolian, colluvial and alluvial regolith (Johnson 2009). Weathering and erosion of these substrates, together with nitrogen fixation from the atmosphere, provides the nutrient capital of Australian landscapes, in some cases supplemented by aerosol inputs from onshore winds and legacies of antecedent marine incursions (McTainsh and Strong 2007). Nutrients

Australian Vegetation, Third Edition, ed. David A. Keith. Published by Cambridge University Press. © Cambridge University Press 2017.
Citation: Brundrett M. *et al.* (2017). Soils and the Below-ground Interactions that Shape Australian Vegetation. In D. A. Keith, ed., *Australian Vegetation*, 3rd edn. Cambridge: Cambridge University Press, pp. 156–81.

are redistributed by geomorphic processes, which result from interactions of topography with climate (e.g. precipitation, wind). Leaching is a key process that can transport dissolved nutrients vertically through the soil profile, and potentially beyond the vertical reach of plant roots.

While climate influences movement and cycling of nutrients it is also the primary regulator of water availability to vegetation at continental scales. Mean precipitation, predominantly as rainfall, varies by about 1.5 orders of magnitude across the Australian continent and variability both within and among years may be extreme, particularly in the lower rainfall zones (Australian Bureau of Meteorology 2016). Topography and geomorphological processes are critical to the lateral and vertical movement of both water and nutrients, making it possible for wetlands and desert to coexist within the same landscape in some of the most arid parts of Australia (Catford *et al.*, Chapter 20, this volume). Groundwater can also be a critical source of water in some Australian ecosystems (Eamus and Froend 2006; Pettit *et al.* 2016). Although rainfall is the ultimate source of water to vegetation, groundwater recharge can be far removed both temporally and spatially from the time and location of plant uptake (Fan 2015).

In this chapter, we focus on four major aspects of the relationships between soils and vegetation in Australia. First, we consider vegetation–soil relationships at landscape scales, focussing on how gradients in edaphic features influence patterns in vegetation composition and structure. We next briefly review the primary mechanisms influencing the availability of nutrients in Australian soils, focussing on key cycling processes and the role of periodic fires. We then summarise how below-ground traits of Australian plants influence their uptake of nutrients and water. Finally, we consider how the expression of these traits along local and continental resource gradients at the plant community scale informs our understanding of the composition, biogeography and ecology of Australian vegetation.

7.2 Soil as a major influence on vegetation patterns

The following soil properties are key to understanding the distribution and structure of vegetation as they influence the productivity, diversity and distribution of plants:

- Soil nutrient availability, which is the capacity of soils to provide plants with essential mineral nutrients, especially phosphorus (P) and nitrogen (N), but also potassium (K), calcium (Ca), magnesium (Mg) and essential trace elements;
- Soil pH, which regulates the availability of nutrients through reduction–oxidation (redox) potential and solubility equilibria: acidity (i.e. pH < 5.5) is characteristic of weathered soils of the tropics and subtropics as well as most forest and heathland soils in Australia; while alkalinity (i.e. pH > 7.5) is more characteristic of arid and semi-arid regions, and of highly calcareous soils;
- Imbalances in the availability or concentration of ions in soil, such as between Mg and Ca;
- Salinity (sodicity), where high concentrations of sodium chloride (NaCl) or other salts can restrict plant growth and survival;
- Concentrations of toxic elements such as aluminium (Al) or nickel (Ni) that may inhibit growth of roots or other organs;
- Depth to groundwater and drainage; and
- Physical properties such as texture, strength and permeability that affect root growth, aeration, water-holding capacity (as a function of clay content and mineralogy) and soil stability.

Vegetation patterns have been strongly linked to soil complexity across Australia. In eastern Australia, landscape mosaics of soils are often associated with patterns of forests, woodlands, heathlands and grasslands (Keith 2004). In an analysis of spatial and floristic data from climatic, geological and vegetation maps of south-east Australia, Keith (2011) found that soils and climate accounted for approximately equal, but essentially non-overlapping portions of variation in plant species composition. The distribution of vegetation complexes across south-west Australia is also strongly associated with soil type within areas of relatively similar climate (Beard 1990; Wardell-Johnson and Williams 1996; Havel and Mattiske 2000; Gibson *et al.* 2004). An example of complex vegetation patterns that closely follow patterns in soils and landforms is provided in Box 7.1.

At local scales, strong edaphic boundaries often result in abrupt changes in vegetation structure and composition, which is observable across a broad range of landscapes in the tropical north and the arid zone as well as the temperate south of the continent. For example, on the Mitchell Plateau of the Kimberley region in tropical north-west Australia, contrasting savanna woodlands with either tussock or hummock grass understories are distributed on mafic and sandstone substrates, respectively (Mucina *et al.* 2013). The structural and floristic features that distinguish mulga shrublands and hummock (*Triodia*) grasslands in arid central Australia have also been related to edaphic characteristics, particularly to differences in clay content, pH and soil depth, although patterns are moderated by effects of fire (Nano and Clarke 2008). Regional patterns of plant species turnover across eastern New South Wales are also strongly predicted by concurrent changes in both lithology and elevation (Di Virgilio *et al.* 2012). Thus, it can be argued that the high taxonomic, structural and functional diversity of Australia's native vegetation is in large part a reflection of the complex mosaics of the underlying substrates on which they grow.

Australian landscapes are old with surfaces that have often been exposed to continuous weathering for many millions of years with the youngest exposed rocks across almost all of Australia being of Triassic age and the oldest of Precambrian origin (Taylor 1994). Even the youngest landscapes in Australia tend to be much older than many landscapes on other continents where glaciation and volcanic activity has been more widespread and frequent. However, there are also many cases where rocks have been covered by more recent deposits of materials resulting from erosion, such as coastal sand dunes and river valleys. In many cases Australian landscapes show long-term impacts of weathering processes due to exposure of rocks and soils to rainfall, the atmosphere, plant root exudates and plant decomposition products, and weathering is typically higher under vegetation than in barren areas (Macphail 2007). Long-term weathering and erosion leads to complex soil patterns which can be linked to vegetation patterns, as shown in Box 7.1.

Landscape age is linked to soil complexity, but also provides time for plants to diversify and specialise to local conditions (Hopper 2009; Fine 2015; Linder and Verboom 2015). Climatic stability is also important in ancient landscapes (Mucina and Wardell-Johnson 2011; Cowling *et al.* 2015). Plant community diversity and species richness often increase with soil age and heterogeneity. This is especially pronounced in south-west Western Australia, within the kwongan shrublands and woodlands.

The mineralogy, texture and depth of soils are strong determinants of vegetation type and plant species composition, presumably due to their influence on nutrient and water availability. Plants exhibit a range of traits and strategies that enable them to occupy a range of edaphic niches across ecosystems. Australia has extensive areas of flat, highly weathered landscapes, resulting in poor drainage and areas with saline groundwater and salt lakes in the lowest parts of the landscape (McArthur 2004; Pate and Verboom 2009). It has also been proposed that dominant plants in ecosystems have long-term impacts on soil profile development, especially in soils where laterite or clay occurs (Pate and Verboom 2009). This theory is based on the impacts of ion transport and root exudates. Impacts of plants on soil structure over long time periods should be expected, but the extent to which this happens requires further investigation. Interactions between soils and vegetation are very complex since soils are themselves the product of underlying geologies, topography, climate and vegetation, and soil materials are often transported by erosion. These interactions are summarised in Figure 7.1 and Table 7.1.

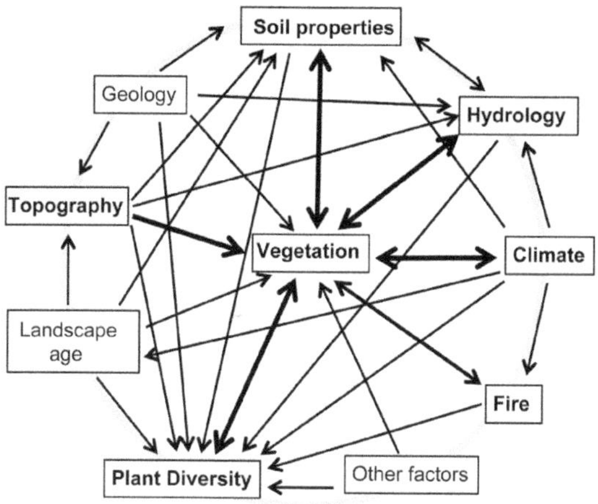

Figure 7.1 Interacting factors linked to soils and landscapes that have been suggested to regulate vegetation patterns in the scientific literature. Factors considered to have a strong influence on vegetation have thicker lines. These factors are explained further in Table 7.1.

7.2.1 Soil nutrients

Catenary sequences may reveal the age of Australian landscapes and the climatic conditions under which soils developed. This is particularly evident on sequentially deposited sandplains, where the relationship between age since deposition, nutrient availability and vegetation is relatively simple because the initial capital of nutrients at the time of sediment deposition is continually depleted by leaching (notwithstanding aerosol inputs) (Myerscough and Carolin 1986; Lambers *et al.* 2013). On lithic substrates, leaching processes interact with wide variation in the mineral composition of rocks and their rates of weathering. Even though the underlying substrates and their soil profiles may be of similar age, soils with contrasting nutrient availability and vegetation may occur across topographically uniform substrate gradients, such as those observable between shale and sandstone in the Sydney Basin (Benson and Howell 1990). In other cases, the soil age nutrient relationship is reversed; for example, where more fertile and older igneous substrates are juxtaposed with younger infertile sands, such as the Tertiary basalt outcrops within the Eurunderee sand mass of the Myall Lakes area in New South Wales (Myerscough and Carolin 1986). Thus, while vegetation structure and diversity are strongly linked to landscape and soil age in the coastal dune systems of south-west Australia (Lambers *et al.* 2010; Laliberté *et al.* 2013) and southern Queensland (Skjemstad *et al.* 1992), more nuanced soil age – vegetation relationships exist in other landscapes across the continent (Bui and Henderson 2013).

Soil P content is correlated with vegetation patterns, with higher total P content in soils under rainforests, wet sclerophyll forests, tussock grasslands and wetlands, and lower in mallee, hummock grasslands, dry sclerophyll forests, heathlands and tall shrublands (Rossel and Bui 2016; Keith and Tozer, Chapter 1, this volume). Comparative studies at local or regional scales in Australia and elsewhere have shown that species richness tends to be inversely related to soil nutrient levels (Heddle and Specht 1975; Specht and Rundel 1990; Keith and Myerscough 1993; Olde Venterink *et al.* 2003; Reynolds *et al.* 2007). Nevertheless, the importance of soil properties relative to other factors in explaining species turnover and shifts in species dominance remains poorly studied and may vary across Australia. For example, distribution and turnover of *Acacia* species across northern Australia is best explained by climate, while soil characteristics such as pH, P, Na and electrical conductivity (EC) better explain the occurrence of *Acacia* species in southern Australia (Bui *et al.* 2014). Present-day gradients in soil chemistry (especially of P, pH and Mg) and climate (winter rainfall) explain around a third of floristic turnover across the South-west Australian Floristic Region (SWAFR) (Jones *et al.* 2016). Other factors such as variation in soil depth or water-holding capacity also interact with climatic gradients, both past and present, in shaping plant distributions (Jones *et al.* 2016). Variations in hydrological connectivity across landscapes, for example between rivers and their surrounding

Table 7.1 **Factors hypothesised to have major roles regulating soil properties and vegetation patterns in Australian ecosystems. Positive or negative interactions between factors are also listed (see Figure 7.1).**

Factor	Details	Interactions – positive	Interactions – negative	Examples and case studies
1. Topographic complexity	Variations in landforms and substrates	3, 4, 6, 7, 8, 9	6	Box 7.2
2. Soil complexity	Spatial variations in soil types	3, 6, 8, 9		Box 7.2
3. Soil properties	Infertility, salinity, clay, hardpans, ferrocrete, laterite	8, 9	8, 9	See text for examples
4. Climate	Rainfall and temperature – seasonality, uniformity and reliability	2, 3, 4, 6, 8	1, 6, 8, 9	Lower rainfall reduces diversity, but so can high rainfall in some cases due to overstorey dominance (see text)
5. Landscape age	Time for increased plant specialisation and diversification	2, 3, 4, 6, 7, 8	1	Precambrian Yilgarn Craton of south-west Western Australia (Taylor 1994)
6. Hydrology	Drainage, flooding, capacity of soils to store water	2, 3, 8, 9	8, 9	Specialised wetland plants, groundwater-dependent vegetation
7. Fire	Fire frequency, intensity, variability	8, 9	8, 9	Miller and Murphy (Chapter 5, this volume)
8. Plant diversity (species richness)	High in shrubland (kwongan) and rainforest for different reasons	9	9	Plant diversity and vegetation patterns are often linked to soil variations (e.g. Verboom and Pate 2013)
9. Vegetation types and complexity	The result of all of the above factors	2, 8	2, 8	See Keith and Tozer (Chapter 1, this volume); Eldridge *et al.* (Chapter 6, this volume); Tozer *et al.* (Chapter 13, this volume)
10. Other factors	Biotic - weeds, pathogens, grazing, pollination; human - pollution, disturbance	8, 8	8, 9	These can be of minor or major importance (not discussed here)

Note interaction numbers refer to rows in this table. See text for additional references.

Box 7.1 Vegetation patterns linked to soils in south-west Australia

Vegetation composition and diversity in the south-west region of Western Australia is strongly dependent on local soil and other landform features (Wardell-Johnson *et al.* 1989; Inions *et al.* 1990; Wardell-Johnson and Williams 1996; Havel and Mattiske 2000). Figure 7.2 shows the complex pattern of vegetation types in an area in the northern jarrah forest of Western Australia, where soil variation patterns are equally complex. The vegetation map for the Western Australian forest bioregion by Mattiske and Havel (1998) closely follows the soil and landscape units described by Churchward *et al.* (1988). A gradient analysis by Fordyce *et al.* (2007) confirmed the strong association between soil properties and vegetation composition and structure in the jarrah forest, but also noted that both were closely linked to topography and hydrology. Thus, the impact of soil patterns on vegetation may be indirect, as both are influenced from the long-term impacts of the same factors, as explained in Figure 7.1.

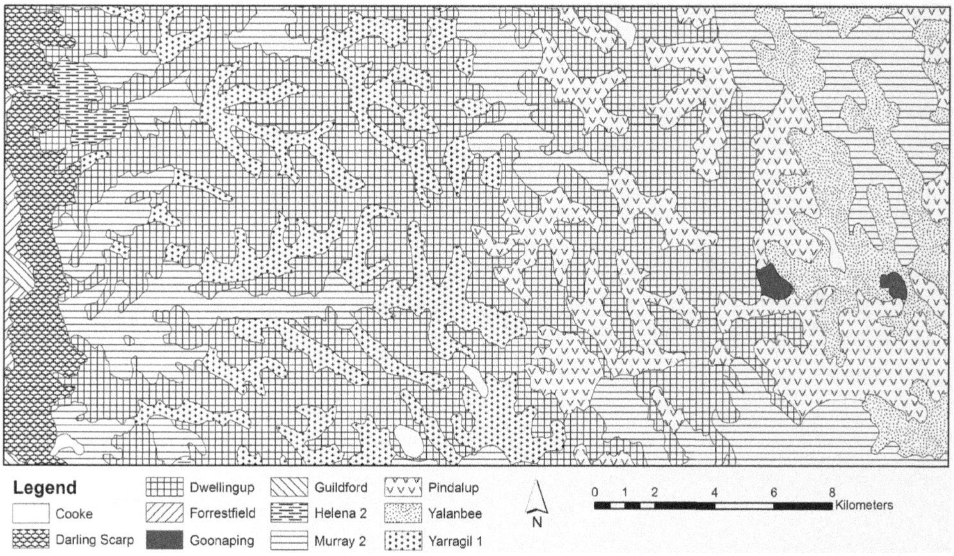

Figure 7.2 Vegetation map for an area of the jarrah forest and adjacent Swan Coastal Plain in Western Australia. This mapping was originally produced by Havel (1975) and updated by Mattiske and Havel (1998).

floodplains, can also drive spatial complexity in soil nutrients (Fellman *et al.* 2013; Leigh *et al.* 2015). Bioavailable soil P and N increase with reduced inundation in the Murray–Darling Basin of eastern Australia; gradients in inundation frequency thus create patterns of both water and nutrient availability that in turn are major contributors to the heterogeneity in composition and productivity of vegetation found in lowland riverine landscapes (Woodward *et al.* 2015).

7.2.2 Soil moisture

Soil hydrology can have a profound direct influence on vegetation structure and composition regardless of nutrient availability. This is most apparent in strongly seasonal environments where trees and larger shrubs rely on subsoil water in the unsaturated zone, from perched watertables or deeper groundwater (see Catford *et al.*, Chapter 20, this volume; Good *et al.*, Chapter 21, this volume). Particular local climatic characteristics, such as the moisture-gathering effects of granite outcrops (see Jones *et al.* 2016) or the deeper soils of valley floors, can also influence species distributions and vegetation structure. Diversity of root architecture and rooting depth of the dominant ecological stratum as well as of understorey species are likely to be important adaptive traits that help explain more fine-scale vegetation patterns (Groom 2004; Groom and Lamont 2015). For example, shrubs and trees can have strongly dimorphic root systems, incorporating a deep taproot coupled with a surface network of lateral roots in surface horizons, which enables them to access water at different soil depths, e.g. *Banksia prionotes* (Dawson and Pate 1996; Burgess *et al.* 2000). Other species may occupy a different niche within the same vegetation type by having a shallow network of surface roots that can readily take up water from small rainfall events. Although several tree species, mostly eucalypts, are recorded to have roots that extend at least to 40 m into the subsoil (Dell *et al.* 1983; Canadell *et al.* 1996), there have been remarkably few direct observations of rooting depth of Australian plant species across a diversity of vegetation and substrate types. Consequently, plant utilisation and thus dependency on shallow versus deeper water sources, is mostly inferred. For example, putative groundwater-dependent ecosystems (GDEs; Eamus and Froend 2006) are typically identified as more productive vegetation within a landscape than might be predicted on the basis of climate and soil nutrients alone. However, the delineation of GDEs is complex and can include surface expressions where groundwater is shallow and discharges naturally to the land, such as spring-fed wetlands, as well as terrestrial expressions such as riparian and floodplain forests and woodlands, where the watertable may be at considerable depth (> 20 m). Nevertheless, rooting depth of the dominant species in terrestrial GDEs is often assumed to be the same as the depth to the saturated zone (Eamus 2009), and changes in depth to groundwater is reflected in turnover of dominant species across a floodplain with increasing distance from a river or stream (Zolfaghar *et al.* 2015).

Diversity of rooting architectures and growth phenology of fine roots in relation to water availability is most strongly developed in plant communities growing in highly seasonal environments. In southern Australia, herbaceous perennial plants, ephemerals, prostrate shrubs and geophytes tend to have shallow root systems and grow and reproduce in winter to early spring when surface soil moisture is generally higher than at other times of the year. Tree species of the jarrah forests of Western Australia (*Eucalyptus marginata* and *Banksia grandis*) also produce prolific fine roots in the wetter winter and spring when soil moisture is high to enhance nutrient availability (Grierson and Adams 1999). However, these same species tend to produce new leaves and stem growth in the dry summer, suggesting they have access also to deeper water sources, although not necessarily from the saturated zone (see Specht and Rundell 1990; Pate and Bell 1999). Thus, patterns in the structure and composition of vegetation across landscapes also reflect underlying differences in relative water availability. The particular below-ground adaptations that different species have to acquire and utilise water for growth is also dependent on the seasonality and reliability of water supply.

7.3 How nutrient availability has shaped Australian vegetation

7.3.1 Nutrient cycling processes

Most nutrients (> 95%) in Australian ecosystems are held in the biomass and bound to organic and inorganic complexes in the soil (Attiwill and Leeper 1987). Australian vegetation depends on the biogeochemical cycling of nutrients via the decomposition of organic matter and mineralisation processes to maintain growth and productivity. The major cycles of both N and P in native forests, woodlands, shrublands and grasslands are reliant on both the above- and below-ground recycling of nutrients held in litterfall. Rates of nutrient cycling in Australian ecosystems tend to increase with rainfall and temperature (Attiwill and Adams 1993) but can become highly episodic in inland and subtropical Australia. However, while the relative amounts and rates of flux may be less, the significance of nutrient cycling increases as both rainfall and growth decline, so that in arid and semi-arid environments, litterfall can account for 90–100% of the external cycles of major nutrients, such as N and P (Attiwill and Adams 1993).

In general, soil nutrient concentrations vary slightly over time due to fluctuations in the rate of nutrient cycling from plant material to humus and mineral soil layers and fluctuations in uptake by plants. Rates are dependent on seasonal changes in temperature, rainfall and disturbance regimes (e.g. Grierson and Adams 2000; Muqaddas *et al.* 2015; Turner and Lambert 2016). When soil nutrient pools are very limited, even small amounts of a nutrient released from plant material (e.g. after a fire, or with a flush in mineralisation on soil rewetting at the start of the rainy season) can significantly increase the total soil concentration of that nutrient at a micro-scale. In southern Australia, there is a predictable spring increase in nutrient availability due to increases in microbial activity, as well as flushes of nutrients in autumn and early winter associated with rewetting of dry soils (Grierson and Adams 2000). In most strongly seasonal ecosystems, it is likely that much of the nutrient requirements of plants are met by uptake during such relatively short periods, rather than by steady-state (constant) absorption. Theory suggests that the competitive advantage of different plant species also depends on the availability of soil nutrients (Grime 1979; Tilman 1988; Fraser *et al.* 2015). Consequently, high temporal fluctuation in the rate of cycling of different nutrients allows more species to coexist at nutrient-limited sites (Tilman 1983; Laliberté *et al.* 2013). In contrast, when soil nutrient pools are large, temporal fluctuations in nutrient cycling rates do not cause a significant reduction in nutrient availability, and a few species that quickly take up nutrients to support fast growth can outcompete slower-growing species resulting in reduced species richness (Hobbie 2015; Kardol *et al.* 2015).

Total nitrogen (N) concentrations of Australian surface soils usually lies between 0.02 and 0.5%, broadly correlating with the organic carbon (C) content of soils, and consequently the rainfall and temperature specific to the region (McKenzie *et al.* 2004). However, in some highly productive cool temperate eucalypt forests and rainforests of south-east Australia, total N concentrations may be as high as 0.9% (Fedrigo *et al.* 2014). Nitrogen is primarily derived from the recycling of organic matter or from additions to the soil through biological N fixation. The majority of N supplied from the soil is in organic form and is released through mineralisation into inorganic forms such as ammonium (NH_4^+) and nitrate (NO_3^-), which can become available for plant uptake. Simple forms of organic N, such as the amino acid glycine, may also be directly utilised by some plants, including the Proteaceae (Schmidt and Stewart 1999). However, the ecological significance of organic N uptake for plant N nutrition remains unknown (Wei *et al.* 2015).

The infertility of many Australian ecosystems is primarily due to soils that have low P content (generally less than 0.02% total P in the top 30 cm; Rossel and Bui 2016) or because what P there is, is also held tightly within mineral oxides such that plant roots are not able to access it (Holford 1997). Soil P availability is defined here as the ability of the soil–plant system to supply P to particular species or vegetation associations over the lifetime of those plants. The main forms of P taken up by plants from the soil solution are the inorganic orthophosphates, $H_2PO_4^-$ and HPO_4^{2-}. When soil pH is less than 7.0, $H_2PO_4^-$ is the predominant form in the soil. Thus, the decomposition of litter and subsequent mineralisation of organic P plays a key role in maintaining P supply to vegetation in many Australian ecosystems (Grierson and Adams 2000; McLaren *et al.* 2015; Turner and Laliberté 2015). The association of particular organic P forms with different vegetation types is largely unknown (but see Adams and Byrne 1989), although studies of Australian agricultural soils have suggested that organic matter content influences simpler forms of organic P (such as phospholipids and RNA), but has relatively little influence on the presence of more complex organic P forms (Moata *et al.* 2016).

Many of the soils of southern and western Australia in particular are characterised by very low concentrations of available P and the growth of many native plant species is negatively impacted by extra P (Ozanne and Specht 1977; Lambers *et al.* 2013). Growth of some species does increase with P supply (e.g. the woodland tree *Eucalyptus tereticornis*; Crous *et al.* 2015), but often at the expense of plant longevity (Handreck 1997) or tolerance to drought (Heddle and Specht 1975). Given that plants take up P from the soil solution, dry soil restricts P mineralisation, P mobility and also P uptake. Consequently, disentangling the impacts of differential water and P availability on vegetation structure and composition can be complex. In many instances, water and P availability are closely related (Pekin *et al.* 2012), even at the continental scale (Bui and Henderson 2013).

7.3.2 Fire impacts on soil nutrients

Fires (Miller and Murphy, Chapter 5, this volume) play an important role in releasing nutrients stored in the biomass, particularly in the nutrient-poor ecosystems of Australia (Raison 1979; Humphreys and Craig 1981; Pekin *et al.* 2012; Huber *et al.* 2013). Burning modifies and redistributes litter and humus and can increase surface soil cation exchange capacity through increased pyrogenic carbon (Hingston *et al.* 1989; Krishnaraj *et al.* 2016). This increases both surface soil nutrient holding capacity and the availability of nutrients, particularly N, for plants because a fraction of organic N remains in the ash residue of burnt plant material and acts as a source for N mineralisation (Raison 1979; Polglase *et al.* 1986; Chambers and Attiwill 1994). However, the net loss of N due to volatilisation during burning also increases with fire intensity (Raison 1979). Nitrogen volatilisation can start at relatively low temperatures (~ 200°C), and based on global estimates, can vary from 907 kg ha^{-1} in temperate forests to as much as 1607 kg ha^{-1} in tropical forests depending on the fire regime (Wan *et al.* 2001). Nonetheless, N supply is often higher after intense fire than before, particularly in nutrient-poor ecosystems, because fire changes nutrient fractions in the soil to forms more easily available to plants (Raison 1979; Humphreys and Craig 1981; Huber *et al.* 2013). For example, labile inorganic N (meaning easily decomposable or available forms that stay in the soil for fairly short periods, usually from a few days to months), in the form of NH_4^+, increased up to 182% directly following fire in a south-west Australian eucalypt forest

(Grove *et al.* 1986). While the increase in labile N is often immediate, soil NH_4^+ has been shown to remain elevated for between 12 and 18 months in sub-alpine plant communities (Huber *et al.* 2013) and wet forest communities (Weston and Attiwill 1990) of south-east Australia.

The colonisation of recently burnt sites by N-fixing species, such as *Acacia pulchella* in the jarrah forests of south-west Australia (Hingston *et al.* 1989) or *Acacia dealbata* in the mountain ash (*E. regnans*) forest of south-east Australia (May and Attiwill 2003), is also thought to have an important role in increasing N availability following fires. In the jarrah forests of south-west Australia, legumes can fix up to 3.3 kg ha^{-1} yr^{-1} of N within the first 2 years following fire (Hingston *et al.* 1982), although these rates decrease to less than 0.5 kg.ha^{-1}.yr^{-1} over the first six years after fire (Hansen *et al.* 1987). A more substantive example comes from the highly productive mountain ash forests of south-east Australia, where N fixation by understorey *Acacia* trees has been estimated to be 50 kg ha^{-1} yr^{-1} for the first 5 years after post-harvest fire (May and Attiwill 2003). Such high rates of N fixation are considered central to N nutrition in the early stages of mountain ash forest growth but become less important as the forest matures (Pfautsch *et al.* 2009).

In contrast to N, the influence of fires on soil P is poorly understood and also more variable due to differences in the phosphate bonding or fixing capacity of soils and immobilisation of P by soil microbes (Rundel 1983; Adams *et al.* 2003). Nonetheless, labile or plant available P can also increase after fire due to conversion of organic P to orthophosphate and reduced acidity from ash formation (Adams *et al.* 1994, 2003; Cade-Menun *et al.* 2000). For example, Grove *et al.* (1986) recorded a 138% increase in labile P in the top 0-3 cm of the soil profile in the south-west Australian jarrah forest directly following fire. However, these increases in P availability are often short lived because phosphates quickly bind to Al, Fe and Mn oxides in acid soils and to Ca minerals in alkaline and neutral soils (Certini 2005).

Beyond the effects of single fires, multiple studies indicate the potential for repeated short-interval fires to influence the balance of available nutrients in Australian ecosystems. In a wet sclerophyll eucalypt forest of south-east Queensland, burning every two years for > 30 years led to significant decreases in soil total and available N and significant increases in available P relative to a long-unburnt control (Toberman *et al.* 2014; Liu *et al.* 2015; Muqaddas *et al.* 2015). This 'decoupling' of soil N and P effect is manifested as significant changes in litter stoichiometry in the two-yearly burnt areas, notably increases in C:N and decreases in N:P ratios, indicating potential for N limitation in the post-fire environment (Toberman *et al.* 2014). In south-east Australia, burning at intervals of 3 to 5 years, repeated over 27 years, has also been associated with decreased soil total N in open forest (Shesthra 2015) as well as with decreased soil N (total and potentially mineralisable) and increased available P in heathland (Adams *et al.* 1994). Nonetheless, long-term repeated burning at 1 to 3-year intervals had no detectable effect on soil N in a Queensland dry sclerophyll forest (Liu *et al.* 2015), nor did four-yearly burning in the above-mentioned wet eucalypt forest (Muqaddas *et al.* 2015), suggesting that repeated fire effects will vary with soil and vegetation types (Liu *et al.* 2015).

At the other end of the fire-interval scale, long periods without fire can also influence the nutrient status of Australian soils. Soil N increased with time since fire to 39 years in dry mixed-species eucalypt forests in eastern Australia (Turner *et al.* 2008), and to 46 years in wet *E. delegatensis* forests in north-west Tasmania (Jones and Davidson 2014). Soil available P did not vary with time since fire, leading to suggestions that premature eucalypt decline in both forest types in the absence of fire could be due to P limitations exacerbated by excess N (Turner *et al.* 2008; Jones and Davidson 2014). This interpretation is consistent with a conceptual model of premature decline of temperate eucalypts in the long-term absence of fire, which nonetheless includes many potentially confounding factors like increased cover of understorey vegetation leading to increased competition for nutrients and water, and to changes in the soil microclimate and microbiota (Close *et al.* 2009). While it remains difficult to isolate the direct causes of eucalypt decline, it seems clear that regular fire is central to maintaining the health of temperate eucalypts, particularly in releasing P for plant uptake – a role of particular importance given the old and highly weathered soils of temperate Australia (Close *et al.* 2009).

7.4 Below-ground traits in Australian plant communities

Plants have root traits that are as important to survival and productivity as their adaptations above-ground to leaves, flowers and stems. These important root adaptations include storage of carbohydrates to allow resprouting after fire or drought, transportation networks to obtain water and nutrients and deep roots to obtain groundwater. As described earlier in this chapter, a range of plant–soil interactions influences the uptake of nutrients by plant roots. This section describes some of the dominant below-ground (root associated) traits exhibited by Australian plants that influence their interactions with the soils in which they grow. We focus primarily on root adaptations considered to enhance nutrient uptake. These adaptations help to shape Australian vegetation communities by influencing the diversity and productivity of plants that are present, because different plant species may have different root adaptations and the effectiveness of these adaptations is likely to vary with soil properties. Mycorrhizal fungi also have adaptations to different soil types and local climates that indirectly benefit plants by increasing their capacity to acquire soil nutrients (Tedersoo *et al.* 2014). There are strong phylogenetic patterns, as mycorrhizal types are consistent within most plant families (Brundrett 2009). Many mycorrhiza types are closely associated with particular vegetation types, while others are widespread across many different vegetation types, as shown in Box 7.2. The two largest Australian plant families, the Fabaceae and Myrtaceae have very complex root strategies which seem to have evolved very rapidly, resulting in more lineages that have switched from arbuscular mycorrhizae (AM) to ectomycorrhizae (EM) roots than occur in other families (Brundrett 2009).

7.4.1 Mycorrhizal roots

Mycorrhizae are symbiotic associations between a fungus and a root or other organ of a living plant that are primarily responsible for nutrient transfer. Mycorrhizal associations function primarily by forming a network of hyphae in the soil to effectively increase the ratio of root surface area to soil volume, thereby reducing the diffusion path for phosphate ions to the root surface and exploring a greater soil volume. These hyphae are also known to exude enzymes and organic compounds that help mobilise insoluble soil nutrients (Smith and Read 2008; Ryan *et al.* 2012). The primitive plants that colonised land over 400 000 years ago had mycorrhizas before they had roots (Brundrett 2002), so symbiotic associations with fungi are one of the primary functions of roots. Different types of mycorrhizal associations vary in their host plants and roles, but are primarily defined by structural characteristics (Table 7.2). A comparison of the number of species of plants with mycorrhizal associations or nonmycorrhizal roots on a global scale and in Australia (Figure 7.3) shows that Australia has disproportionately large numbers of both ectomycorrhizal and nonmycorrhizal species.

Arbuscular mycorrhizae (AM) are the most widespread fungal symbioses and include about 80% of vascular plants, which are present in almost every habitat where plants grow. These involve Glomeromycete (Zygomycete) fungi that colonise roots internally. Arbuscular mycorrhizas consist

Table 7.2 **Summary of defining features, host plants and habitat relationships for types of mycorrhizas (after Brundrett 2008).**

Category	Definition	Role	Host plants	Fungi
Arbuscular mycorrhizae (AM)	Associations formed by Glomeromycotan fungi in plants that usually have arbuscules and often have vesicles	Nutrient acquisition for plant, especially P	Most families of plants (# species)	Glomeromycota
Ectomycorrhiza mycorrhizae (EM)	Associations with a hyphal mantle enclosing short lateral roots and a Hartig net of labyrinthine hyphae that penetrate between root cells	Nutrient acquisition for plant, especially P, N	Certain families or genera (often host AM also)	Most are higher fungi (many basidiomycetes and ascomycetes)
Orchid mycorrhizae	Associations where coils of hyphae (pelotons) penetrate within cells in the plant family Orchidaceae	Nutrient acquisition for plant, especially P, N	Orchidaceae	Most are basidiomycetes in Rhizoctonia alliance
Ericoid mycorrhizae	Coils of hyphae within very thin roots (hair roots) of the Ericaceae	Nutrient acquisition for plant, especially P, N	Ericaceae	Most are ascomycetes
Sub-epidermal	Hyphae in cavities under epidermal cells, only known from the Australian monocot *Thysanotus*	Unknown	*Thysanotus* spp. (Laxmanniaceae)	Unknown

of hyphae that spread through the root cortex where they form arbuscules that consist of highly branched hyphae within root cells. These associations provide plants with nutrients, especially phosphorus, and the fungi with carbohydrates, which are exchanged at the arbuscule interface. These fungi also form propagules and storage structures in the form of vesicles in roots and/or spores in soil.

Ectomycorrhizal associations (EM) are formed primarily by higher fungi in the Basidiomycotina and Ascomycotina, which form mycorrhizas with vascular plants in certain families. Hyphae of ectomycorrhizal fungi form an extracellular interface, with highly branched hyphae growing between epidermal or cortical cells to form a network known as the Hartig net. These associations tend to occur in relatively short and highly branched lateral roots. The majority of plants with EM are trees or large shrubs and the most iconic examples in Australia are gum trees (*Eucalyptus* spp.) and southern beech trees (*Nothofagus* spp.). Many Australian plants in the families Fabaceae and Myrtaceae have both arbuscular mycorrhizas and ectomycorrhizas in the same root system (see mycorrhizas.info/ozplants). For example, roots of *Eucalyptus* spp. seedlings often have dual AM and EM associations, but the EM becomes increasingly dominant over AM as they mature (Chen *et al.* 2000). On a global scale, Australian plant communities tend to have a relatively high diversity of ectomycorrhizal hosts, such as eucalypt trees and many large shrubs (Brundrett 2009).

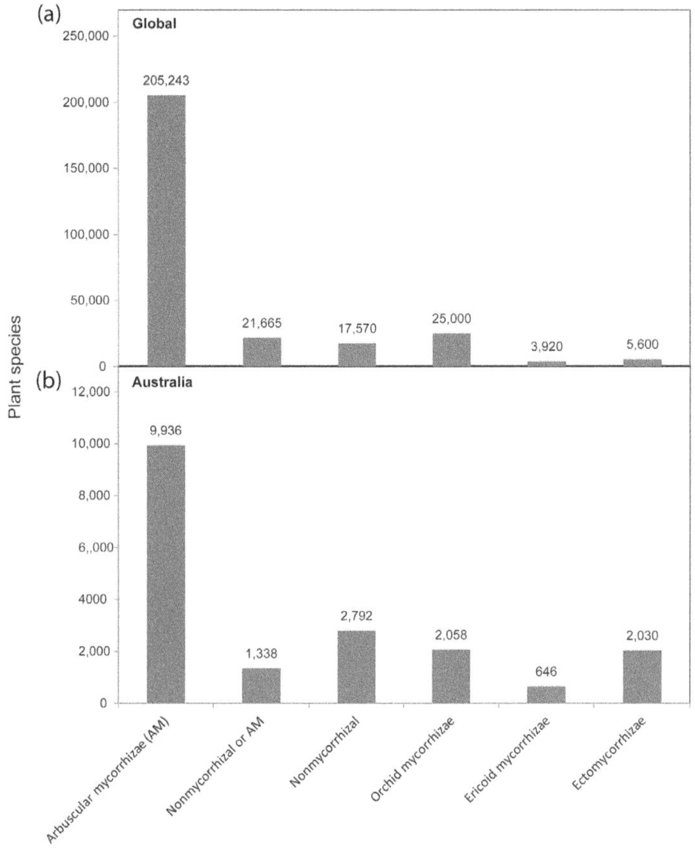

Figure 7.3 **(a)** The relative diversity of different categories of mycorrhizal plants on a global scale. These data are totals of all taxa that can be assigned to each category based on a review of the scientific literature (Brundrett 2009). **(b)** The same classification for Australian plants using diversity data from the *Atlas of Living Australia* (2015).

There are other types of mycorrhizae associated with species in the Orchidaceae and Ericaceae and the monocot genus *Thysanotus*. These types of mycorrhizas are associated with a wide diversity of habitat types, but tend to be most diverse in higher rainfall habitats. Host-specific mycorrhizal associations, such as those found on Ericaceae and orchid species, which are particularly abundant on the nutrient-poor soils of south-west Australia (Groom and Lamont 2015), may also be better at utilising organic nutrient sources when inorganic nutrient sources are limiting (Sommer *et al.* 2012; Nurfadilah *et al.* 2013; Schmidt *et al.* 2013).

7.4.2 Nonmycorrhizal plants

Nonmycorrhizal (NM) plants fail to form mycorrhizae, even when inoculum of these fungi are present, but there are also cases of facultative NM plants which occur in habits where fungal activity is likely to be suppressed by soil or climatic conditions (Brundrett 2009). Facultative mycorrhizal plant

Table 7.3 **Relative diversity of plants with different categories of nonmycorrhizal root traits in Australia and globally (data from Brundrett 2009 and the *Atlas of Living Australia* 2015).**

Root trait category	Australia	Global	Main examples
Cluster and dauciform roots	2 130	6 700	Most species in Proteaceae, Restionaceae, Cyperaceae, etc.
Carnivores	210	600	Sundews (*Drosera* spp.), bladderworts (*Utricularia* spp.), pitcher plants
Parasites	280	4 000	Mistletoes (Loranthaceae) and hemiparasites such as sandalwood (*Santalum* spp.)
Halophytes	330	1 250	Samphires, salt bushes (Amaranthaceae), etc.
Disturbance opportunists	450	12 000	Short-lived weedy plants
Aquatic	450	3 300	Grow partly of fully submerged
Marine	75	320	Seagrasses, mangroves, etc.
Arid	230	5 650	Desert plants
Other	130	6 300	
Total	4 300	38 000	

species in the second category may be mycorrhizal in some habitats and not in others. Plants that lack mycorrhizas tend to have a replacement strategy for nutrient acquisition (see below). Examples of specialised plants that grow in habitats where plant species generally do not have mycorrhizal roots include hydrophytes, halophytes, xerophytes, epiphytes and alpine and subpolar plants (Table 7.3). Aquatic plants and halophytes also include some mycorrhizal species or plants that are facultatively mycorrhizal depending on soil conditions that vary seasonally (Brundrett 2009). Australia has a relatively high taxonomic diversity of NM plants on a global scale, as shown in Table 7.3.

In Australia NM plants with specialised means of nutrition include carnivorous plants, parasitic plants and cluster-rooted plants (including species with proteoid roots) (Table 7.3). The diversity of these plants is higher than expected in Australia, presumably because they are highly competitive in low-nutrient soils (Brundrett 2009). Groom and Lamont (2015) provide detailed explanations of how each of these strategies functions, but in general they all provide access to mineral nutrient sources that are not accessible to other plants. Carnivorous plants have insect traps containing glands producing exudates with the capacity to digest prey and absorb nutrients from them. The Western Australian biodiversity hotspot is the global centre of diversity for carnivorous plants containing about one-quarter of all species of these plants (Brundrett 2009). Parasites are plants with haustorial connections to the vascular tissue of a host plant that is used to supply all of their nutrients and water. Hemiparasites are similar, but also have leaves and roots so are capable of some photosynthesis and water uptake (Groom and Lamont 2015).

Common among species adapted to low P supply, is the capacity to obtain P from sparingly soluble organic and inorganic forms in soils dominated by Fe and Al. One adaptation is the production of large amounts of root exudates, including organic acids and extracellular enzymes, which have strong P solubilisation or mineralisation properties (Grierson and Attiwill 1989; Grierson 1992; Lambers 2010). Both mycorrhizal (Dodd *et al.* 1987; MacFall *et al.* 1991) and nonmycorrhizal roots (Dinkelaker and Marschner 1992) produce phosphatase enzymes, which catalyse the mineralisation of organic P, releasing inorganic P. Root-mediated changes in pH, production of phytosiderophores and phytochelates, and root-mediated microbiological changes may increase nutrient availability in the rhizosphere. For example, *Banksia* spp. have cluster roots that can significantly modify the soil-root environment by lowering pH through the production of organic acids and phosphatase enzymes (Grierson and Attiwill 1989; Grierson 1992; Shane and Lambers 2005). Sedges in the Cyperaceae have highly branched lateral roots similar to cluster roots as well as dauciform roots, swollen lateral roots which function in a similar manner (Shane *et al.* 2006). In addition to root modifications to increase nutrient acquisition, many Australian native species efficiently retain P and other nutrients within the plant, partly through internal recycling and partly due to retention of leaves for up to several years (Lambers *et al.* 2010).

In Australian ecosystems, nonmycorrhizal plants have greater species richness than in most other continents, especially due to the presence of cluster-rooted species, as well as carnivorous and parasitic plants, and these seem to be more often found in relatively low-nutrient soils (Pate *et al.* 2001; Lambers *et al.* 2010; Groom and Lamont 2015).

7.4.3 Nitrogen-fixing associations

Nitrogen-fixing associations are root nodules or specialised roots containing bacteria or cyanobacteria that fix gaseous nitrogen and makes it available to the host plant. These associations only occur in members of the families Fabaceae, Casuarinaceae and Zamiaceae in Australia, but members of these families occur in most Australian plant communities. Nitrogen-fixing nodules are formed by most members of the Fabaceae, which includes Caesalpinoidae, Mimosoidae (acacias), as well as the Faboidae (Stępkowski *et al.* 2012; Barrett *et al.* 2015). Nitrogen-fixing plants are very important both in agriculture and in natural ecosystems in Australia. They are found in most habitats and can be more abundant in recently disturbed or burnt habitats due to their 'hard seeds' that germinate best after heat shock or mechanical scarification (Auld

Box 7.2 The relative dominance and diversity of plants with different nutritional strategies

The relative dominance or diversity of plants with different nutritional adaptations varies at a continental scale. Comparison of the total diversity of plants with mycorrhizal associations or nonmycorrhizal roots on a global scale and in Australia (Figure 7.3) shows that Australia has disproportionately large numbers of both ectomycorrhizal (EM) and nonmycorrhizal (NM) species compared to the rest of the world. These nutritional strategies are both linked to highly infertile soils. On a global scale, Australian plant communities also tend to have a relatively high diversity of ectomycorrhizal hosts, such as eucalypt trees and many large shrubs (Brundrett 2009). The species richness patterns of plants with EM or NM roots, as well as nitrogen-fixing symbiotic associations in Australia are shown in Figure 7.4. Species with cluster roots and parasites using haustoria to feed on other plants are widespread, but carnivorous plants primarily occur in high-rainfall areas. As well as continental patterns, there is

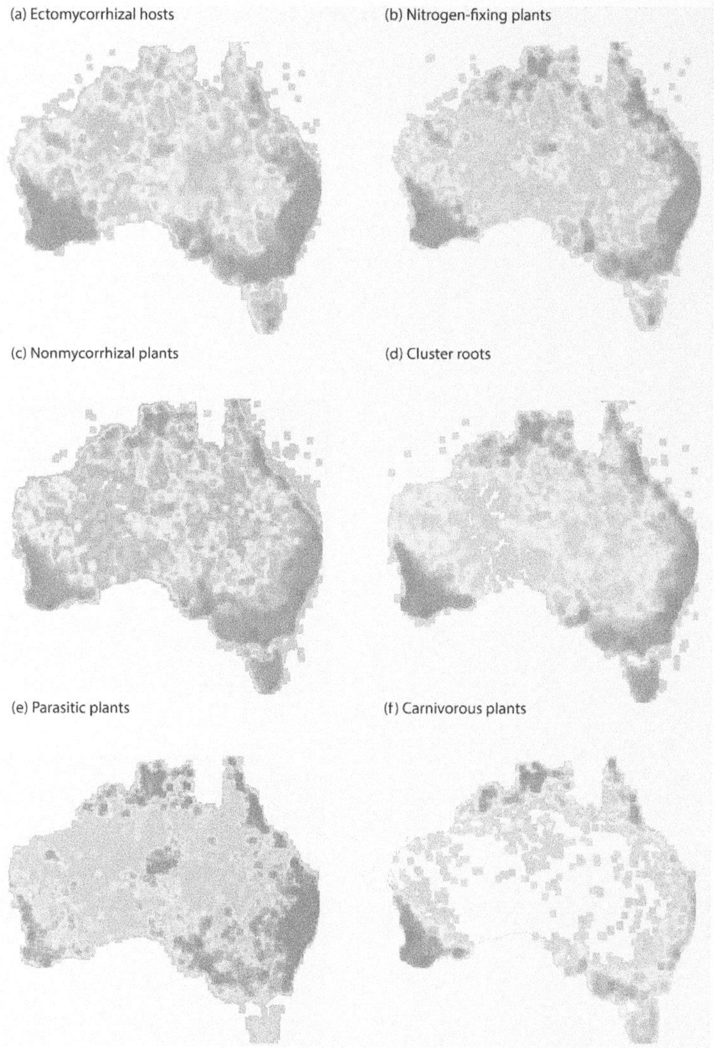

Figure 7.4 Diversity maps for key root strategies: **(a)** ectomycorrhizal plants, **(b)** nitrogen-fixing plants, **(c)** nonmycorrhizal plants, **(d)** plants with cluster roots **(e)** parasitic plants and **(f)** carnivorous plants. These maps show the average count of the number of recorded taxa at different locations using data from the *Atlas of Living Australia* (2015). Maps are produced with the ALA Points to grid function with a 0.1 degree scale and 9 cell moving averages. The highest diversity is shown as red (dark in greyscale), lower diversity as blue (pale in greyscale) and a lack of shading reflecting the absence of records. These maps show general trends only and are affected by limited sampling in remote areas, particularly in central Australia.
Maps: Mark Brundrett. (A black and white version of this figure will appear in some formats. For the colour version, please refer to the plate section.)

Box 7.2 (cont.)

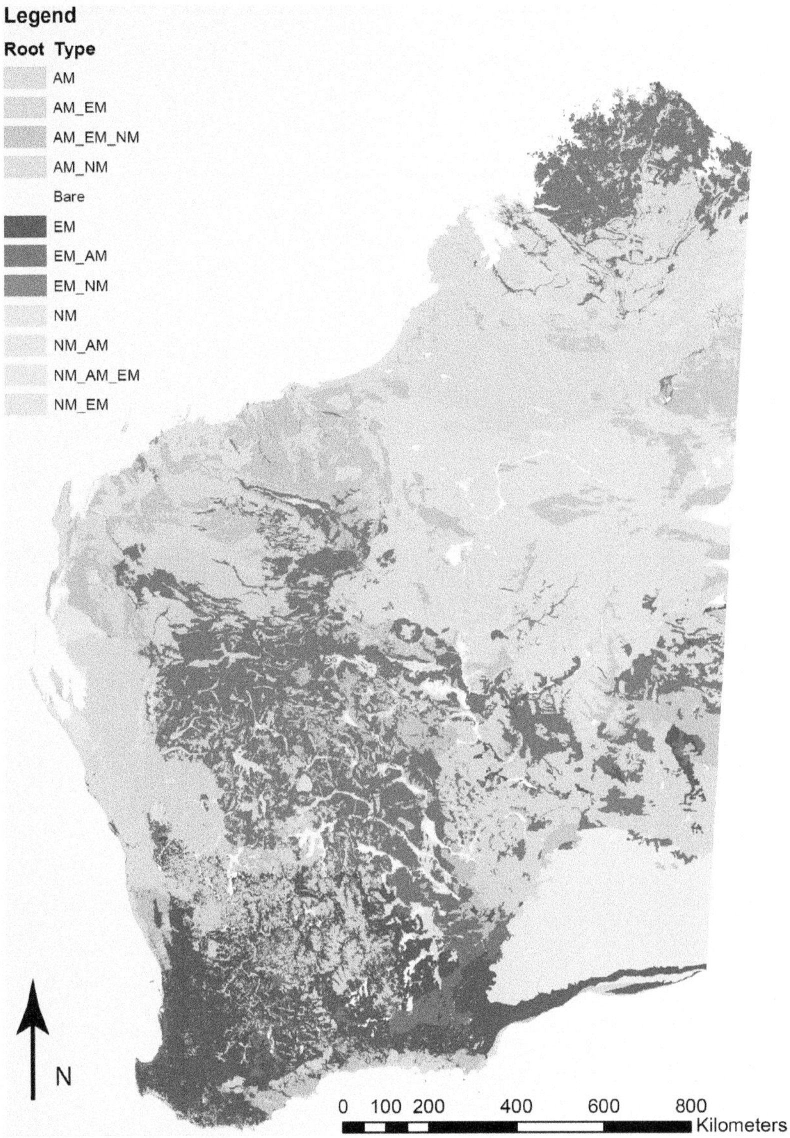

Figure 7.5 The West Australian Vegetation map (Beard *et al.* 2013) with major physiognomic vegetation types reclassified to show below-ground root type adaptations linked to nutrient uptake (AM – arbuscular mycorrhizal, EM – ectomycorrhizal; NM – nonmycorrhizal). This map provides information on root traits of the dominant plants in vegetation types. Many of these types include taxa with traits that differ from those of the dominant taxa. The root status of dominant plants was inferred from existing data and phylogeny. *Source: http://mycorrhizas.info/ozplants; Brundrett (2009). Map: Mark Brundrett. (A black and white version of this figure will appear in some formats. For the colour version, please refer to the plate section.)*

strong variation in trait frequencies along local resource gradients, which can have an important influence on plant diversity. In south-western Australia, for example, the highest diversity and dominance of cluster-rooted NM species tends to be on soils with the lowest P content. But even in these soils they co-occur with mycorrhizal species, which replace them as dominants as soil P increases (Lambers *et al.* 2010, 2014). While Figure 7.4 provides an indication of the diversity of all recorded species, as an example to illustrate dominance patterns, Figure 7.5 is a vegetation map that shows where dominant plants tend to have mycorrhizal or NM roots in Western Australia. These maps show that EM, NM and nitrogen-fixing plants are important in many Australian habitats and that plants with arbuscular mycorrhizal associations are common in most habitats. Thus mycorrhizas are important in almost every habitat, including some of the harshest landscapes on earth. The widespread occurrence of EM hosts has other ecological implications, since their associated fungi include truffles that produce subterranean fruit bodies, which are important food sources for the animals that disperse them (e.g. Claridge 2002).

and Ooi, Chapter 10, this volume). Actinorhizal nitrogen fixation root nodules occur in many but not all of the Casuarinaceae. Studies of actinobacteria from *Allocasuarina* and *Casuarina* show a relatively low diversity of compatible strains that differ from isolates from other families of host plants that occur in other regions of the globe (Simon *et al.* 1999). Some *Allocasuarina* species do not consistently form nodules and this seems to depend both on host species and soil types (Reddell *et al.* 1986; Santi *et al.* 2013). Cycads produce upward facing dichotomously branched 'coralloid roots'. These roots grow close to the soil surface and have green tissue within them containing nitrogen-fixing cyanobacteria (Groom and Lamont 2015).

7.5 Conclusions

- Spatial variation in soil properties, particularly those that influence availability of nutrients and water, are key factors influencing the distribution of Australian plant communities and species. This is evident at continental scales in relationships between vegetation formations and geological substrates and at local scales with species turnover along catenary sequences and across boundaries of geological substrates. Soils and vegetation are also strongly regulated by climate across Australia.
- While climate primarily drives large-scale patterns in Australian vegetation, soil properties that determine the geochemistry and water-holding capacity of soils across landscapes play a key role in explaining the distribution of plant species and the functional composition of vegetation.
- Australian vegetation is often limited by mineral nutrients, especially N and P and most plants have acquired symbiotic associations with soil fungi to help acquire these nutrients. Mycorrhizal plants are dominant in the majority of Australian habitats, especially trees and large shrubs with ectomycorrhizal roots and

plants with arbuscular mycorrhizas. Plants without mycorrhizal roots are also relatively common in many Australian habitats, especially in highly infertile soils and habitats where adverse conditions such as salinity or waterlogging limit plant growth and fungal activity. These plants often have alternative nutrient uptake mechanisms such as cluster roots, or are parasitic or carnivorous. Australia is the world centre of diversity for nonmycorrhizal plants, which tend to be dominant in sites where plant productivity is severely limited by soil conditions.

- Soil-based processes and feedbacks, such as nutrient cycling and responses to fire, are essential for ecosystem sustainability.
- In addition to the factors discussed above, many Australian habitats are severely limited by water availability. Some trees tolerate or avoid severe drought, while others have developed the capacity to utilise groundwater.
- Feedbacks and complex interactions between plants and soils and the relative importance of plant nutrition, hydrology, climate, and history on plant diversity and vegetation structure are still being resolved.

References

Adams, M.A. and Byrne, L.T. (1989) 31 P-NMR analysis of phosphorus compounds in extracts of surface soils from selected Karri (*Eucalyptus diversicolour* F. Muell.) forests. *Soil Biology and Biochemistry* 21, 523–528.

Adams, M.A., Iser, J., Keleher, A.D. and Cheal, D.C. (1994). Nitrogen and phosphorus availability and the role of fire in heathlands at Wilsons Promontory. *Australian Journal of Botany* 42, 269–281.

Adams, M.A., Grierson, P.F. and Burrows, C. (2003) Fires, soils and plant nutrition in south-west Western Australian forests: a focus on spatial and temporal variability. In *Fire in Ecosystems of South-west Western Australia: Impacts and Management*, ed. I. Abbott and N. Burrows. Backhuys Publishers: Leiden.

Atlas of Living Australia (ALA) (2015) *Atlas of Living Australia.* Available at: http://www.ala.org.au/.

Attiwill, P.M. and Adams, M.A. (1993) Nutrient cycling in forests. *New Phytologist* 124, 561–582.

Attiwill, P.M. and Leeper, G.W. (1987) *Forest Soils and Nutrient Cycles.* Melbourne University Press: Melbourne.

Australian Bureau of Meteorology (2016) Available at: http://www.bom.gov.au/.

Barrett, L.G., Bever, J.D., Bissett, A. and Thrall, P.H. (2015) Partner diversity and identity impacts on plant productivity in *Acacia*–rhizobial interactions. *Journal of Ecology* 103, 130–142.

Beard, J.S. (1990) *Plant Life of Western Australia*. Kangaroo Press: Kenthurst.

Beard, J.S., Beeston, G.R., Harvey, J.M. *et al.* (2013) The vegetation of Western Australia at the 1: 3 000 000 scale. Explanatory memoir. *Conservation Science Western Australia* 9, 1–152.

Benson, D.H. and Howell, J. (1990) *Taken for Granted: The Bushland of Sydney and its Suburbs*. Kangaroo Press: Kenthurst.

Brundrett, M.C. (2002) Coevolution of roots and mycorrhizas of land plants. *New Phytologist* 154, 275–304.

Brundrett, M.C. (2008) *Mycorrhizal associations: the web resource.* Available at: mycorrhizas.info (accessed 27 July 2016).

Brundrett, M.C. (2009) Mycorrhizal associations and other means of nutrition of vascular plants: understanding the global diversity of host plants by resolving conflicting information and developing reliable means of diagnosis. *Plant and Soil* 320, 37–77.

Bui, E.N. and Henderson, B.L. (2013) C: N: P stoichiometry in Australian soils with respect to vegetation and environmental factors. *Plant and Soil* 373, 553–568.

Bui, E.N., González-Orozco, C.E. and Miller, J.T. (2014) *Acacia*, climate, and geochemistry in Australia. *Plant and Soil* 381, 161–175.

Burgess, S.S., Pate, J.S., Adams, M.A. and Dawson, T.E. (2000) Seasonal water acquisition and redistribution in the Australian woody phreatophyte, *Banksia prionotes. Annals of Botany* 85, 215–224.

Cade-Menun, B.J., Berch, S.M., Preston, C.M. and Lavkulich, L.M. (2000) Phosphorus forms and related soil chemistry of Podzolic soils on northern Vancouver Island. II. The effects of clear-cutting and burning. *Canadian Journal of Forest Research* 30, 1726–1741.

Canadell, J., Jackson, R.B., Ehleringer, J.B., Mooney, H.A., Sala, O.E. and Schulze, E.D. (1996) Maximum rooting depth of vegetation types at the global scale. *Oecologia* 108, 583–595.

Certini, G. (2005) Effects of fire on properties of forest soils: a review. *Oecologia* 143, 1–10.

Chambers, D.P. and Attiwill, P.M. (1994) The ash-bed effect in *Eucalyptus regnans* forest: chemical, physical and microbiological changes in soil after heating or partial sterilisation. *Australian Journal of Botany* 42, 739–749.

Chen, Y.L., Brundrett, M.C. and Dell, B. (2000) Effects of ectomycorrhizas and vesicular–arbuscular mycorrhizas, alone or in competition, on root colonization and growth of *Eucalyptus globulus* and *E. urophylla*. *New Phytologist* 146, 545–555.

Churchward, H.M. and McArthur, W.M. (1980) Landforms and soils of the Darling System, Western Australia. In *Atlas of Natural Resources, Darling System, Western Australia.* Department of Conservation and Environment: Perth.

Claridge, A.W. (2002) Ecological role of hypogeous ectomycorrhizal fungi in Australian forests and woodlands. In *Diversity and Integration in Mycorrhizas*. Springer: Netherlands, pp. 291–305.

Close, D.C., Davidson, N.J., Johnson, D. *et al.* (2009) Premature decline of Eucalyptus and altered ecosystem processes in the absence of fire in some Australian forests. *The Botanical Review* 75, 191–202.

Cowling, R.M., Potts, A.J., Bradshaw, P.L. *et al.* (2015) Variation in plant diversity in Mediterranean-climate ecosystems: the role of climatic and topographical stability. *Journal of Biogeography* 42, 552–564.

Crous, K.Y., Ósvaldsson, A. and Ellsworth, D.S. (2015) Is phosphorus limiting in a mature *Eucalyptus* woodland? Phosphorus fertilisation stimulates stem growth. *Plant and Soil* 391, 293–305.

Dawson, T.E. and Pate, J.S. (1996) Seasonal water uptake and movement in root systems of Australian phraeatophytic plants of dimorphic root morphology: a stable isotope investigation. *Oecologia* 107, 13–20.

Dell, B., Bartle, J.R. and Tacey, W.H. (1983) Root occupation and root channels of jarrah forest subsoils. *Australian Journal of Botany* 31, 615–627.

Dell, B., Havel, J.J. and Malajczuk, N. (2012) *The Jarrah Forest: A Complex Mediterranean Ecosystem*, ed. B. Dell, J.J. Havekm and N. Malajczuk. Kluwer Academic Publishers: Dordrecht.

Dinkelaker, B. and Marschner, H. (1992) In vivo demonstration of acid phosphatase activity in the rhizosphere of soil-grown plants. *Plant and Soil* 144, 199–205.

Di Virgilio, G., Laffan, S.W. and Ebach, M.C. (2012) Fine-scale quantification of floral and faunal breaks and their geographic correlates, with an example from south-eastern Australia. *Journal of Biogeography* 39, 1862–1876.

Dodd, J.C., Burton, C.C., Burns, R.G. and Jeffries, P. (1987) Phosphatase activity associated with the roots and the rhizosphere of plants infected with vesicular-arbuscular mycorrhizal fungi. *New Phytologist* 107, 163–172.

Eamus, D. (2009) *Identifying Groundwater Dependent Ecosystems: A Guide for Land and Water Managers*. Land and Water Australia: Canberra. Available at: https://www.uts.edu.au/sites/default/files/aglw-groundwater.pdf (accessed 27 July 2016).

Eamus, D. and Froend, R. (2006) Groundwater-dependent ecosystems: the where, what and why of GDEs. *Australian Journal of Botany* 54, 91–96.

Fan, Y. (2015) Groundwater in the Earth's critical zone: relevance to large-scale patterns and processes. *Water Resources Research* 51, 3052–3069.

Fedrigo, M., Kasel, S., Bennett, L.T. *et al.* (2014) Carbon stocks in temperate forests of south-eastern Australia reflect large tree distribution and edaphic conditions. *Forest Ecology and Management* 334, 129–143.

Fellman, J.B., Pettit, N.E., Kalic, J. and Grierson, P.F. (2013) Influence of stream–floodplain biogeochemical linkages on aquatic foodweb structure along a gradient of stream size in a tropical catchment. *Freshwater Science* 32, 217–229.

Fine, P.V. (2015) Ecological and evolutionary drivers of geographic variation in species diversity. *Annual Review of Ecology, Evolution and Systematics* 46, 369–392.

Fraser, L.H., Pither, J., Jentsch, A. *et al.* (2015) Worldwide evidence of a unimodal relationship between productivity and plant species richness. *Science* 349, 302–305.

Gibson, N., Keighery, G.J., Lyons, M.N. and Webb, A. (2004) Terrestrial flora and vegetation of the Western Australian wheatbelt. *Records of the Western Australian Museum Supplement* 67, 139–189.

Grierson, P.F. (1992) Organic acids in the rhizosphere of *Banksia integrifolia* Lf. *Plant and Soil* 144, 259–265.

Grierson, P.F. and Adams, M.A. (1999) Nutrient cycling and growth in forest ecosystems of southwestern Australia: relevance to agricultural landscapes. *Agroforestry Systems* 45, 215–244.

Grierson, P.F. and Adams, M.A. (2000) Plant species affect acid phosphatase, ergosterol and microbial P in a Jarrah (*Eucalyptus marginata* Donn ex Sm.) forest in south-western Australia. *Soil Biology and Biochemistry* 32, 1817–1827.

Grierson, P.F. and Attiwill, P.M. (1989) Chemical characteristics of the proteoid root mat of *Banksia integrifolia* Lf. *Australian Journal of Botany* 37, 137–143.

Grierson, P.F. and Comerford, N.B. (2000) Non-destructive measurement of acid phosphatase activity in the rhizosphere using nitrocellulose membranes and image analysis. *Plant and Soil* 218, 49–57.

Grime, J.P. (1979) Primary strategies in plants. *Transactions of the Botanical Society of Edinburgh* 43, 151–160.

Groom, P.K. (2004) Rooting depth and plant water relations explain species distribution patterns within a sandplain landscape. *Functional Plant Biology* 31, 423–428.

Groom, P.K. and Lamont, B. (2015) *Plant Life of Southwestern Australia: Adaptations for Survival*. De Gruyter Open: Warsaw.

Grove, T.S., O'Connell, A.M. and Dimmock, G.M. (1986) Nutrient changes in surface soils after an intense fire in jarrah (*Eucalyptus marginata* Donn ex Sm.) forest. *Australian Journal of Ecology* 11, 303–317.

Handreck, K.A. (1997) Phosphorus requirements of Australian native plants. *Australian Journal of Soil Research* 35, 241–290.

Hansen, A.P., Pate, J.S., Hansen, A. and Bell, D.T. (1987) Nitrogen economy of post-fire stands of shrub legumes in jarrah (*Eucalyptus marginata* Donn ex Sm.) forest of SW Australia. *Journal of Experimental Botany* 38, 26–41.

Havel, J.J. (1975) *Site-vegetation Mapping in the Northern Jarrah Forest (Darling Range).* Forests Department: Perth.

Havel, J.J. (2000) *Ecology of the forests of south western Australia in relation to climate and landforms.* PhD Thesis, Murdoch University, Perth.

Heddle, E.M. and Specht, R.L. (1975) Dark Island Heath (Ninety-Mile Plain South Australia). VIII. The effect of fertilisers on composition and growth, 1950–1972. *Australian Journal of Botany* 23, 151–164.

Hingston, F.J., Malajczuk, N. and Grove, T.S. (1982) Acetylene reduction (N_2-fixation) by jarrah forest legumes following fire and phosphate application. *Journal of Applied Ecology* 19, 631–645.

Hingston, F.J., O'Connell, A.M. and Grove, T.S. (1989) Nutrient cycling in jarrah forest. In *The Jarrah Forest*, ed. B. Dell and J.J. Havel. Springer: Netherlands.

Hobbie, S.E. (2015) Plant species effects on nutrient cycling: revisiting litter feedbacks. *Trends in Ecology and Evolution* 30, 357–363.

Holford, I.C.R. (1997) Soil phosphorus: its measurement, and its uptake by plants. *Soil Research* 35, 227–240.

Hopper, S.D. (2009) OCBIL theory: towards an integrated understanding of the evolution, ecology and conservation of biodiversity on old, climatically buffered, infertile landscapes. *Plant and Soil* 322, 49–86.

Huber, E., Bell, T.L. and Adams, M.A. (2013) Combustion influences on natural abundance nitrogen isotope ratio in soil and plants following a wildfire in a sub-alpine ecosystem. *Oecologia* 173, 1063–1074.

Humphreys, F.R. and Craig, F.G. (1981) Effects of fire on soil chemical, structural and hydrological properties. *Fire and the Australian Biota*, 177–200.

Johnson, D. (2009) *The Geology of Australia*. Cambridge University Press: Cambridge.

Jones, A.G. and Davidson, N.J. (2014) Altered N, P and C dynamics with absence of fire in *Eucalyptus* forests affected by premature decline. *Austral Ecology* 39, 587–599.

Jones, M.M., Gibson, N., Yates, C. *et al.* (2016) Underestimated effects of climate on plant species turnover in the Southwest Australian Floristic Region. *Journal of Biogeography* 43, 289–300.

Kardol, P., Veen, G.F., Teste, F.P. and Perring, M.P. (2015) Peeking into the black box: a trait-based approach to predicting plant–soil feedback. *New Phytologist* 206, 1–4.

Keith, D.A. (2004) Ocean shores to desert dunes: the native vegetation of NSW and the ACT (Selected Extracts). Department of Environment and Conservation: NSW.

Keith, D.A. and Myerscough, P.J. (1993) Floristics and soil relations of upland swamp vegetation near Sydney. *Australian Journal of Ecology* 18, 325–344.

Krishnaraj, S.J., Baker, T.G., Polglase, P.J. *et al.* (2016) Prescribed fire increases pyrogenic carbon in litter and surface soil in lowland *Eucalyptus* forests of south-eastern Australia. *Forest Ecology and Management* 366, 98–105.

Laliberté, E., Grace, J.B., Huston, M.A. *et al.* (2013) How does pedogenesis drive plant diversity? *Trends in Ecology and Evolution* 28, 331–340.

Lambers, H., Brundrett, M.C., Raven, J.A. and Hopper, S.D. (2010) Plant mineral nutrition in ancient landscapes: high plant species diversity on infertile soils is linked to functional diversity for nutritional strategies. *Plant and Soil* 334, 11–31.

Lambers, H., Ahmedi, I., Berkowitz, O. *et al.* (2013) Phosphorus nutrition of phosphorus-sensitive Australian native plants: threats to plant communities in a global biodiversity hotspot. *Conservation Physiology* 1, 1–21.

Leigh, C., Bush, A., Harrison, E.T. *et al.* (2015) Ecological effects of extreme climatic events on riverine ecosystems: insights from Australia. *Freshwater Biology* 60, 2620–2638.

Linder, H.P. and Verboom, G.A. (2015) The evolution of regional species richness: the history of the southern African flora. *Annual Review of Ecology, Evolution, and Systematics* 46, 393–412.

Liu, X., Chen, C.R., Wang, W.J. *et al.* (2013) Soil environmental factors rather than denitrification gene abundance control N_2O fluxes in a wet sclerophyll forest with different burning frequency. *Soil Biology and Biochemistry* 57, 292–300.

Liu, X., Chen, C., Wang, W., Hughes, J.M. and Lewis, T. (2015) Response of soil denitrifying communities to long-term prescribed burning in two Australian sclerophyll forests. *Geomicrobiology Journal* 32, 577–584.

MacFall, J., Slack, S.A. and Iyer, J. (1991) Effects of *Hebeloma arenosa* and phosphorus fertility on root acid phosphatase activity of red pine (Pinus resinosa) seedlings. *Canadian Journal of Botany* 69, 380–383.

Macphail, M.K. (2007) Australian palaeoclimates: Cretaceous to Tertiary: a review of palaeobotanical and related evidence to the year 2000. *CRC LEME Special Publication 7.*

Mattiske, E.M. and Havel, J.J. (1998) *Vegetation Complexes of the South-west Forest Region of Western Australia.* Department of Conservation and Land Management and Environment Australia: Western Australia.

May, B.M. and Attiwill, P.M. (2003) Nitrogen-fixation by *Acacia dealbata* and changes in soil properties 5 years after mechanical disturbance or slash-burning following timber harvest. *Forest Ecology and Management* 181, 339–355.

McArthur, W.M. (2004) *Reference Soils of South-western Australia*, 2nd edition. Department of Agriculture: Kensington.

McKenzie, N., Jacquier, D., Isbell, R. and Brown, K. (2004) *Australian Soils and Landscapes: An Illustrated Compendium*. CSIRO Publishing: Canberra.

McLaren, T.I., Smernik, R.J., McLaughlin, M.J. *et al.* (2015) Complex forms of soil organic phosphorus: a major component of soil phosphorus. *Environmental Science and Technology* 49, 13238–13245.

McTainsh, G. and Strong, C. (2007) The role of aeolian dust in ecosystems. *Geomorphology* 89, 39–54.

Moata, M.R., Doolette, A.L., Smernik, R.J. *et al.* (2016) Organic phosphorus speciation in Australian red chromosols: stoichiometric control. *Soil Research* 54, 11–19.

Muqaddas, B., Zhou, X., Lewis, T. *et al.* (2015) Long-term frequent prescribed fire decreases surface soil carbon and nitrogen pools in a wet sclerophyll forest of southeast Queensland, Australia. *Science of The Total Environment* 536, 39–47.

Mucina, L. and Wardell-Johnson, G.W. (2011) Landscape age and soil fertility, climatic stability, and fire regime predictability: beyond the OCBIL framework. *Plant and Soil* 34, 1–23.

Mucina, L., Stephenson, G., Daniel, G., van Niekerk, A. and Boonzaaier, I. (2013) *Floristic-Ecological Vegetation Map of the Mitchell Plateau Region (Northern Kimberley, Western Australia).* Vegetation Mapping In The Northern Kimberley, Western Australia. Curtin University: Perth.

Myerscough, P.J. and Carolin, R.C. (1986) The vegetation of the Eurunderee sand mass, headlands and previous islands in the Myall Lakes area, New South Wales. *Cunninghamia* 1, 399–466.

Nano, C.E. and Clarke, P.J. (2008) Variegated desert vegetation: covariation of edaphic and fire variables provides a framework for understanding mulga-spinifex coexistence. *Austral Ecology* 33, 848–862.

Nurfadilah, S., Swarts, N.D., Dixon, K.W. *et al.* (2013) Variation in nutrient-acquisition patterns by mycorrhizal fungi of rare and common orchids explains diversification in a global biodiversity hotspot. *Annals of Botany* 111, 1233–1241.

Olde Venterink, H., Wassen, M.J., Verkroost, A.W.M. and De Ruiter, P.C. (2003) Species richness–productivity patterns differ between N-, P-, and K-limited wetlands. *Ecology* 84, 2191–2199.

Ozanne, P.G. and Specht, R.L. (1981) Mineral nutrition of heathlands: phosphorus toxicity. In *Ecosystems of the World Vol 9B. Heathlands and Related Shrublands. Descriptive Studies*, ed. R.L. Specht. Elsevier Scientific, Amsterdam, pp. 209–213.

Pate, J.S. and Bell, T.L. (1999) Application of the ecosystem mimic concept to the species-rich Banksia woodlands of Western Australia. *Agroforestry Systems* 45, 303–341.

Pate, J.S. and Verboom, W.H. (2009) Contemporary biogenic formation of clay pavements by eucalypts: further support for the phytotarium concept. *Annals of Botany* 103, 673–685.

Pate, J.S., Verboom, W.H. and Galloway, P.D. (2001) Co-occurrence of Proteaceae, laterite and related oligotrophic soils: coincidental associations or causative inter-relationships? *Australian Journal of Botany* 49, 529–560.

Pekin, B.K., Boer, M.M., Wittkuhn, R.S. *et al.* (2012) Plant diversity is linked to nutrient limitation of dominant species in a world biodiversity hotspot. *Journal of Vegetation Science* 23, 745–754.

Pettit, N., Dowe, J. and Dixon, I. (2016) Riparian vegetation of tropical northern Australia. In *Vegetation of Australian Riverine Landscapes: Biology, Ecology and Management*, ed. S. Capon, C. James and M. Reid. CSIRO Publishing: Clayton, pp. 221–238.

Pfautsch, S., Rennenberg, H., Bell, T.L. and Adams, M.A. (2009) Nitrogen uptake by *Eucalyptus regnans* and *Acacia* spp.: preferences, resource overlap and energetic costs. *Tree Physiology* 29, 389–399.

Polglase, P.J., Attiwill, P.M. and Adams, M.A. (1986) Immobilization of soil nitrogen following wildfire in two eucalypt forests of south-eastern Australia. *Acta oecologica. Oecologia plantarum* 7, 261–271.

Raison, R.J. (1979) Modification of the soil environment by vegetation fires, with particular reference to nitrogen transformations: a review. *Plant and Soil* 51, 73–108.

Reddell, P., Bowen, G.D. and Robson, A.D. (1986) Nodulation of Casuarinaceae in relation to host species and soil properties. *Australian Journal of Botany* 34, 435–444.

Reynolds, H.L., Mittelbach, G.G., Darcy-Hall, T.L., Houseman, G.R. and Gross, K.L. (2007) No effect of varying soil resource heterogeneity on plant species richness in a low fertility grassland. *Journal of Ecology* 95, 723–733.

Rossel, R.A.V. and Bui, E.N. (2016) A new detailed map of total phosphorus stocks in Australian soil. *Science of The Total Environment* 542, 1040–1049.

Ryan, M.H., Tibbett, M., Edmonds-Tibbett, T. *et al.* (2012) Carbon trading for phosphorus gain: the balance between rhizosphere carboxylates and arbuscular mycorrhizal symbiosis in plant phosphorus acquisition. *Plant, Cell and Environment* 35, 2170–2180.

Santi, C., Bogusz, D. and Franche, C. (2013) Biological nitrogen fixation in non-legume plants. *Annals of Botany* 111, 743–767.

Schmidt, S. and Stewart, G.R. (1999) Glycine metabolism by plant roots and its occurrence in Australian plant communities. *Functional Plant Biology* 26, 253–264.

Schmidt, S., Raven, J.A. and Paungfoo-Lonhienne, C. (2013) The mixotrophic nature of photosynthetic plants. *Functional Plant Biology* 40, 425–438.

Shane, M.W. and Lambers, H. (2005) Cluster roots: a curiosity in context. *Plant and Soil* 274, 101–125.

Shane, M.W., Cawthray, G.R., Cramer, M.D. *et al.* (2006) Specialized 'dauciform' roots of Cyperaceae are structurally distinct, but functionally analogous with 'cluster' roots. *Plant, Cell and Environment* 29, 1989–1999.

Skjemstad, J.O., Fitzpatrick, R.W., Zarcinas, B.A. and Thompson, C.H. (1992) Genesis of podzols on coastal dunes in southern Queensland. II. Geochemistry and forms of elements as deduced from various soil extraction procedures. *Soil Research* 30, 615–644.

Smith, S.E. and Read, D.J. (2008) *Mycorrhizal Symbiosis*, 3rd edition. Academic Press: Oxford.

Sommer, J., Pausch, J., Brundrett, M.C. *et al.* (2012) Limited carbon and mineral nutrient gain from mycorrhizal fungi by adult Australian orchids. *American Journal of Botany* 99, 1133–1145.

Specht, R.L. and Rundel, P.W. (1990) Sclerophylly and foliar nutrient status of Mediterranean-climate plant communities in southern Australia. *Australian Journal of Botany* 38, 459–474.

Stępkowski, T., Watkin, E., McInnes, A. *et al.* (2012) Distinct *Bradyrhizobium* communities nodulate legumes native to temperate and tropical monsoon Australia. *Molecular phylogenetics and evolution* 63, 265–277.

Taylor, G. (1994) Landscapes of Australia: their nature and evolution. In *History of the Australian Vegetation: Cretaceous to Recent*, ed. R.S. Hill. Cambridge University Press: Cambridge, pp. 60–79.

Tedersoo, L., Bahram, M., Põlme, S. *et al.* (2014) Global diversity and geography of soil fungi. *Science* 346, 1256688.

Tilman, D. (1983) Plant succession and gopher disturbance along an experimental gradient. *Oecologia* 60, 285–292.

Tilman, D. (1988) *Plant strategies and the dynamics and structure of plant communities (No. 26)*. Princeton University Press: New Jersey.

Toberman, H., Chen, C., Lewis, T. and Elser, J.J. (2014) High-frequency fire alters C: N: P stoichiometry in forest litter. *Global Change Biology* 20, 2321–2331.

Turner, B.L. and Laliberté, E. (2015) Soil development and nutrient availability along a 2 million-year coastal dune chronosequence under species-rich Mediterranean shrubland in southwestern Australia. *Ecosystems* 18, 287–309.

Turner, J. and Lambert, M. (2016) Pattern of carbon and nutrient cycling in a small *Eucalyptus* forest catchment, NSW. *Forest Ecology and Management* 372, 258–268.

Turner, J., Lambert, M., Jurskis, V. and Bi, H. (2008) Long term accumulation of nitrogen in soils of dry mixed eucalypt forest in the absence of fire. *Forest Ecology and Management* 256, 1133–1142.

Verboom, W.H. and Pate, J.S. (2013) Exploring the biological dimension to pedogenesis with emphasis on the ecosystems, soils and landscapes of southwestern Australia. *Geoderma* 211, 154–183.

Verboom, W.H. and Pate, J.S. (2015) An integrated approach to mapping and understanding of vegetation: soil systems. *Catena* 126, 134–145.

Wan, S., Hui, D. and Luo, Y. (2001) Fire effects on nitrogen pools and dynamics in terrestrial ecosystems: a meta-analysis. *Ecological Applications* 11, 1349–1365.

Wardell-Johnson, G. and Williams, M. (1996) A floristic survey of the Tingle Mosaic, south-western Australia: applications in land use planning and management. *Journal of the Royal Society of Western Australia* 79, 249–276.

Wei, L., Chen, C. and Yu, S. (2015) Uptake of organic nitrogen and preference for inorganic nitrogen by two Australian native Araucariaceae species. *Plant Ecology and Diversity* 8, 259–264.

Weston, C.J. and Attiwill, P.M. (1990) Effects of fire and harvesting on nitrogen transformations and ionic mobility in soils of *Eucalyptus regnans* forests of south-eastern Australia. *Oecologia* 83, 20–26.

Woodward, K.B., Fellows, C.S., Mitrovic, S.M. and Sheldon, F. (2015) Patterns and bioavailability of soil nutrients and carbon across a gradient of inundation frequencies in a lowland river channel, Murray–Darling Basin, Australia. *Agriculture, Ecosystems and Environment* 205, 1–8.

Zolfaghar, S., Villalobos-Vega, R., Zeppel, M. and Eamus, D. (2015) The hydraulic architecture of *Eucalyptus* trees growing across a gradient of depth-to-groundwater. *Functional Plant Biology* 42, 888–898.

8 Climatic Influence over Vegetation Pattern and Process

MARK K.J. OOI, TONY D. AULD, LINDA J. BEAUMONT AND ROSS A. BRADSTOCK

Climatic influences on vegetation vary in scale from impacts on individual plants, to population- and species-level responses and changes in composition, structure and distribution of vegetation types. Distribution patterns are based on a combination of natural climatic, edaphic, disturbance-driven and biotic factors, all of which can be dramatically affected by climate change. Understanding the impacts of these factors across all levels of biological organisation is a key driver of research questions, both now and for the foreseeable future. In this chapter, we explore the role of climate as a driver of vegetation patterns across Australia and provide an in-depth look at the climatic factors that control many critical plant life cycle and population processes, which in turn influence the distribution of species. We also outline the advances in approaches for modelling future distributions, which need to incorporate climatic niche and life-history responses, as well as account for the "fertilisation effect" of elevated CO_2 on growth and the adaptability of plants, in order to robustly predict species and vegetation distribution as the climate changes.

8.1 Introduction

Climate has often taken centre stage in our attempts to understand and predict vegetation patterns and processes in Australia. More broadly, the question of how climate interacts with vegetation has driven scientific research for centuries. Ancient observations of population-scale processes gave birth to modern ecology, much of it stemming from the work of Theophrastus (371–286 BCE) who investigated the natural history of plants including phenology, seed longevity and growth. Early understanding of large-scale patterns of vegetation in relation to the environment were made possible with the advent of worldwide exploration and the development of methodologies for synthesising environmental measurements and vegetation characteristics. This precipitated a global shift in the study of plants. As a pioneer of this work, Alexander von Humboldt (1769–1859), laid the foundations for the science of biogeography from the late eighteenth century onward, and moved the Linnean-inspired taxonomic focus on species distributions to quantification and extent of vegetation units (Nicholson 2013).

Importantly, Humboldt identified isothermal lines, which could delineate the global pattern of environmental factors and vegetation, and explicitly promoted the study of units of 'socially organised plant life' (Shugart and Woodward 2011). This influenced the first attempts of worldwide vegetation classification, resulting in global

Citation: Ooi, M.K.J. *et al.* (2017). Climatic Influence over Vegetation Pattern and Process. In D. A. Keith, ed., *Australian Vegetation*, 3rd edn. Cambridge: Cambridge University Press, pp. 182–206.

vegetation maps. These included a physiognomic presentation by Grisebach (1872), followed by Schimper (1898) who mapped current-day biomes (references in Küchler and Zonneveld 1988). Perhaps not surprisingly, a direct line from the Humboldtian tradition lead to the first vegetation map of Australia. Ludwig Diels (1874–1945) was awarded funding from the Humboldt Foundation for Nature Research and Travel in 1900 for a botanical expedition of the southern hemisphere (Beard 2001a). Diels' map identified eight vegetation categories in Australia; tropical and temperate rainforests, sclerophyll and savanna forests, savanna, mulga scrub, mallee scrub and desert (Beard 2001b).

From the early development of ecology and biogeography in Australia and abroad, much of the research sought to understand the role of weather (which includes short-term atmospheric changes in temperature, precipitation and humidity, among other things) and climate (long-term averages of daily weather patterns) in determining vegetation patterns and processes. Today, these same questions are driving researchers' attempts to predict the impacts and consequences of anthropogenic climate change, which is causing increases in mean temperature conditions, altering precipitation patterns and intensifying extreme weather events (e.g. storms, droughts and heat waves). Assessing the relationship between climatic factors and vegetation patterns or the control of key life-history processes is essential for understanding the mechanistic impacts of climate change on population function, species distributions and biodiversity (Figure 8.1).

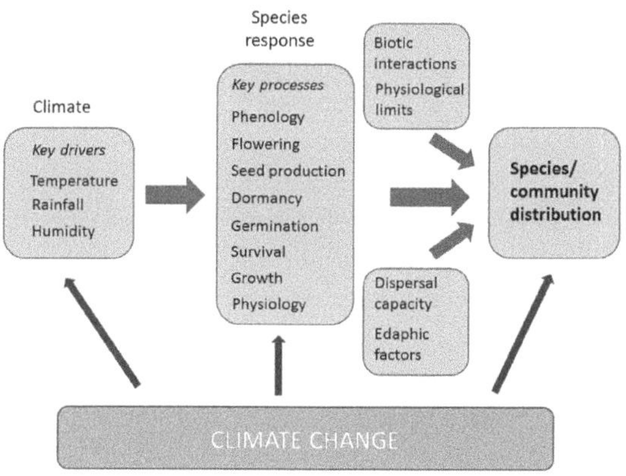

Figure 8.1 Climatic factors drive many key plant population processes, which contribute to determining species and community distribution and ultimately vegetation patterns. Climate change alters the long-term climate means and the variation in associated weather patterns and therefore has an impact on population processes and distribution. Understanding the mechanistic impacts of climate change on population function is therefore essential for robustly modelling and predicting future species distributions.

This chapter explores the role of climate as a driver of vegetation patterns across Australia. It includes an investigation of the relationship between weather conditions and local-scale population processes, which undoubtedly mediate some of the larger-scale patterns observed, and highlights remaining challenges faced to understand and predict future impacts.

8.2 Key climatic drivers of vegetation in Australia

8.2.1 Determinants and classification of Australian vegetation patterns

Climate has long been recognised as the fundamental determinant of vegetation patterns in most Australian classification schemes. Early researchers primarily emphasised water availability, as a function of precipitation, evaporation, soil texture and depth (e.g. Beard 1981; Specht 1981), as the key driver, with other factors often included as secondary determinants. Beadle (1981), for example, emphasised soil fertility (see Brundrett *et al.*, Chapter 7, this volume), while Jackson (1968) highlighted the influence of fire regimes, which are themselves influenced by climate. Paleo-ecological studies also highlighted the central role of climate as a determinant of historical changes in Australian vegetation patterns (Hope *et al.* 2004; Pickett *et al.* 2004; Mooney *et al.*, Chapter 3, this volume), with coupled changes reflecting human activity via fire also often implicated (Lynch *et al.* 2007; Mooney *et al.* 2012).

Several broad climatic elements influence continental-scale vegetation patterns. These include precipitation (quantity and seasonality), temperature, and interactions between these and other abiotic factors (Beadle 1981), which can vary with latitude and altitude. For example, average annual precipitation ranges from very low (<200 mm yr^{-1}) in arid Australia, to extremely high (>3000 mm yr^{-1}) in far north-east Queensland and south-west Tasmania, with Beadle (1981) delineating eight vegetation regions based on rainfall alone. Within these regions, however, interactions with other factors contribute to determining vegetation communities. Central Australia is characterised by very low rainfall but its vegetation ranges from arid shrublands, *Casuarina* woodlands and chenopod shrublands to *Triodia* hummock grasslands (Keith and Tozer, Chapter 1, this volume). These vegetation types reflect interactions between climatic and edaphic drivers, including rainfall seasonality (summer rainfall in the north, aseasonal or winter rainfall in the south), soils (chenopod shrublands are associated with more saline or calcareous soils) and fire (which may influence boundaries between arid shrublands and *Triodia* grasslands) (Bowman *et al.* 2008). Conversely, temperature is critical in determining plant growth rates in the wetter coastal regions, while colder conditions may limit the growth of trees and shrubs in high altitude areas (Venn *et al.*, Chapter 19, this volume).

At the continental scale, a variety of empirical approaches has identified strong links between vegetation patterns, rainfall and temperature. Pickett *et al.* (2004) tested a biome classification system based on plant functional types, water availability and mean temperature in the coolest month (i.e. maximum constraint on growth) using palynological metadata. They found reasonable correspondence between biomes, derived from Carnahan's (1997) national vegetation map, and inferences from the palynological data (although with some important exceptions). Gonzales-Orozco *et al.* (2014) redefined six phytogeographic regions across the continent based on meta-analyses of species turnover. Generally, physiographic factors failed to correspond strongly with the proposed regions, with climate variables found to be the dominant environmental drivers. Division of the phytogeographic regions tended to reflect rainfall gradients across the northern tropical summer and southern temperate winter. While precipitation was a significant driver for northern Australia, below the Tropic of Capricorn high solar radiation and lower temperatures were more important. In a similar vein, Hilbert and Fletcher (2012) related the distribution of ecoregions and major vegetation groups (MVGs, *sensu* DWER 2007) to a suite of potential biophysical influences using an artificial neural network model. Temperature and rainfall were again the predominant determinants of ecoregion and MVG distributions, although in some cases edaphic predictors that influence water availability (e.g. soil depth, texture and hydraulic conductivity) were also influential.

The strong relationship between climatic factors and vegetation patterns, demonstrated over multiple time scales and across a range of empirical approaches, means that it is extremely likely climate change will affect species life cycles, distributions, and community composition. For example, regional-scale analyses of species and community distributions in relation to biophysical variation (e.g. Hilbert and Ostendorf 2001; Austin and van Neil 2011) provide a basis for the projection of trends under future climates. The impact of changes to rainfall, temperature and other variables on these broader vegetation community patterns relate directly to functional responses and persistence at the population scale.

8.3 Influence of climate on plant life cycle and population processes

Climatic factors control many critical plant life cycle and population processes, which in turn influence the distribution of species. For example, fundamental aspects of the life history of plants, including the phenology of flowering and germination, seed dormancy mechanisms and their related seed bank dynamics, are strongly dependent on seasonal temperature changes. Alternatively, climatic variables can provide key resources (e.g. moisture via rainfall) or induce stress (e.g. heat waves or drought) (Rathcke and Lacey 1985; Bell 1999; Merritt *et al.* 2007; Ooi 2012; Baskin and Baskin 2014). Thus, it is unsurprising that plant traits within different climate regions differ in somewhat predictable ways. For example, seasonal patterns of rainfall can drive seedling recruitment and mortality, thereby determining the dominant species in northern monsoon savannas (Williams *et al.* 1999; Russell-Smith and Setterfield 2006). In fire-prone regions, interactions between seasonal temperatures and the fire regime can shift dominance of post-fire recruiting species (Wellington and Noble 1985; Ooi *et al.* 2004; Ooi 2010). Sporadic rainfall events in the vast central arid ecosystems can control germination; thus, germination strategy and subsequent recruitment or mortality of many species is correlated with climatic zones (Mott 1972; Watson *et al.* 1997; Denham and Auld 2004).

Because of the dependence of key life-history processes on climatic variables, it is inevitable that climate change will have an impact on plant population dynamics and species persistence through changes to weather patterns, such as increased frequency or intensity of heat waves or storms (Hughes 2000; Walck *et al.* 2011; Ooi 2012; Parmesan and Hanley 2015). The latest assessment report from the Intergovernmental Panel on Climate Change (Settele *et al.* 2014) highlights the likely impacts on many population processes including pollination, phenology, growth and stress-induced mortality, and the related consequences on species abundance and distribution. In the following sections, we outline the mechanisms by which temperature and rainfall affect the physiology and ecology of plants, and influence their interactions with biotic and other abiotic factors.

8.3.1 Flowering phenology and pollination

One of the most visible responses to annual seasonal cycles is the timing of flowering and fruiting. Phenological studies of Australian plants have historically been hampered by a paucity of long-term data sets. However, recent analyses of herbarium records have highlighted the strong correlation between flowering and temperature, predominantly for species in temperate and alpine ecosystems, but also for some arid zone taxa (e.g. Gallagher *et al.* 2009; Hoffman *et al.* 2010; Rumpff *et al.* 2010; Beaumont *et al.* 2015; Rawal *et al.* 2015). Experimental studies (Day *et al.* 1994; King *et al.* 1996; King 1998) and analysis of observational data (Schneemilch *et al.* 2012; Chambers *et al.* 2013) support the conclusion that temperature is a key driver of the onset or magnitude of flowering across numerous plant families in Australia. Seasonal precipitation can also play a key role in driving trends of phenological change. Unsurprisingly, temperature appears less important in tropical rainforests, where phenology is often coupled with rainfall, responding to the onset of wet or dry seasons, or displays no clear seasonal pattern at all (Franklin and Bach 2006; Inkrot *et al.* 2007).

Flower production is also strongly associated with seasonal temperature. Vernalisation is a requirement for many native Australian species. Cool temperatures are also needed for subsequent floral development, as shown in *Boronia megastigma* (Day *et al.* 1994), *Hardenbergia violacea* (King 1998) and several *Acrotriche* species (Schneemilch *et al.* 2012), and may interact with non-climatic factors such as photoperiod. Similarly, rainfall can influence the magnitude of flowering. In a 3-year study in eastern coastal Australia, Morrison and Myerscough (1989) found that the number of buds produced on wild populations of *Acacia suaveolens* were positively correlated with increasing autumn rainfall, but not with rainfall at any other time during bud formation. This highlights the capacity for plants to regulate maternal investment based on environmental stress experienced by the parent plant.

Increasing temperatures or altered rainfall patterns related to climate change can influence flower production and shift the timing of phenological events. In response to warming, flower production can increase or decrease, with species-specific differences reported. For example, increases in future temperature may reduce the amount of flowering, or increase bud abortion, if cool seasons fail to reach the low temperatures required for anthesis (King *et al.* 2008), or they may increase flowering of some perennial grasses (Hovenden *et al.* 2007). Warming has also widely been reported to change phenology, with altered flowering times potentially decoupling plant–pollinator life cycles or increasing competition for pollinators with formerly asynchronous conspecific neighbours (Peñuelas and Filella 2001; Schneemilch *et al.* 2012). A study of phenological records of Australian plants by Chambers *et al.* (2013) reported that global warming has already resulted in a mean spring advance of 9.7 days per decade across all phenophases, although there are also species-specific examples of warmer temperatures delaying flowering (Peñuelas and Filella 2001; Rawal *et al.* 2015), shortening the time available for ripening.

At present, robust predictions of phenological changes are limited, with a clearer understanding needed of the numerous complex factors including interactions between temperature and other key drivers, such as rainfall or day length, and the identification of constraints to species responses (Beaumont *et al.* 2015). For example, between-species differences in phenological drivers, where some species depend only on temperature while others have an additional photoperiod response, may differentiate the direction and magnitude of phenological trends among species (Venn and Morgan 2007; Saikkonen *et al.* 2012). Higher CO_2 concentrations may also affect phenology; however, experimental evidence to date has provided little support in Australian systems (e.g. Hovenden *et al.* 2008).

8.3.2 Seed production

Seed production, an important determinant of eventual seed bank size and recruitment, can be strongly influenced by interannual variation in rainfall and temperature. The study of climatic drivers of variation in seed supply across Australia has been biased towards semi-arid and arid regions, and northern rainforests and savannas (e.g. Davies 1976; Russell-Smith and Setterfield 2006), at least until relatively recently. In semi-arid and arid plant communities, sporadic rainfall is a driver of population dynamics, with variation in its timing and magnitude a key determinant of seed production. For example, numerous perennial species produce annual seed crops, however larger seed crops are associated with high rainfall within particular seasons. In semi-arid Western Australia and in the Gibson Desert, winter rainfall in the previous year resulted in high seed production for numerous *Acacia* species, whereas summer rainfall was correlated with increased *Hakea lorea* seed production (Davies 1976; Davies and Kenny 2013). Masting events have been identified in arid zone species, such as *Triodia*, with the amount of seed set related to the total rainfall recorded over the previous year (Wright *et al.* 2014).

The limited amount of work regarding the role of seed supply in other ecosystems in Australia is surprising, especially considering the importance, and variability, of the regeneration niche across many of these vegetation types. The lack of study of this life-history stage in, for example, temperate fire-prone regions (but see Morrison and Myerscough 1989) may partly be a result of the strong role played by long-term persistent seed banks where interannual variation in seed production is assumed to be less crucial for population persistence. However, the need to understand climate change impacts has provided an impetus to investigate all aspects of the plant life cycle from a broader range of habitats. Additionally, many of these studies include interactions between climate and elevated CO_2. In a Free Air CO_2 Enrichment (FACE) experiment conducted in lowland grasslands in Tasmania, Hovenden *et al.* (2007) found little overall effect across the community of warmer mean temperatures, increased CO_2 levels or their interaction on reproductive output. However, some positive effects occurred at the species level, where increased CO_2 and a warming × CO_2 interaction resulted in greater seed production for the shrubs, *Hibbertia hirsuta* and *Bossiaea prostrata*, respectively. Interestingly, these effects were inconsistent across the 3 years of the study, suggesting that rainfall variability mediated species responses. Further, warming proceeding severe summer drought reduced seed production for two grass species.

Research on seed fill has been a high priority in the agricultural sector due to the importance of seed yield for food production (e.g. Hampton *et al.* 2013). It is equally important for our understanding of the ecology of natural ecosystems, particularly in the context of predicting the consequences of climate change. Increases in mean temperature and the number and duration of extreme heat wave events are projected across much of Australia. Heat waves, even of relatively short duration, during seed development can have negative effects on seed fill for many agricultural species, such as *Glycine max* (e.g. Spears *et al.* 1997; Hampton *et al.* 2013), and germination performance in natural systems (Gutterman 2000).

8.3.3 Seed dormancy and germination

Ambient temperature and soil moisture (regulated by precipitation) play a crucial role in controlling seed dormancy and germination (Baskin and Baskin 2014). Across most ecosystems, seasonal temperature provides the primary cue regulating the dormancy status and timing of seed germination (Fenner and Thompson 2005). However, while the concept of optimal germination temperature has received considerable study (e.g. Bellairs and Bell 1990; Bell 1999), broader recognition of the role seasonal temperature plays in alleviating dormancy, other than dry after-ripening in arid species, has been slow to gain traction in Australia.

Overcoming physiological dormancy occurs when seeds are subjected to either seasonal temperature stratification (moist cold or warm conditions) or dry after-ripening (dry-stored warm conditions) (see Auld and Ooi, Chapter 10, this volume). For example, a period of dry after-ripening can overcome dormancy in a number of species from semi-arid Western Australia (Commander *et al.* 2009). This is a successful strategy for recruitment in many dry climate regions, because it ensures that seeds do not germinate in response to spring/summer rainfall events, avoiding exposure of seedlings to harsh summers. Instead, dormancy is alleviated only after the hot summer, allowing germination and establishment to occur over the winter months. The rate of dormancy loss increases with after-ripening temperature (e.g. Steadman *et al.* 2003). In contrast, cold stratification can alleviate dormancy and increase germination in Australian alpine systems (e.g. Venn and Morgan 2007; Sommerville *et al.* 2013), while in subtropical regions, warm stratification produces the same result (e.g. Hoyle *et al.* 2008a). These studies emphasise that dormancy-breaking requirements are often related to the climate and environment in which species occur.

In fire-prone regions, cues such as smoke or heat can interact with seasonal temperatures to promote germination post-fire, but within particular seasons. While this is often assumed to be an adaptation to seasonal rainfall environments, such as Mediterranean-climate regions of Australia (Baker *et al.* 2005; Merritt *et al.* 2007), an increasing number of studies have found similar seasonal germination responses in aseasonal rainfall regions (Ooi *et al.* 2006; Thomas *et al.* 2010; Mackenzie *et al.* 2016). These regions nonetheless experience considerable water deficits over the summer months (Tozer *et al.*, Chapter 13, this volume). Seasonal temperatures or heat wave events can also play a role in breaking physical dormancy (also hard seededness; see Auld and Ooi Chapter 10, this volume) in dryland and tropical species (Ooi *et al.* 2009) and, potentially, in gaps or post-fire environments where soil surface temperatures may exceed 50°C (Auld and Bradstock 1996; Santana *et al.* 2010; Ooi *et al.* 2014).

Climate change may alter dormancy and germination characteristics through increases in seasonal temperatures, changes to humidity and extreme weather events (Cochrane *et al.* 2014). For example, there is some evidence suggesting that higher maternal environmental temperature and lower humidity could increase the proportion of physically dormant seeds produced (Tozer and Ooi 2014) or alter their dormancy-breaking temperature thresholds (Hudson *et al.* 2015). Higher temperature thresholds reduce the proportion of the seed bank germinating after any given fire, or restrict germination to high-severity fires, and decouple the relationship between seeds with low thresholds and favourable emergence conditions (Liyanage *et al.* 2016). Conversely, warmer and drier maternal environmental conditions can decrease seed dormancy among some physiologically dormant species (Hoyle *et al.* 2008b). So far, there is less evidence for increased CO_2 directly affecting seed dormancy (e.g. Hovenden *et al.* 2008), however, it may contribute to subsequent seed provisioning. Nevertheless, as shown by these studies, the strong relationship between climatic variables and dormancy means that projected changes to climate will affect recruitment (Ooi 2012), with the persistence of some species potentially reduced and a likely shift in species composition of plant communities.

8.3.4 Seed bank dynamics

Seed bank dynamics are influenced by seed longevity, dormancy, how dormancy is broken, and the variability in cues that promote germination response, all of which are directly or indirectly linked with climate. Changes to climate are likely to alter the timing and success of seedling emergence, because these processes are governed by several interacting event-dependent factors, such as soil warming and available moisture, which are in turn influenced by climatic variation (Walck *et al.* 2011; Ooi 2012). Persistent soil seed banks are common in a range of vegetation types, including in fire-prone regions where post-fire recruitment can be restricted to particular seasons due to temperature (Roche *et al.* 1998; Auld *et al.* 2000; Ooi *et al.* 2004) or seasonal rainfall (Williams *et al.* 2007; Russell-Smith and Lucas 1994), and are often mediated by fire season. In dryland systems, a stochastic element to the timing and magnitude of rainfall drives much of the variation in emergence (Figure 8.2a, b), particularly in the recruitment of ephemeral species (Wright and Clarke 2009). Facelli *et al.* (2005) found that only a portion of the seed bank would respond to any particular rainfall event in arid southern Australia. Community composition of desert annuals varied between years depending on the response to season of rainfall, driven by germination temperature requirements.

A long-lived seed bank ensures that germination is either delayed until the onset of favourable conditions (i.e. during a rainfall event, which provides sufficient moisture for recruitment, or after a fire), or that the seed bank does not germinate all at once (Ooi 2012; Ooi *et al.* 2014; Baskin and Baskin 2014). Seed bank longevity may be compromised by changing climatic conditions. For example, Ooi *et al.* (2009) found that in semi-arid eastern Australia, projected increases in air temperature of 4°C during summer could increase soil temperatures by 10°C, with consequent reductions in seed viability or dormancy. The relationship between higher temperatures and greater loss of dormancy (e.g. Tieu *et al.* 2001; Ooi *et al.* 2014), or greater levels of germination, demonstrates one underlying mechanism by which climate change affects soil seed bank dynamics. Soil temperatures projected for the end of the twenty-first century potentially compromise the bet-hedging ability of several species through increased loss of dormancy (Ooi *et al.* 2009). Kochanek *et al.* (2010) found the seed bank of the semi-arid species, *Wahlenbergia tumidifructa*, was similarly compromised because parent plants exposed to higher temperatures produced seeds of reduced longevity. Such results are likely to vary

Figure 8.2 (a–f) Climatic factors control many plant population processes. For example, the season of rainfall can determine species composition in many arid and semi-arid systems, mediated by mechanisms controlling seed dormancy and germination. Grasses tend to recruit after summer rainfall, as shown by (a) extensive *Enneapogon avenaceus* recruitment in far western New South Wales, while winter rain promotes germination of forbs in the same regions (photo: Tony Auld), as seen by (b) the flush of *Myriocephalus stuartii* (Asteraceae) recruitment (photo: Tony Auld). Both species are physiologically dormant but have warm temperature and cool temperature germination requirements respectively. Climate change also causes mean increases in temperature and alters weather patterns. This can result in changes to the environment necessary to maintain component species, such as in (c) where a reduction in the number of days with cloud cover (and therefore humidity) can affect cloud forest at higher elevations on Lord Howe Island (photo: Tony Auld). Climate change interacts with disturbance regimes such as fire, where alterations to frequency, intensity and season are projected, as seen in (d) an out-of-peak season autumn fire in south-eastern Australia (photo: Mark Ooi). Understanding the response of plants to climate change is complex, with many innovative approaches including Free Air CO_2 Enrichment (FACE) plots, as seen in (e) at the University of Tasmania's TasFACE experiment (photo: Mark Hovenden), and (f) large-scale common garden experiments, seen here at the University of Wollongong, established to understand the plasticity and potential adaptability of species to future changes (photo: Mark Ooi).
(A black and white version of this figure will appear in some formats. For the colour version, please refer to the plate section.)

between species, with Hoyle *et al.* (2008b) finding only a small effect of the maternal environment on seed dormancy levels in *Goodenia fascicularis*, a species from a similar semi-arid habitat.

8.3.5 Seedling recruitment and growth

Rainfall governs seedling recruitment across many of Australia's ecosystems. As discussed above, dormancy, germination mechanisms and seed bank dynamics are adapted to ensure that seedlings emerge into conditions favourable for recruitment. Nonetheless, many species experience high mortality at the seedling stage. Insufficient rainfall can prevent seedlings from establishing, while numerous studies have shown drought to be the main driver of seedling mortality (Lamont *et al.* 1993; Tozer and Bradstock 1997; Russell-Smith and Setterfield 2006).

Projected changes to rainfall patterns, particularly the increasing likelihood of more sporadic rainfall combined with elevated temperatures and associated increases in evaporation rates, will lead to even greater fluctuations in available soil moisture (Stevens *et al.* 2014). As rainfall becomes more erratic, it increases the likelihood of 'false-start' rains, where rainfall levels are enough to initiate germination but not to sustain seedling survival. This will restrict the availability of 'safe sites' for recruitment (Auld and Ooi, Chapter 10, this volume). Furthermore, seed development under warmer maternal conditions may reduce seedling vigour independently of increased CO_2 (Hovenden *et al.* 2008).

8.3.6 Growth and survival of established plants

There is a strong relationship between plant growth (e.g. biomass) and temperature and moisture. Increasing aridity reduces growth; however, temperature effects are dependent on an individual's optimal temperature for growth and the range of temperatures under which it can physiologically function (thermal breadth). Consequently, increased temperature may increase plant growth, but additional warming will induce stress, and eventually death, if the plants' thermal breadth is exceeded. Plant growth is also strongly influenced by edaphic conditions and available nutrients, along with ecological factors such as competition, predation and herbivory.

Changes to temperature and moisture availability under future climatic conditions are highly likely to affect plant growth and survival, especially in species with a narrow thermal breadth. The combination of increased heat stress with increased extent and severity of droughts is likely to rapidly change vegetation structure (e.g. Allen *et al.* 2010; Fensham *et al.* 2015). As projected increases in drought frequency eventuate, the potential for tree recovery will diminish (Allen *et al.* 2015). This form of 'interval squeeze' (see below) may lead to the elimination of susceptible tree and shrub species. Increased concentrations of CO_2 will also affect plant water use and growth, although the extent and magnitude of responses remains somewhat uncertain (see CO_2 section below).

8.3.7 Indirect effects of climate change on vegetation

Climate change can cause several indirect impacts on vegetation communities via interactions with biotic and abiotic factors. Sea-level rise (SLR), a result of melting of Antarctic and Greenland ice sheets and thermal expansion of the oceans, may cause seawater inundation of coastal vegetation, especially during extreme tides and storms. For instance, drier monsoonal conditions and SLR are responsible for loss of vegetation along coastal plains in the Northern Territory (Winn *et al.* 2006). Poleward and elevational range expansion of mangroves, usually at the expense of adjacent saltmarsh communities, has been recorded throughout Australia and the world, coincident with shifts in temperature thresholds and SLR (Saintilan *et al.* 2014). Saltmarshes have been most impacted because upslope areas are often extensively urbanised, reducing the capacity for landward dispersal of species (McKee *et al.* 2012). Projected increases of extreme weather events can result in storm surges and extreme tides, bringing saltwater incursions to communities unable to withstand saline conditions, such as littoral rainforest and freshwater wetlands.

Climate change also interacts with major disturbance regimes that are key drivers of plant life histories (e.g. flood, fire and storms). Modifications to these regimes could affect species survival, reproduction, dispersal and recruitment capacity. The most adverse changes are likely to be where disturbance regimes shift beyond the tolerable thresholds of the component species, reducing their capacity to maintain ecological processes. For example, Enright *et al.* (2015) highlighted how changes to fire regimes may result in 'interval squeeze', where reductions in plant recruitment, growth and survival coincides with more frequent fire under a warming climate. Changes in fire season (Figure 8.2d) may also reduce fire-related recruitment by increasing seedling exposure to desiccating conditions. Similarly, Kingsford (2000) demonstrated that human-altered flooding regimes impact negatively on biodiversity, with further reductions in water flow or changes to seasonality likely to compromise vegetation of freshwater wetlands, riparian forests, woodlands and other communities.

One of the fastest growing areas of research focuses on how vegetation community structure and composition will be altered by synergies between climate change and other threats. The invasive capacity of introduced plants and pathogens, (e.g. Myrtle Rust) may increase if their environmental niches shift or expand (Beaumont *et al.* 2009a, b; Gallagher *et al.* 2010; Leishman *et al.*, Chapter 9, this volume).

Ongoing increases in human populations will interact synergistically with climate change, broadening our ecological footprint and causing greater loss of native vegetation via urbanisation, agriculture and mining. For example, projections for 18 *Banksia* species by Yates *et al.* (2010) indicates that species' range contractions will be the predominant response under future climate scenarios, and interactions with land transformation will likely cause the largest declines. This highlights how interactive effects of climate change will place additional stress on species and communities, increasing the need for management to ameliorate other stresses such as habitat loss and biological invasions.

8.4 Vegetation response to increased CO_2

CO_2 concentrations have now risen from a pre-industrial value of 280 ppm to more than 400 ppm. Concentrations of approximately 600 ppm (depending on scenario) are projected by the end of the twenty-first century (Collins *et al.* 2013). Substantial research has focused on the extent to which plants may benefit from elevated CO_2 (termed the 'fertilisation effect') due to greater water-use efficiency and C assimilation.

In native Australian systems, research has been conducted in FACE facilities (Figure 8.2e), established in Tasmanian grassland, western Sydney woodland and north-east Queensland tropical savanna (TasFACE, EucFACE and OzFACE, respectively), in addition to greenhouse and growth chamber experiments. Perhaps not surprisingly, plant growth responses vary, with increases and decreases reported among co-occurring species, and/or between years. For example, Tooth and Leishman (2013) found that enhanced CO_2 treatments could generate shifts in community composition of Cumberland Plain Woodland in western Sydney by increasing biomass of some grass species and decreasing that of others, but community-level biomass remained unaffected. In Tasmania, Williams *et al.* (2007) also reported a mixed response, with the C_4 grass *Themeda triandra* unaffected by warming or CO_2 increase, while the C_3 grass *Rytidosperma caespitosa* underwent lower population growth in response to higher CO_2.

Studies from around the world suggest that interactions between CO_2 and other environmental variables may explain some variation in species' responses. For example, nutrient scarcity (particularly phosphorus and nitrogen), characteristic of many Australian ecosystems, can severely reduce plant productivity responses to CO_2 (e.g. Reich *et al.* 2014; Crous *et al.* 2015). This suggests that responses to changes in water use, leaf or seed chemistry may be important effects of climate change (Hovenden *et al.* 2010). Water limitation, in particular, may mediate positive effects of CO_2-induced growth, with several studies suggesting this as a key interaction mediating the response in biomass growth and seed production (e.g. Hovenden *et al.* 2007; Fensham *et al.* 2009; Reich *et al.* 2014; Scheiter *et al.* 2014).

Additional experimental studies that incorporate the interactive effects of elevated CO_2, warming and rainfall variation are required to understand species responses to climate change. Incorporating these factors could improve modelled projections of species' potential distributions under future climates. Although traditional species distribution modelling approaches cannot account for the effects of elevated CO_2 on plant growth or other life-history stages, such demographic responses to CO_2, if quantified experimentally, could be incorporated into life-history models (see below).

8.5 Predicting climate change impacts and vegetation distribution in Australia

As climate changes, plant populations may respond by tolerating the change, adapting, migrating to areas with suitable conditions or going extinct. Understanding and predicting these outcomes has become a cornerstone of current ecological research and scaling such responses to predict the future distribution of species and communities is vital for effective biodiversity and landscape management. This presents challenges: to quote Nobel Prize recipient Niels Bohr, '*Prediction is very difficult, especially about the future*' (as quoted by Arthur K. Ellis, 1970, p. 431). Previous sections of this chapter have highlighted the impacts of climate change on particular life-cycle processes. Here, we focus on; (i) advances in approaches for modelling species and vegetation distributions and subsequent ecosystem-level effects, and (ii) assessing the adaptability of plants to future environmental changes.

8.5.1 Modelling responses to climate change

Ecologists have a number of tools at their disposal to explore relationships between environmental conditions and the distribution, abundance, phenology or physical requirements of a species. These include correlative species distribution models (SDMs) and mechanistic (ecophysiological) models. Models that predict responses at the community level or describe the future distribution of broad vegetation patterns are less commonly applied but include community-level SDMs and dynamic global vegetation models (DGVM), respectively.

At their simplest, correlative SDMs mathematically define the environmental conditions within a species' observed distribution (Figure 8.3a). This enables the spatial extent of 'suitable' habitat, as well as its temporal dynamics across past, current or future conditions, to be identified and mapped (Elith and Leathwick 2009). Although SDMs can reveal the potential exposure of a species habitat to climate change, they generally do not incorporate dispersal, life history, phenotypic plasticity or adaptive capacity. The accuracy and bounds of uncertainty of SDM projections depend on a number of factors including the biological and environmental data available, and the algorithms used (Figure 8.3b).

Species- and community-level SDMs project substantial changes to the composition and structure of ecosystems across the continent. For example, alpine and cloud forest communities are projected to migrate to higher elevations and contract in extent (Auld and Leishman 2015; Williams *et al.* 2015) (Figure 8.2c). Many temperate systems are likely to undergo rapid changes. In Tasmania, lowland *Themeda* grassland is expected to contract to

(a)

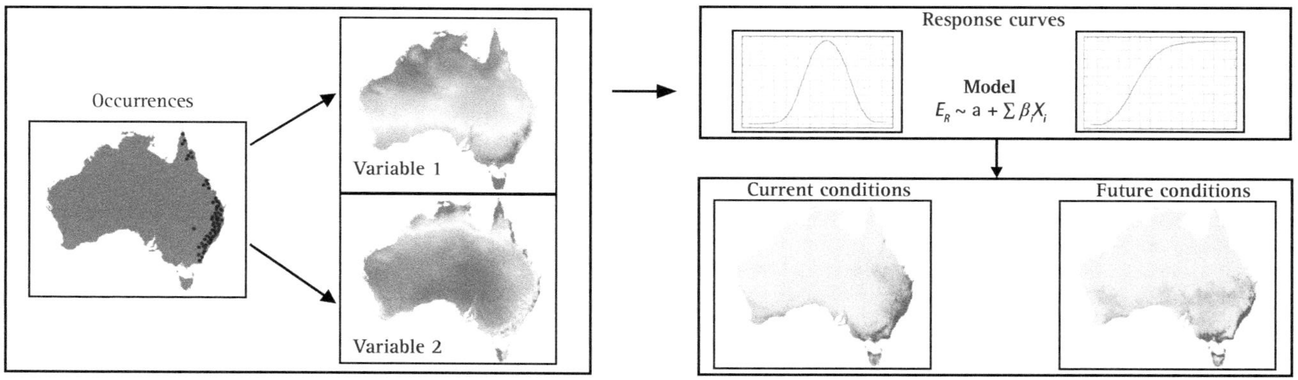

(b)

Biological considerations

- Which predictor variables should be used and at what spatial scale? Predictor variables should reflect ecophysiological responses. Responses to elevated CO_2 are rarely included.
- Species location data are frequently drawn from natural history collections, and may be spatially biased.
- Species characteristics usually excluded or poorly represented, e.g. dispersal, plasticity of responses to environmental change, adaptation potential.
- Biotic interactions are rarely included. The extent to which a species' climate envelope is truncated due to biotic factors is frequently unknown.

Environmental considerations

- Current climate and environmental data: Errors and uncertainty due to data quality, interpolation procedure, spatial resolution.
- Background data: If absence data are unavailable, where will background or pseudo-absence data be drawn from?
- Future climate scenarios: Uncertainty due to emissions of greenhouse gas; computer limitations; understanding of physical processes; parameterisation of models. Which scenarios should be used?

Algorithm considerations

- Species Distribution Model: Numerous algorithms available, but little guidance on what model to use under what circumstances. Models may have similar predictive power but project different maps of distributions.
- Model evaluation: Independent data are rarely available to test predictive power.
- Interpretation of models: What do they actually represent (e.g. 'potential' distribution of species versus distribution of climatically suitable habitat)?
- Model assumptions are frequently violated, e.g. that climate limits species' observed range; environmental envelopes are conserved in space and time.

Figure 8.3 **(a)** Diagrammatic representation of a correlative species distribution model (SDM). Key inputs are occurrence records of the target species (dependent variable) and environmental predictor variables thought to limit its distribution. A model is generated of the relationship between the dependent and predictor variables, which can then be applied to environmental data to assess the extent of suitable habitat under current or future climate scenarios. **(b)** A number of factors can influence accuracy or uncertainty in SDM studies, which can be broadly grouped in biological, environmental, and algorithm considerations. Based on Beaumont *et al.* (2008).

such an extent that, by 2050, there will be < 1% overlap with its current distribution, while only small areas of lowland *Poa* grassland will likely remain suitable by 2100 (Harris *et al.* 2015). Wardle *et al.* (2015) estimated that suitable habitat for arid zone Georgina gidgee woodlands in central Australia may decrease by at least 30% by 2050. Similar declines for mire (wetland) ecosystems are projected, with possible contraction to refugia in the southern parts of its distribution (Keith *et al.* 2014).

Tropical rainforest communities in northern Queensland are very sensitive to climate change, with several forest types likely to be highly stressed by 1°C warming and small declines in precipitation (Hilbert *et al.* 2001). Between 2050 and 2100 most forest types in this area will experience climates that are more appropriate for a structurally different forest type (Hilbert *et al.* 2001), while some lowland vegetation communities will face extirpation (Accad and Neil 2009). Community-level models of highly threatened temperate grassy ecosystems in south-east Australia suggest that climate change may lead to large-scale resorting of species and formation of new communities, with widespread degradation of these ecosystems greatly compromising their capacity to adapt (Prober *et al.* 2012). In topographically varied landscapes such as the Flinders Ranges and southern Mount Lofty Ranges, the distributions of plant species may contract to shared refugia (Guerin and Lowe 2013). SDMs combined with genetic data also suggest that the Southern Flinders Ranges may serve as a possible genetic refugium for *Callistemon teretifolius* (McCallum *et al.* 2014).

Projections indicate that some of the dominant and widespread genera in Australia may undergo substantial range contractions in the future. For instance, changes to current precipitation regimes in south-western Western Australia are likely to drive losses of *Banksia* species, with the greatest declines projected to occur in the transition zone between wet coastal and interior arid areas (Fitzpatrick *et al.* 2008). Climatically suitable habitat for many *Eucalyptus* and *Corymbia* species along the eastern and southern seaboards may be increasingly limited to the continental fringe, although species in the central desert and open woodland regions are likely to lose a greater extent of their climate space than those in eastern Australia (Butt *et al.* 2013).

In general, eucalypts occupy narrow climate envelopes, with the mean annual temperature range of 41% of 819 taxa spanning < 2°C (Hughes *et al.* 1996). However, trials that grow taxa under climatic conditions beyond those they experience in their native range provide insights into their ability to tolerate a broader set of conditions. For example, *Eucalyptus regnans* has been grown at trial sites where mean annual temperature was 5°C warmer than the hottest location in its native range (Booth *et al.* 2015). Yet, substantial changes to the distribution of Australian eucalypts are likely to occur even if only a modest number of taxa are limited by temperature or rainfall (Hughes *et al.* 1996).

Use of empirical approaches for continental-scale projections of future vegetation patterns has developed relatively recently, but several key examples exist. For instance, Hilbert and Fletcher's (2012) models of ecoregions and major vegetation groups project a significant expansion of treeless ecosystems, such as temperate and tropical grassland, savannas and shrublands, deserts and xeric shrublands, and a concomitant decline of temperate and tropical woodland and forested ecoregions.

One of the disadvantages of correlative SDMs is their inability to account for effects of elevated CO_2 on plant growth. In contrast, dynamic global vegetation models (DGVMs) overcome this problem by explicitly modelling the physiology of carbon and water exchange in differing plant functional types. Nonetheless, comparative evaluation of differing DGVMs suggests an inability to accurately predict the broad distribution of contemporary Australian biomes and vegetation types, even though they produced satisfactory estimates of contemporary global carbon uptake by terrestrial ecosystems (Sitch *et al.* 2008). In part, the performance of DGVMs has been constrained by the use of a limited number of plant functional types (typically < 15) to represent global vegetation variation and the fixed approaches to parametrisation of these functional types (van Bodegom *et al.* 2014; Yang *et al.* 2015). Attempts to introduce variation in key attributes of plant functional types using relationships between environmental variation and plant traits (e.g. Pavlick *et al.* 2013; van Bodegom *et al.* 2014) have improved aspects of DGVM performance to some degree. However, the potential for DGVMs to project future vegetation patterns across Australia remains limited, despite the rapid expansion of modelling approaches within this field.

8.5.2 Recent developments and future improvements of model projections

The vast majority of studies projecting potential future distributions of species and ecosystems have used process implicit (i.e. correlative) models, such as SDMs, rather than process explicit (mechanistic) models. Ultimately, although a region may contain suitable climate for the target species, myriad factors will determine whether populations persist or establish. Booth *et al.* (2015) argue that analyses that exclude the dispersal capabilities and intrinsic adaptability of tree species are unlikely to provide reliable predictions of climate change impacts, as they may overestimate the ability of populations to colonise newly suitable habitat and losses among existing populations. Indeed, dispersal ability and barriers to movement are rarely incorporated in a realistic manner into SDMs, although this may be irrelevant for poor dispersers, such as eucalypts, which are unlikely to track climate change (Booth *et al.* 2015). Some studies have attempted to include more realistic dispersal kernels, whereby access to future areas of suitable habitat is constrained by distance to known locations of the species. This approach was used by Summers *et al.* (2012) and Crossman *et al.* (2012) to identify spatial priorities for the conservation of Australian plant species. Alternate approaches (e.g. based on life form or data for individual species) are not always feasible due to computational limitations or lack of data (see review by Bateman *et al.* 2012).

Both correlative and mechanistic models are founded on niche theory, i.e. that a species' distribution is constrained by sets of environmental conditions and interactions with other organisms (Wiens *et al.* 2009). However, mechanistic models aim to realistically quantify the effects of climate and environmental variables on physiological processes. Importantly, these models can account for the 'fertilisation effect' of elevated CO_2 on plant growth, as well as biophysical thresholds and demographic patterns. Although semi-mechanistic models such as CLIMEX (Sutherst and Maywald 1985) have a long history of use in mapping and understanding the invasive potential of alien plant species, there are few examples of the use of mechanistic models to assess future distributions of native Australian plants. Perhaps the main barrier to widespread adoption of these models is their requirement for detailed physiological knowledge of the study species.

Recent attempts to improve the robustness of the projections derived from correlative SDMs have focused on coupling time series of these models with stochastic life-history process models, providing a potentially powerful method for overcoming the limitations of the habitat niche approach (Keith *et al.* 2008). This is particularly useful for species for which life history and niche are well studied. In an example of this approach, Swab *et al.* (2012) modelled *Leucopogon setiger*, a plant species in fire-prone dry sclerophyll forests of south-eastern Australia, incorporating interactions between the fire regime and projected climate change with information on the response of species' key life-history characteristics to fire. They concluded that the species was at greatest risk of decline from climate-related changes in fire regimes, with the loss of bioclimatically suitable habitat a secondary threat. Incorporation of important trait responses into these models is critical for producing robust predictions of future distributions, however, the demographic data required to quantify these traits is often limiting. Nevertheless, generalisations across functional groups can be made to broaden the utility of these studies.

Advances in modelling vegetation distribution at the continental scale have also been made. Kelley and Harrison (2014) incorporated post-fire resprouting into the Land Processes and Exchanges (LPX) DGVM and reported projected increases in tree cover across Australia under two scenarios of future emissions and climate change. They demonstrated that projected tree cover was sensitive to area burned and increased carbon uptake via elevated CO_2, creating the potential for considerable increases in woody plant cover in the dry interior. As Kelley and Harrison (2014) noted, such results are consistent with 'thickening' in drier regions of Australia (e.g. semi-arid woodlands), a phenomenon observed via remote sensing and attributed to increased water-use efficiency and enhanced growth of C_3 woody taxa in response to CO_2 increases (Ukkola *et al.* 2015). Further improvement of the representation of functional trait variations, community assembly mechanisms (Yang *et al.* 2015) and key processes such as resprouting in relation to fire and drought (Zeppel *et al.* 2015; Pausas *et al.* 2016) will enhance the capacity of these models to effectively capture patterns and explore vegetation change. Future consolidation of vegetation data sets and exploration of environmental controls on patterns of variation at the community level will be required to assist with the validation of outputs from these powerful modelling approaches.

8.5.3 Assessing the ability of plants to adapt to climate change

Correlative SDMs assume niche conservatism, meaning that the environmental envelope of the species studied remains the same over time. However, populations of some species may be adapted to local conditions more so than others, and it is possible that plants can persist *in situ* as environmental conditions change if they are able to adjust to the new environment through phenotypic plasticity or adapt via natural selection (Nicotra *et al.* 2010). While the potential for genetic adaptation is understood, evidence of evolutionary responses to climate change is still rare (Merilä and Hendry 2014; Parmesan and Hanley 2015), and there is consensus that the rate of climate change may exceed the capacity of many plant species to adapt. For example, Dalmaris *et al.* (2015) concluded that *Eucalyptus wandoo* had low adaptive capacity under climate change based on a phylogeographic analysis combined with correlative SDMs, showing historical contractions of the species from areas of decreasing rainfall during the last glacial maximum.

Quantifying the variability and phenotypic plasticity of critical life-history traits, in response to specific environmental cues, is essential to robustly predict persistence under climate change (Chevin *et al.* 2010; Long *et al.* 2015). A plastic response in growth, phenology, seed production, dormancy, germination or recruitment, to any given change can buffer a species to new environmental conditions, allowing persistence in the short term, and may assist adaptation in the longer term (Nicotra *et al.* 2010; Parmesan and Hanley 2015). Both lab- and field-based reciprocal transplant experiments provide vital information for assessing plasticity across populations, an important aspect for unravelling how much local adaptation of genotypes can mediate environmental response. For example, Huang *et al.* (2015) examined the relative plastic response in growth between cool- and warm-climate genotypes of *Telopea speciosissima* and found that only the latter produced increased growth under warmer conditions.

Despite the considerable research effort focused on local adaptation and plasticity, there are still large knowledge gaps. In particular, there is a lack of experimental work focused on the variability of early life-history stages surrounding seeds and seedlings, arguably the most critical in mediating persistence (Parmesan and Hanley 2015). There is also a need for more studies investigating variation in plasticity between co-occurring species (and its effect on interactions such as competition), and between populations within species (Nicotra *et al.* 2010). Valladares *et al.* (2014) demonstrated the value of improved understanding in these areas by showing that niche models incorporating observed differences in plasticity among populations produced projections that were more pessimistic than if a uniform level of plasticity was assumed. The number of such studies is limited by the availability of population-level data, highlighting the need for more research to fill this gap.

8.6 Conclusions

Climatic factors can affect vegetation at scales ranging from impacts on individual plants (or at the sub-plant level), to population- and species-level responses, to the distribution of vegetation types. Distribution patterns are based on a combination of natural climatic, edaphic, disturbance-driven and biotic factors, all of which can be dramatically affected by human activity and climate change. A need to understand the impacts of these factors across all levels of biological organisation at multiple spatial and temporal scales will continue to drive research questions for the foreseeable future.

Although a complex set of interactions between climate, water availability, soils and fire have been identified as major drivers of Australian vegetation, a continental-scale analysis is needed to resolve the specific roles of these factors in mediating change in macro-scale vegetation patterns. Furthermore, incorporating fine-scale climate sensitivities of ecological processes into predictions of broad-scale vegetation responses is fundamental to our understanding of future impacts (Shugart and Woodward 2011) (Figure 8.4). Finally, adaptive capacity mediated by phenotypic and genetic diversity,

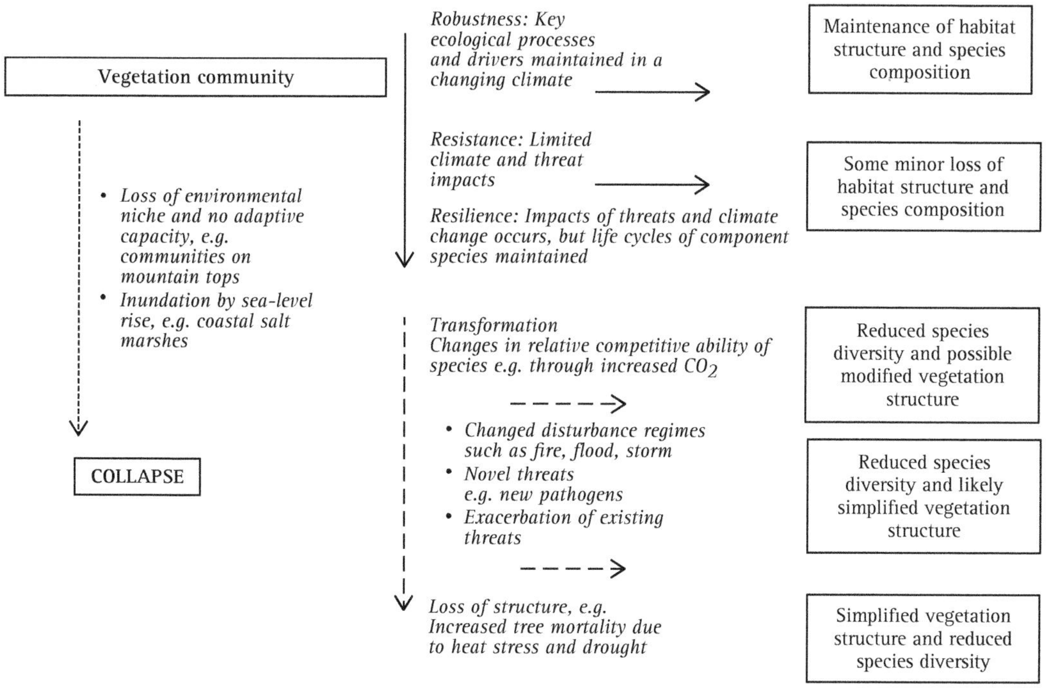

Figure 8.4 Schematic of potential responses of vegetation communities to climate change, via interactions with plant- and population-level processes, based on a framework of community robustness, resistance, resilience, transformation and collapse.

community composition and ecological interactions, are only beginning to be incorporated in modelling studies (Summers *et al.* 2012). New approaches coupling data from provenance trials with SDMs more effectively account for phenotypic plasticity among locally adapted populations (e.g. Benito-Garzón *et al.* 2013, Valladares *et al.* 2014). Measures of contemporary gene flow help to identify environmental barriers to movement, locations of past (or future) climate refugia (Rossetto *et al.* 2012), and to assess the capacity of species to respond to climate change through migration (McCallum *et al.* 2014). As next-generation sequencing becomes more accessible the merging of landscape genomics with SDMs will broaden the scope of scientific inquiry, expanding our understanding of, and ability to forecast, species responses to climate change.

Acknowledgements

Thank you to Adrienne Nicotra, John Baumgartner and Manuel Esperon-Rodríguez for comments on an earlier version of this manuscript.

References

Accad, A. and Neil, D. (2009) Lowland rainforest structural vegetation communities of northeastern Australia: spatial response to predicted climate change. *Proceedings of the 18th World IMACS/MODSIM Congress*, 2028–2034.

Allen, C.D., Macalady, A.K., Chenchouni, H. *et al.* (2010) A global overview of drought- and heat-induced tree mortality reveals emerging climate change risks for forests. *Forest Ecology and Management* 259, 660–684.

Allen, C.D., Breshears, D.D. and McDowell, N.G. (2015) On underestimation of global vulnerability to tree mortality and forest die-off from hotter drought in the Anthropocene. *Ecosphere* 6, 129.

Auld, T.D. and Bradstock, R.A. (1996) Soil temperatures after the passage of a fire: do they influence the germination of buried seeds? *Australian Journal of Ecology* 21, 106–109.

Auld, T.D. and Leishman, M.R. (2015) Ecosystem risk assessment for Gnarled Mossy Cloud Forest, Lord Howe Island, Australia. *Austral Ecology* 40, 364–372.

Auld, T.D., Keith, D.A. and Bradstock, R.A. (2000) Patterns in longevity of soil seed banks in fire-prone communities of south-eastern Australia. *Australian Journal of Botany* 48, 539–548.

Austin, M.P. and Van Niel, K.P. (2011) Improving species distribution models for climate change studies: variable selection and scale. *Journal of Biogeography* 38, 1–8.

Baker, K.S., Steadman, K.J., Plummer, J.A. *et al.* (2005) Dormancy release in Australian fire ephemeral seeds during burial increases germination response to smoke water or heat. *Seed Science Research* 15, 339–348.

Baskin, C.C. and Baskin, J.M. (2014) *Seeds: Ecology, Biogeography and Evolution of Dormancy and Germination.* Academic Press: San Diego.

Bateman, B.L., VanDerWal, J., Williams, S.E. and Johnson, C.N. (2012) Biotic interactions influence the projected distribution of a specialist mammal under climate change. *Diversity and Distributions* 18, 861–872.

Beadle, N.C.W. (1981) *The Vegetation of Australia*. Cambridge University Press: Cambridge.

Beard, J.S. (1981) The history of the phytogeographic region concept in Australia. In *Ecological Biogeography of Australia*, ed. A. Keast. Dr Junk: The Hague pp. 337–353.

Beard, J.S. (2001a) The botanists Diels and Pritzel in Western Australia: a centenary. *Journal of the Royal Society of Western Australia* 84, 143–148.

Beard, J.S. (2001b) A historic vegetation map of Australia. *Austral Ecology* 26, 441–443.

Beaumont, L.J., Hughes, L. and Pitman, A.J. (2008) Why is the choice of future climate scenarios for species distribution modelling important? *Ecology Letters* 11, 1135–1146.

Beaumont, L.J., Gallagher, R.V., Downey, P.O. *et al.* (2009a) Modelling the impact of *Hieracium* spp. on protected areas in Australia under future climates. *Ecography* 32, 757–764.

Beaumont, L.J., Gallagher, R.V., Thuiller, W. *et al.* (2009b) *Different climatic envelopes among invasive populations may lead to underestimations of current and future biological invasions. Diversity and Distributions* 15, 409–420.

Beaumont, L.J., Hartenthaler, T., Keatley, M.R. and Chambers, L.E. (2015) Shifting time: recent changes to the phenology of Australian species. *Climate Research* 63, 203–214.

Bell, D.T. (1999) Turner Review No. 1: the process of germination in Australian species. *Australian Journal of Botany* 47, 475–517.

Bellairs, S.M. and Bell, D.T. (1990) Temperature effects on the seed germination of 10 Kwongan species from Eneabba, Western Australia. *Australian Journal of Botany* 38, 451–458.

Benito-Garzón, M., Ruiz-Benito, P. and Zavala, M.A. (2013) Interspecific differences in tree growth and mortality responses to environmental drivers determine potential species distributional limits in Iberian forests. *Global Ecology and Biogeography* 22,1141–1151.

Booth, T., Broadhurst, L., Pinkard, E. *et al.* (2015) Native forests and climate change: lessons from eucalypts. *Forest Ecology and Management* 347,18–29.

Bowman, D.M.J.S., Boggs, G.S. and Prior, L.D. (2008) Fire maintains an *Acacia aneura* shrubland–*Triodia* grassland mosaic in central Australia. *Journal of Arid Environments* 72, 34–47.

Butt, N., Pollock, L.J. and McAlpine, C.A. (2013) Eucalypts face increasing climate stress. *Ecology and Evolution* 3, 5011–5022.

Carnahan, J.A. (1997) *Australia Natural Vegetation: Australia's Vegetation in the 1780's. Map (1:5 000 000).* Australian Surveying and Land Information Group (AUSLIG): Belconnen.

Chambers, L.E., Altwegg, R., Barbraud, C. *et al.* (2013) Phenological changes in the southern hemisphere. *PLoS One* 8, e75514.

Chevin, L.M., Lande, R. and Mace, G.M. (2010) Adaptation, plasticity, and extinction in a changing environment: towards a predictive theory. *PLoS Biology* 8, e1000357.

Cochrane, A., Hoyle, G.L., Yates, C.J. *et al.* (2014) Predicting the impact of increasing temperatures on seed germination among populations of Western Australian *Banksia* (Proteaceae). *Seed Science Research* 24, 195–205.

Collins, M.R., Knutti, R., Arblaster, J. *et al.* (2013) Long-term Climate Change: Projections, Commitments and Irreversibility. In *Climate Change 2013: The Physical Science Basis. Contribution of Working Group I to the Fifth Assessment Report of the Intergovernmental Panel on Climate Change*, ed. T.F. Stocker, D. Qin, G.-K. Plattner, M. Tignor, S.K. Allen, J. Boschung, A. Nauels, Y. Xia, V. Bex and P.M. Midgley. Cambridge University Press: Cambridge and New York.

Commander, L.E., Merritt, D.J., Rokich, D.P. and Dixon, K.W. (2009) The role of after-ripening in promoting germination of arid zone seeds: a study on six Australian species. *Botanical Journal of the Linnean Society* 161, 411–421.

Crossman, N.D., Bryan, B.A. and Summers, D.M. (2012) Identifying priority areas for reducing species vulnerability to climate change. *Diversity and Distributions* 18, 60–72.

Crous, K.Y., Ósvaldsson, A. and Ellsworth, D.S. (2015) Is phosphorus limiting in a mature *Eucalyptus* woodland? Phosphorus fertilisation stimulates stem growth. *Plant and Soil* 391, 293–305.
Dalmaris, E., Ramalho, C.E., Poot, P. *et al.* (2015) A climate change context for the decline of a foundation tree species in south-western Australia: insights from phylogeography and species distribution modelling. *Annals of Botany* 116, 941–952.
Davies, S.J.J.F. (1976) Studies of the flowering season and fruit production of some arid zone shrubs and trees in Western Australia. *Journal of Ecology* 64, 665–687.
Davies, S.J.J.F. and Kenny, S.A. (2013) The ages and fecundity of some arid zone plants in Western Australia. *The Rangeland Journal* 35, 455–468.
Day, J.S., Loveys, B.R. and Aspinall, D. (1994) Environmental control of flowering of *Boronia megastigma* (Rutaceae) and *Hypocalymma angustifolium* (Myrtaceae). *Australian Journal of Botany* 42, 219–229.
Denham, A.J. and Auld, T.D. (2004) Survival and recruitment of seedlings and suckers of trees and shrubs of the Australian arid zone following habitat management and the outbreak of Rabbit Calicivirus Disease (RCD). *Austral Ecology* 29, 585–599.
DWER (2007) *Australia's native vegetation: a summary of Australia's Major Vegetation Groups, 2007.* Australian Government: Canberra. Available at: https://www.environment.gov.au/resource/australias-native-vegetation-summary-australias-major-vegetation-groups (accessed 28 July 2016).
Elith, J. and Leathwick, J.R. (2009) Species distribution models: ecological explanation and prediction across space and time. *Annual Review of Ecology, Evolution and Systematics* 40, 677–697.
Ellis, A.K. (1970) *Teaching and Learning Elementary Social Studies*. Allyn and Bacon: London.
Enright, N.J., Fontaine, J.B., Bowman, D.M.J.S. *et al.* (2015) Interval squeeze: altered fire regimes and demographic responses interact to threaten woody species persistence as climate changes. *Frontiers in Ecology and the Environment* 13, 265–272.
Facelli, J.M., Chesson, P. and Barnes, N. (2005) Differences in seed biology of annual plants in arid lands: a key ingredient of the storage effect. *Ecology* 86, 2998–3006.
Fenner, M. and Thompson, K. (2005) *The Ecology of Seeds*. Cambridge University Press: Cambridge.
Fensham, R.J., Fairfax, R.J. and Ward, D.P. (2009) Drought-induced tree death in savanna. *Global Change Biology* 15, 380–387.
Fensham, R.J., Fraser, J., MacDermott, H.J. and Firn, J. (2015) Dominant tree species are at risk from exaggerated drought under climate change. *Global Change Biology* 21, 3777–3785.
Fitzpatrick, M.C., Gove, A.D., Sanders, N.J. and Dunn, R.R. (2008) Climate change, plant migration, and range collapse in a global biodiversity hotspot: the *Banksia* (Proteaceae) of Western Australia. *Global Change Biology* 14, 1337–1352.
Franklin, D.C. and Bach, C.S. (2006) Assessing intraspecific phenological synchrony in zoochorous trees from the monsoon forests of northern Australia. *Journal of Tropical Ecology* 22, 419–429.
Gallagher, R.V., Hughes, L. and Leishman, M.R. (2009) Phenological trends among Australian alpine species: using herbarium records to identify climate-change indicators. *Australian Journal of Botany* 57, 1–9.
Gallagher, R.V., Beaumont, L.J., Hughes, L. and Leishman, M.R. (2010) Evidence of climatic niche and biome shifts between native and novel ranges in plant species introduced to Australia. *Journal of Ecology* 98, 790–799.

Gonzales-Orozco, C.E., Ebach, M.C., Laffan, S. *et al.* (2014) Quantifying phytogeographical regions of Australia using geospatial turnover in species composition. *PLoS One* 9, e92558.

Guerin, G.R. and Lowe, A.J. (2013) Multi-species distribution modelling highlights the Adelaide Geosyncline, South Australia, as an important continental-scale arid-zone refugium. *Austral Ecology* 38, 427–435.

Gutterman, Y. (2000) Maternal effects on seeds during development. In *Seeds: The Ecology of Regeneration in Plant Communities*, ed. M. Fenner. CABI: Wallingford, pp. 59–84.

Hampton, J.G., Boelt, B., Rolston, M.P. and Chastain, T.G. (2013) Effects of elevated CO_2 and temperature on seed quality. *Journal of Agricultural Science* 151, 154–162.

Harris, R.M.B., Carter, O., Gilfedder, L. *et al.* (2015) Noah's Ark conservation will not preserve threatened ecological communities under climate change. *PloS One* 10, e0124014.

Hilbert, D.W. and Fletcher, C.S. (2012) *Using artificial neural networks to assess the impacts of future climate change on ecoregions and major vegetation groups in Australia*. CSIRO Climate Adaptation Flagship: Canberra. Available at: https://publications.csiro.au/rpr/download?pid=csiro:EP101610&dsid=DS4 (accessed 28 July 2016).

Hilbert, D.W. and Ostendorf, B. (2001) The utility of empirical, artificial neural network approaches for modelling the distribution of regional to global vegetation in past, present and future climates. *Ecological Modelling* 146, 311–327.

Hilbert, D.W., Ostendorf, B. and Hopkins, M.S. (2001) Sensitivity of tropical forests to climate change in the humid tropics of north Queensland. *Austral Ecology* 26, 590–603.

Hoffman, A.A., Camac, J.S., Williams, R.J. *et al.* (2010) Phenological changes in six Australian subalpine plants in response to experimental warming and year-to-year variation. *Journal of Ecology* 98, 927–937.

Hope, G., Kershaw, A.P., van der Kaars, S. *et al.* (2004) History of vegetation and habitat change in the Austral-Asian region. *Quaternary International* 118–119, 103–126.

Hovenden, M.J. and Williams, A. (2010) The impacts of rising CO_2 concentrations on Australian terrestrial species and ecosystems. *Austral Ecology* 35, 665–684.

Hovenden, M.J., Wills, K.E., Vander Schoor, J.K. *et al.* (2007) Flowering, seed production and seed mass in a species-rich temperate grassland exposed to FACE and warming. *Australian Journal of Botany* 55, 780–794.

Hovenden, M.J., Wills, K.E., Vander Schoor, J.K. *et al.* (2008) Flowering phenology in a species-rich temperate grassland is sensitive to warming but not elevated CO_2. *New Phytologist* 178, 815–822.

Hoyle, G.L., Daws, M.I., Steadman, K.J. and Adkins, S.W. (2008a) Mimicking a semi-arid tropical environment achieves dormancy alleviation for seeds of Australian native Goodeniaceae and Asteraceae. *Annals of Botany* 101, 701–708.

Hoyle, G.L., Steadman, K.J., Daws, M.I. and Adkins, S.W. (2008b) Pre- and post-harvest influences on seed dormancy status of an Australian Goodeniaceae species, *Goodenia fasicularis. Annals of Botany* 102, 93–101.

Huang, G., Rymer, P.D., Duan, H. *et al.* (2015) Elevated temperature is more effective than elevated CO_2 in exposing genotypic variation in *Telopea speciosissima* growth plasticity: implications for woody plant populations under climate change. *Global Change Biology* 21, 3800–3813.

Hudson, A.R., Ayre, D.J. and Ooi, M.K.J. (2015) Physical dormancy in a changing climate. *Seed Science Research* 25, 66–81.

Hughes, L. (2000) Biological consequences of global warming: is the signal already apparent? *Trends in Ecology and Evolution* 15, 56–61.

Hughes, L., Casey, E. and Westoby, M. (1996) Climatic range sizes of *Eucalyptus* species in relation to future climate change. *Global Ecology and Biogeography Letters* 5, 23–29.

Inkrot, D., Sattler, D., Geyer, C. and Morawetz, W. (2007) Flowering and fruiting phenology of *Normanbya normanbyi* (W. Hill) L.H. Bailey (Arecaceae), a palm endemic to the lowland tropical rainforest of north-eastern Australia. *Austral Ecology* 32, 21–28.

Jackson, W.D. (1968) Fire, air, water and earth, an elemental ecology of Tasmania. *Proceedings of the Ecological Society of Australia* 3, 9–16.

Keith, D.A., Akçakaya, H.R., Thuiller, W. *et al.* (2008) Predicting extinction risks under climate change: coupling stochastic population models with dynamic bioclimatic habitat models. *Biology Letters* 4, 560–563.

Keith, D.A., Elith, J. and Simpson, C.C. (2014) Predicting distribution changes of a mire ecosystem under future climates. *Diversity and Distributions* 20, 440–454.

Kelley, S.I. and Harrison, S.P. (2014) Enhanced Australian carbon sink despite increased wildfire during the 21st century. *Environmental Research Letters* 9, 104015.

King, R.W. (1998) Dual control of flower initiation and development by temperature and photoperiod in *Hardenbergia violacea*. *Australian Journal of Botany* 46, 65–74.

King, R.W., Pate, J.S. and Johnstone, J. (1996) Ecotypic differences in the flowering of *Pimelia ferruginea* (Thymelaeceae) in response to cool temperatures. *Australian Journal of Botany* 44, 47–55.

King, R.W., Worrall, R. and Dawson, I.A. (2008) Diversity in environmental controls of flowering in Australian plants. *Scientia Horticulturae* 118, 161–167.

Kingsford, R.T. (2000) Ecological impacts of dams, water diversions and river management on floodplain wetlands in Australia. *Austral Ecology* 25, 109–127.

Kochanek, J., Buckley, Y.M., Probert, R.J. *et al.* (2010) Pre-zygotic parental environment modulates seed longevity. *Austral Ecology* 35, 837–848.

Küchler, A.W. and Zonneveld, I.S. (1988) *Vegetation Mapping*. Kluwer Academic Publishers: Dordrecht.

Lamont, B.B., Witkowski, E.T.F. and Enright, N.J. (1993) Post-fire microsites: safe for seeds, unsafe for seedlings. *Ecology* 74, 501–512.

Liyanage, G.S., Ayre. D.J. and Ooi, M.K.J. (2016) Seedling performance covaries with dormancy thresholds: a case for the maintenance of cryptic seed heteromorphism in a fire-prone ecosystem. *Ecology* 97, 3009–3018.

Long, R.L., Gorecki, M.J., Renton, M. *et al.* (2015) The ecophysiology of seed persistence: a mechanistic view of the journey to germination or demise. *Biological Reviews* 90, 31–59.

Lynch, A.H., Beringer, J., Kershaw, P. *et al.* (2007) Using the paleorecord to evaluate climate and fire interactions in Australia. *Annual Review of Earth and Planetary Sciences* 35, 215–239.

Mackenzie, B.D.E., Auld, T.D., Keith, D.A. *et al.* (2016) The effect of seasonal ambient temperatures on fire-stimulated germination of species with physiological dormancy: a case study using *Boronia* (Rutaceae). *PLoS ONE* 11, e0156142.

McCallum, K.P., Guerin, G.R., Breed, M.F. and Lowe, A.J. (2014) Combining population genetics, species distribution modelling and field assessments to understand a species vulnerability to climate change. *Austral Ecology* 39, 17–28.

McKee, K., Rogers, K. and Saintilan, N. (2012) Response of salt marsh and mangrove wetlands to changes in atmospheric CO2, climate, and sea level. In *Global Change and the Function and Distribution of Wetlands*, ed. B.A. Middleton. Springer: Dordrecht. pp. 63–96.

Merilä, J. and Hendry, A.P. (2014) Climate change, adaptation, and phenotypic plasticity: the problem and the evidence. *Evolutionary Applications* 7, 1–14.

Merritt, D.J., Turner, S.R., Clarke, S. and Dixon, K.W. (2007) Seed dormancy and germination stimulation syndromes for Australian temperate species. *Australian Journal of Botany* 55, 336–344.

Mooney, S.D., Harrison, S.P., Bartlein, P.J. and Stevenson, J. (2012) The prehistory of fire in Australia. In *Flammable Australia: Fire Regimes, Biodiversity and Ecosystems in a Changing World*, ed. R.A. Bradstock, R.J. Williams and A.M. Gill. CSIRO Publishing: Melbourne, pp. 3–25.

Morrison, D.A. and Myerscough, P.J. (1989) Temporal regulation of maternal investment in populations of the perennial legume *Acacia suaveolens*. *Ecology* 70, 1629–1638.

Mott, J.J. (1972) Germination studies on some annual species from an arid region of Western Australia. *Journal of Ecology* 60, 293–304.

Nicholson, M. (2013) Community concepts in plant ecology: from Humboldtian plant geography to the superorganism and beyond. *Web Ecology* 13, 95–102.

Nicotra, A.B., Atkin, O.K., Bonser, S.P. *et al.* (2010) Plant phenotypic plasticity in a changing climate. *Trends in Plant Science* 15, 684–692.

Ooi, M.K.J. (2010) Delayed emergence and post-fire recruitment success: effects of seasonal germination, fire season and dormancy type. *Australian Journal of Botany* 58, 248–256.

Ooi, M.K.J. (2012) Seed bank persistence and climate change. *Seed Science Research* 22, S53–S60.

Ooi, M.K.J., Auld, T.D. and Whelan, R.J. (2004) Delayed post-fire seedling emergence linked to season: a case study using Australian native *Leucopogon* species (Epacridaceae). *Annals of Botany* 174, 183–196.

Ooi, M.K.J., Auld, T.D. and Whelan, R.J. (2006) Dormancy and the fire-centric focus: germination of three *Leucopogon* species (Ericaceae) from south-eastern Australia. *Annals of Botany* 98, 421–430.

Ooi, M.K.J., Auld, T.D. and Denham, A.J. (2009) Climate change and bet-hedging: interactions between increased soil temperatures and seed bank persistence. *Global Change Biology* 15, 2375–2386.

Ooi, M.K.J., Denham, A.D., Santana, V.M. and Auld, T.D. (2014) Temperature thresholds of physically dormant seeds and plant functional response to fire: variation among species and relative impact of climate change. *Ecology and Evolution* 4, 656–671.

Parmesan, C. and Hanley, M.E. (2015) Plants and climate change: complexities and surprises. *Annals of Botany* 116, 849–864.

Pausas, J.G., Pratt, R.B., Keeley, J.E. *et al.* (2016) Towards understanding resprouting at the global scale. *New Phytologist* 209, 945–954.

Pavlick, R., Drewry, D.T., Bohn, K. *et al.* (2013) The Jena Diversity-Dynamic Global Vegetation Model (JeDi-DGVM): a diverse approach to representing terrestrial biogeography and biogeochemistry based on plant functional trade-offs. *Biogeosciences* 10, 4137–4177.

Peñuelas, J. and Filella, I. (2001) Phenology: responses to a warming world. *Science* 294, 793–795.

Pickett, E.J., Harrison, S.P., Hope, G. *et al.* (2004) Pollen-based reconstructions of biome distributions for Australia, Southeast Asia and the Pacific (SEAPAC region) at 0, 6000 and 18 000 14C yr BP. *Journal of Biogeography* 31, 1381–1444.

Prober, S.M., Hilbert, D.W., Ferrier, S. and Dunlop, M. (2012) Combining community-level spatial modelling and expert knowledge to inform climate adaptation in temperate grassy eucalypt woodlands and related grasslands. *Biodiversity and Conservation* 21, 1627–1650.

Rathcke, B. and Lacey, E.P. (1985) Phenological patterns of terrestrial plants. *Annual Review of Ecology and Systematics* 16, 179–214.

Rawal, D.S., Kasel, S., Keatley, M.R. and Nitschke, C.R. (2015) Herbarium records identify sensitivity of flowering phenology of eucalypts to climate: implications for species response to climate change. *Austral Ecology* 40, 117–125.

Reich, P.B., Hobbie, S.E. and Lee, T.D. (2014) Plant growth enhancement by elevated CO_2 eliminated by joint water and nitrogen limitation. *Nature Geoscience* 7, 920–924.

Roche, S., Dixon, K.W. and Pate, J.S. (1998) For everything a season: smoke-induced seed germination and seedling recruitment in a Western Australian *Banksia* woodland. *Austral Ecology* 23, 111–120.

Rossetto, M., Allen, C.B., Thurlby, K.A., Weston, P.H. and Milner, M.L. (2012) Genetic structure and bio-climatic modeling support allopatric over parapatric speciation along a latitudinal gradient. *BMC Evolutionary Biology* 12, 149.

Rumpff, L., Coates, F. and Morgan, J.W. (2010) Biological indicators of climate change: evidence from long-term flowering records of plants along the Victorian coast, Australia. *Australian Journal of Botany* 58, 428–439.

Russell-Smith, J. and Lucas, D.E. (1994) Regeneration of monsoon rainforest in northern Australia: the dormant seed bank. *Journal of Vegetation Science* 5, 161–168.

Russell-Smith, J. and Setterfield, S.A. (2006) Monsoon rain forest seedling dynamics, northern Australia: contrasts with regeneration in eucalypt-dominated savannas. *Journal of Biogeography* 33, 1597–1614.

Saikkonen, K., Taulavuori, K., Hyvönen, T. *et al.* (2012) Climate change-driven species' range shifts filtered by photoperiodism. *Nature Climate Change* 2, 239–242.

Saintilan, N., Wilson, N.C., Rogers, K. *et al.* (2014) Mangrove expansion and salt marsh decline at mangrove poleward limits. *Global Change Biology* 20, 147–157.

Santana, V.M., Bradstock, R.A., Ooi, M.K.J. *et al.* (2010) Effects of soil temperature regimes after fire on seed dormancy and germination in six Australian Fabaceae species. *Australian Journal of Botany* 58, 539–545.

Scheiter, S., Higgins, S.I., Beringer, J. and Hutley, L. (2014) Climate change and long-term fire management impacts on Australian savannas. *New Phytologist* 205, 1211–1226.

Schneemilch, M., Kokkinn, M. and Williams, C.R. (2012) Flowering timing prediction in Australian native understorey species (*Acrotriche* R.Br Ericaceae) using meteorological data. *International Journal of Biometeorology* 56, 95–105.

Settele, J., Scholes, R., Betts, R. *et al.* (2014) Terrestrial and inland water systems. In *Climate Change 2014: Impacts, Adaptation, and Vulnerability. Part A: Global and Sectoral Aspects. Contribution of Working Group II to the Fifth Assessment Report of the Intergovernmental Panel on Climate Change*, ed. C.B. Field, V.R. Barros, D.J. Dokken *et al.* Cambridge University Press, Cambridge, pp. 271–359.

Shugart, H.H. and Woodward, F.I. (2011) *Global Change and the Terrestrial Biosphere: Achievements and Challenges.* Wiley-Blackwell: Chichester.

Sitch, S., Huntingford, C., Geditioney, N. *et al.* (2008) Evaluation of the terrestrial carbon cycle, future plant geography and climate-carbon cycle feedbacks using 5 Dynamic Global Vegetation Models (DGVMs). *Global Change Biology* 14, 1–25.

Sommerville, K.D., Martyn, A.J. and Offord, C.A. (2013) Can seed characteristics or species distribution be used to predict the stratification requirements of herbs in the Australian Alps? *Botanical Journal of the Linnean Society* 172, 187–204.

Spears, J.F., TeKrony, D.M. and Egli, D.B. (1997) Temperature during seed filling and soybean seed germination and vigour. *Seed Science and Technology* 25, 233–244.

Specht, R.L. (1981) Ecophysiological principles determining the biogeography of major vegetation formations in Australia. In *Ecological Biogeography of Australia*, ed. A. Keast. Dr Junk: The Hague, pp. 299–334.

Steadman, K.J., Crawford, A.D. and Gallagher, R.S. (2003) Dormancy release in *Lolium rigidum* seeds is a function of thermal after-ripening time and seed water content. *Functional Plant Biology* 30, 345–352.

Stevens, N., Seal, C.E., Archibald, S. and Bond, W.J. (2014) Increasing temperatures can improve seedling establishment in arid-adapted savanna trees. *Oecologia* 175, 1029–1040.

Summers, D.M., Bryan, B.A., Crossman, N.D. and Meyer, W.S. (2012) Species vulnerability to climate change: impacts on spatial conservation priorities and species representation. *Global Change Biology* 18, 2335–2348.

Sutherst, R.W. and Maywald, G.F. (1985) A computerised system for matching climates in ecology. *Agriculture, Ecosystems and Environment* 13, 281–299.

Swab, R.M., Regan, H.M., Keith, D.A. *et al.* (2012) Niche models tell half the story: spatial context and life-history traits influence species responses to global change. *Journal of Biogeography* 39, 1266–1277.

Thomas, P.B., Morris, E.C., Auld, T.D. and Haigh, A.M. (2010) The interaction of temperature, water availability and fire cues regulates seed germination in a fire-prone landscape. *Oecologia* 162, 293–302.

Tieu, A., Dixon, K.W., Meney, K.A. and Sivasithamparam, K. (2001) The interaction of heat and smoke in the release of seed dormancy in seven species from south-western Western Australia. *Annals of Botany* 88, 259–265.

Tooth, I.M. and Leishman, M.R. (2013) Post-fire resprouting responses of native and exotic grasses from Cumberland Plain Woodland (Sydney, Australia) under elevated carbon dioxide. *Austral Ecology* 38, 1–10.

Tozer, M.G. and Bradstock, R.A. (1997) Factors influencing the establishment of seedlings of the mallee *Eucalyptus luehmanniana* (Myrtaceae). *Australian Journal of Botany* 45, 997–1008.

Tozer, M.G. and Ooi, M.K.J. (2014) Humidity-regulated dormancy onset in the Fabaceae: a conceptual model and its ecological implications for the Australian wattle *Acacia saligna*. *Annals of Botany* 114, 579–590.

Ukkola, A.M., Prentice, I.C., Keenan, T.F. *et al.* (2015) Reduced streamflow in water-stressed climates consistent with CO_2 effects on vegetation. *Nature Climate Change* DOI: 10.1038/NCLIMATE2831

Valladares, F., Matesanz, S., Guilhaumon, F. *et al.* (2014) The effects of phenotypic plasticity and local adaptation on forecasts of species range shifts under climate change. *Ecology Letters* 17, 1351–1364.

Van Bodegom, P.M., Douma, J.C. and Verheijen, L.M. (2014) A fully traits-based approach to modeling global vegetation distribution. *Proceedings of the National Academy of Sciences* 111, 13733–13738.

Venn, S.E. and Morgan, J.W. (2007) Phytomass and phenology of three alpine snowpatch species across a natural snowmelt gradient. *Australian Journal of Botany* 55, 450–456.

Walck, J.L., Hidayati, S.N., Dixon, K.W. *et al.* (2011) Climate change and plant regeneration from seed. *Global Change Biology* 17, 2145–2161.

Wardle, G.M., Greenville, A.C., Frank, A.S.K. *et al.* (2015) Ecosystem risk assessment of Georgina gidgee woodlands in central Australia. *Austral Ecology* 40, 444–459.

Watson, I.W., Westoby, M. and Holm, A.Mc.R. (1997) Demography of two shrub species from an arid grazed ecosystem in Western Australia 1983–93. *Journal of Ecology* 85, 815–832.
Wellington, A.B. and Noble, I.R. (1985) Post-fire recruitment and mortality in a population of the mallee *Eucalyptus incrassate* in semi-arid, south-eastern Australia. *Journal of Ecology* 73, 645–656.
Wiens, J.A., Stralberg, D., Jongsomjit, D. *et al.* (2009) Niches, models, and climate change: assessing the assumptions and uncertainties. *Proceedings of the National Academy of Sciences* 106, 19729–19736.
Williams, R.J., Myers, B.A., Eamus, D. and Duff, G.A. (1999) Reproductive phenology of woody species in a north Australian tropical savanna. *Biotropica* 31, 626–636.
Williams, A.L., Wills, K.E., Janes, J.K. *et al.* (2007) Warming and free-air CO_2 enrichment alter demographics in four co-occurring grassland species. *New Phytologist* 176, 365–374.
Williams, R.J., Wahren, C.-H., Stott, K.A.J. *et al.* (2015) An International Union for the Conservation of Nature Red List ecosystems risk assessment for alpine snow patch herbfields, South-Eastern Australia. *Austral Ecology* 40, 433–443.
Winn, K.O., Saynor, M.J., Eliot, M.J. and Elio, I. (2006) Saltwater intrusion and morphological change at the mouth of the East Alligator River, Northern Territory. *Journal of Coastal Research* 22, 137–149.
Wright, B.R. and Clarke, P.J. (2009) Fire, aridity and seed banks. What does seed bank composition reveal about community processes in fire-prone desert? *Journal of Vegetation Science* 20, 663–674.
Wright, B.R., Zuur, A.F. and Chan, G.C.K. (2014) Proximate causes and possible adaptive functions of mast seeding and barren flower shows in spinifex grasses (*Triodia* spp.) in arid regions of Australia. *The Rangeland Journal* 36, 297–308.
Yang, Y., Zhu, Q., Peng, C. *et al.* (2015) From plant functional types to plant functional traits: a new paradigm in modelling global vegetation dynamics. *Progress in Physical Geography* DOI: 10.1177/0309133315582018.
Yates, C.J., McNeill, A., Elith, J. and Midgley, G.F. (2010) Assessing the impacts of climate change and land transformation on *Banksia* in the south west Australian Floristic Region. *Diversity and Distributions* 16, 187–201.
Zeppel, M.J.B., Harrison, S.P., Adams, H.D. *et al.* (2015) Drought and resprouting plants. *New Phytologist* 206, 583–589.

9 Invasive Plants and Pathogens in Australia

MICHELLE R. LEISHMAN, RACHAEL V. GALLAGHER, JANE A. CATFORD, TONY GRICE, JOHN W. MORGAN AND SAMANTHA SETTERFIELD

Exotic plants and pathogens introduced to Australia have had significant impacts on species, vegetation communities and on ecological processes. There are approximately 2700 exotic plant species that have become naturalised within Australia and of those, around 250 are considered problematic. The sources of introduction have shifted from Europe, Australasia and North America to South America more recently, with the vast majority of introductions associated with the horticultural industry. The proportion of exotic species within the flora increases from northern Australia to the southern areas of the continent, with successful invasion related to high human density and associated disturbance, particularly increased nutrients and high grazing pressure. This chapter describes the main functional types of plant invaders and their impacts across the major regions of Australia. Invasion by exotic woody invaders, plants, perennial grasses, annual herbs and grasses, vines and aquatic plants have caused substantial changes in vegetation structure and composition, with reduced native species diversity and changes in ecosystem processes, such as hydrological and fire regimes. This chapter also describes the impact of two important plant pathogens that have been introduced to Australia – phytophthora (*Phytophthora cinnamomi*) and myrtle rust (*Puccinia psidii*). These pathogens have both resulted in declines in abundance of some susceptible plant species and resultant changes in vegetation composition. Finally, we discuss current weed management approaches in Australia and future research directions.

9.1 Introduction

Invasive plants and plant pathogens are considered major threats to biodiversity globally, resulting in declines of native species (Gilbert and Levine 2013) and substantial changes to vegetation composition and structure, as well as ecosystem function (Vilà *et al.* 2011). Australia has evolved a flora that is phylogenetically distinct due to its isolation since the break-up of Gondwana (Weston and Jordan, Chapter 2, this volume) and thus the establishment of exotic plant species from quite different plant lineages is particularly noticeable in Australian vegetation communities (the 'modern intrusive' element of Keith and Tozer, Chapter 1, this volume). Few, if any, of Australia's biogeographic regions and vegetation communities remain uninvaded by exotic plants (Adair and Groves 1998; Groves *et al.* 2005). However, the success of exotic plants across the continent is not uniform – the proportion of the flora that is exotic generally increases from north to south and varies among vegetation types and with human influence.

Australian Vegetation, Third Edition, ed. David A. Keith. Published by Cambridge University Press. © Cambridge University Press 2017.

Citation: Leishman, M.R. *et al.* (2017). Invasive Plants and Pathogens in Australia. In D. A. Keith, ed., *Australian Vegetation*, 3rd edn. Cambridge: Cambridge University Press, pp. 207–29.

Similarly, there is variation among exotic plant species as to which become successful invaders following introduction, due to a combination of introduction effort, plant traits and environmental conditions.

Exotic invasive plants can have substantial economic and environmental impacts. Economic costs associated with exotic plant invaders have focused on losses in agricultural productivity and direct costs associated with quarantine and control procedures (Sinden *et al.* 2004). It is harder to estimate the economic impacts on native vegetation communities, but it is clear that the cost of removing environmental weeds for ecosystem rehabilitation through national, state and local programmes (e.g. National Landcare Programme; Australian Government 2016) has been considerable.

In this chapter, we describe the mechanisms by which exotic invasive plants affect vegetation communities, the major sources and invasion pathways of these exotic invaders, and the most important functional types. We describe their impacts in the major regions and habitats of Australia: the tropical savannas and woodlands of northern Australia; the woodlands and forests of temperate southern Australia; the arid and semi-arid zones; the alpine zone; and riparian and wetland systems. We discuss the impacts of two invasive plant pathogens on Australian plant species and vegetation communities: phytophthora and myrtle rust. Finally, we examine the management of exotic invasive plants and pathogens and outline future research directions.

9.2 The state of invasive plants in Australia

An astonishing number of plant species (19 012; Diez *et al.* 2009), roughly equal to the number of native species (19 324; Chapman 2009), have been introduced to Australia since European colonisation, more than 30% of which are from plant families that are novel to the continent (Gallagher and Leishman 2014). Yet, despite this enormous number of introductions, only around 2700 species have become naturalised (i.e. formed self-sustaining populations), with 253 viewed as problem invaders due to serious impacts on agriculture or the environment (Rod Randall, pers. comm. 2011; used in Gallagher *et al.* 2015). These 'problem invaders' are generally targeted for management, however the pool of naturalised species provides a large source of potential invaders, particularly as environmental conditions change into the future (O'Donnell *et al.* 2012).

9.2.1 Processes of invasion

9.2.1.1 The introduction-invasion continuum

Richardson *et al.* (2000) conceptualised invasion as a continuum in which species must cross a series of geographic and environmental (abiotic and biotic) barriers to reach a new region (introduction), form self-sustaining populations (naturalisation), and produce offspring in areas distant from sites of introduction (invasion) (Figure 9.1). Notably, the term 'invasive' does not imply any environmental or economic impact (Richardson *et al.* 2000). Factors associated with successful transition through the invasion continuum are introduction effort, traits of the introduced species and environmental conditions in the recipient community (Davis *et al.* 2000).

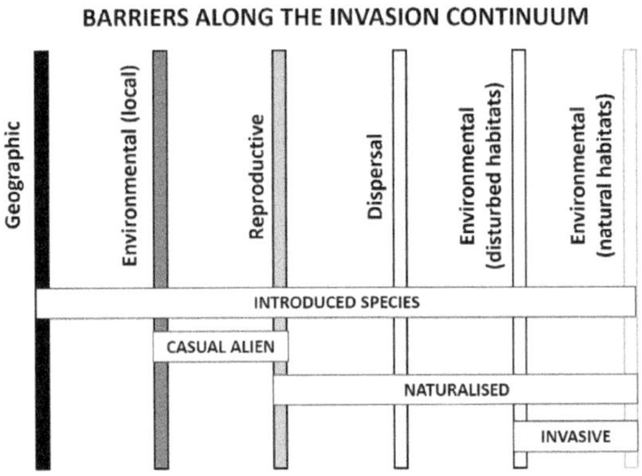

Figure 9.1 The invasion continuum. A schematic representation of major barriers limiting the spread of introduced plants. *Source: Gallagher and Leishman (2015).*

9.2.1.2 Sources and pathways of introduction

The exotic flora of Australia has been introduced from a variety of regions, including South America (35% of the total number of introduced exotic species), Africa (22%), Asia (17%) and Europe (10%) (Gallagher and Leishman 2014). Region of origin has varied through time, with the oldest introductions being from Europe, Australasia and North America and the newest from South America (Phillips *et al.* 2010). One-third of Australia's introduced flora is from the Poaceae, Fabaceae and Asteraceae families (Dodd *et al.* 2015), which are families that dominate the exotic flora in many regions globally.

Exotic plants have been introduced to Australia via accidental (e.g. as a contaminant) and deliberate means. Since early colonial times there have been deliberate introductions of plants, including under government initiatives such as the Commonwealth Plant Introduction Scheme (Cook and Dias 2006). Many species were imported as crop or fodder plants for the developing colony or as horticultural specimens with the aim of transforming the Australian landscape to appear 'more like home' for early settlers (Low 1999). The most striking feature of Australia's exotic flora is the enormous influence of the horticultural sector. Of the ~250 exotic plant species that have become problematic invaders, 70% were introduced for horticulture, 17% for agriculture and 10% were accidental introductions (Randall 2007; Gallagher and Leishman 2014). Twenty-three of the 32 Weeds of National Significance were introduced originally as ornamental plants (Australian Weeds Committee 2012). Several factors contribute to the success of horticultural species as invaders. First, they are planted extensively in areas with large human populations, and there is a clear positive relationship between the availability of horticultural species and their rate of naturalisation (Mulvaney 2001; Pemberton and Lui 2009). Second, many horticultural species have been selected for their characteristics such as fast growth rates and fleshy fruits, which facilitates their invasive success. Finally, many horticultural species are highly suited to environmental conditions associated with human populations, such as nutrient enrichment (Lake and Leishman 2004; Leishman and Thomson 2005).

Successful invaders owe their spread across the landscape to deliberate transport and introduction by humans, long-distance dispersal via wind, water or via frugivorous birds. Native and exotic birds consume the fruits and excrete the seeds of a broad range of exotic plants, potentially resulting in seed dispersal distances of hundreds of metres to kilometres (White and Vivian-Smith 2011), enabling plant introduction into remote and otherwise inaccessible locations (e.g. *Miconia calvescens* in the wet tropics; Hardesty *et al.* 2011). Infestations of exotic fleshy fruited plants are often found clumped under bird roosts or perches: for example, *Olea europaea* subsp. *cuspidata* (African olive) is frequently found under power poles (Cuneo and Leishman 2006). Many of the most problematic and widespread plant invaders in Australia are fleshy fruited bird-dispersed species, including *Ligustrum* species (the privets), *Asparagus* species, *Lantana camara* (lantana), *Cinnamomum camphora* (camphor laurel), *Olea europaea* subsp. *cuspidata* (African olive), *Chrysanthemoides monilifera* (bitou bush and boneseed) and African boxthorn (*Lycium ferocissimum*).

9.2.1.3 Species traits and the environmental context

Certain traits of species are associated with naturalisation and invasion success, particularly those related to reproduction and growth rate. For example, Mason *et al.* (2008) showed that successful invaders produce more seeds in total and for a given seed mass compared to native species, while Leishman *et al.* (2007, 2010) found that successful invaders have leaf-level traits associated with fast-growth strategies. Gallagher *et al.* (2015) reported that invasive species had larger specific leaf areas, flowered for longer and were generally taller than congeneric naturalised species. The role of particular traits is best considered, however, in the context of the environmental conditions of the recipient community. Anthropogenic disturbance typically results in greater availability of resources (e.g. space, light and soil nutrients), which favours exotic plant species with traits enabling colonisation of vacant space and rapid growth in non-limiting conditions (Thomson and Leishman 2004; Leishman and Thomson 2005). Many plant invasions in the Australian context are associated with roadsides (Coulson *et al.* 2014), areas of nutrient enrichment (particularly on low-fertility soils; Lake and Leishman 2004), or high grazing pressure (Close *et al.* 2008).

9.2.2 Functional types of plant invaders

The diverse exotic species that have become successful invaders in Australia can be grouped into broad functional types based on their physical traits and impacts (Table 9.1). Such groupings may be useful for the management and control of weeds and indeed have been recognised in legislation by the listing of invasion by exotic functional groups of species as key threatening processes to biodiversity. The New South Wales *Threatened Species Conservation Act* 1995, for example, lists 'invasion and establishment of exotic vines and scramblers' and 'invasion of native plant communities by exotic perennial grasses' as key threatening processes.

Among the main functional groups of plant invaders there is variation in their introduction, impacts and types of habitats they invade. Invasive woody trees and shrubs have been introduced for horticulture or rehabilitation in most habitats. They have significant impacts due to their size and consequent capacity to transform systems by modifying vegetation structure, with substantial flow-on effects to ecosystem processes and biodiversity. Perennial grass species have been introduced to Australia largely to improve agricultural productivity, particularly in rangelands. Relatively few have proved useful for production, however, and a large proportion of deliberately introduced grasses have become environmental and agricultural weeds (Lonsdale 1994). *Cenchrus ciliaris* (buffel grass) is probably the most widely planted exotic perennial grass and is now widespread across the rangelands (Jackson 2005). Perennial grasses such as *Nassella trichotoma* (serrated tussock) and *N. neesiana* (Chilean needlegrass) are increasingly problematic in temperate grassy woodlands where they outcompete native understorey species (Grice *et al.* 2013). Similarly, exotic annual herbs and grasses replace native perennial grasses in many semi-arid and temperate shrublands and woodlands where grazing pressure from livestock is high (Prober *et al.* 2005). In contrast, exotic vines and scramblers have mainly been introduced for horticulture due to their showy flower displays and rapid growth (Gallagher and Leishman 2014). Harris *et al.* (2007) reported over 180 exotic vine or scrambler species in Australia. Some exotic vines are considered 'transformer' species, particularly in rainforests and riparian vegetation, due to their ability to substantially modify vegetation structure and composition (Figure 9.2a–d). Aquatic weeds have also been introduced as horticultural species, although specifically for the aquarium trade: for example, more than 400 freshwater aquatic plant species have been available for sale in Australia in the last 30 years (Petroeschevsky and Champion 2008). Exotic aquatic invaders enter freshwater systems through the inappropriate dumping of aquarium material (Champion *et al.* 2010) and become established in creeks, rivers, lakes and dams where they can modify water and habitat quality, causing declines in native species.

9.2.3 Plant invasions in major regions of Australia

Invasion by exotic plant species has had a variety of impacts on Australian native vegetation, from the level of individual species through to community-level impacts on vegetation structure and composition, with associated flow-on effects

Table 9.1 **The main functional types of exotic plant invaders in Australian vegetation: Weeds of National Significance and examples of impacts.**

	Weeds of National Significance*	Impacts and examples
Woody trees and shrubs	pond apple (*Annona glabra*), boneseed (*Chrysanthemoides monilifera* subsp. *monilifera*), bitou bush (*Chrysanthemoides monilifera* subsp. *rotundata*), English broom (*Cytisus scoparius*), Mediterranean broom (*Genista linifolia*), Montpellier broom (*Genista monspessulana*), jatropha (*Jatropha gossypifolia*), lantana (*Lantana camara*), African boxthorn (*Lycium ferocissimum*), giant sensitive tree (*Mimosa pigra*), parkinsonia (*Parkinsonia aculeata*), mesquite (*Prosopis* spp.), blackberry (*Rubus fruticosus* aggregate), willows (*Salix* spp.), athel pine (*Tamarix aphylla*), gorse (*Ulex europaeus*), prickly acacia (*Vachella nilotica*)	Reduce recruitment of native species, e.g. African olive, *Olea europaea* subsp. *cuspidata* (Cuneo and Leishman 2013) Reduce habitat suitability for native fauna (Holland-Clift *et al.* 2011) Modify hydrological processes, e.g. athel pine (Low 1999); willows (Stokes and Cunningham 2006)
Perennial grasses	gamba grass (*Andropogon gayanus*), Chilean needle grass (*Nassella neesiana*), serrated tussock (*Nassella trichotoma*), hymenachne (*Hymenachne amplexicaulis*)	Outcompete native understorey species, resulting in declines in the abundance and diversity of those native species (Brooks *et al.* 2010; Kutt and Kemp 2012) Modify habitat for fauna such as nesting seabirds, e.g. Kikuyu, *Pennisetum clandestinum* (Weerheim *et al.* 2003) Alter fire regimes, with significant flow-on consequences, e.g. gamba grass (Rossiter *et al.* 2003)
Vines and scramblers	Madeira vine (*Anredera cordifolia*), the climbing asparaguses (*Asparagus africanus*, *Asparagus plumosus* and *Asparagus scandens*), bridal creeper (*Asparagus asparagoides*), bridal veil (*Asparagus declinatus*), rubber vine (*Cryptostegia grandiflora*), cat's claw vine (*Dolichandra unguis-cati*)	Smother and cause death of host, create canopy gaps that may facilitate further establishment of exotic species (Gallagher *et al.* 2010)
Aquatic weeds	Alligator weed (*Alternanthera philoxeroides*), cabomba (*Cabomba caroliniana*), hymenachne (*Hymenachne amplexicaulis*), water hyacinth (*Eichhornia crassipes*), salvinia (*Salvinia molesta*)	Reduce habitat quality for native species by decreasing dissolved oxygen content and light penetration and changing water temperatures (Ruiz-Avila and Klemm 1996; Perna and Burrows 2005)
Annual herbs and grasses	Fireweed (*Senecio madagascariensis*)	Replace perennial herbs and grasses in heavily grazed and fragmented grassy woodlands in the temperate region (Prober *et al.* 2005)
Opuntioid cacti	*Austrocylindropuntia*, *Cylindropuntia* and *Opuntia* species	Compete with native vegetation, reducing the growth of understorey shrubs, herbs and grasses, particularly in low rainfall environments

* Australian Weeds Committee 2012

Figure 9.2 Some species are considered transformer species as they cause substantial changes in the structure of the communities they invade: (a) and (b) exotic invasive vines can cause dieback and mortality of host trees, opening up the canopy to the invasion of other species (photos: Rachael Gallagher); (c) and (d) African Olive invades grassy woodland, creating a deep shaded canopy that native grasses and herbs are unable to survive beneath (photos: Peter Cuneo); (e) savanna with a native grass understorey (photo: M. Douglas); (f) savanna invaded by the exotic gamba grass, with associated tree mortality due to transformation of fire regimes (photo: Samantha Setterfield).
(A black and white version of this figure will appear in some formats. For the colour version, please refer to the plate section.)

to landscape-level processes, including species interactions, nutrient cycling and hydrological and disturbance regimes. Areas invaded by exotic plants generally have reduced native plant diversity and changed structure and composition (Lake and Leishman 2004; Turner *et al.* 2008; Kutt and Kemp 2012; Cuneo and Leishman 2013). Impacts may be caused by direct mechanisms, such as competition for light, nutrients, space and water (Lenz *et al.* 2003; Lai *et al.* 2015), or indirectly via changes to plant–animal interactions, disturbance regimes or other ecosystem processes (Rossiter *et al.* 2003; White and Vivian-Smith 2011).

National and state biodiversity laws recognise several individual exotic invasive plant species, as well as functional groups of invasive plants, as increasing the risk of extinction of native plants and vegetation communities. In the Commonwealth's *Environment Protection and Biodiversity Conservation Act* 1999, 'invasion of northern Australia by gamba grass and other introduced grasses' and the 'loss and degradation of native plant and animal habitat by invasion of escaped garden plants, including aquatic plants' are listed as key threatening processes to biodiversity. Invasive plant species have been identified as a threat to 341 of the 900 plant species and 103 of the 107 ecological communities listed on the schedules of the New South Wales *Threatened Species Conservation Act* 1995 (as of May 2015; M. Leishman and C. Laws unpublished data).

The suite of species of exotic plant invaders and their impact vary across the major regions and habitats of Australia. In this section we describe the major patterns of exotic plant invasion in each of these regions.

9.2.3.1 Tropical Australia

The vast tropics of northern Australia (where rainfall is >600 mm yr^{-1}) cover ~2 M km^2 and comprise predominately *Eucalyptus*-dominated savanna woodlands and open forests (Williams *et al.*, Chapter 15, this volume), with large areas of tropical wetlands in coastal and near-coastal areas. Rainforests occur in the wet tropics of northern Queensland and in small pockets in Western Australia and the Northern Territory (Woinarski *et al.* 2007; Metcalfe and Green, Chapter 11, this volume). Limited development of intensive agriculture in the tropics of northern Australia means that <5% of the region has undergone landscape transformation (Douglas *et al.* 2011). Although the proportion of invasive plants is also relatively low (5–8% of the total plant species) compared to the Australian average of 12% (Woinarski *et al.* 2007), invasive plants increasingly threaten the region's ecological values. Eighteen of Australia's 32 Weeds of National Significance are invaders of tropical ecosystems. In the past two decades, large areas of tropical Australia's savannas and wetlands have been invaded by ecologically transforming exotic alien plants (Abom *et al.* 2015).

Exotic C_4 grasses are among the greatest threat to Australia's tropical ecosystems. Over 1000 tropical and subtropical grasses have been imported into Australia (Cook and Dias 2006), many as part of programmes aimed at increasing agricultural production. Invasion into relatively undisturbed savannas by the African C_4 grasses *Andropogon gayanus* (gamba grass) and *Cenchrus polystachios* (mission grass) is associated with a decrease in understorey native species richness and abundance, a change in species composition and a shift in the relative abundance of functional groups (Brooks *et al.* 2010). C_4 grass invasions have caused similar declines in species richness and abundance in other tropical systems (Foxcroft *et al.* 2010; Rossi *et al.* 2014).

Invasion by high densities of exotic plants shifts vegetation communities to well-defined alternative states, primarily due to altered fire regimes associated with increased fuel loads (Brooks *et al.* 2010; Setterfield *et al.* 2010). For example, invasion by gamba grass results in a change from short (~0.5 m height), relatively sparse, native grass fuel beds (typically 1–3 t ha^{-1}), to tall (~4 m height), dense fuel beds (up to 30 t ha^{-1}) (Rossiter *et al.* 2003;

Setterfield *et al.* 2010). As a result, fire intensity increases significantly, from 1–3 MW m^{-1} in native grass fires to 16 MW m^{-1} in gamba grass-fuelled fires in the early dry season. The frequent, intense fires associated with gamba grass cause significant tree death, furthering the trajectory from woodland to exotic grassland (Brooks *et al.* 2010) (Figure 9.2e, f).

Invasive exotic grasses are also altering ecosystem structure and function in tropical wetland ecosystems. Significant areas of Queensland and the Northern Territory have been invaded by two transforming aquatic C_4 grass species, *Hymenachne amplexicaulis* and *Urochloa mutica* (Setterfield *et al.* 2013). Both species form monocultures, displacing the diverse mosaic of native grasses and sedges (Douglas and O'Connor 2003). Detailed studies in Kakadu National Park show that areas dominated by *Urochloa mutica* have lower species richness and altered fuel characteristics compared to areas dominated by the native grasses it replaces (Douglas and O'Connor 2003). *Urochloa mutica*-dominated areas have more intense fires, with negative impacts on fire-sensitive woody vegetation, including *Melaleuca* spp. and rainforest.

Large areas of Australia's tropical wetlands have undergone ecosystem transformation from floodplain grassland to impenetrable shrubland following invasion by the woody shrub *Mimosa pigra*. This hard-seeded, thorny, leguminous shrub grows to 6 m in height in northern Australia, which is substantially taller than the 1–2 m it achieves in its native range in tropical America (Setterfield *et al.* 2013). Despite large-scale management efforts, mimosa is estimated to have invaded approximately 140 000 ha in the Northern Territory (Burrows and Lukitsch 2012), although a strong control effort has prevented significant invasion of the World Heritage-listed Kakadu National Park (Adams *et al.* 2015).

Intact tropical rainforests have a higher natural resilience to invasion by exotic plants than most ecosystems due to their closed canopies and high tree densities. To be successful in these environments, invaders need to be capable of germinating and establishing under the canopy or taking advantage of conditions created by tree-falls (Goosem 2008). Nevertheless, a number of exotic plants threaten the integrity of Australia's tropical rainforests. For example, disturbed areas on rainforest edges are ideal for herbs such as the Singapore daisy, *Sphagneticola trilobata*. More threatening than this to rainforest integrity are slower-growing persistent woody weeds such as *Miconia calvescens*, a 4–15 m tall shade-tolerant tree that produces small, many-seeded fleshy fruits (Murphy *et al.* 2008). First discovered in northern Australia in 1996, *Miconia calvescens* is considered a serious threat in tropical Australia because of the ecosystem changes that have occurred following its invasion of the tropical Pacific, where it forms extensive monospecific stands in rainforest (Meyer 1998). A number of vines, such as *Thunbergia* spp., *Mikania micrantha* and *Turbina corymbosa*, also threaten Australia's tropical rainforests by smothering vegetation and altering microclimatic conditions in the understorey (Goosem 2008).

9.2.3.2 Temperate Australia

The Australian humid temperate zone covers the southern edge of the continent: the coast between the New South Wales–Queensland border and Port Lincoln in South Australia, coastal areas between Esperance and Geraldton in Western Australia, and all of Tasmania (Bureau of Meteorology 2012). Environmental conditions across the zone vary widely, resulting in a diverse range of vegetation, including eucalypt woodlands, heath and shrublands and cool temperate rainforest. Although the humid temperate zone covers only ~10% of the continent (812 433 km^2; Bureau of Meteorology 2012), it is the region where Australia's human population is concentrated. High population density and consequent land-use change and urbanisation have contributed to the humid temperate zone being perhaps the most invaded region in Australia.

Few habitats in the humid temperate zone are untouched by weed invasions, with the plant communities most resilient to invasion, such as heathlands (Lamont and Keith, Chapter 14, this volume) and dry sclerophyll forests (Tozer *et al.*, Chapter 13, this volume), characterised by low-nutrient soils. The mosaic of abiotic conditions, such as cool-climate and high-rainfall uplands, coastal dunes and headlands, and summer-droughted Mediterranean climates, provides a variety of potential niches for the establishment and proliferation of introduced plant species. The humid temperate zone hosts diverse introduced, naturalised and invasive plant species, including herbaceous annuals, grasses and sedges, aquatics, and tree and shrub species. In the south-eastern corner of Australia alone, >2500 species are recognised as weeds of cultivation, environmental weeds, or weeds of disturbed areas (Richardson *et al.* 2011). Twenty-three of the 32 Weeds of National Significance have widespread populations in the humid temperate zone and are under active management (Australian Weeds Committee 2012). For example, *Chrysanthemoides monilifera* subsp. *rotundata* (bitou bush) which was introduced for dune stabilisation, has transformed coastal dune environments in eastern Australia by forming impenetrable thickets, and substantial resources are now spent on its control (French *et al.* 2008). Exotic perennial grasses and woody trees and shrubs have arguably had the largest impact in the humid temperate zone, displacing native grasses and shifting vegetation structure.

9.2.3.3 Arid and semi-arid Australia

The climate of Australia's arid and semi-arid zone constrains plant growth but does not preclude plant invasions. Australia's arid and semi-arid zone boasts a wide variety of climate and soil environments, and the exotic flora of arid and semi-arid Australia encompasses a wide variety of growth forms, such as annual and perennial forbs and grasses, succulents, vines, shrubs and trees, and aquatic and semi-aquatic species. No single list exists of exotic plant species in Australia's arid and semi-arid zone. The *Flora of Central Australia* (Jessop 1981), which does not cover all of Australia's arid and semi-arid zones, lists 114 exotic species (about 6% of the region's flora). More than 20% (400 species) of the plant species in western New South Wales are exotic (Cunningham *et al.* 1981).

Many deliberate plant introductions in the arid and semi-arid zone were of species perceived to have potential as pasture plants, and some – such as buffel grass, *Cenchrus ciliaris* (Grice *et al.* 2012) and the forage shrub, *Leucaena leucocephala* – are now important in the cattle industry (Walton 2009). Buffel grass is arguably the most important introduced pasture grass in arid and semi-arid northern Australia, but it is also recognised as a transformer species, particularly in central Australia, where it has naturalised and now dominates large areas. Buffel grass invasion results in loss of native herbaceous and grass species as well as a shift in vegetation structure from hummock to tussock grassland. The South Australian Government formally recognised the environmental problems associated with buffel grass by declaring it a weed in the *Natural Resources Management Act 2004*.

Among the most obvious structural changes caused by invasive species in the arid and semi-arid zone are those associated with invasion of relatively open communities by woody vines, shrubs and trees. Examples include the invasion of Mitchell grasslands (*Astrebla* spp. communities) by *Vachellia nilotica* (Mackey 1998), and invasion of tropical woodlands by *Prosopis* spp. and *Ziziphus mauritiana* (Grice 2009).

It is often difficult to disentangle cause and effect in the changes associated with invasive species in the arid and semi-arid zone. Plant introductions do not take place in isolation of other changes in the invaded systems. For example, a range of factors was involved in the spread of buffel grass in central Queensland. The native woody strata were often cleared in association with the sowing of buffel grass, releasing nutrients that were then sequestered by the grass; fertilisers were also added during the establishment phase. Grazing by cattle favoured the more grazing-tolerant buffel grass over native grass species, and the establishment of buffel grass-dominated areas produced large quantities of buffel grass seeds, facilitating the spread of the species to adjacent areas.

Fire also interacts with plant species' introductions. Plant invaders can alter fire regimes, either promoting fire or suppressing it. High-biomass grasses can contribute to more intense or more frequent fires by providing greater and more contiguous fuel loads, potentially creating a positive feedback fire-grass cycle in which the invasive grass promotes fire and fire promotes the invasive grass (Rossiter *et al.* 2003; Friedel *et al.* 2006). On the other hand, a structural shift from more grassy to more woody vegetation communities associated with invasive shrubs may lead to fewer or lower-intensity fires in another feedback loop in which the new fire regime aids the proliferation of the invasive species that suppresses grassy fuels.

Many of the plant species that invade arid and semi-arid areas are most prolific in mesic parts of the landscape. In central Australia, buffel grass colonises hillslopes that support *Triodia* spp. communities, but it does best in lower parts of the landscape, particularly in riparian zones dominated by, for example, *Eucalyptus camaldulensis*. Many invasive woody species are also most prevalent in riparian zones and run-on areas. In the Burdekin catchment in north Queensland, species such as *Cryptostegia grandiflora* (rubber vine), *Jatropha gossypifolia* (bellyache bush), *Parkinsonia aculeata* (parkinsonia) and *Ziziphus mauritiana* (Indian jujube), all occur at higher frequencies in riparian zones (Lawes *et al.* 2006).

The management of invasive plants in the arid and semi-arid zone faces major challenges. Arid and semi-arid systems are 'boom-and-bust' systems with interannual and other longer-term fluctuations in rainfall and, therefore, soil moisture. Episodic major rainfall events can trigger the dispersal of propagules of invasive species as well as recruitment events. Low human population densities and large areas in the arid and semi-arid zone make the detection of infestations a major challenge and limit the availability of resources for countering plant invasions.

9.2.3.4 Alpine zone

The Australian Alps is one of 11 Australian centres of plant diversity and one of the world's 187 biodiversity hotspots. Alpine regions are characterised by low growing-season temperatures, short growing seasons, and severe temperature extremes – conditions thought to limit the establishment and survival of exotic species maladapted to alpine conditions. High mountains are therefore often considered to be at low risk of plant invasions (Humphries *et al.* 1991), but more than 125 exotic species have been recorded as naturalised above the treeline in the Australian Alps, albeit mostly at low densities (McDougall *et al.* 2005). The observed low number of exotic species in mountains may be related more to a time-lag effect or the relatively low intensity of human activities at higher elevations than to an inherent resistance of mountain ecosystems to plant invasions (Seipel *et al.* 2012). Typically, exotic plants reaching high elevations are not specialised stress-tolerators but have broad climatic tolerances and are capable of growing in a wide elevational range. Patterns of exotic species distributions can be explained by the unidirectional expansion of exotics from anthropogenic sources at low elevations and the progressive dropping out of species with narrow elevational ranges (Alexander *et al.* 2011).

While the diversity of exotic plant species declines with increasing altitude in high-mountain regions in Australia (Johnston and Pickering 2001; McDougall *et al.* 2005), few reliable datasets document changes in the distribution of exotic species over time and hence the vulnerability or otherwise of alpine ecosystems to ongoing invasion. Several species (e.g. *Hypochaeris radicata, Taraxacum officinale, Acetosella vulgaris, Trifolium repens, Cerastium glomeratum, Agrostis capillaris* and *Poa annua*) are long-established (>50–100 yr) and common in the Australian Alps (Morgan and

Carnegie 2009), while others (e.g. *Anthoxanthum odoratum*, *Plantago major*, *Polygonum aviculare* and *Stellaria media*) are restricted to disturbed areas alongside roads and around huts (Godfree *et al.* 2004). It is likely that invasion by exotic plant species is facilitated by increases in tourism, through anthropogenic disturbance, and by climate change, and there is evidence that exotic plant species are expanding their ranges upwards (Johnston and Pickering 2001; Keith McDougall, *unpublished data* 2016). The impacts of invasions on native plant communities are mostly unquantified – but see Wearne and Morgan (2004) for the impacts of invasion by *Cytisus scoparius* in sub-alpine vegetation communities.

Roadsides and walking tracks have acted as dispersal corridors for many new arrivals in the Australian Alps (Morgan and Carnegie 2009), and other species have invaded after fire. For example, bushfires in 2003 triggered the invasion of Victoria's Bogong High Plains of grey willow (*Salix cinerea*) by creating favourable conditions for the germination of its seeds in *Sphagnum* bogs and wetlands.

Many exotic plants now in alpine ecosystems originated in ski resorts, with most deliberately introduced (McDougall *et al.* 2005). Orange hawkweed (*Pilosella aurantiacum*, syn. *Hieracium*), for example, was introduced to the Falls Creek alpine resort as a garden plant (probably in the 1980s). It spread to adjoining sub-alpine vegetation (Morgan 2000), and new populations have been detected at Mount Buller and around Mount Jagungal in Kosciuszko National Park. Orange hawkweed does not require major disturbances to establish in natural vegetation and is highly competitive, making it a significant threat to native biodiversity.

9.2.3.5 Riparian and wetland systems

Riparian zones are among the most highly invaded ecosystems globally (Richardson *et al.* 2007). According to vegetation survey data from 30 000 quadrats across Victoria, 69% of exotic species occur in riparian and wetland ecosystems (Catford and Jansson 2014). Fifteen of the 32 Weeds of National Significance invade riparian or aquatic systems and were introduced to Australia intentionally (Australian Weeds Committee 2012), although only a few exotic plant species have been introduced deliberately to riparian zones (e.g. *Hymenachne amplexicaulis* for ponded pasture; *Salix nigra* for soil stabilisation).

Human activities often concentrate around rivers, providing a ready source of propagules. Riparian systems act as landscape sinks for propagules and plant resources (water, nutrients) and have a large edge-to-core ratio, which makes them particularly vulnerable to species invasions. Humans can act as dispersal vectors by increasing exotic propagule pressure in landscapes through introductions for horticulture and agriculture (Richardson *et al.* 2007). At least half of the riparian Weeds of National Significance benefit from secondary anthropogenic dispersal, e.g. *Alternanthera philoxeroides* can be dispersed in mud attached to machinery or in turf (Parsons and Cuthbertson 1992). Most riparian invaders disperse opportunistically via water.

Riparian habitats are environmentally heterogeneous, providing opportunities for a wide range of species (Catford and Jansson 2014). The flow regime is considered the most important structuring force in riparian ecosystems and hydrological modification through water extraction and flow regulation may alter crucial environmental conditions in riparian zones. Along the regulated River Murray in south-eastern Australia, changes to wetland flooding regimes appear to have inhibited native species and facilitated exotic plants (Catford *et al.* 2011). Trait-based analyses suggest that exotic species are not specifically adapted to the modified conditions, but are likely to indirectly benefit from a decline in competition from native species poorly adapted to the modified conditions (Catford *et al.* 2014).

Of the 170 plant species in the Global Invasive Species Database (ISSG 2016) that invade riparian ecosystems, only two do so exclusively (Catford and Jansson 2014). Only 10 of the 518 exotic species in Victoria that invade riparian scrubs do so exclusively – most species have a much broader range (Catford and Jansson 2014). Thus it seems most likely that exotic species opportunistically invade riparian systems, rather than being riparian specialists. Traits associated with invasive success in riparian environments include vegetative reproduction (Lake and Leishman 2004), high propagule output (Catford and Kyle 2016), ability to disperse by water (Catford and Jansson 2014), rapid growth and resource acquisition (Catford *et al.* 2014), and an ability to cope with burial and inundation. Most exotic plant species that invade riparian environments in south-eastern Australia tend to do so between floods or in higher-elevation areas that are less frequently flooded (Stokes *et al.* 2010; Catford *et al.* 2011).

The high frequency of disturbance and ready availability and dispersal of exotic propagules in riparian zones means that management needs to be long term, sustained and catchment-wide. Instead of species-specific measures, it may be more efficient and effective to facilitate greater ecological resistance to invasion by rejuvenating native plant communities or manipulating environmental conditions by, for example, the design and implementation of environmental flows to favour native species over exotic ones (Catford *et al.* 2011). Similar approaches to building resilience of vegetation communities to exotic invasion have been adopted in non-riparian areas throughout the temperate and tropical regions.

9.3 The state of invasive pathogens in Australia

Pathogens can significantly affect plant individuals, populations and species assemblages (Loo 2009), but few data are available on the introduction of plant pathogens to Australia, or their impacts. Here we focus on two exotic pathogens – phytophthora and myrtle rust – recognised as having a substantial impact on native plants and communities.

9.3.1 Phytophthora

Phytophthora (*Phytophthora cinnamomi*) is a soil-borne pathogen that infects the roots of some woody plant taxa and is considered one of the world's most invasive species (ISSG 2016). There are a number of phytophthora species within Australia, including some considered native to the continent, but here we use the term 'phytophthora' to refer to *Phytophthora cinnamomi* only. Phytophthora was possibly introduced to Australia by European settlers or later through the horticultural industry, but it has been recognised as a major threat to native vegetation since the 1960s (Commonwealth of Australia 2014). There is a range of susceptibility among Australian plant species to phytophthora, with some species not susceptible at all, others being resistant hosts and some species being highly susceptible. Highly susceptible species infected with phytophthora become chlorotic before dying back and collapsing within 3 years (Weste and Marks 1987).

Phytophthora spreads by zoospores that transfer between plants via root-to-root contact or in free water. The movement of soil – via the nursery trade, construction activities, rehabilitation works, vehicles and bushwalkers – is an important introduction pathway (Commonwealth of Australia 2014). Phytophthora now occurs in all states and territories of Australia except the Northern Territory (Table 9.2). It likely infects over 2500 Australian plant species, with temperate heathlands, coastal woodlands and dry eucalypt forests most at risk (Hardham 2005). At the community level, phytophthora typically results in canopy dieback and decline of susceptible species, resulting in shifts in vegetation structure, for example from shrub-dominated to herb-dominated understoreys (McDougall and Summerell 2003). The 'disease caused by *Phytophthora cinnamomi*'

Table 9.2 **Impacts of phytophthora (*Phytophthora cinnamomi*) on vegetation types and species across the states of Australia.**

	Vegetation types most impacted	Species types most impacted
Western Australia	Jarrah (*Eucalyptus marginata*) forests, *Banksia* woodlands and heathlands (Hardy *et al.* 2001).	2284 (40%) of native species are susceptible (Shearer *et al.* 2004)
Victoria	Heathlands, coastal forests and woodlands. Considered at risk are Grampians lowland heath flora, the Brisbane Ranges flora, sclerophyll communities in the eastern Otways, and dry sclerophyll forests on sandy and lateritic soils (Department of Sustainability and Environment 2008).	Structural dominants, Proteaceae, Fabaceae and Ericaceae
Tasmania	Moorlands, heathlands and dry eucalypt forests	*Epacris*, *Phebalium*, *Tetratheca* and *Xanthorrhoea* (Schahinger *et al.* 2003)
South Australia	Woodlands and shrublands in southern regions where annual rainfall is > 400 mm	*Xanthorrhoea*, *Banksia*, *Isopogon*, *Acacia* and *Eucalyptus* species
New South Wales and ACT	Forests, heathlands and woodlands	Proteaceae, Fabaceae and Ericaceae families
Queensland	Notophyll-dominated vegetation, wet coastal forests, Wet Tropics	Proteaceae, Lauraceae, Myrtaceae and Ericaceae families

is recognised as a key threatening process under the Commonwealth *Environment Protection and Biodiversity Conservation Act 1999.*

9.3.1.1 Managing phytophthora

Phytophthora threat abatement plans have been developed nationally and in several states. Although it is unlikely that phytophthora can be eradicated from systems where it is established (Hardy *et al.* 2001), integrated management strategies aim to control its spread and impact (Weste and Marks 1987; Commonwealth of Australia 2014), including through prevention (education, hygiene and access controls) and chemical control (e.g. application of phosphite and fungicides). The application of phosphite (a biodegradable systemic fungicide) inhibits the growth of the pathogen and stimulates host defence (Jackson *et al.* 2000), although it may also fertilise the soil with the potential to cause significant changes in vegetation composition and invasion by exotic plant species, particularly on low-phosphorus soils (Commonwealth of Australia 2014).

9.3.2 Myrtle rust

Myrtle rust (*Puccinia psidii*), a fungus in the order Uredinales, is native to Central and South America. It was introduced accidentally to Australia and first detected in the Central Coast region of New South Wales in 2010 (Carnegie *et al.* 2010). Myrtle rust spores travel by wind or by attaching to animals, and the species therefore has high dispersal capacity. In the few years since its introduction to Australia it has spread along the east and north coast as far north as Melville Island (DPIF 2015) and south to Tasmania (Biosecurity Tasmania 2015), although at the time of writing (April 2016) it has not yet reached climatically suitable areas (Elith *et al.* 2013) in Western Australia or been detected beyond parks or gardens in Victoria or Tasmania. Myrtle rust attacks species in the Myrtaceae family (Carnegie and Lidbetter 2012), with infections occurring mainly on young growing tissues such as leaves, floral buds and fruits (Glen *et al.* 2007). The infection is seen as yellow spore pustules and can result in defoliation and death (Figure 9.3).

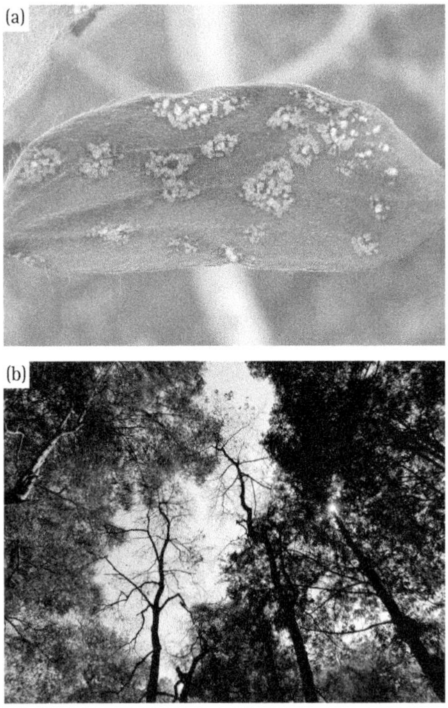

Figure 9.3 (a) Yellow spore pustules on young leaves of *Melaleuca quinquenervia*, indicating infection by myrtle rust (Photo: A Carnegie). (b) *Rhodamnia rubescens* severely affected by myrtle rust in Royal National Park, Sydney. Canopy dieback and mortality of this species results in opening of the canopy which may result in shifts in species assemblages in rainforest vegetation (Photo: L Fernandez). *(A black and white version of this figure will appear in some formats. For the colour version, please refer to the plate section.)*

Australian vegetation is likely to be highly vulnerable to myrtle rust infection because Myrtaceae species are widespread structural dominants (1646 species in 70 genera within Australia, Glen *et al.* 2007). According to inoculation tests and field surveys, 231 Australian species in 51 genera are susceptible to the pathogen (Carnegie and Lidbetter 2012; Pegg *et al.* 2014), with highly susceptible genera including *Agonis*, *Austromyrtus*, *Callistemon*, *Eucalyptus*, *Leptospermum*, *Melaleuca* and *Rhodamnia*. Field studies have also shown that several key rainforest (*Rhodamnia rubescens*, *Rhodomyrtus psidioides*) and coastal wetland (*Melaleuca quinquenervia*) species are highly susceptible (Carnegie *et al.* 2015). Although understanding of the long-term impacts of myrtle rust in Australia is limited, it is possible myrtle rust may limit post-fire regeneration due to the susceptibility of new growth. Mortality of susceptible species may also result in changes in species composition, potentially including invasion by exotic plant species into canopy gaps. More research is needed on potential impacts of myrtle rust and their mitigation. The 'introduction and establishment of Exotic Rust Fungi of the order Pucciniales pathogenic on plants of the family Myrtaceae' has been listed under the New South Wales *Threatened Species Conservation Act 1995* as a key threatening process.

9.4 Managing invasive plants in Australian vegetation

The management of invasive plant species in Australia must be integrated across local, state and federal boundaries using a number of legal and regulatory instruments. A national agenda for weed management is set by the Australian Weeds Strategy (Natural Resource Management Ministerial Council 2006). State governments are typically responsible for post-border risk assessment, declaration and management planning, and local governments and landholders generally fund and perform the majority of weed control and removal actions. In some instances, this is in response to rulings under state and territory noxious weed legislation. Enforcement of noxious weed declarations is the jurisdiction of a variety of government (local, state) and natural resource management bodies. For instance, in New South Wales, around 300 weeds have been declared 'noxious' under the Noxious Weeds Act 1993 and must be controlled on private and public lands (Australian Weeds Committee 2012). Controlling noxious and other weed species in Australia's natural environments has been estimated to cost AUD 20 million per year (Sinden *et al.* 2004).

9.4.1 Risk assessment and regulation of introductions

The feasibility of weed eradication is largely determined by the total area infested, as well as the ease of detection and control (Panetta 2015). Early detection and containment are far more cost-efficient than eradication of a widely established exotic species. Therefore, strong emphasis is placed on preventing the introduction of potential invaders through strict quarantine inspections and Weed Risk Assessment (WRA). The Biosecurity Act 2015 limits the introduction of biological material into Australia and is enforced at all designated international points of entry into Australia, and the importation of novel plant species is limited by WRA protocols. Australia's WRA scheme is recognised as one of the world's most thorough and robust mechanisms for screening potential invaders (Gordon *et al.* 2008). However, the globalisation of trade and increasing movement of people between continents makes the task of managing introductions highly challenging.

9.4.2 Management mechanisms

9.4.2.1 Physical removal

The method used to physically remove established weeds depends largely on the characteristics of the invader. Herbaceous species may be controlled in the short-term through the use of mowing or threshing, however this has the potential to spread seed locally. For woody weeds, such as trees and shrubs like African olive (*Olea europaea* subsp. *cuspidata*) and privet (*Ligustrum* spp.), mechanical removal may need to use heavy machinery, resulting in disturbance of the soil structure and erosion of topsoil, thus providing recruitment opportunities for other weeds. Machinery used to remove weed infestations can spread weeds throughout the landscape if care is not taken to prevent the movement of stem fragments and seeds between sites. In riparian ecosystems, mechanical control of weeds can fragment plant material, leading to further spread throughout water bodies, e.g. *Alternanthera philoxeroides* (Parsons and Cuthbertson 1992).

9.4.2.2 Chemical control

Herbicide application is the most common method used to control weeds in Australia, largely due to high costs of mechanical removal and biological control. Highly problematic weeds, such as *Lantana camara*, have been controlled in some regions through the use of herbicides. However, chemical control is not appropriate in all situations, in particular in riparian zones as aquatic ecosystems are highly sensitive to herbicides. Chemical control, alongside physical removal, is a common method employed by volunteers dedicated to the restoration of local environments, through groups such as Bushcare and Landcare. These groups play an important role in the control of non-native species across Australia.

9.4.2.3 Biological control

For plants, biological control involves the deliberate introduction of invertebrate pests and pathogens to reduce plant growth and reproduction leading to declines in plant density (De Bach 1964). Several successful biocontrol programmes have been implemented in Australia, in some cases leading to long-term control of serious invaders. For example, the cactus moth (*Cactoblastis cactorum*) has had long-lasting effects on invasive prickly pear (*Opuntia* spp.), which advanced at a rate of 1 million hectares per year in the early 1900s (Dodd 1959). More recently, some degree of control of rubber vine (*Cryptostegia grandiflora*) has been achieved through the introduction of a rust fungus that reduces plant vigour and seed output, at least in some parts of its invaded range (Tomley 1995), and a range of other biocontrol agents have been released on bridal creeper (*Asparagus asparagoides*). Nevertheless, biological control of many plant invaders may not be feasible. A major barrier to its use is the significant financial cost of identifying and introducing control agents through field surveys in native ranges, efficacy and host specificity testing, and quarantine procedures, as well as the potential for negative impacts on non-target species (Pemberton 2000).

9.5 Future of invasive plant and pathogen research

There is a continuing need for basic biological information on individual exotic species, their impacts on native vegetation and improved control methods. However, direct and indirect impacts of climate change are increasingly the focus of research on invasive plant species. Exotic invaders may respond more favourably than native species to elevated levels of CO_2, although effects may be species and context specific (Sorte *et al.* 2013). Modelling indicates that across species, both expansions and contractions in weed distributions are likely under changing climatic conditions (O'Donnell *et al.* 2012). Perhaps the most likely impacts of climate change on the spread of invasive plants and pathogens will be through changes in the intensity and timing of natural disturbance regimes, such as fire cycles and tropical cyclones (Leishman and Gallagher 2015), and the interactions between invasive plants and pathogens. Techniques and tools for rapidly responding to weed spread following disturbance are urgently needed, particularly where fires and cyclones are predicted to become more common. Improved understanding of the interactions between multiple stressors – such as disturbance, elevated CO_2 and changing climate – and their impacts on exotic plant invasions is also needed. Given the clear link between human activities and exotic plant success, scientists need methods for better communicating the vast body of knowledge about plant invasions and their impacts on the community in order to engage effectively with industry and the Australian population.

References

Abom, R., Vogler, W. and Schwarzkopf, L. (2015) Mechanisms of the impact of a weed (grader grass, *Themeda quadrivalvis*) on reptile assemblage structure in a tropical savannah. *Biological Conservation* 191, 75–82.

Adair, R.J. and Groves, R.H. (1998) *Impact of Environmental Weeds on Biodiversity: A Review and Development of a Methodology*. Commonwealth of Australia: Canberra.

Adams, V.M., Setterfield, S.A., Douglas, M.M. *et al.* (2015) Measuring benefits of protected area management: trends across realms and research gaps for freshwater systems. *Philosophical Transactions of the Royal Society B: Biological Sciences* 370, 20140274.

Alexander, J.M., Kueffer, C., Daehler, C.C. *et al.* (2011) Assembly of nonnative floras along elevational gradients explained by directional ecological filtering. *Proceedings of the National Academy of Sciences* 108, 656–661.

Australian Government (2016) National Landcare Programme. Available at: www.nrm.gov.au (accessed 20 December 2016).

Australian Weeds Committee (2012) *Weeds of National Significance 2012*. Department of Agriculture, Fisheries and Forestry: Canberra.

Biosecurity Tasmania (2015) *Myrtle rust has been detected in Tasmania*. Department of Primary Industries, Parks, Water and Environment: Tasmania. Available at: http://dpipwe.tas.gov.au/biosecurity/plant-biosecurity/pests-and-diseases/myrtle-rust (accessed 25 February 2015).

Brooks, K.J., Setterfield, S.A. and Douglas, M.M. (2010) Exotic grass invasions: applying a conceptual framework to the dynamics of degradation and restoration in Australia's tropical savannas. *Restoration Ecology* 18, 188–197.

Bureau of Meteorology (2012) *Climate Classification of Australia*. Bureau of Meteorology: Canberra.

Burrows, N. and Lukitsch, B. (2012) Biological control agents are observed on *Mimosa pigra* six and 12 years after their release in the Northern Territory, Australia. In *Developing Solutions to Evolving Weed Problems. Proceedings of the 18th Australasian Weeds Conference*. Weed Society of Victoria: Melbourne, pp. 347–348.

Carnegie, A.J. and Lidbetter, J.R. (2012) Rapidly expanding host range for *Puccinia psidii* sensu lato in Australia. *Australasian Plant Pathology* 41, 13–29.

Carnegie, A.J., Lidbetter, J.R., Walker, J. *et al.* (2010) *Uredo rangelii*, a taxon in the guava rust complex, newly recorded on Myrtaceae in Australia. *Australasian Plant Pathology* 39, 463–466.

Carnegie, A.J., Kathuria, A., Pegg, G.S. *et al.* (2015) Impact of the invasive rust *Puccinia psidii* (myrtle rust) on native Myrtaceae in natural ecosystems in Australia. *Biological Invasions* 18, 127–144.

Catford, J.A. and Jansson, R. (2014) Drowned, buried and carried away: effects of plant traits on the distribution of native and alien species in riparian ecosystems. *New Phytologist* 204, 19–36.

Catford, J.A. and Kyle, G. (2016) Alien plant invasions in Australia's riparian ecosystems. In *Vegetation of Australia's Riverine Landscapes: Biology, Ecology & Management*, ed. S.J. Capon, C. James and M. Reid. CSIRO Publishing: Melbourne, pp. 325–342.

Catford, J.A., Downes, B.J., Gippel, C.J. and Vesk, P.A. (2011) Flow regulation reduces native plant cover and facilitates exotic invasion in riparian wetlands. *Journal of Applied Ecology* 48, 432–442.

Catford, J.A., Morris, M.K., Vesk, P.A. *et al.* (2014) Species and environmental characteristics point to flow regulation and drought as drivers of riparian plant invasion. *Diversity and Distributions* 20, 1084–1096.

Champion, P.D., Clayton, J.S. and Hofstra, D.E. (2010) Nipping aquatic plant invasions in the bud: weed risk assessment and the trade. *Hydrobiologia* 656, 167–172.

Chapman, A.D. (2009) *Numbers of living species in Australia and the world*. 2nd edition. Department of the Environment, Water, Heritage and the Arts: Canberra. Available at: https://www.environment.gov.au/system/files/pages/2ee3f4a1-f130-465b-9c7a-79373680a067/files/nlsaw-2nd-complete.pdf (accessed 28 July 2016).

Close, D.C., Davidson, N.J. and Watson, T. (2008) Health of remnant woodlands in fragments under distinct grazing regimes. *Biological Conservation* 141, 2395–2402.

Commonwealth of Australia (2014) *Threat abatement plan for disease in natural ecosystems caused by* Phytophthora cinnamomi. Department of the Environment: Canberra. Available at: https://www.environment.gov.au/biodiversity/threatened/publications/threat-abatement-plan-disease-natural-ecosystems-caused-phytophthora-cinnamomi (accessed 28 July 2016).

Cook, G.D. and Dias, L. (2006) It was no accident: deliberate plant introductions by Australian Government agencies during the 20th century. *Australian Journal of Botany* 54, 601–625.

Coulson, C., Spooner, P.G., Lunt, I.D. and Watson, S.J. (2014) From the matrix to roadsides and beyond: the role of isolated paddock trees as dispersal points for invasion. *Diversity and Distributions* 20, 137–148.

Cuneo, P. and Leishman, M.R. (2006) African Olive (*Olea europaea* subsp. cuspidata) as an environmental weed in eastern Australia: a review. *Cunninghamia* 9, 545–577.

Cuneo, P. and Leishman, M.R. (2013) Ecological impacts of invasive African olive (*Olea europaea* ssp. cuspidata) in Cumberland Plain Woodland, Sydney, Australia. *Austral Ecology* 38, 103–110.

Cunningham, G.M., Mulham, W.E., Milthorpe, P.L. and Leigh, J.H. (1981) *Plants of Western New South Wales*. NSW Government Printer: Sydney.

Davis, M.A., Grime, J.P. and Thompson, K. (2000) Fluctuating resources in plant communities: a general theory of invasibility. *Journal of Ecology* 88, 528–534.

De Bach, P. (1964) *Biological Control of Insect Pests and Weeds*. Chapman and Hall: London.

Department of Primary Industry and Fisheries (DPIF) (2015) Myrtle rust detected on the Tiwi Islands. Available at: http://www.nt.gov. au/d/index.cfm?Header=Myrtle%20rust (accessed 22 December 2015).

Department of Sustainability and Environment (2008) *Victoria's public land* Phytophthora cinnamomi *management strategy*. Department of Sustainability and Environment: Melbourne.

Diez, J.M., Williams, P.A., Randall, R.P. *et al.* (2009) Learning from failures: testing broad taxonomic hypotheses about plant naturalization. *Ecology Letters* 12, 1174–1183.

Dodd, A.P. (1959) The biological control of prickly pear in Australia. In *Biogeography and Ecology in Australia*. Springer: Netherlands, pp. 565–577.

Dodd, A.J., Burgman, M.A., McCarthy, M.A. and Ainsworth, N. (2015) The changing patterns of plant naturalization in Australia. *Diversity and Distributions* 21, 1038–1050.

Douglas, M.M. and O'Connor, R.A. (2003) Effects of the exotic macrophyte, para grass (*Urochloa mutica*), on benthic and epiphytic macroinvertebrates of a tropical floodplain. *Freshwater Biology* 48, 962–971.

Douglas, M.M., Pusey, B.J., Setterfield, S.A. *et al.* (2011) Northern futures: threats and opportunities for freshwater ecosystems. In *Aquatic Biodiversity in Northern Australia: Patterns, Threats and Future*, ed. B.J. Pusey. Charles Darwin University Press: Darwin, pp. 203–220.

Elith, J., Simpson, J., Hirsch, M. and Burgman, M.A. (2013) Taxonomic uncertainty and decision making for biosecurity: spatial models for myrtle/guava rust. *Australasian Plant Pathology* 42, 43–51.

Foxcroft, L., Richardson, D., Rejmánek, M. and Pyšek, P. (2010) Alien plant invasions in tropical and sub-tropical savannas: patterns, processes and prospects. *Biological Invasions* 12, 3913–3933.

French, K., Ens, E., Gosper, C.R. *et al.* (2008) Management implications of recent research into the effect of bitou bush invasion. *Plant Protection Quarterly* 23, 24.

Friedel, M., Puckey, H., O'Malley, C. *et al.* (2006) *Buffel Grass: Both Friend and Foe. An Evaluation of the Advantages and Disadvantages of Buffel Grass Use, and Recommendations for Future Research*. Desert Knowledge Co-operative Research Centre: Alice Springs.

Gallagher, R.V. and Leishman, M.R. (2014) Invasive plants and invaded ecosystems in Australia: implications for biodiversity. In *Austral Ark: The State of Wildlife in Australia and New Zealand*, ed. A. Stow, N. Maclean and G. Holwell. Cambridge University Press: Cambridge, pp. 105–133.

Gallagher, R.V., Hughes, L., Leishman, M.R. and Wilson, P.D. (2010) Predicted impact of exotic vines on an endangered ecological community under future climate change. *Biological invasions* 12, 4049–4063.

Gallagher, R.V., Randall, R.P. and Leishman, M.R. (2015) Trait differences between naturalized and invasive plant species independent of residence time and phylogeny. *Conservation Biology* 29, 360–369.

Gilbert, B., Levine, J.M. (2013) Plant invasions and extinction debts. *Proceedings of the National Academy of Sciences* 110, 1744–1749.

Glen, M., Alfenas, A.C., Zauza, E.A.V. *et al.* (2007) Puccinia psidii: a threat to the Australian environment and economy: a review. *Australasian Plant Pathology* 36, 1–16.

Godfree, R., Lepschi, B. and Mallinson, D. (2004) Ecological filtering of exotic plants in an Australian sub-alpine environment. *Journal of Vegetation Science* 15, 227–236.

Goosem, S. (2008) Invasive weeds in the Wet Tropics. In *Living in a Dynamic Tropical Forest Landscape*, ed. N.E. Stork and S.M. Turton. Blackwell Publishing: Oxford, pp. 307–321.

Gordon, D.R., Onderdonk, D.A., Fox, A.M. and Stocker, R.K. (2008) Consistent accuracy of the Australian weed risk assessment system across varied geographies. *Diversity and Distributions* 14, 234–242.

Grice, A.C. (2009) *Ziziphus mauritiana* Lam. In *The Biology of Australian Weeds.* Volume 3, ed. F.D. Panetta. R.G. and F.J. Richardson: Melbourne, pp. 294–310.

Grice, A.C., Friedel, M.H., Marshall, N.A. and van Klinken, R.D. (2012) Tackling contentious invasive plant species: a case study of buffel grass in Australia. *Environmental Management* 49, 285–294.

Grice, A.C., Vanderduys, E.P., Perry, J.J. and Cook, G.D. (2013) Patterns and processes of invasive grass impacts on wildlife in Australia. *Wildlife Society Bulletin* 37, 478–485.

Groves, R.H., Lonsdale, M. and Boden, R. (2005) *Jumping the Garden Fence: Invasive Garden Plants in Australia and their Environmental and Agricultural Impacts*. WWF-Australia: Sydney

Hardesty, B.D., Metcalfe, S.S. and Westcott, D.A. (2011) Persistence and spread in a new landscape: dispersal ecology and genetics of *Miconia* invasions in Australia. *Acta Oecologica* 37, 657–665.

Hardham, A.R. (2005) Phytophthora cinnamomi. *Molecular Plant Pathology* 6, 589–604.

Hardy, G.S.J., Barrett, S. and Shearer, B.L. (2001) The future of phosphite as a fungicide to control the soilborne plant pathogen *Phytophthora cinnamomi* in natural ecosystems. *Australasian Plant Pathology* 30, 133–139.

Harris, C.J., Murray, B.R., Hose, G.C. and Hamilton, M.A. (2007) Introduction history and invasion success in exotic vines introduced to Australia. *Diversity and Distributions* 13, 467–475.

Holland-Clift, S., O'Dowd, D.J. and Mac Nally, R. (2011) Impacts of an invasive willow (*Salix × rubens*) on riparian bird assemblages in south-eastern Australia. *Austral Ecology* 36, 511–520.

Humphries, S., Groves, R. and Mitchell, D. (1991) Plant invasions of Australian ecosystems. In *Kowari: Kowari 2. Plant Invasions. The Incidence of Environmental Weeds in Australia*. National Parks and Wildlife Service: Canberra, pp. 1–127.

ISSG (2016) Global Invasive Species Database. Available at: www.issg.org/database

Jackson, J. (2005) Is there a relationship between herbaceous species richness and buffel grass (*Cenchrus ciliaris*)? *Austral Ecology* 30, 505–517.

Jackson, T.J., Burgess, T., Colquhoun, I. and Hardy, G.S. (2000) Action of the fungicide phosphite on *Eucalyptus marginata* inoculated with *Phytophthora cinnamomi*. *Plant Pathology* 49, 147–154.

Jessop, J. (1981) *Flora of Central Australia*. Reed Books: Sydney.

Johnston, F.M. and Pickering, C.M.C. (2001) Alien plants in the Australian Alps. *Mountain Research and Development* 21, 284–291.

Kutt, A.S. and Kemp, J.E. (2012) Native plant diversity in tropical savannas decreases when exotic pasture grass cover increases. *The Rangeland Journal* 34, 183–189.

Lai, H.R., Mayfield, M.M., Gay-des-combes, J.M. *et al.* (2015) Distinct invasion strategies operating within a natural annual plant system. *Ecology Letters* 18, 336–346.

Lake, J.C. and Leishman, M.R. (2004) Invasion success of exotic plants in natural ecosystems: the role of disturbance, plant attributes and freedom from herbivores. *Biological Conservation* 117, 215–226.

Lawes, R.A., Murphy, H.T. and Grice, A.C. (2006) Comparing agglomerative clustering and three weed classification frameworks to assess the invasiveness of alien species across spatial scales. *Diversity and Distributions* 12, 633–644.

Leishman, M.R. and Gallagher, R.V. (2015) Will there be a shift to alien-dominated vegetation assemblages under climate change? *Diversity and Distributions* 21, 848–852.

Leishman, M.R. and Thomson, V.P. (2005) Experimental evidence for the effects of additional water, nutrients and physical disturbance on invasive plants in low fertility Hawkesbury Sandstone soils, Sydney, Australia. *Journal of Ecology* 93, 38–49.

Leishman, M.R., Haslehurst, T., Ares, A. and Baruch, Z. (2007) Leaf trait relationships of native and invasive plants: community and global-scale comparisons. *New Phytologist* 176, 635–643.

Leishman, M.R., Thomson, V.P. and Cooke, J. (2010) Native and exotic invasive plants have fundamentally similar carbon capture strategies. *Journal of Ecology* 98, 28–42.

Lenz, T.I., Moyle-Croft, J.L. and Facelli, J.M. (2003) Direct and indirect effects of exotic annual grasses on species composition of a South Australian grassland. *Austral Ecology* 28, 23–32.

Lonsdale, W.M. (1994) Inviting trouble: introduced pasture species in northern Australia. *Australian Journal of Ecology* 19, 345–354.

Loo, J.A. (2009) Ecological impacts of non-indigenous invasive fungi as forest pathogens. *Biological Invasions* 11, 81–96.

Low, T. (1999) *Feral Future: The Untold Story of Australia's Exotic Invaders*. University of Chicago Press: Chicago.

Mackey A.P. (1998) *Acacia nilotica* ssp. indica (Benth.) Brenan. *The Biology of Australian Weeds* 2, 1–18.

Mason, R.A., Cooke, J., Moles, A.T. and Leishman, M.R. (2008) Reproductive output of native and exotic plants: a global comparison. *Global Ecology and Biogeography* 17, 633–640.

McDougall, K.L. and Summerell, B.A. (2003) The impact of *Phytophthora cinnamomi* on the flora and vegetation of New South Wales: a re-appraisal. In *Phytophthora in Forests and Natural Ecosystems 2nd International IUFRO Working Party Meeting, Albany, Western Australia*, ed. J.A. McComb, G.E. St Hardy and I. Tommerup. Murdoch University: Perth.

McDougall, K.L., Morgan, J.W., Walsh, N.G. and Williams, R.J. (2005) Plant invasions in treeless vegetation of the Australian Alps. *Perspectives in Plant Ecology, Evolution and Systematics* 7, 159–171.

Meyer, J.Y. (1998) Observations on the reproductive biology of *Miconia calvescens* DC (Melastomataceae), an alien invasive tree on the island of Tahiti (South Pacific Ocean) 1. *Biotropica* 30, 609–624.
Morgan, J.W.J. (2000) Orange Hawkweed *Hieracium aurantiacum* L.: a new naturalised species in alpine Australia. The Victorian *Naturalist* 117, 50–51.
Morgan, J.W.J. and Carnegie, V. (2009) Backcountry huts as introduction points for invasion by non-native species into subalpine vegetation. *Arctic, Antarctic and Alpine Research* 41, 238–245.
Mulvaney, M. (2001) The effect of introduction pressure on the naturalization of ornamental woody plants in south-eastern Australia. In *Weed Risk Assessment*, ed. R.H. Groves, F.D. Panetta and J.G. Virtue. CSIRO Publishing: Canberra, pp. 186–193.
Murphy, H.T., Hardesty, B.D., Fletcher, C.S. *et al.* (2008) Predicting dispersal and recruitment of *Miconia calvescens* (Melastomataceae) in Australian tropical rainforests. *Biological Invasions* 10, 925–936.
Natural Resource Management Ministerial Council (2006) *Australian Weeds Strategy: A national strategy for weed management in Australia*. Australian Government Department of the Environment and Water Resources: Canberra ACT.
O'Donnell, J., Gallagher, R.V., Wilson, P.D. *et al.* (2012) Invasion hotspots for non-native plants in Australia under current and future climates. *Global Change Biology* 18, 617–629.
Panetta, F.D. (2015) Weed eradication feasibility: lessons of the 21st century. *Weed Research* 55, 226–238.
Parsons, W.T. and Cuthbertson, E.G. (1992) *Noxious Weeds of Australia*. Inkata Press: Melbourne.
Pegg, G.S., Brawner, J.T. and Lee, D.J. (2014) Screening *Corymbia* populations for resistance to *Puccinia psidii*. *Plant Pathology* 63, 425–436.
Pemberton, R.W. (2000) Predictable risk to native plants in weed biological control. *Oecologia* 125, 489–494.
Pemberton, R.W. and Liu, H. (2009) Marketing time predicts naturalization of horticultural plants. *Ecology* 90, 69–80.
Perna, C. and Burrows, D. (2005) Improved dissolved oxygen status following removal of exotic weed mats in important fish habitat lagoons of the tropical Burdekin River floodplain, Australia. *Marine Pollution Bulletin* 51, 138–148.
Petroeschevsky, A. and Champion, P.D. (2008) Preventing further introduction and spread of aquatic weeds through the ornamental plant trade. In *Sixteenth Australian Weed Conference, Cairns*. Queensland Weed Society: Queensland, pp. 200–302.
Phillips, M.L., Murray, B.M., Leishman, M.R. and Ingram, R. (2010) The naturalization to invasion transition: are there introduction-history correlates of invasiveness in exotic plants of Australia? *Austral Ecology* 35, 695–703.
Prober, S.M., Thiele, K.R., Lunt, I.D. and Koen, T.B. (2005) Restoring ecological function in temperate grassy woodlands: manipulating soil nutrients, exotic annuals and native perennial grasses through carbon supplements and spring burns. *Journal of Applied Ecology* 42, 1073–1085.
Randall, R.P. (2007) *The Introduced Flora of Australia and its Weed Status*. Cooperative Research Centre for Australian Weed Management: Adelaide.
Richardson, D.M., Pyšek, P., Rejmánek, M. *et al.* (2000) Naturalization and invasion of alien plants: concepts and definitions. *Diversity and Distributions* 6, 93–107.

Richardson, D.M., Holmes, P.M., Esler, K.J. *et al.* (2007) Riparian vegetation: degradation, alien plant invasions, and restoration prospects. *Diversity and Distributions* 13, 126–139.
Richardson, F.J., Richardson, R.G. and Shepherd, R.C.H. (2011) *Weeds of the South-east: An Identification Guide for Australia.* CSIRO Publishing: Canberra.
Rossi, R.D., Martins, C.R., Viana, P.L. *et al.* (2014) Impact of invasion by molasses grass (*Melinis minutiflora* P. Beauv.) on native species and on fires in areas of campo-cerrado in Brazil. *Acta Botanica Brasilica* 28, 631–637.
Rossiter, N.A., Setterfield, S.A., Douglas, M.M. and Hutley, L.B. (2003) Testing the grass-fire cycle: alien grass invasion in the tropical savannas of northern Australia. *Diversity and Distributions* 9, 169–176.
Schahinger, R., Rudman, T. and Wardlaw, T.J. (2003) *Conservation of Tasmanian plant species and communities threatened by* Phytophthora cinnamomi. Strategic Regional Plan for Tasmania. Technical Report 03/03. Nature Conservation Branch, Department of Primary Industries, Water and Environment: Hobart.
Seipel, T., Kueffer, C., Rew, L.J. *et al.* (2012) Processes at multiple scales affect richness and similarity of non-native plant species in mountains around the world. *Global Ecology and Biogeography* 21, 236–246.
Setterfield, S.A., Rossiter-Rachor, N.A., Hutley, L.B. *et al.* (2010) Turning up the heat: the impacts of *Andropogon gayanus* (gamba grass) invasion on fire behaviour in northern Australian savannas. *Diversity and Distributions* 16, 854–861.
Setterfield, S.A., Douglas, M.M., Petty, A.M. *et al.* (2013) Weeds of protected areas: floodplain weeds in Australia's Kakadu National Park. In *Plant Invasions in Protected Areas: Patterns, Problems and Challenges*, ed. L.C. Foxcroft, D.M. Richardson, P. Pyšek and P. Genovesi. Springer: Dordrecht, pp. 167–189.
Shearer, B.L., Crane, C.E. and Cochrane, A. (2004) Quantification of the susceptibility of the native flora of the South-West Botanical Province, Western Australia, to *Phytophthora cinnamomi. Australian Journal of Botany* 52, 435–443.
Sinden, J., Jones, R., Hester, S. *et al.* (2004) *The Economic Impact of Weeds in Australia.* CRC for Australian Weed Management: Adelaide.
Sorte, C.J., Ibanez, I., Blumenthal, D.M. *et al.* (2013) Poised to prosper? A cross-system comparison of climate change effects on native and non-native species performance. *Ecology Letters* 16, 261–270.
Stokes, K.E. and Cunningham, S.A. (2006) Predictors of recruitment for willows invading riparian environments in south-east Australia: implications for weed management. *Journal of Applied Ecology* 43, 909–921.
Stokes, K., Ward, K. and Colloff, M. (2010) Alterations in flood frequency increase exotic and native species richness of understorey vegetation in a temperate floodplain eucalypt forest. *Plant Ecology* 211, 219–233.
Thomson, V.P. and Leishman, M.R. (2004) Survival of native plants of Hawkesbury Sandstone communities with additional nutrients: effect of plant age and habitat. *Australian Journal of Botany* 52, 141–147.
Tomley, A.J. (1995) The biology of Australian weeds. 26. *Cryptostegia grandiflora* R.Br. *Plant Protection Quarterly* 10, 122–130.
Turner, P.J., Scott, J.K. and Spafford, H. (2008) The ecological barriers to the recovery of bridal creeper (*Asparagus asparagoides* (L.) Druce) infested sites: impacts on vegetation and the potential increase in other exotic species. *Austral Ecology* 33, 713–722.

Vilà, M., Espinar, J.L., Hejda, M. *et al.* (2011) Ecological impacts of invasive alien plants: a meta-analysis of their effects on species, communities and ecosystems. *Ecology Letters* 14, 702–708.

Walton, C.S. (2009) *Leucaena leucocephala* (Lam.) de Wit. In *The Biology of Australian Weeds,* Volume 3, ed. F.D. Panetta. R.G and F.J. Richardson: Melbourne, pp. 143–156.

Wearne, L.J. and Morgan, J.W. (2004) Community-level changes in Australian subalpine vegetation following invasion by the non-native shrub *Cytisus scoparius*. *Journal of Vegetation Science* 15, 595–604.

Weerheim, M.S., Klomp, N.I., Brunsting, A.M. and Komdeur, J. (2003) Population size, breeding habitat and nest site distribution of little penguins (*Eudyptula minor*) on Montague Island, New South Wales. *Wildlife Research* 30, 151–157.

Weste, G. and Marks, G.C. (1987) The biology of *Phytophthora cinnamomi* in Australasian forests. *Annual Review of Phytopathology* 25, 207–229.

White, E. and Vivian-Smith, G. (2011) Contagious dispersal of seeds of synchronously fruiting species beneath invasive and native fleshy-fruited trees. *Austral Ecology* 36, 195–202.

Woinarski, J., Mackey, B., Nix, H. and Traill, B. (2007) *The Nature of Northern Australia: Natural Values, Ecological Processes and Future Prospects*. Australian National University Press: Canberra.

10 Plant Life Cycles Above- and Below-ground

TONY D. AULD AND MARK K.J. OOI

An understanding of plant life histories and the factors that promote the successful ongoing completion of the life stages of species, delivers insights into both the dynamic nature of Australian vegetation and the processes that maintain plant communities and drive change. We discuss the main elements of plant life history that drive the transition between life stages in seed plants (reproduction, dispersal, seed banks, germination, establishment, growth and survival). We illustrate the variety and pattern in life histories for plant species and the vegetation communities they form, along with predictive plant traits that are associated with these patterns. We identify key elements of the capacity for species to either cope with the inherent variability and changes that occur (robustness and resistance) or to recover from adverse impacts (resilience and rehabilitation). We then present some key examples of ecosystems that illustrate the role that understanding of plant life history processes can play in managing vegetation. In particular, we highlight two issues critical in species persistence: the importance of the dispersal capacity of seeds in the ability of species and vegetation communities to track changes in environmental niches over time; and the role of the soil seed bank as a critical buffer to offset adverse above ground impacts.

10.1 Introduction

The pace and direction of vegetation change is a consequence of turnover in the populations of component plant species. The time scales over which such changes occur may vary from years to centuries, and even millennia, with the time scale of observation determining whether plant communities are perceived as stable, or undergoing cyclical or directional change (Wiens 1989). Thus, at any given location, some species may colonise and/or increase in numbers, while others may decline and even become locally extinct. Colonisation depends on the ability of species to disperse to, and establish at, new sites, while population persistence depends on the ability of plant species to successfully complete their life cycles *in situ* (Figure 10.1). These processes depend on the interplay between plant traits, cohabiting organisms and the prevailing environment (Keddy 1992; Keith and Tozer, Chapter 1, this volume).

Plants display a remarkable diversity of life histories, as well as morphological and physiological traits, that influence rates of survival, growth, reproduction and dispersal (Lavorel *et al.* 2007). In essence, these are variations on a common blueprint that comprises the same basic life stages and transitions between them (Figure 10.1). For instance, most terrestrial plants are characterised largely by the sessile adult stage that is rooted in the ground, with the successful persistence of plants measured by the

Australian Vegetation, Third Edition, ed. David A. Keith. Published by Cambridge University Press. © Cambridge University Press 2017.

Citation: Auld, T.D. and Ooi, M.K.J. (2017). Plant Life Cycles Above- and Below-ground. In D. A. Keith, ed., *Australian Vegetation*, 3rd edn. Cambridge: Cambridge University Press, pp. 230–53.

number and health of these standing individuals (adults in Figure 10.1). However, this is only part of the story. Plants have mobile stages (e.g. pollen and seeds) and many possess storage organs (root-stocks, rhizomes, spores, seeds, tubers, bulbs) that allow persistence, even when there may be few or no plants observable above ground (Pate and Dixon 1981; Bell *et al.* 1993; Bell 1999) (e.g. seed bank in Figure 10.1). As well as the variety of physical forms, the transitions between life stages may be continuous or episodic and triggered by a range of different environmental cues. Seeds and storage organs play a key role in the long-term robustness, resistance and resilience of plant populations (*sensu* IPCC 2001), particularly by mediating escape from stress and disturbance. They also influence persistence under environmental change (Ooi *et al.*, Chapter 8, this volume) as they provide the capacity for long-term survival and avoidance of adverse conditions. Hence, a key challenge for the management of vegetation, is to ensure that strategies developed to maintain persistence of above-ground plants are complemented by actions that also protect the seeds and storage organs, while maintaining the environmental conditions in which these can grow and transition to other life stages.

The variety of plant life histories represented across Australian rainforests, eucalypt woodlands, grasslands, deserts, wetlands and alpine ecosystems represent the alternative evolutionary solutions to biological and physical forces of selection that characterise those environments. The future vegetation of Australia depends on the ability of species to either cope with the inherent environmental variability and directional changes (robustness and resistance) or to recover from adverse impacts (resilience and rehabilitation) (IPCC 2001; Lake 2012; Figures 10.1, 10.3, 10.4). In this chapter, we explore the processes that interact with plant life histories in Australian ecosystems, and highlight plant functional traits that characterise some of the broader patterns of response across species. We focus on components of the life cycle that are important for both persistence *in situ* under current conditions and those that will be critical in determining the likely responses of Australian vegetation to environmental change.

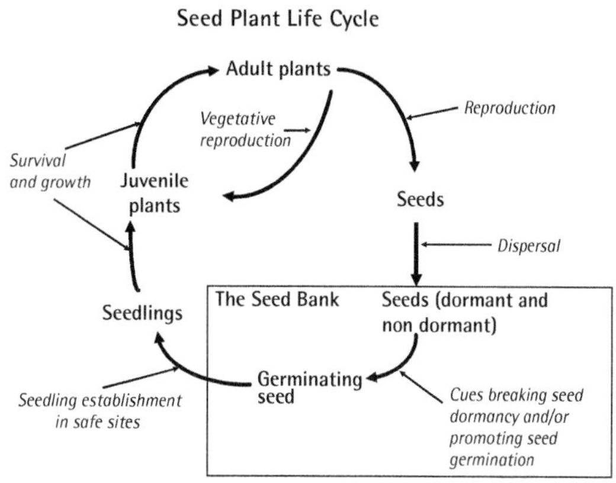

Figure 10.1 Simplified life cycle of a seed plant (gymnosperms and angiosperms). Life-cycle stages in bold. Key factors controlling transitions across life-cycle stages are shown in italics. Rectangular box represents the seed bank.

10.2 Plant life histories

Plants display a variety of life histories. Bryophytes (liverworts, mosses, hornworts) are non-vascular and have haploid gametophytes as the dominant life-cycle stage. Seedless vascular pteridophytes (ferns, club mosses and allies) have a separate growing haploid stage, but the diploid (sporophyte) stage is dominant. Vascular seed plants (gymnosperms and angiosperms) are dominated by the diploid stage with greatly reduced gametophytes enclosed in the sporophyte (leading to the production of seeds). Seed plants have a number of key elements in their life histories that are essential for survival and reproduction (Figure 10.1). Here, focussing on Australian examples, we present a brief overview of the key plant life history processes based on Figure 10.1, including: (i) key elements; (ii) their interaction or interrelatedness; (iii) how they vary among species; (iv) how plant traits might help explain this variation; and (v) their importance to plant population persistence. We also highlight those life-history stages or ecological processes most likely to be affected or important under a changing climate.

10.2.1 Reproduction: flowering, pollination and seed production

Reproduction is needed to produce new offspring in order to replace those plants that die (Harper 1977). Plants may reproduce sexually or asexually (Figure 10.1), with some species combining both of these strategies. Sexual reproduction results in the production of seeds that are a recombination of parental genotypes, are independent of the parent and have some capacity for dispersal. Overall seed production in a species will be determined by flower production, pollination, seed abortion and seed predation. Asexual reproduction does not involve the recombination of genotypes and may vary from independent units including fern spores, bulbils and turions (Brock 1982; Casanova and Brock 1999), again with a capacity for dispersal, to stems produced by stolons, rhizomes or roots that may be dependent upon the parent for a period of time and have limited capacity to move from the parent (Pate and Dixon 1981). Asexual reproduction bypasses the germination and seedling recruitment stages (see Figure 10.1) and may be particularly common in wetlands and among orchids, grasses and sedges.

10.2.1.1 Flowering

In Australia, flowering and seed set are most often seasonally cued via changing temperatures (Entwisle 2014). The reproductive response to different seasonal cues is perhaps best demonstrated using the extensive research conducted on *Acacia*. Many Australian *Acacia* species flower during mid-winter in August, with subsequent seed development and release in October to December (Entwisle 2014). Benson and McDougall (1996) found that this was the dominant pattern for species in the Sydney Basin, however some species flowered in summer (*Acacia elata*, *A. longissima*, *A. parramattensis*), or late summer–autumn (*A. linifolia* and *A. terminalis*), highlighting between-species phenological differences. Variation in flowering and fruiting phenology can also depend on environmental conditions and location. For example, in *Acacia suaveolens*, a shrubby wattle found in coastal eastern Australia from southern Queensland to Tasmania, fruit maturation varies across its distribution from early June in the northern limits of its range to late January in its southern and south-western limits (Morrison 1987). Similar variation also occurs among *Acacia* species in other environments. Friedel *et al.* (1994) found that while fruiting commonly occurred in October in arid central Australia, flowering was more variable and was influenced by soil moisture, temperature and location. Similarly, Davies (1976) and Davies and Kenny (2013) found variations in the timing of seed production in *Acacia* spp. and other arid trees and shrubs, with magnitude related to seasonal rainfall patterns.

The magnitude of seed production also varies over time and space, depending on both biotic and abiotic factors. In general, seed production increases to a maximum and then declines with age of the maternal plant (e.g. Vaughton 1998). Other environmental factors such as rainfall, temperature and disturbances can also cue flowering and affect the magnitude of seed set. Rainfall impacts can occur either independently or in combination with season, e.g. patterns of flowering in the wet tropics (Boulter *et al.* 2006) and masting events in arid *Triodia* species (Wright *et al.* 2014). Weather conditions can influence effective pollination or the magnitude of flowering and therefore the subsequent capacity to produce seeds (Davies 1976; Friedel *et al.* 1993; Williams *et al.* 1999; Davies and Kenny 2013). For example, Zimmer *et al.* (2015) found cone production in *Wollemia nobilis* was reduced under higher spring minimum temperatures (during the pollination period), potentially indicating a negative consequence of heat stress on reproductive processes. Conversely, cone production increased with higher summer rainfall suggesting an effect of resource availability on reproductive output (Zimmer *et al.* 2015). Disturbance-related reproduction cues are well represented in the

Australian flora, with key examples including post-fire flowering (Lamont and Downes 2011), flowering response of annuals to season of flooding (Greet *et al.* 2013) and the positive relationship between reproductive output and canopy gap size (Morgan 1997). Globally, there is a trade-off between seed mass and seed production per individual, with output decreasing as seed mass increases. Seed size is also a strong determinant of both seedling vigour, particularly at the early growth stage (Westoby *et al.* 2002), and the ability to emerge from depth in the soil (Bond *et al.* 1999).

10.2.1.2 Pollination

Wind and animal pollination are common in Australian vegetation, as occurs elsewhere in the world (Regal 1982). A number of important genera that structure Australian plant communities are wind pollinated, including *Casuarina* and *Callitris* (Regal 1982). Animals can have major impacts at several reproductive stages and many plant species are dependent on animals for pollination. While the majority of Australian plants have generalist pollinator relationships, a number have specialised floral structures tied closely to a small number of vertebrate or invertebrate pollinators (Armstrong 1979; Ford *et al.* 1979). For example, a number of species in the Proteaceae (e.g. *Banksia* and *Grevillea*) have large showy nectar-producing inflorescences, and are pollinated by avian honeyeaters (Meliphagideae) and marsupials like the sugar glider (*Petaurus breviceps*) (Ayre and Whelan 1989). Shifts in pollinator assemblages can therefore potentially affect plant fitness. Whelan *et al.* (2009) found that *Grevillea macleayana* populations with a higher frequency of honeyeater visitation displayed greater levels of outcrossing than populations pollinated predominantly by introduced honeybees (*Apis mellifera*). Insect pollinators exhibit relationships of varying specificity with plants (e.g. for terrestrial orchids, see Adams and Lawson 1993). Major animal pollination syndromes vary from food rewards (petals, nectar and pollen) to mimicry (sexual and food). Where there is a strong host-specific relationship, such as in sexual mimicry, persistence of the plant species is dependent upon persistence of the specific pollinator. The strong interaction between plant and insect may imply an interactive relationship, possibly co-evolutionary (*sensu* Janzen 1980), and becomes an important consideration when addressing plant persistence and movement under a changing climate for insects that are important pollinators (Dixon 2009) or those with other ecological roles (Moir *et al.* 2011).

10.2.1.3 Effects of pre-dispersal predators on seed production

Seed losses to predators, either targeted during pre-dispersal predation or untargeted from grazing of undeveloped fruits, can strongly influence eventual seed output. Large-scale observations show that the magnitude of seed predation is not related to latitude (Moles and Westoby 2003) nor strongly to seed mass (Moles *et al.* 2003) and can vary considerably both spatially and temporally within a species (Janzen 1971). However, few studies have linked pre-dispersal seed predation effects with subsequent population dynamics or trait evolution (Kolb *et al.* 2007). In Australia, pre-dispersal seed predators are not well-known, although sufficient work on some groups of plants has identified close plant–insect relationships. In a study of five *Grevillea* species, weevils (*Cydmaea* spp.) caused the largest seed losses in two species and wasps (*Eurytoma* spp.) were the main predators in another three species (Auld and Denham 2001). In the Fabaceae, weevils, wasps and moths consume seeds across a range of different vegetation types (New 1983; Auld 1991). Decoupling of any host-specific relationships between plant and insect seed predator (e.g. caused by mismatched dispersal capacity under shifting climatic envelopes) may increase the number of seeds that escape seed predation. Other pre-dispersal seed predators include vertebrates, such as parrots, and animal grazers (both native and exotic). Seed losses to parrots can be high but they are generally variable in space and time (e.g. Vaughton 1998). Some level of 'escape' from predation may occur where there is predator satiation, for example with masting events after periods of high rainfall in arid and semi-arid vegetation (Robertson 1988; Gutterman 1994; Wright *et al.* 2014). Plant communities in these systems may be relatively resilient to impacts of seed losses by herbivores (although they may be much less resilient to impacts of herbivores on plant survival; see Figure 10.3).

10.2.2 Dispersal

Dispersal is a key element in the dynamics of both species' populations and the extinction/colonisation processes within metapopulations (Hanski 1998), as it is necessary both to maintain local populations (e.g. through rescue effects), recolonise suitable habitat where populations go extinct and to move species to new locations. Australian vegetation is comprised of species with a range of dispersal syndromes (e.g. see Willson *et al.* 1990; Table 10.1). Key modes of species dispersal across different vegetation communities range from unassisted (widespread in many communities), wind (common in arid shrublands, mallee, rainforests, alpine vegetation), ingestion (rainforests, wet sclerophyll forests, wetlands), adherence (wetlands), ants (sclerophyll heaths, woodlands and forests) and water (wetlands) (Westoby *et al.* 1990; Willson *et al.* 1990; Jurado *et al.* 1991; Leishman and Westoby 1992; Butler *et al.* 2007; Thomson *et al.* 2010; Table 10.1). Growth form, seed mass and vegetation type were found to be the best predictors of plant dispersal mechanisms from over 3000 plants in the Sydney region of south-eastern Australia (Thomson *et al.* 2010). Individual species may have multiple dispersal mechanisms. For example, species with initial ballistic dispersal can be dispersed subsequently by ants (Berg 1975; Westoby *et al.* 1990), while He *et al.* (2009) showed that the ant-dispersed shrub *Daviesia triflora* is also likely to be dispersed by emus.

10.2.2.1 Hydrochory (water dispersal)

Dispersal by water is a key component of marine and freshwater ecosystems, and plant species in these habitats have propagules adapted for movement by water (Table 10.1), although in terrestrial systems there are also a number of species that have seeds dispersal locally by water (Murray 1986). Examples include dispersal of seeds or seedlings in coastal sea currents (Murray 1986; Clarke *et al.* 2001; Guja *et al.* 2010) and dispersal of charophyte propagules in inland rivers and lakes (Casanova and Brock 1999). Changes to both the magnitude and timing of flooding and populations of key waterbird dispersal vectors can therefore impact on recruitment processes in plant communities dependent upon these dispersal mechanisms. For example, water extraction leading to reduced flooding over floodplain vegetation may impact on where species recruit, leading to localised transformations, e.g. from semi-permanent wetland to terrestrial vegetation (Bino *et al.* 2015).

10.2.2.2 Anemochory (wind dispersal)

Movement of seeds by wind is common in many open plant communities (arid and semi-arid shrublands, grasslands, alpine vegetation), with the Asteraceae and Myrtaceae major families that utilise this mechanism (Willson *et al.* 1990; Jurado *et al.* 1991; Thomson *et al.* 2010; Table 10.1). Wind dispersal can occur in many other communities, often in combination with secondary dispersal. Wind can move seeds over a range of distances, depending on seed weight and morphology, seed release height, landscape topography, turbulence and weather conditions (Burrows 1986; Tackenberg 2003). A number of rainforest canopy tree and climber species utilise wind to disperse seeds into canopy gaps (Butler *et al.* 2007). In dry forests and heaths, a number of proteaceous taxa have seeds with wings or hairy appendages encased in woody fruits (to ensure they survive the passage of a fire) and are wind dispersed upon release (Lamont *et al.* 1991; Whelan 1995).

10.2.2.3 Zoochory (animal dispersal)

There are a diversity of mechanisms involving seed dispersal by animals (e.g. Willson *et al.* 1989), although our ability to predict dispersal kernels (*sensu* Clark *et al.* 1998) resulting from animal dispersal is poor (Cousens *et al.* 2010). The major animal dispersal mechanisms in Australian vegetation are ingestion (endozoochory) by birds, mammals and occasionally reptiles, adherence (epizoochory) on birds or mammals, and seed movement by ants (myrmecochory; e.g. Berg 1975). Many plant communities have component species with multiple modes of dispersal (Leishman and Westoby 1992; Thomson *et al.* 2010; Table 10.1). Rainforests are dominated by plants with fleshy fruits promoting dispersal by birds, bats and other mammals, but also have species that are wind dispersed, or with no apparent dispersal mechanism (Butler *et al.* 2007). Grasslands and alpine areas have many species that are wind or animal dispersed via attachment, combined with ant dispersal, and species with no specialised dispersal capacity (Willson *et al.* 1990). Dry heathland, sclerophyll woodlands and forests have the world's most diverse ant-dispersed flora (Berg 1975; Buckley 1982; Westoby *et al.* 1990; Gomez and Espadaler 1998), but also have components of their flora with wind, water and other animal dispersal adaptations (Thomson *et al.* 2010; Table 10.1).

10.2.2.4 Dispersal variation within genera and species

Within a plant genus there may also be a range of dispersal strategies. For example, in *Acacia*, many Australian species have a lipid body attached to the seed which is attractive to ants, and the seeds may be moved a small distance (mostly a few metres, Gomez and Espadaler 1998, but see Whitney 2002) by several major seed-dispersing ant genera (O'Dowd and Gill 1986). This is very common in nutrient-poor soils and heaths, woodland, and dry sclerophyll forest along with mallee, arid and semi-arid shrublands (Berg 1975). Other *Acacia* species have brightly coloured arils (red, yellow, orange) and are attractive to birds (although seeds that fall to the ground will also be moved by ants). Bird dispersal generally results in a greater dispersal distance and can be found across a diversity of habitats, from arid ecosystems to rainforests (O'Dowd and Gill 1986). Within a species, there can be a variety of bird dispersal agents as well as other dispersers such as ants and lizards (e.g. in *Acacia ligulata*; Davidson and Morton 1984; Letnic *et al.* 2000; Whitney 2005). In other *Acacia* spp. the fruit does not dehisce and there is no seed aril. In these cases, the fruit or segments of the fruit fall below the parent and these may be dispersed by water (e.g. in heavy rainfall or floods). Species possessing these attributes are associated with watercourses (e.g. *A. stenophylla*; O'Dowd and Gill 1986). Other large Australian genera with a diversity of dispersal types include *Grevillea* (water, ant, bird, wind). Some, such as *Eucalyptus*, may use wind, ants and water but have no specific strong adaptive features across species that favour one or other of these mechanisms.

10.2.2.5 Dispersal and environmental change

Differences in dispersal syndromes across plant communities will have consequences for future plant migration and hence, will influence the relative need to manage vegetation by addressing *in situ* robustness, resistance and resilience, versus migration ability, range shifts or translocation. This is because the shape of seed dispersal curves for species (dispersal kernels) will play a significant role in their ability to migrate under a changing climate (see Ooi *et al.*, Chapter 8, this volume). Two key questions will determine if Australian plant species will be able to maintain their dispersal capacity under environmental change. First, how will changes to climate impact on different dispersal vectors and their interactions with plants? Second, how long is the tail of the dispersal curve (Portnoy and Willson 1993)? This latter question will govern long-distance dispersal ability (e.g. for wind, see Nathan *et al.* 2011)

Table 10.1 **Seed dispersal syndromes across Australian vegetation types and plant genera. Only five main types of dispersal are included. Dispersal syndrome data are estimates of species (%) within each vegetation community except wetlands, where a qualitative estimate is given and water is also an important dispersal vector. Data sourced from Westoby *et al.* (1990); Willson *et al.* (1990); Jurado *et al.* (1991); Leck and Brock (2000); Green *et al.* (2008); Thomson *et al.* (2010); Hutton and Auld (*unpubl.*).**

Vegetation type	Major dispersal types	Wind	Ant	Adhesion	Ingestion	Unassisted	Keystone long-distance dispersal agents
Temperate rainforest	Unassisted and ingestion	12–24	7–11	0–5	33–44	24–35	Flying foxes, pigeons, cassowaries
Cloud forest	Wind and ingestion	44	0	14	42	0	Birds
Coastal heath	Unassisted and ant	6–12	21–47	1–5	4–16	36–62	Emus (where they still persist)
Alpine heath	Unassisted	5–27	2–28	1–13	1–10	37–80	Unknown
Dry sclerophyll forest	Unassisted and ant	6–11	20–44	0–8	8–22	36–48	Unknown
Wet sclerophyll forest	Unassisted	10–25	7-26	0–10	8–31	32–49	Pigeons?
Alpine grassland	Unassisted	12–18	12–15	5–9	2–4	51–61	Unknown
Arid shrubland	Unassisted and wind	38	9	7	7	40	Emus, kangaroos
Mallee	Unassisted and wind	17–43	5–28	4–15	1–13	28–50	Emus, kangaroos
Saltmarsh	Unassisted	6	4	3	12	63	Waders
Wetland	Adhesion and ingestion and water	Low–moderate	Absent	High	High	Low–moderate	Waterbirds

necessary for plant species to move at a rate comparable to global warming (Massot *et al.* 2008). Animal dispersal is the syndrome most likely to be disrupted under environmental change, although there will also be changes to both frequency and intensity of wind and water dispersal and movement of ocean currents that will have consequences for plant migration. Several animal vectors play a key role in the dispersal capacity of Australian vegetation and we might consider such vectors as keystone (*sensu* Mills *et al.* 1993) dispersal agents. Those that allow relatively long-distance dispersal of plant species are important across a range of vegetation types (Table 10.1). Plant communities, either largely lacking long-distance keystone dispersal vectors or where such vectors have been lost, will be most at risk and dependent upon *in situ* survival (robustness, resistance, resilience), or else may rely on assisted translocation to colonise suitable new locations. One of the major dry heath, woodland and forest animal dispersers (ants) has little capacity to move seeds large distances (Gomez and Espalader 1998) (see Text Box 10.1). Hence disruptions to ant–plant communities are not likely to impact on migration but

Box 10.1 Plant dispersal capability

We estimated dispersal capacity for a sample of the vegetation classes in New South Wales (Keith 2004) by estimating the dispersal capability of component species, and hence identifying where the ability of component species to track a changing climate (by southward and upward migration) is likely to be constrained. Species were assigned to the dispersal mode most common for that genus, and dispersal ranges were assumed to fall within the typical dispersal distances for that dispersal syndrome (Keith 2012). We then classified New South Wales vegetation classes (Keith 2004) into five different groupings of dispersal capacity (from not limited to very highly limited) based on the dispersal syndrome that dominates in that vegetation class (cf. Table 10.1). Dispersal distances were estimated to be particularly small (generally in the order of less than 10 m) in plant communities dominated by myrmecochory (Figure 10.2; Table 10.1). The vegetation types most at risk were the sclerophyll woodlands, forests and heaths because a high proportion of their component species have small-scale dispersal capacity via ants. Persistence of these communities through future climates will involve the maintenance of their *in situ* robustness/ resistance and resilience (*sensu* IPCC 2001; Lake 2012; Bernhardt and Leslie 2013), including a capacity to resist change, to recover from disturbance and to adapt to changed conditions into the future. Where threats overwhelm this capacity, active management to move species to new locations will be necessary, as plant communities transform into new assemblages at their existing locations. Many of the vegetation types at risk due to their limited dispersal capacity are confined to geographically

Box 10.1 (cont.)

limited soil types (e.g. on Sydney sandstone). Hence, there may be little capacity for any re-establishment southwards beyond the areas where their soils occur, and maintenance *in situ* may be the only option. As these vegetation communities are highly diverse (Keith 2004), large numbers of endemic species are at considerable risk. Similarly, highly diverse vegetation communities (heaths and woodlands) in south-western Western Australia are also likely to be at risk from a limited dispersal capacity of many component species.

rather may affect local population persistence and distributions. However in such vegetation, there may be rare long-distance movement, for example by emus (Calviño-Cancela *et al.* 2006; He *et al.* 2009). Disruption of these keystone dispersal agents (as has occurred in eastern coastal Australian vegetation) may significantly impact on the capacity of plant species to track changing climate conditions.

For some vegetation communities a future long-distance dispersal potential is somewhat irrelevant, as there is no suitable climate zone into which they can move. Notable examples of such situations occur on mountain summits and include alpine vegetation (key changes predicted include less snow and changes to frequency of fire, Williams *et al.* 2015) and cloud forests (key changes include a reduction of cloud cover and altered frequency of storms, Auld and Leishman 2015).

Figure 10.2 Estimated limitations to dispersal capacity of vegetation classes (*sensu* Keith 2004) in New South Wales. Darker colours represent increased relative dispersal limitation. White areas are cleared of native vegetation.

10.2.3 Seed dormancy and germination

The mechanisms controlling the timing of germination mediate whether or not seedlings emerge into an environment that is favourable for recruitment. Alleviation of seed dormancy is therefore usually related to the climatic and environmental conditions that seeds experience after dispersal (Baskin and Baskin 2014). Seasonal temperature provides the primary cue across most ecosystems around the world and can regulate both the dormancy status and the timing of seed germination (Fenner and Thompson 2005).

Dormancy can be assigned to five classes, outlined by Baskin and Baskin (2014), but is broadly divided into a physical (germination constrained by impermeable seed coat) or physiological (germination constrained by chemical inhibitors) component. Physiological dormancy is the most common type around the world, including many Australian ecosystems (Table 10.2), and is represented by many species with soil seed banks within the Apiaceae, Asteraceae, Dilleniaceae, Ericaceae, Proteaceae and Rutaceae (Merritt *et al.* 2007; Ooi 2007). Quantitative estimates for two vegetation types highlight this (Ooi 2007), with dry sclerophyll forests having species with the following dormancy types (40–45% physical; 35–40% physiological,

Table 10.2 **Seed bank, seed dormancy and factors promoting germination cues across Australian vegetation types. Seed bank types: soil and transient; soil and persistent; canopy and persistent. Canopy and transient is very restricted. Dormancy types: non-dormant; physical dormancy; physiological dormancy; both morphophysiological and morphological dormancy are rare.**

Vegetation type	Seed bank type	Main factors promoting germination	Major seed dormancy types
Temperate rainforest	Soil transient and soil persistent	Temperature, light	Non-dormant and physiological
Cloud forest	Soil transient and soil persistent	Temperature, light	Non-dormant and physiological
Coastal heath	Soil persistent and canopy persistent	Smoke, heat, temperature, moisture	Physical and physiological
Alpine heath	Soil persistent	Temperature, moisture smoke, heat,	Physical and physiological
Dry sclerophyll forest	Soil persistent and canopy persistent	Smoke, heat, temperature, moisture	Physical and physiological
Wet sclerophyll forest	Soil persistent and canopy persistent	Smoke, heat, temperature, moisture	Physical and physiological
Alpine grassland	Soil persistent	Smoke, heat, temperature, moisture	Physical and physiological
Arid shrubland	Soil persistent and canopy persistent	Temperature, moisture, heat, smoke	Physical and physiological
Mallee	Soil persistent and canopy persistent	Smoke, heat, temperature, moisture	Physical and physiological
Saltmarsh	Soil persistent	Temperature, moisture	Non-dormant and physiological
Wetland	Soil persistent	Temperature, moisture	Non-dormant and physiological

15–20% morphophysiological and <5% are either morphological or non-dormant) and semi-arid vegetation having 15–20% physical; 75–80% physiological and <5% are morphophysiological, morphological or non-dormant. In physiologically dormant seeds, while seed coats are permeable, a chemical change within the seed embryo is required (i.e. a change in the balance of the plant hormones abscisic acid and gibberellic acid) before germination can occur (Finch-Savage and Leubner-Metzger 2006). This change occurs when seeds are subjected to either seasonal temperature stratification (moist cold or warm conditions) or dry after-ripening (dry-stored warm conditions). For example, Hoyle *et al.* (2008) found that warm treatments and dry/wet cycles provided the best method to alleviate dormancy of three native forbs from semi-arid tropical Queensland, a combination that mimicked seasonal rainfall conditions in the region. Commander *et al.* (2009) found that a period of dry after-ripening can overcome dormancy of a number of species from semi-arid Western Australia. Work conducted in Australian alpine areas has found that cold stratification alleviates dormancy and increases germination for many physiologically dormant species (Venn and Morgan 2007; Sommerville *et al.* 2013).

Once physiological dormancy is overcome, specific cues may still be required to promote germination (Thompson and Ooi 2010). These can include particular seasonal temperatures, light or smoke (Table 10.2). The distinction between overcoming physiological dormancy and the subsequent promotion of germination by other cues (*sensu* Thompson and Ooi 2010) is often unclear in the Australian literature. This may be due to the fact that temperature can play a dual role, both alleviating physiological dormancy, and promoting germination (via specific external conditions such as seasonal temperatures, diurnal range or amplitude; Table 10.2). In fire-prone regions, additional cues, such as smoke (the main fire-related germination cue for physiologically dormant seeds), may be needed in combination with specific temperatures to ensure germination is promoted in the post-fire environment (Baker *et al.* 2005; Ooi *et al.* 2006; Merritt *et al.* 2007; Thomas *et al.* 2010).

Physically dormant seeds, notably from species within the Fabaceae, Malvaceae and Rhamnaceae have an impermeable seed coat that needs to be broken before water uptake and germination can occur. In Australia, fire-related temperatures generating a heat pulse in the soil are the most recognised cue for breaking physical dormancy (Table 10.2). However, seasonal temperatures play a role in breaking physical dormancy in many arid, semi-arid and tropical species (Baskin *et al.* 2000) and potentially, in canopy gaps or post-fire environments (Auld and Bradstock 1996; Santana *et al.* 2010; Ooi *et al.* 2012).

Non-dormant seeds are common in species with canopy-stored seed banks (e.g. *Banksia* and *Hakea*), pyrogenic flowerers (e.g. *Xanthorrhoea, Telopea* and *Doryanthes*) and in primary rainforest species (Hopkins and Graham 1983; Bellairs and Bell 1990; Denham *et al.* 2011). Although there are no requirements for overcoming dormancy, temperature cues can still restrict germination to particular seasons, often identified as the thermal niche or the optimal germination temperature (Table 10.2). Many *Banksia* species prefer cooler season temperatures for germination, germinating less and at slower rates at higher temperatures, presumably to prevent large numbers of seedlings from emerging as a result of out-of-season summer rainfall in Mediterranean-type climates (e.g. Bellairs and Bell 1990; Cochrane *et al.* 2014) or, more generally, to avoid harsher conditions that occur from greater evapotranspiration in aseasonal rainfall regions (Sonia and Heslehurst 1978; Bradstock and Bedward 1992).

10.2.4 Seed banks

Australian vegetation types display a range of seed bank types (Keith and Tozer, Chapter 1, this volume; Table 10.2). Most have a significant component of species that have persistent soil (sometimes combined with canopy) seed banks and are dominated by seeds with physiological or physical dormancy (Table 10.2). Rainforests and cloud forests in particular, also have component species whose seeds are short-lived (non-dormant) and hence form transient seed banks (Table 10.2). Some species in these forest communities have transient soil seed banks combined with banks of shade-tolerant juvenile plants (*sensu* the Oskars of Silvertown 1987), e.g. *Wollemia nobilis* (Zimmer *et al.* 2014), *Lepidorrhachis mooreana* and *Hedyscepe canterburyana* (Auld *et al.* 2010).

Many plant species have either long-lived standing plants or long-lived stages in their life cycles (such as seeds, bulbs, tubers). The existence of a soil seed bank is one of the key means of plant population persistence, conferring an ability to survive disturbances, recover from adverse conditions, or take advantage of favourable conditions via recruitment of new individuals. While several studies review the role of above-ground mechanisms of plant persistence in Australia, for example bradyspory (Lamont *et al.* 1991) and epicormic resprouting (Gill *et al.* 1981; Clarke *et al.* 2015) in fire-prone systems, we focus here on the role of soil seed banks in population persistence below ground. Two ecosystems, arid and those that are fire prone, provide models for how soil seed banks operate, how they are critically important for the persistence of plant species and how they may be impacted by future climate changes. The principles here are also applicable to many other vegetation types, which include varying proportions of species with similar soil seed bank mechanisms.

10.2.4.1 Arid ecosystems

Vegetation in arid environments is comprised of a range of plant life-history strategies spanning two broad extremes (Noy-Meir 1973; Jurado *et al.* 1991). These are drought persistents versus ephemerals (*sensu* Noy-Meir 1973). Drought persistents include trees, shrubs and subshrubs that are perennial with life spans extending from a few years to several decades to centuries (e.g. Davies and Kenny 2013). They are able to endure droughts as adult plants and take advantage of occasions when rainfall is sufficient to recruit new individuals, either from seed in soil-stored seed banks or vegetatively (Figures 10.1 and 10.3). While some long-lived perennial shrubs and trees maintain persistent seed banks, regular seed production combined with the low predictability of rainfall in arid Australian systems appear to have selected for an increase in the proportion of non-dormant seeds produced and a decrease in seed bank persistence compared to their ephemeral counterparts. Seeds produced by some *Acacia* tree species, including *A. oswaldii*, *A. peuce* and *A. loderi*, showed a high non-dormant fraction, meaning 'false-start' rainfall events could deplete soil seed banks (Auld 1995; Nano *et al.* 2013; Pound *et al.* 2015). Such species may rely on nurse plants, low grazing pressure or high-rainfall conditions (Nano *et al.* 2013) for successful recruitment from the seed bank. Changes to these shrublands by exotic grazers, weed invasion, increased fire and increased tree mortality will impact on structure and species composition (Figure 10.3) and soil seed banks will play a key role in any recovery.

Ephemerals include both true annuals and short-lived perennials and have the general characteristic that they do not persist above ground through dry periods (droughts). The desert 'ephemeral' species that occupy arid landscapes survive by having dormant persistent soil seed banks or tuber/bulbs or rhizomes. Growth in good conditions is triggered by a combination of temperature and rainfall, with many species being able to coexist, but utilising different seasonal cues for germination and/or growth (Gutterman 1994) (Figure 10.4). This supports the inference that plant life histories in arid Australia are shaped by soil moisture availability (Morton *et al.* 2011). Robertson (1988) showed that summer rains tend to promote recruitment and growth of native grasses, while autumn rains favour forbs (e.g. peas, daisies) as well as Mediterranean weeds in arid western New South Wales. In Central Australia, Jurado and Westoby (1992) found that grass species predominantly responded to summer rains, while forbs either were favoured by winter rains or had no seasonal preference. Changes projected for both seasonal temperatures and rainfall patterns in the future are therefore likely to favour some species while others are likely to decline should the required temperature and moisture cues for germination no longer occur (Figure 10.4). This will result in shifts in the ephemeral species composition for arid vegetation, with a likely simplification of diversity where a number of species are adversely affected, unless there is the capacity for dispersal of additional species into the habitat. The capacity of the soil seed bank to buffer against these changes may be compromised and will depend on seed tolerance to lethal soil heating combined with plasticity and adaptive capacity in response to changing seasonal germination cues (Ooi *et al.* 2009; Figure 10.4).

10.2.4.2 Fire-prone ecosystems

Survival in relation to fire is governed by contrasting plant life-history strategies (Whelan 1995; Miller and Murphy, Chapter 5, this volume). There are those species that can persist through a fire and recover by resprouting from dormant buds or

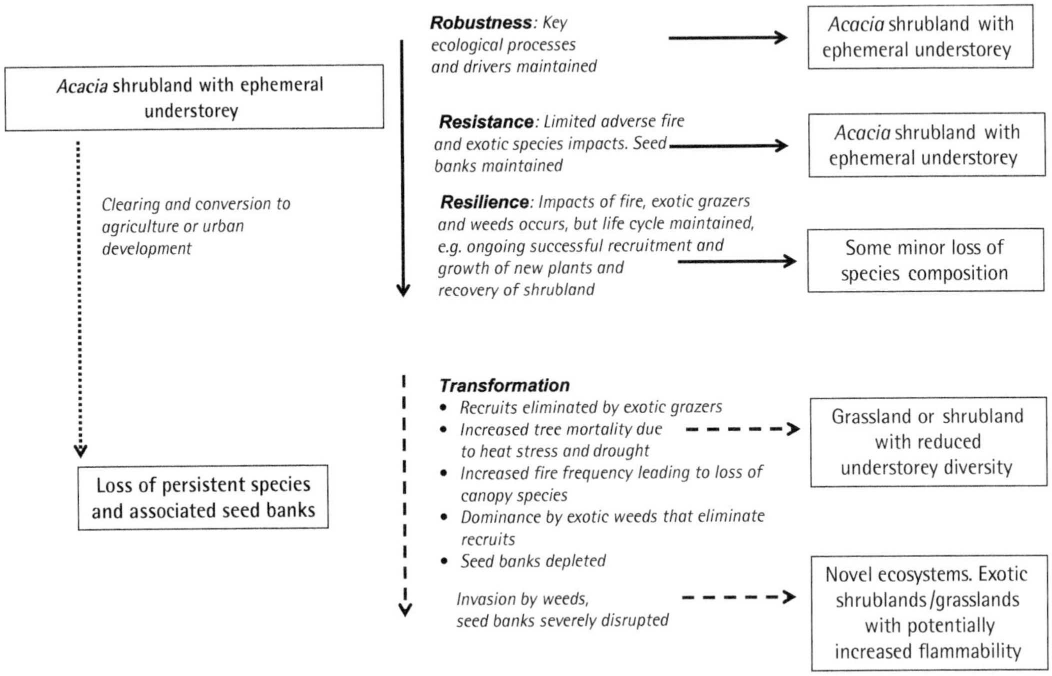

Figure 10.3 Conceptual cause–effect model of drought persistents (*sensu* Noy-Meir 1973) with persistent soil seed banks in long-lived arid and semi-arid *Acacia* shrublands. Boxes represent different potential states of persistence of long-lived plants and associated soil seed bank. Arrows represent transitions between the states with the inferred process indicated by text. Solid lines represent ongoing persistence. Dashed lines represent transformation. Dotted line represents loss of both persistents and soil seed banks. Tree mortality, increased fire frequency and impacts of exotic grazers result in mortality of existing perennial species, decline of soil seed banks, significantly reduced recruitment of palatable structural dominants of the ecosystem and consequent loss of the shrub canopy. Some buffer may be provided by soil seed banks. Invasion by exotic shrubs (weeds) results in a canopy dominated by exotic shrubs such as African boxthorn (*Lycium ferocissimum*) or a flammable grass understorey (buffel grass), with severely depleted native soil seed bank.

meristematic tissues protected from the heat of the fire. These buds or meristematic tissues may be found either on aerial stems (leading to epicormics shoots) or underground in lignotubers, rhizomes, bulbs or tubers (Gill *et al.* 1981). Other species are killed by fire and are totally reliant on the recruitment of new individuals for post-fire recovery. This is usually from a soil- or canopy-stored seed bank, but can occur from dispersal of seeds from unburnt areas. While species with a resprouting capacity are likely to have some resilience to altered fire regimes (Clarke *et al.* 2013), seedling recruitment will still be essential to maintain populations in the longer term (Enright *et al.* 2014). For those species dependent upon post-fire seed germination from a seed bank, the interaction of seed dormancy and fire cues (such as heat and smoke, see Table 10.2) with ambient temperature and rainfall (Thomas *et al.* 2010; Ooi *et al.* 2012) is crucial to recruitment outcomes. These interactions could be sensitive to altered fire regimes under future climates (Bradstock *et al.* 2014). Species persistence will be

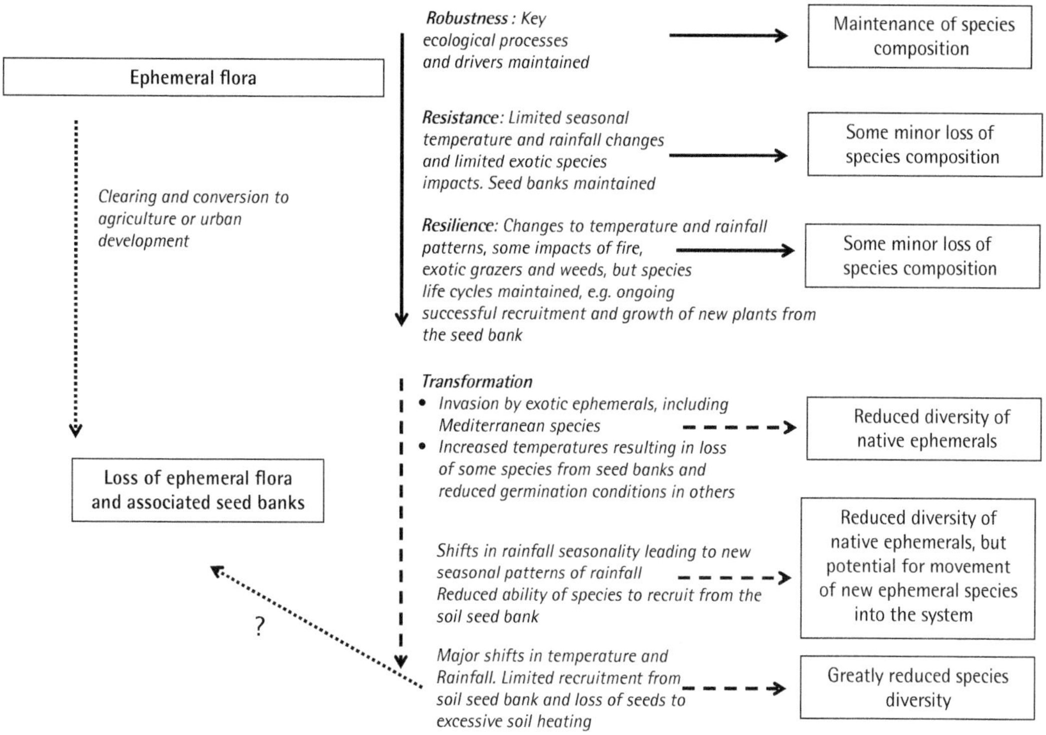

Figure 10.4 Key elements to maintaining arid and semi-arid ephemeral species. Solid lines represent ongoing persistence. Dashed lines represent transformation/simplification. Dotted line represents loss of ephemeral flora. Changes to temperature and seasonality of rainfall will reduce the number of species able to successfully recruit from the soil seed bank.

governed by the ability of soil seed banks to cope with:

- increases in the soil heat pulse generated by fire if future predictions of increased fire severity are realised (heating during a fire may kill seeds, or in physically dormant species, break dormancy or in physiologically dormant species may play an interactive role with smoke in promoting or retarding seed germination);
- changes to seasonal temperature cues (for many species with physiological or morpho-physiological dormancy, seasonal temperature fluctuations and after-ripening influence seed dormancy);
- changes to seasonal rainfall patterns (post-fire recruitment may be diminished if fires occur at times of low rainfall);
- changes to the capacity to maintain a persistent soil seed bank and bet-hedge germination success, particularly in response to 'interval squeeze', *sensu* Enright *et al.* (2015).

10.2.5 Establishment

After germination, the successful establishment of new individuals through seedling emergence and survival is a key requirement for the continued persistence of a plant population at a site.

Establishment is governed by the availability of 'safe sites' (Figure 10.1), and the seedling stage is considered to be the most vulnerable time in the life cycle of plants (Harper 1977). Establishment success varies across species. Larger seed size favours increased establishment success (e.g. Moles and Westoby 2006) as individuals with larger seeds have increased parental resources, greater competitive ability and greater shade tolerance than smaller seeds. Establishment success will also vary in relation to environmental conditions (seasonally and inter-annually), disturbances, and biotic and abiotic conditions. For example, establishment requirements may be tightly bound to particular seasons (e.g. during wet winters in Mediterranean-type climates), or cued to disturbances such as fire, floods and storms (see below), high-rainfall events (Dickman *et al.* 2014), favourable site conditions including light gaps (Hubbell *et al.* 1999), nutrient-rich sites, nurse plants (particularly in arid vegetation, Flores and Jurado 2003), and reduced competition. After establishment, ongoing seedling survival will be influenced by biotic (predators, competitors) and abiotic factors (temperature, moisture, light, nutrients).

10.2.6 Survival and growth

The maintenance and structure of plant communities is dependent upon successful growth and survival of the different life forms of component species (see Figures 10.3 and 10.4). Failure for any one component may seriously impact that structure. Survival and growth are influenced by available abiotic resources and conditions (moisture, temperature, nutrients) and biotic interactions (competition, facilitation). Australian plant communities have a number of mechanisms to enhance the survival of individuals, particularly in response to repeated disturbances, and Bond and Midgley (2001) stress the importance of persistence through sprouting as a life-history strategy. Sprouting may be from a range of organs that store resources such as carbohydrates and possess dormant buds for recovery (e.g. stems, bulbs, rootstocks, lignotubers). The trade-off between plant longevity and reproductive output is widespread in many vegetation communities. Examples include contrasts between long-lived perennial trees and shrubs versus: ephemerals in arid and semi-arid vegetation (Noy-Meir 1973); short-lived and fast-growing gap colonisers in rainforests (Rees *et al.* 2001); fire ephemerals in heaths, woodlands, forests and mallee (Bell *et al.* 1993; Baker *et al.* 2005); and ephemerals in wetlands or floodplain habitats (Casanova and Brock 2000).

Survival of adults alone does not guarantee persistence. If there is no successful seed production or no recruitment, or no recruits ever make it to the adult stage, plants may be thought of as the 'living dead' (*sensu* Janzen 1986). The Australian plant communities most at risk from such a scenario, where long-lived individuals (trees and shrubs) persist but are limited by a failure in one component of the life cycle (Figure 10.1), include a number of arid and semi-arid shrublands across southern Australia (Pickard 1991; Sinclair 2005) (see Figure 10.3). In arid systems, a number of factors influence survival and growth, including rainfall (promoting effect), drought (limiting growth, decreasing survival and eliminating recruitment), fire (generally decreasing survival, while post-fire growth and recruitment are dependent upon rainfall) and grazing pressure (generally limiting growth, decreasing survival and decreasing or eliminating recruitment in palatable species) (Figure 10.3). Other systems may also show a long-lived adult structure but still be at risk, as seen, for example, in tropical savannas when intervals between fires become so short that recruits are killed (Setterfield 2002). Where dominant species are very long lived (across centuries), perception of change may be difficult to record over human time scales, even when there have been serious disruptions to life-history processes such as recruitment (Figure 10.3).

10.3 Key processes that drive persistence of plants in Australian vegetation

A number of important ecological processes are necessary to maintain the persistence and identity of Australian vegetation (Keith and Tozer, Chapter 1, this volume). Here we examine several disturbance regimes that are critical to the life histories of Australian plant species. Globally, ecosystems are subjected to a range of disturbance regimes that impact on the survival and recruitment of plant populations, and hence vegetation composition and structure. Changes to these regimes will have a large impact on the vegetation.

10.3.1 Fire

Fire is a key disturbance across the globe and most ecosystems in Australia experience fire at some time. An understanding of the range of components that make up the fire regime is critical for fully understanding the impact that fire has on population processes, community composition and associated biodiversity (Miller and Murphy, Chapter 5, this volume). Increased fire frequency is arguably the most studied change in fire regimes, and is regarded as one of the most likely processes to impact negatively on vegetation (Bradstock *et al.* 1995; Keith *et al.* 2002), as sufficient time between fires is needed for recruiting seedlings to become mature. Enright *et al.* (2015) (see also Miller and Murphy, Chapter 5, this volume) stress the importance of climate impacts on plant life histories and fire regimes through 'interval squeeze'. This is particularly an issue for plant species in which adults are killed by fire (obligate seeders), but it is also relevant for resprouters, as some successful seedling recruitment will most likely still be important for long-term persistence (Keith 1996). Changes to other aspects of the fire regime, including season and severity, may also have significant effects on persistence and species composition (e.g. Ooi *et al.* 2004; Ooi 2010). Research on these effects is limited in most ecosystems, although some important studies have improved understanding, for example, in tropical savanna (Williams *et al.* 2003).

10.3.2 Water

Water regimes are particularly important in riverine, floodplain and wetland ecosystems. Changes to water flows, in terms of seasonality, magnitude and duration, have occurred as a consequence of rural and urban development in Australia (Kingsford 2000). Water is also a driving force in terrestrial ecosystems, with plant growth and survival, and seed germination and establishment dependent upon available moisture. Consequently, many plants have features that allow them to respond to patterns of rainfall (seasonality, intensity) (e.g. Robertson 1988; Jurado and Westoby 1992) (see 10.2.3 Seed dormancy and germination, above). Mediterranean-climate ecosystems in south-western Australia have little summer rain, and recruitment of seedlings is generally cued to temperature so that germination is confined to the wetter, cooler months (Bell 1999). Comparable vegetation (e.g. heaths, woodlands, mallee) in other parts of Australia exhibits a broader seasonal response with fewer taxa having strong seasonal cues (Ooi 2007). The most striking example of how variation in seasonality and magnitude of rainfall affects plant growth is in arid ecosystems. Here the ephemeral flora and its associated dependent fauna undergo boom-and-bust cycles (Dickman *et al.* 2014). At any one location, the composition of the ephemeral flora may vary considerably over time as species have different temperature and moisture requirements for seed germination. Hence, many species that coexist in soil seed banks are displaced through time in their above-ground life stages (Facelli *et al.* 2005; Figure 10.4).

10.3.3 Storm-driven processes

Gap dynamics play an important role in recruitment and regeneration of trees, shrubs and vines in rainforest and cloud forest vegetation in high-rainfall areas (Metcalfe and Green, Chapter 11, this volume). Recruitment strategies in these systems include shade-tolerant seedling banks (Oskars, *sensu* Silvertown 1987) and rapid germination and growth of seedlings when gaps appear leading to 'gap-phase succession' (Rees *et al.* 2001). Changing frequency and severity of storms that create forest gaps, combined with moisture availability (rainfall patterns), will play a strong role in species persistence within these vegetation types.

References

Adams, P.B. and Lawson, S.D. (1993) Pollination in Australian orchids: a critical assessment of the literature 1882–1992. *Australian Journal of Botany* 41, 553–575.

Armstrong, J.A. (1979) Biotic pollination mechanisms in the Australian flora: a review. *New Zealand Journal of Botany* 17, 467–508.

Auld, T.D. (1991) Patterns of predispersal seed predators in the Fabaceae of the Sydney region, south-eastern Australia. *Australian Journal of Zoology* 39, 519–528.

Auld, T.D. (1995) Soil seedbank patterns of four trees and shrubs from arid Australia. *Journal of Arid Environments* 29, 33–45.

Auld, T.D. and Bradstock, R.A. (1996) Do post-fire soil temperatures influence seed germination? *Australian Journal of Ecology* 21, 106–109.

Auld, T.D. and Denham, A.J. (2001) Predispersal seed predation in shrubs of *Grevillea* (Proteaceae) from south-eastern Australia. *Australian Journal of Botany* 49, 17–21.

Auld, T.D. and Leishman, M.R. (2015) Ecosystem risk assessment for Gnarled Mossy Cloud Forest, Lord Howe Island, Australia. *Austral Ecology* 40, 364–372.

Auld, T.D., Hutton, I., Ooi, M.K.J. and Denham, A.J. (2010) Disruption of recruitment in two endemic palms on Lord Howe Island by invasive rats. *Biological Invasions* 12, 3351–3361.

Ayre, D.J. and Whelan, R.J. (1989) Factors controlling fruit set in hermaphroditic plants: studies with the Australian Proteaceae. *Trends in Ecology and Evolution* 4, 267–272.

Baker, K.S., Steadman, K.J., Plummer, J.A. *et al.* (2005) Dormancy release in Australian fire ephemeral seeds during burial increases germination response to smoke water or heat. *Seed Science Research* 15, 339–348.

Baskin, C.C. and Baskin, J.M. (2014) *Seeds: Ecology, Biogeography and Evolution of Dormancy and Germination*. 2nd edition. Elsevier: San Diego.

Baskin, J.M., Baskin, C.C. and Xiaojie, L. (2000) Taxonomy, anatomy and evolution of physical dormancy in seeds. *Plant Species Biology* 15, 139–152.

Bell, D.T. (1999) Turner Review No. 1: the process of germination in Australian species. *Australian Journal of Botany* 47, 475–517.

Bell, D.T., Plummer, J.A. and Taylor, S.K. (1993) Seed germination ecology in south-western Western Australia. *The Botanical Review* 59, 24–73.

Bellairs, S.M. and Bell, D.T. (1990) Temperature effects on the seed germination of ten Kwongan species from Eneabba, Western Australia. *Australian Journal of Botany* 38, 451–458.

Benson, D. and McDougall, L. (1996) Ecology of Sydney plant species: part 4. Dicotyledon family Fabaceae. *Cunninghamia* 4, 553–752.

Berg, R.Y. (1975) Myrmecochorous plants in Australia and their dispersal by ants. *Australian Journal of Botany* 23, 475–508.

Bernhardt, J.R. and Leslie, H.M. (2013) Resilience to climate change in coastal marine ecosystems. *Annual Review Marine Science* 5, 371–392.

Bino, G., Sisson, S.A., Kingsford, R.T. *et al.* (2015) Developing state and transition models of floodplain vegetation dynamics as a tool for conservation decision-making: a case study of the Macquarie Marshes Ramsar wetland. *Journal of Applied Ecology* 52, 654–664.

Bond, W.J., Honig, M. and Maze, K.E. (1999) Seed size and seedling emergence: an allometric relationship and some ecological implications. *Oecologia* 120, 132–136.

Bond, W.J. and Midgley, J.J. (2001) Ecology of sprouting in woody plants: the persistence niche. *Trends in Ecology and Evolution* 16, 45–51.

Boulter, S.L., Kitching, R.L. and Howlett, B.G. (2006) Family, visitors and the weather: patterns of flowering in tropical rain forests of northern Australia. *Journal of Ecology* 94, 369–382.

Bradstock, R.A. and Bedward, M. (1992) Simulation of the effect of season of fire on post-fire seedling emergence of two Banksia species based on long-term rainfall records. *Australian Journal of Botany* 40, 75–88.

Bradstock, R.A., Keith, D.A. and Auld, T.D. (1995) Fire and conservation: imperatives and constraints on managing for diversity. In *Conserving Biodiversity: Threats and Solutions*, ed. R.A. Bradstock, T.D. Auld, D.A. Keith, R. Kingsford, D. Lunney and D. Sivertsen. Surrey Beatty & Sons: Sydney, pp. 323–333.

Bradstock, R., Penman, T., Boer, M. *et al.* (2014) Divergent responses of fire to recent warming and drying across south-eastern Australia. *Global Change Biology* 20, 1412–1428.

Brock, M.A. (1982) Biology of the salinity tolerant genus *Ruppia* L. in saline lakes in South Australia II. Population ecology and reproductive biology. *Aquatic Botany* 13, 249–268.

Buckley, R.C. (1982) *Ant–Plant Interactions in Australia.* Dr Junk: The Hague.

Burrows, F.M. (1986) The aerial motion of seed, fruits, spores and pollen. In *Seed Dispersal*, ed. D.R. Murray. Academic Press: Sydney, pp. 1–47.

Butler, D.W., Green, R.J., Lamb, D. *et al.* (2007) Biogeography of seed-dispersal syndromes, life-forms and seed sizes among woody rain-forest plants in Australia's subtropics. *Journal of Biogeography* 34, 1736–1750.

Calviño-Cancela, M., Dunn, R.R., van Etten, E.J.B. and Lamont, B.B. (2006) Emus as non-standard seed dispersers and their potential for long-distance dispersal. *Ecography* 29, 632–640.

Casanova, M.T. and Brock, M.A. (1999) Life histories of charophytes from permanent and temporary wetlands in eastern Australia. *Australian Journal of Botany* 47, 383–397.

Casanova, M.T. and Brock, M.A. (2000) How do depth, duration and frequency of flooding influence the establishment of wetland plant communities? *Plant Ecology* 147, 237–250.

Clark, J.S., Fastie, C., Hurtt, G. *et al.* (1998) Reid's paradox of rapid plant migration: dispersal theory and interpretation of paleoecological records. *Bioscience* 48, 13–24.

Clarke, P.J., Kerrigan, R.A. and Westphal, C.J. (2001) Dispersal potential and early growth in 14 tropical mangroves: do early life history traits correlate with patterns of adult distribution? *Journal of Ecology* 89, 648–659.

Clarke, P.J., Lawes, M.J., Midgley, J.J. *et al.* (2013) Resprouting as a key functional trait: how buds, protection and resources drive persistence after fire. *New Phytologist* 197, 19–35.

Clarke, P.J., Lawes, M.J., Murphy, B.B. *et al.* (2015) A synthesis of postfire recovery traits of woody plants in Australian ecosystems. *Science of the Total Environment* 534, 31–42.

Cochrane, A., Hoyle, G.L., Yates, C.J. *et al.* (2014) Predicting the impact of increasing temperatures on seed germination among populations of Western Australian *Banksia* (Proteaceae). *Seed Science Research* 24, 195–205.

Commander, L.E., Merritt, D.J., Rokich, D.P. and Dixon, K.W. (2009) The role of after-ripening in promoting germination of arid zone seeds: a study on six Australian species. *Botanical Journal of the Linnean Society* 161, 411–421.

Cousens, R.D., Hill, J., French, K. and Bishop, I.D. (2010) Towards better prediction of seed dispersal by animals. *Functional Ecology* 24, 1163–1170.

Davidson, D.W. and Morton, S.R. (1984) Dispersal adaptations of some *Acacia* species in the Australian arid zone. *Ecology* 65, 1038–1051.

Davies, S.J.J.F. (1976) Studies of the flowering season and fruit production of some arid zone shrubs and trees in Western Australia. *The Journal of Ecology* 64, 665–687.

Davies, S.J.J.F. and Kenny, S.A. (2013) The ages and fecundity of some arid zone plants in Western Australia. *The Rangeland Journal* 35, 455–468.

Denham, A.J., Whelan, R.J., Auld, T.D. and Denham, R.J. (2011) The coupling of recruitment and disturbance by fire in two resprouting Proteaceae species. *Plant Ecology* 212, 471–481.

Dickman, C., Wardle, G., Foulkes, J. and de Preu, N. (2014) Desert complex environments. In *Biodiversity and Environmental Change: Monitoring, Challenges and Direction*, ed. D. Lindenmayer, E. Burns, N. Thurgate and A. Lowe. CSIRO Publishing: Collingwood, pp. 379–438.

Dixon, K.W. (2009) Pollination and restoration. *Science* 325, 571–573.

Enright, N.J., Fontaine, J.B., Lamont, B.B. *et al.* (2014) Resistance and resilience to changing climate and fire regime depend on plant functional traits. *Journal of Ecology* 102, 1572–1581.

Enright, N.J., Fontaine, J.B., Bowman, D.M.J.S. *et al.* (2015) Interval squeeze: altered fire regimes and demographic responses interact to threaten woody species persistence as climate changes. *Frontiers in Ecology and the Environment* 13, 265–272.

Entwisle, T.J. (2014) *Sprinter and Sprummer. Australia's Changing Seasons.* CSIRO Publishing: Collingwood.

Facelli, J.M., Chesson, P. and Barnes, N. (2005) Differences in seed biology of annual plants in arid lands: a key ingredient of the storage effect. *Ecology* 86, 2998–3006.

Fenner, M. and Thompson, K. (2005) *The Ecology of Seeds*. Cambridge University Press: Cambridge.

Finch-Savage, W.E. and Leubner-Metzger, G. (2006) Seed dormancy and the control of germination. *New Phytologist* 171, 501–523.

Flores, J. and Jurado, E. (2003) Are nurse-protégé interactions more common among plants from arid environments? *Journal of Vegetation Science* 14, 911–916.

Ford, H.A., Paton, D.C. and Forde, N. (1979) Birds as pollinators of Australian plants. *New Zealand Journal of Botany* 17, 509–519.

Gill, A.M., Groves, R.H. and Noble, I.R. (1981) *Fire and the Australian Biota.* Australian Academy of Science: Canberra.

Gomez, C. and Espadaler, X. (1998) Myrmecochorous dispersal distances: a world survey. *Journal of Biogeography* 25, 573–580.

Green, A.J., Jenkins, K.M., Bell, D. *et al.* (2008) The potential role of waterbirds in dispersing invertebrates and plants in arid Australia. *Freshwater Biology* 53, 380–392.

Greet, J., Cousens, R.D. and Webb, J.A. (2013) Seasonal timing of inundation affects riparian plant growth and flowering: implications for riparian vegetation composition. *Plant Ecology* 214, 87–101.

Guja, L.K., Merritt, D.J. and Dixon, K.W. (2010) Buoyancy, salt tolerance and germination of coastal seeds: implications for oceanic hydrochorous dispersal. *Functional Plant Biology* 37, 1175–1186.

Gutterman, Y. (1994) Strategies of seed dispersal and germination in plants inhabiting deserts. *Botanical Review* 60, 373–425.

Hanski, I. (1998) Metapopulation dynamics. *Nature* 396, 41–49.

Harper, J.L. (1977) *Population biology of plants.* Academic Press: London.

He, T., Lamont, B.B., Krauss, S.L. *et al.* (2009) Ants cannot account for interpopulation dispersal of the arillate pea *Daviesia triflora*. *New Phytologist* 181, 725–733.

Hopkins, M.S. and Graham, A.W. (1983) The species composition of soil seed banks beneath lowland tropical rainforests in north Queensland, Australia. *Biotropica* 15, 90–99.

Hoyle, G.L., Daws, M.I., Steadman, K.J. and Adkins, S.W. (2008) Mimicking a semi-arid tropical environment achieves dormancy alleviation for seeds of Australian native Goodeniaceae and Asteraceae. *Annals of Botany* 101, 701–708.

Hubbell, S.P., Foster, R.B., O'Brien, S.T. *et al.* (1999) Light-gap disturbances, recruitment limitation, and tree diversity in a neotropical forest. *Science* 283, 554–557.

IPCC (2001) *Climate Change 2001: Impacts, Adaptation and Vulnerability.* Cambridge University Press: Cambridge.

Janzen, D.H. (1971) Seed predation by animals. *Annual Review of Ecology and Systematics* 2, 465–492.

Janzen, D.H. (1980) When is it coevolution? *Evolution* 34, 611–612.

Janzen, D.H. (1986) Dan Janzen's thoughts from the tropics 5. Blurry catastrophes. *Oikos* 47, 1–2.

Jurado, E. and Westoby, M. (1992) Germination biology of selected central Australian plants. *Australian Journal of Ecology* 17, 341–348.

Jurado, E., Westoby, M. and Nelson, D. (1991) Diaspore weight, dispersal, growth form and perenniality of central Australian plants. *Journal of Ecology* 79, 811–828.

Keddy, P.A. (1992) Assembly and response rules: two goals for predictive community ecology. *Journal of Vegetation Science* 3, 157–164.

Keith, D.A. (1996) Fire-driven extinction in plant populations: a synthesis of theory and review of evidence from Australian vegetation. *Proceedings of Linnean Society New South Wales* 116, 37–78.

Keith, D.A. (2004) *Ocean Shores to Desert Dunes: The Native Vegetation of New South Wales and the ACT.* Department of Environment and Conservation: Sydney.

Keith, D.A. (2012) Functional traits: their roles in understanding and predicting biotic responses to fire regimes from individuals to landscapes. In *Flammable Australia: Fire Regimes, Biodiversity and ecosystems in a Changing World*, ed. R.A. Bradstock, A.M. Gill and R.J. Williams. CSIRO Publishing: Collingwood, pp. 97–125.

Keith, D.A., Williams, J.E. and Woinarski, J.C.W. (2002) Fire management and biodiversity conservation: key approaches and principles. In *Flammable Australia: The Fire Regimes and Biodiversity of a Continent*, ed. R.A. Bradstock, A.M. Gill and J.E. Williams. Cambridge University Press: Cambridge, pp. 401–425.

Kingsford, R.T. (2000) Ecological impacts of dams, water diversions and river management on floodplain wetlands in Australia. *Austral Ecology* 25, 109–127.

Kolb, A., Ehrlen, J. and Eriksson, O. (2007) Ecological and evolutionary consequences of spatial and temporal variation in pre-dispersal seed predation. *Perspectives in Plant Ecology, Evolution and Systematics* 9, 79–100.

Lake, S. (2012) Resistance, resilience and restoration. *Ecological Management and Restoration* 14, 20–24.

Lamont, B.B. and Downes, K.S. (2011) Fire-stimulated flowering among resprouters and geophytes in Australia and South Africa. *Plant Ecology* 212, 2111–2125.

Lamont, B.B., Le Maitre, D.C., Cowling, R.M. and Enright, N.J. (1991) Canopy seed storage in woody plants. *Botanical Review* 57, 277–317.

Lavorel, S., Díaz, S., Cornelissen, J.H.C. *et al.* (2007) Plant functional types: are we getting any closer to the Holy Grail? In *Terrestrial Ecosystems in a Changing World*, ed. J.G. Canadell, D. Pataki and L. Pitelka. Springer-Verlag: Berlin, pp. 149–160.

Leck, M.A. and Brock, M.A. (2000) Ecological and evolutionary trends in wetlands: evidence from seeds and seed banks in New South Wales, Australia and New Jersey, USA. *Plant Species Biology* 15, 97–112.

Leishman, M.R. and Westoby, M. (1992) Classifying plants into groups on the basis of associations of individual traits: evidence from Australian semi-arid woodlands. *Journal of Ecology* 80, 417–424.

Letnic, M., Dickman, C.R. and McNaught, G. (2000) Bet-hedging and germination in the Australian arid zone shrub *Acacia ligulata*. *Austral Ecology* 25, 368–374.

Massot, M., Clobert, J. and Ferriere, R. (2008) Climate warming, dispersal inhibition and extinction risk. *Global Change Biology* 14, 461–469.

Merritt, D.J., Turner, S.R., Clarke, S. and Dixon, K.W. (2007) Seed dormancy and germination stimulation syndromes for Australian temperate species. *Australian Journal of Botany* 55, 336–344.

Mills, L.S., Soulé, M.E. and Doak, D.F. (1993) The keystone species concept in ecology and conservation. *BioScience* 43, 219–224.

Moir, M.L., Vesk, P.A., Brennan, K.E.C. *et al.* (2011) Identifying and managing threatened invertebrates through assessment of coextinction risk. *Conservation Biology* 25, 787–796.

Moles, A.T. and Westoby, M. (2003) Latitude, seed predation and seed mass. *Journal of Biogeography* 30, 105–128.

Moles, A.T. and Westoby, M. (2006) Seed size and plant strategy across the whole life cycle. *Oikos* 113, 91–105.

Moles, A.T., Warton, D.I. and Westoby, M. (2003) Do small-seeded species have higher survival through seed predation than large-seeded species? *Ecology* 84, 3148–3161.

Morgan, J.W. (1997) The effect of grassland gap size on establishment, growth and flowering of the endangered *Ruitdosis leptorrhynchoides* (Asteraceae). *Journal of Applied Ecology* 34, 566–576.

Morrison, D.A. (1987) A review of the biology of *Acacia suaveolens* (Smith) Willd. (Mimosaceae). *Proceedings of Linnean Society New South Wales* 109, 271–292.

Morton, S.R., Stafford Smith, D.M., Dickman, C.R. *et al.* (2011) A fresh framework for the ecology of arid Australia. *Journal of Arid Environments* 75, 313–329.

Murray, D.R. (1986) Seed dispersal by water. In *Seed Dispersal*, ed. D.R. Murray. Academic Press: Sydney, pp. 49–85.

Nano, C.E.M., Bowland, A.E. and Pavey, C.R. (2013) Factors controlling regeneration in a rare desert tree *Acacia peuce*: limits to soil seed bank accumulation in time and space. *Journal of Arid Environments* 90, 114–122.

Nathan, R., Katul, G.G., Bohrer, G. *et al.* (2011) Mechanistic models of seed dispersal by wind. *Theoretical Ecology* 4, 113–132.

New, T.R. (1983) Seed predation of some Australian acacias by weevils (Coleoptera: Curculionidae). *Australian Journal of Zoology* 31, 345–352.

Noy-Meir, I. (1973) Desert ecosystems: environment and producers. *Annual Review of Ecology and Systematics* 4, 25–41.

O'Dowd, D.J. and Gill, A.M. (1986) Seed dispersal syndromes in Australian *Acacia*. In *Seed Dispersal*, ed. D.R. Murray. Academic Press: Sydney, pp. 87–121.

Ooi, M.K.J. (2007) Dormancy classification and potential dormancy-breaking cues for shrub species from fire-prone south-eastern Australia. In *Seed: Biology, Development and Ecology*, ed. S.W. Adkins, S. Ashmore and S.C. Navie. CABI Publishing: Wallingford, pp. 205–216.

Ooi, M.K.J. (2010) Delayed emergence and post-fire recruitment success: effects of seasonal germination, fire season and dormancy type. *Australian Journal of Botany* 58, 248–256.

Ooi, M.K.J., Auld, T.D. and Whelan, R.J. (2004) Delayed post-fire seedling emergence linked to season: a case study with *Leucopogon* species (Epacridaceae). *Plant Ecology* 174, 183–196.

Ooi, M.K.J., Auld, T.D. and Whelan, R.J. (2006) Dormancy and the fire-centric focus: germination of three *Leucopogon* species (Ericaceae) from south-eastern Australia. *Annals of Botany* 98, 421–430.

Ooi, M.K.J., Auld, T.D. and Denham, A.J. (2009) Climate change and bet-hedging: interactions between increased soil temperatures and seed bank persistence. *Global Change Biology* 15, 2375–2396.

Ooi, M.K.J., Auld, T.D and Denham, A.J. (2012) Projected soil temperature increase and seed dormancy response along an altitudinal gradient: implications for seed bank persistence under climate change. *Plant and Soil* 353, 289–303.

Pate, J.S. and Dixon, K.W. (1981) Plants with fleshy underground storage organs: a Western Australian survey. In *The Biology of Australian Plants*, ed. J.S. Pate and A.J. McComb. University of Western Australia Press: Nedlands, pp. 181–215.

Pickard, J. (1991) Sheep and rabbits: the biological chainsaws. *Search* 22, 48–50.

Portnoy, S. and Willson, M.F. (1993) Seed dispersal curves: behavior of the tail of the distribution. *Evolutionary Ecology* 7, 25–44.

Pound, L.M., Ainsley, P.J. and Facelli, J.M. (2015) Dormancy-breaking and germination requirements for seeds of *Acacia papyrocarpa*, *Acacia oswaldii* and *Senna artemisioides* ssp. *coriacea*, three Australian arid zone Fabaceae species. *Australian Journal of Botany* 62, 546–557.

Rees, M., Condit, R., Crawley, M. *et al.* (2001) Long-term studies of vegetation dynamics. *Science* 293, 650–655.

Regal, P.J. (1982) Pollination by wind and animals: ecology of geographic patterns. *Annual Review of Ecology and Systematics* 13, 497–524.

Robertson, G. (1988) Effect of rainfall on biomass, growth and dieback of pastures in an arid grazing system. *Australian Journal of Ecology* 13, 519–528.

Santana, V.M., Bradstock, R.A., Ooi, M.K.J. *et al.* (2010) Effects of soil temperature regimes after fire on seed dormancy and germination in six Australian Fabaceae species. *Australian Journal of Botany* 58, 439–545.

Setterfield, S.A. (2002) Seedling establishment in an Australian tropical savanna: effects of seed supply, soil disturbance and fire. *Journal of Applied Ecology* 39, 949–959.

Silvertown, J. (1987) *Introduction to Plant Population Ecology.* Longman Group: United Kingdom.

Sinclair, R. (2005) Long-term changes in vegetation, gradual and episodic, on the TGB Osborn Vegetation Reserve, Koonamore, South Australia (1926–2002). *Australian Journal of Botany* 53, 283–296.

Sommerville, K.D., Martyn, A.J. and Offord, C.A. (2013) Can seed characteristics or species distribution be used to predict the stratification requirements of herbs in the Australian Alps? *Botanical Journal of the Linnean Society* 172, 187–204.

Sonia, L. and Heslehurst, M.R. (1978) Germination characteristics of some *Banksia* species. *Australian Journal of Ecology* 3, 179–186.

Tackenberg, O. (2003) Modelling long-distance dispersal of plant diaspores by wind. *Ecological Monographs* 73, 173–189.

Thomas, P.B., Morris, E.C. and Auld, T.D. (2010) The interaction of fire-cues, temperature and water availability regulates seed germination in a fire-prone landscape. *Oecologia* 162, 293–302.

Thompson, K. and Ooi, M.K.J. (2010) To germinate or not to germinate: more than just a question of dormancy. *Seed Science Research* 20, 209–211.

Thomson, F.J., Moles, A.T., Auld, T.D. *et al.* (2010) Chasing the unknown: predicting seed dispersal mechanisms from plant traits. *Journal of Ecology* 98, 1310–1318.

Vaughton, G. (1998) Soil seed bank dynamics in the rare obligate seeding shrub, *Grevillea barklyana* (Proteaceae). *Australian Journal of Ecology* 23, 375–384.

Venn, S.E. and Morgan, J.W. (2007) Phytomass and phenology of three alpine snowpatch species across a natural snowmelt gradient. *Australian Journal of Botany* 55, 450–456.

Westoby, M., Rice, B. and Howell, J. (1990) Seed size and plant growth form as factors in dispersal spectra. *Ecology* 71, 1307–1315.

Westoby, M., Falster, D.S., Moles, A.T. *et al.* (2002) Plant ecological strategies: some leading dimensions of variation between species. *Annual Review of Ecology and Systematics* 33, 125–159.

Whelan, R.J. (1995) *The Ecology of Fire.* Cambridge University Press: Cambridge.

Whelan, R.J., Ayre, D.J. and Beynon, F.M. (2009) The birds and the bees: pollinator behaviour and variation in the mating system of the rare shrub *Grevillea macleayana*. *Annals of Botany* 103, 1395–1401.

Whitney, K.D. (2002) Dispersal for distance? *Acacia ligulata* seeds and meat ants *Iridomyrmex viridiaeneus*. *Austral Ecology* 27, 589–595.

Whitney, K.D. (2005) Linking frugivores to the dynamics of a fruit colour polymorphism. *American Journal of Botany* 92, 859–867.

Wiens, J.A. (1989) Spatial scaling in ecology. *Functional Ecology* 3, 385–397.

Williams, P.R., Congdon, R.A., Grice, A.C. and Clarke, P.J. (2003) Fire-related cues break seed dormancy of six legumes of tropical eucalypt savannas in north-eastern Australia. *Austral Ecology* 28, 507–514.

Williams, R.J., Myers, B.A., Eamus, D. and Duff, G.A. (1999) Reproductive phenology of woody species in a north Australian tropical savanna. *Biotropica* 31, 626–636.

Williams, R.J., Wahren, C.H., Stott, K.A.J. *et al.* (2015) An international union for the conservation of nature Red List ecosystems risk assessment for alpine snow patch herbfields, south-eastern Australia. *Austral Ecology* 40, 433–443.

Willson, M.F., Irvine, A.K. and Walsh, N.G. (1989) Vertebrate dispersal syndromes in some Australian and New Zealand plant communities. *Biotropica* 21, 133–147.
Willson, M.F., Rice, B.L. and Westoby, M. (1990) Seed dispersal spectra: a comparison of temperate plant communities. *Journal of Vegetation Science* 1, 547–562.
Wright, B.R., Zuur, A.F. and Khan, G.C.K. (2014) Proximate causes and possible adaptive functions of mast seeding and barren flower shows in spinifex grasses (*Triodia* spp.) in arid regions of Australia. *The Rangeland Journal* 36, 297–308.
Zimmer, H.C., Auld, T.D., Benson, J. and Baker, P.J. (2014) Recruitment bottlenecks in the rare Australian conifer *Wollemia nobilis*. *Biodiversity and Conservation* 23, 203–215.
Zimmer, H.C., Meagher, P.F., Auld, T.D. *et al.* (2015) Year-to-year variation in cone production in *Wollemia nobilis*. *Cunninghamia* 15, 79–85.

PART II Vegetation Types

11 Rainforests and Vine Thickets

DANIEL J. METCALFE AND PETER T. GREEN

In Australia, rainforests and vine thickets are closed-canopy vegetation types that typically occur in a more mesic habitat than the immediately surrounding sclerophyllous vegetation. Formerly widespread across much of the continent (>5 million years ago), they now occur in an arc from the Kimberley region of Western Australia through the northern parts of the Northern Territory, Cape York Peninsula, and down the east coast through central and southern Queensland, to NSW, Victoria and Tasmania. In eastern mainland Australia most rainforest occurs to the east and south of the Great Dividing Range, with pockets of dry rainforest and vine thickets occurring further inland. Rainforest occurs under a diverse set of rainfall regimes, from 500–600 mm per year and 5–6 dry months, to several metres of rain and essentially no dry months per year. Although rainfall is important in shaping landscape patterns and the distribution of rainforests, soil fertility and fire also play a significant role in determining where rainforests occur. Rainforests harbour a disproportionate percentage of Australia's native vascular plant diversity; collectively, rainforest species account for nearly 20% of native species in just 0.3% of the area. About 66% of the pre-European extent of rainforest remains, but some types have been cleared to near-extinction. Rainforest clearing has now largely stopped, and much of what is left is reserved. However, rainforests still face significant threats including the legacy impacts of past fragmentation, invasive species, and future threats from climate change.

11.1 Introduction

The last few decades have seen transformational change in public perceptions of rainforest, from 'bastard scrub' that was once valued in large part because of the comparative ease with which it could be cleared for agriculture (Francis 1951), to an appreciation of its rich biodiversity and natural heritage values (Metcalfe *et al.* 2014). The Tasmanian Wilderness, Gondwana Rainforests of Australia from southern Queensland to southern New South Wales, and the Wet Tropics of Queensland region from Cooktown to Townsville have all been inscribed on the World Heritage List since 1982, 1986 and 1988, respectively. Other World Heritage properties with representation of smaller but important areas of Australian rainforest include Fraser Island, Kakadu National Park, Blue Mountains and the Lord Howe Island Group.

The grandeur that these forests evoke stems partly from the stark contrast they make with other vegetation types for which Australia is internationally known – its broad expanses of desert and eucalypt-dominated forests and woodlands. Consequently,

Australian Vegetation, Third Edition, ed. David A. Keith. Published by Cambridge University Press.

Citation: Metcalfe, D.J. and Green, P.T. (2017). Rainforests and Vine Thickets. In D. A. Keith, ed., *Australian Vegetation*, 3rd edn. Cambridge: Cambridge University Press, pp. 257–80.

in an Australian context the term 'rainforest' in its broadest sense has been used to encompass those exceptional closed-canopy forest types which are typically more mesic than the immediately surrounding forest and woodland vegetation. Although rainfall is important in shaping landscape patterns and the distribution of rainforests, soil fertility and fire also play a significant role in determining where rainforests occur. This observation is supported by the diverse rainfall regimes under which 'rainforests' occur, ranging from 500–600 mm per year and with 5–6 dry months to several metres of rain and essentially no dry months per year.

The rainforests and vine thickets group includes a wide variety of closed forest types ranging from tropical Queensland where they feature large buttressed trees, palms, vines, large-leaved herbaceous plants, ferns, mosses and epiphytes (Webb and Tracey 1994) to cool temperate Tasmania, where structurally simple beech forests dominate with understories of ferns and bryophytes (Busby and Brown 1994). Between these extremes are littoral rainforests on coasts from Victoria to Cape York (Adam 1992), gallery forests along rivers, semi-deciduous vine thickets spanning the monsoon tropics from Cape York to the Kimberley (Russell-Smith 1998), and a range of subtropical and warm temperate forests, mostly within 100 km of the eastern coast but ranging from near sea level to over 1000 m in altitude (Adam 1992).

Rainforest vegetation has been extensively treated in reviews, not least by Webb and Tracey (1994) and Busby and Brown (1994) in the second edition of *Australian Vegetation*. Aspects of the ecology and dynamics of rainforests have been covered in volumes such as Werren and Kershaw (1987) and Bowman (2000) while Stork and Turton (2008) and Kitching *et al.* (2010) address both social and natural history of rainforests and their use. In this chapter, we provide an overview of aspects of rainforest history, ecology, the threats it faces and the resulting conservation efforts, mainly drawing from literature published in the last 25 years.

11.2 Rainforest history

Once distributed widely throughout the continent, rainforests have contracted enormously (Figure 11.1) since Australia began rifting from the rest of Gondwana 65 million years ago (Weston and Jordan, Chapter 2, this volume). With global climatic cycles in the past 2 million years, they advanced and retreated from local refugia (Rossetto 2008; Mooney *et al.*, Chapter 3, this volume). One consequence of the cycle of expansions and contractions has been high levels of speciation and significant species endemism in refugial areas (Weber *et al.* 2013; Costion *et al.* 2014).

As a broad group, the rainforests and vine thickets include many species belonging to the 'primitive' or near-basal flowering plant families: Winteraceae, Eupomatiaceae, Monimiaceae, Lauraceae and Cunoniaceae (Keith 2004; Metcalfe and Ford 2009). Other families typical of the group include: Capparaceae, Celastraceae, Ebenaceae, Elaeocarpaceae, Euphorbiaceae (*sens. lat.*), Fabaceae (*sens. lat.*), Moraceae, Rutaceae, Meliaceae, Myrtaceae, Pittosporaceae, Rubiaceae, Rutaceae, Sapindaceae, Sapotaceae, Sterculiaceae and Verbenaceae (Webb and Tracey 1994). The persistence and later isolation of the Wet Tropics' rainforests has resulted in high levels of speciation in some families, including the primitive Lauraceae (82 spp., 8 genera) as well as the Myrtaceae (124 spp., 33 genera) and Proteaceae (49 spp., 26 genera). Elements of both the tropical and temperate rainforests reflect their Gondwanan origins. However, the proportion of taxa originating from Gondwanan vicariance is greatest in cool temperate rainforests, declines in more northern and lowland rainforests (Sniderman and Jordan 2011), and is lowest in the monsoon forests of more recent origin (Crisp and Cook 2013).

In a continental analysis, Kooyman *et al.* (2013) showed that species richness and phylogenetic diversity trended with moisture and latitude. The richest flora occurs in the northern wet tropics and diversity declines with increasing seasonality across the north, and with increasing latitude. The declines

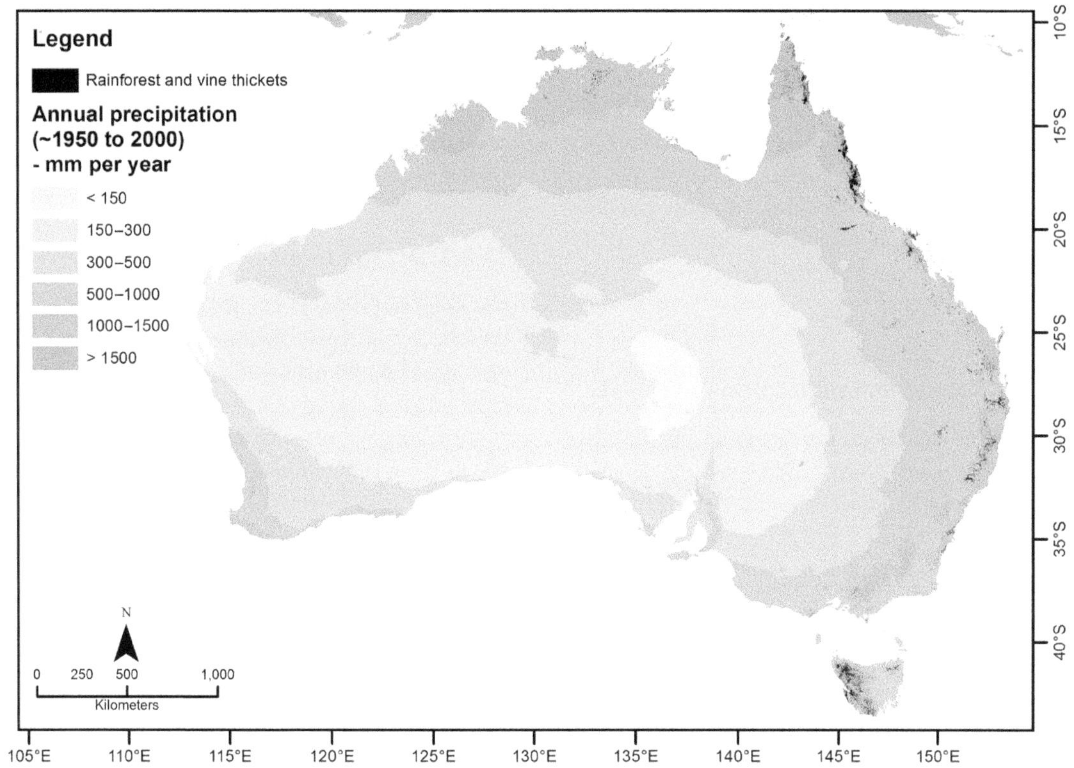

Figure 11.1 Distribution of 'rainforest and vine thicket' (National Vegetation Information System, Major Vegetation Group MVG 1) with rainfall isohyets, showing strong correlation of the distribution of these closed forest types with high annual rainfall.

in both richness and phylogenetic diversity with latitude are driven by temperature not rainfall; the forests in the coolest areas of southern Australia receive high rainfall, yet are very species poor, and also richness declines with altitude (and hence decreasing temperature and increasing rainfall), at least in the subtropics and temperate zones (Floyd 1990). Phylogenetic endemism is highest in the largest areas of moist rainforest – Cape York, the Wet Tropics, the Border Ranges of southern Queensland – northern New South Wales, and western Tasmania.

Anthropogenic modification of rainforest habitat has occurred in two phases. First, Indigenous Australians utilised rainforest habitat as a source of food and materials over 50 000 years (Horsfall 1991; Bridgwater *et al.* 1998; Hill and Baird 2003). Second, there has been a rapid and dramatic period of landscape modification by non-Indigenous Australians over the last *c.* 200 years as extensive timber extraction and clearing for agriculture has fundamentally changed much of the landscape, particularly in the south and east. More recently there have been efforts to protect, restore or replant rainforest (Metcalfe *et al.* 2014).

11.3 Structural and floristic variability

Several structural, physiognomic and floristic classifications have been proposed for Australian rainforests (Adam 1992). The current National Vegetation Information System (Department of the

Environment and Energy 2007) recognises four Major Vegetation Subgroups (MVS) within the Rainforests and Vine Thickets Major Vegetation Group (MVG), 'tropical or subtropical rainforest', 'dry rainforest or vine thickets', 'cool temperate rainforest' and 'warm temperate rainforest'. We separate 'monsoon forests' from other dry rainforest and vine thickets following Keith and Pellow (2015). Characteristic features of different rainforest types are discussed (Table 11.1).

11.3.1 Tropical and subtropical rainforests

Tropical and subtropical rainforests are characterised by high taxonomic diversity and structural complexity (Figure 11.2a). Tropical rainforests primarily occur on Cape York Peninsula and in the Wet Tropics bioregion of north Queensland; the southern limit of most tropical species is around Mackay. Subtropical rainforests have a patchy distribution from Rockhampton through south-east Queensland and northern New South Wales, extending as far south as Narooma.

The rainforests of Cape York are distributed between Torres Strait to the north and the Laura Basin to the south, where a band of sclerophyll vegetation restricts the spread of rainforest species. The largest area of lowland rainforest in Australia occurs around the Lockhart River, and together with the rainforest on the adjacent Iron and McIlwraith Ranges this is the most extensive area of rainforest on Cape York (Metcalfe and Ford 2008). The tropical lowlands have a high proportion of species allied to New Guinea or Asia, while the upland rainforests on the granitic McIlwraith Range have close affinities with the forests of the Wet Tropics, and support a greater proportion of species with Gondwanan ancestry (Hitchcock *et al.* 2013). Species diversity is not especially high at the plot level by tropical standards even in the Wet Tropics rainforests, though they contain a large number of endemic species (674/2141 rainforest species; Metcalfe and Ford 2008). Subtropical rainforest communities (Figure 11.2b) occur within c. 150 km of the coast, and species diversity declines southwards as they grade with warm temperate forests. Relatively large stands remain on mountain and escarpment slopes, but extensive clearing at lower elevations and on coastal sands means that we have already lost the chance to understand much of these communities.

11.3.2 Dry rainforests

Dry rainforest occurs in small, discontinuous inland patches across a wide range of latitudes in Australia, from Chillagoe in north Queensland to the Mitchell River Gorge in Victoria (Keith and Sanders 1990; Cameron 1992; Fensham 1995; Curran *et al.* 2008). These communities come under the rainforest umbrella due to their structural and floristic affinities with mesic rainforest (Webb and Tracey 1981), and because the recruitment dynamics of their component species are not fire-driven. However, they are typically low statured and occur in climates where total annual rainfall is relatively low (often < 800 mm) and sufficiently unreliable to cause seasonal drought stress. Vine thickets (Figure 11.2c) represent extreme forms of dry rainforest in conditions where seasonal drought is more pronounced (McDonald 1996). In Queensland, dry rainforest can even persist up to 500 km inland from the coast and where mean annual rainfall is c. 500 mm, while some stands in New South Wales experience near-freezing minimum temperatures during winter (Fensham 1995; Curran *et al.* 2008). A comprehensive biogeographic treatment of dry rainforest in Australia has not been attempted, but a comparison of typological studies in Queensland (Fensham 1995) and New South Wales (Curran *et al.* 2008) indicates some similarity at the genus level; 15 of 35 tree genera, but only 9 of 47 tree species, listed for New South Wales also occur in Queensland.

One factor allowing dry rainforest to persist in xeric, otherwise fire-prone environments is that they often occur on rocky substrates affording protection from fire (Cameron 1992; Fensham 1995). Dry rainforest species also show adaptations to persist

Table 11.1 **Comparison table of key attributes of the five major rainforest and vine thicket subgroups recognised by the National Vegetation Information System 4.2. Based on a variety of sources including Adam (1992), Webb and Tracey (1994), Russell-Smith (1998), Busby and Brown (1994) and Keith and Pellow (2015).**

Subgroup	Tropical and subtropical rainforest	Dry rainforest and vine thickets	Monsoon Vine Forest	Warm Temperate Rainforest	Cool Temperate Rainforest
Structure	Complex species-rich, multi-layered tree canopies typically exceeding 30 m in height, often with emergent trees and with a variety of compound and simple leaves	Simple low-statured canopies, 10–25 m	Canopy height <2 m to >25 m depending on exposure and seasonal dryness	Tree canopy structurally complex, often with multiple tree canopy layers	Usually a single tree stratum with canopy 8–30 m tall
Plank buttresses	Common	Absent to rare	Absent	Rare	Absent
Canopy leaf retention	Evergreen, but with a minority of deciduous species	Evergreen to deciduous with at least some semi-deciduous species	Almost entirely deciduous or entirely evergreen, depending on dry season soil moisture retention and springs	Evergreen	Evergreen
Leaf sizes of dominant tree species[a]	Notophylls to macrophylls	Mostly notophylls	Mostly notophylls	Microphylls to notophylls	Nanophyll and microphyll
Palms	Free-standing palms often abundant; climbing rattans, *Calamus* spp., common in the tropics	Typically absent (though see emergent palm *Corypha utan* in Figure 11.2c)	Uncommon; if present mainly *Carpentaria* and *Livistona*	Uncommon; if present mainly *Livistona*	Absent (except on montane Lord Howe Island)
Lianes and epiphytes	Often abundant	Lianes abundant, few epiphytes	Lianes abundant, few epiphytes	Not typically abundant	Few vascular epiphytes and lianes
Gymnosperms	*Cycas*, *Bowenia*, *Lepidozamia*, Araucariaceae, *Callitris*, Podocarpaceae	*Cycas*, *Araucaria*, *Callitris*	*Cycas*, *Callitris*	*Araucaria*, *Callitris*, *Podocarpus*	*Arthrotaxis*, Podocarpaceae

Table 11.1 (cont.)

Subgroup	Tropical and subtropical rainforest	Dry rainforest and vine thickets	Monsoon Vine Forest	Warm Temperate Rainforest	Cool Temperate Rainforest
Dominant/typical tree genera[b]	Many codominant species at plot level incl. multiple genera from many families incl. Lauraceae, Myrtaceae, Proteacae, Sapindaceae, Moraceae, Rubiaceae, Elaeocarpaeae, Cunoniaceae, Sterculiaceae	Euphorbiaceae, Moraceae, Myrtaceae, Sapindaceae and Sterculiaceae. *Araucaria* and several *Acacia* species may occur at subtropical latitudes.	Evergreen forests may be dominated by *Allosyncarpia ternata*, semi-deciduous forests typically include many genera from Combretaceae, Euphorbiabeae, Moraceae, Myrtaceae and Rubiaceae	Typical tree genera include *Acmena*, *Ceratopetalum*, *Caldcluvia*, *Schizomeria*, *Elaeocarpus*, *Callicoma*, *Orites*, *Doryphora*, *Endiandra*, *Pittosporum*	*Nothofagus*, *Athrotaxis*, *Phyllocladus*, *Atherosperma*, *Eucryphia*, *Lagarostrobus* and/or *Diselma*, usually one dominant per site
Other notes	Strangling *Ficus* common; terrestrial herbs uncommon	Bryophytes generally absent, ferns generally uncommon. Understorey and ground layer sparse but may include grasses and herbs		Strangling *Ficus* generally absent.	Greatest species richness is in lichens and mosses rather than vascular plants; tree ferns and ground ferns common
Mean annual rainfall (mm)	>1200 mm and up to 8000 mm, with significant dry season cloud stripping in more elevated sites	600–1100 mm, with a pronounced dry period	500–2000 mm, where there is a very long dry season, on lowlands and floodplains or in rocky gorges	1000–2000 mm, predominantly summer rainfall	1200–3500 mm, with persistent low cloud supplementing limited dry season rainfall
Distribution	QLD and NSW, east of the Great Dividing Range	QLD, NSW, VIC Queensland south from Townsville district to northern NSW, becoming confined to the coastal hinterland at latitudes greater than 32°S	NT, WA	NSW, VIC	TAS, VIC, NSW and SE QLD

[a]Leaf sizes according to Webb (1959).
[b]Common families or genera taken from a variety of sources including Webb and Tracey (1994), Russell-Smith (1998), Keith (2004) and Harden *et al.* (2009).

Figure 11.2 (a) Tropical rainforest, Mt Lewis, Far north Queensland (photo: Dan Metcalfe); (b) subtropical rainforest, Bunya Mountains, Queensland (photo: Dan Metcalfe); (c) semi-deciduous vine thicket, Melamen River, Cape York (photo: Dan Metcalfe); (d) monsoon vine thicket, Kimberley, Western Australia (photo: Bruce Webber); (e) warm temperate rainforest, Bola Creek, Royal National Park, New South Wales (photo: David Keith); (f) cool temperate rainforest, Overland Track, Cradle Mountain-Lake St Clair National Park, Tasmania (photo: Travis Sydes).
(A black and white version of this figure will appear in some formats. For the colour version, please refer to the plate section.)

under conditions of seasonal drought. Evergreen species have hydraulic architecture that, during drought, permits them to withstand greater xylem pressure gradients required to maintain water transport to the canopy without embolism. Deciduous species have low wood density and high water storage capacity, but hydraulic architecture that places them at greater risk of embolism (Choat *et al.* 2005). Deciduous species are more common in the tropics and this strategy becomes rarer with increasing latitude (Curran *et al.* 2008), presumably because this strategy is disadvantageous in climates where the onset of the growing season is highly variable (Sobrado 1991), and on low-fertility soils (Orians and Milewski 2007), one or both of which may characterise areas of dry rainforest in southern Australia.

11.3.3 Monsoon rainforests

The monsoon rainforests of northern Australia (Figure 11.2d) are distinguished from the dry rainforests on floristic grounds, but also because they experience a more strongly seasonal drought (Keith and Pellow 2015). In the Northern Territory this formation occurs close to the coast as mostly small, discrete and widely separated patches in a matrix of savanna grassland and woodland. Structural complexity and floristic diversity are highest in a narrow coastal or subcoastal belt. As in the Wet Tropics, diversity is greater than other vegetation formations at higher taxonomic levels (358 species in 250 genera in 95 families; Russell-Smith and Dunlop 1987). Rainforest exists in areas receiving 500–2000 mm rainfall annually, being more concentrated in higher rainfall areas and more isolated in drier areas. The patchiness of these forests is likely a result of fire, combined with the presence of ephemeral and more ancient refugia from which propagules might disperse when circumstances permit the (re)development of rainforest vegetation (Russell-Smith and Dunlop 1987; Bowman 2000). Evergreen rainforests on sandstones with perennial moisture availability show strong similarities to the riverine forests of the Kimberley (Figure 11.2d); vine forests on seasonally dry sandstones, in contrast, are typically almost mono-dominant stands of *Allosyncarpia ternata* or *Xanthostemon* species. On other rock types, vine forests are typically more stunted and thicket-like, coping with very limited soil development and consequently little or no accessible water through the dry season. Lowland vine forests are best developed at perennially moist sites where root systems have access to the water table during the dry season. As dry season water availability declines, canopy height diminishes and the proportion of deciduous species increases. Vine forests also extend onto coastal regions, occupying headlands and beach ridges (Russell-Smith 1998).

Most species in the relatively depauperate Western Australian rainforest flora (Figure 11.2d) also occur in the Northern Territory, and about two-thirds of them are also found in Queensland (Kenneally *et al.* 1991), though one of the best-known and distinctive regional endemics, the boab (*Adansonia gregorii*) occurs at the edge of rainforest and in savanna. Deciduous canopy emergent trees in the semi-evergreen mesophyll vine forests, especially on the Mitchell Plateau, permit the development of a dense understorey with masses of climbers and scandent shrubs develop at sites damaged by fire and cyclones (Kenneally and Beard 1987). Low endemicity, broad species distributions, high dispersability, and the rapid development of rainforest stands (e.g. Russell-Smith and Dunlop 1987) suggest that the rainforests of the Kimberley and Dampier Peninsula are not relicts of an earlier, more widespread vegetation but instead are more recent invasions facilitated by suitable climate and protection from fire (Kenneally *et al.* 1991).

11.3.4 Warm temperate rainforests

Warm temperate rainforests (Figure 11.2e) occur from southern Queensland through New South Wales to east Gippsland, the Strzelecki Ranges and Wilsons Promontory in Victoria. This type of rainforest occurs mostly as small (<100 ha), linear patches in sheltered, moist gullies in hilly terrain, sheltered sides and plateaus of high mountains, but it can also occur on river flats of the coastal plain in east Gippsland. In New South Wales, 'northern' warm temperate rainforest occurs from the Queensland border south to Bateman's Bay and is dominated by *Ceratopetalum apetalum*, *Doryphora sassafras* and *Syzygium smithii* (Keith 2004). The northern types tend to occur on lower-nutrient acidic soils derived from rhyolites, porphyries or granites, are very free draining and probably dry out excessively during rare drought events (J. Dwyer, *pers. comm.*). The 'southern' type occurs from the Shoalhaven River to the Victoria border, and has *Acacia melanoxylon*, *Doryphora sassafras*, *Pittosporum undulatum* and *Syzygium smithii* as canopy dominants. This southern warm temperate rainforest is shorter statured than the northern variant, and has a more open understorey and relatively dense fern layer. In Victoria, some stands of warm temperate rainforest tend towards single-species canopy dominance by *Syzygium smithii*, with various woody vine species. In New South Wales, warm temperate rainforest may contain remnant *Eucalyptus* or *Lophostemon* (Floyd 1990), and sclerophyll emergents also occur in Victorian warm temperate rainforest but mostly within marginal ecotones where there is a history of disturbance (Cameron 1992).

11.3.5 Cool temperate rainforests

Cool temperate rainforests occur from southern Queensland to southern Tasmania, and differ from other types of rainforest in Australia in their relatively low floristic diversity of woody species, and rarity of lifeforms such as woody lianas and vascular epiphytes (Cameron 1992). This rainforest type reaches its greatest areal extent and structural diversity in Tasmania (Figure 11.2f), where the majority of stands occur in the western third and north-eastern portion of the state. High rainfall is widespread in Tasmania and cool temperate rainforest occurs across a range of topographies, aspects and soils, giving rise to a diversity of structural types and floristic associations (Busby and Bridgewater 1977). Tall rainforest is usually dominated by *Nothofagus cunninghamii* with *Atherosperma moschatum* as the main subdominant species (Jarman *et al.* 1987). Lower-stature rainforest (<20 m tall) has uneven, broken canopies and a dense, tangled understorey. The intermediate rainforest type is usually dominated by trees of medium height (20–30 m tall), and with a well-developed shrub layer; it is often referred to as 'horizontal scrub' because of the growth habit of the characteristic species *Anodopetalum biglandulosum*. Open montane rainforest is confined to high altitudes associated with boulder fields or rocky outcrops on the Central Plateau and south to Mt Field. Only callindendrous cool temperate rainforest occurs on mainland Australia; *Nothofagus moorei* is the canopy dominant from southern Queensland to the Hunter River in New South Wales where it usually occurs at high elevations, and *Eucryphia moorei* or *Elaeocarpus holopetalus* dominate between Sydney and the Victorian border (Floyd 1990), while *Nothofagus cunninghamii* dominates in Victoria.

11.4 Ecological processes that influence the distribution and dynamics of rainforest

Rainforest patches are highly dispersed both within and between regions (Figure 11.1). They naturally occur in various sizes within a mosaic of other vegetation types, and can include linear strips and small patches that are below the resolution of vegetation mapping or legislative protection, and also combine with other vegetation types in places to form mixed communities. This dispersion through the landscape is natural, but exacerbated by human activities. Species-specific ability for dispersal within and between such patches helps to define the composition and nature of local plant communities. Very small patches and mixed communities also facilitate seed dispersal between larger rainforest patches. Seed dispersal is discussed below, but the prime determinants of landscape-scale rainforest dynamics and distribution are abiotic.

11.4.1 Precipitation

The distribution of rainforest types across eastern and northern Australia is strongly associated with the amount and seasonality of rainfall (Figure 11.1), the availability of water in the dry season, soil fertility, the incidence of fire, and stochastic destructive events such as tropical cyclones. Classic definitions of rainforest focus on plant communities which receive sufficient precipitation that they

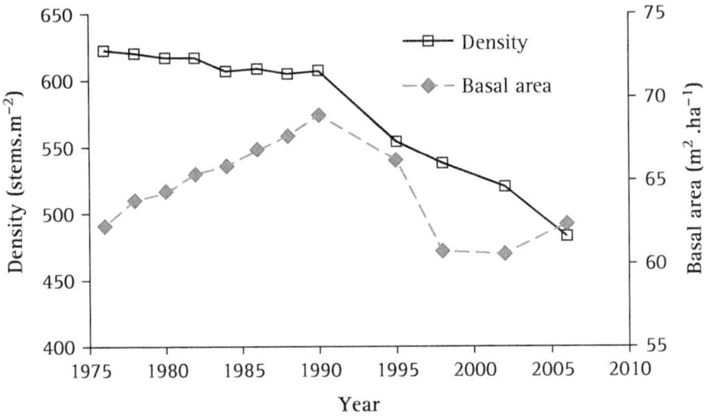

Figure 11.3 Change in basal area (m² ha⁻¹) and stem density at the Curtain Fig permanent tropical rainforest plot (Atherton Tableland, north-east Queensland) between 1975-2006, showing the prolonged impact of a 'dry' wet season in 1992-93 on basal area and stem number. This was the most severe drought in the area for 90 years. Replacement of dead trees only began to occur with recruitment of new saplings during 2002-06.
Redrawn from Metcalfe et al. (2014).

never come under drought stress (e.g. Walter and Breckle 1986). Few if any of these areas exist in Australia, though some areas of the Wet Tropics of Queensland and south-western Tasmania provide optimal conditions for rainforest development. The Great Dividing Range runs close to the coast in the Wet Tropics, so moisture-bearing winds from the Coral Sea must rise to pass over this formation, depositing huge amounts of rain on the eastern flanks. Most of the rain falls in the summer months when monsoonal troughs and cyclonic lows can release metres of rain over a few days or weeks, but importantly rain may fall in any month as the prevailing south-easterlies bring moist air off the sea.

In years when summer rainfall is reduced, physiological drought can have significant and long-lasting effects on community structure. A 30-year time series from a permanent plot on the Atherton Tableland, north-east Queensland, shows steady growth and gradual thinning of trees, until a major drought in 1992–93 caused sustained loss of large trees, with recruitment of new saplings only beginning to compensate drought-related tree mortality more than a decade later (Figure 11.3). In addition to orographic precipitation, up to 50% of winter precipitation may be obtained through cloud stripping, when water-heavy clouds pass through the forest canopies and condensation forms on leaves and branches and runs or drips down to the forest floor (Wallace and McJannet 2008). Deciduous species dominate as dry season precipitation decreases, and evergreen forests are only found in areas where roots can reach permanent ground water, such as in riverine forests.

11.4.2 Edaphic factors

Soil depth, texture and fertility compound the effects of water availability, though modifying rather than determining distribution. Tracey (1969) and Webb (1969) considered soil physical and chemical characteristics under rainforest and adjacent sclerophyll communities between Victoria and the Wet Tropics, and found no consistent pattern at the continental scale. Rather, evaporation, temperature, rainfall, altitude, soil fertility, topography and incidence of fire all contributed to vegetation type. For example, in the higher rainfall zones soil aeration and the tendency to become seasonally inundated was a significant factor, though modified by topography. At a more local level, a detailed assessment of forest differentiation at a single site (Fraser Island) found that more complex vegetation was generally associated with greater soil fertility (Webb 1969), and Beadle (1954) suggested that while rainforest litter was notably higher in phosphorus, this was driving rather than responding to soil fertility. Ash (1988) showed that 38% of rainforest/sclerophyll boundaries in the Wet Tropics coincided with a geological boundary, but Warman *et al.* (2013) concluded that soil attributes themselves did not cause the boundaries; variation in soil characteristics within rainforest or sclerophyll soil types was as great as between them. However, they noted that as sclerophyll woodlands are invaded by rainforest their soils become more like rainforest soils, due to reduced incidence of fire and higher levels of bacterial activity, so similarities between sclerophyll and rainforest soils around boundaries probably reflect a legacy of fluid boundaries driven by climatological trends and fire incidence. Adam (1992) reviewed the effects of high aluminium content in acid soils under warm temperate rainforest, where aluminium accumulation is a marked feature of many species.

11.4.3 Fire

After precipitation, fire is probably the greatest determinant of rainforest distribution, especially of drier types. Fire has long been recognised for its ability to create and maintain rainforest boundaries. The last several decades have seen the development of conceptual models explaining the role of fire in determining rainforest boundaries, and much empirical work that has aided mechanistic understanding. Although early versions of models describing the expansion of rainforest and its fate through the next fire were relatively simple (Hopkins *et al.* 1993), subsequent models have developed a more nuanced understanding of boundary dynamics that is determined by fire severity and return intervals, along with plant traits and fuel type and load (reviewed in Bowman 2000).

With few exceptions stand-replacing fires do not currently occur in structurally mature and undisturbed rainforest in Australia. This has been attributed to a combination of factors that include lack of fuel in the understorey (especially grasses), humid microclimates (Turton and Sexton 1996; Little *et al.* 2012) that impede the spread of fire, and in the case of some stands of warm temperate, cool temperate and monsoonal rainforest, the topographic protection afforded by rocky outcrops, incised gullies and gorges (Cameron 1992; Russell-Smith 1998; Doherty *et al.* 2011; Wood *et al.* 2011). Somewhat surprisingly, leaf-level plant traits appear to play little if any role in the relative pyrophobicity of rainforest (Clarke *et al.* 2014). Rainforest disturbed by frost (Duff and Stocker 1989) or cyclones (Webb 1958; Unwin *et al.* 1988; Bowman and Panton 1994; Bowman *et al.* 1999) may be more susceptible to fire because of increased fuel loads. Although rare, crown fires do occur under extreme weather conditions in temperate forests (e.g. Chesterfield *et al.* 1991). Patch extent, landscape context and fire severity may affect the probability of stand replacement in these systems, and the interpretation of rainforest and adjacent sclerophyll forest as alternative stable states (Warman and Moles 2009; Wood *et al.* 2011). Drying associated with climate change may increase the likelihood of such intense fires, and their impacts (Holz *et al.* 2015).

Under drought or seasonally very dry conditions, fires that start outside rainforest can sometimes penetrate boundaries and burn through extensive areas of rainforest understorey. For example, scorch heights to 12 m have been observed in monsoonal rainforest in the Northern Territory (Fordyce *et al.* 1997), and low 'litter' fires have been recorded in tropical rainforest in Queensland (Edwards and Krockenberger 2006) and temperate rainforest in Tasmania and Victoria (Hill and Read 1982; McMahon 1987). Probably because of their rarity, the impact of fire on the understorey of rainforest has seldom been quantified in Australia.

For a long time, rainforest species and formations were defined in terms of their inability to withstand fire, and their ability to recruit juveniles in the absence of fire (e.g. Francis 1951). Probably the most significant advance in our understanding of the dynamics of rainforest boundaries is the decoupling of sensitivity and recruitment, and the treatment of these as independent rather than co-occurring traits. Contrary to simple models, rainforest species are not uniformly susceptible to fire, and there is widespread evidence across the full range of rainforest types that many species can resprout after fire (monsoonal species – Bowman 1991; Fordyce *et al.* 1997; Russell-Smith *et al.* 1998; tropical species – Stocker 1981; Marrinan *et al.* 2005; Williams *et al.* 2012; subtropical species – Campbell and Clarke 2006; warm temperate – Cameron 1992; Knox and Clarke 2011; cool temperate – Barker 1992). Although the resprouting ability of many rainforest species has been known for some time, it is only relatively recently that this phenomenon has been explicitly recognised as an important stabilising mechanism in rainforest boundary dynamics (Bowman 2000; Clarke *et al.* 2015). Regular fire weakens this stabilising mechanism and rainforest dominance in two ways – through cumulative mortality of larger stems across successive fires, and for the survivors, impact on plant height. Because their crowns are killed by fire, recurrent fire keeps these plants short statured, smothered in more rapidly regrowing vines or reduces their ability to shade out grasses, promoting the persistence of flammable fuel (Williams *et al.* 2012). A small number of rainforest species actively invade wet sclerophyll or grassland communities, and in these only regular fire will dislodge them (Williams *et al.* 2012).

11.4.4 Cyclones

Cyclones are an important driver of the dynamics of the rainforests of the tropics – in 2011 severe Tropical Cyclone Yasi affected around 9000 km^2 of forest and caused the death of c. 3×10^8 trees (Negrón-Juárez *et al.* 2014). Along the east coast the prevailing winds are from the south-east, but tropical cyclones irregularly cross the coast from a variety of directions, when high wind speeds, localised gusting, very high-rainfall events and storm surges may affect the vegetation. The effects of wind and rain are most apparent on the coastal ranges, where repeated canopy disturbance and subsequent growth of vine towers on remaining trees causes a distinctive 'cyclone scrub' (*sensu* Webb 1958; Bruce *et al.* 2008). This is dominated by a few species (e.g. *Merremia peltata*), and under it, there is very limited regeneration of tree and shrub species. Another effect of cyclonic winds is canopy shredding, leading to removal of most of the leaves and exposure of the understorey to high light levels. This exposure can lead to mass recruitment of light-demanding species, and increased light levels may also trigger mass flowering of shrubs and small trees (Hopkins and Graham 1987). Both wind speed and direction and also species-specific characteristics affect the impacts of any given cyclone on individual species and on communities, with some species being more susceptible than others to damage at lower wind speeds (Metcalfe *et al.* 2008). The scale of cyclone disturbance can also affect seedling survival and thus the composition of post-cyclone recruitment (Murphy *et al.* 2014).

The impacts of repeated cyclones on community structure are essentially cumulative. Time series data from lowland rainforest near Cairns, north-east Queensland, show very slow rates of recovery (Figure 11.4). Each event causes major declines in tree density and size, with basal area recovering slowly by growth, and no evidence of net recruitment of new stems after 20 years. While severe cyclones are thought to occur in the region at 50–100 year intervals (Turton 2012), return intervals of 20 years between major cyclones are likely to cause long-term structural changes. Furthermore, in highly disturbed forest patches recruitment may be dominated by invasive species (Murphy *et al.* 2008).

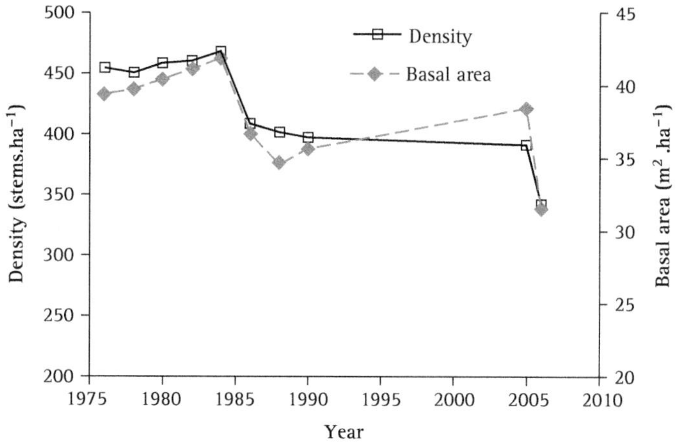

Figure 11.4 Change in basal area (m² ha⁻¹) and stem density at the Woopen Creek permanent tropical rainforest plot (south of Cairns, north-east Queensland) between 1975 and 2006, showing the impacts of successive disturbances associated with Tropical Cyclone Winifred (1986) and Severe Tropical Cyclone Larry (2006).
Redrawn from Metcalfe et al. (2014).

11.4.5 Seed dispersal

Seed dispersal is mediated by vertebrates in 70–90% of tropical rainforest plant species, and the majority of these are endozoochorous (dispersed by ingestion), with most plants serviced by several different animals (Westcott and Dennis 2006). Major groups of dispersers include rodents, flying foxes, cassowaries, pigeons, bowerbirds, riflebirds, orioles, currawongs, honeyeaters and metallic starlings. Dispersal distances range from <100 m to >800 m, with cassowaries and flying foxes regularly transporting seeds over landscape-scale distances (Westcott and Dennis 2006). This dependence on vertebrates in rainforest relative to other vegetation types, where wind, gravity and invertebrates play a greater role, may in part be linked to the near-exclusion of wind from closed-canopy forests and to the mesic conditions which permit investment in water-rich fruits as disperser attractants (Butler *et al.* 2007; Kooyman *et al.* 2013). Wind dispersal may only be a viable alternative for canopy emergents and the lianas that climb them. The production of attractant fruits and consequently the dominance of endozoochory declines in temperate latitudes (Chen *et al.* 2016).

11.5 Major conservation issues and threatening processes now, and in the future

Conservation International has listed the Forests of East Australia, which includes the majority of rainforest to the east of the Great Dividing Range, as their 35th Global Biodiversity Hotspot in recognition of the exceptional degree of vascular plant endemism, combined with high levels of threat (Williams *et al.* 2011). Although iconic examples of Australian rainforest are now protected with World Heritage status, the legacies of past land use and future global environmental change pose ongoing threats to their persistence. In this section, we review the principal threats.

11.5.1 Clearing and fragmentation

Notwithstanding extensive protection in some areas, clearing and consequent fragmentation is still the greatest single threat to the persistence of rainforest across Australia (e.g. Sattler and Williams 1999). Available data suggest that about 34% of the estimated pre-European extent of rainforest has been cleared, but rainforest in some regions has been more heavily cleared than others (Department of the Environment and Energy 2007). For example, the 'Big Scrub' of northern New South Wales was reduced to <1% of its former extent by 1900 (Floyd 1987), and the rainforests of the Illawarra by around 75% (Mills 1986). Subtropical vine thickets from central Queensland to northern New South Wales, have also suffered substantial losses (Keith 2004). Several rainforest formations are now listed under the federal legislation as Endangered or Critically Endangered, and many regional variants are listed under relevant state legislation (SPRAT 2013). Increasing appreciation of the biological and aesthetic values of rainforest and the ecosystem services they provide has resulted in greater legislative protection and almost complete cessation of rainforest clearing in Australia. Those areas still under threat are subject to intense public debate.

In addition to the loss of forested area *per se*, one of the enduring impacts to rainforest of past clearing is habitat fragmentation. Although Australian rainforests occur naturally as archipelagos of patches that vary in size within a non-rainforest matrix, clearing creates edges across which abiotic variables such as light, temperature, humidity and wind speeds differ from the forest interior (Turton and Freiburger 1997; Pohlman *et al.* 2009). These altered environments favour disturbance-adapted plant species, leading to a change in vegetation structure and composition, and invasion by weedy species in both tropical and warm temperate rainforest (Fox *et al.* 1997). Cyclones are an important feature of vegetation dynamics and a potential cause of at least temporal fragmentation in tropical rainforests (see below), and logically it might be expected that small remnants should be more vulnerable to them than continuous forest. However, several studies conducted in the wake of Tropical Cyclone Larry in 2006 found no such effect (Catterall *et al.* 2008; Grimbacher *et al.* 2008, though see Bruce *et al.* 2008).

The effects of fragmentation on population genetic structure have only been studied for a single rainforest tree, *Elaeocarpus grandis* (Rosetto *et al.* 2004a). There was no detectable genetic structure across the species at the southern end of its range in New South Wales, despite populations being isolated for at least a century. The maintenance of gene flow was attributed to the dispersal of fleshy fruits by birds and bats. Another species in the same genus, *E. williamsianus*, is mostly clonal and Rosetto *et al.* (2004b) have postulated that recent fragmentation has caused an increase in vegetative reproduction in that species.

In Australia, numerous studies have documented the effects of fragmentation on populations of small and medium-sized rainforest mammals and birds (Laurance 1997; Moran and Catterall 2014), sometimes leading to community-level shifts in vertebrate assemblages that have been linked empirically to altered patterns of seed input and seed predation in rainforest fragments (Moran *et al.* 2009).

11.5.2 Invasive species

Invasive species, particularly invasive plants, probably pose the greatest threat to the integrity and quality of remnant rainforests. Although undisturbed wet rainforests are generally resistant to invasion by non-native plant species (Rejmánek 1996), several invaders pose serious threats in the tropics because they are shade tolerant and can establish in the dim understorey of even intact forest (e.g. *Miconia calvescens*, Murphy *et al.* 2008; *Miconia nervosa*, Brooks *et al.* 2009; *Cinnamomum camphora*, Neilan *et al.* 2006), or they are vines and colonise rainforest canopies (e.g. *Mikania micrantha, Chromolaena odorata*, Goosem 2008). Exotic vines are a serious threat to subtropical rainforests, especially when disturbed or fragmented by clearing or logging (Leishman *et al.*, Chapter 9, this volume).

Other species such as *Lantana camara* are notorious invaders of rainforest edges and gaps from historical logging or natural disturbances (Gentle and Duggin 1997) or more open types of rainforest (Fensham *et al.* 1994), where they can alter the diversity of native species (Gooden *et al.* 2009) or increase fire risk (Berry *et al.* 2011). Invasive woody shrubs such as *Hyptis suaveolens*, *Senna occidentalis* and *Sida cordifolia*, and the passion vine *Passiflora foetida* are prevalent in monsoonal rainforest of the Northern Territory, where Russell-Smith and Bowman (1992) estimated 22% of patches were 'infested' with weeds. A variety of weeds have been identified as a key threat to the highly endemic rainforest flora on Lord Howe Island (Auld and Hutton 2004). A number of factors promote the invasion of weeds, including altered fire regimes and disturbance by feral animals in monsoon rainforests (Russell-Smith and Bowman 1992), fragmentation (see above), the structural damage caused by cyclones (Goosem 2008; Murphy *et al.* 2008), and dispersal by native seed frugivores (Westcott *et al.* 2008).

Of the myriad of non-native animals that have invaded rainforest in Australia, large vertebrates have the largest and most direct negative impacts on vegetation. For example, Russell-Smith and Bowman (1992) recorded major tree death caused by feral cattle and Asian water buffalo at 20% of monsoon rainforest sites, and recorded severe soil disturbance by feral pigs at 10% of sites. Such disturbance causes significant declines in seedling numbers, at least in the short term (Taylor *et al.* 2011). Congdon and Harrison (2008) regard the feral pig as the highest impact vertebrate pest in rainforest vegetation in the Wet Tropics. Herbivory and environmental degradation caused by feral deer is recognised as a threatening process in New South Wales and Victoria (Keith and Pellow 2005; Peel *et al.* 2005).

Rainforest on Christmas Island is threatened by mutualism between non-native invasive invertebrates, the yellow crazy ant *Anoplolepis gracilipes*, and several non-native honeydew-producing scale insects (O'Dowd *et al.* 2003). This mutualism has permitted the ants to form high-density supercolonies (Abbott 2005). This novel mutualism has manifold negative impacts on native species abundances, interactions among species, and forest structure. While the rainforests of Christmas Island are Asian rather than Australian in character, the loss of biodiversity and ecosystem integrity following invasion by exotic ants indicates the potential impacts these invertebrates may have on the mainland, with yellow crazy ants having recently established in the Wet Tropics (Lach *et al.* 2015).

11.5.3 Disease

Relatively little is known about the extent and magnitude of impacts of disease on rainforest communities in natural settings. The oomycete *Phytophthora cinnamomi* causes significant dieback impacts in the subtropics (Scarlett *et al.* 2015) and temperate rainforests (Podger *et al.* 1990) where it has been introduced but has more limited impacts in tropical rainforests, presumably because it is native to those regions (including New Guinea) and the flora has evolved with it (Metcalfe and Bradford 2008). Nonetheless, there is evidence of significant and prolonged impacts of disease infections at sites such as Garrawalt, north-west of Townsville in north-east Queensland (Figure 11.5). While disease-free sites steadily increase in tree density and size, disease infection causes significant mortality of large trees. Recruitment occurs rapidly in the resulting gaps, but typically with a different suite of species that may take several decades to restore pre-disease forest structure (Figure 11.5).

Myrtle wilt (*Chalara australis*) is a native pathogen and the main cause of death in *Nothofagus cunninghamii* in Tasmania and Victoria (Packham *et al.* 2008). It is most prevalent in forests which also have the highest densities of *Nothofagus*, and infection seems to be through wounds caused by branch fall. The full impact of the exotic pathogen myrtle rust *Puccinia psidii* is yet to be realised,

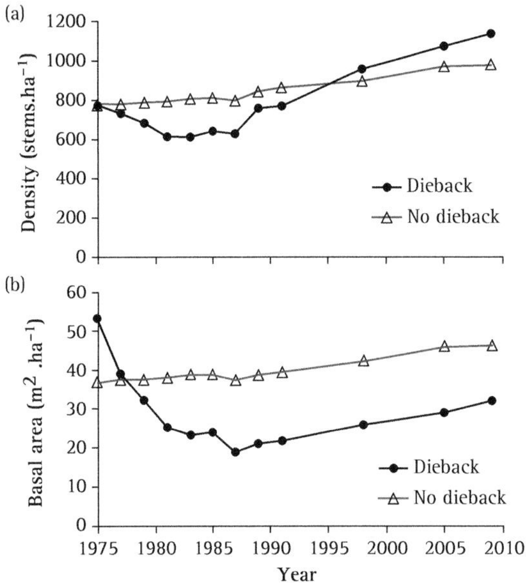

Figure 11.5 Change in basal area ($m^2\ ha^{-1}$) and stem density at the Garrawalt permanent tropical rainforest plot (north-west of Townsville in north-east Queensland) between 1975 and 2009, showing the impact of dieback disease on plot structure, and subsequent recovery. *Redrawn from Metcalfe et al. (2014).*

but the latest observations indicate that highly susceptible species may be at risk of future extinction (Carnegie *et al.* 2016).

11.5.4 Climate change

While the broader impacts of climate change on rainforest vegetation are unknown, modelling projections of suitable habitat suggest significant distributional shifts and an increase in fragmentation (Hilbert *et al.* 2007; VanDerWal *et al.* 2009). Palaeoecology and modelling indicate that similar shifts and contractions took place during Pleistocene climate fluctuations (Hopkins *et al.* 1993; Hilbert *et al.* 2007; Mooney *et al.*, Chapter 2, this volume), and extrapolation of this understanding suggest significant loss of mountain top endemics (Costion *et al.* 2015), changes to patterns of seed dispersal (Mokany *et al.* 2015), and increasing floristic turnover and development of novel communities (Laidlaw *et al.* 2011). Predicted increased frequency of extreme climate events is likely to manifest in more highly damaging tropical cyclones (Webster *et al.* 2005) and greater incidence of droughts leading to fire in rainforest habitat (Holz *et al.* 2015).

11.6 Future research directions

Unlike most regions, the rainforest flora of Australia is extremely well documented and described, and the biogeography and evolution of the flora is fairly well understood. The vast majority of extant species have been formally described and their taxonomy is relatively stable. Between them, two richly illustrated interactive electronic keys cover virtually the entire mainland rainforest flora of Australia; *Australian Tropical Rainforest Plants* (2553 species; https://www.anbg.gov.au/cpbr/cd-keys/rfk/) covers Western Australia, the Northern Territory and northern Queensland, while *Rainforest Plants of Australia* (1139 species; http://rainforests.net.au/product/rainforest-plants-of-australia/) covers Rockhampton to southern Victoria. These keys are the products of decades of research, and facilitate research by making the flora more accessible to non-specialists. Serious effort is also being devoted to genetically barcode entire regional floras. For example, Shapcott *et al.* (2015) have recently barcoded 86% of the known south-east Queensland vascular rainforest flora of 870 species excluding epiphytic orchids and ferns.

Internationally, the research agenda for tropical rainforests over the last two or three decades has had three main foci: documenting the impacts of fragmentation on plant communities; predicting the impacts of climate change; and unravelling the mechanisms that promote species coexistence among ecologically similar species at local spatial scales. Space precludes a comprehensive inventory, synthesis and review of the research effort in Australia, but our impression is that while we know a lot about the impacts of fragmentation in the Australian tropics and subtropics, and have some insights into the possible impacts of climate change, we have comparatively little understanding of the mechanisms promoting species coexistence in Australian rainforests. While some important insights have been derived from relatively small but very long-term permanent plots (Green *et al.* 2014), the lack of large plots in the Australian tropics and subtropics is hampering our ability to understand mechanisms of species coexistence in these rainforests.

In temperate systems lower diversity reduces the primacy of the coexistence question, though may help to explain it. Pressing concerns are centred around minimising anthropogenic impacts through changed fire regimes, spread of diseases and fragmentation, and on understanding how environmental change, particularly climate change, is likely to test the viability of these systems even further.

Acknowledgements

This chapter stemmed from conversations initiated through projects and meetings funded by the Terrestrial Environment Research Network's LTERN facility. Nat Raisbeck-Brown kindly produced Figure 11.1. The manuscript was improved by input from Andrew Ford, John Dwyer, Maurizio Rossetto, Carla Catterall and David Keith.

References

Abbott, K.L. (2005) Supercolonies of the invasive yellow crazy ant, Anoplolepis gracilipes, on an oceanic island: forager activity patterns, density and biomass. *Insectes Sociaux* 52, 266–273.

Adam, P. (1992) *Australian Rainforests*. Oxford University Press: Oxford.

Ash, J. (1988) The location and stability of rainforest boundaries in north-eastern Queensland, Australia. *Journal of Biogeography* 15, 619–630.

Auld, T. and Hutton, I. (2004) Conservation issues for the vascular flora of Lord Howe Island. *Cunninghamia* 8, 490–500.

Barker, P.C.J. (1992) Fire and the stability of a Victorian cool temperate rainforest. In *Victoria's Rainforests: Perspectives on Definition, Classification and Management*, ed. P. Gell and D. Mercer. Monash University: Melbourne, pp. 127–131.

Beadle, N.C.W. (1954) Soil phosphate and the delimitation of plant communities in Eastern Australia. *Ecology* 35, 370–375.

Berry, Z.C., Wevill, K. and Curran, T.J. (2011) The invasive weed *Lantana camara* increases fire risk in dry rainforest by altering fuel beds. *Weed Research* 51, 525–533.

Bowman, D.M.J.S. (1991) Recovery of some northern Australian monsoon forest tree species following fire. *Proceedings of the Royal Society of Queensland* 101, 21–25.

Bowman, D.M.J.S. (2000) *Australian Rainforests: Islands of Green in a Land of Fire*. Cambridge University Press: Cambridge.

Bowman, D.M.J.S. and Panton, W.J. (1994) Fire and cyclone damage to woody vegetation of the north coast of the Northern Territory, Australia. *Australian Geographer* 25, 32–35.

Bowman, D.M.J.S., Panton, W.J. and Head, J. (1999) Abandoned orange-footed scubfowl (*Megapodius reinwardt*) nests and coastal rainforest boundary dynamics during the late Holocene in monsoonal Australia. *Quaternary International* 59, 27–38.

Bridgwater, P.B., Russell-Smith, J. and Cresswell, I.J. (1998) Vegetation science in a cultural landscape: the case of Kakadu National Park. *Phytocoenologia* 28, 67–83.

Brooks, S.J., Panetta, D.F. and Sydes, T.A. (2009) Progress towards the eradication of three melastome shrub species from northern Australia. *Plant Protection Quarterly* 24, 71–78.

Bruce, C., Kroon, F., Sydes, D. and Ford, A. (2008) Impact of severe tropical cyclone Larry on riparian revegetation sites in a Wet Tropics floodplain. *Austral Ecology* 33, 516–524.

Busby, J.R. and Bridgewater, P.B. (1977) Studies in Victorian vegetation, II. A floristic survey of the vegetation associated with *Nothofagus cunninghamii* (Hook.) Oerst. in Victoria and Tasmania. *Proceedings of the Royal Society of Victoria* 89, 173–182.

Busby, J.R. and Brown, M.J. (1994) Southern rainforests. In *Australian Vegetation*, ed. R.H. Groves. Melbourne: Cambridge University Press, pp. 131–155.

Butler, D.W., Green, R.J., Lamb, D., McDonald, W.J.F. and Forster, P.I. (2007) Biogeography of seed-dispersal syndromes, life forms and seed sizes among woody rain-forest plants in Australia's subtropics. *Journal of Biogeography* 34, 1736–1750.

Cameron, D. (1992) A portrait of Victoria's rainforests: distribution, diversity and definition. In *Victoria's Rainforests: Perspectives on Definition, Classification and Management*, ed. P. Gell and D. Mercer. Monash University: Melbourne, pp. 13–96.

Campbell, M.L and Clarke, P.J. (2006) Response of montane wet sclerophyll forest understorey species to fire: evidence from high and low intensity fires. *Proceedings of the Linnaean Society of New South Wales* 127, 63–73.

Carnegie, A.J., Kathuria, A., Pegg, G.S. *et al.* (2016) Impact of the invasive rust *Puccinia psidii* (myrtle rust) on native Myrtaceae in natural ecosystems in Australia. *Biological Invasions* 18, 127–144.

Catterall, C.P., McKenna, S., Kanowski, J. and Piper, S.D. (2008) Do cyclones and forest fragmentation have synergistic effects? A before–after study of rainforest vegetation structure at multiple sites. *Austral Ecology* 33, 471–484.

Chen, S.C., Cornwell, W.K., Zhang, H.X. and Moles, A.T. (2016) Plants show more flesh in the tropics: variation in fruit type along latitudinal and climatic gradients. *Ecography doi.* 10.1111/ecog.02010

Chesterfield, E.A., Taylor, S.J. and Molnar C.D. (1991) Recovery after fire: warm temperate rainforest at Jones Creek, East Gippsland, Victoria. *Australian Forestry* 54, 157–173.

Choat, B., Ball, M.C., Luly, J.G. and Holtum, J.A.M. (2005) Hydraulic architecture of dry rain forest trees from north Queensland, Australia. *Trees* 19, 305–311.

Clarke, P.J., Prior, L.D., Vincent, B. *et al.* (2014) Using a rainforest-flame forest mosaic to test the hypothesis that leaf and litter fuel flammability is under natural selection. *Oecologia* 176, 1123–1133.

Clarke, P.J., Bell, D.M. and Lawes, M.J. (2015) Testing the shifting persistence niche concept: plant resprouting along gradients of disturbance. *American Naturalist* 185, 747–755.

Congdon, B.C. and Harrison, D.A. (2008) Vertebrate pests of the Wet Tropics bioregion: current status and future trends. In *Living in a Dynamic Tropical Forest Landscape*, ed. N.E. Stork and S.M. Turton. Blackwell: Oxford, pp. 322–333.

Costion, C.M., Edwards, W., Ford, A.J. *et al.* (2014) Using phylogenetic diversity to identify ancient rainforest refugia and diversification zones in a biodiversity hotspot. *Diversity and Distributions* 21, 279–289.

Costion, C.M., Simpson, L., Pert, P.L. *et al.* (2015) Will tropical mountaintop plant species survive climate change? Identifying key knowledge gaps using species distribution modelling in Australia. *Biological Conservation* 191, 322–330.

Crisp, M.D. and Cook, L.G. (2013) How was the Australian flora assembled over the last 65 million years? A molecular phylogenetic perspective. *Annual Review of Ecology and Systematics* 44, 303–324.

Curran, T.J., Clarke, P.J. and Bruhl, J.J. (2008) A broad typology of dry rainforests on the western slopes of New South Wales. *Cunninghamia* 10, 381–405.

Department of the Environment and Energy (2007) *NVIS data products 4.2.* Australian Government: Canberra. Available at: http://www.environment.gov.au/land/native-vegetation/national-vegetation-information-system/data-products (accessed 10 May 2016).

Doherty, M.D., Robertson, G., Corcoran, D., Wright, G. (2011) Cool temperate rainforest in the Pilot Wilderness Area, Kosciuszko National Park, New South Wales: distribution, composition and impact of the 2003 fires. *Cunninghamia* 12, 119–127.

Duff, G.A. and Stocker, G.C. (1989) The effects of frosts on rainforest-open forest ecotones in the highlands, north Queensland, Australia. *Proceedings of the Royal Society of Queensland* 100, 49–54.

Edwards, W. and Krockenberger, A. (2006) Seedling mortality due to drought and fire associated with the 2002 El Niño event in a tropical rain forest in north-east Queensland, Australia. *Biotropica* 38, 16–26.

Fensham, R.J. (1995) Floristics and environmental relations of inland dry rainforest in north Queensland, Australia. *Journal of Biogeography* 22, 1047–1063.

Fensham, R.J., Fairfax, R.J. and Cannell, R.J. (1994) The invasion of Lantana camara L in Forty Mile Scrub National Park, north Queensland. *Australian Journal of Ecology* 19, 297–305.

Floyd, A.G. (1987) The status of rainforests in northern New South Wales. In *The Rainforest Legacy*, ed. G. Werren and P. Kershaw. Australian Government Publishing Service: Canberra, pp. 95–117.

(1990) *Australian Rainforests in New South Wales*. Surrey Beatty and Sons: Sydney.

Fordyce, I.R., Eamus, D., Duff, G.A. and Williams, R.J. (1997) The role of seedling age and size in the recovery of *Allosyncarpia ternata* following fire. *Australian Journal of Ecology* 22, 262–269.

Fox, B.J., Taylor, J.E., Fox, M.D. and Williams, C. (1997) Vegetation changes across edges of rainforest remnants. *Biological Conservation* 82, 1–13.

Francis, W.D. (1951) *Australian Rainforest Trees*. Australian Government Publishing Service: Canberra.

Gentle, C.B. and Duggin, J.A. (1997) *Lantana camara* L. invasions in dry rainforest–open forest ecotones: the role of disturbances associated with fire and cattle grazing. *Australian Journal of Ecology* 22, 298–306.

Gooden, B., French, K. and Turner, P.J. (2009) Invasion and management of a woody plant, Lantana camara L., alters vegetation diversity within wet sclerophyll forest in southeastern Australia. *Forest Ecology and Management* 257, 960–967.

Goosem, S. (2008) Invasive weeds in the Wet Tropics. In *Living in a Dynamic Tropical Forest Landscape*, ed. N.E. Stork and S.M. Turton. Blackwell Publishing: Oxford, pp. 307–321.

Green, P.T., Harms, K.E. and Connell, J.H. (2014) Nonrandom, diversifying processes are disproportionately strong in the smallest size classes of a tropical forest. *Proceedings of the National Academy of Sciences of the United States of America* 111, 18649–18654.

Grimbacher, P.S., Catterall, C.P. and Stork, N.E. (2008) Do edge effects increase the susceptibility of rainforest fragments to structural damage resulting from a severe tropical cyclone? *Austral Ecology* 33, 525–531.

Harden, G., McDonald, B. and Williams, J. (2009) *Rainforest Trees and Shrubs a Field Guide to their Identification*. Gwen Harden Publishing: Nambucca.

Hilbert, D.W., Graham, A. and Hopkins, M.S. (2007) Glacial and interglacial refugia within a long-term rainforest refugium: the Wet Tropics Bioregion of NE Queensland, Australia. *Palaeogeography, Palaeoclimatology, Palaeoecology* 251, 104–118.

Hill, R. and Baird, A. (2003) Kuku-Yalanji rainforest Aboriginal people and carbohydrate resource management in the Wet Tropics of Queensland, Australia. *Human Ecology* 31, 27–52.

Hill, R.S. and Read, J. (1982) Rainforest fire in western Tasmania. *Australian Journal of Botany* 30, 583–589.

Hitchcock, P., Kennard, M., Leaver, B. *et al.* (2013) The natural attributes for World Heritage Nomination of Cape York Peninsula, Australia. Department of Sustainability, Environment, Water, Population and Communities: Canberra. Available at: http://www.environment.gov.au/system/files/resources/5ab50983-6bb4-4d87-8298-f1bcf1ab652a/files/sciencepanelreport.pdf (accessed 27 April 2016).

Holz, A., Wood, S.W., Veblen, T.T. and Bowman, M.J.S. (2015) Effects of high-severity fire drove the population collapse of the subalpine Tasmanian endemic conifer *Athrotaxis cupressoides*. *Global Change Biology* 21, 445–458.

Hopkins, M.S. and Graham, A.W. (1987) Gregarious flowering in a lowland tropical rainforest: a possible response to disturbance by Cyclone Winifred. *Australian Journal of Ecology* 12, 25–29.

Hopkins, M.S., Ash, J., Graham, A.W. *et al.* (1993) Charcoal evidence of the spatial extent of the *Eucalyptus* woodland expansions and rainforest contractions in North Queensland during the late Pleistocene. *Journal of Biogeography* 20, 357–372.

Horsfall, N. (1991) The prehistoric occupation of Australian rainforests. In *The Rainforest Legacy*, ed. G. Werren and P. Kershaw. Australian Government Publishing Service: Canberra, pp. 77–84.

Jarman, S.J., Brown, M.J. and Kantvilas, G. (1987) The classification, distribution and conservation status of Tasmanian forests. In *The Rainforest Legacy*, ed. G. Werren and P. Kershaw. Australian Government Publishing Service: Canberra, pp. 9–22.

Keith, D.A. (2004) *Ocean Shores to Desert Dunes: The Native Vegetation of New South Wales and the ACT*. Department of Environment and Conservation: Sydney.

Keith, D.A. and Pellow, B.J. (2005) Effects of Javan rusa deer (*Cervus timorensis*) on native plant species in the Jibbon-Bundeena Area, Royal National Park, New South Wales. *Proceedings of the Linnaean Society of New South Wales* 126, 99–110.

Keith, D.A. and Pellow, B.J. (2015) *Review of Australia's Major Vegetation Classification and Descriptions*. Centre for Ecosystem Science: Sydney.

Keith, D.A. and Sanders, J.M. (1990) Vegetation of the Eden Region, South-eastern Australia: species composition, diversity and structure. *Journal of Vegetation Science* 1, 203–232.

Kenneally, K.F. and Beard, J.S. (1987) Rainforests of Western Australia. In *The Rainforest Legacy*, ed. G. Werren and P. Kershaw. Australian Government Publishing Service: Canberra, pp. 289–304.

Kenneally, K.F., Keighery, G.J. and Hyland, B.P.M. (1991) Floristics and phytogeography of Kimberley rainforests, Western Australia. In *Kimberley Rainforests*, ed. N.L. McKenzie, R.B. Johnston and P.G. Kendrick. Surrey Beatty and Sons: Chipping Norton, pp. 93–131.

Kitching, R.L., Braithwaite, R. and Cavanaugh, J. (eds) (2010) *Remnants of Gondwana: A Natural and Social History of the Gondwanan Rainforests of Australia*. Surrey Beatty: Sydney.

Knox, J.E. and Clarke, P.J. (2011) Fire severity and nutrient availability do not constrain resprouting in forest shrubs. *Plant Ecology* 212, 1967–1978.

Kooyman, R.M., Rossetto, M., Sauquet, H. and Laffan, S.W. (2013) Landscape patterns in rainforest phylogenetic signal: isolated islands of refugia or structured continental distributions? *PLoS ONE* 8, e80685.

Lach, L., Volp, T.M., Greenwood, T.A. and Rose, A. (2015) High invasive ant activity drives predation of a native butterfly larva. *Biotropica* 48, 146–149.

Laidlaw, M.J., McDonald, W.J.F., Hunter, R.J. *et al.* (2011) The potential impacts of climate change on Australian subtropical rainforest. *Australian Journal of Botany* 59, 440–449.

Laurance, W.F. (1997) Responses of mammals to rainforest fragmentation in tropical Queensland: a review and synthesis. *Wildlife Research* 24, 603–612.

Little, J.K., Prior, L.D., Williamson, G.J. *et al.* (2012) Fire weather risk differs across rain forest–savanna boundaries in the humid tropics of north-eastern Australia. *Austral Ecology* 37, 915–925.

Marrinan, M.J., Edwards, W. and Landsberg, J. (2005) Resprouting of saplings following a tropical rainforest fire in north-east Queensland, Australia. *Austral Ecology* 30, 817–826.

McDonald, W.J.F. (1996) Spatial and temporal patterns in the dry season subtropical rainforests of eastern Australia, with particular reference to the vine thickets of central and southern Queensland. PhD Thesis, University of New England, Armidale.

McMahon, A.R.G. (1987) *The effects of the 1982-83 wildfires on some sites of botanical and zoological significance in east and central Gippsland and the westernport catchment*. Department of Conservation, Forests and Lands: Victoria.

Metcalfe, D.J. and Bradford, M.G. (2008) Rain forest recovery from dieback, Queensland, Australia. *Forest Ecology and Management* 256, 2073–2077.

Metcalfe, D.J. and Ford, A.J. (2008) Floristic biodiversity in the Wet Tropics. In *Living in a Dynamic Tropical Forest Landscape*, ed. N. Stork and S. Turton. Blackwell: Oxford, pp. 123–132.

Metcalfe, D.J. and Ford, A.J. (2009) A re-evaluation of Queensland's Wet Tropics based on primitive plants. *Pacific Conservation Biology* 15, 80–86.

Metcalfe, D.J., Bradford, M.G. and Ford, A.J. (2008) Cyclone damage to tropical rain forests: species- and community-level impacts. *Austral Ecology* 33, 432–441.

Metcalfe, D., Liddell, M., Bradford, M. and Green, P.T. (2014) Tropical rainforest of eastern Australia. In *Biodiversity and Environmental Change Monitoring, Challenges and Direction*, ed. D. Lindenmayer, E. Burns, N. Thurgate and A. Lowe. CSIRO Publishing: Melbourne, pp. 111–166.

Mills, K.G. (1986) The Illawarra rainforests: an historical, floristic and environmental study of their distribution and ecology. PhD Thesis. (University of Wollongong: Wollongong) Available at: http://ro.uow.edu.au/theses/1389 (accessed 28 April 2016).

Mokany, K., Prasad, S. and Westcott, D.A. (2015) Impacts of climate change and management responses in tropical forests depend on complex frugivore-mediated seed dispersal. *Global Ecology and Biogeography* 24, 685–694.

Moran, C. and Catterall, C.P. (2014) Responses of seed-dispersing birds to amount of rainforest in the landscape around fragments. *Conservation Biology* 28, 551–560.

Moran, C., Catterall, C.P. and Kanowski, J. (2009) Reduced dispersal of native plant species as a consequence of the reduced abundance of frugivore species in fragmented rain forest. *Biological Conservation* 142, 541–552.

Murphy, H.T., Metcalfe, D.J., Bradford, M.G. *et al.* (2008) Recruitment dynamics of invasive species in rainforest habitats following Cyclone Larry. *Austral Ecology* 33, 495–502.

Murphy, H.T., Metcalfe, D.J., Bradford, M.G. and Ford, A.J. (2014) Community divergence in a tropical forest following a severe cyclone. *Austral Ecology* 39, 696–709.

Negrón-Juárez, R., Chambers, J.Q., Hurtt, G.C. *et al.* (2014) Remote sensing assessment of forest disturbance across complex mountainous terrain: the pattern and severity of impacts of tropical cyclone Yasi on Australian rainforests. *Remote Sensing* 6, 5633–5649.

Neilan, N., Catterall, C.P., Kanowski, J. and McKenna, S. (2006) Do frugivorous birds assist rainforest succession in weed dominated oldfield regrowth of subtropical Australia? *Biological Conservation* 129, 393–407.

O'Dowd, D.J., Green, P.T. and Lake, P.S. (2003) Invasional 'meltdown' on an oceanic island. *Ecology Letters* 6, 812–817.

Orians, G.H. and Milewski, A.V. (2007) Ecology of Australia: the effects of nutrient-poor soils and intense fires. *Biological Reviews* 82, 393–423.

Packham, J.M., Elliott, H.J. and Bashford, R. (2008) Rate of spread of myrtle wilt disease in undisturbed Tasmanian rainforests. *Australian Forestry* 71, 64–69.

Peel, B., Bilney, R.J. and Bilney, R.J. (2005) Observations of the ecological impacts of Sambar *Cervus unicolour* in East Gippsland, Victoria, with reference to destruction of rainforest communities. *The Victorian Naturalist* 122, 189–200.

Podger, F., Pakzer, C. and Wardlaw, T. (1990) A guide to the Tasmanian distribution of *Phytophthora cinnamomi* and its effects on native vegetation. *Tasforests* 2, 13–20.

Pohlman, C.L., Turton, S.M. and Goosem, M. (2009) Temporal variation in microclimatic edge effects near powerlines, highways and streams in Australian tropical rainforest. *Agricultural Forest Meteorology* 149, 84–95.

Rejmánek, M. (1996) Species richness and resistance to invasions. In *Biodiversity and Ecosystem Processes in Tropical Forests*, ed. G. Orians, R. Dirzo and J.H. Cushman. Springer-Verlag: Berlin, pp. 153–172.

Rossetto, M. (2008) From populations to communities: understanding changes in rainforest diversity through the integration of molecular, ecological and environmental data. *Telopea* 12, 47–58.

Rosetto, M., Jones, R. and Hunter, J. (2004a) Genetic effects of rainforest fragmentation in an early successional tree (*Elaeocarpus grandis*). *Heredity* 93, 610–618.

Rosetto, M., Gross, C.L., Jones, R. and Hunter, J. (2004b) The impact of clonality on an endangered tree (*Elaeocarpus williamsianus*) in a fragmented rainforest. *Biological Conservation* 117, 33–39.

Russell-Smith, J. (1998) Monsoon forest. In *Top End Native Plants*, ed. J. Brock. J. Brock: Winnellie, pp. 22–35.

Russell-Smith, J. and Bowman, D.M.J.S. (1992) Conservation of monsoon rainforest isolates in the Northern Territory, Australia. *Biological Conservation* 59, 51–63.

Russell-Smith, J. and Dunlop, C. (1987) The status of monsoon vine forests in the Northern Territory: a perspective. In *The Rainforest Legacy*, ed. G. Werren and P. Kershaw. Australian Government Publishing Service: Canberra, pp. 227–288.

Russell-Smith, J., Ryan, P.G., Klessa, D. *et al.* (1998) Fire regimes, sensitive vegetation and fire management of the sandstone Arnhem Plateau, monsoonal northern Australia. *Journal of Applied Ecology* 35, 829–846.

Sattler, P.S. and Williams, R. (eds) (1999) *The Conservation Status of Queensland's Bioregional Ecosystems*. Queensland Government Environmental Protection Agency: Brisbane.

Scarlett, K., Daniel, R., Shuttleworth, L.A. *et al.* (2015) *Phytophthora* in the Gondwana rainforests of Australia World Heritage Area. *Australian Plant Pathology* 44, 335–348.

Shapcott, A., Forster, P.I., Guymer, G.P. *et al.* (2015) Mapping biodiversity and setting conservation priorities for SE Queensland's rainforests using DNA barcoding. *PLoS ONE* 10, e0122164.

Sniderman, J.M.K. and Jordan, G.J. (2011) Extent and timing of floristic exchange between Australian and Asian rain forests. *Journal of Biogeography* 38, 1445–1455.

Sobrado, M.A. (1991) Cost–benefit relationships in deciduous and evergreen leaves of tropical dry forest species. *Functional Ecology* 5, 608–616.

SPRAT (2013) *Species Profile and Threats Database*. (Department of the Environment: Canberra) Available at: http://www.environment.gov.au/cgi-bin/sprat/public/sprat.pl (accessed 27 April 2016).

Stocker, G.C. (1981) Regeneration of a north Queensland rainforest following felling and burning. *Biotropica* 13, 86–92.

Stork, N.E. and Turton, S.M. (2008) *Living in a Dynamic Tropical Forest Landscape*. Blackwell: Oxford.

Taylor, D.L., Leung, L.K.-P. and Gordon, I.J. (2011) The impact of feral pigs (*Sus scrofa*) on an Australian lowland tropical rainforest. *Wildlife Research* 38, 437–445.

Tracey, J.G. (1969) Edaphic differentiation of some forest types in eastern Australia. I. Soil physical factors. *Journal of Ecology* 57, 805–816.

Turton, S.M. (2012) Securing landscape resilience to tropical cyclones in Australia's Wet Tropics under a changing climate: lessons from Cyclone Larry (and Yasi). *Geographical Research* 43, 140–151.

Turton, S.M. and Freiburger, H.J. (1997) Edge and aspect effects on the microclimate of a small tropical forest remnant on the Atherton Tableland, Northeastern Australia. In *Tropical Forest Remnants, Ecology, Management, and Conservation of Fragmented Communities*, ed. W.F. Laurance and R.O. Bierregard. The University of Chicago Press: Chicago, pp. 45–54.

Turton, S.M. and Sexton, G.J. (1996) Environmental gradients across four rainforest-open forest boundaries in northeastern Queensland. *Australian Journal of Ecology* 21, 245–254.

Unwin, G.L., Applegate, G.B., Stocker, G.C. and Nicholson, D.I. (1988) Initial effects of tropical cyclone 'Winifred' on forests in north Queensland. *Proceedings of the Ecological Society of Australia* 15, 283–296.

VanDerWal, J., Shoo, L.P. and Williams, S.E. (2009) New approaches to understanding late Quaternary climate fluctuations and refugial dynamics in Australian wet tropical rain forests. *Journal of Biogeography* 36, 291–301.

Wallace, J. and McJannet, D. (2008) Modelling interception in coastal and montane rainforests in northern Queensland, Australia. *Journal of Hydrology* 348, 480–495.

Walter, H. and Breckle, S.-W. (1986) *Tropical and Subtropical Zonobiomes*. Springer-Verlag: Berlin and New York.

Warman, L. and Moles, A.T. (2009) Alternative stable states in Australia's Wet Tropics: a theoretical framework for the field data and a field-case for the theory. *Landscape Ecology* 24, 1–13.

Warman, L., Bradford, M.G. and Moles, A.T. (2013) A broad approach to abrupt boundaries: looking beyond the boundary at soil attributes within and across tropical vegetation types. *PLoS ONE* 8, e60789.

Webb, L.J. (1958) Cyclones as an ecological factor in tropical lowland rain-forest, north Queensland. *Australian Journal of Botany* 6, 220–228.
Webb, L.J. (1959) A physiognomic classification of Australian rain forests. *Journal of Ecology* 47, 551–570.
Webb, L.J. (1969) Edaphic differentiation of some forest types in eastern Australia. II. Soil chemical factors. *Journal of Ecology* 57, 817–830.
Webb, L.J. and Tracey, J.G. (1981) Australian rainforests: patterns and change. In *Ecological biogeography in Australia*, ed. A. Keast. Dr. W. Junk: The Hague, pp. 605–694.
Webb, L.J. and Tracey, J.G. (1994) The rainforests of northern Australia. In *Australian Vegetation*, ed. R.H. Groves. Melbourne: Cambridge University Press, pp. 87–130.
Weber, L.C., VanDerWal, J., Schmidt, S. *et al.* (2013) Patterns of rain forest plant endemism in subtropical Australia relate to stable mesic refugia and species dispersal limitations. *Journal of Biogeography* 41, 222–238.
Webster, P.J., Holland, G.J., Curry, J.A. and Chang, H.-R. (2005) Changes in tropical cyclone number, duration, and intensity in a warming environment. *Science* 309, 1844–1846.
Werren, G.L. and Kershaw, A.P. (eds) (1987) *The Rainforest Legacy*. Australian Government Publishing Service: Canberra.
Westcott, D.A. and Dennis, A.J. (2006) Reducing complexity when studying seed dispersal at community scales: a functional classification of vertebrate seed dispersers in tropical forests. *Oecologia* 149, 620–634.
Westcott, D.A., Setter, M., Bradford, M.G. *et al.* (2008) Cassowary dispersal of the invasive pond apple in a tropical rainforest: the contribution of subordinate dispersal modes in invasion. *Diversity and Distributions* 14, 432–439.
Williams, K.J., Ford, A., Rosauer, D.F. *et al.* (2011) Forests of east Australia: the 35th biodiversity hotspot. In *Biodiversity hotspots*, ed. F.E. Zachosm and J.C. Habel. Springer-Verlag: Berlin Heidelberg, pp. 295–310.
Williams, P.R., Parsons, M., Jensen, R. and Tran, C. (2012) Mechanisms of persistence and recruitment in frequently burnt wet tropical eucalypt forests. *Austral Ecology* 37, 268–275.
Wood, S.M., Murphy, B.P. and Bowman, D.M.J.S. (2011) Firescape ecology: how topography determines the contrasting distribution of fire and rainforest in the south-west of the Tasmanian Wilderness World Heritage Area. *Journal of Biogeography* 38, 1807–1820.

12 Wet Sclerophyll Forests

GRANT WARDELL-JOHNSON, JOHN NELDNER AND JAYNE BALMER

Wet sclerophyll forests (WSFs) are dominated by sclerophyllous trees (*Eucalyptus* and closely allied genera; family Myrtaceae) usually taller than 30 m at maturity in areas receiving more than 1 000 mm of annual rainfall. They include the tallest angiosperms, and exhibit biomass amongst the highest in terrestrial ecosystems. Although occupying only 0.75% of Australia's land-cover (4.83 million ha), the environmental, economic and cultural significance of WSF is far out of proportion to their distributional area. They occur discontinuously in a zone of high rainfall from north-eastern Queensland to southern Tasmania, with an outlier in south-western Australia; and also occur in eastern Malesia. In moist or protected margins in eastern Australia (and Malesia), and more extensively in Tasmania, mature eucalypts may occur above a well-developed rainforest; and are sometimes classified separately as mixed forest. This contrasts with WSF in south-western Australia, which includes no rainforest trees. Wet sclerophyll forest grades into rainforest in wetter, more fertile, or less frequently burnt margins; and into dry sclerophyll forests and woodlands in drier, more seasonal, less fertile, more frequently burnt, or cooler sites. Boundaries to adjacent forest types may be sharp, but sometimes unstable in eastern Australia, depending on disturbance regimes. These forests include important refugia under climate change, harbouring a diverse biota, featuring many endemic taxa. However, they are impacted by climate disruption and associated changes in fire regime, and distributions of weed and pest taxa. As a result, there are conservation concerns for the old-growth elements that are unique to these systems.

12.1 Introduction

The iconic wet sclerophyll forests (WSFs) are significant to the Australian identity, having been central to a native forest timber industry, which has contributed to Australia's economic development. These forests occupy only a small fraction of Australia (Figure 12.1), but are disproportionately important; economically, ecologically and culturally. Their aesthetic, scientific and cultural values are increasingly recognised globally, with representation in four World Heritage sites distributed from tropical to temperate latitudes in eastern Australia. Their importance was also central to the 1989 and 2013 extensions of the Tasmanian Wilderness World Heritage Area and the addition of the Greater Blue Mountains to the World Heritage List in 2000. Inclusion of WSF in both of Australia's global biodiversity hotspots and in several national hotspots demonstrates their conservation significance and the serious threats facing them.

Wet sclerophyll forests have an upper tree canopy dominated by species of *Eucalyptus* or closely

Australian Vegetation, Third Edition, ed. David A. Keith. Published by Cambridge University Press.

Citation: Wardell-Johnson, G., Neldner, J. and Balmer, J. (2017). Wet Sclerophyll forests. In D. A. Keith, ed., *Australian Vegetation*, 3rd edn. Cambridge: Cambridge University Press, pp. 281–313.

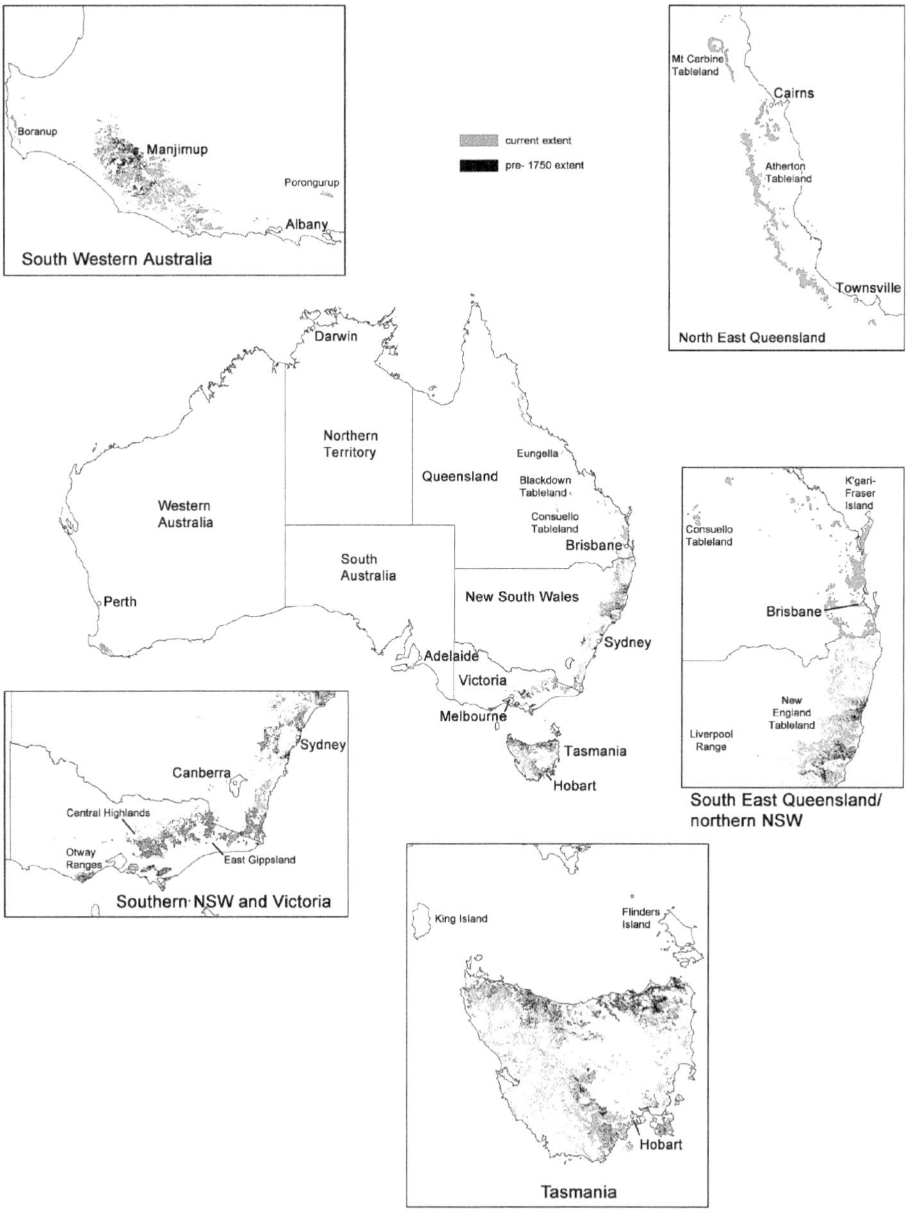

Figure 12.1 Distribution of extant and pre-1750 wet sclerophyll forest (WSF) of Australia. The map and insets show the six regions of WSF in Australia and locations of major cities and place names mentioned in the text. Areas vary according to definition of canopy closure; changes in species identification; data updates; and jurisdiction (see Box 12.2, Table 12.2). The majority of the WSF mapped is Major Vegetation Group 2 (MVG2) in NVIS version 4.0, while for Victoria MVG2 in NVIS version 3.0 is used (Department of Environment and Energy 2007). For Queensland (QLD) the more detailed QLD Herbarium regional ecosystem mapping of Broad Vegetation Group 1 (BVG1) is used. The estimates of mapped area from MPIG (2015) are substantially lower for QLD (155 000 ha) compared to the QLD estimate of 183 886 ha in 2013. The remnant/current extent was derived from a Vegetation Assets States and Transitions (VAST) product from the Commonwealth. The area statistics are derived from this WSF coverage, and total area is larger than figures used to calculate above-ground biomass (AGB) in Table 12.2 (see Commonwealth of Australia 2007).

Table 12.1 **Definitions and areas around wet sclerophyll forest, tall eucalypt forest and related terms.**

Definition	Author	Area	Characteristics
Tall open-forest	Specht (1970)	No estimate provided	Tree canopy > 30 m tall and projective foliage cover of 30–70%.
Tall eucalypt forest	Ashton and Attiwill (1994)	No estimate provided	Tree canopy > 30 m tall and projective foliage cover of 30–70%.
Eucalypt tall open forest	Montreal Process Implementation Group for Australia and National Forest Inventory (2013)	4.9 Mha	Tree canopy > 30 m tall, canopy > 50–80% crown cover.
Eucalypt tall open forests	Commonwealth of Australia (2007)	4.4 M ha (preclear); present 3.6 Mha	Forests of eucalypt trees > 30 m tall. Variable, but most commonly dense broadleaved understorey.
Wet sclerophyll (including mixed) forests	Wardell-Johnson *et al.*, this chapter	6.13 Mha (pre-1750); present 4.83 Mha	Forests of eucalypt canopy, which usually reach at least 30 m tall at maturity; and have a foliage projective cover of between 30 and 70%. The tree/shrub subcanopy may vary from sclerophyllous to rainforest species, or may be sparse.

allied genera (*Corymbia*, *Lophostemon* or *Syncarpia*) in the family Myrtaceae, which usually reach at least 30 m tall at maturity. Although this forest type is usually defined as having a foliage projective cover of between 30 and 70% (Specht 1970; see also Table 12.1), tall woodland (with 5–30% foliage projective cover of eucalypts in the upper canopy) over a dense subcanopy of smaller trees or tree ferns under 30 m tall, is also included in WSF. They usually occur in areas of high rainfall (at least 1 000 mm). These forests share globally significant structural features, of great height and carbon density; and also provide habitat to unique biotic assemblages. Understoreys vary greatly in structure and function depending on factors such as fire history, climate and soils. They include shrubland and grassland components; and others with a significant rainforest element. Wet sclerophyll communities with a well-developed rainforest understorey are sometimes classified separately as mixed forest (*sensu* Gilbert 1959; Lynch and Neldner 2000).

In this chapter, we provide an overview of the biogeography and ecology of some of the world's tallest and most carbon-dense forests, and also a prognosis for their future. We discuss the landscape and ecological context of WSF, and the ecosystem processes driving this formation. We also assess the conservation significance of this ecosystem and its vulnerability to climate disruption, changes to fire frequency and other threats including logging.

12.2 Structural and floristic features

In eastern Australia, WSF (also known as tall open-forest and tall woodland, *sensu* Specht 1970) typically forms an interface between rainforest (Metcalfe and Green, Chapter 11, this volume) and dry sclerophyll forest (DSF; Beadle and Costin 1952; Ashton and Attiwill 1981; Tozer *et al.*, Chapter 13, this volume). Wet sclerophyll forest may transition towards rainforests in response to reduced fire frequency, or towards DSF or wet scrub with increased fire frequency (Gilbert 1959; Jackson 1968; Ash 1988; Stanton *et al.* 2014). Mixed forests no longer occur in south-western Australia; rainforest species such as *Dacrydium*, *Phyllocladus* and *Nothofagus* being lost from the region pre-Pleistocene with the onset of more arid conditions (Hopper 2003). Throughout, DSF tends to be of shorter stature or with a more open canopy, and abuts in drier, more seasonal, less fertile, more frequently burnt, or cooler environments (Tozer *et al.*, Chapter 13, this volume).

The structure of WSF, especially old growth is complex, can be multi-layered and may have a canopy of 70 m or more in height. Wet sclerophyll forest is notable for the hollows and other resources that its trees provide, harbouring diverse arboreal mammal (Braithwaite *et al.* 1983), avifauna (Kikkawa and Dwyer 1992) and invertebrate (Williams 1993) assemblages. Tree hollows suitable for vertebrates generally do not form in trees younger than 120 years of age, with larger hollows taking much longer (Gibbons and Lindenmayer 2001). These forests also provide habitat for ground-dwelling fauna, many species restricted to, or most abundant in WSF (Lunney 2004).

12.2.1 Eucalypts of wet sclerophyll forest

Eucalypts are the dominant tree species in WSF, but they are not as species rich in this formation as they are in DSF, woodland or mallee habitats. Of more than 800 species recognised within *Eucalyptus* and allied genera (i.e. *Eucalyptus*, *Corymbia*, *Angophora*, *Allosyncarpia* and *Stockwellia*; Bayly *et al.* 2013) only about 16 species are confined to WSF (Table 12.2). However, another 50 or more eucalypt species may occur there (Wardell-Johnson *et al.* 1997; Table 12.2). All are shade intolerant, fire regenerators. Some are capable of giant stature, with rapid growth in the first 100 years of life and senescing after c. 500 years (Wood *et al.* 2010). Indeed, the tallest living *E. regnans* (99.6 m) is also the tallest flowering plant (Ashton 1981), with only some conifers (e.g. *Sequoia sempervirens*, *Pseudotsuga menziesii*) of the Pacific Northwest growing taller (Sillett *et al.* 2010). Drivers of gigantism in WSF species include rapid post-fire growth (Wood *et al.* 2010) and strong intraspecific competition among young trees following intense fires and intraspecific and interspecific competition among adult trees (Ashton 1981; Tng *et al.* 2012).

Eucalypts of WSF belong to only a few major taxonomic subgroups within *Eucalyptus*. They are mostly northern blue gums and mahoganies in section *Transversaria*, and southern blue gums in section *Maidenaria* of subgenus *Symphyomyrtus*. Many white mahoganies, peppermints and ashes (including the tallest eucalypts) in subgenus *Monocalyptus* are also in WSF. Several *Corymbia* species may also occur as tall eucalypts, while other dominant myrtaceous taxa in the canopy of WSF include *Syncarpia* and *Lophostemon* (Table 12.2).

12.2.2 Understorey floristics

Generally, the understorey plant species associated with WSF may be either grassy or shrubby (e.g. the grassy and shrubby subformations of Keith 2004). Where the understorey is dominated by rainforest taxa, it is termed mixed forest (*sensu* Gilbert 1959). Characteristic plant species in each of the main structural layers of WSF in the six major regions are summarised in Table 12.2, and examples provided in Figures 12.2 and 12.3. In the tropics and subtropics, the grassy subformation may have a dense grass layer comprising taxa such as *Imperata cylindrica* or *Themeda triandra* with limited shrub cover (Neldner *et al.* 2015). A grassy understorey does not generally occur in the mature WSF of south-eastern or south-western Australia. Dense sedgeland or grasses, such as *Dryopoa dives* or *Poa ensiformis*, may form a ground layer, especially in the early stages of regeneration, but is typically below a taller stratum dominated by shrubs or trees. Similarly, the understorey of WSF dominated by *E. jacksonii* near the south coast of Western Australia may include *Lepidosperma effusum* sedgeland, as a dominant ground layer (Figure 12.3a) beneath a shrub canopy.

Table 12.2 **Characteristic and most frequently occurring vascular plant species of tall eucalypt forests for six regions of Australia. Modified and updated from Ashton and Attiwill (1994), Wardell-Johnson and Williams (1996), Wardell-Johnson *et al.* (2007) and Wood *et al.* (2015). Eucalypt taxa that are largely confined to wet sclerophyll and/or mixed forest are bolded.**

Stratum	Tropics (north-east QLD)	Subtropics (south-east QLD, north-east NSW)	Warm Temperate (central and southern NSW)	Cool Temperate (VIC)	Cool Temperate (TAS)	Warm Temperate (south-west WA)
Tall canopy trees (all Myrtaceae)	***Eucalyptus grandis***, ***E. resinifera***, *E. portuensis*, *Corymbia intermedia*, ***E. pellita***, ***E. cloeziana***, ***Corymbia torelliana***, *Syncarpia glomulifera*	***E. grandis***, ***E. resinifera***, *E. pilularis*, ***E. saligna***, *E. acmenoides*, *E. microcorys*, *Syncarpia glomulifera*, ***S. hillii***, *Lophostemon confertus*, *E.* ***dunnii***, ***E. brunnea***, *E. cameronii*, *E. laevopinea*, *E. campanulata*, *E. obliqua*, *E. propinqua*, *E. dorrigoensis*, *E. nobilis*, *E. siderophloia*	***E. saligna***, *E. pilularis*, *E. paniculata*, *E. fastigata*, *E. cypellocarpa*, *E. elata*, *E. muelleriana*, *E. deanei*, *E. obliqua*, *E. viminalis*, *Corymbia gummifera*, *C. maculata*, *E. globoidea*, *E. longifolia*, *E. paniculata*, *Syncarpia glomerifera*, *E. bicostata*, *E. dalrympleana*, *E. viminalis*	***E. regnans***, *E. viminalis*, *E. obliqua*, *E. cypellocarpa*, *E. delegatensis*, *E. dalrympleana*, *E. nitens*	***E. regnans***, *E. obliqua*, *E. ovata* var. *ovata*, *E. delegatensis* spp. *tasmaniensis*, *E. viminalis* spp. *viminalis*, *E. dalrympleana* spp. *dalrympleana*, *E. subcrenulata*, ***E. johnstonii***, *E. globulus* spp. *globulus*, ***E. brookeriana***, ***E. nitida***, ***E. radiata* var. *radiata***	*Corymbia calophylla*, ***E. diversicolor***, ***E. jacksonii***, *E. guilfoylei*, *E. brevistylis*, *E. patens*, *E. marginata*

Table 12.2 (cont.)

Stratum	Tropics (north-east QLD)	Subtropics (south-east QLD, north-east NSW)	Warm Temperate (central and southern NSW)	Cool Temperate (VIC)	Cool Temperate (TAS)	Warm Temperate (south-west WA)
Subcanopy trees and shrubs	*Syncarpia glomulifera, Rhodomyrtus trineura, Allocasuarina torulosa, Callitris macleayana, Acacia melanoxylon, Alphitonia excelsa, Glochidion harveyanum, Breynia oblongifolia, Neolitsea dealbata, Cryptocarya glaucescens, Endiandra discolour, Notelaea longifolia*	*Syncarpia glomulifera, Allocasuarina torulosa, Lophostemon confertus, Myrsine howittiana, Elaeocarpus reticulatus, Synoum glandulosum, Schizomeria ovata, Callicoma serratifolia, Caldcluvia paniculosa, Cryptocarya rigida, Glochidion ferdinandi, Persoonia media*	*Syncarpia glomulifera, Acacia binervata. A. dealbata, Allocasuarina torulosa, Callicoma serratifolia, Myrsine howittiana, Elaeocarpus reticulatus, Zieria arborescens, Polyscias sambucifolia, Dodonaea triquetra*	*Acacia melanoxylon, Acacia dealbata, Pomaderris aspera, Bedfordia arborescens, Cassinia aculeata, Olearia argophylla,Hedycarya angustifolia Zieria arborescens, Correa lawrenceana, Nematolepis squamea, Prostanthera lasianthos, Polyscias sambucifolia*	*Acacia melanoxylon, A. dealbata, Pomaderris apetala, Nothofagus cunninghamii, Atherosperma moschatum, Pittosporum bicolor, Leptospermum lanigerum, Bedfordia arborescens, Cassinia aculeata, Olearia argophylla* and *O. lirata Zieria arborescens, Correa lawrenceana, Nematolepis squamea, Prostanthera lasianthos, Tasmannia lanceolata, Cenarrhenes nitida*	*Allocasuarina decussata, Agonis flexuosa, Bossiaea aquifolium* subsp. *laidlawiana, Oxylobium lanceolatum, Hovea elliptica, Acacia urophylla, A. pentadenia, Lasiopetalum floribundum, Trymalium odoratissimum, Hibbertia cuneiformis, Choralaena quercifolia, Pimelea clavata*
Palms, tree ferns and cycads (or Pachycauls)	*Archontophoenix alexandrae, Livistona australis*	*Archontophoenix cunninghamiana, Livistona australis, Cyathea australis, C. leichhardtiana, Lepidozamia peroffskyana*	*Livistona australis, Cyathea australis, Macrozamia communis*	*Cyathea australis, Dicksonia antarctica*	*Cyathea australis, Dicksonia antarctica*	*Macrozamia riedlei*

Lianes	*Calamus australis, Melodinus australis, Rubus moluccanus, Smilax australis, Flagellaria indica, Palmeria scandens, Cissus hypoglauca, Hibbertia scandens*	*Marsdenia rostrata, Parsonsia brownii, Smilax australis, Flagellaria indica, Cissus hypoglauca, Pandorea pandorana, Morinda jasminoides, Celastrus australis*	*Marsdenia rostrata, Parsonsia brownii, Smilax australis, Clematis aristata, Cissus hypoglauca, Pandorea pandorana, Morinda jasminoides, Geitonoplesium cymosum*	*Clematis aristata, Parsonsia brownii, Pandorea pandorana*	*Clematis aristata, C. decipiens, Parsonsia brownii, Prionotes cerinthoides*	*Clematis pubescens, Billardiera floribunda, Hardenbergia comptoniana, Chorizema retrorsum, Opercularia volubilis*
Ground ferns	*Pteridium esculentum, Calochlaena dubia, Blechnum cartilagineum, Lygodium microphyllum*	*Calochlaena dubia, Blechnum cartilagineum, Asplenium australasicum*	*Calochlaena dubia, Blechnum cartilagineum, Adiantum hispidulum, Pteridium esculentum*	*Polystichum proliferum, Blechnum nudum, B. wattsii, Histiopteris incisa, Pteridium esculentum, Calochlaena dubia*	*Polystichum proliferum, Blechnum nudum, B. wattsii, Histiopteris incisa, Hypolepis rugosula, Pteridium esculentum, Gleichenia microphylla, Sticherus* spp.	*Adiantum aethiopicum, Pteridium esculentum*
Herbs	*Oplismenus burmannii, Imperata cylindrica, Sorghum nitidum, Microlaena stipoides, Ottochloa nodosa, Dianella caerulea Hardenbergia violacea, Themeda triandra, Gahnia aspera*	*Oplismenus aemulus, Imperata cylindrica, Microlaena stipoides, Themeda triandra, Viola betonicifolia, Dianella caerulea, Hardenbergia violacea, Gahnia aspera*	*Oplismenus aemulus, Imperata cylindrica, Hydrocotyle hirta*	*Australina muelleri, Hydrocotyle hirta, Senecio* spp., *Viola hederacea, Dryopoa dives, Dianella revoluta*	*Australina pusilla, Hydrocotyle hirta, Senecio* spp., *Viola hederacea, Gahnia grandis, Lepidosperma* spp., *Drymophila cyanocarpa, Dianella tasmanica*	*Lepidosperma effusum, Tremandra stelligera, Hybanthus debilissimus, Tetrarrhena laevis, Senecio gilbertii*

Figure 12.2 Wet sclerophyll forest of eastern Australia: (a) *Eucalyptus pilularis* with shrubby understorey, Mapleton National Park, south-east Queensland (photo: Rosemary Niehus); (b) *Syncarpia hillii* and *Lophostemon confertus* with rainforest understorey on Quaternary sand dunes, K'gari–Fraser Island, south-east Queensland (photo: Grant Wardell-Johnson); (c) *Eucalyptus resinifera* and *Syncarpia glomulifera* with grassy understorey, Tumoulin State Forest, north-east Queensland (photo: E. Collins); (d) *Eucalyptus grandis* with rainforest understorey, Mt Tamborine, south-east Queensland (photo: John Neldner); (e) *Eucalyptus regnans* with shrub and fern understorey in the Otway Ranges, Victoria (photo: anon.); (f) *Eucalyptus regnans* (~72 m high) above rainforest understorey (30 m high) of *Nothofagus cunninghamii* and *Atherosperma moschatum*, Styx Valley, Tasmania (photo: anon.); (g) *Eucalyptus sphaerocarpa*, with shrubby understorey, Blackdown Tablelands, south-east Queensland (photo: D. Hopkins); (h) *Eucalyptus campanulata*, with rainforest understorey, New England escarpment, north-east New South Wales (photo: Grant Wardell-Johnson); (i) *Eucalyptus obliqua* and *E. regnans* above *Eucryphia lucida*, Warra, Tasmania (photo: anon.).

Figure 12.3 Wet sclerophyll forest of south-western Australia: (a) *Eucalyptus jacksonii* with dense ground layer of the sedge, *Lepidosperma effusum* near the south coast, south-western Australia (2015). The regeneration and epicormics branches are the result of a high-intensity fire in 1937, the most recent intense fire in the area; (b) Multi-aged, monotypic stand of *Eucalyptus diversicolor* near Manjimup, south-western Australia; (c) *E. diversicolor* at its south-eastern range limits at Nile Creek, south-western Australia; (d) Western outlier of *E. diversicolor* on limestone at Boronup, south-western Australia; (e) Eastern Outlier of *E. diversicolor* in the Porongurup Range, south-western Australia, subjected to intense fires five and 45 years previously (i.e. 1967, 2007).
All photos: Grant Wardell-Johnson.

The shrubby subformation of WSF is widespread throughout all regions. It is characterised by broadleaved (mesophyllous) tall shrubs or trees. Families typical of these forests include many that are strongly associated with mesic habitats (e.g. Lauraceae, Cunoniaceae, Euphorbiaceae, Rubiaceae). However, others include species of less mesic habitats, although with non-sclerophyllous leaves (e.g. Asteraceae, Dilleniaceae, Fabaceae, Rhamnaceae and Rutaceae, and genus *Persoonia* within the family Proteaceae, Table 12.2). Frequent genera represented in the subcanopy of these WSFs include *Acacia*, *Pomaderris*, *Bedfordia*, *Syncarpia*, *Olearia*, *Allocasuarina*, *Bossiaea* or *Trymalium*.

The prominence of non-scleromorphic taxa in WSF in southern regions of Australia contrast with more sclerophyllous taxa in neighbouring dry sclerophyll forests (Wardell-Johnson and Williams 1996). These neighbouring habitats also include much greater alpha and beta species diversity than the WSFs of southern Australia (Wardell-Johnson and Williams 1996). The most plant-diverse of the shrubby subformation of WSF are those characterised by an array of forbs, grasses and ground ferns in subtropical to warm temperate forests (Keith 2004). In eastern Australia, plant diversity reduces in this subformation with increased latitude and elevation.

Mixed forests often have a well-developed understorey of rainforest taxa. For example, in Tasmania, albeit rarely, mixed forest may have a 70+ m tall eucalypt canopy above a 30 m high closed subcanopy of *Nothofagus cunninghamii* and/or *Atherosperma moschatum* (Figure 12.2f). Here, understorey taxa include endemic rainforest shrubs such as *Anodopetalum biglandulosum*, *Anopterus glandulosus* and *Trochocarpa* species, and an array of epiphytic and ground ferns and other Gondwanan taxa. Mixed forest tends to be best developed in old growth (i.e. > 100 years after the most recent intense fire) as younger forest understoreys are not typically dominated by rainforest species. These taxa are in general slow to establish, such that other mesophyllous understorey shrub/tree species tend to dominate the understorey for the first 50–100 years after major disturbance.

12.3 Distribution of wet sclerophyll forests

Wet sclerophyll forest occurs over a relatively narrow ecological range, comprising only 4% of the 123 million hectares of native forest in Australia and around 0.75% of total land cover (MPIGAQ and NFISC 2013). Wet sclerophyll forest is distributed in a discontinuous arc of high-rainfall land from north-eastern Queensland (latitude 16°S) to southern Tasmania (latitude 42°S). A disjunct area of WSF occurs in south-western Australia and also in east Malesia, where *E. deglupta* occurs on deep, moist, well-aerated soils of high fertility (Wardell-Johnson *et al.* 1997). Wet sclerophyll forests occur in tropical to cool temperate environments with mean annual temperatures ranging from 23°C to 5°C (Wood *et al.* 2015).

Area estimates and definitions of WSF vary, with about 4.83 million hectares (6.13 million hectares pre-1750), mapped in the National Vegetation Information System (NVIS; Department of the Environment and Energy (2007)) as 'eucalypt tall open-forest' in the most recent State of the Forests Report (MPIGAQ and NFISC 2013; Table 12.1). In Australia there are six regions supporting WSF: (i) tropical north-east Queensland; (ii) subtropical south-east Queensland and north-east New South Wales; (iii) warm temperate central and southern New South Wales; (iv) cool temperate Victoria and south-east New South Wales; (v) cool temperate Tasmania; and (vi) warm temperate south-west Western Australia (Figure 12.1).

12.3.1 Tropical north-east Queensland

Wet sclerophyll forest occupies about 0.76 million hectares (0.78 million hectares pre-1750) of tropical north-east Queensland. This forest dominated by

E. grandis or *E. resinifera* (Figure 12.2c) includes understoreys ranging from grasslands, to secondary tree layers dominated by *Allocasuarina* or rainforest species, and occurs in a narrow band in the west of the Atherton and Mt Carbine Tablelands (Figure 12.1). Similar forest occurs less extensively on the Eungella Tableland. These forests develop on a wide range of geological substrates and soils, but occur most extensively on dermosols derived from granite and metamorphic uplands and ranges, and less extensively on fertile and deep red ferrosols derived from basalt, or alluvia (Neldner *et al.* 2015).

12.3.2 Subtropical south-east Queensland and north-east New South Wales

This region has the greatest diversity of WSF eucalypts and understorey plants (Keith 2004) and occurs over about 1.95 million hectares (2.65 million hectares pre-1750). Wet sclerophyll forest dominated by *E. grandis* (Figure 12.2d) or *E. saligna* with understoreys of mesophyllous shrubs and ferns occur on alluvial, or metamorphic lowlands and ranges in subtropical south-east Queensland and north-east New South Wales. *Eucalyptus pilularis* (Figure 12.2a), *E. microcorys*, grey gums, mahoganies and ironbarks also form extensive WSF with shrub or grassy understories, typically in drier sites on granitic, sedimentary and metamorphic ranges and Tertiary weathered surfaces in this region (Keith 2004). There are strong altitudinal gradients in species composition, with *E. campanulata* often dominant at elevations of 800–1200 m on the New England escarpment (Figure 12.2h). In south-east Queensland, WSF dominated by *E. pilularis* or *Syncarpia hillii* occurs on the infertile aeric podosols on Quaternary sand dunes of K'gari–Fraser Island (Figures 12.1; 12.2b). Wet sclerophyll forest also occurs on moist outliers west of the main escarpment of the Great Divide, such as on the sedimentary Blackdown Tableland (Figures 12.1; 12.2g), basaltic Consuelo Tableland (Neldner *et al.* 2015) and Liverpool Range (Keith 2004).

12.3.3 Warm temperate central and southern New South Wales

Wet sclerophyll forest occurs over about 0.58 million hectares (0.68 million hectares pre-1750) in warm temperate central and southern New South Wales. This limited range is due to the extensive oligotrophic soils and limited areas of high rainfall in this region. The shrubby subformation dominated by *E. saligna* and *E. pilularis* occurs in sheltered gullies on loamy soils in the north, and is replaced by *E. cypellocarpa*, *E. obliqua* and *E. fastigata* with increasing latitude and altitude. Here, *E. fastigata* sometimes forms pure stands above about 1300 mm annual rainfall. Wet sclerophyll forests of the grassy subformation (Keith 2004) are notable in lowland forests on the mid-south coast, and in marginally wet areas of the tablelands. The former include *Corymbia maculata*, and the latter *E. viminalis* and *E. radiata* (Keith 2004).

12.3.4 Cool temperate Victoria and south-east New South Wales

The cool temperate WSFs of Victoria and southern New South Wales occurs over about 0.63 million hectares (0.72 million hectares pre-1750). They are mostly on deep dermosols derived from Ordovician and Devonian sediments, extrusives, granite substrates and alluvium on sheltered hillsides mostly between 600 and 1300 m above sea level (Burns *et al.* 2015). Some of the most fertile soils are deep red kandosols overlaying felsic intrusives. Forests on these soils are dominated by ashes (e.g. *E. regnans*, Figure 12.2e; *E. delegatensis* and *E. obliqua*, Keith 2004), and many areas of *E. regnans* are monotypic stands. This species is sometimes associated with, or replaced by *E. obliqua* and *E. cypellocarpa* in central and western Victoria, and by *E. fastigata* and *E. denticulata* in eastern Victoria, extending into far south-east New South Wales. At higher altitudes *E. delegatensis*, *E. fraxinoides* and *E. nitens* dominate.

12.3.5 Cool temperate Tasmania

Wet sclerophyll forest is widespread in Tasmania below 900 m elevation (Kirkpatrick *et al.* 1988), and occupies about 1.42 million hectares (1.79 million hectares pre-1750). *Eucalyptus regnans* dominated WSF (Figure 12.2f) occurs on the most fertile well-drained soils, including brown dermosols on Devonian granite and Permian tillite, and red/brown ferrosols on Tertiary basalt and Jurassic dolerite (Grant *et al.* 1995). It may co-occur with, or be replaced by, *E. obliqua* (Figure 12.2i; Figure 12.4a, c) on more rocky or less fertile soils, or on drier, warmer or more fire-prone sites. Above 600 m, *E. delegatensis* is dominant on red/brown ferrosols on dolerite (Grant *et al.* 1995), but also extends to lower elevation where drainage and fertility is poorer. *Eucalyptus nitida* is dominant in western Tasmania where siliceous substrates result in the development of soils such as oxyaquic hydrosols, brown dermosols and brown kurosols (Grant *et al.* 1995). Several taxa including *E. brookeriana*, *E. dalrympleana*, *E. globulus*, *E. johnstonii*, *E. ovata*, *E. subcrenulata* and *E. viminalis* may dominate on more restricted substrates and habitats (Kirkpatrick *et al.* 1988; Kitchener and Harris 2013).

12.3.6 Warm temperate south-western Australia

South-western Australia supports no mixed forest; and WSF occurs in moist, fertile and freely-drained habitats (Churchill 1968; Wardell-Johnson *et al.* 1997) in the highest rainfall and least seasonal area of south-western Australia. Here, WSF characterised by *E. diversicolor* (Figures 12.3b, c, d, e) occupies about 0.168 million hectares (0.204 million hectares pre-1750), largely on granitic terrain between Manjimup and Albany (Wardell-Johnson *et al.* 1997; 2004; 2007; Figure 12.1). *Eucalyptus diversicolor* may occur in pure stands (Figure 12.3b) on red kandosols and dermosols, or with *Corymbia calophylla;* or in drier or more gravelly sites with *E. marginata*. In WSF, along the south-western coast, *E. diversicolor* can occur as a minor component in stands of *E. jacksonii* and *E. guilfoylei* (Plate 13b). Outliers of WSF with *E. diversicolor* occur at Boranup on limestone (Figure 12.3d), and in dermosols in the Porongurup Range (Figures 12.1; 12.3e). Throughout WSF in south-western Australia, the understorey is sclerophyllous, but species poor in comparison with neighbouring vegetation types (Wardell-Johnson *et al.* 2004; 2007). However, it provides important refugial habitat for high rainfall endemic and relictual biota (e.g. *Anthocercis silvicola*, *Moggridgea tingle*, *Metacrinia nicholsii*).

12.4 Ecosystem processes

12.4.1 Water balance

Much of the WSF distribution in Australia occurs where annual rainfall is 1000–1500 mm. In areas supporting WSF in south-western Australia the driest month may have a mean rainfall as low as 15 mm (Churchill 1968). Wet sclerophyll forests at the dry margins of their climatic range are generally associated with gullies, which may receive additional ground water and/or provide shelter from the desiccation of fire-promoting winds. The regional distribution of mixed forest is controlled primarily by high, reliable rainfall, and this type of WSF, together with rainforests, is most common in regions receiving between 1500 and 2000 mm annual rainfall, with at least 50 mm in the driest month (Ashton and Attiwill 1994; Read and Brown 1996).

12.4.2 Disturbance ecology

Catastrophic disturbance, usually fire, is essential for regeneration of dominant canopy trees and many of the other shade-intolerant taxa in WSF (Ashton 1981). In the long-term absence of disturbance, species that are unable to regenerate by resprouting, or to germinate and grow in canopy

Figure 12.4 Shrubby and mixed wet sclerophyll forest, southern Australia. (a) Stand of *Eucalyptus obliqua* with rainforest understorey of *Nothofagus cunninghamii* and *Atherosperma moschatum* in the lower Weld Valley, Tasmania (photo: Rob Blakers); (b) wet sclerophyll forest of *E. jacksonii* and *E. diversicolor* (white trunks in background) at Mt Clare near the south coast, south-western Australia (photo: Grant Wardell-Johnson); (c) *Eucalyptus obliqua*, lower Weld Valley, Tasmania (photo: Rob Blakers).
(A black and white version of this figure will appear in some formats. For the colour version, please refer to the plate section.)

gaps, senesce and disappear from the above-ground vegetation. They are replaced by more shade-tolerant taxa (Ashton 2000). At the end of the life span of the dominant eucalypts the vegetation transitions to rainforest (Jackson 1968). The WSF of south-western Australia is unusual in that climatic conditions, fuel loadings, ignition-sources and lack of topographic barriers ensure that there is no long term (i.e. > 150 years) absence of fire (Burrows and Wardell-Johnson 2003). Here there are no rainforest plants (with the exception of the resprouting shrub *Podocarpus drouynianus*) to replace shade-intolerant taxa.

In common with many Australian ecosystems, the WSF flora has a range of adaptive traits enabling perpetuation through disturbance (Ashton 1981; Miller and Murphy, Chapter 5, this volume). Fires in WSF are often sufficiently intense to kill plants, including canopy trees (Figure 12.5c, d). Therefore post-fire recovery often depends on regeneration from seed. The seed of some species is stored in woody capsules that usually protect the seed from being killed by fire (e.g. Myrtaceae, Casuarinaceae). The fire's heat, or death of the plant, triggers a mass post-fire release of this aerially stored seed (Ashton 1981). Heat and smoke also trigger a mass germination of soil-stored seed including many WSF taxa (e.g. Fabaceae, Rutaceae, Rhamnaceae). Fire clears the understorey vegetation (and in some cases, the canopy), allowing light to reach the ground, enabling seedling growth. The litter, humus and duff are often burnt to expose soil; and the resultant ash bed enhances growth of many WSF species, especially eucalypts (Cremer *et al.* 1978).

Some WSF species are able to survive fire, recovering vegetatively, this capacity varying depending on fire severity, between species and with age and health of plants (Figure 12.5). Some species have capacity to resprout following 100% defoliation from a young age (Wardell-Johnson 2000; Figure 12.5b). The likelihood of a plant being killed by wildfire depends on age or growth stage, with immature and young plants often more susceptible (Taylor *et al.* 2014; Bassett *et al.* 2015). All WSF eucalypts have capacity to resprout from epicormic strands embedded in the xylem and bark, an ancient adaptation in the broader group (Crisp *et al.* 2011). Despite less capacity to resprout from lignotubers, most tall forest trees have thick bark, providing considerable protection, even in the most intense crown fires. Hence, for many WSF eucalypt taxa, a high proportion of the stand resprouts following occasional intense crown fires; and many WSF eucalypts show considerable fire resistance (Figure 12.5a, b, c). Consequently, while some eucalypts are restricted to WSF, many others also occur in more frequently burnt open forest and woodland habitat (Ashton and Attiwill 1994; Wardell-Johnson *et al.* 1997). For example, *E. obliqua* and *E. viminalis* tolerate a wide range of environmental conditions, occurring in a reduced growth form in drier forest types (Wardell-Johnson *et al.* 1997).

A multi-age stand structure arises when fire events initiate recruitment, but fail to kill a portion of the standing trees; or when recruitment occurs during the fire intervals (e.g. following soil disturbance with tree-fall). This gives rise to a post-fire stand structure of older surviving eucalypts and younger, regenerating eucalypts, usually of the same species. Multi-aged forests occur for many WSF types, and are well documented for *E. regnans* and *E. obliqua* in Tasmania (Turner *et al.* 2009), *E. regnans* forests of Victoria (Lindenmayer *et al.* 1991; 1999; McCarthy and Lindenmayer 1998), and *E. diversicolor* (Wardell-Johnson 2000) and *E. jacksonii* (Wardell-Johnson 2000; McCaw and Middleton 2015) forests of south-western Australia.

Despite a capacity to resprout, some WSF eucalypt taxa, including *E. regnans*, *E. delegatensis* and *E. fraxinoides* have a higher probability of being killed by intense fire. Wildfires in these forests can result in large tracts of fire-killed trees, and regeneration of even-aged stands (Wood *et al.* 2010; Bassett *et al.* 2015; Figure 12.5d). These fire-sensitive species are generally in cooler or more mountainous areas. Hence, such areas have larger proportions of even-aged stands. Here, shifts in vegetation type are more likely to occur in response to changing fire regimes or recruitment failures following fire (Bassett *et al.* 2015).

Many rainforest taxa (e.g. *Anodopetalum biglandulosum, Atherosperma moschatum, Eucryphia lucida*) also regenerate by resprouting after the less intense fires more typical of rainforest and mixed forest (Hill and Read 1984; Clarke *et al.* 2015). Species recovering vegetatively after fire may have

Figure 12.5 Impacts of fire on wet sclerophyll forests. (a) Hollow-butting is typical following repeated fires in WSF of *Eucalyptus jacksonii* in south-coastal, south-western Australia. Note the resprouting epicormic shoots (photo: Grant Wardell-Johnson); (b) *Eucalyptus guilfoylei* regeneration following clear-felling near south coast, south-western Australia in 1976, burnt by an intense wildfire in 1991. Note some plants have produced seed (photo: Grant Wardell-Johnson, taken 1992); (c) Dead canopies as signs of an intense fire in *Eucalyptus jacksonii* WSF in 1937 near south coast, south-western Australia (photo: Grant Wardell-Johnson, taken 2014); (d) Large stand of *Eucalyptus delegatensis* that had regenerated following a wildfire in 1939, killed by a subsequent wildfire in 2007. Note seedling regeneration (photo: Lynda Prior). *(A black and white version of this figure will appear in some formats. For the colour version, please refer to the plate section.)*

faster growth rates enabling them to compete with species recruiting from seed. Few rainforest species have soil-stored seed. Further, the post-fire environment may have increased browsing pressure, and surface soil dryness (Knox and Clarke 2012). Therefore, if killed, rainforest species with poor dispersal capacity (e.g. *Eucryphia lucida*, *Nothofagus cunninghamii*), or low capacity to germinate and grow in the post-fire environment (e.g. *Atherosperma moschatum*) may take longer to recolonise.

Without fire to initiate germination of eucalypts (Gilbert 1959; Ashton 1981) rainforest species eventually dominate the understorey, becoming mixed forest, and eventually rainforest. Hence, changes in community properties have been closely linked to the passage of fire, and length of fire-free periods (Webb 1968; Kirkpatrick *et al.* 1988; Ashton and Attiwill 1994; Lynch and Neldner 2000; Russell-Smith *et al.* 2004). However, intense fires at too frequent an interval (i.e. < 150 years) are likely to lead to reduced prevalence of rainforest taxa and loss of mixed forest vegetation. Mixed forest, previously common in Victoria, is now rare, and termed 'ecotone' forest (Petrie *et al.* 1929; Fedrigo *et al.* 2014). Much was replaced by WSF following the 1939 fires (Ashton 1981).

The climate envelope of WSF often overlaps with rainforest along the east coast of Australia (Mackey *et al.* 2002; Tng *et al.* 2012). These forests may occur within a landscape mosaic of less flammable (i.e. rainforest) and more flammable vegetation (i.e. open forest, woodland or savanna; Bowman 2000). This pattern is consistent with a long history of fire in WSF, where particular regimes maintain WSF eucalypts and understorey taxa, often at the expense of fire-sensitive rainforest (Bowman 2000; Wood and Bowman 2012). Studies of boundaries between vegetation types in the Wet Tropics of Queensland (Warman and Moles 2009) and the New England Tableland (Knox and Clarke 2012) show these boundaries to be quite stable in some regions.

Knox and Clarke (2012) provided evidence for environmental and pyric feedbacks between vegetation types, but found no evidence for intense fire triggering a change to an alternate state. There was, for example, a markedly different fire and temperature regime for each vegetation type (rainforest, WSF and DSF) within the regional vegetation mosaic. Importantly, they observed that more recently burnt WSF was more likely to burn with extreme severity than long-unburnt WSF, but that this was not true for rainforest. Fuel loads after 6 years did not differ between forest types or fire severity classes. Temperatures were warmer in more severely burnt forest, enabling greater drying of fuel beds, thereby increasing probability of fire.

Although WSF generate great quantities of flammable leaf litter and coarse woody debris (CWD), they occur in relatively moist (albeit seasonal) environments, and are flammable for less of the year than open forest or woodland habitats. Rainforest in Tasmania is only able to sustain fire at times of the year when the total rainfall in the preceding month has been less than 50 mm (Styger and Kirkpatrick 2015). Wet sclerophyll forests generally become less flammable over time as mesophyllous understorey taxa, especially rainforest species, become more dominant (Jackson 1968; Harrington and Sanderson 1994).

12.4.3 Carbon dynamics

One of the important habitat factors distinguishing WSF from other vegetation such as rainforest and DSF is the immense volumes of carbon that they store within complex structures of tall trees, fallen logs and organic-rich soils (Keith *et al.* 2009; Dean *et al.* 2012c; Fedrigo *et al.* 2014; Box 12.1). For WSF, the stand-level biomass oscillates in the long term with wildfire. With stand-replacement fires the total carbon for an even-aged mixed forest oscillates by around ±24% around the long-term temporal average (Dean *et al.* 2012b). A considerable portion of killed biomass remains on site as CWD.

Box 12.1 Loss of area and carbon from wet sclerophyll forests (WSF)

Wet sclerophyll forests (WSF) occupy around one hundredth of the area of rangeland, but their potential biomass is overall ~17 times more concentrated (rangeland data from Dean *et al.* 2012a). Hence, WSF is significant in carbon management. Estimates of area and volume vary based on differences in definition (e.g. Table 12.1), changes in species recognition, and on data updates (Table 12.1). For example, WSF in south-western Australia occupies 0.168 million hectares or 0.193 million hectares according to NVIS and MPIG and NFISC (2013) respectively, while pre-1750 area is 0.204 million hectares and 0.275 million hectares in NVIS and Beard *et al.* (2013), respectively. Figure 12.6a gives a visual example showing why estimates vary. These ~25% variances suggest change is best understood in relative terms, and based on clearly defined criteria.

The above-ground dry biomass (AGB) for WSF for each Australian State, extant and pre-1750 (Table 12.3) was calculated by overlaying NVIS data using a GIS with a national data layer of potential AGB, as used in the FullCAM model for the National Carbon Accounting System (NCAS; Richards and Brack 2004). Estimates provide a 'pre-1750' area of WSF of 4.44 million hectares, and an extant area (2012) of 3.61 million hectares – a 23% loss in area (Commonwealth of Australia 2007). Nationally, the AGB of WSF is 875 Tg, with the deforested area potentially containing 198 Tg, i.e. a loss of ~23%. New South Wales with the largest area of WSF, contained (and still contains) the highest total above-ground biomass in WSF of any State. Of the Australian states, most WSF has been deforested, or converted (0.41 million hectares) by Tasmania.

In remnant carbon-dense WSFs, reductions in the age class structure and carbon stocks of the dominant stratum are associated with the rise of the pulpwood industry (Bradshaw 2012). Due to land-use change, the remnant eucalypt forests store less carbon and have produced a net carbon emission (Mackey *et al.* 2008). Forest carbon in the 14.5 million hectares of remnant eucalypt forests of south-eastern Australia (including 3.4 million hectares of WSF) have decreased in living biomass, litter and soil pools because of logging by approximately: 15%, 10% and 4% respectively (Roxburgh 2009). Even without climate change, that carbon pool is forecast to take at least ~1500 years to equilibrate under the new biomass regime accompanying logged secondary forest.

For Tasmanian mixed forest logged by clear-fell, burn and sowed silviculture (Figure 12.6b), the net emission across all pools from first logging cycle (including wood products and waste) is 43 (±5)%. The long-term net emission is ~50%, totalling

Box 12.1 (cont.)

20 (±16) Tg from logging of WSF from 1999 to 2009 (Dean *et al.* 2012b). Forecast emissions over ~400 years from conversion of WSF in Tasmania were ~70%. This discounted a change in soil carbon, but included wood-product decomposition (Dean *et al.* 2012b). Independent modelling of similar WSF in Victoria showed a 55% loss (Keith *et al.* 2014). Thus, long-term losses of carbon from WSFs are most likely to be at least 50% upon conversion to a logging regime, a gap in carbon stocks not filled by wood products (Dean *et al.* 2012b). Conservative management of WSF is warranted considering the vast quantities of carbon potentially lost from these forest types, and their critical role as reservoirs of carbon, biodiversity and social values.

Figure 12.6 **(a)** Legacy carbon from an earlier mixed forest – a *Eucalyptus regnans* log spans Cliff Creek in the Styx Valley, Tasmania, mapped as 200 m wide rainforest – typical of the blurred spatiotemporal boundary in carbon dynamics between WSF and rainforest (photo: anon); **(b)** Clear-felling and conversion of old growth to production forest, such as this site in Tasmania reduces long-term carbon stocks. Note the Tasmanian Wilderness World Heritage Area in the background (photo: Perpetua Turner).

While new biomass increases after fire, the CWD decomposes sooner than if it had remained alive, except possibly for the small portion left as charcoal (Dean *et al.* 2012b).

Carbon content varies spatio-temporally within forest type, linked partly to fire frequency; and also with management history (Dean and Wardell-Johnson 2010; Dean *et al.* 2012b). For example, a rainforest understorey of *Nothofagus cunninghamii* in *E. regnans* mixed forest may be replaced by young *E. regnans* in marginally drier conditions and increased fire frequency. This increases the carbon stock once the younger eucalypts reach canopy height (Mackey *et al.* 2008).

A combination of age classes may explain high carbon stocks observed in some areas of the Victorian Central Highlands (Keith *et al.* 2009). This diverse cohort structure is likely maintained by slightly more frequent fire than required to maintain mixed forest. However, frequent intense fire (i.e. < 150 years) will likely see a transition from WSF to other forest types, with lower carbon stocks (Dean and Wardell-Johnson 2010).

Wet sclerophyll forests have an important role in reducing atmospheric carbon through management to maximise storage and sequestration of carbon (Mackey *et al.* 2008; Keith *et al.* 2009; Dean *et al.* 2012a; 2012b). Prescribed burning has been proposed to reduce fuel loads, and thereby increase climate change resilience in both WSF and adjacent habitats (Boer *et al.* 2009). However, prescribed burning in WSF is unlikely to reduce carbon emissions, or risks to infrastructure (Bradstock *et al.* 2012). Rather fire management of 'fuel' levels close to human infrastructure would be more fruitful (Enright and Fontaine 2013). In addition, WSF (with the exception of *E. diversicolor* forest) are only marginally suited to prescribed burning because these forests are often in rugged topography, are generally wet, and only burn under extreme conditions (Lindenmayer 2009). Limiting levels of fragmentation and disturbance, especially logging, is likely to help maintain mesic micro-environments that protect 'fuel' against drying (Wardell-Johnson *et al.* 2011; 2015; Mackey *et al.* 2015); and therefore also minimise loss of carbon stocks (Dean *et al.* 2012b).

12.5 Conservation status

Currently three WSF ecological communities are listed as threatened under the Commonwealth Environment Protection and Biodiversity Protection Act 1999, all in the Sydney Basin Bioregion (blue gum high forest, turpentine-ironbark forest, upland basalt eucalypt forests; Department of the Enviroment and Energy 2016). Further WSF communities are under investigation. In Tasmania, *E. brookeriana* WSF, *E. globulus* WSF and *E. viminalis* WSF are listed as threatened native vegetation at the state level. Several other eucalypts that occur principally in WSF (e.g. *E. jacksonii*, *E. brevistylis*, *E. dunnii*) are also very restricted, though most are currently protected in conservation reserves. The majority of extant WSF in Australia occurs in crown land or leasehold tenure (64% of open forests) with 22% of open forest in nature conservation reserves (MPIGAQ and NFISC 2013). The level of clearing of WSFs varies greatly between regions (Table 12.3) as does the amount in protected areas. While many WSFs may be regarded as adequately protected, there are several threatening processes likely to impact on their condition and resilience.

12.6 Threatening processes

Threats to the ecological integrity of remaining WSF include unsustainable logging, disrupted natural fire regimes, disease and insect outbreaks, and climate change. These threats are frequently cumulative (Lindenmayer *et al.* 2009; 2011), and interactions between these processes and ecological values of WSF are complex (Wardell-Johnson *et al.* 2005; 2011; 2015). These threats affect tall forests globally (Allen *et al.* 2009; 2015; van Mantgem *et al.* 2009), and most forest types in Australia where they have been reviewed extensively (e.g. Wardell-Johnson and Christensen 1992; Wood *et al.* 2015). Eucalypt species in WSF have a particular suite of traits that make them particularly vulnerable to a warmer, drier climate.

12.6.1 Land clearance

The area of WSF has decreased since European occupation in Australia (Table 12.3), as these forests were either mined for timber or cleared for agriculture (Box 12.1). For example, in Queensland the

Table 12.3 **Areas and above-ground biomass (AGB) of wet sclerophyll forest (WSF), with spatial estimates of standard deviation for AGB in brackets. Extant, missing (deforested), and pre-1750 figures are based on Commonwealth of Australia (2007) and Richards and Brack (2004). Details of calculations are provided in Box 12.1.**

State	Area (ha)	Average above-ground biomass (Mg ha^{-1})	Standard deviation %	Above-ground biomass (Tg)	Area loss (%)	Above-ground biomass loss (%)
Extant						
NSW	2 237 000	278(82)	29	622	15	16
QLD	155 100	258(106)	41	40	28	30
TAS	946 000	183(61)	33	173	43	46
VIC	53 700	294(37)	13	16	0.1	0.1
WA	167 500	141(26)	19	24	22	20
Average		231(62)	27		22	23
total	3 559 000			875	23	23
Deforested						
NSW	335 000	306(74)	24	102		
QLD	43 690	278(78)	28	12		
TAS	410 880	193(72)	37	79		
VIC	50 00	277(46)	17	0.014		
WA	36 890	129(21)	16	4.8		
Average		236(58)	25			
Total	826 200			198		
Pre-1750						
NSW	2 571 000	282(81)	29	724		
QLD	198 800	262(100)	38	52		
TAS	1 357 000	186(64)	35	253		
VIC	53 750	294(37)	13	16		
WA	204 410	138(25)	18	28		
Average		233(61)	26			
Total	4 385 000			1073		

percentage remaining of individual WSF regional ecosystems varies from 100% in rugged ranges to less than 30% for ecosystems in alluvial or Tertiary plain situations (Neldner *et al.* 2015). Deforestation has occurred in lowlands for cropping (e.g. sugar cane or horticultural crops), and pastures for dairy or beef cattle. In Tasmania, 7.6% of WSFs were cleared between 1996 and 2011, with most converted to timber plantations for the production of wood and paper (Forest Practices Authority 2012).

In addition, large proportions of remaining WSF in Australia, have been modified, through the impacts of logging, and altered fire regimes that have led to changes in forest structure and composition (RAC 1992; Bradshaw 2012).

12.6.2 Logging

Wet sclerophyll forests have long been targeted for logging because of: (i) the quality of the hardwood timber and associated wood products; (ii) rapid growth rates; and (iii) the high volumes of timber they produce. Recently, forest management has been informed by principals of ecologically sustainable management (ESFM; Lindenmayer *et al.* 2012b), which seeks to ameliorate deleterious effects of logging. Central to ESFM has been the use of silvicultural practices that attempt to mimic aspects of natural disturbances that maintain WSF (Lindenmayer and Franklin 2002; Baker and Read 2011). Forest practices codes which include buffers around streams, and various other regulations, have been implemented to reduce many environmental impacts. However, the degree to which these aims are, or can be achieved; or indeed are relevant has long been debated (e.g. Lindenmayer *et al.* 1991; Wardell-Johnson and Horwitz 2000).

The ecological impact of various silvicultural techniques upon biodiversity and other forest values manifest at a range of spatial and temporal scales (Lindenmayer and Franklin 2002). The impacts of converting large portions of the landscape from old growth to silvicultural regrowth include loss of carbon (Dean *et al.* 2012b; Box 12.1), changes in stand structure and demography (Lindenmayer *et al.* 2011; Taylor *et al.* 2014), as well as fragmentation by logging, clearance and associated road construction. Removal of complex structural components of old growth (e.g. large hollow-bearing trees, large decaying logs and CWD) has deleterious effects on local habitat suitability for a range of fungi, fauna (Lindenmayer and Franklin 2002; Lindenmayer *et al.* 2010a) and plants (Ough and Murphy 1996; Neyland and Jarman 2011). In southern Australia, clear-felling has been a prevalent practice for more than 50 years; and most logged areas now lack large old-growth trees.

Coupe size impacts on the opportunities for the re-establishment of species from adjacent unlogged forests. This is because of a decline in colonising capacity with distance from source populations, and other influences, such as edge effect associated with proximity to mature forest (Wardell-Johnson and Williams 2000; Tabor *et al.* 2007; Baker *et al.* 2013; Fountain-Jones *et al.* 2014). Larger coupes generally have more of their area further from the propagule sources in adjacent forest. The introduction of forest practices codes, which reduced coupe size and their aggregation, may increase opportunity for a greater diversity of species to colonise logged sites. However, this gain is offset by the reduced area of remaining of old growth, and an associated reduction in supply of propagules of rainforest species.

Although rotation periods for clear-fell operations in Tasmania were anticipated to be about 80–90 years for most coupes, marketing opportunities have led to second rotation logging after only 50 years. Repeat disturbances at this frequency have a detrimental impact on the persistence of rainforest species within WSF (Jackson 1968; Ashton 1981). Other threats posed by logging include the introduction of weeds such as *Lantana camara* in the north or *Rubus fruticosus* and *Ulex europaeus* in the south (Wells and Hickey 1999) that outcompete native shrub and herb species, disrupt canopy tree regeneration, and impact on fauna (see Box 12.2). Recognition of the need for retention of structural complexity has led to variable retention silviculture in south-eastern Australian WSF. This is operational in Tasmania and New South Wales, and experimental in Victoria (Baker and Read 2011; Lindenmayer *et al.* 2012b). In south-western Australia, retention of habitat trees has recently increased in intensively managed forests.

Box 12.2 Bell miner associated dieback (BMAD) as a threatening ecological process

Bell miner associated dieback (BMAD) refers to eucalypt forest dieback associated with a dysfunctional interaction between WSF eucalypts and psyllids, one of their invertebrate herbivores (see Eldridge *et al.*, Chapter 6, this volume, for a perspective on the ecological mechanisms) the co-incidence of outbreaks of sap-sucking psyllid insects (Hemiptera: Psylloidea) and colonies of the Bell Miner (*Manorina melanophrys*). This form of dieback was reported in WSF in the 1920s and has recently been listed as a key threatening process in New South Wales (Schedule 3 of the *Threatened Species Conservation Act 1995*). BMAD has been linked to extensive decline of isolated patches of WSF in Victoria, tens of thousands of hectares in north-eastern New South Wales and substantial areas of south-east Queensland (Wardell-Johnson *et al.* 2005).

Bell miners are insectivorous birds (family Meliphagidae), endemic to south-eastern Australia that feed mainly on lerp, a sugary protective covering built by psyllids. They live in colonies and aggressively defend territory, excluding other avian insectivores. Colonies of bell miners are associated with higher abundances of psyllids than sites without bell miners, but a high density of psyllids is not required for bell miners to colonise (Dare *et al.* 2007). Colonisation by bell miners does not necessarily lead to an infestation by psyllids, but the behaviour of bell miners can exacerbate infestation levels. Disturbance leading to formation of a denser understorey, including of the introduced *Lantana camara*, makes previously unsuitable sites attractive for colonisation by bell miners, but a weedy understorey is not required for colonisation (Wardell-Johnson and Lynch 2005; Dare *et al.* 2007).

The management history of WSF is likely to be a significant determinant of the impact of BMAD (Wardell-Johnson *et al.* 2005). Stone (1999) suggested that changes to the overstorey as a result of logging may be sufficient to commence a spiral of dieback on some WSF sites, with this disturbance leading to initial colonisation of psyllids. A case study of Toonunbar in northern New South Wales (Wardell-Johnson and Lynch 2005) illustrates the key role played by disturbance and management history, particularly logging. Fire regimes in shrubby WSF where BMAD is having a pronounced impact are naturally infrequent, although fire regimes differ between vegetation types (Keith and Henderson 2002). Very frequent fire (i.e. < 10 year intervals) may be required to maintain a grassy understorey in some WSF, and a reduction in frequency may lead to higher densities of shrubs in the understorey. It is likely that BMAD, if present in the area, may then also impact on these forests.

Management history may be critical in WSF susceptible to BMAD, with exacerbation of BMAD likely through interactions with a warming climate. Climate change modelling suggests that shrubby WSF (and mixed forest) may also be susceptible to increases in fire frequency and intensity as a result of the climate disruption associated with global warming.

Figure 12.7 **(a)** Bell miner associated dieback in wet sclerophyll forest of *E. saligna* near Toonunbar, NSW, 2006. **(b)** Stable population of bell miners in wet sclerophyll forest of *E. viminalis* near Eden, NSW 2016. Photos: Grant Wardell-Johnson.

12.6.3 Disease and dieback

Stands of WSF undergo periodic attack from diseases and pests. Damage to WSF ecosystems can be widespread, and is sometimes severe. Premature decline of overstorey eucalypts in localised areas across a diverse range of forest types has been reported (Close *et al.* 2009), including some WSF (i.e. *E. delegatensis*, Harvest *et al.* 2008; *E. regnans*, Lindenmayer *et al.* 2012a). In coastal Victoria, New South Wales and southern Queensland, bell miner (*Manorina melanophrys*) associated dieback (BMAD) linked to the build-up of populations of psyllids

(Box 12.2), occurs in *E. saligna*, *E. pilularis* and other shrubby WSFs (Stone 1999; 2005; Wardell-Johnson and Lynch 2005; Wardell-Johnson *et al.* 2005). As well as leading to the death of the canopy trees, BMAD can have long-term impacts on the structure and floristics of WSF and associated fauna (Box 12.2).

Pathogens can have severe impacts on WSF ecosystems. For example, *Armillaria luteobubalinia*, the causal fungus of *Armillaria* root disease is widespread in forests of *E. diversicolor*, and causes scattered mortality and wind-throw (Robinson *et al.* 2008). In Tasmania, regrowth and gully dieback thought to be associated with drought, insect defoliators and infection by *Armillaria* species occurs in regrowth stands of *E. obliqua* and *E. regnans* (Wardlaw 1990). The impact of myrtle rust (*Puccinia psidii*) on WSF species is not yet clear, but has rapidly spread through eastern Australia in the past decade, and has been recorded on flower buds, flowers and fruits of at least 28 native host species within the family Myrtaceae, including members of the genus *Eucalyptus* (Pegg *et al.* 2014). It has killed *Rhodomyrtus psidioides* trees as a result of repeated infections.

12.6.4 Climate disruption

The magnitude and direction of change caused by anthropogenic global warming (IPCC 2007) and deforestation (Steffen *et al.* 2009; Wardell-Johnson *et al.* 2011) varies across the range of WSF (Lindenmayer *et al.* 2010b). There is increased stress in many Australian forest systems (e.g. Calder and Kirkpatrick 2008; Allen *et al.* 2009) as a result of changes already experienced (Steffen *et al.* 2009). Climatic changes are likely to lead to more frequent extreme climate events, including severe drought (Clarke *et al.* 2015). Reduced growth is forecast for eucalypts in WSFs due to reduced water availability (Bowman *et al.* 2014), while a reduction in soil carbon stocks is forecast for WSF in south-western Australia, and for Tasmania due to drought stress and higher fire frequency (Dean and Wardell-Johnson 2010; Wardell-Johnson *et al.*, in press). As rising temperatures lead to increased water stress, this will force major shifts in dominant plants (Allen *et al.* 2015; McDowell and Allen 2015), disadvantaging large trees (Lindenmayer *et al.* 2012c; Bennett *et al.* 2015).

Prior *et al.* (2011) analysed relationships between tree growth and climate using a continental analysis of several thousand permanent inventory plots, and found tree diameter increment decreased with a rise in mean annual temperature and tree size. A predicted increase in mean annual temperature of ~3°C and a decrease in mean annual rainfall of ~40 mm (A2 emissions scenario; IPCC 2000) across the forest estate would likely lead to a decrease in diameter increment of ~17% (Wood *et al.* 2015). These results correspond with physiologically based growth modelling (ABARES 2011). However, this pattern varies between species, so that growth of *E. fraxinoides* is projected to decrease, while growth of *E. delegatensis* is projected to increase with increasing temperature within the limits studied (Wood *et al.* 2015).

Predictions of a general decline in forest growth and productivity across the native forest estate (ABRES 2011; Prior *et al.* 2011) are concerning as they do not account for projected increases in intense fire frequency in these flammable systems (Hennessy *et al.* 2006; Cary *et al.* 2012). Changes in temperature and rainfall are likely to increase the frequency of high-intensity fires in both south-eastern (Hennessy *et al.* 2006) and south-western Australia (Williams *et al.* 2009). Climate modelling has confirmed an increased fire danger index forecast for Tasmania, especially for the southern region which contains the most carbon-dense WSFs and largest trees (Fox-Hughes *et al.* 2014). Rapid (< 150 years) return-times of high-intensity events is most concerning, as this reduces the capacity of WSF species to recover. The dominant eucalypts of WSF usually take at least 100 years to regain their former heights following intense fire (Bradshaw and Rayner 1997).

Abrupt losses of large areas of WSF and progressive losses of structure are underway as a result of the formation of landscape traps (Lindenmayer *et al.* 2009b; 2011; Taylor *et al.* 2014). A demographic shift resulting in replacement of WSF (Ashton 1981) with relatively shorter-lived tree and shrub communities able to persist following more frequent fires is therefore likely during the present century (Lindenmayer *et al.* 2011). This has implications for the persistence of old-growth forest elements of WSF in southern Australia.

12.7 Conclusion

WSFs are globally unique. They occupy a relatively small area of high-rainfall country in Australia, but are disproportionately important for their biodiversity, timber yields and social values. These forests persist in a regime of infrequent intense fires, but are highly vulnerable to warming, drying climates and resultant changed fire regimes. It is likely that the extent of these forests will decrease as a result of interactions associated with anthropogenic climate change. There is also likely to be a decline in the hardwood timber industry as timber is increasingly obtained from diverse sources, and demand to utilise WSF for recreation and nature conservation increases. Paradoxically, as these forests are more widely appreciated, they are being placed under increasing stress by climate disruption. It would be tragic if these once splendid forests were further reduced and degraded, just as society more broadly embraces the multiple values and services that they provide.

Acknowledgements

We thank Jamie Kirkpatrick for valuable discussions on vegetation structure and dynamics, and for comments on an earlier version of the manuscript. We also thank David Keith and two anonymous referees for helpful comments and suggestions. Arnon Accad and Will Smith of Queensland Herbarium produced the map.

References

ABARES (2011) *Potential effects of climate change on forests and forestry in Australia*. Australian Bureau of Agricultural and Resources Economics and Sciences: Canberra.

Allen, C.D., Macalady, A.K., Chenchouni, H. *et al.* (2009) A global overview of drought and heat-induced tree mortality reveals emerging climate change risks for forests. *Forest Ecology and Management* 259, 660–684.

Allen, C.D., Breshears, D.D. and McDowell, N.G. (2015) On underestimation of global vulnerability to tree mortality and forest die-off from hotter drought in the Anthropocene. *Ecosphere* 6, 1–55.

Ash, J. (1988) The location and stability of rainforest boundaries in north-eastern Queensland, Australia. *Journal of Biogeography* 15, 619–630.

Ashton, D.H. (1981) Fire in tall open-forests (wet sclerophyll forests). In *Fire and the Australian Biota*, ed. A.M. Gill, R.H. Groves and I.R. Noble. Australian Academy of Science: Canberra, pp. 121–151.

Ashton, D.H. (2000) The environment and plant ecology of the Hume range, central Victoria. *Proceedings of the Royal Society of Victoria* 112, 185–278.

Ashton, D.H. and Attiwill, P.M. (1994) Tall open-forests. In *Australian Vegetation*, ed. R.H. Groves. Melbourne: Cambridge University Press, pp. 157–196.

Baker, S.C. and Read, S.M. (2011) Variable retention silviculture in Tasmania's wet forests: ecological rationale, adaptive management and synthesis of biodiversity benefits. *Australian Forestry* 74, 218–232.

Baker, S.C., Spies, T.A., Wardlaw, T.J. *et al.* (2013) The harvested side of edges: effect of retained forests on the re-establishment of biodiversity in adjacent harvested areas. *Forest Ecology and Management* 302, 107–121.

Bassett, O.D., Prior, L.D., Slijkerman, C.M. *et al.* (2015) Aerial sowing stopped the loss of alpine ash (*Eucalyptus delegatensis*) forest burnt by three short-interval fires in the Alpine National Park, Victoria, Australia. *Forest Ecology and Management* 342, 39–48.

Bayly, M.J., Rigault, P., Spokevicius, A. *et al.* (2013) Chloroplast genome analysis of Australian eucalypts – Eucalyptus, Corymbia, Angophora, Allosyncarpia and Stockwellia (Myrtaceae). *Molecular Phylogenetics and Evolution* 69, 704–716.

Beadle, N.C.W. and Costin, A.B. (1952) Ecological classification and nomenclature. *Proceedings of the Linnean Society of NSW* 77, 61–82.

Beard, J.S., Beeston, G.R., Harvey, J.M. *et al.* (2013) The vegetation of Western Australia at the 1: 3 000 000 scale. Explanatory memoir. 2nd edition. *Conservation Science Western Australia* 9, 1–152.

Bennett, A.C., McDowell, N.G., Allen, C.D. and Anderson-Teixeira, K.J. (2015) Larger trees suffer most during drought in forests worldwide. *Nature Plants* 1, doi:10.1038/nplants.2015.139

Boer, M.M., Sadler, R.J., Wittkuhn, R.S. *et al.* (2009) Long-term impacts of prescribed burning on regional extent and incidence of wildfires - evidence from 50 years of active fire management in SW Australian forests. *Forest Ecology and Management* 259, 132–142.

Bowman, D.M.J.S. (2000) *Australian Rainforests: Islands of Green in an Ocean of Fire*. Cambridge University Press: Cambridge.

Bowman, D.M.J.S., Williamson, G.J., Keenan, R.J. and Prior, L.D. (2014) A warmer world will reduce tree growth in evergreen broadleaf forests: evidence from Australian temperate and subtropical eucalypt forests. *Global Ecology and Biogeography* 23, 925– 934.

Bradshaw, C.J.A. (2012) Little left to lose: deforestation and forest degradation in Australia since European colonization. *Journal of Plant Ecology* 5, 109–120.

Bradshaw, F.J. and Rayner, M.E. (1997) Age structure of the karri forest: 2. Projections of future forest structure and implications for management. *Australian Forestry* 60, 188–195.

Bradstock, R.A., Boer, M.M., Cary, G.J. *et al.* (2012) Modelling the potential for prescribed burning to mitigate carbon emissions from wildfires in fire-prone forests of Australia. *International Journal of Wildland Fire* 21, 629–639.

Braithwaite, L.W., Dudzinski, M.L. and Turner, J. (1983) Studies on the arboreal marsupial fauna of eucalypt forests being harvested for woodpulp at Eden, N.S.W. II. Relationship between the fauna density, richness and diversity and measured variables of habitat. *Australian Wildlife Research* 10, 231–247.

Burns, E.L., Lindenmayer, D.B., Stein, J. *et al.* (2015) Ecosystem assessment of mountain ash forest in the central highlands of Victoria, south-eastern Australia. *Austral Ecology* 40, 386–399.

Burrows, N. and Wardell-Johnson, G. (2003) Interactions of fire and plants in south-western Australia's forested ecosystems: a review. In *Fire in South-western Australian Ecosystems: Impacts and Management*, ed. N. Burrows and I. Abbott. Backhuys: Leiden, pp. 225–268.

Calder, J.A. and Kirkpatrick, J.B. (2008) Climate change and other factors influencing the decline of the Tasmanian cider gum (*Eucalyptus gunnii*). *Australian Journal of Botany* 56, 684–692.

Cary, G., Bradstock, R., Gill, A.M. and Williams, R. (2012) Global change and fire regimes in Australia. In *Flammable Australia: Fire Regimes, Biodiversity and Ecosystems in a Changing World*, ed. R. Bradstock, A.M. Gill and R. Williams. CSIRO Publishing: Melbourne, pp. 149–180.

Churchill, D.M. (1968) The distribution and prehistory of *Eucalyptus diversicolour* F. Muell, *E. marginata* Donn ex Sm. and *E. calophylla* R. BR in relation to rainfall. *Australian Journal of Botany* 16, 125–151.

Clarke, H., Lucas, C. and Smith, P. (2013) Changes in Australian fire weather between 1973 and 2010. *International Journal of Climatology* 33, 931–944.

Clarke, P.J., Lawes, M.J., Murphy, B.P. *et al.* (2015) A synthesis of postfire recovery traits of woody plants in Australian ecosystems. *Science of The Total Environment* 534, 31–42.

Close, D.C., Davidson, N.J., Johnson, D.W. *et al.* (2009) Premature decline of Eucalyptus and altered ecosystem processes in the absence of fire in some Australian forests. *Botanical Review* 75, 191–202.

Commonwealth of Australia (2007) *Eucalypt Tall Open Forests (MVG2). Australia's Native Vegetation: A Summary of Australia's Major Vegetation Groups, 2007*. Australian Government: Canberra. Available at: https://www.environment.gov.au/system/files/resources/a9897cf2-9d38-4201-bea2-13dadf3af9a8/files/major-veg-summary.pdf (accessed 4 July 2016).

Commonwealth of Australia (2012) *Australia's native vegetation framework. A national framework to guide the ecologically sustainable management of Australia's native vegetation*. Department of Sustainability, Environment, Water, Population and Communities: Canberra. Available at: https://www.environment.gov.au/system/files/resources/76f709dc-ccb3-4645-a18b-063fbbf0a899/files/native-vegetation-framework.pdf (accessed 4 July 2016).

Cremer, K.W., Cromer, R.N. and Florence, R.G. (1978) Stand establishment. In *Eucalypts for Wood Production*, ed. W.E. Hillis and A.G. Brown. CSIRO Publishing: Melbourne, pp. 81–135.

Crisp, M.D., Burrows, G.E., Cook, L. *et al.* (2011) Flammable biomes dominated by eucalypts originated at the Cretaceous–Paleogene boundary. *Nature Communications* 2, 193.

Dare, A.J., McDonald, P.G. and Clarke, M.F. (2007) The ecological context and consequences of colonisation of a site by bell miners (*Manorina melanophrys*). *Wildlife Research* 34, 616–623.

Dean, C. and Wardell-Johnson, G. (2010) Old-growth forests, carbon and climate change: functions and management for tall open-forests in two hotspots of temperate Australia. *Plant Biosystems* 144, 180–193.

Dean, C., Wardell-Johnson, G.W. and Harper, R.J. (2012a) Carbon management of commercial rangelands in Australia: major pools and fluxes. *Agriculture, Ecosystems & Environment* 148, 44–64.

Dean, C., Wardell-Johnson, G.W. and Kirkpatrick, J.B. (2012b) Are there any circumstances in which logging primary wet-eucalypt forest will not add to the global carbon burden? *Agricultural and Forest Meteorology* 161, 156–169.

Dean, C., Fitzgerald, N. and Wardell-Johnson, G. (2012c) Pre-logging carbon accounts in old-growth forests via allometry: an example of mixed forest in Tasmania, Australia. *Plant Biosystems* 146, 223–236.

Department of the Environment and Energy (2007) NVIS data products 4.2. Australian Government: Canberra. Available at: http://www.environment.gov.au/land/native-egetation/national-vegetation-information-system/data-products (accessed 10 May 2016).

Department of the Environment and Energy (2016) *Species Profile and Threats Database: EPBC Act List of Threatened Ecological Communities*. Canberra: Australian Government. Available at: http://www.environment.gov.au/cgi-bin/sprat/public/publiclookupcommunities.pl (accessed 17 June 2016).

Enright, N.J. and Fontaine, J.B. (2013) Climate change and the management of fire-prone vegetation in southwest and southeast Australia. *Geographical Research* 52, 34–44.

Fedrigo, M., Kasel, S., Bennett, L.T. *et al.* (2014) Carbon stocks in temperate forests of south-eastern Australia reflect large tree distribution and edaphic conditions. *Forest Ecology and Management* 334, 129–143.

Forest Practices Authority (2012) *State of the forests Tasmania 2012*. Forest Practices Authority: Hobart. Available at: http://www.fpa.tas.gov.au/__data/assets/pdf_file/0005/111596/State_of_the_forests_Tasmania_2012_report.pdf (accessed 4 July 2016).

Fountain-Jones, N.M., Jordan, G.J., Baker, T.P. *et al.* (2014) Living near the edge: being close to mature forest increases the rate of succession in beetle communities. *Ecological Applications* 25, 800–811.

Fox-Hughes, P., Harris, R., Lee, G. *et al.* (2014) Future fire danger climatology for Tasmania, Australia, using a dynamically downscaled regional climate model. *International Journal of Wildland Fire* 23, 309–321.

Gibbons, P. and Lindenmayer, D.B. (2001) *Tree Hollows and Wildlife Conservation in Australia*. CSIRO Publishing: Melbourne.

Gilbert, J.M. (1959) Forest succession in the Florentine Valley, Tasmania. *Papers and Proceedings of the Royal Society of Tasmania* 93, 129–151.

Grant, J.C., Laffan, M.D., Hill, R.B. and Neilsen, W.A. (1995) Forest soils of Tasmania: a handbook for identification and management. *Forestry Tasmania*: Hobart.

Harrington, G.N. and Sanderson, K.D. (1994) Recent contraction of wet sclerophyll forest in the Wet Tropics of Queensland due to invasion by rainforest. *Pacific Conservation Biology* 1, 319–327.

Harvest, T., Davidson, N.J. and Close, D.C. (2008) Is decline in high altitude eucalypt forests related to rainforest understorey development and altered soil bacteria following the long absence of fire? *Austral Ecology* 33, 880–890.

Hennessy, K., Lucas, C., Nicholls, N. *et al.* (2006) *Climate Change Impacts on Fire Weather in Southeast Australia*. CSIRO Publishing: Canberra.

Hill, R.S. and Read, J. (1984) Post fire regeneration of rainforest and mixed forest in Western Tasmania. *Australian Journal of Botany* 32, 481–493.

Hopper, S.D. (2003) An evolutionary perspective on south-west Western Australian landscapes, biodiversity and fire: a review and management implications. In *Fire in South-western Australian Ecosystems: Impacts and Management*, ed. N. Burrows and I. Abbott. Backhuys: Leiden, pp. 9–35.

IPCC (2000) Chapter 4: An overview of scenarios. In *Special Report on Emissions Scenarios*, ed. N. Nakicenovic and R. Swart. Cambridge University Press: Cambridge, pp. 167–238.

IPCC (2007) *Climate change 2007: the physical basis. Contribution of working group to the fourth assessment report of the Intergovernmental Panel on Climate Change*, ed. S. Soloman, D. Qin, M. Manning *et al.* Cambridge University Press: Cambridge.

Jackson, W.D. (1968) Fire, air, water and earth: an elemental ecology of Tasmania. *Proceedings of the Ecological Society of Australia* 3, 9–16.

Keith, D.A. (2004) *Ocean shores to desert dunes: the native vegetation of New South Wales and the ACT*. NSW Department of Environment and Conservation: Sydney.

Keith, D.A. and Henderson, M.K. (2002) Restoring the prepastoral condition: response to Jurskis. *Austral Ecology* 27, 691–693.

Keith, H., Mackey, B.G. and Lindenmayer, D.B. (2009) Re-evaluation of forest biomass carbon stocks and lessons from the world's most carbon-dense forests. *Proceedings of the National Academy of Sciences* 106, 11635–11640.

Keith, H., Lindenmayer, D., Mackey, B. *et al.* (2014) Managing temperate forests for carbon storage: impacts of logging versus forest protection on carbon stocks. *Ecosphere* 5, 1–34.

Kikkawa, J. and Dwyer, P.D. (1992) Use of scattered resources in rainforest of humid tropical lowlands. *Biotropica* 24, 293–308.

Kirkpatrick, J.B., Peacock, R.J., Cullen, P.J. and Nyland, M.G. (1988) *The Wet Eucalypt Forests of Tasmania*. Tasmanian Conservation Trust: Hobart.

Kitchener, A. and Harris, S. (2013) From forest to fjaeldmark: descriptions of Tasmania's vegetation. 2nd edition. Department of Primary Industries, Parks, Water and Environment: Tasmania. Available at: http://dpipwe.tas.gov.au/conservation/flora-of-tasmania/from-forest-to-fjaedlmark-descriptions-of-tasmanias-vegetation-(edition-2) (accessed 4 July 2016).

Knox, K.J.E. and Clarke, P.J. (2012) Fire severity, feedback effects and resilience to alternative community states in forest assemblages. *Forest Ecology and Management* 265, 47–54.

Lindenmayer, D.B. (2009) Old forests, new perspectives: insights from the mountain ash forests of the Central Highlands of Victoria, south-eastern Australia. *Forest Ecology and Management* 258, 357–365.

Lindenmayer, D.B. and Franklin, J.F. (2002) *Conserving Forest Biodiversity: A Comprehensive Multiscaled Approach.* Island Press: Washington.

Lindenmayer, D.B., Norton, T.W. and Tanton, M.T. (1991) Differences between wildfire and clearfelling on the montane ash forests of Victoria and their implications for fauna dependent on tree hollows. *Australian Forestry* 53, 61–68.

Lindenmayer, D.B., Mackey, B.G., Mullen, I.C. *et al.* (1999) Factors affecting stand structure in forests - are there climatic and topographic determinants? *Forest Ecology and Management* 123, 55–63.

Lindenmayer, D.B., Hunter, M.L., Burton, P.J. and Gibbons, P. (2009) Effects of logging on fire regimes in moist forests. *Conservation Letters* 2, 271–277.

Lindenmayer, D.B., Knight, E., McBurney, L. *et al.* (2010a) Small mammals and retention islands: an experimental study of animal response to alternative logging practices. *Forest Ecology and Management* 260, 2070–2078.

Lindenmayer, D.B., Stefan, W., Burbidge, A.A. *et al.* (2010b) Conservation strategies in response to rapid climate change: Australia as a case study. *Biological Conservation* 143, 1587–1593.

Lindenmayer, D.B., Hobbs, R.J., Likens, G.E. *et al.* (2011) Newly discovered landscape traps produce regime shifts in wet forests. *Proceedings of the National Academy of Sciences* 108, 15887–15891.

Lindenmayer, D.B., Blanchard, W., McBurney, L. *et al.* (2012a) Interacting factors driving a major loss of large trees with cavities in a forest ecosystem. *PLoS ONE* 7, e41864.

Lindenmayer, D.B., Franklin, J.F., Lõhmus, A. *et al.* (2012b) A major shift to retention forestry can help resolve global forest sustainability issues. *Conservation Letters* 5, 421–431.

Lindenmayer, D.B., Laurance, W.F. and Franklin, J.F. (2012c) Global decline in large old trees. *Science*, 338, 1305–1306.

Lunney, D. (2004) *Conservation of Australia's forest fauna*. 2nd edition. Royal Zoological Society of New South Wales: Mosman.

Lynch, A.J.J. and Neldner, V.J. (2000) The problems of placing boundaries on ecological continua: options for a workable national rainforest definition in Australia. *Australian Journal of Botany* 48, 511–530.

Mackey, B., Lindenmayer, D.B., Gill, A.M. *et al.* (2002) *Wildlife, Fire and Future Climate: A Forest Ecosystem Analysis*. CSIRO Publishing: Melbourne.

Mackey, B.G., Keith, H., Berry, S.L. and Lindenmayer, D.B. (2008) *Green Carbon. The Role of Natural Forests in Carbon Storage. Part 1. A Green Carbon Account of Australia's South-eastern Eucalypt Forests, and Policy Implications*. The Australian National University: Canberra.

Mackey, B., Dellasala, D.A., Kormos, C. *et al.* (2015) Policy options for the world's primary forests in multilateral environmental agreements. *Conservation Letters* 8, 139–147.

McCarthy, M.A. and Lindenmayer, D.B. (1998) Multi-aged mountain ash forest, wildlife conservation and timber harvesting. *Forest Ecology and Management* 104, 43–56.

McCaw, W.L. and Middleton, T. (2015) Recovery of tall open eucalypt forest in south-western Australia following complete crown scorch. *Fire Ecology* 11, 95–107.

McDowell, N.G. and Allen, C.D. (2015) Darcy's law predicts widespread forest mortality under climate warming. *Nature Climate Change* 5, 669–672.

Montreal Process Implementation Group for Australia and the National Forest Inventory Steering Committee (MPIGAQ and NFISC) (2013) *Australia's State of the Forests Report (2013)* Five-yearly report Department of Agriculture, ABARES, Canberra.

Neldner, V.J., Niehus, R.E., Wilson, B.A. *et al.* (2015) *The vegetation of Queensland. Descriptions of broad vegetation groups. Version 2.0*. Queensland Herbarium, Department of Science, Information Technology and Innovation: Queensland. Available at: https://publications.qld.gov.au/dataset/vegetation-qld/resource/921fa786-e6d5-4a8a-9b0c-e532d2ce3f32 (accessed 5 July 2016).

Neyland, M.G. and Jarman, S.J. (2011) Early impacts of harvesting and burning disturbances on vegetation communities in the Warra silvicultural systems trial, Tasmania, Australia. *Australian Journal of Botany* 59, 701–712.

Ough, K. and Murphy, A. (1996) The effect of clearfelling logging on tree ferns in Victoria. *Australian Forestry* 59, 178–188.

Pegg, G.S., Giblin, F.R., McTaggart, A.R. *et al.* (2014) *Puccinia psidii* in Queensland, Australia: disease symptoms, distribution and impact. *Plant Pathology* 63, 1005–1021.

Petrie, A.H.K., Jarrett, P.H. and Patton, R.T. (1929) The vegetation of the Blacks' Spur region: a study in the ecology of some Australian mountain eucalyptus forests: I. The mature plant communities. *Journal of Ecology* 17, 223–248.

Prior, L.P., Williamson, G. and Bowman, D.M.J.S. (2011) *Using permanent forestry plots to understand the possible effects of climate change on Australia's production forest estate*. Department of Agriculture, Fisheries and Forestry: Canberra.

Read, J. and Brown, M.J. (1996) Ecology of Australian *Nothofagus* forests. In *The Ecology and Biogeography of Nothofagus Forests*, ed. T.T. Veblen, R.S. Hill and J. Read. Yale University Press: Michigan, pp. 131–181.

Resource Assessment Commission (RAC) (1992) *Forest and Timber Inquiry Final Report Volume 1*. Resource Assessment Commission: Canberra.

Richards, G.P. and Brack, C. (2004) A continental biomass stock and stock change estimation approach for Australia. *Australian Forestry* 67, 284–288.

Robinson, R.M., Mellican, A.E. and Smith, R.H. (2008) Epigeous macrofungal succession in the first five years following a wildfire in karri (*Eucalyptus diversicolour*) regrowth forest in Western Australia. *Austral Ecology* 33, 807–820.

Roxburgh, S.H. (2009) Increase in carbon stocks in pre-1990 eucalypt forests. In *An Analysis of Greenhouse Gas Mitigation and Carbon Biosequestration Opportunities from Rural Land Use*, ed. S. Eady, M. Grundy, M. Battaglia and B. Keating. CSIRO Publishing: St. Lucia, pp. 90–100.

Russell-Smith, J., Stanton, J.P., Whitehead, P.J. and Edwards, A. (2004) Rain forest invasion of eucalypt-dominated woodland savanna, Iron Range, north-eastern Australia: I. Successional processes. *Journal of Biogeography* 31, 1293–1303.

Sillett, S.C., Van Pelt, R., Koch, G.W. *et al.* (2010) Increasing wood production through old age in tall trees. *Forest Ecology and Management* 259, 976–994.

Specht, R.L. (1970) Vegetation. In *The Australian Environment*, ed. G.W. Leeper. CSIRO Publishing and Melbourne University Press: Melbourne, pp. 44–67

Stanton, P., Parsons, M., Stanton, D. and Stott, M. (2014) Fire exclusion and the changing landscape of Queensland's wet tropics bioregion 2. The dynamics of transition forests and implications for management. *Australian Forestry* 77, 58–68.

Steffen, W., Burbidge, A.A., Hughes, L. *et al.* (2009) *Australia's biodiversity and climate change: a strategic assessment of the vulnerability of Australia's biodiversity to climate change*. Australian Government: Canberra. Available at: https://www.environment.gov.au/system/files/resources/eab369d6-76f9-46c8-beb4-aaae8ece112e/files/biodiversity-vulnerability-assessment.pdf (accessed 4 July 2016).

Stone, C. (1999) Assessment and monitoring of decline and dieback of forest eucalypts in relation to ecologically sustainable forest management: a review with a case study. *Australian Forestry* 62, 51–58.

Stone, C. (2005) Bell-miner-associated dieback at the tree crown scale: a multi-trophic process. *Australian Forestry* 68, 237–241.

Styger, J. and Kirkpatrick, J.B. (2015) Less than 50 millimetres of rainfall in the previous month predicts fire in Tasmanian rainforest. *Papers and Proceedings of the Royal Society of Tasmania* 149, 1–5.

Tabor, J., McElhinny, C., Hickey, J. and Wood, J. (2007) Colonisation of clearfelled coupes by rainforest tree species from mature forest edges, Tasmania, Australia. *Forest Ecology and Management* 240, 13–23.

Taylor, C., McCarthy, M.A. and Lindenmayer, D.B. (2014) Nonlinear effects of stand age on fire severity. *Conservation Letters* 7, 355–370.

Tng, D.Y.P., Williamson, G.J., Jordan, G.J. and Bowman, D.M.J.S. (2012) Giant eucalypts: globally unique fire-adapted rain-forest trees? *New Phytologist* 196, 1001–1014.

Turner, P.M., Balmer, J. and Kirkpatrick, J.B. (2009) Stand-replacing wildfires? The incidence of multicohort and single-cohort *Eucalyptus regnans* and *E. obliqua* forests in southern Tasmania. *Forest Ecology and Management* 258, 366–375.

van Mantgem, P.J., Stephenson, N.L., Byrne, J.C. *et al.* (2009) Widespread increase of tree mortality rates in the western United States. *Science* 323, 521–524.

Wardell-Johnson, G. (2000) Responses of locally endemic and regionally distributed eucalypts to moderate and high intensity fire in the Tingle Mosaic, south-western Australia. *Austral Ecology* 25, 409–421.

Wardell-Johnson, G. and Christensen, P. (1992) A review of the effects of disturbance on wildlife of the Karri forest. In *Research on the Impact of Forest Management in South-west Western Australia*. Department of Conservation and Land Management: Como, pp. 33–58.

Wardell-Johnson, G. and Horwitz, P. (2000) The recognition of heterogeneity and restricted endemism in the management of forested ecosystems in south-western Australia. *Australian Forestry* 63, 218–225.

Wardell-Johnson, G. and Lynch, A.J.J. (2005) Landscape processes and eucalypt dieback associated with bell miner habitat in south-eastern Australia. *Australian Forestry* 68, 237–241.

Wardell-Johnson, G. and Williams, M. (1996) A floristic survey of the Tingle Mosaic, south-western Australia. *Journal of the Royal Society of Western Australia* 79, 249–276.

Wardell-Johnson, G. and Williams, M. (2000) Edges and gaps in mature karri forest, southwestern Australia: logging effects on bird species abundance and diversity. *Forest Ecology and Management* 131, 1–21.

Wardell-Johnson, G., Williams, J., Hill, K. and Cummings, R. (1997) Evolutionary biogeography and contemporary distribution of eucalypts. In *Eucalypt Ecology: Individuals to Ecosystems*, ed. J. Williams and J. Woinarski. Cambridge University Press: Cambridge, pp. 92–128.

Wardell-Johnson, G.W., Williamson, M., Mellican, A.E. and Annells, A. (2004) Floristic patterns and disturbance history in karri forest, south-western Australia: 1. Environment and species richness. *Forest Ecology and Management* 199, 449–460.

Wardell-Johnson, G., Stone, C., Recher, H. and Lynch, A.J.J. (2005) A review of eucalypt dieback associated with bell miner habitat in south-eastern Australia. *Australian Forestry* 68, 231–236.

Wardell-Johnson, G.W., Williams, M., Mellican, A. and Annells, A. (2007) Floristic patterns and disturbance history in karri forest, south-western Australia: 2. Origin, growth form and fire response. *Acta Oecologica* 31, 137–150.

Wardell-Johnson, G., Keppel, G. and Sander, J. (2011) Climate change impacts on the terrestrial biodiversity and carbon stocks of Oceania. *Pacific Conservation Biology* 17, 220–240.

Wardell-Johnson, G.W., Calver, M., Burrows, N. and Di Virgilio, G. (2015) Integrating rehabilitation, restoration and conservation for a sustainable jarrah forest future during climate disruption. *Pacific Conservation Biology* 21, 1–11.

Wardell-Johnson, G., Crellin, L., Napier, C. *et al.* (in press) Has canopy height and biomass recovered 78 years after an intense fire in south-western Australia's red tingle (*Eucalyptus jacksonii*) forests? *International Journal of Wild Fire*. (in press).

Wardlaw, T. (1990) Changes in forest health associated with short-term climatic fluctuation. *Tasforests* 2, 107–110.

Warman, L. and Moles, A.T. (2009) Alternative stable states in Australia's Wet Tropics: a theoretical framework for the field data and a field-case for the theory. *Landscape Ecology* 24, 1–13.

Webb, L.J. (1968) Environmental relationships of the structural types of Australian rainforest vegetation. *Ecology* 49, 296–311.

Wells, P. and Hickey, J. (1999) Wet sclerophyll, mixed and swamp forest. In *Vegetation of Tasmania*, ed. J.B. Reid, R.S. Hill, M.J. Brown and M. Hovenden. Australian Biological Resources Study, Environment Australia: Canberra, pp. 224–243.

Williams, G. (1993) *Hidden Rainforests: Subtropical Rainforests and Their Invertebrate Biodiversity*. NSW University Press: Kensington.

Williams, R.J., Bradstock, R.A., Cary, G.J. *et al.* (2009) *Interactions between climate change, fire regimes and biodiversity in Australasia: a preliminary assessment*. Department of Climate Change: Canberra. Available at: https://www.environment.gov.au/system/files/resources/ab711274-004c-46cb-a9f1-09dba68d98c3/files/20100630-climate-fire-biodiversity-pdf.pdf (accessed 4 July 2016).

Wood, S.W. and Bowman, D.M.J.S. (2012) Alternative stable states and the role of fire-vegetation-soil feedbacks in the temperate wilderness of southwest Tasmania. *Landscape Ecology* 27, 13–28.

Wood, S.W., Hua, Q., Allen, K.J. and Bowman, D.J.M.S. (2010) Age and growth of a fire-prone Tasmanian temperate old-growth forest stand dominated by *Eucalyptus regnans*, the world's tallest angiosperm. *Forest Ecology and Management* 260, 438–447.

Wood, S.W., Prior, L.D., Stephens, H.C. and Bowman, D.M.J.S. (2015) Macroecology of Australian tall eucalypt forests: baseline data from a continental-scale permanent plot network. *PLOS One* doi:10.1371/journal.pone.0137811.

13 Biogeography of Australia's Dry Sclerophyll Forests: Drought, Nutrients and Fire

MARK G. TOZER, CHRISTOPHER C. SIMPSON,
ISAAC B. JANSENS AND DAVID A. KEITH

Australia's dry sclerophyll forests are fire-prone ecosystems characterised by a diverse and spectacular sclerophyll flora and occurring on infertile soils. In the east, they are distributed in an arc from south-east Queensland to the Adelaide region and in the eastern half of Tasmania. In the west, they are restricted to the south-west corner on the Darling scarp and Swan coastal plain. The distribution of dry sclerophyll forests has historically been interpreted as reflecting patterns of seasonal drought, but this theory has long been discarded in favour of one of edaphic control (nutrient-poverty). Both theories exemplify classical resource-based models of evolutionary or ecological convergence. Neither, however, can adequately account for patterns in the structure, floristics and dynamics of fire-prone ecosystems. In this chapter, we examine the influence of climate and geology on the distribution of dry sclerophyll forest in terms of both resource constraints and as drivers of fire-regimes. Climate, through its control of primary productivity and fire weather, and soils, which influence flammability via the expression of sclerophylly, underpin biogeographical patterns in the frequency, intensity and seasonality of fires, and thus, potentially, the distribution of dry sclerophyll forests. Fire is a primary focus of management and necessitates finding a delicate balance between fuel reduction and biodiversity conservation. Dry sclerophyll forests are relatively resilient to invasion by alien plant species but many species are susceptible to fungal pathogens such as *Phytophthora cinnamomi*. The most serious threats to dry sclerophyll forests in the future include spread of disease and changes in rainfall patterns and fire regimes associated with anthropogenic climate change.

13.1 Introduction

Australia's dry sclerophyll forests are something of a global anomaly. The most similar sclerophyll communities on other continents are treeless shrublands (Keeley *et al.* 2012), which also occur in Australia (Lamont and Keith, Chapter 14, this volume). The tree-dominated analogues of dry sclerophyll forests are largely coniferous, although a number of closed-canopy broadleaf taxa, such as *Quercus*, may have sclerophyllous foliage.

The leaf trait of sclerophylly is central to the concept of Australia's dry sclerophyll forests. The presence of a well-developed sclerophyll shrub layer is a defining feature of the formation, along with an open canopy of sclerophyll trees, typically eucalypts (Keith 2004). Sclerophyll plants are characterised by hard, stiff or leathery leaves (Schimper 1903)

Australian Vegetation, Third Edition, ed. David A. Keith. Published by Cambridge University Press.

Citation: Tozer, M.G. *et al.* (2017). Biogeography of Australia's Dry Sclerophyll Forests: Drought, Nutrients and Fire. In D. A. Keith, ed., *Australian Vegetation*, 3rd edn. Cambridge: Cambridge University Press, pp. 314–38.

as a consequence of one or more of the following: a high proportion of sclerenchyma within the leaf structure; a thick cuticle; or the presence of silica in the leaf (Beadle 1966).

Lamont *et al.* (2002) characterised scleromorphy as one of the oldest controversies in ecology: whether small, highly lignified leaves with thick cuticles represent an adaptive response to low water availability (Schimper 1903; Oertli *et al.* 1990) or nutrient impoverishment in general, and soil phosphate in particular (Beadle 1954, 1966; Loveless 1962). The soil nutrient hypothesis was dominant when the second edition of *Australian Vegetation* was published (Gill 1994) and the association with soils of extremely low-nutrient status is still considered a defining feature of dry sclerophyll forests, and the primary factor controlling their distribution (Keith 2004).

More recently, Bond and Midgley (1995) and Keeley *et al.* (2012) have expanded on Mutch's (1970) hypothesis in linking sclerophylly and flammability to enhanced fitness through increased opportunities for seedling establishment following fire. The role of fire as an evolutionary driver (as opposed to an emergent property) of fire-prone ecosystems is still debated (Axelrod 1989; Hopper 2009), although there is a growing consensus that the dynamics and distributions of these communities cannot be entirely explained by resource-based models (Mucina and Wardell-Johnson 2011; Keeley *et al.* 2012).

In contrast to the theories above, Orians and Milewski (2007) eschewed adaptive arguments in suggesting sclerophylly is a consequence of the overproduction of lignified tissues (wood, sclerenchyma). They postulated that this occurs under circumstances where light and moisture are (periodically) abundant, but nutrient stocks limit the production of metabolically active tissues. Highly lignified, nutrient-poor tissues result in low rates of herbivory and slow rates of decomposition which, in turn, promote a rapid accumulation of fuel, intense fires and thus further depletion of soil nutrients via volatilisation (Orians and Milewski 2007).

The aim of this chapter is to develop a cohesive climatic, edaphic and pyrophytic characterisation of the dry sclerophyll forest formation. First, we briefly outline the arguments for a reappraisal of the role of climate as a factor influencing the distribution of sclerophyll species. We use a water-balance model to assess the extent to which summer drought can be implicated as a common factor across the distribution of dry sclerophyll forests and evaluate the evidence in support of the drought and nutrient hypotheses at both continental and regional scales. Then we consider how factoring in the effects of recurrent fires improves upon correlative, resource-based models in diagnosing the essential processes that determine the distribution of fire-prone vegetation formations. This is the key conceptual advance of the two decades since the second edition of *Australian Vegetation* and defines the framework within which climate, soils and fire interact in driving convergence in plant traits and community assembly in dry sclerophyll forests.

13.2 Description and distribution

Dry sclerophyll forests are the most widespread of the eucalypt-dominated formations occurring in the wetter coastal and subcoastal regions of southern Australia (Figure 13.1). The formation accounts for an estimated 230 000 square kilometres, or approximately 3% of the vegetation of Australia prior to European settlement. Dry sclerophyll forests span an unusually large range in annual aggregate rainfall, from as little as 500 mm in inland areas to in excess of 1500 mm in parts of the east coast, and in excess of 2000 mm in parts of Queensland and Tasmania (Figure 13.1). Rainfall imposes a fundamental limitation to the accumulation of biomass (Specht 1970) and thus accounts for the transition from dry sclerophyll forest to woodlands and shrublands with increasing aridity. This factor explains the absence of dry sclerophyll forests from central Australia.

Dry sclerophyll forests are also absent from the monsoonal tropics where the rainfall pattern is summer dominated. Eucalypt forests and woodlands in such climates include a major component of C_4 grasses and generally lack a well-developed layer of sclerophyll shrubs, although rocky uplands may support open heathlands interspersed with occasional trees (Lamont

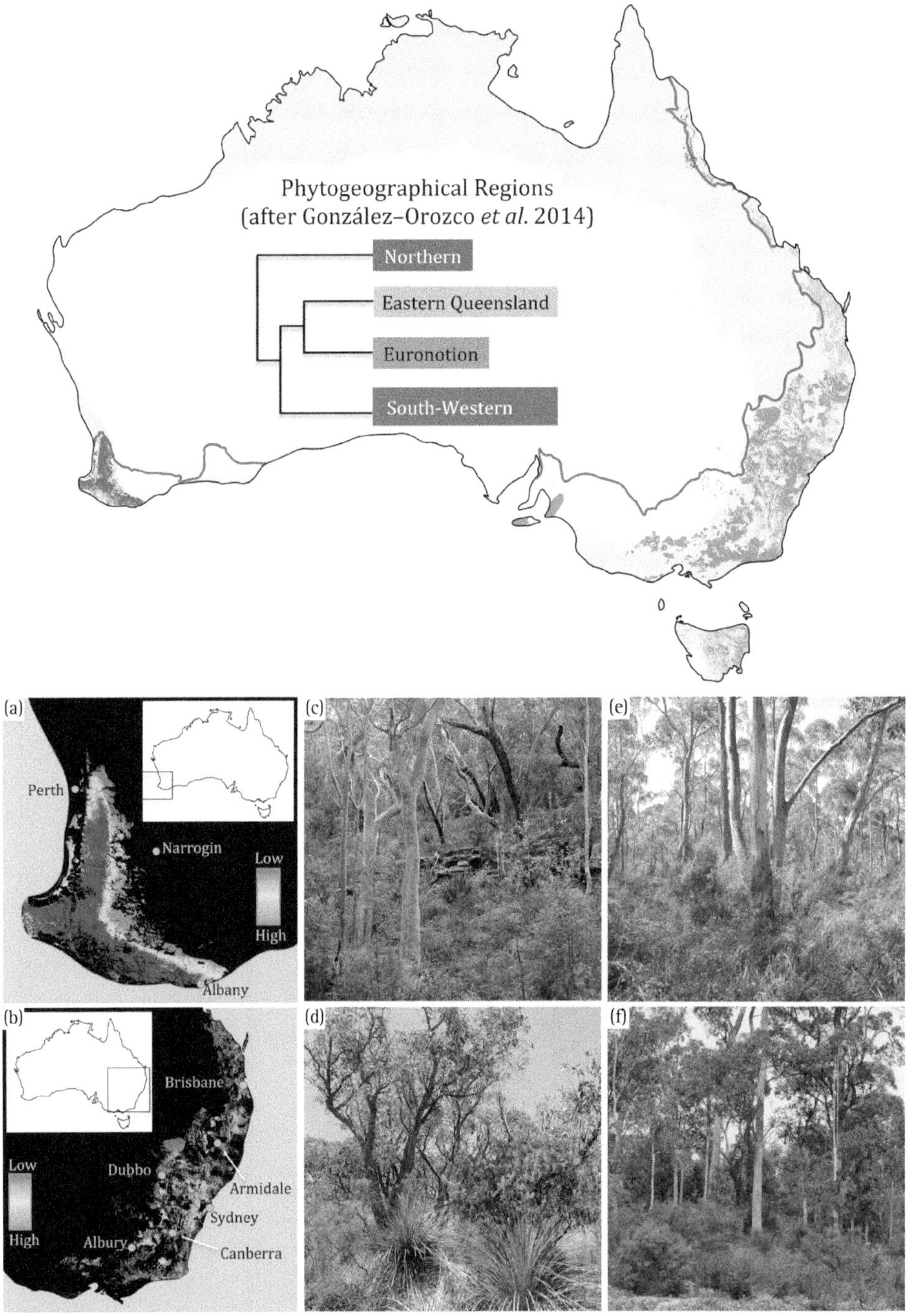

Figure 13.1 The map shows the distribution of dry sclerophyll forests prior to European colonisation compiled from units assigned to Dry Sclerophyll Forest Formation in New South Wales (Keith 2004), Medium Forest and Woodland in Western Australia (Beard *et al.* 2013), and lower-level categories in maps of Queensland (Neldner *et al.* 2015), Victoria, South Australia and Tasmania (Kitchener and Harris 2013) whose descriptions indicated the presence of a well-developed sclerophyll shrub

and Keith, Chapter 14, this volume; Williams *et al.*, Chapter 15, this volume). In southern Australia, dry sclerophyll forests occupy contrasting climates. Winter rainfall and prolonged summer drought dominate in the Mediterranean-climate zone of the south-west, while the south-east is dominated by (often) higher aggregate rainfall distributed more evenly throughout the year. South-eastern Australia contains the world's most extensive expression of sclerophyll vegetation outside Mediterranean-type climates (Keeley *et al.* 2012).

Dry sclerophyll forests broadly encompass Specht's (1970) open forest, low open forest and woodland categories, with the height of the tallest stratum varying in the range 5–30 m and projective foliage cover in the range of 10–70% (e.g. Keith 2004; Kitchener and Harris 2013; Neldner *et al.* 2015). Although Gill (1994) considered Specht's Open Forest category equivalent to dry sclerophyll forests, Specht's Open Forest also included forest with a predominantly herbaceous understorey.

Dry sclerophyll forests are characterised by a continuous but open canopy dominated by one or more species of eucalypts (i.e. *Angophora*, *Corymbia* and *Eucalyptus*; Gill 1994; Ladiges 1997). Understorey shrub species are predominantly from the families Ericaceae, Myrtaceae, Fabaceae, Proteaceae and (less commonly) Dilleniaceae, Euphorbiaceae and Rhamnaceae (Beadle 1966; Keith 2004). The density and diversity of shrub species vary along edaphic and climatic continua. Densely shrubby vegetation with a sparse layer of wiry graminoids is characteristic of highly impoverished, siliceous soils in high-rainfall regions, while open grassy woodlands occupy the other end of the gradients on soils of heavier texture, or in lower rainfall areas (Keith 2004; Prober *et al.*, Chapter 17, this volume).

The subgenus *Eucalyptus* is most widely distributed on soils of low fertility (Noble 1989; Austin *et al.* 1997), although a few species of *Angophora* and *Corymbia* may be locally abundant (Table 13.1). Eucalypts typically occur in associations of two or more species in the south-east (Gill 1994). Monospecific stands are more common in the south-west and either *Eucalyptus marginata* or *E. wandoo* are frequently the dominant species. Shrub species in the genera *Acacia*, *Banksia*, *Daviesia*, *Hakea*, *Hibbertia*, *Leptospermum*, *Leucopogon*, *Pultenaea* and *Xanthorrhoea* are characteristic of dry sclerophyll forest and high levels of local or regional endemism are evident, particularly in the south-west. Disjunctions in the distribution of both overstorey and understorey species are a feature throughout the range of dry sclerophyll forests (Gill 1994). Dry sclerophyll forests account for a higher proportion of the diversity of Australian plant species than expected on an area basis (Keith 2004). On a global level, dry sclerophyll forests fall within Box's (1996) subhumid woodland biome type.

13.3 The relative roles of water and nutrients

Beadle's contention that soil phosphorus is primarily responsible for the delimitation of sclerophyllous and mesomorphic communities has become a central paradigm of Australian ecology (Beadle 1954, 1966;

Figure 13.1 (cont.)

layer beneath a eucalypt overstorey. Areas receiving more than 500 mm in annual rainfall are shaded grey. The red line marks the limits of zones on the south and east coasts in which the annual deficit in water availability does not exceed 800 mm (refer to Box 13.1). (a) and (b) show the forecast probability of habitat conditions being suitable for dry sclerophyll forest in 2070, assuming high population growth and fossil fuel consumption, within the current distribution in, respectively, the south-west and south-east of Australia (see Box 13.1). Examples of dry sclerophyll forest: (c) open forest dominated by *Angophora costata* and *Eucalyptus piperita* in Royal National Park, Sydney (photo: Mark Tozer); (d) woodland dominated by *E. baxteri*, *Banksia marginata*, and *Xanthorrhoea semiplana* in Deep Creek Conservation Park on the Fleurieu Pensinsula, South Australia (photo: David Keith); (e) open forest dominated by *E. ovata* with *Pultenaea juniperina*, *Leptospermum scoparium* and *Lepidosperma laterale* at Longley, Tasmania (photo: David Keith); (f) jarrah forest dominated by *E. marginata* with shrubs of Fabaceae and Proteaceae in the understorey, south-western Australia (photo: Mark Brundrett).
(A black and white version of this figure will appear in some formats. For the colour version, please refer to the plate section.)

Table 13.1 **Summary of dry sclerophyll vegetation, and its landscape characteristics, across Australia based on Beadle's alliances (Beadle 1981). Alliances are sorted by phytogeographical regions (Figure 13.1) after González-Orozco *et al.* (2014).**

Beadle's alliances	Key compositional features	Landscape characteristics
Eastern Queensland Region		
Corymbia intermedia–Eucalyptus acmenoides–E. racemosa–E. eugenioides	Variable understorey from poor to well-developed further from the Tropics including *Lophostemon suaveolens, Grevillea banksii, Allocasuarina littoralis, Xylomelum pyriforme, Xanthorrhoea* spp., *Macrozamia communis, Themeda australis*	1000–1500 mm; sandstones, granites and sandy alluvium; flat to gentle hills between Tropics and central coastal lowlands NSW
Euronotion Region		
C. gummifera–E. racemosa–E. sieberi	One of the most diverse eucalypt alliances with most abundant families being: Orchidaceae, Myrtaceae, Fabaceae, Proteaceae, Cyperaceae, Gramineae, Ericaceae, Asteraceae, Rutaceae, Goodeniaceae, Phormiaceae, Lamiaceae, Dilleniaceae, Xanthorrhoeaceae, Euphorbiaceae, Apiaceae, Restionaceae	750–1500 mm; sandy soils in southern QLD, SA, TAS and NSW, including the Hawkesbury Sandstone plateau
E. baileyana–E. planchoniana–E. bancroftii	Pure to mixed-canopy stands with two shrub layers and a sporadic herbaceous layer: tall shrubs including *Acacia concurrens, Banksia integrifolia, Allocasuarina littoralis, Jacksonia scoparia;* and small shrubs including *Banksia oblongifolia, Bossiaea rhombifolia, Daviesia ulicifolia, Isopogon anemonifolius, Lambertia formosa, Leptospermum trinervium*	1100–1700 mm; sand and gravel soils; coastal lowlands in north-east NSW (<170 m) and south-east QLD tablelands (500–800 m)
E. globoidea	Understorey commonly incl. *Leptospermum obovatum, Banksia serrata, Acacia longifolia, Epacris impressa, Allocasuarina pusilla, Lepidosperma longitudinale, Schoenus imberbis*	Relatively fertile soils on coastal lowlands NSW south of Kempsey
E. radiata	Canopy rarely in pure stands, but associated with *E. acaciiformis, E. dalrympleana, E. youmanii* (north)*; E. piperita, E. sieberi, E. mannifera* (central)*; E. bicostata, E. viminalis, E. mannifera* (south). Understorey consists of *Acacia* spp., *Banksia* spp., *Hakea* spp., *Leucopogon* spp.	1000 mm; North, Central and South Tablelands between 300–1100 m on low fertility soils

Beadle's alliances	Key compositional features	Landscape characteristics
E. mannifera	Understorey incl. *Acacia dealbata, Daviesia latifolia, Lissanthe strigosa, Aristida ramosa, Cymbopogon refractus, Rytidosperma laevis, Dichelachne micrantha, Poa sieberiana, Themeda triandra, Chrysocephalum apiculatum, Acaena ovina*	500–1200 mm; low fertility soils at elevations of 300 m in VIC and 1100 m in Northern Tablelands NSW
E. macrorhyncha–E. rossii	Open woodlands associated with *Callitris* spp. and a dense shrub layer including *Acacia* spp., *Dillwynia* spp., *Astroloma* spp., *Bossiaea* spp., *Epacris* spp., *Leucopogon* spp.	600 mm; extends from NSW to VIC between 200–1000 m on silica-rich granite and sedimentary rock
E. baxteri	Canopy occasionally shared with *E. viminalis* or *E. obliqua*. Common understorey taxa incl. *Astroloma conostephioides, Hibbertia sericea, Xanthorrhoea* spp.	625–750 mm; deep sandy soils with low fertility; restricted to SA and VIC
E. amygdalina	Pure stands with a continuous understorey incl. *Acacia dealbata, A. mucronata, A. myrtifolia, A. stricta, A. verticillata, A. gunnii, Banksia marginata, Beyeria viscosa, Allocasuarina monilifera, Daviesia latifolia*	500–900 mm; eastern TAS up to 800 m on silica-rich sandstones, mudstones and granites
E. dalrympleana (parts)	Canopy stands pure, but often associated with *E. caliginosa*. Typical low tree layer incl. *Banksia integrifolia, Allocasuarina littoralis, Acacia falciformis, A. filicifolia, Exocarpos cupressiformis;* a dense shrub layer incl. *Jacksonia scoparia, Daviesia latifolia, Persoonia cornifolia, Hakea dactyloides, Leucopogon lanceolatus*	800–1150 mm; low fertile, granitic soils from Northern Tablelands NSW to QLD between 950–1250 m
E. obliqua–E. fastigata (parts)	South Australian composition has pure stands of *E. obliqua* which may be associated with *E. viminalis, E. leucoxylon, E. fastigata*, and *E. baxteri*	Only SA conditions in the Mt Lofty Ranges are DSF. 870–1200 mm; low fertile podzolic soils
South Western Region		
E. marginata–E. calophylla	Common understorey taxa incl. *Banksia* spp., *Grevillea* spp., *Hakea* spp., *Isopogon* spp., *Petrophile* spp., *Acacia* spp., *Daviesia* spp., *Hovea* spp., *Eriostemon* spp., *Bossiaea* spp., *Gompholobium* spp., *Styphelia* spp., *Xanthorrhoea* spp.	650–1300 mm; gravel/sand and sandy lateritic soils in south-western WA

Lamont *et al.* 2002). One of the attractions of the soil nutrient hypothesis is that it resolves the paradox of a putative drought-adapted flora distributed primarily on the wetter margins of the continent. Beadle (1966) suggested that sclerophylly was a consequence of a reduction in leaf area associated with the adaptation of species to low soil fertility (primarily low phosphorus). He asserted that the survival of sclerophyll species was related to their ability to tolerate mineral starvation as opposed to drought.

Several lines of evidence suggest that a re-evaluation of the relative importance of drought and nutrients in controlling the distribution of dry sclerophyll forests is warranted. First, ecological convergence (in plant form and vegetation structure) is generally interpreted as a response to energy and moisture constraints imposed by climate patterns at the global and continental scales (Whittaker 1975; Ricklefs 2008; Keeley *et al.* 2012; Keith and Tozer, Chapter 1, this volume). Convergence of leaf traits in Mediterranean-type shrublands, for example, is evident despite considerable variation in soil fertility among these regions at a global scale (di Castri 1981).

Second, a growing body of evidence suggests that the characterisation of climate in aggregate terms (e.g. annual rainfall, average temperatures) is inadequate in representing the interactive effects of energy and water supply as experienced by the plant (Stephenson 1990). Soil moisture deficits are exacerbated by factors such as soil texture, depth and drainage patterns and may arise irrespective of the timing of rainfall if evapotranspirative demand exceeds supply (Keeley *et al.* 2012). Mediterranean-type shrublands in Arizona and South Africa's eastern cape, for example, experience significant soil moisture deficits during summer despite receiving appreciable summer rainfall (van Wilgen 1984; Vankat 1989).

Third, because climatic and edaphic drivers operate at different scales (Keith and Tozer, Chapter 1, this volume) the potential for different drivers to operate via similar mechanisms may be underestimated. Edaphic controls may relate either to the chemical properties of soil and the implications for plant nutrition, or to the effects of texture and depth on water supply. A potential correlation between soil fertility and available moisture capacity is particularly relevant in the case of dry sclerophyll forest. Highly siliceous (quartz-rich) soils are low in nutrients (nitrogen, phosphorus, exchangeable cations and active clays) as a consequence of the leaching of soil nutrients over extended periods of weathering. These coarse-textured freely draining soils also have low available water capacity (Gray and Murphy 1999; Christopherson 2006) and hence are capable of generating significant drought conditions (Sperry and Hacke 2002).

Boundaries between soil derived from different parent materials may therefore equate to significant differences in fertility, moisture availability or both which, in turn, may be reflected in contrasts of vegetation structure and species composition (Brundrett *et al.*, Chapter 7, this volume). A confounding of these factors is evident in forests on both sides of the continent: the most sclerophyllous forest communities are associated with soils that have the lowest fertility and lowest moisture holding capacity (Keith and Sanders 1990; Wardell-Johnson and Williams 1996). Lower water deficits in gullies relative to slopes and ridges may also be attributable to higher water capacities due to deeper soils and/or the concentration of clays and nutrients through illuviation (Gray and Murphy 1999). These gradients are demonstrably amplified by topographic effects on evapotranspiration and translate to spatial patterns that correlate well with the distribution of different vegetation formations.

In the next sections, we use a water-balance model to assess the extent to which the distribution of dry sclerophyll forest can be attributed to summer drought. The key questions are whether: (i) the phenomenon of summer drought is restricted to the dry sclerophyll forest in the south-west of the continent; and (ii) patterns in the distribution of formations along fertility gradients are independent of gradients in water availability.

13.4 Patterns at the continental scale

A regime of spring to summer drought and autumn to winter recharge is a plausible unifying factor underpinning the distribution of dry sclerophyll forests in Australia (Box 13.1). Summer drought in the south-west is a product of low rainfall and high radiance in the summer months and water stored in the soil makes a higher contribution to meeting demand than in the eastern temperate and subtropical regions (Box 13.1; Figure 13.2; Silberstein *et al.* 2001).

Reliable winter rainfall plays a crucial role in recharging groundwater during the period of low evapotranspiration demand. This dependence has implications for climate change, which we return to in the final section. Conversely, summer drought is less severe in the eastern temperate and subtropical regions, but water supply is still insufficient to meet evapotranspiration demand over summer. Although eastern forests are less reliant on stored water during summer, winter recharge is less reliable and the contribution of soil storage to meeting demand is more variable than in the highly seasonal south-west.

Box 13.1 Statistical model of the distribution of dry sclerophyll forest

We developed a distribution model of dry sclerophyll forest to examine the potential role of climatic and edaphic drivers in its continental-scale distribution patterns. The model was based on a set of spatially explicit environmental predictors, themselves derived from spatial models of observational data (Bureau of Meterology n.d.; Dyer 2009). The predictors included components of a water-balance model, temperature variables and soil phosphorus. Water-balance models incorporate the timing of rainfall in relation to seasonal temperature variation, the effects of temperature on evaporation and transpiration and soil available water capacity (AWC). Potential evapotranspiration (PET), a function of temperature, radiation and water vapour pressure, represents evaporative losses (demand) associated with processing energy available for photosynthesis, transpiration to support cooling, nutrient transport and cell metabolism under conditions of unlimited water supply (Stephenson 1990). Actual evapotranspiration (AET), a function of precipitation, radiation and water storage capacity of the soil based on texture and depth, represents the available water supply. AET correlates strongly with net primary production (Rosenzweig 1968; Lieth 1975, 1976). Annual Evapotranspiration deficit (ETD) is the difference between PET and AET. Annual surplus is the amount of rainfall in excess of the requirements for both evapotranspiration and soil moisture recharge. Temperature variables (mean minimum and maximum temperatures) were also included in the model to test for limitations imposed by temperature on plant growth independent of its role in driving water balance. Total soil phosphorus (P) was included in the model as an indicator of soil fertility.

Temperature and rainfall spatial data (6' resolution, i.e. 6 minutes of latitude and longitude) were obtained from the Bureau of Meteorology (n.d). Soil spatial

Box 13.1 (cont.)

data (3' resolution) were obtained from the Commonwealth Scientific and Industrial Research Organisation (n.d.; total phosphorus, available water capacity). Parameters of a water-balance model (9' resolution) were compiled according to Dyer (2009) (AET, ETD, surplus). Water storage was calculated over the estimated depth of the A and B horizons in order to best represent the water availability for understorey species. Deficit estimates may therefore overestimate the severity of drought for species with the potential to source water from greater depths. Nevertheless, patterns in actual evapotranspiration/ water deficit in the Mediterranean-type climate regions suggest the model is adequate for comparative purposes. The graphs in Figure 13.2 show (for indicative locations) the cumulative contributions of rainfall (darkest fill), and soil storage (white fill) to actual evapotranspiration in each month of the year as well as deficit (intermediate fill) in those months where potential evapotranspiration exceeds actual evapotranspiration.

The training data for the model were derived by intersecting 100 000 random points with the distribution of dry sclerophyll forest as described in Figure 13.1. Points were only selected from areas above the 500 mm annual rainfall isohyet to reduce the effect of extreme low rainfall values on model performance. The resulting binary response (presence/absence) data were used to build a boosted regression tree (BRT) model (Elith *et al.* 2008) based on predictors described above. The model was fitted using 10-fold cross-validation with a bag fraction of 0.5, learning rate of 0.01 and tree complexity of 7.

The strongest predictor in the distribution model was ETD (Table 13.2), suggesting that dry sclerophyll forests are restricted by the magnitude of the annual moisture deficit to a zone that includes the south-west, south-east and central east coastal regions but largely excludes the tropics. The upper level of drought tolerance corresponds to a nominal annual deficit of 800 mm. Conversely, dry sclerophyll forest rarely occurs in areas with an annual deficit less than 200 mm and is entirely replaced by rainforests and wet sclerophyll forests in areas of the east coast that feature both high rainfall and soils with high water capacity (no such areas occur in the south-west).

The implications of drought-control on the distribution of dry sclerophyll forest under a changing climate were examined by using the model to predict the future probability of occurrence within the formation's current distribution from projected climate conditions for the year 2070. Under a scenario of rapid economic and population growth based on intensive fossil fuel consumption (scenario A1FI, IPCC 2007), dry sclerophyll forest is predicted to contract by up to 70 km from the north and east of its present distribution in south-west Australia, with substantially larger contractions to the coast predicted for the south-east (Figure 13.1).

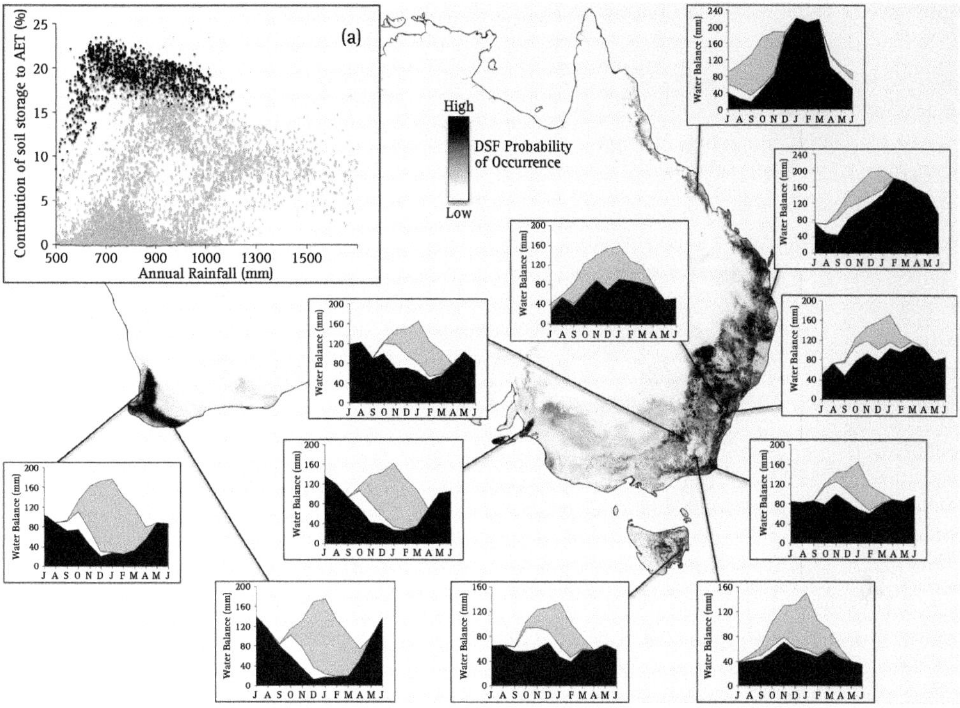

Figure 13.2 Spatial model of the distribution of dry sclerophyll forest (see Box 13.1 for methods). The level of shading indicates the probability of occurrence (white = low probability, black = high probability). Graphs show the contribution of rainfall (black fill) and soil storage (white fill) to actual evapotranspiration (AET) at selected locations throughout the distribution of dry sclerophyll forest. Grey fill indicates moisture deficit. Inset (a) shows the contribution of soil storage to AET as a function of rainfall where black points sample regions with winter-dominated rainfall and grey points sample regions with aseasonal or summer-dominated rainfall.

Table 13.2 **Contributions of model variables to variance explained**

Model	Water balance, temperature, phosphorus	Water balance
Variance explained (%)	60.0	49.9
Area under the receiver operating curve (ROC)	0.954 (±0.001)	0.935 (±0.001)
Variable	Contribution (%)	Contribution (%)
Annual evapotranspiration deficit	40.5	61.7
Annual actual evapotranspiration	13.0	22.3
Average annual maximum temperature	11.5	N/A
Average annual minimum temperature	10.0	N/A
Soil available water capacity	8.8	N/A
Annual moisture surplus	8.5	16.0
Total soil phosphorus	7.7	N/A

Dry sclerophyll forests are replaced by wet sclerophyll forests at temperate latitudes where the summer deficit is weak, or by savanna in the tropics where deficits occur in winter rather than summer, and are more substantial than experienced in the temperate or Mediterranean zones. The limits of the distribution of dry sclerophyll forests in the north-east correspond to a shift in the timing of the drought period towards the winter months. As the summer moisture deficit increases into arid regions in southern Australia, shrub species are progressively replaced by either grass species (in eastern Australia) or chenopods and other xeromorphic shrubs (in western Australia). Grass species are considered to compete more successfully for irregular pulses of rainfall (Stephenson 1990; Coomes and Grubb 2000), while xeromorphs are presumably more drought tolerant.

In comparison to evapotranspiration and water deficit, total phosphorus and soil available water capacity (AWC) are relatively weak predictors of dry sclerophyll forest distribution (Box 13.1). However, continental-scale estimates of these variables may be relatively insensitive to edaphic variation at the local and landscape scales. We address this issue in the following section by examining species turnover at a landscape scale along moisture and nutrient gradients.

13.5 Patterns at landscape and local scales

Both soil fertility and moisture availability potentially influence vegetation composition and structure at the landscape scale. We examined these relationships in a case study of species turnover among vegetation formations in south-east New South Wales (Box 13.2). Dry sclerophyll forest was strongly differentiated from grassy woodland along a gradient of increasing soil fertility inferred from opposing trends in the concentrations of silica and total phosphorus (Figure 13.3a). However, both of these formations were strongly segregated from wet sclerophyll forests and rainforests along a water availability gradient (Figure 13.3a). In comparison to grassy woodlands, these latter mesomorphic communities occupied wetter environments within a similar range of soil fertility. In contrast, dry sclerophyll forests were distinguished from mesomorphic communities by their association with both low-fertility soils and higher water deficit (Figure 13.3a). This suggests that the water availability gradient is at least of equal importance to, and independent of, soil fertility in delimiting a unique environmental niche for dry sclerophyll forest (Figure 13.3a, b). Variation in soil silica concentration was almost completely independent of all of the major parameters of water supply, thus there is no evidence (at least at the landscape scale) to support the hypothesis that high soil silica concentrations exacerbate drought as proposed in the previous section. There is equivocal evidence on whether water availability or nutrients more strongly influence transitions between dry sclerophyll forest and heath.

While no previous studies quantify responses to both soil fertility and water availability gradients, our results largely corroborate earlier studies of soil phosphorus gradients. Beadle (1962) was first to show that dry sclerophyll forests occupied soils with lower total P than rainforests and wet sclerophyll forests. Adam *et al.* (1989) later showed that the number of rainforest species declines geometrically with total soil P along a gradient from rainforest to sclerophyll forest in coastal eastern Australia. Similarly, in the Blue Mountains of New South Wales, Harrington and Humphreys (2004) found total soil P concentrations differed by an order of magnitude (169.9 ± 21.7 cf. 1796 ± 396 mg l^{-1}) at either end of a structural and physiognomic gradient comprising heathland–sclerophyll woodland–dry sclerophyll forest (on soils derived from Triassic quartz sandstone), ending in mesic forest (on soils derived from Tertiary basalt). Wardell-Johnson *et al.* (2007) demonstrated significant correlation

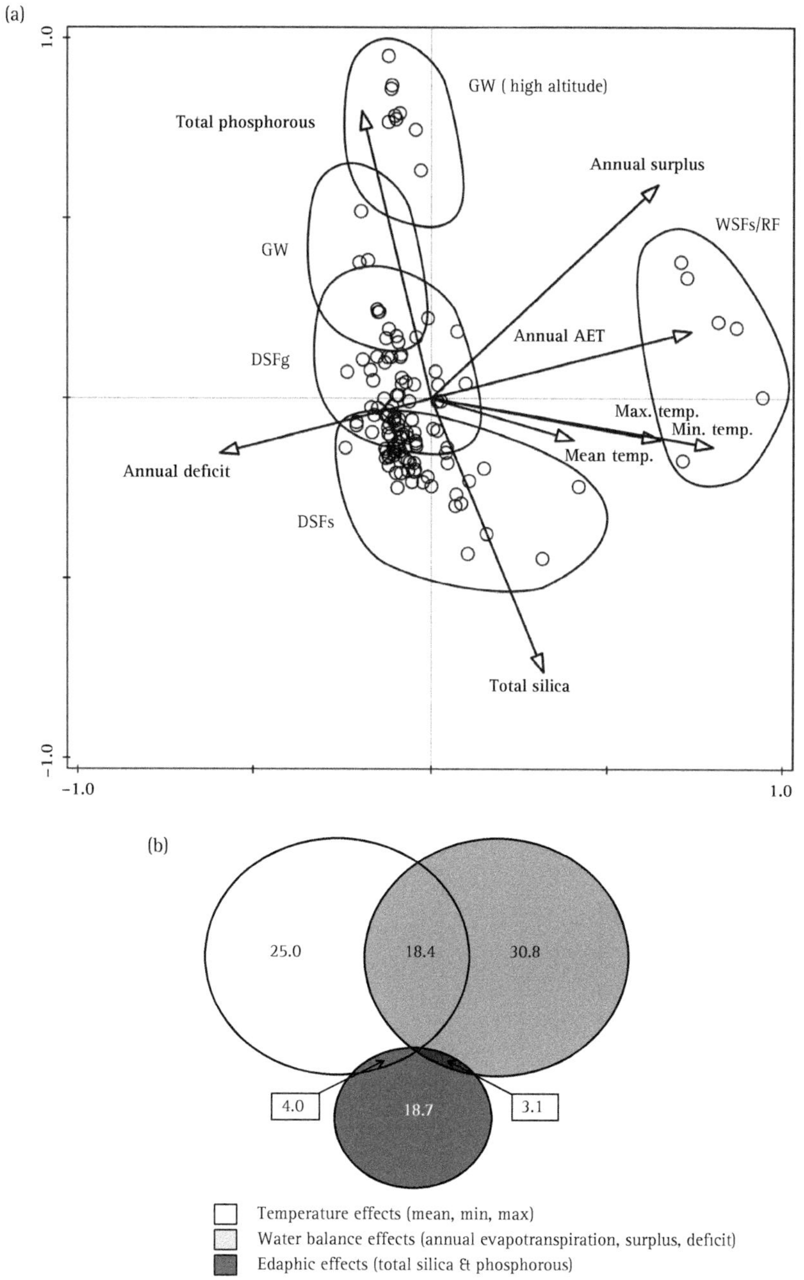

Figure 13.3 (a) Ordination plot derived from canonical correspondence analysis showing the distribution of relevés representing different vegetation formations. Vectors represent variables that explain significant proportions of the floristic variation (see Box 13.2). Formations are: dry sclerophyll forest shrubby and grassy subformations (DSFs/DSFg, *sensu* Keith 2004), grassy woodlands (GW), shrubby wet sclerophyll forests (WSFs) and rainforests (RF). (b) Venn diagram showing portions of variation in floristic composition attributable to temperature, water balance and edaphic effects determined by partial variance analysis. Data are from the Sydney Basin and southern Tablelands of New South Wales (Tozer *et al.* 2010).

between soil phosphorus and species composition (also related to cation exchange capacity and soil carbon) within a south-east Queensland dry sclerophyll forest complex. Conversely, Le Brocque and Buckney (1995) identified a large overlap in total P concentrations between sclerophyll and rainforest vegetation types in the Sydney region, as well as no discernible patterns in available water capacity, sand/silt/gravel ratios, nitrogen and exchangeable cations.

Box 13.2 Soil fertility and water availability gradients analysis across the Sydney Basin and Southern Tablelands

To examine the landscape-scale edaphic and climatic niche of dry sclerophyll forests in relation to other vegetation types, we compiled 135 vegetation survey relevés (0.04 ha quadrats) in which all vascular plant species were recorded and assigned cover/abundance scores according to the method of Poore (1955). These were a subset of a regional dataset for which all relevés had been assigned to compositional vegetation types by hierarchical and non-hierarchical clustering (Tozer *et al.* 2010). Plots were selected from the regional dataset to sample: (i) a range of vegetation formations (rainforest, wet and dry sclerophyll forest, grassy woodland); (ii) a gradient in water availability; and (iii) soils derived from a range of parent materials including Triassic and Permian quartz and lithic sandstones, Triassic shales, Carboniferous granites, Tertiary basalts and Silurian limestones. The sites were revisited to obtain samples of the topsoil (0–8 cm) for chemical analysis. Total phosphorus and silica were estimated by X-ray diffraction (Buhrke *et al.* 1998). Measurements of total phosphorus may overestimate the quantity available to plants because some proportion may be bound up in solid crystals (e.g. phosphorus in apatite) (Gray and Murphy 1999). Available phosphorus was estimated using the Olsen-P, Melich-3 P and diffuse gradient thin-film-P (DGT-P) extraction methods (Mason *et al.* 2010; Rayment and Lyons 2011). Annual evapotranspiration, deficit and surplus were estimated by the procedure described in Box 13.1, which was repeated for the study area at a 30 m grid resolution.

We undertook a series of canonical correspondence analyses in which the effects of temperature, moisture availability and edaphic factors were sequentially partialled out in order to quantify the proportion of floristic variation uniquely attributable to each combination of variables (Šmilauer and Lepš 2014). The ordination from these analyses (Figure 13.3a) shows a continuum from shrub-dominated dry sclerophyll forests through to grassy woodlands along an edaphic gradient of decreasing silica and increasing total phosphorus. Measures of available phosphorus were correlated with total phosphorus but did not explain a statistically significant portion of the floristic variation and hence are not shown in Figure 13.3a. Wet sclerophyll forests and rainforests clustered separately from dry sclerophyll forests and woodlands along a water supply gradient roughly perpendicular to the edaphic gradient.

Partial variance analysis (Figure 13.3b) showed that in combination, the set of water-balance variables accounted for a greater proportion of variation in species composition than either temperature or edaphic effects. A substantial overlap in temperature and water-balance effects is consistent with the role of temperature in driving evapotranspiration but temperature, nevertheless, exerts a strong independent effect. In contrast, total soil phosphorus and silica content explained a smaller proportion of floristic variation among communities with understoreys dominated by either sclerophyll shrubs or herbs, although their influence may be underestimated because the full range of high fertility sites in wet environments could not be sampled.

The marginal contribution of annual rainfall to variance explained was less than half of that contributed by the water-balance model parameters collectively (actual evapotranspiration, deficit, surplus), and its conditional contribution was statistically non-significant. This reinforces the point that annual rainfall is a relatively insensitive measure of the extent to which water-balance limits species' distribution.

All three measures of annual temperature (average minimum, mean and maximum) retain independent and statistically significant conditional contributions in explaining variation in species composition with respect to the water-balance parameters. This suggests temperature exerts a significant control over species distributions independent of its role in driving evapotranspiration.

Heathlands are strongly associated with oligotrophic soils (Lamont and Keith, Chapter 14, this volume) and share many species with dry sclerophyll forests where they co-occur. The tendency for dry sclerophyll forests to occupy sands of Holocene origin while those of Pleistocene age support heath has been attributed to the nutrient poverty of the latter (Myerscough and Carolin 1986; Adam *et al.* 1990). However, several other studies were unable to detect differences in soil phosphorus levels between dry sclerophyll forests and heathlands (Coaldrake and Haydock 1958; Beadle 1966; Harrington and Humphreys 2004). Pidgeon (1940) identified the influence of soil depth on water availability as the factor limiting the development of heathland into dry sclerophyll forest on sandstone soils in the Sydney Basin. Beadle (1954) argued that the capacity of trees to exploit ground water by penetrating crevices negated any potential effects of soil depth, however it is possible that crevice formation is largely associated with joint fracture on, or adjacent to, valley slopes (at least for Triassic Hawkesbury sandstone). If so, then higher water deficits associated with shallow soils on broad ridgelines (possibly alternating with seasonal waterlogging) are a potential mechanism delimiting heath and dry sclerophyll forests in these areas. However, it remains to be investigated whether local variations in soil depth and available water capacity generate significant water deficit gradients.

Laboratory methods suggest that only a small proportion of total P is available for plant uptake. For example, in their study, Harrington and Humphreys (2004) found that basalt soils yielded just over twice the phosphorus available on sandstone (7.67 ± 1.3 cf. 3 ± 0 mg/l), less than 5% of total P, which differed by an order of magnitude between these soil types. Studies that examine available P suggest that it has little influence in delimiting dry sclerophyll forests from other vegetation formations, particularly those exposed to high water deficit (Box 13.2; Harrington and Humphreys 2004). The stronger relationships with total P support the inference that plant root traits allowing uptake of insoluble forms of phosphorus are strongly represented in dry sclerophyll forest floras (Brundrett *et al.*, Chapter 7, this volume). Nonetheless, the causal role for P in discriminating sclerophyll and non-sclerophyll communities requires further investigation.

13.6 Strengths and limitations of resource-based models

The analyses described in Boxes 13.1 and 13.2 suggest that both drought and soil fertility are significant factors determining the distribution and composition of dry sclerophyll forests, although their relative influence varies as a function of scale. At the continental scale, dry sclerophyll forests are associated with areas subject to a regime of summer drought/winter recharge. As water deficit increases, we hypothesise that sclerophyllous shrubs are replaced by more drought-tolerant xeromorphic shrubs such as chenopods or grasses and herbs able to exploit irregular pulses of rainfall (Stephenson 1990; Coomes and Grubb 2000). The transition from dry to wet sclerophyll forests corresponds with declining water deficit, although we postulate a role for edaphic control at local scales.

Patterns at the landscape scale are attributable to gradients in fertility (for a given water supply) although there is also evidence for the delimitation of wet and dry sclerophyll formations on the basis of water availability. The existence of strong gradients in water availability at the local scale remains a possibility, and may explain patterns that are not otherwise attributable to fertility.

Patterns in the distribution of dry sclerophyll forest at the continental and landscape scale are mirrored by trends in the expression of the sclerophyll trait at similar scales. Fonseca *et al.* (2000) surveyed sclerophylly in species assemblages across gradients in mean annual rainfall (estimated range 216–2034 mm) and total soil phosphorus (estimated range 22–2063 $\mu g.g^{-1}$) within a mean annual temperature band of 15–18°C in New South Wales. They approximated the degree of sclerophylly in a two-dimensional trait space based on leaf width and specific leaf area (SLA, leaf area per unit leaf dry mass). On average, they found similar trends in sclerophylly along gradients in rainfall and total soil P. That is, low rainfall and low soil P favoured similar trait combinations although the range of variation in leaf characters within sites exceeded the difference in mean values between sites.

Lamont *et al.* (2002) took an alternative approach by analysing foliar concentrations of phosphorus and nitrogen in restricted phylogenetic groups. These comprised species of *Hakea* occurring in 10 sites over 1300 km in south-western Australia and all species of Proteaceae in 14 sites over 400 km in the western Cape of South Africa. The sites were configured to sample gradients in rainfall. In both regions a strong inverse curvilinear relationship existed between mean leaf mass per area per site (indicative of sclerophylly) and mean annual rainfall but the same was not true for mean foliar P or N. They concluded that role of sclerophylly as a drought adaptation warranted further consideration, although the sampled range of nutrient availability may have been small relative to the sampled rainfall gradients.

In combination, these results corroborate the hydrological and nutritional control of sclerophylly, expressed, respectively, along regional climatic and local-scale edaphic gradients. They support the postulated adaptive significance attributed to sclerophyll leaves, whereby high lignin content reduces herbivory and enhances leaf longevity, thereby increasing nutrient-use efficiency and carbon gain in resource-limited environments (Turner 1994). Although the evolutionary pressure of drought remains a persuasive explanation for leaves that conserve water and resist desiccation (Oertli *et al.* 1990; Lamont *et al.* 2002), Bartlett *et al.* (2012) postulated that sclerophylly has an indirect role in maintaining drought tolerance, which is primarily related to osmotic potential at full turgor. The interplay between drought and nutrients is perhaps best encapsulated by Ordoñez *et al.* (2009) in their meta-analysis of relationships between leaf traits, climate and soil fertility. They concluded that leaf traits are most closely associated with soil fertility while climate governs the distribution of life forms.

One of the difficulties with nutritional poverty as a primary driver of sclerophyllous vegetation is the apparent dynamism between sclerophyllous and non-sclerophyllous vegetation. Positive feedbacks are postulated to have a critical role mediating such transitions (Jackson 1968; Bond *et al.* 2005; Warman and Moles 2009). The mechanisms relate to the stochastic fires, which open the canopy promoting more rapid drying of fine surface fuels, making the vegetation more flammable. Conversely, the absence of such fires promotes mesic microclimates, maintaining the vegetation in a low-flammability state. According to this theory of alternate stable states, moisture status and its effect on fire probability, rather than nutritional poverty, thus emerges as the key factor defining the distribution of sclerophyllous vegetation. Nevertheless, both drought and nutrient impoverishment have important consequences for the dynamics of dry sclerophyll forests under regimes of recurring fire and are integral to the conception of a fire–nutrient–drought nexus explored in the next sections.

13.7 The nutrient–drought–fire nexus

Fire has emerged to challenge the paradigm of climate as the major determinant of plant growth form and the distribution of terrestrial biomes. An evolutionary role for fire in the development of sclerophylly has been postulated, although the relative importance of drought, oligotrophic soils and fire in shaping the evolution of fire-prone communities remains contested (Axelrod 1989; Cowling *et al.* 1996; Hopper *et al.* 2003; Lambers *et al.* 2008; Midgley 2013). Evidence for the role of fire in shaping biomes is compelling. For example, significant discrepancies between the observed and predicted (climate-determined) distributions of tree cover have been attributed to the effects of recurring fires (Bond *et al.* 2005; Bond and Keeley 2005), although the predictions take limited account of edaphic factors.

Fire has been a semi-continuous feature of the Australian environment since the late Cretaceous (Lynch *et al.* 2007; Lamont and He 2012; Miller and Murphy, Chapter 5, this volume). Fire regimes in dry sclerophyll forests are characterised by high-intensity, canopy fires occurring in warmer months at decadal cycles. Plant species may persist through individual fires via a relatively small number of fire responses and life-history syndromes (Keith 1996). Persistence in the face of recurring fires is determined by interactions between demographic processes and elements of the fire regime (frequency, intensity, season, type).

The postulated nexus between fire, nutrients and climate relates to the interactions and feedbacks which, collectively, determine the nature of key elements of the fire regime (Figure 13.4). Under this model, climate and weather influence both fuel production and flammability which, in turn, determine the potential for fire to occur through time (Miller and Murphy, Chapter 5, this volume). Climate also influences the chemical composition of the soil via leaching and weathering. Soil nutrients determine the flammability of plant foliage by promoting the expression of sclerophylly. Fire, in turn, exacerbates nutrient

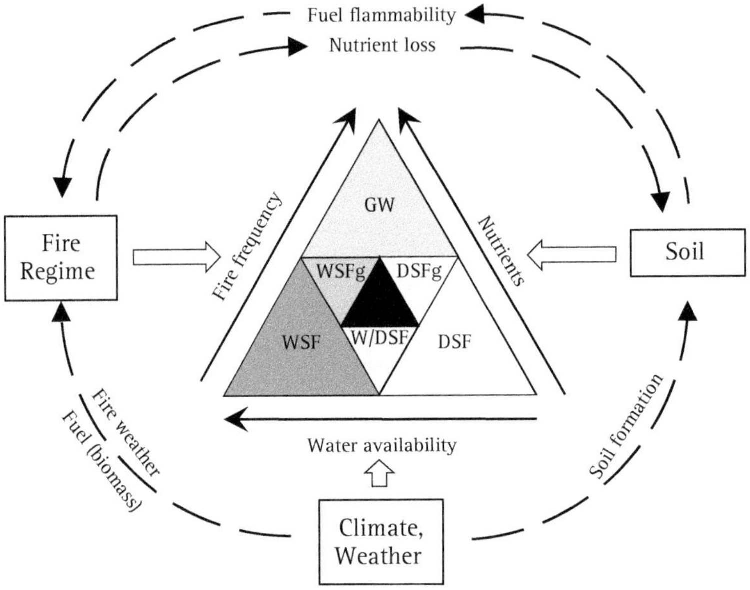

Figure 13.4 Interactions between fire, edaphic and climatic factors in determining the distribution of eucalypt forest and woodland formations in temperate Australia. Dry sclerophyll forests (DSF) occupy areas with the lowest fertility soil where water is seasonally but not strongly limiting and the distribution of fire intervals is predominately intermediate to long ((10)15 to 50 years) in the spectrum typical of temperate Australian forests. Wet sclerophyll forests (WSF) occur in areas where water supply is rarely limiting, soil nutrients are at intermediate to high levels and fire intervals are predominantly long (50 to >100 years). Grassy woodlands (GW) occur on soils of intermediate to high fertility where water is moderately to strongly limiting and fire intervals are short to intermediate ((3)5 to 50 years). Intermediate forms (DSFg and WSFg) support mixtures of shrubs and grasses (see Keith 2004).

poverty through the volatilisation of nutrients contained in plant material (Orians and Milewski 2007).

The position of dry sclerophyll forests in the continuum of Australia's temperate eucalypt forests and woodlands reflects both the physiological constraints imposed by climatic and edaphic drivers and the influence of climate and edaphically related sclerophylly on prevailing fire regimes (Figure 13.4). These climatic and edaphic influences on fire regimes relate to the quantity, moisture content and flammability of available fuel, the occurrence of appropriate weather conditions for fire propagation and the regime of ignitions (Bradstock 2010). Orians and Milewski (2007) hypothesised a central role for oligotrophic soils in this nexus. The central tenet of their model is that plants growing in productive environments where nutrients, but not light, are limiting will produce more carbohydrates than they are able to utilise in metabolic processes or reproduction. One of the consequences of this is the production of highly lignified, nutrient-poor foliage that is subject to low consumption rates by herbivores and microbes, and thus accumulates, providing the fuel for intense fires. Fires, in turn, exacerbate nutrient poverty because nutrients stored in plant tissues are volatilised during fire (Orians and Milewski 2007).

The quantity of fuel available and the potential for fire propagation depend on both climate and weather. Bradstock (2010) demonstrated how biogeographic variation in these factors translates to predictable patterns in fire regimes across Australia. Rainfall determines the quantity of fuel available (biomass production), its composition and continuity. Wetter climates are characterised by abundant woody fuel and continuous *Eucalyptus* litter beds. Decreasing rainfall is associated with a decline in tree cover and increasing discontinuity of *Eucalyptus* litter (Bradstock 2010). As well, moderate productivity combined with increasing severity of annual drought favours herbaceous species over shrubs (Stephenson 1990). Climate determines the prevailing weather (temperature, wind speed, humidity, antecedent rainfall) as well as exerting some control over the type of vegetation present (flammable versus non-flammable). In wetter climates, the potential for fire in forested areas is substantially limited by the probability that weather conditions conducive to fire spread (high temperatures, high wind, low humidity) coincide with times when fuels have dried out sufficiently to burn. Conversely, weather conditions conducive to fire spread become increasingly prevalent (and thus non-limiting) with increasing aridity.

As a consequence of these patterns in fuel accumulation and climate, the potential for fire varies along a continuum from coastal wet sclerophyll forests to semi-arid woodlands depending on whether fire is limited by fuel moisture (availability to burn) or fuel quantity and connectivity (available biomass) (Figure 13.4; Bradstock 2010). Wet and dry sclerophyll forests and temperate grasslands initially accumulate fuel at similar rates but differ in terms of long-term fuel load and availability to burn (i.e. when low fuel moisture and fire weather coincide). Grassy woodlands have the potential to burn more frequently (not fuel moisture-limited) and wet sclerophyll forests less frequently (fuel moisture-limited) than dry sclerophyll forests (Bradstock 2010). Tropical savannas are unique among forests and woodlands in generating sufficient biomass during regular wet seasons to support the potential for annual fires during reliable and prolonged dry seasons. Bradstock's model suggests that fire regimes are capable of driving patterns in the density and distribution of shrub species in woodlands and open forests. Annual fires limit shrub densities in tropical savannas while shrub density and diversity increases in a continuum from grassy woodlands to dry sclerophyll forests as the average interval between fires increases. Maximum development of the shrub understorey occurs under fire frequencies at which recruitment potential is maximised.

13.8 Management

The extensive representation of dry sclerophyll forests in conservation reserves is largely a consequence of the unsuitability of their nutrient-impoverished, free-draining drought-prone soils and steep terrain for agricultural exploitation. As a consequence, rates of clearing in many dry sclerophyll forest landscapes were historically low. Nevertheless, large areas of dry sclerophyll forests occurring on more gently undulating landscapes adjacent to more fertile lands have been converted to pastures and fertilised, or degraded by a range of other practices such as deliberate, high frequency (and low-intensity) burning, periodic grazing, timber-cutting and weed invasion (Keith 2004).

Dry sclerophyll forests are extensively distributed along the most heavily populated fringes of the continent and, as a consequence, have been subject to extensive clearing and fragmentation associated with residential developments in peri-urban areas (Keith 2004). Extensive weed invasion on urban fringes has been facilitated by the enrichment of soils by nutrients carried in stormwater runoff or overflowing sewerage networks (Leishman *et al.* 2004; Leishman and Thomson 2005). Forests on the most infertile soils are, however, generally resistant to weed invasion.

Infestation by the plant pathogen *Phytophthora cinnamomi* is a serious threat throughout the range of dry sclerophyll forests, although its impact is most severe in the south-west and far south-east (Cahill *et al.* 2008). Species affected by the disease are primarily in the families Proteaceae, Ericaceae, Fabaceae and Xanthorrhoeaceae. The indirect effects of changes to the structure and diversity of infested forests have been demonstrated in some fauna species but the full extent of ecological disruption is unknown. Management of the disease is primarily limited to restricting its spread by implementing hygiene protocols and restricting access to infected and vulnerable areas. Eradication of the pathogen is currently impossible and the use of chemicals such as phosphite has been only partially successful in containing its impacts.

Fire is a major focus of management of dry sclerophyll forests, and involves balancing the requirements of a fire-tolerant (but not fire-proof) biota with the need for the protection of human lives and assets. Planned burning is the primary means of achieving either a reduction in fuel to assist in wildfire containment or to promote a desired response in the biota (e.g. creating suitable habitat, initiating recruitment of plant species). However, these goals may be mutually exclusive (Morrison *et al.* 1996). Fuel reduction burns would be required at regular intervals (~5 years) to achieve desirable levels of protection because dry sclerophyll forests typically have relatively rapid rates of biomass production (Bradstock 2010). Such frequent fires over an extended period are likely to lead to reductions in the abundance and diversity of shrub species (Keith 1996). Targeting fuel reduction along the interface of urban areas is one method of minimising the area subjected to fire frequencies undesirable for biodiversity conservation although the required width of such buffers may expand considerably if the frequency and severity of fire weather increases (Price and Bradstock 2012, 2013).

Management systems have been developed which define desirable fire-frequency envelopes based on meta-analyses of plant functional traits and demographic models. These are increasingly being deployed to achieve the conservation of biodiversity in dry sclerophyll forests and other fire-prone vegetation types in Australia (e.g. Bradstock *et al.* 1995; Bradstock and Kenny 2003; Burrows 2008; Edwards and Russell-Smith 2009; Keith 2012). Further research is required to determine whether a focus on fire-interval effects alone will be sufficient to ensure the persistence of all species. Factors specific to individual fires (the event-dependent effects of Bond and van Wilgen 1996: their intensity, antecedent and subsequent weather) or density-dependent feedbacks and competitive interactions may be less amenable to management, but could potentially overwhelm the deterministic effects of fire interval in interrupting and initiating demographic processes.

The feedbacks and interactions between fire frequency, fire severity, fire season and plant response are still poorly understood in dry sclerophyll forests. The fire regimes that promote persistence of dry sclerophyll forest fauna are even more poorly understood (Driscoll *et al.* 2010). Parallel functional response classifications for animal taxa have yet to be incorporated in management frameworks (Keith 2012).

Finally, future declines in annual rainfall projected to occur as a result of greenhouse gas emissions are likely to have significant impacts on the structure, composition and fire regimes of dry sclerophyll forests. Given the nexus between climate, fire regimes and nutrients (Figure 13.4), projected trends of declining rainfall and increasing frequency of fire weather (IPCC 2007) may drive transitions of dry sclerophyll forests to novel communities that resemble grassy or semi-arid woodlands. Tree canopy dieback during prolonged dry spells may be an early symptom of increasing aridity because the hydraulic architecture of *Eucalyptus* species (and hence the ability to avoid vascular embolism) appears to be under genetic rather than environmental control (Pfautsch *et al.* 2016). Declining rainfall has major implications for forests in the Mediterranean-climate regions of Australia such as south-west Australia due to the dependence on winter recharge of groundwater to offset extended summer drought. Nevertheless, projections suggest that the impacts of reduced rainfall are likely to be even more severe in areas where the contribution of groundwater storage is less (Figure 13.1). This is evident in south-eastern Australia where areas predicted to decline in suitability as habitat for dry sclerophyll forest (Figure 13.1) correspond with areas of browning (decreasing growing season integrated leaf area index) detected during the period 1982–2009 (Zhu *et al.* 2016). This trend is counter to a global trend towards greening driven by atmospheric CO_2 enrichment (Zhu *et al.* 2016). Changes in fire activity as a consequence of shifts between fuel dryness-limited and productivity-limited states are likely to compound these changes (Bradstock 2010).

Acknowledgements

We gratefully acknowledge the support and assistance of Jane Elith, Neil MacKenzie, and Dan Tindall in compiling the maps and models presented in this chapter.

References

Adam, P., Stricker, P. and Anderson, D.J. (1989) Species-richness and soil phosphorus in plant communities in coastal New South Wales. *Australian Journal of Ecology* 14, 189–198.

Adam, P., Stricker, P., Wiecek, B.M. and Anderson, D.J. (1990) The vegetation of seacliffs and headlands in New South Wales, Australia. *Australian Journal of Ecology* 15, 515–547.

Austin, M.P., Pausas, J.G. and Noble, I.R. (1997) *Modelling Environmental and Temporal Niches of Eucalypts.* Cambridge University Press: Cambridge.

Axelrod, D.I. (1989) Age and origin of chaparral. In *The California Chaparral: Paradigms Reexamined*, ed. S.C. Keeley. Natural History Museum of Los Angeles County: USA, pp. 7–19.

Bartlett, M.K., Scoffoni, C. and Sack, L. (2012) The determinants of leaf turgor loss point and prediction of drought tolerance of species and biomes: a global meta-analysis. *Ecology Letters* 15, 393–405.

Beadle, N.C.W. (1954) Soil phosphate and the delimitation of plant communities in eastern Australia. *Ecology* 35, 370–375.

Beadle, N.C.W. (1962) Soil phosphate and the delimitation on plant communities in Eastern Australia II. *Ecology* 43, 281–288.

Beadle, N.C.W. (1966) Soil phosphate and its role in molding segments of the Australian flora and vegetation, with special reference to xeromorphy and sclerophylly. *Ecology* 47, 992–1007.

Beadle, N.C.W. (1981) *The Vegetation of Australia*. Cambridge University Press: Cambridge.

Beard, J.S., Beeston, G.R., Harvey, J.M. *et al.* (2013) The vegetation of Western Australia at the 1: 3 000 000 scale. Explanatory memoir. *Conservation Science Western Australia* 9, 1–152.

Bond, W.J and Keeley, J.E. (2005) Fire as a global 'herbivore': the ecology and evolution of flammable ecosystems. *Trends in Ecology and Evolution* 20, 387–394.

Bond, W.J. and Midgley, J.J. (1995) Kill thy neighbour: an individualistic argument for the evolution of flammability. *Oikos* 73, 79–85.

Bond, W.J. and Van Wilgen, B.W. (1996) *Fire and Plants*. Springer Netherlands: Netherlands.

Bond, W.J., Woodward, F.I. and Midgley, G.F. (2005) The global distribution of ecosystems in a world without fire. *New Phytologist* 165, 525–538.

Bowman, D.M.J.S. (2000) *Australian Rainforests: Islands of Green in a Land of Fire.* Cambridge University Press: Cambridge.

Box, E.O. (1996) Plant functional types and climate at the global scale. *Journal of Vegetation Science* 7, 309–320.

Bradstock, R.A. (2010) A biogeographic model of fire regimes in Australia: current and future implications. *Global Ecology and Biogeography* 19, 145–158.

Bradstock, R.A. and Kenny, B.J.(2003) An application of plant functional types to fire management in a conservation reserve in southeastern Australia. *Journal of Vegetation Science* 14, 345–354.

Bradstock, R.A., Keith, D.A., Auld, T.D. *et al.* (1995) Fire and conservation: imperatives and constraints on managing for diversity. In *Conserving Biodiversity: Threats and Solutions*, ed. R.A. Bradstock, D.A. Keith, T.D. Auld, R.T. Kingsford, D. Lunney and D.P. Sivertsen. Surrey Beatty and Sons: Sydney, pp. 323–333.

Buhrke, V.E., Jenkins, R. and Smith, D.K. (1998) *Practical guide for the preparation of specimens for X-ray fluorescence and X-ray diffraction analysis.* Wiley-VCH: New York.

Bureau of Meteorology n.d. Gridded climate data. Canberra: Australian Government. Available at: http://www.bom.gov.au/climate/averages/climatology/gridded-data-info/gridded-climate-data.shtml.

Burrows, N.D. (2008) Linking fire ecology and fire management in south-west Australian forest landscapes. *Forest Ecology and Management* 255, 2394–2406.

Cahill, D.M., Rookes, J.E., Wilson, B.A. *et al.* (2008) Turner Review No. 17. *Phytophthora cinnamomi* and Australia's biodiversity: impacts, predictions and progress towards control. *Australian Journal of Botany* 56, 279–310.

Christopherson, R.W. (2006) *Geosystems: An Introduction to Physical Geography*, 6th edition. Pearson/Prentice Hall: Upper Saddle River.

Coaldrake, J.E. and Haydock, K.P. (1958) Soil phosphate and vegetal pattern in some natural communities of south-eastern Queensland, Australia. *Ecology* 39, 1–5.

Commonwealth Scientific and Industrial Research Organisation (CSIRO) (n.d.) Soil and Landscape Grid of Australia. Canberra: TERN. Available at: http://www.clw.csiro.au/aclep/soilandlandscapegrid/

Coomes, D.A. and Grubb, P.J. (2000) Impacts of root competition in forests and woodlands: a theoretical framework and review of experiments. *Ecological Monographs* 70, 171–207.

Cowling, R.M., Rundel, P.W., Lamont, B.B. *et al.* (1996) Plant diversity in Mediterranean-climate regions. *Trends in Ecology and Evolution* 11, 362–366.

di Castri, F. (1981) Mediterranean-type shrublands of the world. In *Ecosystems of the World*, ed. F di Castri. Elsevier Scientific: New York, pp. 1–52.

Driscoll, D.A., Lindenmayer, D.B., Bennett, A.F. *et al.* (2010) Fire management for biodiversity conservation: key research questions and our capacity to answer them. *Biological Conservation* 143, 1928–1939.

Dyer, J.M. (2009) Assessing topographic patterns in moisture use and stress using a water-balance approach. *Landscape Ecology* 24, 391–403.

Edwards, A.C. and Russell-Smith, J. (2009) Ecological thresholds and the status of fire-sensitive vegetation in western Arnhem Land, northern Australia: implications for management. *International Journal of Wildland Fire* 18, 127–146.

Elith, J., Leathwick, J.R. and Hastie, T. (2008) A working guide to boosted regression trees. *Journal of Animal Ecology* 77, 802–813.

Fonseca, C.R., Overton, J.M., Collins, B. and Westoby, M. (2000) Shifts in trait-combinations along rainfall and phosphorus gradients. *Journal of Ecology* 88, 964–977.

Gill, A.M. (1994) Patterns and processes in open forests of Eucalyptus in southern Australia. In *Australian Vegetation*, ed. R.H. Groves. Melbourne: Cambridge University Press, pp. 197–227.

González-Orozco, C.E., Ebach, M.C., Laffan, S. *et al.* (2014) Quantifying phytogeographical regions of Australia using geospatial turnover in species composition. *PLoS ONE* 9, e92558.

Gray, J.M. and Murphy, B.W. (1999) *Parent Material and Soils: A Guide to the Influence of Parent Material on Soil Distribution in Eastern Australia*. NSW Department of Land and Water Conservation: Sydney.

Harrington, B.A. and Humphreys, G.S. (2004) *Plant accessible phosphorus as a discriminator of native vegetation in the upper Blue Mountains*. The Regional Institute: Gosford. Available at: http://www.regional.org.au/au/asssi/supersoil2004/s11/oral/1840_harringtonb.htm (accessed 8 August 2016).

Hopper, S.D. (2009) OCBIL theory: towards an integrated understanding of the evolution, ecology and conservation of biodiversity on old, climatically buffered, infertile landscapes. *Plant and Soil* 322, 49–86.

Hopper, S.D., Abbott, I. and Burrows, N. (2003) An evolutionary perspective on south-west Western Australian landscapes, biodiversity and fire: a review and management implications. In *Fire in Ecosystems of South-west Western Australia: Impacts and Management*, ed. I. Abbott and N. Burrows. Backhuys Publishers: Leiden, pp. 9–35.

IPCC (2007) *Climate change 2007: synthesis report. Contribution of working groups I, II and III to the fourth assessment report of the Intergovernmental Panel on Climate Change*, ed. R.K. Pachauri and A. Reisinger. IPCC: Geneva.

Jackson, W.D. (1968) Fire, air, water and earth–an elemental ecology of Tasmania. *Proceedings of the Ecological Society of Australia* 3, 9–16.

Keeley, J.E., Bond, W.J., Bradstock, R.A. *et al.* (2012) *Fire in Mediterranean Ecosystems: Ecology, Evolution and Management.* Cambridge University Press: Cambridge.

Keith, D.A. (1996) Fire-driven extinction of plant populations: a synthesis of theory and review of evidence from Australian vegetation. *Proceedings of the Linnean Society of New South Wales* 116, 37–78.

Keith, D.A. (2004) *Ocean shores to desert dunes: the native vegetation of NSW and the ACT.* NSW Department of Environment and Conservation: Hurstville.

Keith, D.A. (2012) Functional traits: their roles in understanding and predicting biotic responses to fire regimes from individuals to landscapes. In *Flammable Australia: Fire Regimes, Biodiversity and Ecosystems in a Changing World*, ed. J.E. Williams, A.M. Gill and R.A. Bradstock. CSIRO Publishing: Collingwood, pp. 97–125.

Keith, D.A. and Sanders, J.M. (1990) Vegetation of the Eden region, south-eastern Australia: species composition, diversity and structure. *Journal of Vegetation Science* 1, 203–232.

Kitchener, A. and Harris, S. (2013) *From Forest To Fjaeldmark: Descriptions of Tasmania's Vegetation*, 2nd edition. Department of Primary Industries, Parks, Water and Environment: Tasmania.

Ladiges, P.Y. (1997) Phylogenetic history and classification of eucalypts. In *Eucalypt Ecology: Individuals to Ecosystems*, ed. J. Williams and J. Woinarski. Cambridge University Press: Cambridge, pp. 16–29.

Lambers, H., Raven, J.A., Shaver, G.R. and Smith, S.E. (2008) Plant nutrient-acquisition strategies change with soil age. *Trends in Ecology and Evolution* 23, 95–103.
Lamont, B.B. and He, T. (2012) Fire-adapted Gondwanan angiosperm floras evolved in the Cretaceous. *BMC evolutionary biology* 12, 223.
Lamont, B.B., Groom, P.K., Cowling, R.M. (2002) High leaf mass per area of related species assemblages may reflect low rainfall and carbon isotope discrimination rather than low phosphorus and nitrogen concentrations. *Functional Ecology* 16, 403–412.
Le Brocque, A.F. and Buckney, R.T. (1995) Vegetation and environmental patterns on soils derived from Hawkesbury Sandstone and Narrabeen substrata in Ku-ring-gai Chase National Park, New South Wales. *Australian Journal of Ecology* 20, 229–238.
Leishman, M.R. and Thomson, V.P. (2005) Experimental evidence for the effects of additional water, nutrients and physical disturbance on invasive plants in low fertility Hawkesbury Sandstone soils, Sydney, Australia. *Journal of Ecology* 93, 38–49.
Leishman, M.R., Hughes, M.T. and Gore, D.B. (2004) Soil phosphorus enhancement below stormwater outlets in urban bushland: spatial and temporal changes and the relationship with invasive plants. *Soil Research* 42, 197–202.
Lieth, H. (1975) Modelling the primary productivity of the world. In *Primary Productivity of the Biosphere*, ed. H. Lieth and R.H. Whittaker. Springer: Berlin, pp. 237–263.
Lieth, H. (1976) The use of correlation models to predict primary productivity from precipitation or evapotranspiration. In *Water and Plant Life*. Springer: Berlin, pp. 392–407.
Loveless, A.R. (1962) Further evidence to support a nutritional interpretation of sclerophylly. *Annals of Botany* 26, 551–561.
Lynch, A.H., Beringer, J., Kershaw, P. *et al.* (2007) Using the paleorecord to evaluate climate and fire interactions in Australia. *Annual Review of Earth Planet Science* 35, 215–239.
Mason, S., McNeill, A., McLaughlin, M.J. and Zhang, H. (2010) Prediction of wheat response to an application of phosphorus under field conditions using diffusive gradients in thin-films (DGT) and extraction methods. *Plant and Soil* 337, 243–258.
Midgley, J.J. (2013) Flammability is not selected for, it emerges. *Australian Journal of Botany* 61, 102–106.
Morrison, D.A., Buckney, R.T., Bewick, B.J. and Cary, G.J. (1996) Conservation conflicts over burning bush in south-eastern Australia. *Biological Conservation* 76, 167–175.
Mucina, L. and Wardell-Johnson, G.W. (2011) Landscape age and soil fertility, climatic stability, and fire regime predictability: beyond the OCBIL framework. *Plant and Soil* 341, 1–23.
Mutch, R.W. (1970) Wildland fires and ecosystems: a hypothesis. *Ecology* 51, 1046–1051.
Myerscough, P.J. and Carolin, R.C. (1986) The vegetation of the Eurunderee sand mass, headlands and previous islands in the Myall Lakes area. New South Wales. *Cunninghamia* 1, 399–466.
Neldner, V.J., Niehus, R.E., Wilson, B.A. *et al.* (2015) *The Vegetation of Queensland. Descriptions of Broad Vegetation Groups. Version 2.0.* Queensland Herbarium, Department of Science, Information Technology and Innovation: Queensland.
Noble, I.R. (1989) Ecological traits of the Eucalyptus L'Hérit subgenera *Monocalyptus* and *Symphyomyrtus*. *Australian Journal of Botany* 37, 207–224.
Oertli, J.J., Lips, S.H. and Agami, M. (1990) The strength of sclerophyllous cells to resist collapse due to negative turgor pressure. *Acta Oecologica* 11, 281–289.

Ordoñez, J.C., Van Bodegom, P.M., Witte, J.P.M. *et al.* (2009) A global study of relationships between leaf traits, climate and soil measures of nutrient fertility. *Global Ecology and Biogeography* 18, 137–149.

Orians, G.H. and Milewski, A.V. (2007) Ecology of Australia: the effects of nutrient-poor soils and intense fires. *Biological Reviews* 82, 393–423.

Pfautsch, S., Harbusch, M., Wesolowski, A. *et al.* (2016) Climate determines vascular traits in the ecologically diverse genus *Eucalyptus*. *Ecology Letters* 19, 240–248.

Pidgeon, I.M. (1940) The ecology of the central coastal area of New South Wales. III. Types of primary succession. *Proceedings of the Linnean Society of NSW* 65, 221–249.

Poore, M.E.D. (1955) The use of phytosociological methods in ecological investigations: I. The Braun-Blanquet system. *The Journal of Ecology* 43, 226–244.

Price, O. and Bradstock, R. (2013) Landscape scale influences of forest area and housing density on house loss in the 2009 Victorian bushfires. *PLoS ONE* 8, e73421.

Price, O.F. and Bradstock, R.A. (2012) The efficacy of fuel treatment in mitigating property loss during wildfires: insights from analysis of the severity of the catastrophic fires in 2009 in Victoria, Australia. *Journal of Environmental Management* 113, 146–157.

Rayment, G.E. and Lyons, D.J. (2011) *Soil Chemical Methods: Australasia.* CSIRO Publishing: Collingwood.

Ricklefs, R.E. (2008) *The Economy of Nature*, 6th edition. Freeman and Co: New York.

Rosenzweig, M.L. (1968) Net primary productivity of terrestrial communities: prediction from climatological data. *American Naturalist* 102, 67–74.

Schimper, A.F.W. (1903) *Plant-Geography upon a Physiological Basis*, translated by W.R. Fisher, ed. P. Groom and I.B. Balfour. Clarendon Press: Oxford.

Silberstein, R., Held, A., Hatton, T. *et al.* (2001) Energy balance of a natural jarrah (*Eucalyptus marginata*) forest in Western Australia: measurements during the spring and summer. *Agricultural and Forest Meteorology* 109, 79–104.

Šmilauer, P. and Lepš, J. (2014) *Multivariate Analysis of Ecological Data Using CANOCO 5*, 2nd edition. Cambridge University Press: Cambridge.

Specht, R.L. (1970) Vegetation. In *The Australian Environment*, ed. G.W. Leeper. CSIRO Publishing and Melbourne University Press: Melbourne, pp. 44–67.

Sperry, J.S. and Hacke, U.G. (2002) Desert shrub water relations with respect to soil characteristics and plant functional type. *Functional Ecology* 16, 367–378.

Stephenson, N.L. (1990) Climatic control of vegetation distribution: the role of the water balance. *American Naturalist* 135, 649–670.

Tozer, M.G., Turner, K., Simpson, C. *et al.* (2010) Native vegetation of southeast NSW: a revised classification and map for the coast and eastern tablelands. *Cunninghamia* 11, 359–406.

Turner, I.M. (1994) Sclerophylly: primarily protective? *Functional Ecology* 8, 669–675.

Van Wilgen, B.W. (1984) Fire climates in the southern and western Cape Province and their potential use in fire control and management. *South African Journal of Science* 80, 358–362.

Vankat, J.L. (1989) Water stress in chaparral shrubs in summer rain versus summer drought climates: whither the Mediterranean type climate paradigm. In *The California Chaparral: Paradigms Reexamined*, ed. S.C. Keeley. Natural History Museum of Los Angeles County: Los Angeles, pp. 117–124.

Wardell-Johnson, G. and Williams, M. (1996) A floristic survey of the Tingle Mosaic, south-western Australia: applications in land use planning and management. *Journal of the Royal Society of Western Australia* 79, 249–276.

Wardell-Johnson, G.W., Lawson, B.E. and Coutts, R.H. (2007) Are regional ecosystems compatible with floristic heterogeneity? A case study from Toohey Forest, south-east Queensland, Australia. *Pacific Conservation Biology* 13, 47–59.

Warman, L. and Moles, A.T. (2009) Alternative stable states in Australia's Wet Tropics: a theoretical framework for the field data and a field-case for the theory. *Landscape Ecology* 24, 1–13.

Whittaker, R.H. (1975) *Communities and Ecosystems*, 2nd edition. Macmillan Publishing Co: New York.

Zhu, Z., Piao, S., Myneni, R.B. *et al.* (2016) Greening of the Earth and its drivers. *Nature Climate Change* 6, 791–795.

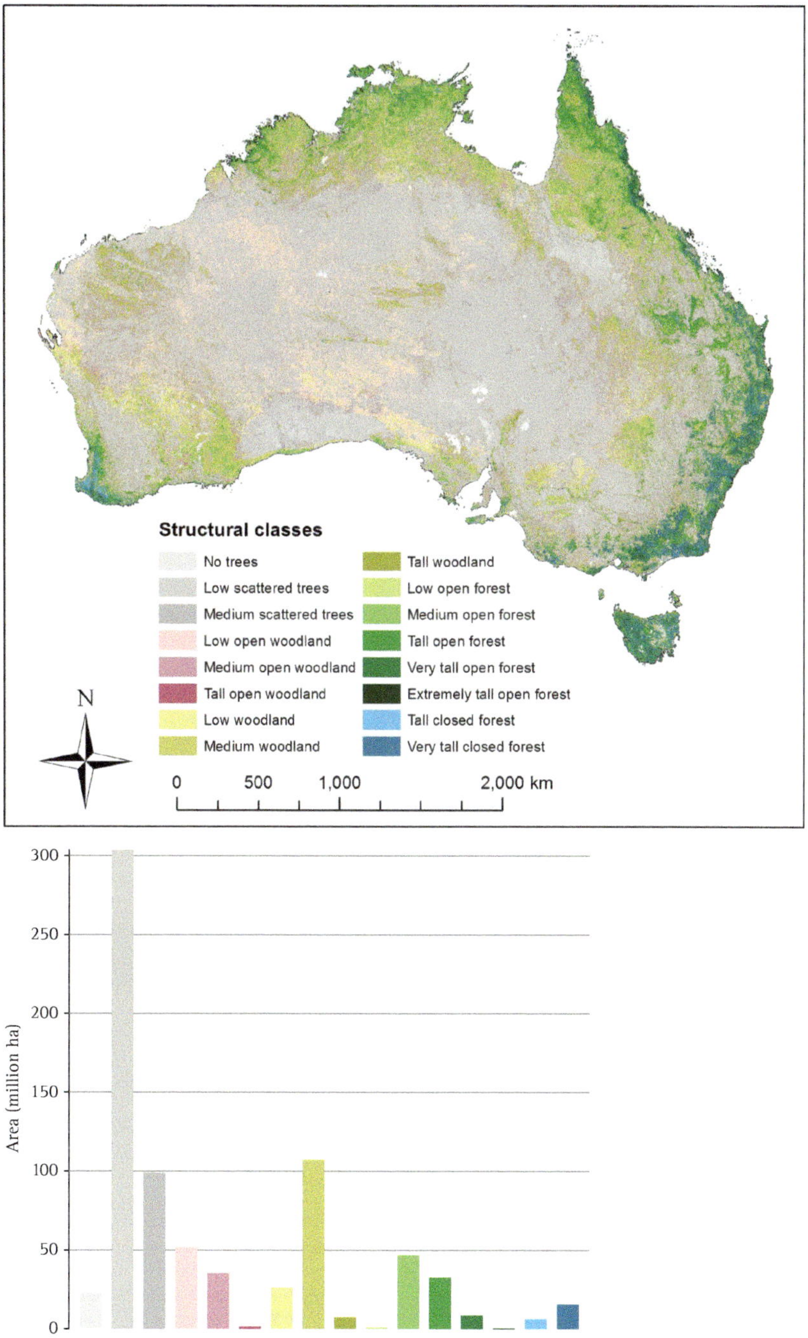

Figure 1.1 Map of vegetation structure of Australia derived from derived from ALOS-1 PALSAR, ICESat/GLAS and Landsat imagery time series for 2003–2009. Histogram shows areas of structural types with a vast core of low open woodland and scattered trees, girt by a semi-continuous band of medium woodland and narrow coastal fringe of forest.
Data from Auscover (2016).

Figure 2.1 Macrofossils and pollen in dated sediments provide independent calibration data for phylogenies and evolutionary narratives. Phylogenetically isolated lineages also provide historical signals. *Photos: Greg Jordan.*

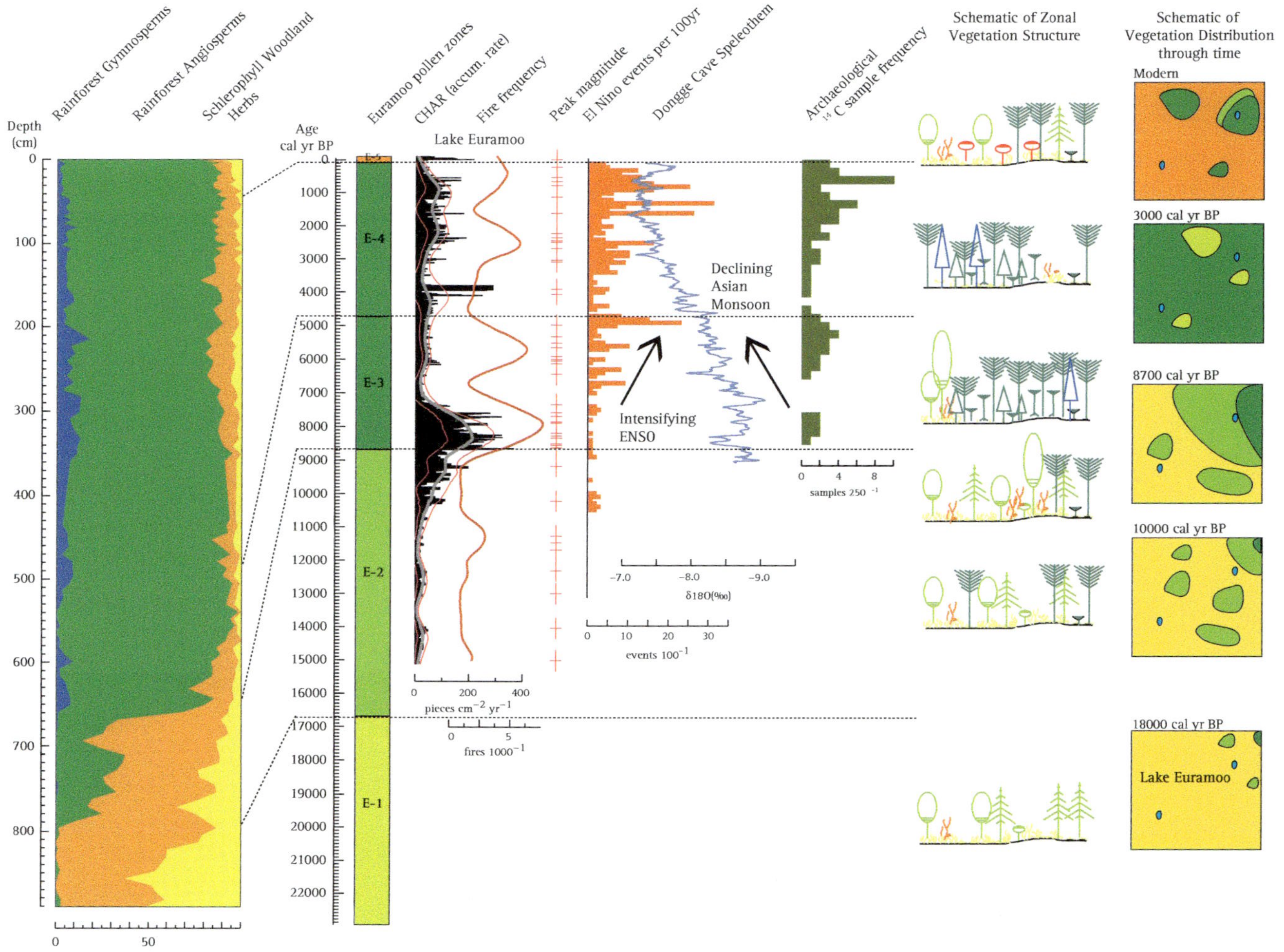

Figure 3.5 A summary diagram of the high-resolution pollen record from Lake Euramoo on the Atherton Tableland in north-eastern Queensland. The reconstruction of fire frequency, based on charcoal accumulation (CHAR), extends back to 15 000 years ago, as prior to that time charcoal abundance is very low. Some of the possible drivers of vegetation and fire change in this landscape are shown for comparison including the total number of strong El Niño events per century interpreted from flood deposits in Laguna Pallcacocha, Ecuador (orange bars; Moy *et al.* 2002) and past precipitation in China from the Dongge Cave speleothem δ180 record (blue line; Wang *et al.* 2005). ***Diagrams: Simon Haberle.***

Figure 4.2 Contemporary Aboriginal burning in the western Great Sandy Desert. In spinifex hummock grasslands with isolated *Melaleuca glomerata* stands, at least four post-fire age classes result from fires ignited by both Martu rangers and Martu families who burn to hunt near Punmu. *Photo: Gareth Catt of Kanyirninpa Jukurrpa.*

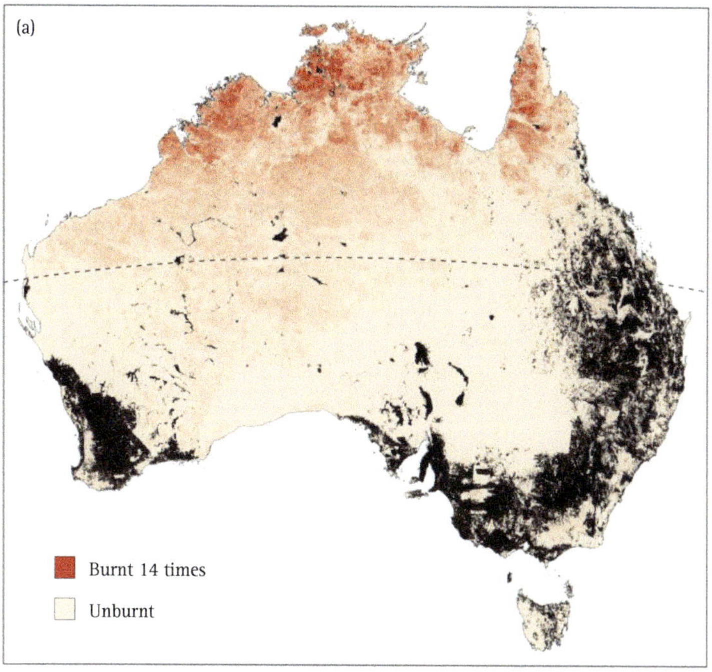

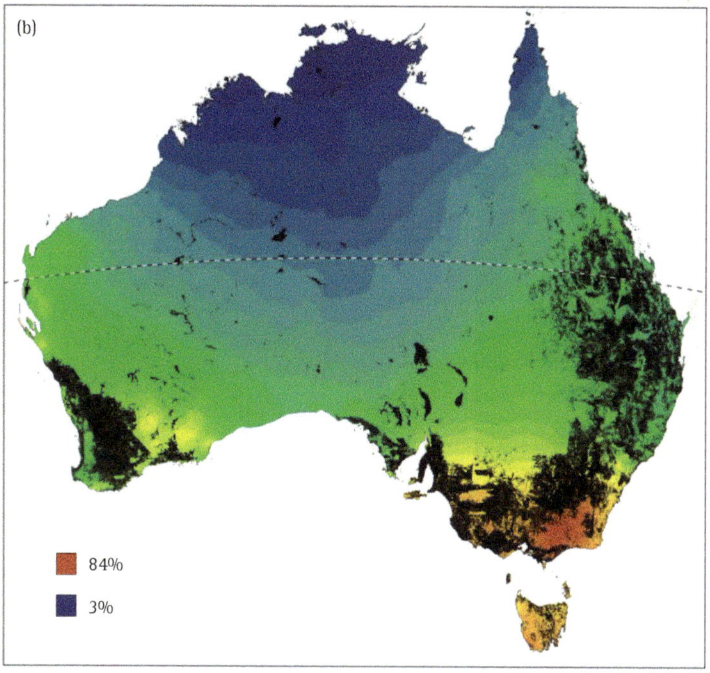

Figure 5.2 Australia-wide variation in: (a) fire frequency (number of times burnt in a 14-year period, 1997–2010), with fire scars derived from Advanced Very High Resolution Radiometer satellite imagery (Russell-Smith *et al.* 2007); and (b) the proportion of burning in the summer and autumn months of December–March, inclusive, inferred from the density of active fire detections by the Moderate-Resolution Imaging Spectroradiometer (MODIS) satellite sensors (Giglio *et al.* 2003). The surface shown in (b) is derived using interpolation with inverse distance weighting. Cleared areas, bare areas and waterbodies are shown in black (National Vegetation Information System 2012). *Maps: Brett Murphy.*

Figure 6.2 (a) Contrasting biomass, soil exposure and species compostion of the vegetation ground-layer in semi-arid woodland either side of experimental herbivore exclosure fence, at Yathong Nature Reserve, central western New South Wales. Feral goats, rabbits and native kangaroos have access to the left hand side of the fence. Note dense growth of *Austrostipa scabra* and undershrubs on right-hand side of fence. (b) Piosphere effect with livestock tracks to water, severe denudation of vegetation, soil compaction and scalding near Bourke, north-west New South Wales. (c) Legacies of over-grazing, including soil compaction, inhibition of ground-layer vegetation and shrub encroachment persist 15 years after livestock removal near Enngonia Bourke, north-west New South Wales. *Photos: David Eldridge.*

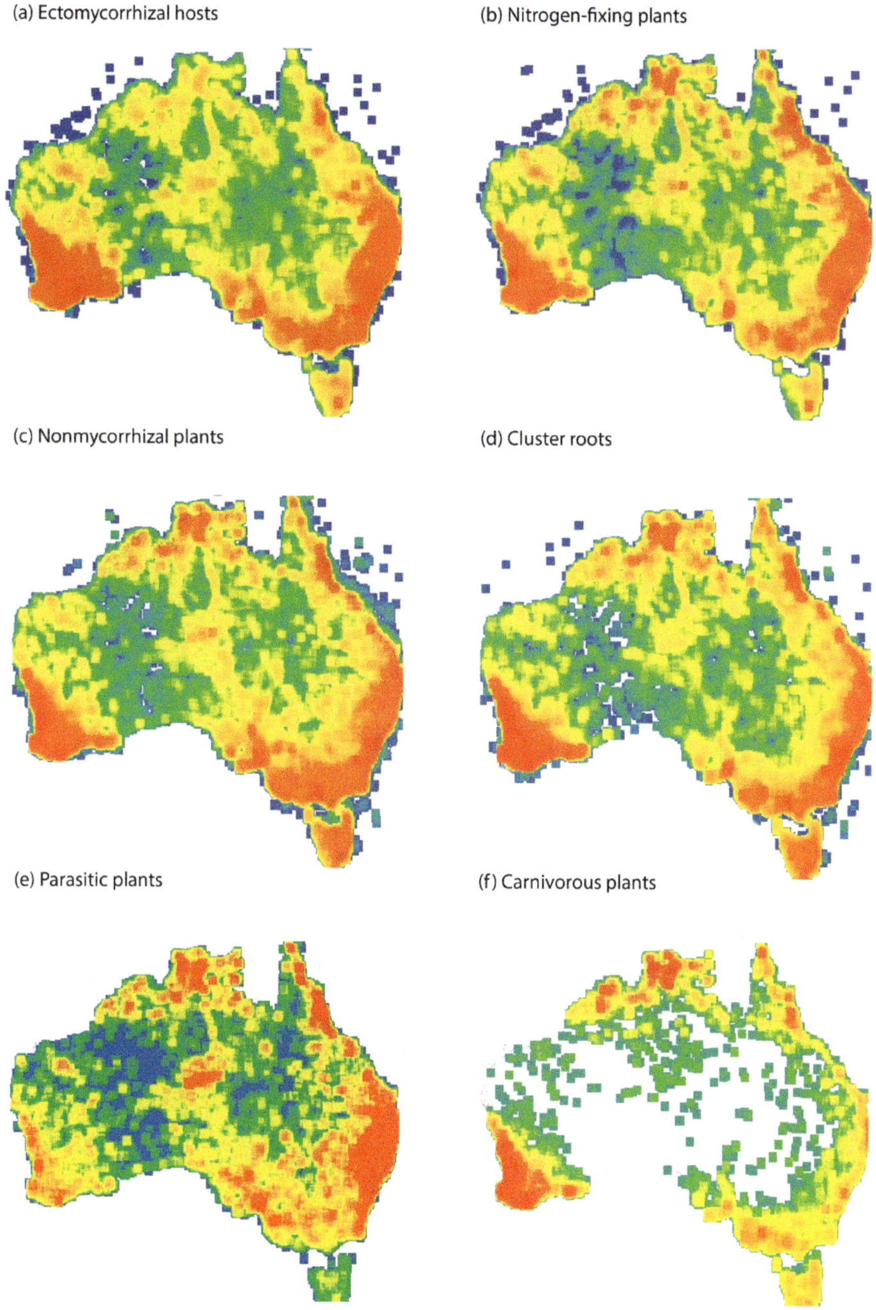

Figure 7.4 Diversity maps for key root strategies: (a) ectomycorrhizal plants, (b) nitrogen-fixing plants, (c) nonmycorrhizal plants, (d) plants with cluster roots (e) parasitic plants and (f) carnivorous plants. These maps show the average count of the number of recorded taxa at different locations using data from the *Atlas of Living Australia* (2015). Maps are produced with the ALA Points to grid function with a 0.1 degree scale and 9 cell moving averages. The highest diversity is shown as red, lower diversity as blue and a lack of shading reflecting the absence of records. These maps show general trends only and are affected by limited sampling in remote areas, particularly in central Australia. *Maps: Mark Brundrett.*

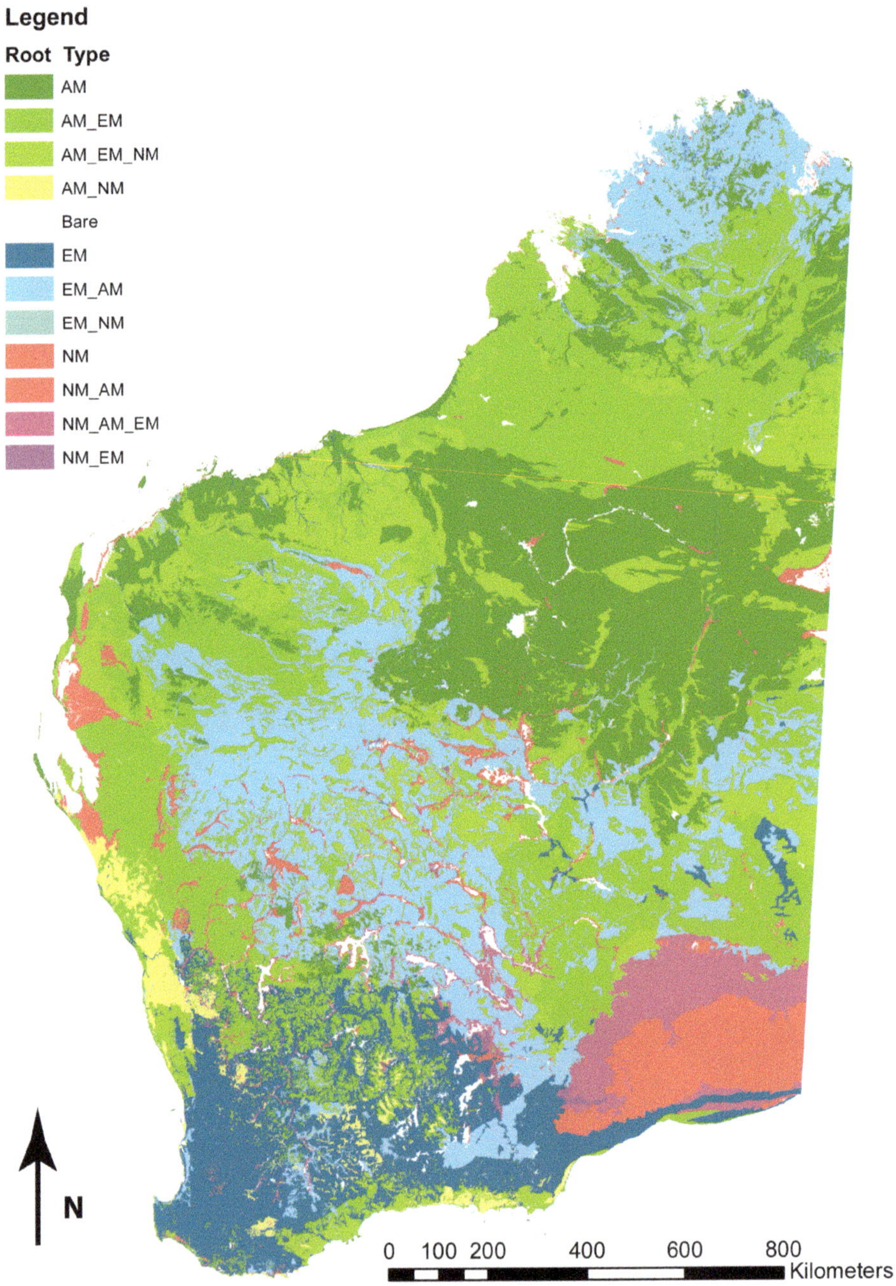

Figure 7.5 The West Australian Vegetation map (Beard *et al.* 2013) with major physiognomic vegetation types reclassified to show below-ground root type adaptations linked to nutrient uptake (AM – arbuscular mycorrhizal, EM – ectomycorrhizal; NM – nonmycorrhizal). This map provides information on root traits of the dominant plants in vegetation types. Many of these types include taxa with traits that differ from those of the dominant taxa. The root status of dominant plants was inferred from existing data and phylogeny. *Source: http://mycorrhizas.info/ozplants; Brundrett (2009). Map: Mark Brundrett.*

Figure 8.2 (a–f) Climatic factors control many plant population processes. For example, the season of rainfall can determine species composition in many arid and semi-arid systems, mediated by mechanisms controlling seed dormancy and germination. Grasses tend to recruit after summer rainfall, as shown by (a) extensive *Enneapogon avenaceus* recruitment in far western New South Wales show, while winter rain promotes germination of forbs in the same regions (photo: Tony Auld), as seen by (b) the flush of *Myriocephalus stuartii* (Asteraceae) recruitment (photo: Tony Auld). Both species are physiologically dormant but have warm temperature and cool temperature germination requirements respectively. Climate change also causes mean increases in temperature and alters weather patterns. This can result in changes to the environment necessary to maintain component species, such as in (c) where a reduction in the number of days with cloud cover (and therefore humidity) can affect cloud forest in the higher elevations of Lord Howe Island (photo: Tony Auld). Climate change interacts with disturbance regimes such as fire, where alterations to frequency, intensity and season are projected, as seen in (d) an out-of-peak season autumn fire in south-eastern Australia (photo: Mark Ooi). Understanding the response of plants to climate change is complex, with many innovative approaches including Free Air CO_2 Enrichment (FACE) plots, as seen in (e) at the University of Tasmania's TasFACE experiment (photo: Mark Hovenden), and (f) large-scale common garden experiments, seen here at the University of Wollongong, established to understand the plasticity and potential adaptability of species to future changes (photo: Mark Ooi).

Figure 9.2 Some species are considered transformer species as they cause substantial changes in the structure of the communities they invade: (a) and (b) exotic invasive vines can cause dieback and mortality of host trees, opening up the canopy to the invasion of other species (photos: Rachael Gallagher); (c) and (d) African Olive invades grassy woodland, creating a deep shaded canopy that native grasses and herbs are unable to survive beneath (photos: Peter Cuneo); (e) savanna with a native grass understorey (photo: M. Douglas); (f) savanna invaded by the exotic gamba grass, with associated tree mortality due to transformation of fire regimes (photo: Samantha Setterfield).

Figure 9.3 (a) Yellow spore pustules on young leaves of *Melaleuca quinquenervia*, indicating infection by myrtle rust (Photo: A Carnegie). (b) *Rhodamnia rubescens* severely affected by myrtle rust in Royal National Park, Sydney. Canopy dieback and mortality of this species results in opening of the canopy which may result in shifts in species assemblages in rainforest vegetation (Photo: L Fernandez).

Figure 11.2 (a) Tropical rainforest, Mt Lewis, Far north Queensland (photo: Dan Metcalfe); (b) subtropical rainforest, Bunya Mountains, Queensland (photo: Dan Metcalfe); (c) semi-deciduous vine thicket, Melamen River, Cape York (photo: Dan Metcalfe). (d) monsoon vine thicket, Kimberley, Western Australia (photo: Bruce Webber); (e) warm temperate rainforest, Bola Creek, Royal National Park, New South Wales (photo: David Keith); (f) cool temperate rainforest, Overland Track, Cradle Mountain-Lake St Clair National Park, Tasmania (photo: Travis Sydes).

Figure 12.4 Shrubby and mixed wet sclerophyll forest, southern Australia. (a) Stand of *Eucalyptus obliqua* with rainforest understorey of *Nothofagus cunninghamii* and *Atherosperma moschatum* in the lower Weld Valley, Tasmania (photo: Rob Blakers); (b) wet sclerophyll forest of *E. jacksonii* and *E. diversicolor* (white trunks in background) at Mt Clare near the south coast, south-western Australia (photo: Grant Wardell-Johnson); (c) *Eucalyptus obliqua*, lower Weld Valley, Tasmania (photo: Rob Blakers).

Figure 12.5 Impacts of fire on wet sclerophyll forests. (a) Hollow-butting is typical following repeated fires in WSF of *Eucalyptus jacksonii* in south-coastal, south-western Australia. Note the resprouting epicormic shoots (photo: Grant Wardell-Johnson); (b) *Eucalyptus guilfoylei* regeneration following clear-felling near south coast, south-western Australia in 1976, burnt by an intense wildfire in 1991. Note some plants have produced seed (photo: Grant Wardell-Johnson, taken 1992). (c) Dead canopies as signs of an intense fire in *Eucalyptus jacksonii* WSF in 1937 near south coast, south-western Australia (photo: Grant Wardell-Johnson, taken 2014). (d) Large stand of *Eucalyptus delegatensis* that had regenerated following a wildfire in 1939, killed by a wildfire in 2007. Note seedling regeneration (photo: Lynda Prior).

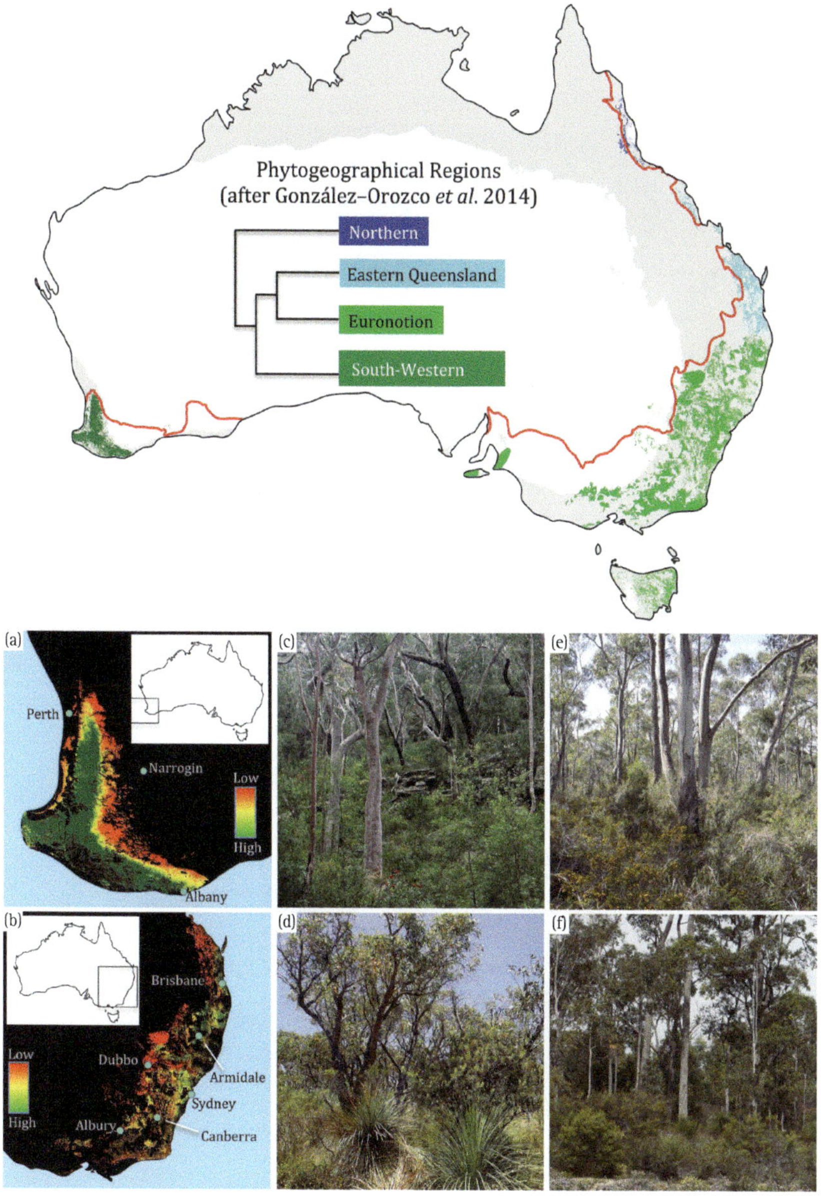

Figure 13.1 The map shows the distribution of dry sclerophyll forests prior to European colonisation compiled from units assigned to Dry Sclerophyll Forest Formation in New South Wales (Keith 2004), Medium Forest and Woodland in Western Australia (Beard *et al.* 2013), and lower-level categories in maps of Queensland (Neldner *et al.* 2015), Victoria, South Australia and Tasmania (Kitchener and Harris 2013) whose descriptions indicated the presence of a well-developed sclerophyll shrub layer beneath a eucalypt overstorey. Areas receiving more than 500 mm in annual rainfall are shaded grey. The red line marks the limits of zones on the south and east coasts in which the annual deficit in water availability does not exceed 800 mm (refer to Box 13.1). (a) and (b) show the forecast probability of habitat conditions being suitable for dry sclerophyll forest in 2070, assuming high population growth and fossil fuel consumption, within the current distribution in, respectively, the south-west and south-east of Australia (see Box 13.1). Examples of dry sclerophyll forest: (c) open forest dominated by *Angophora costata* and *Eucalyptus piperita* in Royal National Park, Sydney (photo: Mark Tozer); (d) woodland dominated by *E. baxteri, Banksia marginata,* and *Xanthorrhoea semiplana* in Deep Creek Conservation Park on the Fleurieu Pensinsula, South Australia (photo: David Keith); (e) open forest dominated by *E. ovata* with *Pultenaea juniperina, Leptospermum scoparium* and *Lepidosperma laterale* at Longley, Tasmania (photo: David Keith); (f) jarrah forest dominated by *E. marginata* with shrubs of Fabaceae and Proteaceae in the understorey, south-western Australia (photo: Mark Brundrett).

Figure 14.3 Impression of leaf transection of *Hakea invaginata* showing adaptations to drought tolerance among sclerophyllous heath plants. Five crypts are bounded by tannin-bearing, photosynthetic tissues to which the sunken stomates are confined. These are surrounded by wedge-shaped lignified fibrebands that replace the palisade beneath the high light, external surfaces and enable the crypts to close to prevent further water loss without the leaf collapsing when turgor is lost. The wedges are topped by a tannin-filled hypodermis (reduces light interception) with a cuticle as thick as the epidermis. The vascular bundles are protected by two fibrebands. Large tannin-filled cells storing water and osmotic substances make up a parenchymatous core. The leaves are ~15 cm long, 2 mm diameter, vertically oriented and survive ~3 years. Section stained with toluidine blue: cellulose shows in green, lignin in blue, tannins various colours including magenta, xylem square red cells. *From painting by Byron Lamont.*

(a)

(b)

Figure 14.7 (a) Scrub-heath in south-western Australia on dune crest on the Eneabba sandplain (see profile, Figure 14.2) in winter 14 years after fire. 113 vascular plant species were recorded in a 40 × 40-m plot at this site (Enright *et al* 2007). The vegetation pictured is dominated by Proteaceae: serotinous banksias – *B. menziesii* in flower (left, 1.2 m tall), *B. hookeriana* in flower (right), *B. attenuata* with cones (mid-centre, also inset: flower head 15 cm tall). Insets: *Daviesia epiphylla* (left); *Banksia attenuata* (right). *Photos: Byron Lamont.* (b) Wet heath (or Restioid heath) in south-eastern Australia on a poorly drained sandstone plateau in Royal National Park (near to Dharawal Nature Reserve; Box 14.1) in early spring 5 years after fire. A total of 75 vascular plant species were recorded in a 20 × 20-m plot at this site (Keith, unpubl. data). The vegetation pictured includes *Xanthorrhoea resinosa* (left foreground), *Banksia ericifolia* (centre) and *Dillwynia floribunda* (yellow flowers throughout). Insets: *Banskia ericifolia* (left); *Dillwynia floribunda* and *Sprengelia incarnata* (right). *Photos: David Keith.*

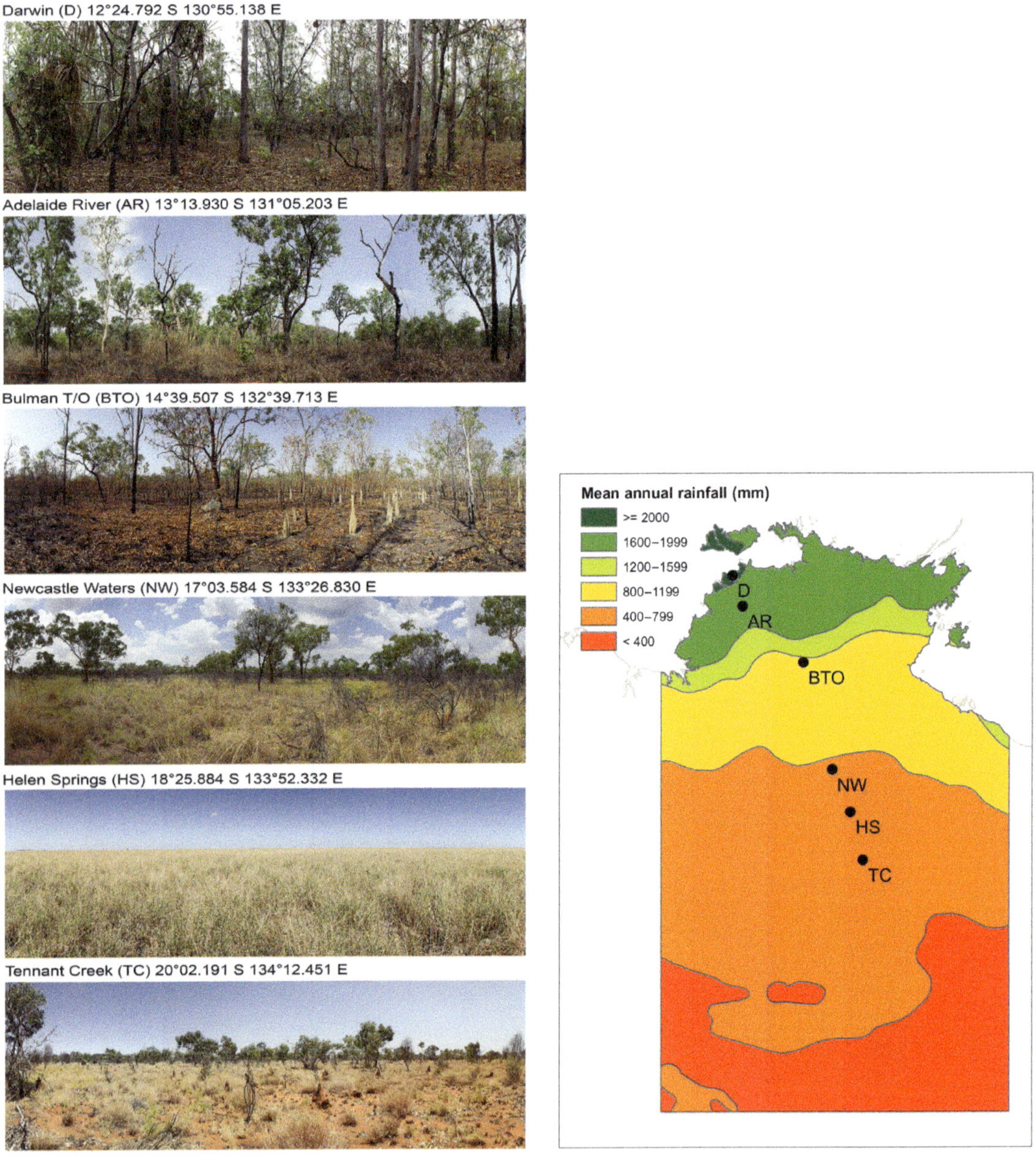

Figure 15.2 Australian savanna landscapes from sites along the North Australian Tropical Transect (NATT) in the Northern Territory. Sites: Open forest of *Eucalyptus tetrodonta* over recently burnt grass layer dominated by annual *Sorghum;* Darwin region (D); woodland of *Corymbia* and *Eucalyptus* over perennial tussock grass layer, Adelaide River (AR); woodland of *Corymbia,* Bullman turnoff on Stuart Highway (BTO); Open woodland of *Corymbia terminalis* over perennial tussock grass layer, Newcastle Waters (NW); Grassland of *Astrebla* and *Dichanthium* spp., Helen Springs (HC); Open woodland of *Eucalyptus brevifolia* with tussock and hummock grass understorey, Tennant Creek (TC). Latitudes and locations given in decimal degrees. *Photos: Adam Liedloff.* Map shows rainfall isohyets, and location of sites illustrated in photographs.

Figure 16.2 Subtropical brigalow forests and eucalypt woodlands. (a) Brigalow forest formed by pure *Acacia harpophylla* (photo: Teresa Eyre); (b) box forest dominated by *Eucalyptus populnea* (photo: Teresa Eyre); (c) silver-leaved ironbark forest dominated by *Eucalyptus melanophloia* (photo: Rod Fensham); (d) landscape dominated by regrowth brigalow (*Acacia harpophylla*) (photo: Don Butler); (e) brigalow (*Acacia harpophylla*) forest degraded by fires fuelled by buffel grass (*Cenchrus ciliaris*) seen flowering in the foreground (photo: Teresa Bracknell).

Figure 17.2 Some examples of temperate eucalypt woodlands (a) *Eucalyptus albens* (white box) grassy woodland near Orange, NSW, (b) *E. viminalis* (manna gum) grassy woodland in a grazing paddock near Bombala, NSW, (c) *E. microcarpa* (grey box) grassy woodland in a Travelling Stock Reserve near Parkes, NSW, (d) *E. loxophleba* subsp. *loxophleba* (York gum) forb-rich woodland in the Korrelocking Nature Reserve, WA, (e) *E. clelandii* (blackbutt) woodland with chenopod understorey near Norseman, WA, (f) *E. salubris* (gimlet) woodland with sclerophyll understorey near Marvel Loch, WA, (g) degraded *E. pauciflora* (snow gum) grassy woodland in grazing paddocks of the Tasmanian Midlands, (h) remnant woodlands in a characteristic temperate agricultural landscape (York, WA). *Photos: a–f, h Suzanne Prober; g, Louise Gilfedder.*

Figure 18.3 Examples of Australian tussock grasslands (a) One of the last relicts of subtropical grassland on the Darling Downs, south-east Queensland, with diversity of C_4 grasses and forbs (including Asteraceae and *Wahlenbergia* species) (photo: Rod Fensham); (b) Mitchell grass (*Astrebla* spp.) is an example of an arid and semi-arid tussock grassland in central Queensland (photo: Rod Fensham); (c) livestock grazing can alter vegetation structure, as shown by grazing exclusion experiments in central Queensland, but research suggests that diversity is generally robust under extensive grazing (photo: Rod Fensham); (d) kangaroo grass (*Themeda triandra*) is the dominant tussock grass of many temperate grasslands in high-rainfall areas including this one at Terrinallum, Victoria (photo: John Morgan); (e) where rainfall is lower, as at Avoca Plain, Victoria, kangaroo grass is replaced by chloridoid grasses (e.g. *Enteropogon acicularis, Chloris truncata, Sporobolus caroli*), annual forbs and chenopods (photo: Paul Foreman).

Figure 19.2 Alpine vegetation (a) Fen, Pretty Valley, Bogong High Plains (photo: Keith McDougall); (b) mosaic bolster heath with invading gymnosperm (*Microcachrys tetragona*) at Newdegate Pass, Mt Field Tasmania. Quadrat is 35 cm × 35 cm (photo: Jamie Kirkpatrick).

Figure 19.3 Alpine vegetation. (a) Feldmark between Mt Northcote and Mt Lee, Snowy Mountains (photo: Keith McDougall). (b) Late-lying snow associated with snowpatch vegetation on the south-eastern side of the Main Range near Mt Lee and Mt Northcote, with Club Lake in the foreground, Snowy Mountains (photo: Susanna Venn). (c) A valley of young megaherbs *Pleurophyllum hookeri* (Macquarie megadaisy) recovering after removal of rabbits on Macquarie Island (photo: Kate Kiefer).

Figure 20.2 Range of common Australian wetland plant species highlighting the diversity of wetland plant growth forms: (a) *Ludwigia peploides* subsp. *montevidensis* and *Myriophyllum* species; (b) outer, darker band of *Eleocharis acuta* and inner band of *Pseudoraphis spinescens* illustrate characteristic zones of wetland vegetation; (c) *Typha orientalis;* (d) exotic *Nymphaea mexicana,* Jerrabomberra Wetlands, ACT; (e) floating *Azolla pinnata* with submerged exotic *Egeria densa,* Euston Lakes, NSW; (f) emergent form of exotic *Sagittaria platyphylla* with an inset showing its submerged form, which is just starting to bolt into the emergent form. *Photos: Jane Catford (a, b, c, f: all taken in River Murray wetlands in Victoria and NSW); Jane Roberts (d, e).*

Figure 20.4 Examples of wetland ecosystem types: (a) tropical coastal floodplain, Northern Territory; (b) spring wetland in central Australia; (c) floodplain wetland of the River Murray, Victoria; (d) monoculture of exotic *Urochloa mutica* (para grass) in Kakadu National Park, Northern Territory; (e) groundwater-dependent wetland in south-west Western Australia; and (f) constructed stormwater treatment wetland in Canberra, Australian Capital Territory. *Photos: Michael Douglas (a, d); Jane Catford (b, d, f); Ray Froend (e).*

Figure 21.4 (a) Upper Fitzroy River in the Kimberly region of north-western Australia. Riparian trees are *Melaleuca argentea* and *Terminalia bursarina.* (b) Back swamp area of Magela Creek floodplain in Kakadu National Park with *Melaleuca leucadendra* trees. (c) Remnant *Eucalyptus coolabah* woodland in the northern Murray–Darling Basin. (d) Inundated *Eucalyptus camaldulensis* in the Barmah–Millewa Forest on the Murray River. (e) *Eucalyptus camaldulensis* dieback in the northern Murray–Darling Basin. (f) Dense regeneration of *Eucalyptus coolabah* in the northern Murray–Darling Basin. (g) Floodplain development for agriculture in the Ord River Scheme, Western Australia. *Photos: (a), (b) Neil Pettit; (c), (d), (f) Megan Good; (e) Rhiannon Smith; (g) Google Earth Pro 7.1.7.2600. (November 28, 2015). Ord River, Kununurra, Australia. 15° 37' 20.30'S, 128° 45' 06.64'W, Eye alt 19.75 km. CNES/Astrium 2016.*

Figure 22.3 Saltmarsh diversity around the Australian coastline varies at a range of spatial scales. At the regional scale, the most diverse communities occur in temperate south-eastern Australia, such as (a) French Island, Western Port Bay, Victoria, where *Tecticornia* spp. form a shrub-dominated zone near the lower intertidal mangrove and herbfields, with mosaics of species occupy the upper intertidal zone (photo: Kerrylee Rogers). Where freshwater contributions from adjacent catchments are high, diversity may be locally limited in temperate zones to those species tolerant of brackish conditions, as seen in (b) with monospecific stands of the rush *Juncus kraussii* on the middle reaches of the Hawkesbury River, NSW (photo: Kerrylee Rogers). In subtropical and tropical Australia, species diversity diminishes and intertidal zones landward of mangrove may be dominated by few species. Saltmarsh in (c) at Ukerebagh Island, Tweed River, NSW, is dominated by *Sporobolus virginicus* (photo: Kerrylee Rogers), and at (d) Moreton Bay, Qld, is dominated by *Sarcocornia quinqueflora* (photo: Catherine Lovelock).

Figure 22.4 Some examples of complex root morphology of mangrove in Australia: (a) cone-shaped pneumatophore roots of *Sonneratia alba* in Darwin Harbour, Northern Territory (photo: Kerrylee Rogers); (b) pneumatophores of dwarf *Avicennia marina* in Corner Inlet, Victoria (photo: Paul Boon); (c) exposed cable roots and pneumatophores of a mature *Avicennia marina* at Carama Inlet, Jervis Bay, NSW (photo: Kerrylee Rogers); and (d) complex stilt roots of *Rhizophora stylosa* at Low Isles, north Queensland (photo: Catherine Lovelock).

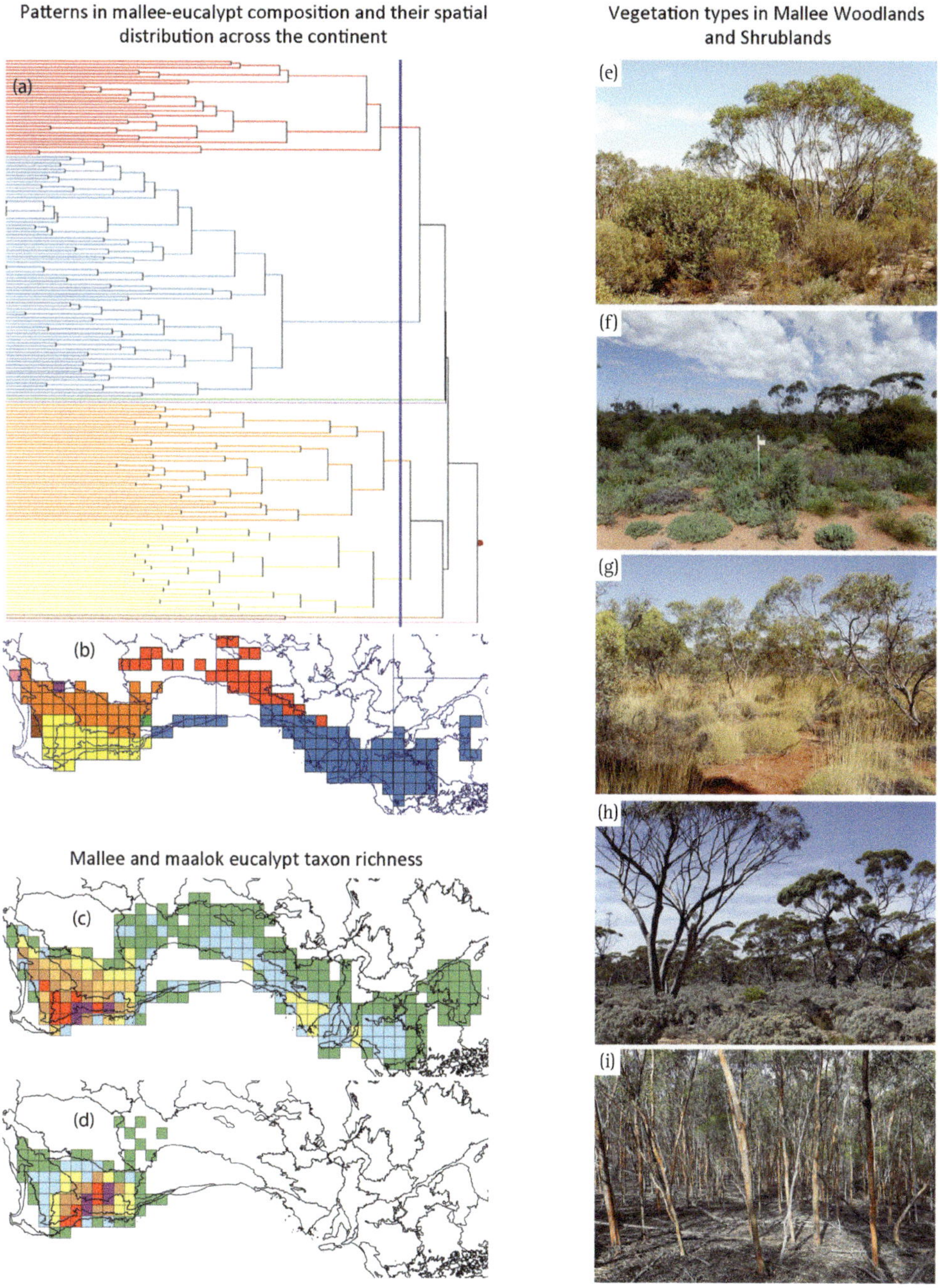

Figure 23.2 Biogeography and spatial variation in richness of terminal *Eucalyptus* taxa with different post-fire responses and major vegetation types in Mallee Woodlands and Shrublands (see Box 23.1).

Figure 24.2 Examples of chenopod shrubland communities. (a) *Atriplex nummularia* (old man saltbush) shrubland on the riverine plain north of Booligal, south-western New South Wales. (b) *Nitraria billardieri* (nitre bush) and *Atriplex vesicaria* (bladder saltbush) with *Autrostipa scabra* on the Olary plain, South Australia. (c) *Maireana pyradimata* on loamy sands, Fowlers Gap, far north-west New South Wales). (d) Dense *Maireana sedifolia* (pearl bluebush) on the dry bed of Lake Mungo, south-west New South Wales. (e) The Paradise total livestock exclosure south of Hay (south-western New South Wales), established 1954 during a wet year, contrasts with an adjoining area that is used to coral travelling stock, where all *Atriplex vesicaria* shrubs have been eliminated. (f) *Maireana sedifolia* persists under sustained grazing, which reduces vegetation cover and compacts soils in the inter-shub spaces, Maralinga, South Australia. *Photos: David Eldridge.*

Figure 25.2 Examples of *Acacia* shrublands/open woodlands in inland Australia: (a) mulga (*Acacia aneura*) shrubland; (b) gidgee (*Acacia georginae*) shrubland; (c) waddy wood (*Acacia peuce*) open woodland; (d) western myall (*Acacia papyrocarpa*) shrubland; (e) bowadga (*Acacia ramulosa*) shrubland; (f) *Acacia undoolyana* shrubland. *All photos: Catherine Nano, except (d) Jen Silcock.*

Figure 26.1 The life forms and reproductive stages of *Triodia basedowii.* (a) Hummock grassland vegetation in the Simpson Desert. Older hummocks in loose rings with darker areas and multiple tufts of dry leaves around the edges. The interplant spaces have an absence of other grasses and forbs and only a biological crust in dry times; (b) large hummock rings in long-unburnt vegetation; (c) a ring of an older plant showing bare area in centre; (d) hummock about 1 m across in full flower; (e) flowering stem; (f) husk (left) and seeds.

Ecological processes within hummock grasslands. (g) Post-fire recruitment of *Triodia basedowii* in the north-east Simpson Desert (SW Queensland). Relatively uniform, smallish hummocks are regenerating after wildfire with bare areas, and some areas of biological soil crust between. Associated shrubs and the crest of a sand dune can be seen towards the back of the image. (h) Tanami Desert 5-years post-fire showing reproductive stems in foreground and woody associations. (i) Cattle grazing in *T. basedowii* hummock grasslands with scattered beefwood, *Grevillea striata* R.Br., Simpson Desert, south-west Queensland. (j) *Triodia* grasslands in the Western Desert with marble gum (*Eucalyptus gongylocarpa*) and *Grevillea juncifolia* Hook. in flower on left. *Photos: Glenda Wardle, except (h) Catherine Nano.*

14 Heathlands and Associated Shrublands

BYRON B. LAMONT AND DAVID A. KEITH

Australia's heathlands are characterised by evergreen shrubs mostly < 2 m tall, with evergreen subshrubs, forbs and graminoids also conspicuous. They occur over a wide range of rainfall/temperature regimes that surround the semi-arid core of Australia, but are best represented under Mediterranean-type climates. Environmental constraints include nutrient-impoverished soils (low concentrations and/or skeletal), hydrological gradients over the landscape often accompanied by strong seasonal drought, and intense recurrent fires. Their evolutionary history can be traced back 100 million years when rainforest clothed much of Australia with local refugia not unlike those conducive to the formation of heaths today. Leaves are typically small, sclerophyllous, long-lived, and highly modified to reduce water loss and heat loads and store nutrients. The heathlands are renowned for their high species diversity at fine scales, landscape turnover and local endemism, with the Proteaceae, Myrtaceae, Fabaceae, Ericaceae, Restionaceae and Cyperaceae accounting for > 50% of biomass and species richness at any site. Most plant species resprout after fire, with some species showing pyrogenic flowering. Most seeds are of limited dispersability, including ant-dispersal, and are either stored in the soil or on the plant and require fire to stimulate their germination or release. Almost all species possess specialised modes of nutrition, especially mycorrhizas and clustered roots. Recent research has focused on understanding mechanisms that explain the high levels of species coexistence: these can be grouped into 11 hypotheses, with the most promising based on small growth forms, large regional species pools and stabilising processes (resource-partitioning) associated with rarity. Land clearing, weed invasion and root pathogens threaten the conservation of Australia's world-heritage heathlands.

14.1 Introduction

In 1770, Joseph Banks and Daniel Solander made plant collections on the shores of Botany Bay, a place within the present-day confines of Sydney, so named to honour one of the world's great botanical discoveries. Some 30 years later, Robert Brown and Peter Good, followed closely by François Péron and Alexander Lesueur, made similarly spectacular collections from their landing places in King George Sound on the south coast of Western Australia. Both collections were remarkable as the first representative inventories of a diverse and novel heathland flora distinguished by vivid if not bizarre floral forms set against harsh and leathery foliage. They were quite unlike the delicate wildflowers present in the meadows of Europe, but somewhat reminiscent of the Cape flora in southern Africa. Although heathlands and associated shrublands account for only about 1.5% of the Australian continent they contain one-quarter of

Australian Vegetation, Third Edition, ed. David A. Keith. Published by Cambridge University Press.

Citation: Lamont, B.B. and Keith, D.A. (2017). Heathlands and Associated Shrublands. In D. A. Keith, ed., *Australian Vegetation*, 3rd edn. Cambridge: Cambridge University Press, pp. 339–67.

its flora. Thus, they deserve special attention from the scientific and conservation points of view.

The distinctive, leathery leaf texture (sclerophylly) of heathland plants, so striking to early European naturalists (Bennett 1834; Diels 1906), provides an insight into the ecological processes that define Australian heathlands. The strong expression of sclerophylly in heathlands relative to other ecosystems is widely attributed to the nutrient poverty of oligotrophic soils and to seasonal drought (Schimper 1903; Beadle 1966; Lamont *et al.* 2002). Feedback between fire and sclerophylly has also been postulated because fires lead to nutrient loss through volatilisation and increased leaching (Orians and Milewski 2007; Tozer *et al.*, Chapter 13, this volume). Sclerophylly is linked to leaf longevity of several years or more, and serves to increase nutrient conservation and resistance to annual drought, heat and herbivory (Witkowski *et al.* 1992). That this relationship holds from tropical to cool temperate climates across the Australian continent suggests a degree of climatic insensitivity. Despite spanning the full latitudinal and altitudinal range on the Australian continent (Figure 14.1), heathlands are best represented in Mediterranean-type climates where they once covered millions of hectares. In arid climates, they are replaced by open shrublands and hummock grasslands (Nano *et al.*, Chapter 25, this volume; Wardle and Nano, Chapter 26, this volume). Conversely, in the most humid climates, local accumulation of soil moisture may cause waterlogging and subsoil anoxia, and heathlands may be replaced by mires. Many Australian forms of mire vegetation retain the sclerophyllous character of heathlands, but also exhibit a range of ecophysiological and

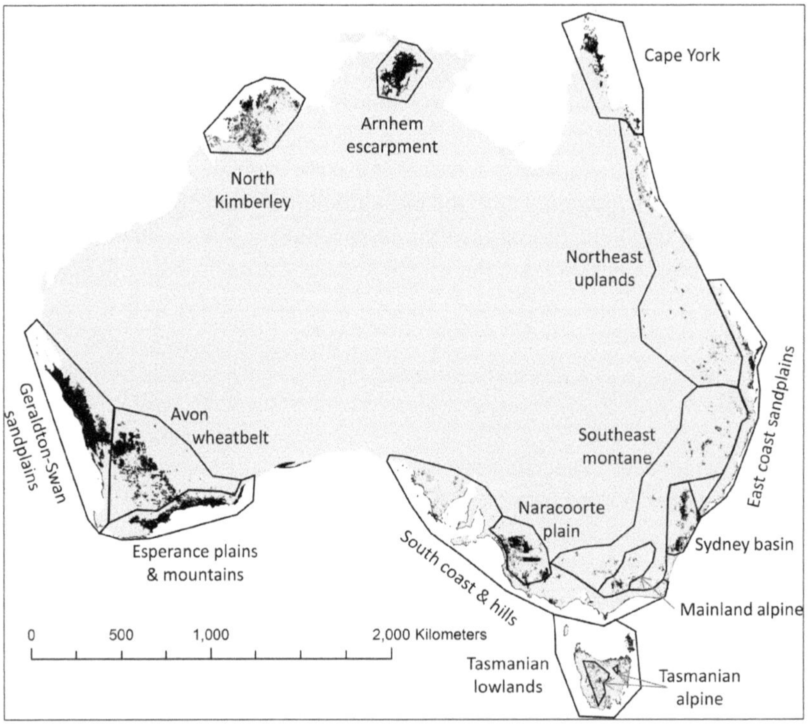

Figure 14.1 Distribution of heathlands and related shrublands in Australia showing 15 biogeographic provinces based on aggregated Australian bioregions. Compiled from data in the National Vegetation Information System (Department of the Environment and Energy 2007; D. Keith).

morphological traits related to oxygen deprivation (Catford and Jansson 2014).

In this chapter, we describe the key features of heathland ecosystems and review the distribution and biogeography of Australian heathlands. Our review highlights the heathlands of south-western and south-eastern Australia as they account for the greatest area and have been most intensively studied. We identify three major environmental drivers that shape selection pressures and the variety of adaptive responses in heathland vegetation: nutrient deprivation; hydrological gradients; and fire regimes. We conclude by discussing the conservation challenges facing these ecosystems by reviewing threatening processes and potential management options.

14.2 Distribution and biogeography

Heathlands are scattered widely around the Australian coast and associated hinterlands and ranges, but are absent from the arid interior and occupy about 1.5% of the continent (Figure 14.1). Fifteen provinces based on groups of related bioregions characterise floristically contrasting heathlands that are geographically distinct and/or differentiated along regional environmental gradients (Table 14.1). Three provinces in south-western Australia account for 75% of the heathland area (Table 14.1). In other regions, they occur as discrete patches within a matrix of eucalypt forest or woodland, although heathlands in rocky tropical landscapes may be interspersed within mosaics of savanna or rainforest (Russell-Smith *et al.* 1998; Mucina and Daniel 2013).

Heathlands occupy remarkably diverse climates that vary from humid tropical north-east Queensland and cool temperate south-west Tasmania, where mean annual rainfall exceeds 3000 mm, to semi-arid south-western Australia receiving as little as 300 mm. Rainfall seasonality varies from a strong winter maximum–summer drought rainfall pattern in the south-west to aseasonal rainfall, though still with summer drought in the mid-latitudes of the south-east, to tropical monsoonal climates with strong summer maxima and winter drought in the north (Table 14.1). Mean monthly minimum and maximum temperatures vary over 13–38°C in the tropics (Katherine) to 3–20°C in cool temperate latitudes (Zeehan) (data from Australian Bureau of Meteorology 2015). For alpine heathlands (see Venn *et al.*, Chapter 19, this volume) these temperature ranges may be as low as -4–19°C (Mt Kosciusko).

14.2.1 Nutrient poverty

Heathlands are associated essentially with oligotrophic landscapes (Specht 1994; Hopper 2009; Mucina and Wardell-Johnson 2011), a feature that distinguishes them from most other ecosystems. Many occur on coarse granitoids, sandstone, lateritic or quartzitic uplands, acid volcanics or (especially) deep leached sands and sands over laterites, although some occur on stony ultramafic or dolerite outcrops or low-lying, deep humic substrates (Table 14.1; Specht 1994). Sometimes it is the restricted access to soil, rather than low-nutrient concentrations, that is the key to nutrient (and water) impoverishment.

The acidity of heathland soils usually varies over pH 5 to 7, but calcareous and serpentine soils supporting distinctive heathland communities may be slightly alkaline. The major nutrients at lowest concentrations compared with other vegetation types are phosphorus (P), nitrogen (N), potassium (K), magnesium (Mg) and calcium (Ca) (Lamont 2003), while limiting trace elements include copper (Cu), zinc (Zn), molybdenum (Mo) and boron (B). P levels range from 1 to 4 mg kg^{-1} in Eneabba heathland (Herath *et al.* 2009) compared with 10 to 25 mg kg^{-1} in heathlands on sand at Myall Lakes (Myerscough and Carolin 1986) and 35 to 75 mg kg^{-1} in wet heathlands on sandstone at Dharawal (Keith and Myerscough 1993). At Eneabba, N varies 3 to 7 mg kg^{-1} on the sandplain and 4–13 mg kg^{-1} on calcareous substrates (Herath *et al.* 2009; Westcott *et al.* 2014). P, followed by N, is most likely to be stored preferentially in seeds (Witkowski

Table 14.1 **Australian heathland provinces (based on bioregions) and features of heathland vegetation within them. Areas of heathland are estimated prior to European land use (Figure 14.1).**

Province	Indicative taxa	Substrates	Rainfall seasonality	Elevation (m)	Estimated area (ha)
Geraldton–Swan plains	*Acacia rostellifera, Anigozanthos manglesii, Melaleuca leuropoma, Stylidium crossocephalum*	Deep sands, laterite, sand over laterite	Winter maximum, summer drought	< 350	3 116 000
Esperance plains and mountains	*Banksia baxteri, Eucalyptus pleurocarpa, Hakea denticulata, Xanthorrhoea platyphylla*	Deep sands, sandstones, granites	Winter maximum, summer drought	< 250, mtns up to 1000	2 018 000
Avon wheatbelt	*Allocasuarina campestris, Melaleuca thyoides, Hakea meisneriana*	Granites, laterite, deep sands	Winter maximum, summer drought	100–350	2 771 000
Naracoorte plain	*Leptospermum myrsinoides, Banksia ornata, Allocasuarina pusilla, Phyllota pleurandroides*	Deep sands	Winter maximum, summer drought	< 200	662 000
South coast and hills	*Allocasuarina paludosa, Epacris impressa, Persoonia juniperina*	Granites, deep sands	Winter maximum, summer drought	< 500	258 000
Tasmanian lowlands	*Allocasuarina monilifera, Leucopogon collinus, Leptospermum glaucescens*	Granites, dolerites	Winter maximum, summer drought	< 700	159 000
Tasmanian alpine	*Abrotanella forsteroides, Microcachrys tetragona, Orites revolutus, Richea scoparia*	Dolerites	Winter maximum	700–1600	69 000
Mainland alpine	*Nematolepis ovatifolia, Orites lancifolius, Prostanthera cuneata*	Granites	Winter maximum	1500–2200	78 000
South-east montane	*Allocasuarina nana, Hakea laevipes, Monotoca scoparia*	Sandstones, acid volcanics	Aseasonal, summer drought	500–1500	108 000
Sydney Basin	*Banksia ericifolia, Eucalyptus stricta, Lambertia formosa, Xanthorrhoea resinosa*	Sandstones	Aseasonal, summer drought	0–1100	114 000
East coast sandplains	*Banksia aemula, Hypolaena fastigiata, Styphelia viridis*	Deep sands	Aseasonal, summer drought	< 100	185 000

Province	Indicative taxa	Substrates	Rainfall seasonality	Elevation (m)	Estimated area (ha)
Cape York	*Asteromyrtus lysicephala, Jacksonia thesioides, Hibbertia banksii, Neofabricia myrtifolia*	Deep sands	Summer maximum, winter drought	< 100	470 000
North-east uplands	*Asteromyrtus lysicephala, Calytrix* spp., *Choriceras tricorne*	Granites, sandstones	Summer maximum, variable drought	100–1000	102 000
Arnhem escarpment	*Asteromyrtus magnifica, Hibbertia brownii, Pityrodia jamesii, Triodia plectrachnoides*	Sandstones	Summer maximum, winter drought	100–400	946 000
North Kimberley	*Acacia gonocarpa, Calytrix achaeta, Calytrix exstipulata*	Sandstones	Summer maximum, winter drought	< 250	235 000
Total					**11 291 000**

and Lamont 1996) and retranslocated efficiently to the developing seedling (Lamont and Groom 2002). However, P is not the only limiting nutrient. In a factorial experiment, Lamont (1995) showed that wild oats (*Avena fatua*) could not survive until P, N *and* K were added to Eneabba sand. Other soils were P *or* N-limited. Heathland plants overcome these limitations by exploiting novel nutrient sources and having efficient uptake mechanisms, for example, through: (i) continual exploration of the soil profile by short-lived fine roots; (ii) extended leaf longevity; (iii) efficient export of leaf nutrients before their abscission; (iv) efficient translocation to the growing points of the plant; (v) preferential accumulation in seeds and efficient retranslocation to the seedlings; and (vi) growth of root systems at the expense of shoots (Groom and Lamont 2015).

Many heathland soils accumulate toxic levels of aluminium (Al), manganese (Mn), silicon (Si) and iron (Fe), either by redeposition in lateritic or acid horizons after leaching or because toxic minerals such as siderite are weathered from rocky substrates (Atwell *et al.* 1999).

14.3 Structure and physiognomy

Heathlands are dominated by sclerophyllous shrubs, typically with co-occurring subshrubs, sedges, forbs and a few grasses (Keith *et al.* 2014). The shrub stratum is commonly up to 1–2 m tall, sometimes with large shrubs to 5 m forming closed thickets, with canopy cover varying from 90% to as low as 10% on rocky sites. In these rocky sites, bryophytes and lichens may be conspicuous. Trees are absent or sparsely scattered emergents, but eucalypts may be common as mallee (multi-stemmed) growth forms. Vegetation size typically varies above and below ground along local hydrological gradients, especially where heathlands grade into peatlands in the Sydney Basin and East coast sandplains (Keith and Myerscough 1993; Myerscough *et al.* 1996) and into low heath in the Geraldton/Swan sandplains (Figure 14.2).

Distinctive physiognomic features of heathland shrubs are their small (leptophyll to nanophyll, Raunkiaer 1934) scleromorphic leaves hardened by high levels of cellulose and lignin in thickened cell

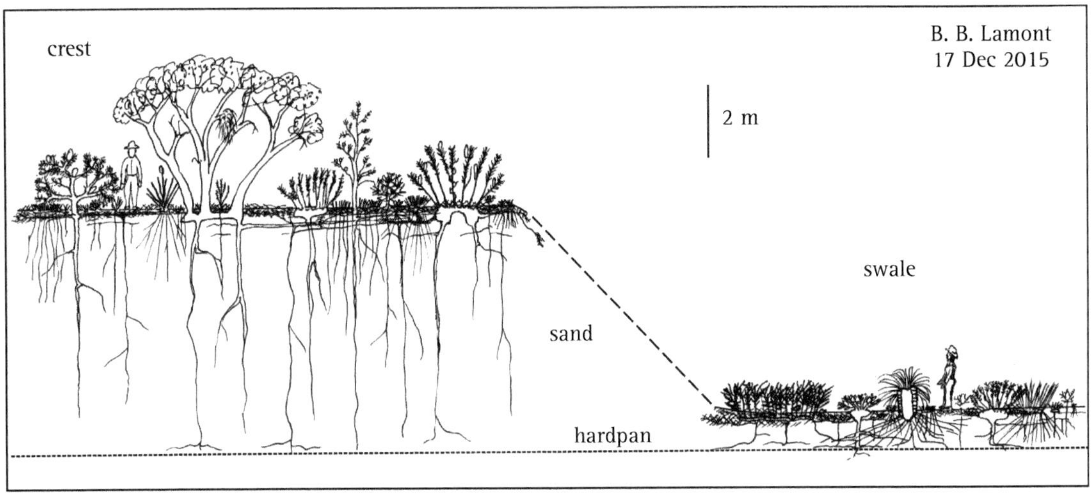

Trait	Dune crest	Swale
Plant stature	taller	shorter
Mallees/tall shrubs	common	rare
Graminoids	less common	more common
Subshrubs	less common	more common
Fire response	more nonsprouters	more resprouters
Leaf size	larger	smaller
Leaf type	broader	narrower
Pubescence	smoother	hairier
Leaflessness	less common	more common
Diallagy (temporary rustic colours over summer-autumn)	no	yes
Root system	deep (up to 5 m to hardpan with larger plants having deeper roots)	shallow (usually < 1 m in vertical plane)

Figure 14.2 Gradients in vegetation and plant traits across dune–swale systems on the Eneabba Plain (20 years after fire).

walls, abundant tannins and sometimes silica granules, and thick cuticles over the epidermis bearing sunken stomates (Groom and Lamont 2015). Leaves are usually thick as well as having a high dry bulk density (Lamont *et al.* 2015). Leafless and phyllodinous species are well represented and even forbs and graminoids may possess sclerophyllous tissues.

14.4 Floristic diversity

Compared with heathland floras of other continents (except the Cape, South Africa and New Caledonia), those within Australia exhibit high diversity and endemism at the species level, with close crowding of many small, perennial species at the local scale, and high species turnover through the landscape (Perry *et al.* 2009; Laliberté *et al.* 2014). Box 14.1 profiles two examples from opposite sides of the continent in which fine-scale species richness is among the highest recorded for shrublands and grasslands worldwide. However, six plant families account for much of the biomass and species-level diversity. These are the Proteaceae (shrubs, mallees), Myrtaceae (shrubs, mallees), Fabaceae (small shrubs), Ericaceae (small shrubs), Restionaceae (wire-rushes) and Cyperaceae (sedges) (Keith *et al.* 2014). Other families, such as Orchidaceae (geophytes) and Stylidiaceae (forbs), are represented by many heathland species in some regions (Hopper and Gioia 2004), while some major families, such as the Asteraceae (shrubs, annuals), Euphorbiaceae (small shrubs), Rubiaceae (shrubs) and Poaceae (grasses), are under-represented in Australian heathlands even though they may be prominent in some regional floras.

Box 14.1 Comparative ecology of heathlands

A comparison of heathlands on either side of the Australian continent shows broad similarities and subtle differences (Table 14.2). Heathlands centred 250 km north of Perth on the Eneabba Plain (Geraldton–Swan Province) occur on deep sands that receive a mean annual rainfall of 500–600 mm, mostly in winter. Those sampled 50 km south of Sydney at Dharawal (Sydney Basin Province) occur on poorly drained peaty depressions over sandstone within a matrix of sclerophyll forest, receiving mean annual rainfall of 1200–1500 mm, with no strong seasonal pattern. The four most speciose families are the same in both and several speciose genera are in common. Both floras are essentially evergreen, while shorter growth forms (< 1.5 m tall) with smaller leaf sizes (< 25 mm^2) are more common at Eneabba. Needle leaves are more strongly represented at Eneabba, whereas strap and ericoid leaves are more frequent at Dharawal. These trait differences may reflect the evolutionary outcomes of stronger seasonal drought at Eneabba.

Plant regeneration is driven by fire in both heathlands. However, serotiny is proportionately more represented at Eneabba, and transient soil seedbanks more so at Dharawal. Pyrogenic flowering occurs in ~10% of resprouting species in both subregions. Lignotubers dominate at Eneabba and rhizomes at Dharawal. Forbs are more common at Dharawal and tufted graminoids at Eneabba. The greater representation of clonal species at Dharawal may reflect limitations on seedling recruitment in dense vegetation that regrows rapidly after fire. Greater use is made of seed dispersal agents at Eneabba, including ants, wind and birds, especially emus. They have a remarkably similar representation of mycorrhizal species, although root clusters are more frequent at Eneabba at the expense of other strategies, except for nitrogen fixation. Some of these differences may reflect phylogenetic legacies, such as the greater radiation of Proteaceae in south-western Australia (Lamont *et al.* 2016).

Box 14.1 (cont.)

Table 14.2 **Comparative plant trait data for heathlands in south-eastern and south-western Australia.**

Trait/attribute	Dharawal (34°14'S 150°54'E) Sydney Basin Province	Eneabba (29°49'S 115°16'E) Geraldton–Swan Province
Sampling	60, 7.5 m^2 plots within 5,000 ha	100, 100 m^2 plots within 105,000 ha
Diversity		
Families/genera/species	59/135/235	49/136/365
Most speciose families	Cyperaceae (11.9%), Proteaceae, Myrtaceae, Fabaceae, Poaceae, Restionaceae, Ericaceae, Orchidaceae, Apiaceae, Dasypogonaceae (2.6%)	Proteaceae (15.6%), Myrtaceae, Fabaceae, Cyperaceae, Haemodoraceae, Ericaceae, Restionaceae, Asparagaceae, Goodeniaceae, Lamiaceae (2.2%)
Most speciose genera	*Schoenus* (3.4%), *Lepidosperma*, *Leptospermum*, *Banksia*, *Acacia*, *Baumea*, *Drosera*, *Epacris*, *Gonocarpus*, *Xyris* (1.7%)	*Banksia* (4.7%), *Acacia*, *Hakea*, *Leucopogon*, *Stylidium*, *Conostylis*, *Verticordia*, *Calytrix*, *Schoenus*, *Grevillea* (2.2%)
Life form (% of species)		
Annual	0.4	4.1
Geophyte	4.7	3.6
Forbs/graminoids (evergreen)	52.3	32.3
Shrubs (evergreen)	40.4	59.7
Trees (evergreen)	2.1	0.3
Plant stature (% of species)		
< 0.5 m	50.2	74.9
0.5–1.5 m	28.1	20.7
1.5–2.5 m	11.5	3.3
2.5–4 m	7.7	0.8
> 4 m	2.6	0.3
Leaf size (% of species)		
Leafless	10.2	11.0
Picophyll	4.7	11.9
Leptophyll	12.8	22.9

Trait/attribute	Dharawal (34°14'S 150°54'E) Sydney Basin Province	Eneabba (29°49'S 115°16'E) Geraldton–Swan Province
Nanophyll	38.3	36.6
Microphyll	24.3	15.4
Mesophyll	9.8	2.0
Leaf type (% of species)		
Elliptic (simple)	20.4	29.0
Compound	0.4	3.3
Dissected	5.1	0
Strap	27.2	16.3
Ericoid	29.8	11.7
Needle	4.7	26.7
Cladode/phyllode	12.3	11.0
Hairy (excluding leafless)	7.8	8.3
Regeneration (% of species)		
Serotiny (canopy-stored seeds)	13.2	20.1
Seed dormancy (soil stored)	72.4	77.7
Non-dormant seeds	14.4	2.2
Resprouters	67.7	73.6
– epicormic or apical buds	4.4	1.6
– lignotuber or basal stems	28.8	62.0
– rhizomes	41.9	12.1
– tubers, corms, bulbs	15.6	6.3
– tillers	9.4	18.0
– pyrogenic flowering	10.0	12.5
Dispersal (% of species)		
Endozoochorous	5.1	10.7
Exozoochorous	1.3	0.3

Box 14.1 (cont.)

Trait/attribute	Dharawal (34°14'S 150°54'E) Sydney Basin Province	Eneabba (29°49'S 115°16'E) Geraldton–Swan Province
Myrmecochorous	20.4	29.7
Anemochorous (winged)	6.8	10.2
Anemochorous (hairs)	4.7	18.4
Anemochorous (microscopic)	9.4	6.0
Unassisted (passive)	52.3	28.6
Nutritional modes (% of species)		
Cluster roots	19.0	27.0
Rhizoids (pteridophytes)	4.3	0
Carnivory	2.6	1.9
Nitrogen fixation	5.5	9.9
Parasitism	1.7	2.5
Mycorrhizas	61.7	61.3

Eneabba plot data and identifications by Vanessa Westcott, Ben Miller, Joe Fontaine and Neal Enright, trait data and collations by Byron Lamont; Dharawal plot data, identifications, trait data and collations by David Keith.

A continental overview of floristic endemism and diversity in Australian heathlands is hampered by lack of a systematic analysis, but virtually all heathland species are endemic to Australia. In south-western Australia the heathland flora displays particularly high levels of local endemism and exclusivity. In heaths of the Eneabba sandplain (Box 14.1), about 65% of 365 species recorded are locally endemic and 95% are endemic to the Geraldton–Swan Province. Hopper and Gioia (2004) identified major centres of local endemism within this region at Lesueur and Stirling Range national parks, with smaller concentrations in the northern Swan Coastal Plain and Fitzgerald River. Other centres of endemism occur in west Arnhem, Sydney sandstone, the mainland Alps and Tasmania (Figure 14.1; Laffan and Crisp 2003), but many of the locally endemic plant species in these heathlands are shared with the sclerophyll woodland matrices in which they are embedded.

Diversity relationships have not been widely examined at the continental scale but richness appears to be highest in the Mediterranean climates of south-western Australia (Lamont *et al.* 1984), followed by humid temperate climates in the south-east (Keith and Myerscough 1993). Both regions have moderate rainfall and temperatures and lithic-derived, often quartzitic, oligotrophic substrates, and also exhibit high rates of species turnover along environmental gradients (Keith and Myerscough 1993; Laliberté *et al.* 2014). Species richness is lower in tropical regions, at high elevations and on skeletal substrates.

14.4.1 Evolutionary history

Heathlands were widely distributed across Australia when rainforests still extended throughout the continent (Weston and Jordan, Chapter 2 this volume). Fossils of sclerophyllous Proteaceae and Ericaceae, with abundant charcoal, provide evidence for heathland species in central Australia 75–65 million years ago (Ma) (Carpenter *et al.* 2015), while fossil Proteaceae and dry-fruited Myrtaceae on the west side of the Eucla Basin date from 40 Ma (Mack and Milne 2015). Some major heathland groups had their origins even earlier, including Restionaceae, arising 100 Ma (Bremer 2002), Haemodoraceae originating 90 Ma (Hopper *et al.* 2009) and three proteoid clades originating 75 Ma (Sauquet *et al.* 2009). Additional heathland taxa appeared 65–50 Ma, including Myrtaceae (*Eucalyptus*, Melaleuceae, Chamelaucieae) (Crisp *et al.* 2011), Proteaceae (*Banksia*, Conosperminae) (Sauquet *et al.* 2009; He *et al.* 2011) and *Actinostrobus* (Cupressaceae; Mao *et al.* 2012). Some of the earliest species in these groups may have begun in sclerophyll woodlands and forests and speciated progressively to form heathlands when growing conditions deteriorated and suitable (infertile) substrates expanded. Heathland lineages continued to emerge 50–20 Ma (Hopper *et al.* 2009; Sauquet *et al.* 2009), with speciation peaking as Australia became drier 20–5 Ma. During glacial maxima of the Pliocene and Pleistocene, heathlands were temporarily displaced at some sites by chenopods, daisies and grasses (Dodson and Macphail 2004) or by shifting sand-sheets and bare rock (Glassford and Semeniuk 1995). The widespread extinction of heath species in south-west Victoria < 1 Ma was attributed to glacial cold and aridity (Sniderman *et al.* 2013), but in south-west Australia prolonged episodic drought may instead have promoted speciation of some heathland lineages (Lamont *et al.* 2016).

14.4.2 Coexistence mechanisms

Research into the mechanisms that maintain local species diversity in heathlands has generated a number of explanatory hypotheses. These have been explored most extensively in the species-rich heathlands of the Eneabba sandplain. No single mechanism unconditionally explains the high levels of species coexistence present there (Table 14.3).

Table 14.3 **Mechanisms sustaining high species diversity in heathlands. Hypotheses specifically applicable to heathlands given in bold.**

	Hypothesis	Mechanism	Support and limitations
1	**Species packing**	Small plant size enables many species to coexist in a small area	High densities of individuals and low number of individuals per species. Does not explain why coexisting individuals belong to different species (Lamont *et al.* 1984; Perry *et al.* 2008)
2	**Resource scarcity**	No one species can dominate when nutrients (and water?) are scarce, hence competitive exclusion is limited	Species richness is inversely related to fertility, congeners tolerate rather than avoid neighbour interactions (Lamont *et al.* 1984)
3	**Seed dispersion**	Contrasting seed dispersal vectors (e.g. ants, post-fire winds) result in disparate spatial dispersion of seedlings, reducing neighbour interactions	Empirically untested, but evidence that density-dependent mortality occurs in post-fire seedlings (Lamont *et al.* 1993; Enright *et al.* 2007)

Table 14.3 (cont.)

	Hypothesis	Mechanism	Support and limitations
4	Fire frequency fluctuations	Disparate fire regimes in different parts of the community or landscape favour different suites of species dependent on their fire responses and seed storage traits.	Recurrent short or long fire intervals favour different suites of species so reduces species diversity at fine scales but increases it at the community and landscape scales provided a mosaic is maintained (Keith and Tozer 2012)
5	**Lottery recruitment**	Recruitment success is a spatial and temporal lottery with post-fire conditions taking turns to favour different species	Variable recruitment success demonstrated for a few congeners, but there are limits to tolerance of variation in fire frequency and observed fluctuations are less than expected (Lamont and Witkowski 1995; Groeneveld *et al.* 2002)
6	Storage effect	Species with regenerative organs or persistent seed banks recruited under favourable antecedent conditions persist through multiple fires	Resprouters persist through multiple fires with stable populations, but not applicable to many species (Chesson and Huntley 1988)
7	Compensatory survival	Survival rates are positively related to rarity (stabilising processes)	Empirical support is limited to habitat and experimental patches, but mechanism unknown (Lamont *et al.* 1993; Lamont *et al.* unpublished)
8	Trait divergence and resource partitioning	Trait divergence promotes coexistence through exploitation of resources from different sources or at different times	Trait trade-offs explain some but not all species diversity in modelled systems (Esther *et al.* 2011)
9	Neutral theory	Random recruitment of competitively equal species	Models based on neutral theory fail to simulate sigmoid species-abundance curves observed in heathlands (Groeneveld *et al.* 2013)
10	Spatial mixing and clumping	Spatial mixing reduces intraspecific competition and maximises facilitation; spatial clumping reduces interspecific competition but promotes self-thinning	Most heterospecific nearest neighbour pairs are explained by stochastic geometry, but conspecific clumping exceeds random expectations (Perry *et al.* 2014)
11	Metacommunity – species pool and immigration	Heathlands are metacommunities whereby immigration into patches from a species-rich regional pool counterbalances local extinctions	Community simulations suggest this mechanism is feasible, but dispersal kernels are poorly known and hypothesis does not explain high initial richness or why the regional pool is large (Esther *et al.* 2008; Perry *et al.* 2009; Laliberté *et al.* 2014)

Four possible coexistence mechanisms are particularly relevant to heathlands (Table 14.3): (i) high species diversity is possible at fine scales because of the prevalence of diminutive or low life forms that can be packed into small spaces; (ii) resource scarcity coupled with divergence in resource acquisition traits is evident in soils with low-nutrient levels and variable soil moisture (Groom and Lamont 2015); (iii) limited but disparate seed dispersal methods may lead to a scattering of microsites bearing different propagules; (iv) fluctuating fire frequencies act like a lottery, favouring different suites of species dependent on their fire responses and seed storage traits (see 14.6 Fire responses for more details). More general models include spatial structuring, functional trade-offs and lottery processes that play minor roles in understanding species co-occurrence. Stabilising processes, involving compensatory responses to increased density and rarity at the recruitment stage, may be important but await rigorous field testing (Table 14.3; Figure 14.4). Ultimately, the issue will only be resolved by understanding the evolutionary and geological history of Australian heathlands.

The size of the species pool at landscape and regional scales involves differential adaptive responses to gradients in hydrology, fertility, other substrate features and climate set against constraints on adaptive variation, phenotypic plasticity and dispersability (Perry *et al.* 2009). Among the most important of these are local hydrological gradients that exist within sandplain dune–swale landscapes (Myerscough *et al.* 1996) and on lithic plateaus (Keith and Myerscough 1993). Heathland plant communities spanning these gradients exhibit extremely high turnover in species composition, physiognomy and functional traits (Figure 14.2), illustrating their profound role in structuring the high species diversity for which heathland landscapes are renowned.

14.5 Plant traits

14.5.1 Life history

There are four life-history types in Australian heathlands: fire-killed (seeders, non(re)sprouters), fire survivors (resprouters), perennial fire avoiders (geophytes) and fire ephemerals. Fire ephemerals, such as short-lived legumes (e.g. *Kennedia*, *Acacia pulchella*), Gyrostemonaceae (e.g. *Tersonia*) and some geophytes (e.g. *Drosera*), are opportunists that take advantage of the increased resource availability after fire but generally persist above ground for only a few years. At the other extreme, resprouters may live for hundreds of years, especially clonal species. Thus, individuals of the rhizomatous *Banksia candolleana*, whose clones may eventually attain a diameter of 10 m, are estimated to reach an age of 1200 years (Merwin *et al.* 2012). In between are non-sprouters that begin to senesce after approximately 40 years, e.g. *Banksia hookeriana* (Enright *et al.* 1996), though invariably they are burnt before this age is reached.

Typically, resprouters make up 40–80% of local heathland floras (Keith *et al.* 2002; Box 14.1) with the majority of genera consisting of a mixture of fire-killed and fire-surviving species. There is conflicting support among the most speciose genera (> 100 species) for the hypothesis that seeders evolve more rapidly than resprouters due to their more rapid generational turnover (Pausas and Keeley 2014). Some speciose genera, such as *Acacia*, *Leucopogon*, *Stylidium*, *Grevillea* and *Baeckea*, are largely fire-killed, but others such as *Eucalyptus*, *Melaleuca*, *Caladenia* and *Hakea*, are mostly resprouters or (rarely) geophytes. This gives support for the alternative proposition that somatic mutations among long-lived resprouters may enable their rate of speciation to keep pace with fire-killed species (Lamont *et al.* 2011).

14.5.2 Leaves

Leaves show many features that suggest adaptations to nutrient poverty, recurrent drought/heat and herbivory/florivory. Most species have small leaf sizes or develop alternative photosynthetic structures (Box 14.1) that reduce water loss. Another feature is the vertical orientation of needle and isobilateral leaves that reduces their heat load. Many heath species have spinescent leaves which, when clumped, protect new axillary growth and cryptic inflorescences and fruits from marsupials, emus and cockatoos. In these cases, sclerophylly serves to protect plant parts other than the leaves themselves, i.e. they have secondary functions (Hanley *et al.* 2007, 2009).

Low availability of both nutrients and water contribute to the strong development of sclerophylly (Lamont *et al.* 2002). The high-density component of scleromorphic leaves seems to be best-associated with low nutrients and the high-thickness component to seasonal drought/heat (Groom *et al.* 2004; Lamont *et al.* 2015). *Hakea*, the most sclerophyllous genus in the world (Figure 14.3), has both properties and several other features that suggest strong selection for conservation of nutrients and water (Lamont *et al.* 2015). Internally, the thick mesophyll serves to store water, osmotic substances (e.g. tannins, that also reduce macropod herbivory, Rafferty *et al.* 2005) and nutrients. The fibrous tissue wrapped around the vascular bundles, sunken stomates and thick, hydrophobic cuticles all restrict transpirational water loss (Figure 14.3). Experiments on heterophyllous species confirm that needle leaves conserve more water but at the expense of photosynthesis per unit mass compared with broad leaves (Groom *et al.* 1994). Other traits that promote resistance to drought and heat include coatings of resin or wax that increase light reflectivity of leaf surfaces, and dense indumenta that also limit surface airflow and hence transpiration (Groom and Lamont 2015).

The primary evolutionary pressure that promotes sclerophylly is nutrient poverty as severe leaching of the substrate, without landscape rejuvenation, during the Cretaceous/Paleogene preceded drought in the Neogene (Beadle 1966; Glassford and Semeniuk 1995; Orians and Milewski 2007; Groom and Lamont 2015). Although uptake mechanisms are efficient, a large fraction of absorbed inorganic nutrients is directed to the seeds (Groom and Lamont 2010)

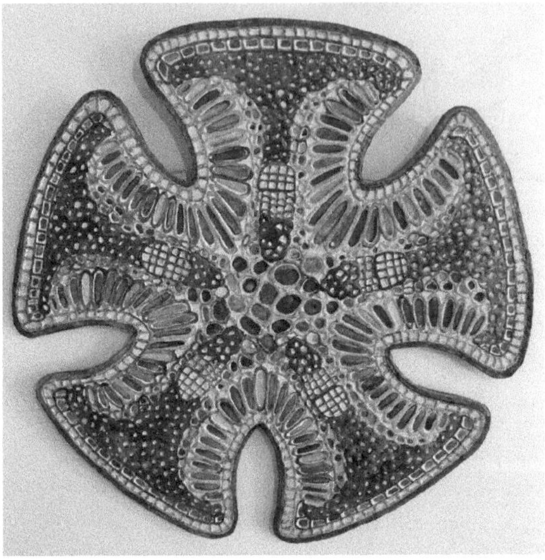

Figure 14.3 Impression of leaf transection of *Hakea invaginata* showing adaptations to drought tolerance among sclerophyllous heath plants. Five crypts are bounded by tannin-bearing, photosynthetic tissues to which the sunken stomates are confined. These are surrounded by wedge-shaped lignified fibrebands that replace the palisade beneath the high light, external surfaces and enable the crypts to close to prevent further water loss without the leaf collapsing when turgor is lost. The wedges are topped by a tannin-filled hypodermis (reduces light interception) with a cuticle as thick as the epidermis. The vascular bundles are protected by two fibrebands. Large tannin-filled cells storing water and osmotic substances make up a parenchymatous core. The leaves are ~15 cm long, 2 mm diameter, vertically oriented and survive ~3 years. Section stained with toluidine blue: cellulose shows in green, lignin in blue, tannins various colours including magenta, xylem square red cells. *From painting by Byron Lamont. (A black and white version of this figure will appear in some formats. For the colour version, please refer to the plate section.)*

because of their critical role in successful recruitment. This creates a shortfall of nutrients essential for photosynthesis (Denton *et al.* 2007), while leaf turnover involves nutrient wastage due to incomplete retranslocation of mobile nutrients, such as P, N and K, and negligible retranslocation of immobile nutrients, such as Ca and Fe. This promotes selection for increased leaf longevity beyond the current growing season (evergreeness). In these circumstances, selective pressure against non-sclerophyllous species is hypothesised to arise via two risks associated with increased longevity: losses to environmental extremes (drought/heat/abrasion) and herbivory. Both problems are mitigated in species with high rates of production of polymers of CH_2O (cellulose, hemicellulose and lignin), that underpins the resistance of sclerophyllous foliage to drought and herbivory but does not carry an associated cost for inorganic nutrients, especially P (Hanley *et al.* 2007; Lamont *et al.* 2015). Tannins, strongly associated with sclerophylly (Read *et al.* 2009), are essentially polymers of CHO, and may also serve to reduce palatability (Parsons *et al.* 2006).

As a result, leaves of many heath species, except geophytes and ephemerals, survive for 12 months or more and may even exceed 5 years (Pate *et al.* 1984; Witkowski *et al.* 1992). As an extreme example, *Banksia petiolaris* has leaves that survive for up to 13 years and their leaf-mass-area increases with age. In turn, mobile nutrients are retranslocated efficiently out of the leaves to the growing points before they finally die (Witkowski and Lamont 1996). Thus, 86% of K and 75% of P are removed from senescent/dead leaves of *B. hookeriana* (Witkowski and Lamont 1996).

14.5.3 Root systems

In temperate heathlands, survival of the first summer drought may be a function of the depth to which the seedling's root system can reach before summer so that species with larger seeds (up to 20 mg) that store more mobile nutrients are more likely to survive (Groom and Lamont 2015). Thus, a 15 cm tall seedling can have a >200 cm long taproot by mid-summer at Eneabba (Enright and Lamont 1992). Drought-resistant mechanisms among young plants with short root systems include physiological tolerance via strong osmotic adjustment or ability to survive several months in a wilted condition (Richards and Lamont 1996). The enhanced survival of young plants in species mixture relative to monoculture can be explained by differentiation of root distributions and thus resource partitioning at the micro-scale (Figure 14.4), hence promoting species diversity (hypothesis 8, Table 14.3).

Three features of adult root systems are notable in heathlands: (i) the predominance of fine root mass in the uppermost 15 cm of the soil profile; (ii) the presence of massive rootstocks (lignotubers) in the uppermost 30 cm; and (iii) the wide horizontal and vertical extent of the major supporting roots (Figures 14.2 and 14.5).

Most temperate heathlands in Australia are dominated by species of Proteaceae and other species that produce mats of root clusters in the uppermost 3–15 cm of soil. In winter, these roots grow into the decomposing litter where N, P, Ca, Mg and other scarce nutrients are most available (Lamont 2003). The clusters of rootlets acidify the rhizosphere and thus accelerate the decomposition process (Brundrett *et al.*, Chapter 7, this volume). The fine roots are attached to laterals that may extend well beyond the plant crown. For the site reported in Figure 14.5, three *Banksia* species accounting for most biomass averaged a height of 1.6 m and width of 1.1 m but had 59 m of laterals on average reaching up to 7.3 m away from the plant with 20 (vertical) sinker roots attached to the laterals attaining a depth > 2.5 m (Lamont and Bergl 1991). These sinkers supplement the role of the taproot in reaching soilwater stored at depth, critical over summer–autumn (Groom and Lamont 2015). In old lignotuberous plants, the

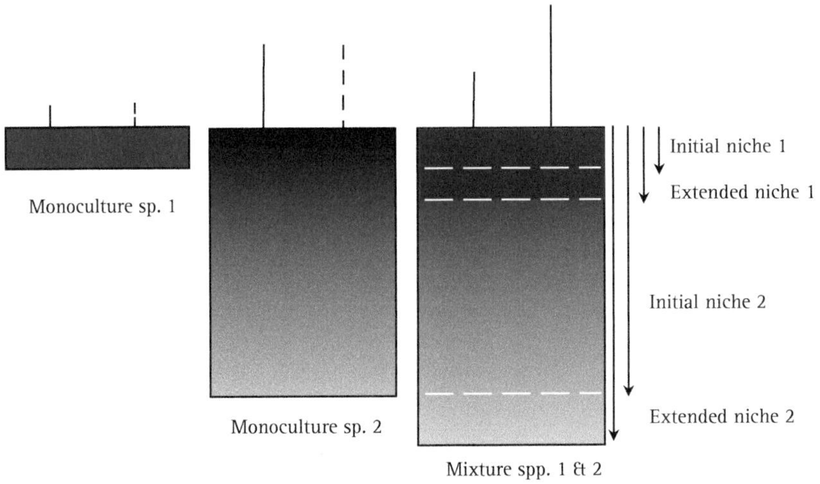

Figure 14.4 Resource-partitioning model that can explain how two post-fire-recruiting heath species of contrasting size (root system distribution) perform better in mixture than monoculture at the within-community scale. Intensity of grey is equivalent to root density. Length of vertical lines corresponds to plant size. Broken vertical line = dead plant. Extended niche = greater volume of soil that can be explored by plants that are not competing for resources in the same part of the profile (and hence the surviving plants reach a greater size) (B. Lamont).

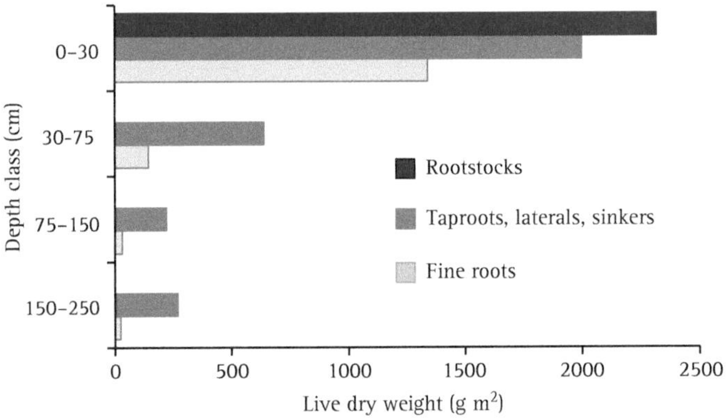

Figure 14.5 Mean distribution of live underground parts through the soil profile for four 8-m² plots in scrub-heath on a dune crest at Eneabba (Figure 14.2). Drawn from data in Low and Lamont (1990).

central rootstock decays and the taproot is lost so that, octopus-like, the plant relies entirely on sinker roots for its dry season water supply.

Depth and width of the root system is a function of plant size among shrubs and mallees (Figure 14.2). Thus, small plants may cease transpiration (and nutrient uptake) in the dry season becoming anabolically inert, even losing chlorophylls under stress until reconstituted in the same leaves during the next wet season. Graminoids and grasstrees have fibrous roots that may sometimes enable shoot growth to continue over the dry season (Lamont and Bergl 1991; Lamont *et al.* 2004). With their root systems confined to the humus-rich surface horizon, geophytes die back in the dry season to a dormant corm or tuber cluster that retains moisture, metabolites and mineral nutrients, enabling rapid return to growth at the start of the wet season, especially after fire (Pate and Dixon 1982).

14.5.4 Specialised modes of nutrition

Lamont (1982) first drew attention to the wide range of nutrient uptake mechanisms in the nutrient-impoverished heathlands of south-western Australia and South Africa, but much more is now known about the mechanisms and species with these traits (Groom and Lamont 2015; Brundrett *et al.*, Chapter 7, this volume). The diverse range of specialised mechanisms represented in heathlands include (Box 14.1) root clusters (especially Proteaceae and Cyperaceae), rhizoids (pteridophytes), carnivory (*Drosera, Utricularia, Cephalotus, Byblis*), nitrogen fixation (Fabaceae, Casuarinaceae), parasitism (*Cassytha, Pilostyles, Cuscuta, Nuytsia, Olax*) and mycorrhizas (many families). A relatively high fraction of Australian heathland species deploy nonmycorrhizal mechanisms (Box 14.1) compared with other plant communities (Brundrett *et al.*, Chapter 7, this volume), and a few have developed multiple specialisations (Figure 14.6).

Figure 14.6 Part of the surface root system of the shrub legume, *Viminaria juncea*, that occurs in waterlogged shrublands in southern Australia. Three nodules are visible plus two root clusters (the larger is 5 cm long) and a vertical pneumatophore arising from the smaller root cluster that assists oxygen uptake (B. Lamont).

14.5.5 Sexual reproduction and dispersal

Some flowering occurs throughout the year in heathlands, but Australian heathlands are renowned for their spectacular displays of flowers in spring (Figure 14.7). This coincides with high demand among insect pollinators and actively breeding bird and mammal pollinators. Most geophytes flower in winter-spring, as do some large shrubs such as *Banksia ornata* and *B. ericifolia* that produce abundant nectar, while those that can still access water at the warmest time of year flower in spring–summer (e.g. *Xylomelum angustifolium*) or summer–autumn (e.g. *Banksia robur*). Irrespective of flowering time, many plants release their seeds via desiccation processes over summer, unless they are retained within serotinous woody fruits (see Fire responses section).

Other special features of flowering in Australian heathlands include: (i) protandry – anthers maturing before the stigma as an outbreeding device (Groom and Lamont 2015); (ii) flower colour change (usually yellow to red) during maturation, that promotes pollination efficiency (Weiss and Lamont 1997); (iii) species with high pollinator-specificity, e.g. bee-pollinated verticordias, wasp-pollinated orchids (Groom and Lamont 2015); (iv) abundance of species pollinated by birds or mammals, rather than wind; (v) presence of poisonous compounds in exposed flowers and protection of concealed flowers by spiny foliage (Hanley *et al.* 2009);

(a)

Figure 14.7 (a) Scrub-heath in south-western Australia on a dune crest on the Eneabba sandplain (see profile, Figure 14.2) in winter 14 years after fire. 113 vascular plant species were recorded in a 40 × 40-m plot at this site (Enright *et al.* 2007). The vegetation pictured is dominated by Proteaceae: serotinous banksias – *B. menziesii* in flower (left, 1.2 m tall), *B. hookeriana* in flower (right), *B. attenuata* with cones (mid-centre, also inset: flower head 15 cm tall). Insets: *Daviesia epiphylla* (left); *Banksia attenuata* (right). *Photos: Byron Lamont.* (b) Wet heath (or Restioid heath) in south-eastern Australia on a poorly drained sandstone plateau in Royal National Park (near to Dharawal Nature Reserve; Box 14.1) in early spring 5 years after fire. A total of 75 vascular plant species were recorded in a 20 × 20-m plot at this site (Keith, unpubl. data). The vegetation pictured includes *Xanthorrhoea resinosa* (left foreground), *Banksia ericifolia* (centre) and *Dillwynia floribunda* (yellow flowers throughout). Insets: *Banskia ericifolia* (left); *Dillwynia floribunda* and *Sprengelia incarnata* (right). *Photos: David Keith. (A black and white version of this figure will appear in some formats. For the colour version, please refer to the plate section.)*

and (vi) occurrence of fire-stimulated flowering among geophytes and some shrubs (Lamont and Downes 2011; Box 14.1).

Heathland plants display a range of dispersal mechanisms, although there is probably higher reliance on ants and passive dispersal (Box 14.1), resulting in short dispersal distances relative to other ecosystems, including heathlands overseas. However, spatial patterns in the distribution of alleles in seeds and seedlings suggest that some species disperse further than expected from their principal vectors alone (He *et al.* 2004, 2009). This has highlighted the importance of low-frequency dispersal events involving vectors such as updrafts, cockatoos and emus.

14.6 Fire responses

Accumulation of highly combustible biomass at ground level, together with periods of hot, dry weather and exposure to lightning strikes associated with regular storms, make heathlands strongly prone to fire (Miller and Murphy, Chapter 5, this volume). Thus, fire plays a crucial role in the dynamics of heathland ecosystems, as a selective agent in the evolution of key adaptive traits, as an instigator of temporal variation in resource availability, and as a mediator of interspecific interactions (Keith *et al.* 2002; He *et al.* 2011; Groom and Lamont 2015). Fire has been present throughout the long history of Australian heathlands, as shown by the presence of charcoal derived from sclerophyll shrubs 75 Ma (Carpenter *et al.* 2015). The ancestors of heathland taxa with key fire-related traits, such as below-ground resprouting, serotiny, fire-enhancing traits, and fire-stimulated flowering and germination, can be traced back to 100 Ma using molecular techniques (Figure 14.8). Among Proteaceae, Petrophileae, a clade of exclusively serotinous taxa, can be traced to 73 Ma, while *Conospermum*, a genus with smoke-stimulated germination, originated 71 Ma (Lamont and He 2012). The origin of clades with fire-adapted traits peaked in the Eocene (usually considered too wet for fire) and especially the Miocene, a time when drought and seasonality (and thus fire) were increasing markedly.

Many species in Australian heathlands exhibit post-fire resprouting from underground bud sources – lignotubers, rhizomes and lateral roots. These include examples of obligate resprouting, in which post-fire recruitment from seeds is negligible, and reproduction is primarily clonal. Many of these species flower profusely but set few seeds, suggesting that deleterious somatic mutations may be responsible for low seed set in resprouters (Lamont and Wiens 2003; El-ahmir *et al.* 2015). Despite their low reproductive effort and slow growth rate compared with fire-killed species (Pate *et al.* 1990; Lamont *et al.* 2011), the fitness advantages of resprouters revolve around their greater tolerance of extreme climates and conditions less favourable for seed set or seedling recruitment. Thus, within *Banksia*, more resprouters occur in the hotter, drier and more fire-prone sandplains of the Geraldton–Swan plains Province than under the milder climates of the southern Esperance plains and mountains Province where seedling recruitment is more likely (Figure 14.1; Lamont and Markey 1995).

Non-sprouters are highly sensitive to fire frequency because successive fires separated by short intervals may reduce the accumulation of seedbanks upon which their populations rely for persistence. For the Eneabba Plain, a minimum fire interval of 7 years is required for reproduction to commence among fire-killed species with soil-stored seeds, extending to 12 years with a 20% reduction in rainfall, while serotinous species require 10 and 14 years, respectively (Enright *et al.* 2015). Resprouters show less short-term response to variations in fire interval or rainfall, but may be more sensitive to successive short fire intervals (Keith *et al.* 2007a). Since adult mortality increases with age, seedling recruitment, even among resprouters, is increasingly required to restore species richness over longer fire intervals.

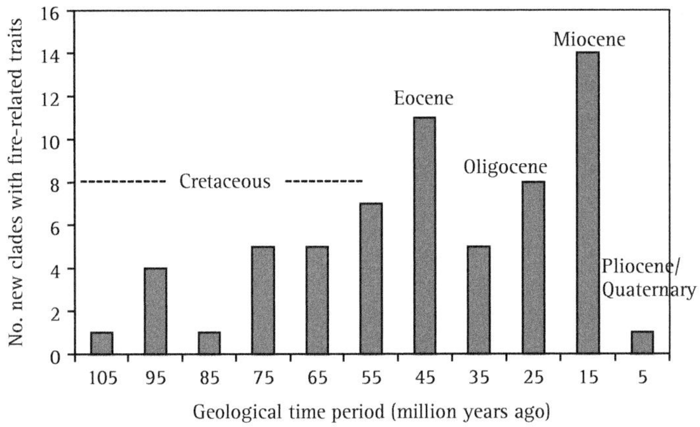

Figure 14.8 Time of origin of clades possessing fire-related traits in Australian heathlands. The 62 records cover fire-enhancing traits, fire-stimulated flowering and germination, below-ground resprouting and serotiny. *Collated from references cited in Groom and Lamont (2015).*

Australia's heathlands have the world's greatest concentration of serotinous species. These belong to Proteaceae, Myrtaceae, Casuarinaceae and Cupressaceae. Heathland plants are generally more strongly serotinous than conspecifics in woodland, as fire is more likely to reach the crown (as plants are shorter) to induce seed release and as seedling recruitment increasingly becomes restricted to the post-fire environment under drier conditions (Cowling and Lamont 1985). Serotiny has two components: (i) seed storage on the plant; and (ii) seed release in response to fire heat (pyriscence) (Lamont *et al.* 1991). It is a specific adaptation to fire because its fitness benefits can only be expressed in the presence of fire (Lamont and Enright 2000). For example, minimum temperatures required to open 50% of follicles varies among banksias from 145°C (*B. tricuspis*) to 390°C (*B. hookeriana*) (Enright and Lamont 1989). Drought is neither as efficient a releaser of seeds as fire nor as effective in seedling recruitment (He *et al.* 2015; Causley *et al.* 2016).

All taxa that release their seeds at maturity remain dormant and viable in the soil, except *Nuytsia*, *Macrozamia*, *Xanthorrhoea* and most geophytes (Lamont and Downes 2011). Thus, there may be 200–1400 seeds m^{-2} in south-western Australian heathlands (Groom and Lamont 2015) and species with soil-stored seeds account for 71–82% of all species at Eneabba (Enright *et al.* 2007). In contrast to the restricted taxonomic spread of serotiny, at least 85 families in Australia's heathlands possess genera with soil seed storage, with some stimulated to germinate by heat shock, some by smoke and some by both (Auld and O'Connell 1988; Keith 1997; Roche *et al.* 1997). The oldest known clade with soil seed storage, that is essentially smoke sensitive, is the graminoid heathland family complex, Restionaceae–Centrolepidaceae–Anarthriaceae, originating 101 Ma, confirming that both heathland species and fire have had a remarkably long history in Australia (Bremer 2002; B. Lamont, unpublished).

Another feature that distinguishes heathlands from other vegetation, especially in south-western Australia, is the retention of dead material on the plant in numerous species of Proteaceae, Cupressaceae, Xanthorrhoeaceae and Cyperaceae

(Low and Lamont 1990). Retention of dead leaves ensures that heat reaches serotinous fruits and stimulates rejuvenation, including pyrogenic flowering (Lamont *et al.* 2004), although it is more common among non-sprouters. Retained dead leaves are also an important source of post-fire N, P and (especially) Ca (Witkowski and Lamont 1996). Some banksias also retain spent flowers on serotinous cones (He *et al.* 2011). These features coevolved with serotiny from 61 Ma and, by igniting during fire, ensure that sufficient heat is received by the follicles to melt the resin sealing their valves (Lamont and Cowling 1984). Heathland fuels generally recover quickly from fire, especially where resprouters dominate their floras (Keith *et al.* 2002; Westcott *et al.* 2014). Accumulation of fine fuels may approach an asymptote as early as 5 years post fire, or may continue for as long as 30 years in heathlands dominated by large fire-killed shrubs (Keith *et al.* 2002).

Fire plays a critical role in mediating interspecific competition in heathland communities that may be dominated by high densities of large fire-killed shrubs. In these systems, a dense shrub canopy, that may develop one to two decades after fire, suppresses standing plants of smaller species, reducing their rates of growth, reproduction and survival (Keith and Bradstock 1994). Slow-growing species, typically woody resprouters, are most susceptible to these effects, especially those with transient or serotinous seedbanks because they have no means of population persistence when standing plants are eliminated by competition (Keith *et al.* 2007b). In contrast, small rapidly maturing obligate seeders evade competition by accumulating persistent soil seedbanks before canopy closure, and these provide a basis for re-establishment of standing plants after the next fire. Thus, fire frequency mediates coexistence in these systems by regulating the density of the serotinous, fire-killed dominants, with fire intervals of < 5 years promoting a switch from dense thicket to open heath and recurrent intervals of 15–25 years promoting replacement of open heath by thicket, both extremes resulting in reduction in species richness at the local scale (Keith and Tozer 2012). These fire-mediated dynamic heathland/thicket mosaics are known from the following provinces (Figure 14.1): Sydney Basin (Keith *et al.* 2007b); Esperance plains (Bond and Ladd 2001); South coast and hills (O'Loughlin *et al.* 2015), and East coast sandplains and Naracoorte plains (Specht and Specht 1989).

14.7 Conservation and management

The major threats to heathland biodiversity include land clearing, changes to fire regimes, eutrophication, plant invasions, plant pathogens and climate change (Keith *et al.* 2014; Keith and Auld, Chapter 27, this volume). Clearing is primarily a threat where heathlands occur on flat land suitable for agriculture, over mineral deposits such as heavy minerals, coal or bauxite, or in urban and peri-urban settings around the coasts where infrastructure related to housing, transport, services, tourism and industry is rapidly expanding. The three south-west Australian and Naracoorte plain provinces (Figure 14.1) have suffered the greatest losses of heathland, with an estimated 4.6 million hectares cleared nationally for agriculture since European settlement (Keith and Auld, Chapter 27, this volume). Heathlands in the East coast sandplains, South coast and hills and Tasmanian lowlands are most at risk from clearing associated with coastal development. Remaining heathlands in landscapes that have been fragmented by clearing may be further threatened by disruption to mating systems (Llorens *et al.* 2012), plant invasions (Milberg and Lamont 1995) and fire exclusion (Parsons and Gosper 2011).

Representation in conservation reserves is the most effective action for protection against land clearing. Heathlands are comparatively well represented in the reserve estate, although given high levels of beta diversity this may be insufficient to protect the full range of heathland biodiversity. Several heathland communities in south-western Australia are listed as threatened under Commonwealth legislation that requires approval for developments that may cause significant impacts.

Keith *et al.* (2014) identified five fire regimes that may threaten heathlands through contrasting mechanisms. The best known of these involves the interruption of plant life-history processes by high frequency fires, which are particularly problematic in the Arnhem and North Kimberley provinces (Russell-Smith *et al.* 1998), as well as temperate coastal heathlands near human sources of ignition (Bradstock *et al.* 1997). The reverse process involving very low fire frequencies occurs in fragmented landscapes, resulting in loss of heath biota through senescence or transition to other vegetation types (Keith *et al.* 2007b; Baker and Catterall 2015). Other potentially detrimental fire regimes include those dominated by low-intensity, cool-season or pre-drought fires in the temperate zone that fail to initiate seedling recruitment, and high-severity or peat fires that are a particular threat to Tasmanian alpine heaths. A range of fire management practices including strategic fuel reduction and suppression operations are implemented to varying degrees to reduce the risks of detrimental fire regimes. However, this is hampered by a dearth of local knowledge about specific life histories and sensitive ecological processes, as well as the effectiveness of mitigation measures.

Soil-borne plant pathogens, notably *Phytophthora cinnamomi*, have had major impacts on heathland biodiversity in the three south-western Australian provinces, South coast and hills and Tasmanian lowlands. This disease kills root tissues in susceptible species belonging to such iconic heathland taxa as Proteaceae, Fabaceae, Ericaceae and *Xanthorrhoea* (Cahill *et al.* 2008; Leishman *et al.*, Chapter 9, this volume). Management responses are limited to quarantine and chemical treatment. Quarantine and hygiene measures are difficult to enforce and potentially undermined by movement of disease propagules by water and mammals. Chemical treatments based on phosphite have proven to be an effective short-term measure to boost plant resistance (Shearer and Fairman 2007), but are expensive to maintain, variably effective in different environments and risk long-term accumulation of phosphate in oligotrophic heathland systems.

Climate change may have direct and indirect impacts on heathland biodiversity. Limited bioclimatic modelling suggests that suitable habitat for some heathland dominants may contract or disappear as the climate warms over the next century (Yates *et al.* 2010). Climate change is also likely to interact with fire regimes, resulting in demographic processes such as fire-interval squeeze (Enright *et al.* 2015) that inflate risks of local extinctions. Other demographic effects include reduced flower production related to declining rainfall in the Naracoorte and south-western Australian provinces (Keith *et al.* 2014), depletion of soil seedbanks and reductions in germinability in response to elevated temperatures (Cochrane *et al.* 2014; Hudson *et al.* 2015). Climate change may also make heathland habitats more suitable for invasions by exotic plants and pathogens (including *Phytophthora*, currently limited by cool temperatures), while some heathlands on coastal sandplains may be vulnerable to encroachment of saline groundwater as sea levels rise. The main climate adaptation strategies relate to management of the threats that interact with climate change, including fire regimes, invasive species, habitat loss and fragmentation (e.g. Keith *et al.* 2008).

Heathlands are charismatic and unique Australian ecosystems with spectacular floral displays, high levels of species diversity, turnover and endemism, and diverse adaptive features as a consequence of a long evolutionary history of nutritional poverty, episodic drought and recurring fire. Although their diversity has been substantially eroded since the first botanical collections were made by Banks and his contemporaries, the remaining heathlands, if effectively conserved and managed, will continue to make a vital contribution to conservation of global biodiversity and ongoing enrichment of Australian outdoor culture.

Acknowledgements

We thank George Perry for commenting on 14.4.2 Coexistence mechanisms, and the reviewers for their perceptive comments that improved the content and presentation. Chris Simpson and Chantel Benbow assisted with preparation of Figure 14.1.

References

Atwell, B.J., P.E. Kriedemann and C.G.N. Turnbull (1999) *Plants in Action*. MacMillan Education Australia: Melbourne.

Auld, T.D. and O'Connell, M.A. (1991) Predicting patterns of post-fire germination in 35 eastern Australian Fabaceae. *Australian Journal of Ecology* 16, 53–70.

Baker, A. and Catterall, C. (2015) Where has all the fire gone? Quantifying the spatial and temporal extent of fire exclusion in Byron Shire, Australia. *Ecological Management and Restoration* 16, 106–113.

Beadle, N.C.W. (1966) Soil phosphate and its role in molding segments of the Australian flora and vegetation, with special reference to xeromorphy and sclerophylly. *Ecology* 47, 992–1007.

Bennett, G. (1834) *Wanderings in New South Wales, Batavia, Pedir coast, Singapore and China: The Journal of a Naturalist*. Richard Bentley: London.

Bond, W.J. and Ladd, P.G. (2001) Dynamics of the overstorey and species richness in Australian heathlands. *Journal of Mediterranean Ecology* 2, 247–257.

Bradstock, R.A., Tozer, M.G. and Keith, D.A. (1997) Effects of high frequency fire on floristic composition and abundance in a fire-prone heathland near Sydney. *Australian Journal of Botany* 45, 641–655.

Bremer, K. (2002) Gondwanan evolution of the grass alliance of families (Poales). *Evolution* 56, 1374–1387.

Cahill, D.M., Rookes, J.E., Wilson, B.A. *et al.* (2008) *Phytophthora cinnamomi* and Australia's biodiversity: impacts predictions and progress towards control. *Australian Journal of Botany* 56, 279–310.

Carpenter, R.J., Macphail, M.K., Jordan, G.J. and Hill, R.S. (2015) Fossil evidence for open, Proteaceae-dominated heathlands and fire in the Late Cretaceous of Australia. *American Journal of Botany* 102, 1–16.

Catford, J.A. and Jansson, R. (2014) Drowned, buried and carried away: effects of plant traits on the distribution of native and alien species in riparian ecosystems. *New Phytologist* 204, 19–36.

Causley, C.L., Fowler, W.M., Lamont, B.B. and He, T. (2016) Fitness benefits of serotiny in fire- and drought-prone environments. *Plant Ecology* 217, 773–779.

Chesson, P.L. and Huntley, N. (1988) Community consequences of life-history traits in a variable environment. *Annals of Zoology Fennici* 25, 5–16.

Cochrane, A., Hoyle, G.L., Yates, C.J. *et al.* (2014) Predicting the impact of increasing temperatures on seed germination among populations of Western Australian *Banksia* (Proteaceae). *Seed Science Research* 24, 195–205.

Cowling, R.M. and Lamont, B.B. (1985) Variation in serotiny of three *Banksia* spp. along a climatic gradient. *Australian Journal of Ecology* 10, 345–350.

Crisp, M.D., Burrows, G.E., Cook, L.G. *et al.* (2011) Flammable biomes dominated by eucalypts originated at the Cretaceous–Palaeogene boundary. *Nature Communications* 2, 193–200.

Denton, M.D., Veneklaas, E.J., Freimoser, F.M. and Lambers, H. (2007) *Banksia* species (Proteaceae) from severely phosphorus-impoverished soils exhibit extreme efficiency in the use and re-mobilization of phosphorus. *Plant, Cell and Environment* 30, 1557–1565.

Department of the Environment and Energy (2007) *NVIS data products 4.2*. Australian Government: Canberra. Available at: http://www.environment.gov.au/land/native-vegetation/national-vegetation-information-system/data-products (accessed 10 May 2016).

Diels, L. (1906) *Die Vegetation der Erde. VII. Die Planzenwelt von west-Australien sudlich des Wandekraises.* W. Engelmann: Leipzig.

Dodson, J.R. and Macphail, M.K. (2004) Palynological evidence for aridity events and vegetation change during the Middle Pliocene, a warm period in southwestern Australia. *Global and Planetary Change* 41, 285–307.

El-ahmir, S.M., Lim, S.L., Lamont, B.B. and He, T. (2015) Seed size, fecundity and postfire regeneration strategy are interdependent in Hakea. *PLoS ONE* 10, e0129027.

Enright, N.J. and Lamont, B.B. (1989) Fire temperatures and follicle-opening requirements in ten *Banksia* species. *Australian Journal of Ecology* 14, 107–113.

Enright, N.J. and Lamont, B.B. (1992) Survival, growth and water relations of Banksia seedlings in a sand mine rehabilitation site and adjacent scrub-heath sites. *Journal of Applied Ecology* 29, 663–671.

Enright, N.J., Lamont, B.B. and Marsula, R. (1996) Canopy seed bank dynamics and optimum fire regime for a highly serotinous shrub, *Banksia hookeriana. Journal of Ecology* 84, 9–17.

Enright, N.J., Mosner, E., Miller, B.P. *et al.* (2007) Soil vs. canopy seed storage and plant species coexistence in species-rich Australian shrublands. *Ecology* 88, 2292–2304.

Enright, N.J., Fontaine, J.B., Bowman, D.M.J.S. *et al.* (2015) Interval squeeze: altered fire regimes and demographic responses interact to threaten woody species persistence as climate changes. *Frontiers in Ecology and the Environment* 13, 265–272.

Esther, A., Groeneveld, J., Enright, N.J. *et al.* (2008) Assessing the importance of seed immigration on coexistence of plant functional types in a species-rich ecosystem. *Ecological Modelling* 213, 402–416.

Esther, A., Groeneveld, J., Enright, N.J. *et al.* (2011) Low-dimensional trade-offs fail to explain richness and structure in species-rich plant communities. *Theoretical Ecology* 4, 495–511.

Glassford, D.K. and Semeniuk, V. (1995) Desert-aeolian origin of late Cenozoic regolith in arid and semi-arid southwestern Australia. *Palaeogeography, Palaeoclimatology, Palaeoecology* 114, 131–166.

Groeneveld, J., Enright, N.J., Lamont, B.B. and Wissel, C. (2002) A spatial model of coexistence among three Banksia species along a habitat gradient in fire-prone shrublands. *Journal of Ecology* 90, 762–774.

Groeneveld, J., Enright, N.J., Lamont, B.B. *et al.* (2013) Species-specific traits plus stabilizing processes best explain coexistence in biodiverse fire-prone plant communities. *PLoS One* 8, e65084.

Groom, P.K. and Lamont, B.B. (2010) Phosphorus accumulation in Proteaceae seeds: a synthesis. *Plant and Soil* 334, 61–72.

Groom, P.G. and Lamont, B.B. (2015) *Plant Life of Southwestern Australia: Adaptations for Survival*. De Gruyter Open: Warsaw.

Groom, P.K., Lamont, B.B. and Kupsky, L. (1994) Contrasting morphology and ecophysiology of co-occurring broad and terete leaves in *Hakea trifurcata* (Proteaceae). *Australian Journal of Botany* 42, 307–320.

Groom, P.K., Lamont, B.B., Leighton, S. *et al.* (2004) Heat damage in sclerophylls is influenced by their leaf properties and plant environment. *Ecoscience* 11, 94–101.

Hanley, M.E., Lamont, B.B., Fairbanks, M.M. and Rafferty, C.M. (2007) Plant structural traits and their role in anti-herbivore defence. *Perspectives in Plant Ecology, Evolution and Systematics* 8, 157–178.

Hanley, M.E., Lamont, B.B. and Armbruster, S.W. (2009) Pollination and plant-defence traits co-vary in Western Australian Hakeas. *New Phytologist* 182, 251–260.

He, T., Krauss, S., Lamont, B.B. *et al.* (2004) Long-distance seed dispersal in a metapopulation of *Banksia hookeriana* inferred by population allocation analysis from AFLP data. *Molecular Ecology* 13, 1099–1109.

He, T., Lamont, B.B., Krauss, S.L. *et al.* (2009) Ants cannot account for interpopulation dispersal of the arillate pea *Daviesia triflora*. *New Phytologist* 181, 725–733.

He, T., Lamont, B.B. and Downes, K.S. (2011) Banksia born to burn. *New Phytologist* 191, 184–196.

He, T., Belcher, C.M., Lamont, B.B. and Lim, S.L. (2015) A 350-million-year legacy of fire adaptation among conifers. *Journal of Ecology* 104, 352–363.

Herath, D.N., Lamont, B.B., Enright, N.J. and Miller, B.P. (2009) Impact of fire on plant-species persistence in post-mine restored and natural shrubland communities in southwestern Australia. *Biological Conservation* 142, 2175–2180.

Hopper, S.D. (2009) OCBIL theory: towards an integrated understanding of the evolution, ecology and conservation of biodiversity on old, climatically buffered, infertile landscapes. *Plant Soil* 322, 49–86.

Hopper, S.D. and Gioia, P. (2004) The Southwest Australian Floristic Region: evolution and conservation of a global hot spot of biodiversity. *Annual Review of Ecology and Systematics* 35, 623–650.

Hopper, S.D., Smith, R.J., Fay, M.F. *et al.* (2009) Molecular phylogenetics of Haemodoraceae in the Greater Cape and Southwest Australian Floristic Regions. *Molecular Phylogenetics and Evolution* 51, 19–30.

Hudson, A.R., Ayre, D.J., Ooi, M.K.J. (2015) Physical dormancy in a changing climate. *Seed Science Research* 25, 66–81.

Keith, D.A. (1997) Combined effects of heat shock, smoke and darkness on germination of *Epacris stuartii* Stapf, an endangered fire-prone Australian shrub. *Oecologia* 112, 340–344.

Keith, D.A. and Bradstock, R.A. (1994) Fire and competition in Australian heath: a conceptual model and field investigations. *Journal of Vegetation and Science* 5, 347–354.

Keith, D.A. and Myerscough, P.J. (1993) Floristics and soil relations of upland swamp vegetation near Sydney. *Australian Journal of Ecology* 18, 325–344.

Keith, D.A. and Tozer, M.G. (2012) Vegetation dynamics in coastal heathlands of the Sydney Basin. *Proceedings of the Linnean Society of New South Wales* 134, B181–B197.

Keith, D.A., McCaw, W.L. and Whelan, R.J. (2002) Fire regimes in Australian heathlands and their effects on plants and animals. In *Flammable Australia: The Fire Regimes and Biodiversity of a Continent*, ed. R.A. Bradstock, J.E. Williams and A.M. Gill. Cambridge University Press: Cambridge, pp. 199–237.

Keith, D.A., Tozer, M.G., Regan, T.J. and Regan, H.M. (2007a) The persistence niche: what makes it and what breaks it for two fire-prone plant species. *Australian Journal of Botany* 55, 273–279.

Keith, D.A., Holman, L., Rodoreda, S. *et al.* (2007b) Plant functional types can predict decade-scale changes in fire-prone vegetation. *Journal of Ecology* 95, 1324–1337.

Keith, D.A., Akçakaya, H.R., Thuiller, W. *et al.* (2008) Predicting extinction risks under climate change: coupling stochastic population models with dynamic bioclimatic habitat models. *Biology Letters* 4, 560–563.

Keith, D.A., Lindenmayer, D.B., Lowe, A. *et al.* (2014) Heathlands. In *Biodiversity and Environmental Change: Monitoring, Challenges and Direction*, ed. D. Lindenmayer, E. Burns, N. Thurgate and A. Lowe. CSIRO Publishing: Melbourne, pp. 215–285.

Laffan, S.W. and Crisp, M.D. (2003) Assessing endemism at multiple spatial scales, with an example from the Australian vascular flora. *Journal of Biogeography* 30, 511–520.

Laliberté, E., Zemunik, G. and Turner, B.L. (2014) Environmental filtering explains variation in plant diversity along resource gradients. *Science* 345, 1602–1605.

Lamont, B.B. (1982) Mechanisms for enhancing nutrient uptake in plants, with special reference to Mediterranean South Africa and Western Australia. *Botanical Review* 48, 597–689.

Lamont, B.B. (1995) Mineral nutrient relations in Mediterranean regions of California, Chile and Australia. In *Ecology and Biogeography of Mediterranean Ecosystems in Chile, California and Australia*, ed. M.T. Arroya, P.H. Zedler and M.D. Fox. Springer Verlag: New York, pp. 211–235.

Lamont, B.B. (2003) Evolutionary convergence and sclerophylly in Mediterranean regions. In *Ecology: An Australian Perspective*, ed. P. Attiwill and B. Wilson. Oxford University Press: Melbourne, pp. 280–294.

Lamont, B.B. and Bergl, S.M. (1991) Water relations, shoot and root architecture, and phenology of three co-occurring *Banksia* species: no evidence for niche differentiation in the pattern of water use. *Oikos* 60, 291–298.

Lamont, B.B. and Cowling, R.M. (1984) Flammable infructescences in *Banksia*: a fruit-opening mechanism. *Australian Journal of Ecology* 9, 295–296.

Lamont, B.B. and Downes, K.S. (2011) Fire-stimulated flowering among resprouters and geophytes in Australia and South Africa. *Plant Ecology* 212, 2111–2125.

Lamont, B.B. and Enright, N.J. (2000) Adaptive advantages of aerial seed banks. *Plant Species Biology* 15, 157–166.

Lamont, B.B. and Groom, P.K. (2002) Green cotyledons of two *Hakea* species control seedling mass and morphology by supplying mineral nutrients rather than organic compounds. *New Phytologist* 153, 101–110.

Lamont, B.B. and He, T. (2012) Fire-adapted Gondwanan angiosperm floras evolved in the Cretaceous. *BMC Evolutionary Biology* 12, 223.

Lamont, B.B. and Markey, A. (1995) Biogeography of fire-killed and resprouting *Banksia* species in southwestern Australia. *Australian Journal of Botany* 43, 283–303.

Lamont, B.B. and Wiens, D. (2003) Are seed set and speciation always low among species that resprout after fire, and why? *Evolutionary Ecology* 17, 277–292.

Lamont, B.B. and Witkowski, E.T.F. (1995) A test for lottery recruitment among four *Banksia* species based on their demography and biological attributes. *Oecologia* 101, 299–308.

Lamont, B.B., Hopkins, A.J. and Hnatiuk, R.J. (1984) The flora: composition, diversity and origins. In *Kwongan: Plant Life of the Sandplain*, ed. J.S. Pate and J.S. Beard. University of Western Australia Press: Perth, pp. 27–50.

Lamont, B.B., Le Maitre, D., Cowling, R.M. and Enright, N.J. (1991) Canopy seed storage in woody plants. *The Botanical Review* 57, 277–317.

Lamont, B.B., Witkowski, E.T.F. and Enright, N.J. (1993) Post-fire litter microsites: safe for seeds, unsafe for seedlings. *Ecology* 74, 501–512.

Lamont, B.B., Groom, P.K. and Cowling, R.M. (2002) High leaf mass per area of related species assemblages may reflect low rainfall and carbon isotope discrimination rather than low phosphorus and nitrogen concentrations. *Functional Ecology* 16, 403–412.

Lamont, B.B., Wittkuhn, R. and Korczynskyj, D. (2004) Ecology and ecophysiology of grasstrees. *Australian Journal of Botany* 52, 561–582.

Lamont, B.B., Enright, N.J. and He, T. (2011) Fitness and evolution of resprouters in relation to fire. *Plant Ecology* 212, 1945–1957.

Lamont, B.B., Groom, P.K., Williams, M. *et al.* (2015) LMA, density and thickness: recognizing different leaf shapes and correcting for their nonlaminarity. *New Phytologist* 207, 942–947.

Lamont, B.B., He, T. and Lim, S.L. (2016) *Hakea*, the world's most sclerophyllous genus, arose in southwest Australian heathland and diversified throughout Australia over the last 12 million years. *Australian Journal of Botany* 64, 77–88.

Llorens, T.M., Byrne, M., Yates, C.J. *et al.* (2012) Evaluating the influence of different aspects of habitat fragmentation on mating patterns and pollen dispersal in the bird-pollinated *Banksia sphaerocarpa* var. *caesia*. *Molecular Ecology* 21, 314–328.

Low, A.B. and Lamont, B.B. (1990) Aerial and below-ground phytomass of *Banksia* scrub-heath at Eneabba, Western Australia. *Australian Journal of Botany* 38, 351–359.

Mack, C.L. and Milne, L.A. (2015) Eocene palynology of the Mulga Rocks deposits, southern Gunbarrel Basin, Western Australia. *Alcheringa* 39, 444–458.

Mao, K., Milne, R.I., Zhang, L. *et al.* (2012) Distribution of living Cupressaceae reflects the breakup of Pangea. *Proceedings of the National Academy of Sciences USA* 109, 7793–7798.

Merwin, L., He, T., Lamont, B.B. *et al.* (2012) Low rate of between-population seed dispersal restricts genetic connectivity and metapopulation dynamics in a clonal shrub. *PLoS ONE* 7, e50974.

Milberg, P. and Lamont, B.B. (1995) Fire enhances weed invasion of roadside vegetation in southwestern Australia. *Biological Conservation* 73, 45–49.

Mucina, L. and Daniel, G. (2013) *Vegetation Mapping in the Northern Kimberley, Western Australia*. Curtin University: Perth.

Mucina, L. and Wardell-Johnson, G.W. (2011) Landscape age and soil fertility, climatic stability, and fire regime predictability: beyond the OCBIL framework. *Plant and Soil* 341, 1–23.

Myerscough, P.J. and Carolin, R.C. (1986) The vegetation of the Eurunderee sand mass, headlands and previous islands in the Myall Lakes area, New South Wales. *Cunninghamia* 1, 399–466.

Myerscough, P.J., Clarke, P.J. and Skelton, N.J. (1996) Plant coexistence in coastal heaths: habitat segregation in the post-fire environment. *Australian Journal of Ecology* 21, 47–54.

O'Loughlin, L.S., Green, P.T. and Morgan, J.W. (2015) The rise and fall of *Leptospermum laevigatum*: plant community change associated with the invasion and senescence of a range-expanding native species. *Applied Vegetation Science* 18, 323–331.

Orians, G.H. and Milewski, A.V. (2007) Ecology of Australia: the effects of nutrient-poor soils and intense fires. *Biological Reviews* 82, 393–423.

Parsons, B.C. and Gosper, C.R. (2011) Contemporary fire regimes in a fragmented and an unfragmented landscape: implications for vegetation structure and persistence of fire-sensitive malleefowl. *International Journal of Wildland Fire* 20, 184–194.

Parsons, M.H., Lamont, B.B., Davies, S.J.J.F. and Kovacs, B.R. (2006) How energy and co-available foods affect the currency of forage for the western grey kangaroo. *Animal Behaviour* 71, 765–772.

Pate, J.S. and Dixon, K.W. (1982) *Tuberous, Cormous and Bulbous Plants: Biology of an Adaptive Strategy in Western Australia*. University of Western Australia Press: Nedlands.

Pate, J.S., Dixon, K.W. and Orshan, G. (1984) Growth and life form characteristics of kwongan species. In *Kwongan: Plant Life of the Sandplain*, ed. J.S. Pate and J.S. Beard. University of Western Australia Press: Nedlands, pp. 84–100.

Pate, J.S., Froend, R.H., Bowen, B.J. *et al.* (1990) Seedling growth and storage characteristics of seeder and resprouter species of Mediterranean-type ecosystems of S.W. Australia. *Annals of Botany* 65, 585–601.

Pausas, J.G. and Keeley, J.E. (2014) Evolutionary ecology of resprouting and seeding in fire-prone ecosystems. *New Phytologist* 204, 55–65.

Perry, G.L.W., Enright, N.J., Miller, B.P. and Lamont, B.B. (2008) Spatial patterns in species-rich Mediterranean-type shrublands of southwestern Australia. *Journal of Vegetation Science* 19, 705–716.

Perry, G.L.W., Enright, N.J., Miller, B.P. *et al.* (2009) Dispersal, edaphic fidelity, and speciation in species-rich Western Australian shrublands: evaluating a neutral model of biodiversity. *Oikos* 118, 1349–1362.

Perry, G.L.W., Miller, B.P., Enright, N.J. and Lamont, B.B. (2014) Stochastic geometry best explains spatial associations among species pairs and plant functional types in species-rich shrublands. *Oikos* 123, 99–110.

Rafferty, C., Lamont, B.B. and Hanley, M.E. (2005) Selective feeding by kangaroos (*Macropus fuliginosus*) on seedlings of *Hakea* species: effects of chemical and physical defences. *Plant Ecology* 177, 201–208.

Raunkiaer, C. (1934) *The Life Forms of Plants and Statistical Plant Geography, being the Collected Papers of C. Raunkiaer.* Oxford University Press: Oxford.

Read, J., Sanson, G.D., Caldwell, E. *et al.* (2009) Correlations between leaf toughness and phenolics among species in contrasting environments of Australia and New Caledonia. *Annals of Botany* 103, 757–767.

Richards, M.B. and Lamont, B.B. (1996) Post-fire mortality and water relations of three congeneric shrub species under extreme water stress: a trade-off with fecundity? *Oecologia* 107, 53–60.

Roche, S., Dixon, K.W. and Pate, J.S. (1997) Seed ageing and smoke: partner cues in the amelioration of seed dormancy in selected Australian native species. *Australian Journal of Botany* 45, 783–815.

Russell-Smith, J., Ryan, P.G., Klessa, D. *et al.* (1998) Fire regimes, fire-sensitive vegetation, and fire management of the sandstone Arnhem Plateau, monsoonal northern Australia. *Journal of Applied Ecology* 35, 829–846.

Sauquet, H., Weston, P.H., Anderson, C.L. *et al.* (2009) Contrasted patterns of hyperdiversification in Mediterranean hotspots. *Proceedings of the National Academy of Sciences USA* 106, 221–225.

Schimper, A.F.W. (1903) *Plant Geography upon a Physiological Basis*. English translation by W.R. Fisher. Clarendon Press: Oxford.

Shearer, B.L. and Fairman, R.G. (2007) Application of phosphite in a high-volume foliar spray delays and reduces the rate of mortality of four *Banksia* species infected with *Phytophthora cinnamomi*. *Australasian Plant Pathology* 36, 358–368.

Specht, R.L. (1994) Heathlands. In *Australian Vegetation*, ed. R.H. Groves. Melbourne: Cambridge University Press, pp. 321–344.

Specht, R.L. and Specht, A. (1989) Species richness of sclerophyll (heathy) plant communities in Australia: the influence of overstorey cover. *Australian Journal of Botany* 37, 337–350.

Sniderman, J.M.K., Jordan, G.J. and Cowling, R.M. (2013) Fossil evidence for a hyperdiverse sclerophyll flora under a non–Mediterranean-type climate. *Proceedings of the National Academy of Sciences* 110, 3423–3428.

Vidal-Russell, R. and Nickrent, D.L. (2008) The first mistletoes: origins of aerial parasitism in Santalales. *Molecular Phylogenetics and Evolution* 47, 523–537.

Weiss, R.W. and Lamont, B.B. (1997) Floral colour change and insect pollination: a dynamic relationship. *Israel Journal of Plant Sciences* 45, 185–199.

Westcott, V.C., Enright, N.J., Miller, B.P. *et al.* (2014) Biomass and litter accumulation patterns in species-rich shrublands for fire hazard assessment. *International Journal of Wildland Fire* 23, 860–871.

Witkowski, E.T.F. and Lamont, B.B. (1996) Allocation of mineral nutrients and carbon between vegetative and reproductive structures in *Banksia hookeriana*. *Oecologia* 105, 38–42.

Witkowski, E.T.F., Lamont, B.B., Walton, C.S. and Radford, S. (1992) Leaf demography, sclerophylly and ecophysiology of two Banksia species with contrasting leaf life spans. *Australian Journal of Botany* 40, 849–862.

Yates, C.J., McNeill, A., Elith, J. and Midgley, G.F. (2010) Assessing the impacts of climate change and land transformation on Banksia in the Southwest Australian Floristic Region. *Diversity and Distributions* 16, 187–201.

15 Australia's Tropical Savannas: Vast, Ancient and Rich Landscapes

RICHARD J. WILLIAMS, GARRY D. COOK, ADAM C. LIEDLOFF AND WILLIAM J. BOND

Tropical savannas are the dominant vegetation type of northern Australia, covering about 25% of the continent. They consist of sparse-to-moderate tree cover over a ground layer dominated by C4 grasses. The climate is monsoonal, wet-dry tropical. Temperatures are high year-round. Rainfall is summer-dominated, with strong continental-scale gradients in mean annual rainfall from the mesic savannas of the near-coastal regions to the arid interior. Soils are generally nutrient-poor. The savannas have evolved over millions of years, are species-rich, and are not derived from closed forest systems by recent human activity. Structure and composition of the vegetation varies strongly with variation in mean annual rainfall and soil type. Tree cover increases with increasing rainfall, but declines as soil clay content increases. Phenology is strongly seasonal, but growth, flowering and fruiting can occur in the tree stratum during the dry season. Fire is frequent and extensive; late dry season fires are more intense than early dry season fires. The ground stratum is relatively insensitive to variation in fire regime, whereas tree abundance and recruitment declines with increasing fire intensity. Savannas account for ca. 3% of Australia's greenhouse gas emissions, and there is considerable potential for managing fire regimes such that emissions are reduced and environmental and social benefits are enhanced.

15.1 Introduction

Tropical savannas are one of the world's most extensive biomes, widely distributed in the wet–dry tropics of Africa, South America, Australia and South-east Asia. Savannas cover about 20% of the Earth's land surface (Lehmann *et al.* 2011, 2014) and account for approximately 30% of global net primary productivity (Grace *et al.* 2006); global fire activity is dominated by savanna fires (van der Werf *et al.* 2008). The savannas of Africa and Asia have played an integral part in early human evolution (Dennell and Roebroeks 2005), and presently most of the world's people live in the tropics.

Australia's tropical savannas occur in a vast arc from Broome in Western Australia, through the Top End of the Northern Territory, to Rockhampton in Queensland (Fox *et al.* 2001; Figure 15.1). They are comparatively intact ecologically, and, despite an apparent structural simplicity, are biodiverse and of high conservation significance. Australia's tropical regions, both wet and wet–dry, are sparsely populated, (< 5% of the total Australian population), albeit with a relatively high proportion of

Australian Vegetation, Third Edition, ed. David A. Keith. Published by Cambridge University Press. © Cambridge University Press 2017.

Citation: Williams, R.J. *et al.* (2017). Australia's Tropical Savannas: Vast, Ancient and Rich Landscapes. In D. A. Keith, ed., *Australian Vegetation*, 3rd edn. Cambridge: Cambridge University Press, pp. 368–88.

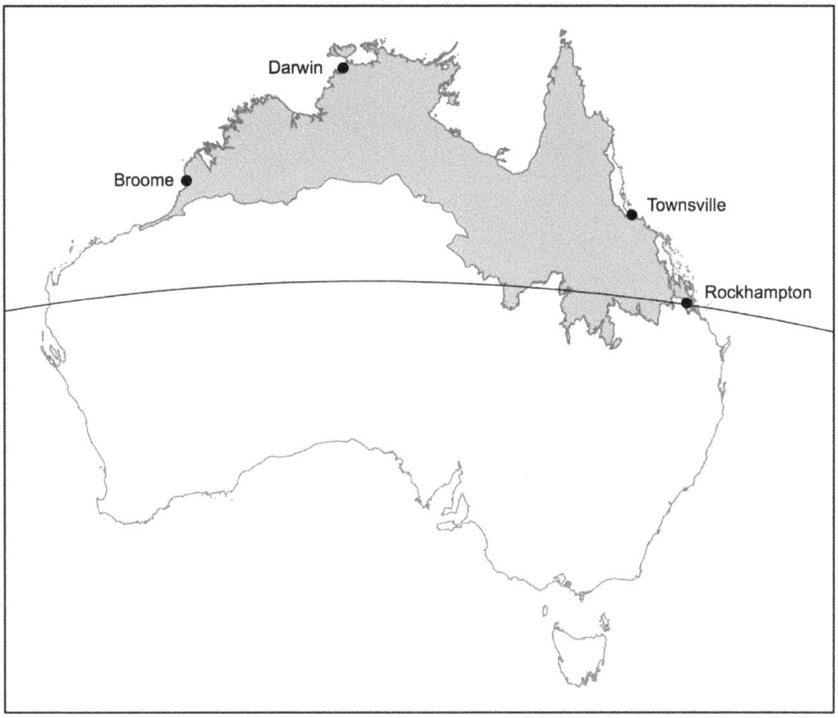

Figure 15.1 The Australian tropical savanna region. The region delimited as savanna follows Fox *et al.* (2001).

Indigenous people (Williams *et al.* 1997a; Woinarski *et al.* 2007). Land uses within the savannas over the past century and a half have centred on livestock grazing, mining and nature conservation. Emerging industries such as carbon farming and customary Indigenous land management have great potential for livelihood development in northern Australia (Woinarski *et al.* 2007). Sustainable management of Australia's savannas depends upon understanding the complex, interacting factors, both past and present, that have shaped the landscapes of northern Australia. In this chapter, we review the vegetation ecology of Australia's savannas. We emphasise the major research fields of the past three decades – the nature and evolution of the savanna biome, climate–soil–disturbance drivers of variation in savanna vegetation, fire–vegetation interactions, carbon dynamics and potential socio-ecological futures for the savannas.

15.2 The global savanna biome

The savanna biome is truly global, occurring across the world where the climate is seasonally wet–dry tropical. Vegetation structure is relatively simple, consisting of an upper stratum dominated by C_3 trees above a ground stratum dominated by C_4 grasses. However, savanna form and function varies substantially, as a consequence of phylogenetic history, variation in rainfall, soil type and disturbance regimes, and the consequent effects on the tree and grass strata (Sankaran *et al.* 2005; Lehmann *et al.* 2011, 2014). In this chapter, we follow the definition of Lehmann *et al.* (2011): a savanna is 'any system with a continuous layer of C_4 grasses, regardless of whether trees are present'. Under this definition treeless, tropical grasslands dominated by C_4 grasses are savannas.

Savannas may coexist with other tropical vegetation formations such as rainforests, scrub thickets, and shrublands. Two enduring questions that ecologists have addressed in detail are: (i) what factors, past and present, separate savannas from other vegetation types, particularly rainforests (e.g. Murphy and Bowman 2012); and (ii) how do interactions between climate, soils and disturbance, especially fire, affect vegetation composition and structure within savannas, especially the coexistence of trees and grasses (e.g. Higgins *et al.* 2000). In this chapter, we explore the second of these questions, drawing on findings from paleoecology, plant physiology, plant demography, biogeography and landscape fire ecology.

Savannas are an ancient biome. Savannas and other grassy ecosystems became prominent from the late Miocene (approx. 8 Ma; Beerling and Osborne 2006; Edwards *et al.* 2010). During the Pliocene, savannas extended from West Africa to north China. During the Pleistocene, open ecosystems such as tropical savannas, underwent cycles of expansion during the glacials, and contractions (as closed forest expanded) during the interglacials. Changes in the concentration of atmospheric CO_2 have been implicated in such large-scale biome shifts during glacial–interglacial periods (Bond and Midgley 2012). The savanna biota is rich, in both plants and animals, and this rich endemic biota is further evidence that, at the biome scale, savannas are not derived from closed forest systems by recent anthropogenic activity, but have evolved over millions of years (Bond 2008; Bowman *et al.* 2010).

Australia's savannas occupy 2 million km^2, or approximately 25% of the continent. They encompass all of the major regions of northern Australia: The Kimberley in the north of Western Australia, the Top End of the Northern Territory, and the Gulf Country and Cape York Peninsula in northern Queensland. Some extensive areas of savanna in north Queensland have been cleared for agriculture (Cook and Dias 2006; Woinarski *et al.* 2007). In north-western Australia, in contrast, there has been relatively little clearing for intensive agriculture; production forestry is rare. Given the extent and relative intactness of Australia's savannas, at least compared with the ecosystems of southern Australia, and savannas elsewhere in the world, Australian savannas have global conservation significance (Woinarski *et al.* 2007).

15.3 Climate and landforms of Australian savannas

The climate of the Australian savanna biome is monsoonal, wet–dry tropical (Suppiah 1992). Temperatures are high year-round (mean annual temperature > 18°C) and not limiting for plant growth. In contrast, rainfall is highly seasonal, with a distinct summer wet season, followed by a winter dry season. Approximately 70–90% of the annual rainfall occurs in 3–4 months over summer, typically between October and March. Winter rainfall may constitute 20–30% of the annual rainfall in parts of north-eastern Queensland, but in the Top End of the Northern Territory and the north of Western Australia, it is < 5%, with some winter months (e.g. June–July in the Top End of the Northern Territory) being virtually rain-free (Cook and Heerdegen 2001).

The strong rainfall seasonality is a direct consequence of the monsoon – a global weather system wherein wind direction reverses seasonally, causing seasonal shifts in precipitation (Suppiah 1992; Cook and Heerdegen 2001). The wet season is dominated by moist, west-north-westerly airflows that are associated with monsoonal periods of rainfall (typically 5–7 days duration), short-duration convective storms and tropical cyclones. The onset of the dry season is associated with a reversal of the wind direction, from west-north-westerly of the monsoon, to south-easterly trade winds, and declines in atmospheric humidity. The transition from the dry season to the wet season (termed 'the build-up') is characterised by increasing atmospheric humidity, and highly localised convective storms. Throughout the build-up and the wet season, there is a high incidence of lightning (Kuleshov 2012), which may ignite fires.

There are strong continental-scale gradients in mean annual precipitation (MAP) in the Australian savannas, as in savannas elsewhere in the world (Lehmann *et al.* 2011). MAP varies from c. 1500–2000 mm in the more humid, coastal regions of Western Australia, the Northern Territory and far north Queensland, to ca. 400 mm in the interior regions of the continent. Williams (1997a) and Fensham (2012) recognised two broad Australian savanna types – humid/monsoonal (MAP 900–2000 mm), and semi-arid/subhumid (MAP 400–900 mm). The highly seasonal rainfall regime, combined with strong gradients in annual rainfall, influences virtually all aspects of savanna vegetation and associated disturbance regimes, particularly fire. Savanna vegetation can also be affected substantially by variation in the timing of the onset and retreat of the monsoon, and by intra-monsoonal periods of drought (Cook and Heerdegen 2001).

Landforms over most of the savanna biome are at elevations of less than 500 m above sea level, with local relief generally < 200 m (Woinarski *et al.* 2005). In both the mesic and semi-arid savannas there are three broad landform classes and associated soils (Williams 1991; Isbell 1996; McKenzie *et al.* 2004). First, and most extensive, are the lowland plains derived from laterised Tertiary surfaces. The main soils are light-textured sands (rudosols and tenosols), and medium-textured loams (kandosols). Second, there are the rocky escarpments, slopes and plateaus of the 'stone country'. These occur in eastern Arnhem Land and parts of the Victoria River District in the Northern Territory, the Kimberley in Western Australia and north-eastern Cape York in Queensland. Soils are typically shallow rudosols, or deep sands on outwash slopes (Williams 1991). Third, there are the 'black-soil plains', where the dominant soils are cracking heavy-textured clays (vertosols), derived from Quaternary–Recent sediments on the floodplains of the major river systems, or *in situ* weathering of pre-Tertiary fine sediments and basalts. Tenosols and kandosols are the most common, although vertosols may be regionally very extensive, such as in Barkly Tablelands of eastern Northern Territory/north-western Queensland.

Soils are nutrient poor (Cook 2003), due to the long history of weathering. Clays generally have higher nutrient status than the lighter-textured sands and loams. Most soils are acidic (pH 5.0–6.5), but may be neutral–mildly alkaline if derived from limestone. Unlike the grassy woodlands of temperate, southern Australia, areas of arable soil in the tropical savannas are generally small and disjunct (Cook 2009).

15.4 Composition, structure, and function in Australian savanna vegetation

Research into the ecology and management of Australia's savannas commenced in earnest at the end of World War II. Research focused on the climate, soils and vegetation of natural savanna landscapes, to support development of cultivated crops and pastures via wholesale transformation of the native landscape with introduced species, particularly grasses (Cook and Dias 2006). Over the last 30–40 years research has been directed towards understanding landscape function and its application in fields as diverse as biodiversity conservation, Indigenous customary management, fire ecology and management, the emerging carbon economy, and sustainable land management (Garnett *et al.* 2008). Hundreds of publications have addressed research questions germane to this quest for sustainable land management. Below, we highlight the key findings from studies on structural and compositional variation in savanna vegetation (including international comparative studies), phenology and growth of vegetation, plant-soil-water relationships, fire regimes, fire–vegetation interactions (especially tree dynamics), faunal ecology, tropical cyclones, and carbon dynamics (Werner 1991; Williams *et al.* 2002; Andersen *et al.* 2003; Russell-Smith *et al.* 2003a, 2009; Williams and Farrer 2005; Fensham 2012).

15.4.1 Variation in structure and composition

Australian savannas encompass a range of structural vegetation types (*sensu* Specht 1981): forests, woodlands, and grasslands (Figure 15.2). Australia's savannas are a matrix or 'sea' that surrounds and abuts a number of other vegetation types, such as rainforest, sedgelands and herblands of riverine and littoral complexes. The Australian savanna vegetation map (Fox *et al.* 2001) shows the savanna biome consists of 22 Interim Bioregionalisation of Australia (IBRA) units (v.5.0; Environment Australia 2000). The 125 vegetation map units are grouped into 26 broad vegetation groups (BVGs). Of these, 23 BVGs are savanna units *sensu stricto* (i.e. forests, woodlands, or shrublands with a grassy understorey, or grasslands). The majority (14 BVGs) are dominated by *Eucalyptus* and/or *Corymbia*; three by *Acacia*; two by *Melaleuca*; one by various pan-tropical deciduous trees and shrubs; one by shrubs; and two are grasslands. Approximately one-third of the BVGs are 'mesic savannas', and the remaining two-thirds are 'semi-arid savannas' (*sensu* Williams *et al.* 1997a; Fensham 2012). The mosaic pattern of closed forest inter-digitating with open savanna, such a distinctive feature of many African savanna landscapes (Bond 2008), is not a feature of Australian savannas.

The tree stratum of Australian savannas varies from c. 10–70% cover, and 5–25 m in height (Fox *et al.* 2001). The tree layer is overwhelmingly dominated by *Eucalyptus* and/or *Corymbia*. Dominant subgenera within *Eucalyptus* are *Blakella* and *Eudesmia;* subgenera typical of the forests and woodlands of southern Australia such as *Symphyomyrtus* and *Monocalyptus* are relatively uncommon. Woodlands dominated by *Melaleuca* with a grassy understorey may occur locally. Pan-tropical genera such as *Erythrophleum*, *Terminalia*, *Buchanania*, *Cochlospermum*, *Brachychiton*, and *Planchonia* are common in the mesic savannas; *Bursaria* and *Gardenia* are common in more subhumid/semi-arid savannas. Woodlands dominated by *Acacia* may be locally common especially in semi-arid savannas (Fox *et al.* 2001; see also Table 8.1 in Fensham 2012). There is a mix of phenological types, based on leaf life span, from fully deciduous, to semi-deciduous and evergreen (Williams *et al.* 1997b).

The dominant grasses are virtually all C_4. There is mix of life forms: tall grasses (0.5–2 m) and short grasses (< 0.25 m), annual species, and perennial tussock-forming species. Common, dominant genera include *Alloteropsis*, *Aristida*, *Astrebla*, *Bothriochloa*, *Chrysopogon*, *Dichanthium*, *Eragrostis*, *Eriachne*, *Heteropogon*, *Schizachyrium*, *Sorghum* and *Themeda*. Tall, annual species of *Sorghum* may dominate in the high-rainfall areas of north-western Australia; tussock species such as *Chrysopogon* are the dominant life form in the semi-arid savannas; the peculiarly Australian hummock-forming genus *Triodia* becomes more common with increasing aridity (MAP < 600 mm), but is also prominent on stone country in high-rainfall areas. Perennial grasses, particularly in the semi-arid savannas, regulate basic landscape function at and near the soil surface by controlling resource distribution, particularly water, as has been shown in the semi-arid grassy woodlands in south-eastern Australia (Ludwig *et al.* 1999).

Much research has addressed how variation in annual precipitation and soil texture affect the composition and structure of Australian savannas, especially along the strong rainfall gradient from the coastal/subcoastal regions of the Top End of the Northern Territory, to the arid interior (the North Australian Tropical Transect, NATT; Cook *et al.* 2002). Vegetation height, tree basal area, and species diversity all decrease with decreasing rainfall along this gradient. Tree basal area varies from 0–20 $m^2\ ha^{-1}$ depending on annual rainfall and soil texture (Figure 15.3). On deep sandy, light-textured soils tree basal area decreases by about 0.7 $m^2\ ha^{-1}$ per 100 mm decrease in annual rainfall, from the wettest (*c.* 1600 mm per annum) to the driest savannas (*c.* 300 mm per annum). On heavier textured loams the rate of change is about 0.5 $m^2\ ha^{-1}$ per 100 mm. On clay soils, trees are generally sparse or absent, and basal area varies from 0 to 3 $m^2\ ha^{-1}$ (Williams *et al.* 1996; Cook *et al.* 2002).

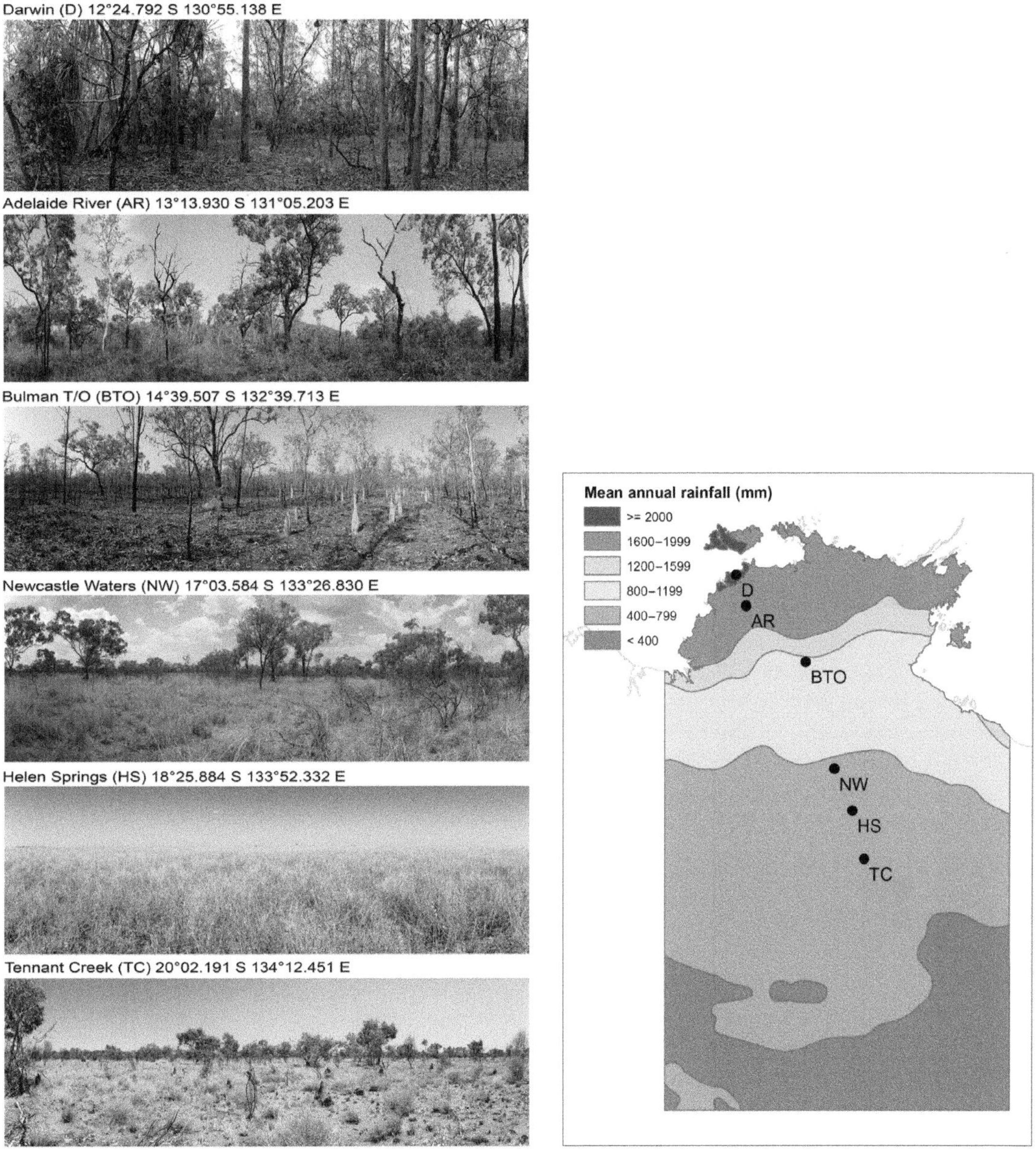

Figure 15.2 Australian savanna landscapes from sites along the North Australian Tropical Transect (NATT) in the Northern Territory. Sites: Open forest of *Eucalyptus tetrodonta* over recently burnt grass layer dominated by annual *Sorghum;* Darwin region (D); woodland of *Corymbia* and *Eucalyptus* over perennial tussock grass layer, Adelaide River (AR); woodland of *Corymbia*, Bullman turnoff on Stuart Highway (BTO); Open woodland of *Corymbia terminalis* over perennial tussock grass layer, Newcastle Waters (NW); Grassland of *Astrebla* and *Dichanthium* spp., Helen Springs (HS); Open woodland of *Eucalyptus brevifolia* with tussock and hummock grass understorey, Tennant Creek (TC). Latitudes and locations given in decimal degrees. *Photos: Adam Liedloff.* Map shows rainfall isohyets, and location of sites illustrated in photographs. *(A black and white version of this figure will appear in some formats. For the colour version, please refer to the plate section.)*

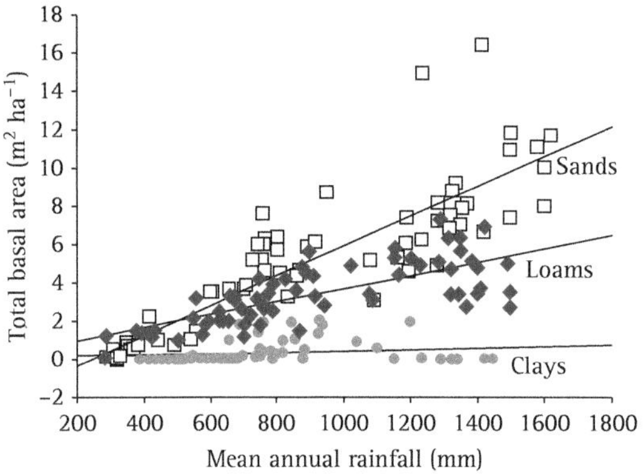

Figure 15.3 Variation in tree basal area (m^2 ha^{-1}) of Australian savannas as a function of mean annual precipitation and soil textural type (□ sands, ◆ loams, ● clays) from 1 ha sample sites in the Northern Territory and north-west Western Australia.
Data sources: Cook et al. (2002); Bond (2008); Lehmann et al. (2014); G.D. Cook and R.J. Williams (unpublished data).

Patterning within the ground stratum also varies with MAP. The size of fetches – patches of bare or sparsely covered ground in between individual perennial grass tussocks – increases with decreasing rainfall (Ludwig *et al.* 1999). Such fetches collect and direct resources to perennial species in the ground stratum, particularly grass tussocks. That fetch size increases with declining MAP indicates that the 'catchment' needed to support a given amount of perennial grass increases as resources decline.

Plant species diversity varies with MAP in complex ways. There is a weak decline in overall diversity with decreasing MAP (Williams *et al.* 1996). The diversity and abundance of deciduous tree species declines with declining MAP (Egan and Williams 1996; Williams *et al.* 1996). Many species are widely distributed; for example open forests dominated by *Eucalyptus miniata/E. tetrodonta* occur on more than 180 000 km^2, or about 10%, of the Northern Territory alone (Woinarski *et al.* 2005). Woinarski *et al.* (2005) showed compositional similarity decreased along the NATT as the difference in average annual rainfall between sites increased. However, plant species composition decreased by only 10 units for every 400 km of geographic separation; it would require ~4000 km separation for a complete turnover of plant species! This suggests that fine-scale factors other than the broad rainfall patterns or soil texture can also strongly influence plant species composition. For example, in the Mitchell grasslands of the Barkly Tableland of the eastern Northern Territory and north-western Queensland, Fisher (2001) found that animal and plant species richness in 100 km^2 paddocks increased by 25–50% if only 5% of the paddock area contained small gravel rises, elevated by *c.* 1 m above the surrounds.

15.4.2 Phenology and growth

Phenology – the periodicity in plant and animal life cycles – is strongly seasonal, in response to the seasonality of precipitation and humidity. This phenological rhythm underpins plant water use, growth and reproduction, ecosystem carbon dynamics, resources for consumers, growth and reproduction of fauna, and savanna fire regimes. Intra- and interannual variation in plant phenology has been most extensively studied in the mesic *Eucalyptus miniata/E. tetrodonta* savannas of the Top End of the Northern Territory. We summarise the major findings below.

15.4.2.1 Grass phenology

Grass phenology is fundamental to savanna ecology: grasses fix substantial amounts of carbon, are a major food source for fauna (vertebrate and invertebrate), and are the dominant constituent of fine fuels. Growth commences with the first rains of the wet season, continuing through the wet season. Globally, *c.* 60% of annual production in savannas can be from the grass stratum (Lloyd *et al.* 2008). Flowering is generally restricted to the wet season (Brennan 1997). Curing (i.e. death of grass shoots) commences at the beginning of the dry season, and the degree of curing increases with the progression of the dry season (Williams *et al.* 1998). Tall, annual grasses (e.g. *Sorghum intrans*) of the mesic savannas of north-west Australia commence curing earlier than their perennial counterparts (e.g. *Chrysopogon*; Gill *et al.* 1996).

15.4.2.2 Tree phenology

Tree canopy cover peaks towards the end of the wet season in all phenological types. Leaf fall in deciduous and semi-deciduous taxa generally commences in the early dry season. In fully deciduous taxa, such as *Terminalia, Buchanania* and *Planchonia*, leaf fall commences soon after the rains end, well before soil moisture falls below wilting point in the top 1 m of the profile (Williams *et al.* 1997b). Fully deciduous trees are also obligately deciduous, as continued watering through the dry season does not prevent leaf fall (Myers *et al.* 1998). Semi-deciduous species, such as *Erythrophleum chlorostachys*, commence leaf fall towards the middle of the dry. The timing of leaf fall may vary from species to species, but peak litterfall is in August (Cook 2003).

Leaf flush generally occurs in the late dry (September–October) often before the first rains and while soil moisture in the top 1 m is below wilting point. Within species, it varies supra-annually, indicating such phenophases may be controlled by variation in atmospheric and/or soil moisture, rather than by photoperiod (Duff *et al.* 1997; Williams *et al.* 1997b). The timing of leaf fall and leaf flush in some evergreen and semi-deciduous taxa (e.g. *Erythrophleum*; dominant eucalypts) can be affected by fire early in the dry season (Williams *et al.* 2003a; see below).

Flowering phenology in trees shows a distinct dry season peak (Setterfield and Williams 1996; Williams *et al.* 1999a). In the mesic savannas of the Top End, most species (34/50) flowered and fruited in each year from 1992 to 1995, and generally in the same month each year. For most taxa, the peak period of flowering and fruiting is the mid-dry season (July/August), and the 'build-up' (October/November); the nadir of reproductive activity is in the late wet season (February/March). The timing of events varies between phenological types. Dominant eucalypts such as *E. miniata* and *E. tetrodonta* flower relatively early in the dry season (May–June). Flowering in fully deciduous taxa, in contrast, generally commences in the mid-dry, after leaf fall has finished. Such flowering can be spectacular: floral buds and flowers on the leafless stems of taxa such as *Brachychiton, Cochlospermum* and *Planchonia* stand out as flashes of red, yellow and white against the background sea of the straw-brown, cured grass layer. Importantly, and in contrast to temperate, southern Australian forests and woodlands, serotiny is absent; i.e. there is no super-annual carryover of seed-bearing fruit in the canopy of any of the common dominant or subdominant tree species. Seed is dispersed by the onset of the wet season, with seedling emergence early in the wet season, which allows the whole of the critical first wet season for further growth.

15.4.2.3 Biomass, water use and growth

The biomass of the grass stratum is generally 2–8 t ha^{-1}, in both semi-arid and mesic savannas, and on light-, medium- and heavy-textured soils. Tree biomass varies strongly with mean annual rainfall. It may be zero on treeless, grassy plains of heavy-textured soils in both semi-arid and mesic savannas, and c. 70 t ha^{-1} on deep sandy loams in the wettest of the mesic savannas (Cook *et al.* 2015).

The carbon stocks of savannas have been studied intensively (Williams and Farrer 2005; Williams *et al.* 2012). There are good predictive relationships between tree basal area and biomass (Cook *et al.* 2005), and individual tree diameter and biomass (Williams *et al.* 2005b). Hence, allometry has been used extensively to estimate carbon stocks. In the mesic *Eucalyptus miniata-E. tetrodonta* savannas at Kapalga in the Northern Territory, Cook *et al.* (2005) estimated stocks of 23 tC ha^{-1} on shallow sandy loam soils, and 43 tC t ha^{-1} on deep sandy loam soils. Similar values have been determined by Chen *et al.* (2003) for this vegetation type on similar soils at nearby Howard Springs. Savanna-wide estimates of stocks have also been calculated, based on these allometric relationships (Figure 15.4; Cook *et al.* 2015).

Water use and carbon assimilation, and hence plant growth, is highly seasonal. Across phenological types, daily maximum leaf conductance, pre-dawn leaf water potential, and rates of photosynthesis and transpiration are greatest in the wet season, and decrease during the dry season (Myers *et al.* 1997; Eamus *et al.* 2001; Eamus and Prior 2001).

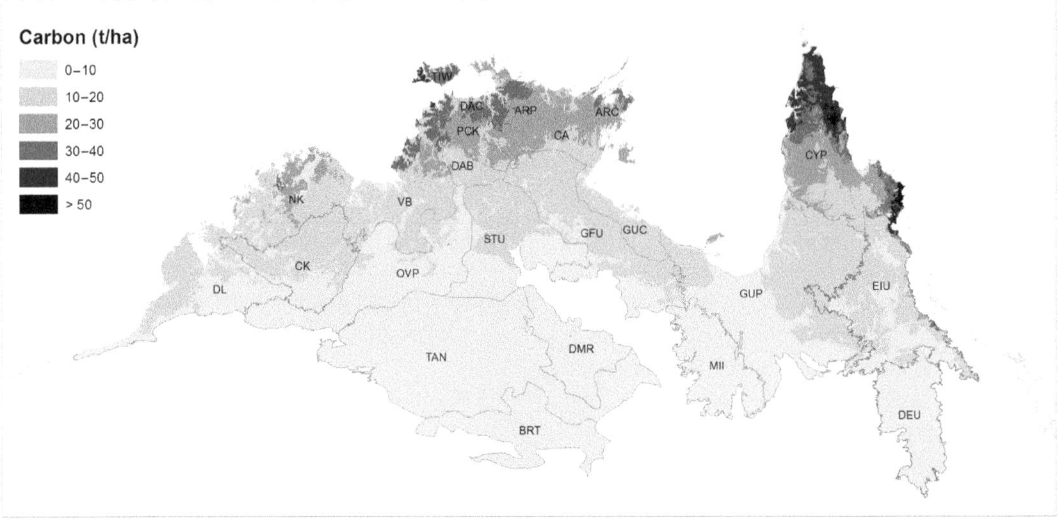

Figure 15.4 Estimated carbon stocks (tC ha^{-1}) of the tree stratum in selected biogeographic regions of Australia's tropical savannas. Bioregion abbreviations as per Environment Australia (2000).
Source: Cook et al. (2015).

These estimates of plant water use have been fundamental to the parameterisation of savanna fire–vegetation–carbon models, such as FLAMES (Liedloff and Cook 2007).

Beringer *et al.* (2007) estimated that, between 2001 and 2005 in a mesic savanna, wet season fluxes accounted for 72% of mean annual net ecosystem productivity (NEP) of 3.6 tC ha^{-1} yr^{-1}. Importantly, transpiration and carbon assimilation can continue throughout the dry season (O'Grady *et al.* 1999; Eamus *et al.* 2001; Beringer *et al.* 2007) due to the ability of dominant, evergreen eucalypts to access deeper-stored soil water during the dry season (Eamus *et al.* 1999; Bowman and Prior 2005).

Growth rates of savanna trees vary between species, but are remarkably similar between mesic and semi-arid savannas. Cook *et al.* (2005) measured growth increments of ca. 0.15 - 0.25 cm yr^{-1}, for various species in a mesic savanna at Kapalga. Along a MAP gradient of 300–1600 mm yr^{-1}, average diameter increments were very similar to the Kapalga estimate: about 0.24 cm yr^{-1} (Cook *et al.* 2015). Such diameter increments, in conjunction with estimates of water use and C-isotope discrimination, can be used to estimate the carbon sequestration potential of savannas. For example, Cook *et al.* (2005) estimated sequestration of c. 2 tC t ha^{-1} yr^{-1} on shallow soils and 3.8 tC t ha^{-1} yr^{-1} on deep soils. These estimates are very similar to estimates for mesic savannas based on mass-balance and FLUX tower measurements (Chen *et al.* 2003; Beringer *et al.* 2007).

15.5 Savanna vegetation dynamics

Savanna vegetation dynamics depend upon complex interactions between 'bottom-up' drivers of ecosystem function such as rainfall and soil nutrients as discussed above, and disturbance, particularly fire, as a 'top-down' regulator of ecosystem state. Disturbance is integral to savanna ecosystems worldwide. Unlike the savannas of Africa, disturbance due to large, native herbivores does not occur in Australian savannas. Fire, however, is pervasive, especially in the mesic savannas. Other landscape-scale agents of disturbance in northern Australia are cyclones, grazing by both small–medium-sized native vertebrates and invertebrates, and grazing by domestic livestock (Andersen and Lonsdale 1990; Woinarski *et al.* 2007). Here, we explicitly focus on fire–vegetation dynamics, and briefly discuss cyclones. Our approach to understanding disturbance is to focus on the disturbance 'regime', the pattern of recurrence of disturbances in time and space across the landscape.

15.5.1 Savanna fire regimes

Fire in the savannas is frequent and extensive, and intervals between fires are relatively short (generally years rather than decades or centuries; Williams *et al.* 2002; Fensham 2012). Fires affect all savanna landscapes and land tenures: pastoral properties, conservation lands, Indigenous lands, mining leases and, increasingly, the urban–savanna interface (Dyer *et al.* 2001; Andersen *et al.* 2003; Russell-Smith *et al.* 2003a). Savanna fire regimes have been studied intensively over the past 30 years. Early interest centred on understanding the use of fire by Indigenous people (e.g. Braithwaite 1991), a topic of continuing scientific and cultural interest (Russell-Smith *et al.* 2009). There have been several short- to medium-term fire experiments in the Northern Territory and Queensland (Andersen *et al.* 2003; Russell-Smith *et al.* 2003b; Williams *et al.* 2003a,b). Regular, remotely sensed monitoring of fire occurrence at the continental scale is now routine (Russell-Smith *et al.* 2007). Managing savanna fire regimes is a fundamental component of land management in the savannas (Russell-Smith *et al.* 2007).

Savanna fire regimes are a direct consequence of the highly seasonal climate and phenology of the grass layer. During the wet season the growth of grasses and other herbs is prolific. During the dry season the grass layer cures, which, combined with consistent, dry winds, provides sufficient dry fuel to carry fire across the landscape, given any ignition. There is, and has been for millennia, a high incidence of lightning in the transition from dry to wet season. Presently, human ignitions are common. In terms of the continental-scale 'four-switch' model of Australian fire regimes and their determinants (Bradstock 2010), all four switches (fuel amount, fuel availability to burn, fire weather conducive to spread, and ignition) are 'on' on an annual basis in the savannas. As a consequence, mean fire frequency is high, increasing from approximately once every 5 years in the arid savannas to once every 1–2 years in the mesic savannas (Figure 15.5).

Prior to the arrival of Aboriginal people, fire was most likely to have occurred during the dry season/wet season transition period, the only period during the year which has both a dry, connected, combustible fine fuel bed, dry winds and an ignition source, namely lightning. With the arrival of Aboriginal people, ignition patterns changed from solely in the late dry season/build-up, to ignitions throughout the dry season, and potentially into the wet season (Braithwaite 1991; Ens *et al.*, Chapter 4, this volume).

Fine fuel loads vary between c. 2–10 t ha^{-1}, which is comparable to savannas elsewhere in the world. Fuel load varies with time since fire, increasing to average equilibrium values of c. 5–8 t ha^{-1} within 3–4 years of the last fire. Fine fuel loads increase with the progression of dry season, due to leaf fall in deciduous species, and rapid, complete turnover of the canopy leaves in the evergreen eucalypts during the late dry season (Williams *et al.* 1997, 1998). Ladder fuels, such as long strands of bark or tall flammable shrubs, which connect the ground stratum to the tree canopy, are very uncommon in Australian savannas. Fire weather becomes increasingly severe with the progression of the dry season, because relative humidity declines and average wind speed during the day increases. Thus, the Forest Fire Danger Index (FFDI) increases from an

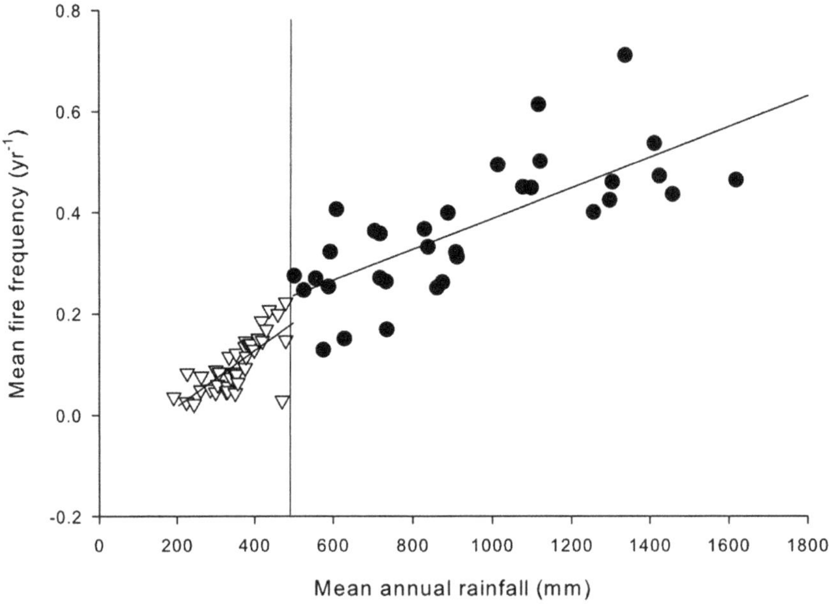

Figure 15.5 The relationship between mean fire frequency and mean annual rainfall. Data from the arid zone, where mean annual rainfall is < 500 mm and aseasonal (triangles), and the tropical savannas, where mean annual rainfall is > 500 mm and seasonal (circles). Fire-frequency data derived from MODIS fire scar mapping 2000–2013 (G. Cook, A. Andersen, A. Edwards, unpublished data).

average monthly maximum of c. 20 in June to c. 40 in September (Gill *et al.* 1996; Williams *et al.* 1998).

There is significant seasonal variation in fire behaviour, due to the seasonal changes in both fuels and fire weather. Late dry season fires are more intense and extensive than early dry season fires. At Kapalga from 1990–1994, mean fire intensity of fronting fires in the annual early dry season (EDS) treatment was c. 2100 kW m^{-1} compared with c. 7700 kW m^{-1} for fires in the annual late dry season (LDS) treatment (Williams *et al.* 1998). There was a similar differential in fire intensity between EDS and LDS fires at nearby Munmarlary over the period 1973–1996 (Russell-Smith *et al.* 2003b). Fires less than 2000 kW m^{-1} generally do not consume the entire fine fuel bed, and the tree canopy is not entirely scorched. In contrast, fires of 4000–5000 kW m^{-1} will consume the whole of the ground stratum, and scorch the whole canopy to heights of c. 20 m. Thus, early dry season fires burn vegetation more patchily than late dry season fires. Flame heights are generally low (0.5–3.0 m), and crown fires do not occur, because there are few ladder fuels, maximum daily FFDI throughout the dry season is generally < 50, and the terrain is relatively flat.

15.5.2 Vegetation, fire and rain interactions

Interactions involving seasonal variation in vegetation growth as affected by inter- and intra-annual patterns of rainfall, and spatial and temporal variation in fire regime components, all have significant implications for vegetation dynamics, and hence the capacity to manage savanna vegetation. Fire impacts on vegetation have been studied extensively through both manipulative and natural experiments, and long-term ground-based and remotely sensed monitoring, in both mesic and semi-arid savannas (Williams *et al.* 2003b). Research into savanna fire regimes, combined with studies on plant demography and plant ecophysiology, has resulted in a predictive understanding of the outcomes of climate-disturbance interactions.

The Kapalga fire experiment demonstrated a high level of resilience of the biota of the mesic savannas to variation in fire regime (Andersen *et al.* 2003). The composition of the grass layer was little affected by fire regime, being more sensitive to variation in annual rainfall, and the cover of the tall annual grass *Sorghum intrans* (Williams *et al.* 2003a). Bowman *et al.* (1988) also concluded that there was little impact of fire regime on the overall floristic composition of the savannas at nearby Munmarlary. Such resilience to burning in the ground vegetation is primarily due to the ground-layer flora being overwhelmingly dominated by resprouters. In the savannas of north-eastern Australia, in contrast, the ground stratum may be sensitive to variation in fire regime. At Cape Cleveland on Cape York, ephemeral forbs increased following late dry season fires, and the dominant grass (*Heteropogon triticeus*) declined in abundance in response to fire intervals of 4 years (P.R. Williams *et al.* 2003).

In contrast to the ground layer, the tree stratum is highly sensitive to variation in fire regime, through fire effects on phenology, recruitment and mortality. At Kapalga, the EDS treatment caused partial canopy scorch in the dominant and subdominant evergreen and semi-deciduous trees, leading to leaf fall and flush within days of scorching. Such post-fire leaf flush is much earlier than the normal peak flush period of September–October (Williams *et al.* 1997b, 2003a). Flowering and fruiting were significantly reduced under the LDS treatment compared with the EDS and 'unburnt' treatments for *Eucalyptus miniata*, *Corymbia porrecta*, and *Terminalia ferdinandiana*. (Williams *et al.* 2003a). For *E. miniata*, dry season fire, whether early or late, can reduce seed output 10-fold, and seedling recruitment 10–100 fold (Setterfield 1997; Williams *et al.* 2003a). Late dry season fires, which coincide with the peak period of the year for both leaf flush and flowering, can dramatically reduce the potential for tree recruitment via seed.

Tree stem survival declines with increasing fire intensity, as does recruitment into the > 3 m height class, particularly in the broadleaf deciduous taxa (Williams *et al.* 2003a). At Kapalga, from 1990 to 1995, the survival of the stems of five of the most common trees – *Eucalyptus miniata*, *E. tetrodonta*, *Corymbia porrecta*, *Erythrophleum chlorostachys* and *Terminalia ferdinandiana* – was significantly higher in the 'unburnt' and EDS treatments than in the LDS treatment. In both EDS and LDS treatments, stem survival varied non-linearly with stem size, and varied between species. For example, in *E. miniata*, stem survival was markedly humped–survival of large stems (> 40 cm diameter at breast height; dbh) was relatively low, and similar to that of saplings; peak survival was in the range of 15 to 30 cm dbh. In contrast, stem survival in the larger size classes was higher in *Eucalyptus tetrodonta* than in *E. miniata*. Survival in the subdominant *Corymbia porrecta*, *Erythrophleum chlorostachys* and *Terminalia ferdinandiana* was substantially lower than that of the dominant eucalypts. In *E. chlorostachys* and *T. ferdinandiana* it was particularly low (< 20%) across virtually all size classes under the LDS treatment. Stem survival in the < 3 m size classes from 1990 to 1995 was very similar across fire regimes (80–90%). However, net recruitment of stems into the > 3 m stratum was significantly higher in 'unburnt' than in EDS treatments, which in turn was significantly higher than the LDS treatment (Williams *et al.* 2003a).

The strong, predictable variation in tree abundance in Australian savannas as a function of annual rainfall, combined with the pronounced fire-mediated variation in tree survival/mortality as a function of fire intensity suggests both bottom-up and top-down control of savanna structure. This enduring question in savanna ecology has been discussed widely (e.g. Bond *et al.* 2005; Sankaran *et al.*

2005; Bond 2008; Lehmann *et al.* 2014; Murphy *et al.* 2015). Determining the relative importance of climate versus disturbance is important because of the global extent of the savannas, the high frequency of fire therein, and the prominent role savannas play in the global carbon cycle. Bond *et al.* (2005) used a modelling approach to 'turn fire off' and found that much of the globe becomes more woody in the absence of fire. In southern African savannas, there is a wide envelope of possible woodiness, with an upper bound set by climate/rainfall, but local expression depends on fire regime (Sankaran *et al.* 2005). Globally, however, there appears to be variation in the relative strength of the two forces (Lehmann *et al.* 2014) with Australian savannas having a lower relative contribution of fire compared to South American and southern African savannas. Using the FLAMES model, Liedloff and Cook (2011) argued that the tree biomass in the savannas of north-western Australia is limited more by water availability than fire. In the savannas of north-eastern Australia, there is also a strong effect of moisture supply on the abundance of trees, with periodic droughts drastically reducing live tree biomass (Fensham and Fairfax 2007; Fensham *et al.* 2009). These findings are consistent with the results of the Bond *et al.* (2005) model in northern Australia: the relative increase in tree biomass is less in northern Australia when fire is turned off, compared with other savannas, and indeed most temperate forest bioregions of the world.

In the global context, a key question about Australian savannas is why they persist at levels of mean annual rainfall up to 2000 mm, and are not replaced by closed forest (Liedloff and Cook 2011). Australian savannas have a longer dry season at higher rainfall than almost anywhere in the world. The length of this dry season appears to be the main determinant of the forest–savanna transition, with eucalypts being able to (i) maintain growth and water use throughout the dry season and (ii) escape the 'fire trap' as saplings more frequently than non-eucalypts (Williams *et al.* 1999b; Cook *et al.* 2002; Bond *et al.* 2012). Fire suppresses the non-eucalypts but does not prevent a relatively high basal area of eucalypts developing at high rainfall with a long dry season. Thus, although there is evidence of a fire-induced recruitment bottleneck in savannas (Prior *et al.* 2010) and tropical eucalypts can escape the 'fire trap' (Bond *et al.* 2012), the mechanism may ultimately be related to drought. The Late Miocene and Pliocene (*c.* 5.3–2.6 Ma) periods coincide with the expansion of aridity across the Australian continent, the contraction of rainforests, expansion of eucalypts and other sclerophyll vegetation types, and the global expansion of savannas and C_4 grasslands. Thus the capacity of savanna eucalypts to escape the 'fire trap' may be a consequence of the evolution, over *c.* 10 million years, of traits that have allowed them to escape the 'drought trap'.

Cyclones, and their interactions with fire, also affect the dynamics of Australian savannas (Cook and Goyens 2008). Major cyclones recur in the savannas, with potentially very extensive and long-lasting impacts. For example, Cyclone Monica (2006), a category 5 cyclone, had peak wind gusts > 300 km hr^{-1}. It felled trees in Arnhem Land over an area of *c.* 7000 km^2, reducing the standing biomass of live trees to near zero in the worst affected areas. The downed woody debris may be consumed by fires, resulting in substantial greenhouse gas emissions. However, there is a compensatory feedback effect with cyclones. Canopy removal, and hence reduction in tree-tree competition, can increase tree growth rates (as measured by diameter increment) from a rate of 2–3 mm yr^{-1}, typical of intact savannas, to 10 mm yr^{-1} where the canopy has been removed (Cook *et al.* 2005).

15.6 Savanna futures

The future of Australian savannas will depend on climate change, land-use change between and within tenures (especially agricultural, conservation and Indigenous lands) and changes in societal norms. There is increasing pressure for intensification of resource use, especially in agriculture. Moreover, Aboriginal people own and manage c. 20% of Australia's savanna biome. Alien plants have established and spread in some areas, affecting fire regimes, nutrient cycling and biodiversity (Rossiter *et al.* 2003; Setterfield *et al.* 2010). Feral animals are widespread (Woinarski *et al.* 2007). Substantial areas of the savanna are subject to inappropriate fire regimes (Russell-Smith *et al.* 2009). Climate change has major implications for savanna vegetation dynamics and fire regimes. The interplay of these ecological, economic and social factors is complex, and beyond the scope of this chapter (but see Whitehead *et al.* 2005; Woinarski *et al.* 2007 for detailed treatments). Here, we focus on how two of these major agents of change – managing fire regimes and global climate change – may influence Australia's savannas in the coming century.

15.6.1 Fire management and savanna vegetation

Managing fire regimes in any biome is complex. It involves understanding the basic biophysical and social drivers of the patterns of recurrence of fire in the landscape; the efficacy of management actions; and the trade-offs between competing objectives (Bond and Archibald 2003; Andersen *et al.* 2005; Bradstock *et al.* 2012a). Managing this complexity has been aided by advances in our understanding of fire ecology, remote sensing of fire regime components, and evaluating management effectiveness (Andersen *et al.* 2003; Williams *et al.* 2003b, 2009; Russell-Smith *et al.* 2009b, 2013; Price *et al.* 2015). This understanding applies across land uses: biodiversity conservation (Andersen *et al.* 2003), livestock grazing (Dyer and Stafford Smith 2003), greenhouse gas abatement and other forms of 'carbon farming' (Cook *et al.* 2010) and customary Indigenous land management (Cook *et al.* 2012).

A fundamental practice in savanna fire management is the use of low-intensity prescribed fires in the early dry season to reduce fuel mass and connectivity across the landscape. This reduces the extent of unplanned, more intense, late dry season fires, thereby reducing area burnt, and fire frequency. Reducing fire frequency, by reducing the extent of late dry season fires, can have multiple environmental benefits. These include reduced tree mortality (Williams *et al.* 2003a); reduced greenhouse gas emissions (Russell-Smith *et al.* 2009; see below) and increased areas of long-unburnt savanna, upon which some key elements of the savanna biota may depend (Andersen *et al.* 2003, 2005). High fire frequency has also been implicated (in association with introduced predators) in the disastrous decline in the abundance of small mammals across northern Australia (Lawes *et al.* 2015; Woinarski *et al.* 2015).

Prescribed burning has relatively high potential for manipulating savanna fire regimes to achieve these various land management objectives because of relatively high 'leverage' (Price *et al.* 2015). Leverage is the unit reduction in unplanned fire area that results from one unit of prior fire at a regional scale over multiple years. Leverage is about 1 in the savannas (compared to c. 0.2–0.3 in temperate eucalypt forests) and thus early dry season fire can supplant late dry season fire without increasing overall fire frequency (Gill *et al.* 2000). Research on the extent to which people can manipulate fire regimes to achieve particular environmental and social outcomes is a priority in fire management for the twenty-first century (Bradstock *et al.* 2012a).

15.6.2 Climate change and savanna vegetation

Climate change may affect the savanna vegetation through direct, indirect and interactive effects. These include increasing atmospheric CO_2 concentrations, increasing temperatures, changes in the moisture and fire regimes, and how people manage savannas.

Increasing atmospheric concentration of CO_2 has the potential to differentially affect trees and grasses within the savanna. Increasing atmospheric CO_2 is likely to benefit trees at the expense of C_4 grasses. In South African savannas, increasing CO_2 has been linked to stimulation of recruitment of trees, and closing of the tree canopy (Wigley *et al.* 2010).

Within the grass stratum, groups such as the Andropogoneae, with the relatively efficient NADP form of C_4 photosynthesis, may decline in abundance, with a concomitant increase in the abundance of groups with the less efficient PCK form of C_4 photosynthesis, such as the Paniceae (Collatz *et al.* 1998). With further increases in CO_2, there may also be a decline in C_4 grasses relative to C_3 grasses (Chamaillé-Jammes and Bond 2010). There are few C_3 grasses in Australian savannas (Wurm 1998) so any increase in the abundance in C_3 grasses is likely to be driven by migration of C_3 taxa into the savannas, not by increases in abundance of C_3 species from within.

Savannas account for about 30% of Australia's terrestrial carbon stocks (Williams *et al.* 2012). The emerging carbon economy will have a major influence on the structure and composition of savannas in the coming century, as fire regimes are managed at landscape scales for carbon benefits. Reducing the incidence of extensive, intense late dry season fires, via early dry season prescribed burning, results in substantial reductions in greenhouse gas emissions (Russell-Smith *et al.* 2009, 2013; Williams *et al.* 2012; Bradstock *et al.* 2012b). This aspect of savanna fire management has been intensively studied, widely validated and increasingly well-resourced by both private and public investments (Russell-Smith *et al.* 2013). Managing to reduce emissions may also bring a number of collateral environmental and social benefits (Cook and Meyer 2009; Russell-Smith *et al.* 2009, 2013; Williams *et al.* 2009; Cook *et al.* 2012). These include enhanced carbon sequestration in trees and soil, improved habitat for rare small mammals, and enhanced livelihoods in remote Aboriginal communities. Such multi-faceted management of fire and vegetation for carbon and other benefits has substantial international application (Russell-Smith *et al.* 2013) and Australian savannas are an ideal place for continued investment in this international experiment in sustainable land management.

In the coming century, the nation must decide how best to manage ecological and social change in the savannas. The ecological research of the past 50 years has provided a robust platform for future research, which will be necessary if we are to understand the complex interactions between the biophysical and socioeconomic environments of this vast, internationally significant biome.

References

Andersen, A.N. and Lonsdale, W.M. (1990) Herbivory by insects in Australian tropical savannas: a review. *Journal of Biogeography* 17, 433–444.

Andersen, A.N., Cook, G.D. and Williams, R.J. (eds) (2003) *Fire in Tropical Savannas. The Kapalga Experiment. Ecological Studies 169*. Springer: New York.

Andersen, A.N., Cook, G.D., Corbett, L.K. *et al.* (2005) Fire frequency and biodiversity conservation in Australian tropical savannas: implications from the Kapalga fire experiment. *Austral Ecology* 30, 155–167.

Beerling, D.J. and Osborne, C.P. (2006) The origin of the savanna biome. *Global Change Biology* 12, 2023–2031.

Beringer, J., Hutley, L.B., Tapper, N.J. and Cernusak, L.A. (2007) Savanna fires and their impact on net ecosystem productivity in North Australia. *Global Change Biology* 13, 990–1004.

Bond, W.J. (2008) What limits trees in C_4 grasslands and savannas? *Annual Review of Ecology, Evolution and Systematics* 39, 641–659.

Bond, W.J. and Archibald, S. (2003) Confronting complexity: fire policy choices in South African savanna parks. *International Journal of Wildland Fire* 12, 381–389.

Bond, W.J. and Midgley, G.F. (2012) Carbon dioxide and the uneasy interactions of trees and savannah grasses. *Proceedings of the Royal Society B* 367, 601–612.

Bond, W.J., Woodward, F.I. and Midgley, G.F. (2005) The global distribution of ecosystems in a world without fire. *New Phytologist* 165, 525–538.

Bond, W.J., Cook, G.D. and Williams, R.J. (2012) Which trees dominate in savannas? The escape hypothesis and eucalypts in northern Australia. *Austral Ecology* 37, 678–685.

Bowman, D.M.J.S. and Prior, L.D. (2005) Turner Review No. 10. Why do evergreen trees dominate the Australian seasonal tropics? *Australian Journal of Botany* 53, 379–399.

Bowman, D.M.J.S., Wilson, B.A. and Hooper, R.J. (1988) Response of *Eucalyptus* forest and woodland to four fire regimes at Munmarlary, Northern Territory, Australia. *Journal of Ecology* 76, 215–232.

Bowman, D.M.J.S., Brown, G.K., Braby, M.F. *et al.* (2010) Biogeography of the Australian monsoon tropics. *Journal of Biogeography* 37, 201–216.

Bradstock, R.A. (2010) A biogeographic model of fire regimes in Australia: current and future implications. *Global Ecology and Biogeography* 19, 145–158.

Bradstock, R.A., Williams, R.J. and Gill, A.M. (2012a) Future fire regimes of Australian ecosystems: new perspectives on enduring questions of management. In *Flammable Australia: Fire Regimes, Biodiversity and Ecosystems in a Changing World*, ed. R.A. Bradstock, R.J. Williams and A.M. Gill. CSIRO Publishing: Melbourne, pp. 307–324.

Bradstock, R., Boer, M., Cary, G. *et al.* (2012b) Modelling the potential of prescribed burning to mitigate carbon emissions in fire-prone, Australian ecosystems. *International Journal of Wildland Fire* 21, 629–639.

Braithwaite, R.W. (1991) Aboriginal fire regimes of monsoonal Australia in the 19th century. *Search* 22, 247–249.

Brennan, K. (1997) *Flowering and Fruiting Phenology of Native Plants in the Alligator Rivers Region with Particular Reference to the Ranger Uranium Mine Lease Area*. Supervising Scientist: Canberra.

Chamaillé-Jammes, S. and Bond, W.J. (2010) Will global change improve grazing quality of grasslands? A call for a deeper understanding of the effects of shifts from C_4 to C_3 grasses for large herbivores. *Oikos* 119, 1857–1861.

Chen, X., Hutley, L.B. and Eamus, D. (2003) Carbon balance of a tropical savanna in northern Australia. *Oecologia* 137, 405–416.

Collatz, G.J., Berry, J.A. and Clark, J.S. (1998) Effects of climate and atmospheric CO_2 partial pressure on the global distribution of C_4 grasses: present, past, and future. *Oecologia* 114, 441–454.

Cook, G.D. (2003) Fuel dynamics, nutrients and atmospheric chemistry. In *Fire in Tropical Savannas: The Kapalga Experiment*, ed. A.N. Andersen, G.D. Cook and R.J. Williams. Springer: New York, pp. 47–58.

Cook, G. (2009) Historical perspectives on land use development in northern Australia: with emphasis on the Northern Territory. In *Northern Australia Land and Water Science Review 2009*. Department of Infrastructure, Transport, Regional Development and Local Government: Canberra. Available at: http://northernaustralia.gov.au/page/water/northern-australia-land-and-water-taskforce (accessed 2 August 2016).

Cook, G.D. and Dias, L. (2006) Turner Review No. 12. It was no accident: deliberate plant introductions by Australian Government agencies during the 20th century. *Australian Journal of Botany* 54, 601–625.

Cook, G.D. and Goyens, C.M.A.C. (2008) The impact of wind on trees in Australian tropical savannas: lessons from Cyclone Monica. *Austral Ecology* 33, 462–470.

Cook, G.D. and Heerdegen, R.G. (2001) Spatial variation in the duration of the rainy season in monsoonal Australia. *International Journal of Climatology* 21, 1723–1732.

Cook, G.D. and Meyer, C.P. (2009) Fire, fuels and greenhouse gases. In *Culture, Ecology and Economy of Savanna Fire Management in Northern Australia: Rekindling the Wurrk Tradition*, ed. J. Russell-Smith, P.J. Whitehead, P.M. Cooke. CSIRO Publishing: Melbourne, pp. 313–327.

Cook, G.D., Williams, R.J., Hutley, L.B. *et al.* (2002) Variation in vegetative water use in the savannas of the North Australian Tropical Transect. *Journal of Vegetation Science* 13, 413–418.

Cook, G.D., Liedloff, A.C., Eager, R.W. *et al.* (2005) The estimation of carbon budgets of frequently burnt tree stands in savannas of northern Australia, using allometric analysis and isotopic discrimination. *Australian Journal of Botany* 53, 621–630.

Cook, G.D., Williams, R.J., Stokes, C.J. *et al.* (2010) Managing sources and sinks of greenhouse gases in Australia's rangelands and tropical Savannas. *Rangeland Ecology and Management* 63, 137–146.

Cook, G.D., Jackson, S., Williams, R.J. (2012) A revolution in northern Australian fire management: recognition of Indigenous knowledge, practice and management. In *Flammable Australia: Fire Regimes, Biodiversity and Ecosystems in a Changing World*, ed. R.A. Bradstock, A.M. Gill and R.J. Williams. CSIRO Publishing: Melbourne, pp. 293–305.

Cook, G.D., Liedloff, A.C., Cuff, N.J. *et al.* (2015) Stocks and dynamics of carbon in trees across a rainfall gradient in a tropical savanna. *Austral Ecology* 40, 845–856.

Dennell, R. and Roebroeks, W. (2005) An Asian perspective on early human dispersal from Africa. *Nature* 438, 1099–1104.

Duff, G.A., Myers, B.A., Williams, R.J. *et al.* (1997) Seasonal patterns in soil moisture, vapour pressure deficit, tree canopy cover and pre-dawn water potential in a northern Australian savanna. *Australian Journal of Botany* 45, 211–224.

Dyer, R. and Stafford Smith, M. (2003) Ecological and economic assessment of prescribed burning impacts in semi-arid pastoral lands of northern Australia. *International Journal of Wildland Fire* 12, 403–413.

Dyer, R., Jacklyn, P., Partridge, I. *et al.* (2001) *Savanna Burning: Understanding and Using Fire in Northern Australia*. Tropical Savannas Cooperative Research Centre: Darwin.

Eamus, D. and Prior, L.D. (2001) Ecophysiology of trees of seasonally dry tropics: comparisons among phenologies. *Advances in Ecological Research* 32, 113–197.

Eamus, D., Myers, B., Duff, G. and Williams, R. (1999) Seasonal changes in photosynthesis of eight savanna tree species. *Tree Physiology* 19, 665–671.

Eamus, D., Hutley, L.B. and O'Grady, A.P. (2001) Daily and seasonal patterns of carbon and water fluxes above a north Australian savanna. *Tree Physiology* 21, 977–988.

Edwards, E.J., Osborne, C.P., Strömberg, C.A.E. *et al.* (2010) The origins of C4 grasslands: integrating evolutionary and ecosystem science. *Science* 328, 587–591.

Egan, J.L. and Williams, R.J. (1996) Lifeform distributions of woodland plant species along a moisture availability gradient in Australia's monsoonal tropics. *Australian Systematic Botany* 9, 205–217.

Environment Australia (2000) *Interim Biogeographic Regionalisation of Australia (IBRA) Version 5.1.*

Fensham, R. (2012) Fire regimes in Australian tropical savanna: perspectives, paradigms and paradoxes. In *Flammable Australia: Fire Regimes, Biodiversity and Ecosystems in a Changing World*, ed. R.A. Bradstock, A.M. Gill and R.J. Williams. CSIRO Publishing: Melbourne, pp. 173–193.

Fensham, R.J. and Fairfax, R.J. (2007) Drought-related tree death of savanna eucalypts: species susceptibility, soil conditions and root architecture. *Journal of Vegetation Science* 18, 71–80.

Fensham, R.J., Fairfax, R.J. and Ward, D.P. (2009) Drought-induced tree death in savanna. *Global Change Biology* 15, 380–387.

Fisher, A. (2001) Biogeography and conservation of Mitchell grasslands in northern Australia. PhD Thesis, Northern Territory University, Darwin.

Fox, I.D., Neldner, V.J., Wilson, G.W. and Bannick, B.W. (2001) *The Vegetation of the Australian Tropical Savannas*. Environmental Protection Agency: Brisbane.

Garnett, S., Woinarski, J., Gerritsen, R. and Duff, G. (2008) *Future Options for North Australia*. Charles Darwin University Press: Darwin.

Gill, A.M., Moore, P.H.R. and Williams, R.J. (1996) Fire weather in the wet–dry tropics of the World Heritage Kakadu National Park, Australia. *Australian Journal of Ecology* 21, 302–308.

Gill, A.M., Ryan, P.G., Moore, P.H.R. and Gibson, M. (2000) Fire regimes of World Heritage Kakadu National Park, Australia. *Austral Ecology* 25, 616–625.

Grace, J., San Jose, J., Meir, P. *et al.* (2006) Productivity and carbon fluxes of tropical savannas. *Journal of Biogeography* 33, 387–400.

Higgins, S.I., Bond, W.J. and Trollope, W.S.W. (2000) Fire, resprouting and variability: a recipe for grass–tree coexistence in savanna. *Journal of Ecology* 88, 213–229.

Isbell, R.F. (1996) *The Australian Soil Classification*. CSIRO: Melbourne.

Kuleshov, Y. (2012) Thunderstorm and lightning climatology of Australia. *Modern Climatology* 11, 85–120.

Lawes, M.J., Murphy, B.P., Fisher, A. *et al.* (2015) Small mammals decline with increasing fire extent in northern Australia: evidence from long-term monitoring in Kakadu National Park. *International Journal of Wildland Fire* 24, 712–722.

Lehmann, C.E.R., Archibald, S.A., Hoffmann, W.A. and Bond, W.J. (2011) Deciphering the distribution of the savanna biome. *New Phytologist* 191, 197–209.

Lehmann, C.E.R., Anderson, M.T., Sankaran, M. *et al.* (2014) Savanna vegetation-fire-climate relationships differ among continents. *Science* 343, 548–552.

Liedloff, A.C. and Cook, G.D. (2007) Modelling the effects of rainfall variability and fire on tree populations in an Australian tropical savanna using the Flames simulation model. *Ecological Modelling* 201, 269–282.

Liedloff, A.C. and Cook, G.D. (2011) The interaction of fire and rainfall variability on tree structure and carbon fluxes in savannas: application of the Flames model. In *Ecosystem Function in Savannas: Measurement and Modeling at Landscape to Global Scales*, ed. M.J. Hill and N.P. Hanan. CRC/Taylor & Francis: Boca Raton, pp. 293–308.

Lloyd J., Bird, M.I., Vellen, L. *et al.* (2008) Contributions of woody and herbaceous vegetation to tropical savanna ecosystem productivity: a quasi-global estimate. *Tree Physiology* 28, 451–468.

Ludwig, J.A., Tongway, D.J., Eager, R.W. *et al.* (1999) Fine-scale vegetation patches decline in size and cover with increasing rainfall in Australian savannas. *Landscape Ecology* 14, 557–566.

McKenzie, N.J., Jacquier, D., Isbell, R.F. and Brown, K. (2004) *Australian Soils and Landscapes: An Illustrated Compendium*. CSIRO Publishing: Melbourne.

Murphy, B.P. and Bowman, D.M.J.S.B. (2012) What controls the distribution of tropical forest and savanna? *Ecology Letters* 15, 748–758.

Murphy, B.P., Liedloff, A.C. and Cook, G.D. (2015) Does fire limit tree biomass in Australian savannas? *International Journal of Wildland Fire* 24, 1–13.

Myers, B.A., Duff, G.A., Eamus, D. *et al.* (1997) Seasonal variation in water relations of trees of differing leaf phenology in a wet–dry tropical savanna near Darwin, northern Australia. *Australian Journal of Botany* 45, 225–240.

Myers, B.A., Williams, R.J., Fordyce, I.R., Duff, G.A. and Eamus, D. (1998) Does irrigation affect leaf phenology in deciduous and evergreen trees of the savannas of northern Australia? *Australian Journal of Ecology* 23, 329–339.

O'Grady, A.P., Eamus, D. and Hutley, L.B. (1999) Transpiration increases during the dry season: patterns of tree water use in eucalypt open forests of northern Australia. *Tree Physiology* 19, 591–597.

Price, O.F., Pausas, J.G., Govender, N. *et al.* (2015) Global patterns in fire leverage: the response of annual area burnt to previous fire. *International Journal of Wildland Fire* 24, 297–306.

Prior, L.D., Williams, R.J. and Bowman, D.M.J.S. (2010) Experimental evidence that fire causes a tree recruitment bottleneck in an Australian tropical savanna. *Journal of Tropical Ecology* 26, 595–603.

Rossiter, N.A., Setterfield, S.A., Douglas, M.M. and Hutley, L.B. (2003) Testing the grass-fire cycle: alien grass invasion in the tropical savannas of northern Australia. *Diversity and Distributions* 9, 169–176.

Russell-Smith, J., Whitehead, P.J., Williams, R.J. and Flannigan, M. (eds) (2003a) Fire and savanna landscapes in northern Australia – regional lessons and global challenges. *International Journal of Wildland Fire* 12, v–ix.

Russell-Smith, J., Whitehead, P.J., Cook, G.D. and Hoare, J.D. (2003b) Response of *Eucalyptus*-dominated savanna to frequent fires: lessons from Munmarlary, 1973–1996. *Ecological Monographs* 73, 349–375.

Russell-Smith, J., Yates, C.P., Whitehead, P.J. *et al.* (2007) Bushfires 'down under': patterns and implications of contemporary Australian landscape burning. *International Journal of Wildland Fire* 16, 361–377.

Russell-Smith, J., Whitehead, P. and Cooke, P. (eds) (2009) *Culture, Ecology and Economy of Fire Management in North Australian Savannas: Rekindling the* Wurrk *Tradition*. CSIRO Publishing: Melbourne.

Russell-Smith, J., Monagle, C., Jacobsohn, M. *et al.* (2013) Can savanna burning projects deliver measurable greenhouse emissions reductions and sustainable livelihood opportunities in fire-prone settings? *Climatic Change* DOI 10.1007/s10584-013-0910-5.

Sankaran, M., Hanan, N.P., Scholes, R.J. *et al.* (2005) Determinants of woody cover in African savanna. *Nature* 438, 846–849.

Setterfield, S.A. (1997) The impact of experimental fire regimes on seed production in two tropical eucalypt species in northern Australia. *Australian Journal of Ecology* 22, 279–287.

Setterfield, S.A. and Williams, R.J. (1996) Patterns of flowering and seed production in Eucalyptus miniata and *Eucalyptus tetrodonta* in a tropical savanna woodland, northern Australia. *Australian Journal of Botany* 44, 107–122.

Setterfield, S.A., Rossiter-Rachor, N.A., Hutley, L.B. *et al.* (2010) Turning up the heat: the impacts of *Andropogon gayanus* (gamba grass) invasion on fire behaviour in northern Australian savannas. *Diversity and Distributions* 16, 854–861.

Specht, R.L. (1981) Ecophysiological principles determining the biogeography of major vegetation formations in Australia. In *Ecological Biogeography of Australia*, ed. A. Keast. Dr Junk: The Hague, pp. 299–334.

Suppiah, R. (1992) The Australian summer monsoon: a review. *Progress in Physical Geography* 16, 283–318.

van der Werf, G.R., Randerson, J.T., Giglio, L. *et al.* (2008) Climate controls on the variability of fires in the tropics and subtropics. *Global Biogeochemical Cycles* 22, GB3028.

Werner, P.A. (ed.) (1991) *Savanna Ecology and Management. Special Edition of the Journal of Biogeography*. Blackwell Scientific Publications: Oxford.

Whitehead, P.J., Russell-Smith, J. and Woinarski, J.C.Z. (2005) Fire, landscape heterogeneity and wildlife management in Australia's tropical savannas: introduction and overview. *Wildlife Research* 32, 369–375.

Wigley, B.J., Bond, W.J. and Hoffman, M. (2010) Thicket expansion in a South African savanna under divergent land use: local vs. global drivers? *Global Change Biology* 16, 964–976.

Williams, M.A.J. (1991) Evolution of the landscape. In *Monsoonal Australia. Landscape, Ecology and Man in the Northern Lowlands*, ed. C.D. Haynes, M.G. Ridpath and M.A.J. Williams. Balkema: Rotterdam, pp. 5–17.

Williams, P.R., Congdon, R.A., Grice, A.C. and Clarke, P.J. (2003) Effect of fire regime on plant abundance in a tropical eucalypt savanna of north-eastern Australia. *Austral Ecology* 28, 327–338.

Williams, R.J. and Farrer, S. (eds) (2005) Carbon accounting, land management, science and policy uncertainty in Australian savanna landscapes. *Australian Journal of Botany* 53, 583–747.

Williams, R.J., Duff, G.A., Bowman, D.M.J.S. and Cook, G.D. (1996) Variation in the composition and structure of tropical savannas as a function of rainfall and soil texture along a large-scale climatic gradient in the Northern Territory, Australia. *Journal of Biogeography* 23, 747–756.

Williams, R.J., Cook, G.D., Ludwig, J.L. and Tongway, D. (1997a) Torch, trees, teeth and tussocks: disturbance in the tropical savannas of the Northern Territory (Australia). In *Frontiers in Ecology. Building the Links*, ed. N. Klomp and I. Lunt. Elsevier: Oxford, pp. 55–66.

Williams, R.J., Myers, B.A., Müller, W.J. *et al.* (1997b) Leaf phenology of woody species in a northern Australian tropical savanna. *Ecology* 78, 2542–2558.

Williams, R.J., Gill, A.M. and Moore, P.H.R. (1998) Seasonal changes in fire behaviour in a tropical savanna in northern Australia. *International Journal of Wildland Fire* 8, 227–239.

Williams, R.J., Myers, B.A., Eamus, D. and Duff, G.A. (1999a) Reproductive phenology of woody species in a north Australian tropical savanna. *Biotropica* 31, 626–636.

Williams, R.J., Cook, G.D., Gill, A.M. and Moore, P.H.R. (1999b) Fire regime, fire intensity and tree survival in a tropical savanna in northern Australia. *Australian Journal of Ecology* 24, 50–59.

Williams, R.J., Griffiths, A.D. and Allan, G.E. (2002) Fire regimes and biodiversity in the wet–dry tropical landscapes of northern Australia. In *The Fire Regimes and Biodiversity of a Continent*, ed. R.A. Bradstock, J.E. Williams and A.M. Gill. Cambridge University Press: Cambridge, pp. 281–304.

Williams, R.J., Müller, W.J., Wahren, C.H. *et al.* (2003a) Fire impacts on savanna vegetation. In *Fire in Tropical Savannas: The Kapalga Fire Experiment*, ed. A.N. Andersen, G.D. Cook and R.J. Williams. Springer-Verlag: New York, pp. 79–106.

Williams, R.J., Woinarski, J.C.Z. and Andersen, A.N. (2003b) Fire experiments in northern Australia: contributions to ecological understanding and biodiversity conservation in tropical savannas. International *Journal of Wildland Fire* 12, 391–402.

Williams, R.J., Zerihun, A., Montagu, K.D. *et al.* (2005b) Allometry for estimating aboveground tree biomass in tropical and subtropical eucalypt woodlands: towards general predictive equations. *Australian Journal of Botany* 53, 607–619.

Williams, R.J., Barrett, D., Cook, G. *et al.* (2009) Landscape-scale fire research in northern Australia: delivering multiple benefits in a changing world. In *Culture, Ecology and Economy of Savanna Fire Management in Northern Australia: Rekindling the Wurrk Tradition*, ed. J. Russell-Smith, P.J. Whitehead, P.M. Cooke. CSIRO Publishing: Melbourne, pp. 181–200.

Williams, R.J., Bradstock, R.A., Barrett, D. *et al.* (2012) Fire regimes and carbon in Australian vegetation. In *Flammable Australia: Fire Regimes, Biodiversity and Ecosystems in a Changing World*, ed. R.A. Bradstock, A.M. Gill and R.J. Williams. CSIRO Publishing: Melbourne, pp. 273–292.

Woinarski, J.C.Z., Williams, R.J., Price, O. and Rankmore, B. (2005) Landscapes without boundaries: wildlife and their environments in northern Australia. *Wildlife Research* 32, 377–388.

Woinarski, J., Mackey, B., Nix, H. and Traill, B. (2007) *The Nature of Northern Australia. Natural Values, Ecological Processes and Future Prospects*. ANU E-Press: Canberra.

Woinarski, J.C.Z., Burbidge, A.A. and Harrison, P.L. (2015) Ongoing unraveling of a continental fauna: decline and extinction of Australian mammals since European settlement. *Proceedings of the National Academy of Sciences of the United States of America* 112, 4531–4540.

Wurm, P. (1998) The population ecology of *Oryza meridionalis* Ng on the South Alligator River floodplain, Kakadu National Park, monsoonal Australia. PhD Thesis, Northern Territory University, Darwin.

16 Brigalow Forests and Associated Eucalypt Woodlands of Subtropical Eastern Australia

ROD J. FENSHAM, ANDREW J.W. BIGGS, DON W. BUTLER AND HARRY J. MACDERMOTT

Brigalow forest and woodlands dominated by box and ironbark *Eucalyptus* occur on deep-soil peneplains in sub-humid climates inland from the subtropical east coast of Australia. The Brigalow occupies clay-rich soils, the ironbark woodlands sandy soils and the box woodlands intermediate texture-contrast soils. In its natural state brigalow has low grass cover and is therefore generally fire retardant. The eucalypt woodlands have grassy understories but flammable fuels can be constrained during drought. Periods of drought can result in high mortality of adult trees and seedling regeneration is initiated in high rainfall periods. The eucalypts are extremely fire resistant. The paradigm that the vegetation has 'thickened' because of livestock grazing and a lack of fire under European management is not supported by the evidence, which instead highlights the importance of rainfall patterns for mortality and regeneration. Grazing can have a substantial impact on floristics where grazing is far more intense than average under commercial pastoralism. Otherwise fire and moderate grazing have a similar effect to drought, allowing for a flush of ephemeral plants because the perennial grasses have been suppressed. The brigalow vegetation has been reduced to about 10% of its former area and the remaining areas are heavily degraded after the proliferation of exotic grasses such as buffel grass (*Cenchrus ciliaris*) which has provided fuel for destructive fires. A suite of rare plant species are threatened by the same process. The *Eucalyptus* woodlands have also been extensively cleared, and buffel grass diminishes floristic diversity in the remnants. Brigalow and *Eucalyptus* vegetation can recover after it has been cleared but livestock grazing will be required to manage the threat of fires that burn with exaggerated intensity because of exotic grasses.

16.1 Introduction

On the eastern side of the Australian continent, distinctive brigalow vegetation and associated eucalypt woodlands occur in a belt of country with a land-use history characterised by very recent and rapid transformation. This area is mostly set back from the coast and spans subhumid and semi-arid subtropical climates. It includes the Desert Uplands, Brigalow Belt North, and Brigalow Belt South biogeographic regions plus the eastern section of the mulga lands (Figure 16.1). These regions have a diversity of landforms from rugged mountain ranges to floodplains. Some of the vegetation types in this region such as dry rainforest (Metcalfe and Green, Chapter 11, this volume), tussock grasslands (Morgan *et al.*, Chapter 18, this volume), floodplain forests and woodlands (Good *et al.*, Chapter 21, this

Australian Vegetation, Third Edition, ed. David A. Keith. Published by Cambridge University Press.

Citation: Fensham, R.J. *et al.* (2017). Brigalow Forests and Associated Eucalypt Woodlands of Subtropical Eastern Australia. In D. A. Keith, ed., *Australian Vegetation*, 3rd edn. Cambridge: Cambridge University Press, pp. 389–409.

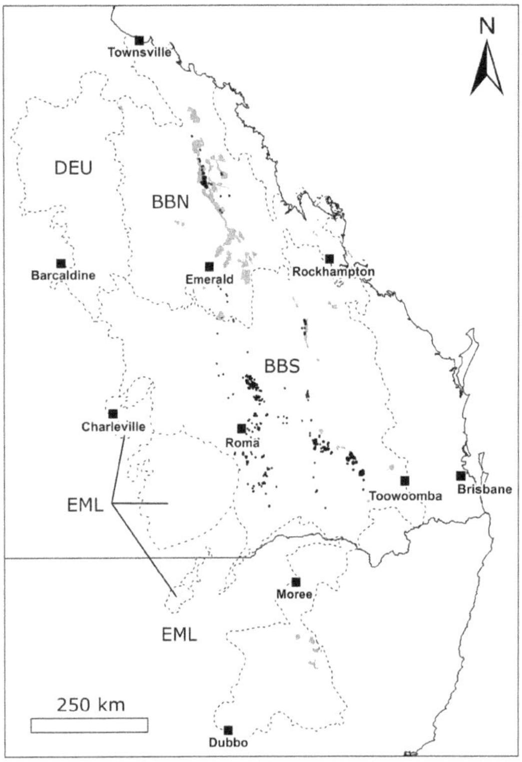

Figure 16.1 The extent of the Desert Uplands (DEU), Brigalow Belt North (BBN), eastern Mulga Lands (EML) and Brigalow Belt South (BBS) bioregions containing the woodlands and forests of brigalow, poplar box, silver-leaved ironbark and/or narrow-leaved ironbark described in this chapter. Active mining leases (grey) and coal seam gas developments (black dots) are indicated (see 16.7.4 Mining).

volume) and wetlands (Catford *et al.*, Chapter 20, this volume) are addressed within other chapters.

The scope of this chapter is the woodlands and forests of the flat and rolling peneplains occurring between the hills and the floodplains, where brigalow (*Acacia harpophylla*), and box and ironbark *Eucalyptus* are the dominant species (Figure 16.2). Of these types the most ecologically and botanically distinctive is the brigalow. Taxonomically, brigalow sits within an informal but distinctive 'Microneurae group' (Pedley 1986) within the section Plurinerves of *Acacia*, which is characterised by globular inflorescences, thick leaves and a thin seed coat ensuring germination can only occur immediately after shedding. Like brigalow, many of the other members of the Microneurae have also preserved their Aboriginal names, probably because of their ecological significance as the dominant species on fertile clay soils throughout north-east and central Australia. Brigalow and myall (*Acacia pendula*) occur in high-rainfall areas, yarran (*Acacia homalophylla*) extends furthest south, gidgee (*Acacia cambagei*) replaces brigalow as the climate becomes more arid, and boree (*Acacia tephrina*) forms an open woodland over grassland. Brigalow is ecologically distinct with a unique combination of characteristics relative to the other major vegetation types within their geographic domain, notably rainforest, eucalypt and mulga dominated vegetation (Table 16.1). The *Eucalyptus* woodlands form a continuum across the continent intermediate between the tropical savannas (Williams *et al.*, Chapter 15, this volume) and temperate woodlands (Prober *et al.*, Chapter 17, this volume) with floristic elements from both.

Grasslands (mostly dominated by *Dichanthium sericeum*) on clay soils and other *Eucalyptus* woodlands on rocky hills occur amidst the brigalow, box and ironbark woodlands. The box woodlands in the eastern Mulga Lands bioregion grade into mulga vegetation (Nano *et al.*, Chapter 25, this volume) as rainfall declines to the west.

16.2 The structure and composition of the vegetation types

16.2.1 Brigalow vegetation

The structure of brigalow vegetation is distinguished from the box and ironbark *Eucalyptus* woodlands by the higher density of trees and the lower coverage of grass and other herbs. This distinction has important implications for the ecology and land use of these systems.

Figure 16.2 Subtropical brigalow forests and eucalypt woodlands. (a) Brigalow forest formed by pure *Acacia harpophylla* (photo: Teresa Eyre); (b) box forest dominated by *Eucalyptus populnea* (photo: Teresa Eyre); (c) silver-leaved ironbark forest dominated by *Eucalyptus melanophloia* (photo: Rod Fensham); (d) landscape dominated by regrowth brigalow (*Acacia harpophylla*) (photo: Don Butler); (e) brigalow (*Acacia harpophylla*) forest degraded by fires fuelled by buffel grass (*Cenchrus ciliaris*) seen flowering in the foreground (photo: Teresa Bracknell).

(A black and white version of this figure will appear in some formats. For the colour version, please refer to the plate section.)

Table 16.1 **Some ecological characteristics of brigalow vegetation relative to rainforest, subtropical eucalypt and mulga dominated vegetation.**

	Brigalow	Rainforest	Eucalypt	Mulga
Tree species	few	many	few	few
Fire retardance	high	high	low	moderate
Tolerance of adults to fire	low	low	high	moderate
Root suckering capacity	high	moderate	none	none
Seedling regeneration	occasional	continuous	occasional	continuous
Tree soil seed bank	no	no	no	yes

Table 16.2 **Mean (standard error) of structural characteristics of undisturbed brigalow forest, poplar box woodland and ironbark woodland from between latitude −22.27° and −25.97°S; and longitude 145.87° and 150.83°E in plots from 1200 m^2 to 4000 m^2 depending on stem densities (N – number of replicate plots). Data from Fensham *et al.* (2003b) and Lucas *et al.* (2010). Stems are >2 m tall.**

Vegetation type	Canopy cover < 5 m high	Canopy cover > 5 m high	Basal area (m^2. ha^{-1})	Biomass (t. ha^{-1})	Stem density (ha^{-1})	N
Brigalow vegetation	10.54 (1.38)	42.91 (4.68)	20.30 (2.72)	131.47 (25.50)	515 (71)	14
Box woodlands	5.22 (2.48)	24.49 (3.57)	12.98 (1.58)	94.17 (9.16)	174 (41)	14
Ironbark woodlands	5.58 (1.83)	28.60 (3.58)	14.29 (1.87)	116.18 (19.79)	142 (32)	11

Brigalow vegetation is dominated by the brigalow tree *Acacia harpophylla*, which can form woodlands but mostly forms forest with crown cover of 50% or more and has relatively high stem densities (Table 16.2). Brigalow can occur in pure stands, or with emergent blackbutt (*Eucalyptus cambageana*) or mountain yapunyah (*Eucalyptus thozetiana*), or in mixtures with belah (*Casuarina cristata*) or yellow-wood (*Terminalia oblongata*), and can grade gradually into dry rainforest with a diverse range of canopy species (Fensham 1995). The narrow-leaved bottle tree (*Brachychiton rupestris*) and ooline (*Cadellia pentastylis*) are particularly distinctive trees associated with brigalow vegetation in some areas. Other associated small tree and shrub species are *Santalum lanceolatum, Citrus glauca, Geijera parviflora, Eremophila deserti* and *E. mitchellii. Carissa ovata* and *Capparis lasiantha* are common low shrubs with multiple flexuous and thorny stems. Semi-succulent subshrubs (chenopods) in Amaranthaceae, such as *Einadia nutans, Enchylaena tomentosa, Rhagodia spinescens* and *Sclerolaena tetracuspis* are common. Other subshrubs include *Abutilon oxycarpum, Sida fibulifera, S. trichopoda* and *Solanum parvifolium.*

The foliage in brigalow canopies is typically dense enough to suppress the biomass of the herbaceous layer beneath them. Common grasses forming a sparse ground layer in undisturbed brigalow include *Ancistrachne uncinulata, Enteropogon acicularis, Paspalidium constrictum, Paspalidium caespitosum* and *Sporobolus caroli.* Common herbs include succulent (e.g. *Zygophyllum apiculatum, Trianthema triquetra*) and non-succulent forms (e.g. *Brunoniella australis, Nyssanthes erecta, Rostellularia adscendens*). Brigalow stands become more open with declining rainfall but there are also considerable variations in structure probably related to soil conditions. There are floristic gradients from the south to the north with examples including the association of narrow-leaved bottle tree (*Brachychiton rupestris*) with brigalow in the south, and yellow-wood (*Terminalia oblongata*) in the north.

Gilgai microrelief is a common feature of soils in brigalow and belah communities and typically takes the form of 'melon-hole' depressions 5–30 m in diameter, with up to 2 m relief (Russell *et al.* 1967). The larger melon-holes form seasonal wetlands where brigalow and other trees are generally excluded (Johnson 1980). These were diverse environments with fine-scaled alternation of wetlands in the hollows and brigalow forest on the hummocks. Brigalow could also form a low diversity 'whipstick' structure of dense small stems less than 10 m tall (Johnson 1964).

The brigalow and the box woodlands were overrun with the central American cactus prickly pear (*Opuntia stricta*) during the early twentieth century. After a huge but futile investment of money and effort the cactus was famously subdued by a biological control, the lava of the moth *Cactoblastis cactorum* (Nix 1994). Prickly pear and a range of other Cactaceae (*Eriocereus martinii, Opuntia tomentosa, O. aurantiaca*) from Central and South America seem well-adapted to the environment of brigalow, where they are widespread. Another notable exotic is the African rubber vine (*Cryptostegia grandiflora*), but none of these exotics transform the brigalow ecosystem to the extent of a suite of high-biomass African perennial grasses (see 16.7.3 Invasion by exotic grasses).

Queensland's regional ecosystem framework preordains bioregion, landzone and dominant plant species as the primary determinants of ecosystem classification (Wilson *et al.* 2002). There has been no analysis of the adequacy of this system for classifying brigalow vegetation communities, but the four bioregions are likely to segment continuous variation related to climate effects, while the various landzones within the bioregions distinguish metamorphic geology where brigalow often occurs in banded patterns with eucalypt forest (landzone 11), fine-grained shales that frequently have a rolling topography, ancient flat clay sheets with gilgais (landzone 4), and floodplains that are regularly inundated (landzone 3). Finally the presence of emergent *Eucalyptus cambageana* or *E. thozetiana* distinguishes edaphic environments that probably underpin distinctive floristic assemblages.

16.2.2 Box woodlands

Box woodlands are dominated by poplar box (*Eucalyptus populnea*) in southern areas and by Reid-river box (*Eucalyptus brownii*) in the north. These species form a broad hybrid swarm in the vicinity of −22°S latitude (Holman *et al.* 2003). Woodlands dominated by *Eucalyptus conica* and *E. microcarpa* occur in similar environmental settings in the south as does *E. microneura* in the north. Other box eucalypts including *E. tectifica* and *E. chlorophylla* extend into the tropics and like the boxes of this chapter also predominate on sodic duplex soils (see 16.4 Soil–vegetation associations). Box eucalypts can also occur as an emergent over brigalow but more typically occur in subtropical savanna woodlands (20–50% canopy cover, Table 16.2) with an understorey formed by a patchy distribution of many perennial grasses (e.g. *Bothriochloa decipiens, Bothriochloa ewartiana, Themeda triandra, Chrysopogon fallax, Paspalidium constrictum* and *Dichanthium sericeum*). Associated trees and shrubs include *Archidendropsis basaltica, Atalaya hemiglauca, Bauhinia carronii, Eremophila mitchellii, Geijera parviflora* and *Grevillea striata.*

16.2.3 Ironbark woodlands

The ironbark woodlands can be segregated based on the colour of the foliage of the dominant *Eucalyptus. Eucalyptus melanophloia* is the most extensive 'silver-leaved' species but this species gradually grades into *Eucalyptus whitei* through the Desert Uplands in the north-west. The green-leaved woodland ironbarks are generally referred to as 'narrow-leaved ironbarks' with *Eucalyptus crebra* being most common, *Eucalyptus xanthoclada* (now regarded as part of *E. crebra*) occurring in some northern areas, and numerous more localised species distributed across the region. The silver-leaved ironbark and narrow-leaved ironbark woodlands have a broad area of overlap where they have very similar structure and floristics. As a broad generalisation, narrow-leaved ironbark woodlands are more common closer to the coast and silver-leaved ironbark woodlands extend further inland. Also, narrow-leaved ironbark woodlands are more likely than silver-leaved ironbark woodlands to extend from the plains to rocky hills. Despite similar structure (Table 16.2) ironbark woodlands are floristically distinct from poplar box woodlands (Fensham 1998), including a different suite of dominant grasses (e.g. *Bothriochloa ewartiana, Themeda triandra, Triodia pungens, Chrysopogon fallax, Heteropogon contortus*). Shrubs strongly associated with ironbark woodlands include *Acacia coriacea, Hakea lorea, Melaleuca nervosa* and *Petalostigma pubescens. Terminalia aridicola* is restricted to northern areas and provides a floristic link to the tropics.

16.3 Climate

Mean annual rainfall in the study region is between 450 mm and 900 mm, with mean monthly maximum winter temperatures between 16.8°C and 25.3°C and mean monthly maximum summer temperatures between 31.1°C and 33.8°C. Frosts occur throughout but are more common in southern areas. Rainfall is highly variable and extended droughts (defined as less than 66% of expected rainfall over 3-year periods) have occurred once since 1900 in the Brigalow Belt North and seven times in the eastern Mulga Lands area. The Brigalow Belt North has fewer extremes in terms of 3-yearly droughts (Figure 16.3) because of the reliability of rainfall in its coastal parts. Extremely wet periods equivalent to 150% of expected rainfall over a 3-year period have occurred a couple of times since 1900. In general, the first half of the twentieth century was drier than the second half, with the 'Federation Drought' culminating in 1903 being the most extreme on record for most regions and the 1950s being a particularly wet decade.

There is a widely held view that there has been a general progression in brigalow vegetation, as well as subtropical box and ironbark forests, from a park-like structure maintained by Aboriginal burning to a much denser structure after more than a century of commercial livestock grazing and associated fire exclusion (Burrows *et al.* 2002; Gammage 2012; see also Fensham 2008). The implications of grasslands converting to forests and woodlands are substantial for the grazing industry because trees generally suppress grass. Quantified studies using aerial photography suggest that 'vegetation thickening' has occurred in *Eucalyptus* woodland (Fensham *et al.* 2003b), albeit at rates much lower than suggested by more intensive but less comprehensive studies (Burrows *et al.* 2002). This thickening has occurred during the relatively wet decades of the second half of the twentieth century (Figure 16.3). Thickening must be considered in association with extensive tree death that occurs during drought (Fensham and Holman 1999; Fensham *et al.* 2015a). Regional contraction of tree cover during drought is balanced against thickening over wetter periods (Fensham *et al.* 2009). A correlative study revealed tree cover change in the Brigalow Belt North and Desert Uplands was related to rainfall history, but not grazing intensity (Fensham *et al.* 2005), and 10 years of monitoring revealed substantial stem death during drought but negligible impacts of fire (Fensham *et al.* 2003a). Long-term and experimental studies confirm that the variable climate of the region is directly reflected in vegetation structure.

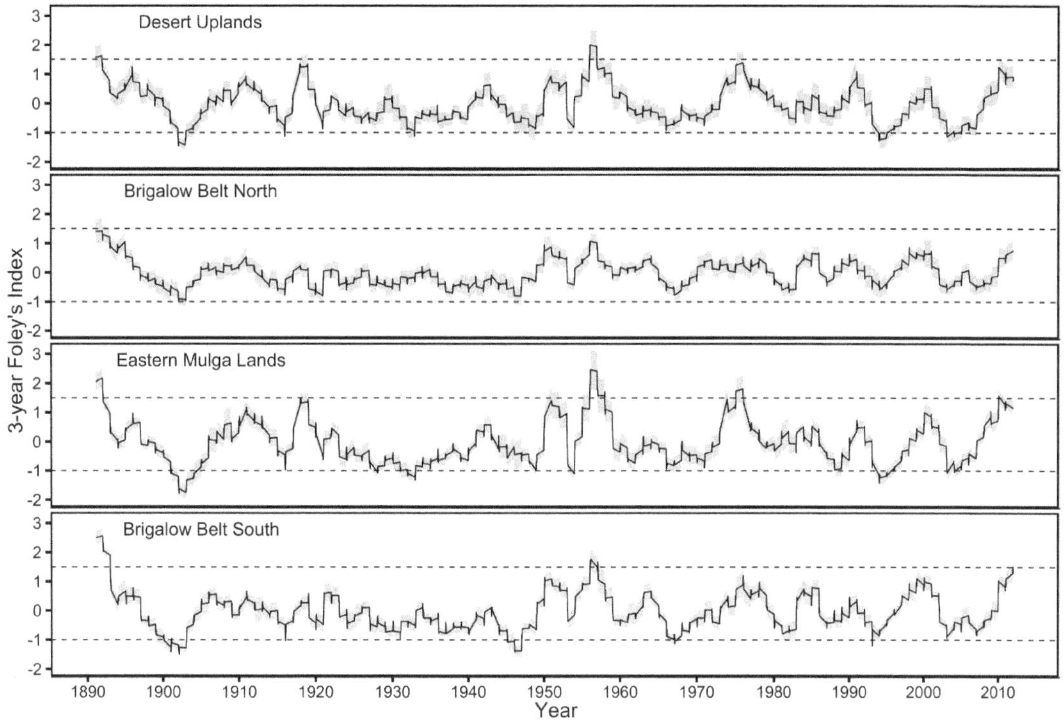

Figure 16.3 Average (and standard deviation) 3-year Foley's rainfall index (3-year prior rainfall less 3-year expected rainfall divided by mean annual rainfall) throughout the period of available rainfall record, from intersections with a one-degree grid in each of the four target regions. Rainfall data is modelled from SILO (Jeffrey *et al.* 2001). When the Foley's Index falls below -1 the previous 3-year rainfall is less than 66.7% of expected. When Foley's Index is above +1 rainfall the previous 3-year rainfall is greater than 150% of expected.

16.4 Soil–vegetation associations

Brigalow forests occur on clay-rich soils on both flat landscapes formed on depositional plains and on undulating erosional landscapes formed on fine-grained sedimentary rocks. Despite the dramatic differences in structure and composition, both grassland and brigalow soils occur on very similar clay soil types, i.e. vertosols and dermosols (Table 16.3) with relatively high fertility. Within the bioregions under consideration, brigalow forest tends to occur on clay-rich soils derived from sedimentary rocks while grasslands are more typically associated with soils derived from basalt (Figure 16.4). However, there were extensive areas where brigalow forest and grassland occurred in a mosaic known as 'patchy plain'. The brigalow soils tend to have lower clay content through the profile, lower pH and higher sodicity and salinity than grasslands, although this difference is not as evident when non-basaltic grasslands, outside of the study bioregions, are included in the analysis. The salinity and sodicity in soils that support brigalow and belah (*Casuarina cristata*) is an important soil characteristic that explains the preponderance of succulent plants in halophytic plant families (i.e. Chenopodiaceae, Aizoaceae, Zygophyllaceae) within brigalow plant communities.

Table 16.3 **Percentage association of Australian Soil Classification Order (Isbell 2002) with vegetation types (see Box 16.1 for methods). Chromosols (CH): non-sodic soils with an abrupt increase in clay content below the top soil; dermosols (DE): well-structured (aggregated) non-cracking clay soils; ferrosols (FE): iron- and clay-rich non-cracking clay soils; kandosols (KA): sandy soils that are massive (non-aggregated) with gradual profile development; kurosols (KU): acid soils with an abrupt increase in clay content below the top soil; tenosols (TE): Sandy soils with little profile development; sodosols (SO): sodic soils with an abrupt increase in clay content below the top soil; vertosols (VE): cracking clay soils.**

	CH	DE	FE	KA	KU	SO	TE	VE	Other	Count of sites
Brigalow Forest types										
Belah/brigalow	3.5	11.7	0.0	0.4	0.0	35.1	0.0	49.3	0.0	282
Brigalow/blackbutt	2.0	8.1	0.0	0.7	1.3	61.7	0.7	25.5	0.0	149
Brigalow	1.4	12.7	0.5	1.1	0.0	14.1	0.2	70.1	0.0	569
Brigalow/mountain yapunyah	11.5	7.7	0.0	3.8	15.4	15.4	3.8	42.3	0.0	26
Grassland	0.2	6.0	1.1	0.1	0.1	1.1	0.2	91.2	0.1	1710
Box woodland	8.1	10.5	0.4	8.8	1.8	52.7	2.3	15.2	0.2	1085
Silver-leaved ironbark woodland	20.3	16.6	1.1	28.3	3.0	15.6	7.3	7.8	0.0	463
Narrow-leaved ironbark woodland	15.0	16.3	1.6	15.5	7.1	29.9	7.8	5.5	1.3	748
Mulga forest	20.0	2.0	0.0	68.0	0.0	2.0	8.0	0.0	0.0	50
Count of sites	332	536	43	386	94	1162	127	2388	14	5082

In the south, brigalow is associated with belah over extensive areas subject to extensive flooding, but also in upland areas. There were no significant differences in the physical soil characteristics between brigalow and belah/brigalow dominated forest (Figure 16.5). Belah/brigalow dominated forest has higher total phosphorus (P) and lower cation exchange capacity than brigalow dominated forest. Brigalow/blackbutt forest generally has soils with lower clay content and surface electrical conductivity (and salinity) than pure brigalow forest (Figure 16.5). Brigalow/mountain yapunyah (*Eucalyptus thozetiana*) typically occurs with pure stands of brigalow on the footslopes of sandstone outcrops where soils have sodic subsoils and acid character.

Brigalow/blackbutt and box woodlands occupy similar edaphic settings, intermediate in many respects between pure brigalow forest and other eucalypt woodlands (Figure 16.5). Both these vegetation types typically occupy Sodosols (Table 16.3) characterised by high subsoil sodicity (ESP; Figure 16.5).

There are few differences between narrow-leaved and silver-leaved ironbark woodlands in terms of their soil characteristics (Figure 16.5). Both occupy a wide range of soil orders but narrow-leaved ironbark are most common on Sodosols while

Box 16.1 Survey data for soils and vegetation.

To describe the association of the vegetation types with soils, data from the Queensland Government soils database (SALI) was used after excluding sites from rocky landscapes. Soil sites in the SALI database include data on dominant flora species, allowing extraction of soil sites for four subtypes of brigalow forest, based on the dominant overstorey species: brigalow (*Acacia harpophylla*); mountain yapunyah (*Eucalyptus thozetiana*); blackbutt (*E. cambageana*); and belah (*Casuarina cristata*). Soil sites for box (*E. brownii, E. populnea*), silver-leaved ironbark (*E. melanophloia, E. whitei*) and narrow-leaved ironbark (*E. crebra, E. xanthoclada*) woodlands were extracted based on their dominant species. Mulga forest (dominated by *Acacia aneura*) and tussock grassland (selected using broad vegetation group 30; a unit in the Queensland Herbarium mapping database; Neldner *et al.* 2015) were also included to provide context with other widespread vegetation types occurring on deep-soil plains. Soil attributes were selected to highlight important differences between vegetation types, graphically prepared and the significance ($P < 0.05$) of individual differences were determined using non-parametric Wilcoxon paired-tests.

silver-leaved ironbark are mostly on kandosols and chromosols (Table 16.3). Silver-leaved ironbark woodlands can also occur on slightly more saline soils such as those derived from some metamorphic substrates. Soils in ironbark woodlands collectively tend to be less clay-rich (sandier), more acidic and have lower electrical conductivity (salinity) than the soils of poplar box woodlands. Mulga (*Acacia aneura*) forest and woodland can occupy similar soil types to the ironbark woodlands, although the mulga soils tend to be more acidic (Figure 16.5).

Surprisingly there is little discrimination between vegetation types in terms of total P and cation exchange capacity (Figure 16.5). P tends to be lower in the ironbark woodlands and mulga collectively than the other types, but differences are not statistically significant. In general, however, soil characteristics are the fundamental determinant of the patterns in vegetation addressed in this chapter. The importance of soil for determining vegetation patterns is illustrated by a spatial analysis demonstrating that apart from mechanical clearing the structure and composition of vegetation has changed little since the initial traverse of the region by the explorer Leichhardt (Fensham 2008). However, there is substantial overlap between vegetation types in many of the soil characters, and although our experience suggests that local vegetation changes often reflect soil and landform changes, the generalisations described here for individual soil characteristics can often be contradicted in local situations. Presumably vegetation pattern can be underpinned by complex interactions between climate, topography and edaphic conditions.

16.5 Fire history

A general lack of herbaceous fuel in the ground layer of brigalow forest means that it does not carry fire as effectively as grassy woodlands, and

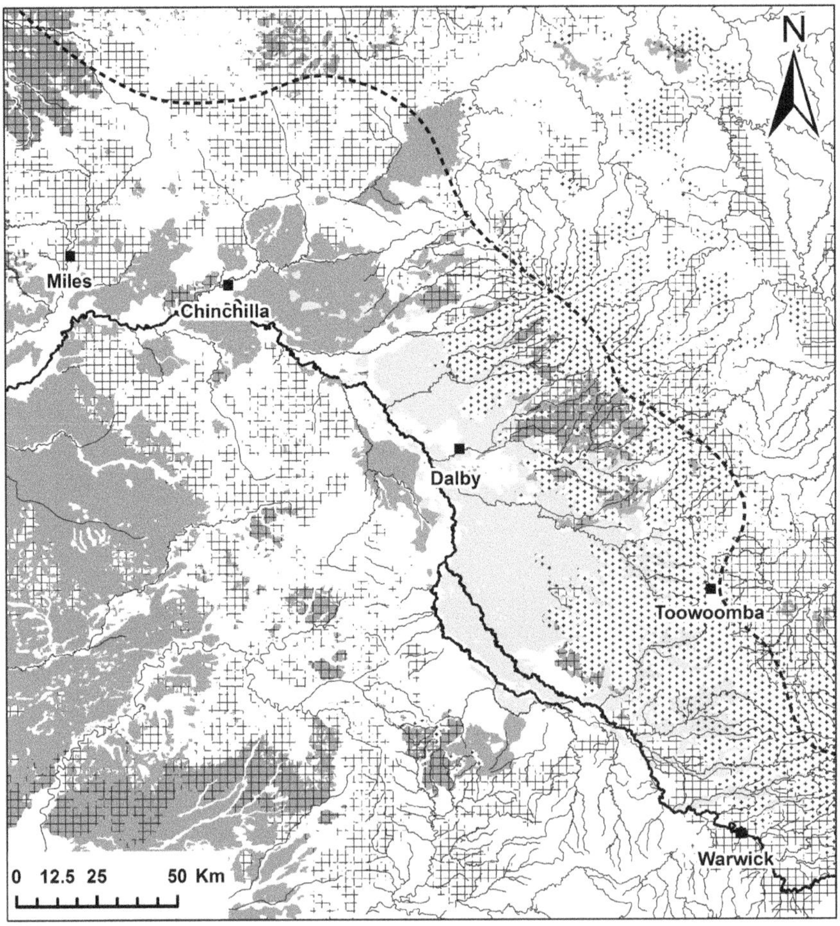

Figure 16.4 The eastern Darling Downs area within the Brigalow Belt South bioregion demonstrating the association of grassland (light grey) on alluvium with basalt geology (stippled) as the primary source material; and brigalow forest (dark grey) on alluvium with sedimentary rocks (hatched) as the source material. The crest of the Great Dividing Range is identified by the dashed line. Grassland is represented as Broad Vegetation Group 30; Brigalow forest as Broad Vegetation Group 25; Basalt as Land Zone 8; and Sedimentary as Land Zone 7, 9 and 10 in the Queensland Herbarium mapping database (Neldner *et al.* 2015).

over the course of millennia fire must have been infrequent in most brigalow ecosystems. Ludwig Leichhardt tracked through the brigalow belt prior to European settlement on two separate occasions, covering approximately 2500 km over a total of 16 months. In all these travels detailed by the expeditioner's journals (Leichhardt 1847; Bunce 1979; Sprod 1989, Sprod 2006) there is only a single mention of fire in brigalow forest from Hovenden Hely, who while searching for their mules in the vicinity of Chinchilla recorded 'In the large scrub we saw no less than about 100 fresh fires of blacks, whom I assume to have been in the scrub hunting' (Sprod 1989, p. 217). This suggests that Aboriginal people did light fires in the brigalow but generally they did not carry. The boundaries of the brigalow

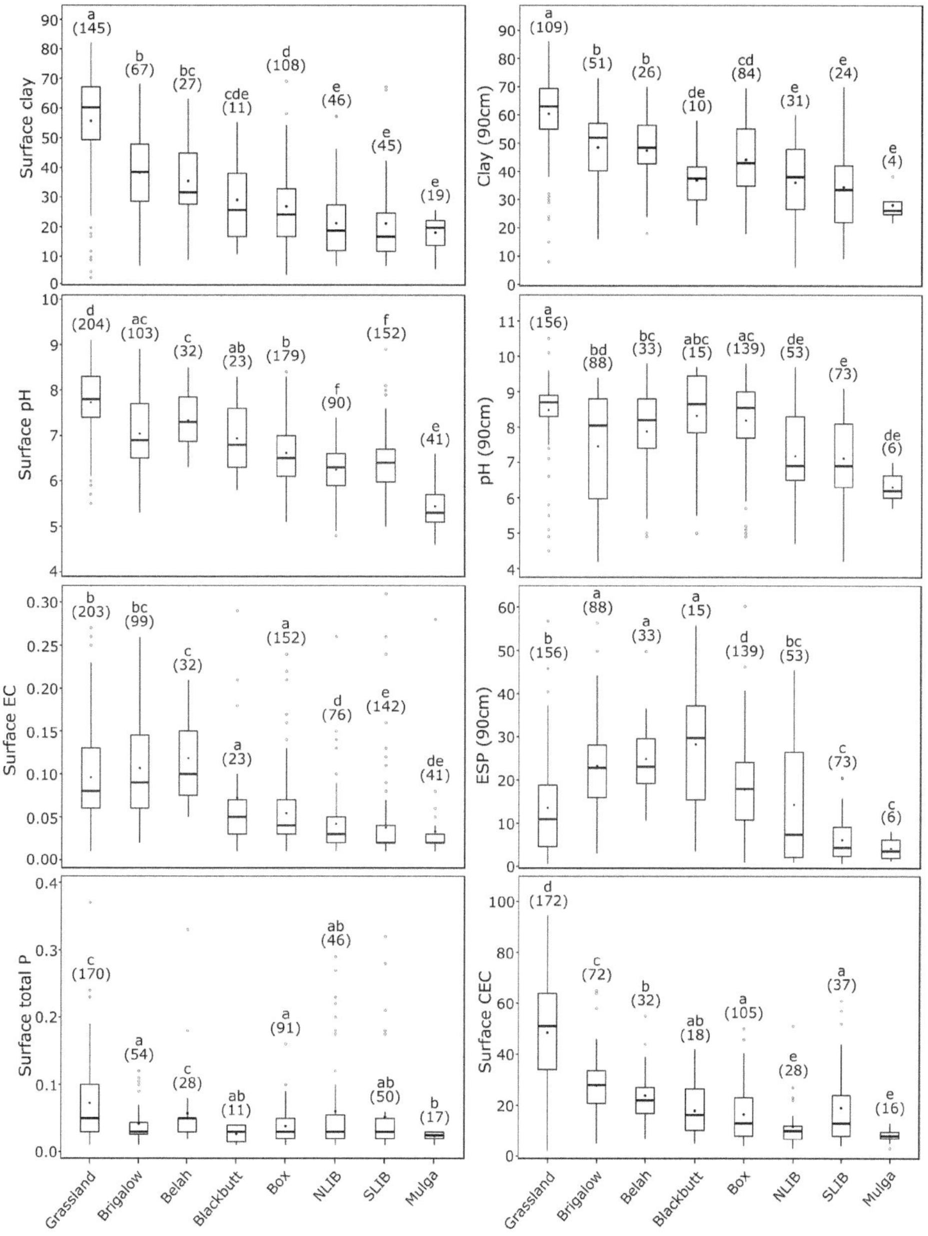

Figure 16.5 Box plots representing selected soil characteristics of the vegetation types represented in this chapter (left to right: Brigalow – brigalow forest; Belah – belah/brigalow forest; Blackbutt – blackbutt-brigalow/blackbutt forest; Box – box woodland; NLIB – narrow-leaved ironbark woodland; SLIB – silver-leaved ironbark woodland) (see Box 16.1 for methods). Other vegetation types within the study area are included for cross-reference (Grassland (far left) – tussock grassland mostly dominated by *Dichanthium sericeum*; Mulga (far right)– forest and woodland dominated by *Acacia aneura*). EC – surface electrical conductivity (mS cm^{-2}); ESP – exchangeable sodium potential (%); CEC – cation exchange capacity (meq/100 g). All other units are percentages. The mean (solid point) absolute median (thick horizontal line), 25–75 percentiles (box) and 10–90 percentiles (thin line) and outliers (hollow point) are indicated.

may have been manipulated by fire particularly in the 'patchy plain' environments where brigalow forest and grassland were intermixed, but the vast majority of the brigalow would have presumably burnt less than once a century, and only after exceptionally wet conditions generated sufficient ground fuel. Periods of high rainfall like the 1950s probably only occurred every few centuries (Figure 16.3).

The general lack of fire in the brigalow forests and woodlands contrasts with the grassy *Eucalyptus* woodlands which would have been regularly burnt. However, these woodlands do not have sufficient fuel to burn every year as do the *Eucalyptus* woodlands in tropical monsoonal environments of northern Australia (Murphy *et al.* 2013; Miller and Murphy, Chapter 5, this volume; Wiliams *et al.*, Chapter 15, this volume). The accumulation of grass fuel is severely restricted by insufficient rainfall in most years. Herbaceous dry matter varies from about 1 t ha^{-1} in dry years to about 5 t ha^{-1} in wet seasons, and these variations greatly exceed variations associated with a range of livestock grazing intensities (O'Reagain and Bushell 2011). Fires in eucalypt woodlands are common after wet spells, when fuel loads are high, and are frequently ignited by lightning strikes.

16.6 Regeneration ecology

Acacia harpophylla is unusual within its genus because its seed has only short-term viability (Johnson 1964). It is a mast seeding species and germinates rapidly and readily in the rare years when seed fall coincides with sustained adequate rainfall. The species can regenerate prodigiously from root suckers after disturbance (Johnson 1964). It may be disturbance events, such as the occasional fire or drought, that foster dense regeneration in places characterised as the 'whipstick' form of brigalow forest (Johnson 1964). However, the relative contribution of sexual and asexual reproduction in brigalow forest over the long term is poorly understood. The size-class distribution of stems in brigalow forest is evenly spread suggesting fairly continuous regeneration (Figure 16.6), probably coinciding with disturbance events to promote suckering and occasional wet years to promote seedlings.

The regeneration ecology of poplar box woodland and ironbark woodland are similar. Seedlings of the dominant *Eucalyptus* species germinate only occasionally in extremely wet years and rapidly develop a lignotuber (Fensham unpublished data). These small trees are extremely resistant to fire (Fensham *et al.* 2003a), although they typically remain stunted in the understorey (Fensham unpublished data) as a cohort of straggly resprouts. Resprouts advance to larger size classes regardless of burning, possibly triggered by release from competition when nearby adult trees die.

Adults of the dominant box and ironbark species are extremely vulnerable to droughts (as measured by extreme 3-year rainfall deficit), with mortality in the vicinity of 25% caused by 1 in 30 year droughts (Fensham *et al.* 2015a). Phasic seedling regeneration, the patchy death of overstorey trees, and recruitment from the resprout cohort when competition is released, results in a spread of size classes in the woodlands (Figure 16.6). Fire sensitivity varies widely among common understorey species: *Senna artemisioides* (37–42% survival), *Dodonaea viscosa* (50–59%), *Geijera parviflora* (43–71%), *Eremophila deserti* (61–69%) and *Eremophila mitchellii* (75–81%) with individual fires (Hodgkinson 1991). Some of the relatively fire-sensitive species such as *Dodonaea viscosa* can recover rapidly with post-fire germination. There is also a suite of *Acacia* species (including *Acacia cowleana, A. cretata, A. elachantha, A. laccata, A. longispicata, A. melleodora, A. tenuissima* and *A. verniciflua*), which have adopted an 'obligate-seeder' strategy germinating *en masse* after fire, producing fresh seed within two years, and then dying after less than 10 years. These species are

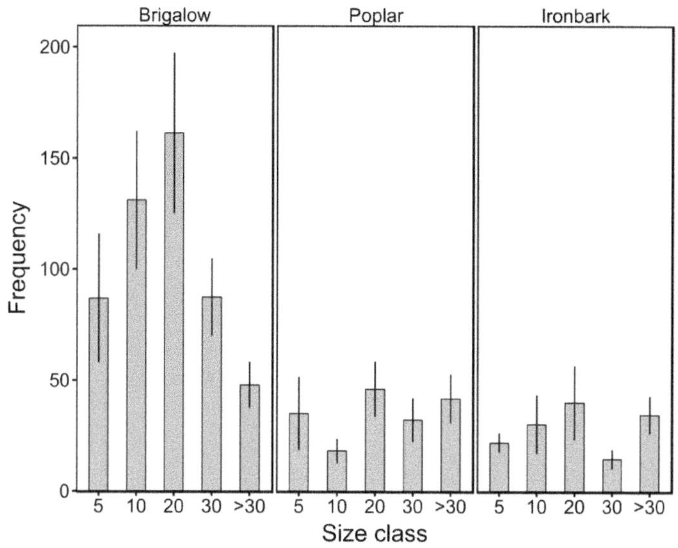

Figure 16.6 The size class (diameter at breast height, cm) distribution of stems (per ha) from brigalow forests, poplar box woodlands and ironbark woodlands (silver-leaved and narrow-leaved species combined) derived from the same data represented in Table 16.2. The bar labels are the maximum value of stem diameters (cm) for each size class, with the exception of >30.

fire sensitive as adults but have a life cycle readily adapted to short fire intervals.

16.7 Land-use history and conservation status of the communities

16.7.1 Land clearing

It has long been recognised that brigalow occupied relatively fertile soils but was relatively worthless for production in its natural state. The strong imperative to clear was curtailed by a lack of suitable means for broad-scale transformation, but the axe was put to use in higher rainfall areas during the first 50 years of settlement. This intense manual effort was rewarded with productive crops and pastures for dairy cattle. However, it wasn't until suitable machinery was developed during World War II that broad-scale clearing of brigalow forest began. The Queensland Government supported the conversion of the brigalow to pastureland by issuing new leases by ballot, with interest-free loans and a proviso that leases could be rescinded unless sufficient brigalow was cleared. This statutory obligation to clear was balanced by other provisions requiring that portions of the leases were conserved. The brigalow forest was levered from the ground by giant chains or cables pulled between two bulldozers, and then the trees were piled into heaps and burnt. No leases were ever rescinded for insufficient or excessive clearing.

From World War II until the 1990s clearing continued at a rate of 1–2% per annum (Figure 16.7) and more than 7 million hectares of brigalow woodlands and forests were felled, perhaps one of the most rapid land conversions in world history (Seabrook *et al.* 2006). The conversion was mostly to develop pastures for cattle production and the development of successful cultivars of exotic pasture grasses was an important enhancement to the clearing. Buffel grass (*Cenchrus ciliaris*) proved particularly successful, providing highly productive

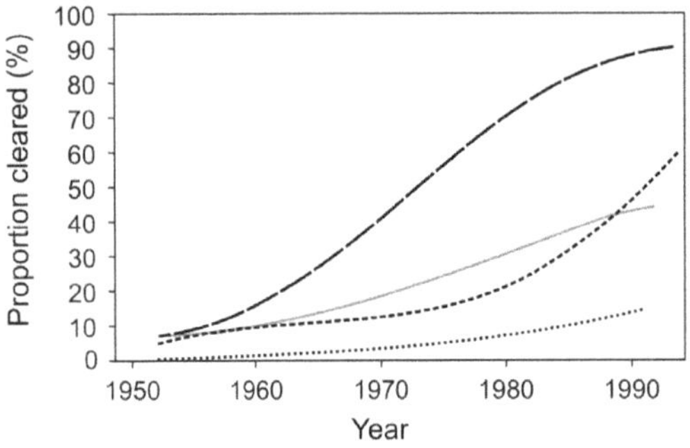

Figure 16.7 Clearing history determined from interpretation of random points from a time series of aerial photography from 1945–1999 within a 125 755 ha area of the Brigalow Belt North and Desert Uplands bioregions (Fensham and Fairfax 2003). Line symbols and the number of sites are brigalow forest (long dashes, 56); Box woodlands (short dashes, 50); Ironbark woodlands (dots, 34) and all vegetation types (grey, 205).

perennial pastures that are resilient to heavy grazing and respond far more rapidly to occasional rainfall than the native species. Only a minority of graziers perceive the capacity of buffel grass to monopolise pasture and exclude other species as a disadvantage (Friedel *et al.* 2011).

Before the clearing of the brigalow, the grazing industry was reliant on the native pastures of the *Eucalyptus* savanna woodland. As the brigalow was developed and the success of buffel grass was realised, effort was extended to clear and intensify the pastures of the *Eucalyptus* woodland. The motivation to clear the woodlands was as much a desire to replace the native pastures with buffel grass, as it was to remove the trees, but the best way to establish exotic pasture was through the soil disturbance and ash bed associated with clearing the trees. In the northern parts of our study region broad-scale clearing of poplar box woodland only commenced as recently as the 1970s, and for the ironbark woodlands on the plains not until the 1980s (Figure 16.7). In the early 2000s the woodlands were being cleared at very substantial rates, about 2% per year in the case of poplar box woodlands (Figure 16.8). The woodlands were following the fate of the brigalow.

In the 1990s, after a protracted process, legislation was implemented in Queensland and New South Wales to restrict clearing on all land tenures. In Queensland the legislation restricted clearing of remnant areas by defining these areas on maps. Despite a wave of panic that resulted in several years with accelerated woodland clearing rates prior to its final implementation in 2005, the legislation has slowed clearing. It is too late to protect a substantial proportion of the brigalow, which has been reduced to 10.0% of the original area, but 41.1% of the poplar box woodlands, 63.5% of the silver-leaved ironbark and 60.8% of the narrow-leaved ironbark woodlands remain uncleared (Figure 16.8). Fragmentation of remnant vegetation is generally less advanced in the north (Desert Uplands and Brigalow Belt North) compared to the south (eastern Mulga Lands and Brigalow Belt South) (Table 16.4), but by far the largest remnant areas of brigalow forest occur in the Brigalow Belt South and Brigalow Belt North (Table 16.4). The Desert Uplands is an important area for preserving remaining box woodlands and silver-leaved ironbark, while the Brigalow Belt North and Brigalow Belt South include the largest areas of remnant narrow-leaved ironbark (Table 16.4).

Despite significant slowing, clearing of remnant vegetation is ongoing in the study area and the effectiveness of legislation will depend on the will of government to maintain effective policies and compliance. Brigalow is listed nationally as a threatened ecological community under the *Environment Protection and Biodiversity Conservation Act 1999* (EPBC), while poplar box woodlands are currently under consideration for listing.

The proportion of brigalow forests and woodlands within conservation reserves is less than 1% in every region (Table 16.4), and the box, silver-leaved ironbark and narrow-leaved ironbark woodlands have only slightly better representation, with

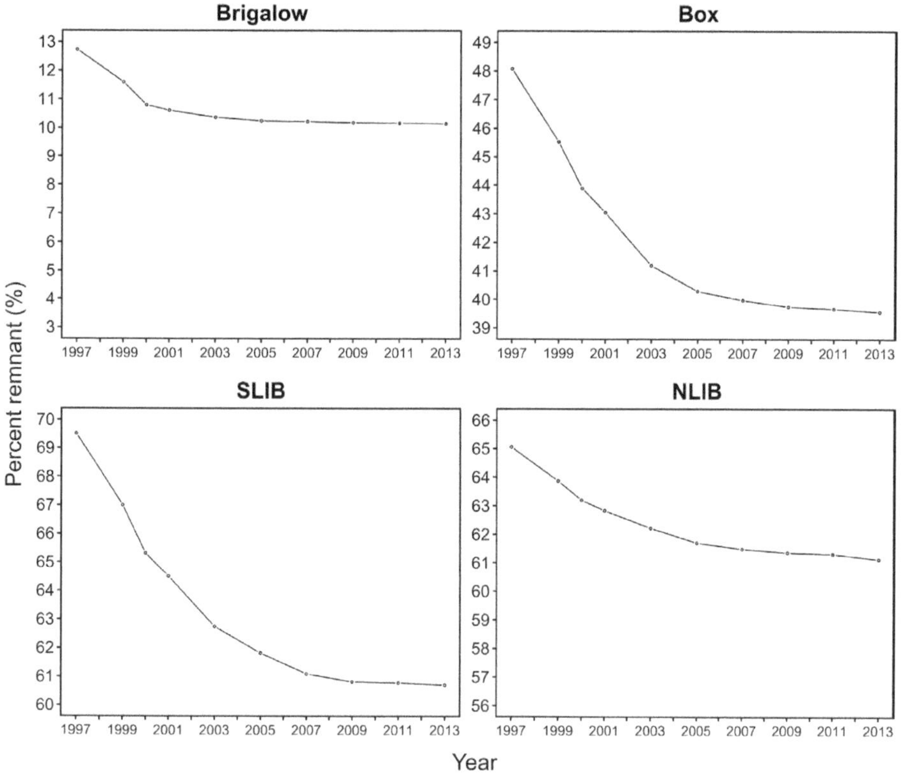

Figure 16.8 Decline in remnant vegetation for brigalow forest, box, silver-leaved ironbark (SLIB), and narrow-leaved ironbark (NLIB) woodlands over the 1997–2013 period when clearing regulations were implemented. The vegetation types are defined by the dominant species of regional ecosystems (Wilson *et al.* 2002) on landzone 3 (recent alluvium), 4 (ancient clay deposits), 5 (ancient sand deposits), 9 (fine-grained sedimentary rocks), 11 (metamorphosed sediments). Where a regional ecosystem is defined with two dominant species the area of this regional ecosystem was subdivided equally to the relevant vegetation types (e.g. a regional ecosystem described as: *Eucalyptus populnea* with *Acacia harpophylla* mid-layer was assigned to 'brigalow' and 'box' vegetation types).

narrow-leaved ironbark woodlands in the Brigalow Belt North having the highest proportion reserved of any vegetation type in any of our study bioregions.

There are 13 listed threatened plant species in the target vegetation types (Table 16.5). The relatively low number of threatened species in the box and silver-leaved ironbark woodlands indicates the lack of locally endemic plant species in these widespread ecosystems that form a continuum with similar ecosystems in the temperate and tropical zones. A systematic survey and assessment of the threat status of the flora in this region is required but the loss and further degradation by fire and exotic grasses poses an imminent threat to all of the listed species restricted to brigalow forest (Table 16.5).

16.7.2 Effects of grazing and fire on floristics

The region has been the subject of extensive livestock grazing since European settlement from the 1830s. This has had a clear impact on botanical composition where it is intense, such as around

Table 16.4 **The mapped area (km^2) and percentage (in brackets) of four vegetation types in 2013 relative to their estimated area before clearing in four regions of Queensland only (Figure 16.1). SLIB – silver-leaved ironbark; NLIB – narrow-leaved ironbark. The percentage and area of remnant vegetation in conservation reserves (including State Forests) is also provided. The attribution of the vegetation types from regional ecosystems is described in the caption of Figure 16.8.**

		Region				
Vegetation type		Brigalow Belt North	Brigalow Belt South	Desert Uplands	Eastern Mulga Lands	Total
Brigalow	Remnant	5853 (12.09)	3597 (6.39)	1470 (46.25)	495 (10.66)	11415 (10.15)
	Reserve	427 (0.88)	423 (0.75)	2 (0.08)	2 (0.05)	855.74 (0.76)
Box	Remnant	17618 (44.54)	15337 (26.27)	18824 (71.68)	6661 (28.31)	58440 (39.56)
	Reserve	1598 (4.04)	2078 (3.56)	37 (0.14)	220 (0.93)	3933 (6.61)
SLIB	Remnant	9771 (40.27)	9932 (42.99)	40528 (83.47)	1889 (29.50)	62120 (60.71)
	Reserve	1349 (5.56)	459 (1.99)	121 (0.25)	44 (0.69)	1973 (1.93)
NLIB	Remnant	20102 (68.78)	12004 (47.87)	4271 (81.99)	2 (95.38)	36379 (61.12)
	Reserve	2231 (7.63)	2450 (9.77)	32 (0.60)	0 (0.00)	4713 (7.92)

water points (McCosker *et al.* 2009). However, in areas more typical of commercial grazing practices, without the benefit of long-term manipulative experiments, impacts are more difficult to distinguish from other sources of variation. Variation in the abundance of the dominant perennial grasses is strongly controlled by seasonal influences (Scanlan *et al.* 1996; O'Reagain and Bushell 2011), which together with variations in soil characteristics are a fundamental determinant of floristic composition overall (Fensham *et al.* 2015b). One study found substantial effects of interannual variation in rainfall, with and without grazing, with a flush of annual herb diversity associated with a diminished perennial grass sward during drought (Fensham *et al.* 2015b). Moderate grazing may have a similar effect. A number of grazing-sensitive species may have been suppressed by a long history of grazing, and the perennial grass *Sehima nervosum* may be an example (McCosker *et al.* 2009).

Individual fires seem to have a similar effect to drought with subsequent rain by precipitating a short-term flush of ephemeral plants that diminish as perennial grasses recover (Williams *et al.* 2003; Fensham *et al.* 2015b). With very long-term fire exclusion, plant species typical of dry rainforest can colonise eucalypt woodland (Fensham and Butler 2004), although this probably happens more rapidly where a seed source of these species is in close vicinity, and as rainfall increases (Fensham 2012).

16.7.3 Invasion by exotic grasses

Land development in subhumid eastern Australia has converted a relatively economically unproductive region into one of Australia's most important areas for beef production. However the demise of the brigalow forest and the clearing of half the eucalypt woodlands over a vast area represents a major environmental cost. The subset of the fauna that survive the removal of trees and shrubs associated with clearing tend to be the same species that have proliferated more generally under agricultural land use and urbanisation (Hannah *et al.* 2007).

Table 16.5 **Threatened plant species listed in State or Federal (*Environment Protection and Biodiversity Conservation Act 1999*) jurisdictions from the target vegetation types for this chapter. Species listed in New South Wales but not threatened in Queensland are excluded. The table is partly informed by Benson *et al.* (2010). SLIB – silver-leaved ironbark; NLIB – narrow-leaved ironbark. Status: E – Endangered; V – Vulnerable.**

Species	Brigalow	Box	SLIB	NLIB	Queensland status	NSW status	EPBC status
Eucalyptus argophloia	X				V		V
Lepidium aschersonii	X					V	V
Solanum elachophyllum	X				E		
Solanum dissectum	X				E		
Solanum johnsonianum	X				E		
Solanum adenophorum	X				E		
Xerothamnella herbacea	X				E		E
Homopholis belsonii	X	X			E		V
Zieria verrucosa	X			X	V		V
Lepidium monoplocoides		X				E	E
Solanum stenopterum		X			V		
Capparis humistrata				X	E		
Androcalva procumbens[#]				X		V	
Cycas cupida				X	V		
Cycas megacarpa				X	E		E
Tylophora linearis				X	E	V	E

[#]syn. *Commersonia procumbens*

In the study region, the noisy miner (*Manorina melanocephala*) and the yellow-throated miner (*M. flavigula*) proliferate in fragmented landscapes and displace other birds by their 'despotic' behaviour (Thomson *et al.* 2015).

Most of the trees are sufficiently widespread to have maintained relatively large populations in remnant vegetation, but the major impacts in terms of plant conservation are associated with the proliferation of buffel grass (*Cenchrus ciliaris*). This species is common throughout the 25.4 million hectares of cleared lands but is also abundant through remnant vegetation. The species has a southern limit in the region around 27°S latitude, and Indian couch (*Bothriochloa pertusa*) is equally pervasive north of 22°S latitude. Other exotic grasses including Rhodes grass (*Chloris gayana*) and green panic (*Megathyrsus maximus*) are also invasive and develop high biomass. When exotic grasses become excessively dominant the diversity of native species is diminished (Fairfax and Fensham 2000) and, in particular, the diversity of perennial grasses declines in the *Eucalyptus* woodlands (Fensham *et al.* 2015b). In brigalow they are particularly destructive because they fuel abnormally intense fires that open up the dense canopy of the forest, creating conditions for accelerated invasion (Butler and Fairfax 2003). A positive feedback between the exotic grass and fire drives negative feedbacks for the tree canopy and reduces the diversity of the understorey (Fairfax and Fensham 2000). This feedback can also be initiated when drought opens up the canopy of brigalow forest, allowing exotic grasses to colonise when rain returns and providing fuel for future fires. Brigalow remnants are under ongoing threat from intense fires fuelled by buffel grass, and this is a particularly potent threat for roadside patches that represent the majority of remnant vegetation in some areas. The process is also a grave concern in many conservation reserves that contain some of the last remaining large remnants. The only feasible option to reduce fuels at large scales in these situations is to incorporate domestic stock grazing. Livestock grazing in reserves seems to be an anathema to the principles of conservation management, but when a region has been subject to wholesale transformation, dramatic side-effects will require radical redress. While comparative data is lacking the regeneration of brigalow does not seem to be suppressed by commercial livestock grazing, as demonstrated by the inevitable recovery of brigalow regrowth in cleared areas spelled from livestock.

16.7.4 Mining

The region includes very significant deposits of coal and coal seam gas. While the mining areas are small compared to the areas cleared for pastoralism, the impacts of open cut mines involve the transformation of the land surface and opportunities for rehabilitation of semi-natural vegetation are much more limited and potentially futile, given that equivalent resources could be used to foster conservation outcomes for remnant and regrowth vegetation with high conservation values. Some active leases for coal mining and coal seam gas extraction (Figure 16.1) include valuable remnants of high conservation significance.

Coal seam gas developments involve extraction from multiple wells spread over large areas. Within an active lease (Figure 16.1), wells are connected by tracks and these linear disturbances provide an opportunity for the spread of exotic species (including buffel grass) into remnant vegetation and may also increase dominance by edge-loving noisy miners. After the abandonment of the well field the disturbances themselves may regrow native woody vegetation, but the consequences of the exotic species may be iniquitous.

16.8 Opportunities for recovery

Trees return as stunted 'regrowth' in both brigalow and eucalypt communities after typical clearing by bulldozers. Generally the trees recover and require re-clearing over cycles of about 10–20 years in order to maintain open pastures. Regrowth extends over the vast proportion of the cleared land (Lucas *et al.* 2014) and could be recovered to semi-natural vegetation with appropriate incentives and ongoing grazing to reduce fuel loads (Fensham and Guymer 2009). Natural brigalow forest has above-ground biomass between 50–150 t ha^{-1} (Fensham and Guymer 2009; Ngugi *et al.* 2011), and stores substantial carbon in dead wood (Dwyer *et al.* 2009). The biomass recovers in about 45 years but recovery of the floristic elements will take longer (Dwyer *et al.* 2009; Ngugi *et al.* 2011; Johnson *et al.* 2016). Incentives may emerge from the carbon market as it seems likely that the recovery of regrowth provides relatively profitable returns on investment compared to plantations (Evans *et al.* 2015). This potential market may provide opportunities for a future change of land use that would allow for the partial recovery of the brigalow forest. Foregoing the option to reclear, managing regrowth to recover should be recognised as a deliberate reforestation effort in the same way as plantation establishment.

References

Benson, J.S., Richards, P., Waller, S. and Allen, C.B. (2010) New South Wales vegetation classification and assessment: part 3 plant communities in the NSW Brigalow Belt South, Nandewar and west New England Bioregions and update of NSW Western Plains and NSW South-western Slopes plant communities, Version 3 of the NSWVCA database. *Cunninghamia* 11, 457–579.

Bunce, D. (1979) *Travels with Dr Leichhardt in Australia.* Oxford University Press: Melbourne.

Burrows, W.H., Henry, B.K., Back, P.V. *et al.* (2002) Growth and carbon stock change in eucalypt woodlands in northeast Australia: ecological and greenhouse sink implications. *Global Change Biology* 8, 769–784.

Butler, D.W. and Fairfax, R.J. (2003) Buffel grass and fire in a gidgee and brigalow woodland: a case study from central Queensland. *Ecological Management and Restoration* 4, 120–124.

Dwyer, J.M., Fensham, R.J., Butler, D.W. and Buckley, Y.M. (2009) Carbon for conservation: assessing the potential for win–win investment in an extensive Australian regrowth ecosystem. *Agriculture, Ecosystems and Environment* 134, 1–7.

Evans, M.C., Carwardine, J., Fensham, R.J. *et al.* (2015) Carbon farming via assisted natural regeneration as a cost-effective mechanism for restoring biodiversity in agricultural landscapes. *Environmental Science and Policy* 50, 114–129.

Fairfax, R.J. and Fensham, R.J. (2000) The effect of exotic pasture development on floristic diversity in central Queensland, Australia. *Biological Conservation* 94, 11–21.

Fensham, R.J. (1995) Floristics and environmental relations of inland dry rainforest in north Queensland, Australia. *Journal of Biogeography* 22, 1047–1063.

Fensham, R.J. (1998) The grassy vegetation of the Darling Downs, South-Eastern Queensland, Australia. Floristics and grazing effects. *Biological Conservation* 84, 301–310.

Fensham, R.J. (2008) Leichhardt's maps: one hundred years of change in vegetation structure in inland Queensland. *Journal of Biogeography* 35, 141–156.

Fensham, R.J. (2012) Fire regimes in Australian tropical savanna: perspectives paradigms and paradoxes. In *Flammable Australia: Fire Regimes, Biodiversity and Ecosystems in a Changing World*, ed. R.A. Bradstock, A.M. Gill and R.J. Williams. CSIRO: Melbourne, pp. 173–193.

Fensham, R.J. and Butler, D.W. (2004) The spatial pattern of dry rainforest colonising unburnt Eucalyptus savanna. *Austral Ecology* 29, 121–128.

Fensham, R.J. and Fairfax, R.J. (2003) A land management history for central Queensland, Australia as determined from land-holder questionnaire and aerial photography. *Journal of Environmental Management* 68, 409–420.

Fensham, R.J. and Guymer, G.P. (2009) Carbon accumulation through ecosystem recovery. *Environmental Science and Policy* 12, 367–372.

Fensham, R.J. and Holman, J.E. (1999) Temporal and spatial patterns in drought-related tree dieback in Australian savanna. *Journal of Applied Ecology* 36, 1035–1050.

Fensham, R.J., Fairfax, R.J., Bowman, D.M.J.S. and Butler, D.W. (2003a) Effects of fire and drought in a tropical eucalypt savanna colonised by rain forest. *Journal of Biogeography* 30, 1405–1414.

Fensham, R.J., Low Choy, S.J., Fairfax, R.J. and Cavallaro, P.C. (2003b) Modelling trends in woody vegetation structure in semi-arid Australia as determined from aerial photography. *Journal of Environmental Management* 68, 421–436.

Fensham, R.J., Fairfax, R.J. and Archer, S. (2005) Rainfall, land-use and woody vegetation cover change in semi-arid Australian savanna. *Journal of Ecology* 93, 596–606.
Fensham, R.J., Fairfax, R.J. and Ward, D.P. (2009) Drought-induced tree death in savanna. *Global Change Biology* 15, 380–387.
Fensham, R.J., Fraser, J., MacDermott, H.J. and Firn, J. (2015a) Dominant tree species are at risk from exaggerated drought under climate change. *Global Change Biology* 21, 3777–3785.
Fensham, R.J., Wang, J. and Kilgour, C. (2015b) The relative impacts of grazing, fire and invasion by buffel grass (*Cenchrus ciliaris*) on the floristic composition of a rangeland savanna ecosystem. *The Rangeland Journal* 37, 227–237.
Friedel, M.H., Grice, A.C., Marshall, N.A. and van Klinken, R.D. (2011) Reducing contention amongst organisations dealing with commercially valuable but invasive plants: the case of buffel grass. *Environmental Science and Policy* 14, 1205–1218.
Gammage, W. (2012) *The Biggest Estate on Earth. How Aborigines Made Australia*. Allen & Unwin: Crows Nest.
Hannah, D., Woinarski, J.C.Z., Catterall, C.P. *et al.* (2007) Impacts of clearing, fragmentation and disturbance on the bird fauna of eucalypt savanna woodlands in central Queensland, Australia. *Austral Ecology* 32, 261–276.
Hodgkinson, K.C. (1991) Shrub recruitment response to intensity and season of fire in a semi-arid woodland. *Journal of Applied Ecology* 28, 60–70.
Holman, J.E., Hughes, J.M. and Fensham, R.J. (2003) A morphological cline in *Eucalyptus*: a genetic perspective. *Molecular Ecology* 12, 3013–3025.
Isbell, R.F. (2002) *The Australian Soil Classification,* revised edition. CSIRO Publishing: Collingwood.
Jeffrey, S.J., Carter, J.O., Moodie, K.B. and Beswick, A.R. (2001) Using spatial interpolation to construct a comprehensive archive of Australian climate data. *Environmental Modelling and Software* 16, 309–330.
Johnson, R.W. (1964) *Ecology and Control of Brigalow in Queensland.* Queensland Department of Primary Industries: Brisbane.
Johnson, R.W. (1980) Studies of a vegetation transect through brigalow (*Acacia harpophylla*) forest in central Queensland. *Australian Journal of Ecology* 5, 287–307.
Johnson, R.W., McDonald, W.J., Fensham, R.J. *et al.* (2016) Changes over 46 years in plant community structure in a cleared brigalow (*Acacia harpophylla*) forest. *Austral Ecology* 14, 644–656.
Leichhardt, L. (1847) *Journal of an Overland Expedition in Australia from Moreton Bay to Port Essington, a Distance of Upwards of 3000 miles, during the Years 1844–1845.* T & W Boone: London.
Lucas, R., Armston, J., Fairfax, R. *et al.* (2010) An evaluation of the ALOS PALSAR L-Band backscatter–above ground biomass relationship Queensland, Australia: impacts of surface moisture condition and vegetation structure. *IEEE Journal of Selected Topics in Applied Earth Observations and Remote Sensing* 3, 576–593.
Lucas, R.M., Clewley, D., Accad, A. *et al.* (2014) Mapping forest growth and degradation stage in the Brigalow Belt Bioregion of Australia through integration of ALOS PALSAR and Landsat-derived foliage projective cover data. *Remote Sensing of Environment* 155, 42–57.
McCosker, J., Rolfe, J. and Fensham, R. (2009) Can bare ground cover serve as a surrogate for plant biodiversity in grazed tropical woodlands? *The Rangeland Journal* 31, 103–109.
Murphy, B.P., Bradstock, R.A., Boer, M.M. *et al.* (2013) Fire regimes of Australia: a pyrogeographic model system. *Journal of Biogeography* 40, 1048–1058.

Neldner, V.J., Niehus, R.E., Wilson, B.A. *et al.* (2015) *The Vegetation of Queensland. Descriptions of Broad Vegetation Groups. Version 2.0.* Queensland Herbarium, Department of Science, Information Technology and Innovation: Brisbane. Available at: https://publications.qld.gov.au/dataset/vegetation-qld/resource/921fa786-e6d5-4a8a-9b0c-e532d2ce3f32 (accessed 3 August 2016).

Ngugi, M.R., Johnson, R.W. and McDonald, W.J.F. (2011) Restoration of ecosystems for biodiversity and carbon sequestration: simulating growth dynamics of brigalow vegetation communities in Australia. *Ecological Modelling* 222, 785–794.

Nix, H. (1994) The brigalow. In *Australian Environmental History*, ed. S. Dovers. Oxford University Press: Oxford.

O'Reagain, P.J. and Bushell, J.J. (2011) *The Wambiana Grazing Trial: Key Learnings for Sustainable and Profitable Management in a Variable Environment.* Queensland Government: Brisbane.

Pedley, L. (1986) Derivation and dispersal of *Acacia* (Leguminosae), with particular reference to Australia, and the recognition of *Senegalia* and *Racosperma*. *Botanical Journal of the Linnean Society* 92, 219–254.

Russell, J.S., Moore, A.W. and Coaldrake, J.E. (1967) Relationships between subtropical semi-arid forest of *Acacia harpophylla* (brigalow) microrelief, and chemical properties of associated gilgai soil. *Australian Journal of Botany* 15, 481–498.

Scanlan, J.C., Pressland, A.J. and Myles, D.J. (1996) Grazing modifies woody and herbaceous component of North Queensland woodlands. *The Rangeland Journal* 18, 47–57.

Seabrook, L., McAlpine, C. and Fensham, R. (2006) Cattle, crops and clearing: regional drivers of landscape change in the Brigalow Belt, Queensland, Australia, 1840–2004. *Landscape and Urban Planning* 78, 373–385.

Sprod, D. (1989) *Proud Intrepid Heart.* Blubberhead Press: Hobart.

(2006) *Leichhardt's Expeditioners. In the Australian Wilderness 1844–1845.* Blubberhead Press: Hobart.

Thomson, J.R., Maron, M., Grey, M.J. *et al.* (2015) Avifaunal disarray: quantifying models of the occurrence and ecological effects of a despotic bird species. *Diversity and Distributions* 21, 451–464.

Williams, P.R., Congdon, R.A., Grice, A.C. and Clarke, P.J. (2003) Effect of fire regime on plant abundance in a tropical eucalypt savanna of north-eastern Australia. *Austral Ecology* 28, 327–338.

Wilson, B.A., Neldner, V.J. and Accad, A. (2002) The extent and status of remnant vegetation in Queensland and its implications for statewide vegetation management and legislation. *The Rangeland Journal* 24, 6–35.

17 Temperate Eucalypt Woodlands

SUZANNE M. PROBER, CARL R. GOSPER, LOUISE GILFEDDER,
TOM D. HARWOOD, KEVIN R. THIELE, KRISTEN J. WILLIAMS
AND COLIN J. YATES

Temperate eucalypt woodlands are the characteristic vegetation in flat to gently hilly landscapes of the intermediate rainfall zones of southern Australia. They comprise well-spaced trees with their crowns generally not touching, with forb-rich, grassy understoreys in fertile, mesic areas and shrubby understoreys in drier areas or on poorer soils. Key processes driving temperate woodland structure, floristic composition and patterns of variability before European colonisation included gradients in aridity and rainfall seasonality, interactions with woodland fauna, soil heterogeneity, and complex disturbance regimes resulting from interactions among fire, marsupial grazing and soil turnover by fossorial mammals. At least two distinct historical fire regimes are recognised: relatively frequent, low intensity fire in the more productive grassy woodlands (possibly once promoted by frequent Aboriginal burning), and infrequent, intense, stand-replacing fires in woodlands dominated by obligate-seeding eucalypts. Due to widespread exploitation for agriculture and other land uses, temperate eucalypt woodlands are today among the most fragmented and degraded vegetation types in Australia. Widespread clearing, livestock grazing, cultivation, fertilisation, exotic species introductions and altered fire regimes have led to exotic invasions, tree dieback, decline of woodland plants and animals, and altered ecosystem and landscape processes. This degradation also limits resilience to climate change, which is projected to cause substantial changes in plant species composition and shifts in woodland distributions. Nevertheless, significant progress has been made in recent decades to understand woodland ecology, establish conservation policy and management options, and restore woodland landscapes. Future promise lies in working with Aboriginal people to further restore temperate woodland biodiversity.

17.1 Introduction

Temperate eucalypt woodlands, with their characteristic wide-crowned eucalypts and low understorey extending over rolling hills and plains, represent the quintessential Australian bush portrayed in the works of colonial landscape artists. These woodlands are widespread on relatively productive soils in the intermediate rainfall zone between the forests of Australia's temperate coastal fringes, and the *Acacia* woodlands, shrublands and hummock grasslands of the drier interior (Figure 17.1a; DEWR 2007).

Temperate eucalypt woodlands were favoured environments for Aboriginal people, offering plentiful

Australian Vegetation, Third Edition, ed. David A. Keith. Published by Cambridge University Press.

Citation: Prober, S.M. *et al.* (2017). Temperate Eucalypt Woodlands. In D. A. Keith, ed., *Australian Vegetation*, 3rd edn. Cambridge: Cambridge University Press, pp. 410–37.

game and a diverse array of plant roots, tubers and fruits that formed the staple of their diets (Ens *et al.*, Chapter 4, this volume). Indeed, it is thought that Aboriginal people played an important role in maintaining the characteristic open grassy structure of some (but not all) woodlands, using fire to promote grasses and tuberous plants such as *Microseris lanceolata* (yam daisy; Hassell 1975; Gott 2005; Prober *et al.* 2016).

European colonisation of the relatively fertile and well-watered temperate eucalypt woodlands began early in the establishment of the Australian pastoral industry. The woodlands have since been widely cleared for cropping and intensive grazing (Figure 17.1a), and uncleared woodlands have typically been degraded through livestock grazing and other uses (Yates and Hobbs 1997; McIntyre *et al.* 2002; Prober and Thiele 2005; Lindenmayer *et al.* 2010). Consequently, many types of temperate eucalypt woodland are now listed as threatened at national and state levels.

Exceptions to this pattern of clearing and degradation are woodlands on poorer soils and steeper topographies in eastern Australia (e.g. the Snowy River area in Victoria and the Tom Gibson Nature Reserve in Tasmania), and the extensive, semi-arid woodlands east of the Western Australian wheatbelt. The latter, known as the Great Western Woodlands, escaped clearing due to inadequate rainfall for cropping and limited availability of fresh groundwater and forage for livestock (Prober *et al.* 2012).

In the agricultural zones, many farms support patches of (typically degraded) woodland, native pasture and/or scattered trees among crops and exotic pastures, and are interspersed by woodland remnants in travelling stock routes, bush cemeteries, state forests, nature reserves, and other public lands. Together, these are highly significant for the persistence of native biodiversity in temperate woodland landscapes (Yates and Hobbs 1997; Prober and Thiele 2005; Dorrough *et al.* 2006; Prober and Smith 2009).

This chapter characterises the natural distribution, floristic and structural variability, and driving ecological processes of temperate eucalypt woodlands, then discusses their current status, threats and conservation opportunities. Nomenclature follows the Australian Plant Census (2015).

17.2 Scope and distribution

We define temperate eucalypt woodlands as grassy and shrubby eucalypt woodlands occurring in temperate latitudes of Australia. Based on earlier circumscriptions (Moore 1970; Yates and Hobbs 1997; Yates *et al.* 2000b), we selected a northern limit of 27.5°S (Figure 17.1a); north of this latitude they grade into subtropical woodlands (Fensham *et al.*, Chapter 16, this volume). Some woodlands within this scope are described in other chapters so detailed discussion of these is excluded here (although they are included in maps and models). These include *Eucalyptus populnea* (poplar box) woodlands in western New South Wales (included with subtropical woodlands given their mostly northerly distribution; Fensham *et al.*, Chapter 16, this volume); sub-alpine woodlands (included in Venn *et al.*, Chapter 19, this volume), and floodplain woodlands dominated by species such as *E. largiflorens* (black box), *E. coolabah* (coolibah) and *E. camaldulensis* (river red gum) (Good *et al.*, Chapter 21, this volume).

Within southern Australia, temperate eucalypt woodlands are prominent in intermediate (250–1200 mm) mean annual rainfall zones, extending to >2000 mm in cool environments (Figure 17.1a). Mean annual temperatures vary substantially, from c. 5 to 23°C based on the pre-defined northern limit (Figure 17.1a). Relatively deep, fertile soils of varied origin (alluvial, aeolian, igneous, sedimentary and metamorphic; Beard 1990; Prober 1996; Lunt and Bennett 2000; Yates *et al.* 2000b) differentiate the occurrence of woodlands from other vegetation types in this broad climatic envelope.

17.3 Structure and composition

17.3.1 Overstorey

Woodlands comprise widely spaced trees with their crowns not touching and projected foliage cover <30% (Yates and Hobbs 1997; Beard *et al.* 2013; noting some authors use higher cover limits,

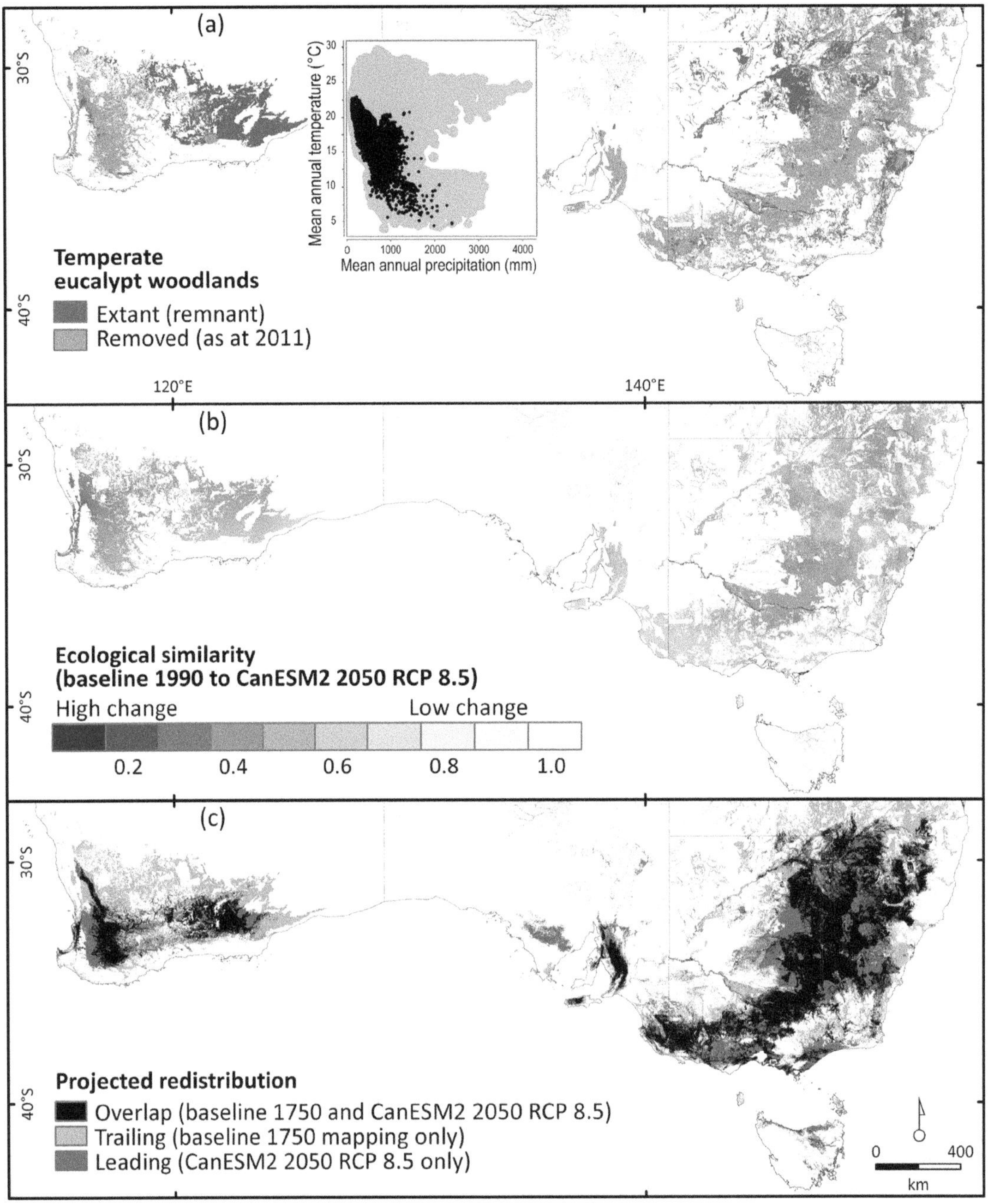

Figure 17.1 (a) Distribution of temperate eucalypt woodlands as mapped in the Native Vegetation Information System (Major Vegetation Groups 5 (eucalypt woodlands) and 11 (eucalypt open woodlands) below 27.5°S, NVIS version 4.1; DSEWPaC 2012), including extant and cleared woodlands. Maps include temperate-latitude eucalypt woodlands discussed in greater detail in Fensham *et al.* (Chapter 16, this volume, poplar box woodlands), Venn *et al.* (Chapter 19, this volume, sub-alpine woodlands)

e.g. 60%, Gillison and Walker 1981). Tree height and density tend to decline with increasing aridity (Yates *et al.* 2000b) though this is complicated by variation in soil fertility and depth. Height and projected foliage cover vary from (rarely) >30 m and 10–30%, respectively, in the highest rainfall zones, to <10 m and <10% in more arid areas or on lighter soils (Yates and Hobbs 1997; Beard *et al.* 2013). Woodlands on heavy clays, such as the Inverell Plateau in New South Wales, can have low tree densities and grade into grasslands; it is likely that tree densities are modified by fire, similar to woodlands elsewhere (Sankaran *et al.* 2005).

Most temperate eucalypt woodlands support one to three dominant eucalypt species at local scales, sometimes more at ecotones. These dominants (e.g. Figure 17.2; Table 17.1) vary widely across the range of the woodlands, particularly in relation to broad-scale aridity gradients and localised soil-topographic patterns. In mainland south-eastern Australia, they often include Box-Ironbark (sect. *Adnataria*) and red gum (sect. *Exsertaria*) species, with a climatic grade from species such as *E. viminalis* (manna gum/white gum) and *E. pauciflora* (snow gum/cabbage gum) in the colder tablelands regions, through *E. melliodora* (yellow box), *E. blakelyi* (Blakely's red gum) and *E. albens* (white box) on the warmer tablelands and inland slopes of the Great Dividing Range, to *E. microcarpa* (grey box), *E. woollsiana* (narrow-leaved grey box) and *E. conica* (fuzzy box) on inland plains (Table 17.1). These are replaced by *E. populnea* (Fensham *et al.* Chapter 16, this volume) and floodplain (Good *et al.* Chapter 21, this volume) woodlands on the extensive, semi-arid inland plains. In rainshadows coastward of the Great Dividing Range, woodlands are commonly dominated by *E. tereticornis* (forest red gum) and *E. moluccana* (coast grey box).

South Australia shares a number of eastern woodland dominants, but has greater prominence of other species from sect. *Adnataria*, particularly *E. odorata* (peppermint box), *E. leucoxylon* (yellow gum), *E. fasciculosa* (pink gum) and *E. porosa* (mallee box) (Paton *et al.* 2000). *Eucalyptus pauciflora*, *E. viminalis* and *E. ovata* (black gum/swamp gum) are important in Tasmania, along with several Tasmanian endemics (Table 17.1).

South-western Australian woodlands in higher rainfall areas (600–1200 mm mean annual rainfall) are dominated by a distinctive suite of species, including *E. rudis* (flooded gum), *Corymbia calophylla* (marri), *E. marginata* (jarrah), *E. gomphocephala* (tuart), *E. wandoo* (wandoo) and *E. accedens* (powderbark wandoo). On the gently undulating, drier (300–600 mm mean annual

Figure 17.1 (cont.)

and Good *et al.* (Chapter 21, this volume, floodplain woodlands). Note that NVIS classifications and mapping are currently under review, with potential for more refined maps and projections as updates become available (DotEE 2017). **Inset**: A climatic envelope estimated for these woodlands. Black circles are a sample of nearly 15,000 locations of the pre-1750 mapped extent of these woodlands. Grey circles are a background sample of nearly 135 000 locations of the other pre-1750 vegetation types throughout Australia. Graphic developed using *Atlas of Living Australia* Spatial Portal (2015). (b) Potential degree of ecological change (in units of ecological similarity, *sensu* Williams *et al.* 2014) in vascular plant composition in the temperate eucalypt woodlands between 1990 and 2050, estimated using generalised dissimilarity modelling under a high emissions (RCP8.5), Canadian Earth System Model (CanESM2; Chylek *et al.* 2011) climate scenario. CanESM2 is a relatively hot climate scenario with moderate rainfall change, representing a moderate scenario on the spectrum of 'Maximum Consensus' futures for Australia (www.climatechangeinaustralia.gov.au). Data included herbarium-derived occurrence records for all plant species Australia-wide, and national spatial layers of environmental variables. See Williams *et al.* (2014) for detailed methodology. (c) Projected distribution of environments supporting temperate eucalypt woodlands in 2050, using the high emissions (RCP8.5) CanESM2 climate scenario. The map indicates projected leading edges (capacity to expand) and trailing edges (potential to be replaced by other vegetation) of temperate eucalypt woodlands, and does not account for clearing. We emphasise this reflects only one of a range of possible future climate scenarios and that projected contractions in Queensland reflect our pre-defined 27.5°S limit. Projections are based on generalised dissimilarity modelling with kernel regression; see Prober *et al.* (2015) for detailed methodology.

Figure 17.2 Some examples of temperate eucalypt woodlands: (a) *Eucalyptus albens* (white box) grassy woodland near Orange, NSW, (b) *E. viminalis* (manna gum) grassy woodland in a grazing paddock near Bombala, NSW, (c) *E. microcarpa* (grey box) grassy woodland in a Travelling Stock Reserve near Parkes, NSW, (d) *E. loxophleba* subsp. *loxophleba* (York gum) forb-rich woodland in the Korrelocking Nature Reserve, WA, (e) *E. clelandiorum* (blackbutt) woodland with chenopod

rainfall) landscapes of the Western Australian wheatbelt, eucalypts from sections *Glandulosae, Bisectae* and *Dumaria* (Nicolle 2015) form complex mosaics, reflecting soil and topographic patterns that result from erosion of the Archaean igneous and metamorphic Yilgarn Craton (Beard 1990; Table 17.1). Some prominent Western Australian wheatbelt woodland species, particularly *E. salmonophloia* (salmon gum), *E. longicornis* (red morrell), *E. salubris* (gimlet) and *E. loxophleba* (York gum), remain common in the uncleared Great Western Woodlands to the east, where a range of other woodland eucalypts also become prominent, including species confined to sporadic alkaline greenstone soils (Table 17.1). The south-western Australian woodlands in turn form distinctive mosaics with other vegetation types, such as sandplain shrublands, mallee and saltmarshes on other parts of the region's soil-topographic mosaics (Beard 1990; Yates *et al.* 2000b; Beard *et al.* 2013).

17.3.2 Understorey

Understoreys in temperate eucalypt woodlands vary from grassy and forb rich in more productive environments to shrubby in drier or less fertile environments (Prober 1996; Clarke 2000; Kirkpatrick and Gilfedder 2000; Prober and Thiele 2004; Wills and Clarke 2008). Grassy woodlands are widespread in the south-east at >400 mm mean annual rainfall, dominated by box and red gum eucalypts and others such as *E. pauciflora, E. viminalis, E. nova-anglica* (New England peppermint) and *E. amygdalina* (black peppermint). Grassy understoreys are poorly documented in Western Australia, and sometimes considered not to occur (Yates *et al.* 2000b). However, evidence from surveys of the least-modified and least weed-invaded remnants suggests that some of the more mesic woodlands are likely to have been grassy or forb rich, particularly *E. loxophleba* subsp. *loxophleba* woodlands above 300 mm mean annual rainfall (Prober and Wiehl 2011), and potentially upland *E. occidentalis* (swamp yate) and *E. rudis* woodlands (Beard 1990).

Grassy understoreys are characterised by a prominent ground layer of perennial native tussock grasses, and diverse native forbs from families such as Asteraceae (daisies) and Fabaceae (legumes) and the orders Liliales and Asparagales (lilies and orchids). Shrubs are generally scattered or patchy, with typical Australian families such as Myrtaceae, Proteaceae and Ericaceae weakly represented (Yates and Hobbs 1997; Prober and Thiele 2005). Western Australia's *Eucalyptus wandoo* woodlands (Yates *et al.* 2000b) and the 'grassy-heathy woodlands' of northern Tasmania (Kirkpatrick *et al.* 1988) often support a dense ground layer of graminoids other than grasses, sometimes giving them a grassy appearance.

Most temperate woodlands in Western Australia (Gosper *et al.* 2016), and woodlands on poorer soils in the south-east (Kirkpatrick *et al.* 1988; Clarke 2000; Wills and Clarke 2008) have shrubby understoreys (as do some poplar box and floodplain woodlands outside the scope of this chapter). Shrubs are typically sclerophyllous, from the families Asteraceae, Ericaceae, Fabaceae, Myrtaceae, Proteaceae, Santalaceae, Scrophulariaceae and Xanthorrhoeaceae, but on alkaline soils with high calcium levels (especially in more arid areas), semi-succulent chenopods may dominate (Clarke 2000; Yates *et al.* 2000b; Gibson *et al.* 2004; Harvey 2014). Annual forbs and short-lived grasses can also occur in shrubby woodlands, especially in mature semi-arid woodlands where shrubs are well-spaced (Gosper *et al.* 2013a).

While most temperate eucalypt woodlands have well-developed understoreys, some have notably sparse understoreys. In mallet woodlands of Western

Figure 17.2 (cont.)

understorey near Norseman, WA, (f) *E. salubris* (gimlet) woodland with sclerophyll understorey near Marvel Loch, WA, (g) degraded *E. pauciflora* (Tasmanian cabbage gum) grassy woodland in grazing paddocks of the Tasmanian Midlands, (h) remnant woodlands in a characteristic temperate agricultural landscape (York, WA).
Photos: a–f, h Suzanne Prober; g, Louise Gilfedder. (A black and white version of this figure will appear in some formats. For the colour version, please refer to the plate section.)

Table 17.1 **Generalised overview of the major Australian temperate eucalypt (*E.* – *Eucalyptus*; *A.* – *Angophora*; *C.* – *Corymbia*) woodland dominants, their distribution and core rainfall and temperature climatic occurrence (see Figure 17.1a). Typical understorey structure and composition is indicated as one or more of grassy (G), graminoid (Gm), sclerophyll shrubby (sS), chenopod shrubby (cS) or bare (B). Note that some temperate woodland types outside the scope of this chapter are included here to place the occurrence of temperate woodlands in context. Information in this table is largely drawn from Kirkpatrick and Gilfedder (2000; Tas), Paton *et al.* (2000; SA), Lunt and Bennett (2000; Vic), Keith (2004; NSW) and Yates *et al.* (2000b; WA). As this treatment is unavoidably superficial, please consult these sources for more detail on woodland plant associations, distribution and environmental correlates.**

Mean annual rainfall	Mean annual temperature	Tasmania	South Australia	Victoria	New South Wales and far south-eastern Queensland	Western Australia
'Wet' >~600 to 2000 mm	'Cool to cold' <~15°C	Sub-alpine woodlands (Venn *et al.* Chapter 19, this volume): *E. coccifera* (snow gum) (sS), *E. delegatensis* (gum-topped stringybark) (G–sS)	Sub-alpine woodlands (Venn *et al.*, Chapter 19, this volume): *E. pauciflora*, *E. niphophila* (snow gums), *E. stellulata* (black Sally) (G–sS) High elevation tablelands: *E. viminalis* (manna gum), *E. dalrympleana* (mountain gum) (G) ---------*E. pauciflora*, *E. stellulata* (G) --------- *E. nova-anglica* (New England peppermint) (G)			
		Highlands, poorly drained flats: *E. gunnii* (cider gum), *E. rodwayi* (swamp peppermint) (G–sS)	Lowlands: ------------------*E. ovata* (swamp gum) (G–Gm) ------------------			
	'Warm to hot' >~15°C			Coastal rainshadow valleys: *E. tereticornis* (forest red gum), *A. floribunda* (rough-barked apple) (G) -------*E. moluccana* (grey box) (G)		Siliceous sands on dunes: *E. gomphocephala* (tuart) (sS) Drainage lines, wetlands, uplands: *E. rudis* (flooded gum) (G–sS) Dry sclerophyll forests and woodlands (Tozer *et al.*, Chapter 13 this volume): *E. marginata* (jarrah)–*C. calophylla* (marri) (sS)

'Inter-mediate' ~400 to 700 mm	'Cool to cold' <~15°C	Lowlands: Fertile soil: *E. viminalis E. pauciflora* (cabbage gum) (G) Infertile soil: *E. pauciflora*, *E. amygdalina* (black peppermint), *E. rubida* (candlebark), *E. tenuiramis* (silver peppermint) (G–Gm–Ss–B) Wet soil: *E. ovata* (black gum) (Gm–sS) Frost hollows: *E. rodwayi* (G)	Lower elevation on ranges: ---------*E. melliodora* (yellow box), *E. blakelyi* (Blakely's red gum) (G) ---*E. bridgesiana* (apple box), *E. polyanthemos* (red box) (G–sS) -------------------*E. goniocalyx* (bundy) (G–sS) ------------------- *E. leucoxylon* (yellow gum), *E. fasciculosa* (pink gum) (G–sS) ---------*E. viminalis* (G)---------	
	'Warm to hot' >~15°C		Slopes of the ranges in topographic mosaics: *E. albens* (white box), *Callitris glaucophylla* (white cypress pine) (G–sS) ---------*E. melliodora* (yellow box), *E. blakelyi* (Blakely's red gum) (G) *E. odorata* (peppermint box) (G) Slopes of ranges and upper floodplains: -----------------*E. microcarpa* (grey box) (G–sS) ----------------- -----*E. woollsiana* (narrow-leaved grey box) (G–sS) ------*E. conica* (fuzzy box) (G–sS)	Slopes; granite-derived, duplex soils: *E. wandoo* (wandoo) (Gm–sS) Laterites: *E. accedens* (powderbark wandoo) (sS) Lateritic breakaways: *E. astringens* and other mallets (B) Drainage lines, wetlands, uplands: *E. occidentalis* (swamp yate) (G–sS)

Table 17.1 (cont.)

Mean annual rainfall	Mean annual temperature	Tasmania	South Australia	Victoria	New South Wales and far south-eastern Queensland	Western Australia
'Semi-arid' 200 to ~500 mm	'Warm to hot' >~15°C		Calcareous soils: *E. porosa* (mallee box) (G)--------------	--------------		Wheatbelt and Great Western Woodlands –
			Floodplain woodlands (Good *et al.*, Chapter 21, this volume):		----*E. coolabah* (coolabah) (G) ----	Heavy soils, valley floors: *E. salmonophloia* (salmon gum), *E. salubris* (gimlet), *E. longicornis* (red morrell) (sS–cS)
			-----------------*E. largiflorens* (black box) (G–cS) -----------------	-----------------	-----------------	
			----------------*E. camaldulensis* (river red gum) (G) ----------------	----------------	----------------	
			Subtropical woodlands (Fensham *et al.*, Chapter 16, this volume):		---------*E. populnea* (poplar box), *C. glaucophylla* (G–sS)	Sandy loams, clay slopes: *E. capillosa* (wheatbelt wandoo) (sS)
						Granitic sandy loams: *E. loxophleba* (York gum) (G–sS–cS)
						Great Western Woodlands only –
						Heavy soils, valley floors: *E. corrugata* (rough-fruited mallee), *E. transcontinentalis* (redwood) (sS–cS)
						Alkaline greenstones: *E. clelandii*, *E. lesouefii*, *E. dundasii* (blackbutts), *E. torquata* (coral gum) (sS–cS)

Table 17.2 **Examples of mean total vascular plant species richness (±se) in temperate eucalypt woodlands indicating a high variability, likely reflecting both ecological (e.g. disturbance history, aridity) and methodological (e.g. plot size, number of times surveyed, season of survey) variation.**

Woodland type	Plot size m^2	Mean species richness	Source
NSW Northern Tablelands	30	~27 ± 2 (low grazing)	McIntyre and Lavorel (1994)
NSW Northern Tablelands	30	~30 ± 3 (moderate grazing)	McIntyre and Lavorel (1994)
NSW Northern Tablelands	30	~25 ± 3 (high grazing)	McIntyre and Lavorel (1994)
NSW *E. albens–E. melliodora*	1000	54.7 ± 1.1	Data from Prober (1996)
NSW *E. albens–E. melliodora*	100	19.8 ± 2.4 (in gaps)	Prober *et al.* (2002)
NSW *E. albens–E. melliodora*	100	29.6 ± 2.1 (under trees)	Prober *et al.* (2002)
NSW grazed derived grassland	400	~43 ± 7	Schultz *et al.* (2014)
Vic *E. melliodora*	128	89.0 ± 1.7	Lunt (1990)
Vic *E. leucoxylon*	64	51.0 ± 3.3	Morgan *et al.* (2011)
Vic *E. leucoxylon*	1024	81.9 ± 3.8	Morgan *et al.* (2011)
WA wheatbelt granite	100	46.2 ± 1.6	Gibson *et al.* (2004)
WA wheatbelt duplex	100	34.8 ± 1.3	Gibson *et al.* (2004)
WA wheatbelt laterite	100	46.5 ± 3.2	Gibson *et al.* (2004)
WA wheatbelt deep sand	100	45.5 ± 2.0	Gibson *et al.* (2004)
WA *E. salubris* (semi-arid)	100	18.5 ± 1.5 (<18 yrs since fire)	Gosper *et al.* (2013a)
WA *E. salubris* (semi-arid)	100	13.6 ± 1.1 (35–120 yrs since fire)	Gosper *et al.* (2013a)
WA *E. salubris* (semi-arid)	100	18.5 ± 2.0 (140->300 yrs since fire)	Gosper *et al.* (2013a)

Australia understorey development is limited, possibly because trees have highly competitive surface roots and can divert rainfall to deep soil layers via canopy interception, stem-flow and pipes in the soil (Verboom and Pate 2006). Similarly, woodlands dominated by *E. amygdalina* and *E. tenuiramis* (silver peppermint) on infertile substrates in Tasmania have high litter cover that may suppress understorey development (Fensham 1992).

Consistent with this variation in understorey types, woodland plant diversity at small scales is highly variable (Table 17.2). Some woodlands, such as those of Western Victoria (89 plant species in 128 m^2, Lunt 1990), are among the most species-rich in the world.

17.3.3 Environmental gradients in woodland understoreys

Plant species of temperate eucalypt woodlands are typically widespread, reflecting the formerly extensive, contiguous distributions of these communities, the gradual nature of climatic gradients and generally low to moderate relief (Lunt and Bennett 2000). Many understorey species and genera are shared between south-eastern and south-western Australia, including shrubs of the genera *Acacia*, *Eremophila*, *Dodonaea* and *Atriplex*, grasses of the genera *Aristida*, *Austrostipa* and *Rytidosperma*, and perennial forbs of the genera *Arthropodium*, *Bulbine*, *Drosera* and *Thysanotus*. Estimates of regional-scale species turnover are not readily available, but Kirkpatrick and Gilfedder (2000) noted low Tasmanian endemicity in Tasmanian dry country woodlands, probably reflecting a continuous distribution across the Bassian Plain (today's Bass Strait) during the last glacial maximum. Similarly, Prober (1996) emphasised the continuity in understorey composition across a 900 km north–south range in grassy *E. albens* woodlands in New South Wales.

Notwithstanding these continuities, strong soil and climate-related patterns in understorey composition are evident across the range of grassy and shrubby woodlands (Kirkpatrick *et al.* 1988; Beard 1990; Gibson *et al.* 2004; Prober and Thiele 2004; Beard *et al.* 2013; Harvey 2014). For example, across a gradient extending 400 km across box woodlands of central New South Wales, mean annual rainfall declines from 800 mm to 400 mm; only 10% of native plant species were recorded across the whole gradient, and there was substantial turnover at the plant family level (Prober and Thiele 2004). Indeed, aridity is a key driver of woodland structure and overstorey species turnover, and of more subtle changes in understorey composition. Increasing aridity corresponds with shifts from mostly perennial to mostly annual forbs, and with turnover in the naturally dominant tussock grasses from *Themeda triandra* (hereafter *Themeda*) and *Poa* spp. in south-eastern (and possibly south-western) Australia at 550–1000+ mm mean annual rainfall, to *Rytidosperma*, *Austrostipa*, *Aristida*, *Enteropogon* and *Neurachne* spp. at <550 mm (e.g. Prober and Thiele 2004). These changes are mirrored on soil productivity gradients within regions of similar climate, sites with poorer and shallower soils having more arid-type understoreys (Prober 1996; Prober *et al.* 2002).

In Western Australia, overstorey and understorey composition broadly covary in relation to soil-topographic mosaics (Beard 1990). In particular, *E. loxophleba* subsp. *loxophleba* woodlands support distinctive grass and forb-rich understoreys, *E. wandoo* woodlands support rich sclerophyllous or graminoid understoreys, while *E. salubris* and *E. salmonophloia* woodlands have more chenopod species. Despite these patterns, the intergrading nature of these mosaics across inland south-western Australia can make it difficult to define distinct woodland communities. For example, Gibson *et al.* (2004) identified few consistent associations between understorey and overstorey taxa in ordination analyses of 379 Western Australian wheatbelt plots. Understoreys can also vary substantially beneath woodlands with the same dominant trees. Harvey (2014), for example, identified distinct shrubby understorey classes across the range of *E. salmonophloia* woodlands, with climate and soil variables each explaining c. 15% of the floristic variation.

17.4 Structuring ecological processes

Climate has an overarching influence on the seasonal and longer-term dynamics of temperate eucalypt woodlands, with disturbance regimes and soil processes playing important roles at finer scales. Aboriginal people may have shaped disturbance regimes through their burning and food gathering activities, especially in the more mesic, grassy woodland types (Hassell 1975; Gott 2005); these are likely to have interacted with marsupial grazing, wildfire and soil turnover by fossorial mammals to form complex disturbance regimes important for maintaining community diversity and heterogeneity across landscapes (Martin 2003; Lunt *et al.* 2012). Interactions with other fauna, such as a distinctive suite of woodland birds, also play an integral role in woodland dynamics.

17.4.1 Climate

As in many ecosystems, climate is a primary driver of plant distributions and seasonal growth patterns in temperate eucalypt woodlands. Winters are typically cold and summers range from mild in elevated (tableland) regions to hot in the drier interior. Rainfall is winter-dominant in most southern and western temperate eucalypt woodlands, but is aseasonal in some areas (south-east Tasmania, central New South Wales, coastal New South Wales, eastern Victoria and in the transitional rainfall zone of south-western Australia), and summer-dominant in northern NSW and southern Queensland where temperate and sub-tropical woodlands merge (Moore 1970; BoM 2015).

Grassy understoreys often include a mix of cool-season C_3 (e.g. *Poa, Rytidosperma* spp., *Austrostipa* spp., *Neurachne alopecuroidea* and *Anthosachne scabra*), and warm-season C_4 (e.g. *Themeda, Aristida* spp. and *Bothriochloa* spp.) grasses, with increasing prominence of C_4 species in summer-rainfall areas (Moore 1970; Whalley *et al.* 1978; Lodge and Whalley 1989; Clarke 2000). This mix allows for active growth in response to rainfall in all but the coldest times of year, and also provides response diversity to differing fire and grazing regimes (e.g. Moore 1970; Lodge and Whalley 1989; Prober *et al.* 2007). Cool-season grasses typically flower and produce seed in spring to summer, while warm-season species produce seed in summer to autumn and can germinate after intermittent heavy rains in these seasons.

Where rainfall is winter-dominant or aseasonal, understorey plant diversity is dominated by cool-season herbaceous forbs and graminoids, many of which avoid the summer moisture deficit by dying back to underground bulbs, tubers or other storage organs (e.g. orchids, lilies and some daisies). Other species remain above-ground and can respond opportunistically to intermittent summer rains (e.g. *Wahlenbergia* spp., Clarke 2000). As rainfall declines inland an increasing understorey component comprises cool-season annuals that survive summers in the seed bank (Whalley *et al.* 1978; Prober and Thiele 2004).

Many cool-season species remain abundant in areas with more summer rainfall, but warm-season C_4 grasses such as *Dichanthium* and *Cymbopogon* spp. become more prominent (Lodge and Whalley 1989) and more forbs, including tuberous species such as *Microseris*, remain active in summer. Some warm-season forbs occur throughout the woodlands, particularly herbaceous legumes such as *Glycine, Desmodium* and *Psoralea* spp. (Whalley *et al.* 1978).

Understorey shrub layers are usually dominated by C_3 species, except where chenopods are prominent (e.g. the C_4 shrubs *Atriplex nummularia, A. vesicaria* and *Maireana sedifolia* dominate understoreys in the northern Great Western Woodlands; Harvey 2014). Substantial turnover in shrub taxa at family and subfamily levels occurs with increasing aridity, for example Dilleniaceae, Ericaceae and Fabaceae subfamily Faboideae are more prominent in mesic areas and Malvaceae, Chenopodiaceae, Scrophulariaceae (*Eremophila*) and Pittosporaceae in semi-arid areas (*Acacia* and *Cassinia* species occur across this gradient; Prober and Thiele 2004; Harvey 2014).

17.4.1.1 Climate variability

Rainfall variability in temperate eucalypt woodlands is mostly low to moderate, usually varying by <50% of mean annual rainfall (BoM 2015). In the north-east Great Western Woodlands, this moderately reliable rainfall, combined with a relatively even seasonal distribution, may facilitate the globally unusual persistence of woodland trees to 20 m tall at as little as 250 mm mean annual rainfall (without fresh ground water); this represents the most arid of the woodlands reviewed here (Prober *et al.* 2012; Figure 17.3).

Nevertheless, interannual rainfall variation still contributes significantly to natural ecological processes in temperate eucalypt woodlands. For example, drought contributes to plant population turnover by increasing mortality in perennials such as *Themeda*, and can promote forb diversity by suppressing the dominant grasses (e.g. Prober *et al.* 2007). Plant strategies to avoid summer dry periods can also facilitate persistence through longer-term droughts, e.g. dormant tubers can persist through several growing seasons (Coates *et al.* 2006; Schultz *et al.* 2014).

Interannual rainfall variation also results in naturally intermittent recruitment of woody and herbaceous species, associated with variation in flowering, seed production, germination and establishment

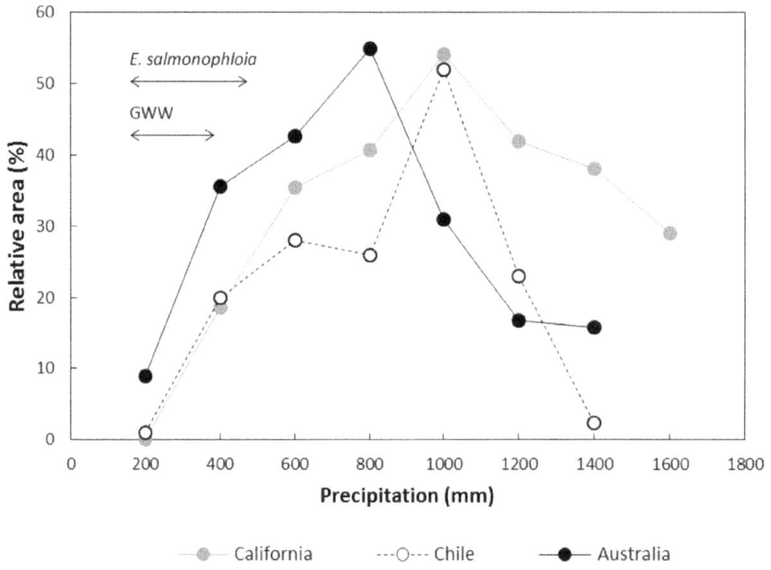

Figure 17.3 Relative area of woodland among all vegetation types by mean annual precipitation within three Mediterranean-climate regions. Vegetation occurrence within each rainfall zone was categorised as forest, woodland, shrubland, scrub, or grassland (see Prober *et al.* 2012). Mean annual rainfall range for the Great Western Woodlands (GWW), and a prominent woodland type (*Eucalyptus salmonophloia*) is also shown.
Figure adapted from Prober et al. (2012).

(Clarke 2000; Vesk and Dorrough 2006). Notably, however, few perennial woodland forbs have long-lived seed banks (e.g. Lunt 1995), suggesting particular vulnerability to long droughts causing excessive mortality of mature plants, as may occur under future climate scenarios (see below).

17.4.2 Soil processes

Temperate eucalypt woodland soils, where little-modified, are typically deep, relatively high in clay content and moderate to rich in most nutrients (McIntyre *et al.* 2002; Prober *et al.* 2002; Martin 2003). In grassy woodlands, the fertile soils promote a dense ground layer, which in turn protects soil surfaces from erosion and exposure to sunlight and raindrop impacts, facilitating water infiltration, biological activity and nutrient cycling (Prober *et al.* 2014). Crusts of bryophytes, lichens and algae, and patchy litter layers increasingly contribute to maintenance of these soil processes in more arid conditions where ground-layers become sparser and shrubs become more frequent (Warren and Eldridge 2001).

While otherwise relatively fertile, topsoils of temperate eucalypt woodlands are characteristically low in available nitrogen and phosphorus (Prober *et al.* 2002; McArthur 2004; Prober and Wiehl 2011). Low available phosphorus is typical of many Australian soils. Nitrogen availability is more variable, but *Themeda*-dominated understoreys typically have extremely low available nitrogen, probably due to uptake by the *Themeda* sward (Prober *et al.* 2002; Prober and Lunt 2009). The low natural availability of nitrogen and phosphorus means that many native woodland plants are sensitive to nutrient enrichment associated with soil disturbance or fertilisation (see below).

Historical records suggest that fossorial mammals such as southern brown bandicoots, numbats and bettongs were abundant in temperate eucalypt woodlands before European colonisation. Although their role in woodland dynamics is poorly characterised owing to their near extinction on mainland Australia (Lunt and Bennett 2000; Martin 2003), it is likely they significantly affected soil processes. Their digging likely influenced biogeochemical cycling, water relations, turnover and redistribution of subterranean plant organs, seed dispersal and plant recruitment (Martin 2003 and references therein; Murphy *et al.* 2005), although in Tasmania where these animals persist, soil compaction is still problematic due to livestock grazing. In some woodlands with abundant tuberous plants, a similar effect may have resulted from extensive digging by Aboriginal people (Gott 2005; Ens *et al.*, Chapter 4, this volume).

At fine scales, woodland trees are important drivers of heterogeneity in soil properties and understorey composition. Topsoils are naturally higher in nutrients, pH and biophysical condition beneath canopies than in gaps, probably due to wide foraging for nutrients by extensive tree roots, protection of soil surfaces, and nutrient enrichment from leaf litter (Prober *et al.* 2002, 2014; Eldridge and Freudenberger 2005; Prober and Wiehl 2011). Lateral and vertical soil water distribution can also be manipulated by woodland trees (Verboom and Pate 2006). In *Themeda–Poa* ecosystems, these patterns are reflected in reduced dominance of *Themeda* and higher forb richness and *Poa* abundance beneath canopies (Prober *et al.* 2002). In the Western Australian wheatbelt, *Eucalyptus salmonophloia* can inhibit understorey and tree recruitment (Yates *et al.* 2000a); whereas in the Great Western Woodlands understorey shrubs appear denser beneath *E. salmonophloia*, potentially due to concentration of soil resources in the semi-arid environment (Figure 17.4).

Figure 17.4 Clustering of shrubs beneath trees in semi-arid *E. salmonophloia* woodlands, perhaps facilitated by concentration of resources in the canopy zone.
Photo: Suzanne Prober.

17.4.3 Fire

Fire can strongly regulate the composition and structure of temperate eucalypt woodlands. Differences in understorey fuel characteristics and post-fire responses of dominant eucalypts are associated with highly distinct fire regimes between grassy and some of the shrubby woodlands (Gosper *et al.* 2016), so we consider these two regimes separately.

17.4.3.1 Fires in woodlands dominated by resprouting eucalypts (grassy woodlands)

Although not well documented, lightning strikes, Aboriginal burning and the ease of surface fire spread through grassy swards, are likely to have led to a regime of relatively frequent, low-intensity fire in the more mesic, grassy woodlands before European colonisation (Hassell 1975; Gott 2005). Many plants in these woodlands recover rapidly by resprouting from epicormic or basal buds (Clarke 2000); for overstorey eucalypts, this in combination with enhanced recruitment after fire (Hobbs and Atkins 1991; Clarke and Davison 2001) is likely to contribute to multi-age stands.

Shrub dynamics are also influenced by fire. Individual fires can encourage establishment of shrubs such as *Cassinia* spp. and hard-seeded legumes (e.g. *Acacia dealbata, Indigofera* spp.) (Leigh and Holgate 1979; Clarke and Davison 2001; Watson *et al.* 2009), but repeated fire promotes open, grassy woodlands (Watson *et al.* 2009). For example on the New South Wales Cumberland Plain, intermediate fire intervals (4–10 year) prevented dense encroachment by the shrub *Bursaria spinosa*, while allowing scattered shrubs to persist (Watson *et al.* 2009). Similarly, lack of fire (and low grazing pressure) can promote dense stands of the fire-intolerant small tree *Callitris glaucophylla* in *E. albens* woodlands (Cohn *et al.* 2013).

In the grassy ground layer, fire influences plant diversity by mediating competition and coexistence along ground-layer productivity gradients (Lunt *et al.* 2012). At the productive end of the gradient, biomass reduction by fire can help maintain diversity in *Themeda–Poa* understoreys. For example, 15 years after cessation of regular burning, 15% of forb species had been lost from experimental plots in cleared *E. albens–E. melliodora* grassy woodlands (Prober *et al.* 2013), whereas geophytic forbs such as *Microseris lanceolata, Tricoryne elatior, Wurmbea dioica* and orchids benefited from fire. The need for burning was lessened when local or seasonal factors reduced *Themeda* dominance or biomass; in particular, drought mimicked fire effects on grass by limiting its growth. Similarly, in long-unburnt woodlands, overstorey trees constrained ground-layer biomass and reduced *Themeda* dominance, potentially helping to maintain forb diversity in the absence of fire (Prober *et al.* 2007, 2013).

Few studies have examined fire effects at the lower end of the productivity gradient in temperate eucalypt woodlands, but a 3-year study of long-grazed *Rytidosperma–Austrostipa* grasslands on the (treeless) Victorian Riverine Plain (Foreman 1996) provides some insight. It indicated no significant change in native plant richness with or without fire, and predicted eventual loss of small native annuals under frequent fire.

The historical timing of fires in temperate grassy eucalypt woodlands is poorly known. Gott (2005) suggested that for south-eastern Australia, historical records place Aboriginal burning in mid to late summer, when edible forbs are dormant. First-hand records in *E. occidentalis* woodlands of south-western Australia suggest Aboriginal burning was undertaken from late summer until the autumn rains (Hassell 1975).

17.4.3.2 Fire in woodlands dominated by obligate-seeding eucalypts

Western Australia has a prominent suite of woodlands in which disturbances (fires, floods or storms) are rare and stand-replacing (Yates *et al.* 1994). Eucalypts in these woodlands are usually obligate seeders (with very thin bark and no lignotuber, e.g. *E. salubris, E. transcontinentalis* and other mallets), or weak to facultative resprouters (e.g. *E. salmonophloia*; Yates *et al.* 1994). Mirroring the dominant eucalypts, the understorey flora has among the lowest proportion of resprouters of all major Australian ecosystems (Gosper *et al.* 2016). Obligate seeders are most frequent, and a substantial minority of taxa have no traits enabling post-fire persistence ('fire avoiders', such as mistletoes and many chenopods). Some of the obligate seeders (e.g. *E. salubris, Callitris columellaris*) can attain ages >350 years (e.g. Gosper *et al.* 2013a), but infrequent fires (or other disturbances) are still important for eventual stand replacement. Consistent with this, contemporary Aboriginal knowledge suggests traditional burning was selective and uncommon in semi-arid woodlands dominated by obligate-seeding eucalypts (Prober *et al.* 2016) and fire regime reconstructions indicate interfire intervals >300–400 years are typical (O'Donnell *et al.* 2011a).

Long fire intervals in these woodlands are made possible by their fire-inhibiting characteristics, which include patchy litter layers interspersed with substantial areas of bare ground or cryptogam crusts, well-spaced shrubs (sometimes including low-flammability chenopods) and sparse grasses

(Figure 17.2e, f; O'Donnell *et al.* 2011a; Gosper *et al.* 2013b). Fires are mostly lightning-ignited during unusually dry years following unusually wet years (O'Donnell *et al.* 2011b). They can spread under very hot and windy conditions; and can be hindered by salt lake systems (O'Donnell *et al.* 2011a).

A fire chronosequence in *E. salubris* woodlands, established using a combination of remotely sensed fire mapping, tree ring analysis and allometric relationships, has allowed characterisation of post-fire development in obligate-seeder woodlands over 400+ year periods (Gosper *et al.* 2013a). Eucalypt seed is released after fire from the canopy, with dense regeneration occurring post-fire (Yates *et al.* 1994; Gosper *et al.* 2013a). Subsequent intense competition in the period ~30–120 years since fire initiates an extended period of self-thinning, after which surviving trees mature and develop an open woodland structure (Figure 17.2f). Mature woodlands thus need >150 years for development (Gosper *et al.* 2013a). Woodland flammability changes with time since fire, but not in the monotonic fashion often assumed: at ~30–120 years since fire, litter and canopy fuel continuity is high and fires potentially more frequent (O'Donnell *et al.* 2011a; Gosper *et al.* 2013b). This forms a flammability bottleneck through which regenerating woodlands must pass to develop into more fire-resistant mature woodlands.

Predictable shifts in plant species richness, composition, structure and associated fauna communities occur in concert with these changes (e.g. Gosper *et al.* 2013a,b). For example, *E. salubris* woodlands exhibit globally unusual 'U'-shaped responses in plant richness and evenness with increasing time since fire. Early post-fire vegetation comprises classic post-fire ephemerals and species with short-distance seed dispersal mechanisms recruiting from local seed banks. As dominant trees and shrubs become increasingly competitive they suppress more species-rich understoreys (Table 17.1), then diversity recovers in long-unburnt vegetation (>170 years) where species with long-distance dispersal traits accumulate (Gosper *et al.* 2013a).

17.4.4 Marsupial herbivory

Marsupial herbivores can modify the structure, composition and fuel characteristics of grassy woodlands but their documented effects are variable, probably due to varying herbivore densities and species (Eldridge *et al.*, Chapter 6, this volume). In woodlands near Melbourne, Robertson (1985) showed that kangaroos fed predominantly on graminoids, thereby promoting forbs; whereas few significant effects on forb diversity and abundance were observed in New South Wales' box woodlands after exclosure from kangaroos for up to 12 years (Prober *et al.* 2013; McIntyre *et al.* 2014). At more extreme, high densities on the other hand, kangaroos can greatly reduce effective establishment of *Themeda*, shrubs and woodland trees, cause mortality of coppicing plants after fire (Leigh and Holgate 1979; Allcock and Hik 2004) and severely degrade vegetation (Meers and Adams 2003). High densities of koalas, although rare, can similarly degrade woodlands through excessive defoliation of eucalypts (Paton *et al.* 2000).

In contrast to kangaroos which prefer herbaceous plants, smaller macropods such as wallabies and pademelons browse on the foliage of woody plant species. Their influence on shrubby woodlands is poorly known, although an absence of these browsing animals led to an increasing density of woody plants in grassy woodlands in Tasmania, potentially also changing woodland flammability (e.g. Kirkpatrick 2004). Marsupials also influence soil turnover and heterogeneity as described under 17.4.2 Soil processes and Eldridge *et al.* (Chapter 6, this volume).

17.4.5 Other interactions with fauna

Animal communities contribute to a wide range of other woodland processes. Although they cannot all be elaborated here, these include pollination of woodland eucalypts and other plants by insects, bats, small marsupials, nectarivorous honeyeaters and lorikeets (Clarke 2000; Wilson 2002); dispersal of seeds such as *Acacia* by ants (Standish *et al.* 2007) and mistletoes (Loranthaceae) by mistletoebirds and painted honeyeaters (Watson 2011); and predation of seed, for example eucalypt seed by ants (Yates *et al.* 1995). Woodland birds also play an important role in regulating the abundance of herbivorous insects, which can in turn influence plant vigour and reproduction (Clarke 2000; Reid and Landsberg 2000; Recher and Majer 2006).

17.5 Conservation challenges

Temperate eucalypt woodlands have been widely modified for agriculture and other land uses. In many areas, more than 90% have been cleared (Figure 17.1, Figure 17.2g, h), leading to their classification as some of the most threatened ecological communities in Australia (Yates and Hobbs 1997; Prober and Thiele 2005; Lindenmayer *et al.* 2010). Only Western Australia's Great Western Woodlands remain relatively intact across extensive landscapes, although still subject to anthropogenic disturbance including broad-scale logging during the early twentieth century, pastoral development, invasion by feral predators, and ongoing mineral exploration and mining (Prober *et al.* 2012). In this section, we summarise the interacting processes contributing to degradation of temperate eucalypt woodlands, then consider current trends and opportunities for their ecological restoration and conservation.

17.5.1 Altered ecosystem processes in agricultural landscapes

Widespread clearing, livestock grazing, soil cultivation and fertilisation for cropping and pasture improvement, exotic species introductions, and altered fire management (Yates and Hobbs 1997; Prober and Thiele 2005; Lunt *et al.* 2007; Prober and Smith 2009; Lindenmayer *et al.* 2010) have each affected woodland biodiversity directly, as well as indirectly through their effects on underlying ecosystem processes (Figure 17.5). Indirect effects can in turn limit the recovery of degraded woodlands, so that simple removal of primary degrading processes (e.g. livestock grazing) is often insufficient for restoring temperate eucalypt woodlands (Yates *et al.* 2000b; McIntyre and Lavorel 2007; Prober and Lunt 2009).

Apart from immense loss of total habitat area, broad-scale clearing has limited landscape connectivity and temporal continuity of resources for fauna (e.g. Radford *et al.* 2005), enhanced nutrients and weed invasion due to edge effects (e.g. Dorrough *et al.* 2006), and altered hydrological processes leading to salinisation in vulnerable parts of woodland landscapes (Yates and Hobbs 1997; Prober and Smith 2009). Clearing and agriculture have also contributed to widespread tree dieback in grassy woodlands by exposing trees to nutrient enrichment and agricultural chemicals, altering the balance between trees and grasses leading to the proliferation of leaf-eating native beetles (e.g. Kirkpatrick and Gilfedder 2000; Reid and Landsberg 2000), and by favouring an ultra-competitive bird, the noisy miner, at the expense of insectivorous birds (Maron *et al.* 2013).

Grazing by domestic livestock directly selects against palatable plant species (see below), as well as affecting plant recruitment, functioning of biological soil crusts, habitat structure for ground-layer fauna, microclimates, soil–water relations and nutrient cycling (e.g. Yates *et al.* 2000c; Warren and Eldridge 2001; Dorrough *et al.* 2004; Lunt *et al.* 2007; Prober *et al.* 2014). Soil water-holding capacity and infiltration can decline substantially following grazing-induced compaction and exposure of the soil surface to raindrop and UV impacts, in turn limiting eucalypt and shrub recruitment (Yates *et al.* 2000c; Skinner *et al.* 2009; Bailey *et al.* 2012; Prober *et al.* 2014, Eldridge *et al.*, Chapter 6, this volume). By contrast, nutrient enrichment and soil disturbance from livestock grazing, fertilisation and cultivation have led to widespread weed invasions (McIntyre and Lavorel 1994; Prober 1996; Yates *et al.* 2000c; Dorrough *et al.* 2004; Lunt *et al.* 2007). Exotic annual grasses

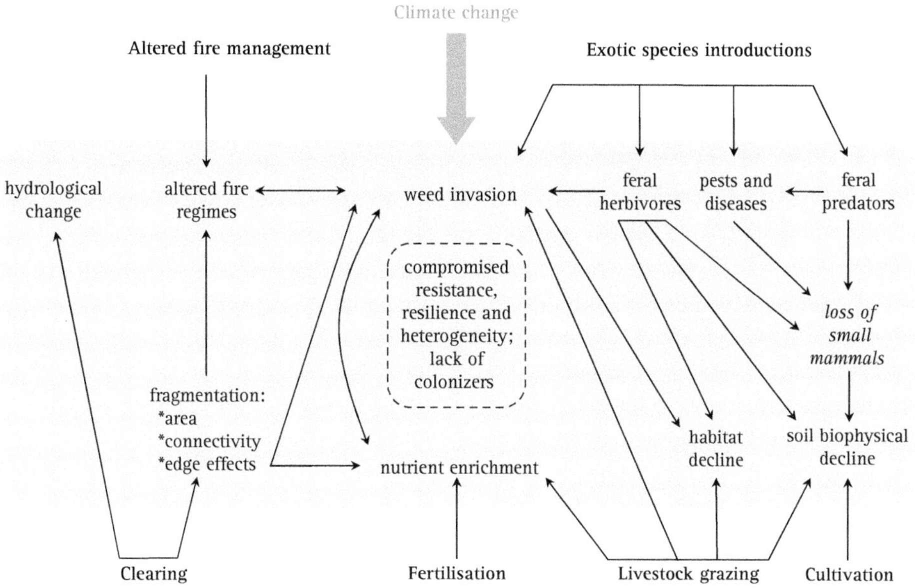

Figure 17.5 Primary and secondary threats to biodiversity of temperate eucalypt woodlands of agricultural landscapes, modified from Prober and Smith (2009). Primary anthropogenic causes of biodiversity decline are indicated in bold text; the shaded area indicates the associated legacy of changes to ecological processes that continue to threaten biodiversity persistence even after primary causes of degradation have ceased. Some changes (central box) emerge as the net outcome of interactions among primary and secondary processes. Due to their complexity, outcomes for biodiversity are not shown, but one example of this (loss of fossorial mammals) is given to emphasise that biodiversity decline itself also impinges on ecosystem processes.

and forbs have become dominant across many grassy and shrubby woodlands, altering soil nitrogen cycling and preventing re-establishment of native plants (Smallbone *et al.* 2007; Prober and Lunt 2009; Prober and Wiehl 2011). Perennial exotics such as *Hyparrhenia hirta* (coolatai grass), *Eragrostis curvula* (love grass), *Nassella neesiana* (Chilean needlegrass), *Phalaris aquatica* (phalaris) and *Hypericum perforatum* (St. John's wort), are also problematic. In particular, C_4 exotic grasses increasingly dominate the grassy swards on the New South Wales North West Slopes (Prober and Thiele 2005). Feral animals such as rabbits, foxes and cats have also dramatically affected woodlands, through grazing and impacts on small vertebrates, in turn altering plant recruitment and soil processes (Martin 2003; Murphy *et al.* 2005).

Compared with possibly frequent, patchy burns facilitated by Aboriginal people in mesic grassy woodlands, fire management currently focuses on fire suppression, and drivers of fire ignition and spread have been altered by landscape fragmentation, weed invasion and livestock grazing. Although not well quantified, observation suggests these have together led to long fire intervals in many woodland landscapes, with regular burning confined largely to occasional rail easements, cemeteries or other specialised sites. In Tasmania, low fire frequencies have been linked to a build-up of leaf litter and litter-derived phenolics on the soil surface, forming a water-repellent layer that limits recruitment (Bailey *et al.* 2012). Conversely, too-frequent fires close to human populations can be problematic, and there are concerns that the frequency of large, intense fires has increased in the fire-sensitive Great Western Woodlands (Prober *et al.* 2012; Gosper *et al.* 2013a,b).

17.5.2 Impacts of altered ecosystem processes on plant species composition

Livestock grazing in *Themeda–Poa* understoreys has led to dramatic shifts in the dominant grasses, with rapid breakdown of the *Themeda–Poa* sward and modification to states dominated by secondary native perennial grasses such as *Rytidosperma, Austrostipa, Aristida* and *Bothriochloa* species and exotic annuals (e.g. Moore 1953; Moore 1970; Lodge and Whalley 1989). In addition, fertilisation and sowing of pasture species has further eliminated native grasses in many sites (McIntyre and Lavorel 2007), as well as reducing eucalypt recruitment (Vesk and Dorrough 2006).

Changes in composition and abundance of subsidiary species are also associated with these transitions. Some native species increase under moderate grazing, notably those from more arid environments (Moore 1970; Prober and Thiele 2004). However, many native species disappear from grazed swards (e.g. Kirkpatrick *et al.* 1988; McIntyre and Lavorel 1994; Prober 1996) and fertilised pastures with high soil phosphorus levels (Dorrough *et al.* 2006).

Grazing-induced changes in the flora of shrubby woodlands and non-*Themeda–Poa* understoreys have been less comprehensively described, but loss of palatable shrubs and invasion by exotic annuals appears common to most woodlands (Yates *et al.* 2000c; Fox 2001). For example, Fox (2001) found that shrub cover declined from 16% to 4% in grazed *E. salmonophloia* woodlands, particularly involving the genera *Acacia, Atriplex, Maireana, Ptilotus, Sclerolaena* and *Santalum*. Other floristic changes in grassy, *Rytidosperma–Austrostipa* dominated communities include reduced native grass diversity (including rapid loss of the subsidiary *Themeda–Poa* component), increased prominence of *Chloris truncata* in south-eastern woodlands, decline in a range of native perennial forbs, and sometimes an increase in native annuals (Moore 1953; Prober and Thiele 2004; Prober and Wiehl 2011).

While much clearing and development in temperate agricultural regions occurred >50 years ago, ongoing clearing, death of isolated paddock trees, and increasing land-use intensity are contributing to ongoing loss of woodlands. In a South Australian agricultural landscape for example, Mogoutnov and Venning (2014) estimated a 36% decline in scattered woodland trees between 1980 and 2008, mostly due to agricultural intensification. Conversely, reduced livestock grazing associated with changing land uses, such as conversion of rural land to residential smallholdings or conservation reserves, has resulted in dense tree and shrub recruitment in some woodlands (Lunt *et al.* 2010). Community recognition of the decline of the biodiversity and functioning of woodland landscapes, and potential for carbon markets, are also contributing to a groundswell of activity towards restoring habitat and connectivity in woodland landscapes (see below).

17.5.3 Linear infrastructure and mining

As well as agriculture, woodlands have been and continue to be subject to a range of other pressures including urbanisation, development of linear infrastructure, mining and climate change. The legacy of historical gold rushes in western Victoria, central New South Wales and the Great Western Woodlands is still seen today in the form of grossly disturbed soil surfaces and multi-stemmed trees that coppiced after widespread harvesting. Mining footprints have increased in recent decades, particularly for coal in northern New South Wales and gold, iron, nickel and rare earths in the Great Western Woodlands. A long-lasting but poorly documented impact of mining and other development is the cumulative effect of extensive networks of roads and exploration tracks, estimated in the Great Western Woodlands to total 150 000 km (Raiter 2016).

In agricultural landscapes more broadly, early development of road, railway and stock route infrastructure left an important legacy of linear remnants in otherwise largely cleared woodland landscapes (Spooner 2005). These remnants are highly significant, but are disappearing due to changing use of stock routes, rail closures, and road widening along major highways and along small country roads (to permit passage of agricultural machinery).

17.5.4 Climate change

Anthropogenic climate change is already impacting ecosystems around the globe, and the temperate eucalypt woodlands are especially vulnerable owing to their highly fragmented and degraded condition. Projections to 2050 using moderate climate change scenarios suggest high potential for change in plant species composition (commonly >50% change in compositional similarity, Figure 17.1b), and south- and coastward shifts in woodland environments, including substantial contractions in south-western Australia (Figure 17.1c; Prober *et al.* 2012). Potential contractions in the south-east under a moderate climate scenario are less dramatic (Figure 17.1c), but substantial compositional change is still expected even where the woodland structure is likely to persist (Figure 17.1b).

The capacity of woodland species to adapt to climate change is not well understood. Adaptive shifts in leaf width are already evident in the widespread shrub *Dodonaea viscosa* (Guerin *et al.* 2012), but equally, evidence is appearing for potentially climate-induced tree mortality. For example rural tree dieback in the Tasmanian Midlands has been linked to reduced annual rainfall and a pronounced shift to a dry autumn since the 1970s (Close and Davidson 2004). Climate change may also induce changes in fire regimes, potentially spurred by invasions of flammable, arid-adapted grasses such as buffel grass (*Cenchrus ciliaris*), threatening persistence of fire-sensitive trees in the Great Western Woodlands (Prober *et al.* 2012).

17.6 Conservation opportunities

Achieving adequate conservation of temperate eucalypt woodlands is challenged by their highly degraded and fragmented state, with few large, relatively intact areas suitable for building a comprehensive, adequate and representative reserve network. Rather, woodland biodiversity is typically dependent on suites of remnants of varying (and often complementary) condition under different land tenures, embedded in multi-use landscapes. Even in the Great Western Woodlands where extensive woodlands remain, secure nature reserves are few, and conservation is challenged by conflicting land uses. Conserving temperate eucalypt woodlands is thus being tackled through a mix of off- and on-reserve approaches (FitzSimons *et al.* 2013; Whitten and Coggan 2013) and collaboration among many landholders and land managers, increasingly including traditional owners. Off-reserve conservation has focused on improved management and restoration of degraded woodlands to maintain and enhance conservation values, revegetation to increase habitat area and support critical ecosystem processes at landscape scales, and development of policy mechanisms and incentives to protect sites and support conservation management (FitzSimons *et al.* 2013; Whitten and Coggan 2013).

17.6.1 Ecological restoration

A first step in managing and restoring the condition of woodland remnants is to limit further degradation caused by changing land use, exotic invasions, and unfavourable fire regimes. Beyond this, a common approach to woodland restoration is fencing to limit livestock grazing. Fencing often leads to positive outcomes for eucalypt recruitment and native understorey diversity, but effectiveness varies with woodland type and degree of initial degradation (e.g. Vesk and Dorrough 2006; Briggs *et al.* 2008; Prober *et al.* 2011). Modifying grazing and fertiliser regimes can also promote native species that are still locally present (Dorrough *et al.* 2004). However, removing degrading factors alone is usually inadequate for restoring original plant composition; indeed no woodlands have been shown to recover to benchmark plant species composition or soil condition through fencing alone (e.g. Prober *et al.* 2011).

Likely reasons for limited recovery are competition with exotic plants and changes in soil processes, seed bank dynamics and establishment conditions. Augmentation of native propagules is typically needed to restore the many plant species that have disappeared from soil seed banks and surrounding landscapes (e.g. Lunt 1995; Standish *et al.* 2007). The widespread persistence of exotic annuals in woodland understoreys is associated with altered soil nitrogen cycling, with seasonally high soil nitrate concentrations promoting exotic annuals at the expense of native plant re-establishment. This cycle can be broken using spring biomass reduction or topsoil removal to reduce weed seed banks, or carbon addition to deplete soil nitrate. Re-establishing dense swards of perennial native tussock grasses, particularly *Themeda*, then becomes feasible, the perennial grasses in turn sequestering soil nitrate and suppressing exotic annuals in the longer term (Prober and Lunt 2009; Prober and Wiehl 2011). Direct drilling and herbicides are additional tools for facilitating re-establishment of native species that restore ecosystem functions (e.g. Gibson-Roy *et al.* 2009).

17.6.2 Revegetation

Another important approach to woodland restoration in the past 20 years has been revegetation (including replanting and natural regeneration) in cleared areas, with multiple goals of promoting biodiversity, storing carbon, production (e.g. sandalwood), and/or ameliorating salinisation. From a biodiversity perspective, recent studies have shown that native plantings can contribute significantly to farm-scale fauna habitat (e.g. Cunningham *et al.* 2008), but there remains a need to improve the diversity and ecological function of plantings. Potential innovations include re-establishing native ground-layer species, reducing soil nutrients and improving soil biophysical processes to optimise persistence of native plants (Gibson-Roy *et al.* 2009; Prober and Lunt 2009; Prober *et al.* 2014), and increasing genotype and species diversity for climate resilience (e.g. Steane *et al.* 2014). Many catchment management organisations now recognise the need to revegetate up to 30% of agricultural landscapes (McIntyre *et al.* 2002), but rates and scales of revegetation need to be substantially increased for this to be achieved.

17.6.3 Policy and incentive structures

On-ground conservation outcomes in temperate eucalypt woodlands often depend on voluntary commitments from land managers and community organisations. A range of policies and market-based instruments have been instigated to facilitate community participation and prevent ongoing biodiversity losses. These include education and motivation programmes, financial incentives schemes such as competitive tenders and environmental stewardship programmes (e.g. Whitten and Coggan 2013), legislative regulations such as clearing controls and threatened ecological community listings, offset schemes when development is permitted, and voluntary legal protection of remnants through covenanting schemes. Bodies such as regional catchment councils and conservation networks also facilitate and coordinate on-ground activities (FitzSimons *et al.* 2013).

While off-reserve conservation measures form a central pillar of the conservation network for temperate eucalypt woodlands, formal reserves are increasingly contributing to conservation of these ecosystems. Many small nature reserves now form a backbone of woodland conservation in the Western Australian wheatbelt (Gibson *et al.* 2004) and the ACT, and State government land acquisition programmes in other states have targeted the purchase of farms to increase representation of the most productive parts of the woodland landscape.

Additionally, reserve networks managed by non-government organisations now include several important temperate eucalypt woodlands.

Finally, Aboriginal people are increasingly re-engaging with land management in temperate eucalypt woodlands, through projects with catchment councils and large-scale corridor programmes such as Slopes to Summit and GondwanaLink. For example, the Ngadju people have recently recovered Native Title over 5 million hectares in the Great Western Woodlands, and are building expertise in conservation planning, weed control and fire management (e.g. Prober *et al.* 2016). Future promise thus lies in working with Aboriginal people to further understand and restore temperate eucalypt woodlands.

Acknowledgements

We thank the many landholders, land managers, policy makers and Traditional Owners who have supported conservation and research in temperate eucalypt woodlands. Preparation of this chapter was supported by the Great Western Woodlands SuperSite of the Australian Government's Terrestrial Ecosystem Research Network.

References

Allcock, K.G. and Hik, D.S. (2004) Survival, growth, and escape from herbivory are determined by habitat and herbivore species for three Australian woodland plants. *Oecologia* 138, 231–241.

Atlas of Living Australia (ALA) (2015) *Atlas of Living Australia Spatial Portal.* Available at: http://www.ala.org.au/ (accessed 1 November 2015).

Australian Plant Census (2015) Available at: https://www.anbg.gov.au/chah/apc/about-APC.html (accessed 1 November 2015).

Bailey, T.G., Davidson, N.J. and Close, D.C. (2012) Understanding the regeneration niche: microsite attributes and recruitment of eucalypts in dry forests. *Forest Ecology and Management* 269, 229–238.

Beard, J.S. (1990) *Plant Life of Western Australia.* Kangaroo Press: Kenthurst.

Beard, J.S., Beeston, G.R., Harvey, J.M. *et al.* (2013) The vegetation of Western Australia at the 1:3,000,000 scale. Explanatory memoir. Second edition. *Conservation Science Western Australia* 9, 1–152.

Briggs, S.V., Taws, N.M., Seddon, J.A. and Vanzella, B. (2008) Condition of fenced and unfenced remnant vegetation in inland catchments in south-eastern Australia. *Australian Journal of Botany* 56, 590–599.

Bureau of Meteorology (BoM) (2015) *Maps of average conditions.* Australian Government: Canberra. Available at: http://www.bom.gov.au/climate/averages/maps.shtml (accessed April 2016).

Chylek, P., Li, J., Dubey, M.K. *et al.* (2011) Observed and model simulated 20th century Arctic temperature variability: Canadian Earth System Model CanESM2. *Atmospheric Chemistry and Physics Discussion* 11, 22893–22907.

Clarke, P.J. (2000) Plant population processes in temperate woodlands in eastern Australia: premises for management. In *Temperate Eucalypt Woodlands in Australia: Biology, Conservation, Management and Restoration*, ed. R.J. Hobbs and C.J. Yates. Surrey Beatty and Sons: Sydney, pp. 248–270.

Clarke, P.J. and Davison, E.A. (2001) Experiments on the mechanism of tree and shrub establishment in temperate grassy woodlands: seedling emergence. *Austral Ecology* 26, 400–412.

Close, D. and Davidson, N.J. (2004) Review of rural tree decline in a changing Australian climate. *Tasforests* 15, 1–18.

Coates, F., Lunt, I.D. and Tremblay, R.L. (2006) Effects of disturbance on population dynamics of the threatened orchid *Prasophyllum correctum* D.L. Jones and implications for grassland management in south-eastern Australia. *Biological Conservation* 129, 59–69.

Cohn, J.S., Lunt, I.D., Bradstock, R.A. *et al.* (2013) Demographic patterns of a widespread long-lived tree are associated with rainfall and disturbances along rainfall gradients in SE Australia. *Ecology and Evolution* 3, 2169–2182.

Cunningham, R., Lindenmayer, D.B., Crane, M. *et al.* (2008) The combined effects of remnant vegetation and tree planting on farmland birds. *Conservation Biology* 22, 742–752.

Department of Sustainability, Environment, Water, Population and Communities (DSEWPaC) (2012) Australia – Estimated Pre 1750 Major Vegetation Groups – NVIS Version 4.1 (Albers 100m analysis product) (Metadata identifier: c9136118-a054-4835-b4d6-1645cba33e10). Australian Government Department of Sustainability, Environment, Water, Population and Communities, Canberra.

Department of the Environmenr and Energy (DotEE) (2017) National Vegetation Information System (BVIS). Available at: www.environment.gov.au/land/native-vegetation/national-vegetation-information-system (accessed 13 January 2017).

Department of the Environment and Water Resources (DEWR) (2007) *Australia's Native Vegetation: A Summary of Australia's Major Vegetation Groups.* Australian Government: Canberra.

Dorrough, J., Yen, A., Turner, V. *et al.* (2004) Livestock grazing management and biodiversity conservation in Australian temperate grassy landscapes. *Australian Journal of Agricultural Research* 55, 279–295.

Dorrough, J., Moxham, C., Turner, V. and Sutter, G. (2006) Soil phosphorus and tree cover modify the effects of livestock grazing on plant species richness in Australian grassy woodland. *Biological Conservation* 130, 394–405.

Eldridge, D.J. and Freudenberger, D. (2005) Ecosystem wicks: woodland trees enhance water infiltration in a fragmented agricultural landscape in eastern Australia. *Austral Ecology* 30, 336–347.

Fensham, R.J. (1992) The management implications of fine fuel dynamics in bushlands surrounding Hobart, Tasmania. *Journal of Environmental Management* 36, 301–320.

FitzSimons, J., Pulsford, I. and Wescott, G. (2013) *Linking Australia's Landscapes. Lessons and Opportunities from Large-scale Conservation Networks.* CSIRO Publishing: Melbourne.

Foreman, P.W. (1996) The ecology of remnant indigenous grasslands of Victoria's northern Riverine Plain. MSc Thesis, La Trobe University, Bundoora.

Fox, P. (2001) Conservation of salmon gum (*Eucalyptus salmonophloia* F. Muell.) woodlands in the Western Australian wheatbelt: a consilience of biology, culture and perception. PhD Thesis, Murdoch University, Perth.

Gibson, N., Keighery, G.J., Lyons, M.N. and Webb, A. (2004) Terrestrial flora and vegetation of the Western Australian wheatbelt. In *Supplement 67: a Biodiversity Survey of the Western Australian Agricultural Zone*, ed. G.J. Keighery, S.A. Halse, M.S. Harvey and N.L. McKenzie. Western Australian Museum: Welshpool, pp. 139–189.

Gibson-Roy, P., Delpratt, J., Moore, G. and Hepworth, G. (2009) Does diversity influence soil nitrate, light availability and productivity in the establishment phase of Australian temperate grassland reconstruction? *Ecological Management and Restoration* 10, 41–50.

Gillison, A.N. and Walker, J. (1981) Woodlands. In *Australian Vegetation*, ed. R.H. Groves. Melbourne: Cambridge University Press, pp. 177–197.

Gosper, C.R., Yates, C.J. and Prober, S.M. (2013a) Floristic diversity in fire-sensitive eucalypt woodlands shows a 'U'-shaped relationship with time since fire. *Journal of Applied Ecology* 50, 1187–1196.

Gosper, C.R., Prober, S.M. and Yates, C.J. (2013b) Multi-century changes in vegetation structure and fuel availability in fire-sensitive eucalypt woodlands. *Forest Ecology and Management* 310, 102–109.

Gosper, C.R., Prober, S.M. and Yates, C.J. (2016) Continental-scale syntheses of Australian pyromes: misclassification of south-western eucalypt woodlands misinforms management. *Journal of Biogeography* 43, 858–861.

Gott, B. (2005) Aboriginal fire management in southeastern Australia: aims and frequency. *Journal of Biogeography* 32, 1203–1208.

Guerin, G.R., Wen, H. and Lowe, A.J. (2012) Leaf morphology shift linked to climate change. *Biology Letters* 8, 882–886.

Harvey, J. (2014) Regional variability in salmon gum (*Eucalyptus salmonophloia*) woodlands of south-western Australia, with particular focus on the Great Western Woodlands. MSc thesis, Curtin University, Perth.

Hassell, E. (1975) *My Dusky Friends: Aboriginal Life, Customs and Legends and Glimpses of Station Life at Jarramungup in the 1880's.* CW Hassell: East Fremantle.

Hobbs, R.J. and Atkins, L. (1991) Interactions between annuals and woody perennials in a Western Australian nature reserve. *Journal of Vegetation Science* 2, 643–654.

Keith, D.A. (2004) *Ocean Shores to Desert Dunes: The Native Vegetation of New South Wales and the ACT.* NSW Department of Environment and Conservation: Sydney.

Kirkpatrick, J.B. (2004) Vegetation change in an urban grassy woodland 1974–2000. *Australian Journal of Botany* 52, 597–608.

Kirkpatrick, J.B. and Gilfedder, L. (2000) Distribution, ecology and conservation status of lowland and subalpine woodlands in Tasmania. In *Temperate Eucalypt Woodlands in Australia: Biology, Conservation, Management and Restoration*, ed. R.J. Hobbs and C.J. Yates. Surrey Beatty and Sons: Sydney, pp. 45–56.

Kirkpatrick, J.B., Gilfedder, L. and Fensham, R. (1988) *City Parks and Cemeteries. Tasmania's Remnant Grasslands and Grassy Woodlands.* Tasmanian Conservation Trust: Hobart.

Leigh, J.H. and Holgate, M.D. (1979) Responses of the understorey of forests and woodlands of the Southern Tablelands to grazing and burning. *Australian Journal of Ecology* 4, 25–45.

Lindenmayer, D., Bennett, A.F. and Hobbs, R. (2010) An overview of the ecology, management and conservation of Australia's temperate woodlands. *Ecological Management and Restoration* 11, 201–209.

Lodge, G.M. and Whalley, R.D.B. (1989) *Native and Natural Pastures of the Northern Slopes and Tablelands of NSW: A Review and Annotated Bibliography.* NSW Agriculture and Fisheries: Sydney.

Lunt, I.D. (1990) Species–area curves and growth-form spectra for some herb-rich woodlands in western Victoria, Australia. *Australian Journal of Ecology* 15, 155–161.

Lunt, I.D. (1995) Seed longevity of six native forbs in a closed *Themeda triandra* grassland. *Australian Journal of Botany* 43, 439–449.

Lunt, I. and Bennett, A.F. (2000) Temperate woodlands in Victoria: distribution, composition and conservation. In *Temperate Eucalypt Woodlands in Australia: Biology, Conservation,*

Management and Restoration, ed. R.J. Hobbs and C.J. Yates. Surrey Beatty and Sons: Sydney, pp. 17–31.

Lunt, I.D., Eldridge, D.J., Morgan, J.W. and Witt, G.B. (2007) A framework to predict the effects of livestock grazing and grazing exclusion on conservation values in natural ecosystems in Australia. *Australian Journal of Botany* 55, 401–415.

Lunt, I.D., Winsemius, L.M., McDonald, S.P. *et al.* (2010) How widespread is woody plant encroachment in temperate Australia? Changes in woody vegetation cover in lowland woodland and coastal ecosystems in Victoria from 1989 to 2005. *Journal of Biogeography* 37, 722–732.

Lunt, I., Prober, S.M. and Morgan, J. (2012) How do fire regimes affect ecosystem structure, function and diversity in grasslands and grassy woodlands of southern Australia? In *Flammable Australia: Fire Regimes, Biodiversity and Ecosystems in a Changing World*, ed. R.A. Bradstock, A.M. Gill and R.J. Williams. CSIRO Publishing: Melbourne, pp. 253–270.

Maron, M., Grey, M.J., Catterall, C.P. *et al.* (2013) Avifaunal disarray due to a single despotic species. *Diversity and Distributions* 19, 1468–1479.

Martin, G. (2003) The role of small ground-foraging mammals in topsoil health and biodiversity: implications to management and restoration. *Ecological Management and Restoration* 4, 114–119.

McArthur, W.M. (2004) *Reference Soils of South-western Australia.* Department of Agriculture: Western Australia.

McIntyre, S. and Lavorel, S. (1994) Predicting richness of native, rare, and exotic plants in response to habitat and disturbance variables across a variegated landscape. *Conservation Biology* 8, 521–531.

McIntyre, S. and Lavorel, S. (2007) A conceptual model of land use effects on the structure and function of herbaceous vegetation. *Agriculture, Ecosystems and Environment* 119, 11–21.

McIntyre, S., McIvor, J.G. and Heard, K.M. (eds) (2002) *Managing and Conserving Grassy Woodlands.* CSIRO Publishing: Melbourne.

McIntyre, S., Cunningham, R.B., Donnelly, C.F. and Manning, A.D. (2014) Restoration of eucalypt grassy woodland: effects of experimental interventions on ground-layer vegetation. *Australian Journal of Botany* 62, 570–579.

Meers, T. and Adams, R. (2003) The impact of grazing by eastern grey kangaroos (*Macropus giganteus*) on vegetation recovery after fire at Reef Hills Regional Park, Victoria. *Ecological Management and Restoration* 4, 126–132.

Mogoutnov, A. and Venning, J. (2014) Remnant tree decline in agricultural regions of South Australia. *Pacific Conservation Biology* 20, 366–375.

Moore, C.W.E. (1953) The vegetation of the south eastern Riverina, New South Wales II. The disclimax communities. *Australian Journal of Botany* 1, 548–567.

Moore, R.M. (1970) South-eastern temperate woodlands and grasslands. In *Australian Grasslands*, ed. R.M. Moore. Australian National University Press: Canberra, pp. 169–190.

Morgan, J.W., Wong, N.K. and Cutler, S.C. (2011) Life-form species–area relationships in a temperate eucalypt woodland community. *Plant Ecology* 212, 1047–1055.

Murphy, M.T., Garkaklis, M.J. and Hardy, G.E.S. (2005) Seed caching by woylies *Bettongia penicillata* can increase sandalwood *Santalum spicatum* regeneration in Western Australia. *Austral Ecology* 30, 747–755.

Nicolle, D. (2015) *Classification of the eucalypts (*Angophora, Corymbia *and* Eucalyptus*). Version 2.* Available at: http://www.dn.com.au/Classification-Of-The-Eucalypts.pdf (accessed 28 July 2015).

O'Donnell, A.J., Boer, M.M., McCaw, W.L. and Grierson, P.F. (2011a) Vegetation and landscape connectivity control wildfire intervals in unmanaged semi-arid shrublands and woodlands in Australia. *Journal of Biogeography* 38, 112–124.

O'Donnell, A.J., Boer, M.M., McCaw, W.L. and Grierson, P.F. (2011b) Climatic anomalies drive wildfire occurrence and extent in semi-arid shrublands and woodlands of southwest Australia. *Ecosphere* 2, 1–15.

Paton, D.C., Prescott, A.M., Davies, R.J.P. and Heard, L.M. (2000) The distribution, status and threats to temperate woodlands in South Australia. In *Temperate Eucalypt Woodlands in Australia: Biology, Conservation, Management and Restoration*, ed. R.J. Hobbs and C.J. Yates. Surrey Beatty and Sons: Sydney, pp. 57–85.

Prober, S.M. (1996) Conservation of the grassy white box woodlands: rangewide floristic variation and implications for reserve design. *Australian Journal of Botany* 44, 57–77.

Prober, S.M. and Lunt, I. (2009) Restoration of *Themeda australis* swards suppresses soil nitrate and enhances ecological resistance to invasion by exotic annuals. *Biological Invasions* 11, 171–181.

Prober, S.M. and Smith, F.P. (2009) Enhancing biodiversity persistence in intensively-used agricultural landscapes: a synthesis of 30 years of research in the Western Australian wheatbelt. *Agriculture, Ecosystems and Environment* 132, 173–191.

Prober, S.M. and Thiele, K.R. (2004) Floristic patterns across an east–west gradient in grassy box woodlands of central New South Wales. *Cunninghamia* 8, 306–325.

Prober, S.M. and Thiele, K.R. (2005) Restoring Australia's temperate grasslands and grassy woodlands: integrating function and diversity. *Ecological Management and Restoration* 6, 16–27.

Prober, S.M. and Wiehl, G. (2011) Resource heterogeneity and persistence of exotic annuals in long-ungrazed Mediterranean-climate woodlands. *Biological Invasions* 13, 2009–2022.

Prober, S.M., Lunt, I. and Thiele, K.R. (2002) Determining reference conditions for management and restoration of temperate grassy woodlands: relationships among trees, topsoils and understorey flora in little-grazed remnants. *Australian Journal of Botany* 50, 687–697.

Prober, S.M., Standish, R.J. and Wiehl, G. (2011) After the fence: vegetation and topsoil condition in grazed, fenced and benchmark eucalypt woodlands of fragmented agricultural landscapes. *Australian Journal of Botany* 59, 369–381.

Prober, S.M., Thiele, K.R. and Lunt, I. (2007) Fire frequency regulates tussock grass composition, structure and resilience in endangered temperate woodlands. *Austral Ecology* 32, 808–824.

Prober, S.M., Thiele, K.R., Rundell, P. *et al.* (2012) Facilitating adaptation of biodiversity to climate change: a conceptual framework applied to the world's largest Mediterranean-climate woodland. *Climatic Change* 110, 227–248.

Prober, S.M., Stol, J., Piper, M. *et al.* (2014) Towards climate-resilient restoration in mesic eucalypt woodlands: characterizing topsoil biophysical condition in different degradation states. *Plant and Soil* 383, 231–244.

Prober, S.M., Thiele, K.R. and Speijers, J. (2013) Management legacies shape decadal-scale responses of plant diversity to experimental disturbance regimes in fragmented grassy woodlands. *Journal of Applied Ecology* 50, 376–386.

Prober, S.M., Williams, K.J., Harwood, T.D. *et al.* (2015) *Helping Biodiversity Adapt: Supporting Adaptation Planning Using a Community-level Modelling Approach.* CSIRO Land and Water Flagship: Canberra. Available at: www.AdaptNRM.org (accessed 4 August 2016).

Prober, S.M., Yuen, E., O'Connor, M. and Schultz, L. (2016) Ngadju kala: Australian Aboriginal fire knowledge in the Great Western Woodlands. *Austral Ecology* 41, 716–732.

Radford, J.Q., Bennett, A.F. and Cheers, G.J. (2005) Landscape-level thresholds of habitat cover for woodland-dependent birds. *Biological Conservation* 124, 317–337.

Raiter, K. (2016) Enigmatic ecological impacts of mining and linear infrastructure development in Australia's Great Western Woodlands. School of Plant Biology, University of Western Australia.

Recher, H.F. and Majer, J.D. (2006) Effects of bird predation on canopy arthropods in wandoo Eucalyptus wandoo woodland. *Austral Ecology* 31, 349–360.

Reid, N. and Landsberg, J. (2000) Tree decline in agricultural landscapes: what we stand to lose. In *Temperate Eucalypt Woodlands in Australia: Biology, Conservation, Management and Restoration*, ed. R.J. Hobbs and C.J. Yates. Surrey Beatty and Sons: Sydney, pp. 127–166.

Robertson, D. (1985) Interrelationships between kangaroos, fire and vegetation dynamics at Gellibrand Hill Park, Victoria. PhD Thesis, University of Melbourne, Melbourne.

Sankaran, M., Hanan, N.P., Scholes, R.J. *et al.* (2005) Determinants of woody cover in African savannas. *Nature* 438, 846–849.

Schultz, N.L., Reid, N., Lodge, G. and Hunter, J.T. (2014) Seasonal and interannual variation in vegetation composition: implications for survey design and data interpretation. *Austral Ecology* 39, 755–766.

Skinner, A.K., Lunt, I.D., Spooner, P. and McIntyre, S. (2009) The effect of soil compaction on germination and early growth of *Eucalyptus albens* and an exotic annual grass. *Austral Ecology* 34, 698–704.

Smallbone, L., Prober, S.M. and Lunt, I.D. (2007) Restoration treatments enhance early establishment of native forbs in a degraded grassy woodland. *Australian Journal of Botany* 55, 818–830.

Spooner, P. (2005) On squatters, settlers and early surveyors: historical development of country road reserves in southern New South Wales. *Australian Geographer* 36, 55–73.

Standish, R.J., Cramer, V.A., Wild, S.L. and Hobbs, R.J. (2007) Seed dispersal and recruitment limitation are barriers to native recolonization of old fields in western Australia. *Journal of Applied Ecology* 44, 435–445.

Steane, D.A., Potts, B.M., McLean, E.H. *et al.* (2014) Genome-wide scans detect adaptation to aridity in a widespread forest tree species. *Molecular Ecology* 23, 2500–2513.

Verboom, W.H. and Pate, J.S. (2006) Bioengineering of soil profiles in semiarid ecosystems: the 'phytotarium' concept. A review. *Plant and Soil* 289, 71–102.

Vesk, P.A. and Dorrough, J.W. (2006) Getting trees on farms the easy way? Lessons from a model of eucalypt regeneration on pastures. *Australian Journal of Botany* 54, 509–519.

Warren, S.D. and Eldridge, D.J. (2001) Biological soil crusts and livestock in arid ecosystems: are they compatible? In *Biological Soil Crusts: Structure, Function, and Management*, ed. J. Belnap and O.L. Lange. Springer-Verlag: Berlin, pp. 401–415.

Watson, D.M. (2011) *Mistletoes of Southern Australia.* CSIRO Publishing: Collingwood.

Watson, P.J., Bradstock, R.A. and Morris, E.C. (2009) Fire frequency influences composition and structure of the shrub layer in an Australian subcoastal temperate grassy woodland. *Austral Ecology* 34, 218–232.

Whalley, R.D.B., Robinson, G.G. and Taylor, J.A. (1978) General effects of management and grazing by domestic livestock on the rangelands of the Northern Tablelands of New South Wales. *Australian Rangelands Journal* 1, 174–190.

Whitten, S.M. and Coggan, A. (2013) Market-based instruments and ecosystem services: opportunity and experience to date. In *Ecosystem Services in Agricultural and Urban Landscapes*, 1st edition, ed. S. Wratten, H. Sandhu, R. Cullen and R. Costanza. John Wiley and Sons: Oxford, pp. 178–192.

Williams, K.J., Prober, S.M., Harwood, T.D. *et al.* (2014) *Implications of Climate Change for Biodiversity: A Community-level Modelling Approach.* CSIRO Land and Water Flagship: Canberra. Available at: www.AdaptNRM.org (accessed 4 August 2016).

Wills, K.E. and Clarke, P.J. (2008) Plant trait-environmental linkages among contrasting landscapes and climate regimes in temperate eucalypt woodlands. *Australian Journal of Botany* 56, 422–432.

Wilson, J. (2002) Flowering ecology of a Box-Ironbark *Eucalyptus* community. PhD Thesis, Deakin University, Melbourne.

Yates, C.J. and Hobbs, R.J. (1997) Temperate eucalypt woodlands: a review of their status, processes threatening their persistence and techniques for restoration. *Australian Journal of Botany* 45, 949–973.

Yates, C.J., Hobbs, R.J. and Bell, R.W. (1994) Landscape-scale disturbances and regeneration in semi-arid woodlands of southwestern Australia. *Pacific Conservation Biology* 1, 214–221.

Yates, C.J., Taplin, R., Hobbs, R.J. and Bell, R.W. (1995) Factors limiting the recruitment of *Eucalyptus salmonophloia* in remnant woodlands. II. Post-dispersal seed predation and soil seed reserves. *Australian Journal of Botany* 43, 145–155.

Yates, C.J., Hobbs, R.J. and Atkins, L. (2000a) Establishment of perennial shrub and tree species in degraded *Eucalyptus salmonophloia* (Salmon Gum) remnant woodlands: effects of restoration treatments. *Restoration Ecology* 8, 135–143.

Yates, C.J., Hobbs, R.J. and True, D.T. (2000b) The distribution and status of eucalypt woodlands in Western Australia. In *Temperate Eucalypt Woodlands in Australia: Biology, Conservation, Management and Restoration*, ed. R.J. Hobbs and C.J. Yates. Surrey Beatty and Sons: Sydney, pp. 86–106.

Yates, C.J., Norton, D.A. and Hobbs, R.J. (2000c) Grazing effects on plant cover, soil and microclimate in fragmented woodlands in south-western Australia: implications for restoration. *Austral Ecology* 25, 36–47.

18 Australian Tussock Grasslands

JOHN W. MORGAN, ROD J. FENSHAM,
ROBERT GODFREE AND PAUL W. FOREMAN

In this review, we highlight the evolution, distribution and ecological features of tussock grasslands throughout Australia. Grassland distribution appears to be largely driven by bottom-up processes (i.e. climate-edaphic factors), although there are some situations where fire has likely played an important role, and derived grasslands are now widespread due to clearing and grazing. The strong relationship of grasslands with fertile, clay soils, developed for agriculture since European arrival, is one of the key reasons why they are now reduced across much of their natural range. Plant diversity in tussock grasslands has long been thought to be promoted by disturbance (e.g. fire), although it is apparent that the paucity of experimental studies at appropriate spatial scales, especially stratified across climate-regulated productivity gradients, hinders our confidence in this prediction. Additionally, the almost complete lack of climate change experiments in tussock grasslands limits our ability to predict its likely impacts on ecosystem composition, structure and function. There have been substantial gains in the preservation of tussock grasslands over the last two decades, and changes in land use offer opportunities for passive restoration. Fragmented tussock grasslands are currently subjected to numerous ecosystem stresses and these will likely have a big impact on future tussock grassland resilience and stability.

18.1 Introduction

A tussock grassland is a plant community in which the dominant species are typically perennial native C_3 or C_4 tussock grasses (either a single species or several genera from the family Poaceae), there are few or no shrubs, and trees are sparse to absent. Tussock grasses are perennial plants that grow as singular plants in clumps, tufts or bunches, rather than forming a carpet or lawn. Associated with the dominant grasses are usually subdominant grass species and a variety of other herbaceous plants, both annual and perennial forbs and monocots. Common plant families in grasslands include Poaceae, Asteraceae, Fabaceae and Cyperaceae.

Natural grasslands dominated by tussock-forming grasses are widespread across southern, eastern and northern Australia, occupying > 55 million hectares (7%) of the continent (Lindenmayer *et al.* 2014). They are one of Australia's 23 major vegetation groups (NVIS Major Vegetation Group 19; Department of the Environment and Energy 2007), extending in a broad arc from central and western Queensland to eastern South Australia, with sporadic occurrences in eastern Tasmania, across the

Australian Vegetation, Third Edition, ed. David A. Keith. Published by Cambridge University Press. © Cambridge University Press 2017.

Citation: Morgan, J.W. *et al.* (2017). Australian Tussock Grasslands. In D. A. Keith, ed., *Australian Vegetation*, 3rd edn. Cambridge: Cambridge University Press, pp. 438–60.

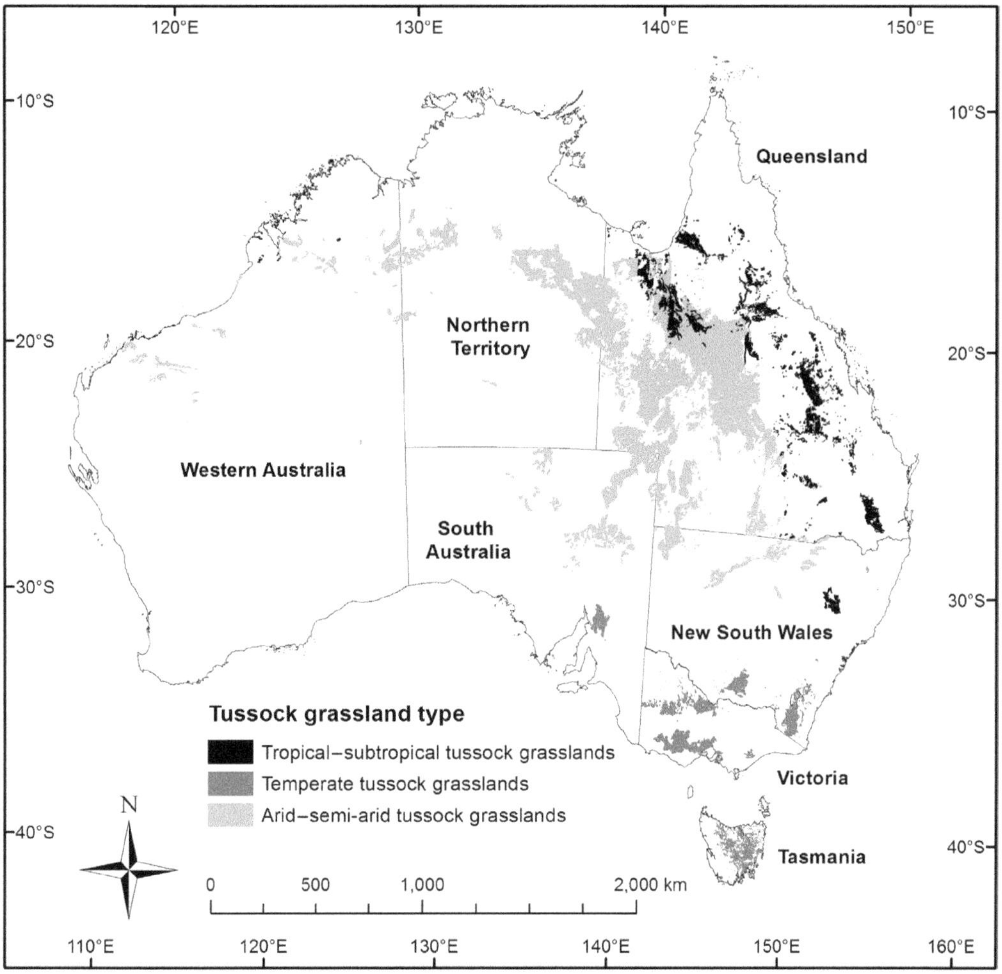

Figure 18.1 Distribution map showing the extent of three major types of tussock grasslands in Australia. Tropical and arid grasslands are composed almost exclusively of grasses with C_4 photosynthetic pathways, while temperate grasslands include mixtures of both C_3 and C_4 grasses. Localised occurrences of tussock grasslands in south-west Western Australia, as well as coastal grasslands, cannot be shown at this scale.

Northern Territory, and the Kimberly and Pilbara regions of Western Australia (Figure 18.1). Tussock grasslands are largely absent from the south-western portion of the continent in climatic zones that support them elsewhere, probably because of the lack of fertile, fine-textured soils. Where fine-textured soils do occur, they are heavily weathered and nutrient poor, supporting the notion that grasslands tend not to occur on lateritic and leached infertile soils (Diels 1906). Nevertheless, localised occurrences of tussock grasslands did exist in the south-west, such as the now extinct *Themeda* grasslands of the northern wheatbelt near Geraldton (O'Connor 2001), as well as occurring on drainage floors and claypans on the Nullarbor, in mosaic with chenopod shrublands (Waddell *et al.* 2010).

The substantial latitudinal distribution (11° to 44°S) generates a wide range of grassland environments. Tussock grasslands generally occur on fertile, fine-textured soils (i.e. clay-rich vertosols) in valley bottoms and plains at low elevations (Kirkpatrick *et al.* 1995; Lunt and Morgan 2002; Keith 2004) and across a wide range of climates: arid to tropical to temperate (Figure 18.1). Tussock grasslands also occur in frost hollows at higher elevations and above the alpine treeline (these are considered further by Venn *et al.*, Chapter 19, this volume). Hummock grasslands, dominated by spiny, hummock-forming spinifex grasses (*Triodia* spp.) are very different from tussock grasslands and are described by Wardle and Nano (Chapter 26, this volume).

18.2 The rise of grasses and grasslands in Australia

In Australia, the contemporary distribution of grasses shows that C_4 grasses are more widespread than C_3 grasses (Hattersley 1983). With decreasing latitude, the dominance and diversity of C_4 species increase, their carbon-concentrating mechanism suppressing the energetically inefficient photorespiration that handicaps C_3 grasses at high temperatures (Osborne 2008). In contrast, C_3 grasses are largely confined to southern temperate climates where photorespiration is naturally limited by lower temperatures (Hartley 1961, 1973). While tussock grasslands are a widespread vegetation type, grassland ecosystems were largely absent from Australia until very recent geological time (Barlow 1981; Jacobs *et al.* 1999; Strömberg 2011).

Grasses originated in Gondwana during the Cretaceous (>65 million years ago (Ma)) (Hsiao *et al.* 1999; Bouchenak-Khelladi *et al.* 2010) with Bremer (2002) speculating, using phylogenetic evidence, that graminoids possibly originated in East Gondwana. In Australia, however, there are almost no early preserved grass macrofossils (Macphail and Hill 2002). The earliest record of Australian grass pollen is from the mid-Eocene (~45 Ma), but it has always been relatively sparse in the Australian fossil record (Macphail and Hill 2002), becoming most abundant within the last 2 million years (White 1986).

Kershaw *et al.* (1994) trace the entry of the Poaceae across Australia using palynological methods. Grass pollen first occurred in what is now north-western Australia, possibly as part of savanna-like vegetation by the mid-Miocene (~15 Ma). Many grasses that are now common in Australia – including the C_4 genera *Themeda*, *Dichanthium* and *Bothriochloa* from the Andropogoneae tribe – appear to have migrated from Asia at this time, when the Australian plate collided with Sundaland, the Indonesian archipelago (Jones 1997). Hartley (1964) showed that the Andropogoneae has major centres of distribution in south-eastern Indonesia and India, and a lesser centre in central eastern Africa. The entry of grasses into more southern and eastern areas of Australia occurred in the late Miocene (~10 Ma) and early Pliocene (~5 Ma) due to increased climate seasonality (Martin 2006). There were rapid evolutionary radiations in grasses around this time (Syme 2012), as well as many large Australian groups such as the sclerophyll taxa *Eucalyptus*, *Banksia* and *Allocasuarina* (Crisp *et al.* 2004). *Austrostipa*, for example, originated and began to rapidly diversify between 25 and 10 Ma (Syme 2012). Increasing taxonomic diversity may have resulted from adaptation to newly derived arid niches created by climatic changes. Molecular data for the C_3 genus *Poa* are also broadly congruent with fossil data that record the appearance of vegetation with a prominent grassland component from the mid-Pliocene (~4 Ma) (Birch *et al.* 2014). Fires were part of the landscape throughout the Miocene, potentially facilitating the spread of grasses and grasslands (Osborne 2008). Increases in the abundance of the major C_4 clades Paniceae and Andropogoneae may have been favoured by these changed conditions, although the reasons for the rise of C_4-dominated grasslands per se are still debated (Keeley and Rundel 2005; Beerling and Osborne 2006; Edwards and Still 2008; Osborne 2008; Edwards and Smith 2010).

18.3 Contemporary Australian tussock grasslands

Tussock grasslands occur across a wide range of climatic domains driven by latitude and distance from the coast, ranging from temperate, to arid, to tropical (Figure 18.1). Gradients in rainfall quantity and seasonality, as well as temperature, drive turnover in composition from C_3 dominance in the temperate south, where pooids and danthonioids predominate (e.g. *Austrostipa, Poa, Rytidosperma, Anthosachne*), to C_4 dominance in the tropical north (Moore 1970; Figure 18.2). The C_4 grasses can be further divided into: (i) the xerophytic chloridoids, typically dominant in warm, arid to semi-arid (mostly inland) regions where there is very low available soil moisture; and (ii) the mesophytic panicoids, mostly found in warm tropical to subtropical regions where available soil moisture is moderate (Twiss 1992). Chloridoids include many genera of the northern inland 'Mitchell Grass Downs' such as *Astrebla, Eragrostis* and *Sporobolus*, while the panicoids include many genera of the higher rainfall inland slopes and plains of the eastern ranges in New South Wales and Queensland (e.g. *Themeda, Bothriochloa, Dichanthium, Digitaria, Panicum*) (Figure 18.2).

Three principle 'natural' tussock grassland types are defined (following the broad classification adopted by Beadle 1981): arid and semi-arid; tropical and subtropical; temperate. Two important anomalies are the heterogeneous assemblage

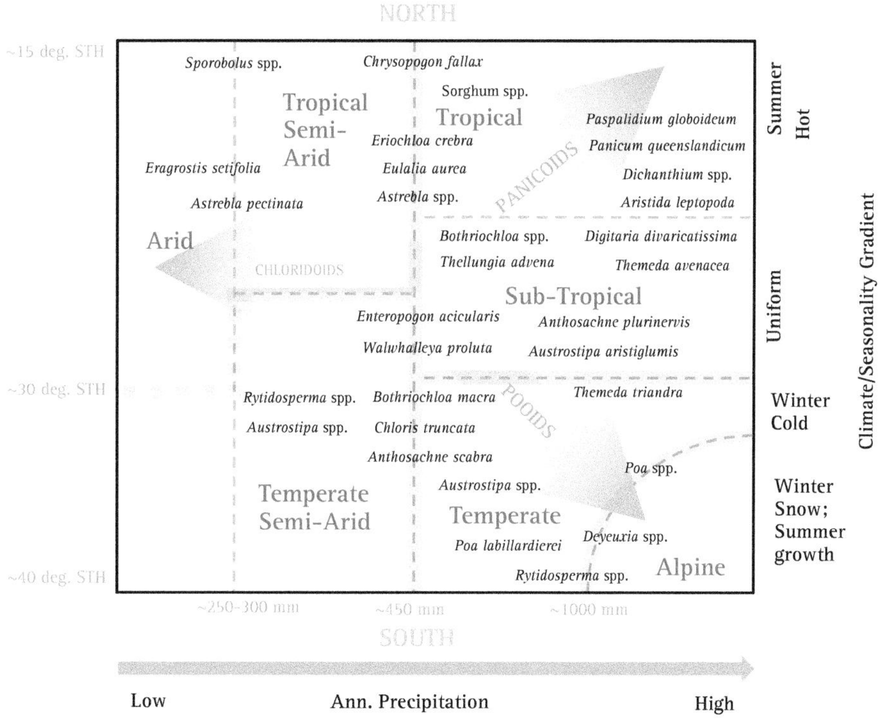

Figure 18.2 Schematic showing broad biogeography and dominant grass taxon gradients between key grassland types in Australia.
Adapted from Beadle (1981).

of coastal grasslands, and the derived grasslands, sometimes referred to as 'disclimax' (Moore 1953, 1970; McDougall 2008) or 'induced' grasslands (Beadle 1981). The latter are dominated by native perennial grasses converted under human agency from another structural form, such as wetland, shrubland, woodland, forest or even rainforest.

18.3.1 Arid and semi-arid grasslands

Arid and semi-arid tussock grasslands are by far the most abundant natural tussock grasslands of the continent, occupying vast expanses of inland eastern Australia and across parts of the north where annual rainfall ranges between 200 and 500 mm (White *et al.* 2014). Their primary occurrence is on cracking clay soil plains (vertosols) derived from fine-textured marine sediments or on alluvial floodplains from throughout western Queensland to the Barkly Tablelands of the Northern Territory. Major outliers are found in the Kimberley and Pilbara regions of Western Australia associated with the Victoria, Ord and Fitzroy Rivers (Bennett *et al.* 2002; Beard *et al.* 2013), on the Stony Plains south and west of Lake Eyre, and along the Darling River floodplains and its tributaries in Queensland and New South Wales (Perry 1970). These 'downs' are characterised by robust, deep-rooted and long-lived, endemic chloridoids (Mitchell grasses, *Astrebla* spp.) that are well-adapted to survive extended droughts and to exploit summer monsoonal storms and floods (Figure 18.3b). Tussock interstices support a variety of ephemeral forbs (from Fabaceae, Asteraceae, Malvaceae and Amaranthaceae) and vegetation patterns are spatially and temporally complex due to the interaction between strongly seasonal rainfall and the heterogeneous microtopography (Fisher *et al.* 2002; White *et al.* 2014). Intense grazing, such as occurs around water points, can alter species composition in arid grasslands (Phelps *et al.* 2002), but recent evidence (e.g. Fensham *et al.* 2014) supports their reputation as robust in the face of extensive grazing typical of commercial pastoralism (Figure 18.3c).

18.3.2 Tropical and subtropical grasslands

Tropical and subtropical grasslands are almost entirely restricted to higher rainfall (500 to >750 mm) inland slopes and plains of Queensland and north-eastern New South Wales. They span from the Liverpool Plains (31.5°S) to the Einasleigh Uplands and Gulf Plains in north Queensland. Soils are cracking clay vertosols formed over young basalt or sometimes fine-grained sedimentary rocks, or recent alluvium along the floodplains of major watercourses (TSSC 2008). These grasslands are relatively mesic because of the summer monsoon moisture penetrating south along the eastern ranges. Panicoids characterise the tropical grasslands, particularly at the generic level (e.g. *Dichanthium, Bothriochloa, Themeda, Panicum*), but in many respects these grasslands are mesic analogues of the arid grasslands, and *Astrebla* can become dominant in dry years where the two systems intergrade (Austin and Williams 1988). In the south, a distinct temperate pooid influence distinguishes the subtropical grasslands with species such as *Rytidosperma bipartitum, Austrostipa aristiglumis* and *Anthosachne plurinervis*, and also one of the few examples of a panicoid with temperate affinities, *Walwhalleya proluta. Digitaria divaricatissima* and *Paspalidium globoideum* are further panicoids characteristic of the tropical grasslands, and others more widespread across the north include *Chrysopogon fallax, Eriochloa crebra* and *Eulalia aurea.* A number of chloridoids are also present in the tropical systems, including *Astrebla* spp., *Sporobolus* spp. and *Thellungia advena.* Because these systems naturally occur on some of Australia's most productive landscapes, they have long been targeted and intensively developed for agriculture and other uses. Recent estimates put the level of decline between 86% in central Queensland and as high as 99% in the Darling Downs (Accad *et al.* 2006; Figure 18.3a). Grazing can also alter species composition in these grasslands, and one of the formerly dominant grasses in the subtropics, *Dichanthium queenslandicum*, is now

Figure 18.3 Examples of Australian tussock grasslands (a) One of the last relicts of subtropical grassland on the Darling Downs, south-east Queensland, with diversity of C_4 grasses and forbs (including Asteraceae and *Wahlenbergia* species) (photo: Rod Fensham); (b) Mitchell grass (*Astrebla* spp.) is an example of an arid and semi-arid tussock grassland in central Queensland (photo: Rod Fensham); (c) livestock grazing can alter vegetation structure, as shown by grazing exclusion experiments in central Queensland, but research suggests that diversity is generally robust under extensive grazing (photo: Rod Fensham); (d) kangaroo grass (*Themeda triandra*) is the dominant tussock grass of many temperate grasslands in high-rainfall areas including this one at Terrinallum, Victoria (photo: John Morgan); (e) where rainfall is lower, as at Avoca Plain, Victoria, kangaroo grass is replaced by chloridoid grasses (e.g. *Enteropogon acicularis*, *Chloris truncata*, *Sporobolus caroli*), annual forbs and chenopods (photo: Paul Foreman).
(A black and white version of this figure will appear in some formats. For the colour version, please refer to the plate section.)

restricted to roadsides and very lightly grazed paddocks (Fensham 1999).

18.3.3 Temperate grasslands

Lowland temperate grasslands occur on the fertile plains, slopes and tablelands of south-eastern Australia, mostly in Victoria and New South Wales, but also in parts of the south-east of South Australia, the ACT, the Tasmanian Midlands, and possibly pockets of the northern wheatbelt of Western Australia (e.g. O'Connor 2001). United by their relatively cold winters and reliable cool-season rainfall, these grasslands are characterised by pooid genera, especially *Rytidosperma* and *Austrostipa*, with *Poa* (such as *P. labillardierei* typical of wetter areas and drainage lines) and the exceptional panicoid, *Themeda triandra*, characteristic of the higher rainfall systems (>500 mm) (Figure 18.3d). Chloridoids like *Enteropogon acicularis* and the temperate-affiliated panicoid *Walwhalleya proluta* generally occur in the lower rainfall regions (300–500 mm) in South Australia, Victoria and New South Wales. Again, these natural grasslands are mostly associated with cracking clay vertosols derived from recent deposits of basalt and alluvium, although in some regions, such as the montane Monaro Tablelands south of Canberra, grasslands also occur on clays derived from a greater range of parent rocks including limestone, fine-grained sedimentary rocks and even granite (Benson 1994). In this latter case, it is thought shrubs and trees are sparse or absent due to influences of cold air drainage. While six of the recognised temperate grassland communities are listed as endangered under the *Environment Protection and Biodiversity Conservation Act 1999*, arguably the *Themeda triandra*-dominated grasslands of higher rainfall areas have been most impacted since European arrival. For instance, the natural grasslands of the Gippsland Plains of eastern Victoria and the northern wheatbelt in Western Australia have been totally lost to agriculture (Lunt 1994; O'Connor 2001). Today the only grassland remaining on the Gippsland Plains has been derived from a related grassy woodland (Williams *et al.* 2015). Fortunately, elsewhere in Tasmania, across the volcanic plains west of Melbourne, and on the southern Tablelands of New South Wales and the ACT, some fragments remain on roadsides, public land and occasionally on private land. The ecologically linked grasslands of the drier semi-arid regions – the Riverina and Victorian Wimmera, and in the Lofty Ranges north from Adelaide in SA – primarily differ in terms of composition and biomass dynamics. *Themeda* is largely replaced by the chloridoids *Enteropogon acicularis*, *Chloris truncata* and *Sporobolus caroli* in the Wimmera and Riverina, and *Lomandra* spp. in the Lofty Ranges of South Australia (Williams *et al.* 2015). The presence of annual forbs, especially from the Asteraceae, and chenopods also distinguish these grasslands (Figure 18.3e).

18.3.4 Coastal grasslands

Grasses and grasslands are well-adapted to thrive in the harsh conditions found on a range of coastal landforms and soil types. Dotted around the Australian coastline, these often highly localised coastal environs support a heterogeneous mix of treeless grassy habitats, united not so much by patterns of climate and soil, but rather coastal exposure, including persistent strong winds and salt spray, storms and king tides, brackish waterlogging and, in some cases, disturbance by nesting or burrowing seabirds. Rocky headlands and near-coastal islands tend to support grasslands dominated by large tussock grasses and sedges. While these grasslands share some species and characteristics of those further inland (e.g. *Themeda triandra*, *Poa labillardierei*), they tend to be dominated by coastal specialists such as *Poa poiformis*, *Poa billardierei*, *Austrostipa stipoides* and *A. flavescens*, along with a range of maritime adapted herbs (Beadle 1981; Keith 2004). Such environs often support seabird breeding colonies that play an important role in dispersing seed, nutrient recycling and fine-scale disturbance. Grasslands also occur on poorly drained, swamp and floodplain-derived brackish clays associated with sheltered estuaries that are vulnerable to waterlogging and inundation (Beadle 1981). These grasslands are distinguished by the presence of brackish wetland influences with species like *Hemarthria uncinata*, *Rytidosperma semiannulare* and rushes, sedges and wetland forbs.

18.3.5 Derived grasslands

18.3.5.1 Pre-European grasslands

A number of probable pre-colonisation derived grasslands occur in eastern Australia, in high rainfall, often mountainous areas suitable for forest growth. Examples include the Bunya Mountains 'grassy balds' in south-east Queensland (Fensham and Fairfax 1996), the north-east Tablelands of New South Wales and in northern Tasmania (Ellis 1985). Annual rainfall is high in all three regions (>700 mm), and grasslands occur within a dynamic eucalypt forest/rainforest mosaic on relatively fertile soils derived from igneous rocks. Two main hypotheses have been proposed to explain the genesis and maintenance of these grasslands. In all areas, climate has been suitable for rainforest dominance throughout the Holocene (Ellis and Thomas 1988; Fensham and Fairfax 1996), before which grassy ecosystems dominated. Thus, grassland patches may represent 'Pleistocene relicts' which have been progressively invaded by forests, except where frequent burning has prevented tree encroachment. The alternative, that grassland patches may have been created in fire-sensitive rainforest more recently by extreme disturbances (e.g. fire, wind, frost), followed by repeated Aboriginal burning, however, seems more compatible with the available evidence (Ellis 1985; Stubbs 2001; Fensham and Fairfax 2007; Butler *et al.* 2014). Much anthropological research suggests localised 'fire-stick farming' (*sensu* Jones 1969) – the deliberate use of fire by Aboriginal people to boost the availability of staple foods – was widespread at the time of European colonisation and important for maintaining the functioning of ecosystems (Bowman 1998; Gott 2005; Gammage 2011; Ens *et al.*, Chapter 4 this volume). Elsewhere, disturbance such as frequent fire has been shown to reduce woody cover below what would be expected climatically (Sankaran *et al.* 2008; Lunt *et al.* 2012), but in Australia there has been insufficient study of century-scale change using the rich historic archives to show how the 'fire-stick' might have had an influence on grass/tree boundaries, especially in southern mesic regions (Jones 1999; Lunt 1998, 2002).

18.3.5.2 Post-European grasslands

Derived grasslands have also arisen across much of southern and eastern Australia in response to European agriculture, particularly on the more fertile slopes, tablelands and plains (Moore 1953; McDougall 2008). Tree clearing, intensive livestock grazing, fertiliser application and exotic species 'pasture improvement' has given rise to homogenised pasturelands structurally similar to herbaceous 'natural' grasslands, but dominated by grazing-tolerant native perennial grasses and disturbance-adapted exotic annuals. Populations of palatable native forbs, such as *Microseris lanceolata* (Gott 1983), have crashed and there has been a widespread decline in ecosystem diversity and resilience (Dorrough 2012). Whatever the origin of grasslands in these areas, they are a very significant economic resource and can also be important habitat for native biodiversity. Some of the most diverse derived grassland remnants are habitat for localised rare or threatened plants and animals, and comprise nationally threatened ecological communities (albeit in a somewhat modified state; see Table 18.1) (Prober and Thiele 2005).

Table 18.1 **Current list of lowland tussock grassland Ecological Communities (EC) listed under the Environment Protection and Biodiversity Conservation (EPBC) Act 1999.**

Type	Group	Status	Date Listed	EPBC EC Name
Natural	Temperate	Critically Endangered	June 2009	Lowland Native Grasslands of Tasmania
		Critically Endangered	June 2008	Natural Temperate Grassland of the Victorian Volcanic Plain
		Critically Endangered	January 2009	Gippsland Red Gum (*Eucalyptus tereticornis* subsp. *mediana*) Grassy Woodland and Associated Native Grassland
		Endangered	July 2000	Natural Temperate Grassland of the Southern Tablelands of NSW and the ACT
	Temperate, semi-arid	Critically Endangered	June 2007	Iron-grass Natural Temperate Grassland of South Australia
		Critically Endangered	September 2012	Natural Grasslands of the Murray Valley Plains
	Coastal	Critically Endangered	February 2015	Natural Damp Grassland of the Victorian Coastal Plains
	Sub-tropical	Critically Endangered	January 2009	Natural grasslands on basalt and fine-textured alluvial plains of northern NSW and southern Queensland
	Tropical	Endangered	January 2009	Natural Grasslands of the Queensland Central Highlands and the northern Fitzroy Basin
Derived	Temperate	Endangered	April 2010	Grey Box (*Eucalyptus microcarpa*) Grassy Woodlands and Derived Native Grasslands of South-eastern Australia
		Critically Endangered	May 2006	White Box–Yellow Box–Blakely's Red Gum Grassy Woodland and Derived Native Grassland

18.4 The role of bottom-up and top-down processes on grassland distributions

Expansive treeless grasslands occurred in many lowland regions in Australia at the time of European settlement across a range of temperatures and rainfall, but climate models alone cannot predict the presence of grassland at a particular locality (Jones 1999). Most grasslands occur on relatively young, fertile soils derived *in situ* from parent material producing clays (vertosols) and where there are distinct wet and dry seasons (Beadle 1981; Keith 2004; Fensham *et al.* 2015). Many early survey maps show sharp grass/tree boundaries which often accord closely with soil and topographic patterns (e.g. Fensham and Fairfax 1997; Lunt 1997; Morcom and Westbrook 1998; Sinclair and Atchison 2012). Thus, 'bottom-up' processes are widely seen to be the primary drivers of treelessness in Australian tussock grasslands. Clay-rich, waterlogged and/or cracking clay soils can limit tree seedling establishment, particularly when combined with low–moderate rainfall and severe drought (Fensham *et al.* 2015). Low winter temperatures, with cold air draining into low-lying areas, frequent frost and even snow in some regions will prohibit tree encroachment in montane regions and at higher latitudes (e.g. Monaro Tablelands and Tasmania). These temperature-related processes, together with high grazing pressure (livestock or marsupials), competitive exclusion in dense grass swards and frequent fire, have all been cited as influencing grass/tree boundaries in Australia (Beadle 1981; Fensham and Kirkpatrick 1992; Lunt 1997; Fensham and Fairfax 2007).

Clay-rich soils, in particular, are effective at limiting the development of woody biomass in Australia where annual rainfall is <700 mm in the north and <400 mm in the south (Fensham *et al.* 2015). The rainfall threshold is higher in the north because the potential evapotranspiration is also higher than in southern regions of the continent. Clay soils exacerbate water stress in water-limited environments because water is less accessible for plants in drying clay soil than in drying sandy soils (i.e. the 'inverse texture effect'; Williams *et al.* 1983). Thus, as drought-induced tree mortality is exaggerated on clay-rich soils (Fensham and Fairfax 2007; Fensham *et al.* 2015), grasses can be episodically favoured over trees, even in more mesic environments.

The role of anthropogenic fire in grassland development in Australia has been emphasised on the basis of the historic archives (e.g. Gammage 2011). While, to date, this has only been supported by ecological research in some limited areas, such as the montane grassy 'balds' of the Bunya Mountains in south-east Queensland (Fensham and Fairfax 1996), a growing body of research supports the view that at the time of European colonisation Aboriginal fire use was skilful and important for maintaining the functioning of ecosystems (Bowman 1998). In the humid north, it has been shown that traditional burning influences the co-occurrence of forest and savanna (Murphy and Bowman 2012), but it remains unclear how frequent fire may have historically influenced the structure and composition of tussock grasslands, especially where traditional burning has long ceased. While further research and innovation is required to unravel the historical role of fire in grasslands, it does not appear to have been a primary factor driving the continental-scale distribution of the major treeless tussock grassland regions in Australia (Lunt *et al.* 2012; White *et al.* 2014). In the context of mesic grasslands of the south-east, Jones (1999, p. 22) concluded that

> The complex history of soils, climate, fire and humans cannot be unravelled until better dated evidence from more sites becomes available. While not wanting to downgrade the importance of Aboriginal fire on the Australian landscape, the evidence shows that rather than grassland being created, existing grassy ecosystems were modified and ecosystem boundaries shifted.

The historic influence of native herbivores on the extent of tussock grasslands in Australia is also mostly unknown (Mack and Thompson 1982; Jacobs *et al.* 1999). Australia has had a distinct marsupial-dominated fauna, lacking placental ungulates (Mack 1989). Browsing and grazing roles, played most conspicuously by ungulates on other continents, were historically filled mostly by large kangaroos and wombats in Australia, with teeth dentition of fossils suggestive of browsing rather than grazing (Prideaux *et al.* 2009). Soft-footed, large, browsing herbivores (at unknown densities) probably came to prominence with the increasing aridity of the continent, and probably coincided with the rise of grasses (Jacobs *et al.* 1999; Prideaux *et al.* 2009). The intervaginal tillering morphology of tussock grasses further supports the notion that grazing was not a strong selection pressure on Australian grasses. A tightly packed cluster of erect tillers is more exposed to grazing by ungulates than are rhizomatous grasses – the common habit of grasses in grassland biomes that have evolved with hard-hooved, congregating ungulates (Mack 1989).

18.5 Vegetation dynamics: disturbance effects on ecosystem function and diversity

In all tussock grasslands in Australia, the tussock grasses themselves make up the vast majority of the biomass, but the flora occupying the inter-tussock spaces comprise the majority of the diversity (Patton 1935; Tremont and McIntyre 1994; Fensham 1998; Bennett *et al.* 2002). Many of the inter-tussock species die back annually in response to seasonal drought and regenerate from buds at or near the soil surface. In highly productive sites, grasslands can rapidly produce abundant plant material which, when dry, provides readily combustible fuel for grassland fires. One of the characteristics of tussock grasslands is that biomass typically recovers quickly after perturbation. For instance, up to 4.6 t ha^{-1} has been recorded within 2 years of burning a *Themeda* grassland in southern Australia (Morgan and Lunt 1999), and 2.5 t ha^{-1} has been recorded within 2.5 years in subtropical *Themeda* grasslands in north-western Australia (Bennett *et al.* 2002). Typically, it is the rate of grass production (i.e. fuel accumulation), combined with climate, landscape fragmentation and patterns of ignition that drive fire regimes (Bond *et al.* 2003). In addition to fuelling fires, accumulated biomass regulates many ecological functions in grasslands, including nutrient cycling (Morgan 2007), plant population dynamics (Morgan 2001), plant diversity (Stuwe and Parsons 1977; Lunt 1997) and animal habitat (Baker-Gabb 1993; Stevens *et al.* 2010). By diminishing biomass, fire regimes and other disturbances can directly and indirectly influence the function, structure and composition of grasslands.

Inferences can be made from observational studies and experiments about the role of disturbance on grassland diversity. For instance, it has long been postulated that frequent fire is an important mechanism to maintain small-scale interstitial native plant diversity in productive temperate *Themeda* grasslands by reducing light competition (Stuwe and Parsons 1977; Lunt and Morgan 2002). Perennial forbs, the life form that contributes much of the native plant diversity in tussock grasslands (Tremont and McIntyre 1994), typically have small and short-lived soil seed banks (Lunt 1995; Morgan 1998), and many are also poor dispersers. Frequent disturbance to the vegetation has therefore been seen as crucial to the maintenance of the perennial 'bud banks' of geophytes (plants with growing buds below ground) and hemicryptophytes (plants with growing buds at the soil surface) in productive grasslands (Morgan 1999; Lunt and Morgan 2002), and their survival is compromised in dense, typically long-unburnt swards (Williams *et al.* 2006). In these grasslands, frequent fire also enhances the otherwise rare opportunities for seedling recruitment (Morgan 2001) by improving flowering and seed production, and creating microsites (Lunt 1993; Morgan 1996, 1997). Long-term studies are needed to assess the relationship between fire frequency and habitat heterogeneity, and to also decouple the effects of fire on grassland dynamics from other drivers, such as climate (Collins 1992; White *et al.* 2014). The little data that do exist highlight that cover, biomass and composition can be highly dynamic between years (White *et al.* 2014).

While disturbance can positively affect the diversity and composition of grasslands, in some regions it can have a relatively benign influence, even where rainfall is moderately high and where there are high fertility soils (e.g. subtropical *Dichanthium* grasslands, Fensham *et al.* submitted). Thus, the paradigm that productive grasslands require disturbance to maintain species diversity does not always apply, perhaps because of spatiotemporal productivity dynamics (Schultz *et al.* 2011). Undisturbed *Dichanthium* grasslands on the Darling Downs, for instance, have substantially less biomass compared to undisturbed *Themeda* grasslands in western Victoria (3.6 t ha^{-1} versus 8.2 t ha^{-1}; McDougall 1989; Fensham *et al.* unpublished). This discrepancy is believed to result mostly from differential net litter accumulation rates because decay is faster on the Darling Downs than in western Victoria.

The role of disturbance in low productivity (i.e. drier or infertile) grasslands has received less attention. Foreman (1996) compared the effects of a single fire against annual burning on the richness of native and exotic species in semi-arid *Rytidosperma* grasslands over three years. He found that burning reduced exotic richness marginally (and non-significantly), but plant richness was not promoted in this lower productivity grassland, presumably because climate-constrained biomass levels were already relatively low. Lewis *et al.* (2008, 2010) found similar outcomes when considering disturbance in arid (*Astrebla*) grasslands. Here, the highly variable climate probably plays an important role in the coexistence of species by negating competitive exclusion and allowing inter-tussock species to persist. Dense grass swards generally do not develop, and recent rainfall events are far more important for sward density than disturbance. Schultz *et al.* (2011), when comparing the effect of disturbance on tussock grassland richness and diversity across a rainfall gradient, found that disturbance had neutral or negative effects on diversity in low rainfall grasslands (<500 mm) where rates of biomass accumulation were slowest.

18.6 Grasslands of the future and the impact of climate change

Climate change is likely to drive significant changes in tussock grasslands because they are particularly sensitive to changes in the dominant processes that regulate them (e.g. rainfall, temperature; Figure 18.2). However, given that climate models alone cannot predict the presence of tussock grassland at a particular locality due to strong edaphic relationships that impose constraints on tussock grassland distribution, it is likely climate change responses will be expressed mainly through changes in traits, composition, structure and function, rather than shifts in distribution (Prober *et al.* 2012). Rapid changes in the populations of forbs have already been noted during prolonged dry periods (Gilfedder and Kirkpatrick 1994), as has the drought-induced collapse of grass populations in semi-arid grasslands (Godfree *et al.* 2011). Regional climate projections suggest that grasslands in eastern Australia will likely face warmer temperatures, reduced water availability (especially in winter), higher evaporation and increases in heavy precipitation events such as storms and periodic flooding (Prober *et al.* 2012). Based on past evidence, these changes are likely to not only slow biomass accumulation rates, potentially reducing the need for disturbance to maintain diversity (Bradstock 2010), but to also drive fundamental demographic changes in grassland assemblages.

Few *in situ* experimental studies have quantified the impact of climate change on tussock grassland dynamics, making predictions about their response largely speculative (Hovenden *et al.* 2006 is an obvious exception). It is expected that rising atmospheric temperatures and reduced winter precipitation will increase the proportion of semi-arid or drought-adapted vegetation in temperate zones. Indeed, a shift away from mesic grasses towards arid-adapted species was observed during the 2000–2009 Millennium Drought in central New South Wales (Godfree 2012), a process facilitated by natural variation in tolerances to soil water availability (Craine *et al.* 2013). If water availability declines, it is also likely that C_4 species, being generally more water-use efficient and having a higher optimum temperature for net assimilation than C_3 species, will potentially come to dominate warmer and drier parts of Australia. An increase in temperature and summer rainfall would likely benefit C_4 species, although whether this would involve migration and establishment of tropical and subtropical taxa or *in situ* selection for cooler-adapted C_4 species such as *Themeda triandra* remains unclear (Bolger *et al.* 2005).

Rising atmospheric CO_2 concentrations will also play a significant role in structuring grassland communities in coming decades. Typical plant responses to elevated CO_2, such as increased photosynthesis, more rapid growth, greater biomass accumulation, lower plant nitrogen concentration and improved water-use efficiency (Ainsworth and Long 2005) are all likely to be ecologically transformative, depending on the specific nature of concurrent changes to temperature and precipitation regimes (Cullen *et al.* 2009). Generally, simulations indicate a 20–30% increase in productivity in C_3 pastures of south-eastern Australia under 550 ppm of CO_2, exceeding that of the C_4 dominated subtropical and subhumid grasslands (Wand *et al.* 1999). However, in southern Australia, such increases are likely to be counteracted by rising temperatures and declining cool-season rainfall (Cullen *et al.* 2009).

The response of Australian tussock grasslands to climate change will also depend on the ability of species and populations to maintain fitness via adaptive evolutionary change. While relatively little is known about the level of inter- and intra-population variation for traits related to growth, water use, productivity and phenology, ecotypic differentiation and genetic variability has been studied in many grass genera, including *Rytidosperma*, *Themeda*, *Digitaria*, *Dichanthium*, *Austrostipa* and *Microlaena* (Groves 1975; Fitzgerald *et al.* 2011; Waters *et al.* 2011; Godfree *et al.* 2013). Given this variability, it is likely that ecophysiological stress arising from climate change will drive selection in traits linked to water acquisition and resource-use efficiency in many grassland species. However, this has not been well studied and remains the single largest knowledge gap relevant to predicting the evolutionary response of tussock grasslands to future climate change.

An interesting feature of plant genomes of relevance to climate change, and that warrants further study, is polyploidy. Polyploidy is a conspicuous feature described in some Australian grasses and grassland forbs (Hayman 1960; Murray and Young 2001; Waters *et al.* 2010; Broadhurst *et al.* 2012). Polyploids usually undergo ecophysiological, morphological and reproductive divergence from lower ploidy parental lines and often occupy distinct geographical ranges and associated habitats. In *Themeda triandra*, for example, diploid cytotypes dominate mesic environments in south-eastern Australia and polyploid cytotypes dominate more arid inland areas (Hayman 1960). It is thought that such distribution patterns increase species' tolerance of stressful or fluctuating environmental conditions and may confer resilience to natural plant populations under climate change.

18.7 Current status of Australian tussock grasslands

The natural grassy plains and grassy ecosystems of eastern and southern Australia were at the centre of the early settlement of inland Australia. Prior to the gold rush, early explorers such as Mitchell, Oxley, Cunningham and Leichhardt traversed the region in search of productive land, especially open grassy plains immediately suitable for livestock or 'ready to receive the plough' (Mitchell 1838). The explorers' reports were loaded with alluring references to beautiful and verdant park-like valleys and plains that were rapidly exploited by pastoralists, heralding dramatic and irrevocable change that continues to this day. Because native grasslands were settled so early, and managed so intensively, there are few records of their original plant composition (Sutton 1916; Patton 1935; Willis 1964; Silcock and Scattini 2007).

Thus, while we know many grasslands have been catastrophically impacted, we can only speculate on details such as change in structure, composition and function (e.g. Fensham and Fairfax 1997; Sinclair and Atchinson 2012). One thing that is clear is that native grasslands in many regions of Australia have been severely depleted by agriculture and urban development; in some cases they have been driven to extinction. In relictual landscapes, such as the Victorian Volcanic Plains, west of Melbourne, and on the Darling Downs in south-east Queensland, it is estimated that 95–99% of the original grasslands have been lost (Kirkpatrick *et al.* 1995; Fensham 1998). In these regions, it is roadsides, travelling stock routes, rail reserves, cemeteries and public lands that support the most intact, species-rich tussock grassland remnants (e.g. Stuwe and Parsons 1977; Fensham 1998). Tussock grasslands in the Pilbara region, dominated by *Themeda triandra*, are listed as threatened in Western Australia as a result of livestock grazing (Kendrick 2001). In less arable, fragmented regions, such as the Riverina, the south-western Tablelands of New South Wales and the central highlands of Queensland, depletion levels have not been as severe. Even though the vast, economically important Mitchell grass grasslands of arid northern Australia have been more resilient to change, they remain poorly reserved across their range (White *et al.* 2014).

18.8 Conservation challenges and opportunities

Many tussock grasslands have been transformed by land-use changes into non-native communities across much of their range and the best examples of the original systems persist in areas that have been least modified (Stuwe and Parsons 1977; Benson 1994; Fensham 1998). One of the most dramatic changes since the publication of the first edition of *Australian Vegetation* in 1981 has been the awakening consciousness of the need for greater regulatory and reservation protection for many native tussock grasslands. No less than 11 grassland communities have been listed under the *Environment Protection and Biodiversity Conservation Act 1999* (most as Critically Endangered) since 2000 (Table 18.1). Not surprisingly, the majority of these occur in the south-east and eastern wheat/sheep belt and most listings were stimulated by seminal work from the 1970s to 1990s when these systems were first declared the 'most endangered ecosystem in Australia' (Kirkpatrick *et al.* 1995). One of the immediate consequences of this change has been a sharp increase in grassland reserve acquisition. For instance, in the Victorian Riverina, since 1997 the Protected Area Network has increased by several orders of magnitude to well over 10 000 ha today (based on Foreman 2010), and similar improvements have occurred in other regions (Fitzsimons 2015). However, due to the scale of clearing, minimum bioregional reservation targets will not be met for any of the listed grasslands without broad-scale restoration.

The expansion of the reserve system is good news, but ongoing conservation management remains very challenging in the context of increasing costs and regulatory complexity. Shifting from traditional pastoral practices to conservation grazing systems, deploying nuanced ecological burning regimes across innumerable, scattered, tiny remnants, and confronting the invasion of an increasing array of new and threatening weeds are some of the issues reserve managers are struggling with today (Williams *et al.* 2015). Although even the best remaining grasslands are characterised by suites of widespread, annual species that do have a negative impact (Lenz *et al.* 2003), it is the invasion of exotic perennial grasses that poses one of the greatest long-term threats to native tussock grasslands, especially in mesic regions. *Eragrostis curvula* (African lovegrass) now has overrun vast areas of the Monaro, *Hyparrhenia hirta* (coolatai grass) is a threat in subtropical grasslands, and perennial C_3 tussock grasses from South America (e.g. *Nassella neesiana*, Chilean needle grass) have become the most important invaders of grasslands on the Victorian Volcanic Plains since the 1990s (Williams *et al.* 2015). Mitchell grass grasslands in Queensland are threatened by the invasion of *Vachellia nilotica* (prickly acacia) and *Cenchrus ciliaris* (buffel grass; White *et al.* 2014).

While agriculture, principally cropping and grazing, has had profoundly negative impacts on native grasslands across much of their range, ironically, it is agricultural land use that also holds great promise for their recovery in the future. Passive restoration of grasslands after the abandonment of cultivation may be a viable restoration option, at least in some parts of the range (Lewis *et al.* 2008; Wong *et al.* 2010). Recovery of old-field or fallow systems following land abandonment is thought to be influenced by the intensity of past land use, with higher intensity leading to slower recovery, or even the establishment of alternate stable states as both biotic and abiotic thresholds are crossed (Wong *et al.* 2010). Recovery is more likely if native species are prevalent in the soil seed bank (Scott and Morgan 2012), or where there is a matrix of remnant grasslands in the vicinity of fallows to provide a source of propagules. Critically, recovery is more likely where perennial exotic grasses do not monopolise fallows and native grasses can reassert dominance (Dorrough *et al.* 2004; Wong *et al.* 2010). Additionally, old fields have shown great promise as areas suitable for grassland reconstruction. Using techniques borrowed from agriculture and horticulture to address barriers that limit seedling regeneration (e.g. elevated soil nutrients, lack of native propagules in the soil seed bank, poor seed dispersal, competition with exotic species), Gibson-Roy *et al.* (2009) and others (e.g. Prober and Lunt 2009) have shown that diverse, functional tussock grasslands can be established (at small scales) on ex-agricultural land as a practical way to reverse the decline of these ecosystems across their range.

18.9 Conclusions

While grasses and grasslands are a relatively new evolutionary feature of the Australian vegetated landscape – emerging under an increasingly drying climate as Australia drifted north over the last 25 to 10 Ma – they have diversified into an array of associations found mostly on fertile, clay-rich soils occupying a wide range of climates. Dominated by herbaceous plants that emerged in the absence of hard-hooved, congregating ungulates, so dominant across the plains of other continents, the structure and function of Australian grasslands is driven by biomass dynamics regulated by floods and drought, in combination with disturbances such as fire and, since the arrival of Europeans, livestock grazing. In southern mesic regions, fire is a critical regulator of the process of competitive exclusion whereby the vigour of dominant species, combined with litter build-up, can rapidly exclude coexisting forbs. It is likely that climate change will mostly drive changes in the way grasslands function rather than shifts in their distribution, although responses are difficult to predict due to complex counteracting processes that have been poorly studied. Because grasslands were intensively exploited very soon after the arrival of Europeans, many systems are now considered among Australia's most endangered ecosystems. While there has been much conservation action in recent decades, including expansion of reserve networks, conservation management and restoration represents a major challenge in the future.

Acknowledgements

Neil Gibson clarified the situation about tussock grasslands in Western Australia. Bob Parsons, David Keith and Suzanne Prober helped improve early versions of the manuscript.

References

Accad, A., Neldner, V.J. and Wilson, B.A. (2006) *Remnant vegetation in Queensland. Analysis of Remnant Vegetation 1997-1999-2000-2001-2003, Including Regional Ecosystem Information.* Queensland Herbarium, Environmental Protection Agency: Brisbane.

Ainsworth, E.A. and Long, S.P. (2005) What have we learned from 15 years of free-air CO_2 enrichment (FACE)? A meta-analytic review of the responses of photosynthesis, canopy properties and plant production to rising CO_2. *New Phytologist* 165, 351–372.

Austin, M.P. and Williams, O.B. (1988) Influence of climate and community composition on the population demography of pasture species in semi-arid Australia. *Vegetatio* 77, 43–49.

Baker-Gabb, D. (1993) *Managing Native Grasslands to Maintain Biodiversity and Conserve the Plains-wanderer.* Royal Australasian Ornithological Union: Melbourne. Available at: http://birdlife.org.au/documents/OTHPUB-Grasslands_Mngmnt.pdf (accessed 4 August 2016).

Barlow, B.A. (1981) Phytogeography of the Australian region. In *Australian Vegetation*, ed. R.H. Groves. Melbourne: Cambridge University Press, pp. 3–35.

Beadle, N.C.W. (1981) *The Vegetation of Australia*. Cambridge University Press: Cambridge.

Beard, J.S., Beeston, G.R., Harvey, J.M. *et al.* (2013) The vegetation of Western Australia at the 1:3 000 000 scale. Explanatory memoir. 2nd edition. *Conservation Science Western Australia* 9, 1–252.

Beerling, D.J. and Osborne, C.P. (2006) The origin of the savanna biome. *Global Change Biology* 12, 2023–2031.

Bennett, L.T., Judd, T.S. and Adams, M.A. (2002) Growth and nutrient content of perennial grasslands following burning in semi-arid, sub-tropical Australia. *Plant Ecology* 164, 185–199.

Benson, J.S. (1994) The native grasslands of the Monaro region: Southern Tablelands of NSW. *Cunninghamia* 3, 609–650.

Birch, J.L., Cantrill, D.J., Walsh, N.G. and Murphy, D.J. (2014) Phylogenetic investigation and divergence dating of *Poa* (Poaceae, tribe Poeae) in the Australasian region. *Botanical Journal of the Linnean Society* 175, 523–552.

Bolger, T.P., Rivelli, A.R. and Garden, D.L. (2005) Drought resistance of native and introduced perennial grasses of south-eastern Australia. *Crop and Pasture Science* 56, 1261–1267.

Bond, W., Midgley, G. and Woodward, F. (2003) What controls South African vegetation: climate or fire? *South African Journal of Botany* 69, 79–91.

Bouchenak-Khelladi, Y., Verboom, G.A., Savolainen, V. and Hodkinson, T.R. (2010) Biogeography of the grasses (Poaceae): a phylogenetic approach to reveal evolutionary history in geographical space and geological time. *Botanical Journal of the Linnean Society* 162, 543–557.

Bowman, D.M.J.S. (1998) Tansley Review No. 101: the impact of Aboriginal landscape burning on the Australian biota. *New Phytologist* 140, 385–410.

Bradstock, R.J. (2010) A biogeographic model of fire regimes in Australia: current and future implications. *Global Ecology and Biogeography* 19, 145–158.

Bremer, K. (2002) Gondwanan evolution of the grass alliance of families (Poales). *Evolution* 57, 1374–1387.

Broadhurst, L.M., Murray, B.G., Forrester, R. and Young, A.G. (2012) Cryptic genetic variability in *Swainsona sericea* (A. Lee) H. Eichler (Fabaceae): lessons for restoration. *Australian Journal of Botany* 60, 429–438.

Butler, D.W., Fensham, R.J., Murphy, B.P. *et al.* (2014) Aborigine-managed forest, savanna and grassland: biome switching in montane eastern Australia. *Journal of Biogeography* 41, 1492–1505.

Collins, S.L. (1992) Fire frequency and community heterogeneity in tallgrass prairie vegetation. *Ecology* 73, 2001–2006.

Craine, J.M., Ocheltree, T.W., Nippert, J.B. *et al.* (2013) Global diversity of drought tolerance and grassland climate-change resilience. *Nature Climate Change* 3, 63–67.

Crisp, M., Cook, L. and Steane, D. (2004) Radiation of the Australian flora: what can comparisons of molecular phylogenies across multiple taxa tell us about the evolution of diversity in present-day communities? *Philosophical Transactions of the Royal Society of London B* 359, 1551–1571.

Cullen, B.R., Johnson, I.R., Eckard, R.J. *et al.* (2009) Climate change effects on pasture systems in south-eastern Australia. *Crop and Pasture Science* 60, 933–942.

Department of the Environment and Energy (2007) NVIS data products 4.2. Australian Government: Canberra. Available at: http://www.environment.gov.au/land/native-vegetation/national-vegetation-information-system/data-products (accessed 10 May 2016).

Department of the Environment and Water Resources (DEWR) (2007) *Australia's Native Vegetation: A Summary of Australia's Major Vegetation Groups.* Australian Government: Canberra.

Diels, L. (1906; trans. 2007) The plant life of Western Australia. *Conservation Science Western Australia* 6, 1–373.

Dorrough, J. (2012) *How do different levels of grazing and fertilisation affect vegetation composition in temperate Australian grassy ecosystems? Systematic review and meta-analysis*. Report to Victorian Department of Sustainability and Environment, Melbourne. Available at http://www.naturalregen.com.au/wp-content/uploads/2012/05/Dorrough_SystematicReview2012.pdf.

Dorrough, J., Yen, A., Turner, V. *et al.* (2004) Livestock grazing management and biodiversity conservation in Australian temperate grassy landscapes. *Australian Journal of Agricultural Research* 55, 1–17.

Edwards, E.J. and Smith, S.A. (2010) Phylogenetic analyses reveal the shady history of C_4 grasses. *PNAS* 107, 2532–2537.

Edwards, E.J. and Still, C.J. (2008) Climate, phylogeny and the ecological distribution of C_4 grasses. *Ecology Letters* 11, 266–276.

Ellis, R.C. (1985) The relationship among eucalypt forest, grassland and rainforest in a highland area of north-eastern Tasmania. *Australian Journal of Ecology* 10, 297–314.

Ellis, R.C. and Thomas, I. (1988) Pre-settlement and post-settlement vegetational change and probable Aboriginal influences in a highland forested area in Tasmania. In *Australia's Ever Changing Forests,* Special Publication No.1, ed. K.J. Frawley and N. Semple. Department of Geography and Oceanography, Australian Defence Force Academy: Campbell, pp. 199–214.

Fensham, R.J. (1998) The grassy vegetation of the Darling Downs, south-eastern Queensland, Australia. Floristics and grazing effects. *Biological Conservation* 84, 301–310.

(1999) Native grassland of the Central Highlands, Queensland, Australia. Floristics, regional context and conservation. *Rangeland Journal* 21, 82–103.

Fensham, R.J. and Fairfax, R.J. (1996) The disappearing grassy balds of the Bunya Mountains, south-eastern Queensland. *Australian Journal of Botany* 44, 543–558.

Fensham, R.J. and Fairfax, R.J. (1997) The use of the land survey record to reconstruct pre-European vegetation patterns in the Darling Downs, Queensland, Australia. *Journal of Biogeography* 24, 827–836.

Fensham, R.J. and Fairfax, R.J. (2007) Drought-related tree death of savanna eucalypts: species susceptibility, soil conditions and root architecture. *Journal of Vegetation Science* 18, 71–80.

Fensham, R.J. and Kirkpatrick, J.B. (1992) The eucalypt forest-grassland/grassy woodland boundary in central Tasmania. *Australian Journal of Botany* 40, 123–138.

Fensham, R.J., Silcock, J.L. and Firn, J. (2014) Managed livestock grazing is compatible with the maintenance of species diversity in semi-desert grasslands. *Ecological Applications* 24, 503–518.

Fensham, R.J., Butler, D.W. and Foley, J. (2015) How does clay constrain woody biomass in drylands? *Global Ecology and Biogeography* 24, 950–958.

Fisher, A., Baker, B. and Woinarski, J. (2002) *Biodiversity Audit. Bioregional case study: Mitchell Grass Downs, Northern Territory. A report to the National Land and Water Audit, Parks and Wildlife Commission of the Northern Territory.* Department of Infrastructure, Planning and Environment: Darwin.

Fitzgerald, T.L., Shapter, F.M., McDonald, S. *et al.* (2011) Genome diversity in wild grasses under environmental stress. *Proceedings of the National Academy of Sciences* 108, 21140–21145.

Fitzsimons, J.A. (2015) Private protected areas in Australia: current status and future directions. *Nature Conservation* 10, 1–23.

Foreman, P.W. (1996) Ecology of native grasslands of Victoria's northern Riverine Plain. Unpublished MSc. Thesis, La Trobe University, Bundoora.

Foreman, P.W. (2010) Recovery of the Northern Plains grassland community – an overview. *Proceedings of the Royal Society of Victoria* 122, 92–99.

Gammage, B. (2011) *The Biggest Estate on Earth: How Aborigines Made Australia.* Allen & Unwin: Crows Nest.

Gibson-Roy, P., Delpratt, J., Moore, G. and Hepworth, G. (2009) Does diversity influence soil nitrate, light availability and productivity in the establishment phase of Australian temperate grassland reconstruction? *Ecological Management and Restoration* 10, 41–50.

Gilfedder, L. and Kirkpatrick, J.B. (1994) Climate, grazing and disturbance, and the population dynamics of *Leucochrysum albicans* at Ross, Tasmania. *Australian Journal of Botany* 42, 417–430.

Godfree, R.C. (2012) Extreme climatic events as drivers of ecosystem change. In *Diversity of Ecosystems*, ed. M. Ali. InTech Publishers: Rijeka, pp. 339–366.

Godfree, R., Lepschi, B., Reside, A., *et al.* (2011) Multiscale topoedaphic heterogeneity increases resilience and resistance of a dominant grassland species to extreme drought and climate change. *Global Change Biology* 17, 943–958.

Godfree, R.C., Robertson, B.C., Gapare, W.J. *et al.* (2013) Nonindigenous plant advantage in native and exotic Australian grasses under experimental drought, warming, and atmospheric CO_2 enrichment. *Biology* 2, 481–513.

Gott, B. (1983) Murnong–*Microseris scapigera*: a study of a staple food of Victorian Aborigines. *Australian Aboriginal Studies* 2, 2–18.

Gott, B. (2005) Aboriginal fire management in south-eastern Australia: aims and frequency. *Journal of Biogeography* 32, 1203–1208.

Groves, R.H. (1975) Growth and development of five populations of Themeda australis in response to temperature. *Australian journal of Botany* 23, 951–963.

Hartley, W. (1961) Studies on the origin, evolution, and distribution of the Gramineae. IV. The genus *Poa* L. *Australian Journal of Botany* 9, 152–161.

Hartley, W. (1964) The distribution of the grasses. In *Grasses and Grasslands*, ed. C. Barnard. MacMillan: London, pp. 29–46.

Hartley, W. (1973) Studies on the origin, evolution, and distribution of the Gramineae. V. The subfamily Festucoideae. *Australian Journal of Botany* 21, 201–234.

Hattersley, P.W. (1983) The distribution of C_3 and C_4 grasses in Australia in relation to climate. *Oecologia* 57, 113–128.

Hayman, D.L. (1960) The distribution and cytology of the chromosome races of *Themeda australis* in southern Australia. *Australian Journal of Botany* 8, 58–68.

Hovenden, M.J., Miglietta, F., Zaldei, A. *et al.* (2006) The TasFACE climate-change impacts experiment: design and performance of combined elevated CO_2 and temperature enhancement in a native Tasmanian grassland. *Australian Journal of Botany* 54, 1–10.

Hsiao, C., Jacobs, S.W.L., Chatterton, N.J. and Asay, K.H. (1999) A molecular phylogeny of the grass family (Poaceae) based on the sequences of nuclear ribosomal DNA (ITS). *Australian Systematic Botany* 11, 667–688.

Jacobs, B.F., Kingston, J.D. and Jacobs, L.L. (1999) The origin of grass-dominated ecosystems. *Annals of the Missouri Botanical Garden* 86, 590–643.

Jones, R. (1969) Fire-stick farming. *Australian Natural History* 16, 224–228.

Jones, R.N. (1997) The biogeography of the grasses and lowland grasslands of south-eastern Australia. *Advances in Nature Conservation* 2, 11–18.

Jones, R.N. (1999) Natural and human influences on the distribution and extent of Victorian lowland grasslands. In *The Great Plains Crash. Proceedings of a Conference on the Grasslands and Grassy Woodlands of Victoria*, ed. R.N. Jones. Indigenous Flora and Fauna Association and Victorian National Parks Association: Melbourne, pp. 19–39. Available at: http://home.vicnet.net.au/~iffa/GPCproc.htm.

Keeley, J.E. and Rundel, P.W. (2005) Fire and the Miocene expansion of C_4 grasslands. *Ecology Letters* 8, 683–690.

Keith, D. (2004) *Ocean Shores to Desert Dunes: The Native Vegetation of NSW and the ACT.* Department of Environment and Conservation: Hurstville.

Kendrick, P. (2001) Pilbara 3 (Pil–Hamersley subregion). In *A Biodiversity Audit of Western Australia's 53 Biogeographical Subregions in 2001*, ed. J. May and N. McKenzie. Western Australian Government Department of Conservation and Land Management: Perth.

Kershaw, A.P., Martin, H.A. and McEwen Mason, J.R.C. (1994) The Neogene: a period of transition. In *Australian Vegetation History*, ed. R.S. Hill. Cambridge University Press: Cambridge.

Kirkpatrick, J.B., McDougall, K. and Hyde, M. (1995) *Australia's Most Threatened Ecosystems: The Southeastern Lowland Native Grasslands.* World Wide Fund for Nature: Sydney.

Lenz, T.I., Moyle-Croft, J.L. and Facelli, J.M. (2003) Direct and indirect effects of exotic annual grasses on species composition of a South Australian grassland. *Austral Ecology* 28, 23–32.

Lewis, T., Clarke, P.J., Reid, N. and Whalley, R.D.P. (2008) Perennial grassland dynamics on fertile plains: is coexistence mediated by disturbance. *Austral Ecology* 33, 128–139.

Lewis, T., Reid, N., Clarke, P.J. and Whalley, R.D.B. (2010) Resilience of a high-conservation-value, semi-arid grassland on fertile clay soils to burning, mowing and ploughing. *Austral Ecology* 35, 464–481.

Lindenmayer, D., Burns, E., Thurgate, N. and Lowe, A. (2014) General overview. In *Biodiversity and Environmental Change: Monitoring, Challenges and Direction*, ed. D. Lindenmayer, E. Burns, N. Thurgate and A. Lowe. CSIRO Publishing: Collingwood, pp. 3–19.

Lunt, I.D. (1993) Variation in flower production of nine grassland species with time since fire, and implications for grassland management and restoration. *Pacific Conservation Biology* 1, 359–366.

Lunt, I.D. (1994) The extinct grasslands of the lowland Gippsland plains. In *Conservation of Lowland Native Grasslands in South-eastern Australia*, ed. K. McDougall and J.B. Kirkpatrick. Worldwide Fund for Nature Australia: Sydney, pp. 9–12.

Lunt, I.D. (1995) Seed longevity of six native forbs in a closed *Themeda triandra* grassland. *Australian Journal of Botany* 43, 439–449.

Lunt, I.D. (1997) Effects of long-term vegetation management on remnant grassy forests and anthropogenic native grasslands in south-eastern Australia. *Biological Conservation* 81, 287–297.

Lunt, I.D. (1998) Two hundred years of land use and vegetation change in a remnant coastal woodland in southern Australia. *Australian Journal of Botany* 46, 629–647.

Lunt, I.D. (2002) Grazed, burnt and cleared: how ecologists have studied century-scale vegetation changes in Australia. *Australian Journal of Botany* 50, 391–407.

Lunt, I.D. and Morgan, J.W. (2002) The role of fire regimes in temperate lowland grasslands of southeastern Australia. In *Flammable Australia: Fire Regimes, Biodiversity and Ecosystems in a Changing World*, ed. R.A. Bradstock, A.M. Gill and R.J. Williams. CSIRO Publishing: Melbourne, pp. 177–196.

Lunt, I.D., Prober, S.M. and Morgan, J.W. (2012) How do fire regimes affect ecosystem structure, function and diversity in grasslands and grassy woodlands of southern Australia. In *Flammable Australia: Fire Regimes, Biodiversity and Ecosystems in a Changing World*, ed. R.A. Bradstock, A.M. Gill and R.J. Williams. CSIRO Publishing: Melbourne, pp. 253–270.

Mack, R.N. (1989) Temperate grasslands vulnerable to plant invasions: characteristics and consequences. In *Biological Invasions: A Global Perspective*, ed. J.A. Drake *et al.* John Wiley and Sons: New York, pp. 155–179.

Mack, R.N. and Thompson, J.N. (1982) Evolution in steppe with few large, hooved mammals. *The American Naturalist* 119, 757–773.

Macphail, M.K. and Hill, R.S. (2002) Paleobotany of the Poaceae. In *Flora of Australia*, Vol. 43, ed. K. Mallett. Australian Biological Resources Study/CSIRO Publishing: Collingwood, pp. 37–70.

Martin, H.A. (2006) Cenozoic climatic change and the development of the arid vegetation in Australia. *Journal of Arid Environments* 66, 533–563.

McDougall, K.L. (1989) *The re-establishment of* Themeda triandra *(Kangaroo Grass): implications for the restoration of grassland. Arthur Rylah Institute for Environmental Research Technical Report Series No. 89.* Department of Conservation, Forests and Lands: East Melbourne.

McDougall, K.L. (2008) Evidence for the natural occurrence of treeless grasslands in the Riverina region of south-eastern Australia. *Australian Journal of Botany* 56, 461–468.

Mitchell, T.L. (1838) *Three Expeditions into the Interior of Eastern Australia; With Descriptions of the Recently Explored Region of Australia Felix, and of the Present Colony of NSW.* T. and W. Boone: London.

Moore, C.W.E. (1953) The vegetation of the south-eastern Riverina, NSW. 2. The disclimax communities. *Australian Journal of Botany* 1, 548–67.

Moore, R.M. (1970) Australian grasslands. In *Australian Grasslands*, ed. R.M. Moore. Australian National University Press: Sydney.

Morcom, L.A. and Westbrooke, M.E. (1998) The pre-settlement vegetation of the western and central Wimmera Plains of Victoria, Australia. *Australian Geographical Studies* 36, 273–288.

Morgan, J.W. (1996) Secondary juvenile period and community recovery of a *Themeda triandra* grassland. *The Victorian Naturalist* 113, 47–57.

Morgan, J.W. (1997) The effect of grassland gap size on establishment, growth and flowering of the endangered *Rutidosis leptorrhynchoides* (Asteraceae). *Journal of Applied Ecology* 34, 566–576.

Morgan, J.W. (1998) Composition and seasonal flux of the soil seed bank of species-rich *Themeda triandra* grasslands in relation to burning history. *Journal of Vegetation Science* 9, 145–156.

Morgan, J.W. (1999) Defining grassland fire events and the response of perennial plants to annual fire in temperate grasslands of south-eastern Australia. *Plant Ecology* 144, 127–144.

Morgan, J.W. (2001) Seedling recruitment patterns over 4 years in an Australian perennial grassland community with different fire histories. *Journal of Ecology* 89, 908–919.

Morgan, J.W. (2007) Relationship between fire frequency and nitrogen limitation on foliage production in a native grassland community in Victoria, Australia. *The Rangeland Journal* 29, 101–105.

Morgan, J.W. and Lunt, I.D. (1999) Effects of time since fire on the tussock dynamics of a dominant grass (*Themeda triandra*) in a temperate Australian grassland. *Biological Conservation* 88, 379–386.

Murphy, B.P. and Bowman, D.M.J.S. (2012) What controls the distribution of tropical forest and savanna? *Ecology Letters* 15, 748–758.

Murray, B.G. and Young, A.G. (2001) Widespread chromosome variation in the endangered grassland forb *Rutidosis leptorrhynchoides* F. Muell.(Asteraceae: Gnaphalieae). *Annals of Botany* 87, 83–90.

O'Connor, M. (2001) The historical ecology of the Greenough Flats, Western Australia. PhD Thesis, Curtin University of Technology.

Osborne, C.P. (2008) Atmosphere, ecology and evolution: what drove the Miocene expansion of C_4 grasslands? *Journal of Ecology* 96, 35–45.

Patton, R.T. (1935) Ecological studies in Victoria. Part IV: Basalt Plains association. *Proceedings of the Royal Society of Victoria* 48, 172–191.

Perry, R.A. (1970) Arid shrublands and grasslands. In *Australian Grasslands*, ed. R.M. Moore. Australian National University Press: Sydney.

Phelps, D.G. and Bosch, O.J.H. (2002) A quantitative state and transition model for the Mitchell grasslands of central western Queensland. *Rangeland Journal* 24, 242–267.

Prideaux, G.J., Ayliffe, L.K., DeSantis, L.R.G. *et al.* (2009) Extinction implications of a chenopod browse diet for a giant Pleistocene kangaroo. *PNAS* 106, 11646–11650.

Prober, S.M. and Lunt, I.D. (2009) Restoration of *Themeda australis* swards suppresses soil nitrate and enhances ecological resistance to invasion by exotic annuals. *Biological Invasions* 11, 171–181.

Prober, S.M. and Thiele, K.R. (2005) Restoring Australia's temperate grasslands and grassy woodlands: integrating function and diversity. *Ecological Management and Restoration* 6, 16–27.

Prober, S.M., Hilbert, D.W., Ferrier, S. *et al.* (2012) Combining community-level spatial modelling and expert knowledge to inform climate adaptation in temperate grassy eucalypt woodlands and related grasslands. *Biodiversity and Conservation* 21, 1627–1650.

Sankaran, M., Ratnam, J. and Hanan, N. (2008) Woody cover in African savannas: the role of resources, fire and herbivory. *Global Ecology and Biogeography* 17, 236–245.

Schultz, N.L., Morgan, J.W. and Lunt, I.D. (2011) Effects of grazing exclusion on plant species richness and phytomass accumulation vary across a regional productivity gradient. *Journal of Vegetation Science* 22, 130–142.

Scott, A.J. and Morgan, J.W. (2012) Recovery of soil and vegetation in semi-arid Australian old fields. *Journal of Arid Environments* 76, 61–71.

Silcock, R.G. and Scattini, W.J. (2007) The original native pasture ecosystems of the eastern and western Darling Downs – can they be restored? *Tropical Grasslands* 41, 154–163.

Sinclair, S.J. and Atchinson, K. (2012) The pre-colonial distribution of grasslands, woodlands and forests on the Werribee plains, Victoria. *Cunninghamia* 12, 213–227.

Stevens, T.A., Evans, M.C., Osborne, W.S. and Sarre, S.D. (2010) Home ranges of, and habitat use by, the grassland earless dragon (*Tympanocryptis pinguicolla*) in remnant native grasslands near Canberra. *Australian Journal of Zoology* 58, 76–84.

Strömberg, C.A.E. (2011) Evolution of grasses and grassland ecosystems. *Annual Review of Earth and Planetary Sciences* 39, 517–544.

Stubbs, B. (2001) The 'grasses' of the Big Scrub District, north-eastern New South Wales: their recent history, spatial distribution and origins. *Australian Geographer* 32, 295–319.

Stuwe, J. and Parsons, R.F. (1977) Themeda australis grasslands on the basalt plains, Victoria: floristics and management effects. *Australian Journal of Ecology* 2, 467–476.

Sutton, C.S. (1916) A sketch of the Keilor Plains flora. *The Victorian Naturalist* 33, 112–143.

Syme, A.E. (2012) Diversification rates in the Australasian endemic grass Austrostipa: 15 million years of constant evolution. *Plant Systematics and Evolution* 298, 221–227.

Threatened Species Scientific Committee (TSSC) (2008) *Approved Conservation Advice (including listing advice) for the Natural Grasslands of the Queensland Central Highlands and the northern Fitzroy Basin Ecological Community*. Department of Environment: Canberra.

Tremont, R.M. and McIntyre, S. (1994) Natural grassy vegetation and native forbs in temperate Australia: structure, dynamics and life histories. *Australian Journal of Botany* 42, 641–658.

Twiss, P.C. (1992) Predicted world distributions of C_3 and C_4 grass phytoliths. In *Phytolith Systematics: Emerging Issues*, ed. G. Rapp and S.C. Mulholland. Advances in Archaeological and Museum Science Vol. 1 Springer Science and Business Media: New York, pp. 113–128.

Waddell, P.A., Gardner, A.K. and Hennig, P. (2010) *An inventory and condition survey of the Western Australian part of the Nullarbor region. Technical Bulletin No. 97*. Department of Agriculture and Food: Western Australia.

Wand, S.J., Midgley, G., Jones, M.H. and Curtis, P.S. (1999) Responses of wild C_4 and C_3 grass (Poaceae) species to elevated atmospheric CO_2 concentration: a meta-analytic test of current theories and perceptions. *Global Change Biology* 5, 723–741.

Waters, C., Murray, B.G., Melville, G. *et al.* (2010) Polyploidy and possible implications for the evolutionary history of some Australian Danthonieae. *Australian Journal of Botany* 58, 23–34.

Waters, C., Melville, G., Coates, D. *et al.* (2011) Variation in morphological traits among and within populations of *Austrodanthonia caespitosa* (Gaudich.) HP Linder and four related species. *Australian Journal of Botany* 59, 324–335.

White, M.E. (1986) *The Greening of Gondwana.* Reed Books: Frenchs Forest.

White, A., Orr, D., Novelly, P. and Bastin, G. (2014) Tussock grasslands. In *Biodiversity and Environmental Change: Monitoring, Challenges and Directions*, ed. D. Lindenmayer, E. Burns, N. Thurgate and A. Lowe. CSIRO Publishing: Melbourne, pp. 479–518.

Williams, J., Prebble, P.E., Williams, W.T. and Hignett, C.T. (1983) The influence of texture, structure and clay mineralogy on the soil moisture characteristic. *Australian Journal of Soil Research* 21, 15–32.

Williams, N.S.G., Morgan, J.W., McCarthy, M.A. and McDonnell, M.J. (2006) Local extinction of grassland plants: the landscape matrix is more important than patch attributes. *Ecology* 87, 3000–3006.

Williams, N.S.G., Marshall, A. and Morgan, J.W. (2015) *Land of Sweeping Plains – Managing and Restoring the Native Grasslands of South-eastern Australia.* CSIRO Publishing: Clayton South.

Willis, J.H. (1964) Vegetation of the basalt plains in western Victoria. *Proceedings of the Royal Society of Victoria* 77, 397–418.

Wong, N.K., Morgan, J.W. and Dorrough, J. (2010) A conceptual model of plant community changes following cessation of cultivation in semi-arid grassland. *Applied Vegetation Science* 13, 389–402.

19 Alpine, Sub-alpine and Sub-Antarctic Vegetation of Australia

SUSANNA VENN, JAMIE KIRKPATRICK, KEITH MCDOUGALL, NEVILLE WALSH, JENNIE WHINAM AND RICHARD J. WILLIAMS

The alpine and sub-alpine vegetation of Australia occupies habitat islands from Namadgi National Park in the Australian Capital Territory to the Southern Range of Tasmania. Macquarie and Heard Islands support Australia's sub-Antarctic vegetation. The high mountain environments on the mainland experience a continental climate, often accompanied by winter snow; whereas, the sub-Antarctic islands and the Tasmanian highlands have maritime climates in which prolonged snow cover on vegetation is restricted. The distributions of plant communities are influenced by water, snow, wind exposure and topography, soils, biotic factors and disturbance. Many plant species are endemic to Tasmania and the Australian mainland and a few endemic to the Australian sub-Antarctic islands. Many species are scleromorphic, hairy, or have cushion, rosette or tussock forms; characteristics that aid in cold climate survival. Some mainland Australian species have fast phenological cycles to make the most of short growing seasons. Cyclical and dynamic vegetation change, plant-plant interactions, recruitment biology and post-fire vegetation responses are well-documented. Current monitoring and research is focused on climate change, alien plant invasions, feral animals and fire.

19.1 Introduction

The high-mountain environments of Australia comprise approximately 11 700 km^2, only 0.15% of the continent. In a global context, Australian mountains are unusual; they are comparatively low, barely exceeding 2000 m at the highest point on the mainland (Mt Kosciuszko, 2228 m), ancient and, on the mainland, soil covered. Except for sub-Antarctic Heard Island, which supports glaciers, there is no nival zone or areas of permanent snow, very few areas of regularly persisting snow and many Tasmanian alpine areas are snow-free during most of winter.

The vegetation is highly diverse and contains many locally endemic plant species. A range of distinctive plant communities occupy regionally and locally diverse environments. The environmental conditions – which can include snow, sometimes short growing seasons, strong winds and regular frosts – have produced a flora with characteristic growth habits, such as prostrate forms, acaulescent rosettes and cushions. Species in the Poaceae, Cyperaceae, Asteraceae, Proteaceae and Ericaceae dominate much of the vegetation, which in Tasmania, is recognised to have international significance for its unusually woody nature, its concentration of palaeoendemics dating to the Cretaceous and its endogenous cyclic dynamics. The sub-Antarctic islands have an oceanic climate. Temperatures are relatively low (maximum

Australian Vegetation, Third Edition, ed. David A. Keith. Published by Cambridge University Press.

Citation: Venn, S. *et al.* (2017). Alpine, Sub-alpine and Sub-Antarctic Vegetation of Australia. In D. A. Keith, ed., *Australian Vegetation*, 3rd edn. Cambridge: Cambridge University Press, pp. 461–89.

temperatures around 7°C), skies cloudy, rainfall frequent and winds strong. Snow can occur at any time of the year, but rarely lies for more than a week on Macquarie Island. Tussock grasses and lush megaherbs are the tallest plants on sub-Antarctic islands, and woody species are absent.

The alpine zone is the area above the treeline. Globally, treeline position is determined by a seasonal mean ground temperature of approximately 6.7°C (Körner and Paulsen 2004). However, in the case of maritime mountains, as in Tasmania, a mean air temperature of the warmest month of 10°C determines the treeline. *Eucalyptus pauciflora* subsp. *niphophila* is the typical species at the abrupt treelines on the mainland. In Tasmania, the treeline species are diverse and the change from sub-alpine forest to alpine heath is typically gradual (Kirkpatrick 1997). Sub-alpine vegetation occurs above the higher altitudinal limits of *Eucalyptus delegatensis* and/or *Gymnoschoenus sphaerocephalus.*

Despite the small extent of Australian alpine, sub-alpine and sub-Antarctic environments, they are highly important for water supplies, hydroelectricity generation, recreation and conservation. The vegetation responds to these competing land uses and climate change, in ways that differ markedly between mainland Australia, Tasmania and the sub-Antarctic islands.

19.2 Distribution, origins and biogeography

19.2.1 Where does alpine, sub-alpine and sub-Antarctic vegetation occur?

Approximately 1600 km^2 of alpine and treeless sub-alpine vegetation on the mainland (McDougall and Walsh 2007) extends from Lake Mountain in Victoria to Namadgi National Park in the ACT, and ranges in elevation from 1000 m on the Baw Baw Plateau to 2228 m at Mt Kosciuszko (Figure 19.1). The Snowy Mountains in New South Wales contains the largest area of alpine vegetation, covering 100 km^2. Further south, alpine vegetation in Victoria occurs in smaller, isolated summit areas around Mt Bogong, Mt Feathertop and Mt Howitt and sub-alpine plains are scattered throughout the high country.

The Tasmanian alpine and sub-alpine treeless vegetation forms more than 20 habitat islands in the eastern alpine zone, more than 30 in the central alpine zone and more than 40 in the western alpine zone, none attaining more than 1650 m in elevation (Kirkpatrick 1986) (Figure 19.1), covering 2153 km^2. The largest alpine areas are the Ben Lomond Plateau, Clumner Bluff, Ironstone Mountain, Wild Dog Tier, Cradle Mountain-Barn Bluff, Mt Field West and Mt Anne.

Macquarie (54°30'S, 158°57'E) and Heard Islands (53°S, 73°30'E) support sub-Antarctic vegetation. Macquarie Island, located roughly halfway between Tasmania and Antarctica, is nearly 34 km long and up to 5 km wide, with an undulating plateau 150 to 300 m high, with the highest peak at 433 m. Heard Island is located in the Indian Ocean sector of the Southern Ocean, 43 km long from north-west to the south-east, and is dominated by Big Ben, a heavily glaciated roughly circular and mildly active volcano with a base diameter of 18–20 km and maximum elevation of 2750 m. Approximately 70% of the island is covered in glaciers and a further 2% by permanent snow (Ruddell 2006).

19.2.2 The origins of alpine environments and the sub-Antarctic islands

The origins of the elevated environments of mainland Australia have recently been interpreted by linking models of plate tectonics with convection currents (Müller *et al.* 2016). The research indicates that around 150 Ma as the eastern edge of the continent collided with the Pacific plate, it was dragged down and moved beneath the Pacific plate to the east. Then, 100 Ma this subduction process ceased, causing the eastern edge of Australia to rebound upwards between 400–600 m. As Australia's drift north increased to 7 cm per year, the continent straddled the edge of the Pacific Superswell upwelling, which lead to the second uplift event in the Snowy Mountains of about 700 m (Müller *et al.*

(a)

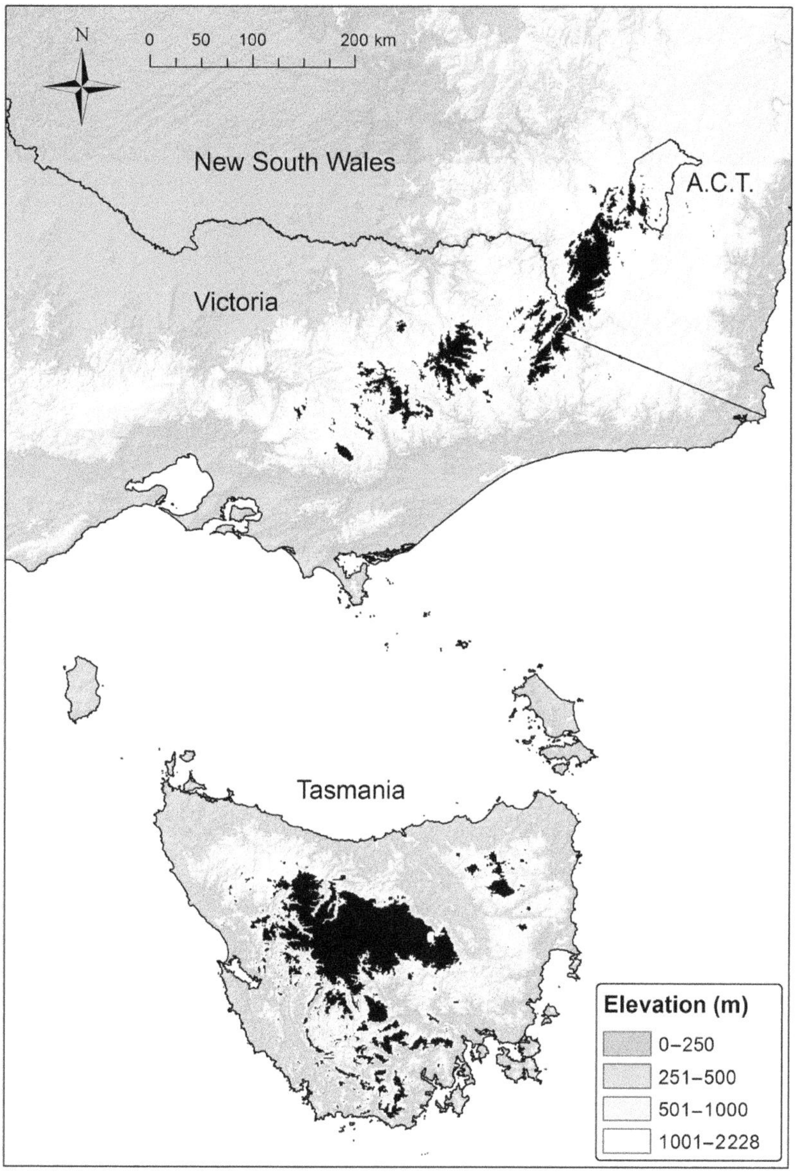

Figure 19.1 (a) Alpine/sub-alpine vegetation in Tasmania and the mainland (shown in black), is represented by the area with an annual mean temperature below 8°C; this is approximately the temperature at the boundary between alpine ash forest and snow gum woodland on the mainland. Elevation is shown in the background and highlights the lower elevation of such vegetation in Tasmania. Macquarie and Heard Islands have annual mean temperatures below 8°C. (b) Generalised distribution of community types on Macquarie Island. Communities such as feldmark may also be scattered within areas dominated by other vegetation types. (c) Sub-Antarctic vegetation on Heard Island occupies between 5 and 10% of the land area and is restricted to lowland areas near the coast (contours interval = 200 m). GIS data from the Australian Antarctic Division are licensed under a Creative Commons Attribution 3.0 Unported License (see also Acknowledgements).

(b)

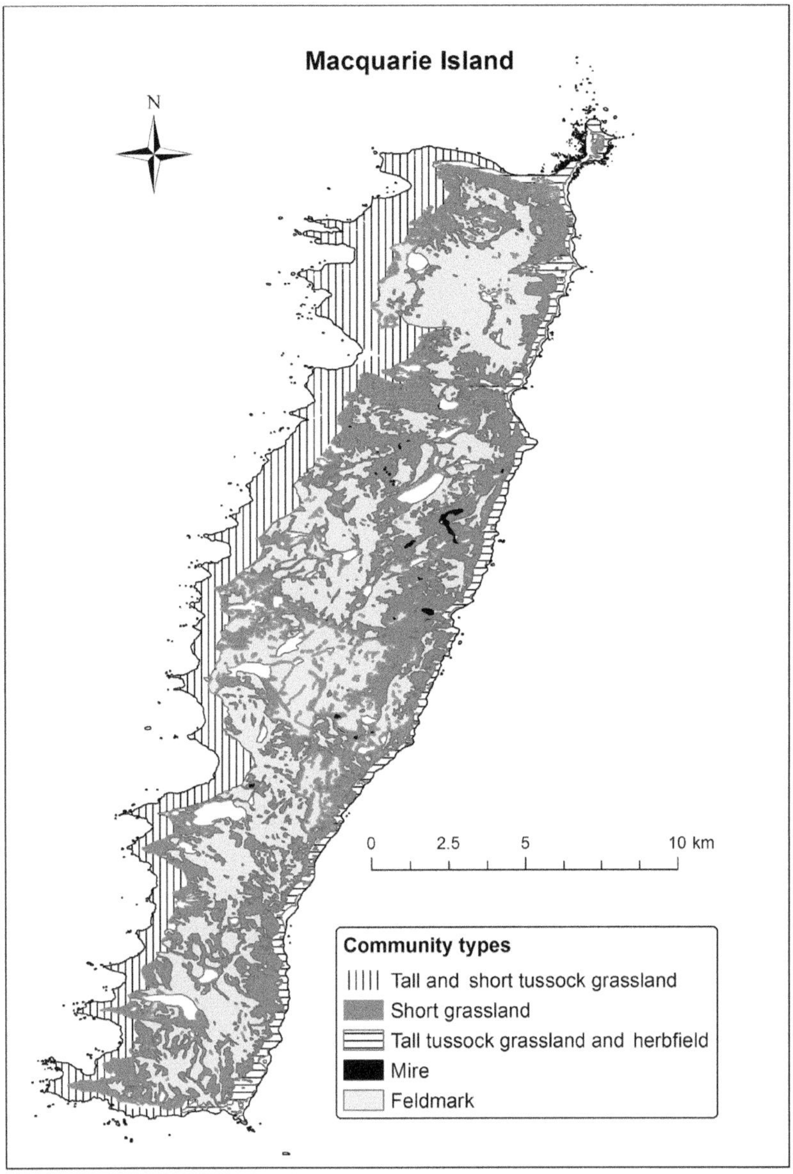

Figure 19.1 (cont.)

(c)

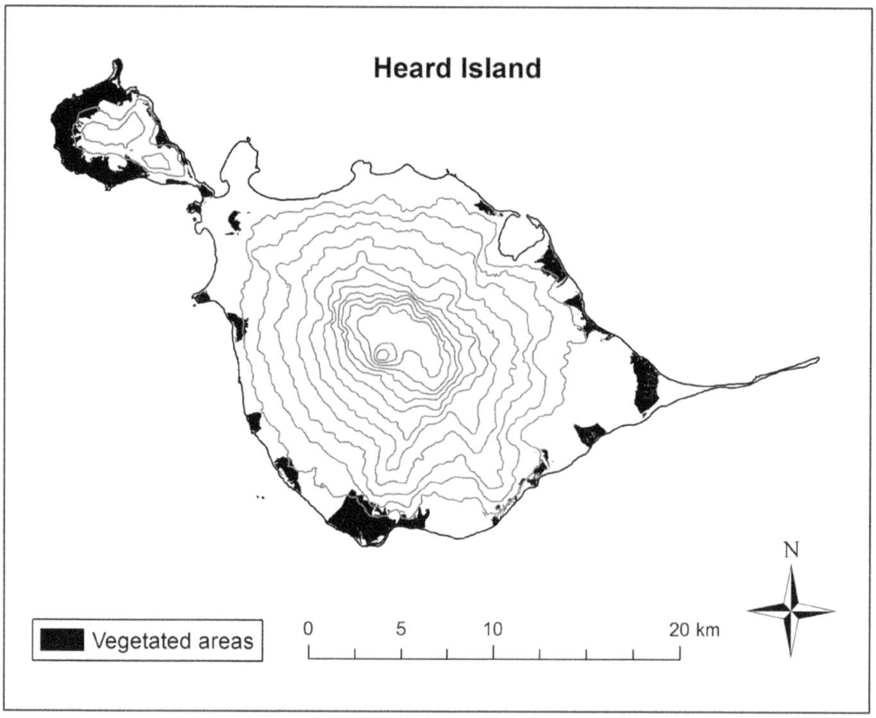

Figure 19.1 (cont.)

2016). The onset of alpine conditions was most likely during the Quaternary. Australian alpine habitats are probably no older than those of New Zealand and New Guinea, and have only experienced glaciation in the late Pleistocene, culminating about 19 000 years ago (Mooney *et al.*, Chapter 3, this volume).

The Pre-Carboniferous quartzose deposits of western Tasmania have been intensely folded in several orogenies to form north–south arcs of mountain ranges, which have been heavily glaciated several times over the last few million years. The alpine mountains of central and eastern Tasmania are in a fault province in which Jurassic dolerite and Permian–Triassic sediments have been exposed to form many high plateaus, most of which have been glaciated and all of which have been subject to periglacial activity, which continues today on fissile rocks at high elevations.

Macquarie Island is the only exposed section of the Macquarie Ridge, which stretches from around 61°S to New Zealand (Selkirk *et al.* 1990). It emerged approximately 600 000–700 000 years ago and continued uplift of the ridge is estimated to be approximately 0.8 mm per year (Adamson *et al.* 1996). Macquarie Island is geologically unique, being the only island composed entirely of oceanic crust, with all rocks formed on or beneath the ocean floor. In contrast, Heard Island is almost entirely igneous in origin. Glaciers on Heard Island are still receding from their last glacial maximum (Bergstrom and Chown 1999).

19.2.3 Vegetation origins

The fossil record from the Australian mainland shows the presence of likely ancestors of the present alpine flora, but does not confirm the existence of alpine vegetation before the late Tertiary (Barlow 1994; Weston and Jordan, Chapter 2, this volume).

Up to 60% of the genera native to the mainland alpine areas probably reached there by expansion from lower elevations. Sub-alpine woodlands are floristically an extension of the same macro-ecosystem, and a reservoir for recolonisation of the alpine habitat (Barlow 1994).

While a general west-to-east dispersal via strong westerly wind and ocean currents (the West Wind Drift) and the Antarctic Circumpolar Current is accepted for some genera, such as *Aciphylla* in New Zealand, evidence is equivocal (Sanmartín *et al.* 2007) and, although rare, east-to-west dispersal may have occurred. Some genera (e.g. *Carex*, *Euphrasia*, *Festuca*, *Isolepis*, *Juncus*, *Luzula*, *Poa*, *Ranunculus*), and species (e.g. *Alchemilla xanthochlora*, *Carex echinata*, *Deschampsia cespitosa*, *Geum urbanum*, *Koeleria macrantha*, *Trisetum spicatum*) have bi-hemispherical distributions. Migratory birds have long been implicated as likely vectors for these otherwise baffling occurrences (Darwin 1859).

In the sub-Antarctic islands, most native plant species have circum-Antarctic distributions. All of the ancestors of sub-Antarctic plant species have established on these islands via long-distance dispersal. Species confined to one island group are not much diverged from close relatives on other land to the north (Clements *et al.* 2007).

Theories of Gondwanan origins for the mainland alpine flora are largely inconsistent with the results of DNA-based phylogenetic analyses, which suggest that the dates of divergence are typically much younger than the tectonic events that led to the ultimate fragmentation of Gondwana between c. 80 and 30 Ma into South America, Australia, New Zealand and Antarctica (McLoughlin 2001; Weston and Jordan, Chapter 2, this volume). Nevertheless, there is evidence that certain families found in Tasmanian alpine vegetation, and the derived genera, radiated *in situ* following isolation (Sanmartín *et al.* 2007; Thomas *et al.* 2014). The element of the Australian plate alpine flora that has been labelled as Gondwanan has floristic affinities at the family level with South Africa (e.g. Proteaceae, Restionaceae) and at a generic level with South America (e.g. *Eucryphia*, *Nothofagus*). Palaeoendemic clades, such as *Bellendena*, *Athrotaxis* and *Isophysis*, are characteristic of the transition zone between alpine and sub-alpine in superhumid parts of Tasmania. These clades appear to have persisted since the Cretaceous (Jordan *et al.* 2016). On Heard Island, *Pringlea antiscorbutica* and six moss species have been identified as potential Tertiary survivors (Bergstrom and Seppelt 1988).

Most vascular plant species in the alpine and treeless sub-alpine zones of mountains in western Tasmania are Tasmanian endemics, with endemism declining to slightly less than half in the eastern Tasmanian mountains (Kirkpatrick and Brown 1984). On the mainland, 30% of taxa are endemic to treeless alpine and sub-alpine vegetation (McDougall and Walsh 2007). Although there is a lower proportion of local endemic alpine/sub-alpine species on the mainland, almost all alpine species within the jurisdiction of Australia are Australian endemics. Many of the plant taxa restricted to one or two alpine areas are obvious neoendemics which have evolved in environmentally/geographically extreme alpine environments (Kirkpatrick 2002).

19.3 Vegetation patterns and plant communities

19.3.1 Community nomenclature

Phytosociological investigation in Australian alpine and sub-alpine areas has taken place at a range of scales employing techniques from descriptive to numerical and from structural to floristic. As a consequence, there are a multitude of classifications and nomenclatures (complicated further by local vernacular and nuance). The structural classification of Specht (1970) has been especially influential; so 'formation names' relating to structure and life form have broad currency. However, many well-known formation names relate to habitat. For example, bog and fen are recognised names globally, but are not used with the same meaning as in the northern hemisphere, where fen refers to mires on minerotrophic sites and bog refers to mires on ombrotrophic sites. In Australia, mires dominated by *Carex gaudichaudiana* or similar taxa are called fen, and those with some *Sphagnum*, not necessarily dominant, are called bog (Kirkpatrick 2016). Feldmark (or fjaeldmark in Tasmania) is an obscure term first used for sparsely vegetated moss communities in the Arctic but now it apparently survives mainly in Australia. Kirkpatrick (1997) distinguishes fjaeldmark from other vegetation types based on the amount of bare ground (>50%), while Costin *et al.* (2000) describe feldmark as being 'characterised by scattered dwarf, prostrate plants', a definition that does not describe Tasmanian fjaeldmark very well. The tall alpine herbfield of Costin *et al.* (2000) includes tall grasslands dominated by *Chionochloa frigida* and some open heathlands, Here we use the term in a narrower sense to refer to communities dominated by forbs (tall only when in flower). In this section we have attempted to preserve local nomenclature while categorising communities firstly by habitat (where such communities are entrenched in vernacular) and then by growth form (Table 19.1).

19.3.2 Distribution of communities

Of the major groupings in Table 19.1, there are examples of most in mainland Australia, Tasmania, Macquarie Island and Heard Island. The exceptions include an absence of shrub and tree communities on sub-Antarctic islands and the absence of cushion-dominated communities on the mainland, although cushion plants occur there. Alpine sedgeland in Tasmania perhaps has analogues on the mainland, but these tend to be highly localised and often at the margins of the alpine area (e.g. the *Carex appressa* community of McDougall and Walsh (2007)). Heathlands with a large range of dominants are the most common vegetation types on the mainland and Tasmania (Kitchener and Harris 2013), whereas feldmark, with an open mosaic of mosses, cushion plants and grasses dominate on sub-Antarctic islands. While mainland Australia and Tasmania have some species and a few communities in common, they share few species and no communities with sub-Antarctic islands.

19.3.3 Environmental gradients and their influence on community distributions

Alpine and sub-alpine vegetation is characterised by its steep environmental gradients, and highly localised and specialised communities with sharp boundaries. There is substantial variation in continentality, which results in variation in snow persistence, which in turn results in strong differences in community composition between mainland Australia and Tasmania (Kirkpatrick and Bridle 1998). Various climatic predictors relating to ambient and soil temperatures are also very important in discriminating communities. Some communities are faithful to particular substrates or topographic positions. Community distribution is associated with strong gradients in soil fertility and depth from the glaciated quartzite mountains of the superhumid south-west of Tasmania, through the dolerite mountains of eastern Tasmania, to the loess top-dressed soil mountains of mainland Australia (Kirkpatrick *et al.* 2014). Organosols and rudosols dominate on the sub-Antarctic islands.

Some communities are a consequence of disturbance or develop only in the long-term absence of disturbance. Marsupial lawns can form under heavy grazing pressure, especially in moist and occasionally inundated areas (Roberts *et al.* 2011). Some communities are associated with recent fire disturbance, while others only develop in long-unburnt areas. Some of the major determinants and resultant vegetation patterns are outlined below.

19.3.3.1 Water and snow

High mountains and the sub-Antarctic islands receive large amounts of precipitation, with some of it falling as snow. Their variable topography,

Table 19.1 **Australian alpine, sub-alpine and sub-Antarctic plant community types, dominant species and habitat characteristics arranged firstly by landscape feature and then by dominant growth form. Local community names (Mainland (M): Costin *et al.* 2000, McDougall and Walsh 2007; Tasmania (T): Kirkpatrick 1997; Macquarie Island (MI): Selkirk *et al.* 1990; Heard Island (H): Bergstrom *et al.* 2002) are generally preserved but may be combined and/or altered where habitat characteristics are similar.**

Local community name	Dominant and characteristic species	Description	Typical habitat
Groundwater and snowmelt water			
Aquatic vegetation (M, T)	M: *Hydrocotyle rivularis, Myriophyllum aquaticum, Nymphoides montana, Ranunculus amphitrichus* T: *Isoetes* spp., *Liparophyllum exiguum, Eleocharis acuta, Ranunculus* spp., *Myriophyllum* spp.	Represented by aquatic species with floating or submerged foliage and a range of herbaceous species, the composition of which is dependent on the period of inundation	Streams, lakes and ponds, usually around (or at) their margins
Bog (M, T) Mire (MI) Wet heath (M)	M: *Sphagnum cristatum, Astelia alpina, Empodisma minus, Epacris paludosa, Richea continentis* T: *Sphagnum cristatum, Astelia alpina, Empodisma minus, Richea gunnii, R. scoparia* MI: *Juncus scheuchzerioides, Hydrocotyle novae-zeelandiae, Montia fontana*, bryophytes	Characterised by moss hummocks and carpets, vegetatively-spreading sedges and rushes and, in Tasmania and the mainland, scleromorphic shrubs. It can be patterned in stripes, regular hummocks and hollows or scattered with small pools	This vegetation occurs where the water table is at or near the surface in valleys and on lower hillslopes. It is extensive on Macquarie Island where it may include fen-like vegetation
Damp herbfields (M)	*Lobelia surrepens, Ranunculus millanii, Hypericum japonicum, Ranunculus pimpinellifolius, Stackhousia pulvinaris*	Dense mats of prostrate forbs	Found on peat at the margins of bogs, wet heath and low velocity streams, and in contour trenches found in some grasslands
Fen (M, T)	*Carex gaudichaudiana*	Species-poor communities of large pools and still water bodies; characterised by emergent tufts of *C. gaudichaudiana*	Fen is associated with periodically or permanently inundated, fertile ground in the lower parts of the alpine zone and the sub-alpine zone

Short alpine herbfield (M, T) Gravelly stream herbfield (M)	M: *Brachyscome stolonifera*, *Celmisia sericophylla*, *Oreobolus pumilio*, *Plantago muelleri*, *P. glacialis*, *Psychrophila introloba* T: *Drosera arcturi*, *Plantago glacialis*, *Gunnera cordifolia*, *Isotoma fluviatilis*, *Psychrophila phylloptera*, *Celmisia saxifraga*	Vegetation dominated by dicotyledonous herbs mostly less than 5 cm tall. Cover is variable, usually consisting of rhizomatous or stoloniferous species. The doughnut-shaped patches of *Oreobolus pumilio* are a feature of short alpine herbfield	Often found on the margins of wetlands and in association with late-lying snow. Distinct forms are found in Tasmania (dominated by *Celmisia saxifraga*) on shallow soils on high elevation dolerite plateaus, and in Victoria (dominated by *C. sericophylla*) on steep rocky streams
Valley grassland (M)	*Poa costiniana*, *P. helmsii*, *P. labillardierei*	Characterised by a high cover of tall tussock grasses, sometimes forming pedestals over a network of creek tributaries	Broad sub-alpine valleys at the margins of fast-flowing waterways; occasionally inundated
Wet mixed herbfield (H)	*Acaena magellanica*, *Azorella selago*, *Deschampsia antarctica*, *Poa cookii*, *Pringlea antiscorbutica*	Species poor but high cover mosaic of cushion plants, low grasses and forbs	On moist moraines and lee slopes at low altitude where the water table is at or close to the surface
Rock or exposed soil			
Feldmark (M, MI) Fjaeldmark (T) Mossy feldmark (H) Open cushion carpet (H)	M: *Epacris microphylla*, *Ewartia nubigena*, *Leucochrysum alpinum* T: Highly variable with no particular marker species. H, MI: *Azorella* spp., *Agrostis magellanica* and mosses (e.g. *Polytrichastrum alpinum*, *Ditrichum* spp., *Racomitrium crispulum*)	Plant cover in these communities is very sparse with fissile rock being a dominant feature	These communities occur in areas exposed to extremely strong winds and little snow cover
Snowpatch feldmark (M)	*Colobanthus nivicola*, *Coprosma niphophila*, *Epilobium tasmanicum*, *Luzula acutifolia* subsp. *nana*	Sparse, low-growing shrubs and forbs with large bare soil or moss-filled gaps	Restricted to steep, sheltered slopes with a very short growing season and much bared soil because of late-lying snow
Podocarp shrubland (M) Coniferous heath (T)	*Podocarpus lawrencei*	*Podocarpus* shrubs form large but often sparse rock-hugging patches. Plant species diversity is generally low	Found in boulder piles and scree of varying size and geology; often on sheltered sites

Table 19.1 (cont.)

Local community name	Dominant and characteristic species	Description	Typical habitat
Cushion plants			
Bolster heath (T) Closed cushion carpet (H)	T: *Abrotanella fosterioides, Donatia novae-zelandiae, Dracophyllum minimum, Mitrasacme archeri, Phyllachne colensoi, Pterygopappus lawrencei* H: *Azorella selago*	Cushions have a tight cluster of short-stemmed shoots that form hard, dense surfaces. They are well-adapted to winter exposure to strong ice-laden winds. In Tasmania, they often form undulating surfaces composed of several species	Associated with constantly moist and poorly drained ground in Tasmania, usually in positions where snow does not persist for more than a few days; on Heard Island, cushion carpets with 100% cover of *A. selago* form on moraine deposits of intermediate exposure
Grasses			
Tussock grasslands (M, T) Tall tussock grassland with megaherbs (MI) Coastal biotic vegetation (H)	M: *Chionochloa frigida, Poa fawcettiae, P. costiniana, P. hiemata, P. hookeri* T: *Deschampsia cespitosa, Poa fawcettiae, P. gunnii, P. costiniana, P. hiemata, P. labillardierei, Rytidosperma* spp. H, MI: *Poa foliosa, P. cookii, Festuca contracta, Agrostis magellanica, Stilbocarpa polaris, Azorella selago*	Widespread communities dominated by tussock-forming grasses but often with a large component of inter-tussock forbs and low shrubs; on Macquarie Island, the megaherb, *Stilbocarpa polaris*, is prominent in gaps	On sub-Antarctic islands, tussock grasslands may be found in many habitats including steep easterly facing coastal slopes and coastal areas of moderate exposure; elsewhere they may occur in a range of habitats but typically where soils are relatively deep and well-drained
Short grassland (MI) Short-turf snowpatch (M) Limestone grassland (M) Marsupial lawn (T)	M: *Poa hothamensis, Rytidosperma nudiflorum* (snowpatch); *Leucochrysum alpinum, Themeda triandra* (limestone) MI: *Agrostis magellanica, Deschampsia chapmanii, Festuca contracta, Luzula crinita, Uncinia* spp. T: *Poa* spp., *Rytidosperma* spp.	Grasslands dominated by grasses that do not form tussocks or where disturbance from animals reduces tussock-forming species to lawns	On Macquarie Island it occurs on sheltered upland areas and beach terraces with a moderate to high water table; on the mainland, it is highly localised on a few limestone valley slopes and gently sloping slopes where snow persists in spring, while in Tasmania the community is the product of heavy marsupial grazing

Forbs			
Herbfield (MI)	*Acaena magellanica*, *Blechnum pennamarina*, *Pleurophyllum hookeri*, *Stilbocarpa polaris*	*Stilbocarpa polaris* can grow taller than a person on Macquarie Island, but most dominant herbs throughout the range of the community are less than 0.5 m tall	Found on sheltered slopes, flats and raised terraces
Tall alpine herbfield (M, T)	M: *Celmisia* spp., *Podolepis robusta*, *Craspedia* spp. T: *Craspedia gracilis*	Vegetation dominated by herbaceous dicotyledonous plants taller than 5 cm, in places mixed with tussock grasses	Occasional on the mainland on deep soils or on the upper margins of areas of late-lying snow; restricted to a few tiny places naturally protected against grazing in Tasmania
Sedges			
Alpine sedgeland (T)	*Astelia alpina*, *Baloskion australe*, *Carpha* spp., *Empodisma minus*, *Gleichenia alpina*, *Isophysis tasmanica*	Characterised by mats of graminoids or rhizomatous ferns in the tallest stratum and the low cover of bare ground	Replaces coniferous and deciduous heath on poorly drained ground after burning
Shrubs			
Coniferous heath (T)	*Diselma archeri*, *Microcachrys tetragona*, *Microstrobos niphophilus*, *Athrotaxis selaginoides*, *A. cupressoides*	Dwarf or mat-forming conifers in a mosaic with a range of shrubs, forbs and grasses common to other heath communities	Found on long-unburnt, well-drained slopes of high-rainfall areas; all the characteristic gymnosperms are fire sensitive
Deciduous heath (T)	*Nothofagus gunnii*	Mat to tall heathland dominated by deciduous shrubs	On sheltered southern slopes of long-unburnt mountains with high precipitation

Table 19.1 (cont.)

Local community name	Dominant and characteristic species	Description	Typical habitat
Heath (T) Open heathland (M) Closed heathland (M)	M: *Grevillea australis, Kunzea muelleri, Nematolepis ovatifolia, Orites lancifolius, Phebalium squamulosum, Prostanthera cuneata* T: *Richea* spp., *Epacris* spp., *Cyathodes* spp., *Orites* spp., *Ozothamnus* spp., *Olearia* spp., *Grevillea australis, Eucryphia milliganii*	Shrub-dominated vegetation varies from short (<30 cm; e.g. *K. muelleri*) to tall (>1 m; e.g. *Orites* spp.) and may be closed or open with abundant lilies, irids, rushes, grasses, forbs, sedges and scleromorphic shrubs in gaps	Occurs in all topographic and edaphic situations but is more common on rocky soils and well-drained peat than on the few areas of deep mineral soils
Mat heath (T)	*Montitega dealbata, Gaultheria depressa, Pentachondra pumila, Cryptandra alpina, Coprosma perpusilla*	Short mats, less than 5 cm tall, consisting of scleromorphic shrubs	Dominates the dry upper slopes of some snow patches and glacier-scraped, exposed flats with shallow well-drained soils
Trees			
Sub-alpine woodland (M, T)	M: Mostly *Eucalyptus pauciflora* subsp. *niphophila* but other subspecies (subsp. *acerina, debeuzevillei, hedraia, pauciflora*) or *E. lacrimans, E. perriniana, E. stellulata* may be locally dominant. T: *Eucalyptus coccifera, E. gunnii, E. subcrenulata*	Dominated by low trees, typically with an open structure; understoreys may be shrub or grass dominated and are often floristically indistinguishable from adjoining treeless communities	Commonly on the warmer side of a distinct line separating vegetation containing trees and where trees are absent; woodland islands above the treeline sometimes occur

including steep and flat terrain, and low evapotranspiration result in steep moisture gradients with a strong seasonal component. On mainland Australia, water availability can be high across the entire landscape during snow melt but hugely divergent during snow-free periods. Tasmanian winters are wet and windy, but too warm for snow to widely accumulate. The lack of persistent winter snow allows native vertebrate herbivores to graze the alpine zone, in contrast to the mainland mountains where native vertebrate herbivores are repelled by deep snow in winter and spring. The rarity of tall alpine herbfield in Tasmania is also related to the presence of vertebrate herbivores.

Where water flow is impeded, bogs and fens (Figure 19.2a) are common. The sub-Antarctic variant is a short mat-forming community with 100% vegetation cover and the vegetation layer often forming a fibrous blanket over very wet peat or water, creating a quaking mire or 'featherbed'. The margins of mires in Tasmania and on the mainland may be occupied by short alpine herbfield, which is also associated with late-lying snow. Communities (locally called bolster heath; Figure 19.2b) comprising several species of cushion plants in Tasmania and cushion communities in the sub-Antarctic are associated with moist, poorly drained ground where snow does not accumulate, making a virtue of their wind-resistant and radiation-absorbing shape. On Heard Island, a pool complex (dominated by *Callitriche antarctica* and with high moss cover), part of the wet mixed herbfield of Scott and Ryan (2004), is prominent.

Towards the dry end of the moisture spectrum on well-drained soils, heathlands (only in Tasmania and mainland Australia) and tussock grasslands are common. They are typically the most species-rich of the communities, containing numerous forbs and tufted grasses in gaps between shrubs and tussocks. Tussock grasslands are typically dominated by *Poa* species. Heathlands are more diverse in their dominance (Table 19.1). On Macquarie Island, the megaherb, *Stilbocarpa polaris*, is prominent in grassland gaps and in herbfield on sheltered slopes.

Figure 19.2 Alpine vegetation (a) Fen, Pretty Valley, Bogong High Plains (photo: Keith McDougall); (b) mosaic bolster heath with invading gymnosperm (*Microcachrys tetragona*) at Newdegate Pass, Mt Field Tasmania. Quadrat is 35 cm × 35 cm (photo: Jamie Kirkpatrick).
(A black and white version of this figure will appear in some formats. For the colour version, please refer to the plate section.)

19.3.3.2 Wind and exposure

Wind exposure is a common trait of alpine and sub-Antarctic ecosystems. Where it is strongest, wind produces characteristically short and wind-pruned plants. The degree of exposure also influences the distribution of plant communities. At high altitude on highly wind-exposed fissile rocks, feldmark forms (Figure 19.3a). Snow is blown off hill tops, saddles and ridges, leaving soils and plants exposed to very low temperatures year-round. Growth forms with adaptations to low temperature and severe wind-pruning for long periods

Figure 19.3 Alpine vegetation. (a) Feldmark between Mt Northcote and Mt Lee, Snowy Mountains (photo: Keith McDougall). (b) Late-lying snow associated with snowpatch vegetation on the south-eastern side of the Main Range near Mt Lee and Mt Northcote, with Club Lake in the foreground, Snowy Mountains (photo: Susanna Venn). (c) A valley of young megaherbs *Pleurophyllum hookeri* (Macquarie megadaisy) recovering after removal of rabbits on Macquarie Island (photo: Kate Kiefer).

(A black and white version of this figure will appear in some formats. For the colour version, please refer to the plate section.)

are well represented (i.e. acaulescent forbs, cushion plants, layering shrubs and bryophytes). In mainland Australia, feldmark is dominated by *Epacris microphylla*. This layering shrub effectively moves across the ridges in the direction of prevailing winds, being pruned on the windward side while sprouting on the leeward side (Barrow *et al.* 1968).

In Tasmania snow accumulates and persists to the north-east whereas on mainland Australia it accumulates and persists to the south-east (Figure 19.3b). The collective term for communities occurring in this situation is 'snowpatch'. Snowpatches comprise a mosaic of communities, arranged according to moisture availability (short alpine herbfield and gravelly stream herbfield in perennially wet parts), soil depth (tussock grassland, tall alpine herbfield or short turf grassland on deeper soils) and the duration of snow persistence (feldmark in areas with long-lasting snow). The spatial complexity of alpine communities on very small areas of lee slopes with persistent snow is similar to that described for Tasmania (Gibson and Kirkpatrick 1985).

19.4 Plant ecology, survival and reproduction in cold environments

19.4.1 The environment experienced by plants

The conditions experienced by plants in cool-climate environments are often very different from ambient conditions. Given sufficient winter snow, low-statured vegetation is protected from frosts and strong winds under the snowpack. Most shrubs are also protected if their branches are weighed down under the snow. Trees, however, are subjected to all kinds of weather, and exposed meristematic regions at the tips of branches make them very vulnerable to frost damage, as evidenced by the many treeless sub-alpine valleys where snow is often shallower and where very low temperatures occur; the lowest on record being −23°C at Charlottes Pass village.

Most alpine and sub-alpine plants in Tasmania are fully exposed to subzero temperatures and strong winds during much of the winter, snow lie being highly localised. The same degree of exposure affects plants in sub-Antarctic vegetation.

Where it occurs, snowmelt generally marks the onset of the growing season, but plants exposed to the sun during the winter can continue to grow, as in the case of cushion plants on Mt Field, Tasmania (Gibson 1990).

Springtime can also present challenges for previously snow-protected plants, especially if the 'end of season climate variability' is accompanied by frosts and additional snowfalls. The length of the growing season in most alpine and sub-alpine regions in snow country is short compared with adjacent low-elevation regions, and in the height of summer many snow country alpine areas can have extreme soil moisture deficits, even where the deepest snowpack accumulated during the previous winter. These patterns of exposure and temperature (largely determined by topography and aspect), growing season length (as determined by snowmelt date) and water-holding capacity of the soil across the landscape drive the patterns in the vegetation and are among the key abiotic factors responsible for the positioning of the many alpine plant communities. Alpine and cool-climate species share many morphological and physiological characteristics that aid their survival: (i) cushion, short-statured, tussock and rosette forms protect growth tips/meristems; (ii) hairy coverings can increase leaf temperatures; (iii) sclerophylly can protect leaves from riming and abrasion, the physical damage caused by wind, and also reduces water loss; (iv) physiological dormancy of seeds, pre-formed flower buds and rapid transition to fruiting synchronise germination and seedling establishment with favourable conditions during the short growing seasons. Several Tasmanian and sub-Antarctic species (e.g. *Diselma archeri* in Tasmania and *Azorella macquariensis* on Macquarie Island) also share the adaptation of withdrawing chlorophyll during winter.

19.4.2 Plant recruitment

Snowmelt is a major cue for the timing of life-cycle events in snow country species. Most species begin intense physiological activity at this time. Pre-formed buds, common among herbaceous species, allow for fast flowering as soon as conditions are suitable (Mark 1970). At landscape scales, the correlation between snowmelt date and phenology is obvious as whole hillsides come into flower. Similarly at smaller scales, such as within snowpatches, many species time their flowering and growth to the regular patterns of snowmelt (Venn and Morgan 2007). Flowering of other species may be cued by day length.

Snowmelt is also a major cue for seed germination from the soil seed bank. A substantial proportion of seeds from the soil seed bank in the Snowy Mountains demonstrated physiological dormancy (Hoyle *et al.* 2013), a strategy to ensure seeds only germinate after a winter cold stratification period, during more favourable conditions in spring. Indeed many, but not all, mainland Australian alpine species require cold stratification, for example seeds of *Aciphylla glacialis* appear to require 15 weeks of very low temperatures (<2°C) (Venn and Morgan 2009a), which is roughly the same duration as an average snow season in Victoria alpine areas. A species' requirement for cold stratification and its germination strategy is difficult to predict, even considering seed mass, length, embryo:seed ratio or species distribution (latitudinal range, altitudinal range and maximum altitude). However, species that require ≥8 weeks of cold stratification have been associated with the presence of an endosperm and either a restricted distribution or being an upland ecotype of a widely distributed species (Sommerville *et al.* 2013). Germination strategies (immediate, delayed or staggered) of alpine plant species from the Snowy Mountains tend to correlate only with other aspects of seed ecology, and cannot be predicted from vegetative traits (Hoyle *et al.* 2015).

Many alpine, sub-alpine and sub-Antarctic species are very long lived and rely on asexual, clonal propagation to escape the many hazards of regeneration from seed in a harsh environment. Despite the constraints on plant growth, a large proportion of the flora is capable of successful seedling recruitment; up to 61 per cent of seedlings tracked across seven Victorian summits survived to become adult plants (Venn and Morgan 2009b). Regular drying of soils below wilting point over summer may be the main cause of seedling and premature adult plant deaths. In 2004, after 50 days of near zero rainfall on the Bogong High Plains, several sub-alpine shrubs were severely desiccated, with all *Epacris glacialis* dying on the west facing slopes of Basalt Hill (Morgan 2004); and high mortality of some common snow grasses, *Poa hiemata* and *P. hothamensis*, was reported after drought in 2007 at various sites on the Bogong High Plains, with no dead tussocks resprouting between October 2007 and January 2009 (Griffin and Hoffmann 2012).

19.4.3 Abiotic and biotic interactions

Future reductions in winter snow across Australian mainland alpine areas due to climate warming and earlier snowmelt (Hennessy *et al.* 2007; Fiddes *et al.* 2015) may induce earlier than usual spring growth, but may also expose plants to extreme low temperature stress at a time when actively growing plants are particularly vulnerable to frost damage. Many mainland alpine species are resistant to freezing; 25 common species tested by Venn *et al.* (2013) from Mt Hotham (1860 m), Mt Buller (1805 m) and King Billy (1696 m) in Victoria could withstand moderate freezing events of −6°C. Shrub species were the life-form group most resistant to freezing, followed by graminoids and forbs. Most shrub species tested were able to withstand −12°C with *Pimelea axiflora* tolerating −18.7°C. An absence of persistent snow over most of Tasmania means that all except 86 ha of Tasmanian alpine vegetation is exposed to the elements for most of the coldest and windiest time of the year. Scleromorphy protects shrubs, cushion plants and graminoids when encased in ice during storms and while being battered with particles of soil and ice in subzero conditions.

Alpine plants can also avoid freezing damage, strong winds, heat stress and desiccation by growing close together and sheltering within neighbouring plant canopies. Plants in a harsh environment are more likely to show facilitative (positive) interactions with each other, whereas under favourable conditions, plants may compete for resources and show negative plant–plant interactions. This switch between facilitative and competitive interactions has been demonstrated experimentally with transplanted seedlings on alpine summits in Victoria (Figure 19.4; Venn *et al.* 2009). Similarly, the removal of close neighbouring plants from adult individuals of *Celmisia pugioniformis* and *Carex breviculmis* reduced their growth and increased mortality throughout a range of tall alpine herbfields on the Bogong High Plains (Jarrad *et al.* 2012). Also, the dominant, prostrate feldmark shrub in the Snowy Mountains, *Epacris microphylla*, acts as a nurse plant providing shelter from abrasive winds and low temperatures. Ballantyne and Pickering (2015) also demonstrated that the shrub assists in the range expansion of more productive but less stress-tolerant species in the community; with twice as

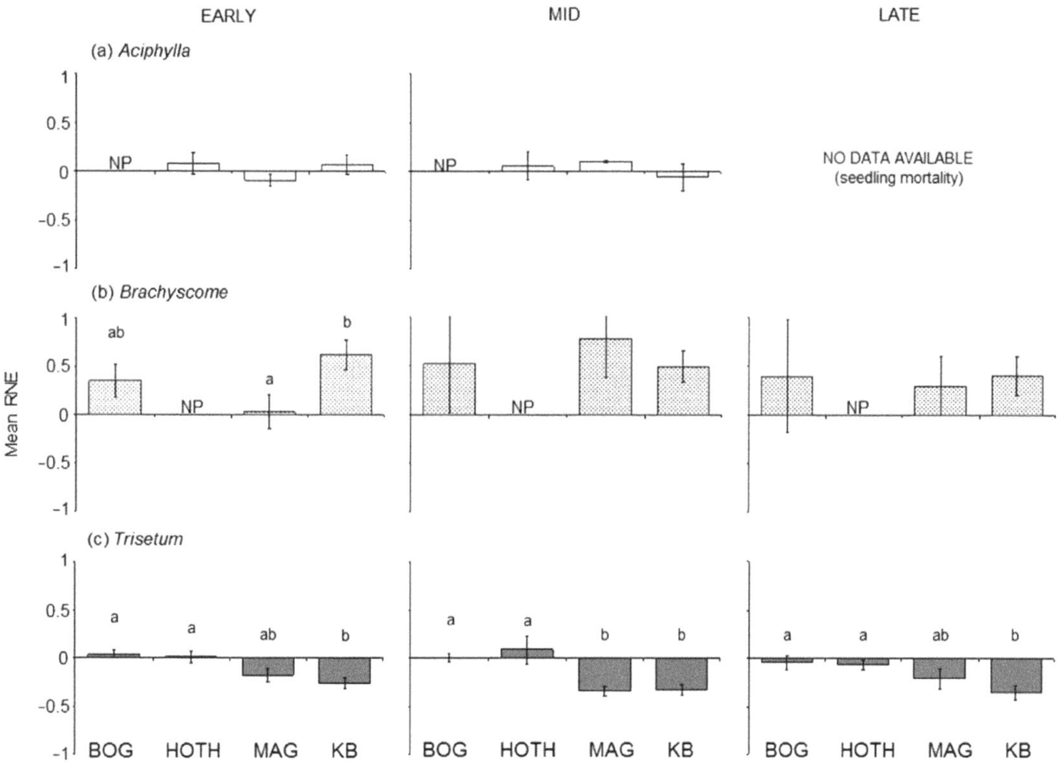

Figure 19.4 Mean relative neighbour effect (RNE) (±se) for (a) *Aciphylla glacialis*, (b) *Brachyscome rigidula*, and (c) *Trisetum spicatum*, for pairs of planted seedlings at King Billy (1696 m) (KB), Mt. Magdala (1725 m) (MAG), Mt. Hotham (1860 m) (HOTH), and Mt. Bogong (1970 m) (BOG), across three sampling intervals (early, mid and late) during the 2005/2006 growing season. Positive values indicate a facilitative interaction and negative values indicate a competitive interaction. Different labels above columns indicate significant differences between sites during that particular sampling interval. NP: *Aciphylla* seedlings were not planted at Mt. Bogong, nor were *Brachyscome* seedlings at Mt. Hotham.
Re-printed with permission from Arctic, Antarctic, and Alpine Research.

many species associated with the shrub than in adjacent, non-shrub-dominated feldmark.

Despite short growing seasons and harsh climatic conditions, alpine, sub-alpine and sub-Antarctic species have relatively high rates of productivity (often measured as biomass accumulation) as do alpine species in other parts of the world (Körner 2002). Peak above-ground phytomass (living and dead plant material) from a snowpatch community on the Bogong High Plains was >1500 g m^{-2} over the 2000/2001 growing season with 140 snow-free days (Venn and Morgan 2007). These levels are higher than that reported in snowbanks in the United States where the growing seasons are very short (Galen and Stanton 1995), but similar to peak phytomass observed in an alpine herbfield in New Zealand (Meurk 1978). Low mean temperatures and cloudy conditions dominate Macquarie Island, but a long sub-Antarctic growing season of 295 days (Jenkin 1972) results in high plant productivity: in grassland (45 m above sea level) dominated by *Poa foliosa* and *Stilbocarpa polaris*, estimated annual above-ground production is 3282 g m^{-2}; and in herbfield (235 m above sea level) dominated by *Pleurophyllum hookeri* estimated annual above-ground production is 489 g m^{-2}. Similar conditions on Heard Island have resulted in rapid colonisation of newly deglaciated land (Scott 1990) and derelict buildings (Whinam *et al.* 2004).

19.4.4 Vegetation dynamics

Ecological processes and vegetation dynamics in alpine and sub-alpine environments must be considered in the context of natural, recurrent disturbance which can influence plant recruitment, successional processes and community structure. Disturbances may be small-scale such as frost heave and wind ablation, medium-scale such as herbivory and seasonal drought, and landscape scale such as fire. The natural disturbance regime may vary considerably both spatially and temporally. Much of our understanding of alpine vegetation dynamics has come from detailed, long-term monitoring of the vegetation, and from experiments that focused on manipulating particular components of the natural disturbance regime, and quantifying effects on patterns of regeneration (Williams *et al.* 2014).

The influence of small- to medium-scale disturbances on shrub-grass dynamics has been well studied in Australian alpine environments. Examples include the cyclical succession in open heathland on the Bogong High Plains (Williams and Ashton 1988), high alpine feldmark in the Snowy Mountains (Barrow *et al.* 1968) and alpine heath on the Central Plateau of Tasmania (McPhail and Kirkpatrick 2015). In open heathland, disturbance of the dead plant litter creates an opportunity for shrub seedlings to establish in the grass sward. If the litter remains fixed to the soil, shrub recruitment is not possible (Williams and Ashton 1987).

Fire is infrequent, but much has been learnt from research that has focused on the major fires that burnt much of the Australian Alps in the last 20 years (Williams *et al.* 2014). The alpine landscape is differentially flammable; heathlands are much more flammable than grasslands (Williams *et al.* 2006) due to inherent differences in slope, fuel mass and fuel bulk. Post-fire regeneration trajectories and rates between the mainland and Tasmania are very different. The alpine heathlands and grasslands of the mainland Alps appear to be capable of rapid regeneration following landscape fires (Wahren *et al.* 2001), and species composition and diversity appear to return to a pre-fire state within 2–5 years for most communities (Wahren *et al.* 2001; Walsh and McDougall 2004; Camac *et al.* 2013; McDougall *et al.* 2015). In Tasmania, there may be a surge of short-lived shrubs after fire, with the dominant species not present in early post-fire succession, presumably because of a lack of sufficient seed dispersal (Kirkpatrick and Bridle 2013). Critically, single fires in Tasmania can eliminate species and whole vegetation types, as well as destroy organic soil (Bridle and Kirkpatrick 1997). Coniferous and deciduous heath dominants can be permanently lost, in contrast to the mainland where most dominant tall heath species regenerate rapidly (Camac *et al.* 2013). An increase in dry lightning strikes in Tasmania since the mid-1990s (Marsden-Smedley 2009) has resulted in losses of fire-susceptible alpine vegetation. In alpine country, fire is the major vegetation conservation concern in Tasmania.

Some alpine communities, such as bogs and bolster heaths, appear to be remarkably stable over time. Clarke *et al.* (2015) found very little change in the floristic composition of bog vegetation in Kosciuszko National Park over 50 years, and attributed an increase in *Sphagnum cristatum* cover over that time to regeneration from fire in 1939, and the cessation of livestock grazing prior in 1958. Similarly, over a 20-year period, McDougall (2007) found little change in the distribution of pools and *Sphagnum* cover in a bog system on the Bogong High Plains. Some cushion/bolster heaths in Tasmania also show remarkable stability. Peat cores showed the continual presence of a cushion heath community at a site at Newdgate Pass for at least 850 years (Gibson and Kirkpatrick 1992). Community composition over time was largely determined by species accumulation rather than species replacement, with little detectable change in total percentage cover of the cushions. However, wind-driven cyclic succession occurs in striped sub-alpine mires on the western Central Plateau. The replacement of swale by ridge is deduced to have a cycle of 170–380 years (Morgan *et al.* 2010).

19.5 Threatening processes and conservation

Many of the threats to treeless alpine and sub-alpine vegetation in Australia are well-documented and monitored by land management agencies. However, many threats have been notoriously difficult to predict and therefore to manage proactively.

In the first edition of *Australian Vegetation*, Costin (1981, p. 374) summarised conservation issues as follows: 'Conflicts in the use of this land have been and are acute, largely because its uses for water harvesting, recreation and for nature conservation are mostly incompatible with grazing, commercial tourism and some hydro-electric developments'. Fire, grazing by cattle and road and trail infrastructure were seen as the greatest threats to protection of the vegetation. Two decades on and the threat from cattle grazing has eased over almost all of this vegetation, the threat from fire (including high fire frequency) and development remains, but the list of threats now includes an increasing range of tourism activities, invasive alien plants, feral animals and climate change, which could act to exacerbate the effects of other threats.

19.5.1 Mammals

Most of the ecological research in the high country in the twentieth century focused on the effects of grazing by sheep and cattle on vegetation (McDougall *et al.* 2012). The obvious negative impact of livestock on alpine vegetation, and the resistance to land-use change by graziers, motivated the establishment and long-term monitoring of livestock exclosures and reference plots. The oldest example of these are the exclosures established in the 1940s on the Bogong High Plains (Carr and Turner 1959). Monitoring plots were established in the Snowy Mountains to document vegetation change after livestock were removed to improve catchment yield (Wimbush and Costin 1979). In Tasmania, a livestock exclosure experiment was established in the 1960s on the Central Plateau (Bridle and Kirkpatrick 1999), and additional plots were established to measure recovery after livestock removal (Bridle *et al.* 2001). Improved understanding from these long-term research plots may have contributed to the removal of livestock from all areas except private land in the alpine and treeless sub-alpine vegetation of mainland Australia and Tasmania, although the stock were removed very late from the Bogong High Plains (2005), and are yet to be moved from the Tasmanian exclosure site. Since then, attention has been focused on feral herbivores such as horses, deer, pigs, goats and rabbits, which have all greatly increased in abundance in the last two decades and all have a detrimental impact on alpine environments.

The 2009 census of horses in the Australian Alps estimated the population at around 7500 but this had grown by 21.6% per annum since the previous census (Dawson 2009). Damage from trampling in wet areas has been reported (Worboys and Pulsford 2013). Pigs are common in some sub-alpine plains, where they dig for roots and tubers in forb-rich grassland. McDougall and Walsh (2002) found that species richness and cover were greatly reduced, and species composition changed, in pig-affected vegetation.

Rabbits now have a negligible effect on vegetation in the Tasmanian high country. It is anticipated that the eradication of rabbits from Macquarie Island will result in a rapid increase in both palatable megaherbs (Figure 19.3c) and *Poa* tussocks, with initial recovery proving promising (Scott and Kirkpatrick 2013; Whinam *et al.* 2014b). Less palatable species that are presently dominant in the landscape (such as *Agrostis magellanica* and *Acaena magellanica*) are likely to decline (Whinam *et al.* 2014b). Rabbits are also present and appear to be increasing in mainland alpine vegetation (S. Venn, *pers. obs.*). European hares in the Snowy Mountains appear to have little impact on tall alpine herbfield, but may affect certain plant species via selective grazing or by spreading seed of alien species (Green and Pickering 2013).

19.5.2 Alien plants and pathogens

Treeless areas and especially the alpine zone in the mountains of mainland Australia were once thought to be impervious to invasions by alien plants (McDougall *et al.* 2005). However, two species, *Acetosella vulgaris* and *Hypochaeris radicata*, have long been present in the alpine zones of mainland Australia, and are regarded as benign because, in the absence of disturbance, they occur as isolated, uncompetitive individuals. These species and many alien species that were known from treeless areas in the twentieth century were largely introduced to the high country during the grazing period of land use, and through the soil conservation activities that followed to repair the damage from grazing. Recent incursions by species from horticulture have shown the vulnerability of treeless vegetation to this threat. Unlike the early plant invaders, *Pilosella* spp. (*Hieracium sensu* Williams and Holland (2007)), *Leucanthemum vulgare* (Benson 2012) and *Cytisus scoparius* (Wearne and Morgan 2004) tend to dominate and displace native species when they become established.

Alien plant species currently at lower altitudes are predicted to invade higher altitudes (Pickering *et al.* 2004). Consistent with that prediction, a recent survey identified 25 new alien species on roadsides in Kosciuszko National Park over a five year period, suggesting a strong upward pressure from lowland areas (Williams *et al.* 2014). However, Petitpierre *et al.* (2016) found that precipitation rather than temperature was likely to be the more important determinant of future alien plant distributions in New South Wales mountains. So, the future distributions of some alien plants may contract rather than expand uphill. On the sub-Antarctic islands, bioturbation and nutrient addition by seals and penguins is associated with most exotic plant invasion (Scott and Kirkpatrick 2005). In Tasmania, the native herbivores eliminate alien plants before they become established (Whinam *et al.* 1994).

While there have been several alarms, there are no cases in which an association between dieback and an alien pathogen has been proven. The dieback of the keystone cushion species on Macquarie Island, *Azorella macquariensis*, appeared to be associated with the presence of a pathogen, although its expression may be secondary (Bergstrom *et al.* 2015), and the dieback not complete (Whinam *et al.* 2014a). In 1994, dead and dying plants were discovered in the Central Plateau Conservation Area of

Tasmania, including the Tasmanian endemic conifer, *Athrotaxis cupressoides*. An American variety of *Phytophthora* was subsequently isolated from plant material (Whinam *et al.* 2001). However, monitoring of sites suggested that it was a short-term disease infection followed by recovery of surviving plants and only localised impacts. In the spring/summer of 2012, mass dieback of *Nematolepis ovatifolia* was observed across the high altitude regions of the Snowy Mountains from Thredbo to the Jagungal Wildnerness Area. *Phytophthora cambivora*, the root-rot fungus, was isolated from two populations of *N. ovatifolia*, in which early symptoms of infection were rapidly followed by plant death. However, after a wet summer, many mildly affected plants recovered, and the suspicious symptoms were in some cases attributed to frost damage from early exposure in the previous winter season (Green 2016).

19.5.3 Climate change

Australian alpine ecosystems are vulnerable to global warming because the alpine and treeless high sub-alpine altitudinal zone is comparatively narrow, with limited scope for the biota to migrate. On the Bogong High Plains, precipitation has declined by 6% and growing season temperatures have increased by approximately 0.4°C since 1979 (Wahren *et al.* 2013). In the Kosciuszko region, both the depth and duration of the snowpack have declined over the past half century (Sánchez-Bayo and Green 2013). In Tasmania there has been warming and drying in the Central Plateau region (Calder and Kirkpatrick 2008), but no indication of warming in much of western Tasmania (Kirkpatrick *et al.* 2002).

Higher temperatures and reduced snow cover and duration are predicted to affect many ecological processes, especially those directly related to the timing and duration of snow cover (Venn *et al.* 2011). On the mainland there have been vegetation changes in snowpatches associated with a decline in snow duration and depth (Pickering *et al.* 2014); range expansions into the alpine/sub-alpine zones (Doherty *et al.* 2015); increases in species richness and species abundance at small scales (Venn *et al.* 2012); and increases in the abundance of native woody species (Wahren *et al.* 2013; Venn *et al.* 2014; Camac *et al.* 2015). At Cradle Mountain, Tasmania, dieback of *Nothofagus gunnii* has been associated with increasing and extreme temperatures (Visoiu and Whinam 2015) and increasing dryness and warmth is associated with dieback of *Eucalyptus gunnii* in sub-alpine vegetation on the eastern Central Plateau (Calder and Kirkpatrick 2008). More intense rainfall on Macquarie Island has been associated with changes in the abundance of herbs and mosses independent of rabbit grazing (Scott and Kirkpatrick 2013), and glacial retreat is allowing the expansion of vegetation on Heard Island (Scott 1990).

Systematic, long-term experimental and monitoring studies of climate change effects on Australian alpine vegetation have been a feature of ecological research in the Australian Alps over the last 10 years. The Global Observational Research Initiative in Alpine Environments (GLORIA) study in the Snowy Mountains has revealed a net increase of six species and a 12% increase in species richness between 2004 and 2011 (Venn *et al.* 2012). Shrub species showed the largest increases in cover at the lower elevation summits, whereas graminoid species showed large increases across all sites. The International Tundra Experiment (ITEX) on the Bogong High Plains revealed that 8 years of experimental warming of 1.2°C over the growing season induced flowering 2–10 days earlier, but some species responses were idiosyncratic, and some were unaffected (Jarrad *et al.* 2009; Hoffmann *et al.* 2010). Transplant experiments have shown that some species e.g. *Poa hiemata*, have an inherent capacity to adapt to higher temperatures, whereas others, such as *Craspedia lamicola*, had little capacity to adapt, and population-level changes to altered temperature regimes were accounted for by plastic changes (Byars and Hoffmann 2009). However, plant diversity did not decline during this period, and the floristic composition of vegetation between

2004 and 2010 was similar to that of the controls and within the spatial and temporal range of variation in nearby long-term monitoring plots (Wahren *et al.* 2013). The latter finding illustrates the importance of long-term monitoring for interpreting ecological change, so that the magnitude and direction of change in the shorter-term experiments such as ITEX can be seen in the context of longer-term historical variability (Williams *et al.* 2014).

19.5.4 Tourism

Direct and indirect threats from tourism are numerous and extensive (e.g. horse riding, skiing, mountain biking). Ski tourism impacts include snowmobile activities within and surrounding the resort areas, large water consumption and treatment within resort areas, snowmaking (Pickering and Buckley 2010) and weed invasions (Pickering *et al.* 2007), as well as direct removal of taller shrubby vegetation. Ski village development in alpine Tasmania is restrained, because of the unreliability of winter and spring snow. Damage from bushwalking in the western and central alpine zones of Tasmania is widespread, visually severe and growing, even where walker numbers are declining (Dixon *et al.* 2004), because of the low carrying capacity of unhardened tracks (Whinam and Chilcott 2003). In New South Wales, there is evidence that bushwalking and mountain biking can impact vegetation structure and soil compaction (Pickering *et al.* 2011). Bushwalkers are also frequent vectors of exotic plant seeds via their clothing (Mount and Pickering 2009), as are vehicles (Pickering and Hill 2007). In the sub-Antarctic, expeditioners have been shown to be the major vector of seed introduction, with 'hook and loop' closures identified as particularly effective transporters of seed (Whinam *et al.* 2005).

19.5.5 Management and conservation

Despite the growing range and complexity of threats in alpine and sub-alpine vegetation, management is feasible because the majority of the land is within protected areas. In mainland Australia the largest are Alpine National Park in Victoria, Kosciuszko National Park in New South Wales and Namadgi National Park in the ACT, covered by the Australian Alps National Parks Memorandum of Understanding. A large proportion of the Tasmanian alpine and sub-alpine treeless vegetation is in the Tasmanian Wilderness World Heritage Area, which includes the full range of Tasmanian communities and broad vegetation types. Ben Lomond National Park includes the largest alpine area in Tasmania. Tasmanian alpine vegetation and its component species have recognised World Heritage values under several criteria (Balmer *et al.* 2004). Both Macquarie Island and the Heard Island and the McDonald Islands (HIMI) group are also recognised as internationally significant through their listing as World Heritage Area properties. The HIMI Territory is managed by the Australian Government via the Australian Antarctic Division, while Macquarie Island is managed by the Tasmanian Parks and Wildlife Service.

The rarity of, and threats to, many species and communities in alpine, sub-alpine and sub-Antarctic areas are recognised in their listing as threatened under State and Commonwealth legislation. Most recently, mainland Australian snowpatch herbfield ecosystems were assessed as being eligible for listing as endangered using International Union for the Conservation of Nature (IUCN) Red List criteria (Williams *et al.* 2015). Feldmark is listed as an endangered ecological community under the New South Wales *Threatened Species Conservation Act 1995* and the *Caltha introloba* Herbland Community (equivalent to the short alpine herbfield) is listed under the Victorian *Flora and Fauna Guarantee Act 1988*. *Sphagnum* bog communities are listed as threatened in State legislation (Victoria and New South Wales) and in the Commonwealth *Environmental Protection and Biodiversity Conservation Act 1999* for their sensitivity to a changing climate.

With continued funding for scientific research, monitoring and appropriate management of Australia's alpine, sub-alpine and sub-Antarctic vegetation, there is hope that these areas will remain a significant and distinctive component of Australian vegetation, and to continue to be appreciated and valued by the people of Australia and the rest of the world.

Acknowledgements

The authors would like to thank Bob Parsons for comments on an early version of the manuscript and two anonymous referees for their helpful suggestions. We would like to thank David Smith, Australian Antarctic Division for providing spatial data for Heard and Macquarie Islands: http://data.aad.gov.au/aadc/metadata/metadata_redirect.cfm?md=/AMD/AU/heard_vegetation_gis; http://data.aad.gov.au/aadc/metadata/metadata_redirect.cfm?md=/AMD/AU/MIvegmap. Heard Island vegetation cover is from mapping by J. Scott and U. Harris (Heard Island: Baseline Data for Monitoring Long-term Change. Vegetation Mapping from Orthophotos). Macquarie Island vegetation types were derived from mapping by P. Selkirk, D. Adamson and U. Harris (1999; Macquarie Island Vegetation and Drainage Structure Data Set).

References

Adamson, D., Selkirk, P., Price, D. *et al.* (1996) Pleistocene uplift and palaeoenvironments of Macquarie Island: evidence from palaeobeaches and sedimentary deposits. *Papers and Proceedings of the Royal Society of Tasmania* 130, 25–32.

Ballantyne, M. and Pickering, C.M. (2015) Shrub facilitation is an important driver of alpine plant community diversity and functional composition. *Biodiversity and Conservation* 24, 1859–1875.

Balmer, J., Whinam, J., Kelman, J. *et al.* (2004) *A review of the floristic values of the Tasmanian Wilderness World Heritage Area. Nature Conservation Report 2004/3.* Department of Primary Industries, Water and Environment: Hobart.

Barlow, B.A. (1994) Phytogeography of the Australian region. In *Australian Vegetation*, ed. R.H. Groves. Melbourne: Cambridge University Press, pp. 3–35.

Barrow, M., Costin, A. and Lake, P. (1968) Cyclical changes in an Australian fjaeldmark community. *The Journal of Ecology* 56, 89–96.

Benson, J. (2012) Ox-eye daisy: an expanding weed on the tablelands. *Nature New South Wales* 56, 24–25.

Bergstrom, D.M. and Chown, S.L. (1999) Life at the front: history, ecology and change on southern ocean islands. *Trends in Ecology and Evolution* 14, 472–477.

Bergstrom, D. and Seppelt, R. (1988) The moss flora of Heard Island: revised checklist, annotations and phytogeographical considerations. *Lindbergia* 14, 184–190.

Bergstrom, D.M., Whinam, J. and Belbin, L. (2002) A classification of sub-Antarctic Heard Island vegetation. *Arctic, Antarctic, and Alpine Research* 34, 169–177.

Bergstrom, D.M., Bricher, P.K., Raymond, B. *et al.* (2015) Rapid collapse of a sub-Antarctic alpine ecosystem: the role of climate and pathogens. *Journal of Applied Ecology* 52, 774–783.

Bridle, K. and Kirkpatrick, J. (1997) Local environmental correlates of variability in the organic soils of moorland and alpine vegetation, Mt Sprent, Tasmania. *Australian Journal of Ecology* 22, 196–205.

Bridle, K. and Kirkpatrick, J.B. (1999) Comparative effects of stock and wild vertebrate herbivore grazing on treeless subalpine vegetation, eastern central plateau, Tasmania. *Australian Journal of Botany* 47, 817–834.

Bridle, K., Kirkpatrick, J., Cullen, P. and Shepherd, R. (2001) Recovery in alpine heath and grassland following burning and grazing, eastern Central Plateau, Tasmania, Australia. *Arctic, Antarctic, and Alpine Research* 33, 348–356.

Byars, S.G. and Hoffmann, A.A. (2009) Lack of strong local adaptation in the alpine forb *Craspedia lamicola* in southeastern Australia. *International Journal of Plant Sciences* 170, 906–917.

Calder, J.A. and Kirkpatrick, J.B. (2008) Climate change and other factors influencing the decline of the Tasmanian cider gum (*Eucalyptus gunnii*). *Australian Journal of Botany* 56, 684–692.

Camac, J.S., Williams, R.J., Wahren, C.-H. *et al.* (2013) Post-fire regeneration in alpine heathland: does fire severity matter? *Austral Ecology* 38, 199–207.

Camac, J.S., Williams, R.J., Wahren, C.-H. *et al.* (2015) Modeling rates of life form cover change in burned and unburned alpine heathland subject to experimental warming. *Oecologia* 178, 615–628.

Carr, S.G. and Turner, J. (1959) The ecology of the Bogong High Plains. II. Fencing experiments in grassland C. *Australian Journal of Botany* 7, 34–63.

Clarke, P.J., Keith, D.A., Vincent, B.E. and Letten, A.D. (2015) Post-grazing and post-fire vegetation dynamics: long-term changes in mountain bogs reveal community resilience. *Journal of Vegetation Science* 26, 278–290.

Clements, M.A., Mackenzie, A.M., Copson, G.R. *et al.* (2007) Biology and molecular phylogenetics of *Nematoceras sulcatum*, a second endemic orchid species from sub-Antarctic Macquarie Island. *Polar Biology* 30, 859–869.

Costin, A. (1981) Alpine and sub-alpine vegetation. In *Australian Vegetation*, ed. R.H. Groves. Melbourne: Cambridge University Press, pp. 361–376.

Costin, A.B., Gray, M., Totterdell, C.J. and Wimbush, D.J. (2000) *Kosciuszko Alpine Flora.* CSIRO: Melbourne.

Darwin, C. (1859) *On the Origin of Species by Means of Natural Selection.* Murray: London.

Dawson, M. (2009) Aerial survey of feral horses in the Australian Alps. (Australian Alps Liaison Committee) Available at: https://theaustralianalps.files.wordpress.com/2013/12/2009feralhorsealpssurvey.pdf (accessed 8 August 2016).

Dixon, G., Hawes, M. and McPherson, G. (2004) Monitoring and modelling walking track impacts in the Tasmanian Wilderness World Heritage Area, Australia. *Journal of Environmental Management* 71, 305–320.

Doherty, M., Wright, G. and McDougall, K.L. (2015) The flora of Kosciuszko National Park, New South Wales: summary and overview. *Cunninghamia* 15, 13–68.

Fiddes, S.L., Pezza, A.B. and Barras, V. (2015) A new perspective on Australian snow. *Atmospheric Science Letters* 16, 246–252.

Galen, C. and Stanton, M.L. (1995) Responses of snowbed plant species to changes in growing season length. *Ecology* 76, 1546–1557.

Gibson, N. (1990) The environments and primary production of cushion species at Mt Field and Mt Wellington, Tasmania. *Australian Journal of Botany* 38, 229–243.

Gibson, N. and Kirkpatrick, J.B. (1985) Vegetation associated with localised snow accumulation at Mount Field West, Tasmania. *Australian Journal of Ecology* 10, 91–99.

Gibson, N. and Kirkpatrick, J.B. (1992) Dynamics of a Tasmanian cushion heath community. *Journal of Vegetation Science* 3, 647–654.

Green, K. (2016) Dieback of *Nematolepis ovatifolia* (Rutaceae), an endemic shrub in the alpine-subalpine heaths of the Snowy Mountains, is facilitated by climate change. *Cunninghamia* 16, 1–9.
Green, K. and Pickering, C. (2013) Limited effect of hare grazing and short-term climatic variations on the most common alpine vegetation community in the Snowy Mountains, Australia. *Plant Ecology and Diversity* 6, 511–522.
Griffin, P.C. and Hoffmann, A.A. (2012) Mortality of Australian alpine grasses (*Poa* spp.) after drought: species differences and ecological patterns. *Journal of Plant Ecology* 5, 121–133.
Hennessy, K., Whetton, P., Walsh, K. *et al.* (2007) Climate change effects on snow conditions in mainland Australia and adaptation at ski resorts through snowmaking. *Climate Research* 35, 255–270.
Hoffmann, A.A., Camac, J.S., Williams, R.J. *et al.* (2010) Phenological changes in six Australian subalpine plants in response to experimental warming and year-to-year variation. *Journal of Ecology* 98, 927–937.
Hoyle, G.L., Steadman, K.J., Good, R.B. *et al.* (2015) Seed germination strategies: an evolutionary trajectory independent of vegetative functional traits. *Frontiers in Plant Science* 6, 731.
Hoyle, G.L., Venn, S.E., Steadman, K.J. *et al.* (2013) Soil warming increases plant species richness but decreases germination from the alpine soil seed bank. *Global Change Biology* 19, 1549–1561.
Jarrad, F.C., Wahren, C.-H., Williams, R.J. and Burgman, M.A. (2009) Impacts of experimental warming and fire on phenology of subalpine open-heath species. *Australian Journal of Botany* 56, 617–629.
Jarrad, F.C., Walshe, T., Chee, Y.E. and Burgman, M.A. (2012) A test of biotic interactions among two alpine plant species in Australia. *Austral Ecology* 37, 90–96.
Jenkin, J.F. (1972) Studies on plant growth in a sub-Antarctic environment. PhD Thesis, University of Melbourne, Melbourne.
Jordan, G., Harrison, P., Worth, J. *et al.* (2016) Palaeoendemic plants provide evidence for persistence of open, well-watered vegetation since the Cretaceous. *Global Ecology and Biogeography* 25, 127–140.
Kirkpatrick, J.B. (1986) Tasmanian alpine biogeography and ecology and interpretation of the past. In *Flora and Fauna of Alpine Australasia: Ages and Origins*, ed. B.A. Barlow. CSIRO Publishing: Melbourne, pp. 229–242.
Kirkpatrick, J.B. (1997) *Alpine Tasmania: An Illustrated Guide to the Flora and Vegetation.* Oxford University Press: Melbourne.
Kirkpatrick, J.B. (2002) Factors influencing the spatial restriction of vascular plant species in the alpine archipelagos of Australia. In *Mountain Biodiversity: A Global Assessment*, ed. C. Körner and E. Spehn. Parthenon Publishing: London, pp. 155–164.
Kirkpatrick, J.B. (2016) Pattern and process in riparian vegetation of the treeless high country of Australia. In *Vegetation of Australian Riverine Landscapes*, ed. S. Capon, C. James and M. Reid. CSIRO Publishing: Collingwood, pp. 201–220.
Kirkpatrick, J.B. and Bridle, K.L. (1998) Environmental relationships of floristic variation in the alpine vegetation of southeast Australia. *Journal of Vegetation Science* 9, 251–260.
Kirkpatrick, J.B. and Bridle, K.L. (2013) Natural and cultural histories of fire differ between Tasmanian and mainland Australian alpine vegetation. *Australian Journal of Botany* 61, 465–474.
Kirkpatrick, J.B. and Brown, M. (1984) A numerical analysis of Tasmanian higher plant endemism. *Botanical Journal of the Linnean Society* 88, 165–183.

Kirkpatrick, J.B., Bridle, K. and Lynch, A. (2002) Changes in alpine vegetation related to geomorphological processes and climatic change on Hill One, Southern Range, Tasmania, 1989–2000. *Australian Journal of Botany* 50, 753–759.
Kirkpatrick, J.B., Green, K., Bridle, K.L. and Venn, S.E. (2014) Patterns of variation in Australian alpine soils and their relationships to parent material, vegetation formation, climate and topography. *Catena* 121, 186–194.
Kitchener, A. and Harris, S. (2013) *From Forest to Fjaeldmark: Descriptions of Tasmania's Vegetation*, 2nd edition. Department of Primary Industries, Water and Environment: Tasmania.
Körner, C. (2002) Alpine plants: stressed or adapted? In *Physiological Plant Ecology: 39th Symposium of the British Ecological Society*. Cambridge University Press: Cambridge, p. 297.
Körner, C. and Paulsen, J. (2004) A world-wide study of high altitude treeline temperatures. *Journal of Biogeography* 31, 713–732.
Mark, A. (1970) Floral initiation and development in New Zealand alpine plants. *New Zealand Journal of Botany* 8, 67–75.
Marsden-Smedley, J.B. (2009) *Planned burning in Tasmania: operational guidelines and review of current knowledge,* ed. PaWS Fire Management Section. Department of Primary Industries, Water and the Environment: Hobart.
McDougall, K.L. (2007) Grazing and fire in two subalpine peatlands. *Australian Journal of Botany* 55, 42–47.
McDougall, K.L. and Walsh, N.G. (2002) The flora of Nungar Plain, a treeless sub-alpine frost hollow in Kosciuszko National Park. *Cunninghamia* 7, 601–610.
McDougall, K.L. and Walsh, N.G. (2007) Treeless vegetation of the Australian Alps. *Cunninghamia* 10, 1–57.
McDougall, K.L., Morgan, J.W., Walsh, N.G. and Williams, R.J. (2005) Plant invasions in treeless vegetation of the Australian Alps. *Perspectives in Plant Ecology, Evolution and Systematics* 7, 159–171.
McDougall, K.L., Williams, R.J. and Wahren, C.-H. (2012) The ecology of domestic grazing in the Australian Alps. In *Grazing Ecology: Vegetation and Soil Impact*, ed. J. Arevalo. Nova Science Publishers Inc.: New York, pp. 27–40.
McDougall, K.L., Walsh, N.G. and Wright, G.T. (2015) Recovery of treeless subalpine vegetation in Kosciuszko National Park after the landscape-scale fire of 2003. *Australian Journal of Botany* 63, 597–607.
McLoughlin, S. (2001) The breakup history of Gondwana and its impact on pre-Cenozoic floristic provincialism. *Australian Journal of Botany* 49, 271–300.
McPhail, D. and Kirkpatrick, J. (2015) Patch mosaic cyclic succession associated with the growth and senescence of an alpine shrub. *Australian Journal of Botany* 64, 111–119.
Meurk, C.D. (1978) Phytomass and primary productivity in central Otago, New Zealand. *New Zealand Journal of Botany* 1, 27–50.
Morgan, J.W. (2004) Drought-related dieback in four subalpine shrub species, Bogong High Plains, Victoria. *Cunninghamia* 8, 326–330.
Morgan, S., Kirkpatrick, J.B. and di Folco, M.-B. (2010) Wind-controlled linear patterning and cyclic succession in Tasmanian Sphagnum mires. *Journal of Ecology* 98, 583–591.
Mount, A. and Pickering, C.M. (2009) Testing the capacity of clothing to act as a vector for non-native seed in protected areas. *Journal of Environmental Management* 91, 168–179.
Müller, R.D., Flament, N., Matthews, K.J. *et al.* (2016) Formation of Australian continental margin highlands driven by plate–mantle interaction. *Earth and Planetary Science Letters* 441, 60–70.
Petitpierre, B., MacDougall, K., Seipel, T. *et al.* (2016) Will climate change increase the risk of plant invasions into mountains? *Ecological Applications* 26, 530–544.

Pickering, C.M. and Buckley, R.C. (2010) Climate response by the ski industry: the shortcomings of snowmaking for Australian resorts. *Ambio* 39, 430–438.
Pickering, C.M. and Hill, W. (2007) Roadside weeds of the snowy mountains, Australia. *Mountain Research and Development* 27, 359–367.
Pickering, C.M., Good, R. and Green, K. (2004) *Potential Effects of Global Warming on the Biota of the Australian Alps.* Australian Greenhouse Office: Canberra.
Pickering, C.M., Bear, R. and Hill, W. (2007) Indirect impacts of nature based tourism and recreation: the association between infrastructure and the diversity of exotic plants in Kosciuszko National Park, Australia. *Journal of Ecotourism* 6, 146–157.
Pickering, C.M., Rossi, S. and Barros, A. (2011) Assessing the impacts of mountain biking and hiking on subalpine grassland in Australia using an experimental protocol. *Journal of Environmental Management* 92, 3049–3057.
Pickering, C.M., Green, K., Barros, A.A. and Venn, S. (2014) A resurvey of late-lying snowpatches reveals changes in both species and functional composition across snowmelt zones. *Alpine Botany* 124, 93–103.
Roberts, C., Kirkpatrick, J. and McQuillan, P. (2011) Tasmanian lentic wetland lawns are maintained by grazing rather than inundation. *Austral Ecology* 36, 303–309.
Ruddell, A. (2006) An inventory of present glaciers on Heard Island and their historical variation. *Heard Island: Southern Ocean Sentinel*, 28–51.
Sánchez-Bayo, F. and Green, K. (2013) Australian snowpack disappearing under the influence of global warming and solar activity. *Arctic, Antarctic, and Alpine Research* 45, 107–118.
Sanmartín, I., Wanntorp, L. and Winkworth, R.C. (2007) West Wind Drift revisited: testing for directional dispersal in the southern hemisphere using event-based tree fitting. *Journal of Biogeography* 34, 398–416.
Scott, J. (1990) Changes in vegetation on Heard Island 1947–1987. In *Antarctic Ecosystems: Ecological Change and Conservation*, ed. R. Kerry and G. Hempel. Springer-Verlag: Berlin, pp. 61–76.
Scott, J. and Kirkpatrick, J. (2005) Changes in sub-Antarctic Heard Island vegetation at sites occupied by *Poa annua*, 1987–2000. *Arctic, Antarctic, and Alpine Research* 37, 366–371.
Scott, J. and Kirkpatrick, J. (2013) Changes in the cover of plant species associated with climate change and grazing pressure on the Macquarie Island coastal slopes, 1980–2009. *Polar Biology* 36, 127–136.
Scott, J. and Ryan, U. (2004) Heard Island vegetation mapping from orthophotos derived from non-metric photography. (Australian Antarctic Data Centre–SnoWhite Metadata) Available at: http://www.aad.gov.au/default.asp (accessed 8 August 2016)
Selkirk, P.M., Seppelt, R.D. and Selkirk, D.R. (1990) *Sub-Antarctic Macquarie Island: Environment and Biology.* Cambridge University Press: Cambridge.
Sommerville, K.D., Martyn, A.J. and Offord, C.A. (2013) Can seed characteristics or species distribution be used to predict the stratification requirements of herbs in the Australian Alps? *Botanical Journal of the Linnean Society* 172, 187–204.
Specht, R. (1970) Vegetation. In *The Australian Environment*, ed. G.W. Leeper. CSIRO Publishing and Melbourne University Press: Melbourne, pp. 44–67.
Thomas, N., Bruhl, J.J., Ford, A. and Weston, P.H. (2014) Molecular dating of Winteraceae reveals a complex biogeographical history involving both ancient Gondwanan vicariance and long-distance dispersal. *Journal of Biogeography* 41, 894–904.
Venn, S.E. and Morgan, J.W. (2007) Phytomass and phenology of three alpine snowpatch species across a natural snowmelt gradient. *Australian Journal of Botany* 55, 450–456.

Venn, S.E. and Morgan, J.W. (2009a) Germination characteristics of Mountains Celery *Aciphylla glacialis* (F. Muell.) Benth. (Apiaceae). *The Victorian Naturalist* 126, 4–12.

Venn, S.E. and Morgan, J.W. (2009b) Patterns in alpine seedling emergence and establishment across a stress gradient of mountain summits in south-eastern Australia. *Plant Ecology and Diversity* 2, 5–6.

Venn, S.E., Morgan, J.W. and Green, P.T. (2009) Do facilitative interactions with neighboring plants assist the growth of seedlings at high altitudes in alpine Australia? *Arctic, Antarctic and Alpine Research* 41, 381–387.

Venn, S.E., Green, K., Pickering, C.M. and Morgan, J.W. (2011) Using plant functional traits to explain community composition across a strong environmental filter in Australian alpine snowpatches. *Plant Ecology* 212, 1491–1499.

Venn, S.E., Pickering, C.M. and Green, K. (2012) Short-term variation in species richness across an altitudinal gradient of alpine summits. *Biodiversity and Conservation* 21, 3157–3186.

Venn, S.E., Morgan, J.W. and Lord, J.M. (2013) Foliar freezing resistance of Australian alpine plants over the growing season. *Austral Ecology* 38, 152–161.

Venn, S.E., Pickering, C. and Green, K. (2014) Spatial and temporal functional changes in alpine summit vegetation are driven by increases in shrubs and graminoids. *AoB Plants* 6, plu008.

Visoiu, M. and Whinam, J. (2015) Extreme weather conditions correspond with localised vegetation death at Cradle Mountain, Tasmania. *Ecological Management and Restoration* 16, 76–78.

Wahren, C-.H., Camac, J., Jarrad, F. *et al.* (2013) Experimental warming and long-term vegetation dynamics in an alpine heathland. *Australian Journal of Botany* 61, 36–51.

Wahren, C.-H., Papst, W.A. and Williams, R.J. (2001) Early post-fire regeneration in subalpine heathland and grassland in the Victorian Alpine National Park, south-eastern Australia. *Austral Ecology* 26, 670–679.

Walsh, N.G. and McDougall, K.L. (2004) Progress in the recovery of the flora of treeless subalpine vegetation in Kosciuszko National Park after the 2003 fires. *Cunninghamia* 8, 439–452.

Wearne, L.J. and Morgan, J.W. (2004) Community-level changes in Australian subalpine vegetation following invasion by the non-native shrub Cytisus scoparius. *Journal of Vegetation Science* 15, 595–604.

Whinam, J. and Chilcott, N.M. (2003) Impacts after four years of experimental trampling on alpine/sub-alpine environments in western Tasmania. *Journal of Environmental Management* 67, 339–351.

Whinam, J., Cannell, E., Kirkpatrick, J. and Comfort, M. (1994) Studies on the potential impact of recreational horseriding on some alpine environments of the Central Plateau, Tasmania. *Journal of Environmental Management* 40, 103–117.

Whinam, J., Chilcott, N. and Rudman, T. (2001) Impacts of dieback at Pine Lake, Tasmania. *Papers and Proceedings of the Royal Society of Tasmania* 135, 41–50.

Whinam, J., Selkirk, P., Downing, A. and Hull, B. (2004) Return of the megaherbs: plant colonisation of derelict ANARE station buildings on sub-Antarctic Heard Island. *Polar Record* 40, 235–243.

Whinam, J., Chilcott, N. and Bergstrom, D. (2005) Sub-Antarctic hitchhikers: expeditioners as vectors for the introduction of alien organisms. *Biological Conservation* 121, 207–219.

Whinam, J., Abdul-Rahman, J., Visoiu, M. *et al.* (2014a) Spatial and temporal variation in damage and dieback in a threatened sub-Antarctic cushion species. *Australian Journal of Botany* 62, 10–21.

Whinam, J., Fitzgerald, N., Visoiu, M. and Copson, G. (2014b) Thirty years of vegetation dynamics in response to a fluctuating rabbit population on sub-Antarctic Macquarie Island. *Ecological Management and Restoration* 15, 41–51.

Williams, N. and Holland, K. (2007) The ecology and invasion history of hawkweeds (*Hieracium species*) in Australia. *Plant Protection Quarterly* 22, 76–80.

Williams, R., Wahren, C.H., Stott, K.A.J. *et al.* (2015) An International Union for the Conservation of Nature Red List ecosystems risk assessment for alpine snow patch herb-fields, south-eastern Australia. *Austral Ecology* 40, 433–443.

Williams, R.J. and Ashton, D.H. (1987) The composition, structure and distribution of heath-land and grassland communities in the sub-alpine tract of the Bogong High Plains, Victoria. *Australian Journal of Ecology* 12, 137–140.

Williams, R.J. and Ashton, D.H. (1988) Cyclical patterns of regeneration in the subalpine heathland communities on the Bogong High Plains, Victoria. *Australian Journal of Botany* 36, 605–619.

Williams, R.J., Papst, W., McDougall, K. *et al.* (2014) Alpine ecosystems. In *Biodiversity and Environmental Change: Monitoring, Challenges and Direction*, ed. E. Burns, D. Lindenmayer, A. Lowe and N. Thurgate. CSIRO Publishing: Collingwood, pp. 167–212.

Williams, R.J., Wahren, C.-H., Bradstock, R.A. and Müller, W.J. (2006) Does alpine grazing reduce blazing? A landscape test of a widely held hypothesis. *Austral Ecology* 31, 925–936.

Wimbush, D. and Costin, A.B. (1979) Trends in vegetation at Kosciusko. 1. Grazing trials in the subalpine zone, 1957–1971. *Australian Journal of Botany* 27, 741–787.

Worboys, G. and Pulsford, I. (2013) Observations of Pest Horse Impacts in the Australian Alps. Available at: www.mountains-wcpa.org (accessed 8 August 2016).

20 Wetland Vegetation of Inland Australia

JANE A. CATFORD, JANE ROBERTS, SAMANTHA J. CAPON, RAY H. FROEND, SARAS M. WINDECKER AND MICHAEL M. DOUGLAS

Wetlands fed by rainwater, surface flows and groundwater occur throughout Australia, even in arid areas. This chapter focuses on temporary wetlands and permanently wet systems that are dominated by non-woody macrophytes. We use nine case studies that span Australia to illustrate their biogeography, dynamics and key threats of Australian Wetlands. The type and distribution of wetland vegetation, from the annually flooded wetlands of northern Australia, to saline lakes of arid and semi-arid Australia, to groundwater-dependent systems of the southwest, to bogs and fens of the Alps and Tasmania, reflects hydrology, climate and geomorphology. Wetland plants have developed a range of adaptations and life histories to tolerate the dynamic water regimes characteristic of Australian wetlands, and can be grouped into seven categories that reflect these adaptations. Waterbirds and water can connect spatially isolated systems, and seedbanks that last for decades allow species to disperse through time. The water regime is a strong driver of species composition and abundance, thus hydrological modification through water extraction, flow regulation or reductions in rainfall is a significant threat to wetland flora, and arguably the principal threat for Australian wetland vegetation. The displacement of native macrophytes by exotic and terrestrial species is both a symptom and cause of ecological change, with exotic plants often being better adapted to modified flooding and fire regimes, livestock grazing and eutrophication than natives. Introduced livestock and feral fauna eat, trample and uproot native plants, and degrade their habitat. These types of threats are expected to intensify, increasing the challenge for wetland management and policy.

20.1 Introduction

Wetlands are among the most variable and productive of Earth's ecosystems. They are highly valuable to humans and of crucial importance for ecosystem health. Being at the interface of land and water, wetlands are home and host to numerous taxa, from phytoplankton to macroinvertebrates, to birds and mammals. Wetlands occur throughout Australia, varying from permanently wet to almost permanently dry. Fed by rainwater, rivers, surface flows and groundwater, the boundaries of many wetlands shift seasonally, or in response to longer climatic cycles, as inundation varies. Waterbirds and water can connect seemingly isolated systems, allowing plants and their propagules to disperse widely. The diversity of these systems is mirrored by the diversity of the flora that inhabits them.

This chapter explores the types of wetlands that occur in Australia, using case studies to illustrate

Australian Vegetation, Third Edition, ed. David A. Keith. Published by Cambridge University Press.

Citation: Catford, J.A. *et al.* (2017). Wetland Vegetation of Inland Australia. In D. A. Keith, ed., *Australian Vegetation*, 3rd edn. Cambridge: Cambridge University Press, pp. 490–515.

their biogeography and dynamics, and the key drivers that shape their vegetation. Over half of Australia's wetlands have been destroyed since European settlement (Bennett 1997); we outline the main drivers of this loss, as well as some management and policy initiatives designed to arrest and ameliorate it. We provide an overview of the major factors and processes affecting wetland vegetation, and give examples of wetland plant adaptations.

We focus on temporary and permanent wetlands that are dominated by non-woody macrophytes (or hydrophytes, i.e. plants adapted to living in water, saturated soil or very moist soil) and can be characterised by water regimes involving periodic or permanent inundation or waterlogging. The high variability of rainfall and runoff in Australia means that some Australian wetlands may only flood once every several years, and possibly less often in ephemeral arid zone river systems (Capon 2003). Wetlands fed by groundwater are buffered against this variability but, being usually quite small and relatively few in number, they account for only a small percentage of Australia's wetland area. Although much of the material discussed in this chapter can apply to vegetation in flowing waters, we focus on vegetation of lentic (standing or slow flowing water) systems.

20.2 Drivers of wetland vegetation

Numerous abiotic, biotic and human factors affect the environmental conditions, dispersal opportunities and biotic interactions that shape wetland vegetation (Boulton *et al.* 2014). The influence of these interacting factors determines the biogeography and ecology of wetland vegetation (Figure 20.1).

Hydrology is the key force structuring wetland and riparian vegetation. Hydrology refers to the distribution and quantity of water in the landscape. Hydrology at the wetland level is characterised by the water regime, which encapsulates the duration, frequency, magnitude, timing (seasonality), rate of change and variability of wetting and drying. Wetlands may experience extremes in physiochemical conditions as a result of changing flooding conditions. Because of different hydrological conditions associated with wetland bathymetry (underwater topography), wetland vegetation typically forms zones (or mosaics) demarcated by subtle changes in water depth or inundation duration (Figure 20.2). The species that occur in each zone usually have physiological, morphological or life-history adaptations suited to those particular hydrological and environmental conditions (Brock and Casanova 1997).

Wetland hydrology is integrally linked with geomorphology (the physical form of the catchment, wetland and associated waterways) and hydraulics (flow dynamics and velocity), and these three factors interact closely to affect other key drivers. Together with water quality (including nutrient fluxes, sediment loads and ionic composition) and soil and sediment characteristics (lithology), these factors determine the key environmental conditions affecting wetland vegetation (Figure 20.1). Plants and animals may affect the abiotic characteristics of a site by altering its physical or chemical characteristics. For example, fallen trees affect hydraulics and channel form, and benthivorous fish disturb the sediment, uprooting plants and increasing turbidity.

Wetland biota may tolerate changing hydrological conditions, but many complete their life cycles during periods in which environmental conditions are suitable. The ability to colonise, grow and reproduce quickly during such times, and persist in a dormant state when conditions are unfavourable, is particularly important under boom-and-bust conditions (Bunn *et al.* 2006). Dispersal in time and space is therefore crucial for species persistence. Plants can effectively disperse through time by storing propagules (or diaspores: seeds, spores and vegetative fragments that can develop into an individual plant) in seed banks in soil, litter and plant canopies. In desert wetlands that may only be inundated a few times per century, seeds must be able to survive intervening dry periods, and germinate when appropriate flooding conditions arise (Brock *et al.* 2003). Like terrestrial plants, wetland plants disperse by a range of mechanisms, including via wind, animal and self-propulsion, and many also spread clonally through above-ground stolons and below-ground

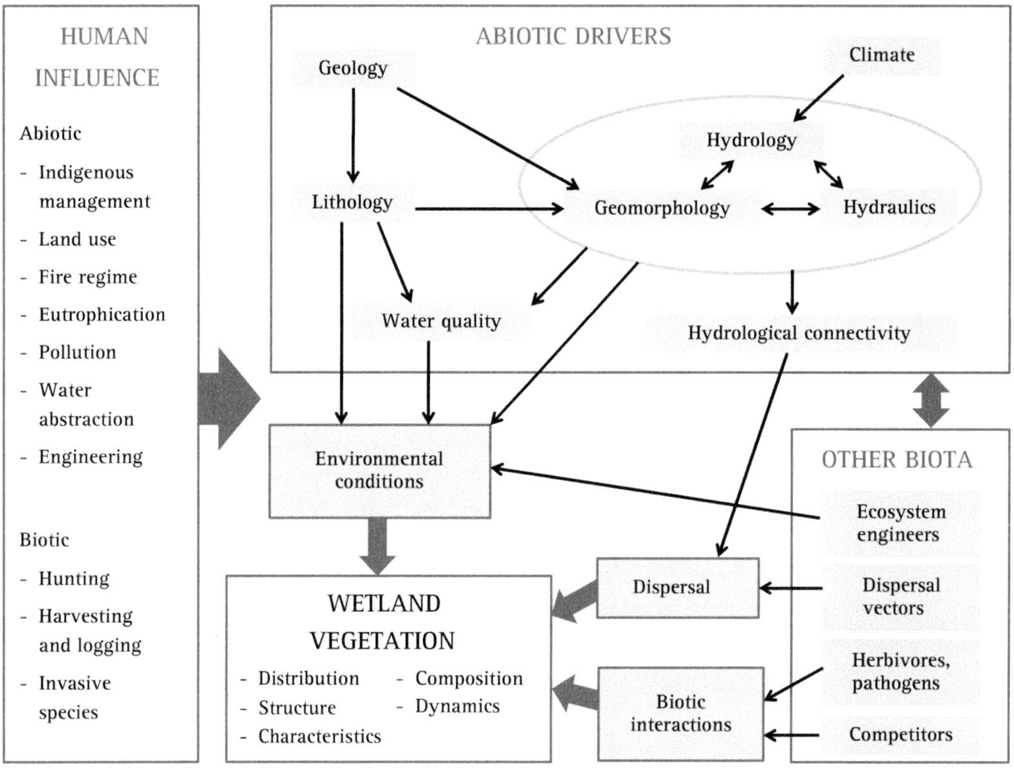

Figure 20.1 Key abiotic, biotic and human factors that shape the biogeography and ecology of wetland vegetation in Australia. Arrows indicate major causal relationships (e.g. climate affects hydrology; lithology affects geomorphology; hydrology, geomorphology and hydraulics collectively affect water quality and hydrological connectivity). Feedbacks (not shown to avoid clutter) occur among and within the abiotic and biotic drivers, between vegetation and abiotic drivers (e.g. vegetation can modify hydraulics, water quality, microclimates) and between external factors (e.g. bushfires, land-use practices) and the abiotic and biotic components of wetlands. Humans affect wetland vegetation directly and indirectly by altering environmental conditions, dispersal opportunities and biotic interactions.

rhizomes. However, a large fraction of wetland plants use water as a means of dispersal (hydrochory), and some may be dispersed by fruit-eating fish (ichthyochory; though there appears to be no evidence of ichthyochory in Australia: Jane Roberts, pers. comm. 2015) (Catford and Jansson 2014). Water-borne dispersal enables propagules to reach sites that are hydrologically connected and likely to have suitable environmental conditions. Hydrological connectivity refers to permanent and temporary connections among wetlands and other (mostly aquatic) systems; it includes lateral, longitudinal (in the cases of riverine wetlands) and vertical (surface to groundwater) connections (Figure 20.1).

Waterbirds are effective in dispersing propagules over long distances to hydrologically disconnected systems. Such long-distance dispersal of many wetland plant species facilitates species and population mixing among wetlands, and contributes to the wide geographic distributions of some species, spanning continents and beyond. Wetlands that are isolated, such as small discharge springs fed by groundwater,

Figure 20.2 Range of common Australian wetland plant species highlighting the diversity of wetland plant growth forms: (a) *Ludwigia peploides* subsp. *montevidensis* and *Myriophyllum* species; (b) outer, darker band of *Eleocharis acuta* and inner band of *Pseudoraphis spinescens* illustrate characteristic zones of wetland vegetation; (c) *Typha orientalis*; (d) exotic *Nymphaea mexicana*, Jerrabomberra Wetlands, ACT; (e) floating *Azolla pinnata* with submerged exotic *Egeria densa*, Euston Lakes, NSW; (f) emergent form of exotic *Sagittaria platyphylla* with an inset showing its submerged form, which is just starting to bolt into the emergent form.
Photos: Jane Catford (a, b, c, f: all taken in River Murray wetlands in Victoria and NSW); Jane Roberts (d, e). (A black and white version of this figure will appear in some formats. For the colour version, please refer to the plate section.)

offer fewer opportunities for long-distance dispersal and levels of endemism are subsequently higher than in more connected systems, like floodplain wetlands.

Like any vegetation, wetland vegetation is affected by interactions with other trophic levels and with other types of vegetation. Terrestrial vegetation may compete with (or facilitate) wetland vegetation under drier conditions, and herbivores may modify the abundance of certain species. As discussed below, humans affect wetland vegetation directly and indirectly by altering environmental conditions, dispersal opportunities and biotic interactions. Owing to the interplay of so many factors, the dynamics of wetlands can be quite diverse, yet there are commonalities as the case studies below illustrate.

20.3 Ecology and floristics of wetland plants

20.3.1 Adaptations of wetland plants

Wetland plants have adapted to specific flooding and drying cycles, and this is particularly evident in regeneration and dispersal.

Germination and recruitment directly affects plant community composition. Water regimes can influence plant phenology (the timing of periodic events, such as flowering and seed set) by providing cues for germination and dispersal, as well as suitable conditions for seedling growth (Pettit and Froend 2001). Flooding can break dormancy by causing a sudden increase in salinity after drought in arid or semi-arid systems (Nielsen *et al.* 2007).

Although not all wetland plants disperse by water and hydrochorous species rarely disperse exclusively by water, water is important for dispersal in two major ways – as a vector and as a trigger for dispersal. Attributes of propagules that facilitate hydrochory include: buoyancy; corky or spongy tissue of low density; waxy, cuticularised epidermis; and surfaces with furrows, pits or hairs that can enhance floating ability (Catford and Jansson 2014). In climates with predictable seasonal patterns, some riparian plants time the release of propagules to coincide with a particular phase of river hydrographs. This adaptation increases the likelihood that propagules will be transported and deposited on recently flooded soils, providing conditions favourable for germination and recruitment. If the seasonal flow patterns are changed, for example by river regulation, native plants may no longer release propagules at optimal times. The summer–autumn timing of peak flows in regulated rivers in Victoria coincides with the seed release of exotic species, whereas native riparian species release seeds in winter-spring when flows were historically at their highest (Greet *et al.* 2012).

20.3.2 Functional types of wetland plants

Macrophytes can be classified into groups based on their adaptations to, and tolerance of, flooding. The most commonly used wetland plant classification scheme in Australia is that developed by Brock and Casanova (1997). This hierarchical scheme classifies wetland plant species into three main groups, Submerged, Amphibious and Terrestrial (with seven subgroups) based on their life history and response to flooding (Table 20.1). Casanova (2011) updated the scheme, but we prefer the relative simplicity of the original one.

The terrestrial group comprises two subgroups, terrestrial-dry and terrestrial-damp. Terrestrial-dry are unable to tolerate flooding, so occur above the high water mark (terrestrial-dry species, Table 20.1; Figure 20.2b). Terrestrial-damp species are found around wetlands, reflecting their need for damp soil and their tolerance of infrequent inundation (see also species described in Good *et al.*, Chapter 21, this volume).

Plants in the Amphibious group are either tolerators (plants that tolerate fluctuations in water level) or responders (plants that respond to them). Amphibious tolerator-emergents are often the

Table 20.1 **Hierarchical classification scheme used to categorise wetland species based on their response to water regime. Classification was devised by Brock and Casanova (1997) based on field and germination trials. [a]Exotic taxa; [b]genera that contain both exotic and native species**

Primary category	Secondary category	Description	Example species
Terrestrial	Dry species: Tdr	Species which germinate, grow and reproduce where there is no surface water and the water table is below the soil surface	*Hypochaeris* spp.[a], *Microlaena stipoides* var. *stipoides*, *Austrodanthonia* spp., *Euchiton* spp., *Mollugo verticillata*, *Lactuca* spp.[a], *Chenopodium* spp.
Terrestrial	Damp species: Tda	Species which germinate, grow and reproduce on saturated soil	*Gratiola peruviana*, *Persicaria prostrata*, *Rorippa palustris*, *Rumex crispus*[a], *Medicago* spp.[a], *Alternanthera denticulata*
Amphibious fluctuation tolerators	Emergent species: ATe	Species which germinate in damp or flooded conditions, which tolerate variation in water level, and which grow with their basal portions under water and reproduce out of water	*Juncus ingens*, *Phragmites australis*, *Eleocharis* spp., *Cyperus eragrostis*[a], *Carex* spp.[b], *Persicaria hydropiper*, *Typha* spp.[b]
Amphibious fluctuation tolerators	Low-growing species: ATl	Species which germinate in damp or flooded conditions, which tolerate variation in water level, which are low-growing and tolerate complete submersion when water levels rise	*Stellaria angustifolia*, *Centipeda cunninghamii*, *Isotoma fluviatilis* subsp. *australis*, *Callitriche stagnalis*[a], *Rorippa nasturtium-aquaticum*[a], *Hydrocotyle verticillata*
Amphibious fluctuation responders	Morphologically plastic species: ARp	Species which germinate in flooded conditions, grow in both flooded and damp conditions, reproduce above the surface of the water, and which have morphological plasticity (e.g. heterophylly) in response to water-level variation	*Myriophyllum* spp.[b], *Potamogeton* spp., *Sagittaria platyphylla*[a], *Crassula helmsii*, *Isolepis* spp.
Amphibious fluctuation responders	Species with floating leaves: ARf	Species which germinate in flooded conditions, grow in both flooded and damp conditions, reproduce above the surface of the water, and which have floating leaves when inundated	*Nymphoides crenata*, *Azolla* spp., *Lemna disperma*, *Ricciocarpus natans*, *Ottelia ovalifolia*, *Eichhornia crassipes*[a], *Salvinia molesta*[a]
Submerged: S	–	Species which germinate, grow and reproduce underwater	*Chara* spp., *Nitella* spp., *Vallisneria* spp., *Isoetes muelleri*, *Ruppia* spp., *Egeria densa*[a]

dominant and most conspicuous type of wetland vegetation (Figure 20.2b, c). These species are able to tolerate fluctuations in water levels by ensuring that a sufficient portion (roughly two-thirds of their above-ground biomass) is above the water, enabling respiration. Amphibious tolerator low-growing species generally contribute a relatively small fraction to the plant biomass in a wetland.

There are two types of amphibious responders: plastic and floating (Figure 20.2a, d, e, f). Plastic species change their morphology and form, and may have heterophyllous leaves, which differ whether submerged or emergent (e.g. *Sagittaria platyphylla*, Figure 20.2f). Waterlilies are a classic example of floating responders, as their long internodes or petioles allow leaves to float to the surface of the water even if its depth fluctuates (Brock 2003). Floating responders includes some of the smallest wetland plants (*Spirodela*, *Lemna* and *Wolffia* species; Figure 20.2e) and invasive exotic plants (Figure 20.2d), including *Eichhornia crassipes* (water hyacinth) and *Salvinia molesta* (salvinia, see Leishman *et al.*, Chapter 9, this volume).

Submerged plants grow below the surface of the water and have adaptations that enable them to germinate, grow and reproduce entirely underwater. Submerged plants cannot tolerate exposure to air. As well as angiosperms like *Vallisneria* species, many of the submerged taxa are charophytes (freshwater green algae), e.g. *Chara* and *Nitella* species.

20.4 Trends in Australian wetland vegetation science

Wetland plant ecology and wetland vegetation science in Australia were in their infancy prior to the 1970s, mirroring global trends (Figure 20.3). Early study of Australian wetland plants was principally concerned with species and habitat descriptions (including taxonomy), and aquatic weed management in irrigation channels.

The main emphasis on emergent macrophytes in the 1970s and 1980s reflected scientific and management priorities at that time. Emergent macrophytes were recognised as drivers of lake ecology and appealed to management interests because of their use as a 'green' alternative to treating wastewater in constructed wetlands (Brix 1987). Australian research, on *Eleocharis sphacelata* and *Typha domingensis*, for example, also contributed to international interest in macrophyte adaptations to anoxic and hypoxic substrates (e.g. Sorrell *et al.* 1997).

Research on salinity and water regime dominated wetland vegetation research from the 1990s onwards. Salinity experiments in the 1990s built on research from the 1970s and 1980s by asking more complex questions, such as identifying tolerances of macrophytes using concentration gradients in growth trials. Experiments in the 2000s considered interactions between salinity and drying, explored geographic variability and interpreted findings in terms of biodiversity (Goodman *et al.* 2011).

Investigations into water regimes have progressed from assessing the response of few species to a single-hydrologic variable (e.g. emergent macrophytes along a depth gradient, Rea and Ganf 1994), to comparing multiple species, to the interaction of the whole water regime with salinity, sedimentation or plant community composition (e.g. Catford *et al.* 2011). In parallel, perspective has shifted from short-term questions (such as within-year patterns of growth, resource allocation and reproduction) to establishing the role of water regime throughout the plant life cycle and over large spatial scales. Years of research have revealed that wetland plants employ many strategies to cope with variable hydrology. More recently, scientists have built on the body of knowledge of altered water regimes to test theoretical questions, for example of invasion ecology (Catford *et al.* 2011).

Australia has made a significant contribution to wetland ecology through seed bank studies. Initially seed bank studies were descriptive, and intent on characterising seed density and species richness,

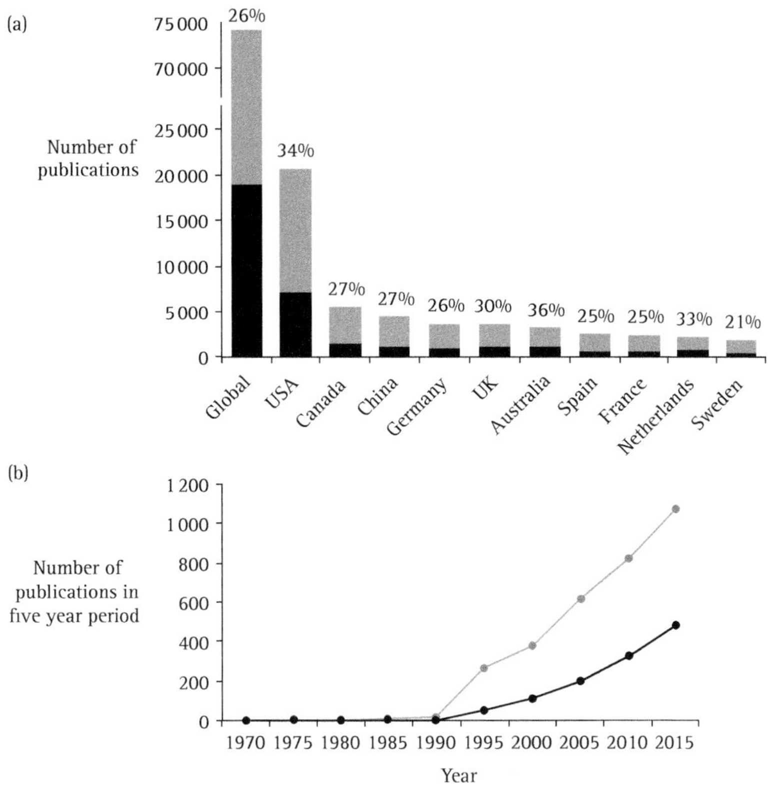

Figure 20.3 **(a)** Number of publications in ISI Web of Science on wetland vegetation globally and for the 10 countries with the greatest number of publications [grey bars are total publications. Note the disjointed *y*-axis; black bars are publications with applied perspective; % indicates percentage of total studies that have an applied perspective]; year of first publication per country, followed by year of first applied publication in parentheses: global 1900 (1968), USA 1900 (1981), Canada 1928 (1983), China 1991 (1995), Germany 1980 (1991), England 1973 (1975), Australia 1974 (1990), Spain 1985 (1992), France 1972 (1990), Netherlands 1973 (1984), Sweden 1972 (1992). **(b)** Number of publications over five year periods focussing on wetland vegetation in Australia. Grey line indicates total publications (<16 per five year interval until 1991–1995 period); black line indicates publication with applied perspective (0 before 1990); number of 2015 publications assumed to be the same as 2014. Search terms on Web of Science (15 October 2015): [TOPIC: ((Wetland* OR Bog* OR Marsh* OR Swamp* OR Fen* OR Mire* OR Riparian* OR Lake* OR Littoral* OR Spring*) AND (Vegetation OR Flora OR Floristic* OR Plant* OR Macrophyt* OR Hydrophyt*)); RESEARCH AREAS: (Environmental sciences ecology OR Plant sciences OR Marine freshwater biology OR Water resources OR Physical geography OR Biodiversity conservation); Timespan: All years.; Search language=Auto] *plus* [TOPIC: ((manag* OR restor* OR rehabilitat* OR applied OR policy OR policies))] *plus* [Country: Australia].

understanding spatial and temporal variability, and establishing the role of hydrology as a driver of this variability. They have since become a powerful tool allowing comparison among wetlands, evaluation of antecedent conditions, scenario testing, impact assessment, and a means of establishing recovery potential and resilience (Dawson *et al.* 2017). One outcome of seedbank research in Australia led to the development of a macrophyte classification system (Brock and Casanova 1997).

Consistent with global trends and terrestrial plant ecology, wetland science has seen growing emphasis on characterising species based on their morphology, responses to environmental stimuli (e.g. flooding) and effects on ecosystem function and services (Catford and Jansson 2014). This shift is motivated by the greater generalisation that trait-based classifications can offer to ecology and management: understanding plant responses and effects based on species characteristics rather than species identity allows findings gathered in one study and one region to be applied elsewhere.

The focus on salinity and water regime has necessarily resulted in fewer studies dedicated to other environmental factors: for example, effects of fire and temperature on wetland vegetation are poorly understood outside of alpine and sub-alpine areas. The short-term effects of native herbivores (mainly waterbirds such as black swans and magpie geese), exotic livestock (cattle, horses, pigs) and exotic benthivorous fish (common carp) on wetland vegetation and seed banks may be locally known, but the cumulative and differential long-term effects of biota on floristic composition and structure need further study. Plant–plant interactions, factors that promote dominance or facilitate invasion, and the nursery effects of structurally dominant species (James *et al.* 2015) have been investigated only rarely, and consequently their importance in structuring wetland plant communities is likely underappreciated.

20.5 Case studies of wetland types

Wetlands are found throughout Australia, with their type and distribution reflecting their hydrology, climate (particularly rainfall and evaporation) and topographic and morphological features (Figure 20.1; Brock 2003). Northern Australia has wet summers, southern coastal parts of temperate Australia have wet winters, and the arid and semi-arid interior, which makes up two-thirds of the continent, has highly unpredictable rainfall. If fed by rainfall, surface runoff or river flows, wetland hydrology reflects these climatic patterns. Flooding is highly predictable and seasonal in the tropics, whereas the saline and episodic wetlands that dominate arid and semi-arid regions can be characterised by their (wet) boom and (dry) bust cycles. Seasonal or intermittent wetlands are more common near the coast, and wetlands that are more persistent often occur where rainfall exceeds 1000 mm annually (Brock 2003; Boulton *et al.* 2014). In contrast to wetlands fed by surface water, many groundwater-fed systems naturally have a permanent supply of water. Springs fed by groundwater can seemingly occur in unlikely places, providing rare oases in the middle of the desert, but their distribution reflects the location of cracks and fissures in the aquitards that confine groundwater or where sediments that form the aquifer project above the ground (Fensham and Fairfax 2003). Wetlands flooded by rivers are necessarily found along river channels. The distribution of wetlands is changing as surface and groundwater are used and regulated by humans, as climatic patterns change, and as wetlands are in-filled and their catchments modified. The nine case studies presented below illustrate some of this diversity.

20.5.1 Floodplain wetlands of the wet–dry tropics

One of the most striking features of northern Australia is the extensive areas of floodplain associated with many of its large river systems (Figure 20.4). When in flood, the whole expanse of a river floodplain can be inundated; wetlands are not just restricted to lower elevation areas of the floodplain. In the lower rainfall regions, such as the Gulf of Carpentaria and the Kimberley, floodplains may only be inundated for a few weeks of the year and contain very few vascular plants. In contrast, wetlands in the higher rainfall regions may be inundated for up to half of the year and support expansive areas of aquatic macrophytes, grasses

Figure 20.4 Examples of wetland ecosystem types: (a) tropical coastal floodplain, Northern Territory; (b) spring wetland in central Australia; (c) floodplain wetland of the River Murray, Victoria; (d) monoculture of exotic *Urochloa mutica* (para grass) in Kakadu National Park, Northern Territory; (e) groundwater-dependent wetland in south-west Western Australia; and (f) constructed stormwater treatment wetland in Canberra, Australian Capital Territory.

Photos: Michael Douglas (a, d); Jane Catford (b, d, f); Ray Froend (e). (A black and white version of this figure will appear in some formats. For the colour version, please refer to the plate section.)

and sedges (Pettit *et al.* 2011). These latter systems have been the focus of much of the past research on tropical floodplain wetlands, particularly in the Alligator Rivers Region in Kakadu National Park (Finlayson 2005). Formed only 1,500–6,000 years ago, these tropical floodplains are relatively recent geomorphological features and contain a flora that is common to many other tropical regions of the world with less than 25% of the plant species endemic to northern Australia (Cowie *et al.* 2000).

The pronounced seasonality of the rainfall, and associated patterns of annual flooding and drying, is the primary driver of vegetation dynamics in the region. Local variation in vegetation community composition reflects microtopography and local differences in inundation (Finlayson 2005). As found in tropical South American floodplains, the predictability of the timing and magnitude of floods that drive wetland plant dynamics is positively correlated with fish species richness, bird population stability and riparian forest production in Australian floodplains (Jardine *et al.* 2015). A few species dominate large areas of Australian tropical floodplains; for example, *Oryza* species (wild rice), *Eleocharis* species (spike-rush), *Hymenachne acutigluma* (native hymenachne), *Pseudoraphis spinescens* (water couch) and water lilies (*Nymphaea* species and *Nymphoides* species) (Finlayson 2005). Although these systems are relatively low in plant species richness compared with the surrounding savanna landscape, they are hotspots of primary production and support an abundance of fauna, particularly fish and waterbirds (Finlayson *et al.* 2006). Recent research on Australia's tropical river ecosystems has highlighted the importance of primary production in floodplain wetlands, particularly by epiphytic algae, for supporting aquatic ecosystems of the floodplain, main river channel and estuary (Jardine *et al.* 2012).

In contrast to many floodplain systems in more populated regions, the hydrology of Australia's tropical floodplains remains largely intact. However, they have been listed as one of Australia's 10 most vulnerable habitats due to exotic plant invasions, which transform the structure and function of these systems, and the risk of sea-level rise (Laurance *et al.* 2011). Invasive perennial grasses such as *Urochloa mutica* (para grass, Figure 20.4d) and *Hymenachne amplexicaulis* (olive hymenachne) were introduced for cattle grazing and have since spread long distances (assisted by floods and waterbirds) and now cover increasingly large areas of some floodplains, including those in Kakadu National Park (Setterfield *et al.* 2013). Although successfully managed in Kakadu, the South American woody shrub *Mimosa pigra* (mimosa) is a serious weed on several floodplains in the Northern Territory (Setterfield *et al.* 2013). Their low elevation (most are <2 m above sea level) makes these coastal floodplains highly susceptible to sea-level rise (Catford *et al.* 2013), which has already transformed areas of freshwater wetland to saline swamp, including more than 17 000 ha on the Mary River in the Northern Territory (Mulrennan and Woodroffe 1998).

20.5.2 Desert floodplain wetlands in arid and semi-arid Australia

Desert floodplains occur throughout inland Australia but are particularly well-developed in the Channel Country of the eastern Lake Eyre Basin and, to a slightly lesser extent, in the lowlands of the western Murray–Darling Basin (Capon *et al.* 2016). Characterised by extremely low topographic gradients with mean annual rainfall below 250 mm, these eastern desert floodplains typically encompass complex channel networks that distribute floodwaters over vast areas during periods of high river flows. In contrast, the ephemeral floodplains of Australia's central and western deserts are usually associated with poorly defined drainage systems and are typically inundated by flashy overland flows. In either case, the defining characteristic of desert floodplains is their highly variable and unpredictable patterns of wetting and drying, which have an overriding influence on the composition and structure of their vegetation (Capon 2005).

Desert floodplain wetlands encompass the whole floodplain and are not restricted to depressions within the floodplain. They support a diverse range of short grass, sedge and forb associations that shift in composition and structure both temporally, in response to wetting and drying, and spatially in relation to flood history (Capon 2005). Plant species inhabiting these environments tend to be widely distributed, many having cosmopolitan distributions extending beyond the arid zone (Box *et al.* 2008). Common families represented include Asteraceae, Amaranthaceae (including chenopods), Cyperaceae, Euphorbiaceae, Fabaceae, Goodeniaceae, Malvaceae, Nyctaginaceae, Poaceae and Polygonaceae. Many species maintain large, persistent soil seed banks, which enable plants to avoid unfavourable conditions (e.g. drought) and re-establish when suitable conditions arise, typically following the drawdown of floodwaters (Capon 2007). In the arid zone, woody vegetation is mostly limited to watercourses and their banks or along flood strandlines. Further to the east however, large areas of semi-arid floodplain woodlands also occur (Good *et al.*, Chapter 21, this volume). Low-lying swampy areas also frequently support wetland shrublands (e.g. of *Duma florulenta*, *Chenopodium auricomum*).

Water resource development, of both surface and ground waters, probably represent the greatest threat to Australia's desert floodplains, governed as they are by hydrology. Land use is predominantly livestock grazing. Although overgrazing may have significant effects on vegetation during dry periods (e.g. severe vegetation cover reduction, soil compaction), cattle have limited access to vegetation when it is at its most productive and diverse (i.e. following flooding) due to surface instability. Recent intensification of mining exploration and extraction also pose a threat to desert floodplain vegetation as a result of infrastructure development, water use and the disposal of excess ground water. Climate change has the potential to further influence vegetation, especially where water regimes are altered.

20.5.3 Inland saline lakes

Saline lakes and wetlands, including lakes, claypans and rain-filled depressions, are a major feature of inland Australia, accounting for over 80% of total wetland area on the continent (Timms and Boulton 2001; Eldridge *et al.*, Chapter 24, this volume). Additionally, many freshwater lakes become saline as they dry. Most saline lakes occur in Western Australia and South Australia (e.g. Lake Eyre) but there are also many in the eastern mainland states. Most are ephemeral or temporary, but some wetlands, especially in southern Australia, can be seasonal or permanent. Saline lakes also vary in wave action and turbidity (Timms and Boulton 2001).

Vegetation composition in saline lakes is driven by salinity, water regime and turbidity (Porter *et al.* 2007). Species richness typically declines with increasing salinity and is usually greater in temporary, rather than permanent, lakes. However, only a few algal and angiosperm species are capable of persisting under both high levels of salinity and widely fluctuating water levels. Nevertheless, even highly saline lakes can support very productive submerged vegetation (Brock 1994). Rapid growth of submerged plants can occur under conditions of low turbidity when salinities are below ~60 g l^{-1} (Timms and Boulton 2001). Charophytes (e.g. *Nitella* species and *Lamprothamnium* species) are typically well represented and aquatic forbs are usually dominated by species of *Ruppia* and *Lepilaena*. Lakes are often fringed by salt-tolerant samphires (e.g. *Halosarcia* species, *Tecticornia* species), succulents (e.g. *Gunniopsis* species) and sedges (e.g. *Cyperus* species). Forbs and grasses may colonise some ephemeral and temporary saline lakes during dry periods, but some lakes will form salt crusts when dry and remain free of vegetation (Porter *et al.* 2007).

Threats to inland saline lakes include dryland salinisation, which has been a widespread result of altered hydrology (from water extraction and river regulation) and vegetation clearing in southern Australia, particularly south-west Western Australia

(Halse *et al.* 2003). Dryland salinity, along with a drying climate, has the potential to significantly alter the salinity and water regimes of inland saline lakes in much of Australia's south. Increased nutrient loads in runoff from surrounding farms may also pose a threat (Timms 2005). Extraction of materials from saline lakes, as well as further changes to hydrology resulting from mining infrastructure and water resource development, are also likely to have significant effects (Timms 2005).

20.5.4 Groundwater-dependent springs of the Great Artesian Basin

Spring wetlands are mostly permanent wetlands fed by groundwater from underlying aquifers and are classified as 'outcrop' or 'discharge' types. Spring wetlands occur throughout Australia but are particularly distinctive in arid environments where, as in other countries, they are one of the few sources of permanent water (Figure 20.4b; Fensham and Fairfax 2003). Spring wetlands can occur in areas where sediments that form the aquifer project above the ground (outcrop springs). These include springs in aquifer recharge areas where rates of rainfall-derived inflow exceed throughflow and where water-bearing sediments are close to the ground surface, which often occurs at the edges of an aquifer. Groundwater that feeds outcrop springs can have a short residence time (i.e. water is not in the aquifer for long), so these types of springs may be ephemeral. Discharge springs occur where water flows upwards through confining beds (aquitard) via faults or conduits in the overlying sediments (Fensham and Fairfax 2003). Discharge springs may sustain small permanent wetlands because groundwater feeding these systems has a long residence time in the aquifer and thus shows minimal fluctuations in flow rates.

The Great Artesian Basin is an aquifer system that underlies 22% of Australia in the north-east, extending over parts of Queensland, New South Wales, South Australia and Northern Territory (Powell *et al.* 2015). Water percolates through the sandstone aquifer generally in a south-westerly direction, with sandstone intake beds located mainly along the eastern margin in north Queensland (Powell *et al.* 2015). Springs in the Great Artesian Basin are often referred to as mound springs, but only a small fraction of spring wetlands have visible mounds associated with them. Mounds may be formed by upwelling of subsoil in the artesian water, from vegetation-derived peat accumulation, from the expansion of surface clays and from accretion of aeolian sand or calcium carbonate (Fensham and Fairfax 2003). Unlike spring wetlands in arid areas of Western Australia and Northern Territory, which are thought to have greater connectivity, the spring wetlands of the Great Artesian Basin have high levels endemism in both flora and fauna and are thus of high conservation value. Of 325 native vascular plants recorded in the Great Artesian Basin spring wetlands, at least eight appear to be endemic, with presumably little opportunity for dispersal outside of these isolated and environmentally unique systems. These include *Myriophyllum artesium*, *Eriocaulon carsonii* and related subspecies, and *Sporobolus pamelae*, along with species from the families Poaceae, Cyperaceae and Scrophulariaceae. At least 30 exotic plant species have also been recorded in these wetlands (Fensham and Price 2004).

Spring wetlands in the Great Artesian Basin are severely threatened by groundwater extraction and aquifer drawdown. Between 1880 and 1990s, water levels were drawn down by as much as 100 m in parts of Queensland. As a result, 89% of spring-groups have become inactive throughout much of Queensland over the last century (Fensham and Fairfax 2003). The discharge springs are most affected by groundwater extraction and 87% of springs in the discharge area have become partly or completely inactive due to groundwater drawdown, as opposed to only 8% in the recharge area. Of the springs that are still active, 26% have been severely degraded through wetland excavation, which is intended to increase access for livestock. Other threats include exotic species used for ponded pasture (e.g. *Hymenachne amplexicaulis*, *Urochloa mutica*, *Echinochloa polystachya*), livestock trampling, exotic animals (e.g. pig rooting) and fire (Fensham and Fairfax 2003). To protect the remaining spring wetlands, no more bores should be developed near springs, bores close to springs should be capped, and spring excavation and use of exotic ponded pasture should be prohibited (Fensham and Fairfax 2003). Greater protection of spring wetlands, especially in discharge areas, through the reserve system would greatly improve the conservation status of these unique systems.

20.5.5 Freshwater wetlands on coastal floodplains of south-eastern Australia

Coastal freshwater wetlands occur around the coast of Australia on sandy loam soils of waterlogged or periodically inundated alluvial flats. Primarily fed by rivers, coastal freshwater wetlands are similar in vegetation structure to inland floodplain wetlands, but tend to be located at elevations less than 20 m above sea level, have sandier soils, higher rainfall, and periodic saline water input (Whitehead *et al.* 1990). Rogers *et al.* (Chapter 22, this volume) review halophytic vegetation, including coastal mangroves and saltmarshes.

Plant communities in coastal floodplain wetlands are largely determined by frequency, duration and depth of inundation, but are being increasingly affected by human activities that include livestock grazing, catchment land clearing and pollution (Pressey and Middleton 1982). Freshwater wetlands in coastal plains typically have low cover of woody species, though will occasionally have scattered trees (e.g. *Casuarina glauca*, *Melaleuca ericifolia*) that are common in neighbouring swamp floodplain forests (de Jong 2000). Wetlands that lack surface water most of the time are typically dominated by dense grassland, sedgeland, or rushland under 0.5 m tall. Common species include *Paspalum distichum*, *Leersia hexandra*, *Pseudoraphis spinescens* and *Carex appressa*. Wetlands with more regular cycles of inundation and drying may support taller emergent species, such as *Bolboschoenus* species and *Schoenoplectus* species over 1 m tall. Common herbs in these areas include *Hydrocharis dubia*, *Philydrum lanuginosum*, *Ludwigia peploides*, *Marsilea mutica* and *Myriophyllum* species. As water levels become deeper or more permanent, these systems begin to develop more floating vegetation such as *Azolla* species, *Lemna* species, *Hydrilla verticillata*, *Ceratophyllum demersum*, *Nymphoides indica*, *Ottelia ovalifolia* and *Potamogeton* species.

Land development is a particular threat for freshwater wetlands along the coast because of the high value of the land, and because high organic matter in the sediments of freshwater wetlands make them ideal for agricultural development, including exotic pasture. Koo Wee Rup Swamp, historically located along the coast between Melbourne and Gippsland, was once one of the largest freshwater wetlands in Victoria. Extending over 40 000 ha, this swamp was drained for agriculture in 1876, destroying dense stands of swamp paperbark (*Melaleuca* species) and giant bulrush (*Typha* species) (Yugovic and Mitchell 2006). Coastal wetlands filter overland flow, improving water quality before it enters the ocean. The degradation and loss of wetlands can result in increased sediment and nutrient concentrations in runoff, which may lead to algal blooms, increased turbidity, and the degradation of marine ecosystems (Johnson *et al.* 1999). Coastal wetlands are themselves threatened by eutrophication, especially by nitrogen and phosphorus that originates from urban and agricultural areas, resulting in the displacement of native species by exotic species.

In coastal areas, where most of the human population resides, many wetlands were either 'reclaimed', via drainage or infilling, or regulated by control structures (e.g. barrages) to permit urban and agricultural development. This led to the fragmentation, degradation and loss of a large proportion of Australian coastal wetlands, which continue to be very poorly researched ecosystems. In south-western Western Australia, around 70% of coastal wetlands are estimated to have been lost since European settlement (Davis *et al.* 2015). In many coastal regions, wetland drainage and land-use change have also led to the exposure of underlying sulfidic sediments (i.e. acid sulfate soils) with a wide range of detrimental outcomes including diminished plant growth (White *et al.* 1997).

20.5.6 Billabongs and lakes of the southern Murray–Darling Basin floodplain

The floodplains of the Murray, Murrumbidgee and Goulburn Rivers are inset with two major types of wetlands that differ in geomorphology: billabongs and lakes (see Good *et al.*, Chapter 21, this volume for descriptions of forests and woodlands on drier parts of these floodplains). These fill, drawdown and dry in response to river inundation cycles, which vary across the floodplain resulting in a mosaic of water regimes, ranging from near permanent to episodic.

Billabongs are former river channels, typically small (surface area <10 ha), mostly clay-based, and very abundant, numbering thousands per floodplain. Though billabongs may be referred to as oxbow lakes, billabongs can take many shapes and forms, and often simply feature as depressions in the floodplain (Figure 20.4c). The diversity of wetland plants within a billabong is strongly influenced by wetland form: deeper wetlands have littoral vegetation on their slopes and submerged herbland on their bed when flooded, whereas shallower wetlands may have sedgeland throughout. Floodplain lakes are deflation basins with a clay or sand lunette on one margin, typically large (surface area 100–600 ha), and are gilgai, clay or sand-based (if not in-filled by sediment): they number tens to hundreds per river system (Pressey 1986). Their compact round shape and uniform floor result in low environmental heterogeneity, so individual lakes typically have a single dominant vegetation type.

The mosaic of water regimes occurring across a floodplain means floodplain wetlands may carry a complete spectrum of growth forms, such as river red gum (*Eucalyptus camaldulensis)* woodland, lignum (*Duma florulenta*) or chenopod shrubland, sedgeland, various types of grassland, and various types of aquatic herbland, and the number of species is correspondingly large. The species pool is a mix of Australian families and species (Myrtaceae and Amaranthaceae for trees and shrubs), cosmopolitan or widely distributed species (e.g. *Phragmites australis, Potamogeton crispus*), Australian aquatic and amphibious representatives of cosmopolitan or widely distributed families (Cyperaceae, Poaceae, Typhaceae, Juncaceae, Juncaginaceae, Haloragaceae, Ranunculaceae, Elatinaceae, Polygonaceae), and a few species with restricted distributions (e.g. *Amphibromus fluitans, Myriophyllum porcatum*). Although the species composition of floodplain wetlands is highly variable, wetlands with similar water regimes have functionally similar plant communities (Casanova 2011).

Over the last 200 years, floodplains of the southern Murray–Darling Basin have been dramatically altered through vegetation clearing, cultivation, livestock grazing, sediment fluxes, altered water quality, river regulation, salinisation and acidification. The widespread decline of submerged macrophytes and high abundance and richness of exotic species exemplify this (Catford *et al.* 2011). River regulation is a major (but not sole) factor determining the abundance of exotic species. In billabongs, 20–40% of species may be exotic, and most of these are short-lived terrestrial species such as pasture grasses or agricultural weeds. Their presence is a legacy of past agricultural enterprises (Lunt *et al.* 2012) but their abundance is boosted by the generally drier conditions that river regulation imposes. These conditions have the dual effect of disadvantaging the already established native amphibious species and of providing conditions suitable for exotic terrestrial species, but not native terrestrial species (Catford *et al.* 2011). Regulation disrupts the match between native species phenology and floods, and seasonal flow inversion (high flows in summer) favours the dispersal and establishment of exotic species generally (Greet *et al.* 2012). Thus, river regulation exacerbates the threat of invasive exotics such as amphibious *Nymphaea mexicana* and *Sagittaria platyphylla* (Figure 20.4d, f).

Submerged macrophyte abundance and diversity has been adversely affected by: (i) increased sediment flux from eroded uplands, which has increased turbidity and reduced photic depth; (ii) exotic common carp (*Cyprinus carpio*) that can dislodge macrophytes through their benthivorous feeding strategy, a strategy not used by native fish; and (iii) the prolonged Millennium Drought in the early twenty-first century. Tolerance to such pressures requires capacity to grow under low light, bulky rhizomes and relatively long-lived seed (6–9 years) that can aid recolonisation (Brock 2011). Only the robust perennial, *Vallisneria australis*, is known to have these characteristics.

20.5.7 Bogs and fens of the Australian Alps and Tasmania

The Alpine Sphagnum Bogs and Associated Fens ecological community predominantly occurs in the Australian Alp Bioregion, which spans 375 km from Mount Baw Baw in Victoria to southern New South Wales and the western margin of Australian Capital Territory. Similar communities are found in Tasmania, and typically occur above the treeline in alpine, sub-alpine, and montane environments. There is no strict altitude requirement, however, since local climatic conditions can enable these communities to flourish at lower altitudes and below the climatic treeline. Alpine bogs and fens shelter endangered plants, such as the bogong eyebright (*Euphrasia eichleri*), and are generally less than 1 ha in size, with such restricted distributions contributing to their eligibility for listing as threatened under the *Environmental Protection and Biodiversity Conservation Act 1999*. For a broader review of alpine vegetation, see Venn *et al.*, Chapter 19, this volume).

Alpine bogs and fens are both peatland ecosystems. Peat is terrestrial soil composed of at least 20% organic matter, which is at least 30 cm deep. Peatlands occur in areas with higher water inflow than outflow, or where there is a large supply of groundwater. Waterlogged soil produces ionically reducing, anaerobic conditions that inhibit activity of microbial decomposers, allowing carbon-rich undecomposed plant remains to accumulate. Accumulation of plant detritus into peat can produce small dams, which further modify hydrology in favour of continuing peat formation. Dynamic growth of peatlands can produce a mosaic of heterogeneous habitat that supports a range of vegetation communities (Camill 1999).

Although alpine bogs and fens are both characterised by presence of peat soils, and often co-occur in the landscape, they differ in nutrient levels and the plant communities they support. Fens are primarily fed by surface flow and groundwater, which is nutrient rich, and support sedge communities dominated by species such as *Carex gaudichaudiana*. Sphagnum mosses are generally absent from these areas as they are outcompeted by plants that are better able to exploit high nutrient environments. Bogs, in contrast, have very low-nutrient levels because they mostly rely on rainwater (though streams can run through 'valley bogs'). Bog soils are generally more acidic and support sphagnum mosses such as *Sphagnum cristatum* and *Sphagnum novozelandicum* and sclerophyllous shrubs up to 1 m tall such as *Richea* species, *Epacris glacialis*, and *Grevillea australis*.

Alpine bogs and fens face a number of threats related to their restricted geographical extent and small patch size. There are more than 11 000 bogs and fens throughout the Australian Alps Bioregion, the high altitude areas of south-eastern Australia's Great Dividing Range. Their small size and patchy distribution make them difficult to protect against trampling from livestock and feral animals. Exotic ungulates (e.g. cattle, sheep, horses, deer, pigs) trample bog and fen vegetation and also create incisions in peat soil that result in wetland erosion and drainage.

Presence of water is vital for the persistence of hydrophytic bog and fen vegetation, and so threats that impact water supply or peat structural integrity endanger these largely endemic communities (Wahren *et al.* 1999). Climate change is predicted to increase occurrence and severity of drought and fire, which damage peat. Fires can destroy carbon-rich peat, changing the physical structure of bog and fen soils and inhibiting their ability to capture and store water. Models predict that alpine bogs and fens around the world are among the ecosystems most threatened by climate change (Schuur *et al.* 2008).

20.5.8 Groundwater-dependent wetlands of the south-west

Characteristics of freshwater wetland vegetation communities in south-western Australia are largely determined by wetland water source, sediment type and duration of inundation (DEC 2012). There is a high number of permanently and seasonally inundated, shallow (<3 m) wetlands in the region for which groundwater is the dominant source of water, either from local and regional superficial groundwater or from perched groundwater held by an impeding sediment layer that is recharged directly by rainfall.

Despite having a hot, dry Mediterranean summer, the south-west is home to flora that represents the world centre for diversity in a number of wetland-associated plant families (e.g. Droseraceae, Juncaginaceae, Centrolepidaceae, Restionaceae and Hydatellaceae). The vegetation communities include a range of endemic wetland genera (e.g. *Reedia, Cephalotus, Tribonanthes, Epiblema, Schoenolaena, Cosmelia, Euchilopsis, Acidonia, Homalospermum, Pericalymma* and *Taxandria*) and high levels of local and regional endemism. The southern Swan Coastal Plain alone has approximately 445 wetland-associated plant taxa, 74% of which are endemic to Western Australia and 3% endemic to the southern Swan Coastal Plain (DEC 2012).

The dominant growth forms in Swan Coastal Plain wetlands appear to reflect the seasonally dry Mediterranean climate. The flora is rich in shrubs, but herbs and sedges make up 77% of taxa (Figure 20.4e). Although 78% of species are perennial, 22% of these regrow their above-ground biomass from underground storage organs each year (DEC 2012), which is likely an adaptation to seasonal drought and disturbance such as fire.

Seasonally waterlogged wetlands are typically maintained via perched groundwater and have the greatest diversity (DEC 2012). This floristic diversity reflects spatial and temporal patterns in waterlogging (e.g. duration), sediment stratigraphy (e.g. distribution of organic fraction) and composition of impeding layer (e.g. ironstone, calcrete or granite). These systems are dry in summer, with their high plant diversity not becoming apparent until late winter and spring when the sedgelands and herblands regrow from their underground storages. They typically support a mosaic of plant assemblages, especially woodlands (dominated by *Eucalyptus rudis* and *Melaleuca* species), shrublands (dominated by a diverse suite of species of Myrtaceae e.g. *Calothamnus lateralis, Melaleuca teretifolia, M. viminea, Astartea* species, *Pericalymma ellipticum, Kunzea* species, *Hypocalymma angustifolium*, and some Proteaceae), sedgelands, herblands and only rarely grasslands.

Seasonally waterlogged and perched groundwater-dependent wetlands were once extensive in the south-west, particularly on the Swan Coastal Plain, but due to land development and hydrological changes, many are now only represented by remnant portions (DEC 2012). Groundwater levels on the Swan Coastal Plain have been declining since the 1970s due to changes in climate, land use and groundwater extraction. Declining groundwater levels have affected vegetation in groundwater-dependent wetlands in a number of ways, including: reduced surface water levels and in some cases drying and loss of wetland habitat; peat fires; declining health and death of some groundwater-dependent vegetation; changes in flow from springs/seeps; and acidification of groundwater and wetlands (Sommer and Froend 2014). With changing hydrological conditions, many permanent/semi-permanent wetlands are developing vegetation characteristics of seasonally waterlogged wetlands, while formerly waterlogged areas are being encroached by dryland species. Longer periods of exposed, dry sediments have also lead to increased damage from illegal vehicle access and disturbance of sediment surfaces. The waterlogged fringes of more permanent wetlands have also been extensively reclaimed as water levels gradually fall and urbanisation continues.

20.5.9 Artificial wetlands

Artificial wetlands are waterbodies that are wholly or largely anthropogenic in origin, in contrast to natural wetlands that have been formed solely by hydrological and/or geomorphic processes over time. They are typically used for livestock watering (farm dams, bore-fed wetlands), water storage (irrigation water storages, weirpools, upland reservoirs and lowland storages), amenity (urban lakes and ponds, Figure 20.4f), water treatment (waste water treatment ponds, ponds and meadows as part of water-sensitive urban design) or production (e.g. rice fields). Artificial wetlands differ from natural wetlands in being younger, and having differing combinations of size, shape and water regime: reservoirs and weirpools are considerably deeper; storages and amenity lakes are permanent; weirpools and amenity lakes have stable water levels; rice fields provide shallow clear water in spring–summer in a semi-arid region.

A wetland assemblage develops (or is planted) that is typically dominated by native species (Dugdale *et al.* 2009) but distinct from its regional equivalents (Badman 1999). Riverine storages and amenity lakes are notable exceptions to native dominance; these are often dominated by floating or submerged macrophytes, including highly invasive exotic species like *Egeria densa*, *Cabomba caroliniana* and *Nymphaea mexicana* (Figure 20.2d; Dugdale *et al.* 2009). Characteristics of artificial wetlands that may constrain the development of native vegetation assemblages are: high turbidity, which excludes submerged plants; water deeper than 1.5 m with steep vertical banks, which precludes the development of a littoral zone; permanent water or extreme fluctuations, which preclude successful plant establishment and persistence (Casanova *et al.* 1997).

Hydrological connectivity is variable. Many farm dams and bore-fed wetlands are hydrologically isolated, and their plant assemblages have developed independently of hydrochory. Some, notably meadows in urban contexts, are only intermittently connected. The continuous hydrological connectivity of riverine storages, such as Lake Mulwala on the River Murray, imposes a continual threat through the downstream dispersal of vegetative fragments of undesirable species. Although utilitarian in intent, artificial wetlands may develop ecological value as habitat for fauna, such as waterbirds: an example is the Ramsar-listed Western Treatment Plant at Werribee, Victoria.

20.5.10 Synopsis

The nine case studies presented illustrate commonalities in the forces that structure wetland vegetation across Australia, as well as the processes that threaten them (Figure 20.1). The primacy of hydrology and wetland water regimes for determining the distribution and characteristics of wetland vegetation is particularly apparent. It is the variation in each of these structuring forces, and interactions among them, that give the plant communities of individual wetlands their own character. The case studies illustrate diverse expression of character among different wetland ecosystems across the continent.

20.6 Threats

Practically all wetland systems in Australia, with the notable exception of tropical floodplain wetlands and some springs, are threatened by changes in water supply and water regimes (Keith and Auld, Chapter 27, this volume). Such changes are caused by the extraction and regulation of surface and groundwater, with modification of surface water being particularly pronounced in south-eastern Australia and groundwater extraction affecting wetlands in central and western Australia. Depending on the system in question, the chief threat may involve an increase in water persistence (i.e. for water storage; affecting wetlands that are naturally ephemeral or intermittent) or a decrease (i.e. for water extraction; affecting wetlands that are naturally always wet, like bogs, fens and some springs). Increased water persistence (e.g. around dams and weir pools) can lead to tree death and shifts towards species favoured by stable water levels, e.g. *Typha* species (Blanch *et al.* 2000). In contrast, reductions in flooding can lead to wetland loss or replacement of native macrophytes by woody species (Bren 1992) and exotic terrestrial herbs (Catford *et al.* 2014).

Climate change also threatens wetland hydrology, as illustrated in alpine bogs threatened by declining rainfall and runoff, and in coastal systems at risk of saltwater intrusions from rising sea levels. Overall, climate change has the capacity to reduce native wetland plant diversity, promote exotic invasions, accelerate wetland fragmentation and homogenise wetland ecosystems at local and landscape scales (Capon *et al.* 2013).

Exotic species are threats throughout Australia, and are both a symptom (e.g. in response to flow regulation along the River Murray, see case studies) and a cause of changes in native vegetation (e.g. exotic pastures that outcompete native vegetation, Fensham and Fairfax 2003). Fifteen of the 32 Weeds of National Significance invade riparian or aquatic systems and were intentionally introduced (Leishman *et al.*, Chapter 9, this volume). Ponded pasture and other exotic pasture species continue to be planted, and new varieties that will likely increase their invasion success continue to be developed (Driscoll *et al.* 2014). Exotic animals also threaten native plants directly through grazing, trampling and uprooting (e.g. ungulates, carp) and indirectly, for example, through soil compaction or increased gully erosion (Eldridge *et al.*, Chapter 6, this volume).

Many wetlands have been degraded by declines in water quality, particularly in intensive agricultural and urban landscapes. Changes in the land use of wetlands and their catchments (e.g. livestock access to alpine and Murray–Darling Basin wetlands) for agriculture, urbanisation and increasingly mining, have led to sedimentation, pollution and eutrophication, salinisation, acidification, and changes to hydrology and connectivity among wetlands.

20.7 Management and policy

20.7.1 Policy relating to wetland vegetation

The first national policy on wetland management was released in 1997 (The Wetlands Policy of the Commonwealth Government of Australia). While all states and territories have legislation to protect the environment and conserve natural resources (e.g. environment protection, land-use planning, protected areas, water and vegetation management), state-level policies rarely address wetlands and wetland vegetation directly. In 2013, the National Wetlands Policy for Australia was initiated and is designed to improve alignment of Australian Government, state/territory and community action and investment, and provide consistency in decision-making across government agencies (WetlandCare Australia 2013).

20.7.2 Who is responsible for the management of wetland vegetation?

The management of wetlands and associated vegetation in Australia is primarily the responsibility of landowners and land managers. Individual states and territories have the legislative and policy responsibility for natural resource management (Finlay-Jones 1997). The Australian Government administers the Ramsar Convention on Wetlands of International Importance by providing national wetland policy, leadership and direction. It also implements the *Environment Protection and Biodiversity Conservation Act 1999* (EPBC Act), and develops programmes to improve the management of wetlands. The EPBC Act specifies that any activities within or outside designated Ramsar wetlands that are likely to have a significant impact on the ecological character of a listed wetland, including its vegetation, must be assessed prior to approval. Managers may also need approval from the relevant agencies under state legislation for activities that potentially impact Ramsar-listed and non-Ramsar-listed wetlands. Indigenous management of wetland vegetation continues in much of the country and has been revived in some places as a result of Native Title (Ens *et al.*, Chapter 4, this volume).

20.7.3 Management approaches

Like all plant communities, wetland vegetation is shaped by the composition of the regional species pool, local dispersal, environmental conditions and interactions with other biota (Figure 20.1; Catford and Jansson 2014). Alteration to any one of these four elements (for example, through exotic species introduction, reduced connectivity among wetlands and flow regulation) can alter wetland vegetation. Effective management requires a comprehensive understanding of the factors and processes that influence wetland vegetation so that the chief threats to wetland vegetation can be identified and targeted (Figure 20.1).

Threats to wetland vegetation along the River Murray include cattle grazing, eutrophication, drought, human activities around towns and farms, and flow regulation. Of these, flow regulation was found to be primarily responsible for an observed decline in native vegetation, and particularly native amphibious vegetation (Catford *et al.* 2014). By altering historical water regimes, and reducing flood magnitude in particular, flow regulation inhibits native plants that are poorly adapted to the modified flooding regimes. With less competition from native plants, exotic plant species (most of which are terrestrial) have increased in abundance. Environmental flows can reduce some of these negative effects, but it is rarely possible to provide the full spectrum of needs for wetland plants, and also service other wetland needs. For example, environmental flows are commonly used to extend the duration of natural flood events because flood duration is crucial for fish spawning and waterbird breeding, but floods extending into the summer months favour competitive emergent macrophytes such as *Typha* species or *Juncus ingens* and can lead to perverse outcomes (Box 20.1).

Once a threat is identified, varied approaches can be used to ameliorate it. For example, for systems with an insufficient supply of propagules, seeds can be sown directly or hydrological connectivity can be re-established among wetlands to facilitate dispersal. When various management options are available and the most effective one unknown, modelling can aid decision-making by predicting the likely response of plant communities to different management actions. Based on ecological understanding and empirical data, a state and transition model has been developed for the Ramsar-listed Macquarie Marshes to model effects of flood frequency, distance to river and fire frequency on vegetation dynamics (Bino *et al.* 2015).

Modelling is a valuable tool for anticipating ecological responses to management actions, but can also be used to anticipate effects of environmental conditions not previously experienced (Catford *et al.* 2013). Novel ecosystems that arise from global environmental change and exotic species invasions present new challenges for environmental policy and management. In cases where maintenance or restoration of historical vegetation is unrealistic, wetland managers may instead focus on the maintenance of key ecosystem services or functions (e.g. water filtering, carbon sequestration, flood

Box 20.1 Perverse outcomes of environmental watery delivery

Environmental water is an accepted means of rehabilitating stressed wetlands and facilitating fish, frog and waterbird breeding. Flow delivery can be a balancing act between needs and risks. Environmental flows delivered to wetlands can sometimes have perverse effects that jeopardise their intended aims.

Since the early 1990s, environmental flows have been delivered to the nationally important Booligal Swamp and the terminal wetlands of Merrimajeel and Muggabah Creeks in the southern Murray–Darling Basin. The environmental flows are delivered in spring–summer to support colonial-nesting waterbird breeding. For the first 10 years of this management action, water for consumptive use (stock, domestic, irrigation) was also delivered at this time, and rejected irrigation deliveries (i.e. requested irrigation water not extracted from the creeks) occurred during spring–summer too. Although waterbird recruitment benefits from summer flows, these creeks historically (i.e. 'naturally') experienced low or no flows in summer. As a result of unseasonal high flows, vegetation productivity increased in the effluent channels, blocking water passage and inadvertently limiting the delivery of environmental water to the wetlands, as well as water for human use.

The current water management regime, developed since the early 2000s, mitigates these effects by delivering more consumptive water during colder months, and through structures that stop unseasonal flows. However, delivery of environmental flows into Booligal Swamp for late breeding birds, such as glossy ibis (*Plegadis falcinellus*; Driver *et al.* 2010), will continue to alter the timing of flows beyond the

Figure 20.5 Lake Merrimajeel, February 2014.
Photo: Patrick Driver.

historic range, and promote overgrowth of several emergent macrophytes (e.g. *Typha* species) and lignum (*Duma florulenta*). Manual 'creek cleaning' to remove this vegetation and associated silt often results in further growth of shrubs, which appear to respond favourably to mechanical disturbance. The timing of flow delivery is therefore a critical consideration in this situation for meeting both ecological and socio-economic objectives.

attenuation). This could involve facilitating the establishment of salt-tolerant species for wetlands affected by secondary (human-induced) salinisation (Sim *et al.* 2006).

20.8 Conclusion

Wetlands and wetland vegetation are of critical ecological, social, cultural and economic value, exemplified by their likely role as refugia and stepping stones for dispersal in a changing climate (Capon *et al.* 2013). Over half of Australia's wetlands have been lost in the last 200 years (Bennett 1997). Those remaining are severely degraded yet they are facing increasing threat from rising water demand, land-use change, mining, climate change, agricultural development in new regions, rising sea levels and associated saltwater intrusion, and the introduction and spread of invasive species. Expansion of irrigated agriculture in northern Australia would increase pressure on freshwater ecosystems in the north (Douglas *et al.* 2011) and continued extraction of groundwater will exacerbate wetland degradation in central and western Australia. Wetland vegetation is resilient by its very nature with plant physiological and life-history adaptations to cope with a highly variable environment. However, wetland ecosystems can only withstand so much. Effective protection of wetlands requires: (i) greater inclusion in the National Reserve System; and (ii) increased policy initiatives that protect wetlands and the water on which they rely. The ongoing loss and degradation of wetland ecosystems throughout the Great Artesian and Murray–Darling Basins provide sobering lessons for other parts of the country.

Acknowledgements

We thank David Keith for inviting us to write this chapter, Patrick Driver for the idea and information for Box 20.1, Margaret Brock and Greg Keighery for feedback on the chapter plan, and Margaret Brock and an anonymous reviewer for comments on a chapter draft. JAC and SW acknowledge financial support from the ARC Centre of Excellence for Environmental Decisions.

References

Badman, F.J. (1999) *The Lake Eyre South Study: Vegetation.* Royal Geographical Society of South Australia: Adelaide.

Bennett, J. (1997) Valuing wetlands. In *Wetlands in a Dry Land: Understanding for Management*, ed. W.D. Williams. Environment Australia and Department of the Environment: Canberra, pp. 283–288.

Bino, G., Sisson, S.A., Kingsford, R.T. *et al.* (2015) Developing state and transition models of floodplain vegetation dynamics as a tool for conservation decision-making: a case study of the Macquarie Marshes Ramsar wetland. *Journal of Applied Ecology* 52, 654–664.

Blanch, S.J., Walker, K.F. and Ganf, G.G. (2000) Water regimes and littoral plants in four weir pools of the River Murray, Australia. *Regulated Rivers: Research and Management* 16, 445–456.

Boulton, A., Brock, M., Robson, B. *et al.* (2014) *Australian Freshwater Ecology: Processes and Management.* John Wiley & Sons: Chichester.

Box, J., Duguid, A., Read, R. *et al.* (2008) Central Australian waterbodies: the importance of permanence in a desert landscape. *Journal of Arid Environments* 72, 1395–1413.

Bren, L. (1992) Tree invasion of an intermittent wetland in relation to changes in the flooding frequency of the River Murray, Australia. *Australian Journal of Ecology* 17, 395–408.

Brix, H. (1987) Treatment of wastewater in the rhizosphere of wetland plants-the root-zone method. *Water Science and Technology* 19, 107–118.

Brock, M.A. (1994) Aquatic vegetation of inland wetlands. In *Australian Vegetation*, ed. R.H. Groves. Melbourne: Cambridge University Press, pp. 437–466.

Brock, M.A. (2003) Australian wetland plants and wetlands in the landscape: conservation of diversity and future management. *Aquatic Ecosystem Health and Management* 6, 29–40.

Brock, M.A. (2011) Persistence of seed banks in Australian temporary wetlands. *Freshwater Biology* 56, 1312–1327.

Brock, M.A. and Casanova, M.T. (1997) Plant life at the edge of wetlands: ecological responses to wetting and drying patterns. In *Frontiers in Ecology: Building the Links*, ed. N. Klomp and I. Lunt. Elsevier Science: Oxford, pp. 181–192.

Brock, M.A., Nielsen, D.L., Shiel, R.J. *et al.* (2003) Drought and aquatic community resilience: the role of eggs and seeds in sediments of temporary wetlands. *Freshwater Biology* 48, 1207–1218.

Bunn, S.E., Thoms, M.C., Hamilton, S.K. and Capon, S.J. (2006) Flow variability in dryland rivers: boom, bust and the bits in between. *River Research and Applications* 22, 179–186.

Capon, S., Chambers, L., Mac Nally, R. *et al.* (2013) Riparian ecosystems in the 21st century: hotspots for climate change adaptation? *Ecosystems* 16, 359–381.

Capon, S.J. (2003) Plant community responses to wetting and drying in a large arid floodplain. *River Research and Applications* 19, 509–520.

Capon, S.J. (2005) Flood variability and spatial variation in plant community composition and structure on a large arid floodplain. *Journal of Arid Environments* 60, 283–302.

Capon, S.J. (2007) Effects of flooding on seedling emergence from the soil seed bank of a large desert floodplain. *Wetlands* 27, 904–914.

Capon, S.J., Porter, J. and James, C.S.(2016) Vegetation of Australia's desert river landscapes. In *Vegetation of Australia's Riverine Landscapes*, ed. S.J. Capon, M. Reid and C.S. James. CSIRO Publishing: Melbourne, pp. 239–258.

Casanova, M.T. (2011) Using water plant functional groups to investigate environmental water requirements. *Freshwater Biology* 56, 2637–2652.

Casanova, M.T., Douglas-Hill, A., Brock, M.A. *et al.* (1997) Farm ponds in New South Wales, Australia: relationship between macrophyte and phytoplankton abundances. *Marine and Freshwater Research* 48, 353–360.

Catford, J.A. and Jansson, R. (2014) Drowned, buried and carried away: effects of plant traits on the distribution of native and alien species in riparian ecosystems. *New Phytologist* 204, 19–36.

Catford, J.A., Downes, B.J., Gippel, C.J. and Vesk, P.A. (2011) Flow regulation reduces native plant cover and facilitates exotic invasion in riparian wetlands. *Journal of Applied Ecology* 48, 432–442.

Catford, J.A., Morris, W.K., Vesk, P.A. *et al.* (2014) Species and environmental characteristics point to flow regulation and drought as drivers of riparian plant invasion. *Diversity and Distributions* 20, 1084–1096.

Catford, J.A., Naiman, R.J., Chambers, L.E. *et al.* (2013) Predicting novel riparian ecosystems in a changing climate. *Ecosystems* 16, 382–400.

Cowie, I.D., Short, P.S. and Osterkamp-Madsen, M. (2000) *Floodplain Flora: A Flora of the Coastal Floodplains of the Northern Territory, Australia.* Australian Biological Resources Study: Canberra.

Davis, J., O'Grady, A.P., Dale, A. *et al.* (2015) When trends intersect: the challenge of protecting freshwater ecosystems under multiple land use and hydrological intensification scenarios. *Science of The Total Environment* 534, 65–78.

Dawson, S.K., Kingsford, R.T., Berney, P. *et al.* (2017) Frequent inundation helps counteract land use impacts on wetland propagule banks. *Applied Vegetation Science.* [DOI:10.111/avsc.12295]

DEC (2012) *A Guide to Managing and Restoring Wetlands in Western Australia.* Department of Environment and Conservation: Perth.

De Jong, N. (2000) Woody plant restoration and natural regeneration in wet meadow at Coomonderry Swamp on the south coast of New South Wales. *Marine and Freshwater Research* 51, 81–89.

Douglas, M.M., Jackson, S., Setterfield, S.A. *et al.* (2011) Northern futures: threats and opportunities for freshwater ecosystems. In *Aquatic Biodiversity in Northern Australia: Patterns, Threats and Future*, ed. B.J. Pusey. Charles Darwin University Press: Darwin, pp. 203–220.

Driscoll, D.A., Catford, J.A., Barney, J.N. *et al.* (2014) New pasture plants intensify invasive species risk. *Proceedings of the National Academy of Sciences* 111, 16622–16627.

Driver, P.D., Chowdhury, S., Hameed, T. *et al.* (2010) Ecosystem response models for lower Calare (Lachlan River) floodplain wetlands: managing wetland biota and climate change modelling. In *Ecosystem Response Modelling in the Murray–Darling Basin.* CSIRO Publishing: Melbourne, pp. 183–196.

Dugdale, T.M., Clements, D., Hunt, T.D. and Saglioco, J.L. (2009) Utilising waterbody draw-down to suppress aquatic weeds. In *Weeds Society of Victoria Fourth Biennial Conference.* Weeds Society of Victoria: Victoria.

Fensham, R.J. and Fairfax, R.J. (2003) Spring wetlands of the Great Artesian Basin, Queensland, Australia. *Wetlands Ecology and Management* 11, 343–362.

Fensham, R.J. and Price, R.J. (2004) Ranking spring wetlands in the Great Artesian Basin of Australia using endemicity and isolation of plant species. *Biological Conservation* 119, 41–50.

Finlay-Jones, J. (1997) Aspects of wetland law and policy in Australia. *Wetlands Ecology and Management* 5, 37–54.

Finlayson, C.M. (2005) Plant ecology of Australia's tropical floodplain wetlands: a review. *Annals of Botany* 96, 541–555.

Finlayson, C.M., Lowry, J., Bellio, M.G. *et al.* (2006) Biodiversity of the wetlands of the Kakadu Region, northern Australia. *Aquatic Sciences* 68, 374–399.

Goodman, A.M., Ganf, G.G., Maier, H.R., Dandy, G.C. (2011) The effect of inundation and salinity on the germination of seed banks from wetlands in South Australia. *Aquatic Botany* 94, 102–106.

Greet, J., Cousens, R.D. and Webb, J.A. (2012) Flow regulation affects temporal patterns of riverine plant seed dispersal: potential implications for plant recruitment. *Freshwater Biology* 57, 2568–2579.

Halse, S., Ruprecht, J. and Pinder, A. (2003) Salinisation and prospects for biodiversity in rivers and wetlands of south-west Western Australia. *Australian Journal of Botany* 51, 673–688.

James, C.S., Capon, S.J. and Quinn, G.P. (2015) Nurse plant effects of a dominant shrub (*Duma florulenta*) on understorey vegetation in a large, semi-arid wetland in relation to flood frequency and drying. *Journal of Vegetation Science* 26, 985–994.

Jardine, T.D., Pettit, N.E., Warfe, D.M. *et al.* (2012) Consumer–resource coupling in wet–dry tropical rivers. *Journal of Animal Ecology* 81, 310–322.

Jardine, T.D., Bond, N.R., Burford, M.A. *et al.* (2015) Does flood rhythm drive ecosystem responses in tropical riverscapes? *Ecology* 96, 684–692.

Johnson, A.K.L., Ebert, S.P. and Murray, A.E. (1999) Distribution of coastal freshwater wetlands and riparian forests in the Herbert River catchment and implications for management of catchments adjacent the Great Barrier Reef Marine Park. *Environmental Conservation* 26(3), 229–235.

Laurance, W.F., Dell, B., Turton, S.M. *et al.* (2011) The 10 Australian ecosystems most vulnerable to tipping points. *Biological Conservation* 144, 1472–1480.

Lunt, I.D., Jansen, A. and Binns, D.L. (2012) Effects of flood timing and livestock grazing on exotic annual plants in riverine floodplains. *Journal of Applied Ecology* 49, 1131–1139.

Mulrennan, M.E. and Woodroffe, C.D. (1998) Saltwater intrusion into the coastal plains of the Lower Mary River, Northern Territory, Australia. *Journal of Environmental Management* 54, 169–188.

Nielsen, D.L., Brock, M.A., Petrie, R. and Crossle, K. (2007) The impact of salinity pulses on the emergence of plant and zooplankton from wetland seed and egg banks. *Freshwater Biology* 52, 784–795.

Pettit, N.E. and Froend, R.H. (2001) Variability in flood disturbance and the impact on riparian tree recruitment in two contrasting river systems. *Wetlands Ecology and Management* 9, 13–25.

Pettit, N., Townsend, S., Dixon, I. and Wilson, D. (2011) Plant communities of aquatic and riverine habitats. In *Aquatic Biodiversity in Northern Australia: Patterns, Threats and Future*, ed. B.J. Pusey. Charles Darwin University Press: Darwin, pp. 37–50.

Porter, J.L., Kingsford, R.T. and Brock, M.A. (2007) Seed banks in arid wetlands with contrasting flooding, salinity and turbidity regimes. *Plant Ecology* 188, 215–234.

Powell, O., Silcock, J. and Fensham, R. (2015) Oases to oblivion: the rapid demise of springs in the south-eastern Great Artesian Basin, Australia. *Groundwater* 53, 171–178.

Pressey, R.L. (1986) *Wetlands of the River Murray below Lake Hume.* River Murray Commission: Canberra.

Pressey, R.L. and Middleton, M.J. (1982) Impacts of Flood Mitigation Works on Coastal Wetlands in New South Wales. *Wetlands (Australia)* 2, 27–44.

Rea, N. and Ganf, G.G. (1994) Water depth changes and biomass allocation in two contrasting macrophytes. *Australian Journal of Marine and Freshwater Research* 45, 1459–1468.

Schuur, E.A.G., Bockheim, J., Canadell, J.G. *et al.* (2008) Vulnerability of permafrost carbon to climate change: implications for the global carbon cycle. *BioScience* 58(8), 701–714. doi:10.1641/B580807.

Setterfield, S.A., Douglas, M.M., Petty, A.M. *et al.* (2013) Invasive plants in the floodplains of Australia's Kakadu National Park. In *Plant Invasions in Protected Areas*, ed. L.C. Foxcroft, P. Pyšek, D.M. Richardson and P. Genovesi. Springer: Berlin, pp. 167–189.

Sim, L.L., Chambers, J.M. and Davis, J.A. (2006) Ecological regime shifts in salinised wetland systems. I. Salinity thresholds for the loss of submerged macrophytes. *Hydrobiologia* 573, 89–107.

Sommer, B. and Froend, R. (2014) Phreatophytic vegetation responses to groundwater depth in a drying Mediterranean-type landscape. *Journal of Vegetation Science* 25, 1045–1055.

Sorrell, B.K., Brix, H. and Orr, P.T. (1997) Eleocharis sphacelata: internal gas transport pathways and modelling of aeration by pressurized flow and diffusion. *New Phytologist* 136, 433–442.

Timms, B. and Boulton, A. (2001) Typology of arid zone floodplain wetlands of the Paroo River (inland Australia) and the influence of water regime, turbidity, and salinity on their aquatic invertebrate assemblages. *Archiv für Hydrobiologie* 153, 1–27.

Timms, B.V. (2005) Salt lakes in Australia: present problems and prognosis for the future. *Hydrobiologia* 552, 1–15.

Wahren, C., Williams, R. and Papst, W. (1999) Alpine and subalpine wetland vegetation on the Bogong High Plains, south-eastern Australia. *Australian Journal of Botany* 47, 165–188.

WetlandCare Australia (2013) Australian national wetlands policy – background information. Available at www.wetlandcare.com.au/index.php/download_file/view/770/89/ (accessed 8 August 2016).

White, I., Melville, M., Wilson, B. and Sammut, J. (1997) Reducing acidic discharges from coastal wetlands in eastern Australia. *Wetlands Ecology and Management* 5, 55–72.

Whitehead, P.J., Wilson, B.A. and Bowman, D.M.J.S. (1990) Conservation of coastal wetlands of the Northern Territory of Australia: The Mary River floodplain. *Biological Conservation* 52, 85–111.

Yugovic, J. and Mitchell, S. (2006) Ecological review of the Koo-Wee-Rup swamp and associated grasslands. *The Victorian Naturalist* 123, 323–334.

21 Forests and Woodlands of Australia's Rivers and Floodplains

MEGAN GOOD, RHIANNON SMITH AND NEIL PETTIT

Australia's floodplain and riparian environments are dynamic, boom-bust systems shaped by spatial and temporal variation in water availability. Plant communities occupying riparian and floodplain environments broadly follow predictable patterns in terms of structure and diversity according to water availability (both surface and groundwater) in the landscape. Dense, often structurally complex and diverse woodlands and forests occupy positions close to the river, while open woodland communities with large old trees and very little midstorey occupy the floodplains. Recruitment dynamics of floodplain species are dictated by the timing, duration, frequency, depth, magnitude and rate of rise and fall of floodwaters. However, regulation of river systems and transformation of floodplain environments by urban and agricultural activity has resulted in a disconnection between rivers and their floodplains, to the detriment of a variety of ecological functions required to maintain floodplain and riparian ecosystems, as well as instream habitats. Many billions of dollars in revenue emanates from agricultural industries that utilise the rich fertile soils laid down by successive flood events, and competing interests for water highlight ongoing tensions over the management of these environments. The future of Australia's floodplain ecosystems relies on a sound understanding of the requirements of the species that occupy these environments and balancing environmental and human water demands in an environmentally, economically and socially sustainable context.

Land of plenty or land of want, where the grey Companions dance, Feast or famine, or hope or fear, and in all things land of chance, Where Nature pampers or Nature slays, in her ruthless, red, romance.

Banjo Patterson's 'Australian Scenery: The Plains'

21.1 Introduction

Australia's inland rivers and floodplains are part of our cultural identity; they feature in Dreaming stories, art, poetry and folklore. Forests and woodlands of these river systems are adapted to utilise various water sources during prolonged droughts and withstand flooding at other times.

Riparian zones and floodplains are the interface between terrestrial and aquatic systems. Gradients

Australian Vegetation, Third Edition, ed. David A. Keith. Published by Cambridge University Press. © Cambridge University Press 2017.

Citation: Good, M., Smith, R. and Pettit, N. (2017). Forests and Woodlands of Australia's Rivers and Floodplains. In D. A. Keith, ed., *Australian Vegetation*, 3rd edn. Cambridge: Cambridge University Press, pp. 516–43.

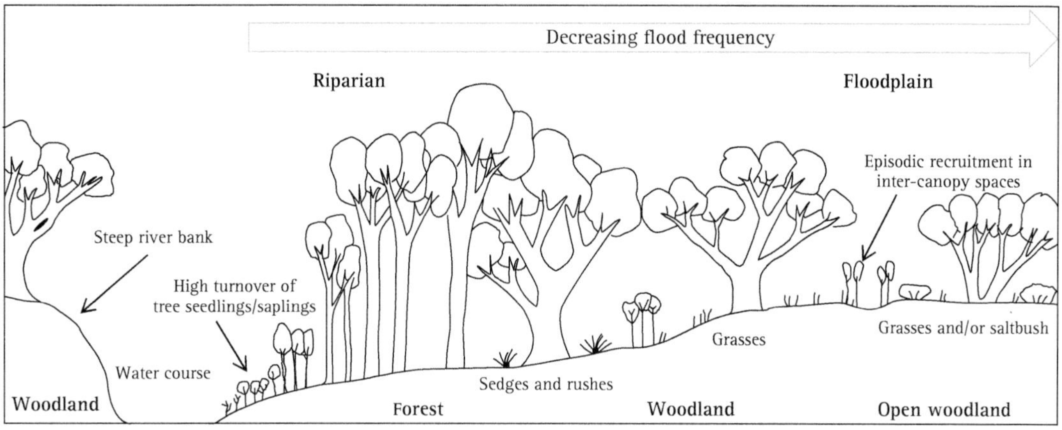

Figure 21.1 Profile diagram of a typical Australian riparian and floodplain landscape with broad vegetation types and typical understorey life forms.
Note: The vertical incline is exaggerated to include more riparian and floodplain vegetation types and trees and other vegetation elements are therefore, not to scale.

of change from terrestrial to aquatic can be steep or gradual and this is reflected in environmental and biological processes and vegetation community composition (Figure 21.1). In general, woodlands and forests that are close to the river are dense, structurally complex and biologically diverse while open woodland communities with large old trees and very little midstorey occupy the floodplains (Figure 21.1). The vegetation within the riparian zone performs important ecological functions contributing to in-stream processes, bank stability and water movement (Bunn 1993). Trees on the river edge provide shade that moderates in-stream primary productivity by reducing light and water temperature, while their roots and fallen branches provide habitat for biota (Bayly and Williams 1981; Bunn 1993). Riparian vegetation also provides important subsidies of food and nutrients to the river (Nakano and Murakami 2001). In contrast, floodplains can be expansive areas often with very little topographic relief with composition, structure and function of vegetation communities dictated by the timing and duration of floods (Junk *et al.* 1989).

In this chapter, we will explore the current knowledge of woody riparian and floodplain vegetation in Australia, from the productive tropics of northern Australia to the highly variable arid interior. We will focus on floodplain and riparian vegetation with an upper stratum of trees or shrubs, as distinct from treeless wetland ecosystems which are covered by Catford *et al.* (Chapter 20, this volume). Our focus will be on the woody component of these systems, but understorey species in common with treeless wetlands are described in the following section on community composition. We will describe riparian and floodplain vegetation in eight broad zones adapted from Australia's 12 major drainage basins (AWRC 1976; Figure 21.2).

21.2 The floodplain and riverine environment

Australia is a dry continent with very little runoff into rivers and streams due to high rates of evaporation and efficient water uptake by native vegetation. Rivers and streams are characterised by extremely variable flows (Kennard *et al.* 2010) and reliable base flows are confined to rivers in higher rainfall regions or areas with groundwater influence. Much of the continent contains no rivers at all or rivers that flow only occasionally (Figure 21.2). In the south-east of the country, including Tasmania, rivers

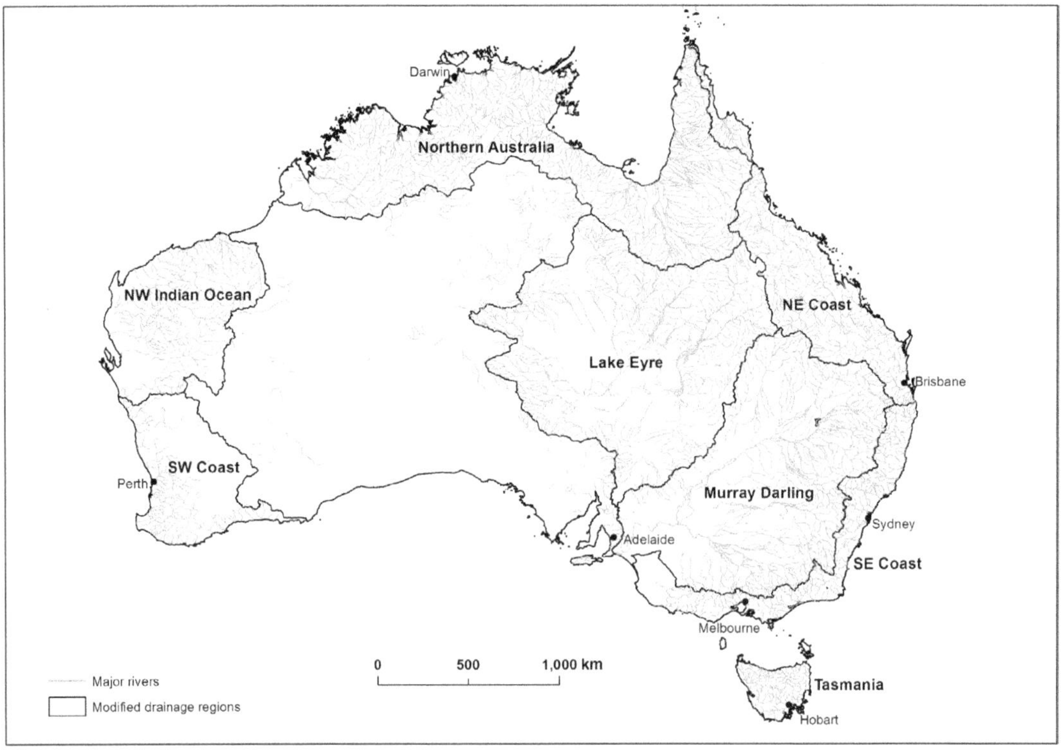

Figure 21.2 Eight floodplain and riparian vegetation zones based on modified drainage regions (AWRC 1976): (1) Northern Australia; (2) North-east Coast; (3) South-east Coast; (4) Murray–Darling; (5) Lake Eyre; (6) South-west Coast; (7) North-west Coast Indian Ocean and (8) Tasmania. Major rivers (light grey lines) and capital cities are included.

are generally perennial and have reliable base flows due to snowmelt and high rainfall in the ranges.

Floodplains form in most climatic regions across Australia where regular floods overtop river banks and spread out across adjoining areas of low relief. In most catchments, floods can be the result of rain falling many tens to thousands of kilometres away. For example, in the Channel Country of western Queensland and north-west New South Wales, large volumes of water move through the region after rain associated with monsoon events in northern Queensland (Gibling *et al.* 1998). Similarly, the Darling River in western New South Wales receives much of its water via rivers that flow from the western side of Great Dividing Range.

In eastern Australia, the Great Dividing Range is the source of most river systems; rain and snow falls on the mountains and water flows down either side. West flowing rivers are usually long, and form part of three main drainage basins: (i) Gulf of Carpentaria basin (the eastern section of the Northern Australia zone, Figure 21.2); (ii) Lake Eyre Basin which is endorheic (does not reach the ocean but instead flows into a lake system); and (iii) Murray–Darling Basin. Rivers that flow east from the Great Divide make up the North-east Coast and South-east Coast drainage basins (Figure 21.2), where rivers are generally fast flowing and cover short distances through higher rainfall areas, due to their proximity to the coast.

The Murray–Darling Basin covers an area of more than 1 million km^2 and incorporates many complex floodplain wetland systems, including the Lowbidgee Floodplain, Menindee Lakes, Darling Anabranch Lakes and Macquarie Marshes (MDBA 2016). Some rivers, such as the Gwydir, Narran and Lachlan, do not reach the Murray River in average years, instead dissipating into terminal wetland systems such as the Gwydir Wetlands, Narran Lakes and Great Cumbung Swamp (MDBC 2006). Much of the Murray–Darling Basin naturally experiences prolonged drought and highly variable hydrology, with many lowland rivers described as having variable, unpredictable and highly to extremely intermittent flows (Kennard *et al.* 2010). On average, only 6% of mean annual rainfall in the Murray–Darling Basin is transformed into runoff, with decreases in runoff predicted as a result of climate change (Chiew *et al.* 2008).

Catchments in the semi-arid Lake Eyre Basin have extremely low gradients and little topographic relief, resulting in a vast area of floodplain (30% of the catchment) and a complex series of channels (Walker *et al.* 1997). These features provide important drought refugia for biota, are sites of high productivity in a dry landscape, and support a productive cattle industry.

Rivers in northern Australia drain west and north into the Timor Sea and Gulf of Carpentaria (wet–dry tropics), and north-east into the Coral Sea (wet tropics). In these rivers, stream flow is strongly seasonal with high interannual variability (Kennard *et al.* 2010). Disconnected pools are sustained to varying degrees by subsurface and local groundwater flow in the dry season. The only perennial rivers (with some sustained base flow) are those that are closely connected to large groundwater aquifers, such as the Daly and Roper rivers in the Northern Territory.

West Australian river systems cover a wide climatic range. In the south-west coast zone (Figure 21.2), the climate is Mediterranean, with wet winters and extended summer drought. These rivers are characterised by their distinctive geology, with areas of ancient, low-relief drainage in the inland headwaters section, higher relief rejuvenated drainage in the middle sections and sedimentary soils in the coastal sections. In the north-west Indian Ocean zone (Figure 21.2), the climate is semi-arid to arid with highly episodic rainfall and stream flows.

21.2.1 Geomorphology and soils

Floodplains are complex fluvial environments consisting of a range of different habitats formed from thousands to millions of years of sediment deposition by successive flood events (Nanson and Croke 1992). Floodplains are formed when suspended sediments settle out of floodwaters, forming alluvial floodplain soils (Boulton *et al.* 2014). Clays are deposited by very slow moving floodwaters, while course sands or gravel may be deposited by fast moving, high-energy floodwaters (Nanson and Croke 1992).

No two flood events are the same, so clays are often interspersed with layers of sand or gravel, depending on the origin of floodwaters, the velocity of water movement and catchment lithology (Page *et al.* 2003). Soils close to river channels often have higher silt and sand content than the surrounding floodplain (Butler 1950) and generally have higher organic matter content, particularly in the surface soil (Stace *et al.* 1968). In inland Australia, active floodplains generally consist of highly fertile clays (black or grey vertosols) that form deep cracks when dry and swell on wetting (Stace *et al.* 1968).

A number of structures may exist within floodplain landscapes, including belts of undulating land with coarse-textured soils associated with prior stream formations, sandy lenses and depressed zones known as 'flood runners' that are important conductors of water away from rivers and streams during floods. These slight differences in elevation across the floodplain result in complex drainage patterns and localised runoff, differing water availability and flood frequencies. Floodplain soils may also form gilgai structures which can produce repeated differences in elevation of a metre or more over a short distance.

21.3 Composition, structure and distribution of floodplain and riparian vegetation

Knowledge of the composition and distribution of floodplain and riparian vegetation is far from complete. Research efforts have generally focused on temperate zones, leaving tropical and arid floodplains under-represented in the published literature (Thoms *et al.* 2016). Temperate inland and coastal river systems can be highly modified and so our knowledge of these plant communities is also hindered. We describe some of the key features of common Australian floodplain and riparian woodlands, forests and scrubs. Common species found in these vegetation communities, along with habitat and climatic factors can be found in Table 21.1.

21.3.1 Northern Australia

In the wetter parts of northern Australia, including the Alligator Rivers region and Arnhem Land, the inundated floodplain is dominated by aquatic grasses and sedges. Scattered areas of trees of *Melaleuca leucadendra* and *M. cajuputi* form an overstorey in some areas of the floodplain while *M. viridiflora* and *M. dealbata* occur on the margins, as they are generally tolerant of drier conditions (Cowie *et al.* 2000). In the deeper, permanent billabongs, floating and submerged plants are prominent, including *Nymphaea* spp., *Nymphoides indica*, *Utricularia* spp., *Ceratophyllum demersum* and *Najas tenuifolia*. In the dry season, the large aquatic grasses dieback and there is a suite of short-lived terrestrial herbaceous plants that take advantage of the dry conditions and complete their life cycle within the dry season.

On flat alluvial floodplains in drier regions, such as rivers draining into the Gulf of Carpentaria, open woodlands of *Eucalyptus microtheca* occur (Neldner *et al.* 2014). Other species present can include *E. chlorophylla*, *Corymbia bella* and *Grevillea striata*. Grasses generally dominate the understorey. *E. microtheca* can occur in low-lying areas of the floodplains with *Excoecaria parvifolia* and a ground storey of *Dichanthium* spp. Riparian forests occur along most rivers in the wet–dry tropics and consist of mixed forest with dominants including *E. camaldulensis*, *Melaleuca leucadendra*, *M. argentea*, *Pandanus spiralis*, *Terminalia platyphylla*, *Lophostemon grandiflorus*, *Nauclea orientalis*, *Barringtonia acutangula* and *Ficus* spp. (Dowe 2008).

21.3.2 North-east Coast

In the wet tropics region of north-east Queensland, *Melaleuca leucadendra/fluviatilis* (paperbark) forest is the most common riparian element on well-drained alluvial soils. Co-occurring species include trees such as *Eucalyptus tereticornis*, *Tristaniopsis exiliflora* and *Xanthostemon chrysanthus*, as well as some vine forest elements of the rainforest (Dowe 2008). *Melaleuca quinquenervia* occurs from Cape York to northern New South Wales, east of the Great Dividing Range, usually near the coast. This species is generally found near streams and fringing tidal estuaries, but has only limited salt tolerance, growing best in fresh water swamps.

21.3.3 Murray–Darling Basin

In riparian zones and floodplain depressions, *E. camaldulensis* subsp. *camaldulensis* or subsp. *acuta* (river red gum) is the dominant canopy species, forming open forests and woodlands (Roberts and Ludwig 1990; Keith 2004). In the most frequently flooded areas with the highest moisture availability, reeds, sedges, rushes and forbs dominate the understorey. In less frequently flooded

Table 21.1 **Common woody vegetation types found in floodplain and riparian environments in Australia and examples of associated understorey species. Note: Only a selection of vegetation types are included for areas with high habitat diversity, such as Tasmania and the South-east Coast. Flood frequencies are estimates only.**

Zone	Climate	Vegetation type	Landscape position	Flood frequency	Upper stratum	Midstorey	Groundstorey	Key references
Northern Australia	Tropical	*Melaleuca* forests	Floodplain, back swamps	Annual	*Melaleuca leucadendra, M. cajuputi, M. viridiflora*	*Pandanus aquaticus*	*Hymenachne acutigluma, Oryza* spp., *Pseudoraphis spinescens*	Cowie *et al.* (2000)
		Melaleuca woodlands	Coastal floodplain	Annual	*Melaleuca leucadendra, M. cajuputi*	*Pandanus aquaticus, Barringtonia acutangula*	*Pseudoraphis spinescens, Hymenachne acutigluma, Nymphaea* spp.	Finlayson (2005)
		Coolibah (*E. microtheca*) woodlands	Upper floodplain	Annual	*Eucalyptus microtheca, Corymbia bella*	*Excoecaria parvifolia, Barringtonia acutangula*	*Dichanthium* spp., *Bothriochloa decipiens, Themeda arguens*	Pettit *et al.* (2012)
		River red gum/paperbark forests	Riparian	Annual	*Eucalyptus camaldulensis, Melaleuca leucadendra, M. fluviatilis, M. argentea*	*Terminalis* spp., *Pandanus aquaticus*	*Mnesithea rottboellioides, Lomandra longifolia, Bothriochloa bladhii*	Dowe (2008)
North-west Indian Ocean	Semi-arid/arid	Smooth-barked coolibah (*E. victrix*) woodlands	Floodplain and riparian	20–30 yrs	*Eucalyptus victrix, Acacia aneura*	*Senna artemisioides* subsp. *oligophylla*	*Eriachne flaccida, Chrysopogon fallax, Cyperus* spp.	Florentine (1999); Lyon (2015)
South-western Australia	Mediterranean	Flooded gum (*E. rudis*) woodlands	Riparian and floodplain	1–10 yrs	*Eucalyptus rudis, Melaleuca rhaphiophylla*	*M. rhaphiophylla, Banksia seminuda*	*Carex appressa, Chorizandra enodis, Baumea* spp.	Hancock *et al.* (1996)

Table 21.1 (cont.)

Zone	Climate	Vegetation type	Landscape position	Flood frequency	Upper stratum	Midstorey	Groundstorey	Key references
Lake Eyre Basin	Arid	Coolibah (*E. coolabah*) and river red gum woodlands	Floodplain, riparian, fringing wetlands	3–10 yrs	*E. coolabah*, *E. camaldulensis*	*Duma florulenta*, *Santalum lanceolatum*, *Eremophila bignoniiflora*	*Paspalidium jubiflorum*, *Eulalia aurea*	Boyland (1984); Pettit (2002)
	Arid	Lignum shrublands	Lower floodplain (fringing wetlands)	3–10 yrs	*Duma florulenta*, *Acacia stenophylla*	*Chenopodium* spp.	*Leptochloa digitata*, *Paspalidium jubiflorum*, *Sporobolus mitchellii*	Craig *et al.* (1991)
Murray–Darling Basin	Semi-arid/ subtropical	Coolibah woodlands	Upper floodplain	1 in 10 yrs	*Eucalyptus coolabah* subsp. *coolabah*, *E. largiflorens*, *E. camaldulensis*	*Acacia stenophylla*, *Duma florulenta*, *Atriplex* spp.	*Paspalidium jubiflorum*, *Einadia nutans*, *Sclerolaena* spp.	Good *et al.* (2012)
	Semi-arid/ temperate	Black box woodlands	Upper floodplain	1 in 10 yrs	*Eucalyptus largiflorens*, *E. camaldulensis*	*Atriplex* spp., *Chenopodium* spp.	*Lachnagrostis filiformis*, *Alternanthera denticulata*, *Sclerolaena tricuspis*,	Roberts (2004)
	Semi-arid/ temperate	River red gum forest	Lower floodplain and riparian	1 in 3 yrs	*Eucalyptus camaldulensis*, *E. largiflorens*, *E. coolabah*	*Duma florulenta*, *Acacia stenophylla*, *Acacia salicina*	*Eleocharis acuta*, *Carex tereticaulis*, *Paspalidium jubiflorum*	Roberts (2004)
	Semi-arid/ temperate	Myall shrublands	Upper floodplain	Rarely	*Acacia pendula*, *Casuarina cristata*, *Alectryon oleifolius*	*Rhagodia spinescens*, *Maireana* spp., *Enchylaena tomentosa*	*Dichanthium sericeum*, *Enteropogon acicularis*, *Austrostipa aristiglumis*	Roberts (2004)

	Semi-arid/ subtropical	Brigalow shrublands	Upper floodplain	Rarely	*Acacia harpophylla, Casuarina cristata, Eucalyptus populnea*	*Geijera parviflora, Apophyllum anomalum, Citrus glauca*	*Sclerolaena* spp., *Austrostipa* spp., *Marsilea drummondii*	Roberts (2004)
	Temperate	River oak scrub	Riparian	N/A	*Casuarina cunninghamiana*	*Acacia mearnsii, A. floribunda*	*Lomandra longifolia, Carex appressa*	Woolfrey and Ladd (2001)
South-east Coast	Subtropical	Coastal floodplain forest	River terraces and alluvial flats	Periodically inundated	*Eucalyptus tereticornis, E. siderophloia, Corymbia intermedia*	*Alphitonia excelsa, Breynia oblongifolia, Acacia concurrens*	*Imperata cylindrica, Themeda triandra, Cyanthillium cinereum*	Keith and Scott (2005)
	Temperate	Coastal swamp sclerophyll forests	Alluvial flats on floodplain margin	Waterlogged/ periodically inundated	*Eucalyptus robusta, M. quinquenervia, E. botryoides*	*Acacia longifolia, Dodonaea triquetra, Leptospermum polygalifolium*	*Gahnia clarkei, Pteridium esculentum, Dianella caerulea*	Keith and Scott (2005)
	Temperate	Swamp oak forest	Low-lying flats, lake and estuarine margins	Waterlogged/ periodically inundated	*Casuarina glauca, Syzygium smithii, Melaleuca ericifolia*	*Parsonsia straminea, Geitonoplesium cymosum*	*Centella asiatica, Carex appressa, Lomandra longifolia*	Keith and Scott (2005)
	Temperate	Lowland rainforest	Riparian and alluvial flats	Periodically inundated	*Syzygium smithii, Alphitonia excelsa, Elaeocarpus obovatus*	*Cissus* spp., *Breynia oblongifolia*	*Commelina cyanea, Dianella caerulea, Lomandra longifolia*	Keith and Scott (2005)
	Temperate	River flat Eucalypt forest	River terraces and flats	Periodically inundated	*Eucalyptus tereticornis, E. amplifolia, Angophora floribunda*	*Melaleuca* spp., *Bursaria spinosa, Solanum prinophyllum*	*Microlaena stipoides, Entolasia marginata, Dichondra repens*	Keith and Scott (2005)

Table 21.1 (cont.)

Zone	Climate	Vegetation type	Landscape position	Flood frequency	Upper stratum	Midstorey	Groundstorey	Key references
North-east Coast	Tropical	Gallery forests	Riparian	N/A	*Melaleuca leucadendra, Eucalyptus tereticornis*	*Tristaniopsis exiliflora, Xanthostemon chrysanthus*	Vine thicket understorey, ferns and bryophytes	Dowe (2008)
	Subtropical/ tropical	Gallery rainforests	Riparian	N/A	*Alphitonia excelsa, Archontophoenix cunninghamii, Castanospermum australe, Ficus coronata*	*Tristaniopsis laurina, Breynia oblongifolia, Maclura cochinchinensis, Notelaea longifolia*	*Commelina cyanea, Lomandra longifolia, Pseuderanthemum variabile*	Keith and Scott (2005)
	Subtropical/ tropical	*Melaleuca* swamps	Fresh water swamps and fringing tidal estuaries	Periodically inundated	*Melaleuca quinquenervia, Eucalyptus tereticornis, Lophostemon suaveolens*	*Banksia* spp., *Glochidion* spp., *Petalostigma pubescens*	*Gahnia sieberiana*, grasses and sedges	Keith and Scott (2005)
Tasmania[a]	Temperate	Riparian rainforests and forests	Riparian	N/A	*Eucalyptus delegatensis, E. viminalis, E. regnans.*	*Pomaderris apetala, Acacia dealbata*	Mixed grasses, sedges, ferns	Daley and Kirkpatrick (2004)
	Temperate	Estuarine forest			*Melaleuca ericifolia, Eucalyptus amygdalina, E. ovata*	*Melaleuca squarrosa, Leptospermum scoparium*	*Lomandra longifolia, Juncus kraussii*	Daley and Kirkpatrick (2004)
	Temperate	Mixed eucalypt woodlands			*Eucalyptus viminalis, E. amygdalina*	*Leptospermum lanigerum*	*Poa* spp., *Lepidosperma inops*	Daley and Kirkpatrick (2004)

[a]Only three examples from 21 vegetation types described in Daley and Kirkpatrick (2004)

sites, the understorey is predominantly composed of grasses, forbs and chenopods. A shrub layer of *Duma florulenta, Acacia salicina* and *A. stenophylla* is occasionally present.

Eucalyptus coolabah subsp. *coolabah* (coolibah) occurs on floodplains in the north of the Murray–Darling Basin and is replaced by *E. largiflorens* (black box) south of Bourke, New South Wales (with some overlap). *E. coolabah* subsp. *coolabah* generally has a grassy understorey dominated by C_4 grasses in higher rainfall areas and *Duma florulenta* (lignum) shrubs in frequently flooded sites (Hunter 2005; Good *et al.* 2012), whereas *E. largiflorens* woodlands often have a chenopod-dominated understorey. Both *E. coolabah* subsp. *coolabah* and *E. largiflorens* form open woodlands (interspersed with patches of dense regeneration) with tree density ranging from ~25 trees per hectare in older stands to ~3000 trees per hectare following episodic recruitment (Good *et al.* 2011; Good 2012).

Acacia shrublands dominated by *A. pendula* (myall) and *A. harpophylla* (brigalow) occur in slightly elevated positions and across old lowland river floodplains where flooding is infrequent. *A. harpophylla* occurs throughout central Queensland and northern New South Wales (Fensham *et al.*, Chapter 16, this volume), while *A. pendula* occurs throughout New South Wales and southern central Qld. *A. harpophylla* can be found within riparian zones of some river systems, including the Comet River east of Emerald, in diverse, mixed subtropical shrub assemblages commonly found across northern Australia. *A. pendula* is found across the ancient floodplains of the southern Murray–Darling Basin.

Casuarina cunninghamiana (river she oak) grows densely along the river edge in the headwaters of the Great Dividing Range. This is a widely distributed obligate riparian tree occurring from southern New South Wales to the Daly River in the Northern Territory, but does not occur in areas with less than 500 mm mean annual rainfall, which means it does not extend far along west flowing rivers (Woolfrey and Ladd 2001). *C. cunninghamiana* forms a dense forest or scrub with an understorey of rushes and sedges and some co-occurring *Acacia* species (Table 21.1; Beadle 1981; Keith 2004).

21.3.4 South-east Coast

The south-east coast of Australia contains a diverse array of floodplain and riparian habitats, and considering its relatively small area, contains a large proportion of the country's Ramsar-listed wetlands (Boon *et al.* 2016). These areas have largely been cleared and converted to agricultural or urban land since European occupation (Ladd 1978; Lunt 1997; Keith and Scott 2005). Along the New South Wales coast, floodplain and riparian vegetation types tend to be distributed according to soil moisture, fertility and salinity tolerance (Keith and Scott 2005). Lowland rainforests once occurred in riparian zones of major rivers and on floodplains where soil moisture was abundant (Keith and Scott 2005). The more elevated, less frequently inundated parts of floodplains of coastal New South Wales once supported tall open eucalypt forests, with the dominant tree species changing from north to south of the Hunter region (Keith and Scott 2005). *Melaleuca* spp. and *E. robusta* forests occur on sandier and less well-drained soils, particularly in depressions associated with coastal sandplains (Keith and Scott 2005). Riverine forests dominated by *Casuarina cunninghamiana* and/or *Tristaniopsis laurina* dominate the fast-flowing streams of the coastal lowlands from the south-east of Queensland to Victoria (Keith 2004).

Along the southern coast, temperate rainforests occur along streams that flood regularly. Various floodplain and riparian eucalypt species (*E. ovata, E. camaldulensis, E. tereticornis,* and *E. radiata*) are found in riparian zones, creek lines, terraces and floodplain depressions as forests or woodlands, depending on moisture availability. The forests have a diverse midstorey of shrubs (*Acacia* and *Leptospermum* spp.) and an understorey of sedges, grasses, herbs and ferns. Riparian thickets occur in regularly flooded stream beds and floodplain terraces where *Melaleuca squarrosa* and *Leptospermum lanigerum* are frequently dominant, with an understorey of ferns and sedges.

21.3.5 Tasmania

The characteristics of native riparian vegetation in high-rainfall areas of Tasmania closely resemble the adjacent dominant vegetation types in structure and species composition. Daley and Kirkpatrick (2004) described and mapped 21 different riparian plant communities for Tasmanian rivers. There are distinct riparian communities at higher elevations such as grassy heath, and grassy woodlands of *Eucalyptus pauciflora* and *E. delegatensis*, widespread areas of closed scrub of *Leptospermum lanigerum* and *Melaleuca squarrosa* and woodlands of mixed *Eucalyptus* spp. and *Acacia* spp., as well as estuarine forest of *Melaleuca ericifolia, Lomandra longifolia* and *Juncus kraussii*. Two rare shrub species (*Gynatrix pulchella* and *Phebalium daviesii*) are restricted to the riparian zone and another 76 species are predominantly found along Tasmanian rivers. Riparian trees include *Eucalyptus ovata, Lagarostrobos franklinii* (Huon pine), *Acradenia frankliniae* and *Callitris oblonga*. Shrub species commonly, though not exclusively found in riparian habitats include *Pomaderris apetala, Leptospermum lanigerum, Acacia dealbata, A. melanoxylon* and *Coprosma quadrifida*.

21.3.6 South-west Coast

Eucalyptus rudis (closely related to *E. camaldulensis*) dominates woodlands along watercourses in south-west Western Australia. There is generally a midstorey of *Melaleuca rhaphiophylla* fringing the area nearest the water edge with rushes and sedges such as *Juncus* spp. and *Baumea* spp. as a groundstorey (Hancock *et al.* 1996). Further east, headwater zones of west and south flowing rivers, where most native vegetation has been cleared for cereal cropping, are now affected by secondary salinisation. Only streams that are in uncleared, mid-catchment areas are unaffected. Salt-tolerant species now predominate in these areas such as *Casuarina obesa* and *Melaleuca hamulosa*, and where stream salinity is particularly high, samphire, bluebush and chenopod species predominate. In some areas of better soils and drainage, such as the Beaufort River flats, areas of high species diversity persist, including diverse myrtaceous shrublands with many *Melaleuca* species.

21.3.7 North-west Indian Ocean

In the semi-arid rangelands (Pilbara, Gascoyne region, WA), rivers have highly intermittent flows and are dry most of the time. Springs and river pools with high water permanence support *Melaleuca argentea, E. camaldulensis* subsp. *refulgens* and *Acacia ampliceps*, with an understorey of reeds, rushes and sedges (Florentine 1999; Lyon 2015). *E. victrix* woodlands occur on floodplain areas with fine-textured soils, while *E. camaldulensis* occupies riparian areas. Thickets of *Duma florulenta* (lignum) can be found on red clays in run-on zones, often along with an overstorey of *E. camaldulensis*. The Fortescue Marsh forms a large endorheic floodplain that is episodically inundated and is dominated structurally by salt-tolerant samphire species including *Muellerolimon salicorniaceum* with some *Tecticornia* spp. (Lyon 2015), and a fringing woodland of *E. victrix*.

21.3.8 Lake Eyre Basin

The riparian vegetation fringing the permanent waterholes of Cooper Creek and the Diamantina and Georgina Rivers consists of an open woodland of *Eucalyptus coolabah* and *E. camaldulensis*. There is a sparse shrub understorey of chenopods including several *Sclerolaena* spp., *Chenopodium auricomum* and perennial grasses. *E. camaldulensis* occurs closer to the water's edge, with a midstorey of *Melaleuca linariifolia* (Boyland 1984; Pettit 2002; Capon 2005). *E. coolabah*, with a midstorey of *Bauhinia gilva*, *Acacia stenophylla*, *Santalum lanceolatum* and *Eremophila bignoniiflora* occur on the floodplain side of the riparian strip. Most of the floodplain areas in the Lake Eyre Basin do not support woody vegetation and herbaceous plants predominate, with community composition influenced by flooding frequency (Capon 2005).

21.4 Vegetation dynamics on floodplains

The vegetation and structure of floodplain landscapes changes in a shifting mosaic across large temporal and spatial scales (Thoms and Parsons 2011). For example, riparian vegetation along high-energy streams can be completely restructured following a single large flood event (Woolfrey and Ladd 2001; Keith 2004), whereas woody vegetation on extensive, semi-arid plains slowly shifts in response to century-long flood regimes through patchy tree mortality and regeneration (Thoms and Parsons 2011; Good *et al.* 2014). Fire and waterlogging are important drivers of vegetation dynamics in northern Australian floodplains and riparian zones (Bowman and Wilson 1986; Edwards *et al.* 2003; Petty and Douglas 2010; Warfe *et al.* 2011). In coastal floodplains the interactions among tidal, groundwater and surface water levels drive long-term vegetation change (Boon *et al.* 2016). Despite these differences, there are some similarities in the life history of dominant woody floodplain plants (Figure 21.3) and broad patterns of vegetation change tend to reflect the interaction between hydrological regimes and tree life cycles.

21.4.1 Phenology and dispersal

Many riparian and floodplain tree species flower and set seed in response to large rainfall events or floods. In the tropics, seed production and release is often annual due to reliable wet seasons (Pettit and Froend 2001b; Figure 21.3). In areas with less reliable rainfall, seed production in eucalypt trees is very strongly related to frequent rainfall (Travers and Eldridge 2013). Floodplain and riparian trees generally do not have a long-lived soil seed bank, possibly due to frequent soil disturbance – from flooding and self-mulching clay soils (Greet *et al.* 2012). Canopy seed storage (serotiny) is not a common trait among these species. This means that seed fall must coincide with optimal germination conditions (following floods or high rainfall). Woolfrey and Ladd (2001) reported mast seeding in *Casuarina cunninghamiana* and this seed production strategy may be common among riparian and floodplain species. Unfortunately, the mechanisms and timing of seed fall in floodplain and riparian eucalypts is not well understood and requires further investigation.

Seeds of floodplain and riparian trees are generally short lived and often dispersed by river and flood water (Pettit and Froend 2001b; Figure 21.3). For example, riparian trees in temperate regions (*E. camaldulensis* and *Callistemon sieberi*) produce and release short-lived seeds during spring, when unregulated river flows peak (Greet *et al.* 2012). If the timing of floods are optimal (coinciding with seed fall), this mode of dispersal results in seeds travelling further from mature trees, avoiding competition, and being deposited in productive and/ or protected microsites (Pettit and Froend 2001b). Alternatively, seed released following rainfall will be dispersed by gravity under or near mature tree canopies.

21.4.2 Tree recruitment

Successful tree recruitment only occurs when seedlings survive to reproductive maturity. There are many abiotic and biotic factors that reduce this likelihood along the way (Figure 21.3). Floodplain eucalypt seeds can germinate *en masse* following

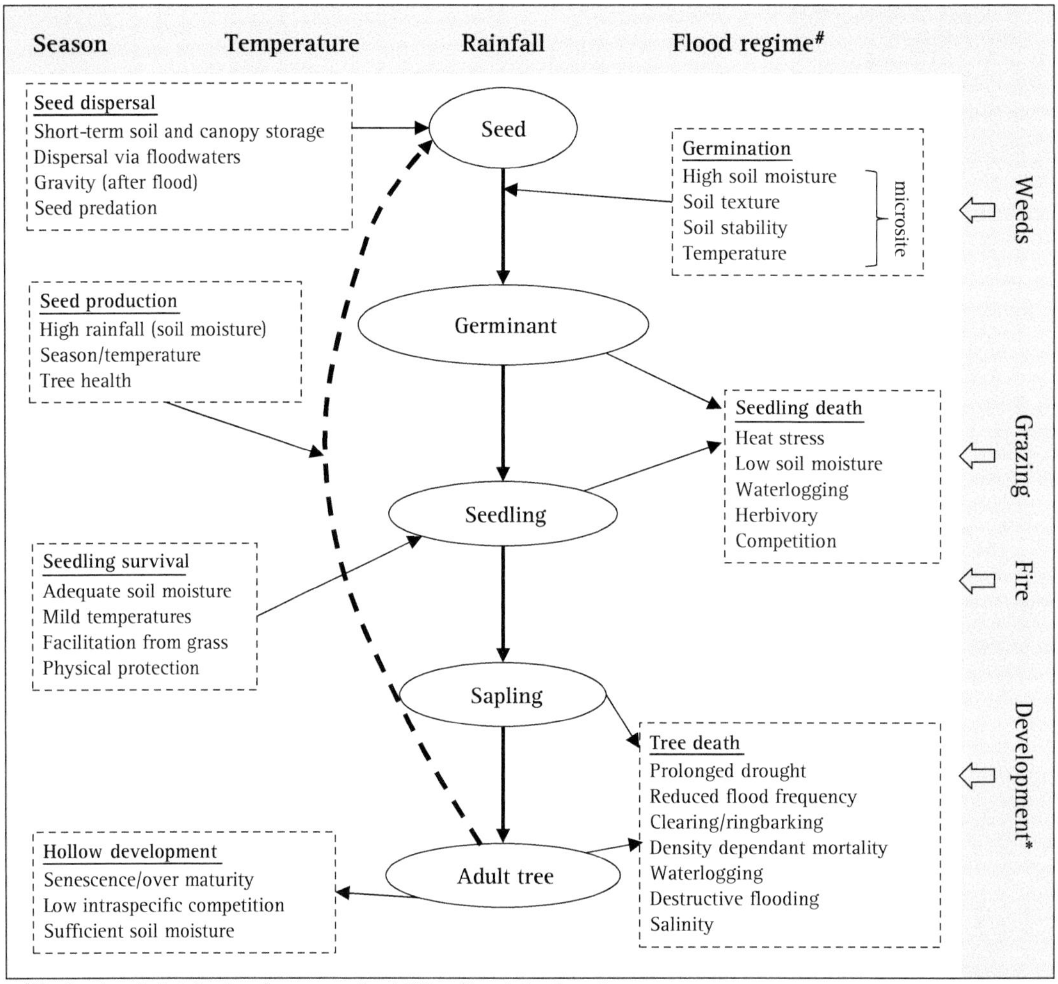

Flood regime: timing, duration, frequency and variability of inundation for a given area

*Development: direct and indirect European modifications including clearing, thinning, salinity, altered flood regimes which mostly affect the life-span of adult trees.

Figure 21.3 General model of floodplain tree life cycle. Environmental conditions that affect most life stages and transitions between life stages are shown at the top and exogenous factors are shown on the right side of the diagram (their position indicates which transitions they have the most impact upon).

flood recession and moisture is the key stimulus for germination in woody floodplain species (Boland *et al.* 1980; Jensen *et al.* 2008). As such, most floodplain woodlands tend to recruit episodically following floods (Roberts 1993; Pettit and Froend 2001a). This is particularly evident in the arid climatic zones where tree regeneration is unlikely without the high soil moisture levels that are only available after a flood (Roberts 1993; Pettit 2002).

Survival of seedlings depends on sufficient follow-up rainfall, protection from herbivory, subsequent flooding and low levels of competition from adult trees (Fox *et al.* 2004; Jensen *et al.* 2008; Good *et al.* 2014). Follow-up rainfall is vital for the survival of *E. camaldulensis* and *E. largiflorens* germinants in the southern Murray–Darling Basin (Dexter 1968; Jensen *et al.* 2008). In the arid Pilbara region of Western Australia, low rainfall resulted in zero survival of *E. victrix* seedlings over 2 years (Fox *et al.* 2004). Good *et al.* (2014) found that herbaceous vegetation facilitates *E. coolabah* seedling survival by reducing water and heat stress. In low-lying areas, long periods of inundation will also prevent seedling establishment. Clearly, seedling survival is a strong bottleneck for floodplain tree regeneration, particularly in arid regions, and this tends to result in a pattern of rare and patchy regeneration in the landscape.

Riparian species, particularly those found east of the Great Dividing Range, such as *Casuarina cunninghamiana* in south-east Australia, recruit more regularly due to their proximity to a somewhat dependable water source. However, floods and droughts tend to increase and decrease recruitment success respectively (Woolfrey and Ladd 2001). Substrate is also very important for *C. cunninghamiana* recruitment success, with seedlings surviving best on a cobble substrate (Woolfrey and Ladd 2001; Figure 21.3). Similar patterns of riparian tree establishment were found on the Ord and Blackwood rivers in north and south Western Australia respectively (Pettit and Froend 2001a). Although flooding increases seedling germination, subsequent high-energy floods following rainfall events kill seedlings and saplings, thus preventing establishment of trees close to the river (Pettit and Froend 2001a).

Tropical floodplain trees are not likely to be limited by soil moisture but may require fires or some disturbance to the adult trees to initiate a recruitment event (Franklin *et al.* 2007). Several *Melaleuca* species that occur in Northern Australia were found to be closely associated with soil type and flood frequency (Franklin *et al.* 2007), and this probably reflects their tolerance of both drought and waterlogging.

21.4.3 Tree persistence

Once established, many floodplain trees use ground water where it is of sufficient quality, and at a depth that can be accessed by the tree roots (Thorburn *et al.* 1994; Cunningham *et al.* 2011). For example, river red gums may preferentially use saline groundwater as it is more reliable than the adjacent creek flows (Thorburn and Walker 1994). River red gums can also switch between water sources, reduce their canopy extent or density, and reduce growth rates in response to dry conditions (Mensforth *et al.* 1994; Doody *et al.* 2015). Floodplain trees have various root adaptations to take advantage of different water sources: 'sinker' roots access deep soil water in times of drought (Colloff 2014), aerenchymatous roots allow river red gums to survive during prolonged inundation (Sena Gomez and Kozlowski 1980), while 'feeder' roots in shallow soil enable uptake of moisture from rainfall events between floods (Colloff 2014).

In semi-arid regions such as the Lake Eyre Basin and North-west Indian Ocean regions, floods are essential for riparian tree survival (Pettit 2002; Capon 2005; Wen *et al.* 2009). Indeed, plant production is much higher along streams than in neighbouring upland areas due to greater access to water and flooding (Smith and Reid 2013). This is evidenced by the taller, denser vegetation along streamlines in these regions and the unique vegetation communities such as *E. camaldulensis* and *Melaleuca* spp. that do not occur outside these riverine zones (Figure 21.4a).

21.4.4 Tree death

Long periods without flooding or adequate rainfall can significantly reduce the life span of floodplain and riparian trees. Changes to groundwater, drought and altered flood regimes have resulted in large-scale tree dieback (Figure 21.4e) in many floodplain woodlands (Cunningham *et al.* 2007, 2011; Wen *et al.* 2009). Indeed, changes to flood regimes have been

Figure 21.4 (a) Upper Fitzroy River in the Kimberly region of north-western Australia. Riparian trees are *Melaleuca argentea* and *Terminalia bursarina.* (b) Back swamp area of Magela Creek floodplain in Kakadu National Park with *Melaleuca leucadendra* trees. (c) Remnant *Eucalyptus coolabah* woodland in the northern Murray–Darling Basin. (d) Inundated *Eucalyptus camaldulensis* in the Barmah–Millewa Forest on the Murray River. (e) *Eucalyptus camaldulensis* dieback in the northern Murray–Darling Basin. (f) Dense regeneration of *Eucalyptus coolabah* in the northern Murray–Darling Basin. (g) Floodplain development for agriculture in the Ord River Scheme, Western Australia.
Photos: (a), (b) Neil Pettit; (c), (d), (f) Megan Good; (e) Rhiannon Smith; (g) Google Earth Pro 7.1.7.2600. (November 28, 2015). Ord River, Kununurra, Australia. 15° 37' 20.30'S, 128° 45' 06.64'W, Eye alt 19.75 km. CNES/Astrium 2016.)
(A black and white version of this figure will appear in some formats. For the colour version, please refer to the plate section.)

cited as the main cause of *E. camaldulensis* dieback along the Murray River (Fensham *et al.* 2009; Horner *et al.* 2009; Doody *et al.* 2014). Flooding can also kill riparian trees; the physical force of high-energy flows damaging trees closest to the river (Pettit and Froend 2001a; Woolfrey and Ladd 2001).

Another cause of tree death in woodlands is density dependant mortality. Following dense episodic recruitment trees generally self-thin over time as slightly larger trees outcompete the smaller trees in an even-aged stand (Mohler *et al.* 1978; White 1981; Baldwin *et al.* 2000; Wiegand *et al.* 2008). Good (2012) demonstrated this relationship in densely regenerating *E. coolabah* subsp. *coolabah* stands in northern New South Wales (Figure 21.4f) and high rates of mortality were reported in dense *E. camaldulensis* stands on the Murray River (Horner *et al.* 2009; Box 21.1).

Box 21.1 Eucalypt woodland dynamics in the Murray–Darling Basin: management and conservation issues

River red gum, black box and coolibah dominate woodlands across the Murray–Darling Basin. These ecosystems are relatively well studied, but uncertainty remains about the causes and management of dense regeneration and mortality of these species.

Dense regeneration of floodplain eucalypts occurred during large floods in the 1950s and 1970s, resulting in patchy dense stands of coolibah and black box on the northern Murray–Darling Basin floodplains (Maher 1995; Good *et al.* 2012; Figure 21.5), and river red gum in the southern Murray–Darling Basin forests (Dexter 1968). This was encouraged in river red gum forests that were valuable for timber production (Dexter 1968). However, concerns have been raised about impacts on biodiversity arising from a shift from forests dominated by large hollow-bearing trees to those dominated by dense regeneration (Horner *et al.* 2010; Mac Nally *et al.* 2011).

These changes in forest and woodland structure are unsurprising given large old trees were removed for timber and agricultural production, reducing competition and providing space for dense recruitment (Mac Nally *et al.* 2011). Recently, reduced flooding, increased salinity and altered groundwater depth have resulted in widespread dieback in regenerating river red gum forests (Kingsford 2000; Cunningham *et al.* 2007; Mac Nally *et al.* 2011). On the upper floodplains, coolibah and black box are still being cleared at high rates (Keith *et al.* 2009). Increased inter-canopy gaps allow larger areas of dense regeneration, resulting in a structural shift from woodlands and forests dominated by large old trees, to dense, even-aged stands.

When considering patterns of tree death and regeneration, a landscape-scale approach is required. Episodic regeneration is characteristic of floodplain and riparian trees.

Box 21.1 (cont.)

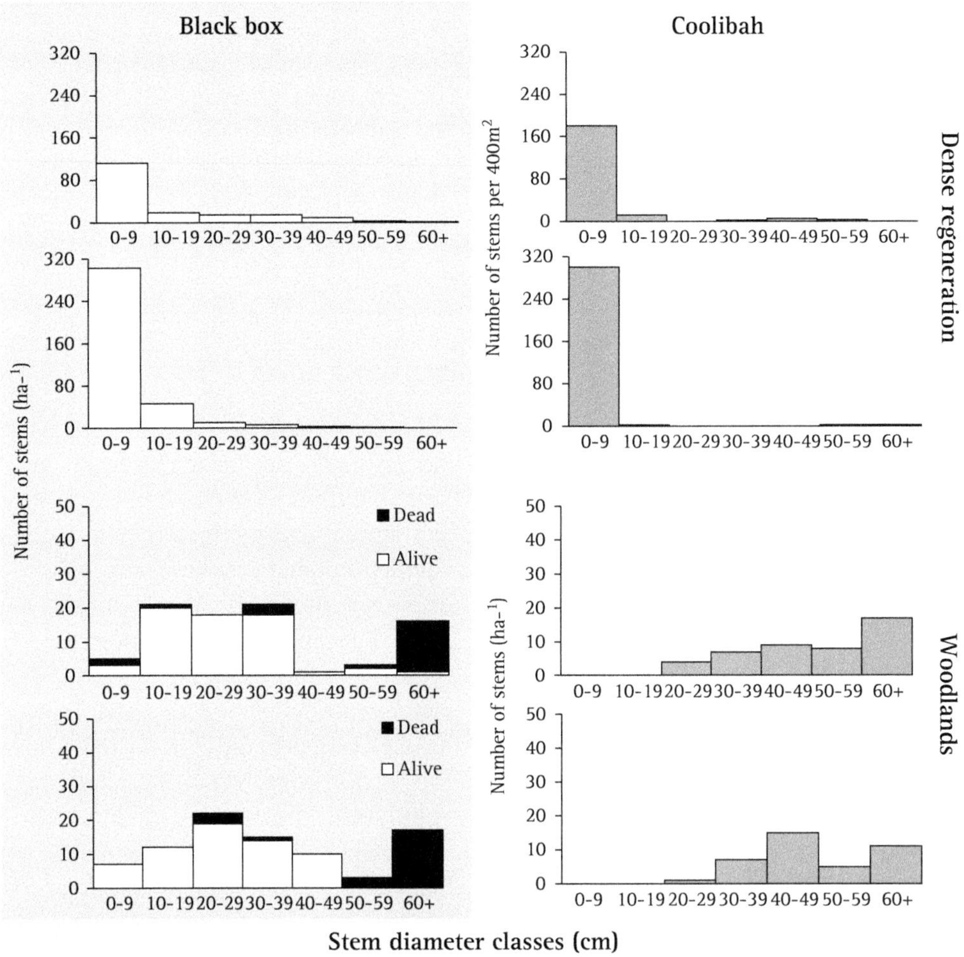

Figure 21.5 Examples of tree size frequency distributions of two floodplain woodland types: *Eucalyptus largiflorens* (left) and *E. coolabah* subsp. *coolabah* (right) in two different structural states: dense regeneration (top) and woodland (bottom).

In a survey of coolibah woodlands on the northern Murray–Darling Basin floodplains, Good (2012) suggested that remnant stands (Figure 21.4c) are the product of historical episodes of dense regeneration and subsequent self-thinning (Figure 21.5). A similar pattern was found in black box woodlands of the southern Murray–Darling Basin (Figure 21.5; unpublished data). At the landscape-scale, patches of dense regeneration (Figure 21.4f) fill the regeneration gap in mature woodlands where recruitment in the 0 – 20 cm diameter class is lacking. However, self-thinning of dense stands and development of mature woodlands on upper floodplains may take several decades.

Dense, structurally homogenous stands of river red gum forest along the Murray River suffer high mortality and low growth rates (Horner *et al.* 2009), and this could be prevented by mechanically thinning dense river red gum stands (Horner *et al.* 2010). A combination of well-timed environmental floods and artificial thinning may be required to encourage structural heterogeneity and habitat complexity, but interventions must not damage understorey plants or fauna. Further, patches of dense regeneration must be retained to increase forest and woodland heterogeneity at the landscape scale. The situation in upper floodplain woodlands is less clear, and Good *et al.* (2011) highlights problems with the New South Wales laws surrounding thinning and clearing of dense coolibah and black box regeneration. In all cases, more research into conditions that promote a sustainable and dynamic coexistence of mature and regenerating woodlands will ensure long-term persistence of these valuable ecosystems.

21.4.5 Herbaceous plant community dynamics

This is a brief description of herbaceous vegetation dynamics and more detail can be found in Catford *et al.* (Chapter 20, this volume). In some floodplain woodlands, a flood will bring about a shift towards a more aquatic assemblage of plant species in the understorey. The herbaceous element of floodplain woodlands can overlap substantially with treeless wetland assemblages and also, at the drier end of the scale, grasslands and savannas (Capon 2005). Indeed, Good *et al.* (2012) found that the understorey of coolibah woodlands in northern New South Wales were mostly indistinguishable from adjoining native grasslands and open savannas. Similarly, in floodplain forests dominated by *Melaleuca* spp. in Northern Australia, the understorey community is largely uncoupled from the overstorey (Franklin *et al.* 2007). The woody overstorey can exert a strong control over understorey biomass and in very dense coolibah woodlands and river red gum forests this can lead to situations with high volume of leaf litter and very low biomass of herbaceous species (Good *et al.* 2011).

21.5 Conservation issues

Humans have been using and shaping riverine landscapes in Australia for tens of thousands of years (Reid *et al.* 2016; Ens *et al.*, Chapter 4, this volume). Pre-European populations of Aboriginal people were largest in productive regions such as inland floodplains and coastal regions (Reid *et al.* 2016). This long phase of management by Aboriginal people ended abruptly when European squatters moved into riverine areas (see Reid *et al.* 2016 for a comprehensive review). Floodplain and riparian areas were attractive to Europeans for their timber and livestock grazing potential, and are today worth many billions of dollars annually in agricultural exports (MDBA 2012). As such, floodplain and riparian ecosystems are vulnerable to a range of competing land uses, leading to tensions relating to their sustainable management.

21.5.1 River regulation and water extraction

River regulation is possibly the biggest conservation issue for many of Australia's floodplains, especially in the Murray–Darling Basin (Kingsford 2000; Keith and Auld, Chapter 27, this volume). The ecological consequences of fundamentally altering the flows of rivers and the timing, duration and extent of flooding are so wide reaching that they cannot be adequately covered here (see Kingsford 2000 for more detail).

Even with abundant evidence of the effect of river regulation on floodplain, riparian, wetland and in-stream ecosystems (Kingsford 2000; Jenkins *et al.* 2005) pressure remains for irrigation development on unregulated river systems. The push for water extraction from dryland rivers in central Australia is especially concerning, given the highly variable flow regimes and ecology of dependant ecosystems in these regions (Kingsford *et al.* 1998). Tropical and subtropical river systems are also facing further development (Kingsford 2000). Although these systems may appear to have abundant water during wet seasons, floodplain ecosystems rely heavily on the connectivity that flooding provides (Warfe *et al.* 2011).

As mentioned in section 21.4, riparian floristics and reproduction are closely related to many aspects of flood regimes so that river regulation is likely to have a significant effect on tree recruitment (Figure 21.5). *Eucalyptus camaldulensis* dieback has also been linked to Murray River regulation and water extraction (Cunningham *et al.* 2007, 2011; Horner *et al.* 2009). Conversely, altered flow regimes have resulted in the encroachment of *E. camaldulensis* into grassy wetlands that were not inhabited by trees prior to the reduction in flood frequency and duration (Bren 1992). Indeed, the areas most affected by river regulation and water extraction are generally those that rely on the most frequent inundation (Kingsford 2000).

21.5.2 Clearing, thinning and logging

The fertility and flatness of floodplain landscapes means they have been very heavily modified for agricultural production, originally for livestock grazing and dairy farming and more recently for cropping (Reid *et al.* 2016). Cropping has resulted in high rates of clearing, especially in the Murray–Darling Basin (Cox *et al.* 2001; Bedward *et al.* 2007; Keith *et al.* 2009). Coastal floodplains were severely cleared and modified (over 99% of trees cleared) soon after European occupation, and very few examples of coastal floodplain vegetation remain (Ladd 1978; Lunt 1997; Keith and Scott 2005). In recent years, clearing rates have been highest in southern Queensland and northern New South Wales (Bradshaw 2012). Native vegetation protection policy and enforcement in these states are lacking (McAlpine *et al.* 2002; Fensham 2008). Indeed, recent changes to vegetation protection laws in Queensland resulted in a dramatic increase in clearing from 2012 to 2014 (Queensland Department of Science, Information Technology and Innovation 2015). The area of *E. coolabah* subsp. *coolabah* woodlands in northern New South Wales had been reduced by 61% of the pre-European extent, and clearing rates increased in recent years (1.7% per year) (Keith *et al.* 2009). These statistics highlight that clearing of native vegetation is very much an ongoing and highly political issue.

21.5.3 Dryland salinity

Dryland salinity is a widespread problem in Australia. The Australian Dryland Salinity Assessment 2000 (NLWRA 2000) estimated that the highest risk areas for salinity, based on groundwater depth and land-use practices, are found in south-west Western Australia, south-east South Australia, and western Victoria. These regions are already experiencing salinity problems caused by broad-scale clearing of native perennial vegetation and increased deep drainage of soil water.

The extensively cleared and cropped wheat belt of south-west Western Australia has been dramatically affected by salinity (George *et al.* 1997; Cramer and Hobbs 2002), with 36% of the total divertible water resources in this region now considered brackish and saline (Mayer *et al.* 2005). Native vegetation is severely affected by salinity and as saline groundwater approaches the soil surface, the combination of waterlogging and salinity causes death of vegetation that is sensitive to these conditions (George *et al.* 1997). Increased soil salinity has resulted in reduced plant species richness, diversity and altered community composition, including decline in riparian trees, shrubs and perennial herbs, and a switch to species that have greater salt tolerance, such as Amaranthaceae (including chenopod) species and exotic grasses including *Hordeum* spp. and *Distichium* spp. (Lymbery *et al.* 2003). Salinity has also caused dieback of *E. largiflorens* forests on the lower Murray River's Chowilla floodplain (Jolly *et al.* 1993) and *E. camaldulensis* forests along the Murray River are experiencing varying levels of dieback due to, among other things, saline groundwater (Slavich *et al.* 1999).

The causes and consequences of dryland salinity have been well studied, but management options are either largely unattainable (landscape-scale revegetation of native perennial vegetation) or have unknown consequences for downstream processes (deep drainage options; Pannell and Ewing 2006). Retention of existing native vegetation and allocation of water to environmental flows in heavily regulated river systems may prevent the spread of salinity in Australia. In a future changing climate, drier conditions may reduce salinity effects as the saline groundwater is lowered. However, as rainfall decreases, the flushing effect of higher rainfall is reduced, possibly leading to increased salinity in some areas (Pettit *et al.* 2015).

21.5.4 Livestock grazing

Livestock grazing is known to have a strong negative influence on riparian and floodplain vegetation (Fleischner 1994; Jansen and Robertson 2001; Lunt *et al.* 2007; Pettit *et al.* 2012). In areas where cattle congregate, there may be a stronger but relatively localised effect on riparian vegetation (Pettit 2002). Riparian condition declines with increasing grazing intensity (Jansen and Robertson 2001) but recovery of riparian vegetation following grazing exclusion is less predictable (Lunt *et al.* 2007). In the Tasmanian Midlands, Leonard and Kirkpatrick (2004) found that grazing intensity was associated with increased exotic species richness. In the same region, Fensham and Kirkpatrick (1989) found that grazing could also have a positive effect on native species richness in fertile sites but not in unproductive sites, a pattern that was found in Victorian grassy ecosystems (Schultz *et al.* 2011) and globally (Proulx and Mazumder 1998). However, species richness is only one measure of ecosystem health; livestock grazing has a homogenising effect on landscape-scale diversity (Schultz *et al.* 2014), as well as negative effects on ecosystem structure and function (Eldridge *et al.* 2016 and Chapter 6, this volume).

21.5.5 Weeds

Non-native invasive plant species have fundamentally altered many floodplain and riparian ecosystems in Australia. Generally, weeds are associated with other disturbances, especially cattle grazing, cropping and river regulation; these conditions reduce the ability of native plant communities to compete with invading species while also directly providing bare soil and resources to allow non-native species to establish. Once established, weeds can change nutrient cycling, species composition, structure, habitat values and fire regimes in affected ecosystems (Leishman *et al.*, Chapter 9, this volume).

Weed invasions generally affect the herbaceous component of floodplain and riparian woodlands, rather than the structure and function of the woody overstorey. However, the overstorey may be affected indirectly, for example, where the invasion of the herbaceous *Phyla canescens* (lippia) into wetlands and floodplain woodlands in the northern Murray–Darling Basin may affect the survival of river red gums (Taylor and Ganf 2005; Price *et al.* 2011). Various management methods have been implemented to control lippia, but the most effective has been the increase in summer flooding (Price *et al.* 2010). Increased inundation duration has been recommended as a management tool for invasive floodplain and riparian species more broadly (Catford *et al.* 2011).

21.5.6 Altered fire regimes

The effects and incidence of fire in Australian floodplains and riparian vegetation has not been well studied. Between 1996 and 2000, 25% of the floodplains in Kakadu National Park were burnt (Edwards *et al.* 2003). Differences in the timing of fires can have a large effect on riparian community structure (Douglas *et al.* 2003; Petty and Douglas 2010, Douglas and Setterfield 2015). Increased burning of floodplains and associated rainforests and *Melaleuca* forests has been reported for Northern Australia (Russell-Smith *et al.* 1997) and has been implicated in the decline of *Melaleuca* forest on the floodplain. There is also evidence of positive feedback between fires and grass invasion in other northern Australian habitats (Rossiter *et al.* 2003). For example, invasive *Urochloa mutica* increases fire frequency and intensity by modifying bushfire fuels. The change in fire regime reduces areas of fire-sensitive vegetation, which promotes further spread.

21.6 Conclusions and future research

Riparian zones and floodplains are an extremely important feature of Australian landscapes as a complex mosaic of terrestrial and aquatic habitats. In the dry Australian climate, riparian zones provide linear hotspots of plant diversity and productivity. The plant communities that occupy this habitat are driven, to a large extent, by seasonality and variability of river flows and flood regimes (Naiman *et al.* 2008; Warfe *et al.* 2011). With increasing competition for water resources to support industrial and agricultural expansion, and growing urban populations in Australia (Cresswell *et al.* 2009), a thorough understanding of the ecological links between water and riparian systems is critical (Warfe *et al.* 2011).

Although there have been many recent advances in our understanding of floodplain and riparian ecosystems, there is an urgent need for further research into landscape-scale floodplain woodland dynamics. Surprisingly, there is still very little known about the mechanisms of floodplain tree flowering, seed production and, to a lesser degree, seed dispersal and germination. There also needs to be further research into the effects of changes to groundwater and flood regimes on the longevity of floodplain trees. A better understanding of floodplain and riparian vegetation dynamics could increase the efficiency with which we manage our finite water resources across the continent.

References

Australian Water Resources Council (AWRC) (1976) *Review of Australia's Water Resources 1975*. Australian Government Publishing Service: Canberra

Baldwin, V.C., Peterson, K.D., Clark III, A. *et al.* (2000) The effects of spacing and thinning on stand and tree characteristics of 38-year-old Loblolly Pine. *Forest Ecology and Management* 137, 91–102.

Bayly, I.A.E. and Williams, W.D. (1981) *Inland Waters and their Ecology*. Longman Cheshire: Melbourne.

Beadle, N.C.W. (1981) *The Vegetation of Australia*. Cambridge University Press: Cambridge.

Bedward, M., Simpson, C.C., Ellis, M.V. and Metcalfe, L.M. (2007) Patterns and determinants of historical woodland clearing in central-western New South Wales, Australia. *Geographical Research* 45, 348–357.

Boland, D.J., Brooker, M.I.H., Turnbull, J.W. and Kleinig, D.A. (1980) *Eucalyptus Seed*. CSIRO Publishing: Canberra.

Boon, P., Keith, D. and Raulings, E. (2016) Vegetation of coastal floodplains and wetlands in south-eastern Australia. In *Vegetation of Australian Riverine Landscapes: Biology, Ecology and Management*, ed. S. Capon, M. Reid and C. James. CSIRO Publishing: Melbourne, pp. 145–176.

Boulton, A.J., Brock, M.A., Robson, B.J. *et al.* (2014) *Australian Freshwater Ecology: Processes and Management*. Wiley-Blackwell: West Sussex.

Bowman, D.M.J.S. and Wilson, B.A. (1986) Wetland vegetation pattern on the Adelaide River floodplain, Northern Territory, Australia. *Proceedings of the Royal Society of Queensland* 97, 69–77.

Boyland, D.E. (1984) *Vegetation Survey of Queensland: South Western Queensland*. Queensland Department of Primary Industries: Brisbane.

Bradshaw, C.J.A. (2012) Little left to lose: deforestation and forest degradation in Australia since European colonization. *Journal of Plant Ecology* 5, 109–120.

Bren, L.J. (1992) Tree invasion of an intermittent wetland in relation to changes in the flooding frequency of the River Murray, Australia. *Austral Ecology* 17, 395–408.

Bunn, S.E. (1993) Riparian-stream linkages: research needs for the protection of in-stream values. *Australian Biologist* 6, 46–51.

Butler, B.E. (1950) A theory of prior streams as a causal factor of soil occurrence in the Riverine Plain of south-eastern Australia. *Crop and Pasture Science* 1, 231–252.

Capon, S.J. (2005) Flood variability and spatial variation in plant community composition and structure on a large arid floodplain. *Journal of Arid Environments* 60, 283–302.

Catford, J.A., Downes, B.J., Gippel, C.J. and Vesk, P.A. (2011) Flow regulation reduces native plant cover and facilitates exotic invasion in riparian wetlands. *Journal of Applied Ecology* 48, 432–442.

Chiew, F.H.S., Vaze, J., Viney, N.R. *et al.* (2008) *Rainfall-runoff modelling across the Murray–Darling Basin. A report to the Australian Government from the CSIRO Murray–Darling Basin Sustainable Yields Project*. CSIRO: Australia.

Colloff, M. (2014) *Flooded Forest and Desert Creek*. CSIRO Publishing: Collingwood.

Cowie, I.D., Short, P.S. and Osterkamp-Madsen, M. (2000) *Floodplain Flora: A Flora of the Coastal Floodplains of the Northern Territory, Australia*. Australian Biological Resources Study, Government Printers: Canberra.

Cox, S.J., Sivertsen, D.P. and Bedward, M. (2001) Clearing of native woody vegetation in the New South Wales northern wheatbelt: extent, rate of loss and implications for biodiversity conservation. *Cunninghamia* 7, 101–155.

Craig, A.E., Walker, K.F. and Boulton, A.J. (1991) Effects of edaphic factors and flood frequency on the abundance of lignum (*Muehlenbeckia florulenta* Meissner) (Polygonaceae) on the river Murray floodplain, South Australia. *Australian Journal of Botany* 39, 431–443.

Cramer, V.A. and Hobbs, R.J. (2002) Ecological consequences of altered hydrological regimes in fragmented ecosystems in southern Australia: impacts and possible management responses. *Austral Ecology* 27, 546–564.

Cresswell, R., Petheram, C., Harrington, G. *et al.* (2009) Final report to the Northern Australia Land and Water Taskforce. In *Northern Australia Land and Water Science Review*, ed. P. Stone. SIRO: Canberra.

Cunningham, S.C., Read, J., Baker, P.J. and Mac Nally, R. (2007) Quantitative assessment of stand condition and its relationship to physiological stress in stands of *Eucalyptus camaldulensis* (Myrtaceae). *Australian Journal of Botany* 55, 692–699.

Cunningham, S.C., Thomson, J.R., Mac Nally, R. *et al.* (2011) Groundwater change forecasts widespread forest dieback across an extensive floodplain system. *Freshwater Biology* 56, 1494–1508.

Daley, E.A. and Kirkpatrick, J.B. (2004) Native riparian vegetation in Tasmania. *Cunninghamiana* 8, 409–430.

Dexter, B.D. (1968) Flooding and regeneration of river red gum, Eucalyptus camaldulensis, Dehn. In *The Commonwealth Forestry Conference*. Forests Commission Victoria: Melbourne.

Doody, T.M., Benger, S.N., Pritchard, J.L. and Overton, I.C. (2014) Ecological response of *Eucalyptus camaldulensis* (river red gum) to extended drought and flooding along the River Murray, South Australia (1997–2011) and implications for environmental flow management. *Marine and Freshwater Research* 65, 1082–1093.

Doody, T.M., Colloff, M.J., Davies, M. *et al.* (2015) Quantifying water requirements of riparian river red gum (*Eucalyptus camaldulensis*) in the Murray–Darling Basin, Australia: implications for the management of environmental flows. *Ecohydrology* 8, 1471–1487.

Douglas, M. and Setterfield, S.A. (2015) The effects of fire on riparian vegetation: landscape-scale experiment in tropical Australia. *Freshwater Science* (in press).

Douglas, M., Townsend, S.A. and Lake, P.S. (2003) Streams. In *Fire in Tropical Savanna: The Kapalga Experiment*, ed. A.N. Andersen, G.D. Cook and R.J. Williams. Springer: New York, pp. 59–78.

Dowe, J.L. (2008) Report 4: distribution and ecological preferences of riparian vegetation in northern Australia. In *A compendium of ecological information on Australia's northern tropical rivers. Sub-project 1 of Australia's tropical rivers: an integrated data assessment and analysis (DET18). A report to Land, Water Australia*, ed. G.P. Luckas and C.M. Finlayson. National Centre for Tropical Wetland Research: Townsville.

Edwards, A., Kennett, R., Price, O. *et al.* (2003) Monitoring the impacts of fire regimes on vegetation in northern Australia: an example from Kakadu National Park. *International Journal of Wild Fire* 12, 427–440.

Eldridge, D.J., Poore, A.G.B., Ruiz-Colmenero, M. *et al.* (2016) Ecosystem structure, function and composition in rangelands are negatively affected by livestock grazing. *Ecological Applications* 26, 1273–1283.

Fensham, R.J. (2008) A protocol for assessing applications to selectively clear vegetation in Australia. *Land Use Policy* 25, 249–258.

Fensham, R.J. and Kirkpatrick, J.B. (1989) The conservation of original vegetation remnants in the midlands, Tasmania. *Papers and Proceedings of the Royal Society of Tasmania* 123, 229–246.

Fensham, R.J., Fairfax, R.J. and Ward, D.P. (2009) Drought-induced tree death in savanna. *Global Change Biology* 15, 380–387.

Finlayson, C.M. (2005) Plant ecology of Australia's tropical floodplain wetlands: A review. *Annals of Botany* 96, 541–555.

Fleischner, T.L. (1994) Ecological costs of livestock grazing in western North America. *Conservation Biology* 8, 629–644.

Florentine, S.K. (1999) Ecology of *Eucalyptus victrix* in grassland in the floodplain of the Fortescue River. PhD Thesis, Curtin University of Technology, Perth.

Fox, J.E.D., Florentine, S.K. and Westbrooke, M.E. (2004) Observations on survival and early growth of natural regeneration in floodplain coolibah *Eucalyptus victrix* (Myrtaceae) in the Pilbara, Western Australia. *The Rangeland Journal* 26, 150–160.

Franklin, D.C., Brocklehurst, P.S., Lynch, D. and Bowman, D.M.J.S. (2007) Niche differentiation and regeneration in the seasonally flooded *Melaleuca* forests of northern Australia. *Journal of Tropical Ecology* 23, 457–467.

George, R., McFarlane, D. and Nulsen, B. (1997) Salinity threatens the viability of agriculture and ecosystems in Western Australia. *Hydrogeology Journal* 5, 6–21.

Gibling, M.R., Nanson, G.C. and Maroulis, J.C. (1998) Anastomosing river sedimentation in the Channel Country of central Australia. *Sedimentology* 45, 595–619.

Good, M.K. (2012) Plant community patterns and population dynamics of coolibah woodlands. PhD Thesis, University of New England, Armidale.

Good, M.K., Price, J.N., Clarke, P.J. and Reid, N. (2011) Densely regenerating coolibah (*Eucalyptus coolabah*) woodlands are more species rich than surrounding derived grasslands in floodplains of eastern Australia. *Australian Journal of Botany* 59, 468–479.

Good, M.K., Price, J.N., Clarke, P.J. and Reid, N. (2012) Dense regeneration of floodplain *Eucalyptus coolabah*: invasive scrub or passive restoration of an endangered woodland community? *The Rangeland Journal* 34, 219–230.

Good, M.K., Clarke, P.J., Price, J.N. and Reid, N. (2014) Seasonality and facilitation drive tree establishment in a semi-arid floodplain savanna. *Oecologia* 175, 261–271.

Greet, J., Cousens, R.D. and Webb, J.A. (2012) Flow regulation affects temporal patterns of riverine plant seed dispersal: potential implications for plant recruitment. *Freshwater Biology* 57, 2568–2579.

Hancock, C.N., Ladd, P.G. and Froend, R.H. (1996) Biodiversity and management of riparian vegetation in Western Australia. *Forest Ecology and Management* 85, 239–250.

Horner, G.J., Baker, P.J., Mac Nally, R. *et al.* (2009) Mortality of developing floodplain forests subjected to a drying climate and water extraction. *Global Change Biology* 15, 2176–2186.

Horner, G.J., Baker, P.J., Mac Nally, R. *et al.* (2010) Forest structure, habitat and carbon benefits from thinning floodplain forests: managing early stand density makes a difference. *Forest Ecology and Management* 259, 286–293.

Hunter, J.T. (2005) Vegetation of Culgoa National Park, central northern New South Wales. *Cunninghamia* 9, 275–284.

Jansen, A. and Robertson, A.I. (2001) Relationships between livestock management and the ecological condition of riparian habitats along an Australian floodplain river. *Journal of Applied Ecology* 38, 63–75.
Jenkins, K.M., Boulton, A.J. and Ryder, D.S. (2005) A common parched future? Research and management of Australian arid zone floodplain wetlands. *Hydrobiologia* 552, 57–73.
Jensen, A.E., Walker, K.F. and Paton, D.C. (2008) The role of seedbanks in restoration of floodplain woodlands. *River Research and Applications* 24, 632–649.
Jolly, I.D., Walker, G.R. and Thorburn, P.J. (1993) Salt accumulation in semi-arid floodplain soils with implications for forest health. *Journal of Hydrology* 150, 589–614.
Junk, W.J., Bayley, P.B. and Sparks, R.E. (1989) The flood pulse concept in riverplain systems. In *International large River Symposium*. Vol. 106, ed. D.P. Dodge. Fisheries and Aquatic Sciences Special Publication: Canada.
Keith, D.A. (2004) *Ocean Shores to Desert Dunes: The Native Vegetation of New South Wales and the ACT*. Department of Environment and Conservation: Hurstville.
Keith, D.A. and Scott, J. (2005) Native vegetation of coastal floodplains: a diagnosis of the major plant communities in New South Wales. *Pacific Conservation Biology* 11, 81–104.
Keith, D.A., Orscheg, C., Simpson, C.C. *et al.* (2009) A new approach and case study for estimating extent and rates of habitat loss for ecological communities. *Biological Conservation* 142, 1469–1479.
Kennard, M.J., Pusey, B.J., Olden, J.D. *et al.* (2010) Classification of natural flow regimes in Australia to support environmental flow management. *Freshwater Biology* 55, 171–193.
Kingsford, R.T. (2000) Ecological impacts of dams, water diversions and river management on floodplain wetlands in Australia. *Austral Ecology* 25, 109–127.
Kingsford, R.T., Boulton, A.J. and Puckridge, J.T. (1998) Challenges in managing dryland rivers crossing political boundaries: lessons from Cooper Creek and the Paroo River, central Australia. *Aquatic Conservation: Marine Freshwater Ecosystems* 8, 361–378.
Ladd, P. (1978) Vegetation history at Lake Curlip in lowland eastern Victoria, from 5200 B.P. to present. *Australian Journal of Botany* 26, 393–414.
Leonard, S.W.J. and Kirkpatrick, J.B.A. (2004) Effects of grazing management and environmental factors on native grassland and grassy woodland, Northern Midlands, Tasmania. *Australian Journal of Botany* 52, 529–542.
Lunt, I.D. (1997) The distribution and environmental relationships of native grasslands on the lowland Gippsland Plain, Victoria: an historical study. *Australian Geographical Studies* 35, 140–152.
Lunt, I.D., Eldridge, D.J., Morgan, J.W. and Witt, G.B. (2007) Turner review no. 13. A framework to predict the effects of livestock grazing and grazing exclusion on conservation values in natural ecosystems in Australia. *Australian Journal of Botany* 55, 401–415.
Lymbery, A.J., Doupé, R.G. and Pettit, N.E. (2003) The effect of salinisation on riparian vegetation in experimental catchments on the Collie River, Western Australia. *Australian Journal of Botany* 51, 667–672.
Lyon, M.N. (2015) The riparian flora and plant communities of the Pilbara region of Western Australia. *Records of the Western Australian Museum, Supplement* 78, 485–513.
Mac Nally, R., Cunningham, S.C., Baker, P.J. *et al.* (2011) Dynamics of Murray–Darling floodplain forests under multiple stressors: the past, present, and future of an Australian icon. *Water Resources Research* 47, 1–11.

Maher, M. (1995) *A thin line: should densities of coolibah and black box be controlled in the Western Division of New South Wales? Report to the Western Lands Commission*. Department of Conservation and Land Management: Dubbo.

Mayer, X.M., Ruprecht, J.K. and Bari, M.A. (2005) *Stream salinity status and trends in south-west Western Australia*. Department of Environment Western Australia: Perth.

McAlpine, C.A., Fensham, R.J. and Temple-Smith, D.E. (2002) Biodiversity conservation and vegetation clearing in Queensland: principles and thresholds. *The Rangeland Journal* 24, 38–55.

MDBA (2012) *Sustainable rivers audit 2: the ecological health of rivers in the Murray–Darling Basin at the end of the Millennium Drought (2008–2010)*. Murray–Darling Basin Authority: Canberra.

MDBA (2016) MDBA website. Murray–Darling Basin Authority: Canberra. Available at: http://www.mdba.gov.au.

Mensforth, L.J., Thorburn, P.J., Tyerman, S.D. and Walker, G.R. (1994) Sources of water used by riparian *Eucalyptus camaldulensis* overlying highly saline groundwater. *Oecologia* 100, 21–28.

Mohler, C.L., Marks, P.L. and Sprugel, D.G. (1978) Stand structure and allometry of trees during self-thinning of pure stands. *Journal of Ecology* 66, 599–614.

Naiman, R.J., Latterell, J.J., Pettit, N.E. and Olden, J.D. (2008) Flow variability and the biophysical vitality of river systems. *Comptes Rendus Geosciences* 340, 629–643.

Nakano, S. and Murakami, M. (2001) Reciprocal subsidies: dynamic interdependence between terrestrial and aquatic food webs. *Proceedings of the National Academy of Science U.S.A.* 98, 166–170.

Nanson, G.C. and Croke, J.C. (1992) A genetic classification of floodplains. *Geomorphology* 4, 459–486.

National Land and Water Resources Audit (NLWRA) (2000) Australian dryland salinity assessment 2000: extent, impacts, processes, monitoring and management options. Land and Water Australia: Canberra.

Neldner, V.J., Niehus, R.E., Wilson, B.A. *et al.* (2014) *The Vegetation of Queensland. Descriptions of Broad Vegetation Groups*. Queensland Herbarium, Department of Science, Information Technology, Innovation and the Arts: Brisbane.

Page, K.J., Nanson, G.C. and Frazier, P.S. (2003) Floodplain formation and sediment stratigraphy resulting from oblique accretion on the Murrumbidgee River, Australia. *Journal of Sedimentary Research* 73, 5–14.

Pannell, D.J. and Ewing, M.A. (2006) Managing secondary dryland salinity: options and challenges. *Agricultural Water Management* 80, 41–56.

Pettit, N.E. (2002) Riparian vegetation of a permanent waterhole on Cooper Creek, south-western Queensland. *Proceedings of the Royal Society of Qld* 110, 15–25.

Pettit, N.E., Bayliss, P., Davies, P.M. *et al.* (2011) Seasonal contrasts in carbon resources and ecological processes on a tropical floodplain. *Freshwater Biology* 56, 1047–1064.

Pettit, N.E. and Froend, R.H. (2001a) Variability in flood disturbance and the impact on riparian tree recruitment in two contrasting river systems. *Wetlands Ecology and Management* 9, 13–25.

Pettit, N.E. and Froend, R.H. (2001b) Availability of seed for recruitment of riparian vegetation: a comparison of a tropical and temperate river ecosystem in Australia. *Australian Journal of Botany* 49, 515–528.

Pettit, N.E., Jardine, T.D., Hamilton, S.K. *et al.* (2012) Seasonal changes in water quality and macrophytes and the impact of cattle on tropical floodplain waterholes. *Marine and Freshwater Research* 63, 788–800.
Pettit, N.E., Naiman, R.J., Fry, J.M. *et al.* (2015) Environmental change: prospects for conservation and agriculture in a south west Australian biodiversity hotspot. *Ecology and Society* 20, 10.
Petty, A. and Douglas, M. (2010) Scale relationships and linkages between woody vegetation communities along a large tropical floodplain river, north Australia. *Journal of Tropical Ecology* 26, 79–92.
Price, J.N., Gross, C.L. and Whalley, W. (2010) Prolonged summer flooding switched dominance from the invasive weed Lippia (*Phyla canescens*) to native species in one small, ephemeral wetland. *Ecological Management and Restoration* 11, 61–63.
Price, J.N., Macdonald, M.J., Gross, C.L. *et al.* (2011) Vegetative reproduction facilitates early expansion of Phyla canescens in a semi-arid floodplain. *Biological Invasions* 13, 285–289.
Proulx, M. and Mazumder, A. (1998) Reversal of grazing impact on plant species richness in nutrient poor vs. nutrient-rich ecosystems. *Ecology* 79, 2581–2592.
Queensland Department of Science, Information Technology and Innovation (2015) *Land cover change in Queensland* 2012–13 *and* 2013–14: *a Statewide Landcover and Trees Study (SLATS) report*. DSITI: Brisbane.
Reid, M., Bickford, S., Gell, P., Kenyon, C. (2016) A history of Australia's riverine habitats and vegetation. In *Vegetation of Australian Riverine Landscapes: Biology, Ecology and Management*, ed. S. Capon, M. Reid and C. James. CSIRO Publishing: Melbourne, pp. 45–64.
Roberts, J. (1993) Regeneration and growth of coolibah, *Eucalyptus coolabah* subsp. *arida*, a riparian tree, in the Cooper Creek region of South Australia. *Australian Journal of Ecology* 18, 345–350.
Roberts, J. (2004) *Floodplain Forests and Woodlands in the Southern Murray–Darling Basin*. Roberts J: Canberra.
Roberts, J. and Ludwig, J.A. (1990) Riparian habitats on the Chowilla floodplain of the River Murray, South Australia. *Wetlands (Australia)* 9, 13–19.
Rossiter, N.A., Setterfield, S.A., Douglas, M.M. and Hutley, L.B. (2003) Testing the grass-fire cycle: alien grass invasion in the tropical savannas of northern Australia. *Diversity and Distributions* 9, 169–176.
Russell-Smith, J.M., Ryan, P.G. and Durieu, R. (1997) A LANDSAT MSS-derived fire history of Kakadu National Park, monsoonal northern Australia, 1980-94: seasonal extent, frequency and patchiness. *Journal of Applied Ecology* 34, 748–766.
Schultz, N.L., Morgan, J.W. and Lunt, I.D. (2011) Effects of grazing exclusion on plant species richness and phytomass accumulation vary across a regional productivity gradient. *Journal of Vegetation Science* 22, 130–142.
Schultz, N.L., Reid, N., Lodge, G. and Hunter, J.T. (2014) Broad-scale patterns in plant diversity vary between land uses in a variegated temperate Australian agricultural landscape. *Austral Ecology* 39, 855–863.
Slavich, P., Walker, G., Jolly, I. *et al.* (1999) Dynamics of *Eucalyptus largiflorens* growth and water use in response to modified watertable and flooding regimes on a saline floodplain. *Agricultural Water Management* 39, 245–264.
Sena Gomez, A.R. and Kozlowski, T.T. (1980) Effects of flooding on *Eucalyptus camaldulensis* and *Eucalyptus globulus* seedlings. *Oecologia* 46, 139–142.

Smith, R. and Reid, N. (2013) Carbon storage value of native vegetation on a subhumid–semi-arid floodplain. *Crop and Pasture Science* 64, 1209–1216.
Stace, H.C.T., Hubble, G.D., Brewer, R. *et al.* (1968) *A Handbook of Australian Soils*. Rellim Technical Publications for CSIRO and International Society of Soil Science: Glenside.
Taylor, B. and Ganf, G.G. (2005) Comparative ecology of two co-occurring floodplain plants: the native *Sporobolus mitchellii* and the exotic *Phyla canescens*. *Marine and Freshwater Research* 56, 431–440.
Thoms, M. and Parsons, M. (2011) Patterns of vegetation community distribution in a large, semi-arid floodplain landscape. *River Systems* 19, 271–282.
Thoms, M., Parsons, M. and Southwell, M. (2016) The physical template of Australia's rivers. In *Vegetation of Australian Riverine Landscapes: Biology, Ecology and Management*, ed. S. Capon, M. Reid and C. James. CSIRO Publishing: Melbourne, pp. 7–25.
Thorburn, P.J. and Walker, G.R. (1994) Variations in steam water uptake by *Eucalyptus camaldulensis* with differing access to stream water. *Oecologia* 100, 293–301.
Thorburn, P.J., Mensforth, L.J. and Walker, G.R. (1994) Reliance of creek-side river red gums on creek water. *Australian Journal of Marine and Freshwater Research* 45, 1439–1443.
Travers, S.K. and Eldridge, D.J. (2013) Increased rainfall frequency triggers an increase in litter fall rates of reproductive structures in an arid eucalypt woodland. *Austral Ecology* 38, 820–830.
Walker, K.F., Puckridge, J.T. and Blanch, S.J. (1997) Irrigation development on Cooper Creek, central Australia: prospects for a regulated economy in a boom and bust ecology. *Aquatic Conservation: Marine and Freshwater Ecosystems* 7, 63–73.
Warfe, D.M., Pettit, N.E., Davies, P.M. *et al.* (2011) The 'wet–dry' in the wet–dry tropics drive river ecosystem structure and processes in northern Australia. *Freshwater Biology* 56, 2169–2195.
Wen, L., Ling, J., Saintilan, N. and Rogers, K. (2009) An investigation of the hydrological requirements of river red gum (*Eucalyptus camaldulensis*) forest, using classification and regression tree modelling. *Ecohydrology* 2, 143–155.
White, J. (1981) The allometric interpretation of the self-thinning rule. *Journal of Theoretical Biology* 89, 475–500.
Wiegand, K., Saltz, D., Ward, D. and Levin, S.A. (2008) The role of size inequality in self-thinning: a pattern-oriented simulation model for arid savannas. *Ecological Modelling* 210, 431–445.
Woolfrey, A.R. and Ladd, P.G. (2001) Habitat preference and reproductive traits of a major Australian riparian tree species (*Casuarina cunninghamiana*). *Australian Journal of Botany* 49, 705–715.

22 Coastal Halophytic Vegetation

KERRYLEE ROGERS, PAUL BOON, CATHERINE LOVELOCK AND NEIL SAINTILAN

Coastal halophytes are plants of the coastal zone that are capable of growing and reproducing under saline conditions. Mostly they occur in areas subject to tidal inundation, though may also occur on headlands where salt spray elevates soil salinity. Mangrove and saltmarsh are the primary plant communities occupying the saline intertidal zone of Australia. Mangrove diversity is greatest in the tropics and saltmarsh diversity greatest in temperate zones. Coastal halophytes are most extensive where tidal conditions facilitate the development of a sedimentary substrate for plant establishment and growth. Over time, strong zonation can develop within the tidal frame. Where mangrove and saltmarsh co-occur saltmarsh is typically positioned higher in the tidal frame and thus is subject to less frequent inundation. Seagrass beds often adjoin mangrove in the lower intertidal and sub-tidal zones where environmental conditions are suitable. Mangrove and saltmarsh plants exhibit unique adaptations that enable them to establish and grow within the harsh conditions created by tidal inundation, conditions typified by recurrent waterlogging, elevated salinity and anaerobic substrate. Common adaptations include mechanisms to exclude, extrude or store salt ions, and unique root morphologies that improve stability and create aerated rhizospheres around the roots. Coastal halophyte communities have responded over geological timescales to fluctuating sea levels by shifting their distribution and by accumulating sediments. More recently, coastal halophytes have increasingly succumbed to habitat change and exploitation, both globally and in Australia, with large areas of former mangrove or saltmarsh lost to development. A shift to ecosystem-based management and policy aims to protect increasingly valued ecosystem services provided by coastal halophytes by reversing their loss and degradation. However, losses of coastal halophyte communities are likely to continue in the 21st century due to the effect of climate change on sea-levels, higher air and sea temperatures, and a suite of related impacts on primary productivity. Adaptation to climate change should aim to limit further losses of coastal halophyte vegetation, and urgently improve or restore ecosystem services provided by these highly valuable coastal plant communities.

22.1 Introduction

Halophytes that live in saline environments are distinguished from glycophytes, which are incapable of growing under saline conditions. This distinction is complex: are conditions saline, brackish or near fresh; what is the ionic composition of the water; are halophytes restricted to saline environments only or can they grow elsewhere? Because of this complexity, attempts to define halophytes have used a range of

Australian Vegetation, Third Edition, ed. David A. Keith. Published by Cambridge University Press.

Citation: Rogers, K. *et al.* (2017). Coastal Halophytic Vegetation. In D. A. Keith, ed., *Australian Vegetation*, 3rd edn. Cambridge: Cambridge University Press, pp. 544–69.

criteria, from salinity thresholds (e.g. Gorham 1993), edaphic characteristics (e.g. Waisel 1972), chemical characteristics of the plant material itself, or possession of specific physiological mechanisms that promote salt tolerance, such as salt secretion (e.g. Breckle 2002). For this chapter we adopt a pragmatic definition of a halophyte (after Flowers and Colmer 2008): plants able to complete their life cycle in a salt concentration of at least 200 mM NaCl (approx. 12 g l^{-1} NaCl, or roughly one-third ocean-water salinity) under conditions that might be expected to occur in the natural environment. We focus on angiosperms, yet recognise that there are many marine algae that meet the definition of being halophytes.

The exact number of halophytic plant species in Australia is largely unknown for a number of reasons. First, it is dependent upon the tolerance threshold used to differentiate halophytes from the salt sensitive glycophytes. In global analyses ~80 mM of sodium chloride has been used as a salinity threshold between the two plant groups, while other assessments use sodium chloride concentrations much closer to seawater resulting in fewer species identified as halophytic plants (Flowers *et al.* 2010). Second, the salt tolerance of many species still remains unknown (Boon 2012). This is particularly apparent at the upper limit of saltmarsh, where salinity conditions are presumed to be low, but for which there is very little published data. For this reason, we focus on those plants known to occur in association with tidal inundation along low-energy coastlines, as well as some that occur in non-tidal, but still saline, settings. Few angiosperms have adapted to the combination of high salinity and waterlogged sediments (Barrett-Lennard 2003; Kozlowski 1997). While they are a minor component of the global flora, they form distinctive and ecologically important habitats. Mangroves, for example, have colonised such environments through the evolution of specialised anatomical, physiological and reproductive adaptations (Clough *et al.* 1982; Saenger 1982) and provide numerous ecosystem services.

In this chapter, we review the ecology, biogeography and geomorphology of Australian coastal halophytic vegetation, focussing on mangrove and saltmarsh. The relationship between halophytic vegetation, sea level and coastal geomorphology, both current and past, is a unifying theme. We then examine climate change and other human influences, such as coastal development, as pervasive and ongoing threats, and discuss governance and management needed to mitigate them. Finally, we identify knowledge gaps and research needed to advance our understanding of ecosystem services provided by halophytic plant communities associated with the Australian coastline.

22.2 Tidal relationships of halophytic communities

Tidal halophytic plants often grow along coastlines where hydrodynamic energy is low, in conditions that enable sediments to accumulate and provide a suitable substrate in which seedlings can establish and plants develop. The occurrence of halophytes along the Australian coastline varies with latitude (Figure 22.1a); saltmarsh species diversity increases with increasing latitude, in contrast to mangrove species, which are predominantly tropical in distribution (Saenger *et al.* 1977; Adam 1996). The coastline of Australia is somewhat atypical in supporting well-developed sympatric mangrove and saltmarsh communities. In the northern parts of the continent, floristically diverse and extensive mangrove stands are seaward of often floristically poor but extensive saltmarshes and saline mudflats. In the southern parts of the continent, mangroves are less diverse but similarly are seaward of saltmarshes that are not only extensive but often floristically rich. A similar juxtaposition is found on parts of the coastlines of Brazil, Gulf of Mexico, Florida, China and the east coast of Africa (Saintilan *et al.* 2014).

Tidally influenced plant communities often exhibit strong zonation; seagrass is restricted to the subtidal and lower regions of the intertidal zone, while mangrove or saltmarsh occur throughout the upper portion of the intertidal zone

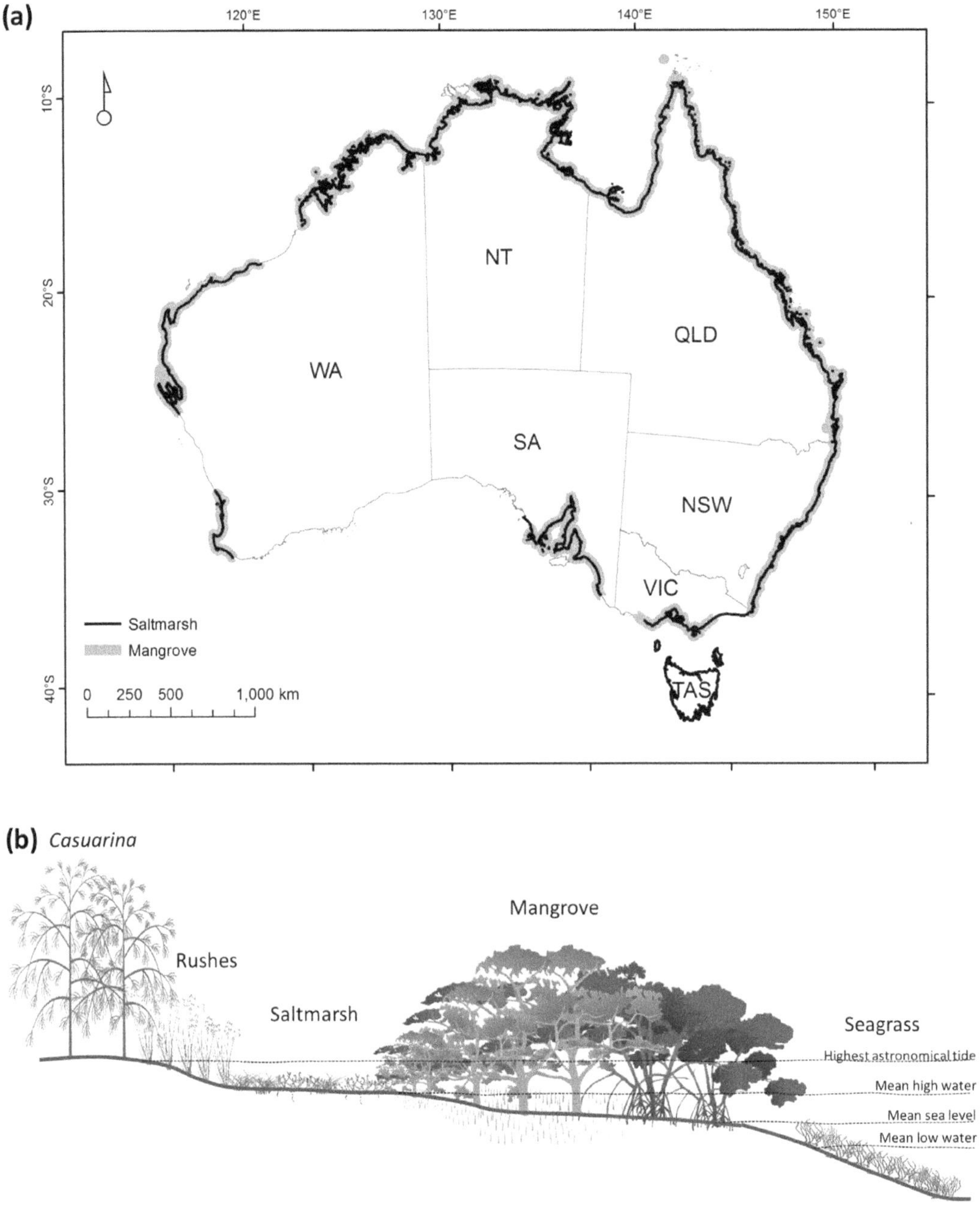

Figure 22.1 Mangrove and saltmarsh halophyte distribution: (a) around the coastline of Australia; and (b) across a tidal gradient, here shown for subtropical and temperate Australia.

above mean sea level (Fig. 22.1b). Seagrasses are ancient flowering plants that exhibit unique adaptations enabling them to complete their life cycle within sheltered coastal and offshore reef environments where salinity conditions are more stable (Sainty *et al.* 2012). Most species grow subtidally, but some may occur lower in the intertidal zone, such as eelgrass (*Zostera* spp.), while *Ruppia* and *Lepilaena* species may occur within pools located within saltmarsh. Where saltmarsh and mangrove co-occur, mangrove forest typically occupies lower elevations (to mean high tide), and saltmarsh occupies high elevations within the tidal frame, usually extending to roughly the limit of the highest astronomic tide. Any simple zonation patterns with increasing distance from the sea are, however, complicated by spatial patterning along the sides of tidal creeks and other waterways that traverse tidal marshes and mangroves. Broad ecotones occur in gently sloping settings and tidal halophytes may merge with 'brackish' ecosystems that may consist of non-woody taxa, such as the sea rush *Juncus kraussii*, as well as trees and shrubs, such as swamp oak *Casuarina glauca*, or paperbarks *Melaleuca* spp. In arid areas, halophytic vegetation may merge with hypersaline mudflats and cyanobacterial mats, sometimes called sabka (Lovelock *et al.* 2010a; Duggan *et al.* 2014). In other areas, where rainfall is high and the elevation gradient over the intertidal is low, freshwater and brackish water wetlands may dominate behind the uppermost levels of the saltmarsh (Catford *et al.*, Chapter 20, this volume).

22.3 Distribution and biogeography

Although there is no single definition of coastal saltmarsh that adequately covers the floristic and structural complexity of this community, mangroves have been defined as trees, shrubs, palms or ferns generally exceeding 0.5 m and which normally grow in the intertidal zone of marine and estuarine environments (Duke 2006). The term 'mangrove' can be used to refer to both the individual tree and to the plant community. For the latter, Macnae (1966) proposed the word 'mangal' be used, but this terminology has not been uniformly adopted by wetland ecologists and managers.

Mangrove forests occupy approximately 18% of the Australian coastline (Duke 2006), with an estimated area of 978 000 ha. Australia has the second largest extent in the world, behind Indonesia (3 113 000 ha) and similar to that of Brazil (963 000 ha) (Giri *et al.* 2011). Six mangrove biogeographic regions have been recognised on a global scale, divided between the Atlantic–East Pacific (AEP) and the Indo-West Pacific (IWP) provinces. Mangrove species diversity is significantly greater in the IWP province, and while Indonesia has the highest mangrove species diversity, Australian species diversity is high on global standards (Duke *et al.* 1998a; Duke 2006). Australia's mangrove flora includes 41 species, drawn from 22 genera and 19 plant families, and this constitutes 57% of the mangrove species in the world. Although mangroves are overwhelmingly dicotylednous angiosperms, Australian mangroves also include one fern (*Acrostichum*, in the family Pteridaceae) and one palm (*Nypa*, in the family Arecaceae).

Despite the high diversity of mangroves in the IWP, mangroves in the region have not evolved and dispersed uniformly (Saenger 1998). Fossil records indicate a much broader distribution of the mangrove palm, *Nypa* than the present-day extant populations of the genus (Plaziat *et al.* 2001), and fossilised *Nypa* pollen has been documented from the Eocene in Tasmania (Pole and Macphail 1996). Unusual occurrences and genetic discontinuities in the current distribution indicate that mangrove evolution was complex (Duke *et al.* 1998b; Duke 2002). Debate about mangrove evolution (Duke 1995; Saenger 1998; Ellison *et al.* 1999) is perhaps unsurprising as mangroves occur in many different plant families, including the Avicenniaceae, Combretaceae, Euphorbiaceae, Meliaceae, Rhizophoraceae, Sonneratiaceae and Sterculiaceae.

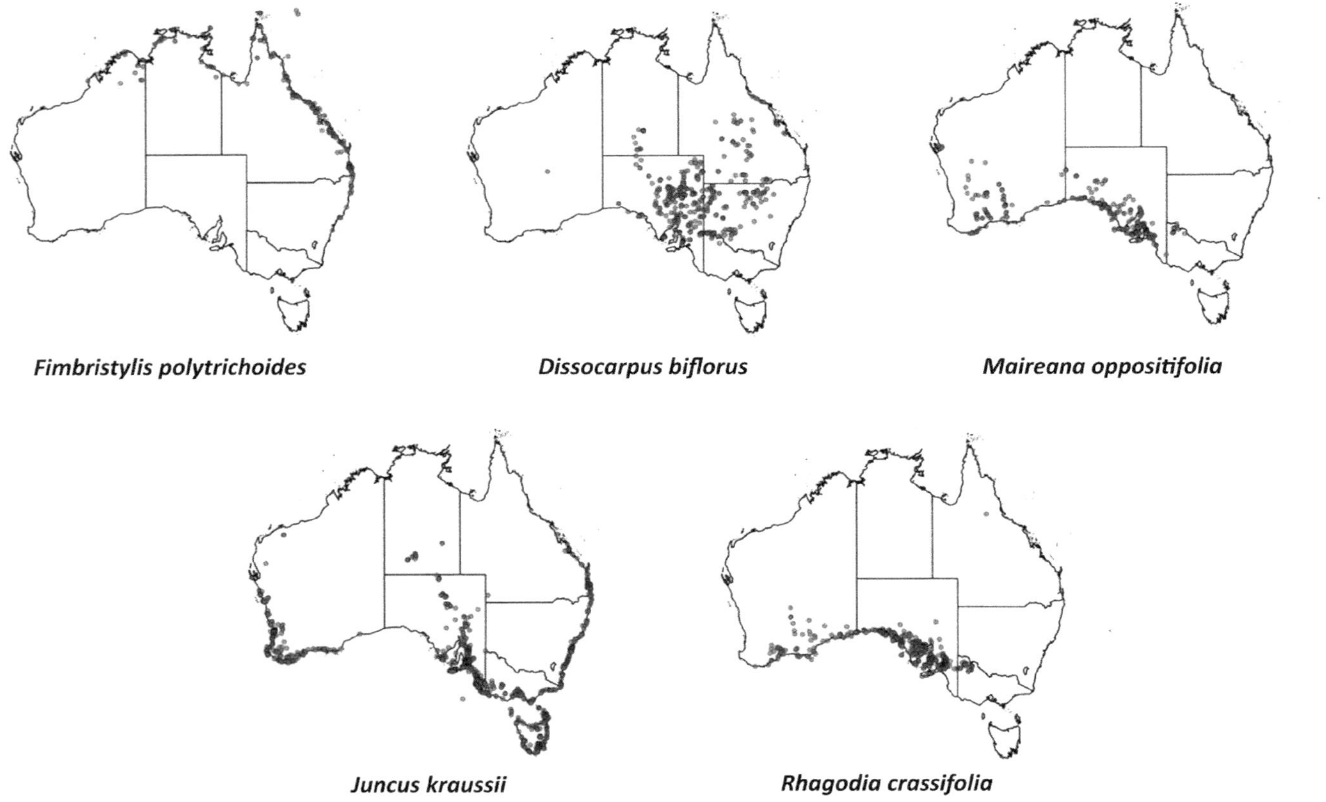

Figure 22.2 Distribution patterns of selected saltmarsh species, being illustrative of predominantly inland distribution (*D. biflorus*), predominantly coastal southern and northern (*J. kraussii* and *F. polytrichoides*, respectively) and arid zone chenopods (*M. oppositifolia* and *R. crassifolia*). *Data and base-maps derived from the Australian Virtual Herbarium (n.d.).*

As mangroves are predominantly tropical in distribution, species diversity is greatest along the wet tropical coastlines of the Indonesian–New Guinea Archipelago and of Northern Australia. Diversity diminishes with increasing latitude, with mangrove distribution broadly corresponding to shorelines where the average sea surface temperature exceeds 20°C (Duke *et al.* 1998a). There is only one species of mangrove, *Avicennia marina* at the highest latitudinal limit (i.e. Corner Inlet, 38°54′25″S); this species has the greatest latitudinal distribution of any mangrove globally (Saintilan *et al.* 2014). The truncated latitudinal distribution of mangroves on the western coasts of continents in the southern hemisphere suggests that aridity may be an important determinant of distribution in these regions (Ball and Sobrado 1998). This pattern is particularly evident along the coastline of north-western Australia, where species diversity declines with increasing aridity (Semeniuk 1983).

Unlike mangrove, the status of many saltmarsh species as either native or introduced to Australia is unknown and still questioned among botanists (Adam 1990; Boon *et al.* 2015). Further complexity arises from occurrence of saltmarsh in both coastal and inland settings, the former with a wide suite of environmental conditions and the latter most often associated with endorheic basins (De Deckker 1983). Saltmarsh taxa may be predominantly inland with some coastal representation (e.g. *Dissocarpus biflorus* in Figure 22.2), predominantly coastal though with inland occurrence (e.g. *Samolus repens*, some species of *Fibristylus*), and more rarely exclusively coastal (e.g. *Austrostipa stipoides, Limonium solanderi*). The conditions required for growth, that is, saline habitat periodically freshened while competition with salt-intolerant vegetation is excluded, are similar between upper intertidal environments and salt lakes. Birds migrating between inland and coastal habitats provide an effective dispersal mechanism between coastal saltmarsh and their inland counterparts. Additionally, salt spray, sometimes combined with groundwater seepage, can facilitate the growth of halophytic plant communities on cliffs, bluffs and rock shelves well above tidal reach.

Notwithstanding the transcontinental distribution of many species and the possible non-native status of some species, clear biogeographic patterns in saltmarsh taxa are noticeable at continental and subcontinental scales. The primary biogeographic division of Australia's saltmarsh flora is between northern and southern bioregions, separated at 23° latitude (Saintilan 2009). The two halves of the continent have less than 25% of species in common. Species occurring primarily in the northern regions include *Tecticornia australasica, Fimbristylis polytrichoides, Portulaca bicolour, Fimbristylis ferruginea, Batis agrillicola, Xerochloa imberbis* and *Sesuvium portulacastrum*. Species occurring primarily in the southern bioregions include *Samolus repens, Atriplex cinerea, Apium prostratum, Centaurium erythraea, Myoporum insulare, Parapholis incurva* and *Cotula coronopifolia* (Saintilan 2009).

Three bioregions in each of the northern and southern continental divisions have distinct saltmarsh floras (Figure 22.3). In both divisions, about two-thirds of the flora is shared with the remainder of the division. Within the southern division, the Gawler, Nullarbor and Hampton bioregions, all largely characterised by the absence of estuaries, because of the arid climate or the cliffed coastline. This would suggest that saltmarsh occurrences are either open coast or (as is more likely) cliff top. For example, no estuarine saltmarshes are recorded on the Great Australian Bight between Fowler's Bay, SA (33°14'S, 124°02'E) and Point Dempster WA (32°11'S, 133°28'E). Sedges and rushes are largely absent from these bioregions and the characterising species are of the Amaranthaceae family (see for example the distribution of *Maireana oppositifolia* in Figure 22.2). In the northern division, species found consistently in humid areas include *Tecticornia australasica*, the sedges *Fimbristylis ferruginea*, and *Fimbristylis polytrichoides* and *Batis agrillicola*. Species found predominantly in the more arid Pilbara, Dampier and Kimberly coastlines

Figure 22.3 Saltmarsh diversity around the Australian coastline varies at a range of spatial scales. At the regional scale, the most diverse communities occur in temperate south-eastern Australia, such as (a) French Island, Western Port Bay, Victoria, where *Tecticornia* spp. form a shrub-dominated zone near the lower intertidal mangrove and herbfields, with mosaics of species occupying the upper intertidal zone (photo: Kerrylee Rogers). Where freshwater contributions from adjacent catchments are high, diversity may be locally limited in temperate zones to those species tolerant of brackish conditions, as seen in (b) with monospecific stands of the rush *Juncus kraussii* on the middle reaches of the Hawkesbury River, NSW (photo: Kerrylee Rogers). In subtropical and tropical Australia, species diversity diminishes and intertidal zones landward of mangrove may be dominated by few species. Saltmarsh in (c) at Ukerebagh Island, Tweed River, NSW, is dominated by *Sporobolus virginicus* (photo: Kerrylee Rogers), and at (d) Moreton Bay, Qld, is dominated by *Sarcocornia quinqueflora* (photo: Catherine Lovelock).
(A black and white version of this figure will appear in some formats. For the colour version, please refer to the plate section.)

are *Tecticornia auriculata*, *Frankenia ambita* and *Neobassia astrocarpa*.

In general, there is little relationship between the intertidal area available to saltmarsh plants and saltmarsh floristic diversity. Northern Australia supports a low diversity of saltmarsh, commonly dominated by grasses including *Sporobolus virginicus*, in spite of the large intertidal area available for saltmarsh colonisation. To some degree, this may be related to climatic constraints, with hypersaline conditions developing during the dry season in many parts of Northern Australia and prolonged inundation in the wet season (Duggan *et al.* 2014). Wide intertidal areas will be devoid of saltmarsh in these situations. Notwithstanding, Tasmania, Victoria, South Australia and New South Wales contain less than 2.5% of the total saltmarsh/saltpan area yet together support 90% of the saltmarsh flora of Australia (Saintilan 2009).

22.4 Ecology, physiology and adaptations

Mangrove more typically grow at mid-tidal elevations where frequent tidal inundation destabilises sediments, reduces soil aeration and exposes soil to seawater salinities. Consequently, mangroves have evolved a number of distinctive adaptations for countering saline and anaerobic soil conditions. Salt tolerance in particular is associated with mechanisms to exclude, extrude, or store salt ions, which is supported by efficient use of water during photosynthesis (Ball 1988; Saenger 2002). Some mangrove species exclude salt more effectively than others, but approximately 80–99% of salt is excluded at the roots (Clough *et al.* 1982; Saenger 2002), and frequent tidal inundation is essential to flush the excluded salt ions and limit the development of hypersaline conditions around the roots. The remaining salt ions that are not excluded are subsequently stored in the vacuoles of cells, with some species extruding salt through salt glands in leaves. Storage of salt ions within leaves is typically accompanied by succulence to dilute salt concentrations; succulence is likely to be an adaptive response to salinity that is absent under freshwater conditions (Saenger 2002).

The high metabolic cost of salt tolerance limits the use of water in mangroves. Many mangroves therefore display traits such as xeromorphy, steep leaf angles and low transpiration rates that promote water-use efficiency and minimise water loss during photosynthesis (Ball 1988). Many mangroves are also strongly dependent on fresh water sources, with evidence for preferential uptake of fresh water sources when available (Santini *et al.* 2015).

Mangroves exhibit a range of root morphologies that improve stability and oxygen transport in anaerobic soils. In *Avicennia marina*, root systems typically consist of laterally spreading cable roots with smaller vertical anchor roots, fine nutritive roots and pneumatophores that increase root aeration, and trap and stabilise sediments (Figure 22.4). Despite generally being restricted to the upper 1–2 m of sediment, the complexity of the mangrove root system is such that the ratio of above-ground to below-ground biomass is lower than in other forests (Saintilan 1997a,b). Above-ground root morphology is equally complex among different mangrove species, with a range of aerial structures present in various genera including pneumatophores – conical root extensions from the cable root system (typical of *Avicennia* and *Sonneratia*); knee roots – cable roots that extend above the soil surface before protruding downward (typical of *Bruguiera*); stilt roots – roots protruding from the trunk before extending into the substrate (typical of *Rhizophora* or *Ceriops*); buttress roots – flattened roots extending laterally from the main trunk (typical of *Heritiera littoralis*); and aerial roots – that descend from branches or the trunk (typical of *Rhizophora* and *Avicennia*). Lenticels on roots facilitate gas exchange by regulating oxygen availability and pressure during tidal inundation and subsequent shifts in pressure as tides recede (Allaway *et al.* 2001). Leakage of oxygen from roots of some species may improve soil oxidation and enhance nutrient availability (Reef *et al.* 2010), though not in all cases (Boon 2006).

Mangrove reproduction takes place in an environment where tidal inundation transports seeds, yet might also limit seedling establishment. Many

Figure 22.4 Some examples of complex root morphology of mangrove in Australia: (a) cone-shaped pneumatophore roots of *Sonneratia alba* in Darwin Harbour, Northern Territory (photo: Kerrylee Rogers); (b) pneumatophores of dwarf *Avicennia marina* in Corner Inlet, Victoria (photo: Paul Boon); (c) exposed cable roots and pneumatophores of a mature *Avicennia marina* at Carama Inlet, Jervis Bay, NSW (photo: Kerrylee Rogers); and (d) complex stilt roots of *Rhizophora stylosa* at Low Isles, north Queensland (photo: Catherine Lovelock).
(A black and white version of this figure will appear in some formats. For the colour version, please refer to the plate section.)

mangrove species possess viviparous seeds, where the seed or propagule germinates while attached to the parent tree to the extent that the embryo ruptures the pericarp, or possess cryptoviviparous seeds, which germinate without the pericarp rupturing. How either is an adaptation to inundation, salinity, low-nutrient availability or waterlogging is not clear, although the peculiar timing of germination of mangrove seeds does improve seed anchoring, establishment and nutrient availability (Saenger 1982). Mature seedling or propagule production typically occurs in the later summer months of February to March, coinciding with the end of the rainy season in the tropics (Duke *et al.* 1984). Studies on *Avicennia marina* in southern New South Wales show it flowers in late summer too (Clarke and Myerscough 1993). Many mangrove propagules are buoyant, aiding dispersal on tides well beyond the parent tree; however the dispersal distance appears to be variable. The vast majority of *Avicenna marina* propagules were reportedly stranded within 50 m of the parent tree in south-eastern Australia (Clarke 1993) and *Ceriops tagal* propagules were stranded within 3 m of the parent tree in Northern Australia (McGuinness 1996), although long-distance dispersal is presumably possible. Some seeds are not buoyant or viviparous (e.g. *Sonneratia*), implying that vivipary and seed buoyancy are not always required as an adaptation to the coastal environment.

Whereas mangroves are often limited to the mid-tide zone, coastal saltmarsh can extend to the very highest levels of tidal inundation, the level set under normal conditions by the highest astronomical tide. In the most landward areas, soil salinities can be highly variable: tidal inundation in some locations can be so rare that hypersalinity often develops due to evaporation, but in others periodic freshening by runoff from local catchments or from adjacent floodplain rivers can result in recurrent periods of low salinity. Plants occupying these environments have strategies for tolerating the highly variable salinity and flooding regimes (Clarke and Hannon 1969). In the upper intertidal environment, nitrogen limitation may also be an important control on plant growth (Feller *et al.* 2003), as it is for many coastal marine systems. Strategies for the fixation and preservation of nitrogen in Australian saltmarshes include plant–plant interactions with nearby nitrogen fixers *Casuarina glauca* (Wei *et al.* 2015).

Given the wide range in plant morphologies occurring in saltmarsh, it is not surprising that saltmarsh plants exhibit a very wide range of salt tolerances and adaptations to living in environments that are variously saline or fresh and inundated or desiccated (Morrisey 1995). In hypersaline areas of Victoria only a few angiosperm species can exist and the wetland is typically a low shrubland dominated by succulent Amaranthaceae such as *Tecticornia pergranulata* and *T. halocnemoides* (Boon *et al.* 2015). Elsewhere, extremely high soil salinities may favour the growth of resilient cyanobacterial mats, particularly along the arid coastline of Western Australia (Paling and McComb 1994; Lovelock *et al.* 2010a). Conversely, the rear parts of coastal wetlands are often subject to localised freshwater flows, and may support a wide range of salt-tolerant taxa such as *Bolboschoenus caldwellii* and *Schoenoplectus pungens* and semi-aquatic species tolerant of at least moderate salinity. In other places, rushland/sedgeland vegetation, typically dominated by *Juncus kraussii*, occasionally with *Phragmites australis* or species of Cyperaceae with components of small halophytic herbs, occur in coastal areas where freshwater flows augment otherwise saline environments (Boon *et al.* 2015). Although limited in extent these habitats are some of the most carbon-rich sediments in the intertidal zone (Lovelock *et al.* 2014; Saintilan *et al.* 2013).

Temperature is a critical determinant of competitive interactions between mangrove and saltmarsh. The winnowing of mangrove species with increasing latitude is thought to be primarily an outcome of different cold tolerance among species (Hutchings and Saenger 1987). On the Gulf Coast of the United States, periodic heavy frost is the primary factor that limits encroachment of *Avicennia germinans* into marsh habitats. In eastern Australia, *Avicennia marina* has a greater tolerance of mild frost, possibly resulting from the higher incidence of mild and lower incidence of heavy frost compared with southern United States (Stuart *et al.* 2007), and in this context frost is a less important limitation in competition with saltmarsh. However, warming may increase the vigour of mangrove growth at higher latitudes, and this may explain the proliferation of mangrove at the expense of saltmarsh at higher latitudes across several continents, Australia included (Saintilan *et al.* 2014). Conversely, the high inverse correlation between minimum monthly temperature and the diversity of saltmarsh species ($r^2 = -0.78$; Saintilan 2009) suggests that warmer temperatures may limit the germination and growth of some saltmarsh species. Limited experimental research on temperature effects on germination of saltmarsh species provides some preliminary support (Greenwood and MacFarlane 2006).

22.5 Zonation

The distribution of coastal halophytes at local scales is strongly influenced by the effect of tidal inundation on soil salinity, anoxia and nutrient availability, and this often leads to conspicuous zonation in those places where more than one species of halophyte coexist. As noted earlier, mangroves and saltmarshes share the intertidal zone over most of the Australian continent, with mangrove typically in the lower portion and saltmarsh adjacent in the upper portion of the tidal frame. The introduced *Spartina* and native *Phragmites australis* may colonise mudflats seaward of mangrove. Only in Tasmania do saltmarshes grow without competition from mangrove.

The earliest ecological descriptions conceptualised succession from mudflats to mangrove (Curtiss 1888; Davis 1940) and from mangrove to saltmarsh in Clementsian terms (Pidgeon 1940). In this model, the colonising mangroves built sediment elevation and were thereby eventually replaced by saltmarsh, which took advantage of the less frequent tidal inundation. According to this explanation, continuing accretion and increasing substrate elevation creates conditions favourable for species replacement, while 'land building' by pioneer species continued at seaward margins. This sequence concords with several stratigraphic studies in south-eastern Australia (e.g. Saintilan and Hashimoto 1999), although can be equally interpreted as a response to geomorphic evolution of the intertidal substrate over the Holocene as sea level stabilised.

Traditional successional hypotheses for the zonation observed in coastal vegetation communities have been challenged on a number of fronts. First, extensive peats beneath mangrove (presumed to be from pioneer species) have been found in some instances to indicate the inverse pattern of colonisation, with mangrove now occupying areas that were previously landward freshwater environments (Scholl 1964). Second, a considerable body of work in Northern Australia on the spatial pattern of mangrove found species distributions that could not be explained by successional hypotheses (Thom *et al.* 1975; Stoddart 1980; Woodroffe 1987). For example, on the Ord River, Western Australia, the distribution of *Rhizophora*, a so-called pioneer species, did not reflect its role as a 'land builder', as the species occurred at both the seaward margin and at the landward fringe where inundation frequency appeared to create soil salinities close to seawater (Thom *et al.* 1975). The occurrence of mangrove on eroding shorelines at King Sound in north-western Australia similarly indicates an absence of successional seaward advancement of mangrove (Semeniuk 1980).

In recent decades, a trend has been observed of saltmarsh being replaced by mangrove or mangrove invading into areas from which they were formerly absent (Saintilan and Hashimoto 1999; Saintilan *et al.* 2014). Relative sea-level rise (Rogers *et al.* 2006) and altered rainfall patterns (Eslami-Andargoli *et al.* 2009), both factors that prevent desiccation of mangrove propagules and reduce soil salinities within saltmarshes, are often proposed as the explanation. Disaggregating these factors is difficult because of the compounding effects of the El Niño Southern Oscillation on rainfall and on sea level, and their subsequent co-correlation in south-eastern Australia (Rogers *et al.* 2014).

The expansion of mangroves into coastal saltmarsh has been reported for five continents; purported mechanisms favour geomorphic control on zonation, with expansion of mangrove influenced by interactions between hydroperiod, sedimentation, elevation and salinity, while nutrients play a role in some settings (Saintilan *et al.* 2014). In Tasmania, where mangroves are absent and saltmarsh occupies a greater portion of the intertidal zone, conversion of shrubland to herbfield is reported to be consistent with changes in relative sea level (Prahalad *et al.* 2011; Saintilan and Rogers 2013).

Landward expansion of mangrove is occurring at the same time as seaward mangrove expansion as sediments reach mean sea level (e.g. Mitchell *et al.* 2007). In addition, as forests mature they follow trends of increasing tree size, increasing carbon stores in sediments and declining nutrient availability (Lovelock *et al.* 2010b), but successional turnover in species composition is not necessarily observed. Because of all these factors, support is growing for the notion that the zonation of coastal halophytic vegetation reflects opportunistic responses to geomorphic changes. These involve complex interactions among tidal inundation and soil salinity, substrate subsidence and compaction, changes to freshwater discharge, and sediment deposition and erosion, the latter mediated by sediment availability and changes in sea level, both eustatic and storm-surge related.

Trans-specific zonation patterns can be relatively easy to distinguish along a mangrove tidal profile, particularly on low-energy open coasts. Macnae (1966), for example, identified parallel zones in north-eastern Australia that were characterised by first a landward *Avicennia* zone; *Ceriops* thickets, *Brugiera* forests; *Rhizophora* forests, and finally a seaward fringe of *Avicennia* or *Sonneratia*. In a large study of 17 estuaries in northern Australia, zonation was affected by the biogeographic influence of temperature and rainfall on species occurrence within estuaries and the distinctive tolerance of species to environmental variables that varied along the length of an estuary (Bunt 1996). The influence of tides diminished up-river, and some mangrove species, such as *Avicennia marina*, *Sonneratia alba* and *Rhizophora stylosa*, preferentially occupied the downstream and more saline reaches, where tidal effects were greater. Some species, however, occurred at typically lower salinities in the upstream reaches of coastal rivers, such as *Heritiera littoralis* and *Aegiceras corniculatum* (Duke 2006). Confounding this patterning is the fact that mangroves are often long-lived trees, and zonation patterns can persist for decades, even when environmental conditions that led to their establishment may have changed (Lovelock *et al.* 2011).

Like mangrove, saltmarsh zonation is commonly characterised by distinct growth forms that correlate with gradients of salinity and inundation, such as low grasslands (e.g. *Distichlis distichophylla*, *Sporobolus virginicus*), herbfields (*Samolus repens*, *Wilsonia backhousei*), rhizomatous or stoloniferous sedges (e.g. *Gahnia filum*, *Juncus* spp. such as *J. kraussii*), large clumped grasses (e.g. *Austrostipa stipoides*) and shrubs (e.g. *Atriplex paludosa* subsp. *paludosa*, *Tecticornia arbuscula*, *Tecticornia halocnemoides* and *Tecticornia pergranulata*) (Kirkpatrick and Glasby 1981). These environmental gradients broadly correspond to position in the landscape, with some distinction in species composition between 'upper' and 'lower' saltmarsh, brackish and fringing marsh (Sainty *et al.* 2012). Within subtropical settings, saltmarshes are predominantly grasslands and herbfields, with sedges dominating brackish environments. In temperate Australia, shrubs of the genus *Tecticornia* occupy mid-intertidal environments giving way to herbfields at upper intertidal elevations, which may be particularly diverse (Saintilan 2009).

With the development of elevation models using light detection and ranging (Lidar) technology, there has been increasing interest in modelling zonation, using elevation as a proxy for environmental gradients. In Darwin Harbour, inundation regime was identified as the most important factor influencing the distribution of *Rhizophora stylosa*, *Sonneratia alba* and *Ceriops tagal*, followed by salinity (Crase *et al.* 2013). Similar correlations between species zonation and inundation regimes have also been established for saltmarsh at Botany Bay (Hickey and Bruce 2010). Clarke (2014) cautioned that the inter-correlated nature of abiotic and biotic variables is poorly accounted for in statistical models; hence, identifying causal factors of zonation is problematic in the absence of experimental manipulation.

22.6 Geomorphic evolution and geomorphological settings

Mangrove and coastal saltmarsh generally occupy low-energy intertidal shorelines above mean sea level, and the influence of sea level on their distribution is profound (Adam 1990; Lovelock *et al.* 2015b; Woodroffe *et al.* 2015). Saltmarsh on headlands are associated with land receiving sufficient salt spray to encourage halophytic plant growth and establishment; in contrast, non-tidally influenced mangrove in Australia are likely to be relict populations where tidal exchange has been lost during the Quaternary period. Sedimentary enclosure is the prevailing explanation for relict inland mangroves at Lake MacLeod (Ellison and Simmonds 2003) and Mandora Marsh, Western Australia (Beard 1967), whereas post-glacial uplift of a shore terrace may explain the occurrence of an elevated *Brugiera* forest on Christmas Island (Woodroffe 1988).

Closure of barriers at estuary entrances along wave-dominated coastlines can significantly reduce tidal exchange between saltmarshes and the ocean. Estuary entrances may close seasonally, as is the case in south-west Australia (Hodgkin and Hesp 1998), or more intermittently, as is the case in many parts of south-eastern Australia. Salinity of inundating waters can fluctuate from hypersaline conditions when closed and hydrological flows from catchments are low, to that of sea waters when tidal exchange with the ocean is not impeded, to freshwater conditions when estuary entrances are closed and rainfall/runoff from catchments is substantial.

Peats derived from halophytic material are useful paleo-indicators of sea level because mangrove and saltmarsh occurrence and zonation are very sensitive to tides. Substantial evidence of mangrove and saltmarsh habitat evolution during the Holocene, the most recent period of sea-level rise, has been extracted from sedimentary cores from the South Alligator River, Northern Territory (Woodroffe *et al.* 1993).

In Australia, sea level over this period was characterised by a low stand at approximately 120 m below present mean sea level (PMSL) at 18 000–20 000 yr BP (Before Present), rising to approximately present level (or up to 2 m above PMSL) about 7000 yr BP. After this period sea level remained relatively stable, although there remains some conjecture about when present levels were attained following highstands that may have been up to 2 m above PMSL (Lewis *et al.* 2013). Three phases of estuarine infill are recognisable from paleo-distributions of halophytic vegetation. The transgressive phase (~8000 yr BP) was dominated by landward transgression of mangrove as the sea drowned the valley of the South Alligator River. Mangrove would not have been able to accumulate sediments at rates equivalent to the high rate of sea-level rise and was restricted to the narrow intertidal margins of the flooded valley. The deceleration in sea-level rise that started approximately 7000 years ago marked the commencement of the 'big swamp' phase in this part of northern Australia. During this period, sediments were able to infill the valley to depths of 10–15 m, creating an expansive intertidal substrate that mangroves colonised. This phase was marked in sediment cores by a dominance of *Rhizophora* pollen, with smaller quantities of pollen from species typical of high intertidal zones such as *Sonneratia*, *Avicennia* and floodplain sedges. The high rate of infill slowed because of reduced tidal inundation, and during the sinuous/cuspate phase, from 5300 years BP, the mangrove dominance gave way to grasses and sedges on saline mudflats and fresher backwater swamps.

The sequence of infill of the tidally dominated South Alligator estuary is probably typical of the tide-dominated estuarine valleys of northern, western and eastern Australia, with similar Holocene evolution described for the Adelaide, Daly, Mary, Ord, King, Richmond and Hawkesbury Rivers, and Lizard Island (Thom *et al.* 1975; Chappell 1993; Woodroffe *et al.* 1993; Saintilan and Hashimoto 1999; Hashimoto *et al.* 2006; Proske and Haberle 2012). The dominance of mangrove on coastal floodplains in the mid-Holocene has led to the development of large carbon pools in the soils of mangrove, saltmarsh and coastal floodplains. Some of the carbon in these pools has been implicated in the generation of pyrites and acid sulfate soils (Sammut *et al.* 1995, 1996; Lin and Melville 2010).

As intertidal halophytes share the intertidal zone, they also share common geomorphic units for much of their distribution along the Australian coastline. Mangrove development along the tide-dominated coasts typical of northern Australia is generally restricted to two primary geomorphological settings: mangrove may occupy the intertidal zone of tide-dominated estuaries, or deltaic environments where the degree of valley infill is greater and mangrove development tends to be most extensive (Figure 22.5a, b). Where rainfall is high, mangrove may occupy the entire intertidal zone and merge with rainforest and woodlands in the hinterland. Saltmarsh, however, may develop in the upper intertidal zone where mangrove expansion is limited by hypersaline conditions. This occurs around Townsville in Northern Queensland. In contrast, cyanobacterial mats occupy the upper intertidal zone on the arid coastline of north-western Australia, where infrequent tidal inundation and high evaporative demand generate hypersaline conditions and likely severe nutrient limitation (Lovelock *et al.* 2010a).

The wave-dominated rocky coastlines of southern Australia do not provide an ideal substrate for extensive development of coastal halophytes but, there are some places where mangrove occurs on rock platforms or other rocky substrates. Examples include Crackneck Point (New South Wales), basalt at The Jawbone Marine Sanctuary, granite at Millers Landing, sedimentary rocks south of Hastings in Western Port (all in Victoria). These rocky shores are often broken up by extensive beach ridges and sand barriers. As a consequence, the tidally restricted estuarine valleys are more lagoon-like (Figure 22.5c–e). Mangrove may also grow in carbonate geomorphological settings on the Great Barrier Reef (Buckley 1983), typically located in low-energy conditions behind fringing coral reefs (Figure 22.5i).

Within wave-dominated estuaries in south-eastern and south-western Australia, intertidal substrates on which mangrove and saltmarsh establish are most extensive on fluvial and flood-tide deltas. Saltmarsh dominates and mangrove tend to be absent from intermittent estuaries where the estuary entrance is usually closed. The primary cause of exclusion is unknown, but may relate to variable salinity and inundation dynamics associated with transitions between opening and closing. Along the New South Wales coastline, training of estuary entrances using seawalls to reinstate daily tidal flushing appears to promote establishment of mangrove along the shorelines (e.g. Lake Illawarra). Some colonisation and growth of coastal halophytes may occur along low-energy shorelines of drowned river valleys where the geomorphology reduces wave energy and elevates the influence of tides on substrates (e.g. the Hawkesbury River, Figure 22.5f). Tides are even more important along coasts where wave energy is diminished, such as coastal embayments (e.g. Western Port, Figure 22.5g) and the open coast where freshwater inputs from catchments are diminished (Figure 22.5h).

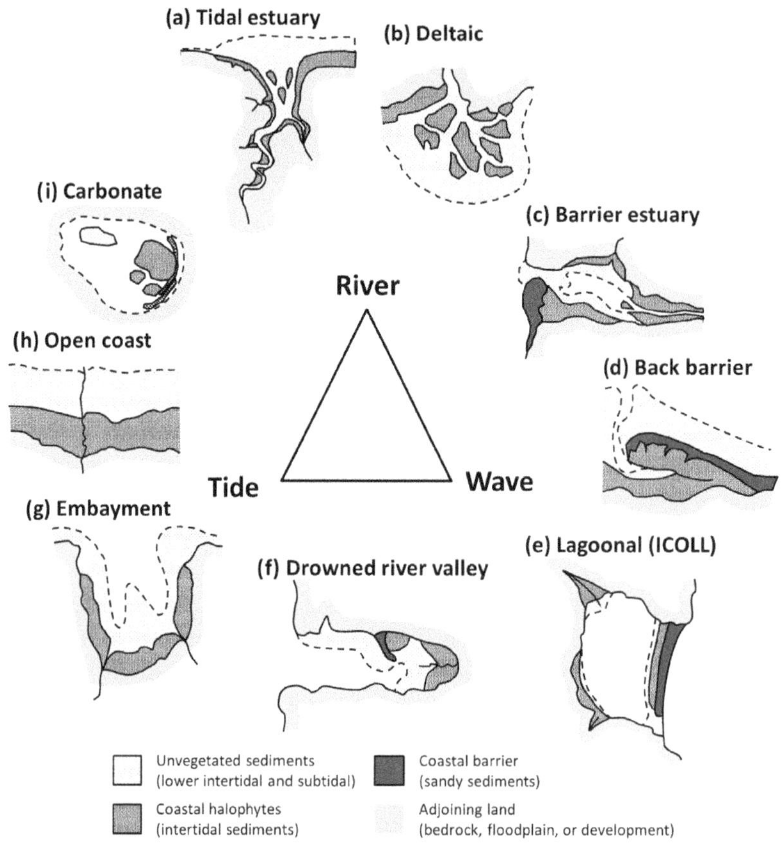

Figure 22.5 Geomorphological settings in which mangrove and saltmarsh are found in Australia. These settings are differentiated on the basis of the influence of wave, tide and river power on sedimentary dynamics within estuarine environments. Conceptualisation of the geomorphic settings of mangrove and saltmarsh in Australia builds upon the ternary classification of estuaries by Boyd *et al.* (1992) and Dalrymple *et al.* (1992); and its application to estuaries of Australia (Harris *et al.* 2002; Heap *et al.* 2004); and integrates the classical geomorphological settings of mangrove (adapted from Thom 1982; Woodroffe 1992) and saltmarsh (Woodroffe 2003).

22.7 Sedimentation and morphodynamic feedbacks

Coastal halophytic vegetation is strongly influenced by sea level. Over extended time periods, eustatic and relative sea level are not stable and as a result intertidal plant communities are dependent upon the supply and accumulation of material to maintain their position within the tidal frame. Once vegetation becomes established in the intertidal zone, further sedimentation is facilitated by organic matter contributions from *in situ* and neighbouring vegetation. Recent analyses using carbon isotopes suggests that *in situ* vegetation is the primary source of organic matter in mangrove and saltmarsh sediments in south-eastern Australia (Saintilan *et al.* 2013).

Once established, the vegetation baffles tides, decreasing flow velocities and enabling entrained sediments to be deposited and trapped (Furukawa *et al.* 1997; Mazda *et al.* 1997; Rogers and Woodroffe 2014). For mangrove in tropical Northern

Australia, the mineral component of sediments may be derived from the considerable mud reserves located offshore and transported onshore by waves, wind and tide (Wolanski 2006). Critical in the deposition process are the drag force of the sediment surface and the complex array of aerial roots, both of which facilitate sedimentation of suspended particles (Mazda *et al.* 1997). Empirical studies emphasise the efficiencies of the aerial components of mangrove in trapping sediments in tropical (Furukawa *et al.* 1997) and subtropical settings (Adame *et al.* 2010). In a microtidal temperate estuary in New South Wales, tidal flow velocity declined from the river through the mudflat and mangrove to the upper intertidal saltmarsh, and varied with distance from the river and suspended particulate matter concentrations (Howe *et al.* 2005). Intertidal halophytic vegetation, therefore, facilitates sediment deposition and stability in a number of ways: the fine mat of surface roots and deeper rhizomes bind sediment and prevent erosion; emergent shoots and roots decrease current velocities and promote deposition of fine particles; and plants add organic matter to sediments via primary production, thus building up the sediment surface.

Saltmarshes in the northern hemisphere commonly exhibit strong relationships between sediment supply, distance to the source of sediment (such as tidal creeks), elevation and deposition (Allen 2000). On the basis of these patterns, Adam (1990) proposed that sediment accumulation was inversely proportional to the time the sediment surface was inundated, known as the hydroperiod. This hypothesis has not been tested in Australia, but analyses of vertical accretion, surface elevation changes and subsidence using surface elevation tables and feldspar marker horizons (i.e. the SET-MH approach) across the intertidal zone consistently showed higher rates of sedimentation in lower intertidal mangrove than upper intertidal saltmarsh in a number of sites in south-eastern Australia (Rogers *et al.* 2006). In most instances, the magnitude of vertical accretion, subsidence and surface elevation increase was mediated by several variables, including sea level (Rogers *et al.* 2005, 2014; Lovelock *et al.* 2015a), turbidity (Lovelock *et al.* 2015a), climatic factors (Rogers *et al.* 2006; Rogers *et al.* 2014), and the effect of climate on groundwater availability (Rogers and Saintilan 2009). Storms were found to have only limited influence on surface elevation trends in vegetated intertidal habitats in south-eastern Australia (Rogers *et al.* 2013b), in contrast with trends reported in North America (Cahoon 2006). In much of south-eastern and south-western Australia, mangrove and saltmarsh occur within barrier estuaries where they are afforded some protection from storm surges.

Established relationships between sedimentation, sediment supply and distance from sediment sources have been the foundation for models of mangrove and saltmarsh evolution in Australia (Akumu *et al.* 2011; Traill *et al.* 2011; McKee *et al.* 2012; Rogers *et al.* 2012) and globally (Fagherazzi *et al.* 2012). With increasing concern about the vulnerability of intertidal vegetation to sea-level rise, models have demonstrated limitations to self-regulation of elevation, with projections indicating that relative sea-level rise will likely lead to declines in mangrove and saltmarsh health and extent unless the rates of sediment supply and accumulation increase (Rogers *et al.* 2013a; Lovelock *et al.* 2015b).

22.8 Current status and trends in extent

The extent of coastal halophytic vegetation is declining across the world, due largely to conversion of halophytic habitats to alternative land uses within the coastal zone (Gedan *et al.* 2009; Polidoro *et al.* 2010). The trend is most rapid in developing countries, but Australia has not been immune to such losses. One of the best indications of declines in extent of halophytic vegetation comes from comparisons of current extents with the estimates of coastal wetland extent prior to extensive drainage of wetlands in the 1970s. In New South Wales, mangrove and saltmarsh is estimated to have declined from 337 km^2 prior to 1969 (Goodrick 1970), to approximately 179 km^2

following the last statewide survey (OzCoasts 2009). Similarly, declines in mangrove and saltmarsh extent in Victoria is estimated to be in the order of 5–20% since the mid-nineteenth century across different sectors of the coast, with some regions having lost as much as 50% of their coastal saline wetland since European colonisation in the early nineteenth century (Sinclair and Boon 2012). In Queensland, there has been widespread conversion of saline habitat to ponded pasture for beef production and other agricultural uses including sugar cane (Holmes *et al.* 2013). In South Australia, anecdotal evidence of mangrove and saltmarsh distribution at the time of European settlement indicates significant habitat loss, particularly around Gulf St Vincent (Fotheringham and Coleman 2008) where approximately 80% of tidal saltmarsh and 100% of saltwater tea tree (*Melaleuca* spp.), reed beds and native coastal grasslands have been lost to urbanisation (Edyvane 1999).

The losses of mangrove and saltmarsh in Australia are largely attributed to land infilling and drainage for coastal developments, solar evaporative ponds for salt production, aquaculture, agriculture and flood mitigation purposes. In New South Wales, Williams and Watford (1997) estimated that approximately 4200 structures prevent tidal flow across coastal floodplains. These facilitated conversion of coastal halophytic vegetation to grazing land or other land uses. Similar impacts on coastal halophytic vegetation have occurred on floodplains across all of the eastern states of Australia (Sinclair and Boon 2012; Boon *et al.* 2015). Flow-on effects from drainage includes the conversion of highly efficient carbon sinks to greenhouse gas sources (Poffenbarger *et al.* 2011), discharge of acid into estuarine waters from pyritic soils associated with halophytes (Sammut *et al.* 1995), habitat fragmentation (Laegdsgaard 2006) and eutrophication (Lovelock *et al.* 2009). Greater governance and community awareness of the value of mangrove and saltmarsh have limited the impact of threatening processes on mangrove and saltmarsh, though resource exploration and infrastructure development remain significant threats.

Despite mounting evidence of the ecosystem services provided to society by coastal communities of halophytic plants (Barbier *et al.* 2011), smaller scale threats are still ongoing and difficult to manage. Vehicle and foot access to adjacent waterways have had a profound effect on saltmarsh extent in urbanised areas. For example, recreational vehicle use on saltmarsh was responsible for the loss of more than 2 ha of restricted saltmarsh along the Georges River in Sydney (Kelleway 2005), and destruction of large stands of the Vulnerable *Wilsonia backhousei* (listed under the New South Wales *Threatened Species Conservation Act 1995*) on the Parramatta River (Kelleway *et al.* 2007). Ongoing disturbance such as mowing, dumping of litter, discharge of stormwater and pollution (Laegdsgaard 2006), trampling by livestock and grazing have reduced the cover and susceptibility of halophyte communities to displacement by specialised exotic species. *Juncus acutus*, of Mediterranean origin, is known to outcompete the native rush *Juncus kraussii*, particularly in highly urbanised estuaries, such as Homebush Bay (Paul and Young 2006). Infestations of the invasive common cord grass (*Spartina* x *townsendii* and *Spartina anglica*), originally introduced to stabilise mudflats and limit erosion of coastal saltmarshes in Tasmania and Victoria, have required significant management efforts (Kriwoken and Hedge 2000). Weed invasions particularly associated with salt-tolerant agricultural plants (e.g. Tall wheat grass, *Lophopyrum ponticum*) are often unrecognised as a major threat to coastal halophytic plant communities (Boon *et al.* 2015; Leishman *et al.*, Chapter 9 this volume).

22.9 Climate change: sea level rise, changing rainfall patterns, increasing temperature, increasing carbon dioxide

Elevated atmospheric carbon dioxide concentration and its effect on temperature, eustatic sea level, storms and rainfall patterns is an ongoing and poorly managed threat. Observed mangrove expansion and saltmarsh contraction has been linked to multiple factors associated with climate change: sea-level rise (Rogers *et al.* 2006); altered rainfall patterns (Eslami-Andargoli *et al.* 2009); and elevated carbon dioxide concentrations (Reef and Lovelock 2014; Saintilan and Rogers 2015). An explanation of the changes in vegetation lies in the unique niche of coastal halophytes, defined closely by (largely unknown) tolerances to inundation and salinity. Alterations to these processes, facilitated by climatic drivers, induce changes in the extent of mangrove and saltmarsh.

Adaptation to sea-level rise is essential for the long-term future of coastal halophytes and their provision of ecosystem services. In this regard Northern Australia is fortunate to have long stretches of coastline and vast coastal floodplains that have not yet been exposed to significant development pressures, and where mangrove and saltmarsh can retreat from rising seas into a largely undeveloped hinterland. In temperate and subtropical areas of Australia, where coastal development pressures are markedly greater and have longer historical legacies, constraints on shoreward retreat are more significant.

Models are important tools for planning the adaptation of halophytic communities and their ecosystem services along the developed coastlines of Australia (Rogers *et al.* 2016). Simple approaches that employ bath-tub modelling techniques provide a first-order indication of the vulnerability of coastal ecosystems to sea-level rise (e.g. DCC 2009). They, however, do not account well for the many internal processes that alter elevations of the land relative to water levels, such as sedimentation, plant productivity and root expansion, autocompaction and organic matter decomposition (Rogers *et al.* 2012). More complex models account for internal processes and for possible management interventions, allowing the adaptive capacity of coastal halophytic communities to be evaluated. Under low rates of sea-level rise, these models predict existing mangrove and saltmarsh communities to shift across the intertidal zone to higher elevations while maintaining their position within an altered tidal frame (Traill *et al.* 2011; Rogers *et al.* 2013a). Under high rates of sea-level rise, however, the resilience of halophytic communities may become exhausted, resulting in significant declines in extent. The models also predict declines in extent where there are physical barriers to landward retreat, such as seawalls, levee banks, floodgates, roads and buildings are not major impediments to retreat (Rogers *et al.* 2013a). This suggests that climate adaptation requires management intervention to reduce barriers in many areas of south-eastern Australia that have been turned over to urban, industrial or agricultural land uses.

22.10 Governance and management

Governance and management of halophytic vegetation in Australia is implemented within both marine and terrestrial environments governance structures across all levels of government (federal, state and local) (Rogers *et al.* 2016). While this approach has successfully reduced large-scale reclamation and flood mitigation works, it has led to a complex system of conservation and management that varies significantly between localities. Many legislative and policy instruments are generated on the basis of geographically defined boundaries that are usually delineated as polygons on maps. For example in New South Wales, coastal wetlands are protected under State Environmental Planning Policy No. 14 (SEPP14), a planning instrument under the New South Wales Environmental Planning and Assessment Act (1979), which affords protection from threatening processes to mapped coastal halophytic vegetation outside the Sydney–Newcastle–Wollongong area. However, the mapping of protected wetlands is based on aerial photography captured in the early 1980s, and may thus fail to protect ecological communities that migrate beyond existing conservation and management boundaries in the context of a changing climate and other drivers (Rogers *et al.* 2013a; Mills *et al.* 2015). More recently, floristic based approaches under the New South Wales *Fisheries Management Act 1994* (amended 2010) and *Environmental Protection and Biodiversity Conservation Act 1999* enable protection of listed saltmarsh communities without the accuracy limitations and temporal constraints imposed by statutory maps.

Climate change may also prompt diversification of conservation priorities and management approaches from traditional measures to achieve comprehensive, adequate and representative conservation of ecosystems (National Biodiversity Strategy Review Task Group 2010), to off-reserve maintenance of the ecosystem services provided by coastal halophytic vegetation. For example, in New South Wales, programmes have been established in selected sites to control expansion of mangrove into adjacent saltmarsh, and in Victoria regional Catchment Management Authorities run a process of single-bid auctions to determine reimbursements to landholders for managing saltmarsh on private land for a minimum of five years according to a pre-existing management plan (Rogers *et al.* 2016). These programmes emphasise the ecological contribution of saltmarsh to fisheries (Mazumder *et al.* 2011) and waterbirds (Spencer *et al.* 2009).

Increasing evidence suggests that mangrove contribute to mitigating the effects of climate change by sequestering carbon from the atmosphere within soils and biomass at a faster rate than saltmarsh (Lovelock *et al.* 2014; Saintilan *et al.* 2013). Balancing the various ecosystem services provided by mangrove and saltmarsh, particularly in the context of a changing climate, might require prioritisation of ecosystem values for conservation and management. Projects and proposals to reinstate tidal exchange across coastal floodplains demonstrate the ecosystem services gained by conversion of low-lying coastal floodplain from marginal agricultural or grazing land, to healthy halophytic communities. Key services include improvements to carbon sequestration, fisheries, biodiversity, reductions in flooding, and improved nutrient cycling (e.g. Creighton *et al.* 2015).

References

Adam, P. (1990) *Saltmarsh Ecology*. Cambridge University Press: Cambridge.

Adam, P. (1996) Saltmarsh. In *State of the Marine Environment Report for Australia. Technical annexe 1*, ed. L.P. Zann and P. Kailola. Department of Environment, Sport and Territory: Canberra, pp. 97–106.

Adame, M.F., Neil, D., Wright, S.F. and Lovelock, C.E. (2010) Sedimentation within and among mangrove forests along a gradient of geomorphological settings. *Estuarine, Coastal and Shelf Science* 86, 21–30.

Akumu, C., Pathirana, S., Baban, S. and Bucher, D. (2011) Examining the potential impacts of sea level rise on coastal wetlands in north-eastern NSW, Australia. *Journal of Coastal Conservation* 15, 15–22.

Allaway, W., Curran, M., Hollington, L. *et al.* (2001) Gas space and oxygen exchange in roots of *Avicennia marina* (Forssk.) Vierh. var. *australasica* (Walp.) Moldenke ex N. C. Duke, the grey mangrove. *Wetlands Ecology and Management* 9, 221–228.

Allen, J.R.L. (2000) Morphodynamics of Holocene salt marshes: a review sketch from the Atlantic and Southern North Sea coasts of Europe. *Quaternary Science Reviews* 19, 1155–1231.

Australian Virtual Herbarium (n.d.) Available at: http://avh.chah.org.au.

Ball, M. (1988) Ecophysiology of mangroves. *Trees* 2, 129–142.

Ball, M.C. and Sobrado, M. (1998) Ecophysiology of mangroves: challenges in linking physiological proceses with patterns in forest structure. In *Physiological Plant Ecology*, ed. M. Press, J. Scholes and M. Barker. Springer: New York, pp. 331–346.

Barbier, E.B., Hacker, S.D., Kennedy, C. *et al.* (2011) The value of estuarine and coastal ecosystem services. *Ecological Monographs* 81, 169–193.

Barrett-Lennard, E.G. (2003) The interaction between waterlogging and salinity in higher plants: causes, consequences and implications. *Plant and Soil* 253, 35–54.

Beard, J. (1967) An inland occurrence of mangrove. *Western Australian Naturalist* 10, 112–115.

Boon, P. (2006) Biogeochemistry, ecology and management of hydrologically dynamic wetlands. In *The Ecology of Freshwater and Estuarine Wetlands*, ed. D.P. Batzer and R.R. Sharitz. University of California Press: Berkeley, pp. 115–176.

Boon, P.I. (2012) Coastal wetlands of temperate eastern Australia: will Cinderella ever go to the ball? *Marine and Freshwater Research* 63, 845–855.

Boon, P.I., Allen, T., Carr, G. *et al.* (2015) Coastal wetlands of Victoria, south-eastern Australia: providing the inventory and condition information needed for their effective management and conservation. *Aquatic Conservation: Marine and Freshwater Ecosystems* 25, 454–479.

Boyd, R., Dalrymple, R. and Zaitlin, B.A. (1992) Classification of clastic coastal depositional environments. *Sedimentary Geology* 80, 139–150.

Breckle, S.-W. (2002) Salinity, halophytes and salt affected natural ecosystems. In *Salinity: Environment–Plants–Molecules*. Springer: Netherlands, pp. 53–77.

Buckley, R. (1983) Biogeography of mangroves on northern Great Barrier Reef Islands. *Bulletin of Marine Science* 33, 767–776.

Bunt, J.S. (1996) Mangrove zonation: an examination of data from seventeen riverine estuaries in tropical Australia. *Annals of Botany* 78, 333–341.

Cahoon, D. (2006) A review of major storm impacts on coastal wetland elevations. *Estuaries and Coasts* 29, 889–898.

Chappell, J. (1993) Contrasting Holocene sedimentary geologies of lower Daly River, northern Australia, and lower Sepik-Ramu, Papua New Guinea. *Sedimentary Geology* 83, 339–358.

Clarke, L.D. and Hannon, N.J. (1969) The mangrove swamp and salt marsh communities of the Sydney district: II. The Holocoenotic complex with particular reference to physiography. *Journal of Ecology* 57, 213–234.

Clarke, P.J. (1993) Dispersal of grey mangrove (*Avicennia marina*) propagules in south-eastern Australia. *Aquatic Botany* 45, 195–204.

Clarke, P.J. (2014) Seeking global generality: a critique for mangrove modellers. *Marine and Freshwater Research* 65, 930–933.

Clarke, P.J. and Myerscough, P. (1993) The intertidal distribution of the grey mangrove (*Avicennia marina*) in southern Australia: the effects of physical conditions, interspecific competition, and predation on propagule establishment and survival. *Australian Journal of Ecology* 18, 307–315.

Clough, B., Andrews, T. and Cowan, I. (1982) Physiological processes in mangroves. In *Mangrove Ecosystems in Australia: Structure, Function and Management*, ed. B.F. Clough. National University Press: Canberra, pp. 193–210.

Crase, B., Liedloff, A., Vesk, P.A. *et al.* (2013) Hydroperiod is the main driver of the spatial pattern of dominance in mangrove communities. *Global Ecology and Biogeography* 22, 806–817.

Creighton, C., Boon, P.I., Brookes, J.D. and Sheaves, M. (2015) Repairing Australia's estuaries for improved fisheries production: what benefits, at what cost? *Marine and Freshwater Research* 66, 493–507.

Curtiss, A. (1888) How the mangrove forms islands. *Garden and Forest* 1, 100.

Dalrymple, R.W., Zaitlin, B.A. and Boyd, R. (1992) Estuarine facies models; conceptual basis and stratigraphic implications. *Journal of Sedimentary Research* 62, 1130–1146.

Davis, J.H. (1940) *The Ecology and Geologic Role of Mangroves in Florida*. Carnegie Institution of Washington: Washington.

DCC (2009) *Climate Change risks to Australia's Coast: A First Pass National Assessment*. Department of Climate Change: Canberra.

De Deckker, P. (1983) Australian salt lakes: their history, chemistry, and biota: a review. *Hydrobiologia* 105, 231–244.

Duggan, M., Connolly, R., Whittle, M. *et al.* (2014) Effects of freshwater flow extremes on intertidal biota of a wet–dry tropical estuary. *Marine Ecology Progress Series* 502, 11–23.

Duke, N. (1995) Genetic diversity, distributional barriers and rafting continents: more thoughts on the evolution of mangroves. *Hydrobiologia* 295, 167–181.

Duke, N., Bunt, J. and Williams, W. (1984) Observations on the floral and vegetative phenologies of north-eastern Australian mangroves. *Australian Journal of Botany* 32, 87–99.

Duke, N.C. (2002) Global distribution and genetic discontinuities of mangroves - emerging patterns in the evolution of *Rhizophora*. *Trees* 16, 65–79.

Duke, N.C. (2006) *Australia's Mangrove*. University of Queensland: Brisbane.

Duke, N.C., Ball, M.C. and Ellison, J.C. (1998a) Factors influencing biodiversity and distributional gradients in mangroves. *Global Ecology and Biogeography Letters* 7, 27–47.

Duke, N.C., Benzie, J.A.H., Goodall, J.A. and Ballment, E.R. (1998b) Genetic structure and evolution of species in the mangrove genus *Avicennia* (Avicenniaceae) in the Indo-west Pacific. *Evolution* 52, 1612–1626.

Edyvane, K.S. (1999) Coastal and marine wetlands in Gulf St. Vincent, South Australia: understanding their loss and degradation. *Wetlands Ecology and Management* 7, 83–104.

Ellison, A.M., Farnsworth, E.J. and Merkt, R.E. (1999) Origins of mangrove ecosystems and the mangrove biodiversity anomaly. *Global Ecology and Biogeography* 8, 95–115.

Ellison, J.C. and Simmonds, S. (2003) Structure and productivity of inland mangrove stands at Lake MacLeod, Western Australia. *Journal of the Royal Society of Western Australia* 86, 25–30.

Eslami-Andargoli, L., Dale, P., Sipe, N. and Chaseling, J. (2009) Mangrove expansion and rainfall patterns in Moreton Bay, southeast Queensland, Australia. *Estuarine, Coastal and Shelf Science* 85, 292–298.

Fagherazzi, S., Kirwan, M.L., Mudd, S.M. *et al.* (2012) Numerical models of salt marsh evolution: ecological, geomorphic, and climatic factors. *Reviews of Geophysics* 50, RG1002.

Feller, I.C., McKee, K.L., Whigham, D.F. and O'Neill, J.P. (2003) Nitrogen versus phosphorus limitation across an ecotonal gradient in a mangrove forest. *Biogeochemistry* 62, 145–175.

Flowers, T.J. and Colmer, T.D. (2008) Salinity tolerance in halophytes. *New Phytologist* 179, 945–963.

Flowers, T.J., Galal, H.K. and Bromham, L. (2010) Evolution of halophytes: multiple origins of salt tolerance in land plants. *Functional Plant Biology* 37, 604–612.

Fotheringham, D. and Coleman, P. (2008) Salt marshes in SA. In *Natural History of Gulf St Vincent*, ed. S. Shepherd. Royal Society of South Australia: Adelaide.

Furukawa, K., Wolanski, E. and Mueller, H. (1997) Currents and sediment transport in mangrove forests. *Estuarine, Coastal and Shelf Science* 44, 301–310.

Gedan, K.B., Silliman, B.R. and Bertness, M.D. (2009) Centuries of human-driven change in salt marsh ecosystems. *Annual Review of Marine Science* 1, 117–141.
Giri, C., Ochieng, E. and Tieszen, L.L. *et al.* (2011) Status and distribution of mangrove forests of the world using earth observation satellite data. *Global Ecology and Biogeography* 20, 154–159.
Goodrick, G.N. (1970) *A Survey of Wetlands of Coastal New South Wales*. CSIRO Division of Wildlife Research: Canberra.
Gorham, J. (1993) Genetics and physiology of enhanced K+/Na+ discrimination. In *Genetic Aspects of Plant Mineral Nutrition*, ed. P.J. Randall, E. Delhaize, R.A. Richards and R. Munns. Kluwer Academic Publishers: Canberra, pp. 151–158.
Greenwood, M.E. and MacFarlane, G.R. (2006) Effects of salinity and temperature on the germination of *Phragmites australis*, *Juncus kraussii*, and *Juncus acutus*: implications for estuarine restoration initiatives. *Wetlands* 26, 854–861.
Harris, P.T., Heap, A.D., Bryce, S.M., *et al.* (2002) Classification of Australian clastic coastal depositional environments based upon a quantitative analysis of wave, tidal, and river power. *Journal of Sedimentary Research* 72, 858–870.
Hashimoto, T.R., Saintilan, N. and Haberle, S.G. (2006) Mid-Holocene development of mangrove communities featuring Rhizophoraceae and geomorphic change in the Richmond River estuary, New South Wales, Australia. *Geographical Research* 44, 63–76.
Heap, A.D., Bryce, S. and Ryan, D.A. (2004) Facies evolution of Holocene estuaries and deltas: a large-sample statistical study from Australia. *Sedimentary Geology* 168, 1–17.
Hickey, D. and Bruce, E. (2010) Examining tidal inundation and salt marsh vegetation distribution patterns using spatial analysis (Botany Bay, Australia). *Journal of Coastal Research* 26, 94–102.
Hodgkin, E.P. and Hesp, P. (1998) Estuaries to salt lakes: Holocene transformation of the estuarine ecosystems of south-western Australia. *Marine and Freshwater Research* 49, 183–201.
Holmes, G., McGrath, C., Larsen, J. *et al.* (2013) *Coastal ecosystems management. Case study: regional planning*. Great Barrier Reef Marine Park Authority: Townsville.
Howe, A., Rodrigues, J. and MacFarlane, G. (2005) Vegetation-sediment-flow interactions in estuarine wetlands. In *MODSIM05: International Congress on Modelling and Simulation: Advances And Applications for Management and Decision Making, 2005*. Modelling and Simulation Society of Australia and New Zealand: Melbourne.
Hutchings, P. and Saenger, P. (1987) *Ecology of Mangroves*. Queensland University Press: St Lucia.
Kelleway, J. (2005) Ecological impacts of recreational vehicle use on saltmarshes of the Georges River, Sydney. *Wetlands (Australia)* 22, 52–66.
Kelleway, J., Williams, R., Allen, C. *et al.* (2007) *An assessment of the saltmarsh of the Parramatta River and Sydney Harbour*. NSW Department Primary Industries: Cronulla.
Kirkpatrick, J.B. and Glasby, J. (1981) *Saltmarsh in Tasmania: distribution, community, composition and conservation. Occasional paper no. 8*. Department of Geography, University of Tasmania: Hobart.
Kozlowski, T.T. (1997) Responses of woody plants to flooding and salinity. *Tree Physiology Monograph* 17, 490.
Kriwoken, L.K. and Hedge, P. (2000) Exotic species and estuaries: managing Spartina anglica in Tasmania, Australia. Ocean and Coastal Management 43, 573–584.
Laegdsgaard, P. (2006) Ecology, disturbance and restoration of coastal saltmarsh in Australia: a review. *Wetlands Ecology and Management* 14, 379–399.

Lewis, S.E., Sloss, C.R., Murray-Wallace, C.V. *et al.* (2013) Post-glacial sea-level changes around the Australian margin: a review. *Quaternary Science Reviews* 74, 115–138.

Lin, C. and Melville, M. (2010) Mangrove soil: a potential contamination source to estuarine ecosystems of Australia. *Wetlands (Australia)* 11, 68–75.

Lovelock, C., Adame, M., Bennion, V. *et al.* (2014) Contemporary rates of carbon sequestration through vertical accretion of sediments in mangrove forests and saltmarshes of south east Queensland, Australia. *Estuaries and Coasts* 37, 763–771.

Lovelock, C.E., Ball, M.C., Martin, K.C. and Feller, I.C. (2009) Nutrient enrichment increases mortality of mangroves. *PLoS ONE* 4, e5600.

Lovelock, C., Grinham, A., Adame, M. and Penrose, H. (2010a) Elemental composition and productivity of cyanobacterial mats in an arid zone estuary in north Western Australia. *Wetlands Ecology and Management* 18, 37–47.

Lovelock, C., Sorrell, B., Hancock, N. *et al.* (2010b) Mangrove forest and soil development on a rapidly accreting shore in New Zealand. *Ecosystems* 13, 437–451.

Lovelock, C.E., Feller, I.C., Adame, M.F. *et al.* (2011) Intense storms and the delivery of materials that relieve nutrient limitations in mangroves of an arid zone estuary. *Functional Plant Biology* 38, 514–522.

Lovelock, C.E., Adame, M.F., Bennion, V. *et al.* (2015a) Sea level and turbidity controls on mangrove soil surface elevation change. *Estuarine, Coastal and Shelf Science* 153, 1–9.

Lovelock, C.E., Cahoon, D.R., Friess, D.A. *et al.* (2015b) The vulnerability of Indo-Pacific mangrove forests to sea-level rise. *Nature* 526, 559–563.

Lucas, R.M., Milne, A.K., Mitchell, A. *et al.* (2000) Use of stereo aerial photography for assessing changes in the extent and height of mangrove canopies in tropical Australia. *Wetlands Ecology and Management* 10, 159–173.

Macnae, W. (1966) Mangroves in eastern and southern Australia. *Australian Journal of Botany* 14, 67–104.

Mazda, Y., Wolanski, E., King, B. *et al.* (1997) Drag force due to vegetation in mangrove swamps. *Mangroves and Salt Marshes* 1, 193–199.

Mazumder, D., Saintilan, N., Williams, J.R., Szymczak, R. (2011) Trophic importance of a temperate intertidal wetland to resident and itinerant taxa: evidence from multiple stable isotope analyses. *Marine and Freshwater Research* 62, 11–19.

McGuinness, K.A. (1996) Dispersal, establishment and survival of *Ceriops tagal* propagules in a north Australian mangrove forest. *Oecologia* 109, 80–87.

McKee, K., Rogers, K. and Saintilan, N. (2012) Response of salt marsh and mangrove wetlands to changes in atmospheric CO2, climate, and sea level. In *Global Change and the Function and Distribution of Wetlands*, ed. B.A. Middleton. Springer: Dordrecht. pp. 63–96.

Mills, M., Leon, J.X., Saunders, M.I. *et al.* (2015) Reconciling development and conservation under coastal squeeze from rising sea-level. *Conservation Letters* doi:10.1111/conl.12213

Mitchell, A.L., Lucas, R.M., Donnelly, B.E. *et al.* (2007) A new map of mangroves for Kakadu National Park, northern Australia, based on stereo aerial photography. *Aquatic Conservation: Marine and Freshwater Ecosystems* 17, 446–467.

Morrisey, D. (1995) Saltmarshes. In *Coastal Marine Ecology of Temperate Australia*, ed. A.J. Underwood and M.G. Chapman. University of New South Wales Press: Sydney, pp. 205–220.

National Biodiversity Strategy Review Task Group (2010) *Australia's Biodiversity Conservation Strategy 2010–2030*. Australian Government: Canberra.

OzCoasts (2009) *National Intertidal/Subtidal Benthic Habitat.* Australian Government: Canberra.

Paling, E.I. and McComb, A.J. (1994) Cyanobacterial mats: a possible nitrogen source for arid-coast mangroves. *International Journal of Ecology and Environmental Sciences* 20, 47–54.
Paul, S., Young, R. (2006) Experimental control of exotic spiny rush, Juncus acutus from Sydney Olympic Park: I. Juncus mortality and re-growth. *Wetlands (Australia)* 23, 1–13.
Pidgeon, I.M. (1940) The ecology of the central coastal area of New South Wales. III. Types of primary succession. *Proceedings of the Linnean Society of New South Wales* 65, 221–249.
Plaziat, J.-C., Cavagnetto, C., Koeniguer, J.-C. and Baltzer, F. (2001) History and biogeography of the mangrove ecosystem, based on a critical reassessment of the paleontological record. *Wetlands Ecology and Management* 9, 161–180.
Poffenbarger, H., Needelman, B. and Megonigal, J. (2011) Salinity influence on methane emissions from tidal marshes. *Wetlands* 31, 831–842.
Pole, M.S. and Macphail, M.K. (1996) Eocene *Nypa* from Regatta Point, Tasmania. *Review of Palaeobotany and Palynology* 92, 55–67.
Polidoro, B.A., Carpenter, K.E., Collins, L. *et al.* (2010) The loss of species: mangrove extinction risk and geographic areas of global concern. *PLoS ONE* 5, e10095.
Prahalad, V.N., Kirkpatrick, J.B. and Mount, R.E. (2011) Tasmanian coastal saltmarsh community transitions associated with climate change and relative sea level rise 1975-2009. *Australian Journal of Botany* 59, 741–748.
Proske, U. and Haberle, S.G. (2012) Island ecosystem and biodiversity dynamics in north-eastern Australia during the Holocene: unravelling short-term impacts and long-term drivers. *Holocene* 22, 1097–1111.
Reef, R. and Lovelock, C.E. (2014) Historical analysis of mangrove leaf traits throughout the 19th and 20th centuries reveals differential responses to increases in atmospheric CO_2. *Global Ecology and Biogeography* 23, 1209–1214.
Reef, R., Feller, I.C. and Lovelock, C.E. (2010) Nutrition of mangroves. *Tree Physiology* 30, 1148–1160.
Rogers, K. and Saintilan, N. (2009) Relationships between surface elevation and groundwater in mangrove forests of southeast Australia. *Journal of Coastal Research* 24, 63–69.
Rogers, K. and Woodroffe, C.D. (2014) Tidal flats and salt marshes. In *Coastal Environments and Global Change*, ed. G. Masselink and R. Gehrels. Wiley and Sons: Oxford, pp. 227–250.
Rogers, K., Saintilan, N. and Heijnis, H. (2005) Mangrove encroachment of salt marsh in Western Port Bay, Victoria: the role of sedimentation, subsidence and sea level rise. *Estuaries* 28, 551–559.
Rogers, K., Wilton, K.M. and Saintilan, N. (2006) Vegetation change and surface elevation dynamics in estuarine wetlands of southeast Australia. *Estuarine, Coastal and Shelf Science* 66, 559–569.
Rogers, K., Saintilan, N. and Copeland, C. (2012) Modelling wetland surface elevation and its application to forecasting the effects of sea-level rise on estuarine wetlands. *Ecological Modelling* 244, 148–157.
Rogers, K., Saintilan, N. and Copeland, C. (2013a) Managed retreat of saline coastal wetlands: challenges and opportunities identified from the Hunter River estuary, Australia. *Estuaries and Coasts* 37, 67–78.
Rogers, K., Saintilan, N., Howe, A.J. and Rodríguez, J.F. (2013b) Sedimentation, elevation and marsh evolution in a south-eastern Australian estuary during changing climatic conditions. *Estuarine, Coastal and Shelf Science* 133, 172–181.
Rogers, K., Saintilan, N. and Woodroffe, C.D. (2014) Surface elevation change and vegetation distribution dynamics in a subtropical coastal wetland: implications for coastal wetland response to climate change. *Estuarine, Coastal and Shelf Science* 149, 46–56.

Rogers, K., Boon, P.I., Branigan, S. *et al.* (2016) The state of legislation and policy protecting Australia's mangrove and salt marsh and their ecosystem services. *Marine Policy* 72, 139–155.

Saenger, P. (1982) Morphological, anatomical and reproductive adaptations of Australian mangroves. In *Mangrove Ecosystems in Australia: Structure, Function and Management*, ed. B.F. Clough. Australian National University Press: Canberra, pp. 153–191.

Saenger, P. (1998) Mangrove vegetation: an evolutionary perspective. *Marine and Freshwater Research* 49, 277–286.

Saenger, P. (2002) *Mangrove Ecology, Silviculture and Conservation.* Kluwer Academic Press: Dordrecht.

Saenger, P., Specht, M.M., Specht, R.L. and Chapman, V.J.(1977) Mangal and coastal saltmarsh communities in Australasia. In *Ecosystems of the World I: Wet Coastal Ecosystems*, ed. V.J. Chapman. Elsevier: Amsterdam.

Saintilan, N. (1997a) Above- and below-ground biomass of mangroves in a sub-tropical estuary. *Marine and Freshwater Research* 48, 601–604.

Saintilan, N. (1997b) Above- and below-ground biomasses of two species of mangrove on the Hawkesbury River estuary, New South Wales. *Marine and Freshwater Research* 48, 147–152.

Saintilan, N. (2009) Biogeography of Australian saltmarsh plants. *Austral Ecology* 34, 929–937.

Saintilan, N. and Hashimoto, T.R. (1999) Mangrove-saltmarsh dynamics on a bay-head delta in the Hawkesbury River estuary, New South Wales, Australia. *Hydrobiologia* 413, 95–102.

Saintilan, N., Rogers, K. (2013) The significance and vulnerability of Australian saltmarshes: implications for management in a changing climate. *Marine and Freshwater Research* 64, 66–79.

Saintilan, N. and Rogers, K. (2015) Woody plant encroachment of grasslands: a comparison of terrestrial and wetland settings. *New Phytologist* 205, 1062–1070.

Saintilan, N., Rogers, K., Mazumder, D. and Woodroffe, C. (2013) Allochthonous and autochthonous contributions to carbon accumulation and carbon store in south-eastern Australian coastal wetlands. *Estuarine, Coastal and Shelf Science* 128, 84–92.

Saintilan, N., Wilson, N.C., Rogers, K. *et al.* (2014) Mangrove expansion and salt marsh decline at mangrove poleward limits. *Global Change Biology* 20, 147–157.

Sainty, G., Hosking, J., Carr, G.W. and Adam, P. (2012) *Estuary Plans and What's Happening to Them in South-east Australia.* Sainty and Associates Pty Ltd: China.

Sammut, J., Melville, M.D., Callinan, R.B. and Fraser, G.C. (1995) Estuarine acidification: impacts on aquatic biota of draining acid sulphate soils. *Australian Geographical Studies* 33, 89–100.

Sammut, J., White, I. and Melville, M. (1996) Acidification of an estuarine tributary in eastern Australia due to drainage of acid sulfate soils. *Marine and Freshwater Research* 47, 669–684.

Santini, N.S., Reef, R., Lockington, D.A. and Lovelock, C.E. (2015) The use of fresh and saline water sources by the mangrove *Avicennia marina. Hydrobiologia* 745, 59–68.

Scholl, D.W. (1964) Recent sedimentary record in mangrove swamps and rise in sea level over the south-western coast of Florida: part 1. *Marine Geology* 1, 344–366.

Semeniuk, V. (1980) Mangrove zonation along an eroding coastline in King Sound, north-western Australia. *The Journal of Ecology* 68, 789–812.

Semeniuk, V. (1983) Mangrove distribution in north-western Australia in relationship to regional and local freshwater seepage. *Vegetatio* 53, 11–31.

Sinclair, S. and Boon, P.I. (2012) Changes in the area of coastal marsh in Victoria since the mid 19th century. *Cunninghamia* 12, 153–176.

Spencer, J.A., Monamy, V. and Breitfuss, M. (2009) Saltmarsh as habitat for birds and other vertebrates. In *Australian Saltmarsh Ecology*, ed. N. Saintilan. CSIRO Publishing: Collingwood.

Stoddart, D.R. (1980) Mangroves as successional stages, inner reefs of the northern Great Barrier Reef. *Journal of Biogeography* 7, 269–284.

Stuart, S.A., Choat, B., Martin, K.C. *et al.* (2007) The role of freezing in setting the latitudinal limits of mangrove forests. *New Phytologist* 173, 576–583.

Thom, B. (1982) Mangrove ecology: a geomorphological perspective. In *Mangrove Ecosystems in Australia*. Australian National University Press: Canberra, pp. 3–17.

Thom, B.G., Wright, L.D. and Coleman, J.M. (1975) Mangrove ecology and deltaic-estuarine geomorphology: Cambridge Gulf-Ord River, Western Australia. *Journal of Ecology* 63, 203–232.

Traill, L.W., Perhans, K., Lovelock, C.E. *et al.* (2011) Managing for change: wetland transitions under sea-level rise and outcomes for threatened species. *Diversity and Distributions* 17, 1225–1233.

Waisel, Y. (1972) *Biology of Halophytes.* Academic Press: California.

Wei, L., Lockington, D.A., Yu, S. and Lovelock, C.E. (2015) Nitrogen sharing and water source partitioning co-occur in estuarine wetlands. *Functional Plant Biology* 42, 410–417.

Williams, R.J. and Watford, F.A. (1997) *Change in the Distribution of Mangrove and Saltmarsh in Berowra and Marramarra Creeks. 1941–1992.* NSW Fisheries Office of Conservation, Fisheries Research Institute: Cronulla.

Wolanski, E. (2006) The evolution time scale of macro-tidal estuaries: examples from the Pacific Rim. *Estuarine, Coastal and Shelf Science* 66, 544–549.

Woodroffe, C. (1992) Mangrove sediments and geomorphology. In *Tropical Mangrove Ecosystems*, ed. A.I. Robertson and D.M. Alongi. American Geophysical Union: Washington, pp. 7–41.

Woodroffe, C.D. (1987) *Pacific Island Mangroves: Distribution and Environmental Settings.* University of Hawaii Press: Hawaii.

Woodroffe, C.D. (1988) Relict mangrove stand on Last Interglacial terrace, Christmas Island, Indian Ocean. *Journal of Tropical Ecology* 4, 1–17.

Woodroffe, C.D. (2003) *Coasts: Form, Process and Evolution.* Cambridge University Press: New York.

Woodroffe, C.D., Mulrennan, M.E. and Chappell, J. (1993) Estuarine infill and coastal progradation, southern van Diemen Gulf, northern Australia. *Sedimentary Geology* 83, 257–275.

Woodroffe, C.D., Rogers, K., McKee, K.L. *et al.* (2015) Mangrove sedimentation and response to relative sea-level rise. *Annual Review of Marine Science* 8, 243–266.

23 Mallee Woodlands and Shrublands: The Mallee, Muruk/Muert and Maalok Vegetation of Southern Australia

COLIN J. YATES, CARL R. GOSPER, STEPHEN D. HOPPER, DAVID A. KEITH, SUZANNE M. PROBER AND MARK G. TOZER

Mallee Woodlands and Shrublands are a characteristic element of southern Australian Mediterranean type climate landscapes, being most extensive in intermediate to low rainfall zones. Mallee plant communities in south-east and southern central Australia, and muruk plant communities in south-west Australia are dominated by *Eucalyptus* spp. with multiple stems and a large subterranean lignotuber from which plants sprout after disturbances. Maalok plant communities in southwest Australia are dominated by *Eucalyptus* spp. with single stems and no lignotuber, relying on seed for regeneration after disturbance. Understoreys in these communities show a continental-scale pattern of decreasing diversity and cover of sclerophyllous shrubs and increasing cover of *Triodia* hummock grass moving towards the arid interior. At regional scales, edaphic factors affecting soil texture and pH exert a strong influence on floristic composition and structure. The diversity and endemism of dominant eucalypts in muruk and maalok are substantially higher in south-west Australia than mallee in central southern and south-east Australia, with both historical-geological and ecological factors playing a role. Fire regimes, grazing regimes and climate variation interact to affect the dynamics of mallee, muruk and maalok plant communities, but the balance of fire versus grazing as the dominant control is dependent on land tenure and land use. In higher rainfall areas there has been widespread clearing and fragmentation of mallee, muruk and maalok plant communities, but in areas less suitable for agriculture there are still major tracts of relatively undisturbed vegetation which include some of the largest conservation reserves in southern Australia.

23.1 Introduction

Vegetation classified and mapped as Mallee Woodlands and Shrublands occurs extensively throughout the Mediterranean-type and semi-arid climates of southern Australia where they are most prominent in areas receiving between 600 mm to c. 150–200 mm mean annual rainfall (Figure 23.1). Typically, but not always, these plant communities are dominated by long-lived *Eucalyptus* spp., 2 to 6 m high, with numerous aerial stems, a narrow canopy zone and a large subterranean lignotuber from which plants sprout after disturbances; a growth form in eucalypts generally referred to as 'mallee' (Box 23.1; Figure 23.2e–h). In south-west Australia, *Eucalyptus* spp. known as maalok (marlocks or mallets) can coexist in landscape mosaics with the more extensive mallee communities (Beadle 1981; Beard 1990). Maalok eucalypts are short in stature (5–7 m), have single stems, no lignotuber

Australian Vegetation, Third Edition, ed. David A. Keith. Published by Cambridge University Press. © Cambridge University Press 2017.

Citation: Yates, C. J. *et al.* (2017). Mallee Woodlands and Shrublands: The Mallee, Muruk/Muert and Maalok Vegetation of Southern Australia. In D. A. Keith, ed., *Australian Vegetation*, 3rd edn. Cambridge: Cambridge University Press, pp. 570–98.

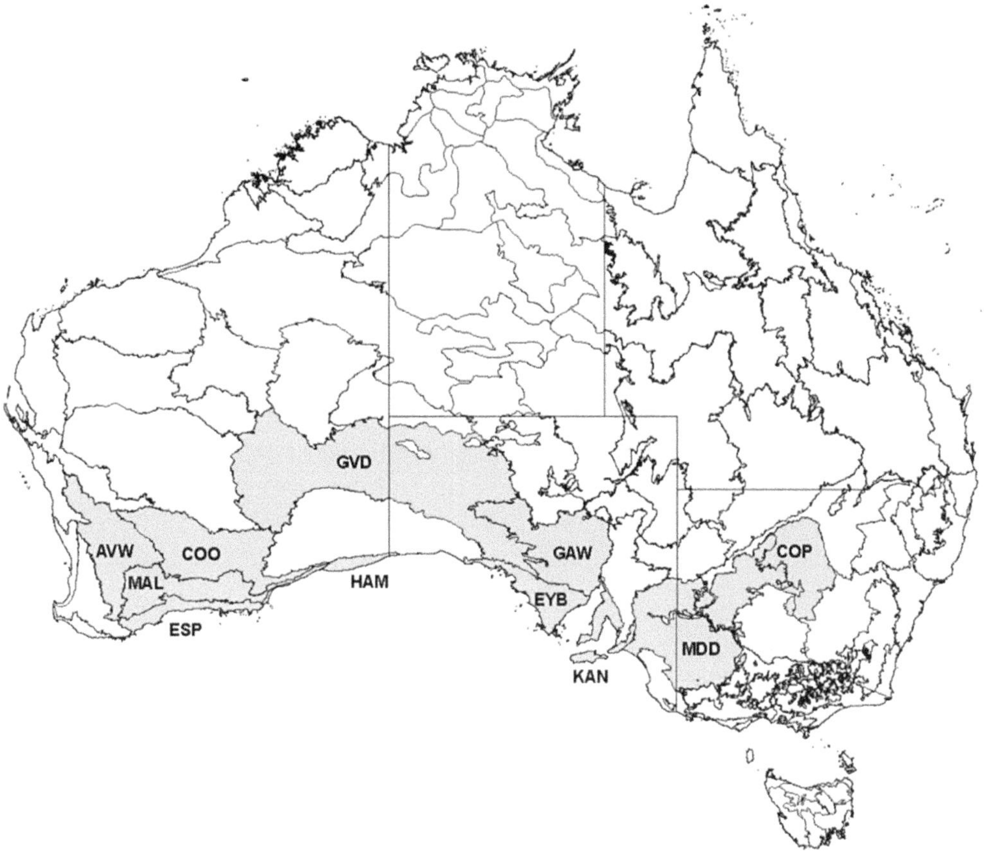

Figure 23.1 Bioregions (shaded) in which Mallee Woodlands and Shrublands (DEWR 2007) are widespread showing the study area over which the analyses in Figure 23.2 were conducted. AVW = Avon Wheatbelt, COO = Coolgardie, COP = Cobar Peneplain, ESP = Esperance Plains, EYB = Eyre Yorke Block, GAW = Gawler, GVD = Great Victoria Desert, HAM = Hampton, KAN = Kanmantoo, MAL = mallee, MDD = Murray–Darling Depression.

and regenerate from seed to form distinctive low forests, woodlands and shrublands (Figure 23.3i).

Aboriginal people have lived in Mallee Woodlands and Shrublands for as long as they have occupied southern Australia (Foulkes and Armstrong 2000; Veth and O'Connor 2013). The word *mallee* is Koori (south-east Australian) in origin (Table 23.1). Correct pronunciation was 'mahlee' according to Henry Fenton, a *Wergaia Dyakkalbaliak* man from Albacutya in the 1890s (Clark 1990: 340). Early encounters with mallee vegetation by eastern Australian explorers Mitchell, Sturt and Eyre did not record the word mallee. Eyre (1845) instead used 'gum scrub'. Two years later, however, Ham's (1847) published map of Australia Felix had the words 'mallee scrub' in two places across *Wergaia* land in north-west Victoria, with one crescentic annotation reading 'Impenetrable Scrub "Eucalyptus Dumosa" termed "mallee" by the Aborigines'. The specific source of this information was not given, but undoubtedly came from personal communication with Chief Protector of Aborigines George Augustus Robinson, who travelled through Wergaia country in April 1845 noting Aboriginal language, collecting herbarium specimens, and also saw and recorded in field notes 'mallee (dumosa)' a year later near Lake Beal Beal in April 1846 (Clark 2000: 23). Ham's adoption of Robinson's 'mallee scrub' was immediately taken up and first

Box 23.1 Variation in richness of terminal *Eucalyptus* taxa

Figure 23.2 includes (a) agglomerative cluster analysis of mallee eucalypt taxa composition and (b) the spatial distribution of clusters with colours in the dendrogram matching colours on the map. The thicker vertical line in the dendrogram (shown in blue in the plate section) shows the level of dissimilarity at which clusters were separated. (c) Cell richness of taxa with a multi-stemmed lignotuber resprouter habit (mallee eucalypts). Cell colour (refer to plate section) shows richness increasing from green (< 10 taxa), blue (11–20 taxa), yellow (21–30 taxa), light orange (31–40 taxa), orange (41–50 taxa), red (51–60 taxa) to purple (> 60 taxa). (d) Cell richness of taxa with a single-stemmed, non-lignotuberous, non-resprouting habit (maalok eucalypts). Cell colour shows richness increasing from green (< 5 taxa), blue (6–10 taxa), yellow (11–15 taxa), orange (16–20 taxa), red (21–25 taxa) to purple (>26 taxa). Maalok eucalypts are restricted to south-west Australia. Data were downloaded from the *Atlas of Living Australia* (2015), and taxa were allocated to one of four post-fire response types (combination resprouter, lignotuber resprouter, stem resprouter or non-resprouter) following Nicolle (2006). After taxonomic and geospatial checking, lignotuber resprouter analyses (using Biodiverse software; Laffan *et al.* 2010) included 281 taxa and 66 500 records, and non-resprouter analyses 82 taxa and 6283 records. Analyses were conducted on cells of 0.75° latitude and longitude. In the agglomerative cluster analysis, the Sorenson dissimilarity metric was used with the link average cluster linkage method (Laffan *et al.* 2010), with cells with less than either 5 taxa and/or 20 records omitted. Analyses made no distinction based on the vegetation type in which taxa occurred. Representative examples of Mallee Woodlands and Shrublands vegetation types are shown: (e) heathy mallee/muruk (photo: Carl Gosper); (f) shrubby mallee (photo: Mark Tozer); (g) *Triodia* mallee (photo: Carl Gosper); (h) chenopod mallee (photo: Suzanne Prober); and (i) maalok (photo: Cristina Ramalho, DPaW).

published in an anonymous article in the *South Australian* newspaper (Anon. 1847). Ham's (1847) and Robinson's use of 'mallee' to refer both to a species of eucalypt and to scrub vegetation has continued, with all multi-stemmed eucalypts that resprout from a lignotuber now called mallees and the Mallee Woodlands and Shrublands they form recognised as one of Australia's Major Vegetation Groups.

Among Noongars of south-west Australia, multi-stemmed resprouter eucalypt scrub is known as *muruk* (*moorook*) by Ballardong people (Wheatbelt NRM n.d.) or the similar *muert* by Esperance Noongars (von Brandenstein 1988). In addition, there are several colonial references (cited in Brooker and Hopper 1991) to use of the term *marlock/maalok* for a dense thicket or impenetrable

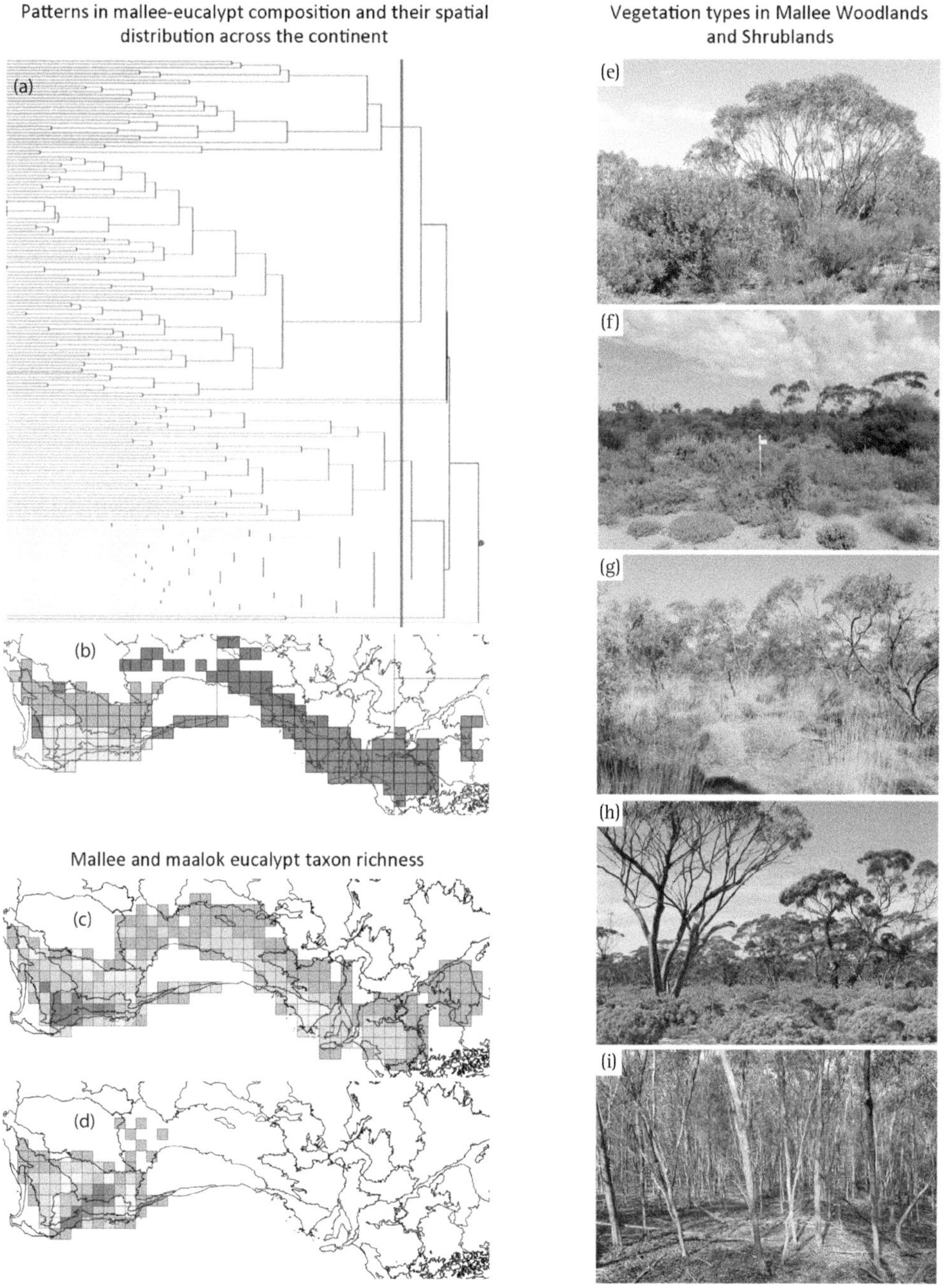

Figure 23.2 Biogeography and spatial variation in richness of terminal *Eucalyptus* taxa with different post-fire responses and major vegetation types in Mallee Woodlands and Shrublands (see Box 23.1).
(A black and white version of this figure will appear in some formats. For the colour version, please refer to the plate section.)

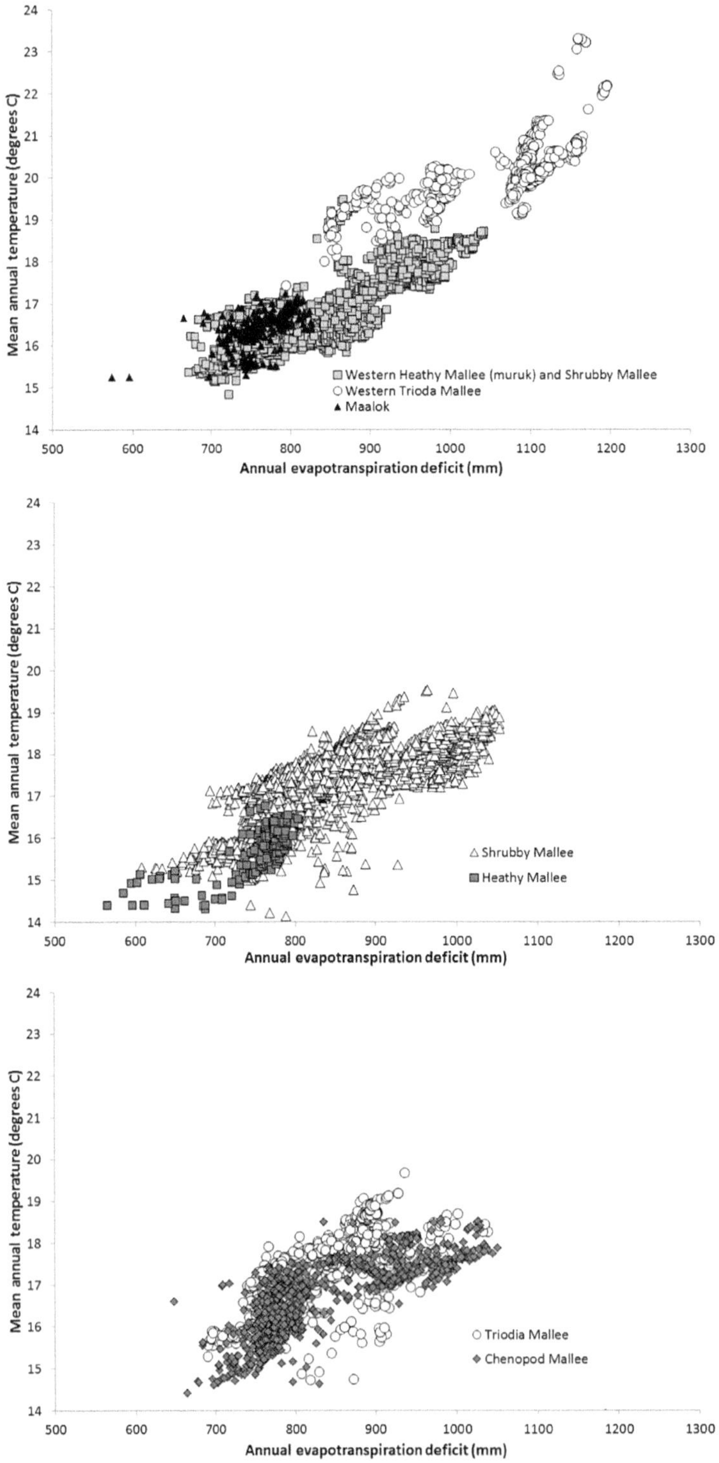

Figure 23.3 Climatic characterisation of Mallee Woodlands and Shrublands. Individual points were sampled randomly from within the distribution of major vegetation types interpreted from National Vegetation Inventory System (Department of the Environment and Energy 2007) and intersected with spatial layers representing mean average annual temperature for the period 1961–1990 and annual evapotranspiration deficit. Gridded climate data were sourced from the Bureau of Meteorology (n.d.). Western Heathy and Shrubby mallee were not separated. Annual evapotranspiration deficit was estimated using a water-balance model (Dyer 2009) based on monthly estimates of precipitation and potential evapotranspiration as inputs and incorporating a soil storage component comprising the available water capacity to 1 m depth (CSIRO n.d.). The model was implemented at the continental scale at a pixel resolution of 270 metres using ArcGIS v.10 (Dyer 2009). Water-balance parameters (actual evapotranspiration, soil storage, surplus and deficit) were calculated using a daily time-step and summed to give monthly and yearly totals.

Table 23.1 **Aboriginal language names for selected mallees, maaloks and moorts. Ngadju words spelt according to Wangka Maya Pilbara Aboriginal Language Centre (2009), Wheatbelt NRM (undated) or according to references cited.**

Latin name	Common term	Aboriginal names (language group: *name*)	Comments
	Thicket	Jardwadjali: *Mallee*	South-east, White (2006)
	Mallee scrub	Esperance Noongar: *Muruk (Moorook)* Noongar: *Muert*	South-west, Wheatbelt NRM (undated) von Brandenstein (1988)
	Thick bush scrub	Noongar: *Nganjima*	South-west, Wheatbelt NRM (undated)
	Dense thicket, impenetrable scrub, Marlock	Noongar: *Marlock/Maalok*	South-west, cited in Brooker and Hopper (1991); typically associated with non-lignotuberous eucalypts such as *E. platypus* (moort)
	Mallee tree/mallee habit	Ballardong Noongar: *Warilj* Yued Noongar: *Woonert*	South-west, Douglas (1976) South-west, Rooney (2011) Place names such as Wyalcatchem, Wyalgima Hill, Warraling Hill and Warralakin are derived from these
	(Non-resprouting) 'mallee'	Noongar: *Moort*	South-west, Wheatbelt NRM (undated); applied to non-resprouting eucalypts such as *E. platypus*, although recorded as 'mallee'
	Mallee saplings traded for spear shafts	Maligundidj: *Jerrum or Kuyen/Queyun*	South-east, Stone (1911); Clark (1990); Hercus (1992)
	Mallee root (used for water)	Wemba Wemba, Baraba Baraba: *Wiya* Esperance Noongar: *Wianyurt* Noongar: *Kudjat*	South-east, Stone (1911); Hercus (1992) von Brandenstein (1988) South-west, Wheatbelt NRM (undated)
Eucalyptus albida	White mallee	Noongar: *Durditj*	South-west, Wheatbelt NRM (undated)
E. celastroides	Snap and rattle	Ngadju: *Nalari*	South-west, makes good hot coals for cooking (Prober *et al.* 2013)

Table 23.1 (cont.)

Latin name	Common term	Aboriginal names (language group: *name*)	Comments
E. caesia	Caesia (commonly used Gungurru is a misinterpretation)	Noongar: *Biant burra*	South-west, Bates (nd~1904-06); Rye and Hopper (1982)
E. doratoxylon	Spearwood mallee	Noongar: *Kiitjmaund* (Geitch-Gmunt)	South-west, Hassell (1975); Bindon and Chadwick (2011); ('spears all together')
E. dumosa	Mallee	Wergaia: *Mallee*	South-east, Ham (1847); Hercus (1992); White (2006)
E. macrocarpa	Mottlecah	Noongar: *Martilgarrang*	South-west, Moore (1884); Douglas (1976). Derived from mara (hand), dilbi (leaf) and garrang (angry), presumably meaning the angry-spirited mallee with hand-sized leaves.
		Balwunga	Moore (1884), derived from bal (he, she it, or leave alone) and wangka (talk), meaning 'leave it alone, it talks (to spirits)'
E. oleosa	Giant mallee, red mallee	Ngadju: *Kutan*	South-west, water obtained from roots; Wangka Maya Pilbara Aboriginal Language Centre (2009); Prober *et al.* (2013).
		Noongar: *Birdidjit*	Wheatbelt NRM (undated)
E. petrensis	Straggly mallee	Noongar Whadjuk: *Koodjat*	South-west, wood is used to make doarks (sticks for knocking the tops off grass trees [*Xanthorrhoea preissii*]), kitjs (spears), wannas (digging sticks), and in recent times, didgeridoos (Collard 2011)
E. pileata	Capped mallee	Ngadju: *Pungal*	South-west, favoured hardwood for spears; stems support 'flathead' bardi grub that is good for chest infections (Prober *et al.* 2013)

scrub. Early colonists familiar with Noongar language associated this term with non-lignotuberous species such as *E. platypus* (*moort*) indicating a clear distinction of these low forests, woodlands and shrublands comprised of small *maalok* trees from lignotuberous *muruk*/*muert* (mallee shrublands).

In this chapter, we build on previous reviews of Australia's Mallee Woodlands and Shrublands (Noble and Bradstock 1989; Noble *et al.* 1990; Parsons 1994). We introduce the term muruk to describe the floristically diverse sclerophyllous shrublands dominated by mallee eucalypts in south-west Australia, thereby distinguishing them from the mallee shrublands of central southern and south-east Australia. In addition we include south-west Australia's maalok vegetation which has been largely ignored in previous treatments of Australia's vegetation. However, we exclude sclerophyll heathlands in humid climates that may superficially resemble mallee shrublands when they contain sparse emergent mallee eucalypts (Lamont and Keith, Chapter 14, this volume). In the period since previous reviews there have been significant advances in the survey, description and inventory of *Eucalyptus* diversity and traits in Mallee Woodlands and Shrublands allowing a continental-scale synthesis of patterns and possible causes. Additionally, there have been significant advances in our understanding of the processes and temporal scales underlying vegetation dynamics. Our aim is to review and synthesise this information, highlighting similarities and dissimilarities in the evolution, ecology and dynamics of major plant communities within the Mallee Woodlands and Shrublands and their implications for conservation.

23.2 Distribution and biogeography

23.2.1 Floristic and structural variability and their broad ecological correlates

The ecological description of Mallee Woodlands and Shrublands at the continental scale (Beadle 1981), together with the more regionally focused work of Beard (1990) and Beard *et al.* (2013) in south-west Australia, and Sparrow (1989), Kinnear *et al.* (2000), Keith (2004) and Haslem *et al.* (2010) in south-east Australia, provides the basis for a general understanding of higher level patterns in vegetation structure and composition.

At the continental scale the composition and structure of Mallee Woodlands and Shrublands is strongly influenced by climate with sclerophyllous shrub diversity and cover decreasing, and *Triodia* hummock grass cover increasing moving towards the arid interior (Beadle 1981; Beard 1990; Gibson *et al.* 2014; Figure 23.3). At regional scales, edaphic factors affecting soil texture and pH may exert a strong influence on floristic composition and structure (Sparrow 1989; Keith 2004; Gibson *et al.* 2014), but biogeographic history is also important, particularly in south-west Australia (Burgman 1988).

A floristic classification of Mallee Woodlands and Shrublands in south-east Australia (Haslem *et al.* 2010) provides a framework for describing high-level patterns of variation in mallee eucalypt-dominated vegetation, and is applicable at the continental scale as follows: (i) heathy mallee including muruk and its south-east counterpart mallee, with a relatively dense sclerophyll shrub understorey dominated by species from temperate plant genera (Figure 23.2e); (ii) shrubby mallee with a relatively open semi-sclerophyll shrub understorey dominated by species from arid plant genera (Figure 23.2f); (iii) *Triodia* mallee with a hummock grass understorey and sparse semi-sclerophyll shrubs from arid plant genera (Figure 23.2g); and (iv) chenopod mallee with a semi-succulent chenopod shrub and tussock grass understorey (Figure 23.2h). A fifth, solely south-west Australian group, maalok vegetation dominated by maalok eucalypts with no understorey or a sparse sclerophyll shrub understorey, may be added to Haslem's *et al.* (2010) classification (Figure 23.2i). Contrasts in the floristic composition of these entities are particularly noticeable for heathy mallee and *Triodia* mallee which occur at the two ends of the climate gradient (Figure 23.4). Variation in composition among major vegetation

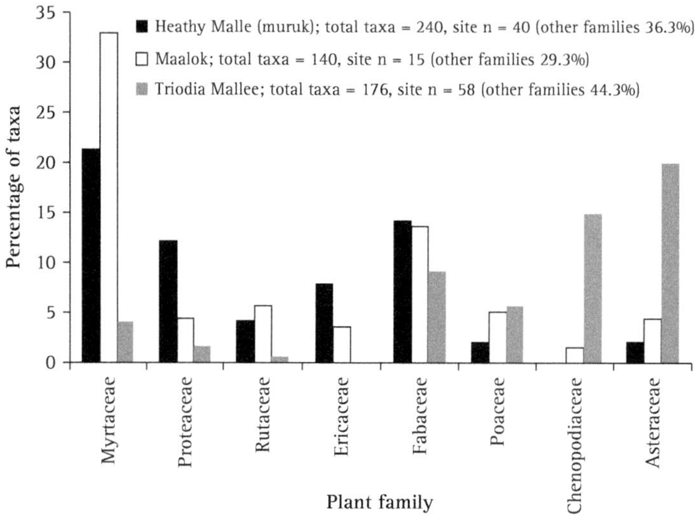

Figure 23.4 Comparative plant family composition of heathy mallee (muruk), *Triodia* mallee and maalok. A global list of taxa from each source was used, excluding exotics. Heathy mallee data are from the Mallee bioregion (Gosper *et al.* 2012a,b); *Triodia* mallee data from the Murray–Darling Depression bioregion (Keith and Tozer unpubl. data); and maalok data from the Mallee and Esperance Plains bioregions (Gibson *et al.* 2004; Markey *et al.* 2012).

types and their ecological correlates are discussed further below and summarised in Table 23.2.

In relatively mesic (receiving > 300 mm mean annual rainfall, MAR) south-west Australia, shrub understoreys of heathy mallee (muruk) and maalok plant communities are dominated by Myrtaceae, Fabaceae, Proteaceae, Ericaceae and Rutaceae, although understoreys in maalok tend to be sparse or even absent. Both heathy mallee (muruk) and maalok vegetation occur on the deeply weathered and lateritised landforms and soils of the ancient Yilgarn Plateau and Albany–Fraser Orogen which occupy most of the region. Heathy mallee (muruk) plant communities are most extensive on mid-slope acid yellow duplex soils where sand transported downslope from lateritic sandplains has accumulated as a thin sheet over the highly weathered pallid zone clays of the laterite profile (Beard 1990; Gibson *et al.* 2004; McArthur 2004). The heathy mallee (muruk) communities are notable for their diversity of mallee eucalypts, with many species having narrow geographic ranges and multiple species coexisting at any point. For example at Lake Magenta (*c.* 350 mm MAR) in the Mallee bioregion, average species richness of mallee eucalypts in 10 m × 10 m plots is 4 species with a range of 2–7 species (Gosper, Prober and Yates unpublished data). Maalok plant communities occur in depressions on mid- and lower-slope positions where there are acid heavy brown and grey clay soils (e.g. *E. platypus*) (Beard 1990; McArthur 2004), but are also found on clay soils of greenstone ranges (e.g. *E. cernua*), younger calcareous sands on coastal dunes (e.g. *E. utilis*) and near-coastal islands (e.g. *E. conferruminata*).

The most extensive occurrences of heathy mallee (muruk) and maalok are in the Mallee bioregion. Heathy mallee (muruk) is widespread but less extensive in the Avon Wheatbelt and Mallee bioregions where duplex soils are often restricted to a narrow band below lateritic breakaways (Beard 1990). In the Avon Wheatbelt lateritic breakaways also support plant communities dominated by non-resprouting

Table 23.2 **Overview of high-level floristic variation for major vegetation types in Mallee Woodlands and Shrublands and their broad ecological correlates.**

	Heathy mallee		Shrubby mallee	Triodia mallee	Chenopod mallee	Maalok
	(muruk)	(mallee)				
Eucalypt post-disturbance response	Lignotuber resprouter		Lignotuber resprouter	Lignotuber resprouter	Lignotuber resprouter	Non-resprouting (obligate seeder)
Widespread and/or characteristic canopy dominants (*E.* = *Eucalyptus*)	*E. celastroides*, *E. cylindriflora*, *E. dissimulata*, *E. flocktoniae*, *E. incrassata*, *E. leptopoda*, *E. phaenophylla*, *E. tenera*, *E. uncinata*, many others	*Callitris verrucosa* *E. costata* subsp. *murrayana*, *E. diversifolia*, *E. dumosa*, *E. leptophylla*	*E. celastroides* *E. dumosa* *E. gracilis*, *E. socialis*,	*E. costata*, *E. dumosa*, *E. oleosa*, *E. socialis*, *E. youngiana*	*E. gracilis*, *E. oleosa* subsp. *oleosa*	*E. cernua*, *E. conferruminata*, *E. alipes*, *E. platypus*, *E. utilis*
Understorey type and widespread and/or characteristic genera	Relatively dense sclerophyll shrubs of temperate affinities e.g. *Acacia*, *Cryptandra*, *Daviesia*, *Grevillea*, *Hakea*, *Hibbertia*, *Leucopogon*, *Leptospermum*, *Melaleuca*, *Phebalium*		Relatively open semi-sclerophyll shrubs of arid affinities e.g. *Acacia*, *Dodonaea*, *Eremophila*, *Senna*	Relatively open hummock grass and sparse sclerophyll shrubs and herbs of arid affinities e.g. *Acacia*, *Beyeria*, *Triodia*, *Vittadinia*	Relatively open semi-succulent chenopod shrubs and tussock grasses e.g. *Austrostipa*, *Atriplex*, *Cratystylis*, *Maireana*, *Olearia*, *Rytidosperma*, *Zygophyllum*	Absent or sparse sclerophyll shrubs of temperate affinities e.g. *Acacia*, *Boronia*, *Gastrolobium*, *Melaleuca*
Distribution	south-west	south-east	south-west/central/south-east	south-west/central/south-east	south-west/central/south-east	south-west
Mean annual rainfall	300–600 mm		200–300 mm	150–300 mm	200–300 mm	300–600 mm

eucalypts known as mallet woodlands (e.g. *E. astringens* and *E.gardneri*, see Prober *et al.*, Chapter 17, this volume). Throughout the Mallee and Avon Wheatbelt bioregions, scattered emergent individuals or clumps of mallee eucalypts occur in sclerophyllous heaths and shrublands (kwongkan) on the sandier and more ironstone gravelly soils of the lateritic sandplains. Some mallee heaths are ecotonal at the margins of muruk and kwongkan (sclerophyllous shrublands) and are best viewed as a complex continuum, but others are unique in so far as the mallee eucalypts (e.g. *E. pleurocarpa*) are largely restricted to mallee heaths, which floristically and structurally have more in common with kwongkan. These mallee heaths are particularly extensive in the Esperance Plains bioregion (Beard *et al.* 2013) and are discussed further by Lamont and Keith (Chapter 14, this volume).

At the more arid end (< 300 mm MAR) of their distribution in south-west Australia, heathy mallee (muruk) grades into shrubby mallee and forms a mosaic with *Eucalyptus* woodlands in the Coolgardie bioregion (Beard *et al.* 2013; see Prober *et al.*, Chapter 17, this volume). Beyond the Yilgarn Plateau towards the increasingly arid interior, chenopod mallee on finer-textured alkaline soils in the Hampton bioregion, and *Triodia* mallee on red sands in the Great Victoria Desert bioregion become more common and form a biogeographic connection with shrubby mallee, chenopod mallee and *Triodia* mallee in central southern and south-east Australia.

In south-east Australia, Mallee Woodlands and Shrublands are most extensive in the Murray–Darling Depression, Kanmantoo and Eyre Yorke Block bioregions where these communities occur on extensive sand deposits and dune fields which have alluvial (red sand) or marine (yellow sand) origins with subsequent aeolian redeposition during the Pleistocene (Bowler 1982; McLaren 2009). In more arid (< 300 MAR) areas to the north, *Triodia* mallee, shrubby mallee and chenopod mallee are most extensive on alkaline red sands (Haslem *et al.* 2010; Gibson *et al.* 2015). Within this region local variation in soil texture affects the distribution of the major mallee vegetation types, with *Triodia* mallee most extensive on the lighter-textured sandy soils, and shrubby and chenopod mallee most extensive on heavier-textured sandy loams. In more mesic (> 300mm MAR) areas to the south, heathy mallee is most extensive on coarser acidic yellow sands (Haslem *et al.* 2010; Gibson *et al.* 2015).

The most mesic occurrences of Mallee Woodlands and Shrublands in south-east Australia are along the southern coast of South Australia, such as on Kangaroo Island, the southern Eyre Peninsula and the Ngarkat area. The understoreys of these heathy mallee communities have structural and floristic similarities to similarly mesic areas of south-west Australia's muruk (Sparrow 1989; Kinnear *et al.* 2000; Davies *et al.* 2013). South Australia is also the core of the distribution of the NVIS Major Vegetation Subgroup 'mallee with a tussock grass understorey' (*Austrostipa* spp., *Rytidosperma* spp.) (Department of the Environment and Energy 2007). This vegetation occurs on finer-textured soils and may be closely aligned to chenopod mallee, but has been extensively cleared for agriculture and/or heavily grazed, making interpretation of original understorey composition difficult (Sparrow 1989).

Patterns of plant species diversity in Mallee Woodlands and Shrublands are generally not well documented, but in south-west Australia's heathy mallee (muruk) beta diversity can be high. For example, Burgman (1988) showed that sites more than 15 km apart differ substantially in floristic composition, with both soil characteristics and geographic distance being important correlates. Comparable published studies of beta diversity for heathy mallee in south-east Australia and other major mallee vegetation types are not available. Similarly there are no published studies of beta diversity in maalok.

23.2.2 Distribution of eucalypt diversity

The distribution and evolution of eucalypt diversity in Mallee Woodlands and Shrublands is best understood in terms of a centre of species richness and endemism extending northwards and eastwards from the south coast of Western Australia (Box 23.1; Figure 23.2c, d). The resprouter mallee growth habit has arisen independently on many occasions in all *Eucalyptus* subgenera across a range of habitats and climates, presumably because the lignotuber enables plants to survive different stresses including frost, drought, insect defoliation and fire. However, there has been major development of the lignotuberous growth habit in subgenus *Symphyomyrtus* Sections *Bisectae* and *Dumaria* in south-west Australia (Figure 23.2c; Brooker 2000). Spatial analysis of mallee eucalypt diversity in Mallee Woodlands and Shrublands indicates four continental-scale compositional groups: two south-west groups, a central southern group and a south-east group (Figure 23.2a, b). One of the south-west groups occurs on the south coast of Western Australia centred on the Mallee and Esperance Plains bioregions, with the other group more northerly and centred on the Coolgardie and northern Avon Wheatbelt bioregions. Significantly, the central southern (mainly from the Great Victoria Desert bioregion) and south-east groups are compositionally more similar with one another than they are with the two south-west groups. Furthermore, the south-east group has an outlying occurrence in the Hampton bioregion in WA in addition to eastern Australian bioregions, clearly showing a strong biogeographical connection in mallee eucalypt composition across the coastal Nullarbor gap.

In contrast to the resprouter mallee growth habit, the obligate-seeder maalok growth habit is restricted to south-west Australian lineages of subgenus *Symphyomyrtus* Sections *Bisectae* and *Dumaria* with greatest species richness in the Mallee and Coolgardie bioregions (Figure 23.2d). Significantly, some of these species are sister taxa to species with a mallee habit and distinguishable only by the presence/absence of a lignotuber, with few or no other morphological character differences (Nicolle 2006). Whether this represents a gain or loss of a lignotuber is unknown and the cause of this high level of character reversibility is a particularly interesting question in the evolution of *Eucalyptus* diversity requiring further investigation with molecular phylogenetic studies of sister species and subspecies.

23.2.3 The role of environmental history in the distribution and evolution of eucalypt diversity

The south-west, central southern and south-east groups of mallee eucalypts occur in landscapes of vastly different ages and histories (see above). It has been postulated that mallee eucalypts from south-west Australia colonised the extensive sand sheets in central southern Australia and south-east Australia relatively recently (Hill 1989). These latter land systems were formed during arid phases of the Pleistocene when climatic thresholds for stabilised vegetation were exceeded (Byrne *et al.* 2008), and are young compared to the Yilgarn Plateau and Albany–Fraser Orogen, where mallee eucalypts occur in the south-west.

Phylogeographic investigations of mallee eucalypt species lineages which span the continent are required to rigorously test this hypothesis. To date, such investigations are restricted to *Eucalyptus* series *Subulatae*, a large group which includes mallee and woodland tree taxa distributed across southern Australia (Nicolle 2008). Patterns of variation in chloroplast DNA across the series do not support rapid and extensive expansion from the south-west to the south-east, but show distinct western, central and eastern clades which diverged 1.25, 1.23 and 0.95 Ma, with species persisting through Pleistocene climate changes in multiple localised refugia (Byrne *et al.* 2008; Nicolle 2008).

Mallee and maalok eucalypts in south-west Australia, like other woody plant groups, have diversified and contribute to the region harbouring one of the world's richest extra-tropical floras (Hopper and Gioia 2004). Why such diversity exists in south-west Australia has been an enduring and intriguing question for biogeography (Hooker 1860;

Diels 1906; Burbidge 1960; Hopper and Gioia 2004). Contemporary ecological factors play an important role in the distribution of plant species, but alone do not appear to adequately explain south-west Australia's spectacular diversity (Burgman 1988; Hopper and Gioia 2004), which has been influenced by environmental history and its impact on the balance of speciation and extinction (Hopper and Gioia 2004; Hopper 2009; Cowling *et al.* 2015). Recent evidence indicates that relatively buffered Pleistocene climate and near-quiescent Cenozoic topography in the south-west may have reduced the intensity of extinctions associated with Pleistocene climate cycles and promoted the persistence of species-rich pre-Pliocene clades (Hopper 2009; Sniderman *et al.* 2013; Cowling *et al.* 2015). The emerging view is that old, climatically buffered, infertile landscapes of the south-west have had more time for species to accumulate than landscapes that have been subject to greater topographic and climate fluctuations (Hopper 2009; Sniderman *et al.* 2013; Cowling *et al.* 2015), and this seems consistent with the distribution of *Eucalyptus* diversity across Australia's Mallee Woodlands and Shrublands.

23.3 Vegetation dynamics: the influence of fire and climate

Fire regimes, grazing regimes and climate variability influence the dynamics of Mallee Woodlands and Shrublands through their influence on the establishment, growth and survival of plant species. For example, fire and rainfall influence seed germination and seedling establishment, while fire, herbivory and drought have selective effects on plant survival (Keith and Tozer 2012). High rainfall and drought can also influence fuel dynamics and flammability.

Throughout the Mallee Woodlands and Shrublands, days of extreme fire weather occur regularly (Bradstock and Cohn 2002a; McCaw and Hanstrum 2003), and are often associated with distinctive weather patterns such as pre-frontal troughs which bring atmospheric instability, dry lightning storms and strong winds during the hot, dry summer months (McCaw and Hanstrum 2003). As a consequence recurrent fire is a primary agent of disturbance in less-modified landscapes and was likely influenced for millennia by Aboriginal people to varying although poorly known degrees. There is a paucity of fire science information for maalok, and chenopod mallee, so much of the following discussion is most relevant to heathy mallee, shrubby mallee and *Triodia* mallee.

23.3.1 Vegetation structure, the nature of fuels and fire behaviour

The quantity and properties of vegetation fuels, including mass, flammability and spatial continuity determine whether and how vegetation will burn. Like other semi-arid ecosystems, heathy mallee, shrubby mallee and *Triodia* mallee are characteristically patchy at a scale that limits lateral connectivity and fire spread. The multi-stemmed eucalypts and understorey shrubs or hummock grasses, the most flammable elements of the vegetation, occur as individual plants or small aggregations with significant gaps between plant canopies (Figure 23.2e–g). Decorticating ribbons of bark are sometimes suspended in the mallee eucalypts, and together with leaf litter, accumulate at the base of plants, forming a discontinuous layer with patches of bare ground (Bradstock and Gill 1993; Cruz *et al.* 2013). The properties of fuel in maalok have not been measured, but differ qualitatively from mallee vegetation types. Generally speaking the single-stemmed eucalypts occur at high densities with less distance between crowns, the community has little or no understorey and a more continuous layer of leaf litter (Figure 23.2i).

In heathy mallee, shrubby mallee and *Triodia* mallee, the mass and continuity of most primary fuel sources increases rapidly in the first

20–30 years after fire, but does not increase substantially thereafter (Haslem *et al.* 2011; Parsons and Gosper 2011). Despite this, quantitative investigations of fire behaviour in heathy mallee and shrubby mallee show that wind speed and fuel moisture content are more important for sustained fire propagation than fuel age (Cruz *et al.* 2013). Wind is particularly important for fire propagation by inclining flames and increasing their lengths to bridge fuel gaps. However, the effect of wind speed on fire sustainability is restricted to the low range of fuel moisture contents (~10% or lower) and higher wind speeds will not offset the damping effects of high litter fuel moisture content (Cruz *et al.* 2013).

Fire hazard within Mallee Woodlands and Shrublands is related to the rainfall gradient from coastal regions to the interior. Fire hazard is independent of time since fire at the arid end of the rainfall gradient and becomes more linearly dependent on time since fire at the wetter end (O'Donnell *et al.* 2011; Gibson *et al.* 2015). This highlights the importance of fuel type. In wetter environments, heathy mallee is commonly dominated by perennial woody shrub fuels that take a number of years to accumulate eventually becoming more laterally continuous. In contrast, in arid environments shrubs and hummock grasses are sparser in shrubby mallee and *Triodia* mallee and significant lateral discontinuities remain even when they are fully grown. Instead, lateral continuity is periodically provided by flushes of ephemeral tussock grasses (mostly *Austrostipa* spp.), which depend on antecedent rain. Hence major fires in semi-arid shrubby and *Triodia* mallee are often associated with 'speargrass years' following above average rainfall (Noble and Vines 1993) and depend less on accumulation of woody fuels with time since fire (Gibson *et al.* 2015).

The fire environment in Mallee Woodlands and Shrublands means that fires are often large (> 10 000 ha), intense, and occur on a multi-decadal basis (Bradstock and Cohn 2002a; O'Donnell *et al.* 2011; Avitabile *et al.* 2013). Quantifying fire return intervals is challenging because the incidence of fire is driven by processes operating on decadal time scales and the historical fire record is relatively short (O'Donnell *et al.* 2011; Avitabile *et al.* 2013). Nevertheless, modelled estimates of fire intervals derived from the fire record vary across Mallee Woodlands and Shrublands, and are influenced by average annual rainfall, understorey and soil type. In south-east Australia higher annual rainfall was consistently associated with shorter fire intervals (Gibson *et al.* 2015). However, at sites along the gradient receiving similar amounts of rainfall the interaction between soil and vegetation type influences the length of fire intervals. Higher fertility sands are associated with shorter fire intervals in *Triodia* mallee, whereas lower fertility sands were associated with shorter fire intervals in heathy mallee (Gibson *et al.* 2015). In south-west Australia the modelled mean fire interval for heathy mallee and shrubby mallee varies from *c.* 40 to 65 years (O'Donnell *et al.* 2011; Parsons and Gosper 2011).

23.3.2 Plant species functional traits and fire responses

As is the case for other fire-prone ecosystems, most plant species in Mallee Woodlands and Shrublands possess traits that confer resilience to periodic burning (Gosper *et al.* 2012a; Clarke *et al.* 2015), either through resprouting or post-fire recruitment from seed. Mallee eucalypts are resilient to multiple disturbances such as fire, drought or frost which kill the canopy. However, burning does increase adult mortality compared to interfire periods (Wellington and Noble 1985). Furthermore, the resprouting capacity of mallee eucalypts can be reduced or even exhausted (leading to plant death) if multiple disturbances over a short period recurrently remove the canopy with the period between disturbances and season affecting resprouting capacity (Noble 2001; Noble and Diggle 2013).

For many woody species in heathy mallee, shrubby mallee and *Triodia* mallee recruitment from seed occurs predominantly after fire, because fire stimulates germination of seed and seedling recruitment is likely to be more successful during reduced competition and/or predator and herbivore satiation in the immediate post-fire period (Bradstock and Cohn 2002a). Rates of recruitment after fire can be quite variable, and in resprouting species, including mallee eucalypts, recruitment success is often very low and partly determined by favourable rainfall following the fire during the seedling establishment phase (Wellington and Noble 1985; Yates *et al.* 2003). Maalok eucalypts are killed by fire and rely on canopy-stored seeds and post-fire germination for population persistence. As a consequence their population dynamics will be particularly sensitive to variation in both fire intervals and fire events (Bond and van Wilgen 1996), as well as post-fire soil moisture, which may be one reason why maalok is restricted to areas with lower evapotranspiration deficits (Figure 23.3). There are no published investigations of post-fire vegetation dynamics or fire-response traits in maalok.

Although Mallee Woodlands and Shrublands are dominated by resprouting mallee eucalypts they have a lower overall proportion of resprouting woody taxa (R+; ~38%) than the other major Australian ecosystems evaluated by Clarke *et al.* (2015). The ability to recruit from seed after fire is strongly represented in Mallee Woodlands and Shrublands, with the highest proportion of woody taxa with post-fire recruitment (S+; ~93%) of all major Australian ecosystems (Clarke *et al.* 2015). When resprouting and post-fire recruitment traits are considered in combination, Mallee Woodlands and Shrublands overall are dominated by obligate seeders (R-S+ ~57%) and facultative resprouters (R+S+ ~35%, including the structurally dominant mallee eucalypts), with both combinations represented in many genera (Pausas and Bradstock 2007; Gosper *et al.* 2012a; Clarke *et al.* 2015). Obligate resprouting (R+S- ~3%) and fire-avoiding (R-S- ~4%) trait combinations are very poorly represented among woody taxa (Clarke *et al.* 2015), although in the latter case this may reflect a lack of sampling and attribution of fire responses in taxa from chenopod mallee, as many chenopods are fire avoiders (Bradstock and Cohn 2002a). In south-east Australia, trait combinations vary along the rainfall gradient, with less arid environments supporting greater representation of facultative resprouters, and obligate seeders predominating in more arid regions (Pausas and Bradstock 2007).

23.3.3 Post-fire dynamics of mallee vegetation

Several authors have suggested that in common with other fire-prone vegetation types, the post-fire recovery of Mallee Woodlands and Shrublands is best described by the initial floristic composition model (Egler 1954; Bradstock and Cohn 2002a; Gosper *et al.* 2012b). This model postulates that all plant species are present in the above-ground vegetation shortly after fire, with species richness and diversity declining with increasing time since fire due to competition and differences in plant longevity, with 'disappearing' species retreating to the soil seed bank. In heathy mallee, successional changes with time since fire consistent with the initial floristic composition model are expressed in the cover of plant functional types (PFTs) affecting survival, persistence and growth, but less clearly expressed in species richness (Gosper *et al.* 2012a,b). The most pronounced changes were increasing cover of resprouting serotinous trees (mallee eucalypts) and decreasing cover of short-lived post-fire ephemerals with time since fire (Gosper *et al.* 2012a).

The structure of heathy mallee, shrubby mallee and Triodia mallee changes over time scales of a century or more after fire, often in non-linear ways and this affects habitat resources used by fauna (Haslem *et al.* 2011; Gosper *et al.* 2012b). Long-unburnt vegetation has wider and taller tree crowns, but fewer stems in the dominant mallee eucalypts and more tree hollows, while in hummock grass understoreys, *Triodia* and litter may reach peak cover at an intermediate time since fire of ~20–80 yrs (Noble 2001; Haslem *et al.* 2011; Gosper *et al.* 2012b). Occasional fire may therefore be important in maintaining the vigour and flammability of *Triodia* mallee communities. Species of birds, mammals and lizards respond to these changes in vegetation structure, showing distinct preferences for vegetation at specific times since fire (Kelly *et al.* 2011; Nimmo *et al.* 2012; Watson *et al.* 2012; Taylor *et al.* 2013). There is no evidence for structural senescence in long-unburnt heathy mallee (Gosper *et al.* 2012b). Indeed, if unburnt for very long (> 100 yrs) periods, some mallee vegetation types may develop a low 'woodland'-like structure, through gradual attrition of subdominant stems in mallee eucalypts, leading to development of a monopodial habit (Noble 2001; Gosper *et al.* 2012b).

The duration of fire intervals will affect the dynamics and composition of Mallee Woodlands and Shrublands because the competitive dominance of species and their regeneration capacity (size of the seed bank for recruitment and carbohydrate and bud reserves for resprouting) vary with time since fire (Keith and Bradstock 1994; Keith *et al.* 2007; Gosper *et al.* 2013). Although most PFTs in Mallee Woodlands and Shrublands are vulnerable to population decline under some scenarios of fire intervals and competing PFTs, bradysporous obligate seeders are usually regarded as particularly vulnerable to extremes of fire interval (Bond and van Wilgen 1996). The juvenile period and plant longevity in key obligate seeders inform minimum and maximum fire-interval thresholds for planning ecological fire management (Bradstock and Kenny 2003; Gosper *et al.* 2013). The temporal development of habitat resources for fauna also provide a means for defining minimum and maximum fire intervals, and species reliant on habitat resources that take decades to develop after vegetation is burnt will be particularly vulnerable to very short fire intervals (Clarke 2008; Haslem *et al.* 2011).

Juvenile periods, seed bank accumulation and plant longevity vary among populations and with site productivity across the small number of plant species and sites for which they have been quantified in heathy mallee, shrubby mallee and *Triodia* mallee. Consequently, generalisation of appropriate minimum fire intervals beyond the few location-specific studies (which suggested a minimum interval > ~20–25 yrs; Bradstock 1989; Gosper *et al.* 2013) to other occurrences is fraught. However, extremely short fire intervals are likely to drive local declines or extinctions of interval-sensitive species. For example, a 10-year fire interval eliminated *Callitris verrucosa* in *Triodia* mallee (Bradstock 1989). Similarly, short intervals (< 10 years) are likely to drive local declines in a number of birds, mammals and reptiles (Kelly *et al.* 2011; Nimmo *et al.* 2012; Watson *et al.* 2012). While theoretically very long fire intervals could result in the loss of some plant species through senescence, there is little evidence to support this. Experimental investigations show that very few plant species are lost through senescence when *Triodia* mallee

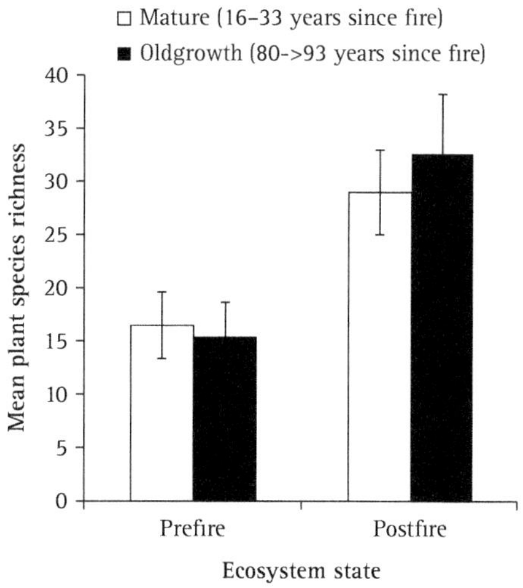

Figure 23.5 Plant species richness before and after fire in Triodia mallee with contrasting times since a previous fire. Mature vegetation was burnt 16 to 33 years previously and old growth 80 to > 93 years previously. The study was conducted at Tarawi Nature Reserve (Scotia region), south-west New South Wales (Keith and Tozer 2012).

remains unburnt for almost a century (Figure 23.5). Furthermore, a number of animal species in *Triodia* mallee, particularly birds, show a preference for older vegetation (> 35 years since fire; Kelly *et al.* 2011; Nimmo *et al.* 2012; Watson *et al.* 2012). In landscapes where Mallee Woodlands and Shrublands are extensive and vegetation is contiguous such as large conservation reserves in south-east Australia and the Great Western Woodlands in south-west Australia, fires can and do occur on a massive scale, sometimes exceeding 100 000 ha in extent (Noble and Vines 1993; Bradstock and Cohn 2002a; Parsons and Gosper 2011). The effects of fires of this scale on biodiversity have been poorly documented. Variation in meso- and micro-scale topography and fuels, and temporal variations in fire weather (diurnally and with wind variability) may add considerable patchiness to recruitment and recovery conditions within the bounds of such fires, but measurements are needed. In *Triodia* mallee a negative feedback between *Callitris verrucosa* maturity and flammability exists such that mature *Callitris* stands suppress fire spread and may escape combustion (Bradstock and Cohn 2002b). This patchiness may contribute to the persistence of the relatively slowly-maturing and obligate-seeding *Callitris* itself.

23.4 Interactions between mallee vegetation and vertebrate herbivores

Although fire is a dominant process in Mallee Woodlands and Shrublands, grazing and browsing by vertebrate herbivores can also influence the dynamics and assembly of plant communities. The current understanding of the role of native herbivores in these communities is limited, but in more intensively used landscapes, livestock grazing can replace fire as the dominant influence on ecosystem structure and composition (Eldridge *et al.*, Chapter 6, this volume). The balance of fire versus grazing as the dominant control of dynamics in Mallee Woodlands and Shrublands is thus dependent on land tenure and land use.

In mallee landscapes managed for conservation or vested as unallocated crown land, native macropods (*Macropus rufus*, *M. fuliginosus*, *M. giganteus* and *M. robustus*) and feral herbivores (mainly rabbits *Oryctolagus cuniculus* and goats *Capra hircus*) may be present. Much of what is known about interactions between these vertebrate herbivores and Mallee Woodlands and Shrublands is derived from herbivore exclusion studies in the *Triodia* mallee of south-east Australia, and thus may not represent the full range of responses. These studies show that a major effect of vertebrate herbivory on mallee vegetation is through the limitation of seedling recruitment, especially for palatable woody species after fire (Cohn and Bradstock 2000; Keith and Tozer 2012).

The magnitude of the effect is event-dependent, with grazing impacts greatest for smaller fires and varying among years (Cohn and Bradstock 2000; Keith and Tozer 2012). The impacts of herbivory on seedlings may also vary markedly between plant species (Table 23.3). Highly palatable *Austrostipa* spp. are well-known as a preferred food source under a range of conditions (Noble 1989), but the consumption of other species is temporally and spatially variable, and many plant taxa rarely show signs of grazing/browsing when exposed to vertebrate herbivores (Table 23.3). The nature of plant herbivore defences has not been explored in Mallee Woodlands and Shrublands, but few species display the extent of physical defences observable in plants from some other semi-arid regions (e.g. Cooper and Owen-Smith 1986). Given the differential outcomes of herbivory and the apparently limited occurrence of physical defensive traits, chemical defences are likely to be important (Carmona *et al.* 2011).

In general, goats appear to have the most diverse diet in *Triodia* mallee, followed by rabbits, with macropods preferring grasses (Wilson 1990), but browsing more on shrub seedlings

Table 23.3 **Classification of selected plant taxa from *Triodia* mallee in Tarawi Nature Reserve (Murray-Darling Depression bioregion, New South Wales) based on the degree of leaf and branch removal from open fenced plots relative to fenced controls (Keith and Tozer unpubl. data). High – biomass consistently removed when exposed to vertebrate herbivores. Intermediate – biomass removal observed at some times and places. Low – biomass removal rarely observed.**

High	Intermediate	Low
Austrostipa	*Acacia*	Asteraceae
Callitris	*Amphipogon*	Solanaceae
Hakea	*Chenopodium*	*Beyeria*
	Eutaxia	*Codonocarpus*
	Haloragis	*Eremophila*
	Maireana	*Eucalyptus*
	Triodia	*Myoporum*
	Westringia	*Sclerolaena*
		Senna
		Zygophyllum

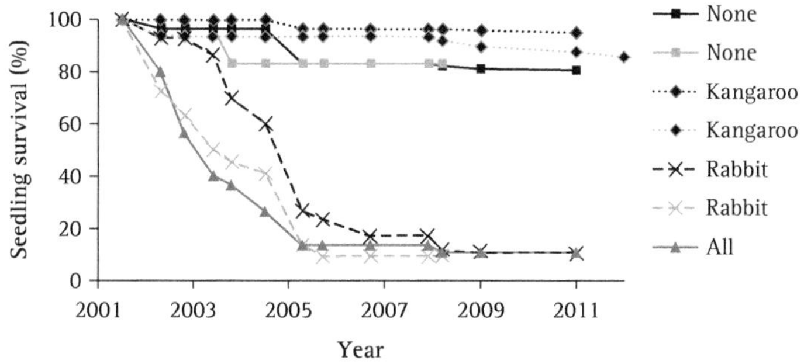

Figure 23.6 Survival of *Callitris verrucosa* seedlings emerging after fire in 1997 in plots at two *Triodia* mallee sites. Fences were erected to limit access to different subsets of the vertebrate herbivores present (kangaroos, rabbits, goats): kangaroos only, rabbits only and no vertebrate herbivores (Keith and Tozer unpubl. data). The study was conducted at Tarawi Nature Reserve (Scotia region) in south-west New South Wales (Keith and Tozer 2012).

when grass availability is limited. Exclosures allowing selective entry of macropods and rabbits in *Triodia* mallee implicate rabbits as the major cause of seedling herbivory in *Acacia* species (Cohn and Bradstock 2000). Similarly, in *Triodia* mallee rabbits and/or goats were responsible for major seedling mortality in post-fire *C. verrucosa* seedling populations, while macropods had no detectable effect relative to total herbivore exclusion (Figure 23.6).

Rangeland sheep grazing for wool production has been extensive and pervasive in south-east Australian mallee landscapes for more than 180 years (Wilson 1990). Stocking rates were probably highest in mallee with a tussock grass understorey and chenopod mallee, which are likely to be more productive and nutritionally suitable for sheep than other types of mallee in drier climates or on poorer soils (Cheal and Parkes 1989). This grazing has had a profound effect with ongoing legacies that continue to affect species composition and community assembly. Davies *et al.* (2013) show substantial differences in plant species composition between mallee vegetation intensively grazed by sheep and cattle, and mallee excluded from livestock in roadside and quarry reserves. Differences in native species richness were even more marked in soil seedbanks than standing vegetation, and differences between standing vegetation and seedbanks were greater in grazed than ungrazed vegetation. The results suggest a cumulative impact on the plant community, whereby reductions in fecundity associated with grazing flow through to depletion of soil seedbanks, which in turn is likely to reduce the diversity of the next generation of standing plants. Conversely, grazing seems to promote the diversity of short-lived introduced plant species (Davies *et al.* 2013).

Temporary or prolonged cessation of livestock grazing in shrubby mallee may result in encroachment by woody shrub species of *Dodonaea*, *Eremophila* and *Senna*, especially if stocking rates have been high. Several landscape-wide episodes of shrub encroachment have occurred in south-east Australia's semi-arid zone in response to either destocking during drought or major La Niña rainfall events that promote high rates of shrub recruitment (Eldridge *et al.* 2013; Eldridge *et al.* Chapter 6, this volume). The resulting increase in shrub cover in shrubby mallee of south-west New South Wales had positive effects on soil water infiltration and nutrient retention. However, there was evidence that some of these processes of ecosystem repair were curtailed by continued livestock grazing leading to long-term declines in productivity. Eldridge *et al.* (2013) suggest that this may be due to dispersion of plant litter by livestock, direct browsing of shrubs (reducing their water interception capacity), a reduction of facilitation interactions among plants or indirect impacts on biocrusts. Other ecosystem-level effects of overgrazing may operate through their adverse impacts on activity of soil invertebrates, which contribute to cycling of nutrients and accumulation of soil carbon (Eldridge *et al.* 2012).

23.5 People in the Mallee Woodlands and Shrublands

23.5.1 Aboriginal perceptions and etymology

Aboriginal perceptions of Mallee Woodlands and Shrublands are not widely documented. Permanent occupation of these semi-arid landscapes was feasible due to the persistence of water in soaks, *gnammas* (rock holes), solution cavities and mallee (e.g. *E. oleosa*) roots, augmented by access to water in neighbouring environments such as riverine wetlands of the Murray or the short coastal rivers of south-west Australia (Eyre 1845; Moore 1884; Maclaren 1912; Hassell 1975; Bindon 1997; Noble and Kimber 1997; Bayly 1999; Foulkes and Armstrong 2000; Prober *et al.* 2013). Indeed, open areas persisting today around soaks in the south-east Australian mallee may be a result of heavy Aboriginal use of these areas including removing timber for firewood (Foulkes and Armstrong 2000). Food resources were varied, including possum, macropods, malleefowl (*Leipoa ocellata*) and emu (*Dromaius novaehollandiae*) eggs, reptiles and plant foods (Ens *et al.*, Chapter 4, this volume).

Among Aboriginal language groups, the mallee eucalypt growth habit had several names, as did individual species (Table 23.1), but not all Aboriginal names for mallee eucalypts are documented. In both south-west and south-east Australia, town and place names are derived from Aboriginal words for the mallee growth habit or particular mallee eucalypt species.

In a contemporary study in the Norseman region of south-west Australia, Ngadju people recognised that mallees come in a wide variety, and in particular they distinguished grassy mallee woodland (*Triodia* mallee) from scrubby mallee vegetation (heathy mallee and shrubby mallee) (Prober *et al.* 2013). Grassy mallee was viewed and managed more like grassland, including frequent ground-layer burning (seeking to avoid burning the mallee eucalypts). Coastal and sandplain mallee scrubs were generally perceived as more inhospitable places prone to bushfires (although still actively burnt to benefit bush tucker). Consistent with this, Ethel Hassell (who settled in the Jerramungup region of south-west Australia in the 1880s) and other colonial ethnographers recorded avoidance of dense 'moort' thickets by Noongar people, as a place where *warra wirn* (bad spirits) linger (Hassell 1975).

Mallee eucalypts are also culturally and practically significant. South coast Noongars still use the inland distribution of *djalyirark* (*Eucalyptus pleurocarpa*, Tallerack) as a marker of their inland boundary. The Esperance Land Claim corporation has named itself for this mallee, and other mallee eucalypt species have many recorded uses by Aboriginal people (Table 23.1).

23.5.2 Pastoralism and agricultural development

Early European settlement of the Mallee Woodlands and Shrublands began in the 1840s and was focused on livestock grazing for wool and meat production (Cooke 2006). Land clearing for cereal growing began in the 1880s with periods of major expansion following the two World Wars (Harris 1990). While agricultural development created food and fibre, and generated substantial wealth, it came with economic and social costs for governments and some individuals, especially on the marginal fringe where climate variability, sandy infertile soils and salinity caused major environmental problems. As a consequence by the mid-1980s clearing in Mallee Woodlands and Shrublands had largely ceased.

Of the estimated 387 230 km^2 pre-European extent of Mallee Woodlands and Shrublands, approximately 30% has been cleared for agriculture (DEWR 2007). This transformation has not been uniform across the continent. Bioregions with notably large vegetation losses include many of the wetter parts of the Mallee Woodlands and Shrublands, including north-western Victoria and south-east South Australia, the Eyre and Yorke peninsulas in South Australia and the southern coast of Western Australia (Table 23.4; Figure 23.1), where cereal cropping and sheep grazing are the main agricultural activities. The Mallee Woodlands and Shrublands in bioregions at the arid end of their distribution have experienced very low rates of land clearing (Table 23.4; Figure 23.1), but pastoralism has had an impact in some areas (see above; Eldridge *et al.* 2013).

23.5.3 Conservation efforts

Despite extensive clearing of Mallee Woodlands and Shrublands in the past, and continuing utilisation of some areas for pastoralism, major tracts of vegetation remain relatively undisturbed, including some of the largest conservation reserves in southern Australia. Approximately 26% of the mapped pre-European extent of Mallee Woodlands and Shrublands is in the National Reserve System (NRS), but because of the pattern of land use, reservation has been opportunistic and representation varies greatly among bioregions where Mallee Woodlands and Shrublands occur (Table 23.4). The high beta diversity and rapid geographical replacement of plant taxa in south-west Australia's muruk and maalok communities presents special challenges for a representative system of conservation reserves (Burgman 1988).

Furthermore, many reserves with Mallee Woodlands and Shrublands occur as isolated remnants embedded in disturbed landscapes ranging from heavily cleared and fragmented agricultural landscapes to more continuously vegetated

Table 23.4 **Representation in the National Reserve System (NRS) and extent of native vegetation clearance in the bioregions in which Mallee Woodlands and Shrublands are widespread. Bioregions are ordered by increasing representation in the NRS (IUCN categories I-VI; Government of Australia 2014; extant vegetation from DEWR (2007).**

Bioregion	NRS (% of bioregion area)	Extant vegetation (% of bioregion area)
Cobar Peneplain	2.8	> 70
Avon Wheatbelt	2.8	11–30
Coolgardie	12.5	> 70
Hampton	14.6	> 70
Eyre Yorke Block	14.7	51–70
Murray–Darling Depression	15.2	51–70
Gawler	15.8	> 70
Mallee	18.0	51–70
Kanmantoo	20.6	31–50
Esperance Plains	28.9	51–70
Great Victoria Desert	30.6	> 70

rangeland landscapes grazed by livestock. As a consequence, Mallee Woodlands and Shrublands are subject to the potentially detrimental effects of habitat fragmentation, inappropriate fire regimes and threats such as exotic invasions and species loss due to grazing.

Conserving the diversity of Mallee Woodlands and Shrublands in the context of fragmentation legacies requires a landscape ecology approach that protects and manages remnant vegetation, increases ecological connectivity, where appropriate, and integrates nature conservation and management of threatening processes on land across tenures. Government-managed national parks or nature reserves continue to be central to the conservation enterprise with private sector reserves, individual landholder initiatives and landscape-scale connectivity approaches such as Gondwana Link in south-west Australia playing an increasingly important role.

Furthermore, because fire regimes, grazing regimes and climate variation interact to affect the dynamics of Mallee Woodlands and Shrublands, conservation efforts must be flexible enough to manage their influences. Fire planning and management in the Mallee Woodlands and Shrublands has historically focused on hazard reduction, but increasingly is focused on using fire to manage the risks of adverse ecological outcomes, such as loss or decline of native species. Planning systems that use prescribed fire strategically to maintain future fire management options and achieve management targets for fire frequency and spatial variability are being applied to some conservation reserves (e.g. Tarawi Nature Reserve; Willson 1999). Management targets are set as ranges bounded by thresholds rather than optima in fire regime parameters (Keith *et al.* 2002). The timing of seed bank accumulation for the slowest-maturing obligate-seeding plant species (Bradstock and Cohn 2002a; Gosper *et al.* 2013) and development of critical habitat resources for fauna (Haslem *et al.* 2011), provide an empirical basis for setting thresholds and ranges for fire return intervals (Keith *et al.* 2002).

Acceptable ranges in interrelated spatial characteristics of the fire regime such as fire intensity, patchiness and size are less well understood, but evidence is emerging that vegetation recovering from smaller fires (< 200 ha) is subject to heavier grazing pressure affecting recovery (Keith *et al.* unpubl. data), while concern has been expressed that large (> 10 000 ha) high-intensity fires homogenise vegetation structure and habitat with potentially detrimental effects on poorly dispersed fauna (Clarke 2008).

Planning approaches that set ranges and thresholds in fire regime parameters, together with an understanding of risks and behaviour of unplanned fires, inform fire managers where planned fires should be implemented in the landscape and when. These strategies aim to manage risks and keep options open to maintain sufficient proportions of mid-successional vegetation (11–40 years since fire) and long-unburnt vegetation (> 50 years) for the conservation of reptile, bird and mammal diversity (Willson 1999; Taylor *et al.* 2013; Kelly *et al.* 2011; Nimmo *et al.* 2012).

Unplanned fire adds an extra layer of complexity for managers seeking to implement specific fire management objectives. Mallee Woodlands and Shrublands exhibit weak relationships between fuel age and fire in low rainfall environments and relatively rapid accumulation of shrub fuels in higher rainfall environments (Gibson *et al.* 2015). Unplanned fires can therefore occur in quite young fuel ages, creating relatively short fire intervals or reducing large areas to a single time since fire state. Many reserves also occur in multiple-use landscapes where there may be conflicting management goals and 'good neighbour' relationships to take into account. Planning approaches need to be flexible enough to manage these risks strategically (Willson 1999; Keith *et al.* 2002).

Additionally, because introduced vertebrate herbivores have such a big influence on vegetation cover and composition, efforts to control goats and rabbits will continue to be a priority for the management of Mallee Woodlands and Shrublands and a particular challenge for multiple-use landscapes. Moreover, integrated management of herbivores and fire in a manner responsive to interannual variation in rainfall will be critical to conservation outcomes under a synergistic regime of threats to mallee biota (Keith and Tozer 2012).

Finally, the South-west Native Title Settlement recognising Noongar people as the Traditional Owners covering *c.* 200 000 km^2 of south-west Australia and Ngadju people's recent recovery of Native Title over *c.* 50 000 km^2 in the Great Western Woodlands will provide opportunities for traditional owners to work on country and undertake conservation management in muruk and maalok vegetation. Management partnerships with traditional owners are well established in mallee regions of south-east Australia. Mungo National Park, a large mallee reserve in south-west New South Wales, has had a joint management agreement since 2001, providing for majority Indigenous representation on a joint management committee, employment and training of traditional owners in reserve management positions and tourist operations.

23.6 Conclusions

There have been significant advances in understanding the biogeography of Mallee Woodlands and Shrublands, but there is still much to be learnt about the evolution and underlying causes for patterns of diversity, particularly on the south coast of Western Australia where resprouting mallee eucalypt species richness is highest and non-resprouting maalok eucalypt species have evolved. Advances in understanding fire science and ecology provide a strong empirical foundation for ecological fire management. Translating this into planning and operations presents many challenges and uncertainties, but also many opportunities for learning about ecological fire management.

References

Anonymous (1847) The great serpent of the north. *South Australian* Friday 23 April 1847, 4.

Atlas of Living Australia (ALA) (2015) *Atlas of Living Australia.* Available at: http://www.ala.org.au/.

Avitabile, S., Callister, K., Kelly, L.T. *et al.* (2013) Systematic fire mapping is critical for fire ecology, planning and management: a case study in the semi-arid Murray Mallee, south eastern Australia. *Landscape Urban Plan* 117, 81–91.

Bates, D.M. (nd ~1904-06) *Ngailbaitch of Korrlup, 50 miles NW of Jerramungup.* [Manuscript or typescript vocabularies] Section XII. Part 2. Language: grammar and vocabularies. Southwestern district. Part 2.B.7 (b), folio 40, 230–255.

Bayly, I.A.E. (1999) Review of how indigenous people managed for water in desert regions of Australia. *Journal of the Royal Society Western Australia* 82, 17–25.

Beadle, N.C.W. (1981) *The Vegetation of Australia.* Cambridge University Press: Cambridge.

Beard, J.S. (1990) *Plant Life of Western Australia.* Kangaroo Press: Kenthurst.

Beard, J.S., Beeston, G.R., Harvey, J.M. *et al.* (2013) The vegetation of Western Australia at the 1:3 000 000 scale. Explanatory memoir, second edition. *Conservation Science Western Australia* 9, 1–252.

Bindon, P.R. (1997) Aboriginal people and granite domes. *Journal of the Royal Society of Western Australia* 80, 173–179.

Bindon, P. and Chadwick, R. (compilers) (2011) *A Nyoongar Wordlist from the South-west of Western Australia.* 2nd edition. Western Australian Museum: Perth.

Bond, W.J. and van Wilgen, B.W. (1996) *Fire and Plants.* Chapman and Hall: London.

Bowler, J.M. (1982) Aridity in the late Tertiary and Quaternary of Australia. In *Evolution of the Flora and Fauna of Arid Australia*, ed. W.R. Barker and P.J.M. Greenslade. Peacock Publications: Frewville, pp. 35–45.

Bradstock, R.A. (1989) Dynamics of a perennial understorey. In *Mediterranean Landscapes in Australia: Mallee Ecosystems and their Management*, ed. J.C. Noble and R.A. Bradstock. CSIRO Publishing: Melbourne, pp. 141–154.

Bradstock, R.A. and Cohn, J.S. (2002a) Fire regimes and biodiversity in semi-arid mallee ecosystems. In *Flammable Australia: Fire Regimes, Biodiversity and Ecosystems in a Changing World*, ed. R.A. Bradstock, A.M. Gill and R.J. Williams. CSIRO Publishing: Melbourne, pp. 238–258.

Bradstock, R.A. and Cohn, J.S. (2002b) Demographic characteristics of mallee pine (*Callitris verrucosa*) in fire-prone mallee communities of central New South Wales. *Australian Journal of Botany* 50, 653–665.

Bradstock, R.A. and Gill, A.M. (1993) Fire in semiarid, mallee shrublands: size of flames from discrete fuel arrays and their role in the spread of fire. *International Journal of Wildland Fire* 3, 3–12.

Bradstock, R.A. and Kenny, B.J. (2003) An application of plant functional types to fire management in a conservation reserve in southeastern Australia. *Journal of Vegetation Science* 14, 345–354.

Brooker, M.I.H. (2000) A new classification of the genus *Eucalyptus* L'Her. (Myrtaceae). *Australian Systematic Botany* 13, 79–148.

Brooker, M.I.H. and Hopper, S.D. (1991) A taxonomic revision of *Eucalyptus wandoo, E. redunca* and allied species (E. series Levispermae Maiden - Myrtaceae) in Western Australia. *Nuytsia* 8, 1–189.

Burbidge, N.T. (1960) The phytogeography of Australia. *Australian Journal of Botany* 8, 75–211.

Bureau of Meteorology (n.d.) *Gridded climate data*. Canberra: Australian Government. Available at: http://www.bom.gov.au/climate/averages/climatology/gridded-data-info/gridded-climate-data.shtml

Burgman, M.A. (1988) Spatial analysis of vegetation patterns in southern Western Australia: implications for reserve design. *Australian Journal of Ecology* 13, 415–429.

Byrne, M., Yeates, D.K., Joseph, L. *et al.* (2008) Birth of a biome: insights into the assembly and maintenance of the Australian arid zone biota. *Molecular Ecology* 17, 4398–4417.

Carmona, D., Lajeunesse, M.J. and Johnson, M.T.J. (2011) Plant traits that predict resistance to herbivores. *Functional Ecology* 25, 358–367.

Cheal, D.C. and Parkes, D.M. (1989) Mallee vegetation in Victoria. In *Mediterranean Landscapes in Australia: Mallee Ecosystems and Their Management*, ed. J.C. Noble and R.A. Bradstock. CSIRO Publishing: Melbourne, pp. 125–140.

Clark, I.D. (1990) Aboriginal languages and clans: an historical atlas of western and central Victoria, 1800–1900. *Monash Publications Geography* 37, 1–448.

Clark, I.D., ed. (2000) *The Journals of George Augustus Robinson, Chief Protector, Port Phillip Aboriginal Protectorate, Volume 5 25 October 1845 – 9 June 1849*. Self-published by the editor.

Clarke, M.F. (2008) Catering for the needs of fauna in fire management: science or just wishful thinking? *Wildlife Research* 35, 385–394.

Clarke, P.J., Lawes, M.J., Murphy, B.P. *et al.* (2015) A synthesis of postfire recovery traits of woody plants in Australian ecosystems. *Science of the Total Environment* 534, 31–42.

Cohn, J.S. and Bradstock, R.A. (2000) Factors affecting post-fire seedling establishment of selected mallee understorey species. *Australian Journal of Botany* 48, 59–70.

Collard, N. (2011) *Plants and People in Mooro country: Nyungar Plant Use in Yellagonga Regional Park*, 2nd edition. City of Joondalup: Joondalup. Available at: http://www.joondalup.wa.gov.au/Files/PlantsandPeopleinMooroCountry.pdf (accessed 10 November 2015).

Cooke, J.W. (2006) European settlement in the Victorian mallee: a brief overview. *Proceedings of the Royal Society Victoria* 118, 295–303.

Cooper, S.M. and Owen-Smith, N. (1986) Effects of plant spinescence on large mammalian herbivores. *Oecologia* 68, 446–455.

Cowling, R.M., Potts, A.J., Bradshaw, P.L. *et al.* (2015) Variation in plant diversity in Mediterranean-climate ecosystems: the role of climatic and topographical stability. *Journal of Biogeography* 42, 552–564.

Cruz, M.G., McCaw, W.L., Anderson, W.R. and Gould, J.S.(2013) Fire behaviour modelling in semi-arid mallee-heath shrublands of southern Australia. *Environmental Modelling and Software* 40, 21–34.

CSRIO (n.d.) Available at: http://www.asris.csiro.au.

Davies, R.J.P., Whalen, M.A., Mackay, D.A. *et al.* (2013) Does soil seed bank diversity limit post-fire regeneration in small, fragmented, long-unburnt remnants of fire adapted vegetation? *Biological Conservation* 158, 287–295.

Department of the Environment and Energy (2007) *NVIS data products 4.2*. Australian Government: Canberra. Available at: http://www.environment.gov.au/land/native-vegetation/national-vegetation-information-system/data-products (accessed 10 May 2016).

Diels, L. (1906) *Die Pflanzenwelt von West-Australien Südlich des Wendekreises.* Vegetation der Erde VII: Leipzig.

Douglas, W.H. (1976) *The Aboriginal Languages of the South-west of Australia.* Australian Institute of Aboriginal Studies: Canberra.

Dyer, J.M. (2009) Assessing topographic patterns in moisture use and stress using a water-balance approach. *Landscape Ecology* 24, 391–403.

Egler, F.E. (1954) Vegetation science concepts. I. Initial floristic composition, a factor in old-field vegetation development. *Vegetatio* 4, 412–418.

Eldridge, D.J., Huang, N. and Hayward, M.W. (2012) Soil disturbance by invertebrates in a semi-arid eucalypt woodland: effects of grazing exclusion, faunal reintroductions, landscape and patch characteristics. *Proceedings of the Linnaean Society of New South Wales* 134, A11–A18.

Eldridge, D.J., Soliveres, S., Bowker, M.A. and Val, J. (2013) Grazing dampens the positive effects of shrub encroachment on ecosystem functions in a semi-arid woodland. *Journal of Applied Ecology* 50, 1028–1038.

Eyre, E.J. (1845) *Journals of expeditions of discovery into central Australia, and overland from Adelaide to King George's Sound, in the Years 1840-1, 2 Vols.* T. and W. Boone: London.

Foulkes, J. and Armstrong, D. (2000) Land-use history. In *A Biological Survey of the Murray Mallee, South Australia*, ed. J.N. Foulkes and J.S. Gillen. Department for Environment and Heritage and Department for Transport, Urban Planning and the Arts: Adelaide, pp. 19–26.

Gibson, N., Keighery, G.J., Lyons, M.N. and Webb, A. (2004) Terrestrial flora and vegetation of the Western Australian wheatbelt. In *A Biodiversity Survey of the Western Australian Agricultural Zone*, ed. G.J. Keighery, S.A. Halse, M.S. Harvey and N.L. McKenzie. Records of the Western Australian Museum: Perth, pp. 139–189.

Gibson, R.K., Bradstock, R.A., Penman, T.D. *et al.* (2014) Changing dominance of key plant species across a Mediterranean climate region; implications for fuel types and future fire regimes. *Plant Ecology* 215, 83–95.

Gibson, R.K., Bradstock, R.A., Penman, T.D. *et al.* (2015) Climatic, vegetation and edaphic influences on the probability of fire across Mediterranean woodlands of southeastern Australia. *Journal of Biogeography* 42, 1750–1760.

Gosper, C.R., Yates, C.J. and Prober, S.M. (2012a) Changes in plant species and functional composition with time since fire in two Mediterranean climate plant communities. *Journal of Vegetation Science* 23, 1071–1081.

Gosper, C.R., Yates, C.J., Prober, S.M. and Parsons, B.C. (2012b) Contrasting changes in vegetation structure and diversity with time since fire in two Australian Mediterranean-climate plant communities. *Austral Ecology* 37, 164–174.

Gosper, C.R., Prober, S.M. and Yates, C.J. (2013) Estimating fire interval bounds using vital attributes: implications of uncertainty and among-population variability. *Ecological Applications* 23, 924–935.

Government of Australia (2014) Collaborative Australian Protected Areas Database 2014. Available at: https://www.environment.gov.au/land/nrs/science/capad/2014.

Ham, T. (1847) *A Map of Australia Felix.* Thomas Ham: Melbourne.

Harris, C.R. (1990) The history of mallee land use: Aboriginal and European. In *The Mallee Lands: A Conservation Perspective*, ed. J.C. Noble, P.J. Joss and G.K. Jones. CSIRO Publishing: East Melbourne, pp. 147–151.

Haslem, A., Callister, K.E., Avitabile, S.C. *et al.* (2010) A framework for mapping vegetation over broad spatial extents: a technique to aid land management across jurisdictional boundaries. *Landscape and Urban Planning* 97, 296–305.

Haslem, A., Kelly, L.T., Nimmo, D.G. *et al.* (2011) Habitat or fuel? Implications of long-term, post-fire for the development of key resources for fauna and fire. *Journal of Applied Ecology* 48, 247–256.

Hassell, E. (1975) *My Dusky Friends. Aboriginal Life, Customs and Legends and Glimpses of Station Life at Jarramungup in the 1880's.* C.W. Hassell: East Fremantle.

Hercus, L.A. (1992) *WembaWemba dictionary.* AIATSIS: Canberra.

Hill, K.D. (1989) Mallee eucalypt communities: their classification and biogeography. In *Mediterranean Landscapes in Australia: Mallee Ecosystems and Their Management*, ed. J.C. Noble and R.A. Bradstock. CSIRO Publishing: Melbourne, pp. 93–108.

Hooker, J.D. (1860) *The Botany of the Antarctic Voyage of H. M. Discovery Ships* Erebus *and* Terror *in the Years 1839–1843. Part III. Flora Tasmaniae. Vol. I. Dicotyledones.* Lovell Reeve: London.

Hopper, S.D. (2009) OCBIL theory: towards an integrated understanding of the evolution, ecology and conservation of biodiversity on old, climatically buffered, infertile landscapes. *Plant Soil* 322, 49–86.

Hopper, S.D. and Gioia, P. (2004) The southwest Australian floristic region: evolution and conservation of a global hot spot of biodiversity. *Annual Review of Ecology, Evolution and Systematics* 35, 623–650.

Keith, D. (2004) *Ocean Shores to Desert Dunes. The Native Vegetation of New South Wales and the ACT.* Department of Environment and Conservation: Hurstville.

Keith, D.A. and Bradstock, R.A. (1994) Fire and competition in Australian heath: a conceptual model and field investigations. *Journal of Vegetation Science* 5, 347–554.

Keith, D.A. and Tozer, M.G. (2012) The influence of fire, herbivores and rainfall on vegetation dynamics in the mallee: a long-term experiment. *Proceedings of the Linnaean Society of New South Wales* 134, A39–A54.

Keith, D.A., Williams, J.E. and Woinarski, J.C.Z. (2002) Fire management and biodiversity conservation: key approaches and principles. In *Flammable Australia: Fire Regimes, Biodiversity and Ecosystems in a Changing World*, ed. R.A. Bradstock, A.M. Gill and R.J. Williams. CSIRO Publishing: Melbourne, pp. 401–425.

Keith, D.A., Holman, S., Rodoreta, S. *et al.* (2007) Plant functional types can predict decade-scale changes in fire-prone vegetation. *Journal of Ecology* 95, 1324–1337.

Kelly, L.T., Nimmo, D.G., Spence-Bailey, L.M. *et al.* (2011) Influence of fire history on small mammal distributions: insights from a 100-year post-fire chronosequence. *Diversity and Distributions* 17, 462–473.

Kinnear, S., Gillen, J., Smith, F. *et al.* (2000) Vegetation. In *A Biological Survey of the Murray Mallee, South Australia*, ed. J.N. Foulkes and J.S. Gillen. Department for Environment and Heritage and Department for Transport, Urban Planning and the Arts: Adelaide, pp. 35–142.

Laffan, S.W., Lubarsky, E. and Rosauer, D.F. (2010) Biodiverse: a tool for the spatial analysis of biological and other diversity. *Ecography* 33, 643–647.

MacLaren, M. (1912) IV. Notes on desert-water in Western Australia. 'Gnamma holes' and 'night wells'. *Geological Magazine* (Decade V) 9, 301–304.

Markey, A., Kern, S. and Gibson, N. (2012) Floristic communities of the Ravensthorpe Range, Western Australia. *Conservation Science Western Australia* 8, 187–239.

McArthur, W.M. (2004) *Reference Soils of South-western Australia.* Australian Society of Soil Science, WA Branch Inc. and Department of Agriculture and Food: Perth.

McCaw, L. and Hanstrum, B. (2003) Fire environment of Mediterranean south-west Western Australia. In *Fire in Ecosystems of South-west Western Australia: Impacts and Management*, ed. I. Abbott and N. Burrows. Backhuys Publishing: Leiden, pp. 87–106.

McLaren, S. (2009) Revised stratigraphy of the Blanchetown clay, Murray Basin: age constraints on the evolution of paleo Lake Bungunnia. *Australian Journal of Earth Sciences* 56, 259–270.

Moore, G.F. (1884; facsimile 1978) *Diary of Ten Years of an Early Settler in Western Australia and Also a Descriptive Vocabulary of the Language of the Aborigines.* University of Western Australia Press: Nedlands.

Nicolle, D. (2006) A classification and census of regenerative strategies in the eucalypts (Angophora, Corymbia and Eucalyptus - Myrtaceae), with special reference to the obligate seeders. *Australian Journal of Botany* 54, 391–407.

Nicolle, D. (2008) Systematic studies of the southern Australian mallees *(Eucalyptus Series Subulatae - Myrtaceae)*. PhD Thesis, Flinders University, Adelaide.

Nimmo, D.G., Kelly, L.T., Spence-Bailey, L.M. *et al.* (2012) Predicting the century-long, post-fire responses of reptiles. *Global Ecology and Biogeography* 21, 1062–1073.

Noble, J.C. (1989) Fire studies in mallee (*Eucalyptus* spp.) communities of western New South Wales: the effects of fires applied in different seasons on herbage productivity and their implications for management. *Australian Journal of Ecology* 14, 169–187.

Noble, J.C. (2001) Lignotubers and meristem dependence in mallee (*Eucalyptus* spp.) coppicing after fire. *Australian Journal of Botany* 49, 31–41.

Noble, J.C. and Bradstock, R.A. (eds) (1989) *Mediterranean Landscapes in Australia: Mallee Ecosystems and their Management.* CSIRO Publishing: Melbourne.

Noble, J.C. and Diggle, P.J. (2013) Population biology of coppicing plants: survival of mallee (*Eucalyptus* spp.) populations exposed to contrasting fire and cutting regimes. *Australian Journal of Botany* 61, 552–557.

Noble, J.C. and Kimber, R.G. (1997) On the ethno-ecology of mallee root-water. *Aboriginal History* 21, 170–202.

Noble, J.C. and Vines, R.G. (1993) Fire studies in mallee (*Eucalyptus* spp.) communities of western New South Wales: grass fuel dynamics and associated weather patterns. *Rangeland Journal* 15, 270–297.

Noble, J.C., Joss, P.J. and Jones, G.K. (eds) (1990) *The Mallee Lands: A Conservation Perspective.* CSIRO Publishing: East Melbourne.

O'Donnell, A.J., Boer, M.M., McCaw, W.L. and Grierson, P.F. (2011) Vegetation and landscape connectivity control wildfire intervals in unmanaged semi-arid shrublands and woodlands in Australia. *Journal of Biogeography* 38, 112–124.

Parsons, B.C. and Gosper, C.R. (2011) Contemporary fire regimes in a fragmented and an unfragmented landscape: implications for vegetation structure and persistence of the fire-sensitive malleefowl. *International Journal of Wildland Fire* 20, 184–194.

Parsons, R.F. (1994) Eucalyptus scrubs and shrublands. In *Australian Vegetation*, ed. R.H. Groves. Melbourne: Cambridge University Press, pp. 291–319.

Pausas, J.G. and Bradstock, R.A. (2007) Fire persistence traits of plants along a productivity and disturbance gradient in Mediterranean shrublands of SE Australia. *Global Ecology and Biogeography* 16, 330–340.

Prober, S.M., Yuen, E., O'Connor, M. and Schultz, L. (2013) *Ngadju Kala: Ngadju fire knowledge and contemporary fire management in the Great Western Woodlands*. CSIRO Ecosystem Sciences: Floreat. Available at: https://publications.csiro.au/rpr/pub?list=SEA&pid=csiro:EP135694 (accessed 10 November 2015).

Rooney, B. (2011) *The Nyoongar Legacy. The Naming of the Land and the Language of the People.* Batchelor Press: Batchelor.

Rye, B.L. and Hopper, S.D. (1982) Misapplication of the Aboriginal name 'Gungurru' to *Eucalyptus caesia* Benth. and notes on the species' distribution. *Journal of the Royal Society of Western Australia* 65, 93–95.

Sniderman, J.M.K., Jordan, G.J. and Cowling, R.M. (2013) Fossil evidence for a hyperdiverse sclerophyll flora under a non–Mediterranean-type climate. *Proceedings of the Natural Academy of Sciences* 110, 3423–3428.

Sparrow, A. (1989) Mallee vegetation in South Australia. In *Mediterranean Landscapes in Australia: Mallee Ecosystems and their Management*, ed. J.C. Noble and R.A. Bradstock. CSIRO Publishing: Melbourne, pp. 109–124.

Stone, A.C. (1911) The Aborigines of Lake Boga, Victoria. *Proceedings of the Royal Society of Victoria* 23, 433–468.

Taylor, R.S., Watson, S.J., Bennett, A.F. and Clarke, M.F. (2013) Which fire management strategies benefit biodiversity? A landscape perspective case study using birds in mallee ecosystems of south-eastern Australia. *Biological Conservation* 159, 248–256.

Veth, P. and O'Connor, S. (2013) The past 50 000 years: an archaeological view. In *The Cambridge History of Australia*, ed. A. Bashford and S. Macintyre. Cambridge University Press: Cambridge, pp. 17–42.

von Brandenstein, C.G. (1988) *Nyungar Anew: Phonology, Text Samples and Etymological and Historical 1500-word Vocabulary of an Artificially re-created Aboriginal Language in the South-west of Australia*. Australian National University: Canberra.

Wangka Maya Pilbara Aboriginal Language Centre (2009) *Ngajumaya dictionary. English-Ngajumaya Wordlist and Topical Wordlists, 2009 edition.* Wangka Maya Pilbara Aboriginal Language Centre in agreement with Karlkurla Language and Cultural Aboriginal Corporation: South Headland.

Watson, S.J., Taylor, R.S., Nimmo, D.G. *et al.* (2012) Effects of time-since-fire on birds: how informative are generalized fire-response curves for conservation management? *Ecological Applications* 22, 685–696.

Wellington, A.B. and Noble, I.R. (1985) Post-fire recruitment and mortality in a population of the mallee *Eucalyptus incrassata* in semi-arid, south-east Australia. *Journal of Ecology* 73, 645–656.

Wheatbelt NRM (n.d.) *Nyungar Budjara Wangany: NRM wordlist and language collection. Booklet of the Avon Catchment region.* Wheatbelt Natural Resource Management: Northam. Available at: http://www.wheatbeltnrm.org.au/knowledge/nyungar-budjara-wangany-nyungar-nrm-wordlist-language-collection-booklet-avon-catchment (accessed 10 November 2015).

White, M. (2006) The mallee vegetation of north western Victoria. *Proceedings of the Royal Society Victoria* 118, 229–244.

Wilson, A.D. (1990) The effect of grazing on Australian ecosystems. *Proceedings of the Ecological Society of Australia* 16, 235–244.

Willson, A. (1999) *Tarawi Nature Reserve Fire Management Plan.* NSW National Parks and Wildlife Service: Hurstville.

Yates, C.J., Hopper, S.D., Brown, A. and van Leeuwen, S. (2003) Impact of two wildfires on endemic granite outcrop vegetation in Western Australia. *Journal of Vegetation Science* 14, 185–194.

24 The Chenopod Shrublands

DAVID J. ELDRIDGE, SAMANTHA K. TRAVERS, AGUSTÍN F. FACELLI, JOSÉ M. FACELLI AND DAVID A. KEITH

Chenopod shrubland communities are among the most extensively studied plant communities in Australia. They occupy large areas of Mediterranean southern Australia that are dominated by winter rainfall, and support a rich array of perennial and annual shrubs, grasses and forbs. The community dominants, *Atriplex* and *Maireana* are important for providing landscape stability, forage production for livestock during droughts, and habitat for animals and understorey plant species. The landscape-level distribution of the main chenopod shrub species is driven largely by soil depth and texture, which control the availability of soil moisture. Differences in microtopography and site history determine distribution of individual shrubs. Chenopod shrubs are facilitated by longer-lived perennial plants such as *Acacia*, and themselves act as nurse plants for annual species by altering soil physical and chemical properties beneath their canopies. The dominant shrubs *Atriplex* and *Maireana* spp. are relatively resistant to low levels of livestock grazing, and the groundstorey plant response to grazing varies markedly with plant community type. Substantial overgrazing since European settlement has resulted in large changes, generally reductions, in the health, density and distribution of chenopod shrublands, and the current community status is poor. State-and-Transition models suggest that recovery of the community dominants after overgrazing is protracted and costly. Forecast reductions in winter rainfall over large areas of the range of the chenopod shrublands will place more stress on the community dominants and likely accelerate changes in plant community composition.

24.1 Introduction

The chenopod shrublands, known colloquially as the Saltbush and Bluebush Country, occur over large areas of southern Australia that are dominated by winter rainfall (Graetz and Wilson 1984). These communities support a rich suite of annual and perennial species of the family Amaranthaceae (formerly Chenopodiaceae). The key components of the chenopod shrublands are relatively short (<2 m tall) and moderately long-lived woody plants *Atriplex* (30 years) and *Maireana* species (~150 years; Crisp 1978). They are also common components of open woodlands dominated by longer-lived perennial species of *Acacia*, *Eucalyptus* and *Casuarina*.

The most widespread and economically important genera include *Sclerolaena*, *Atriplex*, *Maireana*, *Chenopodium*, *Rhagodia* and *Dysphania* (Leigh 1994). The chenopod shrublands are among the most extensively studied types of Australian vegetation, particularly during the second half of the twentieth century. This is probably because these communities

Australian Vegetation, Third Edition, ed. David A. Keith. Published by Cambridge University Press. © Cambridge University Press 2017.

Citation: Eldridge, D.J. *et al.* (2017). The Chenopod Shrublands. In D. A. Keith, ed., *Australian Vegetation*, 3rd edn. Cambridge: Cambridge University Press, pp. 599–625.

support a highly productive pastoral industry and are widely distributed over large areas of southern Australia. While the understorey plants provide most production for livestock, the community dominants, the shrubs themselves, and the trees, where they occur, provide the main structural elements in these low shrublands; play critical roles in nutrient cycling and hydrology; and support an extensive understorey community of native forbs and grasses. They also provide valuable forage during droughts, and protect the soil against wind and water erosion. A substantial amount of research on the chenopod shrublands focused on the effects of grazing. More recent research has aimed to understand the ecological relationships among the major community components.

24.2 Distribution and biogeography

24.2.1 Continental-scale distribution

The chenopod shrub-steppe occupies about 6% of the land area of Australia (434 000 km^2; Graetz and Wilson 1984). It is restricted to semi-arid Mediterranean environments in southern Australia, south of the Tropic of Capricorn, in rainfall areas ranging from about 350 mm on the south-eastern edge of the western Riverina in New South Wales to less than 200 mm on the western border of Western Australia (Graetz 1973; Graetz and Wilson 1984; Leigh 1994). These environments are characterised by hot summers, winter maximum rainfall, and large differences in rainfall from east to west and north to south. Inter-decadal boom–bust cycles are characteristic of the climate, with years of significantly high rainfall between periods of prolonged drought conditions.

Chenopod shrublands are distributed along inland riverine plains, claypans, aeolian sandplains, limestone plateaus and gibber plains (Graetz and Wilson 1984; Keith 2004; Beard *et al.* 2013). Soils range from grey cracking clays, to calcareous loamy sands and deep red sands, to desert loams and brown gibber soils (Graetz and Wilson 1984). On the western edge of their distribution, the chenopod shrublands often occur as an understorey community within *Acacia* woodlands (Nano *et al.*, Chapter 25, this volume) or *Eucalyptus* woodlands (Prober *et al.*, Chapter 17, this volume). Different assemblages of subsucculent chenopod shrublands occur on these contrasting substrates (Table 24.1) and collectively occur across more than 360 000 km^2 of southern Australia, with just over half occurring in South Australia. The most extensive types occur on gibber plains and associated depressions, and the limestone plains of the Nullarbor (Figure 24.1). Succulent samphire vegetation occupies an additional 40 000 km^2 on saline flats and around salt lakes, mostly in South and Western Australia (Table 24.1; Figure 24.1).

24.2.2 Chenopod shrublands on other continents

Worldwide, the chenopod component of the Amaranthaceae family contains about 1250 species from 100 genera, 60% of these genera are found between the Canary Islands and Central Asia (Kadereit *et al.* 2005). They also occur in temperate parts of Africa and the Americas. Australia is the second most diverse hotspot for the chenopods, with 32 genera, 20 that are endemic to Australia; and 302 species, 279 of which are endemic (Kadereit *et al.* 2005). Australia has quite favourable conditions for chenopods, with expansive arid and semi-arid regions covering more than 75%, and saline soils that cover more than 60% of the continent, that are often associated with hydrological features such as salt pans, estuarine plains and coastal lagoons (Northcote and Skene 1972).

24.2.3 Evolutionary biogeography of chenopods

Several hypotheses have been proposed to explain the evolutionary history of chenopod taxa in Australia, from cosmopolitan ancestors which inhabited Cretaceous coastlines (Burbidge 1960), to autochthonous chenopods from Gondwana. More recent evidence suggests that the

Table 24.1 **Summary of the main chenopod community types in Australia.**

Community	Community dominants	Geomorphology	Soils	General distribution	Specific locations	References
Clayplain Saltbush Shrublands	*Atriplex vesicaria, A. nummularia, Maireana aphylla, Rhagodia spinescens, Sclerolaena muricata, Nitraria billardierei*. Open shrubland.	Extensive drainage plains and depressions; associated levees; dry lake beds; areas of run-on.	Deep, heavy-textured, generally clay soils, mainly of Tertiary alluvial origin.	Murray–Darling Basin, mainly Riverina area; originally 27 000 km^2, now c. 20 000 km^2	Hay Plain (NSW), Ned's Corner (Vic)	Zimmer (1946); Parr-Smith and Calder (1979); Williams (1979); Graetz and Wilson (1984); Mitchell and Wilcox (1988); Leigh (1994); Keith (2004)
Sandplain Bluebush Shrublands	*Maireana sedifolia, M. pyramidata* with localised dominance of *Atriplex vesicaria* and scattered grasses (*Austrostipa scabra, Aristida* spp.). Open shrubland.	Aeolian plains often underlain with limestone.	Calcareous red sandy loams over calcrete subsoils.	Lower Darling south-west NSW, to Gawler district of central southern SA; 38 500 km^2	Menindee (NSW), Koonamore, Whyalla (SA),	Osborn *et al.* (1932); Crocker and Wood (1947); Carrodus (1962); Leigh (1994); Keith (2004)
Gibber Chenopod Shrublands	*Atriplex vesicaria, A. leptocarpa, A. lindleyi, Maireana aphylla*, and various *Sclerolaena* species. Low sparse vegetation.	Upland stony ranges, footslopes and pediplains with gilgai pattern.	Brown and red desert loam and brown gibber soils.	South-west Qld, north-west NSW, central and north-east SA; 132 000 km^2	Barrier Ranges (NSW), Olary, Yunta (SA)	Valentine and Nagorcka (1979); Parr-Smith (1982); Eldridge (1988); Leigh (1994); Keith (2004)

Table 24.1 (cont.)

Community	Community dominants	Geomorphology	Soils	General distribution	Specific locations	References
Nullarbor Chenopod Shrublands	*Maireana sedifolia* on the rises, with *Atriplex vesicaria*, *A. acutibractea*, *A. cryptocarpa* and *Cratystylis conocephala* on the slopes and drainage tracts. Grasses (*Austrostipa* and *Austrodanthonia*) and various forbs among shrubs. Open shrubland.	Gently undulating limestone plains and low rises with scattered drainage tracts and clay depressions.	Calcareous shallow loams and earths derived from limestone.	Great Australian Bight; 109 000 km^2	Nullarbor Plain (SA, WA)	Beard (1975); Beard *et al.* (2013)
Mulga Chenopod Shrublands	*Atriplex vesicaria*, *A. lindleyi*, *Maireana sedifolia*, *M. carnosa*, *M. enchylaenoides*, *M. georgei* open shrubland with scattered emergents of *Acacia aneura*, *A. sclerosperma* and other acacias.	Sandplains and sandy alluvial depressions.	Calcareous red sandy loams.	Scattered through the Great Victoria Desert, Yalgoo and Murchison regions of WA; 56 800 km^2	East of Shark Bay (WA)	Beard *et al.* (2013)
Arid Samphire Shrublands	Dominated by *Tecticornia halocnemoides*, various other *Tecticornia* spp., *Disphyma crassifolium*, *Frankenia* spp.	Drainage plains and depressions; groundwater discharge basins and associated levees.	Saline depressions and edges of perennial and ephemeral hypersaline lakes.	Scattered throughout inland Australia, mainly south of the tropics, from the WA coast to western NSW; 40 000 km^2	Fortescue marshes, Pilbara (WA), Lake Eyre, Lake Frome (SA), Neds Corner (SA), Nanya (NSW)	Beadle (1981); Keith (2004); Beard *et al.* (2013)

ancestral chenopods originated after the break-up of Gondwana, and that at least nine different ancestors arrived in Australia independently via long-distance dispersal or slow migration (Kadereit *et al.* 2014). Thus, it is possible that chenopods colonised Australia from several directions, each with a different origin (Kadereit *et al.* 2004).

The distinct lack of fossilised evidence from Australia makes it difficult to ascertain with certainty when chenopods arrived in Australia. Australian chenopods first appear in palynological records in the Oligocene (~35 Ma) in the La Trobe Valley depression, at a time where inland freshwater lakes were present in Australia (Parr-Smith 1982; Martin 1994). On the south-eastern mainland of Australia, during the Oligo-Miocene transgression of the Murray Basin, chenopods were abundant, likely as halophytic shrubs. Chenopods were also relatively abundant in the Lake Eyre Basin in the early-to-mid-Pliocene (6–3 Ma; Martin 1994). They appear to have increased in the Capricorn basin (QLD) in the late Miocene (10 Ma) to Late Pliocene (~3 Ma), but it is not known whether these plants were saltmarshes or whether they represented more widespread arid conditions (Hekel 1972; Martin 1994). The genera *Chenopodium* and *Rhagodia* appear to have arrived during the Pliocene (4.7–2.9 Ma) but their geographic origin is unknown (Kadereit *et al.* 2004). It is possible they dispersed from the Americas, but very limited opportunities exist for sampling palynological records from Antarctica *in situ* to confirm this (Martin 1994). *Atriplex spongiosa* appears to be the youngest Australian chenopod lineage, arriving in the late Pliocene (Kadereit *et al.* 2004).

Pollen records for the Quaternary (2.6 Ma to present) are somewhat unreliable for chenopods, as this was a very dry period for Australia and, like many desert plants, they respond to dry conditions by ceasing flower production and reducing in density. However, the predominantly dry climates Australia experienced during the Quaternary have shaped the diversity of contemporary chenopod species. For example, the genus *Atriplex* contains over 250 species, 70% of which are endemic to Australia. There is an array of *Atriplex* variants, most of which evolved rapidly during the Quaternary, within distinctive geographic or edaphic (varying with soil type) niches (Kadereit *et al.* 2004).

24.3 Plant community structure and composition

Structurally, the saltbushes (*Atriplex* species) generally occur in monospecific stands of shrubs < 1.5 m tall (*Atriplex vesicaria*) with projected foliage cover between 10 and 30% (Specht 1970). *Atriplex nummularia* can attain heights of 3 m. The bluebushes (*Maireana sedifolia*, *M. pyramidata*, *M. astrotricha*) occur in mixed stands with *Atriplex vesicaria* and generally have a foliage cover < 10% (Leigh 1994). Shrubs are shorter, and have lower foliage cover on gibber substrates. Figure 24.2 shows a variety of structural forms.

Six broader community types can be identified (Figure 24.1). The Clayplain Saltbush Shrublands occur on extensive plains with clay soils, and are most strongly developed in the Riverina area of New South Wales. Sandplain Bluebush Shrublands are characterised by mixed stands of *Maireana* species, and occur over large areas of South Australia and south-west New South Wales on highly calcareous aeolian loams soils. Gibber Chenopod Shrublands occur as low, sparsely vegetated bands on upland stony footslopes and pediplains in far western New South Wales, south-western Queensland and north-eastern South Australia. Nullarbor Chenopod Shrublands are common over large areas of the Nullarbor Plain on gently undulating plains underlain by limestone. Mulga Chenopod Shrublands are widely scattered in loamy depressions on sandplains and alluvial corridors. A sixth type, Arid Samphire Shrublands, dominated by *Tecticornia halocnemoides* and associated halophytes, occurs in hypersaline drainage basins and depressions south of the tropics, from western New South Wales to the coast of Western Australia.

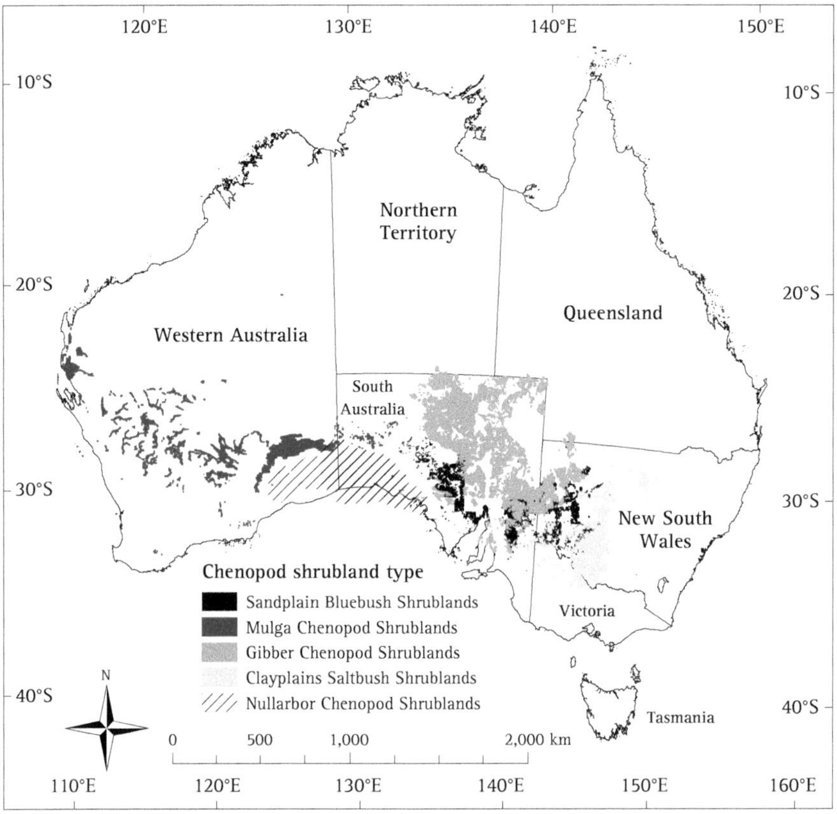

Figure 24.1 Distribution of the chenopod shrublands in Australia based on interpretation of state agency vegetation maps held in NVIS 4.2 (Department of the Environment and Energy 2007).
Note Mulga Chenopod Shrublands only found in Western Australia, Clayplain Saltbush Shrublands only found in south-eastern Australia. Arid Samphire Shrublands not shown.

24.3.1 Vascular plant community composition

The community dominants occur in association with a rich suite of annual and short-lived perennial plants. Notable among these are 'copperburrs' (*Sclerolaena* spp.) and annual saltbushes (*Atriplex* spp.), which often dominate once the shrubs have been removed by overgrazing (see Boxes 24.1 and 24.2), a rich community of generally winter-growing daisies (Asteraceae), annual and perennial grasses (Poaceae) and assorted members of the Brassicaceae, Fabaceae (subfamily Faboideae), Malvaceae, and Portulacaceae families to name but a few.

We analysed data on plant species composition and shrub density from three different chenopod communities (Clayplains Saltbush, Sandplain Bluebush, Gibber Chenopod Shrublands; Table 24.1) for 115 sites in western New South Wales collected between 1989 and 2002 as part of the New South Wales Government's Rangeland Assessment Program (Eldridge and Koen 2003). We found very little

Figure 24.2 Examples of chenopod shrubland communities. (a) *Atriplex nummularia* (old man saltbush) shrubland on the riverine plain north of Booligal, south-western New South Wales. (b) *Nitraria billardieri* (nitre bush) and *Atriplex vesicaria* (bladder saltbush) with *Autrostipa scabra* on the Olary plain, South Australia. (c) *Maireana pyradimata* on loamy sands, Fowlers Gap, far north-west New South Wales. (d) Dense *Maireana sedifolia* (pearl bluebush) on the dry bed of Lake Mungo, south-west New South Wales. (e) The Paradise total livestock exclosure south of Hay (south-western New South Wales), established 1954 during a wet year, contrasts with an adjoining area that is used to coral travelling stock, where all *Atriplex vesicaria* shrubs have been eliminated. (f) *Maireana sedifolia* persists under sustained grazing, which reduces vegetation cover and compacts soils in the inter-shub spaces, Maralinga, South Australia.
All photos: David Eldridge. (A black and white version of this figure will appear in some formats. For the colour version, please refer to the plate section.)

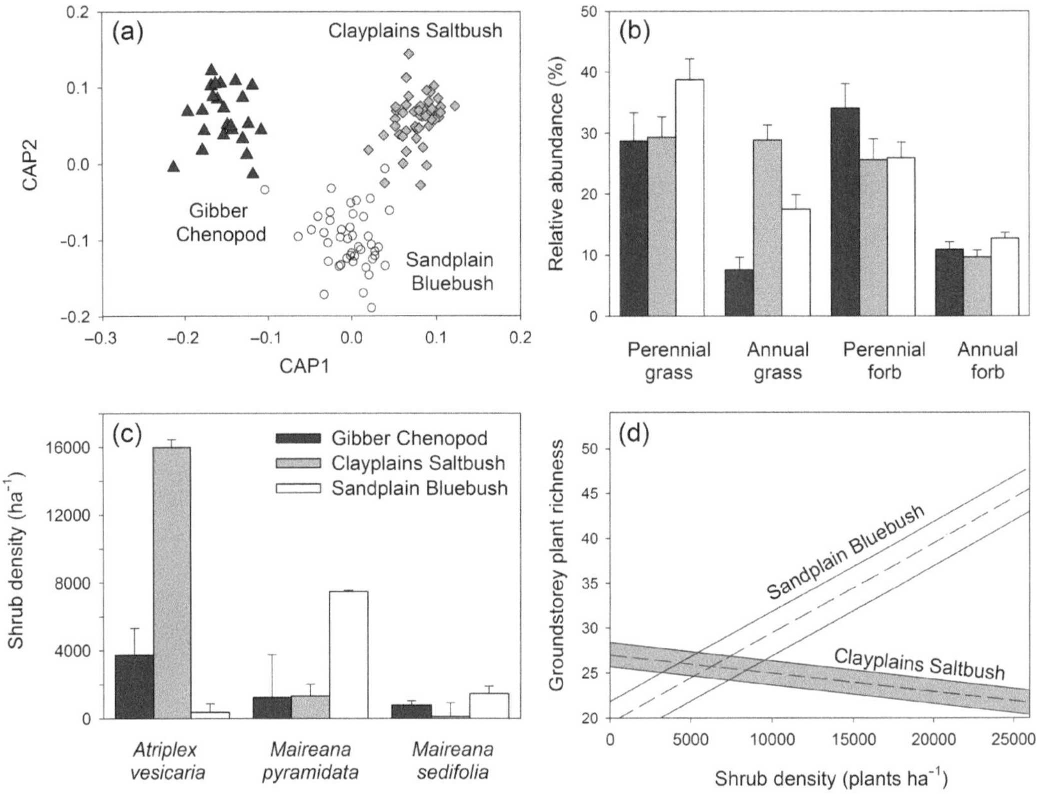

Figure 24.3 (a) Canonical analysis of principal co-ordinates (CAP) ordination plot (based on the biomass of plant species) for three chenopod shrubland communities. Each point represents a sample of the estimated species biomass at a site averaged over 4 to 13 years of sampling between 1990 and 2002. N = 24 for Gibber Chenopod Shrublands, N = 48 for Clayplains Saltbush Shrublands, N = 43 for Sandplain Bluebush Shrublands. (b) Relative abundance (% ± SE) of groundstorey growth forms, perennial and annual grasses and forbs, averaged over sites within each of the three chenopod communities (based on the same data used in (a)). (c) Average density (mean ± SE shrubs ha^{-1}) of three chenopod species, *Atriplex vesicaria, Maireana pyramidata* and *Maireana sedifolia*, within each of our three chenopod communities. (d) Modelled mean number of groundstorey species (±95% CI) in relation to shrub density for Clayplains Saltbush and Sandplain Bluebush communities.

overlap in the species composition of the three communities (Figure 24.3a). Groundstorey plant composition at the Sandplain Bluebush sites was slightly more variable than that at the Clayplains Saltbush and Gibber Chenopod sites (Figure 24.3a). Growth forms of groundstorey species differed markedly among the three communities (Figure 24.3b). Clayplains Saltbush was co-dominated by perennial and annual grasses, while the slightly coarser textured soils of the sandplain bluebush community supported a greater proportion of perennial grasses (Figure 24.3b). The Gibber Chenopod Shrublands were dominated by perennial forbs and had relatively few annuals compared with the other two communities, likely due to their more arid climates. The proportion of annual grass varied markedly among these three communities but the proportion of annual forbs was similar (Figure 24.3b).

Shrub composition and density varied greatly among the three communities, with *Atriplex vesicaria* characterised by densities of 16 000 shrubs per hectare in Clayplains Saltbush Shrublands (Figure 24.3c). The Sandplain Bluebush sites were dominated by *Maireana pyramidata*, but included a variable and low density of *Atriplex vesicaria*.

Chenopod shrubs have been shown to have a strong legacy effect on the soil long after their removal. For example, removal of the canopy has been shown to enhance plant richness and biomass (Tupper and Müller 1985; Weedon and Facelli 2008). However, the relationship between groundstorey plant community composition and shrub density for the New South Wales study suggests that the mechanisms are more complex (Figure 24.3d). Overall, there were no differences in average groundstorey plant richness between the Clayplains Saltbush and Sandplain Bluebush communities. However, plant richness increased substantially with increases in shrub density in Sandplain Bluebush but declined slightly in Clayplains Saltbush (Figure 24.3d). These results may reflect substantial differences in soil type and surface stability provided by the shrubs. For example, bluebush communities may support more groundstorey species than saltbush because it is easier for plants to extract moisture from coarse-textured than fine-textured soils (Noy-Meir 1973; Eldridge 1988).

24.3.2 Non-vascular plant community composition

Biological soil crusts (biocrusts), complex associations between mosses, liverworts, lichens, cyanobacteria and other microscopic organisms, are common over much of Mediterranean Australia and are prominent within chenopod shrublands. Consequently, extensive surveys of biocrust communities are closely aligned with communities dominated by *Atriplex* and *Maireana* spp. (e.g. Rogers and Lange 1971; Eldridge 1998). Calcareous loamy soils, with pH levels > 8.5, particularly in the Sandplain Bluebush Shrublands support a rich suite of crusts with up to 80% cover. The most common lichens of these plant communities are shown in Table 24.2. Studies in the Great Victoria Desert on plains with Sandplain Bluebush Shrublands demonstrated that lichen richness increased with time since disturbance and explained almost 50% of the variation in species richness (Eldridge 1998). Soil crust and lichen cover has also been extensively studied at Koonamore (Hall *et al.* 1964; Sinclair 2005) mostly in relation to the impacts of grazing (see 24.6.3 Changes in plant community composition with grazing).

24.3.3 Adaptive traits of dominant plants

Chenopod plants have evolved a range of different strategies and adaptations to living in dry Mediterranean environments (Anderson 1982). A comparison of the adaptations of three wide-ranging species is shown in Table 24.3. For example, *Atriplex* is a salt excreter while *Maireana* is a salt accumulator. The bracteoles surrounding the seeds of *Atriplex* species contain high levels of sodium chloride, (NaCl; Beadle 1952). High levels of NaCl suppress cytokinin (Anderson 1982), which is essential for cell division, and shoot and root growth. Thus, the salt must be leached before the plants will germinate, with estimates of 100 mm of rainfall required to leach 95% of the NaCl (Charley 1959). In fact, there is a strong positive relationship between NaCl concentration in the bracteoles and concentrations of sodium chloride that prevent germination (Beadle 1952). *Tecticornia*, another widespread genus in the chenopod shrublands is also a salt accumulator, and like *Maireana* spp. has succulent leaves.

Table 24.2 **Responses of broad lichen groups to herbivore activity in chenopod shrublands.**

Lichen group	Key taxa	Response to herbivore activity	Location	References
Foliose lichens	*Xanthoparmelia* spp.,	Substantial decline	Whyalla (SA)	Rogers and Lange (1971); Rogers (1972)
	X. semiviridus, *X. convoluta*		Maralinga (SA)	Eldridge (1998) Eldridge and Ferris (1999)
Squamulose lichens	*Psora decipiens*, *P. crenata*, *P. crystallifera*, *Endocarpon simplicatum*	Slight decline	Whyalla (SA)	Rogers (1972)
			Maralinga (SA)	Eldridge and Ferris (1999)
Crustose lichens	*Diploschistes* spp., *Lecidea* spp.	Slight decline	Maralinga (SA)	Eldridge and Ferris (1999) Eldridge (1998)
Gelatinous lichens	*Heppia despreauxii*, *Collema coccophorum*	No change	Whyalla (SA)	Rogers (1972)
		Increase	Western Riverina, Mildura (NSW)	Eldridge (1996)
Cyanobacteria	*Nostoc*, *Scytonema*, *Microcoleus* spp.	No change	Arid South Australia	Rogers (1989)

Water stress during seed set has been shown to induce hard seededness in *Atriplex holocarpa* (Osmond *et al.* 1980). This form of bet-hedging would ensure that sufficient seeds are retained for periods with conditions that would support higher seedling survival. *Atriplex* species also possess traits that enhance water-use efficiency. For example, Sinclair and Thomas (1970) showed that some species of *Atriplex* have leaves that are highly reflective, due to the dense epidermal structure and salt crystals on their leaves. The dense white indumentum of whorl-branched hairs on the surface of *Maireana sedifolia* leaves probably serves the same purpose. Stomatal sensitivity to light, which could restrict growth and competitive of ability in shaded environments, may explain the tendency of *Atriplex vesicaria* to be restricted to open habitats. Traits that allow annual plants to persist in chenopod shrublands may include cycling dormancy and germination requirements that reduce the risk of germinating in periods when there is insufficient soil moisture (e.g. *Carrichtera annua*; Facelli and Chesson 2008). Finally, three physical traits make chenopod shrubs susceptible to grazing: absence of spines (with the exception of *Rhagodia* branchlets and *Sclerolaena* fruits), the limited ability to produce viable apical or lateral buds, and the ease with which primordia can be removed (Osborn *et al.* 1932; Graetz 1973; see 24.6.3 Changes in plant community composition with grazing).

Table 24.3 **Adaptations of three common chenopod species with C_4 photosynthetic pathways, differences in their adaptations to dryland environments and their functions.**

Species	Adaptation	Function
Atriplex vesicaria	Fan-like shallow roots	Interception of surface water
	Ability to reduce stomatal opening	Reduces evapotranspiration by up to 50%
	Leaf drop and root shedding during dry periods	Reduces evapotranspiration and water use
	Rain-dependent flowering	Allows response at any time of the year
	Bracteoles that contain salt	Ensures minimum moisture levels (~50 mm rainfall) for germination that maximise establishment
	Seeds germinate at low temperatures	Maximises the chances of successful seedling establishment
	Long seed viability ~ 3–5 years	Maximises the chances of successful seedling establishment
	Thick seed coat that inhibits light penetration	Maximises the chances of successful seedling establishment
	Relatively short life span (~25 years) for woody plant	Persists through seedbank rather than mature plants
Maireana pyramidata	Deep roots	Greater access to moisture
	Grows in sandy soils	Competitive advantage over *Atriplex* spp.
	Can withstand high temperatures	Competitive advantage over *Atriplex* spp.
	Resilient to defoliation	Tolerates moderate herbivory
	Semi-succulent leaves	Tolerates moderate soil salt concentrations
	Retains leaves in drought	Rapid post-drought recovery
	Flowers in autumn and winter	Flower and fruit production avoid greatest drought stress
	Short-lived seeds	Germinates readily after rain
	No germination inhibitors	Germinates readily after rain
	Long lived (~ 150 years)	Invests substantially in mature plants
Atriplex nummularia	Varied root system; deep tap and lateral roots	Increases scavenging ability for water and nutrients
	Extensive fibrous roots	Allows it to stabilise surface soils
	Can use saline water	Increased competitive capacity
	Can grow on clay or sandy soils	Increased competitive capacity

24.4 Soil and geomorphological relationships: from plant to landscape scale

24.4.1 Landscape-scale relationships

At landscape scales, the distribution and community composition of chenopod shrublands is driven by physical soil factors, principally soil texture and soil depth, which control the availability of soil moisture (Table 24.1). For example, in the western Riverina, soil moisture is difficult to extract from the clay-rich soils (Noy-Meir 1973) and infiltration rarely exceeds 30 cm into the soil. On these soils *Atriplex vesicaria* has a competitive advantage over *Maireana* spp. because its roots are confined to the top 30 cm of the soil and it can extract water from soils with lower moisture levels than *Maireana* spp. (Carrodus and Specht 1965). The distribution of these dominant species therefore is driven by the extent to which soil texture drives changes in soil water-holding capacity (Osmond *et al.* 1980).

The distribution and longevity of different chenopod species in the saltbush bluebush shrublands are correlated with the depth to which water accesses the subsoil and the presence of hard limestone horizons (Table 24.1). In northern South Australia for example, Carrodus and Specht (1965) showed that *Atriplex vesicaria* was confined to areas with shallow water penetration, but *Maireana sedifolia* and *Maireana pyramidata* were found in soils with an indurated limestone hardpan because they have extensive lateral and tap roots that extend 2 m into the soil and can break through the limestone (Crocker and Wood 1947; Carrodus 1962). This may account for the fact that these plants are extremely long lived relative to other chenopods (Osborn *et al.* 1932; Table 24.3). Finally at the landscape scale, areas that receive supplementary run-on water will likely support more deeply-rooted chenopod species such as *Atriplex nummularia* and *Chenopodium nitrariaceum* (Table 24.1).

A special case of landscape-level plant distribution occurs in the Gibber Chenopod Shrublands, which are characterised by a linear gilgai microrelief (Table 24.1). These systems comprise two distinct micro-regions; shelf and depression. The shelf is typically covered by a mantle of small quartz or silcrete stones (gibbers) protecting the surface against erosion (Eldridge 1988). The long, linear depressions are dominated by brown and grey clays with a high water-holding capacity, and markedly greater (up to 10-times) infiltration than the shelf (Valentine and Nagorcka 1979). Large amounts of runoff generated from low (<0.5%) slopes of the shelf areas are redistributed to the depressions, even after rain events as low as 5 mm. This supplementary water supports a high cover and diversity of chenopod species including *Atriplex vesicaria*, *Maireana pyramidata*, *Maireana sedifolia*, and many perennial grasses (Eldridge 1988). Soil with high levels of moisture, or that may be flooded (e.g. on the Riverine Plain) may support *Maireana aphylla*, which has a high requirement for soil moisture.

24.4.2 Patterns at the individual plant scale

The contemporary distribution of both perennial and annual chenopod species is complex, and relates to subtle differences in microtopography, soil water availability and site history. Much of what we know today about what drives plant distribution in chenopod shrublands comes from comprehensive studies in the early 1970s and 1980s from the Riverina region of New South Wales (*Atriplex vesicaria*) and northern South Australia (*Maireana sedifolia* with *Atriplex vesicaria*). A number of trends have emerged from these studies. First, the intensity of plant-induced nutrient patterns declines with shrub age, i.e. soil nutrients in older stands tend to be more randomly distributed (Anderson 1970; Malik *et al.* 1976). Seedlings tend to aggregate around older plants (Eldridge *et al.* 1992), changing the distribution from random to clumped. However, this clumping is only maintained until reduced soil moisture and competition lead to self-thinning (Slatyer 1973) and the development of more regular patterning. Second, the spatial patterning of juvenile shrubs alters surface microtopography by trapping organic matter and sediment. This enhanced microtopography also drives the local distribution of biennial and annual species of *Atriplex* and *Sclerolaena*. For example, differences in water penetration and retention, runoff and ponding explained differences in the distribution of *Atriplex lindleyi*, an annual chenopod species that grows on gilgaid surfaces on heavy clay soils (Malik and Andersen 1971). The contemporary distribution of this annual chenopod species may have been influenced by the previous distribution of *Atriplex vesicaria*, which has been eliminated from some areas by overgrazing. Third, as soil moisture declines, plants growing on existing shrub hummocks tend to survive longer than those in the depressions, because it is easier for plants to extract water from sandy than clay soils (Noy-Meir 1973). Finally, the ratio of soil sodium and potassium (Na+K) to calcium (Ca) is highly positively correlated with the density of juvenile *Atriplex vesicaria* plants but negatively correlated with the density of mature shrubs (Anderson 1970).

Chenopod shrubs also have higher levels of salt beneath their canopies than in the interspaces, but only in the top 15 cm (Sharma and Tongway 1973). The opposite is true at greater depths. This pattern of salt concentrations beneath the canopy of shrubs is a by-product of uptake and depletion of salts in the subsoil and the deposition and decomposition of leaf litter that is high in sodium chloride because *Atriplex vesicaria* accumulates and excretes salt through glands on its leaf surfaces (Sharma and Tongway 1973). Relocation of NaCl from the subsoil and deposition on the surface has important implications for plant–plant interactions in chenopod shrublands and competition between understorey plants (see 24.5.1 and 24.5.2 Plant–plant interactions). The even greater salt concentrations beneath *Atriplex nummularia* canopies is probably due to their greater size and deeper roots, which allow them to access salt from more of the soil profile, as well as their greater rate of biomass production. These differences are thought to persist at least 2 years after shrub removal (Rixon 1970).

The spatial patterning around chenopod shrubs is likely reinforced by their value as ecosystem wicks, i.e. their tendency to allow higher rates of infiltration compared with the interspaces. For example, Dunkerley (2000) showed that infiltration rates adjacent to the canopies of *Maireana* spp. were up to five-times greater than those in the interspace. Similarly, the soil directly beneath chenopod shrubs may have up to twice the levels of organic carbon and total nitrogen than the interspaces (Rixon 1970), and most of this is in the top 5 cm (Charley and Cowling 1968). These differences are due largely to more moisture and greater levels of organic matter and decomposition beneath the shrubs. Greater levels of moisture and nutrients can extend a substantial distance from the canopy, probably due to greater root mass and associated soil organic matter. Thus, shrubs growing close together effectively have no interspace, enabling greater water infiltration.

24.5 Major ecological processes

24.5.1 Plant–plant interactions: chenopod shrubs as protégé species of taller shrubs

Substantial research, predominantly from arid South Australia, has examined the ecology, threats and challenges to the persistence of long-lived perennial plants such as *Acacia papyrocarpa*, which is an important structural element of chenopod shrublands across a large area of its range (Facelli and Brock 2000; Sinclair 2005). Along with *Atriplex*, *Maireana* and an assortment of other long-lived trees such as *Casuarina* and *Eucalyptus* spp., these plants play a critical role in plant–plant interactions in the chenopod shrublands. Many of the facilitatory effects of these long-lived species on smaller chenopod subshrubs are due to the greater levels of soil physical and chemical constituents (e.g. organic matter, nitrogen, phosphorus and other nutrients) beneath their canopies and greater shade, which enhances soil moisture, reduces the evaporative demand during summer and protects plants from high radiation loads (Facelli and Brock 2000). Lower levels of light and temperature beneath the canopy may even facilitate the invasion of exotic plants (Lentz and Facelli 2003).

The role of *Acacia papyrocarpa* as a nurse plant depends on the particular species growing beneath its canopy. Competitive and facilitative effects have been shown to be strongly species specific. For example, the bird-dispersed subshrubs *Rhagodia spinescens* and *Enchylaena tomentosa* are found almost exclusively under the canopies of large trees (Facelli and Brock 2000; Hastwell and Facelli 2003; Soliveres *et al.* 2012). They also produce more biomass when grown under the canopy (Prider and Facelli 2004). *Atriplex vesicaria*, however, shows the opposite response, with greater cover in the interspaces, particularly during dry times (Prider and Facelli 2004). The reasons for this species-level segregation are complex, but likely relate to greater levels of shade, improved soil macroporosity, finer soil texture and therefore greater infiltration of water over longer periods (Prider and Facelli 2004), favouring plants such as *Enchylaena tomentosa*, which has greater water needs than *Atriplex* (Prider and Facelli 2004). The provision of shade may become more important as water becomes more limiting. Physical amelioration of the subcanopy environment likely explains the difference in plant community composition, rather than specific plant traits or dispersal mechanisms. For example, some species resident beneath tall canopy plants are bird dispersed (e.g. *Enchylaena tomentosa, Rhagodia spinescens, Einadia nutans*), but others are not (e.g. *Zygophyllum* spp., *Sisymbrium irio*) (Facelli and Brock 2000; Soliveres *et al.* 2012).

24.5.2 Plant–plant interactions: chenopod shrubs as nurse plants

Limited research, mainly on *Atriplex vesicaria* and *Maireana* species, demonstrates that these smaller, shorter-lived perennial chenopods also create gradients in facilitation and competition beneath their canopies (e.g. Facelli and Temby 2002). Trade-offs between shading and competition, and the positive effects of the canopy on soil modification can result in both positive and negative effects of shrubs on annual plant composition. This sometimes results in lower emergence and diversity of annual species under *Maireana sedifolia* canopies than the interspaces (Weedon and Facelli 2008). However, the extent to which chenopod shrubs will facilitate or compete with understorey plants is strongly seasonally dependent. While annual plants may be more abundant under *Maireana sedifolia* canopies during dry periods, this effect wanes under conditions of average or above average rainfall (Emmerson and Facelli 1996). Differences may also relate to canopy morphology, which affects plant susceptibility to herbivory. The canopy of *Maireana sedifolia* is dense, often reaching the soil surface (Facelli and Temby 2002), and provides a physical barrier against herbivory (associational resistance). Conversely, groundstorey plants growing beneath *Atriplex vesicaria* shrubs may be more susceptible to herbivory because they are associated with a palatable shrub (associational susceptibility). Thus the benefits of being associated with a nitrogen-rich soil environment may be offset by greater herbivory.

Variation in species-specific effects may be related, in part, to differences in environmental conditions among different nurse species. For example, soil nitrogen is greater beneath *Atriplex vesicaria* than in the interspaces, making it more conducive to plant growth, whereas soil beneath *Maireana sedifolia* has more available phosphorus. Facilitatory effects may even extend beyond the edge of the canopy (Facelli and Temby 2002), possibly due to less competition at the edges, or the greater levels of soil water in the annular zone surrounding these shrubs (Dunkerley 2000).

24.6 Drivers of plant community composition

24.6.1 The influence of herbivore activity

Numerous studies have examined the net effects of herbivores, particularly domestic livestock, on community composition and the structure of the key plant species from the chenopod shrublands (e.g. Graetz and Wilson 1984; Leigh 1994). These studies have focused on the effects of livestock (generally sheep) on the health and vigour of the main structural elements; *Atriplex vesicaria, Maireana sedifolia* and *Maireana pyramidata* or the health of the soil, often using distance from water as a surrogate for grazing intensity (the piosphere effect, Lange 1969). The piosphere effect, which describes gradients in herbivore activity emanating from livestock watering points, is well-developed in chenopod shrublands, and characterised by marked changes in soil quality (e.g. soil nutrients, infiltration and compaction), shrub density, exotic plant cover, bare soil cover, mass of livestock dung, and plant and lichen composition with increasing proximity to water (Lange 1969; Rogers and Lange 1971; Barker 1979; Andrew and Lange 1986a, b). We draw upon this extensive body of research to describe the main effects of livestock activity, both herbivory and soil disturbance, on the structure, composition and function of chenopod shrublands, their key structural elements and their constituent understorey species.

The main role of chenopod shrubs is to provide structural stability in the shrubland system, and this role becomes increasingly important for soil stability as the proportion of fine material in the soil declines (Tupper and Müller 1985). The important elements, *Atriplex vesicaria, Maireana sedifolia, Maireana pyramidata*, and to a lesser extent *Atriplex nummularia* and *Maireana astrotricha*, are susceptible to herbivore activity, but to varying degrees (Graetz and Wilson 1984; see Eldridge *et al.*, Chapter 6, this volume). The net effect of herbivore activity on the constituent species depends on the intensity of herbivore activity, the plant species that replace them, and the stability of the soil surface (Graetz and Wilson 1984). While the main focus of this section is on vertebrate herbivores, we acknowledge the potentially large role of invertebrates in chenopod shrublands. For example, widespread reports of *Atriplex vesicaria* dieback in the 1970s and 1980s were thought to result from caterpillar attack (Clift *et al.* 1987). The relationship between vertebrate and invertebrate herbivory on *Atriplex vesicaria* is complex, but moderately grazed plants are thought to recover better from caterpillar attack than ungrazed plants (Clift *et al.* 1987).

24.6.2 Models of community response to livestock and climate

We use state and transition models to describe the response of two chenopod shrubland community types; the Clayplains Saltbush community (Box 24.1) and the Sandplain Bluebush community (Box 24.2) to livestock activity.

Of the original focal shrub species inhabiting the chenopod shrublands, *Atriplex nummularia* is probably the most sensitive to grazing, with extensive stands virtually eliminated from the Riverina by livestock and human activity by the end of the nineteenth century (Moore 1953). Of the other focal species, low to moderate livestock activity has either little effect, or a slight stimulatory effect on many chenopod shrubs (Wilson and Graetz 1979). *Atriplex vesicaria* plants will generally tolerate intermittent heavy grazing (e.g. Osborn *et al.* 1932; Squires 1981). Sustained herbivory will kill the plants if they are completely defoliated (Boxes 24.1 and 24.2, Transition 1; Wilson and Graetz 1979), leading to cascading effects on community composition and

Box 24.1 Dynamics of *Atriplex vesicaria* on fine-textured soils

State I: community dominated by *Atriplex vesicaria* (cover > ~20%; Westoby *et al.* 1989) with mixed age structure, intact soil seed bank, and variable ephemerals and perennial grasses dependent on seasonal conditions.

State II: variable cover of *Atriplex vesicaria* (5–10%; Eldridge *et al.* 1990) without seedlings, but sustains an intact soil seedbank; dominance by short-lived chenopod forbs (e.g. *Sclerolaena, Atriplex, Maireana* spp.), perennial grasses and annual grasses. Highly transient between states I and III over periods of a few years (Westoby *et al.* 1989; Hunt 1992).

State III: isolated and moribund *Atriplex vesicaria* shrubs with no soil seedbank; dominated by annual grasses (e.g. *Hordeum* sp.) and forbs, particularly *Sclerolaena* spp. May also support derived grasslands (*Austrodanthonia* spp.) in some eastern areas (e.g. western Riverina; Williams 1961; Westoby *et al.* 1989); variable cover of erosion, particularly when associated with prior stream levees. A highly unstable state, generally moving rapidly to state IV.

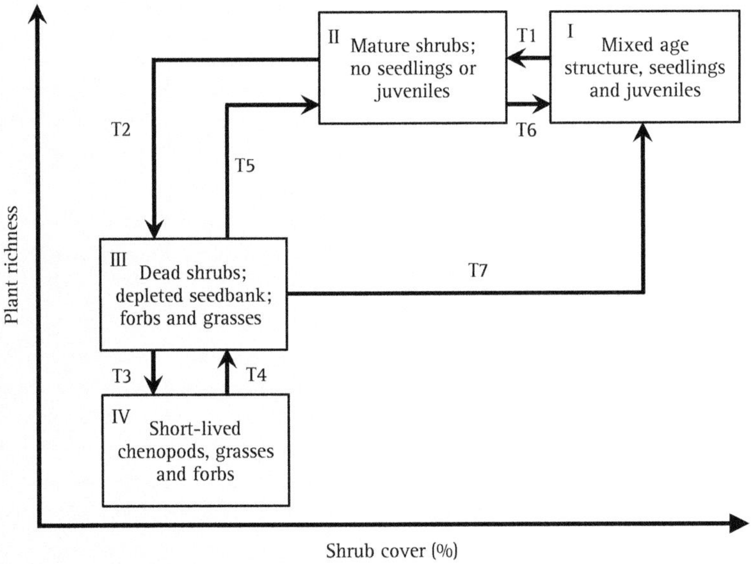

Figure 24.4 State and transition model for Clayplains Saltbush Shrublands.

State IV: low diversity, low resilience community dominated by short-lived *Sclerolaena* spp. and exotic annual grasses; sometimes supports grazing-resistant increaser shrubs such as *Nitraria billardierei* or *Lycium ferocissimum* (Noble 1991); extensive areas of wind and water sheeting resulting from removal of biocrusts (Eldridge 1996).

Transitions (see Figure 24.4)

T1: Moderate sustained livestock activity over 3–5 years reduces shrub density and cover, particularly after complete defoliation (Leigh and Mulham 1971). Effects of defoliation are greatest during droughts when more palatable plants have been consumed. Sheep actively select for female flowers (Walsh *et al.* 2005) thereby reducing seed set. Seed removal by invertebrates exacerbates seedbank depletion (Briese 1982). Insect herbivory can result in extensive die-back (Clift *et al.* 1987) but plants may recover (T6). The return interval of fire in chenopod shrublands is long, but high levels of fuel following La Niña events may result in substantial saltbush death through wildfire (Lay 1976).

T2: Continued herbivore activity (herbivory and soil disturbance), coupled with drought. Episodic recruitment of *Atriplex vesicaria* occurs in all years under variable rainfalls (e.g. Eldridge *et al.* 1990; Eldridge and Westoby 1991; Watson *et al.* 1997). Mortality of germinants increases with intensity and duration of droughts and widespread death occurs if droughts exceed ~8 months while plants are juveniles (<3 years old; Westoby *et al.* 1989).

T3: Continued overgrazing coupled with prolonged drought.

T4 and T5: Complete destocking for up to 3–5 years coupled with restoration techniques using saltbush seeding (Stanley 1978; Rhodes 1987). Sufficient rainfall is required for germination, establishment and growth (Hunt 1992). Probability of success is low for both T4 and T5 given the low frequency of threshold rainfall to sustain seedlings for 3 years to adult establishment (Hunt 1992). Transition T4 requires significant financial investment to attain substantial soil restoration through mechanical methods (pitting, furrowing, water ponding).

T6: Rainfall to germinate and establish seedlings, and protection from grazing to stimulate bush growth and seed production (Hunt 1992). This transition is likely to be slow until adult plants increase in vigour and produce seeds to support the soil seed bank (Hunt 1992).

T7: A financially costly and ecologically risky strategy requiring soil engineering based on physical disturbance, seeding and sometimes, fertilisation (e.g. Stanley 1978). However, engineering has been shown to be successful on some soil types where high-rainfall years coincide with low levels of livestock grazing (Stanley 1978; Rhodes 1987).

Box 24.2 Dynamics of mixed *Atriplex vesicaria–Maireana* spp. community on coarse-textured soils

State I: mixed community of *Atriplex vesicaria, Maireana sedifolia, M. pyramidata, M. astrotricha* with a diverse range of biennial chenopods; variable plant age classes and high productivity of grasses.

State II: community initially dominated by multiple species of *Maireana* spp. with few if any, generally senescent *Atriplex vesicaria*. Continued herbivory leads to selective removal of *M. sedifolia* and dominance by *M. pyramidata* (Eldridge *et al.* 1990; Tupper and Müller 1985). Total shrub cover is generally greater than that in State I due to the dominance by larger *Maireana* spp. shrubs (Eldridge *et al.* 1990) and mixed populations of smaller chenopod sub-shrubs (Leigh 1994).

State III: dominated by grazing-tolerant perennial grasses (e.g. *Aristida, Austrostipa* spp.), variable annual forbs with some bare soil, particularly in drought years. Low cover but variable density of perennial shrubs (generally *M. pyramidata*; Tupper and Müller 1985).

State IV: extensive areas of bare soil dominated by annual grasses and forbs, often exotic, with a few scattered and heavily grazed perennial shrubs (*M. pyramidata*).

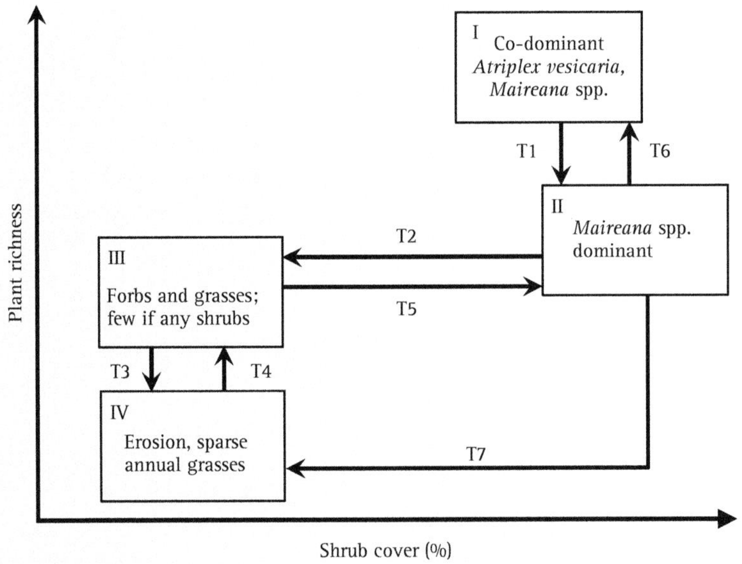

Figure 24.5 State and transition model for Sandplain Bluebush Shrublands.

Transitions (see Figure 24.5)

T1: Overgrazing, drought and removal of *Atriplex vesicaria* plants by herbivory (Wood 1936; Eldridge *et al.* 1990, 1992).

T2: Overgrazing, and sustained physical disturbance by herbivores (Wood 1936).

T3 and T4: Both of these transitions occur regularly, depending on seasonal conditions, oscillating between bare soil and a community dominated by annual forbs and grasses (T4). Effects are likely driven more by rainfall and seasonal conditions than grazing (Stanley and Milthorpe 1978; Eldridge *et al.* 1990).

T5: Unlikely without engineering and physical restoration such as furrowing or pitting, and addition of shrub seed combined with destocking (Stanley 1978). Recovery is likely to occur only during high-rainfall years (Stanley and Milthorpe 1978), irrespective of herbivore activity.

T6: May occur naturally, but dependent on size and viability of remaining *Atriplex vesicaria* seedbank (Burbidge 1945) and proximity to mature *Atriplex* plants.

T7: Can occur over a period of about 5 years due to sustained herbivore activity (both herbivory and soil disturbance) during droughts (Wood 1936; Stanley and Milthorpe 1978).

structure. Graetz (1986), however, posits that low levels of grazing or grazing exclusion may be detrimental to *Atriplex vesicaria* by limiting recruitment of new seedlings, potentially as a result of competition with older shrubs (though see Eldridge *et al.* 1990).

Mature plants of *Maireana sedifolia*, which typically occur with *Atriplex vesicaria* throughout its range (Table 24.1), are more resistant to moderate levels of herbivore activity. Extremely high levels of activity, however, will trigger the replacement of *Maireana sedifolia* by *Maireana pyramidata* (Crisp 1978; Box 24.2, Transition 2). Recovery from overgrazing may be slow, due to depleted soil seed banks or infrequent episodic recruitment (Eldridge *et al.* 1990; Watson *et al.* 1997; Sinclair 2005; Watson *et al.* 2007; Box 24.2, Transition 6). Recruitment of *Atriplex vesicaria* and *Maireana* spp. generally occurs following autumn of spring rains (Burbidge 1945; Box 24.2, Transition 5). In some areas, high stocking rates have been shown to increase the densities of *Maireana sedifolia* and *Maireana pyramidata* (Ratcliffe 1936; Graetz and Ludwig 1978; Eldridge and Westoby 1991), potentially due to reduced competition between seedlings and adult plants, or increases in the cover of bare soil, which is critical for seedling establishment (Box 24.2, Transition 7). Grazers have been shown to have little effect on the survival of *Atriplex vesicaria* seedlings and juveniles (Eldridge *et al.* 1990; Hunt 1992).

The effects of herbivore activity are exacerbated during dry times when livestock trampling suppresses regrowth (Leigh and Mulham 1971). Studies from the Koonamore exclosure in north-eastern South Australia (Hall *et al.* 1964) demonstrate that recovery of *Atriplex vesicaria* is protracted, even with complete removal of herbivores (see Box 24.2, Transitions 3 and 4).

Recovery to full cover and density of *Atriplex vesicaria* plants may take many years (Sinclair 2005; Box 24.2, Transitions 5 and 6).

Maireana pyramidata is relatively unpalatable, even to European rabbits (*Oryctolagus cuniculus*), but ring barking by rabbits during dry times (Tupper and Müller 1985) can kill shrubs (Box 24.2, Transition 2). Increases in the relative dominance of *Maireana pyramidata* in mixed *Maireana* spp. communities across its distribution since the mid-twentieth century have been well documented (Graetz and Ludwig 1978; Tupper and Müller 1985; Sinclair 2005). Complete removal of *Maireana pyramidata*, however, can lead to the dominance of annual forbs and grasses, and increased areas of wind erosion (Box 24.2, Transition 3). However, as indicated earlier, increases in the density of *Maireana pyramidata* are generally associated with increased richness of understorey plant species.

24.6.3 Changes in plant community composition with grazing

Long-term studies have shown that rainfall may have a greater influence on the community composition and survival of both shrubs and their understorey species than grazing (Graetz 1980; Eldridge *et al.* 1990; Eldridge and Westoby 1991; Watson *et al.* 2007). Most chenopod shrubs are generally only grazed during dry times when the more nutritious herbaceous plants are either rare or unavailable (Leigh 1994). The groundstorey component is less affected by grazing than the shrubs themselves because they are largely driven by rainfall rather than grazing intensity (e.g. Graetz 1980; Graetz and Wilson 1984; Westoby *et al.* 1989). However, grazing may bias the sex ratio of mature *Atriplex vesicaria* plants by selecting for male plants, which are less palatable than female plants (Williams *et al.* 1979; Walsh *et al.* 2005).

The exact response of groundstorey plant communities to grazing is strongly dependent on community type. In general, two trends emerge across the range of chenopod shrublands. First, chenopod shrubs typically comprise < 25% of the total diet of domestic livestock even though they may make up a substantial proportion of the total biomass (Squires 1981; Graetz 1986). Second, the most degraded states (Boxes 24.1 and 24.2, States III and IV) are typically dominated by short-lived, grazing insensitive annual forbs and grasses such as *Sclerolaena* spp. (Leigh and Mulham 1971; Leigh and Noble 1972) across all communities. While it is not possible to document the changes in plant community composition across all chenopod community types, we examine changes in an *Atriplex vesicaria* shrubland in the western Riverina as an example (Box 24.1).

Overgrazing by livestock in chenopod shrublands in the western Riverina leads to replacement of *Atriplex vesicaria* by native perennial grasses such as *Rytidosperma caespitosa*, *Chloris truncata*, *Enteropogon acicularis*, *Panicum decompositum* and *Austrostipa scabra* (Williams 1968; Box 24.1, Transition 2). Reductions in *Atriplex vesicaria* density (e.g. Box 24.1, Transition 1) may stimulate increased densities of other chenopod subshrubs such as *Maireana turbinata* and *Maireana georgei*. Continued herbivory and surface destabilisation eventually leads to a dominance by native forbs such as annual saltbushes (*Atriplex* spp.), 'copperburrs' (*Sclerolaena* spp., *Dissocarpus paradoxus*), native and exotic grasses (*Sporobolus caroli*, *Hordeum leporinum*) and exotic forbs (e.g. *Medicago* spp.) (Leigh and Mulham 1971; Leigh and Noble 1972; Box 24.1, Transitions 2 and 3). The replacement of *Atriplex vesicaria* by derived perennial grasslands, a phenomenon that has been reported for the western Riverina (Box 24.1, Transition 2), is not apparent in *Atriplex vesicaria* communities on gibber or sandplains where the resulting vegetation is generally annual forbs and grasses (Box 24.2, Transition 2 and 3). On fine-textured soils, grazing may also favour *Maireana aphylla* shrubs in low-lying areas.

24.6.4 Effects of fire

In general, chenopod shrublands are not adapted to regular fires (Waddell *et al.* 2010). *Atriplex vesicaria* will not survive fires of any intensity (Graetz and Wilson 1984; Box 24.1, Transition 1) but *Maireana sedifolia*, if burned, may survive and resprout, as long as the fire intensity is low. Recovery however is very slow (Waddell *et al.* 2010). Where fires are intense and shrubs are killed, they are generally replaced by perennial grasses, particularly in the Sandplain Bluebush Shrublands and Nullarbor Chenopod Shrublands.

24.7 Current status, conservation and threats

24.7.1 Conservation of chenopod shrublands

Chenopod shrublands mainly occur on leasehold and freehold lands and less than 15% occurs in reserves or protected areas. Relatively large areas of chenopod shrublands are reserved in a number of national parks, reserves and protected areas including Nullarbor Regional Reserve, Gawler Ranges National Park, Chowilla Regional Reserve, Kati-Thanda Lake Eyre National Park in South Australia; Willandra World Heritage Area, Willandra National Park, Oolambeyan National Park and Lachlan Valley Nature Reserve in New South Wales, and Great Victoria Desert Nature Reserve in Western Australia. Chenopod shrublands are home to a number of threatened species. For example, the thick-billed grasswren (*Amytornis textilis modestus*) inhabits almost exclusively, treeless, dense, chenopod shrublands in north-eastern South Australia, but once occurred across a large area spanning western New South Wales, northern South Australia and southern Northern Territory (Garnett and Crowley 2000). Due to their range decline, they have been listed as Endangered in New South Wales, and are nationally and internationally recognised as a Vulnerable species. The vulnerable plant species, ridged noon-flower (*Sarcozona bicarinata*) and sandalwood (*Santalum spicatum*) are found in the Nullarbor Regional Reserve, which also contains one of the world's largest karst areas. *Rhagodia acicularis*, which is endemic to south-west Western Australia, is listed as threatened due to its extremely limited range.

24.7.2 Current status and threats

The current status of chenopod shrublands is poor, with widespread loss of shrubs reported over the last half of the twentieth century. For example, significant reductions of up to 75% on some properties have been recorded over large areas of pastoral South Australia (Lay 1979). There have been few attempts to document the extent of degradation of chenopod shrublands. Newman and Condon (1969) estimated that only about 50% of chenopod shrublands were in pre-European condition based on the health of individual shrubs, community composition of associated plants and lack of soil degradation. A more recent study showed that 60% of the total area of *Atriplex vesicaria* and *Maireana* spp. in New South Wales comprised mainly scattered stands, with only about 8% occurring as dense stands (Graham 1992). In some areas chenopod shrublands are expanding due to increased salinity due to poor land management, at the expense of other vegetation types.

Vegetation clearance for grazing, and more recently irrigated cropping, is a major threat to some chenopod shrublands, but the rate of clearing has been disproportionate across the different shrubland types. Clearing has been almost negligible in the most arid and remote locations (e.g. Gibber Chenopod Shrublands in northern South Australia and Nullarbor Chenopod Shrublands). However, substantial losses of Clayplains Saltbush Shrublands (~25%; Keith 2004) have occurred on the Hay Plain, especially in recent years with expansion of irrigated cropping. Some areas of Sandplain Bluebush Shrublands in western

New South Wales in the Wentworth-Kinchega area have effectively been replaced by ephemeral grassland and forbland due to historical overgrazing.

24.8 Conclusions

Chenopod shrublands have been the subject of considerable study for more than 50 years. Despite this, their current condition is poor, their distribution has contracted substantially and the prognosis for recovery, particularly in the face of declining winter rainfall over the next century in southern Australia, is poor. Two broad themes arise from this chapter. First, while the dominant shrubs themselves are important for ecosystem stability and pastoral use, they provide a substantial role in facilitating groundstorey components of the plant community. Second, the evidence suggests that recovery of the full species component of chenopod shrublands following overgrazing is unlikely, particularly given climatic predictions for southern Australia. Our focus should be on protecting remaining intact stands of healthy functional shrublands, particularly from overgrazing by livestock but also through clearing for irrigated agriculture. As some of these areas occur on highly productive soils, the challenge will be to balance increasing demand for agricultural land with the needs of sustaining this iconic vegetation community and its biodiversity.

References

Anderson, D.J. (1970) Analysis of pattern in *Atriplex vesicaria* area communities from the Riverine Plain of New South Wales. In *The Biology of Atriplex*, ed. R. Jones. CSIRO Australia: Canberra, pp. 63–68.

Anderson, D.J. (1982) Environmentally adaptive traits in arid zone plants. In *Evolution of the Flora and Fauna of Arid Australia*, ed. W.R. Barker and P.J.M. Greenslade. Peacock Publications: Adelaide, pp. 133–140.

Andrew, M.H. and Lange, R.T. (1986a) Development of a new piosphere in arid chenopod shrubland grazed by sheep. 1. Changes to the soil surface. *Australian Journal of Ecology* 11, 395–409.

Andrew, M.H. and Lange, R.T. (1986b) Development of a new piosphere in arid chenopod shrubland grazed by sheep. 2. Changes to the vegetation. *Australian Journal of Ecology* 11, 411–424.

Barker, S. (1979) Shrub population dynamics under grazing: within paddock studies. In *Studies of the Australian Arid Zone. IV. Chenopod Shrublands*, ed. R.D. Graetz and K.M.W. Howes. CSIRO Australia: Perth, pp. 83–106.

Beadle, N.C.W. (1952) Studies in halophytes. I. The germination of the seed and establishment of the seedlings in five species of *Atriplex* in Australia. *Ecology* 33, 49–62.

Beadle, N.C.W. (1981) *The Vegetation of Australia*. Cambridge University Press: Cambridge.

Beard, J.S. (1975) *Vegetation Survey of Western Australia: Nullarbor, 1:1 000 000 Vegetation Series, Explanatory Notes to Sheet 4.* University of Western Australia Press: Perth.

Beard, J.S., Beetson, G.R., Harvey, J.M. *et al.* (2013) *The Vegetation of Western Australia at 1:3 000 000 scale. Explanatory memoir*, 2nd edition. Science Division, Department of Parks and Wildlife: Western Australia.

Briese, D.T. (1982) Damage to saltbush by the coccid *Pulvinaria maskelli* Ollif, and the role played by an attendant ant. *Journal of the Australian Entomological Society* 21, 293–294.

Burbidge, N.T. (1945) Germination studies of Australian Chenopodiaceae with special reference to the conditions necessary for regeneration. I. *Atriplex vesicarium* Heward. *Transactions of the Royal Society of South Australia* 69, 73–84.

Burbidge, N.T. (1960) The phytogeography of the Australian region. *Australian Journal of Botany* 8, 75–211.

Carrodus, B.B. (1962) Some aspects of the ecology of arid South Australia: the relative distribution of *Atriplex vesicaria* Heward ex Benth. and *Kochia sedifolia* s.v.M. PhD Thesis, University of Adelaide, Adelaide.

Carrodus, B.B. and Specht, R.L. (1965) Factors affecting the relative distribution of *Atriplex vesicaria* and *Kochia sedifolia* (Chenopodiaceae) in the arid zone of South Australia. *Australian Journal of Botany* 89, 41–60.

Charley, J.L. (1959) Soil salinity-vegetation patterns in western New South Wales. PhD Dissertation, University of New England, Armidale.

Charley, J.L. and Cowling, S.W. (1968) Changes in soil nutrient status resulting from overgrazing and their consequences in plant communities of semi-arid areas. *Proceedings of the Ecological Society of Australia* 3, 28–38.

Clift, D.K., Semple, W.S. and Prior, J.C. (1987) A survey of bladder saltbush (*Atriplex vesicaria* Heward ex Benth.) dieback on the riverine plain of south-eastern Australia from the late 1970s to 1983. *Australian Rangeland Journal* 9, 39–48.

Crisp, M.D. (1978) Demography and survival under grazing of three Australian semi-desert shrubs. *Oikos* 30, 520–528.

Crocker, R.L. and Wood, J.G. (1947) Some historical influences on the development of the South Australian vegetation communities and their bearing on concepts and classification in ecology. *Transactions of the Royal Society of South Australia* 65, 44–66.

Department of the Environment and Energy (2007) *NVIS data products 4.2.* Australian Government: Canberra. Available at: http://www.environment.gov.au/land/native-vegetation/national-vegetation-information-system/data-products (accessed 10 May 2016).

Dunkerley, D.L. (2000) Hydrological effects of dryland shrubs: defining spatial extent of modified soil water uptake rates at an Australian desert site. *Journal of Arid Environments* 45, 159–172.

Eldridge, D.J. (1988) Soil-landform and vegetation relations in the chenopod shrublands of western New South Wales. *Earth Science Reviews* 25, 493–499.

Eldridge, D.J. (1996) Distribution and floristics of terricolous lichens in soil crusts in arid and semi-arid New South Wales, Australia. *Australian Journal of Botany* 44, 581–599.

Eldridge, D.J. (1998) Soil crust lichens and mosses on calcrete-dominant soils at Maralinga. *Journal of the Adelaide Botanic Gardens* 18, 9–24.

Eldridge, D.J. and Ferris, J.M. (1999) Recovery of populations of the soil lichen *Psora crenata* after disturbance in arid South Australia. *The Rangeland Journal* 21, 194–198.

Eldridge, D.J. and Koen, T.B. (2003) Detecting environmental change in eastern Australia: rangeland health in the semi-arid woodlands. *Science of the Total Environment* 310, 211–219.

Eldridge, D.J. and Westoby, M. (1991) Recruitment and survival in *Atriplex vesicaria* populations in western New South Wales, 1977–1987. *Australian Journal of Ecology* 16, 309–314.

Eldridge, D.J., Westoby, M. and Stanley, R.J. (1990) Population dynamics of the arid zone shrubs *Atriplex vesicaria, Maireana astrotricha* and *Maireana pyramidata* under grazing, 1980–1987. *Journal of Applied Ecology* 27, 502–512.

Eldridge, D.J., Westoby, M. and Holbrook, K.G. (1992) Soil surface characteristics, microtopography and proximity to mature shrubs: effect of survival on several cohorts of *Atriplex vesicaria* seedlings. *Journal of Ecology* 78, 357–364.

Emmerson, L.M. and Facelli, J.M. (1996) Bluebush mound heights: an indicator of grazing regime? In *Conference Papers, the Australian Rangeland Society, 9th Biennial Conference*, ed. L.P. Hunt and R. Sinclair. Australian Rangeland Society: Port Augusta, pp. 227–228.

Facelli, J.M. and Brock, D.J. (2000) Patch dynamics in arid lands: localized effects of *Acacia papyrocarpa* on soils and vegetation of open woodlands of South Australia. *Ecography* 23, 479–491.

Facelli, J.M. and Chesson, P. (2008) Cyclic dormancy, temperature and water availability control the germination of *Carrichtera annua*, an invasive species in chenopod shrublands. *Austral Ecology* 33, 324–328.

Facelli, J.M. and Temby, A.M. (2002) Multiple effects of shrubs on annual plant communities in arid lands of South Australia. *Austral Ecology* 27, 422–432.

Garnett, S. and Crowley, G.M. (2000) *The Action Plan for Australian Birds*. Environment Australia: Canberra. Available at: https://www.environment.gov.au/resource/action-plan-australian-birds-2000 (accessed 12 August 2016).

Graetz, R.D. (1973) Biological characteristics of Australian *Acacia* and chenopodiaceous shrublands relevant to their pastoral use. In *Arid Shrublands*, ed. D.N. Hyder. Society for Range Management: Denver, pp. 33–39.

Graetz, R.D. (1980) The influence of grazing by cattle and sheep on the population dynamics of bladder saltbush (*Atriplex vesicaria*) and white top (*Danthonia caespitosa*). *Australian Rangeland Journal* 2, 208–213.

Graetz, R.D. (1986) A comparative study of sheep grazing a semi-arid saltbush pasture in two condition classes. *Australian Rangelands Journal* 8, 46–56.

Graetz, R.D. and Ludwig, J.L. (1978) A method for the analysis of piosphere data applicable to range assessment. *Australian Rangeland Journal* 1, 126–136.

Graetz, R.D. and Wilson, A.D. (1984) Saltbush and bluebush. In *Management of Australia's Rangelands*, ed. G.N. Harrington, A.D. Wilson and M.D. Young. CSIRO Publishing: Melbourne, pp. 209–222.

Graham, O.P. (1992) Survey of land degradation in New South Wales Australia. *Environmental Management* 16, 205–223.

Hall, E.A.A., Specht, R.L. and Eadley, C.M. (1964) Regeneration of the vegetation on Koonamore Vegetation Reserve 1926-62. *Australian Journal of Botany* 12, 205–264.

Hastwell, G. and Facelli, J.M. (2003) Facilitation in a pulsed environment: differing effects of facilitation on growth and survivorship of the chenopod shrub *Enchylaena tomentosa*. *Journal of Ecology* 91, 941–950.

Hekel, H. (1972) Pollen and spore assemblages from Queensland Tertiary sediments. *Geological Survey of Queensland, Palaeontological Papers* 30, 1–31.

Hunt, L.P. (1992) Piospheres and the state and transition model of vegetation change in chenopod shrublands. In *Proceedings of the 7th Australian Rangeland Society Conference, Cobar*. Australian Rangeland Society: Cobar, pp. 37–45.

Kadereit, G., Gotzek, D., Jacobs, S. and Freitag, H. (2005) Origin and age of Australian Chenopodiaceae. *Organisms Diversity and Evolution* 5, 59–80.

Kadereit, G., Lauterbach, M., Pirie, M.D. *et al.* (2014) When do different C-4 leaf anatomies indicate independent C-4 origins? Parallel evolution of C-4 leaf types in Camphorosmeae (Chenopodiaceae). *Journal of Experimental Botany* 65, 3499–3511.

Keith, D. (2004) *Ocean Shores to Desert Dunes. The Native Vegetation of New South Wales and the ACT.* Department of Environment and Conservation: Hurstville.

Lange, R.T. (1969) The piosphere: sheep track and dung patterns. *Journal of Range Management* 22, 396–400.

Lay, B.G. (1976) Fire in the pastoral country. *Journal of Agriculture South Australia* 79, 9–14.

Lay, B.G. (1979) Shrub population dynamics under grazing: a long term study. In *Studies of the Australian Arid Zone. IV. Chenopod shrublands*, ed. R.D. Graetz and K.M.W. Howes. CSIRO: Perth, pp. 107–124.

Leigh, J.H. (1994) Chenopod shrublands. In *Australian Vegetation*, ed. R.H. Groves. Melbourne: Cambridge University Press, pp. 345–367.

Leigh, J.H. and Mulham, W.E. (1971) The effect of defoliation on the persistence of *Atriplex vesicaria*. *Australian Journal of Agricultural Research* 22, 239–244.

Leigh, J.H. and Noble, J.C. (1972) *Riverine Plain of New South Wales: Its Pastoral and Irrigation Development.* CSIRO Australia: Canberra.

Lenz, T.I. and Facelli, J.M. (2003) Shade facilitates an invasive stem succulent in a chenopod shrubland in South Australia. *Austral Ecology* 28, 480–490.

Malik, A.R. and Andersen, D.J. (1971) An analysis of *Atriplex inflata* populations from western New South Wales. *Australian Journal of Botany* 19, 191–202.

Malik, A.R., Andersen, D.J. and Myerscough, P.J. (1976) Studies on structure in plant communities. VII. Field and experimental analysis of *Atriplex vesicaria* populations from the Riverine plain of New South Wales. *Australian Journal of Botany* 24, 265–280.

Martin, H.A. (1994) Australian Tertiary phytogeography: evidence from palynology. In *History of the Australian Vegetation: Cretaceous to Recent*, ed. R.S. Hill. Cambridge University Press: Cambridge, pp. 104–142.

Mitchell, A.A. and Wilcox, D.G. (1988) *Arid Shrubland Plants of Western Australia.* University of Western Australia Press with the Western Australian Department of Agriculture: Perth.

Moore, C.W.E. (1953) The vegetation of the south-eastern Riverina, New South Wales. I. The climax communities. *Australian Journal of Botany* 1, 485–547.

Newman, J.C. and Condon, R.W. (1969) Land-use and present condition. In *Arid Lands of Australia*, ed. R.O. Slatyer and R.A. Perry. ANU Press: Canberra, pp. 105–132.

Noble, J.C. (1991) On ratites and their interactions with plants. *Revista Chilena de Historia Natural* 64, 85–118.

Northcote, K.H. and Skene, J.K.M. (1972) *Australian Soils with Saline and Sodic Properties.* CSIRO: Melbourne.

Noy-Meir, I. (1973) Desert ecosystems: environment and producers. *Annual Review of Ecology and Systematics* 4, 25–51.

Osborn, T.G.B., Wood, J.G. and Paltridge, T.B. (1932) On the growth and reaction to grazing of the perennial saltbush, *Atriplex vesicarium*: an ecological study of the biotic factor. *Proceedings of The Linnean Society of New South Wales* 57, 377–402.

Osmond, C.B., Bjorkman, O. and Anderson, D.J. (1980) *Physiological Processes in Plant Ecology: Toward a Synthesis with Atriplex.* Springer-Verlag: Berlin-New York.

Parr-Smith, G.A. (1982) Biogeograpghy and evolution in Australian species of *Atriplex*. In *Evolution of the Flora and Fauna of Arid Australia*, ed. W.R. Barker and P.J.M. Greenslade. Peacock Publications: Frewville, pp. 219–299.

Parr-Smith, GA. and Caulder, D.M. (1979) *Atriplex visecaria* Heward ex Benth. and related species. In *Studies of the Australian Arid Zone. IV. Chenopod Shrublands*, ed. R.D. Graetz and K.M.W. Howes. CSIRO Australia: Perth, pp. 17–28.

Prider, J. and Facelli, J. (2004) Interactive effects of drought and shade on three arid zone chenopod shrubs with contrasting distributions in relation to tree canopies. *Functional Ecology* 18, 67–76.

Ratcliffe, F.N. (1936) Sand drift in the arid pastoral areas of South Australia. Australian CSIRO pamphlet No. 64.

Rhodes, D.W. (1987) Vegetation response and productivity, under water ponding. *Journal of the Soil Conservation Service of NSW* 43, 62–67.

Rixon, A.J. (1970) Cycling of nutrients in a grazed *Atriplex vesicaria* community. In *The Biology of Atriplex*, ed. R. Jones. CSIRO Australia: Melbourne, pp. 87–95.

Rogers, R.W. (1972) Soil surface lichens in arid and sub-arid south-eastern Australia. III. The relationship between distribution and environment. *Australian Journal of Botany* 20, 301–316.

Rogers, R.W. (1989) Blue-green algae in southern Australian rangeland soils. *Australian Rangeland Journal* 11, 67–73.

Rogers, R.W. and Lange, R.T. (1971) Lichen populations on arid soil crusts around sheep watering places in South Australia. *Oikos* 22, 93–100.

Sharma, M.L. and Tongway, D.J. (1973) Plant induced soil salinity patterns in two saltbush (*Atriplex* spp.) communities. *Journal of Range Management* 26, 121–125.

Sinclair, R. (2005) Long-term changes in vegetation, gradual and episodic, on the TGB Osborn Vegetation Reserve, Koonamore, South Australia (1926-2002). *Australian Journal of Botany* 53, 283–296.

Sinclair, R. and Thomas, D.A. (1970) Optical properties of leaves of some species in arid South Australia. *Australian Journal of Botany* 18, 261–273.

Slatyer, R.O. (1973) Structure and function of Australian arid shrublands. In *Arid Shrublands: Proceedings of the Third Workshop of the United States/Australia Rangelands Panel*, ed. D.N. Hyder. Society for Range Management: Denver, pp. 66–73.

Soliveres, S., Eldridge, D.J., Hemmings, F. and Maestre, F.T. (2012) Nurse plant effects on plant species richness in drylands: the role of grazing, rainfall and species specificity. *Perspectives in Plant Ecology, Evolution and Systematics* 14, 402–414.

Specht, R.L. (1970) Vegetation. In *The Australian Environment*, ed. G.W. Leeper. CSIRO Australia and Melbourne University Press: Melbourne, pp. 44–67.

Squires, V.R. (1981) *Livestock Management in the Arid Zone.* Inkata Press: Melbourne.

Stanley, R.J. (1978) Establishment of chenopod shrubs by tyne pitting on hardpan soils in western New South Wales, Australia. In *Proceedings of the First International Rangeland Congress*, ed. D.N. Hyder. Society for Range Management: Denver, pp. 639–642.

Stanley, R.J. and Milthorpe, P.L. (1978) The effects of climate and grazing by rabbits on survival and growth of black bluebush (*Maireana pyramidata*) seedlings at Menindee in western New South Wales. In *Working papers, Australian Rangeland Society Conference, Broken Hill*. Australian Rangelands Society: Broken Hill, pp. 129–138.

Tupper, G.J. and Müller, W.J. (1985) Preliminary observations on the effect of removal of black bluebush (*Maireana pyramidata*) and pearl bluebush (*Maireana sedifolia*) on shrub regeneration, herbage production and erosional potential. *Australian Rangeland Journal* 7, 103–106.

Valentine, I. and Nagorcka, B.N. (1979) Contour patterning in *Atriplex vesicaria* communities. In *Studies of the Australian Arid Zone IV. Chenopod Shrublands*, ed. R.D. Graetz and K.M.W. Howes. CSIRO Australia: Melbourne, pp. 121–146.

Waddell, P.A., Gardner, A.K. and Hennig, P. (2010) *An Inventory and Condition Survey of the Western Australian Part of the Nullarbor Region.* Department of Agriculture and Food: Western Australia.

Walsh, D., Andrew, M., Sinclair, R. and Coleman, D. (2005) Evidence for palatability variation between sex phenotypes of bladder saltbush from small-plot grazing trials. *The Rangeland Journal* 27, 89–96.

Watson, I.W., Westoby, M. and Holm, A.M. (1997) Demography of two shrub species from an arid grazed ecosystem in Western Australia 1983–1993. *Journal of Ecology* 85, 815–832.

Watson, I.W., Thomas, P.W.E. and Fletcher, W.J. (2007) The first assessment, using a rangeland monitoring system, of change in shrub and tree populations across the arid shrublands of Western Australia. *The Rangeland Journal* 29, 25–37.

Weedon, J.T. and Facelli, J.M. (2008) Desert shrubs have negative or neutral effects on annuals at two levels of water availability in arid lands of South Australia. *Journal of Ecology* 96, 1230–1237.

Westoby, M., Walker, B. and Noy-Meir, I. (1989) Opportunistic management for rangelands not at equilibrium. *Journal of Range Management* 42, 266–274.

Williams, D.G. (1979) The comparative ecology of two perennial chenopods. In *Studies of the Australian Arid Zone IV. Chenopod Shrublands*, ed. R.D. Graetz and K.M.W. Howes. CSIRO Australia: Melbourne, pp. 29–40.

Williams, O.B. (1968) Studies in the ecology of the riverine plain. III. Phenology of a *Danthonia caespitosa* Gaudich grassland. *Australian Journal of Agricultural Research* 12, 247–259.

Williams, O.B. (1968) Studies in the ecology of the riverine plain. IV. Basal area and density changes of *Danthonia caespitosa* Gaudich. In a natural pasture grazed by sheep. *Australian Journal of Botany* 16, 565–578.

Wilson, A.D. and Graetz, R.D. (1979) Management of the semi-arid and arid rangelands of Australia. In *Management of Semi-arid Ecosystems*, ed. B.H. Walker. Elsevier: Amsterdam, pp. 83–111.

Wood, J.G. (1936) Regeneration of the vegetation on the Koonamore Vegetation Reserve, 1926–1936. *Transactions of the Royal Society of South Australia* 60, 96–111.

Zimmer, W.J. (1946) *The Flora of the Far North-west of Victoria.* Victorian Forests Commission: Victoria.

25 Arid Shrublands and Open Woodlands of Inland Australia

CATHERINE NANO, PETER JOBSON AND GLENDA M. WARDLE

Arid shrublands comprise a wide array of mainly Acacia-dominated habitats, and along with Triodia hummock grasslands, they epitomise the physical and ecological character of Australia's deserts and rangelands. They include some of the most widespread vegetation communities in Australia as well as some of the most localised, and together they support a high diversity of distinctive arid biota. Underscoring the dynamics of arid shrubland ecology are the low, but highly-erratic delivery of rainfall in space and time; the redistribution of limiting soil resources; a diversity of biotic interactions, as well as significant top-down pressure from herbivore and fire consumption. Uniting the floristic components of this almost continent-wide vegetation grouping is a spectrum of well-known drought-avoidance and drought-tolerance traits; but increasingly, research is pointing to a high level of specialisation, particularly at early-life phases, that underpins coexistence and ecosystem stability. Mirroring global rangeland trends, Australia's arid shrublands have been extensively utilised for pastoral production, and the impacts of recurrent drought, overgrazing, pest herbivores, shrub proliferation and clearing, spread of pasture grasses and changed fire regimes, collectively manifest as significant land management challenges, some of which will be exacerbated by climate change. Increasing opportunities for land use diversification in inland Australia will potentially facilitate improved shrubland management and conservation in the upcoming decades. For this to occur though, adaptive management frameworks need to be strengthened through ongoing investment in rangeland monitoring systems that provide accurate reporting and feedback on land condition and biodiversity trends.

25.1 Introduction

Arid and semi-arid shrublands and woodlands (hereafter referred to as arid shrublands) comprise a wide array of mainly *Acacia*-dominated plant communities, and they collectively occupy a vast portion of inland Australia ('the arid zone'). Along with *Triodia* hummock grasslands (Wardle and Nano, Chapter 26, this volume), these shrublands epitomise the physical and ecological character of Australia's deserts and rangelands. Arid shrublands equate with a compilation of two major vegetation groups: '*Acacia* Shrublands' (MVG 16) and '*Acacia* Open Woodlands' (MVG 13) (Department of the Environment and Energy 2016a). The strictly arid waddy wood woodlands (*Acacia peuce* F.Muell.) are also included, as are the arid and semi-arid occurrences of gidgee woodlands (*Acacia cambagei* R.T.Baker, *Acacia georginae* F.M.Bailey or *Acacia calcicola* Forde & Isling). These plant communities are characterised by mainly single-species canopies, though occasionally other shrubs or emergent trees are associates. Other non-*Acacia* vegetation types form part of this broad grouping. Prominent examples are black oak and

Australian Vegetation, Third Edition, ed. David A. Keith. Published by Cambridge University Press. © Cambridge University Press 2017.

Citation: Nano, C., Jobson, P. and Wardle, G.M. (2017). Arid Shrublands and Open Woodlands of Inland Australia. In D. A. Keith, ed., *Australian Vegetation*, 3rd edn. Cambridge: Cambridge University Press, pp. 626–50.

belah (*Casuarina pauper* F.Muell. ex L.A.S. Johnson and *C. cristata* Miq.) and mixed-species woodlands e.g. ironwood (*Acacia estrophiolata* F.Muell.)–whitewood (*Atalaya hemiglauca* (F.Muell.) F.Muell. ex Benth.) and shrublands e.g. *Senna* and *Eremophila*. Included in this grouping are some of the most widespread vegetation communities in Australia as well as some of the most localised; together they support a high diversity of distinctive arid biota, and are habitat for multiple threatened flora and fauna species.

Research on Australia's arid shrublands began for economic reasons but has also contributed substantially to fundamental ecology. The advent of land system mapping in the late 1950s (e.g. see Mabbutt *et al.* 1963) provided the first formal circumscription of arid vegetation types and their environmental correlates. Other significant early contributions include numerous floristic surveys (e.g. Blake 1938); the bioclimatic analysis of *Acacia aneura* F.Muell. ex Benth. by Nix and Austin (1973), and the delineation of the botanical districts of Western Australia by Beard (1975). Fortunately, while much of the early work was undertaken to quantify pastoral land-use capability, it also laid the foundations for future ecological research and, ultimately, conservation management. Presently, by far the bulk of information relates to mulga (*Acacia aneura*) due to its ubiquity, long-exploited pastoral value and ecological importance, though ongoing work on a selection of other prominent canopy-forming species: *Acacia papyrocarpa* Benth.; *Acacia carneorum* Maiden (syn *A. carnei*); *Acacia oswaldii* F.Muell.; *Acacia peuce*; *Acacia georginae*; *Dodonaea viscosa* s.l. Jacq.; *A. peuce; A. georginae* and *Casuarina pauper* contributes significantly to our increasing knowledge of aridland ecology.

On the global stage, the early turning point in desert ecology came with the influential review 'Desert ecosystems: environment and producers' by Noy-Meir (1973) that outlined the now famous 'pulse-and-reserve' model (K. Bridges and M. Westoby unpublished data). This described a simple and direct relationship between rare large rainfall events and pulses of plant productivity, and it was specifically designed to account for the low, discontinuous and stochastic nature of rainfall in deserts. Importantly, this conceptualisation of episodic dynamics was used in equal measure to explain both system stability and dysfunction in deserts (Westoby *et al.* 1989). By now 'desertification' – defined as severe degradation and loss of productivity due to overgrazing – had emerged as a focal theme of arid/semi-arid rangeland ecology (Ward 2009). In particular, the complex issue of shrub proliferation ('bush encroachment') was identified as a principle threat to dryland pastoral productivity globally (Eldridge *et al.* 2013). Importantly, the need to redress rangeland degradation forced a refinement of the 'water-control' paradigm because it was by now apparent that both the physical environment and biotic interactions (consumers and plant competition) can strongly mediate the rainfall–productivity relationship, and that boundaries between shrublands and more open vegetation are potentially dynamic.

For Australia's arid/semi-arid rangelands, the combined effects of recurrent severe drought, overgrazing, pest herbivores, and (in some places) shrub proliferation and wildfire, generated concerns about the condition of the rangelands, and from the 1980s onwards, improved management became a key research focus (McKeon *et al.* 2003). Arid research centres and field sites were established across the continent (e.g. Charleville Pastoral Laboratory; Alice Springs CSIRO; The Mulga Research Centre at Kalgoorlie; Koonamore Station in South Australia, and Fowlers Gap and Lake Mere Stations in western New South Wales), resulting in the generation of an impressive amount of information on grazing impacts, drought tolerance, recruitment dynamics, grass–tree competition, herbage productivity and floristics across multiple shrubland types. Currently, land-use impacts and plant population recovery dynamics, fire effects, and the novel issues of exotic grass invasion and climate change are among the major research directions in arid shrubland ecology.

In this chapter we describe the distribution, biogeography and floristic and structural attributes of Australia's arid shrublands. We further provide an overview of their ecology and land use impacts, and a summary of ideas for their improved management and conservation.

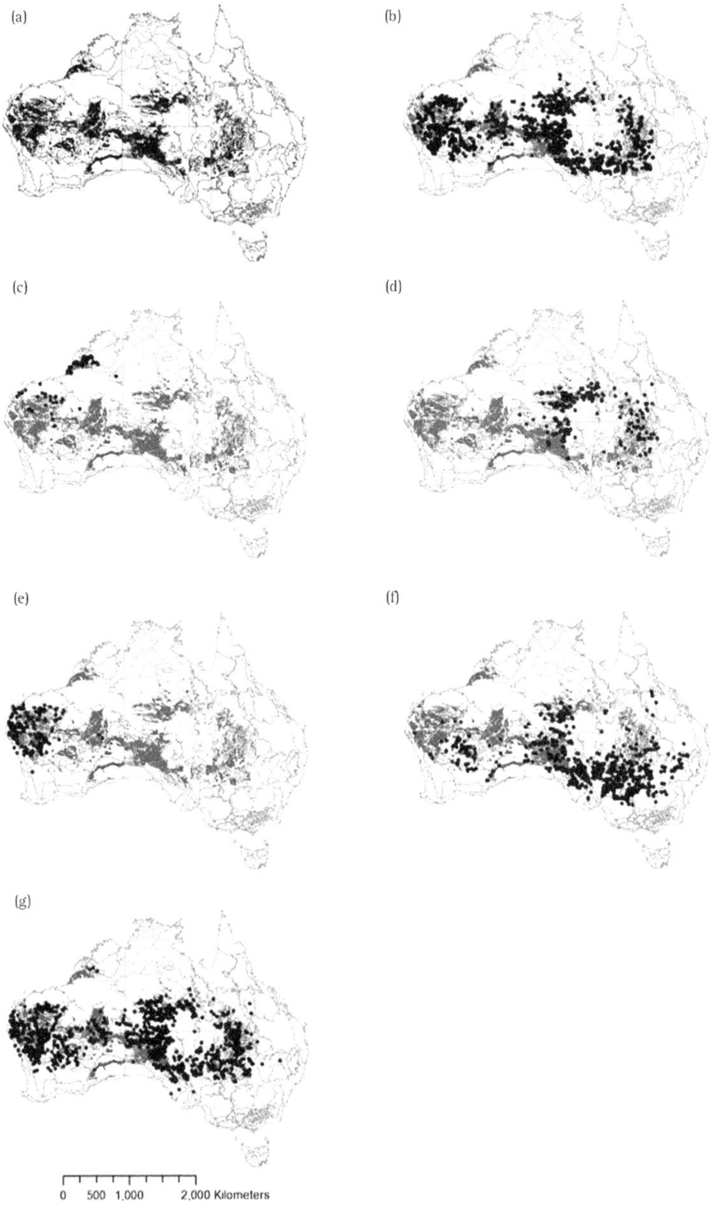

Figure 25.1 Distribution of *Acacia* shrublands/open woodlands in inland Australia. (a) Distribution of NVIS Major Vegetation Groups (MVGs) 16 (*Acacia* shrublands) and 13 (*Acacia* open woodlands) (grey shading). (b) Distribution of mulga shrubland dominants within MVGs 16 and 13 (black circles). (c) Distribution of pindan shrubland dominants within MVGs 16 and 13 (black circles). (d) Distribution of gidgee shrubland dominants within MVGs 16 and 13 (black circles). (e) Distribution of Pilbara shrubland dominants within MVGs 16 and 13 (black circles). (f) Distribution of trans-continental calcareous and solonised soils shrubland dominants within MVGs 16 and 13 (black circles). (g) Distribution of transcontinental sands and alluvial soils shrubland dominants within MVGs 16 and 13 (black circles). Bioregional (IBRA7) boundaries shown in grey outline. All Data sourced from *Atlas of Living Australia* (2015).

Table 25.1 **Summary of arid shrubland types, grouped according to dominant canopy species, soil association and distribution within inland Australia.**

Vegetation grouping	Dominant species	Soil	Distribution
Mulga	*Acacia aneura* and allies (see Table 25.2)	Sands, mixed sand and loams	Pilbara, Gascoyne, Murchison, Goldfields and arid WA; arid NT; arid SA; western NSW; western Qld
Pindan	*Acacia tumida* *A. eriopoda*	Sand	Southern Kimberly (Dampierland)
Gidgee	*Acacia calcicola* *A. georginae* *A. cambagei*	Calcareous soils and clays	Southern NT, northern SA Qld/NT border Eastern arid NT; north-eastern SA; western Qld; north-western NSW
Pilbara–Mid-West	*Acacia xiphophylla* *A. sclerosperma*	Sand	Pilbara and Murchison WA
Transcontinental calcareous and solonised soils	*Acacia peuce*	Calcareous clays	Simpson Desert
	A. pendula		Western (slopes and plains) NSW; southern and central western Qld
	A. oswaldii		Murchison eastwards in WA, southern NT, SA, north-west NSW, central and southern Qld
	Casuarina pauper/ cristata	Calcareous sands	Yalgoo Bioregion WA, SA, western NSW, south-west Qld
	A. papyrocarpa		Yalgoo and Goldfields WA to northern Flinders Ranges SA
	A. carneorum	Solonised (salt) sands	Far west NSW and eastern SA (Broken Hill Complex Bioregion)
	A. loderi		Far west NSW and eastern SA (Broken Hill Complex Bioregion)
Transcontinental sands and alluvial soils	*Acacia ramulosa*	Sands	Yalgoo Bioregion WA, southern NT, SA, north-west NSW
	A. brachystachya		Murchison eastwards in WA, southern NT, SA, north-west NSW, south-western Qld
	A. resinimarginea		WA endemic: Goldfields
	Allocasuarina decaisneana		Arid WA; arid NT and northern arid SA
	A. ligulata *A. sibirica*	Mixed sand, loams and salty sands	Pilbara, Murchison, Goldfields WA, arid NT, arid SA, western NSW, south-west Qld
	A. stenophylla	Alluvial	Arid NT; Murray Riverland and arid SA; western NSW, western Qld, northern Vic
	A. kempeana	Rocky skeletal (central Australia)	Pilbara, Murchison, Goldfields WA, arid NT, arid SA, western NSW, south-west Qld

25.2 Distribution and biogeography

Arid shrublands predominate in inland Australia in areas receiving between 250 mm and 350 mm annual rainfall. Combined, they have a west–east transcontinental distribution, stretching discontinuously from the Yalgoo bioregion in Western Australia; across the Great Victoria Desert and Nullarbor Plain; the southern Northern Territory; the Gawler bioregion and the northern Flinders Lofty block (South Australia), into the Broken Hill Complex of New South Wales; and then ending in the Mulga Lands bioregion of south-central Queensland and western New South Wales (Figure 25.1a; Table 25.1). In Western Australia, arid shrublands continue northwards to include the Murchison and Pilbara bioregions. Arid shrublands in these areas occur in a variety of situations, but they are best developed on extensive plains and dunefields, on low ranges and hills, and in valleys.

The diversity of arid shrublands can be summarised into seven broad groups (Figure 25.1; Table 25.1). The first three groups below are based on taxonomic phylogenies, the following three on broad regional distribution and soil type, and the last one comprises mixed-canopy shrublands. Three of the groups are geographically restricted, while mulga and three others have a transcontinental distribution. Affinities with six broad substrate types are also apparent: sand; calcareous; sand dunes; mixed sand and loams and salty sands; rocky skeletal; and alluvial.

25.2.1 Mulga

Mulga – characterised by shrubs and trees in the *A. aneura* species complex (*sensu* Maslin and Reid 2012) – is the most widespread of the arid shrubland types (Figures 25.1b and 25.2a). Mulga occupies close to 20% of the Australian continent (over 150 million hectares); occurring in a discontinuous belt between latitudes 20°S and 31°S, and longitudinally from near-coastal Western Australia through to south-central Queensland. Rainfall is 200–500 mm/year and is not strongly seasonal (Nix and Austin 1973). Across its continental extent, mulga is known from an extraordinary array of landforms including rocky hills/mountains, red earth and alluvial plains, sandplains and dune fields. Mulga soils are nutrient-poor, with especially-low phosphorous and nitrogen (Keith and Tozer, Chapter 1, this volume). They are light textured (with a hard coherent subsoil on plains), and are neutral to slightly acidic (Maslin and Reid 2012). On gently sloping red earth plains, mulga forms visually-striking parallel bands ('groves and inter-groves') but elsewhere it occurs as continuous stands. Less uniform, but equally striking, are the mosaics of mulga-*Triodia* hummock grassland habitats throughout the central and western deserts (Wardle and Nano, Chapter 26, this volume).

The morphological variation within the *A. aneura* species complex is notorious. Of the 12 species described by Maslin and Reid (2012), five are endemic/mainly restricted to Western Australia; two by contrast, occur in all mainland states except Victoria, and the remaining five occur in Western Australia, the Northern Territory and South Australia (with two of these having outliers in Queensland) (Table 25.2). Outside Western Australia, the taxonomy is unresolved and entities currently recognised outside that state are based on the taxonomy set down by Pedley (2001), which contradicts Maslin and Reid (*q.v.*), and which is now regarded as unsatisfactory.

25.2.2 Pindan

Pindan shrubland (*Acacia tumida* F.Muell. ex Benth.–*A. eriopoda* Maiden and Blakely) of the south-western Kimberley is restricted to a single bioregion (Dampierland; Figure 25.1c) and geographically isolated from the more southern arid shrublands of Western Australia. Pindan occurs in dense thickets on red earth sandplains and it is characterised by emergent trees (e.g. *Corymbia greeniana* (D.J.Carr and S.G.M.Carr) K.D.Hill and L.A.S.Johnson and *Eucalyptus miniata* A.Cunn. ex Schauer) and by a *Triodia* spp. understorey that fuels regular fire. These shrublands blur the line between shrubland and hummock grassland. *Acacia ancistrocarpa* Maiden and Blakely shrublands to the east of pindan are functionally and structurally similar.

Figure 25.2 Examples of *Acacia* shrublands/open woodlands in inland Australia: (a) mulga (*Acacia aneura*) shrubland; (b) gidgee (*Acacia georginae*) shrubland; (c) waddy wood (*Acacia peuce*) open woodland; (d) western myall (*Acacia papyrocarpa*) shrubland; (e) bowadga (*Acacia ramulosa*) shrubland; (f) *Acacia undoolyana* shrubland.
All photos: Catherine Nano, except (d) Jen Silcock. (A black and white version of this figure will appear in some formats. For the colour version, please refer to the plate section.)

Table 25.2 **Current taxonomic status (2016) of *Acacia aneura* and its sister taxa as accepted by the Australian Plant Census.**

Taxon Name	Taxonomy System	Distribution
Acacia aneura (syn. *Acacia aneura var. aneura*, *Acacia aneura* var. *intermedia* [WA only])	Maslin and Reid (2012)	Pilbara, Gascoyne, Murchison, Goldfields and arid WA; arid NT; arid SA; western NSW; western Qld
Acacia aneura var. *intermedia*	Pedley (2001)	Arid NT; arid SA; western NSW; western Qld
Acacia aneura var. *major*	Pedley (2001)	Arid NT; arid SA; western NSW; western Qld
Acacia aptaneura (syn. *Acacia aneura* var. *pilbarana*, *Acacia aneura* var. *tenuis*)	Maslin and Reid (2012)	Pilbara, Murchison, Goldfields and arid WA; arid NT; arid SA; far western NSW; western Qld
Acacia ayersiana (syn. *Acacia ayersiana* var. *latifolia*)	Maslin and Reid (2012)	Pilbara, Murchison, Goldfields and arid WA; arid NT; arid SA; far western NSW; western Qld
Acacia caesaneura (syn. *Acacia aneura* var. *argentea*)	Maslin and Reid (2012)	WA endemic: Murchison
Acacia craspedocarpa	Maslin and Reid (2012)	WA endemic: Murchison
Acacia fuscaneura (syn. *Acacia aneura* var. *fulginea*)	Maslin and Reid (2012)	WA endemic: Pilbara, Gascoyne and Murchison
Acacia incurvaneura (syn. *Acacia aneura* var. *microcarpa*)	Maslin and Reid (2012)	Pilbara, Murchison, Goldfields and arid WA; southern arid NT; northern arid SA
Acacia macraneura (syn. *Acacia aneura* var. *macrocarpa*)	Maslin and Reid (2012)	Southern Pilbara and Gascoyne WA, disjunct - (Mt Agnes) extreme north-west SA
Acacia minyura	Maslin and Reid (2012)	Pilbara, Gascoyne, Murchison and arid WA; arid NT; arid SA; disjunct – central western Qld (Winton–Longreach–Aramac)
Acacia mulganeura (syn. *Acacia minyura* variant – Pedley 2001)	Maslin and Reid (2012)	Pilbara, Gascoyne, Murchison and arid WA; southern arid NT, northern arid SA
Acacia paraneura	Maslin and Reid (2012)	Southern Pilbara, Gascoyne, Murchison and arid WA; arid NT; northern arid SA; far western Qld
Acacia pteraneura (syn. *Acacia aneura* var. *conifera*)	Maslin and Reid (2012)	Pilbara, Murchison, Goldfields and arid WA; arid NT; northern arid SA

25.2.3 Gidgee

Gidgee communities (*A. cambagei*, *A. georginae* and *A. calcicola*) share an evolutionary lineage (Cowan 2001), and occur on heavy-texture calcareous plains and alluvial soils (Figure 25.2b). In dune fields, gidgee is patchily distributed in clay swales between dunes (Wardle *et al.* 2015). Combined, gidgee communities form a natural bio-zone of western Queensland, arid/semi-arid southern Northern Territory and northern South Australia (Figure 25.1d; Beadle 1981).

25.2.4 Pilbara and mid-west (Gascoyne/Murchison)

These shrublands form a natural group (Figure 25.1e). The two bioregions are characterised by ranges of deep iron ore deposits and deposition of alluvial soils through erosion and weathering. Within this group, *Acacia xiphophylla* E.Pritz. (snakewood) and *Acacia sclerosperma* F.Muell. shrublands are common on sandy loam soils. Notably, the 12 Western Australian species of the *A. aneura* complex (Maslin and Reid 2012) are found in this region, with two being Murchison endemics. The mid-west is also characterised by high speciation and endemism in *Eremophila*, and members of this genus are often are co-dominants of low mixed shrublands.

25.2.5 Transcontinental arid shrublands on calcareous and solonised soils

A complex of shrublands occurs across the continent (Figure 25.1f). *Acacia peuce* (Figure 25.2c) occurs on calcareous gibber plain as three disjunct woodland populations on the edge of the Simpson Desert. *Acacia papyrocarpa* (Figure 25.2d) and *A. carneorum* both occur on calcareous sand dunes and ridges, whereas *Acacia loderi* Maiden is restricted to salt derived or solonised soils on flats. Interestingly, *A. papyrocarpa* and *A. loderi* are sister species and without fruit can be difficult to differentiate (Whibley and Symon 1992). As currently circumscribed, the two species are geographically separated by the lower Flinders Ranges and by their soil affinities. *Acacia oswaldii* is widespread and scattered, occurring mainly in calcareous sands or loam. *Acacia pendula* A.Cunn. & G.Don and *Casuarina pauper/cristata* form shrublands (with *Alectryon oleifolius* (Desf.) S.T.Reynolds) along floodplains of rivers and major streams; they are always in patchy, but often discrete communities. *Acacia pendula* and *C. cristata* occur along the western slopes and plains of the Great Dividing Range of central western Queensland and New South Wales. *Casuarina pauper* has a more southern distribution spanning from Western Australia (Kalgoorlie region) eastwards until it is replaced by *C. cristata* in northern New South Wales. *Casuarina pauper* and *C. cristata* are sister species, and were previously recognised as two subspecies of a broader taxon (Wilson and Johnson 1990).

25.2.6 Transcontinental arid shrublands on sands or alluvial soils

Characteristic shrublands on sandy soils include: bowadga (*Acacia ramulosa* W.Fitzg.) (Figure 25.2e), turpentine mulga (*Acacia brachystachya* Benth. syn. *Acacia cibaria*), bastard mulga (*Acacia sibricia* S.Moore syn. *A. stowardii*) and sandhill wattle (*Acacia ligulata* A.Cunn. ex Benth.). The latter three are very widespread and often morphologically variable, and their distribution extends into the north-western arid tropics where they can form extensive stands. Throughout the ranges of *A. aneura*, *A. ramulosa* and *A. brachystachya*, it is not uncommon to find all three species co-occurring (Pedley 2001). *Acacia ligulata* regularly occurs on sands associated with salty lakes or waterways. *Acacia resinimarginea* W.Fitzg. also forms shrublands on sandy soils within the Goldfields region.

Acacia stenophylla A.Cunn. ex Benth. woodland occurs on alluvial soils, being sporadic along ephemeral arid waterways to becoming more extensive along permanent rivers and major streams. Its distribution extends from the Gulf of Carpentaria coast to the Murray River mouth in South Australia along arid eastern Australia. *Acacia kempeana* F.Muell. (witchetty bush) forms pure stands in central Australia, predominantly on rocky shallow soils over sandstone, granite or dolomite, but also on alluvial soils.

25.2.7 Mixed-canopy shrublands

Some shrublands have two or more canopy dominants (Table 25.3). There are occurrences of pure stands of some species (e.g. *Atalaya hemiglauca* or *Acacia excelsa* Benth.), but these are comparatively rare. These shrublands occur on a variety of soils including sands, loamy sands and calcareous soils. Prominent species that form part of mixed canopies/subcanopies are *Senna artemisioides* s.l., *Eremophila scoparia* (R.Br.) F.Muell., *Santalum lanceolatum* R.Br. and *Dodonaea viscosa*. There is a distinct lack of regional endemism in this group. The only pairings of related taxa are *Acacia excelsa* and *A. estrophiolata* and members of *Eremophila* sect. *Eriocalyx* (*Eremophila latrobei–E. forrestii* F.Muell. species alliance and *E. margarethae* S.Moore and allied species) (Chinnock 2007).

Table 25.3 **Commonly encountered mixed assemblage shrublands of arid Australia. Height classes: Tall > 3 m; Medium 1-3 m; Low < 1 m.**

Mixed-species association	Height class	Distribution
Acacia estrophiolata–Atalaya hemiglauca–Grevillea striata	Tall	Arid NT, northern SA
Acacia excelsa–Callitris glaucophylla	Tall	Central southern Qld; western NSW
Acacia pruinocarpa–Acacia aneura	Tall	Pilbara and arid WA; arid NT
Hakea divaricata/eyreana–Atalaya hemiglauca–Acacia estrophiolata	Tall	Arid NT and arid northern and north-eastern SA; 'Channel Country' Qld
Alectryon oleifolius–Flindersia maculosa–Acacia aneura	Tall	Western NSW; south-western Qld
Acacia acuminata–A. quadrimarginea–Allocasuarina huegeliana	Tall and Medium	Southern Murchison and Goldfields, WA
Acacia aneura–Hakea leucoptera–Casuarina pauper	Tall and Medium	Far western NSW; western Qld
Acacia tetragonophylla–Acacia victoriae	Medium	Arid WA; arid NT; northern arid SA; far western NSW; arid western Qld
Eremophila longifolia–Acacia aneura–Dodonaea viscosa	Medium	Arid WA; arid NT; northern arid SA; far western NSW; arid western Qld
Senna artemisioides–Dodonaea viscosa	Medium and Low	Arid NT; arid SA; Western NSW; central southern Qld
Eremophila latrobei–Senna artemisioides	Medium and low	Arid WA; arid NT
Eremophila compacta–Acacia aneura	Medium and low	Murchison, WA
Eremophila citrina–E. latrobei–Acacia aneura	Medium and low	Murchison, WA
Eremophila margarethae–E. latrobei–Acacia aneura/sclerosperma	Medium and low	Murchison and Goldfields, WA
Acacia stellaticeps–Maireana spp.	Low	Pilbara, extending to Kimberley
Eremophila freelingii–Senna artemisioides	Low	Arid NT; arid SA

25.3 Structure and floristics

Arid shrublands characteristically have three structural layers: a canopy of tall 3–6 (–12) m shrubs/low trees, a 1–2 m shrub layer and a ground layer of grasses and forbs. Emergent trees, low trees, mistletoes, vines and subshrubs variously contribute to structural complexity among vegetation types and sites. Canopy dominants are usually long lived, and individuals can be widely spaced, with a gnarled and spreading habit (e.g. species in the gidgee group, Wardle *et al.* 2015). However, for mulga specifically, tall (12 m) stands with >70% canopy cover occur in the high-rainfall portion of the Mulga Lands bioregion (Fensham *et al.* 2012a). The understorey of arid shrublands is typified by shrubs (*Eremophila, Senna, Maireana*), tussock grasses (*Eragrostis, Aristida*) and multiple herbs within the families Asteraceae and Amaranthaceae. It is usually discontinuous, with a concentration of species directly under tree canopies.

Soil differences are strongly correlated with floristic variability in the understorey. For example in the case of *A. loderi* shrubland on solonised sand, the understorey is dominated by chenopods (*Maireana* and *Atriplex*) whereas for the closely allied *A. papyrocarpa*, on calcareous sands, the understorey is a mixture of *Eremophila, Senna* and other low-stature acacias (e.g. *A. ramulosa*). Similarly, the understorey of mulga varies substantially with changes in soil texture, pH and stoniness (Keith 2004; Fensham *et al.* 2012a). The association of *Aristida holathera* Domin with sandy loams and of *Aristida contorta* F.Muell. with red earths illustrates the soil–understorey relationship well.

Within the understorey flora, there is high arid zone endemism and speciation, particularly in the families Asteraceae, Scrophulariaceae, Lamiaceae and Amaranthaceae. Within the Asteraceae tribe Gnaphalieae (Orchard 2015), approximately 10 monotypic genera are confined to arid shrublands, and a further 10 genera are speciose within them. In the Lamiaceae, the subfamily Prostantheroideae (*Newcastelia* and associated genera) has strong representation in these habitats, while two large genera (*Eremophila* and *Ptilotus*) have their highest level of speciation in arid shrublands. Floristically interesting are the species within genera normally associated with rainforests: *Ventilago* (Rhamnaceae); *Flindersia* and *Geijera* (Rutaceae); and *Atalaya* and *Alectryon* (Sapindaceae). These species often look incongruous in the landscape and many additionally occur, or they have sister taxa in mesic climates.

25.4 Ecological interactions

Mirroring global research trends, Australian desert ecology has in recent years undergone a shift away from the broad-sweeping concept of the arid zone as a single entity, uniformly governed by chronically low and unpredictable resources (Morton *et al.* 2011). While most current ecological frameworks (Holmgren *et al.* 2006; Morton *et al.* 2011) do broadly align with the 'episodic world view' (*sensu* Watson *et al.* 1997) of the pulse-and-reserve model, additional emphasis is collectively placed on a wide range of edaphic, biotic and disturbance factors and their potential interactions.

25.4.1 Rainfall and drought

One of the major refinements of the pulse-and-reserve model is the greater emphasis on regional-scale variability in rainfall stochasticity and seasonality (Schwinning and Sala 2004). Where shrublands predominate in inland Australia, rainfall is skewed towards large rare events, and stochasticity is pronounced, especially in the north (Morton *et al.* 2011). Given these climate parameters, theory predicts that primary productivity should be highly episodic, but that species coexistence will be maintained via subtle differences in responsiveness to timing and amount of soil moisture, especially among annual/ephemeral species with a high allocation to seed production ('drought evaders') (Morton *et al.* 2011). Consistent with this, the link between community-scale productivity responses and high-rainfall bouts is now well established throughout the arid zone (Clarke *et al.* 2005; Fensham *et al.* 2012a; Nano and Pavey 2013; Wardle *et al.* 2015). Furthermore, it has been demonstrated that herbs show differential winter (in forbs) versus summer (in grasses) germination tendencies, and there is a 'hierarchy' of responses across individual pulses involving a gradient from rapid-responding shallow-rooted, short-lived species (*Dactyloctenium radulans* (R.Br.) P.Beauv.), to a lagged response in 'high-demand' perennials (Clarke *et al.* 2005; Nano and Pavey 2013). Projected increases in summer-rainfall dominance under future climates (CSIRO and Bureau of Meteorology 2015) may therefore act to especially disadvantage the winter-growing component of the herbaceous flora.

The classic life-history profile of arid woody species is characterised by drought-tolerant, long-lived adults (>200 years) and rare, highly episodic recruitment, and certainly this profile is reported from arid Australia (e.g. *Acacia papyrocarpa*, *A. peuce*, *A. georginae*) (Lange and Purdie 1976; Raghu *et al.* 2013, Wardle *et al.* 2015). For most species, however, uncertainty prevails as to the necessary rates of recruitment for adult replacement and the relative contribution to population dynamics of continuous (slow gradual increases) versus episodic recruitment (Watson *et al.* 1997; Denham and Auld 2004). Available evidence points to a range of early life-history traits that potentially determine seedling recruitment rates. Among species, there is high variability in seedcrop size (e.g. large in *A. ligulata*, small in *A. carneorum*); dormancy levels (e.g. high in *A. papyrocarpa*, negligible in *C. pauper*); seedbank longevity (e.g. long lived in *A. ligulata*, short lived in *A. oswaldii*) and seasonal propensities and speed of germination (e.g. slower rates under cool temperatures in *Senna artemisioides*) (Auld 1995a; Gilpin *et al.* 2014; Pound *et al.* 2015). Ultimately though, desiccation represents the most significant hurdle to juvenile establishment, because rainfall is often sufficient for germination but insufficient for longer-term seedling survival. For species capable of vegetative reproduction, sucker production may be rare in the absence of root disturbance (e.g. *A. oleifolius*) or it may be on an annual basis (e.g. *A. carneorum*), making rainfall a poor predictor of population dynamics (Denham and Auld 2004).

Once established, the risk of drought-death in shrubs and trees evidently declines with age and growth, although soil texture and plant traits, e.g. resprouting ability and root architecture, reduce predictability (Watson *et al.* 1997; Raghu *et al.* 2013; Norman *et al.* 2014). Monitoring data mainly imply incremental mortality in adult populations, e.g. *Eremophila forrestii* F.Muell., *E. maitlandii* Benth., *A. peuce* (Watson *et al.* 1997; Sinclair 2005; Raghu *et al.* 2013), although for *A. aneura*, it is proposed that episodic drought mortality may stabilise tree density over the long-term (Fensham *et al.* 2012a). Climate change projections for the Australian arid zone include more frequent and severe drought, leading to expectations for higher mortality rates in slow-growing species such as *A. aneura* and *A. georginae* (Fensham *et al.* 2012a; Wardle *et al.* 2015).

25.4.2 Modifying effects of the physical environment

The physical environment is a major modifier of plant rainfall responses and community dynamics in arid/semi-arid Australia. Geology, landform, soil texture and chemistry, and relief can all strongly affect species distributions and productivity via their regulating effect on available moisture and nutrients (Beadle 1981; Morton *et al.* 2011). Most particularly, the semi-arid phenomenon of banded vegetation (e.g. mulga 'grove-intergrove' alterations) is one of the clearest illustrations of how rainfall attributes (amount and intensity) and the environment interact to determine dryland vegetation patterning (Tongway and Ludwig 2001). This phenomenon is restricted to medium-texture, gently sloping plains; and arises from the downslope transfer of resources from patches of bare impenetrable soil to vegetated sinks via sheet-flow during intense rain events. The redistribution of limited resources significantly increases ecosystem rainfall-use efficiency and it underpins wooded patch development under otherwise suboptimal climate conditions (Moreno-de las Heras *et al.* 2012). Modelling shows that either a decrease or increase in rainfall intensity (as predicted under climate change) would result in lower rates of soil moisture infiltration in vegetated bands and a loss of plant cover from the landscape (Siteur *et al.* 2014). This is consistent with the findings of Eamus *et al.* (2013) that *A. aneura* largely relies on soil moisture in the upper soil profile rather than on deeper stores of water.

25.4.3 Competition and facilitation

Positive and negative interactions between plants are now understood as critical elements of arid vegetation dynamics. The suppression of recruiting shrub seedlings during wet years by adult grasses is considered a vital underpinning of the grass-shrub balance in arid/semi-arid rangelands in Australia and globally (Harrington 1991; Ward 2009). Relatedly, negative shrub effects on herbaceous biomass are documented for Australian arid shrublands, though site and grazing effects can outweigh this interaction (Page *et al.* 2000; Fensham *et al.* 2012b; Eldridge *et al.* 2013). By contrast, positive interactions between shrubs and their understorey species may result from multiple shrub canopy effects including: (i) higher subcanopy soil fertility (through resource capture and nitrogen fixation); (ii) increased shading and improved microclimate; (iii) increased soil moisture through canopy interception of rainfall and stem-flow; and (iv) positive litter effects including reduced rainfall runoff, and enhanced water penetration and organic matter content (Eldridge *et al.*, Chapter 6, this volume). Similar effects have been noted in eucalypt woodlands (Prober *et al.*, Chapter 17, this volume). This 'fertile island' effect, combined with the provision of perch sites for seed-dispersing birds and branch habitat for mistletoes, facilitates recruitment and survival of understorey plants, including those possibly less well-adapted to stressful conditions (Ward 2009; Eldridge and Solivares 2014). For banded vegetation, theory is extended to include the concept of positive self-reinforcing feedbacks: an increase in plant biomass results in higher moisture infiltration and nutrient capture, leading to enhanced plant growth and grove maintenance. In this situation competition effects are counterbalanced by enhanced resource access (Siteur *et al.* 2014).

25.4.4 Plant–animal interactions

Traditionally, plant–animal interactions have been viewed as unimportant in the structuring of arid vegetation because animals themselves are constrained by intermittent resources and, particularly in Australia, by low plant palatability and nutrient status (Ward 2009; Morton *et al.* 2011). However, for arid Australia (and deserts more generally), available evidence now points to a diversity of important plant–animal relationships.

Detritivores (termites *Isoptera*) have a major impact on Australian deserts through their role as nutrient recyclers, and they are predicted to be highly abundant in shrublands due to the availability of woody debris (Morton *et al.* 2011; Wardle *et al.* 2015). Birds and rodents are major consumers of seed (especially when they are irrupting following high rainfall, Wardle *et al.* 2015), although their direct impacts on community dynamics are untested. Similarly, rapid seed removal by ants has been noted for a range of arid shrub species – *Dodonaea viscosa, Acacia papyrocarpa, A. aneura* – (Harrington and Driver 1995; Ireland and Andrew 1995; Wright and Zurr 2014), but actual consumption rates are not well quantified. Notably, Wright and Zurr (2014) speculate that seedling recruitment in *A. aneura* is strictly limited to the aftermath of pulse events when seed crops are large enough to satiate ant predators. Macropods and insects contribute to total grazing pressure – grazing by the former is associated with high plant species richness in mulga, while leaf-stripping by grasshoppers may increase mortality in drought-stressed trees (Fensham *et al.* 2011; 2012a).

Apart from their role as consumers, ants, along with birds and potentially some large reptiles, are likely to be important seed dispersal agents of arid shrubland species (Davidson and Morton 1984). Within *Acacia*, the different aril types are suggestive of specialised seed dispersal syndromes. Certain species (*A. papyrocarpa, A. aneura, A. kempeana*) have a small white aril, indicative of ant dispersal while others have coloured arils and/or fruit displays that likely promote bird dispersal (including *A. oswaldii, A. ligulata, A. carneorum*) (O'Dowd and Gill 1986). Notably though, certain species (e.g. *A. ligulata*) are attractive to birds and ants, with benefits potentially arising from a portion of the seed crop having its dormancy broken via bird ingestion (Letnic *et al.* 2000). For arid plants, animal-mediated dispersal away from the immediate parent neighbourhood potentially offers many benefits, including an ability to track suitable climates (Pascov *et al.* 2015; Franklin *et al.* 2016); and interestingly, ant dispersal distances of some Australian arid shrub species (*A. ligulata* and *Acacia karina* Maslin & Buscumb) far exceed those recorded from other desert regions (Whitney 2002; Pascov *et al.* 2015). Short-distance (<10m) ant dispersal is, however, also well documented, and it is proposed that the resulting clumped distribution offers resource and/or fire-protection advantages to shrubs (Harrington and Driver 1995). Similarly, seed dispersal by birds occurs over long and short distances. For example, the emu (*Dromaius novaehollandiae*) potentially transports seed of a wide range of species (e.g. *C. pauper, Enchylaena tomentosa* R.Br. *Eremophila* spp., *Santalum acuminatum* (R.Br.) A.DC.) over many kilometres (Dunstan *et al.* 2013); while the specialist mistletoebird (*Dicaeum hirundinaceum*) apparently results in highly localised dispersal (e.g. between shrub canopies) of mistletoe seed (*Lysiana* and *Amyema*) (Start 2015).

Knowledge of plant–pollinator interactions in Australian deserts is presently fragmentary. Lacking are the well-known obligate mutualisms and specialised vertebrate syndromes of other deserts involving conspicuous high-reward flowers and moth or bird/bat visitors (see Franklin *et al.* 2016). Instead, wind- and insect-pollination (predominantly solitary bees) syndromes are more likely to predominate in arid Australia (Keighery 1982), although recent research points to pollination facilitation by small mammal floral visitors during resource pulses (Popic *et al.* 2016). For insect pollinators, generalist relationships dominate over specialist relationships, and a relatively low number of floral visitors are effective pollen transporters/pollinators (Popic *et al.* 2013; Gilpin *et al.* 2014). Interestingly, Gilpin *et al.* (2014) linked the low reproductive success of *A. carneorum* to the poor pollinator services provided by its native flower visitors. By contrast, the consistently large seed crops of *A. ligulata* were explained in part by the pollination efficiency of the introduced *Apis mellifera.* Significant knowledge gaps persist particularly in relation to land use, climate change and *A. mellifera* impacts on the native pollinator assemblages of arid shrublands.

The creation of soil pits as nutrient sinks and 'safe sites' for seedling establishment is another advantage that desert plants derive from animals (mammals and reptiles). Native fauna have been shown to be more effective than exotic rabbits in this role, indicating that important plant services have been compromised by the high rates of fossorial mammal extinction from the Australian deserts (James *et al.* 2009).

25.4.5 Fire

Broadly, arid shrublands are 'fire sensitive', reflecting the inability of most canopy species to withstand short fire intervals due to their long maturation times (e.g. *A. aneura* and allies), low-to-variable resprouting ability (e.g. *A. kempeana*) and their thin, non-protective bark (e.g. *C. pauper*, *A. stenophylla*, *A. estrophiolata*) (Russell-Smith *et al.* 2012; Schubert *et al.* 2016). At the continent scale, these habitats typically burn infrequently (20–100 (>100) years) and at low intensity (0–100 kW m^{-1}) as a consequence of low annual net productivity (0.4 t C ha^{-1} yr^{-1}) (Murphy *et al.* 2013; Clarke *et al.* 2015). At finer scales though, the risk and importance of fire varies greatly among and within the different plant communities, depending on their spatial context, with the relatively fire-prone pindan shrublands of Western Australia and the non-flammable mulga woodlands of central-south Queensland characterising the extremes of the gradient (Fensham *et al.* 2012a). In the flammable landscapes of central Australia and Western Australia (where *Triodia* dominates) shrublands are usually closely associated with local sites that experience low fire activity ('fire refuges') characterised by high terrain roughness and soil that is edaphically unsuitable for *Triodia* dominance (Nano and Clarke 2010; Russell-Smith *et al.* 2012). For mulga in particular, modern fire regimes (characterised by unplanned, rare, large-scale events) are implicated in patch contraction, but support for this remains equivocal (Bowman *et al.* 2008; Ward *et al.* 2014). Current mulga-*Triodia* grassland boundary frameworks vary in their emphasis on endogenous process i.e. the dual negative feedbacks of woody cover on fire activity, and fire activity on woody cover (Murphy *et al.* 2013) versus edaphic control (Nano and Clarke 2010); and it is probable that boundaries vary in nature from 'meta-stable' (potentially dynamic) to stationary, depending on their spatial context. Notably, for central Australia, fire activity is projected to decline under climate change due to diminished herbage growth with lower rainfall – a situation that could potentially favour wooded habitats over grassland (King *et al.* 2013).

For some arid shrubs, occasional fire may be important for population persistence e.g. in the rare *Acacia unguicula* R.S.Cowan & Maslin (Department of Environment and Conservation 2009). More generally though, interval-recruiting species dominate over strict fire-recruiting species in arid shrublands (Clarke *et al.* 2015); and while for some species, recruitment can be greatly enhanced by fire (e.g. *D. viscosa*, *A. aneura;* Hodgkinson and Oxley 1990), there can be significant recruitment between fires in response to rainfall pulses. There is little to suggest, therefore, that arid shrublands are fire-dependent (Dalgleish *et al.* 2015). Moreover, for mulga, functional and structural diversity is highest in the oldest stands, likely due to recolonisation (mistletoes especially) and positive canopy effects (Nano and Clarke 2008; Leavesley *et al.* 2010; Start 2015). Bird assemblages show similar relationships, with core (specialist) mulga species characterising the oldest stands, and generalists the regenerating stands (Leavesley *et al.* 2010). Fire dynamics are poorly known in arid shrublands other than mulga, making it difficult to predict the impacts of novel disturbances such as exotic grass invasion or climate change.

25.5 Human impacts

Under the pulse-and-reserve model, desert systems become destabilised if they are subjected to over-exploitation (Noy-Meir 1973). Since the inception of this model, *desertification* has become a central theme of arid/semi-arid rangelands. Globally, livestock overgrazing is the primary cause of desertification, and climate change is expected to worsen its effects (Ward 2009). In Australia, desertification is referred to as *land degradation* (Pickup 1998), and it is broadly ascribed to the failure of managers to adjust livestock (sheep and cattle) densities to match seasonal conditions (McKeon *et al.* 2003). Its history can be summarised as a series of major episodes dating from the 1890s (in western New South

Wales) through to the 1980s (Gascoyne, western New South Wales, Gawler region, central Australia and Queensland) (McKeon *et al.* 2003); though the process is viewed as ongoing in certain contexts (Pringle *et al.* 2006; Bastin *et al.* 2008). For shrublands specifically, other significant current management issues include, shrub encroachment, fodder harvesting and canopy clearing for pasture, and spread of exotic pasture-grass – all of which must now be approached in the context of the changing climate.

25.5.1 Overgrazing

Grazing impacts are predicted to be worse in areas closest to water tanks and bores where animals tend to congregate ('the piosphere' concept; Lange 1969) and in paddocks intensively grazed for long periods; but site characteristics and past rainfall can modify or mitigate grazing-gradient effects, as has been shown in mulga (Fensham *et al.* 2012a). Biotic and abiotic attributes can also vary in their response to grazing gradients. For example, in *C. pauper* woodland, vascular plant diversity and richness is highest under intermediate grazing intensity, while negative soil impacts are most pronounced under heavy grazing (Eldridge *et al.* 2011). The issue is further complicated by the uncontrolled grazing patterns of native macropods and feral herbivores (rabbits, goats, horses, camels) (McKeon *et al.* 2003; Fensham *et al.* 2011).

The suppression of palatable perennial grasses – notably *Monachather paradoxus* Steud., *Thyridolepis mitchelliana* (Nees) S.T.Blake, and *Themeda triandra* Forssk. – is an often-cited symptom of irreversible rangeland degradation. These ('decreaser') species die under heavy grazing mainly because they under-allocate to their root systems and lose their ability to survive drought (McKeon *et al.* 2003). Notably, a recent examination of plant grazing responses in mulga in southern Queensland demonstrated the ability of palatable grasses to recover with grazing protection, and to persist at the landscape scale under grazing – a result that draws into question the prevailing view of irreversible change in heavily utilised rangelands (Fensham *et al.* 2011). Other herbaceous lifeforms (annual grasses and annual/perennial herbs) appear to have far less predictable responses to livestock grazing, although species-level responses can be pronounced.

Grazing by rabbits and livestock has long been considered a primary limitation on shrub and tree recruitment in arid/semi-arid Australia (Lange and Purdie 1976), although grazing and rainfall effects are notoriously hard to disentangle. Available evidence indicates that for palatable species that reproduce vegetatively, e.g. *A. carneorum*, *S. acuminatum* and *C. pauper*, grazing (particularly by rabbits, and even at low density following biological control) can significantly reduce the likelihood of new suckers surviving to the juvenile stage (Denham and Auld 2004). For palatable species reliant on seedlings, e.g. *A. papyrocarpa* and *A. oswaldii*, the combination of reduced grazing and high rainfall may be necessary for population recovery due to the overriding desiccation risk for seedlings (Auld 1995b; Denham and Auld 2004; Sinclair 2005; Watson *et al.* 2007; Davies *et al.* 2015). Potentially, grazing limits recruitment to high-rainfall phases (when there is predator satiation) or to areas under protective canopies (nurse plants); but much work is required to more fully understand how it interacts with rainfall, site effects, and species palatability and regeneration traits to determine recruitment rates. At high density, large browsers can limit sapling height-growth and maturation (e.g. in *A. aneura*, *A. peuce*, *A. ligulata*, *A. ramulosa*, *A. hemiglauca*) and cause damage to canopies that compromises rainfall interception and litter retention (see Fensham *et al.* 2012a; Brim Box *et al.* 2016). Increased mortality under prolonged heavy sheep grazing is also known to occur, e.g. in *Eremophila maitlandii* (Watson *et al.* 1997).

Livestock grazing can negatively impact the abiotic components of rangeland habitats (Eldridge *et al.*, Chapter 6, this volume). Overgrazing can lead to erosion scalds (particularly in texture-contrast soils), gullies (sheeting by water on loams), and sand drift and dust storms (sands and sandy loams) (Condon 2002; Aubault *et al.* 2015). Trampling disaggregates the soil and reduces organic matter levels, biocrust and litter cover. Compaction increases soil bulk density, causing reduced infiltration and groundwater recharge (Tongway *et al.* 2003; Eldridge *et al.* 2013). For banded vegetation specifically, modelling shows that grazing can cause an abrupt transition to 'leaky' landscapes characterised by increased hydrological connectivity (i.e. mean flowlength between bare patches) and reduced rainfall-use efficiency (Moreno-de las Heras *et al.* 2012). Overall, negative abiotic impacts are predicted to cause muted primary productivity responses in high-rainfall years (Bastin *et al.* 2008). Generally, these effects are difficult to reverse, though there are documented examples of remarkable erosion recovery with reduced stocking over long time frames (Condon 2002).

25.5.2 Shrub encroachment

The thickening of native shrubs in grassy habitat ('shrub/bush encroachment') is a global phenomenon of arid/semi-arid rangelands (Ward 2009; Eldridge and Solivares 2014). The purported causes are many, but there is now some agreement that it likely results from long-term fluctuations in rainfall and atmospheric CO_2, with the added or interacting effects of overgrazing, reduced competition between grasses and woody seedlings, and changed fire regimes (e.g. in Australia due to the cessation of Aboriginal burning) (Fensham *et al.* 2012a, Eldridge *et al.* 2013). In eastern Australia, encroachment (involving mainly *Dodonaea viscosa*, *Eremophila*, *Senna*, but also *Acacia aneura*) has long been equated with land degradation, because of the perception that it suppresses palatable grasses and increases erosion (Harrington 1991). Consequently there has been considerable allocation of management resources to control shrubs ('thinning'), backed by legislation in New South Wales, and again in Queensland since 2013; resulting in the removal of vast amounts of woody biomass from the landscape (Eldridge and Solivares 2014).

It is somewhat staggering, therefore, that the causes, extent and consequences of shrub encroachment have only recently been investigated in any detail in the Australian context. Recent shrub encroachment models support the proposition that grazing per se is more likely to be the cause of the negative effects on ecosystem functioning commonly attributed to encroachment; and that, particularly on loamy versus sandy soils, heavy grazing dampens positive shrub effects, both in relation to plant–plant interactions and ecosystem function (Eldridge *et al.* 2013). Overall, there is now well-founded argument that the management of total grazing pressure over the pulse–interpulse cycle is a far better strategy than the removal of shrubs (Eldridge *et al.* 2013; Eldridge and Solivares 2014). This includes the avoidance of overgrazing during wet years to ensure that grasses maintain their competitive advantage over recruiting shrubs (Ward 2009).

25.5.3 Fodder harvesting and clearing for pasture

In the Mulga Lands bioregion, large tracts of mulga have been cleared for pastoral reasons (Fensham *et al.* 2012b). Extensive tree fodder harvesting in mulga (pushing or pulling trees) underpins the sustainability of the pastoral industry during dry phases in the Mulga Lands bioregion (Fensham *et al.* 2012a, b). However, this practice, if contained, is considered ecologically sustainable because livestock preferentially switch back to herbage when it is again available and trees are allowed to regenerate (see McKeon *et al.* 2003). The clearing of mulga for transition to pasture has also occurred on a broad scale in this bioregion. While the process does not result in substantial changes in plant diversity or composition, the effects of tree removal on nutrient cycling and fauna are likely to be pronounced. Clearing impacts in arid shrublands are potentially reversible through passive management if the process is not accompanied by burning or exotic pasture establishment (Fensham *et al.* 2012a, b).

25.5.4 Exotic pasture grass

The exotic C_4 pasture grass, buffel grass (*Cenchrus ciliaris* L.), poses a significant threat to the biodiversity of inland Australia (Marshall *et al.* 2012). This species has been actively sown since the 1920s as a livestock pasture grass in Western Australia and central Queensland, and from the 1960s for pasture and dust suppression in central Australia. Buffel grass additionally invades from drainage systems where it has naturalised (Clarke *et al.* 2005). Where it proliferates – principally in alluvial woodland, but also in stony witchetty bush and mulga and in cleared gidgee – it suppresses the diversity of winter and summer growing herbs, especially perennial tussock grasses (Clarke *et al.* 2005). Equally concerning, buffel grass produces large volumes of standing dead matter, and invaded areas burn hotter and more frequently than uninvaded areas, leading to the suppression of resprouter (e.g. *Acacia estrophiolata, Hakea divariacata* L.A.S.Johnson) and fire-killed (e.g. *A. aneura*) canopy species. Buffel grass additionally carries fire into shrublands (e.g. *Acacia undoolyana* G.J.Leach, Figure 25.2f) from creeks that previously functioned as natural fire breaks (Marshall *et al.* 2012).

Under current climate conditions, a vast portion of inland Australia is highly suitable for buffel grass invasion (Martin *et al.* 2015). As the climate shifts, habitat suitability should decrease over much of the interior, but increase in the southern bioregions (Yalgoo, Gawler Range and Broken Hill complex) (Martin *et al.* 2015). Management of this species is expensive, and the most effective means of control is the prevention of its unintentional spread (Marshall *et al.* 2012). Problematically, buffel grass is still sanctioned by governments in most states except South Australia, where it was declared a weed in 2015, reflecting its ability to transform landscapes. Importantly though, conflict among stakeholders may be overstated and there is scope for a national approach that goes some way towards accommodating competing interests (Grice *et al.* 2012).

25.6 Conservation

25.6.1 Threatened species

Threatened plant species are associated with arid shrublands throughout inland Australia. They occur as small and restricted populations within more widespread habitat (e.g. *Acacia carneorum, Acacia ammophila* Pedley, *Acacia unguicula, Acacia woodmaniorum* Maslin and Buscomb), or as single-canopy dominants (e.g. *Acacia peuce, Acacia undoolyana, Acacia latzii* Maslin, *Acacia pickardii* Tindale). Many threatened species are now included in recovery plans (Department of the Environment and Energy 2016b), and in spite of their remoteness, they receive considerable conservation attention as well as legislative protection (Raghu *et al.* 2013). The value of single-species conservation is under constant re-evaluation, but for many arid species it can be justified on multiple grounds:

- high extinction risk, e.g. *A. unguicula, A. woodmaniorum*
- short range endemics (restricted distribution), e.g. *A. unguicula, A. undoolyana, A. woodmaniorum*
- umbrella species (conservation actions benefit other threatened species), e.g. *A. peuce, A. undoolyana*
- keystone species (having a disproportionate role in the ecosystem), e.g. *A. peuce, A. pickardii*
- culturally important to Indigenous people, e.g. *A. peuce.*

Threats associated with high grazing pressure and climate change are common to most species, while more regionally-specific threats include: inappropriate fire (*A. undoolyana*); clearing (*A. ammophila, A. carneorum*); mining (*A. pickardii, A. carneorum, A. woodmaniorum*), and buffel grass (*A. undoolyana*). Two species, *A. carneorum* and *A. pickardii*, are highly clonal and for these, populations may comprise few genetically distinct individuals which increases the extinction risk (Department of the Environment and Energy 2016b). In most cases, there is only rudimentary knowledge of species life histories, requiring an adaptive management framework for their conservation (Raghu *et al.* 2013).

25.6.2 Threatened communities

Habitat-level conservation is in its infancy for arid shrublands. One approach to this issue is to identify and protect whole ecological communities at risk of large detrimental changes. For listed communities in Australia, all community components and biological processes are protected by legislation which can be a more efficient use of conservation funding than the single-species approach (Office of Environment and Heritage 2016). No arid shrubland communities have yet been assessed for listing under national legislation, although semi-arid Weeping myall woodlands (*Acacia pendula*) are currently listed as endangered. Similarly, the related brigalow–gidgee woodland/shrubland is listed as Endangered in New South Wales, and it forms part of the nationally endangered brigalow (*Acacia harpophylla* F.Muell. ex Benth. dominant and codominant) (Office of Environment and Heritage 2014; Fensham *et al.*, Chapter 16, this volume). So far, two arid shrubland communities have been assessed using the IUCN Red List of Ecosystems criteria (Keith *et al.* 2013): *A. loderi* shrublands (Endangered in New South Wales; Office of Environment and Heritage 2016), and Georgina gidgee woodlands (Nationally Vulnerable status recommended; Wardle *et al.* 2015). These communities are threatened by a range of pastoral-related issues (heavy total grazing pressure, inappropriate fire regimes, exotic species, and changed hydrology), and they require improved management and the redress of life-history knowledge gaps for their recovery. Potentially, various other arid shrublands meet threatened community listing criteria, and the process is a useful vehicle for highlighting the poor condition of many remote vegetation types (Wardle *et al.* 2015).

25.7 Future directions

Due to a range of economic and social forces, Australia's arid/semi-arid rangelands are increasingly becoming 'contested ground' for the competing land uses options, notably animal production, biodiversity conservation and, most recently, carbon farming (conserving or replenishing carbon stocks) (Dean *et al.* 2015). Presently, landscapes supporting arid shrublands are not comprehensively reserved or conserved, and there is otherwise an overrepresentation of rocky/mountainous arid landscapes of little pastoral value in the National Reserve System. Improved representation in protected areas would be a major advance in the conservation of these ecologically important habitats, and it is likely that Indigenous Protected Areas and privately owned conservation reserves will comprise a significant proportion of the reserve estate in future years (Bastin *et al.* 2008). In the meantime, or in lieu of such progress, the few remaining examples of arid shrublands in water-remote areas can make a significant contribution to biodiversity conservation, if they are appropriately managed and left undeveloped (Bastin *et al.* 2008).

The restoration of degraded land will potentially be an important component of the sustainable use of the rangelands in upcoming decades. In Australia, the recent introduction of a price on carbon provides a financial incentive for landholders to retain/increase woody biomass on their land by way of sequestering carbon (Fensham *et al.* 2012a, b; Eldridge *et al.* 2013). This may in turn trigger a shift in the otherwise negative 'grazing-centric' view of encroachment, especially given that the process occurs passively (without cost); and over long time periods, it may be a viable alternative or supplement to grazing income (Eldridge *et al.* 2013; Dean *et al.* 2015). Most likely though, pastoral production will remain the dominant land use in the foreseeable future where arid shrublands occur. With the modern-day benefit of seasonal climate forecasting systems, managers are far better placed to avoid the mistakes of the past. In this context, several risk minimisation strategies have been framed for the sustainable use of the rangelands: conservative stocking rates; management of total grazing pressure, early destocking response to drought, planned resting to aid pasture recovery, and monitoring for compliance with legislation (McKeon *et al.* 2003; Bastin *et al.* 2008). Furthermore, recent research is providing evidence for differential and context-dependent responses of the different arid shrubland types. Increasingly, therefore, managers should be better equipped to identify the specific conditions where grazing is most closely associated with degradation, and to explore alternative management in those cases. For this to occur though, adaptive management frameworks need to be strengthened through ongoing investment in rangeland monitoring systems that provide accurate reporting and feedback on land condition and biodiversity trends in arid shrublands.

Acknowledgements

We are most grateful for the extremely helpful and insightful comments from the editor, David Keith, and from Tony Auld and Rod Fensham that greatly improved this chapter. We wish to acknowledge the vast amounts of work on arid shrublands carried out by the multiple authors referred to in this chapter as well as many others whose work could not be covered here. We dedicate this chapter to the memory of Peter Clarke, whose enthusiasm for arid Australia and plant ecology, more generally, was always inspiring.

References

Atlas of Living Australia (ALA) (2015) *Atlas of Living Australia.* Available at: http://www.ala.org.au/.

Aubault, H., Webb, N.P., Strong, C.L. *et al.* (2015) Grazing impacts on the susceptibility of rangelands to wind erosion: the effects of stocking rate, stocking strategy and land condition. *Aeolian Research* 17, 89–99.

Auld, T.D. (1995a) Soil seedbank patterns of four trees and shrubs from arid Australia. *Journal of Arid Environments* 29, 33–45.

Auld, T.D. (1995b) Seedling survival under grazing in the arid perennial *Acacia oswaldii.* *Biological Conservation* 72, 27–32.

Bastin, G. The ACRIS Management Committee, Rangelands (2008) *Taking the Pulse.* ACRIS Management Committee by the National Land and Water Resources Audit: Canberra.

Beadle, N.C.W. (1981) *The Vegetation of Australia*. Cambridge University Press: Cambridge.

Beard, J.S. (1975) *Vegetation survey of Western Australia.* University of Western Australia: Perth.

Blake, S.T. (1938) The plant communities of western Queensland and their relationships, with special reference to the grazing history. *Proceedings of the Royal Society of Queensland* 49, 156–204.

Bowman, D.J.M.S., Boggs, G.S. and Prior, L.D. (2008) Fire maintains an *Acacia aneura* shrubland–*Triodia* grassland mosaic in central Australia. *Journal of Arid Environments* 72, 34–47.

Brim Box, J., Nano, C.E.M., McBurnie, G. *et al.* (2016) The impact of feral camels (*Camelus dromedarius*) on woody vegetation in arid Australia. *The Rangeland Journal* 38, 181–190.

Chinnock, R.J. (2007) *Eremophila and allied genera: a monograph of the plant family Myoporaceae.* Rosenberg Publishing: Sydney.

Clarke, P.J., Latz, P.K. and Albrecht, D.E. (2005) Long-term changes in semi-arid vegetation: invasion of an exotic perennial grass has larger effects than rainfall variability. *Journal of Vegetation Science* 16, 237–248.

Clarke, P.J., Lawes, M.J., Murphy, B.P. *et al.* (2015) A synthesis of postfire recovery traits of woody plants in Australian ecosystems. *Science of the Total Environment* 534, 31–42.

Condon, D. (2002) *Out of the West: Historical Perspectives on the Western Division of New South Wales.* Rangeland Management Action Plan: Yowie Bay.

Cowan, R.S. (2001) *Acacia cambagei. Flora of Australia online.* (Australian Biological Resources Study: Canberra) Available at: http://www.environment.gov.au/biodiversity/abrs/online-resources/flora/main/index.html (accessed 6 June 2016).

CSIRO and Bureau of Meteorology (2015) *Climate change in Australia information for Australia's natural resource management regions: technical report.* CSIRO and Bureau of Meteorology: Australia.

Dalgleish, S.A., van Etten, E., Stock, W.D. and Knuckey, K. (2015) Fuel dynamics and vegetation recovery after fire in a semiarid Australian shrubland. *International Journal of Wildland Fire* 24, 613–623.

Davidson, D.W. and Morton, S.R. (1984) Dispersal adaptations of some *Acacia* species in the Australian arid zone. *Ecology* 65, 1038–1051.

Davies, S.J.J.F., Kenny, S.A. and Walsh, T.F.M. (2015) The population dynamics of some arid zone plants during 46 years of grazing on Mileura Station, Western Australia. *The Rangeland Journal* 37, 367–377.

Dean, C., Kirkpatrick, J.B., Harper, R.J. and Eldridge, D.J. (2015) Optimising carbon sequestration in arid and semiarid rangelands. *Ecological Engineering* 74, 148–163.

Denham, A.J. and Auld, T.D. (2004) Survival and recruitment of seedlings and suckers of trees and shrubs of the Australian arid zone following habitat management and the outbreak of Rabbit Calicivirus Disease (RCD). *Austral Ecology* 29, 585–599.

Department of Environment and Conservation (2009) *Acacia unguicula interim recovery plan* 2009–2014. *Interim recovery plan no. 288.* Department of Environment and Conservation: Western Australia.

Department of the Environment and Energy (2016a) *NVIS data products 4.2*. Australian Government: Canberra. Available at: http://www.environment.gov.au/land/native-vegetation/national-vegetation-information-system/data-products (accessed 10 May 2016).

Department of the Environment and Energy (2016b) *Species profile and threats database*. Department of the Environment: Canberra. Available at: http://www.environment.gov.au/sprat (accessed 3 June 2016).

Dunstan, H., Florentine, S.K., Calviño-Cancela, M. *et al.* (2013) Dietary characteristics of emus (*Dromaius novaehollandiae*) in semi-arid New South Wales, Australia, and dispersal and germination of ingested seeds. *Emu* 113, 168–176.

Eamus, D., Cleverly, J., Boulain, N. *et al.* (2013) Carbon and water fluxes in an arid zone *Acacia* savanna woodland: an analyses of seasonal patterns and responses to rainfall events. *Agricultural and forest meteorology* 182, 225–238.

Eldridge, D.J. and Soliveres, S. (2014) Are shrubs really a sign of declining ecosystem function? Disentangling the myths and truths of woody encroachment in Australia. *Australian Journal of Botany* 62, 594–608.

Eldridge, D.J., Val, J. and James, A.I. (2011) Abiotic effects predominate under prolonged live-stock-induced disturbance. *Austral Ecology* 36, 367–377.

Eldridge, D.J., Soliveres, S., Bowker, M.A. and Val, J. (2013) Grazing dampens the positive effects of shrub encroachment on ecosystem functions in a semi-arid woodland. *Journal of Applied Ecology* 50, 1028–1038.

Fensham, R.J., Silcock, J.L. and Dwyer, J.M. (2011) Plant species richness responses to grazing protection and degradation history in a low productivity landscape. *Journal of Vegetation Science* 22, 997–1008.

Fensham, R.F., Fairfax, R.J. and Dwyer, J.M. (2012a) Potential aboveground biomass in drought-prone forest used for rangeland pastoralism. *Ecological Applications* 22, 894–908.

Fensham, R.J., Dwyer, J.M., Eyre, T.J. *et al.* (2012b) The effect of clearing on plant composition in mulga (*Acacia aneura*) dry forest, Australia. *Austral Ecology* 37, 183–192.

Franklin, K.A., Sommers, P.N., Aslan, C.E. *et al.* (2016) Plant biotic interactions in the Sonoran Desert: current knowledge and future research perspectives. *International Journal of Plant Sciences* 177, 217–234.

Gilpin, A., Ayre, D.J. and Denham, A.J. (2014) Can the pollination biology and floral ontogeny of the threatened *Acacia carneorum* explain its lack of reproductive success? *Ecological Research* 29, 225–235.

Grice, A.C., Friedel, M.H., Marshall, N.A. and Van Klinken, R.D. (2012) Tackling contentious invasive plant species: a case study of buffel grass in Australia. *Environmental Management* 49, 285–294.

Harrington, G.N. (1991) Effects of soil moisture on shrub seedling survival in semi-arid grassland. *Ecology* 72, 1138–1149.

Harrington, G.N. and Driver, M.A. (1995) The effect of fire and ants on the seed-bank of a shrub in a semi-arid grassland. *Austral Ecology* 20, 538–547.

Hodgkinson, K.C. and Oxley, R.E. (1990) Influence of fire and edaphic factors on germination of the arid zone shrubs *Acacia aneura*, *Cassia nemophila* and *Dodonaea viscosa*. *Australian Journal of Botany* 38, 269–279.

Holmgren, M., Stapp, P., Dickman, C.R. *et al.* (2006) Extreme climatic events shape arid and semiarid ecosystems. *Frontiers in Ecology and the Environment* 4, 87–95.

Ireland, C.T. and Andrew, M.H. (1995) Ants remove virtually all western myall (*Acacia papyrocarpa* Benth.) seeds at Middleback, South Australia. *Austral Ecology* 20, 565–570.

James, A.L., Eldridge, D.J. and Hill, B.M. (2009) Foraging animals create fertile patches in an Australian desert shrubland. *Ecography* 32, 723–732.

Keighery, G.J. (1982) Pollination syndromes and breeding systems of Western Australian arid zone plants. In *Evolution of the Flora and Fauna of Arid Australia*, ed. W.R. Barker and P.J.M. Greenslade. Peacock Publications: Frewville, pp. 167–172.

Keith, D.A. (2004) *Ocean Shores to Desert Dunes: The Native Vegetation of New South Wales and the ACT.* Department of Environment and Conservation: Hurstville.

Keith, D.A., Rodríguez, J.P., Rodríguez-Clark, K.M. *et al.* (2013) Scientific foundations for an IUCN Red List of Ecosystems. *PLoS ONE* 8(5), e62111.

King, K.J., Cary, G.J., Bradstock, R.A. and Marsden-Smedley, J.B. (2013) Contrasting fire responses to climate and management: insights from two Australian ecosystems. *Global Change* 19, 1223–1235.

Lange, R.T. (1969) The piosphere: sheep track and dung patterns. *Journal of Range Management* 22, 396–400.

Lange, R. and Purdie, R. (1976) Western Myall (*Acacia sowdenii*), its survival prospects and management needs. *The Rangeland Journal* 1, 64–69.

Leavesley, A.J., Cary, G.J., Edwards, G.P. and Gill, A.M. (2010) The effect of fire on birds of mulga woodland in arid central Australia. *International Journal of Wildland Fire* 19, 949–960.

Letnic, M., Dickman, C.R. and McNaught, G. (2000) Bet-hedging and germination in the Australian arid zone shrub *Acacia ligulata*. *Austral Ecology* 25, 368–374.

Mabbutt, J.A., Speck, N.H., Wright, R.L. *et al.* (1963) Land systems of the Wiluna-Meekatharra area, Western Australia. In *General Report on Lands of the Wiluna-Meekatharra Area, Western Australia 1958*, ed. J.A. Mabbutt, W.H. Litchfield, N.H. Speck, J. Sofoulis, D.G. Wilcox, J.M. Arnold, M. Brookfield and R.L. Wright. CSIRO Australia: Melbourne.

Marshall, V.M., Lewis, M.M. and Ostendorf, B. (2012) Buffel grass (*Cenchrus ciliaris*) as an invader and threat to biodiversity in arid environments: a review. *Journal of Arid Environments* 78, 1–12.

Martin, T.G., Murphy, H., Liedloff, A. *et al.* (2015) Buffel grass and climate change: a framework for projecting invasive species distributions when data are scarce. *Biological Invasions* 17, 3197–3210.

Maslin, B.R. and Reid, J.E. (2012) A taxonomic revision of Mulga (*Acacia aneura* and its close relatives: Fabaceae) in Western Australia. *Nuytsia* 22, 129–267.

McKeon, G.M., Crimp, S.J., Cunningham, G.M. *et al.* (2003) Historical degradation episodes in Australia: global climate and economic forces and their interaction with rangeland grazing systems. In *Pasture Degradation and Recovery in Australia's Rangelands: Learning*

from History, ed. G.M. McKeon, W.B. Hall, B.K. Henry, G.S. Stone and I.W. Watson. NRSc Publishing: Brisbane, pp. 27–86.
Moreno-de las Heras, M., Saco, P.M., Willgoose, G.R. and Tongway, D.J. (2012) Variations in hydrological connectivity of Australian semiarid landscapes indicate abrupt changes in rainfall-use efficiency of vegetation. *Journal of Geophysical Research: Biogeosciences* 117, G03009.
Morton, S.R., Stafford Smith, D.M., Dickman, C.R. *et al.* (2011) A fresh framework for the ecology of arid Australia. *Journal of Arid Environments* 75, 313–329.
Murphy, B.P., Bradstock, R.A., Boer, M.M. *et al.* (2013) Fire regimes of Australia: a pyrogeographic model system. *Journal of Biogeography* 40, 1048–1058.
Nano, C.E.M. and Clarke, P.E. (2008) Variegated desert vegetation: covariation of edaphic and fire variables provides a framework for understanding mulga-spinifex coexistence. *Austral Ecology* 33, 848–862.
Nano, C.E.M. and Clarke, P.J. (2010) Woody-grass ratios in a grassy arid system are limited by multi-causal interactions of abiotic constraint, competition and fire. *Oecologia* 162, 719–732.
Nano, C.E.M. and Pavey, C.R. (2013) Refining the 'pulse-reserve' model for arid central Australia: seasonal rainfall, soil moisture and plant productivity in sand ridge and stony plain habitats of the Simpson Desert. *Austral Ecology* 38, 741–753.
Nix, H. and Austin, M. (1973) Mulga: a bioclimatic analysis. *Tropical Grasslands* 7, 9–21.
Norman, P., Denham, R. and Calvert, M.J. (2014) Life histories of two arid zone shrubs change with differences in habitat, grazing and climate. *The Rangeland Journal* 36, 249–257.
Noy-Meir, I. (1973) Desert ecosystems: environment and producers. *Annual Review of Ecology and Systematics* 4, 25–51.
O'Dowd, D.J. and Gill, M. (1986) Seed dispersal syndromes in Australian acacias. In *Seed Dispersal*, ed. D.R. Murray. Academic Press: Australia, pp. 87–121.
Office of Environment and Heritage (2014) *Brigalow-Gidgee woodland/shrubland in the Mulga Lands and Darling Riverine Plains Bioregions profile.* Available at: http://www.environment.nsw.gov.au/threatenedspeciesApp/ / (accessed 14 June 2016).
Office of Environment and Heritage (2016) *Acacia loderi* shrublands profile. Available at: http://www.environment.nsw.gov.au/threatenedspeciesApp/ (accessed 12 April 2016).
Orchard, A.E. (2015) Synoptic classification of Australian Asteraceae. Asteraceae 1. In *Flora of Australia*. CSIRO Publishing: Melbourne, pp. 1–17.
Page, M., Beeton, R.J.S. and Mott, J.J. (2000) Grass response to shrub removal in two semi-arid vegetation communities. *The Rangeland Journal* 22, 220–234.
Pascov, C.M., Nevill, P.G., Elliott, C.P. *et al.* (2015) The critical role of ants in the extensive dispersal of Acacia seeds revealed by genetic parentage assignment. *Oecologia* 179, 1123–1134.
Pedley, L. (2001) Acacia aneura. In *Flora of Australia, 11B*. CSIRO Publishing: Melbourne, pp. 314–322.
Pickup, G. (1998) Desertification and climate change: the Australian perspective. *Climate Research* 11, 51–63.
Popic, T.J., Wardle, G.M. and Davila, Y.C. (2013) Flower-visitor networks only partially predict the function of pollen transport by bees. *Austral Ecology* 38, 76–86.

Popic, T.J., Davila, Y.C. and Wardle, G.M. (2016) Cheater or mutualist? Novel florivory interaction between nectar rich *Crotalaria cunninghamii* and small mammals. *Austral Ecology* 41, 396–404.

Pound, L.M., Ainsley, P.J. and Facelli, J.M. (2015) Dormancy-breaking and germination requirements for seeds of *Acacia papyrocarpa*, *Acacia oswaldii* and *Senna artemisioides* ssp. × *coriacea*, three Australian arid zone Fabaceae species. *Australian Journal of Botany* 62, 546–557.

Pringle, H.J.R., Watson, I.W. and Tinley, K.L. (2006) Rangeland improvement or ongoing degradation: reconciling apparent contradictions from the arid rangelands of Western Australia. *Landscape Ecology* 21, 1267–1279.

Raghu, S., Nano, C.E.M. and Pavey, C.R. (2013) A demographic framework for the adaptive management of the endangered arid zone tree species *Acacia peuce*. *Australian Journal of Botany* 61, 89–101.

Russell-Smith, J., Gardener, M.R., Brock, C. *et al.* (2012) Fire persistence traits can be used to predict vegetation response to changing fire regimes at expansive landscape scales: an Australian example. *Journal of Biogeography* 39, 1657–1668.

Schubert, A.T., Nano, C.E., Clarke, P.J. and Lawes, M.J. (2016) Evidence for bark thickness as a fire-resistance trait from desert to savanna in fire-prone inland Australia. *Plant Ecology* 217, 1–4.

Schwinning, S. and Sala, O.E. (2004) Hierarchy of responses to resource pulses in arid and semi-arid ecosystems. *Oecologia* 141, 211–220.

Sinclair, R. (2005) Long-term changes in vegetation, gradual and episodic, on the TGB Osborn Vegetation Reserve, Koonamore, South Australia (1926–2002). *Australian Journal of Botany* 53, 283–296.

Siteur, K., Eppinga, M.B., Karssenberg, D. *et al.* (2014) How will increases in rainfall intensity affect semiarid ecosystems? *Water Resources Research* 50, 5980–6001.

Start, A.N. (2015) The mistletoe flora of southern Western Australia, with a particular reference to host relationships and fire. *Australian Journal of Botany* 63, 636–646.

Tongway, D.J. and Ludwig, J.A. (2001) Theories on the origins, maintenance, dynamics, and functioning of banded landscapes. In *Banded Vegetation Patterning in Arid and Semiarid Environments. Ecological Processes and Consequences for Management*, ed. D.J. Tongway, C. Valentin and J. Seghieri. Springer: New York, pp. 20–31.

Tongway, D.J., Sparrow, A.D. and Friedel, M.H. (2003) Degradation and recovery processes in arid grazing lands of central Australia. Part 1: soil and land resources. *Journal of Arid Environments* 55, 301–326.

Ward, B.G., Bragg, T.B. and Hayes, B.A. (2014) Relationship between fire-return interval and mulga (*Acacia aneura*) regeneration in the Gibson Desert and Gascoyne–Murchison regions of Western Australia. *International Journal of Wildland Fire* 23, 394–402.

Ward, D. (2009) *The Biology of Deserts.* Oxford University Press: Oxford.

Wardle, G.M., Greenville, A.C., Frank, A.S.K. *et al.* (2015) Ecosystem risk assessment of Georgina gidgee woodlands in central Australia. *Austral Ecology* 40, 444–459.

Watson, I.W., Westoby, M. and Holm, A.M. (1997) Demography of two shrub species from an arid grazed ecosystem in Western Australia 1983–93. *Journal of Ecology* 85, 815–832.

Watson, I.W., Thomas, P.W.E. and Fletcher, W.J. (2007) The first assessment, using a rangeland monitoring system, of change in shrub and tree populations across the arid shrublands of Western Australia. *The Rangeland Journal* 29, 25–37.

Westoby, M., Walker, B. and Noy-Meir, I. (1989) Opportunistic management for rangelands not at equilibrium. *Journal of Rangeland Management* 42, 266–274.

Whibley, D.J.E. and Symon, D.E. (1992) *Acacias of South Australia*. 2nd edition. South Australian Printers: Adelaide.

Whitney, K.D. (2002) Dispersal for distance? *Acacia ligulata* seeds and meat ants *Iridomyrmex viridiaeneus*. *Austral Ecology* 27, 589–595.

Wilson, K.L. and Johnson, L.A.S. (1990) Casuarinaceae. In *Flora of New South Wales*. Vol. 1, ed. G.J. Harden. New South Wales University Press: Sydney, pp. 507–517.

Wright, B.R. and Zuur, A.F. (2014) Seedbank dynamics after masting in mulga (*Acacia aptaneura*): implications for post-fire regeneration. *Journal of Arid Environments* 107, 10–17.

26 Hummock Grasslands: *Triodia*-dominated Grasslands in Arid Australia

GLENDA M. WARDLE AND CATHERINE NANO

Hummock grasslands dominated by *Triodia* are an iconic part of the Australian outback. They cover nearly a third of the continent, are structurally and compositionally diverse, yet still relatively poorly known at fine scales. Across their extent they form associations with trees and shrubs as well as other plants, biological crusts and diverse faunal communities. Aridity underpins the key processes of water and heat stress that shape the ecology of hummock grasslands, but to varying degrees across the monsoonal northern areas, arid central regions and southern Mediterranean regions. Their rich ecological interactions respond dynamically in a pulse-reserve event-driven system, and form spatially variable associations in response to topography, substrate and disturbance. Ongoing changes in land use and the climate will continue to influence hummock grasslands at an accelerating rate into the future, while our understanding of this vast system is only beginning to develop. People have largely ignored this vast biome in comparison to the highly exploited forests and temperate grasslands of Australia. That will not continue and the coming decades are going to be times of rapid change in terms of economic, social and environmental pressure on this nationally important vegetation community.

26.1 Introduction

This chapter describes the distribution and ecology of hummock grasslands dominated by the Australian endemic genus *Triodia* R.Br. (Poaceae). Other grassland types are described by Morgan *et al.* (Chapter 18, this volume) and Venn *et al.* (Chapter 19, this volume). Savanna woodlands in northern Australia (Williams *et al.*, Chapter 15, this volume) and mallee woodlands in southern semi-arid regions (Yates *et al.*, Chapter 23, this volume) may include *Triodia*, beneath a woody canopy of eucalypts.

Grasslands are globally important, covering 40% of the world's terrestrial surface area (World Resources Institute 2000, based on IGBP data). Humans rely on grasslands for food production and livestock grazing, and refer to them as grazing lands or rangelands (Neely *et al.* 2009). Consequently, natural grasslands have declined as vast areas of the world's most productive grasslands in the higher rainfall areas have been converted to arable land or exotic pastures (Dixon *et al.* 2014), leading to increasing pressure for grazing on the remaining lower rainfall areas to feed the growing human population. This pressure to increase productivity underpins the interest in the ecology of grasslands.

Defining natural grasslands as a vegetation unit is complex, partly because of human influences. A second reason is that grasslands intergrade at fine-to-regional scales with areas of low shrub or tree

Australian Vegetation, Third Edition, ed. David A. Keith. Published by Cambridge University Press.

Citation: Wardle, G.M. and Nano, C. (2017). Hummock Grasslands: Triodia-dominated Grasslands in Arid Australia. In D. A. Keith, ed., *Australian Vegetation*, 3rd edn. Cambridge: Cambridge University Press, pp. 651–74.

cover (Nano *et al.*, Chapter 25, this volume) and may overlap functionally. The term 'desert complex environments' encompasses the broad mosaic of grassland, shrubland and woodland associations in the deserts and semi-arid regions of Australia (Dickman *et al.* 2014). However, fine-scale mapping of small woodland units is needed to characterise these interrelations and particularly the threats to endangered ecological communities that remain 'hidden' within these vast regions (Wardle *et al.* 2015). Finally, grasslands are subject to frequent disturbances such as fire that lead to dynamic changes in the cover and extent of the boundaries with adjacent vegetation units (Bowman 2008; Nano and Clarke 2008, 2010). So, despite the familiarity of grasslands, there is much we need to learn and appreciate about Australian grasslands.

26.1.1 The growth form of hummock grasses

Hummock grasses in the genus *Triodia* are xerophytic perennials with sharply pointed leaves clustered on persistent stolons (Beadle 1981). The plant forms low (< 1.5 m) dense hemispherical dome shapes that in some species expand outwards from the centre to form rings (Figure 26.1). An annulus of leaves continues to grow, leaving a central area with flattened rhizomes in a mat-like growth form, sometimes with a mounded, mostly bare area of soil at the centre (Dickman *et al.* 2014). A hummock, or mound, is therefore an apt description of both the plant and the grassland. The growth forms of some other *Triodia* species are open and tussock-like, with more continuous cover – the so-called 'soft spinifex'. Hummock grasses dominate the drier inland and northern savanna bioregions of Australia (Figure 26.2), but the distinctive growth form is relatively unknown outside Australia. The combination of low-nutrient soils and aridity, common across inland Australia, led to the evolution of persistent, sclerophyllous leaves that accumulate as a dense but aerated packing of dry fuel (Beard 1981, 1984; Morton *et al.* 2011), which together with the presence of resins in the leaves (Gamage *et al.* 2012), promotes fire. Given their similar fire ecology, sclerophyllous foliage and the discrete clumps of plants, separated by bare areas; hummocks have been compared to sclerophyll shrubs (Rice and Westoby 1999; Crisp *et al.* 2015). Beadle (1981) also compared the cane-like stems of *Zygochloa paradoxa* (R.Br.) S.T.Blake to shrubs, a hummock-forming species that dominates dune crests within arid areas such as the Simpson Desert.

26.1.2 Names for *Triodia*

Triodia species are commonly called spinifex, however, *Spinifex* L. (Poaceae) is a distinct genus of coastal sand dunes. There are 69 species in the endemic radiation of *Triodia* including the former genus, *Plectrachne*, based on evidence from molecular analyses (Mant *et al.* 2000; Toon *et al.* 2015) and comprehensive sampling (Crisp *et al.* 2015). Recently, new species in the Pilbara (WA) were proposed for the *Triodia basedowii* E.Pritz species complex (Anderson *et al.* 2016). Despite strong support for monophyly, unambiguous morphological features are not available for diagnosing *Triodia sens. lat.* Instead, these grasses can be most reliably recognised by their highly developed scleromorphy and xeromorphy, tough, needle-like leaves resulting from the in-rolling of the leaf blades (at least in dry conditions) and crypsis of stomata in deep grooves, in combination with reproductive characters (Jacobs 2004; Crisp *et al.* 2015). Hereafter, we will use *Triodia* or spinifex to refer to the species that form the dominant cover layer in hummock grasslands.

26.2 Indigenous knowledge and uses

Indigenous peoples value spinifex hummock grasslands as culturally significant country (Ens *et al.*, Chapter 4, this volume). This is recognised through Native Title over the Indigenous estate (33% of Australia), including large areas of hummock grasslands, and 69 Indigenous Protected Areas in the National Reserve System (Australian Government n.d.).

Figure 26.1 The life forms and reproductive stages of *Triodia basedowii*. (a) Hummock grassland vegetation in the Simpson Desert. Older hummocks in loose rings with darker areas and multiple tufts of dry leaves around the edges. The interplant spaces have an absence of other grasses and forbs and only a biological crust in dry times; (b) large hummock rings in long-unburnt vegetation; (c) a ring of an older plant showing bare area in centre; (d) hummock about 1 m across in full flower; (e) flowering stem; (f) husk (left) and seeds.
Ecological processes within hummock grasslands. (g) Post-fire recruitment of *Triodia basedowii* in the north-east Simpson Desert (SW Queensland). Relatively uniform, smallish hummocks are regenerating after wildfire with bare areas, and some areas of biological soil crust between. Associated shrubs and the crest of a sand dune can be seen towards the back of the image.
(h) Tanami Desert 5-years post-fire showing reproductive stems in foreground and woody associations. (i) Cattle grazing in *T. basedowii* hummock grasslands with scattered beefwood, *Grevillea striata* R.Br., Simpson Desert, south-west Queensland. (j) *Triodia* grasslands in the Western Desert with marble gum (*Eucalyptus gongylocarpa*) and *Grevillea juncifolia* Hook. in flower on left.
Photos: Glenda Wardle, except (h) Catherine Nano. (A black and white version of this figure will appear in some formats. For the colour version, please refer to the plate section.)

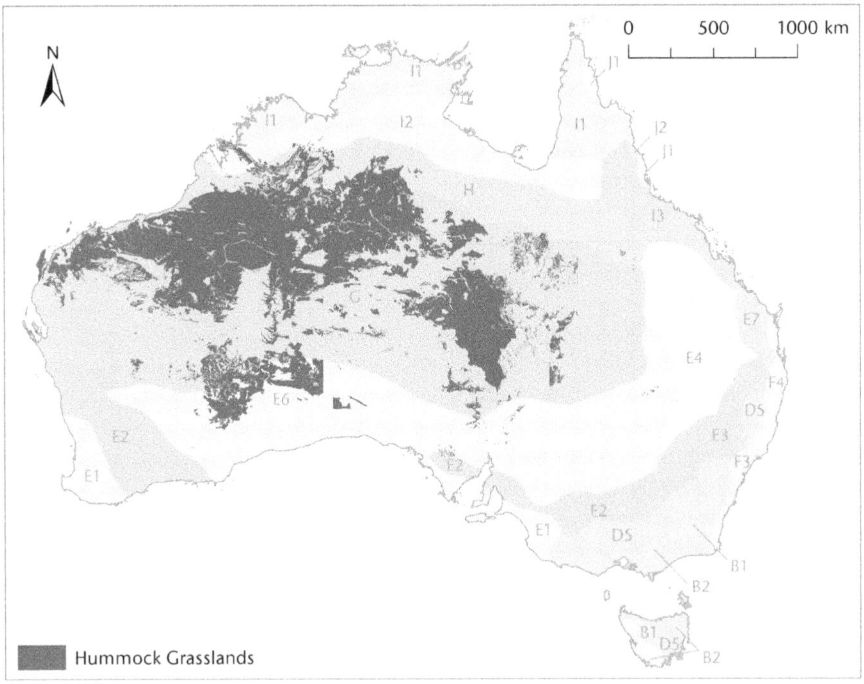

Figure 26.2 Distribution of hummock grasslands, as mapped by the Department of the Environment and Water Resources (2007) and in relation to the agro-climatic classes of Hutchinson *et al.* (2005). The three Agro-climate classes that largely coincide with hummock grasslands are: Class G Desert, Class H Semi-arid with some growth in warm-season and Class E6 Semi-arid climate that is too dry to support field crops. Source: Dickman *et al.* (2014), Fig. 10.3 in ***Biodiversity and Environmental Change: Monitoring, Challenges and Direction.*** (Ed. D. Lindenmayer, E. Burns, N. Thurgate and A. Lowe) (CSIRO Publishing: Melbourne) Created by Eleanor Dormontt.

Indigenous knowledge about the uses of *Triodia* products for food, shelter, medicines, building materials, and to construct tools and ornaments (reviewed by Pitman and Wallis 2012) has been accumulated over tens of thousands of years (Mulvaney and Kamminga 1999). However, interest in expanding the exploitation of spinifex materials for economic return has recently increased (Gamage *et al.* 2012). The fibres of spinifex are a valuable material similar to hemp or flax and used in weaving, rope-making and building. The other significant material is resin, a hydrocarbon secretion of terpenoid and/or phenolic compounds produced by leaves, particularly in soft-type species. Spinifex resin has desirable properties such as a high melting temperature, toughness and malleability, and was traditionally used as an adhesive substance to attach parts of tools or weaponry together; and for water-proofing storage vessels (Pitman and Wallis 2012).

Traditional uses of spinifex included lighting fires in the vegetation for hunting, to produce smoke signals for communication, or burning individual plants to provide light or heat. The important ecological consequences of traditional burning will be discussed below in the section on ecological processes.

26.3 Adaptations to arid conditions and biome shifts

The morphology and anatomy of the xeromorphic leaf characteristics of *Triodia* species appear to have adaptive significance for arid environments. The plants are perennial and evergreen but many dieback to dry, yellowish tissue during extended dry periods, only to rapidly green up after good rains. The leaf anatomy is distinctive and adapted to drought (Toon *et al.* 2015), with the long cylindrical leaf lamina typically rolled inwards and stomata sunken deeply into pits or grooves (Crisp and Cook 2013). The encryption of stomata originated in *Triodia* by c. 7 Ma (Toon *et al.* 2015). A further distinction is the use of a 4-carbon molecule (C_4) photosynthetic pathway (McWilliam and Mison 1974), an adaptation to declining levels of atmospheric CO_2 (Ehleringer *et al.* 1991), and accentuated by lower leaf stomatal density typical of arid environments. C_4 plants have increased water-use efficiency and higher photosynthetic capacity at high light levels compared to typical C_3 plants (Sage and Monson 1999). Similar to other C_4 plants, *Triodia* exhibits the *Kranz anatomy*, or wreath-arrangement of vascular bundles, with a ring of cells containing an alternate form of chloroplasts that give the plants an advantage in conditions of drought, high temperature, and nitrogen or CO_2 limitation. The ideas behind this adaptive hypothesis were recently tested in a comparative phylogenetic study of 117 genera of grasses with 15 lineages of C_4 plants. From the higher number of transitions to arid habitats in C_4 plants, Osborne and Freckleton (2009) concluded that the selection for the C_4 pathway arose in open habitats with high irradiance and is a pre-adaptation (exaptation) to xeric conditions.

There are two major types of leaves in *Triodia* based on their anatomy and presence of resin. The 'hard type' is the ancestral form characterised by amphistomatous leaves (stomata on both adaxial and abaxial surfaces) (Toon *et al.* 2015). These leaves also have well-developed sclerenchyma and typically lack resin (Jacobs 2004). Typical species are *Triodia basedowii*, widespread across the central arid areas and *Triodia irritans* R.Br., with a southerly distribution. The second group, found in two lineages, have epistomatous leaves that lack abaxial stomata, and contain high levels of resin (Toon *et al.* 2015). This group are described as 'soft type' due to their higher palatability to livestock (Mant *et al.* 2000) and being less prickly to touch. Common examples include the widespread *Triodia pungens* R.Br., *Triodia bynoei* (C.E.Hubb.) Lazarides, from northern Australia and the more restricted but locally abundant, *Triodia schinzii* (Henrard) Lazarides.

Modern phylogenetic approaches capable of resolving the timing and origin of phenotypic trait variation across lineages have confirmed that *Triodia* arose in the Eremaean biome in parallel with Australia's drying climate in the mid-Miocene, and that species later shifted into the monsoonal savanna regions to the north over the last 10 Myr (Toon *et al.* 2015). The evolution of the leaf anatomy therefore appears to have occurred after these shifts and was most likely a response to the change in predictability of rainfall moving from long dry periods punctuated by infrequent rainfall in the arid zone to seasonal rainfall, alternating with drought in the northern biome (Toon *et al.* 2015).

26.4 Distribution and biogeography

Hummock grasslands are designated as a major vegetation group of Australia, MVG 20 (Department of the Environment and Water Resources 2007), covering approximately 1.37 million km^2. Smyth *et al.* (2012), based on a broader interpretation that includes mallee and savanna woodlands with *Triodia* understoreys, estimated that they cover 2.82 million km^2, or just over 30% of the continent. This makes them the largest biome in Australia (Figure 26.2). This vast distribution falls under many large Natural Resource Management jurisdictions, including, for example, Desert Channels,

Territory NRM, South Australian Arid Lands and Alinytjara Wilurara. The boundaries vary by definition (Smyth *et al.* 2012), but effectively hummock grasslands span 37 IBRA regions (Interim Biogeographic Regionalisation for Australia, version 7.0; Department of Sustainability, Environment, Water, Population and Communities 2012).

Australia's hummock grasslands are one of the world's largest biomes with a dominant grassland component. Despite being globally distinctive, and meeting the definition, they were surprisingly omitted from the maps and inventory in a recent global classification of grassland types based on the International Vegetation Classification divisions (IVC) integrated into the spatially explicit Terrestrial Ecoregions of the World (TEOW) (Dixon *et al.* 2014).

The historical biogeography of hummock grasslands reflects the changing aridity of the Australian continent (Weston and Jordan, Chapter 2, this volume) and Quaternary vegetation history (Mooney *et al.*, Chapter 3, this volume). By the late Palaeocene (53–65 million years ago (Ma)) north-western Australia was experiencing arid conditions and this spread to inland areas towards the mid-Miocene (5.3–23.5 Ma) (Martin 2006). Aridity intensified throughout the Pliocene (1.75–5.3 Ma), with fully arid landforms appearing 1–4 million years ago (Byrne *et al.* 2008). Therefore, the Australian arid zone, characterised by variable, low rainfall (150–350 mm per year) and a moisture index of less than 0.4 (mean annual rainfall divided by evaporation), is younger in evolutionary terms than the mesic habitats (Byrne *et al.* 2008). While *Triodia* is estimated to have differentiated from its relatives in the mid-Miocene (Toon *et al.* 2015), hummock grasslands were present by the Pleistocene, and diversified into assemblages of arid-adapted organisms (Byrne *et al.* 2008). Speciation continues, and Crisp and Cook (2013) report that of all major Australian plant groups, *Triodia* continues to exhibit one of the steepest rates of divergence, except perhaps for *Acacia*.

26.5 Species distributions and floristic variability

The factors that determine the distribution of *Triodia* species remain poorly understood. Some species appear to be generalists with a wide distribution, while others have small isolated populations (Griffin 1984; Armstrong and Garnett 2011) or disjunct distributions (Table 26.1).

Early accounts of spinifex grasslands for the whole continent described five alliances of 'spinifex country' (Beadle and Costin 1952; Beadle 1981) but these were not quantitatively circumscribed or mapped. They do, however, serve to provide an overview. The following alliances were recognised: (i) the *Triodia* (formerly *Plectrachne*) *schinzii* (feathertop spinifex) alliance on sand in the north-west arid regions; (ii) the *T. pungens* (gummy or soft spinifex) alliance, on sand or stony areas across the arid region mainly north of the Tropic of Capricorn; (iii) the *T. basedowii* (hard spinifex) alliance, usually on sand, mainly south of the Tropic of Capricorn from south-west Queensland to Western Australia; (iv) the *T. irritans* (porcupine grass) alliance, usually in rocky areas and mainly in South Australia; and (v) the *Zygochloa* paradoxa (sandhill cane grass) alliance on sand dunes in low rainfall areas of the Simpson Desert (Beadle 1981).

As a comparison, Griffin (1990) examined the mosaic of vegetation associations of three widespread species, *T. basedowii*, *T. pungens* and *T. schinzii* in central Australia. Across 65 sites, plant composition and vertical structure were sampled using 1000 wheel-points at 1 m spacing along a transect. Vegetation data were analysed by ordination and classification techniques to determine whether stands were dominated by a single species or mixed species, and if distinct alliances were associated with soils and climate characteristics. Overall, 24% of the sites contained admixtures of the *Triodia* species, confirming that they are not necessarily mutually distinct in their requirements. According to Griffin (1990), *T. pungens* dominates

Table 26.1 **Patterns of distribution and abundance of representative *Triodia* species with their occurrence in Australia by state or region. Information derived from Lazarides (1997).**

Distribution type	Species	Occurrence by state or location
Widespread and common	*T. basedowii*	NSW, NT, Qld, SA, WA
	T. bitextura	NT, Qld, WA
	T. brizoides	NT, Qld, WA
	T. longiceps	NT, Qld, SA, WA
	T. microstachya	NT, Qld, WA
	T. pungens	NT, Qld, SA, WA
	T. scariosa	NSW, NT, Qld, SA, Vic, WA
	T. schinzii	NT, Qld, SA, WA
	T. stenostachya	NT, Qld, WA
	T. melvillei	NT, SA, WA
	T. epactia	NT, WA
Restricted but abundant	*T. molesta, T. triaristata, T. inutilis, T. intermedia* and *T. danthonioides*	
Rare everywhere	*T. aeria*	Mt House Range, WA
	T. aristiglumis	Kakadu National Park, NT
	T. biflora	Hamersley Range, WA
	T. bunglensis	Bungle Bungle National Park, WA
	T. fitzgeraldii	Dillon Springs, Carr Boyd Range, WA
	T. prona	Thompsons Springs, Carr Boyd Range, WA
	T. vella	Darling Downs, Qld
Disjunct distributions	*T. procera*	Coastal mainland and islands of the Northern Territory
	T. claytonii Lazarides (formerly *T. mollis*)	Mitchell Plateau, WA, and in northern Qld

areas with > 350 mm annual rainfall in northern regions, while hard *T. basedowii* dominates drier (< 300 mm) southern regions. In the presence of dunes with increased water-holding capacity, *T. schinzii* extends its range from wetter northern areas to more arid southern regions. As non-spinifex species are poorly associated with these species, they do not distinguish the alliances (Grigg *et al.* 2008). Interestingly, the assemblages in the *T. basedowii* alliance sites were sufficiently distinct from the other two, even though few species were restricted to this alliance.

26.5.1 Case study: Western Australia

The vegetation types in Western Australia with a substantial or dominant component of hummock grasses are outlined in Table 26.2. Spinifex country, as it is commonly called, is the most widespread vegetation type in Western Australia and dominates the Eremaean Botanical Province. In the northern area, between latitudes 18-23°S, hummock grasslands occur on almost all substrate types to the exclusion of other components, while south of 23°S hummock grasslands are confined to

Table 26.2 **Description of associations of hummock grassland vegetation types (VT) of Western Australia with *Triodia* as a dominant layer, and estimates of area in hectares. (From Beard *et al.* 2013).**

VT	Description of associations in Western Australia	Area in ha
30	Grasslands, curly spinifex savanna woodland or low trees: *Triodia bitextura* with *Eucalyptus phoenicea, E. brevifolia, Corymbia ferruginea, C. dichromophloia*	304 980 485
34	Tree-steppe: desert oak (*Allocasuarina decaisneana*) with soft spinifex (*Triodia pungens*)	1 826 433
35	Low tree-steppe: hummock grassland (*Triodia* spp.) with scattered bloodwood (*Corymbia dichromophloia*) and snappy gum (*Eucalyptus brevifolia*)	16 053 836
36	Sparse low tree-steppe: hummock grassland (*Triodia* spp.) with sparse eucalypts e.g. bloodwood (*Corymbia dichromophloia*) and snappy gum (*Eucalyptus brevifolia*)	1 751 714
37	Tree-and-shrub steppe: hummock grassland (*Triodia* spp.) with scattered eucalypts (*Eucalyptus gongylocarpa*) over wattle scrub (*Acacia* spp.) or mallee	12 536 719
38	Shrub-steppe: hummock grassland (*Triodia* spp.) with scattered shrubs (*Acacia* spp., *Grevillea* spp.) or mallee (*Eucalyptus* spp.)	25 315 525
39	Sparse shrub-steppe: hummock grassland (*Triodia* spp.) with sparse shrubs (*Acacia* spp.)	7 906 238
40	Grass-steppe: hummock grassland (*Triodia* spp.)	2 527 282
41	Spinifex complexes: hummock grassland (*Triodia* spp.) with scattered low trees over dwarf shrubs or mixed short grass and spinifex mixed species	1 597 775
104	Mosaic 104. Woodland/Open mallee steppe	1 092 784
110	Mosaic 110. Scrub or very open scrub/Grass-steppe	712 124
111	Mosaic 111. Pindan/Tall bunch-grass low tree savanna	261 130
112	Mosaic 112. Curly spinifex low tree savanna/Sparse low tree-steppe	602 455
113	Mosaic 113. High bunch-grass savanna woodland/Curly spinifex savanna	970 179
114	Mosaic 114. Curly spinifex or short-grass low tree savanna/Grass-steppe	393 229
115	Mosaic 115. Short bunch-grass low tree savanna/Low tree-steppe	519 761
116	Mosaic 116. Short bunch-grass savanna/Grass-steppe	958 198
117	Mosaic 117. Sparse low tree-steppe/Sparse shrub-steppe	27 688 259
118	Mosaic 118. Low tree-steppe/Scrub	52 794
119	Mosaic 119. Low woodland or open low woodland/Saltbush and bluebush	172 534

arid sandplains and dunefields (Beard *et al.* 2013). Scattered trees and shrubs are almost always present and distinguish the mapped types of spinifex grasslands. The vegetation types depicted in the 1: 3 000 000-scale vegetation map of Western Australia (Beard *et al.* 2013) provide an example of how the structure, composition and general distribution of associations that include hummock grasslands vary. The main associations were identified by consistency of the dominant plant species based on three features: (i) physiognomy of dominant stratum; (ii) density of strata (projective foliage cover); and

(iii) dominant or diagnostic plant species (Beard and Webb 1974).

26.6 Structural variability

26.6.1 Woody associations

Hummock grasses can either dominate, or occur in association with emergent trees (*Acacia*, *Corymbia*, *Eucalyptus*), small trees and shrubs, e.g. *Acacia*, *Atalaya*, *Crotalaria*, *Eremophila*, *Grevillea*, *Hakea*, *Owenia*, or mallee eucalypts. It is an interesting question to consider what level of shrub cover is consistent with a description of a vegetation unit as 'hummock grassland'. As with many ecological patterns, the scale of the survey will influence whether an area is considered to be homogenous grass cover, or dominated by grasses with scattered trees. Alternatively, it may be a mosaic of vegetation units that are floristically and physiognomically quite distinct, yet function as an interconnected complex ecologically (Wardle *et al.* 2015).

26.6.1.1 Associated species of forbs and grasses

Understandably, vegetation descriptions focus on perennial species (Beard *et al.* 2013) and rely on snapshot surveys, or remotely acquired imagery, to generate the spatial patterns. For example, to track change over large scales in grassland sites, the Western Australian Monitoring System (WARMS) recorded only the presence of grasses, shrubs and trees, rather than herbaceous biomass (Watson *et al.* 2007). Data reflecting the full dynamic of vegetation change are scarce, but see Dickman *et al.* (2014) for a review of available information for desert complex environments. The herbaceous plants, annuals, ephemerals and seed bank diversity contribute substantially to the diversity, productivity and ecological function of most arid zone vegetation types but are typically omitted from assessments used for management and planning. This component is also quite critical for assessing the resilience of ecological interactions to global climate change (Popic and Wardle 2011).

26.6.1.2 Case study: AusPlots

To redress the knowledge gap for a continental-scale plot-based inventory for vegetation and soils across the rangelands of Australia a surveillance monitoring programme called AusPlots was initiated (White *et al.* 2012). Site selection was stratified by bioregion, vegetation and other factors, and a standardised protocol for voucher samples and data collection was used to survey vegetation composition, structure and soils, in approximately 500 1-ha plots, with many of these located in hummock grasslands. An example of the range of cover values and species richness is provided for six plots in the Simpson Strezlecki Desert bioregion (Table 26.3). Cover of *Triodia basedowii* was around 30% for Plots 1, 4 and 8 where it was the dominant growth form, and around 8-9% where it occurred with blue mallee, *Eucalyptus gamophylla*, or *Acacia ligulata* (Plots 2 and 6). Vegetation with Georgina gidgee (*Acacia georginae*) as the dominant woody cover also had cover of spinifex at around 5% (Plot 3) (AusPlots n.d.). AusPlots will be resurveyed to track changes over time and the data are available in the public domain to analyse patterns in vegetation composition and structure in relation to environmental drivers.

26.7 Major environmental factors

The physical factors that most shape Australia's hummock grasslands are low and unpredictable rainfall, punctuated with episodic deluges that flood clay pans or low-lying areas, and nutrient-poor soils that are nonetheless patchy across the landscape (Figure 26.3).

Hummock grasslands occur across a wide geographic area with climatic conditions strongly influenced by extreme years; both droughts and flooding rain events that occur over decadal scales (Dickman

Table 26.3 **Vegetation cover (%) for dominant plant species in the Terrestrial Ecosystem Research Network (TERN) AusPlots located on Ethabuka and Cravens Peak reserves owned by Bush Heritage Australia and located in the Simpson Strezlecki Desert (SSD) Bioregion. Each 1 ha plot was surveyed at 1010 points located 1 m apart and the presence of each species was recorded at each stratum present in April 2014. Data available from AusPlots (n.d), courtesy of Emrys Leitch and Ben Sparrow.**

Plot name	Number of species present in 1 ha plot	Dominant species recorded	Common name	Dominant species, approximate % cover
QDASSD0001	36	*Triodia basedowii*	Hard spinifex	31.87
		Grevillea stenobotrya	Rattle-pod grevillea	1.25
QDASSD0002	20	*Triodia basedowii*	Hard spinifex	8.22
		Acacia ligulata	Dune wattle	3.07
		Eucalyptus gamophylla	Blue mallee	3.07
		Eragrostis eriopoda	A love grass	2.87
		Eremophila longifolia	Berrigan	1.29
QDASSD0003	26	*Acacia georginae*	Georgina gidgee	6.44
		Triodia basedowii	Hard spinifex	4.95
		Enchylaena tomentosa var. tomentosa	Ruby saltbush	3.56
QDASSD0004	29	*Triodia basedowii*	Hard spinifex	32.48
QDASSD0006	28	*Triodia basedowii*	Hard spinifex	9.21
		Aristida holathera var. *holathera*	Erect kerosene grass	2.57
		Halgania cyanea	Mallee blue-flower	1.09
QDASSD0008	23	*Triodia basedowii*	Hard spinifex	29.11
		Grevillea stenobotrya	Rattle-pod grevillea	5.94
		Petalostylis cassioides	Butterfly bush	1.78

et al. 2014). This variability is driven by El Niño and La Niña ENSO events and the Indian Ocean Dipole warming and cooling events (Holmgren *et al.* 2006). Importantly, based on records since 1890 there was a long-term drying event until the mid-1970s that overlapped with high grazing pressure, but coincidentally, before much research on this biome was initiated (Smyth *et al.* 2012).

Aridity is therefore key to understanding both the present climate envelope for this vegetation, and the projected future climate. Spinifex grasslands dominate in the drier arid to semi-arid parts of the continent where mean annual rainfall is low and typically ranges from 150 mm to 350 mm. But some *Triodia* species, and the grassy understorey they form extend into the monsoonal rain regions to the north or into the Mediterranean climate to the south, which has more predictable dry summers replenished by annual rainfalls in late autumn, winter and early spring (Smyth *et al.* 2012). Temperatures, also, are extreme in central Australia with cool nights as low as -7.5°C in July (mid-winter), and hot days, leading to a daily range averaging 37.7°C, based on 30-years of data (Smyth *et al.* 2012). The summer

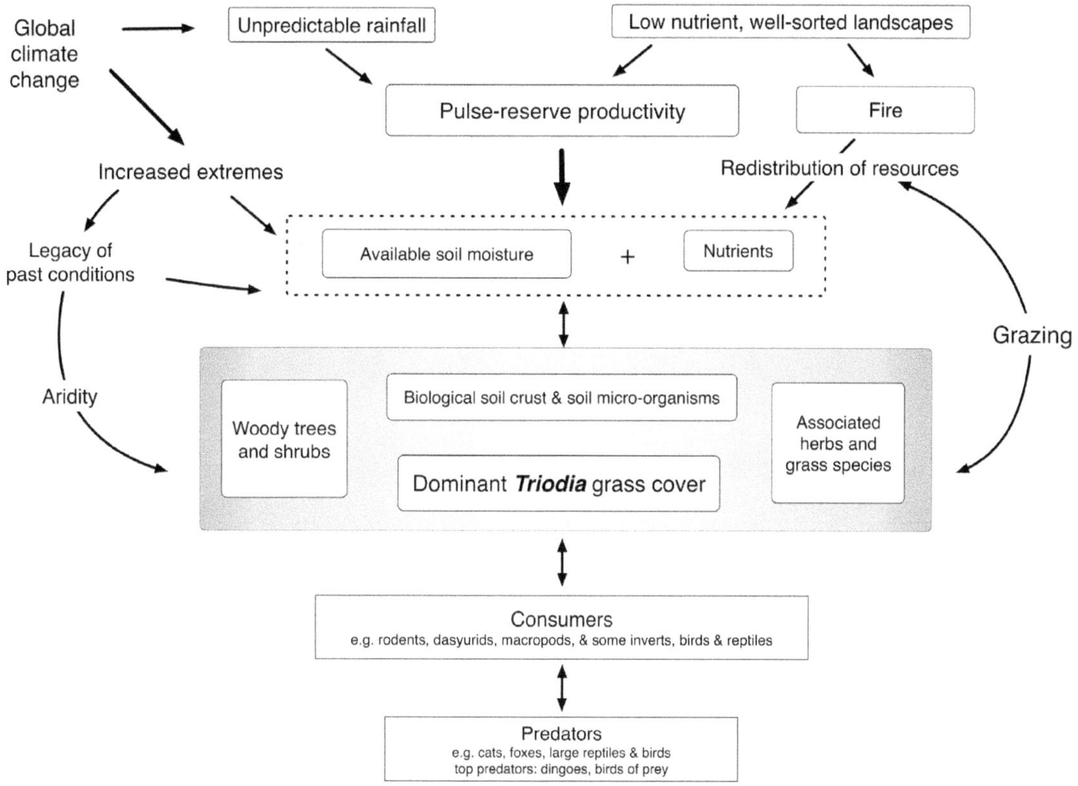

Figure 26.3 Conceptual framework of the ecological interactions that operate in hummock grasslands. Features of the physical environment such as unpredictable rainfall and low nutrients lead to pulse-reserve productivity, which along with fire (the other important ecological process), redistribute resources. The vegetation, dominated by *Triodia*, interacts with associated herbaceous and woody plant species and the biological soil crust, which in turn support the higher trophic levels of consumers and predators. Global climate change is predicted to increase extremes in rainfall and temperature with flow-on effects on available soil moisture, operating partly through legacies of longer droughts or increased floods. Grazing by large herbivores, including domestic livestock, directly influence vegetation structure and composition by selectively removing palatable species, and indirectly, through the redistribution of nutrients and trampling disturbance of soils. *Source: Wardle unpublished, ecological drivers based on ideas from Morton et al. 2011 and biological interactions from Dickman et al. (2014).*

(January) daily temperature range is slightly lower at 34.7°C but the maximum often exceeds 40°C. Therefore, both water and heat stress play a role in limiting available water for plant growth, and *Triodia* has been able to solve this problem with many of its distinctive leaf and physiological characteristics, already discussed.

Hummock grasslands occur in a wide range of substrates and topographic situations. Despite the relatively low relief generally across the inland Australian landscape, uplands and hills are important. *Triodia* species grow on deep sands, red clayey sands and gravels overlying sandstone, limestone, shales, skeletal soils on rock outcrops, and salt pan fringes (Griffin 1984; Lazarides 1997), extending from the arid inland to coastal areas. Soil textures interact with water availability. Noy-Meir (1973) defined the inverse texture effect based on the expectation that a higher proportion of rainfall would be captured at depth by sandy soils than by clay soils in hot arid environments (English *et al.* 2005). However, the lower water-holding capacity of sandy soils means a greater depth is needed to store an equivalent amount of water than in finer textured soils (Grigg 2009).

Digital information on soil attributes across Australia at a 90 m grid spacing and at 6 depth levels is available from Soils to Satellites (CSIRO n.d.). Soils of the hummock grasslands are generally lower than those of arid shrublands and tussock grasslands in bulk density, clay and silt, but slightly elevated in sand and available water capacity. Soil pH is generally acidic at surface levels (pH 4.9-6) but less so at depths down to 200 cm. Soil nutrients are low, with organic carbon (< 1%), nitrogen (< 0.05%) and large areas with total phosphorus (< 0.02%) or (0.02-0.05%). The particularly low levels for phosphorus and nitrogen are well-known (Winkworth 1967; Specht and Specht 1999) but the soils are also deficient in micronutrients needed for both plant and animal life (Orians and Milewski 2007). Despite these low-nutrient levels, heterogeneity is still important. Phosphorus availability for hummock grasses (*Triodia* spp.) is lower on sand dunes (2.8 ppm) compared to river floodplains (18.8 ppm) (Charley and Cowling 1968). At the plant scale, soils beneath spinifex hummocks have lower levels of phosphorus than surrounding soils, but slightly increased levels of nitrogen (Rice *et al.* 1994).

The widespread nutrient poverty and intense fires common in Australia have influenced the biota in distinctive ways. The sclerophylly that arises with low-nutrient conditions leads to retention of leaves, high lignification and slow breakdown of material leading to accumulation of fuel. However, plants growing on nutrient-poor substrates can nonetheless photosynthesise relatively rapidly under mild, sunny climates when soil moisture is adequate leading to carbohydrate-rich, nutrient-poor tissues. The Nutrient-Poverty/Intense-Fire Theory (Orians and Milewski 2007) links these processes. Fire also consumes plant tissues in a similar way to herbivory (Bond and Keely 2005) and is a significant evolutionary force.

26.8 Ecological processes

26.8.1 Pulse-reserve productivity

Hummock grasslands experience extremes of climatic conditions and decadal patterns characterised by long dry periods punctuated by irregular large rainfall episodes that initiate pulses of plant growth and reproduction that in turn underpin system-reserves of carbon and energy (Wardle *et al.* 2013). The influential pulse-reserve rainfall–productivity model (Noy-Meir 1973) captures this general relationship between annual total precipitation and pulses of productivity in arid systems. However, it does not fully account for the variability in magnitude and timing of pulses (van Etten 2009) and the full range of system-wide responses (Nano and Pavey 2013). In particular, we need to understand how rainfall connectivity or closely-timed rain events spur certain species to respond without flooding rains (Fernandez *et al.* 2007) and how soil textures influence outcomes (Nano and Pavey 2013). Typically the ecological responses to increased productivity involve changes to plant cover by rapid growth of perennials and changes to composition by germination and recruitment of seeds. Increased productivity also stimulates biological soil crusts and microorganisms, and influences trophic interactions among consumers and predators (Figure 26.3). New work suggests vegetation patterns characterised by bare circles emerge by self-organisation as a consequence of positive biomass-water feedbacks associated with water runoff and biomass-dependent infiltration (Getzin *et al.* 2016). Big rain years potentially stimulate population responses in invertebrates (Kwok *et al.* 2015); birds (Tischler *et al.* 2013), fish (Kerezsy *et al.* 2013), mammals (Greenville *et al.* 2013; Pavey and Nano 2013) and alter ecosystem processes, such as litterfall (Travers *et al.* 2013); reviewed in (Wardle *et al.* 2013). But big rains also fuel fires.

26.8.2 Fire dynamics of *Triodia* over space and time

Fire is an important ecological process in the hummock grasslands but our knowledge of historic fire regimes and how they have shaped the biota is as patchy as the fires themselves. Aboriginal people use fire for cultural reasons, hunting and other practical purposes and have extensive knowledge for managing the landscape to meet these needs (Mulvaney and Kamminga 1999). Fire is also managed for protecting life and property on pastoral lands and for conservation of biodiversity in the National Reserve System, although given the vast areas involved, and low population base, it is unclear how effective these efforts are (Penman *et al.* 2011). In the hummock grasslands, however, lightning strikes remain a key ignition source. Reconstruction of fire over the last glacial period (70 000 years) from charcoal deposits showed variability on all time scales, but provided no support for changes in biomass burnt at 50 ± 10 thousand years ago, around when Aboriginal people arrived (Mooney *et al.* 2011). More detailed site-level analysis by Stevenson and Hope (2005) reported changes in vegetation over that period were independent of changes in fire or human activity. Caution is needed in taking these broad-scale patterns down to regions, as the available historical data are mostly from outside the arid regions where hummock grasslands are found. The dominant role of climate in changing fire regimes is not so surprising, given that the paleo-fire record shows consistent response to climate changes, with increased fires in warmer climates (Mooney *et al.* 2011) and the importance of fire weather in the extent of fire spread (Cheney and Sullivan 2008).

In hummock grasslands of the more arid regions, widespread fire events occur following high-rainfall events (Figure 26.1h). The reason for this is that although *Triodia*, and associated grasses such as *Aristida*, are highly flammable, the interspaces between hummocks are typically bare, retarding the spread of fires, until rain stimulates recruitment of ephemerals that fill the inter-hummock spaces (Nano *et al.* 2012). Fires can carry with only minimal biomass between hummocks, especially if wind-driven (Gill *et al.* 1999), and are frequently stopped by relatively bare dune crests only to traverse 100 m breaks in other conditions (Wardle pers. obsv.). Fire behaviour in grasslands is complex, with spot ignitions and line fires leading to different rates of spread depending on wind changes and updrafts (Cheney and Sullivan 2008). Fuel accumulation rates vary within and between hummock grass alliances and it is the accumulated substantial rainfalls that drive the increase rather than simply time-since-last-fire (Turner *et al.* 2008; Greenville *et al.* 2009; Nano *et al.* 2012). In central Australia, extensive fires occurred in 1976, 1982–84, 1990–01, 2000–02, and 2011–12 (Greenville *et al.* 2009; Nano *et al.* 2012) and 27% of arid or semi-arid Australia burnt at least once between 1998 and 2004 (Turner *et al.* 2008). Modern approaches for mapping fires have improved our understanding of the variability in spatial extent and patchiness of fires (e.g. in the Western Deserts, mean = 63.9 ha, range 5–6005 ha) (Burrows *et al.* 2006). Fire return intervals also vary, with frequent fires possible, but typical values in recent decades are 7 years (Tanami Desert), up to 26 years (Simpson Desert), or as long as 100 years for long-unburnt spinifex in the driest parts of the Simpson Desert (Allan and Southgate 2002; Wright and Clarke 2007; Greenville *et al.* 2009; Kimber and Friedel 2015). The persistence of long-unburnt areas teaches us that landscapes should not be treated uniformly in managing fire and that we need to retain these areas as they provide valuable habitat refuges for wildlife.

26.8.3 Regeneration in *Triodia*

Hummock grasses respond to fire either by resprouting, seedling recruitment or both (Burbidge 1944; Suijdendorp 1981; Casson and Fox 1987). Species, and populations, vary in these responses depending on the environmental conditions and severity of fire (Rice and Westoby 1999) and most likely genetic differences too. In a survey of 116 sites across Australia less than two years post-fire, Rice and Westoby (1999) recorded three fire-killed species (*T. basedowii*, *T. marginata* and *T. melvillei*); five resprouting species (*T. bynoei*, *T. bunicola* (syn. *T. scariosa*), *T. intermedia*, *T. molesta* and *T. spicata*) and a further seven species exhibited both responses across the sites examined. Resprouting species were more likely to have an open growth habit but no other traits, nor site characteristics, were consistent across the response types. Interestingly, three *Triodia* species that differ in their fire responses have coexisted under similar fire regimes since the Pleistocene in the Kimberly region of Western Australia (Armstrong 2011). *Triodia basedowii* has only recruited from seed after wildfire in the Simpson Desert long-term study sites, but sometimes seedling recruitment is delayed, or there is a cluster of seedlings within an area previously covered by a hummock, which may be mistaken for resprouting in some circumstances (Wardle pers. obsv.).

The patterns of seed production and seedling recruitment in *Triodia* are complex, and still being explored. *Triodia* species have three inflorescence types: spike, raceme and panicle and the spikelet itself may contain one, two, few or many fertile florets, which usually determine its shape and size (Lazarides 1997). Many have noticed the less-than-reliable seed production that seems at odds with fire-killed species (Westoby *et al.* 1988; Farley *et al.* 2013; Dickman *et al.* 2014). The intermittent production of large, synchronised seed crops among plant populations, known as mast seeding, has recently been described for *Triodia* spp. (Wright *et al.* 2014). Based on herbarium and commercial seed collections, pooled across seven species, it was found that 43% of specimens contained completely aborted inflorescences (0% seed fill), while seed set ranged from 2 to 69% in the remaining specimens (Wright *et al.* 2014). Pre-dispersal predation by insects was also recorded in 42% of specimens, further limiting seed available for recruitment.

Seed dormancy, and persistence in soil seed banks are important aspects of recruitment for fire-killed species (Bogusiak *et al.* 1990; Kinloch and Friedel 2005). Florets and seeds of *Triodia* species exhibit physiological dormancy (Erickson 2016) but the levels of dormancy and germination responses vary among species. Removal of the floret, dry after-ripening over 12-24 months storage, and the smoke-derived stimulant karrikinolide all increase germination to varying levels among the seven species tested (Erickson 2016). These findings are being used to assist in restoration of perennial grasses after clearing of vegetation for mining in the Pilbara, Western Australia. Stimulation of germination by fire cues was also recorded for three *Triodia* species from Queensland (Gamage *et al.* 2014) and *T. schinzii* in central Australia (Wright and Clarke 2008).

26.9 Ecological interactions

Hummock grasses exist as part of a web of interacting processes that shape the relationships among the plants, consumers and how they cope in a changing environment (Figure 26.3). There are also important feedbacks among ecosystem engineers, such as termites, ants and the microbes of the biological soil crusts that recycle nutrients, change infiltration patterns and moderate the availability of resources to other components of the system. Below we highlight just a few of these to illustrate the depth of knowledge required to fully understand the function of hummock grasslands.

26.9.1 Grassland–shrubland boundaries

Globally, the boundaries between grasslands and shrublands can be maintained by climate, soil, fire, herbivory or combinations of these factors and are intriguing as grasslands may occur in climatic regions that could in principle support woody vegetation, see review by Bond (2008). Hummock grasslands share boundaries with mulga and other acacia woodlands that appear to be controlled by both dynamic and static landscape features (Bowman *et al.* 1994; Nano and Clarke 2008, 2010; Nano *et al.*, Chapter 25, this volume)

26.9.2 Pollinator networks of associated species

A relatively unknown diversity of invertebrate pollinators interacts with the many plant species associated with *Triodia* grasslands (Popic and Wardle 2011). The rich carbohydrate rewards of nectar (Morton *et al.* 2011) and pollen attract many visitors to flowers especially in years of high rainfall as experienced in central Australia in 2010–2011 (Wardle *et al.* 2013). In a study in *T. basedowii* grasslands in the Simpson Desert, a total of 436 invertebrate species were recorded from 432 hours of net sampling and 1296 pan trap-days (Popic *et al.* 2013). Composition was dominated by Hymenoptera, including wasps (180 spp., 41%), native bees (72 spp., 17%) and ants (26 spp., 6%); as well as Diptera (46 spp., 11%), Hemiptera (45 spp., 10%), Coleoptera (38 spp., 9%) and various Lepidoptera, Orthoptera and Thysanoptera. Overall, 109 bee species, all native and solitary were discovered, with 33 of these being new species. Both species richness and abundance were positively related to the number of flowering species, which in turn was predicted by the previous 9-month cumulative rainfall. Plant-invertebrate networks were highly variable over space and time, and were generally nested in structure, with more specialised species interacting with generalist species (Popic *et al.* 2013).

26.9.3 Termites as ecosystem engineers

Termites take advantage of the large amounts of indigestible biomass that develops on the infertile soils in hummock grasslands (Morton *et al.* 2011). As detritivores, they survive on matter too protein poor and ligneous for vertebrate herbivores and benefit from the economies of scale of colonial life, with a slow and seasonal metabolism. Termites in hummock grasslands tend to be below ground in 'pavements' and not mound-forming, as they are in northern Australia. Their activities move soils and provide food for vertebrate consumers, such as lizards, that in turn, burrow, dig and alter soil properties (Eldridge *et al.*, Chapter 6, this volume). Few studies have quantified their abundance but they make a substantial contribution to breaking down and recycling spinifex tissues (Orians and Milewski 2007), rivalled only by fire (Morton *et al.* 2011).

26.9.4 Ants and other invertebrates

Ants are a diverse group in hummock grasslands and ecologically important as granivores, consumers of plant exudates, and as prey for vertebrates (Andersen 2000). Data from a 6-year survey in the Simpson Desert show ants were the most abundant invertebrate taxon, comprising 83% of all invertebrates captured, from a total of 174 381 invertebrates and 32 orders (Kwok *et al.* 2015). Invertebrate abundance and composition were highly dynamic and responses to environmental change were taxon specific, but six taxa (Acarina, ants, Araneae, Coleoptera, Collembola and Thysanura) showed a positive relationship with the cover of non-*Triodia* vegetation. Certain taxa also showed positive relationships with other indicators of vegetative productivity (seeding, flowering).

26.9.5 Biological soil crusts and soil microorganisms

Biological soil crusts consisting of cyanobacteria, lichens and mosses occur on sand plains in the hummock grasslands and are functionally important (Tongway and Smith 1989), as they bind and stabilise soil particles (Eldridge 2001), improve water infiltration, fix atmospheric nitrogen and reduce water and wind erosion (Eldridge and Green 1994). The intimate functional relationships between plant composition and microbiotic crusts remains under-appreciated in hummock grasslands but is an important part of the productivity of these communities.

26.10 Major conservation issues and threatening processes

The remoteness of the hummock grasslands from high-density human populations has prevented large-scale vegetation clearing (< 1% since 1788 (Dickman *et al.* 2014)), but it also means that there are few people and resources to manage them effectively (Stafford Smith and Cribb 2009). For example, the Australian Government designated 23 Conservation Management Zones, based on ecology and threat characteristics, with a single large Arid Shrubland and Desert zone covering 48% of the continent (or 22% of the National Reserve System). The design of the zonation implies that the drivers within larger-scale regions with lower land-use intensity are more uniform and less complex. However, this claim is in stark contrast to the recommendation from Smyth *et al.* (2012) that large and diverse drivers of this biome prevented it from being managed as a single unit. Below we outline how pastoralism, mining, groundwater extraction and climate change pose risks for the biodiversity of hummock grasslands.

26.10.1 Pastoralism, ferals, exotic grasses and altered fire regimes

Pastoral production dominates in the northern and eastern arid regions of hummock grasslands, (Bastin and ACRIS Management Committee 2008) with cattle in the drier regions (Figure 26.1i) and sheep or mixtures elsewhere (Smyth *et al.* 2012). Generally, artificial water sources (bores, dams) promote concentrations of grazing pressure and environmental degradation through soil erosion, trampling of soil crusts and changes to vegetation and fauna (Frank *et al.* 2013, 2014), but degradation is less clear in short-term studies (e.g. Fensham *et al.* 2014). Kangaroos and feral herbivores such as camels or goats increase with additional waterpoints or large rains events, so it is important to consider total grazing pressure in any assessment of impact (Frank *et al.* 2015). Pastoralism has also led to the spread of weeds such as the exotic pastures grass, buffel grass (*Cenchrus ciliaris* L.), which poses a threat to arid ecosystems through accumulated dead material that fuels altered fire regimes (Nano *et al.*, Chapter 25, this volume).

Since European settlement, burning has been used to improve pasture value and prevent large-scale fires but the implications of these altered fire regimes for the ecosystem are not well understood. Managing fire for biodiversity is complex. Our knowledge of fire behaviour, fire weather and the seasonality, frequency and intensity of fires is steadily progressing (Bradstock *et al.* 2012; Nano *et al.* 2012) but there remains uncertainty about pre-European historic fire regimes and how global climate change will play out on this important process. The intertwining of cultural and ecological knowledge is advancing in the northern savannas (Russell-Smith *et al.* 2009), but elsewhere more work is needed to understand the myriad plant and animal responses to fire and how they vary. Fire management will remain a challenge in hummock grasslands given their remoteness and vastness, but no intervention may be the most precautionary approach until the evidence is attained to manage better. As mentioned, recent research emphasises the importance of retaining long-unburnt patches for native fauna protection against feral predators. For hummock grassland plants, however, there are no clear examples of where too much or too little fire is a threatening process.

26.10.2 Mining, clearing and restoration

The continued extraction of groundwater for pastoralism or mining is becoming unsustainable and can result in lowering of water tables, decreased environmental flows to groundwater-dependent ecosystems, movement of saline water into aquifers, land subsidence, and a decline in the amount of recharge to the base flow of river systems (Smyth *et al.* 2012). Loss of groundwater in turn threatens plants, stygofauna and humans (Dickman *et al.* 2014). As rainfall is critical to the recharge of groundwater, any shift in the frequency or size of rainfall events will exacerbate these losses from land use.

An emerging threat is the potential to commercially harvest *Triodia* from natural grasslands for biomaterials. Research is underway to evaluate the material properties of *Triodia* biomass and resin (Gamage *et al.* 2012; Pitman and Wallis 2012). Aboriginal people traditionally valued and used *Triodia* in many ways and the new efforts to incorporate novel income streams as part of desert living by partnering with commercial operations are clearly important socially. What is uncertain is how the vegetation can be harvested in a sustainable manner. Mining industry requirements for restoration of disturbed areas in Western Australia have increased efforts to understand how to revegetate large areas of *Triodia* but as already mentioned, seedling recruitment can be unpredictable, and we need to know more about how dormancy and dispersal operate (Erickson *et al.* 2016).

26.10.3 Global climate change

Nationally, climate change impacts on the largest relatively intact biome in Australia are important. Economically, the assets of the hummock grass biome support major industries, including pastoralism, mining and ecotourism, worth billions to the national economy (Smyth *et al.* 2012). Present biodiversity of the biome is well-adapted to aridity, but increasing aridity together with fire may impact biodiversity on many levels via physiological and phenological stress. Based on bioclimatic modelling, and with full acknowledgement of the uncertainties inherent in the models, it is predicted that areas of hummock grasslands will become hotter and more arid by 2070 (Smyth *et al.* 2012). These environmental shifts will vary between regions, with more rainfall in the northern Pilbara but decreases elsewhere, temperatures will increase by 3–7°C with effects more pronounced in summer. Such changes will lead to compositional and structural changes in ecological communities and push many taxa past their ecophysiological thresholds. Plants, reptiles and snails are expected to be most impacted by climate change, mammals moderately and birds and frogs to experience the lowest impacts (Smyth *et al.* 2012; Pavey 2014). Planning for these shifts needs to account for the losses of familiar communities and reassembly of novel ones and for the considerable landscape heterogeneity that – for some species at least – may present refugia options, such as in the MacDonnell Ranges (Reside *et al.* 2013).

26.11 Recommendations for future work

There is a strong and growing foundation of knowledge about hummock grasslands but the gaps are just as striking. Where do they occur and why? How do they function and persist or reform diverse assemblages in a highly event-driven environment? What level of disturbance would lead to a transformation and loss of the defining composition and functional components of the ecosystem, or ecosystem collapse? These are some of the more compelling questions to address. Future studies that would be fruitful to answer these are, firstly, to undertake fine-scale mapping and classification across the extent to provide a richer baseline to detect change over time and space. Fortunately, there are many opportunities for integrating new technologies such as Lidar, drones, and remotely sensed data to construct a real-time, fine-scale assessment of vegetation composition and structural change over space and time. Second, better tracking of changes in phenology (Figure 2.12j), reproductive success of *Triodia*, and intermittent recruitment opportunities for shrubs, and other plant species can be tracked with a series of phenocams that continuously record levels of vegetation greenness, structural changes and reproduction. Detailed experimental studies are needed to probe further into the causes of mast seeding and responses to fire, as well as how species such as termites exploit the resources provided by hummock grasses. Third, we need to know more about the importance of groundwater to some parts of this vegetation system, and the fauna it supports, as there is increased pressure to extract water to support pastoral and mining development in remote areas. Fourth, integrating plot-based data with broad-scale, remote sensing datasets and experimental studies on ecosystem function would improve the long-term forecasting of likely responses to ecosystem change. Finally, a formal ecosystem risk assessment would be useful to understand how ecosystem collapse might eventuate and what could be done to mitigate against the increasing pressure to develop economic return from these areas.

Acknowledgements

The following people are thanked for information or data that contributed to this chapter: Chris Dickman, Aaron Greenville, Emrys Leitch, Vuong Nguyen, Ben Sparrow, Bobby Tamayo and Stephen van Leeuwen. Previous discussions with Margaret Friedel, Barbara Rice (deceased), Mark Westoby, Peter Clarke (deceased) and Steve Morton were influential to the formation of ideas. We would also like to thank David Keith for valuable guidance and comments on earlier drafts of this chapter. The work was supported by ARC Discovery Grants (GMW) and NCRIS funding to support the Long Term Ecological Research Network (LTERN). We would like to acknowledge the traditional owners of the lands across the entire distribution of the spinifex hummock grasslands for their custodianship and care of this natural and cultural heritage.

References

Allan, G.E. and Southgate, R.I. (2002) Fire regimes in the *Spinifex* landscapes of Australia. In *Flammable Australia: Fire Regimes, Biodiversity and Ecosystems in a Changing World*, ed. R.A. Bradstock, A.M. Gill and R.J. Williams. CSIRO Publishing: Melbourne, pp. 145–176.

Andersen, A. (2000) *Ants of Northern Australia: A Guide to the Monsoonal Fauna.* CSIRO Publishing: Melbourne.

Anderson, B.M., Barrett, M.D., Krauss, S.L. and Thiele, K. (2016) Untangling a species complex of arid zone grasses (*Triodia*) reveals patterns congruent with co-occurring animals. *Molecular Phylogenetics and Evolution* 101, 1–21.

Armstrong, G. (2011) Evidence for the equal resilience of *Triodia* spp. (Poaceae), from different functional groups, to frequent fire dating back to the late Pleistocene. *Heredity* 107, 558–564.

Armstrong, G. and Garnett, S.T. (2011) Landscape partitioning among *Triodia* spp. (Poaceae) in the fire prone Kimberley, north-west Australia. *Austral Ecology* 36, 1–9.

AusPlots (n.d.) Available at: http://www.ausplots.org/

Australian Government (n.d) National Reserve System. Available at: https://www.environment.gov.au/indigenous/ipa/index.html (accessed 2 November 2015).

Bastin, G.N. ACRIS (Australian Collaborative Rangelands Information System) Management Committee) (2008) *Taking the Pulse.* ACRIS Management Committee and the National Land and Water Resources Audit: Canberra.

Beadle, N.C.W. (1981) *The Vegetation of Australia*. Cambridge University Press: Cambridge.

Beadle, N.C.W. and Costin, A.B. (1952) Ecological classification and nomenclature. *Proceedings of the Linnean Society of New South Wales* 77, 61–82.

Beard, J.S. (1984) The vegetation of the Australian arid zone. In *Arid Australia*, ed. H.G. Cogger and E.E. Cameron. Australian Museum: Sydney, pp. 113–119.

Beard, J.S. and Webb, M.J. (1974) *The Vegetation Survey of Western Australia: Its Aims, Objects and Methods Part I of Great Sandy Desert 1:1 000 000 Vegetation Series: Explanatory Notes to Sheet 2*. University of Western Australia Press: Nedlands.

Beard, J.S., Beeston, G.R., Harvey, J.M. *et al.* (2013) The vegetation of Western Australia at the 1:3 000 000 scale. Explanatory memoir, 2nd edition. *Conservation Science Western Australia* 9, 1–152. Available at: http://www.dpaw.wa.gov.au/cswajournal (accessed 15 August 2016).

Bogusiak, A., Rice, B., Westoby, M. and Friedel, M. (1990) Seedling emergence of hummock grasses in relation to the effects of fire. *The Rangeland Journal* 12, 25–28.

Bond, W.J. (2008) What limits trees in C_4 grasslands and savannas? *Annual Review of Ecology, Evolution, and Systematics* 39, 641–659.

Bond, W.J. and Keely, J.E. (2005) Fire as a global 'herbivore': the ecology and evolution of flammable ecosystems. *Trends in Ecology and Evolution* 20, 387–394.

Bowman, D.M.J.S., Latz, P.K. and Panton, W.J. (1994) Pattern and change in an *Acacia-aneura* shrubland and *Triodia* hummock grassland mosaic on rolling hills in central Australia. *Australian Journal of Botany* 43, 25–37.

Bradstock, R.A., Gill, A.M. and Williams, R.J. (eds) (2012) *Flammable Australia: Fire Regimes, Biodiversity and Ecosystems in a Changing World.* CSIRO Publishing: Melbourne.

Burbidge, N.T. (1944) Ecological succession observed during regeneration of *Triodia pungens* R. Br. after burning. *Journal of the Royal Society of Western Australia* 28, 149–156.

Burrows, N.D., Burbidge, A.A., Fuller, P.J. and Behn, G. (2006) Evidence of altered fire regimes in the Western Desert Region of Australia. *Conservation Science Western Australia* 5, 272–284.

Byrne, M., Yeates, D.K., Joseph, L. *et al.* (2008) Birth of a biome: insights into the assembly and maintenance of the Australian arid zone biota. *Molecular Ecology* 17, 4398–4417.

Casson, N.E. and Fox, J.E.D. (1987) The post-fire regeneration responses of *Triodia wiseana* and *T. basedowii. The Australian Rangeland Journal* 9, 53–55.

Charley, J.L. and Cowling, R.W. (1968) Changes in soil nutrient status resulting from overgrazing and their consequences in plant communities of some arid areas. *Proceedings of the Ecological Society of Australia* 3, 28–38.

Cheney, P. and Sullivan, A. (2008) *Grassfires: Fuel, Weather and Fire Behaviour*, 2nd edition. CSIRO Publishing: Melbourne.

Crisp, M.D. and Cook, L.G. (2013) How was the Australian flora assembled over the last 65 million years? A molecular phylogenetic perspective. *Annual Review of Ecology, Evolution, and Systematics* 44, 303–324.

Crisp, M.D., Mant, J., Toon, A. and Cook, L.G. (2015) Australian spinifex grasses: new names in *Triodia* for *Monodia* and *Symplectrodia*. *Phytotaxa* 230, 293–296.

CSIRO (n.d.) Available at: http://www.clw.csiro.au/aclep/soilandlandscapegrid/.

Department of the Environment and Water Resources (2007) *Australia's Native Vegetation: A Summary of Australia's Major Vegetation Groups, Version 3.0*. Australian Government: Canberra.

Department of Sustainability, Environment, Water, Population and Communities (2012) *Interim Biogeographic Regionalisation for Australia*, Version 7 (IBRA7). Australian Government:

Canberra. Available at: https://www.environment.gov.au/land/nrs/science/ibra#ibra (accessed 15 August 2016).

Dickman, C.R., Wardle, M., Foulkes, J. and de Preu, N. (2014) Desert complex environments. In *Biodiversity and Environmental Change: Monitoring, Challenges and Direction*, ed. D. Lindenmayer, E. Burns, N. Thurgate, A. Lowe. CSIRO Publishing: Melbourne, pp. 379–438.

Dixon, A.P., Faber-Langendoen, D., Josse, C. *et al.* (2014) Distribution mapping of world grassland types. *Journal of Biogeography* 41, 2003–2019.

Ehleringer, J.R., Phillips, S.L., Schuster, W.S.F. and Sandquist, D.R. (1991) Differential utilization of summer rains by desert plants. *Oecologia* 88, 430–434.

Eldridge, D.J. (2001) Biological soil crusts of Australia. In *Biological Soil Crusts: Structure, Function and Management*, ed. J. Belnap and O. Lange. Springer-Verlag: Berlin, pp. 119–132.

Eldridge, D.J. and Green, R.S.B. (1994) Microbiotic soil crusts: a review of their roles in soil and ecological processes in the rangelands of Australia. *Australian Journal of Soil Research* 32, 389–415.

English, N.B., Weltzin, J.F., Fravolini, A. *et al.* (2005) The influence of soil texture and vegetation on soil moisture under rainout shelters in a semi-desert grassland. *Journal of Arid Environments* 63, 324–343.

Erickson, T.E., Shackelford, N., Dixon, K.W. *et al.* (2016) Overcoming physiological dormancy in seeds of *Triodia* (Poaceae) to improve restoration in the arid zone. *Restoration Ecology* doi:10.1111/rec.12357.

Farley, G.F., Bellairs, S.M. and Adkins, S.W. (2013) Germination of selected Australian native grass species, with potential for minesite rehabilitation. *Australian Journal of Botany* 61, 283–290.

Fensham, R.J., Silcock, J.L. and Firn, J. (2014) Managed livestock grazing is compatible with the maintenance of plant diversity in semidesert grasslands. *Ecological Applications* 24, 503–517.

Fernandez, R.J. (2007) On the frequent lack of response of plants to rainfall events in arid areas. *Journal of Arid Environments* 68, 688–691.

Frank, A.S.K., Dickman, C.R., Wardle, G.M. and Greenville, A.C. (2013) Interactions of grazing history, cattle removal and time since rain drive divergent short-term responses by desert biota. *Plos ONE* 8, e68466.

Frank, A.S.K., Wardle, G.M., Dickman, C.R. and Greenville, A.C. (2014) Habitat- and rainfall-dependent biodiversity responses to cattle removal in an arid woodland–grassland environment. *Ecological Applications* 24, 2013–2028.

Frank, A.S.K., Wardle, G.M., Greenville, A.C. and Dickman, C.R. (2015) Cattle removal in arid Australia benefits kangaroos in high quality habitat but does not affect camels. *The Rangeland Journal* 38, 73–84.

Gamage, H.K., Mondal, S., Wallis, L.A. *et al.* (2012) Indigenous and modern biomaterials derived from *Triodia* ('spinifex') grasslands in Australia. *Australian Journal of Botany* 60, 114–127.

Gamage, H.K., Memmott, P., Firn, J. and Schmidt, S. (2014) Harvesting as an alternative to burning for managing spinifex grasslands in Australia. *Advances in Ecology 2014*, Article ID 430431.

Getzin, S., Yizhaq, H., Bell, B. *et al.* (2016) Discovery of fairy circles in Australia supports self-organization theory. *Proceedings of the National Academy of Sciences of the United States of America* 113, 3551–3556.

Gill, A.M. (2000) *Fire-pulses in the heart of Australia: fire regimes and fire management in central Australia. Report to Environment Australia and CSIRO.* Centre for Plant Biodiversity Research: Canberra.

Gill, A.M., York, A. and Woinarski, J.C.Z. (1999) *Australia's Biodiversity. Responses to Fire: Plants, Birds and Invertebrates*. Department of Environment and Heritage: Canberra.

Greenville, A.C., Dickman, C.R., Wardle, G.M. and Letnic, M. (2009) The fire history of an arid grassland: the influence of antecedent rainfall and ENSO. *International Journal of Wildland Fire* 18, 631–639.

Greenville, A.C., Wardle, G.M. and Dickman, C.R. (2012) Extreme climatic events drive mammal irruptions: regression analysis of 100-year trends in desert rainfall and temperature. *Ecology and Evolution* 2, 2645–2658.

Greenville, A.C., Wardle, G.M. and Dickman, C.R. (2013) Extreme rainfall events predict irruptions of rat plagues in central Australia. *Austral Ecology* 38, 754–764.

Griffin, G.F. (1984) Hummock grasslands. In *Management of Australia's Rangelands*, ed. G.N. Harrington, A.D. Wilson and M.D. Young. CSIRO Publishing: Melbourne, pp. 271–284.

Griffin, G.F. (1990) Characteristics of three *Spinifex* alliances in central Australia. *Journal of Vegetation Science* 1, 435–444.

Grigg, A. (2009) An ecophysiological approach to determine problems associated with mine-site rehabilitation: a case study in the Great Sandy Desert, north-western Australia. PhD Thesis, University of Western Australia, Perth.

Grigg, A.M., Veneklaas, E. and Lambers, H. (2008) Water relations and mineral nutrition of *Triodia* grasses on desert dunes and interdunes. *Australian Journal of Botany* 56, 408–421.

Holmgren, M., Stapp, P., Dickman, C.R. *et al.* (2006) Extreme climatic events shape arid and semiarid ecosystems. *Frontiers in Ecology and the Environment* 4, 87–95.

Jacobs, S.W.L. (1992) *Spinifex* (*Triodia*, *Plectrachne*, *Symplectrodia* and *Monodia*: Poaceae) in Australia. In *Desertified Grasslands: Their Biology and Management*, ed. G.P. Chapman. Academic Press: London, pp. 47–62.

Jacobs, S.W.L. (2004) The tribe Triodieae (Chloridoideae: Gramineae). *Telopea* 10, 701–703.

Kerezsy, A., Balcombe, S., Tischler, M. and Arthington, A. (2013) Fish movement strategies in an ephemeral river in the Simpson Desert, Australia. *Austral Ecology* 38, 798–808.

Kimber, R.G. and Friedel, M.H. (2015) Challenging the concept of Aboriginal mosaic fire practices in the Lake Eyre Basin. *The Rangeland Journal* 37, 623–630.

Kinloch, J.E. and Friedel, M.H. (2005) Soil seed reserves in arid grazing lands of central Australia. Part 1: seed bank and vegetation dynamics. *Journal of Arid Environments* 60, 133–161.

Kwok, A.B.C., Wardle, G.M., Greenville, A.C. and Dickman, C.R. (2015) Long-term patterns of invertebrate abundance and relationships to environmental predictors in arid Australia. *Austral Ecology* 41, 480–491.

Lazarides, M. (1997) A revision of *Triodia* including Plectrachne (Poaceae, Eragrostideae, Triodiinae). *Australian Systematic Botany* 10, 381–489.

Mant, J.G., Bayer, R.J., Crisp, M.D. and Trueman, J.W.H. (2000) A phylogeny of Triodieae (Poaceae: Chloridoideae) based on the ITS region of nrDNA: testing conflict between anatomical and inflorescence characters. In *Grasses: Systematics and Evolution*, ed. S.W.L. Jacobs and J. Everett. CSIRO Publishing: Melbourne, pp. 213–217.

Martin, H.A. (2006) Cenozoic climate change and the development of the arid vegetation in Australia. *Journal of Arid Environments* 66, 533–563.

McWilliam, J.R. and Mison, K. (1974) Significance of the C_4 pathway in *Triodia irritans* (Spinifex), a grass adapted to arid environments. *Australian Journal of Plant Physiology* 1, 171–175.

Mooney, S.D., Harrison, S.P., Bartlein, P.J. *et al.* (2011) Late Quaternary fire regimes in Australasia. *Quaternary Science Reviews* 30, 28–46.

Morton, S.R., Stafford Smith, D.M., Dickman, C.R. *et al.* (2011) A fresh framework for the ecology of arid Australia. *Journal of Arid Environments* 75, 313–329.

Mulvaney, D.J. and Kamminga, J. (1999) *Prehistory of Australia*. Allen and Unwin: St Leonards.

Nano, C.E.M. and Clarke, P.J. (2008) Variegated desert vegetation: covariation of edaphic and fire variables provides a framework for understanding mulga-spinifex coexistence. *Austral Ecology* 33, 848–862.

Nano, C.E.M. and Clarke, P.J. (2010) Woody-grass ratios in a grassy arid system are limited by multi-causal interactions of abiotic constraint, competition and fire. *Oecologia* 162, 719–732.

Nano, C.E.M. and Pavey, C.R. (2013) Refining the 'pulse-reserve' model for arid central Australia: seasonal rainfall, soil moisture, and plant productivity in sand ridge and stony plain habitats of the Simpson Desert, Australia. *Austral Ecology* 38, 741–753.

Nano, C.E.M., Clarke, P.J. and Pavey, C.R. (2012) Fire regimes in arid hummock grasslands and *Acacia* shrublands. In *Flammable Australia: Fire Regimes, Biodiversity and Ecosystems in a Changing World*, ed. R.A. Bradstock, A.M. Gill and R.J. Williams. CSIRO Publishing: Melbourne, pp. 195–214.

Neely, C., Bunning, S. and Wilkes, A. (eds) (2009) *Review of evidence on drylands pastoral systems and climate change: implications and opportunities for mitigation and adaptation.* Land Tenure and Management Unit (NRLA) Land and Water Division, Food and Agriculture Organization of the United Nations (FAO): Rome.

Noy-Meir, I. (1973) Desert ecosystems: environment and producers. *Annual Review of Ecology and Systematics* 4, 25–51.

Orians, G.H. and Milewski, A.V. (2007) Ecology of Australia: the effects of nutrient-poor soils and intense fires. *Biological Reviews* 82, 393–423.

Osborne, C.S. and Freckleton, R.P. (2009) Ecological selection pressures for C_4 photosynthesis in the grasses. *Proceedings of the Royal Society B* 276, 1753–1760.

Pavey, C.R. (2014) *Australian Rangelands and Climate Change: Native Species.* Ninti One Limited and CSIRO Publishing: Alice Springs.

Pavey, C.R. and Nano, C.E.M. (2013) Changes in richness and abundance of rodents and native predators in response to extreme rainfall in arid Australia. *Austral Ecology* 38, 777–785.

Penman, T.D., Christie, F.J., Andersen, A.N. *et al.* (2011) Prescribed burning: how can it work to conserve the things we value? *International Journal of Wildland Fire* 20, 731–733.

Pitman, H.T. and Wallis, L.A. (2012) The point of spinifex: Aboriginal uses of spinifex grasses in Australia. *Ethnobotany Research and Applications* 10, 109–131.

Popic, T.J. and Wardle, G.M. (2011) Extremes: understanding flower–visitor interactions in a changing climate. In *Wildlife and Climate Change: Towards Robust Conservation Strategies for Australian Fauna*, ed. C. Dickman and D. Lunney. Royal Zoological Society of New South Wales: Mosman, pp. 99–106.

Popic, T.J., Wardle, G.M. and Davila, Y.C. (2013) Plant-floral visitor networks only partially predict the function of pollen transport by bees. *Austral Ecology* 38, 76–86.

Reside, A.E., VanDerWal, J.J., Phillips, B.L. *et al.* (2013) *Climate Change Refugia for Terrestrial Biodiversity*. National Climate Change Adaptation Research Facility: Gold Coast. Available at: https://www.nccarf.edu.au/publications/climate-change-refugia-terrestrial-biodiversity (accessed 15 August 2016).

Rice, B. and Westoby, M. (1999) Regeneration after fire in *Triodia* R.Br. *Austral Ecology* 24, 563–572.
Rice, B., Westoby, M., Griffin, G. and Friedel, M. (1994) Effects of supplementary soil nutrients on hummock grasses. *Australian Journal of Botany* 42, 687–703.
Russell-Smith, J., Whitehead, P. and Cooke, P. (2009) *Culture, Ecology and Economy of Fire Management in North Australian Savannas: Rekindling the Wurrk Tradition.* CSIRO Publishing: Melbourne.
Sage, R.F. and Monson, R.K. (eds) (1999) *C_4 Plant Biology*. Academic Press: San Diego.
Smyth, A.K., Hilbert, D.W., Ferrier, S. *et al.* (2012) *The Implications of Climate Change for Biodiversity Conservation and the National Reserve System: Hummock Grasslands Biome.* CSIRO Climate Adaptation Flagship: Canberra.
Specht, R.L. and Specht, A. (1999) *Australian Plant Communities: Dynamics of Structure, Growth and Biodiversity*. Oxford University Press: South Melbourne.
Stafford Smith, M. and Cribb, J. (2009) *Dry Times. Blueprint for a Red Land.* CSIRO Publishing: Collingwood.
Stevenson, J. and Hope, G.S. (2005) A comparison of late Quaternary forest changes in New Caledonia and northeastern Australia. *Quaternary Research* 64, 372–383.
Suijendorp, H. (1981) Responses of the hummock grasslands of north-western Australia to fire. In *Fire and the Australian Biota*, ed. A.M. Gill, R.H. Groves and I.R. Noble. Australian Academy of Science: Canberra, pp. 417–424.
Tischler, M., Dickman, C.R. and Wardle, G.M. (2013) Bird functional group responses to rainfall across four vegetation types in the Simpson Desert, central Australia. *Austral Ecology* 38, 809–819.
Tongway, D.J. and Smith, E.L. (1989) Soil surface features as indicators of rangeland site productivity. *Australian Rangeland Journal* 11, 15–20.
Toon, A., Crisp, M.D., Gamage, H.J. *et al.* (2015) Key innovation or adaptive change? A test of leaf traits using Triodiinae in Australia. *Scientific Reports* 5, 1–12.
Travers, S.K. and Eldridge, D.J. (2013) Increased rainfall frequency triggers an increase in litter fall rates of reproductive structures in an arid eucalypt woodland. *Austral Ecology* 38, 820–830.
Turner, D., Ostendorf, B. and Lewis, M. (2008) An introduction to patterns of fire in arid and semi-arid Australia, 1998–2004. *The Rangeland Journal* 30, 95–107.
Van Etten, E.J.B. (2009) Interannual rainfall variability of arid Australia: greater than elsewhere? *Australian Geographer* 40, 109–120.
Watson, I.W., Novelly, P.E. and Thomas, P.W.E. (2007) Monitoring changes in pastoral rangelands: the Western Australian Rangeland Monitoring System (WARMS). *The Rangeland Journal* 29, 191–205.
Wardle, G.M., Pavey, C.R. and Dickman, C.R. (2013) Greening of arid Australia: new insights from extreme years. *Austral Ecology* 38, 731–740.
Wardle, G.M., Greenville, A.C., Frank, A.S.K. *et al.* (2015) Ecosystem risk assessment of Georgina gidgee woodlands in central Australia. *Austral Ecology* 40, 444–459.
White, A., Sparrow, B., Leitch, E. *et al.* (2012) AusPlots rangelands survey protocols manual. Available at: http://www.tern.org.au/AusPlots-Rangelands-Survey-Protocols-Manual-pg23944.html (accessed 15 August 2016).
Winkworth, R.E. (1967) The composition of several arid spinifex grasslands of central Australia in relation to rainfall, soil water relations and nutrients. *Australian Journal of Botany* 15, 107–130.

World Resources Institute (2000) *World Resources 2000-2001. People and Ecosystems: The Fraying Web of Life*. Elsevier: Amsterdam.

Wright, B.R. and Clarke, P. (2007) Fire regime (recency, interval and season) changes the composition of spinifex (*Triodia* spp.)-dominated desert dunes. *Australian Journal of Botany* 55, 709–724.

Wright, B.R. and Clarke, P. (2008) Relationships between soil temperature and properties of fire in feathertop spinifex (*Triodia schinzii* (Henrard) Lazarides) sandridge desert in central Australia. *The Rangeland Journal* 30, 317–325.

Wright, R., Zuur., A.F. and Chan, G.C.K. (2014) Proximate causes and possible adaptive functions of mast seeding and barren flower shows in spinifex grasses (*Triodia* spp.) in arid regions of Australia. *Rangeland Journal* 36, 297–308.

PART III Conclusion

27 Conservation of Australian Vegetation

DAVID A. KEITH AND TONY D. AULD

Australian vegetation is undergoing its greatest transformation since desert shrublands and grasslands replaced extensive forests across the continent during the Miocene aridification 10-15 million years ago. The current wave of change is due entirely to post-industrial human activity, resulting in simplification and homogenisation of native vegetation or its replacement by introduced crops, pastures and urban or industrial infrastructure. Consequently, the unique character of Australian vegetation is being rapidly eroded. Vegetation types in humid and subhumid climates on flat land with productive soils and wetlands are disproportionately affected. We review the principal threatening processes involved in the transformation: vegetation clearing and fragmentation; direct harvesting; changes to fire and flood regimes; eutrophication of soils and water; disruption of biological interactions including overgrazing and invasions of exotic plants and pathogens. Scientific capacity to assess risks and monitor trends in Australia's vegetation and associated biodiversity is rapidly developing, but there are lags and enduring inconsistencies in implementation and reporting. In recent decades, action has begun to redress the losses and degradation of Australian vegetation. Considerable advances have been made by: i) increasing the area and representativeness of the protected area network; ii) adopting laws and policies to promote conservation outside protected areas; iii) developing and implementing strategies for conservation management and threat abatement; and iv) ecological restoration. Despite this good progress, recent acceleration of human population growth and anthropogenic climate change continue to force an irreversible decline in vegetation diversity and to exacerbate the impacts of other threats. Australian ecosystems will continue to become markedly less diverse and more similar to other parts of the world unless more progressive and effective conservation action is resourced and implemented.

27.1 Introduction

Often known as the 'red continent', the majority of the Australian land mass is desert. Much of this desert became arid 10–15 Ma during the Miocene, as extensive forest mosaics with both mesophyll and sclerophyll elements were replaced by open and more fire-prone xeromorphic shrublands and grasslands, a transformation that involved many plant extinctions (Weston and Jordan, Chapter 2 this volume). As a consequence, much of Australia's relic rainforest is characterised by 'deep diversity', with many co-occurring plant families and orders that are each represented by comparatively few species (Adam 1994; Keith and Tozer, Chapter 1, this volume). In contrast, the Miocene aridification,

Australian Vegetation, Third Edition, ed. David A. Keith. Published by Cambridge University Press.

Citation: Keith, D.A. and Auld, T.D. (2017). Conservation of *Australian Vegetation*. In D. A. Keith, ed., *Australian Vegetation*, 3rd edn. Cambridge: Cambridge University Press, pp. 677–710.

along with subsequent Pleistocene climatic fluctuations, fuelled a massive radiation of species into newly arid and seasonally dry regions. Unlike the deep diversity of rainforests, the majority of species within Australian sclerophyll and xeromorphic plant communities come from relatively few species-rich genera and families. Changes in fire regimes were intimately linked with climatic dynamism and plant evolution within Australian sclerophyllous and xeromorphic forests, woodlands, shrublands and grasslands (Crisp *et al.* 2011).

Some of the stocks for this radiation were relictual taxa, such as *Banksia* (Cardillo and Pratt 2013) and *Eucalyptus* (Thornhill *et al.* 2015), whose progenitors occupied Australia before it rifted from other southern land masses. The contemporary native vegetation of Australia also includes derivatives of a large contingent of arrivals from transoceanic dispersal and northern incursions as Australia drew closer to Asia. Many of these migrations occurred long ago and were followed by substantial diversification, as is the case for *Acacia* (Weston and Jordan, Chapter 2, this volume) and *Triodia* (Toon *et al.* 2015), the iconic wattles and spinifex that dominate much of Australia's arid zone. The backdrop for diversification under climatic dynamism is a landscape mosaic comprising extensive old land surfaces with impoverished soils and smaller areas of rejuvenated soils (Hopper 2009; Brundrett *et al.*, Chapter 7, this volume), with topographic relief defining local climatic refugia, primarily in the east of the continent.

The outcome of these evolutionary processes is a rich tapestry of vegetation in which a large core of arid shrublands and grasslands is mostly surrounded by a belt of forests, savanna and temperate woodlands, heathlands, grasslands and herbfields. Both humid and arid landscapes are peppered throughout by wetlands that vary from peaty mires to ephemeral salt lakes that are more commonly dry or hypersaline than wet. While the overall diversity of Australia's flora is unexceptional on a world scale (Mutke and Barthlott 2005), the levels of endemism are extreme, with more than 90% of species found nowhere else. This endemism, together with distinctive growth forms such as open-canopy trees and hummock grasses, contribute a degree of uniqueness that has been influential on Australia's human population since the first migrants arrived some 50 000 years ago. Aboriginal lifestyles, beliefs and subsistence became essentially defined by their concept of 'country' and the regional variations in landscapes, vegetation and fauna that define it (Ens *et al.*, Chapter 4, this volume). In a somewhat similar vein, Australian vegetation, soils and climate became part of a socio-economic system and cultural identity shaped by the post-industrial arrivals of European people and subsequently others. For many, the Australian environment contrasts markedly with their places of origin and now features heavily in contemporary Australian art, sport and music, and profoundly influences economic activity and production of the nation. Australian vegetation is deeply valued, both for its intrinsic qualities and for its utilitarian assets.

While the pre-historical activities of Aboriginal people influenced Australian ecosystems to a largely unknown degree (Ens *et al.*, Chapter 4, this volume; Mooney *et al.*, Chapter 3, this volume), the post-industrial impact of humans has been profound. In eking out a modern existence, humans are transforming Australian vegetation in ways that erode its diversity and uniqueness. The most obvious transformations are those that involve clearance of native vegetation and its replacement with crops, pastures or built infrastructure. Some of the remaining vegetation is subject to industrial overexploitation of resources such as water, soil, wood and forage. Less direct, though generally more pervasive processes of transformation involve invasive species, changes to water quality, soils and fire regimes, and most recently anthropogenic climate change. Responses designed to arrest the decline of biodiversity began in the late-nineteenth century and gathered momentum in the second half of the twentieth century, with development of protected area networks, regulatory legislation and policies, conservation incentives, threat abatement and ecological restoration.

In this chapter, we first review the major threats to Australian biodiversity, using vegetation formations (Keith and Tozer, Chapter 1, this volume) to characterise variation in the incidence and impact of threatening processes across the continent, as well as the varied robustness, resistance and resilience of different ecosystem types. Second, we consider the capacity to assess risks to biodiversity and inform conservation decisions, focussing on methods and data that are currently available and those that are potentially applicable but not yet applied. Third, we review the diverse planning and management responses to conservation challenges, evaluating progress and problems in their application. We conclude by considering how a changing climate might interact with the suite of other threatening processes, novel threats and the implementation of conservation planning and management.

27.2 Key threats to Australian vegetation

Threats to biodiversity are the proximate activities or processes that cause destruction or degradation of biodiversity. Salafsky *et al.* (2008) noted that threats are almost always related to human causes, and may be characterised by their temporal status (historical, ongoing or potential future). They distinguished threats from 'stresses', which they defined as the ecological symptoms that result from a threat (declines in diversity, distribution, function, etc.), and 'contributing factors' or chains of factors that may be ultimate causes and ongoing drivers of threats, such as aspects of the social, economic, political, institutional or cultural environment.

'Key threatening processes' to biodiversity may be listed under Australian legislation and in some states (Auld and Keith 2009). These statutory lists enable threats that affect a wide range of species and ecosystems to be identified and reviewed, and also facilitate strategic policy and management responses through 'Threat Abatement Plans' (e.g. Commonwealth of Australia 2014). They also play a minor contextual role in environmental regulatory instruments established under the legislation (Preston and Adam 2004).

To review the threats to different vegetation types across Australia, we adapted a simple threat classification based on ecological mechanisms and outcomes (Auld and Keith 2009). We reviewed six main themes: vegetation clearing and degradation; direct harvest; changes to disturbance regimes (fires and floods); eutrophication of soils and water; disruptions to biological interactions; and climate change. Different threats often interact to produce additive or synergistic impacts, and these interactions are most complex and least certain when they involve climate change. Within each mechanistic theme, we reviewed the most pervasive or severe types of threats, their causes and spatiotemporal patterns, and infer the susceptibility and exposure of different vegetation types to them (Figure 27.1).

27.2.1 Vegetation clearing and degradation

Clearing is the most severe of threatening processes (Figure 27.1) because it involves removal and replacement of native vegetation, its component plants and dependent animals. More than 83 million hectares (12%) of Australia's native vegetation has been cleared since European settlement (based on Department of the Environment and Energy 2007). The transformation is ongoing, with 17 million hectares of detectable woody native vegetation (at least 2 m tall and with at least 20% woody cover) cleared during 1972–2014 (Evans 2016). Queensland and New South Wales have the highest clearing rates, which abated after 2006 and then accelerated after 2011 (Figure 27.2; Evans 2016). Recent rates of clearing in more open or non-woody vegetation have not been quantified, but are likely to involve an additional loss of a similar magnitude to that recorded for detectable woody native vegetation. The principle causes of clearing are intensification of agriculture, mining, urban and industrial development and transport corridors driven by an increasing human population (Evans 2016).

Exposure to clearing varies markedly both between vegetation types and temporally. Currently, temperate tussock grasslands, temperate subhumid and semi-arid woodlands and subtropical savannas are most

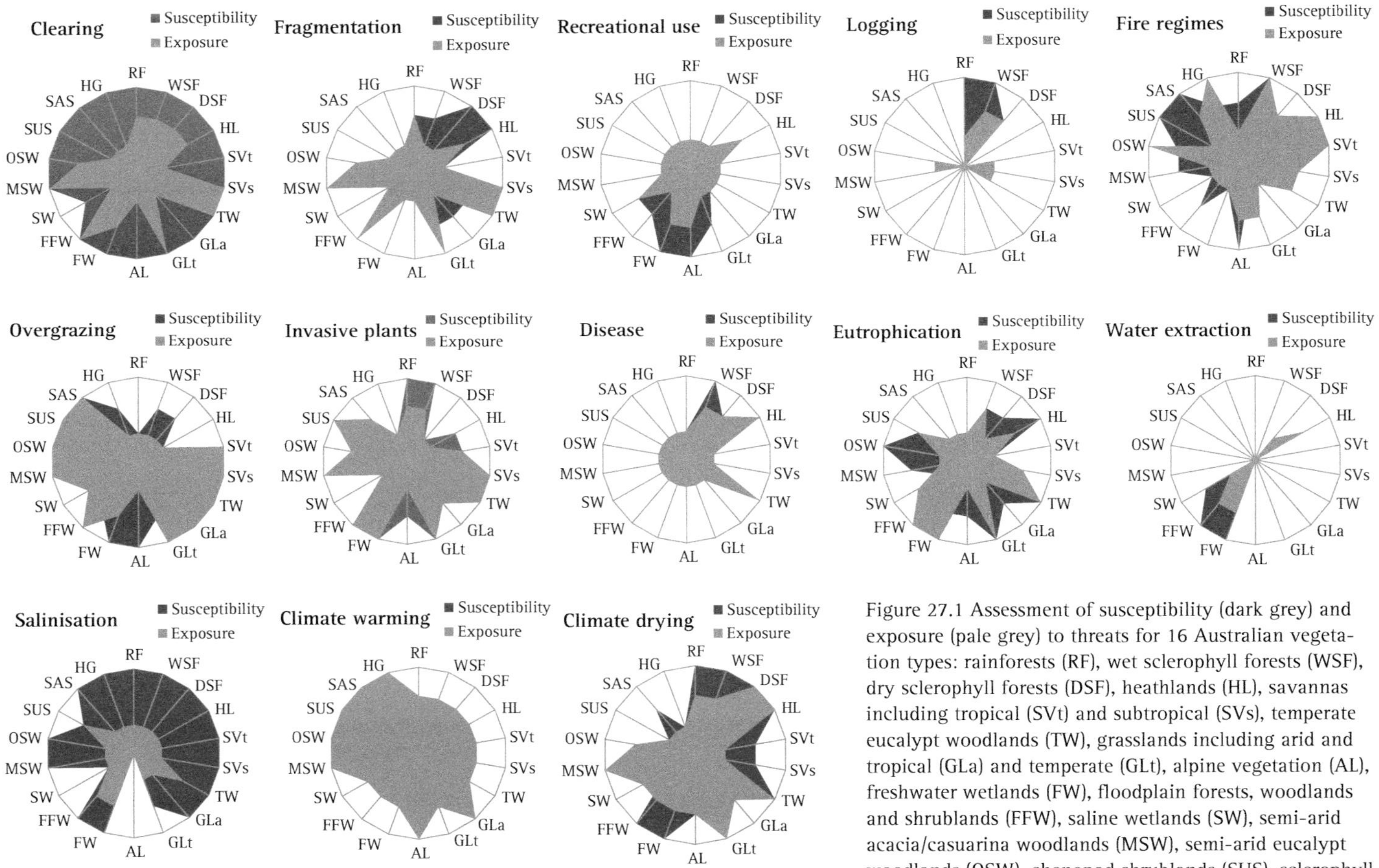

Figure 27.1 Assessment of susceptibility (dark grey) and exposure (pale grey) to threats for 16 Australian vegetation types: rainforests (RF), wet sclerophyll forests (WSF), dry sclerophyll forests (DSF), heathlands (HL), savannas including tropical (SVt) and subtropical (SVs), temperate eucalypt woodlands (TW), grasslands including arid and tropical (GLa) and temperate (GLt), alpine vegetation (AL), freshwater wetlands (FW), floodplain forests, woodlands and shrublands (FFW), saline wetlands (SW), semi-arid acacia/casuarina woodlands (MSW), semi-arid eucalypt woodlands (OSW), chenopod shrublands (SUS), sclerophyll arid shrublands (SAS), hummock grasslands (HG). See Keith and Tozer (Chapter 1, this volume) for descriptions.

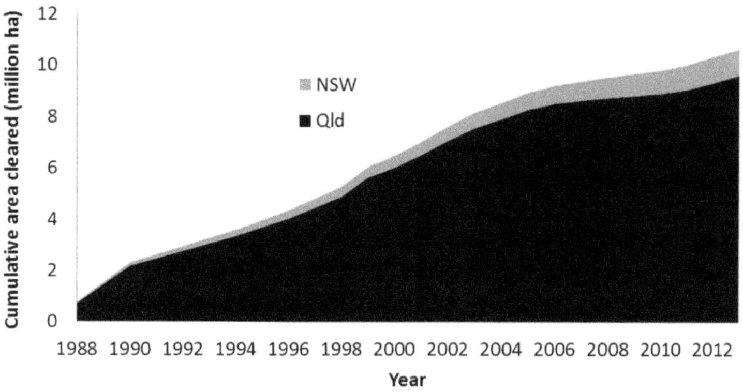

Figure 27.2 Cumulative area of detectable woody vegetation (forest and woodland > 2 m tall and > 20% woody cover) cleared as estimated by SLATS programmes for Queensland and New South Wales from 1988 to 2012. The SLATS method is unable to detect clearing of 'non-woody' vegetation, including open woodlands, grasslands and open shrublands. Hence total clearing rates of native vegetation are likely to be greater than the data show.
Sources: DSITI (2015) and OEH (2014).

exposed. This is only partly reflected in continental-scale analysis of clearing at the level of vegetation formations, which highlights major depletion of temperate eucalypt woodlands and significant reductions in the temperate distributions of other forests, woodlands and heathlands (Figure 27.3a). The averaging effects of such broad-scale analysis tend to mask more extreme impacts, however. For example the extensive arid tussock grasslands, which remain largely intact across 98% of their distribution, contrast markedly with temperate tussock grasslands, which have been reduced to 16% of their former range (Figure 27.3b), indicating that some major types of grassland have undergone much larger declines in diversity than their overall average estimate suggests (11%, Figure 27.3a). Finer scales of analysis reveal even more extreme losses, with Natural Temperate Grassland of the Victorian Volcanic Plain reduced to just 8% of their former range (Threatened Species Scientific Committee 2008). Similar patterns can be seen within other formations such as the Semi-arid Acacia/Casuarina Woodlands, including arid and semi-arid myall woodlands, in which losses to clearing vary from 0.1% to 88% cleared, respectively (Figure 27.3c).

Depending on the nature of land-use intensification, some native species may remain or re-establish, but the outcome is essentially ecosystem transformation, resulting in a novel system that is less ecologically diverse, structurally simpler and composed of both native and exotic plant species with particular traits that promote opportunistic colonisation or high resilience to disturbance (Hobbs *et al.* 2006).

Beyond the immediate temporal and spatial footprint of vegetation clearing, lagged ecological impacts are often associated with fragmentation of the remaining vegetation. These may include extinctions (termed the extinction debt), and occur via mechanisms such as edge effects, and disruptions to pollination, dispersal, trophic networks and disturbance regimes (Saunders *et al.* 1991; Cunningham 2000; Ross *et al.* 2002). Thus, even where clearing has apparently ceased, the realisation of impacts on vegetation and its dependent fauna may be ongoing for decades (Saunders 1989; Laurance 1997; McAlpine *et al.* 2002). Woodlands and grasslands of temperate and subtropical regions that have already suffered high levels of clearing

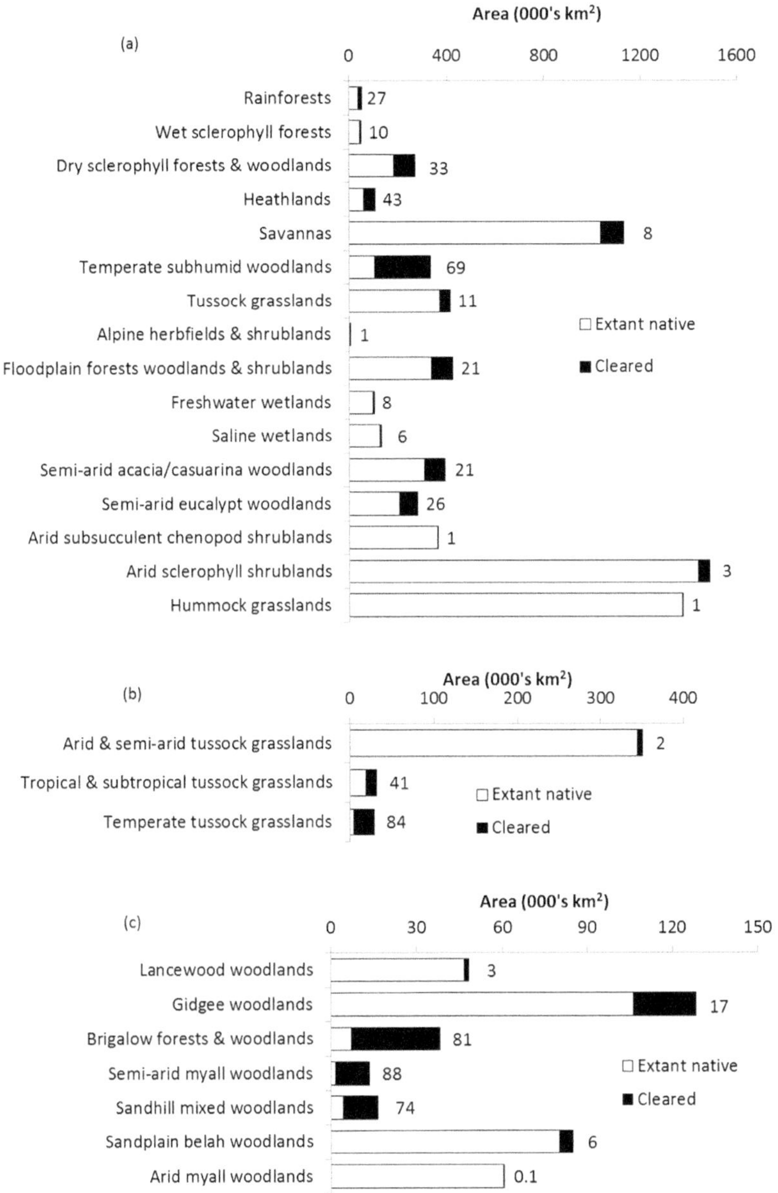

Figure 27.3 (a) Estimated area of extant and cleared native vegetation for each of 16 Australian vegetation formations described by Keith and Tozer (Chapter 1, this volume). Numbers to the right of the bars are estimated percentage of historic (pre-clearing) distribution that has been cleared since European settlement. (b) Variation in clearing impacts among different types of tussock grassland (see Morgan *et al.*, Chapter 18, this volume). (c) Variation in clearing impacts among different types of semi-arid acacia/casuarina woodland (see Fensham *et al.*, Chapter 16, this volume; Nano and Wardle, Chapter 25, this volume). Estimates based on interpretation of state agency vegetation maps held in NVIS 4.2 (Department of the Environment and Energy 2007).

are most severely affected by fragmentation (Figure 27.1). Sclerophyllous vegetation types are susceptible because their plants generally have poor dispersal abilities, although the exposure to fragmentation varies across their distribution (Figure 27.1).

In Australia, as in other regions, vegetation clearing has a distinctive historical spatiotemporal signature, which initiated in coastal nodes where Europeans began to arrive after 1788; Aboriginal people did not clear native vegetation to any appreciable degree. Attention soon focused on fertile coastal floodplains, where rainforests and wet sclerophyll forests were cut for local and export timber and replaced by pastures and crops. Soon after European explorers breached the Blue Mountains and charted the inland rivers, squatters cleared grassy woodlands and grasslands by axe and horse-drawn ploughs. Between 1820 and 1880, extensive areas of temperate woodlands and grasslands were cleared in south-eastern Australia, including the Mt Lofty Ranges and Tasmanian midlands. Waves of agricultural intensification spread across southern Australia and eastern Queensland with soldier settlement programmes after World Wars I and II. By 1925, 94 000 km^2 of farmland had been taken up in these schemes, primarily in south-western Australia, the tablelands and slopes of New South Wales (NSW), riverland of South Australia and western Victoria (Australian Bureau of Statistics 1925). Statistical records of cropping, sown pasture and ringbarking indicated that by 1920, the majority of woodlands on the New South Wales western slopes of the Great Divide had been cleared (Bedward *et al.* 2007), and Victorian and South Australian woodlands were similarly depleted (Bradshaw 2012).

After World War II, clearing of woodlands and heathlands in the south-western Australian wheatbelt accelerated with soldier settlement, which was later extended to civilians until 1969 (Jarvis 1979). The post-World War II period also saw advances in machinery, making it economically viable to clear hitherto undeveloped brigalow woodlands on rich soils in central and southern Queensland and northern New South Wales, with more than 7 million hectares cleared between the mid-1950s and the 1990s (Fensham *et al.*, Chapter 16, this volume). Floodplain woodlands of the upper Darling River in New South Wales and Queensland had a similar but slightly later chronology of clearing as new machinery enabled reliable access through cracking clay soils (Keith *et al.* 2009). With the exception of the Atherton Tableland and some wet tropical lowlands, clearing of tropical forests and savannas is relatively recent, as agricultural schemes exploited alluvial plains of the Ord, Daly, Mary and other tropical rivers onwards from the 1970s (Australian Greenhouse Office 2000). Further intensification seems likely in the coming decades (Australian Government 2015).

Urban expansion and infill has accelerated since the 1960s particularly on the coast as Australians became more numerous and affluent. Long-standing hotspots include Sydney's Cumberland Plain, Melbourne's western basalt plains, the Brisbane valley and Gold Coast, the Adelaide Hills and the southern Swan Coastal Plain. In these areas, native vegetation cover has declined to less than 10% of its extent at settlement, and ongoing clearing is fuelled by high demand for land. More recent expansion of holiday settlements is eroding the area of heathlands, littoral rainforests and coastal forests in north-east and south-east Queensland, the north and central coast of New South Wales, the Otway coast of Victoria, and the northern and southern sandplains of south-western Australia.

Mining activity first affected woodlands in the Goldfields of New South Wales and Victoria in the 1860s, but contemporary impacts are most severe where the mineral resources are associated with geographically restricted substrates that support distinctive vegetation. This includes coastal sandplains, bauxite-rich laterites, limestones, banded ironstone formations and ultramafic outcrops (e.g. Gibson *et al.* 2015). Locally extensive mining operations are associated with coal in the Sydney Basin, central Queensland and Gippsland in Victoria and iron ore in the Pilbara and Yilgarn areas of Western Australia. In contrast, less extensive clearing associated with mineral exploration and coal seam gas infrastructure extends over substantially larger areas.

An emerging threat to native vegetation is degradation associated with recreational overuse. This is intensifying as the human population becomes larger and more socio-economically focused on outdoor recreation. The most susceptible vegetation types are those on erodible substrates (e.g. peaty or sandy soils) and/or with slow plant growth rates, including vegetation formations in extreme cold, dry or oligotrophic environments. Exposure depends on the spatial proximity to population centres and outdoor recreation resources. Thus alpine vegetation is highly susceptible because of its sensitive peaty soils, slow-growing plants and increasing exposure as winter skiing and summer high-country hiking activities become popular and more concentrated within a contracting climatic zone (Good 1992; Venn *et al.*, Chapter 19, this volume). Similarly, demand for access to the coast is increasing exposure of susceptible beach strands and coastal saltmarshes to off-road vehicle use (Rogers *et al.*, Chapter 22, this volume), while some coastal heathlands on erodible substrates are subject to intense visitor pressures near major tourist centres (Keith *et al.* 2014). Increasing visitor pressures pose major challenges to protected area managers, not only through physical damage to vegetation and soils, but through spread of invasive plants and pathogens (Keith *et al.* 2012) and increased risks of wildfire ignitions (Collins *et al.* 2015).

27.2.2 Direct harvesting

Plants are directly harvested from native vegetation primarily for timber products, horticultural stock, cut flowers and the emerging bush tucker trade. In Australia, native forest logging differs from clearing in its effects on vegetation because regeneration is passively facilitated or actively promoted. The principal effects on vegetation are: (i) changes in structure and dominance, as densities of trees with large canopies and hollow-bearing trunks are reduced (Lindenmayer and Franklin 2002); and (ii) dominance shifts towards species with high recruitment rates in open (post-logging) conditions (Ough 2001). Susceptibility to these impacts depends partly on the silvicultural management employed, which varies from clear-fell and re-seeding to various selective and variable retention designs (Scott *et al.* 2015). Interactions with other threats may also be important, for example, clearing for road operations, altered fire regimes and spread of pathogens.

Relatively few of Australia's vegetation formations are exposed to logging, with wet sclerophyll forests having the greatest exposure, along with some dry sclerophyll forests and floodplain forests dominated by river red gum (Figure 27.1). New logging of previously unexploited native forests has declined markedly since the 1990s in response to declining demand (especially for paper products) met increasingly from plantations. Like fragmentation, however, logging may involve long legacies, including weed and pathogen invasion, stasis of forest structural forms with low habitat suitability, and changes to fire regimes (Lindenmayer *et al.* 2011). Consequently, threats to biodiversity will persist for decades in rainforests and wet sclerophyll forests even where logging has largely ceased.

Other forms of direct harvesting include seed harvest for revegetation, illegal collection or orchids and other ornamentals and cut flowers. While seed harvest for revegetation has increased in recent decades due to regulatory requirements for rehabilitation, as well as landcare and other rural initiatives, some of the material is sourced from seed 'orchards' (Gibson-Roy *et al.* 2007). The other threats are most relevant to individual species, particularly ground orchids, while a proportion of the demand for cut flowers and bush tucker is met from plantations, rather than native vegetation.

27.2.3 Changes to disturbance regimes

Disturbance regimes such as fires, floods, and storms are important for driving plant population processes within vegetation communities (Auld and Ooi, Chapter 10, this volume).

27.2.3.1 Fire regimes

Fire is a major driver of vegetation patterns across the Australian landscape (Miller and Murphy, Chapter 5, this volume). The interactions between a changing climate (increased aridity) and fire have shaped the late Miocene vegetation shift from rainforest to sclerophyll-dominated vegetation, as well as Quaternary fluctuations (Mooney *et al.*, Chapter 3, this volume). Changes to ignition patterns and extensive fragmentation of vegetation since European arrival have dramatically altered both fire regimes and their impacts on biodiversity. Today, particular fire regimes associated with biodiversity loss, such as those with frequent fire recurrence, are listed as Key Threatening Processes in some jurisdictions (Auld and Keith 2009). Fire regimes influence community structure, composition and diversity through diverse mechanisms (Keith 2012) and may therefore pose threats to most Australian vegetation types (Figure 27.1), even those that are rarely flammable (Keith and Tozer, Chapter 1, this volume).

Interactions between fire regimes and other processes may generate synergistic threats to vegetation and its dependent biota. Increasing aridity with global climate change may lead to more widespread fires that are of high severity in some systems. The influence of climate change on fire regimes is likely to be exacerbated by increased frequency and changed spatial patterns of wildfire ignition, which are closely associated with the increasing density and movement of human populations (Collins *et al.* 2015). Holz *et al.* (2014) highlight how increasing aridity may make a landscape more fire-prone with the resultant decline and loss of relatively fire-sensitive taxa such as *Athrotaxis cupressoides*. A key challenge will be to identify species and vegetation most at risk from changes to fire regimes. For example, Enright *et al.* (2015) link species life-history processes with fire impacts to highlight the risk of 'interval squeeze', where the ability of species to persist is compromised through reduced plant recruitment, growth and survival in response to more frequent fire under a warming climate. Fire refugia may also be compromised where whole landscapes are burnt and prior clearing and fragmentation limit the ability of species to return to isolated fragments. This may be a particular concern for fauna, including species that play key roles in maintaining plant population processes (Auld and Ooi, Chapter 8, this volume).

27.2.3.2 Water regimes: surface flows and groundwater

Hydrological threats to native vegetation include changes to water regimes resulting from disruption of river and overland flows by dams, weirs, diversions, drains, levees and other floodplain infrastructure, or from extraction of groundwater. Wetland, riparian and floodplain vegetation are among the most susceptible systems to these threats (Figure 27.1), as the ecophysiological traits and life cycles of their component plant species are closely tied to the water regime (Catford *et al.*, Chapter 20, this volume; Good *et al.*, Chapter 21, this volume). Exposure to hydrological change varies from major inland and coastal river systems, where there is intense demand for irrigation or potable water, to cool temperate moorlands, ephemeral arid streams and clay pans, where there are negligible levels of water extraction (Figure 27.1).

Water-dependent systems vary greatly in geomorphic form and function, vegetation structure, species composition and ecohydrological function (Catford *et al.*, Chapter 20, this volume). Groundwater-dependent ecosystems are still being identified across Australia and their susceptibility to hydrological disruption is poorly understood (Eamus *et al.* 2015). However, some exhibit strong sensitivities. Many mound springs of the Great

Artesian Basin have dried with extraction of bore water (Fensham and Fairfax 2003), while declines of *Banksia* woodlands associated with lowering of the water table around bores on the Swan Coastal Plain indicate that some susceptible groundwater-dependent ecosystems are not wetlands (Froend and Sommer 2010). While water extraction is the principal contributing factor, underground mining is also emerging as an important threat to groundwater-dependent systems (Krogh 2007).

The spatiotemporal pattern of water exploitation across Australia contrasts with that of vegetation clearing. In the nineteenth century, water was extracted from local wetlands and rivers to supply population centres with potable water. Industrial-scale intensification began in the early twentieth century when farmers organised drainage unions or trusts that planned major public works to construct co-ordinated drainage networks on the east coast floodplains, systematically replacing wetlands with farmlands (Keith and Scott 2005). Around the same time, the first major dams were constructed to service growing city populations. The first diversions from the Murray and Murrumbidgee Rivers were established in the nineteenth century, but river regulation accelerated during 1922–1935 with construction of 10 locks and weirs to support navigation on the Murray. Construction of major dams occurred in the Murray–Darling during the 1920s to 1930s and was accompanied by irrigation infrastructure in the Riverina and northern Victoria. However, the most intensive development of water infrastructure was from 1950 until 1980, when numerous dams were constructed in metropolitan catchments near major cities, the Snowy Mountains scheme, other Murray–Darling tributaries and the Tasmanian hydro-electric scheme (Kingsford 2016). Conversion of lakes to storage bodies and thousands of small dams, weirs, floodplain drains, levees and other structures contribute to anthropogenic alteration of river flows (Kingsford 2016).

The major impact of alteration to water regimes is deprivation of water to wetland, riparian and floodplain vegetation, resulting in mortality of plant species with high water-dependency (Cunningham *et al.* 2011), shifts in composition of riparian vegetation (Doupé and Petitt 2002), contraction of wetlands and their replacement by dryland vegetation (Kingsford and Thomas 2004; Brock *et al.* 2006). The loss of wetland functions as ecological refugia and hydrological regulators has consequences for whole landscapes (Mac Nally *et al.* 2011). At local scales, impoundment of river water and conversion of natural lakes to storages has reduced hydrological variability and drowned wetland, riparian and dryland ecosystems (Kingsford *et al.* 2004). River regulation also reduces hydrological connectivity, interrupting propagule dispersal and fluxes of nutrients and organic material (Steinfeld and Kingsford 2013).

27.2.4 Pollution of water and soils

27.2.4.1 Nutrient enrichment

Pollution sometimes involves leakages or spills of toxins, but more commonly involves insidious enrichment of nutrients in water and soils by outflow or runoff from developed rural, urban or industrial areas. In aquatic systems, nutrient enrichment promotes eutrophication, a process whereby explosive plant and algal growth leads to hypoxia due to the increased activity of microorganisms engaged in the decomposition of plant matter. In terrestrial systems, the most susceptible vegetation occurs on oligotrophic substrates although naturally fertile depositional landforms may rapidly accumulate ecologically harmful nutrient levels (Figure 27.1). Nutrient enrichment became a more serious threat to Australian biodiversity from the 1930s when use of agricultural fertilisers was subsidised and became more widespread. Dependence of industrial agriculture on fertilisers is rapidly increasing, with use of nitrogen–phosphorus–potassium (NPK) fertilisers by Australian agriculture more than doubling between 1985 and 2005 (Ryan 2010). Nitrogen enrichment through air pollution and atmospheric deposition is a globally recognised ecosystem stressor within and around industrialised regions (Simkin *et al.* 2015).

Nutrient enrichment has profound effects on heathlands, dry sclerophyll forests, grassy woodlands and some wetlands, resulting in replacement of native plant taxa by fast-growing exotic species (Leishman *et al.*, Chapter 9, this volume). Sclerophyllous plant taxa are sensitive to phosphorus toxicity (Shane *et al.* 2004), and tend not to increase rates of growth and fecundity to the extent that certain exotic species can when nutrient supply is elevated, thus making them weak competitors in those circumstances. There is also evidence that nutrient enrichment reduces species asynchrony in growth, a potentially important mechanism sustaining coexistence and stable productivity (Hautier *et al.* 2014). As well as oligotrophic systems, the effects of nutrient enrichment can also be profound and long-lasting in mesotrophic vegetation such as grassy woodlands and grasslands, where native plant taxa are particularly sensitive to soil nitrogen and unable to compete with exotic ephemeral and perennial grasses (Prober *et al.*, Chapter 17, this volume). Vegetation types with high levels of exposure to nutrient enrichment include those adjacent to or downslope from built infrastructure, and those in temperate rural districts where fertilisers are routinely applied (Figure 27.1).

27.2.4.2 Salinisation

Salinisation is associated with clearing of deep-rooted native woody vegetation and replacement by shallow-rooted crops and pastures (Cramer and Hobbs 2002). This reduces evapotranspirative transfer of water from the soil profile to the atmosphere, causing the water table to rise and bringing soluble salts to the upper soil horizons and surface. Extensive saline groundwater reserves originated from Miocene marine transgressions as well as the accumulation of marine aerosols in regions where little runoff of precipitation occurs. Once near the surface, salt is transported laterally by surface water flow and accumulates in depressions and drainage lines. Glycophytic vegetation in these environments is unable to tolerate high levels of salt and is replaced by halophytes or, in the most extreme cases, hypersaline pans (Cramer and Hobbs 2002). All glycophytic vegetation types are susceptible to salinisation, although temperate eucalypt woodlands, wetlands and riparian forests, woodlands and shrublands are most exposed (Figure 27.1), particularly those in the subdued terrain of the south-western Australian wheatbelt and lower Murray–Darling basin.

27.2.5 Disruption of biological interactions

27.2.5.1 Plant and pathogen invasions

Since European settlement, extensive clearing and fragmentation of vegetation in many areas combined with changes to nutrients and disturbance regimes have provided opportunities for exotic species to invade native plant communities (Leishman *et al.*, Chapter 9, this volume). These include both accidental and deliberate introductions such as garden ornamentals or crop species (Leishman *et al.*, Chapter 9, this volume). The rate of introductions of new species that may become invasive has increased with the globalisation of human enterprise.

Depending on their life form, invaders have the potential to reduce species diversity and transform the structure and function of native vegetation. Shrubs such as bitou (*Chrysanthemoides monilifera*) and Scotch broom (*Cystitis scoparia*), for example, can dominate heaths and woodland vegetation and competitively exclude many native species, while vines and scramblers can smother rainforest and wet sclerophyll trees (Leishman *et al.*, Chapter 9, this volume). Other species may transform vegetation by altering the disturbance regimes. Gamba grass (*Andropogon gayanus*), a widespread introduced grass in tropical savannas, has changed fire regimes, particularly fire severity (Setterfield *et al.* 2010), while buffel grass (*Cenchrus ciliaris*) has a similar effect in arid shrublands (Miller *et al.* 2010;

Nano *et al.*, Chapter 25, this volume). Impacts of other exotic plants may be more subtle. Mediterranean forbs and grasses in arid shrublands, for example, may be largely absent above-ground during droughts or after summer rain, but after cool-season rains they can dominate the ephemeral component of these vegetation communities.

Several introduced pathogens are major threats to some Australian plant communities and are recognised as key threats to biodiversity under Commonwealth or state legislation. *Phytophthora cinnamomi* transforms heathlands, temperate woodlands and dry sclerophyll forests into simpler communities (Figure 27.1), particularly in south-western Western Australia (Cahill *et al.* 2008). Myrtle rust (*Puccinia psidii*) is a more recent introduction that selectively infects members of the Myrtaceae and may cause shifts in species composition and structure within forested vegetation communities (Leishman *et al.*, Chapter 9, this volume).

27.2.5.2 Overgrazing and soil degradation

High densities of herbivores threaten native vegetation by selectively removing palatable plant species or reducing their abundance, thereby changing vegetation structure and composition and reducing diversity. Indirect effects include increases in abundance of unpalatable plant species due to release from competition or opening of gaps enabling colonisation. Overgrazing may also cause soil degradation where herbivores disrupt soil surface integrity and soil crusts, resulting in erosion, compaction and/or loss of nutrients, reduced water infiltration (Eldridge *et al.*, Chapter 6, this volume) or degradation of wetlands and waterholes (e.g. Brim Box *et al.* 2016). These impacts can leave long legacies or produce regime shifts, as for example, scalds persist in semi-arid rangelands from episodes of overgrazing during 1860–1890 (Beadle 1948, 1981). Plants have evolved a range of physical and chemical defences to herbivores (Díaz *et al.* 2006; Eldridge *et al.*, Chapter 6, this volume), although these vary greatly between species, and within species depending on life stage, environmental factors and levels of grazing pressure (Vesk *et al.* 2004). For example, many woody species are quite resilient to browsing in their mature phases, but highly susceptible at seedling and juvenile stages (Auld and Ooi, Chapter 10, this volume).

Recent evidence suggests that trophic cascades may initiate overgrazing in some situations, especially in arid systems. The postulated mechanism involves decline of carnivore populations (either native or introduced), release from predation, irruption of herbivore populations and consequent overgrazing effects on vegetation (Letnic *et al.* 2012). Evidence of effects of predators on herbivores and of herbivores on vegetation is well established (Colman *et al.* 2014), while a link between carnivores and vegetation is still under investigation.

Few Australian vegetation types exhibit high levels of resistance and resilience to high densities of herbivores (Figure 27.1), with impacts on vegetation and soils commonly evident even at moderate levels of grazing (Eldridge *et al.* 2016). The vegetation types most susceptible to overgrazing are those in which component plants have slow growth rates and/or few chemical or physical defences against herbivores, and those with erodible or compactable soils (Lunt *et al.* 2007). Tropical grasslands and high-productivity temperate grasslands appear to be the most resilient vegetation types to herbivore activity due to their consolidated clay soils and dominance by tussock grasses with protected buds and rapid regrowth rates, although the response of interstitial herbs is not well researched, especially in the tropics (Morgan *et al.*, Chapter 18, this volume). The alpine, subpolar, arid and semi-arid vegetation formations are among the most sensitive to overgrazing because of their low productivity and erodible soils (Figure 27.1; Venn *et al.*, Chapter 19, this volume; Nono *et al.*, Chapter 25, this volume).

The herbivores associated with overgrazing impacts include invasive species (rabbits, goats, deer, camels), domestic livestock (sheep, cattle), and more rarely, native macropods. There are high levels of exposure to various combinations of these species in savannas, temperate grassy systems and throughout the semi-arid and arid zones (Figure 27.1). Notably, exposure of alpine and subpolar vegetation to overgrazing was very high prior to the 1960s, and sequentially declined with exclusion of livestock from Kosciuszko and later the Victorian highlands, and the eradication of rabbits from Macquarie Island (Venn *et al.*, Chapter 19, this volume).

27.2.6 The changing climate

Climate change is the most pervasive threat to Australian vegetation. Sensitivities and exposure to climatic warming and drying vary between vegetation types (Figure 27.1). Historically, shifts in climate and associated aridity have been a major driver of vegetation change in Australia (Martin 2006; Mooney *et al.*, Chapter 3, this volume). Consequently, rapid changes to climate are likely to lead to abrupt shifts in vegetation distribution patterns, structure, function and composition. Plant communities face environmental change on multiple fronts across multiple scales including changes to temperature, rainfall, available soil moisture, CO_2, sea levels, fire, flood and storm regimes, loss and fragmentation of habitat, nutrient enrichment and salinisation. Thermal tolerance, response to elevated CO_2 and other processes are expressions of climate sensitivities through flowering, fruiting, dispersal, germination, growth and survival. These are being altered by a changing climate and its interaction with other threats (fire, hydrological change and introduced species) that ultimately mediate plant population persistence (Ooi *et al.*, Chapter 8, this volume).

Under future climate scenarios many species are predicted to move outside their current realised climatic niches (Shoo *et al.* 2013; Ooi *et al.*, Chapter 8, this volume). However, evidence from invasive exotic species clearly demonstrates that many species may have a broader potential niche than their current realised niche, with factors such as competition, predation and dispersal limiting the current realised niche (e.g. Beaumont *et al.* 2014). Nonetheless, Dunlop and Brown (2008) highlight the reality that some ecological communities will not persist *in situ* without management.

Vegetation could change relatively rapidly when extreme events cause widespread mortality of established vegetation, creating opportunities for replacement by novel vegetation. This may change functional relationships, for example, by leading to short-term increases in fire risk. Tree or shrub mortality induced by heat stress and/or prolonged drought is one emerging mechanism whereby plant communities may be transformed into simpler assemblages and habitats. Drought-induced tree and shrub mortality has been observed globally in many vegetation types and climate zones, including in parts of arid and semi-arid Australia (Allen *et al.* 2010, 2015; Cunningham *et al.* 2011; Fensham *et al.* 2012) and also poses risks to rainforest, wet and dry sclerophyll forests, woodlands, alpine vegetation and arid shrublands. This has consequences for the flora and fauna dependent upon trees as habitat, along with potential exposure of understorey species through changes to fire regimes and increases in solar radiation. A second mechanism for widespread tree mortality is attack from pathogens (Rizzo and Garbelotto 2003; Carnegie *et al.* 2016) or pests, often in combination with drought stress (Anderegg *et al.* 2015), and it is increasingly important to prevent the introduction of exotic species into Australia to minimise this risk (Leishman *et al.*, Chapter 9, this volume).

27.3 Assessing and monitoring the risks

Internationally, Australia has a comparatively strong statutory and scientific infrastructure to evaluate and report on the status of native vegetation and the risks to the biodiversity it supports. The assessment and 'Red listing' of threatened species, ecological communities and key threatening processes is mandated by law nationally and in some state jurisdictions (Keith 2009; Nicholson *et al.* 2015). The remaining subnational jurisdictions maintain non-statutory lists.

Australia, with some European countries and South Africa, has led the development of Red listing processes and their implementation in conservation policies and law, and they have proved to be workable and effective tools for conservation planning and management (Keith 2015; Keith *et al.* 2015).

Importantly, Australian laws provide for listing threatened biodiversity at two levels of organisation: species and ecological communities. The latter are assemblages of species that co-occur in space and time; essentially the same concept as habitat types and ecosystems, which are listed in other countries and internationally (Keith *et al.* 2015). Current listings include species assemblages defined on the basis of co-occurring plants, vertebrates or invertebrates; although in practice the majority of listings reflect vegetation types defined by co-occurring plants that structure the habitats of other species.

Lists of threatened species and ecological communities in Australia support a number of functions including: (i) identification of key issues for environmental impact assessment and development approval processes under environmental laws; (ii) prioritisation of resources for conservation action within government, industry and community sectors; and (iii) monitoring the status of biodiversity through time as mandated by the Convention on Biological Diversity and Australian legislation.

The current listings under national legislation include 1794 threatened species (including 1299 plants) and 78 ecological communities (https://www.environment.gov.au/epbc/about/epbc-act-lists#species, accessed 7 May 2016). A large majority of species and ecological communities are yet to be assessed for listing, and hence the actual numbers of threatened species and communities are considerably larger. Assessing the Red List status of all species is an aspirational goal, but the large numbers of species impose formidable logistic constraints. On the other hand, comprehensive systematic Red List assessments of ecosystems (or ecological communities) are more tractable and have already been implemented in some countries (Keith *et al.* 2015). If undertaken in Australia, such an assessment could provide a powerful basis for biodiversity monitoring and reporting, inform priorities for statutory listing and guide conservation planning (Keith 2015). One of the requisites for this is a comprehensive ecological typology which, for terrestrial ecosystems, could be constructed from a number of data sources, including the National Vegetation Inventory System (Keith and Pellow 2015).

Australia is relatively well equipped to monitor and report on the status of its biodiversity generally, and its vegetation in particular, although there are significant thematic gaps in data coverage. A network of long-term plot-based field studies is a critical component of the scientific infrastructure needed to support biodiversity monitoring (Lindenmayer *et al.* 2014). These produce valuable time series of vegetation structure, species composition and other indicators that provide unique insights into on-ground trends. A core set of monitoring networks is centrally co-ordinated through Australia's Long Term Ecological Research Network (http://www.tern.org.au/Long-Term-Ecological-Research-Network-LTERN-pg17872.html) and a larger suite of long-term studies are currently active. However, the current infrastructure does not yet represent the range of ecosystems across the Australian continent (Sparrow *et al.* 2014).

The remote sensing data infrastructure is also rapidly expanding with national data sets for time series of vegetation height and fractional cover, vegetation indices, leaf area index and several other variables (http://www.auscover.org.au/data). The Australian Government uses satellite imagery to monitor deforestation rates, primarily for carbon accounting (Commonwealth of Australia 2015). Queensland and more recently New South Wales government agencies have implemented Statewide Landcover and Trees Study (SLATS) programmes, based on repeatable satellite image processing algorithms to monitor rates of woody vegetation clearing (DSITI 2015). These programmes provide valuable publically available sources of information on trends in the extent of Australian forests and woodlands (Figure 27.2) that ought to be extended to other Australian states and non-woody vegetation.

In combination, time series data from field monitoring plots and remote sensing, as well as turnover in lists of threatened species and ecological communities (Butchart *et al.* 2007), have the potential to provide a powerful quantitative capability for monitoring and assessing trends in Australian vegetation. Australia's statutory framework based on periodic State of Environment reports has thus far relied largely upon qualitative appraisals of trends and is yet to synthesise these data to exploit their capability in a repeatable and rigorous manner to report quantitative trends (DSEWPaC 2011).

27.4 Planning and management responses

27.4.1 Protected areas

Protected areas are the most effective and secure means of mitigating threats related to vegetation clearing. They may also be effective in mitigating other threats, depending on their degree of statutory protection, management strategies and resources available to implement them. The IUCN Protected Area Management Categories (Dudley 2008) distinguish those in which nature conservation is the overriding management objective (categories I–IV) from those in which conservation management occurs alongside sustainable resource use (categories V–VI). Some classifications of protected area, for example, permit mining on the surface or underground that may lead to irreversible loss or degradation of vegetation and soils (Wawryk 2014). Most Australian protected areas mitigate threats related to overgrazing by excluding domestic livestock. Policy changes can rapidly erode the effectiveness of such management, however, leaving prolonged legacies of overgrazing (Eldridge *et al.*, Chapter 6, this volume). For example temporary permits for livestock grazing were issued for five national parks in Queensland during drought when vegetation and soils are most susceptible to grazing impacts (Dickson 2013). Many protected areas employ management strategies to control invasive species, regulate visitor activity and implement fire regimes for conservation. The effectiveness of these strategies are highly dependent on resourcing and recent downward trends in budgets and staffing, as well as pressures to develop protected areas, suggest risks of increased impacts into the future (Craigie *et al.* 2015).

The National Reserve System (NRS) is Australia's network of terrestrial protected areas. These vary considerably in tenure, level of statutory protection and management. As of 2014 (CAPAD 2014), the terrestrial NRS included 70.5 million hectares (9.17% of Australia) within IUCN categories I–IV, and a further 67 million hectares (8.71%) in categories V–VI, which are mostly managed by Aboriginal people in the tropics and arid zone. Approximately 48% of the reserve estate is managed by government agencies, 40% by Aboriginal communities, 7% is managed jointly and the remaining 5% is in private ownership (CAPAD 2014).

Like most other parts of the world, development of Australia's reserve system has focused on scenic landscapes, avoiding areas of high primary resource value ('worthless lands' hypothesis, Hall and Shultis 1991). The resulting networks of reserves are biased towards landscapes with low rainfall, steep terrain and infertile soils (Pressey *et al.* 2000; Mendel and Kirkpatrick 2002). Policies to redress this historical bias have developed relatively recently with Australia's adoption of international targets (Watson *et al.* 2016). The goal of the NRS is to develop and effectively manage a comprehensive, adequate and representative system of protected areas, as the primary means for securing long-term protection for terrestrial biodiversity (Australian Government 2009).

Representation of vegetation in Australian protected areas can be assessed in a basic way by calculating the proportion of area reserved in each of Australia's 89 bioregions (Australian Government 2013). In 2014, protected areas covered at least 15% of 42 bioregions, and less than 5% of 15 bioregions (Figure 27.4). A retrospective analysis of protected areas in 1997 produces corresponding estimates of 19 bioregions with more than 15% representation and 41 bioregions with less than 5% representation, suggesting a considerable improvement in comprehensiveness of the reserve system. The major improvements are related to establishment of large reserves (IUCN category VI) managed by Indigenous people in central and northern Australia (Craigie *et al.* 2015), increasing the total extent of the reserve system from 58.5 to 136.9 million hectares. Private reserves managed by non-government organisations funded by philanthropic sources have also increased in recent years (Fitzsimmons 2015).

Arguably, with the extent of the reserve network doubling in less than two decades, the additional area could have been deployed more strategically to improve reservation of under-represented bioregions (Pressey 1994) and to alleviate threats more effectively (Visconti *et al.* 2010). A large majority of the additional reserve area and much of the improvement in representation was in arid environments that are currently exposed to low levels of threat (Figure 27.1). In contrast, bioregions containing vegetation such as temperate grasslands and woodlands with the highest exposure and susceptibility to severe threats (Figure 27.1) remain poorly represented in reserves (DEWR 2007). Larger gaps are apparent at finer resolution because bioregions provide only a coarse means of assessing representation. For example, of 419 biogeographic subregions, only 151 have more than 15% representation in reserves, 163 have less than 5% representation and 34 have no reserves (CAPAD 2014). Nonetheless, protection of small remaining areas of native vegetation in intensively developed bioregions represents important conservation progress. It should also be remembered that placement of large new reserves in arid regions reflects cultural objectives as well as those for biodiversity conservation.

The adequacy of reserve networks, including their capacity to sustain biota and ecological functionality in the long term, has received much less attention than comprehensiveness and representativeness. Klein *et al.* (2009) examined a number of continental-scale conservation planning scenarios in which reserve networks were selected and designed to represent biodiversity while also: (i) building resilience to external threats by adopting subcatchments as planning units; (ii) protecting drought refugia by targeting sites that maintained productivity through dry years; (iii) protecting evolutionary refugia by targeting sites with relictual taxa and local endemism; and (iv) promoting connectivity of waterways. Biodiversity representation was based on 1763 vegetation types (intersection of 62 Major Vegetation Subgroups (DEWR 2007) with 85 bioregions), 151 environmental domains (Mackey *et al.* 2008), modelled distributions of 563 bird species and 1222 listed threatened species. The results showed that relatively small adjustments to reserve design (changed priorities for 10% of selected subcatchments) could achieve targets for drought and ecological refugia as well as biodiversity representation, and thus improve the long-term ecological and evolutionary resilience of the protected area network for little additional cost. They also showed that a more radical reconfiguration of reserves was required to achieve full waterway connectivity at almost double the acquisition cost. However, 30% more river connections could be achieved with less than 1% additional cost, indicating the large connectivity benefits of low-cost reserve design alternatives.

27.4.2 Conservation policies and regulatory instruments

Australia has a relatively well-developed suite of environmental laws, policies, regulatory instruments and funding programmes to conserve native

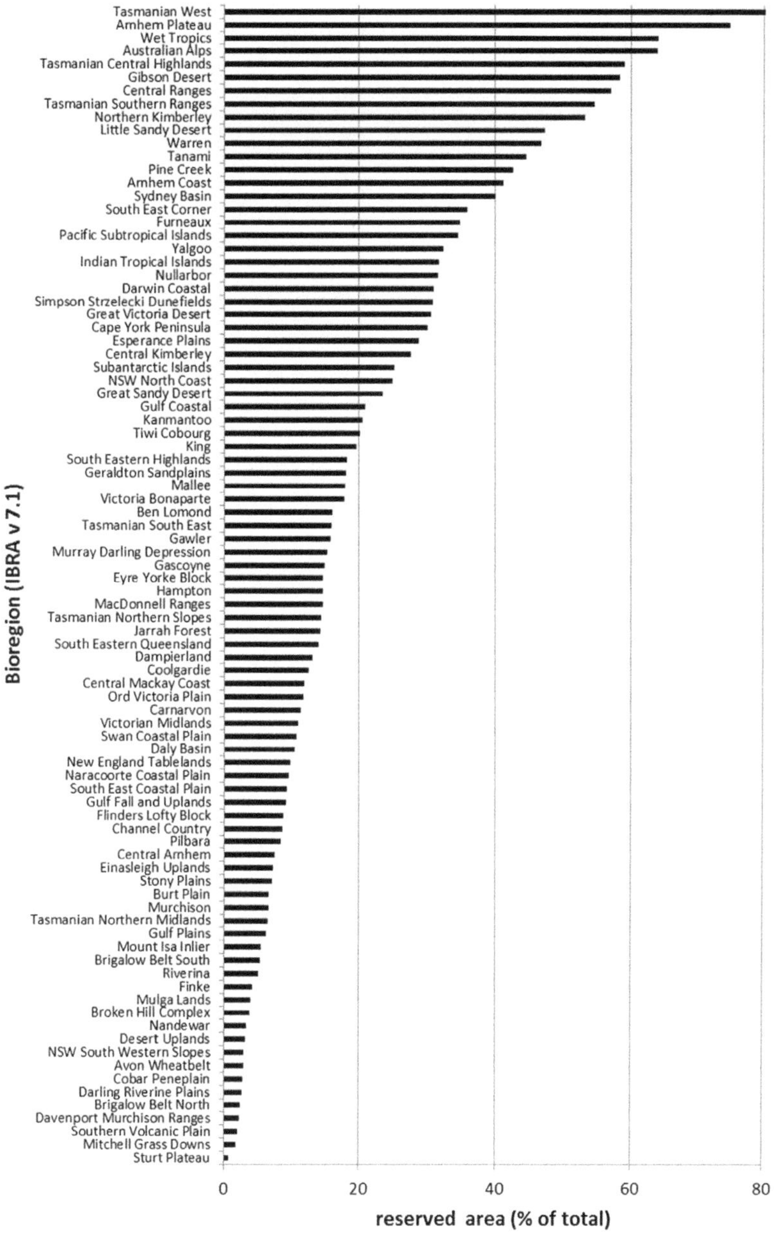

Figure 27.4 Representation of Australia's 88 bioregions (IBRA 7.1) in the National Reserve System (IUCN categories I–VI). *Source: CAPAD (2014).*

vegetation and associated biodiversity outside the protected area network. These are particularly important for the most extensively depleted and degraded Australian vegetation types, such as temperate grasslands and grassy woodlands, which are primarily distributed in intensive agricultural regions and urban/industrial centres.

Three levels of government have different roles in policy implementation. The *Commonwealth Environment Protection and Biodiversity Conservation Act 1999* requires environmental assessments to accompany applications for development approvals. These assessments must consider the potential impacts of developments on 'matters of national environmental significance', including threatened species and ecological communities listed on the schedules of the legislation, as well as RAMSAR wetlands, World Heritage Areas, areas listed on the National Estate, Commonwealth land and habitat for migratory species. State and Territory governments also have environmental legislation that requires assessments for a broader suite of development proposals and environmental assets, however the strength of this legislation varies greatly between jurisdictions. Both Commonwealth and state legislation provide for recovery planning to identify and implement co-ordinated measures designed to improve the status of listed species and ecological communities. Some jurisdictions, notably South Australia, Queensland and New South Wales also administer legislation to regulate clearing of native vegetation and extraction of water in rural areas. Local governments have an important role as approval authorities and are also responsible for land zoning, which defines permissible land uses in different parts of the landscape. This has a major *a priori* influence on the outcomes of conservation and development decisions. Importantly, laws provide for public involvement through submissions on listings and development applications, as well as the ability for proponents and objectors to challenge decisions in court.

Emerging policy instruments include voluntary conservation covenants and offset policies. Covenants enable landholders to make voluntary formal commitments about the conservation and management of native vegetation on their land. These are attached to the land title and thus binding on current and future owners of the property. Offset policies require developers to 'offset' unavoidable impacts on biodiversity through protection and management of biodiversity elsewhere. Key principles of this approach include: (i) the mitigation hierarchy, in which offsets are only implemented to compensate residual impacts after all others are avoided or minimised to the fullest extent possible; (ii) like for like, whereby offsets must be for the same type of vegetation as that lost; (iii) additionality, i.e. offset actions and their outcomes must be additional to those that would have occurred if the development had not taken place; (iv) perpetuity, whereby offset gains are secure into the future; and (v) no net loss, whereby the gains from offsets must at least balance the losses of biodiversity caused by the development (Maron *et al.* 2016a). At present, these principles are not fully implemented in government policy and operations. In most jurisdictions offsets are implemented semi-formally through conditions of development approval, while other jurisdictions such as New South Wales are establishing biobanking trusts that maintain registers of properties that are eligible for offset investments. Finally, governments provide support to landholders and community groups for management and restoration of native vegetation through funding programmes such as 'landcare' and advisory bodies.

Many of the decisions and actions resulting from environmental policies and regulatory instruments address the impacts of vegetation clearing, forestry operations, overgrazing and control of invasive species, but how effective are they in conserving and mitigating threats to Australian vegetation? Answering this question would require analysis of performance against relevant counterfactual scenarios. For example, clearing regulations prevent a fraction of vegetation losses that would have occurred if they had not been in place because some proponents decide not to apply for permits and some applications are refused. These need to be balanced against clearing exemptions and approvals and illegal clearing rates over time. Although acceleration of clearing rates after weakening of clearing regulations in Queensland and New South Wales provide some insights into the effectiveness of the regulations when they are in place (Figure 27.2), simple examination of clearing rates alone cannot provide an adequate evaluation of performance, which requires analysis of other components that are difficult to estimate and not currently reported.

A range of challenges and limitations apply to each of the policy and regulatory tools described above. A key problem is the increasingly narrow focus of environmental assessments on listed threatened species to the neglect of other elements of biodiversity. Encouraged by the structure of environmental regulations, such an approach fails to address ongoing erosion of the diversity and functionality of Australian vegetation and its associated biota, discourages preventative action and ultimately increases the cost of conserving species and communities that were not addressed earlier. Thus major biomes, such as tropical lowland rainforests, brigalow woodlands and inland floodplains, only trigger significant regulatory action after they became fragmented and degraded.

Other challenges involve regulatory burden and the costs of compliance. As the general public and land owner both benefit from conservation of vegetation, how should the costs be shared? This question is made more complex by the interplay between short-term economic benefits of vegetation clearing and the broader and longer-term benefits of vegetation conservation. Another dimension to the problem is the level of government resourcing allocated to administer the regulations, which varies between jurisdictions and over time, but is always less than required to support adequate monitoring, compliance and enforcement. One of the responses has been to minimise regulatory burden by targeting regulatory processes in ways that reduce costs and increase certainty for developers. Examples of problems emerging from this approach can be found in Strategic Assessments and Regional Forest Agreements. These define fixed conditions for development and timber extraction, respectively, over decadal time periods, but lack the flexibility to incorporate new information about the status of biodiversity as knowledge improves during the term of the agreement. Another example is exemptions from clearing permits applying to 'invasive native scrub' in New South Wales. These exemptions override provisions to protect listed species and ecological communities. Thus, where dominants of threatened ecological communities are listed as 'invasive native scrub', the regulations fail to protect valuable areas of natural regeneration of these communities.

Biodiversity offsets are a particularly challenging policy initiative that has generated considerable debate in the scientific community and civil society. Concerns include fundamental ethical objections, the balance of social equity, issues of governance, and technical difficulties quantifying biodiversity losses and gains (Maron *et al.* 2016a). Technical challenges for vegetation offsets include: the definition of adequate baselines and counterfactual scenarios to determine additionality; lags in ecological responses to offset actions; the limited technology and track record in restoring degraded ecosystems to a level that compensates losses; and the development of metrics that accurately represent and value the features lost and gained. In essence, the policies need to achieve no net loss from certain losses and uncertain gains. Governance of offset policies is critical to determining whether they contribute genuinely to conservation of Australian vegetation or simply accelerate its decline by subverting the mitigation hierarchy (Maron *et al.* 2015).

27.4.3 Threat abatement and conservation management

While protected areas effectively mitigate vegetation clearing, most other threats (Figure 27.1) operate across land tenures. The Commonwealth and some state legislation (e.g. New South Wales and Victoria) provide for listing of Key Threatening Processes. Although the listings remain somewhat incomplete (Auld and Keith 2009), they offer a systematic means of reviewing, planning and funding conservation action through threat abatement plans and threat amelioration.

Threats posed by invasive exotic plants and pathogens are managed nationally and locally (Leishman *et al.*, Chapter 9, this volume). National biosecurity measures aim to minimise the introduction of potentially invasive exotic plants and animals into Australia. Site-based management includes preventative measures such as control of water runoff, nutrient enrichment and vehicle access and remedial measures such as chemical control and physical removal of weeds. These measures are usually implemented by land owners, sometimes in collaboration with community groups. Some of the local action is co-ordinated within planning frameworks developed by regional authorities, and the community may receive support for implementation from government funded programmes. Threat Abatement Plans enable strategic approaches, such as developing best practice for managing invasive species and targeting control at the sites considered to be of highest priority for threat control, e.g. where there are threatened species most impacted by the threat (Australian Government 2014). Management success is variable and inconsistently reported at local levels, although it is clear that outcomes at both local and landscape scales would benefit from greater investment.

There is increasing pressure on the remaining Australian vegetation from recreational use, including walking tracks, cycling, horse riding, camping facilities, lodges and 4WD touring. Accommodating these diverse needs of an increasing population will be challenging and requires better understanding of sensitivities of different vegetation and soil types to a range of different facilities and levels of use (e.g. Ballantyne *et al.* 2016). Evidence-based planning and design of minimal impact visitor facilities is still in its infancy, but is most advanced in management of protected areas (especially Wilderness and World Heritage Areas).

Management of altered water regimes for conservation of Australia's riparian and wetland vegetation has focused on the implementation of environmental flows and, to a lesser degree, modification of river and floodplain infrastructure including dams, weirs, levees, channels and diversions (Catford *et al.*, Chapter 20, this volume). Early signs suggest that wetland vegetation responds positively to restoring both the frequency and magnitude of natural flow regimes, although responses may depend on legacies from other threats as well as the duration of flow disruption (Dawson *et al.* 2017).

Fire management is one of the most complex, challenging and uncertain aspects of threat abatement for conservation of Australian vegetation. This is due to the highly variable level and nature of sensitivities among different vegetation types and species, the interactions between fires and other threats, conflicts between conservation and other fire management objectives and the probabilistic influence of management on fire regime outcomes (Miller and Murphy, Chapter 5, this volume). Nonetheless, there have been considerable advances in scientific understanding of responses of some plant species and communities to different components of the fire regime. Versatile management frameworks are developing in parallel with this work, incorporating adaptive principles and management thresholds to define actions required to keep fire regimes within the limits of variation required to maintain the persistence of species and communities (Keith *et al.* 2002; Cook *et al.* 2016). While implementation of such strategies is improving, there is a particular need to avoid simplistic applications of fire management that may be appropriate in one vegetation type to other vegetation that may occur across contrasting ecosystems, landscapes and climates.

Conserving vegetation under a changing climate will require a range of mitigation and adaptation strategies to deal with complex interactive threats and uncertainties in both climate predictions and species responses. These strategies need to promote *in situ* resilience of vegetation and enhance species' ability to track suitable climates and habitats. This requires multiple strategies to: (i) maintain and manage existing vegetation and ecological processes; (ii) enhance connectivity for native species while limiting the spread of detrimental exotic species; and (iii) rehabilitate degraded vegetation and establish plant communities at new locations to compensate for transformation within their extant distributions. Conserving phenotypic and genetic diversity of plant species is fundamental to maximising their adaptive capacity *in situ* (Williams *et al.* 2008; Dawson *et al.* 2011). While maintaining adaptive capacity is more tractable than rehabilitation, which involves greater costs, technological limitations and management risks, it is likely that all three approaches will be necessary to avoid significant loss of diversity in vegetation and associated biota under future climates. This is particularly so for mesic vegetation types (rainforests and cloud forests, wet sclerophyll forests, wetlands and alpine vegetation) that are more susceptible to drying, potentially bereft of climatically suitable areas in the future and virtually impossible to reconstruct in another location.

27.4.4 Ecological restoration

Given widespread depletion and degradation of native vegetation, especially across southern Australia, ecological restoration is widely seen as a priority by Australian governments and civil society, especially in the context of climate change (Hobbs and Harris 2001). Plans and commitments for vegetation restoration commonly arise from conditions of development approvals, recovery plans for threatened species and ecological communities, biodiversity offsets for development, sustainable farm management and community landcare projects. National funding programmes for vegetation restoration have been in place since the 1980s.

Restoration encompasses different approaches that vary in levels of intervention and applicability depending on the initial state of the ecosystem (Clewell and Aronson 2013). Recently, Australia was the first country to adopt national standards for ecological restoration (McDonald *et al.* 2016). These standards articulate several critical principles for restoration projects, including a universal goal to achieve full recovery of six major ecosystem attributes relative to appropriate locally indigenous reference ecosystems.

Understanding the ecosystem properties and ecological processes that influence restoration outcomes for a variety of degraded states is critical to setting specific restoration objectives, selecting appropriate restoration approaches and sites, and monitoring restoration outcomes (McDonald *et al.* 2016). Although restoration activities are widespread in Australia, the few programmes that have been systematically evaluated according to principles adopted in national standards demonstrate the technical difficulties and long time frames in meeting fundamental goals (Box 27.1).

Box 27.1 Evaluating the success of ecological restoration

Although restoration activities are widespread in Australia, few projects are designed, documented, monitored and communicated in ways that contribute reliably and substantially to the developing knowledge base. We reviewed a small sample of 10 well-documented studies that evaluated

Box 27.1 (cont.)

ecological outcomes of restoration consistent with principles in the Australian Standard (McDonald *et al.* 2016) across a range of Australian vegetation types over decadal time frames (Table 27.1). We asked where restoration is most successful, what governs success, how long it takes, and which techniques produce best results. All 10 studies employed the reference concept to establish a restoration goal and measure progress towards it in both restored and untreated vegetation, largely consistent with the Australian standards (McDonald *et al.* 2016). Reference states were represented by multiple samples of non-degraded vegetation of the same type within the same area, and of vegetation on-site prior to degradation. Most studies had a retrospective design based on space-for-time substitution, but some were replicated in space and time, and some involved manipulative experimentation.

Although the studies spanned a wide range of Australian vegetation types, environments, degradation processes and restoration approaches, the qualitative ecological outcomes of restoration were remarkably similar and suggest that full ecological restoration to reference states is rarely achieved within decadal-scale time frames if at all. The following generalisations can be drawn from Table 27.1.

- Restoration involves substantial ecological lags, with most structural attributes restored more readily and more rapidly than compositional attributes, although some structural attributes develop slowly and may take more than 100 years to reach reference states;
- Restoration success is more likely when the initial state has experienced only low levels of degradation – outcomes appear particularly sensitive to site selection;
- Disruption of the soil profile, either by mining, repeated ploughing, fertiliser application or hydrological change, increases the severity of degradation and hence reduces the likelihood and rate of successful restoration to reference states;
- Floodplain vegetation, rainforests and wet sclerophyll forests exhibit stronger restoration responses than dry sclerophyll forests, savanna, temperate eucalypt woodlands and grasslands, with the latter group showing limited compositional convergence over time and a tendency to form novel ecological communities that lack key features of reference states.

Almost all of the studies identified one or more ecological barriers that disrupt, inhibit or redirect progress of restoration towards reference states. These include limited dispersal of native plant propagules, depletion of native soil seed banks, unsuitable soil conditions or microclimate for seedling recruitment, lack of facilitators such as mycorrhizae or nurse plants, overabundance of herbivores or pathogens, and competition from exotic plants.

Table 27.1 **Studies of ecological restoration outcomes in different vegetation types across Australia.**

Vegetation formation	Disturbance history	Initial state	Reference state	Restoration management	Restoration outcomes	Reference
Tropical rainforests, Atherton Tableland Qld	Clearing and livestock grazing	Exotic pasture	Adjacent old-growth rainforest	Unassisted natural regeneration at some sites, diverse tree planting and follow-up maintenance at others	After 40 years, some structural features of regrowth and restored vegetation resembled reference states, but convergence of composition and floristic features was slower, reaching less than 50% of reference state values after 25–60 years. Convergence of planted vegetation was faster than unassisted regeneration.	Shoo *et al.* (2016)
Tropical savanna, Cape York Qld	Bauxite opencut mining	Bare soil	Adjacent savanna forest	Surface ripping, redistribution of topsoil, fertilisation and sowing of native seed with enrichment plantings after several years	After 23 years, restored vegetation remained substantially different to reference states in both structure and composition. Eleven tree species that accounted for > 97% of canopy volume in reference states accounted for < 22% of canopy volume in restored vegetation.	Gould (2012)
Wet and dry sclerophyll forests, West Gippsland Vic	Clearing, pasture improvement, livestock grazing	Exotic pasture	Adjacent remnant forests	Diverse tree and shrub plantings (ecological plantings)	After 26 years, structural complexity of restored vegetation converged with reference states but species composition and richness had not. Restored vegetation was intermediate in composition between untreated pasture and reference states but markedly different from both.	Munro *et al.* (2009)

Table 27.1 (cont.)

Vegetation formation	Disturbance history	Initial state	Reference state	Restoration management	Restoration outcomes	Reference
Dry sclerophyll forests, Myall Lakes NSW	Mineral sand extraction, redistribution of soil profile	Bare sand	Pre-mining forest and adjacent unmined forest	Redistribution of mined sand, seed broadcast of sterile annual grasses, followed by sowing of native species seed, selective tree planting after several years	After 15 years, restored vegetation exhibited decreasing resemblance with pre-mining reference states and no trend in resemblance with adjacent unmined states. Restored vegetation had greater abundance of exotic species, overabundance of some native species, and underrepresentation of other native species relative to reference states.	Buckney and Morrison (1992)
Dry sclerophyll forests, northern jarrah WA	Bauxite opencut mining	Bare soil	Adjacent unmined forest	Surface ripping, redistribution of topsoil, fertilisation and seed sowing of native species, with enrichment plantings after several years	After 14 years, restored vegetation had not become increasingly similar in composition, species density or cover to unmined reference states. Seeded treatments produced greater diversity of native species and lower richness of exotic species than unseeded treatments.	Norman *et al.* (2006)
Temperate eucalypt woodlands, Cumberland Plain NSW	Pasture improvement, livestock grazing	Exotic pasture	Adjacent remnant woodland fragments	Slashing and chemical treatment of exotic pasture, diverse tree and shrub plantings within weed-mat discs, follow-up slashing after 12 months, livestock removal	After 20 years, restored vegetation exhibited increasing resemblance to reference states in some structural features (tree height, cover). Species composition changed over time, but resemblance to reference states did not increase. Restored vegetation included greater abundance of more exotic species and lacked a significant component of the native flora present in reference states.	Wilkins *et al.* (2003); Nichols *et al.* (2010); Keith (unpubl. data)

Temperate eucalypt woodlands, Kellerberin district WA	Clearing, cropping and fertiliser application for 1-4 years	Cropped field	Adjacent remnant woodland	Unassisted natural regeneration	After 45 years, restored vegetation at one of three sites accumulated similar numbers and abundance of native species in standing vegetation and the soil seedbank relative to reference states, but richness and abundance remained low at two sites.	Standish *et al.* (2007)
Temperate eucalypt woodlands, Narrogin-Meredin-Dowerin district WA	Clearing, pasture improvement, livestock grazing	Exotic/ native pasture/ woodland	Local remnant woodlands ungrazed >50 yrs	Livestock exclusion	After 22 years, fenced vegetation had marginally greater resemblance to reference states than unfenced vegetation, reflecting greater richness of native groundcover species and lower cover of exotic species, while there was no trend with time since fencing and soil nutrients showed little response.	Prober *et al.* (2011)
Temperate eucalypt woodlands, central west NSW	Clearing, fertiliser application, livestock grazing	Exotic/ native pasture	Adjacent remnant woodland	Livestock exclusion, application of carbohydrate (sucrose), spring burning, sowing of *Themeda*	Within 3 years, plots treated with carbohydrate, burning and sowing treatments showed increases of native grass cover, reduced cover of exotics and reduced soil nitrogen (to similar levels cf. reference states), but effects were weaker or absent when one or more of the treatments were not applied.	Prober and Lunt (2009)
Floodplain forests, woodlands and shrublands, Macquarie marshes NSW	Clearing, cropping (1-23 yrs), livestock grazing, flood mitigation infrastructure	Drained and cropped field	Periodically flooded forests and shrublands nearby	Reinstatement of floodplain connectivity (levee breaching), environmental flows, cessation of cropping and livestock grazing	After 4 years of flood reinstatement, abundance of plant functional types in restored vegetation resembled reference states for sites exposed to 1 year of cropping, but remained different at sites cropped more than twice.	Dawson *et al.* (2017)

Although restoration technologies have advanced in recent decades, the outcomes remain highly uncertain and involve significant risks and ecological lags. In contrast, protection and management of native vegetation that is presently in, or close to, reference states produces outcomes that are more certainly positive, more immediate and more cost-effective. Conservation and planning approaches that apply the mitigation hierarchy (first avoid, then minimise impacts on native vegetation, and offset/compensate damage as a last resort) will therefore always be more successful and cost-effective in arresting ongoing declines of biodiversity than those that rely on restoration or reconstruction of (even moderately) degraded vegetation to replace further losses of vegetation in reference condition. Recent trends in government policy to increase resourcing for restoration while weakening protection of native vegetation are therefore likely to accelerate, rather than reduce erosion of Australia's native vegetation (e.g. Maron *et al.* 2016b).

Despite substantial risks of failure and ecological lags, a restoration imperative exists for ecosystems that are already highly depleted and degraded throughout their geographic range. The keys to improving restoration outcomes are: (i) selecting sites that are most likely to respond to management actions; (ii) correctly diagnosing potential restoration barriers; and (iii) designing and implementing management to overcome the barriers. Restoration outcomes can sometimes be improved by applying multiple strategies to remove restoration barriers, including soil remediation (e.g. ripping, denitrification), introduction of native species (planting, seeding), control of exotic plants (physical, chemical methods) and herbivores (fencing, culling), or implementation of appropriate fire regimes (lighting or suppressing fires).

27.5 Conclusions

Australian vegetation is undergoing its greatest upheaval since the encroaching deserts swept aside ancient forests more than 10 million years ago. Within a few human generations, millions of square kilometres of forests, woodlands, heathlands and grasslands have been transformed into croplands, pastures and urban landscapes, while other extensive areas of native vegetation are undergoing degradation, simplification and homogenisation. Australian vegetation is losing its unique character.

Most of the threatening processes driving these changes are well understood, though precise mechanisms of some are yet to be resolved. More challenging, is the development and effective implementation of management and policy strategies to redress the ongoing erosion of diversity. The chapters in this book recount substantial scientific advances to meet these challenges. In an international context, Australia is well equipped to arrest the decline with developing networks of protected areas, sophisticated legislative and policy apparatus, strong research support for management and growing community awareness. Yet new and more powerful agents of biodiversity erosion are emerging, including the changing climate, accelerating human population growth and a variable and unpredictable socioeconomic ethos. These will accelerate the current trajectory to a simpler and more cosmopolitan Australian vegetation of the future unless the capacity to protect and manage its current diversity is more effectively resourced and implemented.

Acknowledgements

We are grateful to Chris Simpson for assistance with spatial analyses and Mark Tozer for comments on the manuscript.

References

Adam, P. (1994) *Australian Rainforests.* Oxford University Press: Oxford.

Allen, C.D., Macalady, A.K., Chenchouniet, H. *et al.* (2010) A global overview of drought and heat-induced tree mortality reveals emerging climate change risks for forests. *Forest Ecology and Management* 259, 660–884.

Allen, C.D., Breshears, D.D. and McDowell, N.G. (2015) On underestimation of global vulnerability to tree mortality and forest die-off from hotter drought in the Anthropocene. *Ecosphere* 6, 1–55.

Anderegg, W.R.L., Hicke, J.A., Fisher, R.A. *et al.* (2015) Tree mortality from drought, insects, and their interactions in a changing climate. *New Phytologist* 208, 674–683.

Auld, T.D. and Keith, D.A. (2009) Dealing with threats: integrating science and management. *Ecological Management and Restoration* 10, S79–S87.

Australian Bureau of Statistics (1925) *Settlement of returned soldiers 1914-18. Special article 1301.0 year book Australia, 1925.* Australian Government: Canberra. Available at: http://www.abs.gov.au/AUSSTATS/abs@.nsf/featurearticlesbyCatalogue/72BB159FA215052FCA2569DE0020331D?OpenDocument (accessed 19 August 2016).

Australian Government (2009) *Australia's strategy for the National Reserve System 2009–2030.* Australian Government: Canberra.

Australian Government (2013) *Interim biogeographic regionalisation for Australia (IBRA) v7.* Australian Government: Canberra. Available at: https://www.environment.gov.au/land/nrs/science/ibra/australias-bioregions-maps (accessed 2 May 2016).

Australian Government (2014) *Threat abatement plan for disease in natural ecosystems caused by Phytophthora cinnamomi.* Department of the Environment: Canberra.

Australian Government (2015) *Our north, our future. White paper on developing northern Australia.* Australian Government: Canberra.

Australian Greenhouse Office (2000) *Land clearing: a social history. National carbon accounting system technical report no. 4.* Commonwealth of Australia: Canberra.

Ballantyne, M., Treby, D.L., Quarmby, J. and Pickering, C.M. (2016) Comparing the impacts of different types of recreational trails on grey box grassy woodland vegetation: lessons for conservation and management. *Australian Journal of Botany* 64, 246–259.

Beadle, N.C.W. (1948) *The vegetation and pastures of western New South Wales, with special reference to soil erosion.* Department of Conservation of NSW: Sydney.

Beadle, N.C.W. (1981) *The Vegetation of Australia.* Cambridge University Press: Melbourne.

Beaumont, L.J., Gallagher, R.V., Leishman, M.R. *et al.* (2014) How can knowledge of the climate niche inform the weed risk assessment process? A case study of *Chrysanthemoides monilifera* in Australia. *Diversity and Distributions* 20, 613–625.

Bedward, M., Simpson, C.C., Ellis, M.V. and Metcalfe, L.M. (2007) Patterns and determinants of historical woodland clearing in central-western New South Wales, Australia. *Geographical Research* 45, 348–357.

Bradshaw, C.J.A. (2012) Little left to lose: deforestation and forest degradation in Australia since European colonization. *Journal of Plant Ecology* 5, 109–120.

Brim-Box, J., McBurnie, G., Strehlow, K. *et al.* (2016) The impact of feral camels (*Camelus dromedarius*) on remote waterholes in central Australia. *The Rangeland Journal* 38, 191–200.

Brock, M.A., Capon, S.J. and Porter, J.L. (2006) Disturbance of plant communities dependent on desert rivers. In *Ecology of Desert Rivers*, ed. R.T. Kingsford. Cambridge University Press: Cambridge, pp. 203–247.

Buckney, R.T. and Morrison, D.A. (1992) Temporal trends in plant species composition on mined sand dunes in Myall Lakes National Park, Australia. *Australian Journal of Ecology* 17, 241–254.

Butchart, S.H.M., Akçakaya, H.R., Chanson, J. *et al.* (2007) Improvements to the Red List Index. *PLoS ONE* 2, e140.

Cahill, D.M., Rookes, J.E., Wilson, B.A. *et al.* (2008) *Phytophthora cinnamomi* and Australia's biodiversity: impacts predictions and progress towards control. *Australian Journal of Botany* 56, 279–310.

CAPAD (2014) *Collaborative Australian Protected Area Database. National Summary 2014.* Department of the Environment and Energy: Canberra. Available at: https://www.environment.gov.au/land/nrs/science/capad/2014 (accessed 9 May 2016).

Cardillo, M. and Pratt, R. (2013) Evolution of a hotspot genus: geographic variation in speciation and extinction rates in *Banksia* (Proteaceae). *BMC Evolutionary Biology* 13, 155.

Carnegie, A.J., Kathuria, A., Pegg, G.S. *et al.* (2016) Impact of the invasive rust *Puccinia psidii* (myrtle rust) on native Myrtaceae in natural ecosystems in Australia. *Biological Invasions* 18, 127–144.

Clewell, A.F. and Aronson, J. (2013) *Ecological Restoration: Principles, Values and Structure of an Emerging Profession*, 2nd edition. Cambridge University Press: Cambridge.

Collins, K.M., Price, O.F., Penman, T.D. *et al.* (2015) Spatial patterns of wildfire ignitions in south-eastern Australia. *International Journal of Wildland Fire* 24, 1098–1108.

Colman, N.J., Gordon, C.E., Crowther, M.S. *et al.* (2014) Lethal control of an apex predator has unintended cascading effects on forest mammal assemblages. *Proceedings of the Royal Society B* 281, 20133094.

Commonwealth of Australia (2014) *Threat abatement plan for disease in natural ecosystems caused by* Phytophthora cinnamomi. Commonwealth of Australia: Canberra.

Commonwealth of Australia (2015) *Australian land use, land use-change and forestry emissions projections.* Commonwealth of Australia: Canberra.

Cook, C.N., de Bie, K., Keith, D.A. and Addison, P.F.E. (2016) Decision triggers are a critical part of evidence-based conservation. *Biological Conservation* 195, 46–51.

Craigie, I.D., Grech, A., Pressey, R.L. *et al.* (2015) Terrestrial protected areas of Australia. In *Austral Ark: The State of Wildlife in Australia and New Zealand*, ed. A. Stow, N. Maclean and G.I. Holwell. Cambridge University Press: Cambridge, pp. 560–581.

Cramer, V.A. and Hobbs, R.J. (2002) Ecological consequences of altered hydrological regimes in fragmented ecosystems in southern Australia: impacts and possible management responses. *Austral Ecology* 27, 546–564.

Crisp, M.D., Burrows, G.E., Cook, L.G. *et al.* (2011) Flammable biomes dominated by eucalypts originated at the Cretaceous–Palaeogene boundary. *Nature Communications* 2, 193.

Cunningham, S.A. (2000) Depressed pollination in habitat fragments causes low fruit set. *Proceedings of the Royal Society B* 267, 1149–1152.

Cunningham, S.C., Thomson, J.R., Mac Nally, R. *et al.* (2011) Groundwater change forecasts widespread forest dieback across an extensive floodplain system. *Freshwater Biology* 56, 1494–1508.

Dawson, T.P., Jackson, S.T., House, J.I. *et al.* (2011) Beyond predictions: biodiversity conservation in a changing climate. *Science* 332, 53–58.

Dawson, S.K., Kingsford, R.T., Berney, P. *et al.* (2017) Contrasting influences of inundation and land use on the rate of floodplain restoration. *Aquatic Conservation*, in press.

Department of the Environment and Energy (2007) *NVIS data products 4.2*. Australian Government: Canberra. Available at: http://www.environment.gov.au/land/native-vegetation/national-vegetation-information-system/data-products (accessed 10 May 2016).

Diaz, S., Lavorel, S., McIntyre, S. *et al.* (2006) Plant trait responses to grazing: a global synthesis. *Global Change Biology* 13, 313–341.

Dickson, S. (2013) *National parks emergency grazing measures to end. Media statement.* Minister for National Parks, Recreation, Sport and Racing: Brisbane. Available at: http://statements.qld.gov.au/Statement/2013/10/29/national-parks-emergency-grazing-measures-to-end (accessed 19 August 2016).

Doupé, R.G. and Petitt, N.E. (2002) Ecological perspectives on regulation and water allocation for the Ord River, Western Australia. *River Research and Applications* 18, 307–320.

DSEWPaC (2011) *Peer review of the Australia State of the Environment 2011 report.* Australian Government Department of Sustainability, Environment, Water, Population and Communities on behalf of the State of the Environment 2011 Committee: Canberra. Available at: https://www.environment.gov.au/system/files/pages/ba3942af-f815-43d9-a0f3-dd26c19d83cd/files/soe2011-supplementary-peer-review-australia-state-environment-2011-report.pdf (accessed 19 August 2016).

DSITI (2015) *Land cover change in Queensland 2012–13 and 2013–14: a Statewide Landcover and Trees Study (SLATS) report*. Queensland Department of Science, Information Technology and Innovation: Brisbane.

Dudley, N. (2008) *Guidelines for applying protected area management categories.* International Union for Conservation of Nature and Natural Resources: Gland.

Dunlop, M. and Brown, P.R. (2008) *Implications of climate change for Australia's National Reserve System: a preliminary assessment.* Department of Climate Change: Canberra, Australia. Available at: https://www.environment.gov.au/system/files/resources/917bb661-b626-44bb-bd52-325645ae7c49/files/nrs-report.pdf (accessed 19 August 2016).

Eamus, D., Zolfaghar, S., Villalobos-Vega, R. *et al.* (2015) Groundwater-dependent ecosystems: recent insights from satellite and field based studies. *Hydrology and Earth System Sciences* 19, 4229–4256.

Eldridge, D.J., Poore, A.G.B., Ruiz-Colmenero, M. *et al.* (2016) Ecosystem structure, function and composition in rangelands are negatively affected by livestock grazing. *Ecological Applications* 26, 1273–1283.

Enright, N.J., Fontaine, J.B., Bowman, D.M.J.S. *et al.* (2015) Interval squeeze: altered fire regimes and demographic responses interact to threaten woody species persistence as climate changes. *Frontiers in Ecology and the Environment* 13, 265–272.

Evans, M.C. (2016) Deforestation in Australia: drivers, trends and policy responses. *Pacific Conservation Biology* 22, 130–150.

Fensham, R.J. and Fairfax, R.J. (2003) Spring wetlands of the Great Artesian Basin, Queensland, Australia. *Wetlands Ecology and Management* 11, 343–362.

Fensham, R.J., Fairfax, R.J. and Dwyer, J.M. (2012) Potential aboveground biomass in drought-prone forest used for rangeland pastoralism. *Ecological Applications* 22, 894–908.

Fitzsimons, J.S. (2015) Private protected areas in Australia: current status and future directions. *Nature Conservation* 10, 1–23.

Froend, R.H. and Sommer, B. (2010) Phreatophytic vegetation response to climatic and abstraction-induced GW drawdown: examples of long-term spatial and temporal variability in community response. *Ecological Engineering* 36, 1191–1200.

Gibson, N., Coates, D., van Leeuwen, S. and Yates, C. (2015) Hot dry and ancient: banded iron formations of Western Australia. In *Geossistemas ferruginosos do Brasil. Áreas prioritárias para conservação da diversidade geológica e biológica, patrimônio cultural e serviços ambientais*, ed. F.F. do Carmo and L.H.Y. Kamino. 3i Editora Ltda: Belo Horizonte, pp. 361–391.

Gibson-Roy, P., Delpratt, J. and Moore, G. (2007) Restoring the Victorian Western (Basalt) Plains grassland. 1. Laboratory trials of viability and germination, and the implications for direct seeding. *Ecological Management and Restoration* 8, 114–122.

Good, R.B. (1992) *Kosciusko Heritage*. National Parks and Wildlife Service: Sydney.

Gould, S.F. (2012) Comparison of post-mining rehabilitation with reference ecosystems in Monsoonal Eucalypt Woodlands, Northern Australia. *Restoration Ecology* 20, 250–259.

Hall, C.M. and Shultis, J. (1991) Railways, tourism and worthless lands: the establishment of national parks in Australia, Canada, New Zealand and the United States. *Australian Canadian Studies* 8, 57–74.

Hautier, Y., Seabloom, E.W., Borer, E.T. *et al.* (2014) Eutrophication weakens stabilizing effects of diversity in natural grasslands. *Nature*. 508, 521–525.

Hobbs, R.J. and Harris, J.A. (2001) Restoration ecology: repairing the Earth's ecosystems in the new millennium. *Restoration Ecology* 9, 239–246.

Hobbs, R.J., Arico, S., Aronson, J. *et al.* (2006) Novel ecosystems: theoretical and management aspects of the new ecological world order. *Global Ecology and Biogeography* 15, 1–7.

Holz, A., Wood, S.W., Veblen, T.T. and Bowman, D.M.J.S. (2014) Effects of high-severity fire drove the population collapse of the subalpine Tasmanian endemic conifer *Athrotaxis cupressoides*. *Global Change Biology* 21, 445–458.

Hopper, S.D. (2009) OCBIL theory: towards an integrated understanding of the evolution, ecology and conservation of biodiversity on old, climatically buffered, infertile landscapes. *Plant Soil* 322, 49–86.

Jarvis, N.T. (1979) *Western Australia: An Atlas of Human Endeavour: 1829–1979*. Western Australian Government: Perth.

Keith, D.A. (2009) The interpretation, assessment and conservation of ecological communities and ecosystems. *Ecological Management and Restoration* 10, S3–S15.

Keith, D.A. (2012) Functional traits: their roles in understanding and predicting biotic responses to fire regimes. In *Flammable Australia: Fire Regimes, Biodiversity and Ecosystems in a Changing World*, ed. R.A. Bradstock, A.M. Gill and R.J. Williams. CSIRO Publishing: Melbourne, pp. 97–125.

Keith, D.A. (2015) Assessing and managing risks to ecosystem biodiversity. *Austral Ecology* 40, 337–346.

Keith, D.A. and Pellow, B.J. (2015) *Review of Australia's Major Vegetation Classification and Descriptions*. Centre for Ecosystem Science, UNSW: Sydney.

Keith, D.A. and Scott, J. (2005) Native vegetation of coastal floodplains: a diagnosis of the major plant communities in New South Wales. *Pacific Conservation Biology* 11, 81–104.

Keith, D.A., Williams, J.E. and Woinarski, J.C.W. (2002) Biodiversity conservation: principles and approaches for fire management. In *Flammable Australia: Fire Regimes, Biodiversity and*

Ecosystems in a Changing World, ed. R.A. Bradstock, A.M. Gill and R.J. Williams. CSIRO Publishing: Melbourne, pp. 401–425.

Keith, D.A., Orscheg, C., Simpson, C.C. *et al.* (2009) A new approach and case study for estimating extent and rates of habitat loss for ecological communities. *Biological Conservation* 142, 1469–1479.

Keith, D.A., McDougall, K.L., Simpson, C.C. and Walsh, J.L. (2012) Spatial analysis of risks posed by root rot pathogen, *Phytophthora cinnamomi*: implications for disease management. *Proceedings of the Linnean Society of New South Wales* 134, B14–B179.

Keith, D.A., Lindenmayer, D.B., Lowe, A. *et al.* (2014). Heathlands. In *Biodiversity and Environmental Change: Monitoring, Challenges and Direction*, ed. D. Lindenmayer, E. Burns, N. Thurgate and A. Lowe. CSIRO Publishing: Melbourne, pp. 215–285.

Keith, D.A., Rodríguez, J.P., Brooks, T.M. *et al.* (2015) The IUCN Red List of Ecosystems: motivations, challenges and applications. *Conservation Letters* 8, 214–226.

Kingsford, R.T. (2016) Flow alteration and its effect on Australian riverine vegetation. In *Vegetation of Australian Riverine Landscapes*, ed. S. Capon, C. James and M. Reid. CSIRO Publishing: Melbourne, pp. 261–271.

Kingsford, R.T. and Thomas, R.F. (2004) Destruction of wetlands and waterbird populations by dams and irrigation on the Murrumbidgee River in arid Australia. *Environmental Management* 34, 383–396.

Kingsford, R.T., Jenkins, K.M. and Porter, J.L. (2004) Imposed hydrological stability on lakes in arid Australia and effect on waterbirds. *Ecology* 85, 2478–2492.

Klein, C., Wilson, K., Watts, M. *et al.* (2009) Incorporating ecological and evolutionary processes into continental-scale conservation planning *Ecological Applications* 19, 206–217.

Krogh, M. (2007) Management of longwall coal mining impacts in Sydney's southern drinking water catchments. *Australasian Journal of Environmental Management* 14, 155–165.

Laurance, W.F. (1997) Responses of mammals to rainforest fragmentation in tropical Queensland: a review and synthesis. *Wildlife Research* 24, 603–612.

Letnic, M., Ritchie, E.G. and Dickman, C.R. (2012) Top predators as biodiversity regulators: the dingo *Canis lupus dingo* as a case study. *Biological Reviews* 87, 390–413.

Lindenmayer, D.B. and Franklin, J.F. (2002) *Conserving Forest Biodiversity: A Comprehensive Multi-scaled Approach.* Island Press: Washington DC.

Lindenmayer, D.B., Hobbs, R.J., Likens, G.E. *et al.* (2011) Newly discovered landscape traps produce regime shifts in wet forests. *Proceedings of the National Academy of Science* 108, 15887–15891.

Lindenmayer, D.B., Burns, E., Thurgate, N. and Lowe, A. (2014) *Biodiversity and Environmental Change: Monitoring, Challenges and Direction.* CSIRO Publishing: Melbourne.

Lunt, I.D., Eldridge, D.J., Morgan, J.W. and Witt, G.B. (2007) A framework to predict the effects of livestock grazing and grazing exclusion on conservation values in natural ecosystems in Australia. *Australian Journal of Botany* 55, 401–415.

Mac Nally, R., Cunningham, S.C., Baker, P.J. *et al.* (2011) Dynamics of Murray–Darling floodplain forests under multiple stressors: the past, present, and future of an Australian icon. *Water Resources Research* 47, doi:10.1029/2011WR010383.

Mackey, B.G., Berry, S.L. and Brown, T. (2008) Reconciling approaches to biogeographic regionalization: a systematic and generic framework examined with a case study of the Australian continent. *Journal of Biogeography* 35, 213–229.

Maron, M., Gordon, A., Mackey, B.G. *et al.* (2015) Stop misuse of biodiversity offsets. *Nature* 523, 401–403.

Maron, M., Ives, C.D., Kujala, H. *et al.* (2016a) Taming a wicked problem: resolving controversies in biodiversity offsetting. *BioScience* 66, 489–498.

Maron, M., Laurance, W., Pressey, R. *et al.* (2016b) Queensland land clearing is undermining Australia's environmental progress. The Conversation. Available at: http://theconversation.com/queensland-land-clearing-is-undermining-australias-environmental-progress-54882 (accessed 15 May 2016).

Martin, H.A. (2006) Cenozoic climatic change and the development of the arid vegetation in Australia. *Journal of Arid Environments* 66, 533–563.

McAlpine, C.A., Fensham, R.J. and Temple-Smith, D.E. (2002) Biodiversity conservation and vegetation clearing in Queensland: principles and thresholds. *The Rangeland Journal* 24, 36–55.

McDonald, T., Jonson, J. and Dixon, K.W. (2016) National standards for the practice of ecological restoration in Australia. *Restoration Ecology* S1, 1–34.

Mendel, L.C. and Kirkpatrick, J.B. (2002) Historical progress of biodiversity conservation in the protected-area system of Tasmania, Australia. *Conservation Biology* 16, 1520–1529.

Miller, G., Friedal, M., Adam, P. and Chewings, V. (2010) Ecological impacts of buffel grass (*Cenchrus ciliaris* L.) invasion in central Australia: does field evidence support a fire-invasion feedback? *The Rangeland Journal* 32, 353–365.

Munro, N.T., Fischer, J., Wood, J. and Lindenmayer, D.B. (2009) Revegetation in agricultural areas: the development of structural complexity and floristic diversity. *Ecological Applications* 19, 1197–1210.

Mutke, J. and Barthlott, W. (2005) Patterns of vascular plant diversity at continental to global scales. *Biologiske Skrifter* 55, 521–531.

Nichols, P.W.B., Morris, E.C. and Keith, D.A. (2010) Testing a facilitation model for ecosystem restoration: does tree planting restore ground layer species in a grassy woodland? *Austral Ecology* 35, 888–897.

Nicholson, E., Regan, T.J., Auld, T.D. *et al.* (2015) Towards consistency, rigour and compatibility of risk assessments for ecosystems and ecological communities. *Austral Ecology* 40, 338–363.

Norman, M.A., Koch, J.M., Grant, C.D. *et al.* (2006) Vegetation succession after bauxite mining in western Australia. *Restoration Ecology* 14, 278–288.

OEH (2014) *NSW report on native vegetation.* NSW Office of Environment and Heritage: Sydney.

Ough, K. (2001) Regeneration of wet forest flora a decade after clear-felling or wildfire – is there a difference? *Australian Journal of Botany* 49, 645–664.

Pressey, R.L. (1994) Ad hoc reservations: forward or backward steps in developing representative reserve systems? *Conservation Biology* 8, 662–668.

Pressey, R.L., Hager, T.C., Ryan, K.M. *et al.* (2000) Using abiotic data for conservation assessments over extensive regions: quantitative methods applied across New South Wales, Australia. *Biological Conservation* 96, 55–82.

Preston, B.J. and Adam, P. (2004) Describing and listing threatened ecological communities under the *Threatened Species Conservation Act 1995* (NSW). *Environmental Planning and Law Journal* 21, 250–263.

Prober, S.M. and Lunt, I.D. (2009) Restoration of *Themeda australis* swards suppresses soil nitrate and enhances ecological resistance to invasion by exotic annuals. *Biological Invasions* 11, 171–181.

Prober, S.M., Standish, R.J. and Wiehl, G. (2011) After the fence: vegetation and topsoil condition in grazed, fenced and benchmark eucalypt woodlands of fragmented agricultural landscapes. *Australian Journal of Botany* 59, 369–381.

Rizzo, D.M. and Garbelotto, M. (2003) Sudden oak death: endangering California and Oregon forest ecosystems. *Frontiers in Ecology and Environment* 1, 197–204.

Ross, K.A., Fox, B.J. and Fox, M.D. (2002) Changes to plant species richness in forest fragments: fragment age, disturbance and fire history may be as important as area. *Journal of Biogeography* 29, 749–765.

Ryan, T. (2010) *The Australian Fertilizer Industry: Values and Issues. Australian Fertilizer Industry Conference.* Australian Fertilizer Industry Association: Canberra.

Salafsky, N., Salzer, D., Stattersfield, A.J. *et al.* (2008) A standard lexicon for biodiversity conservation: unified classifications of threats and actions. *Conservation Biology* 22, 897–911.

Saunders, D.A. (1989) Changes in the avifauna of a region, district and remnant as a result of fragmentation of native vegetation: the wheatbelt of Western Australia. A case study. *Biological Conservation* 50, 99–135.

Saunders, D.A., Hobbs, R.J. and Margules, C.R. (1991) Biological consequences of ecosystem fragmentation: a review. *Conservation Biology* 5, 18–32.

Scott, R.E., Neyland, M.G. and Hovenden, M.J. (2015) Variable-retention harvesting in Tasmania: regeneration success? *Australian Forestry* 78, 232–242.

Setterfield, S.A., Rossiter-Rachor, N.A., Hutley, L.B. *et al.* (2010) Turning up the heat: the impacts of *Andropogon gayanus* (gamba grass) invasion on fire behaviour in northern Australian savannas. *Diversity and Distributions* 16, 854–861.

Shane, M.W., McCully, M.E. and Lambers, H. (2004) Tissue and cellular phosphorus storage during development of phosphorus toxicity in *Hakea prostrata* (Proteaceae). *Journal of Experimental Botany* 55, 1033–1044.

Shoo, L.P., Hoffmann, A.A., Garnett, S. *et al.* (2013) Making decisions to conserve species under climate change. *Climatic Change* 119, 239–246.

Shoo, L.P., Freebody, K., Kanowski, J. and Catterall, C.P. (2016) Slow recovery of tropical old-field rainforest regrowth and the value and limitations of active restoration. *Conservation Biology* 30, 121–132.

Simkin, S.M., Allen, E.B., Bowman, W.D. *et al.* (2015) Conditional vulnerability of plant diversity to atmospheric nitrogen deposition across the United States. *Proceedings of the National Academy of Science* 113, 4086–4091.

Sparrow, B., Dormott, E., Thurgate, N. *et al.* (2014) Our capacity to tell an Australian ecological story. In *Biodiversity and Environmental Change: Monitoring, Challenges and Direction*, ed. D. Lindenmayer, E. Burns, N. Thurgate and A. Lowe. CSIRO Publishing: Melbourne, pp. 49–81.

Standish, R.J., Cramer, V.A., Wild, S.L. and Hobbs, R.J. (2007) Seed dispersal and recruitment limitation are barriers to native recolonization of old-fields in western Australia. *Journal of Applied Ecology* 44, 435–445.

Steinfeld, C.M.M. and Kingsford, R.T. (2013) Disconnecting the floodplain: earthworks and their ecological effect on a dryland floodplain in the Murray–Darling Basin, Australia. *River Research and Applications* 29, 206–218.

Thornhill, A.H., Ho, S.Y.W., Külheim, C. *et al.* (2015) Interpreting the modern distribution of Myrtaceae using a dated molecular phylogeny. *Molecular Phylogenetics and Evolution* 93, 29–43.

Threatened Species Scientific Committee (2008) *Commonwealth listing advice on Natural Temperate Grassland of the Victorian Volcanic Plain.* Australian Government: Canberra.

Toon, A., Crisp, M.D., Gamage, H. *et al.* (2015) Key innovation or adaptive change? A test of leaf traits using Triodiinae in Australia. *Scientific Reports* 5, 1–12.

Vesk, P.A., Leishman, M.R. and Westoby, M. (2004) Simple traits do not predict grazing response in Australian dry shrublands and woodlands. *Journal of Applied Ecology* 41, 22–31.

Visconti, P., Pressey, R.L., Bode, M. *et al.* (2010) Habitat vulnerability in conservation planning – when it matters and how much. *Conservation Letters* 3, 404–414.

Watson, J.E.M., Darling, E.S., Venter, O. *et al.* (2016) Bolder science needed now for protected areas. *Conservation Biology* 30, 243–248.

Wawryk, A. (2014) Conservation and access to land for mining in protected areas: the conflict over mining in South Australia's Arkaroola Wilderness Sanctuary. *Journal of Environmental Law* 26, 291–317.

Wilkins, S., Keith, D.A. and Adam, P. (2003) Measuring success: evaluating the restoration of a grassy eucalypt woodland on the Cumberland plain, Sydney, Australia. *Restoration Ecology* 11, 489–503.

Williams, S.E., Shoo, L.P., Isaac, J.L. *et al.* (2008) Towards an integrated framework for assessing the vulnerability of species to climate change. *PLoS Biology* 6, e325.

INDEX

Printed by Integrated Books International,
United States of America